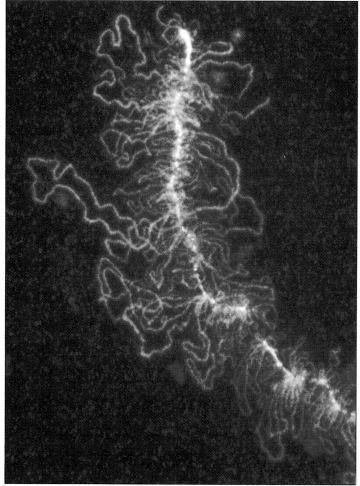

Active transcription units on a newt chromosome
Ooctyes from amphibians such as Notophthalmus viridescens *have lampbrush chromosomes where the active RNA-synthesizing genes loop out. The DNA axis of these loops is stained with a white dye. The red stain is from an antibody that binds to RNA-binding proteins. Chapter 11. (Photograph courtesy of M. B. Roth and J. Gall.)*

Induction of a secondary body axis in *Xenopus*
The normal Xenopus tadpole is seen on top. When mRNA for the activin protein is injected into a single blastomere on the ventral side of a 32-stage embryo, it induces a new axis to form. Chapter 8. (Photograph courtesy of G. Thomsen, M. Whitman, and D. A. Melton.)

Neural migration pathways in insects
Neural axons in insect embryos migrate in very specific patterns. Neurons derived from a common precursor (shown here in the same color) produce axons that selectively migrate along with other axons. The Q1 axon, for instance, travels until it meets the dMP2 axon and then travels with it, while the axon from the G neuron continues moving in a straight line until it meets the P1 axon. Chapter 17. (Photograph courtesy of C. Goodman.)

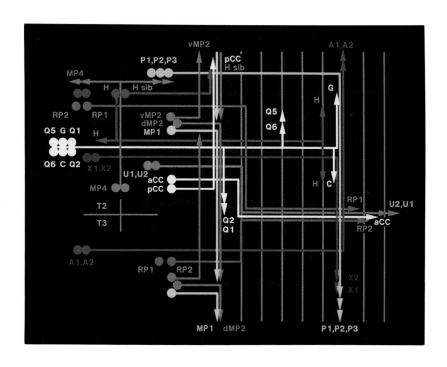

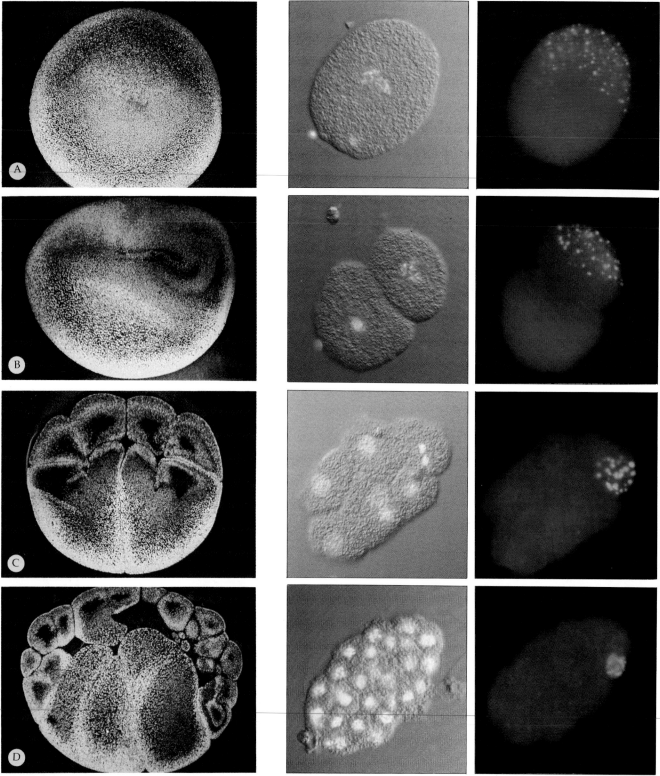

Cytoplasmic rearrangements in the frog, *Xenopus laevis*

(A) The unfertilized Xenopus laevis *egg is radially symmetric. (B) Cytoplasmic movements are seen as the egg starts cleaving 90 minutes after fertilization. The cytoplasm of the future dorsal side (right) differs from that of the future ventral side (left). These differences can be seen throughout embryonic cleavage (C,D) and result in the positioning of dorsal morphogenic determinants in the side of the embryo opposite the point of sperm entry. Chapters 2, 4, and 8. (Photographs courtesy of M. V. Danilchik.)*

Progressive cytoplasmic localization

The segregation of certain cytoplasmic granules ("P-granules") is seen to progress into the posteriormost cells of the Caenorhabditis elegans *embryo. These cells generate the sperm and egg of the nematode. When the pronuclei meet during fertilization, the P-granules move to the posterior portion of the cell. This movement continues until they are found only in the P-cell that gives rise to the gametes. The left-hand column is stained to show the position of the nuclei, while the right-hand column is stained to show the P-granules. Chapter 7. (Photographs courtesy of S. Strome.)*

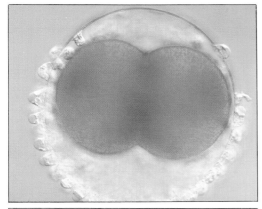
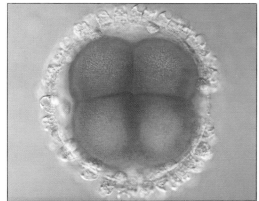
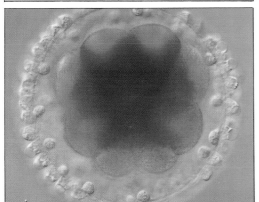
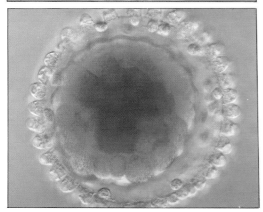

Cytoplasmic localization in tunicate embryos
Cleavage separates regions of cytoplasm into particular cells. The yellow crescent of the Styela embryo becomes localized into a small group of cells that will generate the larval musculature. This figure shows the 2-, 4-, 16-, and 64-cell stages. Chapters 3 and 7. (Photographs courtesy of J. R. Whittaker.)

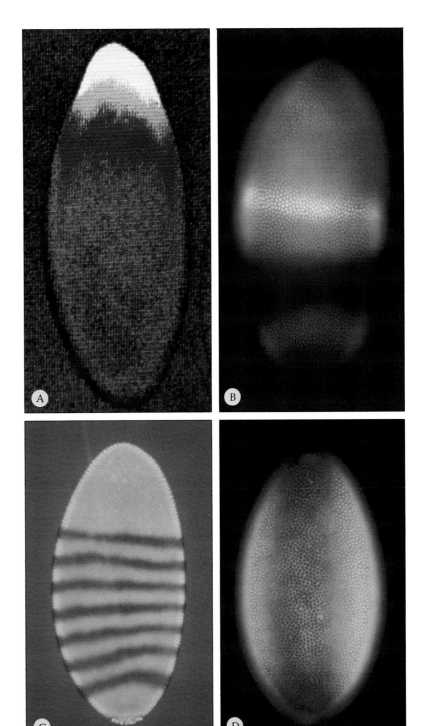

Pattern formation in *Drosophila*
(A) The anterior–posterior axis is specified by cytoplasmic mRNAs and proteins. The gradient of bicoid protein is especially important. High concentrations of this protein (yellow through red) cause head and thorax formation by activating the hunchback gene. (B) Proteins from "gap" genes such as hunchback interact to define the domains of the insect body. Here hunchback (orange) and Krüppel (green) proteins overlap to form a boundary (yellow). (C) These interactions activate the transcription of pair-rule genes (seen as dark bands) that divide the embryo into segments along the anterior–posterior axis. (D) At the same time, the dorsal–ventral axis is forming. The crucial event in the specification of the dorsal–ventral axis is the translocation of the cytoplasmic dorsal protein into the nuclei at the ventral side of the embryo; elsewhere in the embryo, dorsal protein remains in the cytoplasm. White circles are nuclei; dark circles in the ventral midline are nuclei that have taken up the dorsal protein. This view is from the ventral surface. Chapters 7 and 18. (Photographs courtesy of (A) W. Driever and C. Nüsslein-Volhard; (B and D) C. Rushlow and M. Levine; and (C) T. Karr.)

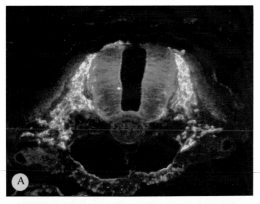

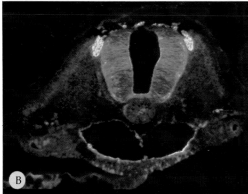

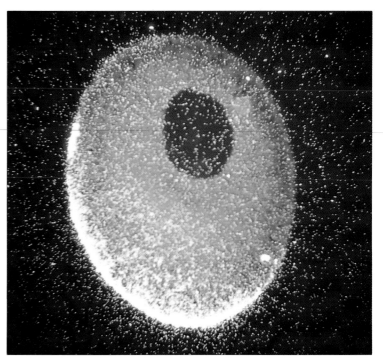

Migration of chick neural crest cells

Chick neural crest cells can be followed in their migration by staining the cells with a fluorescently tagged monoclonal antibody. The neural crest cells (stained green) are found to migrate through the anterior (A) but not the posterior (B) regions of the somite tissue. This specific pattern of neural crest cell migration plays a role in determining the placement of peripheral neurons. Chapter 5. (Photographs courtesy of M. Bronner-Fraser.)

Localization of a specific mRNA in a region of the egg

The vg1 mRNA, which encodes a growth factor-like protein, is found by in situ hybridization to reside solely in the vegetal region of the Xenopus egg. The white crescent at the bottom of the egg is due to the radioactivity of the probe recognizing the mRNA; the rest of the egg is green due to staining with Giemsa dye. Chapters 7, 8, and 14. (Photograph courtesy of D. A. Melton.)

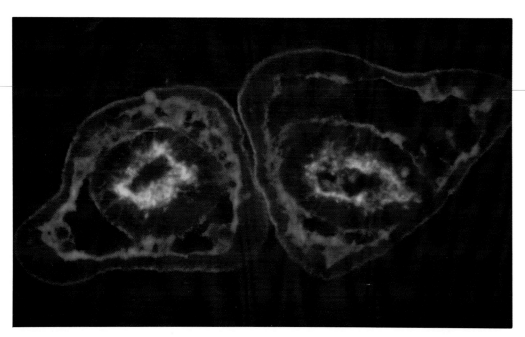

Tissue-specific differences in cell membranes

Not all cell membranes are alike. The ability to change position in the embryo is often seen to be a consequence of changes at the cell surface. Here a particular protein (stained red by a fluorescently labeled monoclonal antibody) is expressed on the cell surface of only those cells destined to become the skeleton of the sea urchin larva. The expression of this protein may be responsible for the migration of these cells into the center of the embryo. Chapters 4 and 15. (Photograph courtesy of G. Wessell.)

Control of development by the environment

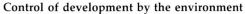

Caterpillars of Nemoria arizonaria *that hatch in the spring eat oak flowers and develop a cuticle that mimics the flowers. Caterpillars of the same species that hatch in the summer (after the flowers are gone) eat oak leaves; these caterpillars develop cuticle that resembles the oak twigs. Chemicals in the leaves appear to modify cuticle development. Chapter 19. (Photographs courtesy of E. Greene.)*

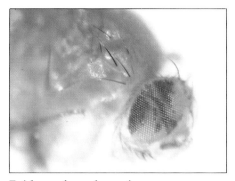

Evidence for a dynamic genome

The mosaic eye in this Drosophila *was created by the spontaneous removal of a piece of DNA (a transposon) that had been in a pigment-forming gene in the cells that generate the eye. White areas contain cells that retain the transposon, while the red cells are able to form pigment because their pigment-forming genes are no longer mutated by containing the transposon. Chapter 10. (Photograph courtesy of G. M. Rubin.)*

Effect of retinoic acid on limb regeneration

Retinoic acid causes regenerating cells to "forget" their original position. Regenerating salamander wrist tissue will usually form only a wrist. After treatment with retinoic acid, however, the regenerating wrist tissue (here from a darkly pigmented salamander) regenerates an entire forearm (lower right limb) when grafted to the hindlimb of a differently pigmented animal. Chapter 17. (Photograph courtesy of K. Crawford.)

Gynandromorph moth

A sexual mosaic (gynandromorph) of an Io moth, divided bilaterally into a rose-brown female half and a smaller-winged, yellow male half. Such sex mosaics are caused when an X chromosome is lost from a nucleus during an early mitotic division. Chapter 21. (Photograph by T. R. Manley; courtesy of The Journal of Heredity.)

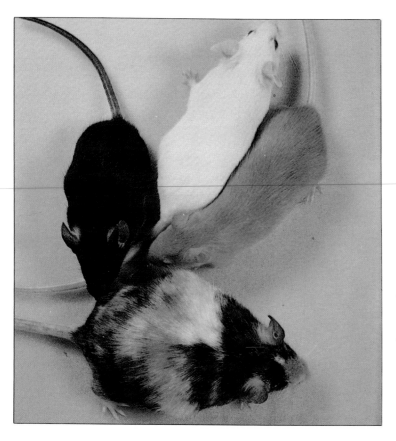

A mouse with six parents

The multicolored mouse was formed by mixing together the cells from three 4-cell stage embryos: an embryo from two black mice; an embryo from two white mice; and an embryo from two brown mice. Instead of forming a three-headed monster, the embryos regulated to form one (normal-size) mouse with contributions from each of the three embryos. That each of the three embryos also produced germ line cells was shown by mating the mouse with a recessive (white) mouse; this mating produced offspring of all three different colors. Chapters 3 and 8. (Photograph courtesy of C. Markert and The Journal of Heredity.)

Developmental Biology

Third Edition

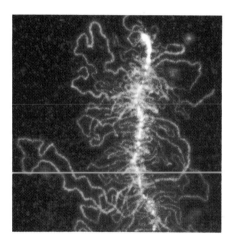

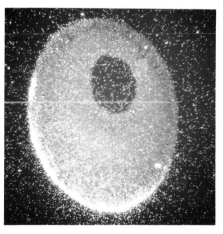

SCOTT F. GILBERT
SWARTHMORE COLLEGE

Sinauer Associates, Inc. • Publishers

Sunderland, Massachusetts

THE COVER

Section of an 8-cell embryo of the frog *Xenopus laevis* stained for the presence of yolk platelets. The colors reflect differences in the concentration of yolk; the heaviest concentration is in the vegetal (bottom) portion of the embryo. The differences in yolk concentration between the right and left sides of the picture reflect movements of cytoplasmic material. This cytoplasmic rearrangement creates the dorsal (back) region of the embryo. (Photograph courtesy of M. V. Danilchik.)

THE TITLE PAGE

The amphibian oocyte: DNA, RNA, and protein. (Left) The yolk protein in the amphibian oocyte is concentrated in the vegetal cytoplasm of the egg, away from the large nucleus. (Center) The lampbrush chromosomes of the oocyte transcribe prodigious amounts of RNA to sustain the metabolism of this enormous cell and to prepare the egg for developing into an embryo. (Right) Some of the messages made in the oocyte become localized to specific areas of the cell, where their protein products will become active in the embryo. (Photographs courtesy of (left) M. V. Danilchik, (center) J. Gall, and (right) D. A. Melton.)

THE PART OPENERS
Formation of the neural tube in the developing chick embryo

Part I. Looking down on the dorsal surface of the 1–2 day chick embryo, one sees the formation of the neural tube in the center of the back. The anterior–posterior axis is first seen by the extension of the primitive streak. The cells migrating through the anterior portion of this streak form the notochord, which extends the length of this axis. Later the cells above the notochord fold inward to produce the neural tube as the embryo becomes progressively more structured. (Photographs courtesy of R. Nagele.)

Part II. During the formation of the neural tube, different genes are expressed in different tissues. The cells that are to become the neural tube gain certain proteins while they lose others. Here, the cells forming the neural tube are seen to lose a cytoskeletal protein whose expression is maintained in the epidermis and primitive gut tissue. (Photographs courtesy of G. Grunwald.)

Part III. Structural features of neural tube formation seen by the scanning electron microscope. A flat area of cells can be seen to lengthen and form a plate. Local contractions then cause this neural plate to fold inward to form the neural tube. (Photographs courtesy of K. Tosney.)

Credits for chapter-opening quotes appear on page 859.

Library of Congress Cataloging-in-Publication Data
Gilbert, Scott F., 1949–
   Developmental biology / Scott F. Gilbert.—3rd ed.
     p.   cm.
   Includes bibliographical references and index.
   ISBN 0-87893-245-3
   1. Developmental biology.   2. Embryology.   I. Title.
QH491.G54   1991
574.3—dc20   90-43141
                                  CIP

Printed in U.S.A.
6 5 4 3 2 1

To Daniel, Sarah, and David

Contents

I. PATTERNS OF DEVELOPMENT

II. MECHANISMS OF CELLULAR DIFFERENTIATION

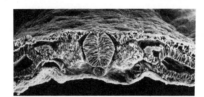

III. CELL INTERACTIONS IN DEVELOPMENT

Preface

This is the third edition of this book to come out since 1985, and if you compare the current volume with the one from six years ago, it is obvious that developmental biology is in a state of profound and exciting intellectual ferment. Much of this richness has resulted from returning to old problems with new techniques. During the past three years, we have seen an astonishing reevaluation of old experiments and concepts as new methods derived from molecular biology and immunology are applied to problems whose solutions have evaded us for almost a century. In fact, the last half of 1990 has been a molecular reprise of classical embryology. The bicoid protein was shown to be an anterior insect morphogen, confirming a hypothesis of Boveri at the turn of the century and the transplantation experiments Klaus Sander performed in the 1950s; a new form of activin has been postulated to be the natural inducer of the dorsal mesoderm, thereby fulfilling a quest started by Spemann's students in 1924; the products of the murine *Steel* and *White spotting* loci were found to interact to bring about the development of germ cells, blood cells, and melanoblasts, completing the embryological investigations initiated in Russell's laboratory during the 1960s; and the major testis determination gene of the Y chromosome has tentatively been identified as the *SRY* locus, thereby answering a question whose origins precede recorded history. Over 2000 years ago, Confucius wrote, "To attain the new by practicing and cherishing the old; And, having obtained the new, to reanimate the old; This is the way of the teacher." Confucius' depiction of a teacher's role seems particularly apt as we try to learn and teach the developmental biology of the 1990s.

There is no single "Developmental Biology," and there is no single way to teach it. The student of development is not confined to any hierarchical level of organization: the transcription of globin genes and the emergence of salamander gills are equally valid problems for the developmental biologist. Nor is the developmental biologist confined to any particular group of organisms or to any system within an organism. Developmental biology includes and integrates them all. No structure originates except by development, and developmental biology is nothing less than the study of every cellular molecule, cell, tissue, organ, and organism as a function of time—all of which means developmental biology is probably the broadest and most inclusive of all biological disciplines.

Because of this breadth, the study of development is bridging together all the areas of biology. It plays the crucial role in relating genotype to phenotype, and interacts with molecular biology at one level and with evolution on another. The two greatest changes between the second and third editions of this book involve the molecular regulation of gene expression during development and the developmental processes that constrain and facilitate evolution. In fact, new research into the relationships between development and evolution (again, going back to old problems) necessitated the inclusion of a new chapter. Thus, the presentation of the cyclical developmental sequence of the organism from fertilization through gametogenesis is now bounded by two termini dealing with the evolution of animal development.

Developmental biology is also a most peculiar science. It is a science of becoming, not of being, and it denies the hegemony of the adult." When most people think of a dog, they only think of the mature hound. For the developmental biologist, dog denotes not only the adult of the species, but also the zygote, gastrula, fetus, and puppy. The adult is merely one state, the product of a long series of embryonic interactions that created it. The interesting problems, to a developmental biologist, are those involving how that organism is generated from a single cell. We know only the smallest portion about development, and interpreting the mechanisms of development from what we do know is analogous to describing the life in the ocean by looking at tidepools. Much of development remains to be known. The problems—How does a single cell give rise to the diverse cell types of the adult body? How do the neurons make their specific connections with other neurons and with peripheral tissues? How do organs form? How do the egg and sperm interact to create a new organism?—are among the most important in all science.

This text attempts to keep pace with the evolution of the field and to make its experiments and conclusions accessible to undergraduates. Although the book does not assume a background more advanced than that obtained from most good introductory biology courses, a previous course in genetics or cell biology is certainly recommended, if only for a deeper appreciation of the studies presented. As before, I have tried to order the chapters in a manner that allows the student to have a conceptual framework from which to understand each successive chapter. The text attempts to present developmental biology as a dynamic, unfinished, human endeavor based on repeated observations and controlled experiments. There are numerous citations so interested individuals can read the original publications.

Elements present in the first two editions of this book have been strengthened. Modern developmental biology demands that both the organismal and molecular approaches to biology be respected (indeed, celebrated). The most interesting areas of the field are probably those where these approaches intersect, and I have tried to integrate them. I have also attempted to blur some of the artificial boundaries that have divided developmental biology from its sibling disciplines of evolutionary biology and genetics. As before, I have tried to increase this book's flexibility by including "Sidelights and Speculations" sections. Some of these sections take a concept from the text and relate it to other areas of biology. Others go into detail or explore concepts that are too new or controversial to be included in the main body of the text.

My friends who study plant development have kidded and chided me for writing a book that is all meat and no vegetables. This is not an accident. I have studied (and even published in) plant developmental biology, and I feel strongly that it deserves its own course and its own book. I could not do the field justice without adding an equivalent number of pages to the existing text, and this book is long enough. If there is anyone who wants to write a plant counterpart to this book, please get in touch with Andy Sinauer, Sinauer Associates, Sunderland, Massachusetts 01375.

A final caveat: At the Marine Biology Laboratory at Woods Hole, Massachusetts (scene of some of the most important embryological investigations in the United States), there hangs a handwritten banner that reads, "Study Nature, Not Books." This sign hangs over the main entrance to the *library* as a reminder that we come to study organisms, not the history of how they are studied. This book is successful to the point that it stimulates one to relinquish it and to study the organism.

Acknowledgments

This book, like previous editions, has benefited enormously from the comments and criticisms of several investigators who took the time to read drafts of these chapters. These stalwarts include: K. Alitalo, D. Boettiger, M. Bronner-Fraser, S. Carroll, M. V. Danilchik, B. A. Edgar, A. Ellison, C. Ettensohn, J. Fallon, S. E. Fraser, G. Gilbert, L. Gilbert, G. Guild, R. Grainger, G. Grunwald, B. Hall, J. Hardin, A. Jacobson, L. A. Jaffe, R. E. Keller, K. J. Kemphues, D. Kirk, J. Lash, W. Loomis, A. Lopo, B. J. Meyer, C. Phillips, R. Raff, S. Rashootin, M. Saha, L. Saxén, R. Schultz, and H. Weintraub. Moreover, this book could not have been completed without the considerable effort from the scores of researchers who sent me the photographs that appear herein, and it has received substantial improvement from those scientists who sent me critiques of the second edition and preprints for the third.

This edition, like its earlier incarnations, owes a great deal to the suggestions and criticisms of the students in my developmental biology and developmental genetics classes. The extremely supportive staff and faculty of Swarthmore College have also played major roles in producing this book, and I believe that every member of the biology department has helped me when my chapters entered into their respective fields. Thanks also to science librarians E. Horikawa and M. Spencer for putting up with me and for keeping recent volumes from being sent to the bindery while I was writing the book. I would also like to thank Dr. Lauri Saxén and his laboratory at the University of Helsinki for their hospitality during this year and for their stimulating scientific discussions.

This is a third edition published by the same individuals who worked on the first two editions—a rare occurrence in contemporary publishing! Andy Sinauer has somehow managed to gather most of the same remarkable people around this project once again, and it has been a privilege to work with them. My thanks to him and to editor Carol Wigg, production coordinators Joe Vesely and Janice Holabird, artist John Woolsey, designer Rodelinde Albrecht, copy editor Chris Decker, and indexer Alina Lopo. Moreover, this book could never have been completed were it not for the encouragement of my wife, Anne Raunio, who, as an obstetrician, knows the joys of developmental biology, and who felt that I, too, could go through labor a third time, if only with a book. My thanks to you all.

SCOTT F. GILBERT

I

Patterns of development

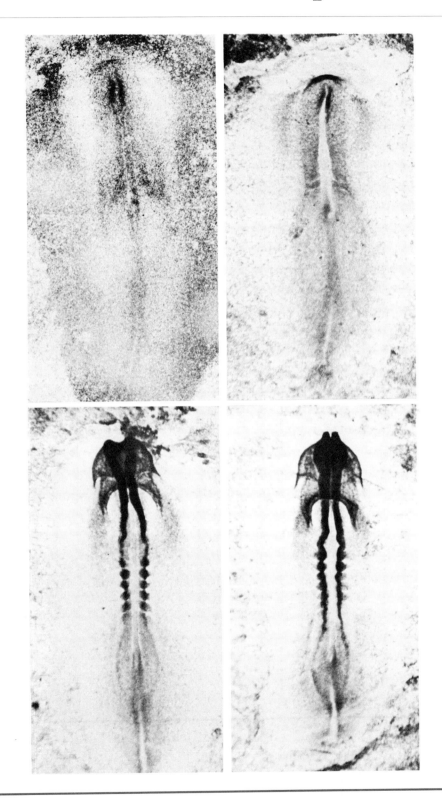

1

An introduction to animal development

I remember perfectly well the intense satisfaction and delight with which I had listened, by the hour, to Bach's fugues . . . and it has often occurred to me that the pleasure derived from musical compositions of this kind . . . is exactly the same as in most of my problems of morphology—that you have the theme in one of the old masters' works followed out in all its endless variations, always reappearing and always reminding you of the unity in variety.

—THOMAS HUXLEY (1895)

Happy is the person who is able to discern the causes of things.

—VIRGIL (37 B.C.)

According to Aristotle, the first major embryologist known to history, science begins with wonder. "It is owing to wonder that people began to philosophize, and wonder remains the beginning of knowledge." The development of animals has been a source of wonder throughout human history, and it has constantly stimulated individuals to seek the causes for such remarkable, yet commonplace, phenomena. The simple procedure of cracking open a chick egg on each successive day of its three-week incubation provides a remarkable experience—a thin band of cells is seen to give rise to an entire bird. Aristotle performed this experiment and noted the formation of the major organs. Anyone can wonder at this phenomenon, but it is the scientist who seeks to discover how development actually occurs. Rather than dissipating wonder, new understanding increases it.

Multicellular organisms on earth do not spring forth fully formed. Rather, they arise by a relatively slow process of progressive change that we call DEVELOPMENT. In nearly all cases, the development of a multicellular organism begins with a single cell—the fertilized egg, or ZYGOTE— which divides mitotically to produce all the cells of the body. The study of animal development has traditionally been called EMBRYOLOGY, referring to the fact that between the stages of the fertilized egg and birth, the developing organism is known as an EMBRYO. But development does not stop at birth, or even at adulthood. Most organisms never cease developing. Each day we replace over a gram of skin cells (the older cells being sloughed off as we move), and our bone marrow sustains the development of millions of new erythrocytes every minute of our lives. Therefore, in recent years it has become customary to speak of DEVELOPMENTAL BIOLOGY as the discipline that involves studies of embryonic and other developmental processes.

Developmental biology is one of the most exciting and fast-growing fields of biology. Part of its excitement comes from its subject matter, for we are just beginning to understand the molecular mechanisms of animal development. Another part of the excitement comes from the unifying role that developmental biology is beginning to assume in the biological sciences. Developmental biology is creating a framework that integrates molecular biology, physiology, cell biology, anatomy, cancer research, immunology, and even evolutionary and ecological studies. The study of development has become essential for understanding any other area of biology.

Principal features of development

Development accomplishes two major functions. It generates cellular diversity and order within each generation, and it assures the continuity of life from one generation to the next. The first function involves the production and organization of all the diverse types of cells in the body. A single cell, the fertilized egg, gives rise to muscle cells, skin cells, neurons, lymphocytes, blood cells, and all the other cell types. The generation of cellular diversity is called DIFFERENTIATION; the processes that organize the differentiated cells into tissues and organs are called MORPHOGENESIS (creation of form and structure) and GROWTH (increase in size). The second major function of development is REPRODUCTION: the continued generation of new individuals of the species.

The major features of animal development are illustrated in Figure 1. The life of a new individual is initiated by the fusion of genetic material from the two GAMETES—the sperm and the egg. This fusion, called FERTILIZATION, stimulates the egg to begin development. The subsequent sequence of stages is collectively called EMBRYOGENESIS. Throughout the animal kingdom an incredible variety of embryonic types exists, but most patterns of embryogenesis comprise variations on four themes.

1 Immediately following fertilization, cleavage occurs. CLEAVAGE is a series of extremely rapid mitotic divisions wherein the enormous volume of zygote cytoplasm is divided into numerous smaller cells. These cells are called BLASTOMERES, and by the end of cleavage they generally form a sphere known as a BLASTULA.
2 After the rate of mitotic division has slowed down, the blastomeres undergo dramatic movements wherein they change their positions relative to one another. This series of extensive cell rearrangements is called GASTRULATION. As a result of gastrulation, the typical embryo contains three cell regions, called GERM LAYERS.* The outer layer—the ECTODERM—produces the cells of

*From the Latin germen, meaning bud or sprout; the same root as germination. The names of the three germ layers are from the Greek: Ectoderm from ektos (outside) plus derma (skin); mesoderm from mesos (middle), and endoderm from endon (within).

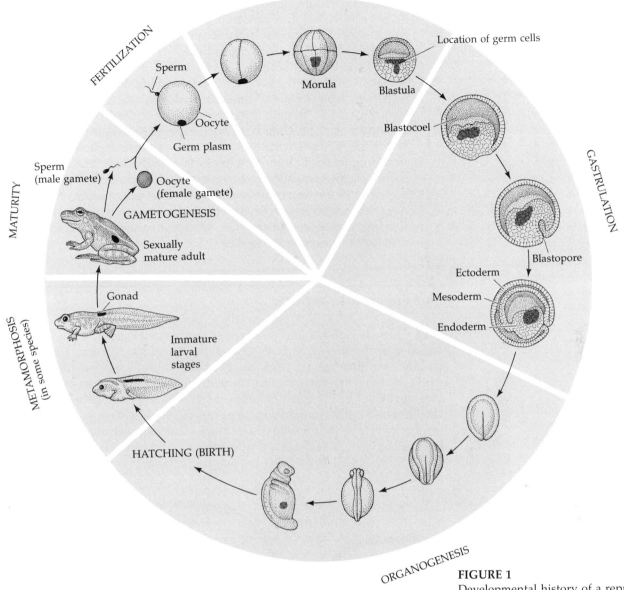

FERTILIZATION

Sperm

Oocyte

Germ plasm

Location of germ cells

Morula

Blastula

Blastocoel

GASTRULATION

Blastopore

Ectoderm

Mesoderm

Endoderm

Sperm
(male gamete)

MATURITY

Oocyte
(female gamete)

GAMETOGENESIS

Sexually
mature adult

Gonad

METAMORPHOSIS
(in some species)

Immature
larval
stages

HATCHING (BIRTH)

ORGANOGENESIS

FIGURE 1
Developmental history of a representative organism, the frog. The stages
from fertilization through hatching (or
birth) are known collectively as embryogenesis. The region of the embryo set
aside for producing germ cells is
shown in color. Gametogenesis, which
is complete in the sexually mature
adult, begins at different times during
development, depending upon the
species.

the epidermis and the nervous system; the inner layer—the
ENDODERM—produces the lining of the digestive tube and its associated organs (pancreas, liver, and so on); and the middle layer
—the MESODERM—gives rise to several organs (heart, kidney,
gonads), connective tissues (bone, muscles, tendons), and the
blood cells.

3 Once the three germ layers are established, the cells interact with
one another and rearrange themselves to produce the bodily organs. This process is called ORGANOGENESIS. (In vertebrates, organogenesis is initiated when a series of cellular interactions
causes the mid-dorsal ectodermal cells to form the neural tube.
This tube will become the brain and spinal cord.) Many organs
contain cells from more than one germ layer, and it is not unusual for the outside of an organ to be derived from one layer
and the inside from another. Also during organogenesis certain

cells undergo long migrations from their place of origin to their final location. These migrating cells include the precursors of blood cells, lymph cells, pigment cells, and gametes.

4 As seen in Figure 1, a portion of egg cytoplasm gives rise to the precursors of the gametes. These cells are called GERM CELLS, and they are set aside for their reproductive function. All the other cells of the body are called SOMATIC CELLS. This separation of somatic cells (which give rise to the individual body) and germ cells (which contribute to the formation of a new generation) is typically one of the first differentiations to occur during animal development. The germ cells eventually migrate to the gonads, where they differentiate into gametes—sex cells capable of participating in fertilization to create a new individual. The development of gametes, called GAMETOGENESIS, is usually not completed until the organism has become physically mature. At maturity, the gametes may be released and undergo fertilization to begin a new life. Meanwhile, the adult organism eventually undergoes senescence and dies.

Our eukaryotic heritage

Organisms are divided into two major groups, their classification depending on whether their cells possess a nuclear envelope. On one side of this classification line are the PROKARYOTES (Gk. *karyon*, meaning "nucleus"), which include the bacteria and the blue-green algae. These cells lack a true nucleus. On the other side are the EUKARYOTES, which include protists, animals, plants, and fungi. Cells of eukaryotes have a well-formed nuclear envelope surrounding their chromosomes. This fundamental difference between eukaryotes and prokaryotes influences how they arrange

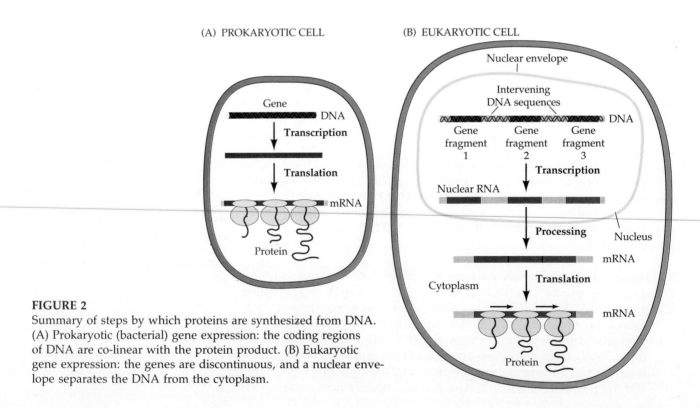

FIGURE 2
Summary of steps by which proteins are synthesized from DNA.
(A) Prokaryotic (bacterial) gene expression: the coding regions of DNA are co-linear with the protein product. (B) Eukaryotic gene expression: the genes are discontinuous, and a nuclear envelope separates the DNA from the cytoplasm.

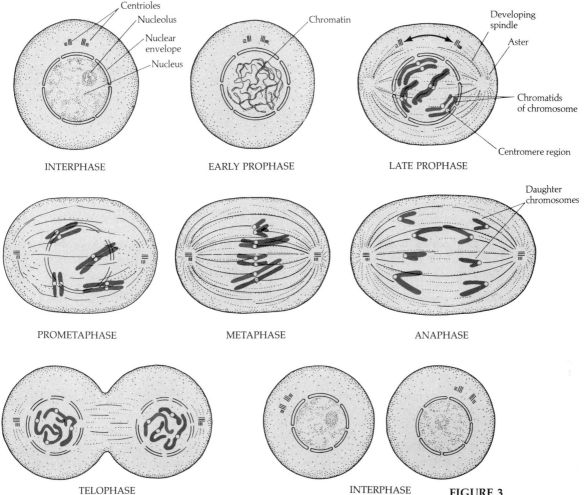

INTERPHASE   EARLY PROPHASE   LATE PROPHASE

PROMETAPHASE   METAPHASE   ANAPHASE

TELOPHASE   INTERPHASE

FIGURE 3
Diagrammatic representation of mitosis in animal cells. During interphase the DNA is doubled in preparation for cell division. During prophase the nuclear envelope breaks down and a spindle forms between the two centrioles. At metaphase the chromosomes align at the equator of the cell, and as anaphase begins the duplicated chromosomes (called chromatids) are separated. At telophase the chromosomes reach the mitotic poles and the cell begins to pinch in. At each pole are the same number and type of chromosomes as were present in the cell before it divided.

and utilize their genetic information. In both groups, the inherited information needed for development and metabolism is encoded in the deoxyribonucleic acid (DNA) sequences of the chromosomes. The prokaryotic chromosome is generally a small, circular, double helix of DNA, having approximately a million base pairs. The eukaryotic cell usually has several chromosomes, and the simplest eukaryotic protists have over ten times the amount of DNA found in the most complex prokaryotes. Moreover, the structure of a eukaryotic gene is more complex than that of prokaryotic genes. The amino acid sequence of a prokaryotic protein is a direct reflection of the DNA sequence in the chromosome. The protein-coding DNA of a eukaryotic gene, however, is usually divided up such that the complete amino acid sequence of a protein is derived from discontinuous segments of DNA (Figure 2). The intervening DNA often contains sequences that are involved with regulating the time and place that the gene is activated.

Eukaryotic chromosomes also are very different from prokaryotic chromosomes. Eukaryotic DNA is wrapped around specific proteins called HISTONES, which can organize the DNA into compact structures; in bacteria, there are no histones. Moreover, eukaryotic cells undergo mitosis wherein the nuclear envelope breaks down and the replicated chromosomes are equally divided between the daughter cells (Figure 3). In prokaryotes, cell division is not mitotic; no mitotic spindle develops, and there is no nuclear envelope to break down. Rather, the daughter chromosomes remain attached to adjacent points on the cell membrane. These attachment points are separated by the growth of the cell membrane between them, eventually placing the chromosomes into different daughter cells.

Prokaryotes and eukaryotes have different mechanisms of gene regulation. In both prokaryotes and eukaryotes, DNA is transcribed by enzymes called RNA polymerases to make RNA. When mRNA is produced in prokaryotes, it is immediately translated into a protein while the other end of it is still being transcribed from the DNA (Figure 4). Thus, transcription and translation are simultaneous and coordinated events in prokaryotes. But the existence of the nuclear envelope in eukaryotes provides the opportunity for an entirely new type of cell regulation. The ribosomes, which are responsible for translation, are on one side of the nuclear envelope, whereas the DNA and the RNA polymerases needed for transcription are on the other side. In between transcription and translation, the transcribed RNA must be processed so that it can pass through the nuclear envelope. By regulating which mRNAs can pass into the cytoplasm, the cell is able to select which of the newly synthesized messages will be translated. Thus, a new level of complexity has been added, one that we shall see is extremely important for the developing organism.

Development among the unicellular eukaryotes

All multicellular eukaryotic organisms have evolved from unicellular protists. It is in these protists that the basic features of development first appeared. Simple eukaryotes give us our first examples of the nucleus directing morphogenesis, the use of the cell surface to mediate cooperation between individual cells, and the first occurrences of sexual reproduction.

Control of developmental morphogenesis in *Acetabularia*

At the turn of the century, it had not yet been proved that the nucleus contained hereditary or developmental information. Some of the best evidence for this theory came from studies in which unicellular organisms were fragmented into nucleate and anucleate pieces (reviewed in Wilson,

FIGURE 4
Concurrent transcription and translation in prokaryotes. A portion of *Escherichia coli* DNA runs horizontally across this electron micrograph. Messenger RNA transcripts can be seen on both sides. Ribosomes have attached to the mRNA and are synthesizing proteins (which cannot be seen). The mRNA can be seen increasing in length from left to right, indicating the direction of transcription. (Courtesy of O. L. Miller, Jr.)

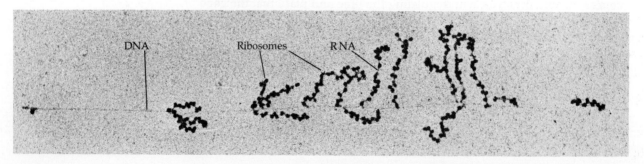

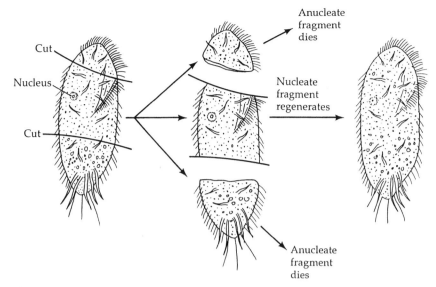

FIGURE 5
Regeneration of the nucleate fragment of the unicellular protist *Stylonychia*. The anucleate fragments survive for a time, but finally die.

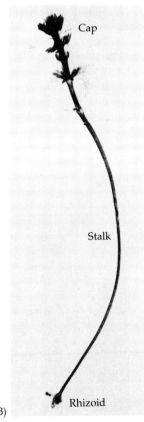

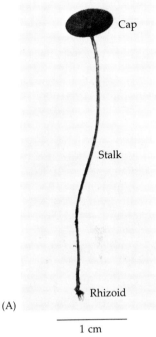

(A)

1 cm

(B)

1896). When various protozoans were cut into numerous fragments, nearly all the pieces died. However, the fragments containing nuclei were able to live and to regenerate entire, complex, cellular structures (Figure 5).

Nuclear control of cell morphogenesis and the interaction of nucleus and cytoplasm are beautifully demonstrated by studies of *Acetabularia*. This enormous single cell (1–2 inches) consists of three parts: a cap, a stalk, and a rhizoid (Figure 6). The rhizoid is located at the base of the cell and holds it to the substrate. The single nucleus of the cell resides within the rhizoid. The size of *Acetabularia* and the location of its nucleus allow investigators to remove the nucleus from one cell and replace it with a nucleus from another cell. In the 1930s, J. Hämmerling took advantage of these unique features and transferred nuclei from one species—*A. mediterranea*—into the enucleated rhizoid of another species—*A. crenulata*. As Figure 6 shows, these two species have very different cap structures. Hämmerling found that when the nucleus from one species was transplanted into the stalk of another species, the newly formed cap eventually assumed the form associated with the donor nucleus (Figure 7). Thus, the nucleus was seen to control *Acetabularia* development.

The formation of a cap is a complex morphogenic event involving the synthesis of numerous proteins, the products of which must be placed in a certain portion of the cell. The transplanted nucleus does indeed direct the synthesis of its species-specific cap, but it takes several weeks to do so. Moreover, if the original nucleus is removed at an early time in *Acetabularia* development, before the cap is formed, a normal cap is formed weeks later, even though the organism will eventually die. This suggests that (1) the nucleus contains information specifying the type of cap pro-

FIGURE 6
Two species of the giant unicellular protist *Acetabularia*. (A) *Acetabularia mediterranea* cell. (B) *Acetabularia crenulata* cell. The rhizoid contains the nucleus. (Courtesy of H. Harris.)

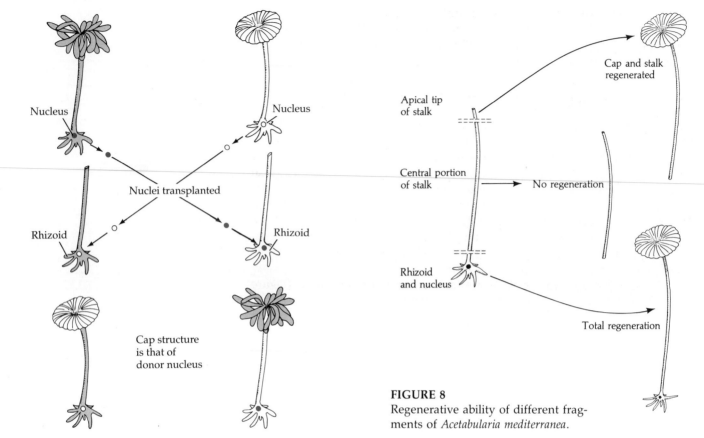

FIGURE 8
Regenerative ability of different fragments of *Acetabularia mediterranea*.

FIGURE 7
Effect of exchanging nuclei between two species of *Acetabularia*; nuclei were transplanted into enucleated rhizoid fragments. *A. crenulata* structures are shown in color; *A. mediterranea* structures are white.

duced (i.e., it contains the genetic information that specifies the proteins responsible for the production of a certain type of cap), and (2) the information enters the cytoplasm long before cap production occurs. The information in the cytoplasm is not used for several weeks.

One hypothesis proposed to explain these observations was that the nucleus synthesizes a stable mRNA, that lies dormant in the cytoplasm until the time of cap formation. More recent evidence based on an observation that Hämmerling published in 1934 supports this hypothesis. Hämmerling fractionated young *Acetabularia* into several parts (Figure 8). The portion with the nucleus formed a cap as expected; so did the apical tip of the stalk. The remaining portion of the stalk did not form a cap. Thus, Hämmerling postulated (nearly 30 years before the existence of mRNA was known) that the instructions for the cap formation were somehow stored at the tip. More recent studies (Kloppstech and Schweiger, 1975) have shown that nucleus-derived mRNA does accumulate at this region. Ribonuclease, an enzyme that cleaves RNA, completely inhibits cap formation when added to the seawater in which *Acetabularia* is growing. In enucleated cells, this effect is permanent; once the RNA is destroyed, no cap formation can occur. In nucleated cells, however, the cap can form after the ribonuclease is washed away—presumably because new mRNA is being made by the nucleus. Garcia and Dazy (1986) have also shown that protein synthesis is especially active in the apex of these cells.

It is obvious from the preceding discussion that nuclear transcription plays an important role in the formation of the *Acetabularia* cap. But note that the cytoplasm also plays an essential role in cap formation. The mRNAs are not translated for weeks, even though they are in the cytoplasm. Something in the cytoplasm controls when the message is utilized.

Hence, the expression of the cap is controlled, not only by nuclear transcription, but also by cytoplasmic factors influencing translation. In this unicellular organism, "development" is controlled at both the transcriptional and the translational levels.

Differentiation in the amoeboflagellate *Naegleria*

One of the most remarkable cases of protist "differentiation" is that of *Naegleria gruberi*. This organism occupies a special place in protist taxonomy because it can change its form from that of an amoeba to that of a flagellate (Figure 9). During most of its life cycle, *N. gruberi* is a typical amoeba, feeding on soil bacteria and dividing by fission. However, when the bacteria are diluted (either by rainwater or by water added by an experimenter), each *N. gruberi* rapidly develops a streamlined body shape and two long anterior flagella. Thus, instead of having several differentiated cell types in one organism, this one cell has different cell structures and biochemistry at different times of its life.

Differentiation into the flagellate form occurs within an hour (Figure 10). During this time the amoeba has to create centrioles to serve as the basal bodies (microtubule organizing centers) of the flagella, as well as to create the flagella themselves. Flagella are made from many proteins, the most abundant of which is TUBULIN. The tubulin molecules are organized

FIGURE 9
Transformation from amoeboid to flagellated state in *Naegleria gruberi*. Top row stained with Lugol's iodine. Bottom row stained with fluorescent antibody to the tubulin protein of the microtubules. (A) 0 minutes. (B) 25 minutes, showing new tubulin synthesis. (C) 70 minutes, showing emergence of visible flagella. (D) 120 minutes, showing mature flagella and streamlined body shape. (From Walsh, 1984; courtesy of C. Walsh.)

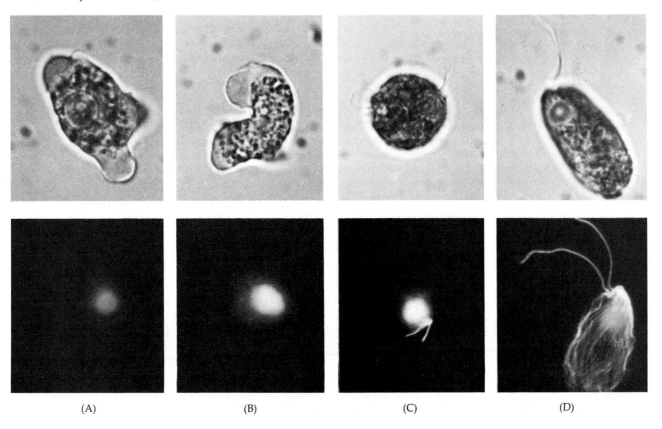

(A) (B) (C) (D)

FIGURE 10

Differentiation of the flagellate phenotype in *Naegleria*. Amoebae that had been growing in the bacteria-enriched medium are washed free of bacteria at time zero. By 80 minutes, nearly the entire population has developed flagella. (After Fulton, 1977.)

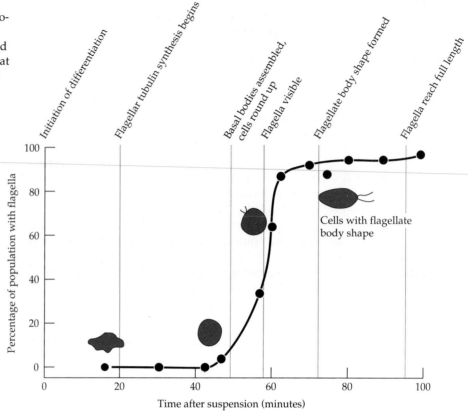

into microtubules, and the microtubules are further organized into an arrangement that will permit flagellar movement. Fulton and Walsh (1980) showed that the tubulin for the *Naegleria* flagella does not exist in the amoeba stage. It is made *de novo* ("from scratch"), starting with new transcription from the nucleus. To show this, the investigators manipulated transcription at various stages with actinomycin D, an antibiotic drug that selectively inhibits RNA synthesis. When added before the dilution of the food supply, this antibiotic prevents tubulin synthesis. However, if the actinomycin D is added 20 minutes after dilution, tubulin is still made at the normal time (about 30 minutes later). Therefore, it appears that the mRNA for tubulin is made during the first 20 minutes after dilution and is used shortly thereafter. This interpretation was confirmed when it was shown that mRNA extracted from amoebae does not contain any detectable messages for flagellar tubulin, whereas mRNA extracted from differentiating cells contains a great many such messages (Walsh, 1984).

Here, then, is an excellent example of the TRANSCRIPTIONAL CONTROL of a development process: the *Naegleria* nucleus responds to environmental changes by synthesizing the mRNA for flagellar tubulin. We also see another process that remains extremely important in the development of all other animals and plants, namely, the assembly of tubulin molecules to produce a flagellum. This arrangement, whereby tubulin is polymerized into microtubules and the microtubules assembled into an ordered array, is seen throughout nature. In mammals it is evident in the sperm flagellum and in the cilia of the spinal cord and respiratory tract. This POSTTRANSLATIONAL CONTROL, whereby proteins interact with one another to generate organized structures within the cell, will be discussed more fully later. We see, then, that development in unicellular eukaryotes can be controlled at the transcriptional, translational, and posttranslational levels.

FIGURE 11

Sex in bacteria. Some bacterial cells are covered with numerous appendages (pili) and are capable of transmitting genes to a recipient cell (lacking pili) through a sex pilus. In this figure, the sex pilus is highlighted by viral particles that bind specifically to that structure. (Photograph courtesy of C. C. Brinton, Jr. and J. Carnahan.)

The origins of sexual reproduction

Sexual reproduction is one other invention of the protists that has had a profound effect on more complex organisms. It should be noted that sex and reproduction are two distinct and separable processes. *Reproduction* involves the creation of new individuals. *Sex* involves the combining of genes from two different individuals into new arrangements. Reproduction in the absence of sex is characteristic of those organisms that reproduce by fission; there is no sorting of genes when an amoeba divides or when a hydra buds off cells to form a new colony. Sex without reproduction is also common among unicellular organisms. Bacteria are able to transmit genes from one individual to another by means of sex pili (Figure 11). This transmission is separate from reproduction. Protists are also able to reassort genes without reproduction. Paramecia, for instance, reproduce by fission, but sex is accomplished by CONJUGATION (Figure 12). When two paramecia join together, they link their oral apparatuses and form a cytoplasmic connection. Each macronucleus (which controls the metabo-

FIGURE 12

Conjugation in paramecia. Two paramecia can exchange genetic material, leaving each with genes that differ from those with which they started. (After Strickberger, 1985.)

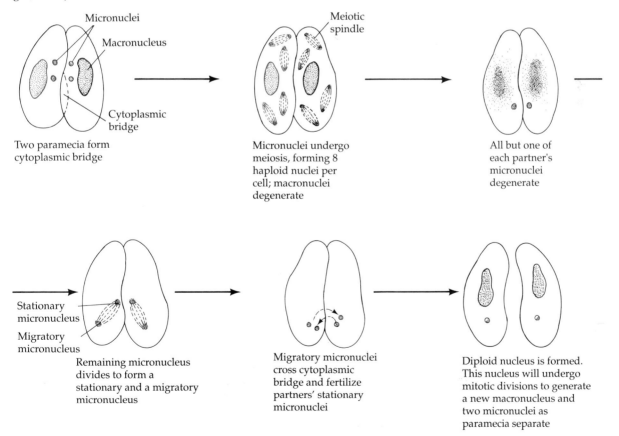

Micronuclei
Macronucleus
Cytoplasmic bridge

Two paramecia form cytoplasmic bridge

Meiotic spindle

Micronuclei undergo meiosis, forming 8 haploid nuclei per cell; macronuclei degenerate

All but one of each partner's micronuclei degenerate

Stationary micronucleus
Migratory micronucleus

Remaining micronucleus divides to form a stationary and a migratory micronucleus

Migratory micronuclei cross cytoplasmic bridge and fertilize partners' stationary micronuclei

Diploid nucleus is formed. This nucleus will undergo mitotic divisions to generate a new macronucleus and two micronuclei as paramecia separate

lism of the organism) degenerates while each micronucleus undergoes meiosis followed by mitosis to produce eight haploid micronuclei, of which all but one degenerate. The remaining micronucleus divides once more to form a stationary micronucleus and a migratory micronucleus. Each migratory micronucleus crosses the cytoplasmic bridge and fuses with ("fertilizes") the stationary micronucleus, thereby creating a new diploid nucleus in each cell. This diploid nucleus then divides to give rise to a new micronucleus and a new macronucleus as the two partners disengage. No reproduction has occurred, only sex.

The union of these two distinct processes, sex and reproduction, into SEXUAL REPRODUCTION, is seen in unicellular eukaryotes. Figure 13 shows the life cycle of *Chlamydomonas*. This organism is usually haploid, having just one copy of each chromosome (like a mammalian gamete). The individuals of each species, however, are divided into two mating groups: *plus* and *minus*. When these meet, they join their cytoplasms together and their nuclei fuse to form a diploid zygote. This zygote is the only diploid cell in the life cycle, and it eventually undergoes meiosis to form four new *Chlamydomonas* cells. Here is sexual reproduction, for chromosomes are

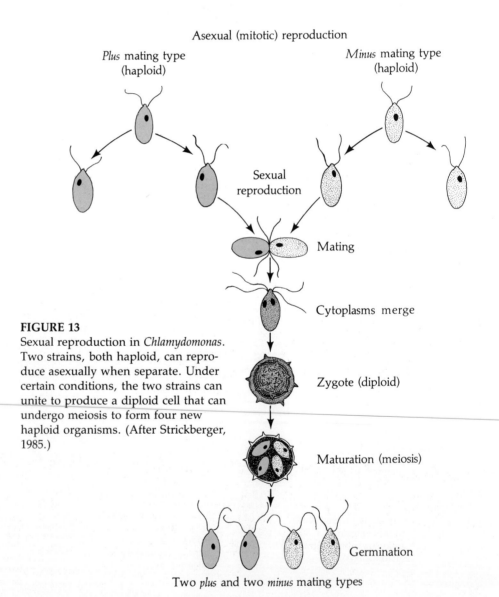

FIGURE 13
Sexual reproduction in *Chlamydomonas*. Two strains, both haploid, can reproduce asexually when separate. Under certain conditions, the two strains can unite to produce a diploid cell that can undergo meiosis to form four new haploid organisms. (After Strickberger, 1985.)

Asexual (mitotic) reproduction

Plus mating type (haploid)

Minus mating type (haploid)

Sexual reproduction

Mating

Cytoplasms merge

Zygote (diploid)

Maturation (meiosis)

Germination

Two *plus* and two *minus* mating types

reassorted during the meiotic divisions and more individuals are formed. Note that in this protist type of sexual reproduction, the gametes are morphologically identical—the distinction between sperm and egg has not yet been made.

In evolving sexual reproduction, two important advances had to be achieved. The first is the mechanism of MEIOSIS (Figure 14), whereby the diploid complement of chromosomes is reduced to the haploid state. (This will be discussed in detail in Chapter 22.) The other advance is the mechanism whereby the two different mating types recognize each other. Recognition occurs first on the flagellar membranes (Figure 15; Goodenough and Weiss, 1975; Bergman et al., 1975). The agglutination of flagella enables specific regions of the cell membranes to come together. These specialized sectors contain mating type-specific components that enable the cytoplasms to fuse. Following agglutination, the *plus* individuals initiate the fusion by extending a "fertilization tube." This tube contacts and fuses with a specific site on the *minus* individual. Interestingly, the mechanism used to extend this tube—the polymerization of the protein actin—is also used to extend processes of sea urchin eggs and sperm. In the next chapter, we shall see that the recognition and fusion of sperm and egg occur in a manner amazingly similar to that of these protists.

Unicellular eukaryotes appear to have the basic elements of the developmental processes that characterize the more complex organisms of

FIGURE 14
Summary of meiosis. The DNA and associated proteins replicate during interphase. During prophase the nuclear envelope breaks down and homologous chromosomes (each chromosome being double, with the chromatids joined at the centromere) align in pairs. Chromosomal rearrangements can occur between the four homologous chromatids at this time. After the first metaphase, the two original homologous chromosomes are segregated into different cells. During the second division the centromere splits, thereby leaving each new cell with one copy of each chromosome.

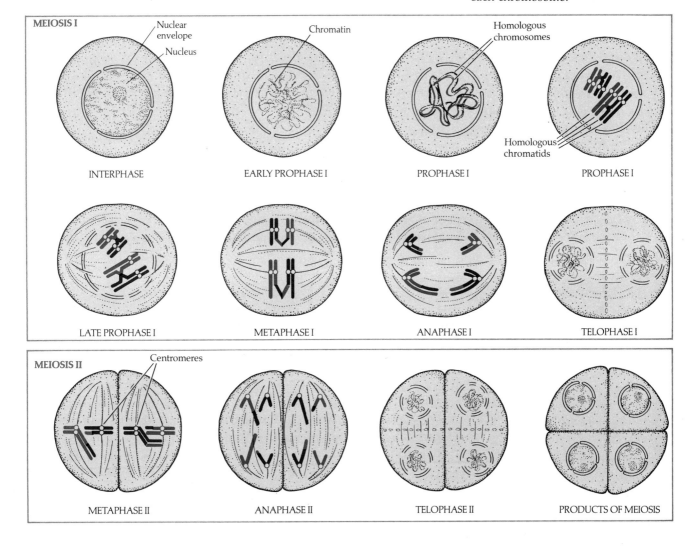

FIGURE 15

Two-step recognition in mating *Chlamydomonas*. (A) Scanning electron micrograph (×7000) of mating pair. The interacting flagella twist about each other, adhering at the tips (arrows). (B) Transmission electron micrograph (×20,000) of a cytoplasmic bridge connecting the two organisms. The microfilaments extend from the donor (lower) cell to the recipient (upper) cell. (From Goodenough and Weiss, 1975; and Bergman et al., 1975; by permission of U. Goodenough.)

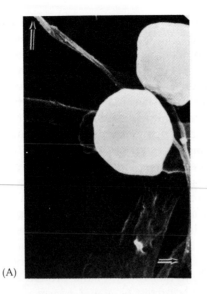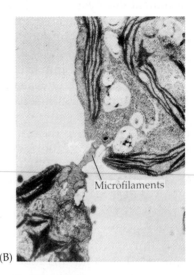

(A) (B)

Microfilaments

other phyla: cellular synthesis is controlled by transcriptional, translational, and posttranslational regulation; a mechanism for processing RNA through the nuclear membrane exists; the structures of individual genes and chromosomes are as they will be throughout eukaryotic evolution; mitosis and meiosis are perfected; and sexual reproduction exists, involving cooperation between individual cells. Such intercellular cooperation becomes even more important with the evolution of multicellular organisms.

Colonial eukaryotes: The evolution of differentiation

One of evolution's most important experiments was the creation of multicellular organisms. There appear to be several paths by which single cells evolved multicellular arrangements; we will discuss only two of them.* The first path involves the orderly division of the reproductive cell and the subsequent differentiation of its progeny into different cell types. This path to multicellularity can be seen in a remarkable series of multicellular organisms collectively referred to as the family Volvocaceae, or the "Volvocaceans."

The Volvocaceans

The simpler organisms among the Volvocaceans are collections of numerous *Chlamydomonas*-like cells; but the more advanced members of this group have developed a second, very different, cell type. A single organism of the Volvocacean genus *Oltmannsiella* contains four *Chlamydomonas*-like cells in a row, embedded in a gelatinous matrix. In the genus *Gonium* (Figure 16), a single cell divides to produce a flat plate of 4 to 16 cells, each with its own flagellum. In a related genus—*Pandorina*—the 16 cells form a sphere; and in *Eudorina*, the sphere contains 32 or 64 cells arranged in a regular pattern. In these organisms, then, a very important developmental principle has been worked out: the ordered division of one cell to generate a number of cells that are organized in a predictable fashion. Like

*A fuller discussion of this transition from unicellularity to multicellularity will be found in Chapter 23.

most animal embryos, the cell divisions by which a single Volvocacean cell divides to produce an organism of 4 to 64 cells occur in very rapid sequence and in the absence of cell growth.

The next two genera of the Volvocacean series exhibit another important principle of development: the differentiation of cell types within an individual organism. The reproductive cells become differentiated from the somatic cells. In all the Volvocacean genera mentioned above, every cell can, and normally does, produce a complete new organism by mitosis (Figure 17A,B). In the genera *Pleodorina* and *Volvox*, however, relatively few cells can reproduce. In *Pleodorina californica*, the cells in the anterior region are restricted to a somatic function. Only those cells on the posterior side can reproduce. In *P. californica*, a colony usually has 128 or 64 cells, and the ratio of the number of somatic cells to the number of reproductive

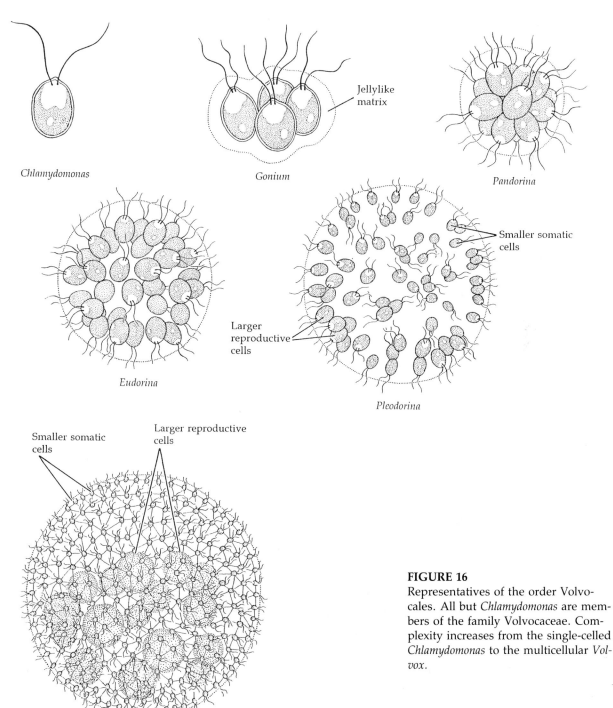

Chlamydomonas

Jellylike matrix

Gonium

Pandorina

Eudorina

Smaller somatic cells

Larger reproductive cells

Pleodorina

Smaller somatic cells

Larger reproductive cells

FIGURE 16
Representatives of the order Volvocales. All but *Chlamydomonas* are members of the family Volvocaceae. Complexity increases from the single-celled *Chlamydomonas* to the multicellular *Volvox*.

cells is usually 3:5. Thus, a 128-cell colony has 48 somatic cells, and a 64-cell colony has 24.

In *Volvox*, almost all the cells are somatic, and very few of the cells are able to produce new individuals. In some species of *Volvox*, reproductive cells, as in *Pleodorina*, are derived from cells that originally look and function like somatic cells before they enlarge and divide to form new progeny. However, in other members of the genus, such as *V. carteri*, there is a complete division of labor: the reproductive cells that will create the next generation are set aside during the division of the reproductive cells that are forming a new individual. The reproductive cells never develop functional flagella and never contribute to motility or other somatic functions of the individual; they are entirely specialized for reproduction. Thus, although the simpler Volvocaceans may be thought of as colonial organisms (because each cell is capable of independent existence and of perpetuating the species), in *V. carteri* we have a truly multicellular organism with two distinct and interdependent cell types (somatic and reproductive), both of which are required for perpetuation of the species (Figure 17C). Although not all animals set aside the reproductive cells from the somatic cells (and plants hardly ever do), this separation of germ (reproductive) cells from somatic cells early in development is characteristic of many animal phyla and will be discussed in more detail in Chapter 7.

Although all of the Volvocaceans, like their unicellular relative *Chlamydomonas*, reproduce predominantly by asexual means (as described earlier) they are also capable of sexual reproduction. This involves the production and fusion of haploid gametes. In many species of *Chlamydomonas*, including the one illustrated in Figure 13, sexual reproduction is ISOGAMOUS, since the haploid gametes that meet are similar in size, structure, and motility. However, in other species of *Chlamydomonas*—as well as many species of colonial Volvocaceans—swimming gametes of very different sizes are produced by the different mating types. This is called HETEROGAMY. But the larger Volvocaceans have evolved a specialized form of heterogamy, called OOGAMY. This involves the production of large, relatively immotile eggs by one mating type and small, motile sperm by the other (see Sidelights and Speculations). Thus, the Volvocaceans include the simplest organisms that have distinguishable male and female members of the species and that have distinct developmental pathways for the production of eggs or sperm. In all the Volvocaceans the fertilization reaction resembles that of *Chlamydomonas* in that it results in the production of a dormant diploid zygote that is capable of surviving harsh environmental conditions. When conditions allow the zygote to germinate, it first undergoes meiosis to produce haploid offspring of the two different mating types in equal numbers.

FIGURE 17
Asexual reproduction in Volvocaceans. (A) Mature colony of *Eudorina elegans*. (B) Each of the *E. elegans* cells divides and produces a new colony. (C) Mature *Volvox carteri*. Most of the cells are incapable of reproduction. Germ cells (gonidia) have begun dividing into new organisms. (A and B after Hartmann, 1921; C from Kirk et al., 1982, courtesy of D. Kirk.)

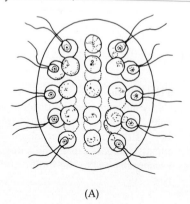

(A)

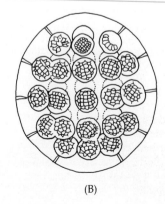

(B)

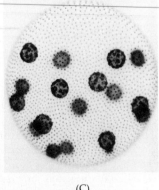

(C)

Sex and individuality in Volvox

Death and differentiation

Simple as it is, *Volvox* shares many features that characterize the life cycles and developmental histories of much more complex organisms, including ourselves. As already mentioned, *Volvox* is among the simplest organisms to exhibit a division of labor between two completely different cell types. And, as a consequence of this, it is among the simplest organisms to include death as a regular, genetically programmed, part of its life history.

Unicellular organisms that reproduce by simple cell division, such as amoebae, are potentially immortal. The amoeba that you see today under the microscope has no dead ancestors! When an amoeba divides, neither of the two resulting cells can be considered either as ancestor or offspring; they are siblings. Death comes to an amoeba only if it is eaten or meets with a fatal accident; and when it does, the dead cell leaves no offspring.

Death becomes an essential part of life, however, for any multicellular organism that establishes a division of labor between somatic (body) cells and germ (reproductive) cells. Consider the life history of *Volvox carteri* when it is

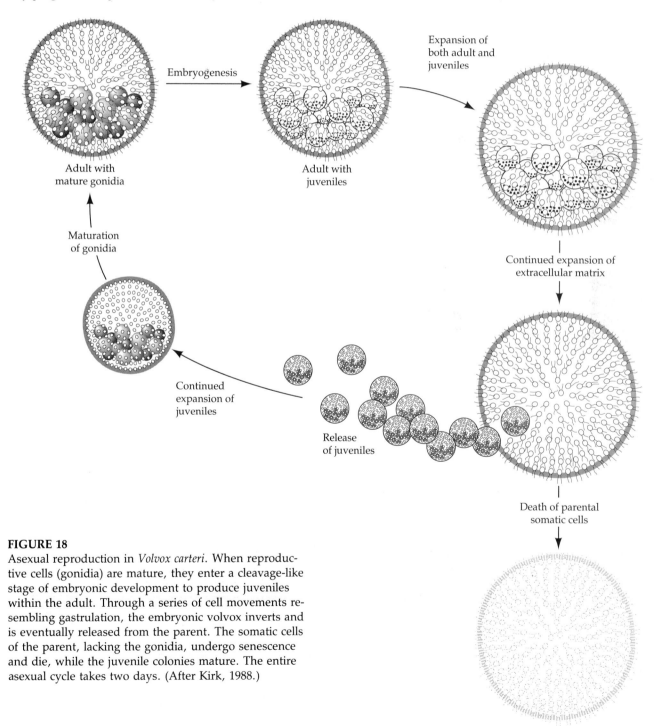

FIGURE 18

Asexual reproduction in *Volvox carteri*. When reproductive cells (gonidia) are mature, they enter a cleavage-like stage of embryonic development to produce juveniles within the adult. Through a series of cell movements resembling gastrulation, the embryonic volvox inverts and is eventually released from the parent. The somatic cells of the parent, lacking the gonidia, undergo senescence and die, while the juvenile colonies mature. The entire asexual cycle takes two days. (After Kirk, 1988.)

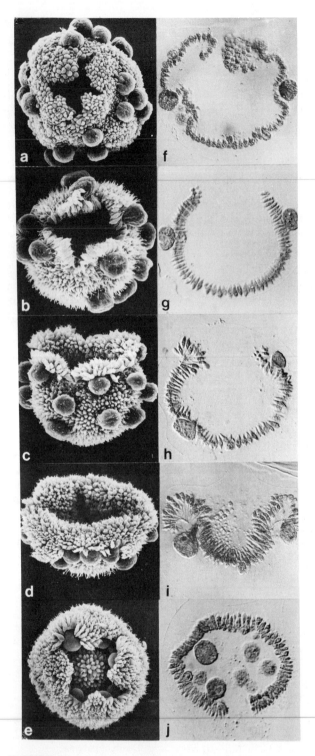

reproducing asexually (Figure 18). Each asexual adult is a spheroid containing some 2000 small biflagellated somatic cells along its periphery, and about 16 large asexual reproductive cells, called GONIDIA, toward one end of the interior. When mature, each gonidium divides rapidly 11 or 12 times. Certain of these divisions are asymmetric, and produce the 16 large cells that will become a new set of gonidia. At the end of cleavage, all of the cells that will be present in an adult have been produced from each gonidium. But the embryo is "inside-out": its gonidia are on the outside and the flagella of its somatic cells are pointing toward the interior of the hollow sphere of cells. This predicament is corrected by a process called "inversion" in which the embryo turns itself right-side out by a set of cell movements that somewhat resemble the gastrulation movements of animal embryos (Figure 19). Clusters of bottle-shaped cells open a hole at one end of the embryo by producing tension on the interconnected cell sheet (Figure 20). The embryo inverts through this hole and then closes it up. About a day after this is done, the juvenile colonies are enzymatically released from the parent and swim away.

What happens to the somatic cells of the "parent" *Volvox* now that its young have "left home"? Having produced offspring and being incapable of further reproduction, these somatic cells die. Actually, they commit suicide, synthesizing a set of proteins that cause the death and dissolution of the cells that make these proteins (Pommerville and Kochert, 1982). Moreover, in this death, the cells release for the use of others—including their own offspring—all the nutrients that they had stored during life. "Thus emerges," notes David Kirk, "one of the great

FIGURE 19
Inversion of asexually produced embryos of *Volvox carteri*. A–E are scanning electron micrographs of whole embryos. F–J are sagittal sections through the center of the embryo and are visualized by differential interference microscopy. Before inversion, the embryo is a hollow sphere of connected cells. When cells change their shape, a hole, the phialopore, opens at the apex of the embryo (A, B, F, G). Cells then curl around and rejoin at the bottom (C–E, H–J). (From Kirk et al., 1982; photographs courtesy of D. Kirk.)

FIGURE 20
"Bottle cells" near the opening of the phialopore. These cells remain tightly connected through cytoplasmic bridges near their elongated apexes, thereby creating the tension that causes the curvature of the interconnected cell sheet. (From Kirk et al., 1982; photograph courtesy of D. Kirk.)

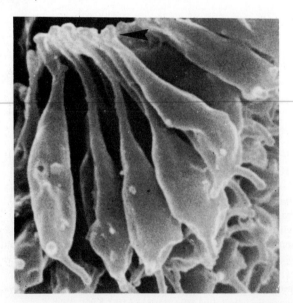

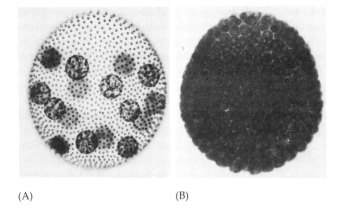

(A)   (B)

FIGURE 21
Mutation of a single gene (called *somatic regulator A*) abolishes programmed cell death in *V. carteri*. The newly hatched volvox carrying this mutation (A) is indistinguishable from the wild-type spheroid. However, shortly before the time when the somatic cells of the wild-type spheroids begin to die, the somatic cells of this mutant redifferentiate as gonidia (B). Eventually, every cell of the mutant will divide to form a new spheroid that will repeat this potentially immortal developmental cycle.

themes of life on planet Earth: 'Some die that others may live'."

A specific gene has been identified in *Volvox carteri* that plays a central role in regulating cell death (Kirk, 1988). In laboratory strains possessing mutations of this gene, somatic cells abandon their suicidal ways, gain the ability to reproduce asexually, and become potentially immortal (Figure 21). The fact that such mutants are never found in nature, however, indicates that cell death most likely plays an important role in the survival of *V. carteri* under natural conditions.

Enter sex

Although *V. carteri* reproduces asexually much of the time, in nature it reproduces sexually once each year. When it does, one generation of individuals passes away, and a new and genetically different generation is produced. The naturalist Joseph Wood Krutch (1956) put it more poetically.

> The amoeba and the paramecium are potentially immortal. . . . But for Volvox, death seems to be as inevitable as it is in a mouse or in a man. Volvox must die as Leeuwenhoek saw it die because it had children and is no longer needed. When its time comes it drops quietly to the bottom and joins its ancestors. As Hegner, the Johns Hopkins zoologist, once

wrote, "This is the first advent of inevitable natural death in the animal kingdom and all for the sake of sex." And he asked: "Is it worth it?"

For *Volvox carteri*, it most assuredly is worth it. *V. carteri* lives in shallow temporary ponds that fill with spring rains, but that dry out in the heat of late summer. During most of that time, *V. carteri* swims about, reproducing asexually. But these asexual volvoxes would die in minutes once the pond dried out. *V. carteri* is able to survive by turning sexual shortly before the pond dries up, thereby producing dormant zygotes that survive the heat and

FIGURE 22
Sexual reproduction in *Volvox carteri*. Males and females are indistinguishable in their asexual phase. When the sexual inducer protein is present, the gonidia of both mating types undergo a modified embryogenesis that leads to the formation of eggs in the females and sperm in the males. When the gametes are mature, sperm packets (containing 64 or 128 sperm each) are released and swim to the females. Upon reaching the female, the sperm packet breaks up into individual sperm, which can fertilize the eggs. The resulting zygote has tough cell walls that can resist drying, heat, and cold. When spring rains cause the zygote to germinate, it undergoes meiosis to produce haploid males and females that reproduce asexually until heat induces the sexual cycle again.

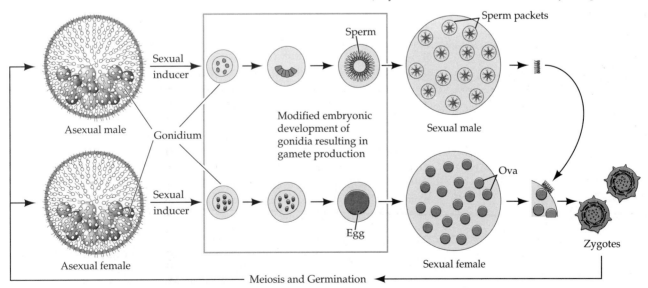

drought of late summer and the cold of winter. When rain fills the ponds in spring, the zygotes break their dormancy and hatch out a new generation of individuals to reproduce asexually until the pond is about to dry up once more. How do these simple organisms predict the coming of adverse conditions with sufficient accuracy to produce a sexual generation just in time, year after year?

The stimulus for switching from the asexual to the sexual mode of reproduction in *V. carteri* is known to be a 30,000-Da SEXUAL INDUCER protein. This protein is so powerful that concentrations as low as 6×10^{-17} M cause gonidia to undergo a modified pattern of embryonic development that results in the production of eggs or sperm, depending on the genetic sex of the individual. The sperm are released and swim to a female where they fertilize eggs to produce the dormant zygotes (Figure 22).

What, then, is the source of this sexual inducer protein? A few years ago, Kirk and Kirk (1986) discovered that the sexual cycle could be initiated by heating the dishes of *Volvox* to temperatures that might be expected in a shallow pond in late summer. When this was done, the somatic cells of the asexual volvoxes produced the sexual inducer protein. Since the amount of sexual inducer protein secreted by one individual is sufficient to initiate sexual development in over 500,000,000 asexual volvoxes, a single inducing volvox can convert the entire pond to sexuality. This discovery explained an observation made nearly 80 years ago that "in the full blaze of Nebraska sunlight, *Volvox* is able to appear, multiply, and riot in sexual reproduction in pools of rainwater of scarcely a fortnight's duration" (Powers, 1908). Thus, in temporary ponds formed by spring rains and dried up by summer's heat, *Volvox* has found a means of survival: it uses the heat to induce the formation of sexual individuals, whose mating produces zygotes capable of surviving conditions that kill the adult organism. Here, too, we see that development is critically linked to the ecosystem wherein the organism has adapted to survive.

Differentiation and morphogenesis in *Dictyostelium*

The life cycle of **Dictyostelium.** Another type of multicellular organization derived from unicellular organisms is found in *Dictyostelium discoideum*.* The life cycle of this fascinating and unusual organism is illustrated in Figure 23. In its vegetative cycle, solitary haploid amoebae (called myxamoebae to distinguish them from the true amoeba species) live on decaying logs, eating bacteria and reproducing by binary fission. When they have exhausted their food supply, tens of thousands of these amoebae join together to form moving streams of cells that converge at a central point. Here they form a conical mound, which eventually absorbs all the streaming cells and bends over to produce the migrating slug. The slug (often given the more dignified titles of pseudoplasmodium or grex) is usually 2–4 mm long and is encased in a slimy sheath. The grex begins migration (if the environment is dark and moist) with its anterior tip slightly raised. When migration ceases, these anterior cells, representing 15–20 percent of the entire cellular population, form a tubed stalk. The stalk begins as some of the central anterior cells begin secreting an extracellular coat and form a hollow tube. More prestalk cells enter the tube, and as they exit they become part of the bottom of the tube. As the prestalk cells differentiate, they form vacuoles and enlarge, lifting the mass of prespore cells (Williams, 1991). The stalk cells die, but the posterior cells, elevated above the stalk, become spore cells. These spore cells disperse, each one becoming a new myxamoeba. In addition to this asexual cycle, there is a possibility for sex in *Dictyostelium*. Two cells can fuse to create a giant cell, which digests all the other cells of the aggregate. When it has eaten all its neighbors, it encysts itself in a thick wall and undergoes meiotic and mitotic divisions; eventually new myxamoebae are liberated.

Dictyostelium has been a wonderful experimental organism for developmental biologists, for it is an organism wherein initially identical cells are differentiated into one of two alternative cell types, spore or stalk. It

---

*This organism is colloquially called a "cellular slime mold." This is a misnomer, since "*Dictyostelium* is neither slimy nor a mold and is as unrelated to *Physarum* (the 'true' slime mold) as it is to *Acanthamoeba* or *Neurospora*. It is certainly not a filamentous fungi and is just an amoebal organism." (W. Loomis, personal communication.)

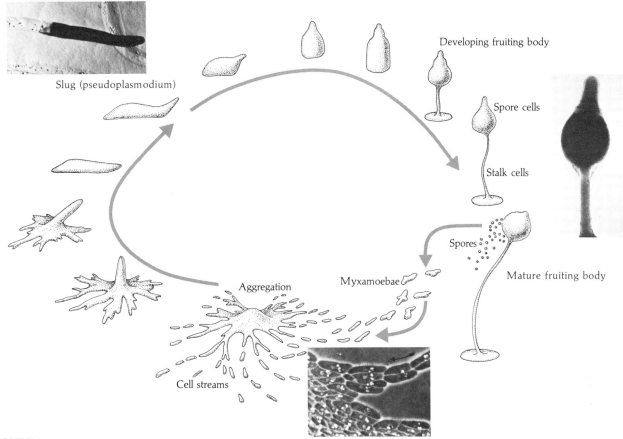

FIGURE 23

Life history of *Dictyostelium discoideum*. Haploid spores give rise to myxamoebae, which can reproduce asexually to form more haploid myxamoebae. As the food supply diminishes, aggregation occurs at central points and a migrating pseudo-plasmodium is formed. Eventually it stops moving and forms a fruiting body that releases more spores. (Photographs courtesy of K. Raper.)

is also an organism wherein individual cells come together to form a cohesive structure composed of differentiated cell types, akin to tissue formation in more complex organisms. The aggregation of thousands of amoebae into a single organism is an incredible feat of organization, and invites experimentation on the mechanisms involved.

Aggregation of **Dictyostelium** *cells.* The first question is, What causes these cells to aggregate? Time-lapse microcinematography has shown that no directed movement occurs during the first 4–5 hours following nutrient starvation. During the next 5 hours, however, the cells are seen moving at about 20 micrometers per minute for 100 seconds. This movement ceases for about 4 minutes, then resumes. Although the movement is directed toward a central point, it is not a simple radial movement. Rather, cells join with each other to form streams; the streams converge into larger streams; and eventually all streams merge at the center. Bonner (1947) and Shaffer (1953) showed that this movement was due to CHEMOTAXIS; that is, the cells were guided to aggregation centers by a soluble substance. This substance was later identified as cyclic adenosine 3′,5′-monophos-phate (cAMP) (Konijn et al., 1967; Bonner et al., 1969), the chemical structure of which is shown in Figure 24A.

Aggregation is initiated as each of the cells begins to synthesize cyclic

AMP. There are no "dominant" cells that begin the secretion or that control the others. Rather, the sites of aggregation are determined by the distribution of amoebae (Keller and Segal, 1970; Tyson and Murray, 1989). Neighboring cells respond to the cAMP in two ways: they initiate a movement toward the cAMP pulse, and they release cAMP of their own (Robertson et al., 1972; Shaffer, 1975). After this, the cell is unresponsive to further cAMP pulses for several minutes. The result is a rotating wave of cAMP that is propagated throughout the population of cells (Figure 24B). As each wave arrives, the cells take another step toward the center.*

The differentiation of individual amoebae into either stalk (somatic) or spore (reproductive) cells is a complex matter. Raper (1940) and Bonner (1957) have demonstrated that in all cases the anterior cells become stalk, while the remaining cells are destined to form spores. Surgically removing the anterior part of the slug does not change this situation. The new anterior cells (which had been destined to produce spores) will now become the stalk (Raper, 1940). Somehow a decision is made so that whichever cells are anterior become stalk cells and whichever are posterior become spores. This ability of cells to change their developmental fates according to their location within the whole organism is called REGULATION. We will see this phenomenon in many embryos, including those of mammals.

***Differentiation in* Dictyostelium.** This differentiation into stalk cell or spore cell reflects one of the major phenomena of embryogenesis: the cell's selection of a "developmental pathway." Cells are often seen to select a particular developmental fate when alternatives are available. A given bone marrow cell, for instance, can become a lymphocyte or a blood cell; a particular embryonic cell can become either an epidermal skin cell or a neuron. In *Dictyostelium*, we see a simple, dichotomous decision, because

*The biochemistry of this reaction involves a receptor that binds cAMP. When this binding occurs, specific gene transcription takes place, motility towards the source of the cAMP is initiated, and adenyl cyclase enzymes (that synthesize cAMP from ATP) are activated. The newly formed cAMP activates its own receptors as well as those of its neighbors. The cells in the area remain insensitive to new waves of cAMP until the bound cAMP is removed from the receptors by another cell-surface enzyme, phosphodiesterase (Johnson et al., 1989). This set of reactions creates an oscillator of the type that will be discussed in Chapter 17. The mathematics of such oscillation reactions predict the rotating spiral form that such aggregating colonies take. Interestingly, the same mathematical formulas predict the behavior of certain chemical reactions and the formation of new stars in rotating spiral galaxies (Tyson and Murray, 1989).

FIGURE 24
Chemotaxis due to waves of cyclic adenosine monophosphate (cAMP). (A) Chemical structure of cAMP. (B) Visualization of the cAMP "waves" in the medium. Central cells secrete cAMP at regular intervals, and each secretion diffuses outward as a concentric wave. Waves are charted by saturating filter paper with radioactive cAMP and placing it on an aggregating colony. The cAMP from the secreting cells dilutes the radioactive cAMP. When the radioactivity on the paper is recorded (by placing it over X-ray film), the regions of high cAMP concentration in the culture appear lighter than those of low cAMP concentration. (B from Tomchick and Devreotes, 1981; courtesy of P. Devreotes.)

(A)

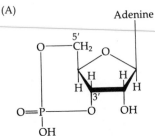

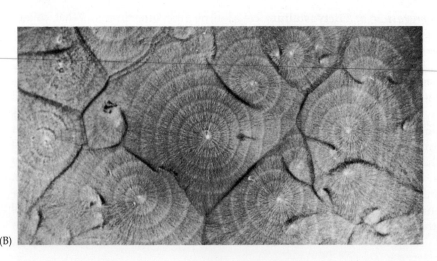

(B)

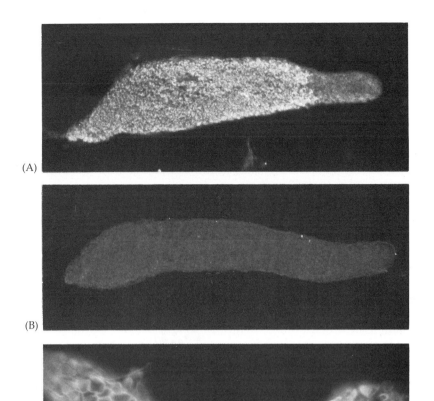

(A)

(B)

(C)

FIGURE 25

Chemicals controlling differentiation in *Dictyostelium*. A and B show the effects of placing *Dictyostelium* slugs into a medium containing enzymes that destroy extracellular cyclic AMP. (A) Control grex stained for the presence of a prespore-specific protein (white regions). (B) Similar grex stained after the treatment with cAMP-degrading enzymes. No prespore-specific product is seen. (C) is a higher magnification of a slug treated with DIF in the absence of ammonia. The stain used here binds to the cellulose wall of the stalk cells. All the cells of the grex have become stalk cells. (A and B from Wang et al., 1988a; C from Wang and Schaap, 1989; photographs courtesy of the authors.)

only two cell types are possible. How is it that a given cell becomes a stalk cell or a spore cell? Although the details are not fully known, a cell's fate appears to be regulated by certain diffusible molecules. One of these, DIFFERENTIATION-INDUCING FACTOR (DIF), appears to be necessary for stalk cell differentiation. This factor, like the sex-inducing factor of *Volvox*, is effective at very low concentrations ($10^{-10} M$); and, like the *Volvox* protein, it appears to induce the differentiation of a particular type of cell. When added to isolated amoebae or even to prespore (posterior) cells, it causes them to form stalk cells. This low molecular weight lipid is genetically regulated, for there are mutant strains of *Dictyostelium* that form only spore precursors and no stalk cells. When DIF is added to these mutant cultures, stalk cells are able to differentiate (Kay and Jermyn, 1983; Morris et al., 1987) and new prestalk-specific mRNAs are seen in the cell cytoplasm (Williams et al., 1987).

While DIF stimulates amoebae to become prestalk cells, the differentiation of prespore cells is most likely controlled by the continuing pulses of cyclic AMP. High concentrations of cyclic AMP initiate the expression of prespore-specific mRNAs in aggregated amoebae. Moreover, when slugs are placed in a medium containing an enzyme that destroys extracellular cAMP, the prespore cells lose their differentiated characteristics (Figure 25; Schaap and van Driel, 1985; Wang et al., 1988a,b).

How the grex knows which end is up

If all the amoebae of the grex start out being equal, how can some cells become spore cells and other cells become stalk cells? How can those cells in the posterior of the slug receive high levels of cAMP and differentiate into spore cells, while those equivalent cells in the anterior obtain relatively lower levels of cAMP and become stalk cells? The answer may lie in the observation that all the original cells are not equal. Those cells that become starved while in the early portion of their cell cycle tend to form the anterior portion of the slug, while those amoebae starved towards the end of their cell cycles tend to form the posterior (McDonald and Durston, 1984; Weijer et al., 1984). This work has been confirmed and extended by Ohmori and Maeda (1987) who showed that those cells starved in the latter part of the cell cycle responded differently to cAMP and showed much higher adhesivity than those cells starved just after mitosis. Williams and co-workers (1989) have found that prespore and prestalk cells can be distinguished in the early aggregates, and that they are randomly distributed. These cells sort out from each other to form prespore and prestalk regions as the grex begins to form.

One would expect that DIF, the putative stalk-forming morphogen, would be most highly concentrated in the anterior end of the grex, where the prestalk cells are to form. This, however, is not the case. DIF is found throughout the slug and may actually be at higher levels in the posterior of the grex (Brookman et al., 1987). Why should it be active only in the anterior end? Wang and Schaap (1989) have shown that ammonia is an inhibitor of DIF activity. They suggest that as the slug stops migrating, the excretion of ammonia into the air is greatest at the anterior tip. This would allow DIF to convert the anterior cells into prespore cells, while it would still be inhibited from acting

on posterior cells. Alternatively, Loomis (1990) has proposed that the differential metabolism of cAMP causes one population of cells to become cells that produce DIF but are themselves insensitive to it (the prespore cells) whereas a second population that does not synthesize DIF is nevertheless sensitive to it (the prestalk cells). A model summarizing some of the data in this field is shown in Figure 26.

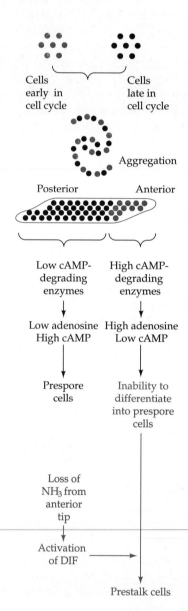

FIGURE 26
Hypothetical model for the differentiation of amoebae into either stalk or spore cells in *Dictyostelium*. Initial differences in the cell cycle are reflected in different enzymatic activities and in the sorting out of cells to the anterior or posterior of the grex. These differences in cAMP-degrading enzymes become reflected in the different concentrations of cAMP found in these regions. This difference in cAMP levels may create differences in the ability of cells to synthesize or respond to DIF. This is a simplified model, and other molecules have also been implicated in facilitating this developmental decision.

***Cell adhesion molecules in* Dictyostelium.** How do these individual cells stick together to form a cohesive organism? This is the same problem that embryonic cells face, and the solution that evolved in the protists is the same one used by embryos: developmentally regulated CELL ADHESION MOLECULES.

While growing mitotically on bacteria, *Dictyostelium* cells are not mutually adhesive. However, once cell division stops, the cells become increasingly adhesive, reaching a plateau of maximum cohesiveness around eight hours after starvation. This initial cell-to-cell adhesion is mediated by a 24,000-Da glycoprotein that is absent in growing cells but is seen shortly thereafter (Figure 27; Knecht et al., 1987; Loomis, 1988). This protein is synthesized from newly transcribed mRNA and it becomes localized in the cell membranes of the myxamoebae. If these cells are treated with antibodies that bind to and mask this protein, the cells will not stick to each other, and all subsequent development ceases.

Once this initial aggregation has occurred, it is stabilized by a second cell adhesion molecule. This 80,000-Da glycoprotein is also synthesized during the aggregation phase. If it is defective or absent in the cells, small slugs will form, and their fruiting bodies will only be about one-third the normal size. Thus, the second cell adhesion system seems to be needed for retaining a large enough number of cells to form large fruiting bodies (Müller and Gerisch, 1978; Loomis, 1988). In addition, a third cell adhesion system is activated late in development, while the slug is migrating. This protein or group of proteins may exist only on prespore cells and may be responsible for separating the prespore cells from the prestalk cells (Loomis, personal communication). Thus, *Dictyostelium* has evolved three developmentally regulated systems of cell–cell adhesion that are necessary for the morphogenesis of individual cells into a coherent organism. As we shall see in subsequent chapters, metazoan cells also use a variety of cell adhesion molecules to form the tissues and organs of the embryo.

Dictyostelium is a "part-time multicellular organism" that does not form many cell types (Kay et al., 1989), and the more complex multicellular organisms do not form from the aggregation of formerly independent cells. However, many of the principles found in the development of this "simple" organism are the same as those found in embryos of the more complex phyla. The ability of individual cells to sense a chemical gradient (as in the amoeba's response to cAMP) is very important for cell migration and morphogenesis during animal development. The role of cell surface proteins for cell cohesiveness is found throughout the animal kingdom, and differentiation-inducing molecules are just beginning to be isolated in metazoan organisms.

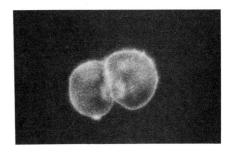

FIGURE 27
Dictyostelium cells synthesize an adhesive, 24,000-Da glycoprotein shortly after nutrient starvation. *Dictyostelium* cells were stained with a fluorescent antibody that binds to the 24,000-Da glycoprotein and were then observed under ultraviolet light. This protein is not seen on amoebae that have just stopped dividing. However, as shown here, ten hours after cell division has ceased, individual amoebae are seen to have this protein in their cell membranes and are capable of adhering together. (Photograph courtesy of W. Loomis.)

SIDELIGHTS & SPECULATIONS

Evidence and antibodies

Biology, like any other science, does not deal with Facts. Rather, it deals with evidence. There are several types of evidence that will be presented in this book, and they are not all equivalent in strength. As an example, we will use the analysis of cell adhesion in *Dictyostelium*. The first, and weakest, type of evidence is *correlative* evidence. Here, correlations are made between two or more events, and there is an inference that one event caused the other. As we have seen, fluorescently labeled antibodies to a certain 24,000-Da glycoprotein do not label dividing, vegetative, cells; but they do find this protein in myxamoeba cell membranes soon after the cells stop dividing and become competent to aggregate (Figure 27). Thus, there is a correlation between the presence of this cell membrane glycoprotein and the ability to aggregate.

Correlative evidence gives a starting point to investigations, and one cannot say with certainty that one event caused the other based solely upon correlations. It could be equally as probable that cell adhesion caused the cell to synthesize this new glycoprotein or that both cell adhesion and the synthesis of the 24,000-Da glycoprotein were separate events initiated by the same underlying cause. The simultaneous occurrence of the two events could even be coincidental, the events having no relationship to each other.*

How, then, does one get beyond mere correlation? In the study of cell adhesion in *Dictyostelium*, the next step

*In a tongue-in-cheek article spoofing such correlative inferences, Sies (1988) demonstrated a remarkably good correlation between the number of storks seen in West Germany from 1965 to 1980 and the number of babies born during those same years.

was to use those same antibodies to block the adhesion of myxamoebae. Based on a technique pioneered by Gerisch's laboratory (Beug et al., 1970), Knecht and co-workers took the antibodies that bound this 24,000-Da glycoprotein and isolated their antigen-binding sites (the portions of the antibody molecule that actually recognize the antigen). This was necessary because the whole antibody molecule contains two antigen-binding sites and would therefore artificially crosslink and agglutinate the myxamoebae. When these antigen-binding fragments (called F_{ab}'s) were added to the aggregation-competent cells, the cells could not aggregate. The antibody fragments had inhibited the cells' adhering together, presumably by binding to the 24,000-Da glycoprotein and blocking its function. This type of evidence is called "loss-of-function" evidence. While stronger than correlative evidence, it still does not make other inferences impossible. For instance, perhaps the antibodies killed the cell (as might have been the case if the 24,000-Da glycoprotein were a critical transport channel). This would also stop the cells from adhering. Or perhaps the 24,000-Da glycoprotein had nothing to do with adhesion itself but is necessary for the real adhesive molecule to function (such as in stabilizing membrane proteins in general). Blocking this glycoprotein would similarly cause the inhibition of cell aggregation. Thus, loss-of-function

evidence must be bolstered by many controls demonstrating that the antibodies specifically knock out the particular function and nothing else.

The strongest type of evidence is the "gain-of-function" evidence. Here, the initiation of the first event causes the second event to happen even in instances where neither event usually occurs. Silva and Klein (1990) and Faix and co-workers (1990) have recently obtained such evidence to show that the 80,000-Da glycoprotein is an adhesive molecule. They isolated the gene for the 80,000-Da protein and modified the gene in a way that would cause it to be expressed all the time. They then placed it back into well-fed, vegetatively growing myxamoebae which do not usually express this protein and which are not usually able to adhere to each other. The presence of this protein on the cell membrane of these dividing cells was confirmed by antibody labeling. Moreover, such cells now adhered to one another even in the vegetative stages when they normally would not. Thus, they had gained an adhesive function solely upon expressing this particular glycoprotein on their cell surfaces. This gain-of-function evidence is more convincing than other types of analyses, and similar experiments have recently been performed on mammalian cells (see Chapter 15) to demonstrate the presence of particular cell adhesion molecules in the developing embryo.

Developmental patterns among metazoans

Since the remainder of this book concerns the development of METAZOANS—multicellular organisms that pass through embryonic stages of development—we will present an overview of their developmental patterns. The last chapter of the text will discuss these patterns in more detail. Figure 28 illustrates the major evolutionary trends of metazoan development. The most striking observation is that life has not evolved in a straight line but rather there are several evolutionary paths. We can see that most of the species of metazoans belong to one of two major branches of animals: protostomes and deuterostomes.

The Porifera

The colonial protozoans are thought to have given rise to two groups of metazoans, both of which pass through embryonic stages of development. One of these groups is the Porifera (sponges). These animals develop in a manner so different from that of any other animal group that some taxonomists do not consider them metazoans at all. A sponge has three types of somatic cells, but one of these, the ARCHEOCYTE, can differentiate into all the other cell types in the body. The cells of a sponge passed through a sieve can regenerate new sponges from individual cells. Moreover, such reaggregation is species-specific, and if individual sponge cells from different species are mixed together, each of the sponges that reforms contains cells from only one species (Wilson, 1907). It is generally thought that the motile archeocytes collect cells from their own species and not from others (Turner, 1978). Sponges contain no mesoderm, so there are

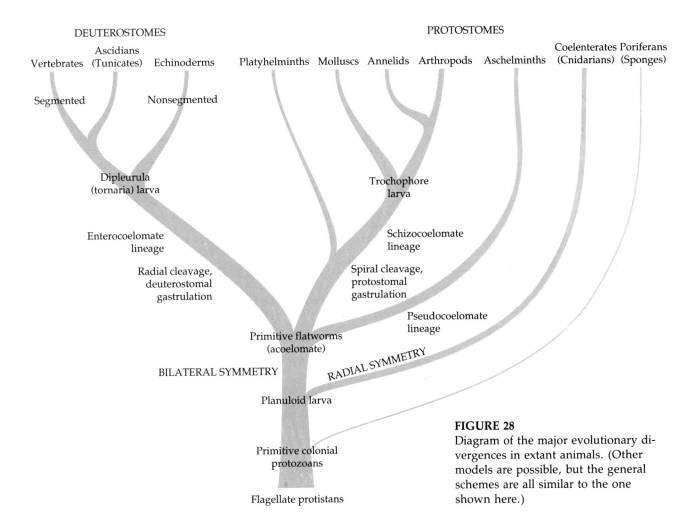

DEUTEROSTOMES

PROTOSTOMES

Vertebrates
Ascidians (Tunicates)
Echinoderms
Platyhelminths
Molluscs
Annelids
Arthropods
Aschelminths
Coelenterates (Cnidarians)
Poriferans (Sponges)

Segmented

Nonsegmented

Dipleurula (tornaria) larva

Trochophore larva

Enterocoelomate lineage

Schizocoelomate lineage

Radial cleavage, deuterostomal gastrulation

Spiral cleavage, protostomal gastrulation

Pseudocoelomate lineage

Primitive flatworms (acoelomate)

BILATERAL SYMMETRY

RADIAL SYMMETRY

Planuloid larva

Primitive colonial protozoans

Flagellate protistans

FIGURE 28
Diagram of the major evolutionary divergences in extant animals. (Other models are possible, but the general schemes are all similar to the one shown here.)

no true organ systems in the Porifera; nor do they have a digestive tube or circulatory system, nerves, or muscles. Thus, even though they pass through an embryonic and a larval stage, they are very unlike most metazoans.

Protostomes and deuterostomes

The other group of metazoans arising from the colonial protists is characterized by the presence of three germ layers during development. Some members of this group constitute the RADIATA, so called because they have radial symmetry, like that of a tube or a wheel. The Radiata include the cnidarians (jellyfish, corals, and hydra) and ctenophores (comb jellies). In these animals, the mesoderm is rudimentary, consisting of sparsely scattered cells in a gelatinous matrix. Most metazoans, however, are bilateral species (BILATERIA) and are classified as either protostomes or deuterostomes. All Bilateria are thought to have descended from a primitive type of flatworm. These flatworms were the first to have a true mesoderm (although it was not hollowed out to form a body cavity), and they are thought to have resembled the larvae of certain contemporary coelenterates. The differences in the two major divisions of the Bilateria are illustrated in Figure 29. PROTOSTOMES (Gk. meaning "mouth first"), which include the mollusc, arthropod, and worm phyla, are so called because during gastrulation the mouth regions are formed first. The body cavity

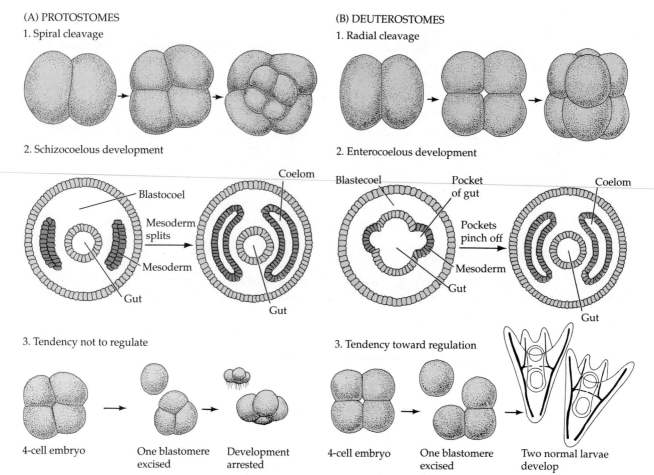

(A) PROTOSTOMES

1. Spiral cleavage

2. Schizocoelous development

Blastocoel

Mesoderm splits

Mesoderm

Gut

Coelom

Gut

3. Tendency not to regulate

4-cell embryo → One blastomere excised → Development arrested

(B) DEUTEROSTOMES

1. Radial cleavage

2. Enterocoelous development

Blastecoel

Pocket of gut

Pockets pinch off

Mesoderm

Gut

Coelom

Gut

3. Tendency toward regulation

4-cell embryo → One blastomere excised → Two normal larvae develop

FIGURE 29

Principal tendencies of the deutero-stomes and protostomes. Exceptions to all these general tendencies have evolved secondarily in certain members of each group. (Mammals and birds, for instance, do not have a strictly enterocoelous formation of the body cavity.)

of these animals forms from the hollowing out of a previously solid cord of mesodermal cells.

The other great division of the animal kingdom is the DEUTEROSTOME lineage. Phyla in this division include chordates and echinoderms. Although it may seem strange to classify humans and wolverines in the same group as starfish and sea urchins, certain embryological features stress this kinship. First, in deuterostomes (Gk. meaning "mouth second"), the mouth region is formed after the anal region. Also, whereas protostomes form their body cavity by hollowing out a solid mesodermal block (schizocoelous formation of the body cavity), most deuterostomes form their body cavities from mesodermal pouches extending from the gut (enterocoelous formation of the body cavity).

In addition, protostomes and deuterostomes differ in the way they undergo cleavage. Most deuterostomes undergo cleavage such that the blastomeres are perpendicular or parallel to each other. This is called RADIAL CLEAVAGE. Protostomes, on the other hand, form blastulae composed of cells that are at acute angles to the polar axis of the embryo. Thus, they are said to undergo SPIRAL CLEAVAGE. Furthermore, the cleavage stage blastomeres of most deuterostomes have a greater ability to regulate development than do those of protostomes. If a single blastomere is removed from a 4-cell sea urchin or human embryo, that blastomere will develop into an entire organism, and the remaining three-quarters of the embryo can also develop normally. However, if the same operation were performed on a snail or worm embryo, both the single blastomere and the remaining ones would develop into partial embryos—each lacking what was formed from the other.

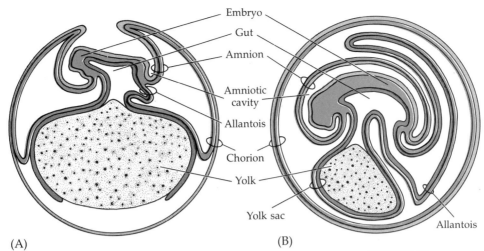

Embryo
Gut
Amnion
Amniotic cavity
Allantois
Chorion
Yolk
Yolk sac
Allantois

(A)   (B)

The evolution of organisms depends upon inherited changes in their development. One of the greatest evolutionary advances—the amniote egg—occurred among the deuterostomes. This type of egg, exemplified by that of a chicken (Figure 30), first appeared in reptiles about 255 million years ago. The amniote egg liberated vertebrates from any dependence upon water during their development; they were free to roam on land, far from existing ponds. Whereas amphibians must return to water to breed and to enable their eggs to develop, the amniote egg carries its own water and food supplies. The egg is fertilized internally and contains YOLK to nourish the developing embryo. Moreover, it contains two sacs: the AMNION, which contains the fluid bathing the embryo, and the ALLANTOIS, in which waste materials from embryonic metabolism collect. The entire structure is encased in a shell that allows the diffusion of oxygen but is hard enough to protect the embryo from environmental assaults. A similar development of egg casings enabled arthropods to be the first terrestrial invertebrates. Thus, the final crossing of the boundary between water and land occurred with the modification of the earliest stage in development, the egg.

Developmental biology provides an endless assortment of fascinating animals and problems. It is the author's task to select a small sample of them to illustrate the major principles of animal development. Yet this sample is an incredibly small collection. Right now, we are merely observing the small tidepool within our reach while the whole ocean of developmental principles lies before us. Our study of animal development will begin, then, with the early stages of animal embryogenesis: fertilization, cleavage, gastrulation, and the establishment of the body plan. Although an attempt has been made to survey the important variations throughout the animal kingdom, a certain deuterostome chauvinism may be apparent.

FIGURE 30
Diagram of the amniote egg of the chick, showing the development of membranes enfolding the embryo. (A) 3-day incubation. (B) 7-day incubation. The origin of the membranes will be detailed in Chapter 6. Note that the yolk is eventually surrounded by the membranes derived from the embryo (to absorb nutrients) and that other membranes extend from the embryo to the shell (where they will exchange oxygen and carbon dioxide and obtain calcium from the shell). The ectoderm eventually covers the outside of the embryo and extends to the shell. The endoderm becomes the gut and encircles the yolk while the mesoderm (color) extends to cover these layers and provide the blood vessels to bring the extraembryonic material into the embryo.

LITERATURE CITED

Bergman, K., Goodenough, U. W., Goodenough, D. A., Jawitz, J. and Martin, H. 1975. Gametic differentiation in *Chlamydomonas reinhardtii*. II. Flagellar membranes and the agglutination reaction. *J. Cell Biol.* 67: 606–622.

Beug, H., Gerisch, G., Kempff, S., Riedel, V. and Cremer, G. 1970. Specific inhibition of cell contact formation in *Dictyostelium* by univalent antibodies. *Exp. Cell Res.* 63: 147–158.

Bonner, J. T. 1947. Evidence for the formation of cell aggregates by chemotaxis in the development of the slime mold *Dictyostelium discoideum*. *J. Exp. Zool.* 106: 1–26.

Bonner, J. T. 1957. A theory of the control of differentiation in the cellular slime molds. *Q. Rev. Biol.* 32: 232–246.

Bonner, J. T., Berkley, D. S., Hall, E. M., Konijn, T. M., Mason, J. W., O'Keefe, G. and Wolfe, P. B. 1969. Acrasin, acrasinase and the sensitivity to acrasin in *Dictyostelium discoideum*. *Dev. Biol.* 20: 72–87.

Brookman, J. J., Jermyn, K. A. and Kay, R. R. 1987. Nature and distribution of the morphogen DIF in the *Dictyostelium* slug. *Development* 100: 119–124.

Faix, J., Gerisch, G. and Noegel, A. A. 1990. Constitutive overexpression of the contact A glycoprotein enables growth-phase cells of *Dictyostelium discoideum* to aggregate. *EMBO J.* 9: 2709–2716.

Fulton, C. 1977. Cell differentiation in *Naegleria gruberi*. *Annu. Rev. Microbiol.* 31: 597–629.

Fulton, C. and Walsh, C. 1980. Cell differentiation and flagellar elongation in *Naegleria gruberi*: Dependence on transcription and translation. *J. Cell Biol.* 85: 346–360.

Garcia, E. and Dazy, A.-C. 1986. Spatial distribution of poly(A)$^+$ RNA and protein synthesis in *Acetabularia mediterranea*. *Biol. Cell* 58: 23–29.

Goodenough, U. W. and Weiss, R. L. 1975. Gametic differentiation in *Chlamydomonas reinhardtii*. III. Cell wall lysis and microfilament associated mating structure activation in wild-type and mutant strains. *J. Cell Biol.* 67: 623–637.

Hämmerling, J. 1934. Über formbildended Substanzen bei *Acetabularia mediterranea*, ihre räumliche und zeitliche Verteilung und ihre Herkunft. *Wilhelm Roux Arch. Entwicklungsmech. Org.* 131: 1–81.

Hartmann, M. 1921. Die dauernd agame Zucht von *Eudorina elegans*, experimentelle Beiträge zum Befruchtungs–und Todproblem. *Arch. Protistk.* 43: 223–286.

Johnson, R. L., Gundersen, R., Lilly, P., Pitt, G. S., Pupillo, M., Sun, T. J., Vaughan, R. A. and Devreotes, P. N. 1989. G-protein-linked signal transduction systems control development in *Dictyostelium*. *Development* [Suppl.]: 75–80.

Kay, R. R. and Jermyn, K. A. 1983. A possible morphogen controlling differentiation in *Dictyostelium*. *Nature* 303: 242–244.

Kay, R. R., Berks, M. and Traynor, D. 1989. Morphogen hunting in *Dictyostelium*. *Development* [Suppl.]: 81–90.

Keller, E. F. and Segal, L. A. 1970. Initiation of slime mold aggregation viewed as an instability. *J. Theoret. Biol.* 26: 399–415.

Kirk, D. L. 1988. The ontogeny and phylogeny of cellular differentiation in *Volvox*. *Trends Genet.* 4: 32–36.

Kirk, D. L. and Kirk, M. M. 1986. Heat shock elicits production of sexual inducer in *Volvox*. *Science* 231: 51–54.

Kirk, D. L., Viamontes, G. I., Green, K. J. and Bryant, J. L. Jr. 1982. Integrated morphogenetic behavior of cell sheets: Volvox as a model. *In* S. Subtelny and P. B. Green (eds.) *Developmental Order: Its Origin and Regulation.* Alan R. Liss, New York, pp. 247–274.

Kloppstech, K. and Schweiger, H. G. 1975. Polyadenylated RNA from *Acetabularia*. *Differentiation* 4: 115–123.

Knecht, D. A., Fuller, D. and Loomis, W. F. 1987. Surface glycoprotein gp24 involved in early adhesion of *Dictyostelium discoideum*. *Dev. Biol.* 121: 277–283.

Konijn, T. M., van der Meene, J. G. C., Bonner, J. T. and Barkley, D. S. 1967. The acrasin activity of adenosine-3′,5′-cyclic phosphate. *Proc. Natl. Acad. Sci. USA* 58: 1152–1154.

Krutch, J. W. 1956. *The Great Chain of Life.* Houghton Mifflin, Boston, pp. 28–29.

Loomis, W. F. 1988. Cell–cell adhesion in *Dictyostelium discoideum*. *Dev. Genet.* 9: 549–559.

Loomis, W. F. 1990. Essential genes for development of *Dictyostelium*. *Prog. Mol. Subcell. Biol.* 11: 159–183.

McDonald, S. A. and Durston, A. J. 1984. The cell cycle and sorting out behaviour in *Dictyostelium discoideum*. *J. Cell Sci.* 66: 196–204.

Morris, H. R., Taylor, G. W., Masento, M. S., Jermyn, K. A. and Kay, R. R. 1987. Chemical structure of the morphogen differentiation-inducing factor from *Dictyostelium discoideum*. *Nature* 328: 811–814.

Müller, K. and Gerisch, G. 1978. A specific glycoprotein as the target of adhesion blocking Fab in aggregating *Dictyostelium* cells. *Nature* 274: 445–447.

Ohmori, T. and Maeda, Y. 1987. The developmental fate of *Dictyostelium discoideum* cells depends greatly on the cell-cycle position at the onset of starvation. *Cell Differ.* 22: 11–18.

Pommerville, J. and Kochert, G. 1982. Effects of senescence on somatic cell physiology in the green alga *Volvox carteri*. *Exp. Cell Res.* 14: 39–45.

Powers, J. H. 1908. Further studies on *Volvox*, with description of three new species. *Trans. Am. Micros. Soc.* 28: 141–175.

Raper, K. B. 1940. Pseudoplasmodium formation and organization in *Dictyostelium discoideum*. *J. Elisha Mitchell Sci. Soc.* 56: 241–282.

Robertson, A., Drage, D. J. and Cohen, M. H. 1972. Control of aggregation in *Dictyostelium discoideum* by an external periodic pulse of cyclic adenosine monophosphate. *Science* 175: 333–335.

Schaap, P. and van Driel, R. 1985. The induction of post–aggregative differentiation in *Dictyostelium discoideum* by cAMP. Evidence for involvement of the cell surface cAMP receptor. *Exp. Cell Res.* 159: 388–398.

Shaffer, B. M. 1953. Aggregation in cellular slime molds: In vitro isolation of acrasin. *Nature* 171: 975.

Shaffer, B. M. 1975. Secretion of cyclic AMP induced by cyclic AMP in the cellular slime mold *Dictyostelium discoideum*. *Nature* 255: 549–552.

Sies, H. 1988. A new parameter for sex education. *Nature* 332: 495.

Silva, A. M. da and Klein, C. 1990. Cell adhesion transformed *D. discoideum* cells:

Expression of gp80 and its biochemical characterization. *Dev. Biol.* 140: 139–148.

Strickberger, M. W. 1985. *Genetics*, 3rd Ed. Macmillan, New York.

Tomchick, K. J. and Devreotes, P. N. 1981. Adenosine 3′,5′ monophosphate waves in *Dictyostelium discoideum*: A demonstration by isotope dilution-fluorography. *Science* 212: 443–446.

Turner, R. S. Jr. 1978. Sponge cell adhesions. *In* D. R. Garrod (ed.), *Specificity of Embryological Interactions.* Chapman and Hall, London, pp. 199–232.

Tyson, J. J. and Murray, J. D. 1989. Cyclic AMP waves during aggregation of *Dictyostelium* amoebae. *Development* 106: 421–426.

Walsh, C. 1984. Synthesis and assembly of the cytoskeleton of *Naegleria gruberi* flagellates. *J. Cell Biol.* 98: 449–456.

Wang, M. and Schaap, P. 1989. Ammonia depletion and DIF trigger stalk cell differentiation in intact *Dictyostelium discoideum* slugs. *Development* 105: 569–574.

Wang, M., van Driel, R. and Schaap, P. 1988a. Cyclic AMP-phosphodiesterase induces dedifferentiation of prespore cells in *Dictyostelium discoideum* slugs: Evidence that cyclic AMP is the morphogenetic signal for prespore differentiation. *Development* 103: 611–618.

Wang, M., Aerts, R. J., Spek, W. and Schaap, P. 1988b. Cell cycle phase in *Dictyostelium discoideum* is correlated with the expression of cyclic AMP production, detection and degradation. Involvement of cAMP signalling in cell sorting. *Dev. Biol.* 125: 410–416.

Weijer, C. J., Duschl, G. and David, C. N. 1984. Dependence of cell-type proportioning and sorting on cell cycle phase in *Dictyostelium discoideum*. *Exp. Cell Res.* 70: 133–145.

Williams, J. G. 1991. Cell movement and regionally localized differentiation in *Dictyostelium* morphogenesis. Forty-Ninth Annual Symposium of the Society for Developmental Biology. In press.

Williams, J. G., Ceccarelli, A., McRobbie, S., Mahbubani, H., Kay, R. R., Early, A., Berks, M. and Jermyn, K. A. 1987. Direct induction of *Dictyostelium* pre-stalk gene expression of DIF provides evidence that DIF is a morphogen. *Cell* 49: 185–192.

Williams, J. G., Duffy, K. T., Lane, D. P., McRobbie, S. J., Harwood, A. J., Traynor, D., Kay, R. R. and Jermyn, K. A. 1989. Origins of the prestalk-prespore pattern in *Dictyostelium* development. *Cell* 59: 1157–1163.

Wilson, E. B. 1896. *The Cell in Development and Inheritance.* Macmillan, New York.

Wilson, H. V. 1907. On some phenomena of coalescence and regeneration in sponges. *J. Exp. Zool.* 5: 245–258.

2

Fertilization: Beginning a new organism

Urge and urge and urge,
Always the procreant urge of the world.
Out of the dimness opposite equals advance,
Always substance and increase, always sex,
Always a knit of identity, always distinction,
Always a breed of life.

—WALT WHITMAN (1855)

The final aim of all love intrigues, be they comic or tragic, is really
of more importance than all other ends in human life. What it
turns upon is nothing less than the composition of the next
generation.

—A. SCHOPENHAUER
(QUOTED BY C. DARWIN, 1871)

Fertilization is the process whereby two sex cells (GAMETES) fuse together to create a new individual with genetic potentials derived from both parents. Fertilization, then, accomplishes two separate activities: sex (the combining of genes derived from the two parents) and reproduction (the creation of new organisms). Thus, the first function of fertilization involves the transmission of genes from parent to offspring, and the second function of fertilization is to initiate in the egg cytoplasm those reactions that permit development to proceed.

Although the actual details of fertilization vary enormously from species to species, the events of conception generally consist of four major activities:

- *Contact and recognition between sperm and egg.* In most instances, this is the quality control step, insuring that the sperm and egg are of the same species.
- *Regulation of sperm entry into the egg.* This is the quantity control step. Only one sperm can ultimately fertilize the egg. All other sperm must be eliminated.
- *Fusion of the genetic material of sperm and egg.*
- *Activation of egg metabolism to start development.*

Structure of the gametes

As we shall soon see, a complex dialogue exists between egg and sperm. The egg is able to activate the sperm metabolism that is essential for fertilization, and the sperm reciprocates by activating the egg metabolism needed for the onset of development. But before proceeding to investigate each of these features of fertilization, we must first discuss the structures of the sperm and egg, the two cell types specialized for fertilization.

Sperm

It is only within the past century that the sperm's role in fertilization has been known. Anton van Leeuwenhoek, the Dutch microscopist who codiscovered sperm in 1678, first believed them to be parasitic animals living within the semen (hence the term *spermatozoa*, meaning "sperm animals"). He originally assumed they had nothing at all to do with reproducing the organism in which they were found, but he later came to believe that each sperm contained a preformed animal. Leeuwenhoek (1685) wrote that sperm were seeds (both *sperma* and *semen* mean "seed") and that the female merely provided the nutrient soil into which the seeds were planted. In this, he was returning to a notion of procreation promulgated by Aristotle 2,000 years earlier. Try as he could, Leeuwenhoek was continually disappointed in his inability to find the preformed seed within the spermatozoa. Nicolas Hartsoeker, the other codiscoverer of sperm, drew what he also hoped to find: a preformed human ("homunculus") within the human sperm (Figure 1). This belief that the sperm contained the entire embryonic organism never gained much acceptance as it implied an enormous waste of potential life.

In a series of experiments performed in the late 1700s, Lazzaro Spallanzani demonstrated that filtered toad semen devoid of sperm would not fertilize eggs. He concluded, however, that the viscous fluid retained by the filter paper, and not the sperm, was the agent of fertilization. He, too, felt that the spermatic animals were clearly parasitic. Yet, this was the first evidence suggesting their importance.

The combination of better microscopic lenses and the cell theory led to a new appreciation of spermatic function. In 1824, J. L. Prevost and J. B. Dumas claimed that sperm were not parasites but rather were the active agents of fertilization. They noted the universal existence of sperm in sexually mature males and their absence in immature and aged individuals. These observations, coupled with the known absence of spermatozoa in the sterile mule, convinced them that "there exists an intimate relation between their presence in the organs and the fecundating capacity of the animal." They proposed that the sperm actually entered the egg and contributed materially to the next generation.

These claims were largely disregarded until the 1840s when A. von Kolliker described the formation of sperm from testicular cells. He concluded that sperm were greatly modified cells that came from the testes of adult males. He ridiculed the idea that the semen could be normal while they served to support an enormous number of parasites. Even so, von Kolliker denied that there was any physical contact between sperm and egg. He believed that the sperm excited the egg to develop, much like a magnet communicated its presence to iron. It was only in 1876 that Oscar Hertwig and Herman Fol independently demonstrated the penetration of the egg by the sperm and the union of the cells' nuclei. Hertwig had sought an organism suitable for detailed microscopic observations, and he

FIGURE 1
The human infant preformed in the sperm as depicted by Nicolas Hartsoeker (1694).

found that the Mediterranean sea urchin, *Toxopneustes lividus*, was perfect. Not only was it common throughout the region and sexually mature throughout most of the year, but its eggs were available in large numbers and were transparent even at high magnifications. After mixing together sperm and egg suspensions, Hertwig repeatedly observed a sperm entering an egg and saw the two nuclei unite. He also noted that only one sperm was seen to enter each egg and that all the nuclei of the embryo were derived from the fused nucleus created at fertilization. Fol made similar observations and detailed the mechanism of sperm entry. Fertilization was at last recognized as the union of sperm and egg, and the union of sea urchin egg and sperm remains the best-studied example of fertilization.

Each sperm is known to consist of a haploid nucleus, a propulsion system to move the nucleus, and a sac of enzymes that enable the nucleus to enter the egg. Most of the cytoplasm of the sperm has been eliminated during the sperm's maturation, leaving only certain organelles that have been modified for their spermatic function (Figure 2). During the course of sperm maturation, the haploid nucleus becomes very streamlined and its DNA becomes tightly compressed. In front of this compressed haploid nucleus lies the ACROSOMAL VESICLE, which is derived from the Golgi apparatus and contains enzymes that digest proteins and complex sugars. Thus, it can be considered to be a modified lysosome. These stored enzymes are used to lyse the outer coverings of the egg. In many species, such as sea urchins, a region of globular actin molecules lies between the nucleus and the acrosomal vesicle. These proteins are used to extend a fingerlike process during the early stages of fertilization. In these species, recognition between sperm and egg involves molecules on this ACROSOMAL PROCESS. Together, the acrosome and nucleus constitute the head of the sperm.

The means by which sperm are propelled varies according to how the species has adapted to environmental conditions. In some species (such as the parasitic roundworm *Ascaris*), the sperm travel by the amoeboid motion of LAMELLIPODIA—local extensions of the cell membrane. In most species, however, each sperm is able to travel long distances by whipping its FLAGELLUM.

Flagella are complex structures. The major motor portion of the flagellum is called the AXONEME. An axoneme is formed by the microtubules emanating from the centriole at the base of the sperm nucleus (Figures 2 and 3). The core of the axoneme consists of two central microtubules surrounded by a row of nine doublet microtubules. Actually, only one microtubule of each doublet is complete, having 13 protofilaments; the other is C-shaped and has only 11 protofilaments (Figure 3C). A three-dimensional model of a complete microtubule is shown in Figure 3B. Here one can see the 13 interconnected protofilaments, which are made exclusively of the dimeric protein tubulin.

Another protein, DYNEIN, is attached to the microtubules (Figure 3B). This protein can hydrolyze molecules of adenosine triphosphate (ATP) and can convert the released chemical energy into the mechanical energy that propels the sperm (Ogawa et al., 1977). The importance of dynein can be seen in those individuals with the genetic syndrome called Kartagener's triad. These individuals lack dynein on all their ciliated and flagellated cells; therefore, these structures are rendered immotile. Males with this disease are sterile (immotile sperm), are susceptible to bronchial infections (immotile respiratory cilia), and have a 50 percent chance of having their heart on the right-hand side of their body (Afzelius, 1976).

The "9 + 2" microtubule arrangement with the dynein arms has been

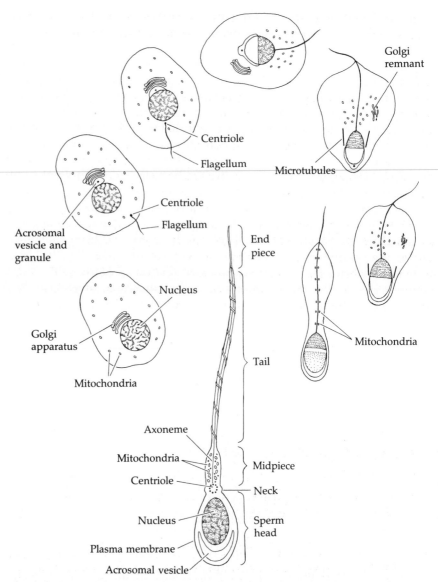

FIGURE 2

The modification of a germ cell to form a mammalian sperm. The centriole produces a long flagellum at what will be the posterior end of the sperm, and the Golgi apparatus forms the acrosomal vesicle at the future anterior end. The mitochondria (hollow dots) collect about the flagellum near the base of the haploid nucleus and become incorporated into the midpiece of the sperm. The remaining cytoplasm is jettisoned, and the nucleus condenses. The size of the mature sperm has been enlarged relative to the other figures. (After Clermont and Leblond, 1955.)

conserved throughout the eukaryotic kingdoms, suggesting that this arrangement is extremely well suited for transmitting energy for movement. The energy to whip the flagellum and thereby propel the sperm comes from rings of mitochondria located in the neck region of the sperm (Figure 2).

In many species (notably mammals), a layer of dense fibers has interposed itself between the mitochondrial sheath and axoneme. This fiber layer stiffens the sperm tail. Because the thickness of this layer decreases toward the tip, the fibers probably serve to prevent the sperm head from being whipped around too suddenly. Thus, the sperm has undergone extensive modification for the transmission of its nucleus to the egg.

Egg

All the material necessary for the beginning of growth and development must be stored in the mature egg (OVUM). Therefore, whereas the sperm has eliminated most of its cytoplasm, the developing egg not only conserves its material but is actively involved in accumulating more. It either synthesizes or absorbs proteins, such as yolk, that act as food reservoirs for the developing embryo. Thus, birds' eggs are enormous single cells that have become swollen with their accumulated yolk. Even eggs with relatively sparse yolk are comparatively large. The volume of a sea urchin egg is about 2×10^5 cubic micrometers, more than 10,000 times the volume of the sperm. The diagram of the sea urchin egg and sperm in Figure 4 shows their relative sizes. The figure also shows the various components of the mature oocyte. So, while sperm and egg have equal haploid nuclear components, the egg also has an enormous cytoplasmic storehouse that it has accumulated during its maturation. This cytoplasmic trove includes proteins, ribosomes, and transfer RNA (tRNA), messenger RNA, and morphogenic factors.

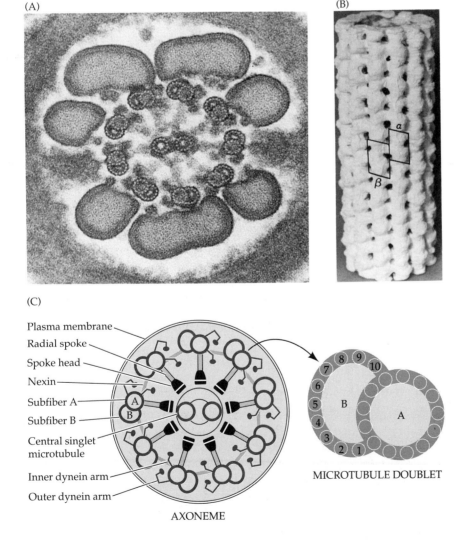

(A)

(B)

(C)

Plasma membrane
Radial spoke
Spoke head
Nexin
Subfiber A
Subfiber B
Central singlet microtubule
Inner dynein arm
Outer dynein arm

AXONEME

MICROTUBULE DOUBLET

FIGURE 3
The motile apparatus of the sperm. (A) Cross section of the flagellum of a mammalian spermatozoon showing the central axoneme and the external fibers. (B) A three-dimensional model of the "A" microtubule. The α and β-tubulin subunits are similar but not identical, and the microtubule can change size by polymerizing or depolymerizing tubulin subunits at either end. (C) Interpretive diagram of the "9 + 2" arrangement of the microtubules and other flagellar components. One microtubule doublet has been "magnified." The first ("A") portion of the doublet is a normal microtubule containing 13 protofilaments; the second ("B") portion of the doublet contains only 11 (occasionally 10) protofilaments. (A courtesy of D. M. Phillips; B from Amos and Klug, 1974; photograph courtesy of the authors; C after Voet and Voet, 1990.)

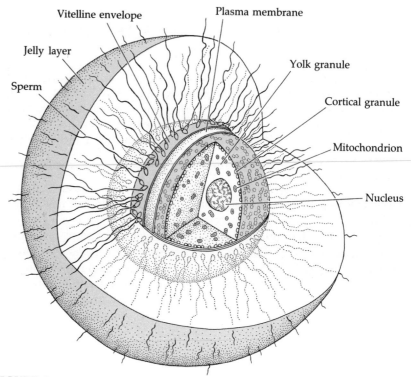

FIGURE 4
Structure of the sea urchin egg during fertilization (After Epel, 1977.)

- *Proteins.* It will be a long while before the embryo is able to feed itself or obtain food from its mother. The early embryonic cells need some storable supply of energy and amino acids. In many species, this is accomplished by accumulating yolk proteins in the egg. Many of the yolk proteins are made in other organs (liver, fat body) and travel to the egg.
- *Ribosomes and tRNA.* As we will soon see, there is a burst of protein synthesis soon after fertilization. This protein synthesis is accomplished by ribosomes and tRNA, which preexist in the egg. The developing egg (called an OOCYTE while completing meiosis) has special mechanisms to synthesize ribosomes, and certain amphibian oocytes produce as many as 10^{12} ribosomes during their meiotic prophase.
- *Messenger RNA.* In most organisms, the messages for proteins made during early development are already packaged in the oocyte. It is estimated that the eggs of sea urchins contain 25,000–50,000 different types of mRNAs. This mRNA, however, remains dormant until after fertilization.
- *Morphogenic factors.* These are molecules that direct the differentiation of cells into certain cell types. They appear to be localized in different regions of the egg and become segregated into different cells during cleavage (Chapters 7 and 8).

Within this enormous volume of cytoplasm resides a large nucleus. In some species (sea urchins, for example) the nucleus is already haploid at the time of fertilization. In other species (including many worms and most mammals) the egg nucleus is still diploid and the sperm enters before the meiotic divisions are completed. The stage of the egg nucleus at the time of sperm entry is illustrated in Figure 5.

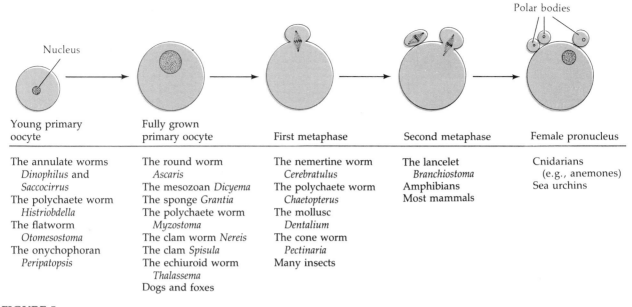

Young primary oocyte	Fully grown primary oocyte	First metaphase	Second metaphase	Female pronucleus
The annulate worms *Dinophilus* and *Saccocirrus*	The round worm *Ascaris*	The nemertine worm *Cerebratulus*	The lancelet *Branchiostoma*	Cnidarians (e.g., anemones)
The polychaete worm *Histriobdella*	The mesozoan *Dicyema*	The polychaete worm *Chaetopterus*	Amphibians	Sea urchins
The flatworm *Otomesostoma*	The sponge *Grantia*	The mollusc *Dentalium*	Most mammals	
The onychophoran *Peripatopsis*	The polychaete worm *Myzostoma*	The cone worm *Pectinaria*		
	The clam worm *Nereis*	Many insects		
	The clam *Spisula*			
	The echiuroid worm *Thalassema*			
	Dogs and foxes			

FIGURE 5

Stages of egg maturation at the time of sperm entry in different animals. (After Austin, 1965.)

Enclosing the cytoplasm is the egg PLASMA MEMBRANE. This membrane must regulate the flow of certain ions during fertilization and must be capable of fusing with the sperm plasma membrane. Above the plasma membrane is the VITELLINE ENVELOPE (Figure 6). This glycoprotein membrane is essential for the species-specific binding of sperm. In mammals, the vitelline envelope is very thick and is called the ZONA PELLUCIDA. The mammalian egg is also surrounded by a layer of cells, the CUMULUS CELLS (Figure 7). These represent ovarian follicular cells, which were nurturing

FIGURE 6

The sea urchin egg cell surface. (A) Scanning electron micrograph of an egg before fertilization. The plasma membrane is exposed where the vitelline envelope has been torn. (B) Transmission electron micrograph of an unfertilized egg showing microvilli and plasma membrane, which is closely covered by the vitelline envelope. A cortical granule lies directly beneath the plasma membrane of the egg. (From Schroeder, 1979; photographs courtesy of T. E. Schroeder.)

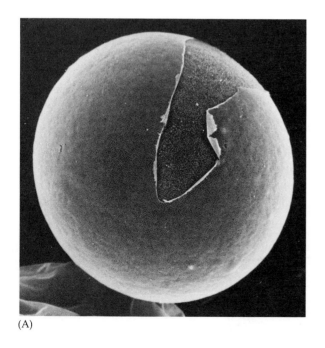

(A)

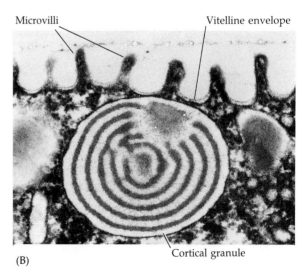

(B)

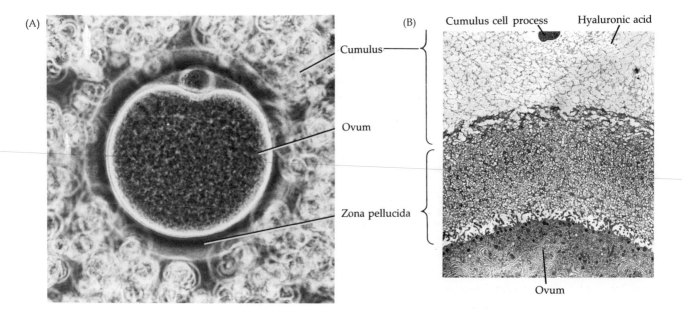

(A)

(B) Cumulus cell process Hyaluronic acid

Cumulus

Ovum

Zona pellucida

Ovum

FIGURE 7

Mammalian eggs immediately before fertilization. (A) The hamster egg, or ovum, is encased in the zona pellucida. This, in turn, is surrounded by the cells of the cumulus. A polar body cell, produced during meiosis, is also within the zona pellucida. (B) Electron micrograph of a hamster ovum fixed in a manner that preserves the structure of the zona pellucida and hyaluronic acid matrices that surround the egg. A process from a cumulus cell can be seen within the hyaluronic acid matrix. (A courtesy of R. Yanagimachi; B from Yudin et al., 1988, courtesy of G. Cherr.)

the egg at the time of its release from the ovary. Sperm have to get past these cells, also, to fertilize the egg.*

Lying immediately beneath the plasma membrane of the sea urchin egg is the cortex. The cytoplasm in this region is more gel-like than the internal cytoplasm and contains high concentrations of globular actin molecules. During fertilization, these actin molecules polymerize to form long cables of actin known as MICROFILAMENTS. Microfilaments are necessary for cell division, and they also are used to extend the egg surface into the microvilli, which aid sperm entry into the cell (Figures 6 and 20). Also within this cortex are the CORTICAL GRANULES (Figures 4 and 6). These membrane structures are homologous to the acrosomal vesicle of the sperm, being Golgi-derived organelles containing proteolytic enzymes. However, whereas each sperm contains one acrosomal vesicle, each sea urchin egg contains approximately 15,000 cortical granules. Moreover, in addition to containing the digestive enzymes, the cortical granules also contain MUCOPOLYSACCHARIDES and HYALINE PROTEIN. The enzymes and mucopolysaccharides are active in preventing other sperm from entering the egg after the first sperm has entered, and the hyaline protein sur-

*In some species, the extracellular coverings of the mammalian eggs are divided into three regions: the zona pellucida, the corona radiata, and the cumulus. The corona radiata is then used to refer to those follicle cells that are immediately adjacent to the zona pellucida, and the term cumulus (or cumulus oophorus) is then reserved for the more diffuse cells that are further from the egg and zona. In many mammals (for example, the hamster, shown in Figure 7B), these two layers are not readily distinguished, and in other species (such as sheep and cattle), these cellular layers are probably shed before fertilization (see Talbot, 1985).

rounds the early embryo and provides support for the cleavage-stage blastomeres.

Many types of eggs also secrete an egg jelly outside their vitelline envelope. This glycoprotein meshwork can have numerous functions. Most often, though, it is used to either attract or activate sperm. The egg, then, is a cell specialized for receiving sperm and initiating development.

Recognition of sperm and egg: Action at a distance

Many marine organisms release their gametes into the local environment. This environment may be as small as a tidepool or as large as the ocean. Moreover, this environment is shared with other species that may shed their sex cells at the same time. These organisms are faced with two problems: (1) How can sperm and eggs meet in such a dilute concentration? (2) What mechanism prevents starfish sperm from trying to fertilize sea urchin eggs? Two major mechanisms have evolved to solve these difficulties: species-specific attraction of sperm and species-specific sperm activation.

Sperm attraction

Species-specific sperm attraction (a type of chemotaxis) has been documented in numerous species, including cnidarians, molluscs, echinoderms, and urochordates (Miller, 1985). In 1978 Miller demonstrated that the eggs of the cnidarian *Orthopyxis caliculata* not only secrete a chemotactic factor but also regulate the timing of its release. Developing oocytes at various stages in their maturation were fixed onto microscope slides, and sperm were released at a certain distance from the eggs. Miller found that when sperm were added to eggs that had not yet completed their second meiotic division, there was no attraction of sperm to eggs. However, after the second meiotic division was finished and the eggs were ready to be fertilized, the sperm migrated toward them. Thus, the oocyte controlled not only the type of sperm it would attract but also the time at which it would attract them.

The mechanisms for chemotaxis are different in other species (reviewed in Metz, 1978). Recently, a chemotactic molecule, RESACT, a 14-amino acid peptide, has been isolated from the egg jelly of the sea urchin *Arbacia punctulata* (Ward et al., 1985). Resact diffuses readily in seawater and has profound effects at very low concentrations when it is added to a suspension of *Arbacia* sperm (Figure 8). When a drop of seawater containing *Arbacia* sperm is placed on a microscope slide, the sperm generally

FIGURE 8
Sperm chemotaxis in *Arbacia*. One nanoliter of a 10-nM solution of resact is injected into a 20-μl drop of sperm suspension. The position of the micropipette is indicated in A. (A) One-second photographic exposure showing sperm swimming in tight circles before addition of resact. (B–D) Similar one-second exposures showing migration of sperm to the center of the resact gradient 20 seconds, 40 seconds, and 90 seconds after injection. (From Ward et al., 1985; photographs courtesy of V. D. Vacquier.)

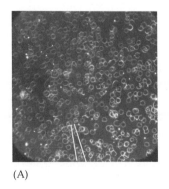

(A)

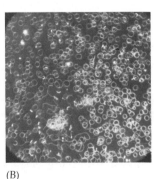

(B)

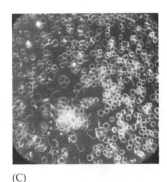

(C)

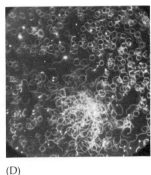

(D)

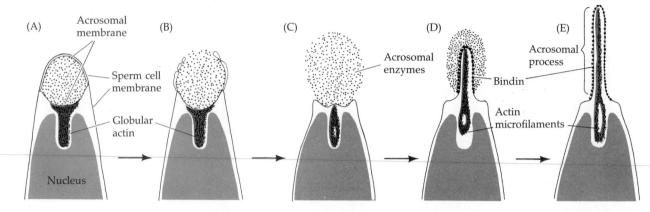

FIGURE 9

Acrosome reaction in sea urchin sperm. (A–C) The portion of the acrosomal membrane lying directly beneath the sperm cell membrane fuses with the cell membrane to release the contents of the acrosomal vesicle. (D,E) As the actin molecules assemble to produce microfilaments, the acrosomal process is extended outward. Actual photographs of the acrosome reaction in sea urchin sperm are shown below. (After Saunders, 1970; photographs courtesy of G. L. Decker and W. J. Lennarz.)

swim in circles about 50 μm in diameter. Within seconds of the introduction of a minute amount of resact into the drop, sperm migrate into the region of the injection and congregate there. As resact continues to diffuse from the area of injection, more sperm are recruited into the growing cluster. Resact is specific for *Arbacia punctulata* and does not attract sperm of other species. *Arbacia* sperm bind resact to receptors in their cell membranes (Ramarao and Garbers, 1985; Bentley et al., 1986) and can swim up a concentration gradient of this compound until they reach the egg.

The acrosome reaction

A second interaction between sperm and egg involves the activation of sperm by egg jelly. In most marine invertebrates, this acrosome reaction has two components: the fusion of the acrosomal vesicle with the sperm plasma membrane (which results in the release of the contents of the acrosomal vesicle) and the extension of the acrosomal process (Colwin and Colwin, 1963; Figure 9). The acrosome reaction can be initiated by soluble egg jelly, by the egg jelly surrounding the egg, or even by contact with the egg itself in certain species. It can also be activated artificially by increasing the calcium concentration of seawater.

In sea urchins, contact with egg jelly causes the fusion of the acrosomal vesicle and release of protein-digesting enzymes that can digest a path through the jelly coat to the egg surface (Dan, 1967; Franklin, 1970; Levine et al., 1978). The sequence of these events is outlined in Figure 9. The acrosome reaction is thought to be initiated by a sulfated polysaccharide in the egg jelly that allows calcium and sodium ions to enter the sperm head and causes potassium and hydrogen ions to leave (SeGall and Lennarz, 1979; Schackmann and Shapiro, 1981). The fusion of the acrosomal vesicle is caused by the calcium-mediated fusion of the acrosomal membrane with the adjacent sperm plasma membrane (Figures 9 and 10). This is essentially an exocytosis reaction wherein a vesicle is brought to the cell surface and fuses with the cell membrane, thereby releasing its contents. Such exocytotic reactions are seen in the release of insulin from pancreatic cells and in the release of neurotransmitters from synaptic terminals. In all cases, there is a calcium-mediated fusion between the secretory vesicle and the cell membrane.

The second part of the acrosome reaction involves the extension of the acrosomal process (Figure 9). This protrusion arises from the polymerization of globular actin molecules into actin filaments, a reaction that appears to be dependent upon the release of hydrogen ions from the sperm head (Tilney et al., 1978; Schackmann et al., 1978). Until activation,

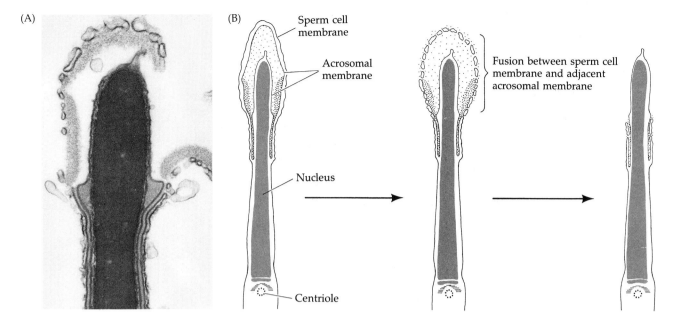

(A)

(B)
Sperm cell
membrane

Acrosomal
membrane

Fusion between sperm cell
membrane and adjacent
acrosomal membrane

Nucleus

Centriole

a regulatory protein may be responsible for blocking actin polymerization, and the rise in intracellular pH accompanying activation might subsequently interfere with this function. In addition to causing the extension of the acrosomal process, this significant increase in pH is also responsible for activating the dynein ATPase in the neck of the sperm. This activation causes a rapid utilization of ATP and a 50 percent increase in mitochondrial respiration. The energy generated is used primarily for flagellar motility (Tombes and Shapiro, 1985). A possible scheme for the events of the acrosome reaction is presented in Figure 11.

The egg jelly factors that initiate the acrosomal reactions of sea urchins are often highly specific. The sperm of sea urchins *Arbacia punctulata* and *Strongylocentrotus drobachiensis* will react only with jelly of their own eggs. However, *S. purpuratus* sperm can also be activated by *Lytechinus variegatus* (but not *A. punctulata*) egg jelly (Summers and Hylander, 1975). Therefore, egg jelly may provide species-specific recognition in some species but not in others.

FIGURE 10

Acrosome reaction in hamster sperm. (A) Transmission electron micrograph of hamster sperm undergoing acrosomal reaction. The membrane can be seen to form vesicles. (B) Interpretive diagram of electron micrographs showing the fusion of the acrosomal and cell membranes in the sperm head. (A from Meizel, 1984; courtesy of S. Meizel. B after Yanagimachi and Noda, 1970.)

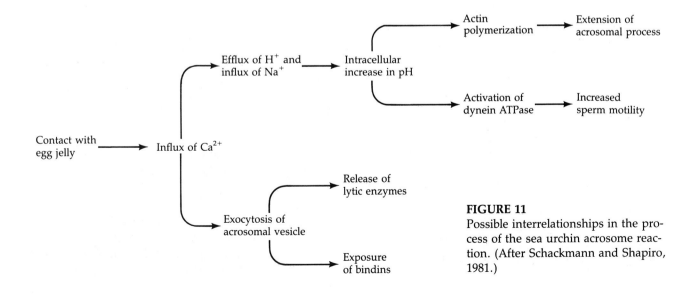

Contact with
egg jelly → Influx of Ca^{2+}

Efflux of H^+ and influx of Na^+ → Intracellular increase in pH

Actin polymerization → Extension of acrosomal process

Activation of dynein ATPase → Increased sperm motility

Exocytosis of acrosomal vesicle

Release of lytic enzymes

Exposure of bindins

FIGURE 11

Possible interrelationships in the process of the sea urchin acrosome reaction. (After Schackmann and Shapiro, 1981.)

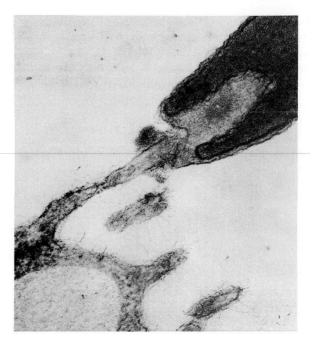

FIGURE 12
Contact of a sea urchin sperm acrosomal process with an egg microvillus. (From Epel, 1977; photograph courtesy of F. D. Collins and D. Epel.)

FIGURE 13
Species-specific agglutination of dejellied eggs by bindin. (A) Agglutination was promoted by adding 212 μg bindin to a plastic well containing 0.25 ml of a 2 percent (volume:volume) suspension of eggs. After 2–5 minutes of gentle shaking, the wells were photographed. (The interpretive diagram is based on photographs of Glabe and Vacquier, 1977.) (B) Fluorescence photomicrograph of *S. purpuratus* eggs bound together by fluorescein-labeled *S. purpuratus* bindin particles. Bindin particles were invariably present at the places where two eggs came together. (From Glabe and Lennarz, 1979; photograph courtesy of the authors.)

Recognition of sperm and egg: Contact of gametes

Species-specific recognition in sea urchins

Once the sea urchin sperm has penetrated the egg jelly, the acrosomal process of the sperm contacts the outer layer of the egg's vitelline envelope (Figure 12). A major species-specific recognition step occurs at this point. The acrosomal protein mediating this recognition is called BINDIN. In 1977 Vacquier and his co-workers isolated this nonsoluble 30,500-Da protein

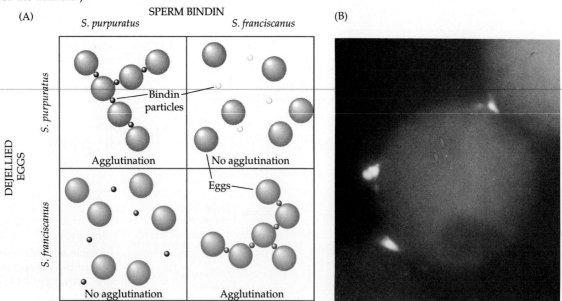

from the acrosome of *Strongylocentrotus purpuratus*. This protein is capable of binding to dejellied eggs or to the isolated vitelline envelopes from *S. purpuratus* (Vacquier and Moy, 1977; Figure 13). Further, its interaction with eggs is species-specific (Glabe and Vacquier, 1977; Glabe and Lennarz, 1979); bindin isolated from the acrosomes of *S. purpuratus* agglutinates its own dejellied eggs, but not those of the closely related species *S. franciscanus*.

In 1979 Moy and Vacquier demonstrated that bindin is located specifically on the acrosomal process, exactly where it should be for sperm–egg recognition. The immunochemical staining procedure they used is illustrated in Figure 14A. Antibodies to purified sea urchin bindin were made by injecting the bindin into rabbits. This rabbit antibindin was then bound to sea urchin sperm that had undergone the acrosome reaction. After any unbound antibody was washed off, the sperm were treated with swine antibodies that could bind to rabbit antibodies. These swine antibodies had been covalently linked to peroxidase enzymes. In such a fashion, peroxidase molecules were placed wherever bindin was present. Peroxidase catalyzes the formation of a dark precipitate from diaminobenzidine and hydrogen peroxide. When these two substrates were added to the treated sperm, the precipitate was found to cover the acrosomal process (Figure 14B). Moreover, bindin could not be detected on the sperm surface until after the acrosomal reaction, and then it was found at the sperm–egg junction (Figure 14C).

Biochemical studies have shown that the bindins of closely related sea urchin species are indeed different. This finding implies the existence of species-specific bindin receptors on the vitelline envelope. Such receptors were also suggested by the experiments of Vacquier and Payne (1973), who saturated sea urchin eggs with sperm. As seen in Figure 15, sperm binding does not occur over the entire egg surface. Even at saturating numbers of sperm (approximately 1500) there appears to be room on the ovum for more sperm heads, which implies there is a limiting number of sperm binding sites. A large glycoprotein complex from the vitelline envelopes of sea urchin eggs has been isolated and shown to bind radioactive bindin in a species-specific manner (Glabe and Vacquier, 1978; Rossignol et al., 1984). This glycoprotein is also able to compete with eggs for the

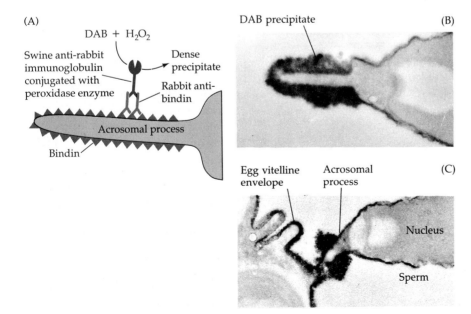

(A)

DAB + H_2O_2

Swine anti-rabbit immunoglobulin conjugated with peroxidase enzyme

Dense precipitate

Rabbit anti-bindin

Acrosomal process

Bindin

DAB precipitate (B)

Egg vitelline envelope   Acrosomal process (C)

Nucleus

Sperm

FIGURE 14
Localization of bindin on the acrosomal process. (A) Immunochemical localization technique places a rabbit antibody wherever bindin is exposed. A swine antibody that reacts against rabbit antibody can then bind to this first antibody. The swine antibody has been covalently linked with a molecule of peroxidase, an enzyme that can catalyze the reaction of diaminobenzidine (DAB) and peroxide to form an electron-dense precipitate. Thus, this precipitate will form only where bindin is present. (B) Localization of bindin to the acrosomal process after the acrosome reaction (×133,200). (C) Localization of bindin to the acrosomal process at the junction of the sperm and the egg. (B and C from Moy and Vacquier, 1979; photographs courtesy of V. D. Vacquier.)

FIGURE 15
Scanning electron micrograph of sea urchin sperm bound to the vitelline envelope of an egg. (Photograph courtesy of C. Glabe, L. Perez, and W. J. Lennarz.)

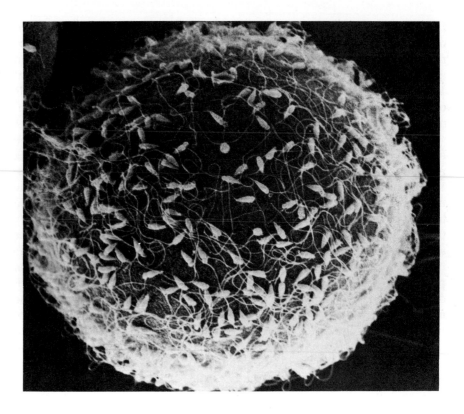

sperm of the same species. That is, if *S. purpuratus* sperm is mixed with the bindin receptor from *S. purpuratus* vitelline envelopes, the sperm bind to it and will not fertilize the eggs. The isolated bindin receptor from *S. purpuratus*, however, does not interfere with the fertilization of other related sea urchins. Thus, species-specific recognition of sea urchin gametes occurs at the levels of acrosome activation and sperm adhesion to the vitelline envelope.

Gamete recognition and binding in mammals

Capacitation. So far we have been focusing our discussion on those organisms for which fertilization takes place externally. In mammalian species, fertilization is internal, and the fertilization process has been adapted to this environment. Indeed, investigators have concluded that the reproductive tract of female mammals plays a very active role in the fertilization process. Newly ejaculated mammalian sperm are unable to undergo the acrosomal reaction without residing for some amount of time in the female reproductive tract. This requirement for CAPACITATION varies from species to species (Gwatkin, 1976) and can be mimicked in vitro by incubating sperm in tissue culture media or in fluid from the oviducts. The molecular changes that account for capacitation are still unknown (Saling, 1989). The concentration of cholesterol in the sperm plasma membrane is seen to be lowered during sperm capacitation in several species (Davis, 1981) and particular proteins or carbohydrates are seen to be lost (Lopez et al., 1985; Wilson and Oliphant, 1987; Poirier and Jackson, 1981). It is possible that these changes in membrane lipids allow for greater mobility of membrane proteins and that the moieties lost during capacitation had been blocking zona-binding proteins. However, it is still uncertain to what extent each of these mechanisms causes capacitation of the sperm.

Primary binding of sperm to zona pellucida.   To reach the egg, the mammalian sperm must first pass through the extracellular matrices that surround the egg. The innermost layer, adjacent to the egg, is the zona pellucida. This glycoprotein shell is made during oogenesis and is present on all mammalian eggs at the time of fertilization. The outer layer is composed of cumulus cells and their hyaluronate-rich matrix. These cumulus cells are the remnants of the granulosa cells of the ovarian follicle, which are shed along with the egg during ovulation. In some species, the cumulus cells are not present at the time of fertilization and they do not constitute a barrier to sperm even when they are present (Talbot et al., 1985). Capacitated sperm pass freely through the cumulus layer to the zona, while noncapacitated sperm (and sperm that have prematurely undergone the acrosome reaction) are held up in the matrix (Austin, 1960; Corselli and Talbot, 1987). In the best-studied mammals—mice and hamsters—the acrosome reaction occurs *after* the sperm has bound to the zona pellucida (Saling et al., 1979; Florman and Storey, 1982; Cherr et al., 1986).

The zona pellucida in mammals plays a role analogous to that of the vitelline envelope. The binding of sperm to the zona is relatively, but not absolutely, species-specific (species specificity should not be a major problem when fertilization occurs internally), and the binding of mouse sperm to the mouse zona can be inhibited by first incubating the mouse sperm with zona glycoproteins. Bleil and Wassarman (1980, 1986, 1988) have isolated from the zona pellucida an 83,000-Da glycoprotein (ZP3) that is the active competitor in this inhibition assay (Figure 16). The other two zona glycoproteins, ZP1 and ZP2, failed to compete for sperm binding. Thus, there is a specific glycoprotein in the zona pellucida to which the mouse sperm bind. Moreover, this same protein has been shown to initiate the acrosome reaction after sperm have bound to it. The mouse sperm can thereby concentrate its proteolytic enzymes directly at the point of attachment at the zona pellucida.

The molecular mechanism by which the zona pellucida and the mammalian sperm recognize each other is presently being studied. The current hypothesis of mammalian gamete binding postulates a set of proteins on the sperm capable of recognizing a specific carbohydrate and protein regions on the egg zona ZP3 (Wassarman, 1987; Saling, 1989). Florman and co-workers (1984; Florman and Wassarman, 1985) have shown that a critical moiety for sperm–zona recognition in the mouse is a carbohydrate

FIGURE 16
Binding of sperm to the zona pellucida. (A) Inhibition assay showing the specific decrease of mouse sperm binding to zonae pellucidae when sperm and zonae are incubated with increasingly large amounts of glycoprotein ZP3. The importance of the carbohydrate portion of ZP3 is also indicated by this figure. (B) Binding of radioactively labeled ZP3 to capacitated mouse sperm. (A after Bleil and Wassarman, 1980, and Florman and Wassarman, 1985; B from Bleil and Wassarman, 1986; courtesy of the authors.)

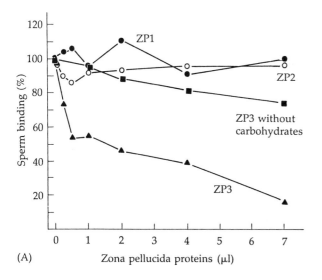

(A)

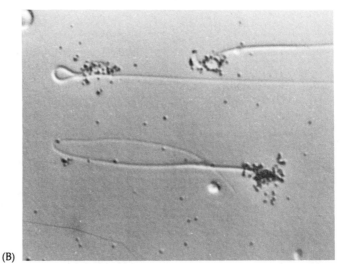

(B)

group on the ZP3 glycoprotein. Removal of these carbohydrate groups abolishes the ability of ZP3 to compete for sperm binding. Bleil and Wassarman (1980) have further shown that the critical sugar is the terminal galactose on these carbohydrate chains. If this terminal galactose is removed or chemically modified, sperm-binding activity is lost.

But what is the sperm component that is recognizing the zona glycoproteins? There are at least three mouse sperm proteins that are capable of binding to the zona pellucida. The first appears to be a protein that specifically binds to the galactose residues of ZP3. Bleil and Wassarman (1990) isolated this protein by binding ZP3 covalently to beads and passing the proteins isolated from mouse sperm membranes over them. Most of the proteins passed through the column, but one, a 56,000-Da peptide, bound to the ZP3-coated beads. It did not bind to ZP2-coated beads in a similar experiment. This protein was found to be exposed in the sperm membrane and bound to galactose residues, strongly suggesting that it is the sperm receptor for the ZP3 carbohydrate.

The second sperm protein that appears to be important for sperm–zona binding is a sperm cell membrane glycosyltransferase enzyme. Shur's laboratory (Shur and Hall, 1982a,b; Lopez et al., 1985) has shown that this receptor for the zona is an enzyme on the outside of the sperm head that recognizes the sugar N-acetylglucosamine. This enzyme, N-acetylglucosamine:galactosyltransferase, is embedded in the sperm plasma membrane, directly above the acrosome, with its active site pointed outward. The enzymatic function of this glycosyltransferase protein would be to catalyze the addition of a galactose sugar onto a carbohydrate chain terminating with an N-acetylglucosamine sugar. However, there are no activated galactose residues in the female reproductive tract. Thus, although the enzyme can bind to the N-acetylglucosamine residues of the zona proteins exactly as any enzyme would bind to a substrate, it cannot catalyze the reaction because the second reactant is not present. Therefore, the enzymes (on the sperm) remain bound to their substrate (on the egg). If this hypothesis were correct, one would expect that sperm–egg binding could be inhibited either by inhibiting the enzyme or by adding the second reactant, activated galactose. This is exactly what Shur and co-workers found to be the case. Either inhibiting the binding of the glycosyltransferase to N-acetylglucosamine residues (with antibodies or modifying proteins) or completing the reaction (with UDP-galactose) stopped the sperm from binding to the egg. Moreover, the galactosyltransferase has been purified from sperm cell membranes and found to compete with mouse sperm for binding to the zona pellucida (Shur and Neely, 1988).

A third sperm protein that binds to the mouse zona pellucida appears to be a protease. Early research into mammalian fertilization had established that protease inhibitors blocked fertilization, but it had been assumed that these inhibitors arrested the digestion of the zona by the already-attached sperm. In 1981, Saling found that these protease inhibitors actually stopped the binding of the capacitated sperm to the zona. Once the sperm had attached, trypsin inhibitors could not block the movement of sperm through the zona nor the fusion of sperm and egg. Moreover, incubation of the egg with these inhibitors did not inhibit sperm binding, but incubating the sperm with the inhibitors did. This trypsin-inhibitor sensitive protein has not yet been isolated, but its characteristics distinguish it from known proteases (Benau and Storey, 1987).

It is possible that all these adhesion systems (and perhaps more) are operating to keep the wiggling sperm attached to the zona. Thousands of sites might be needed to keep these two cells from coming apart. Figure 17 attempts to integrate these various models of sperm–zona adhesion.

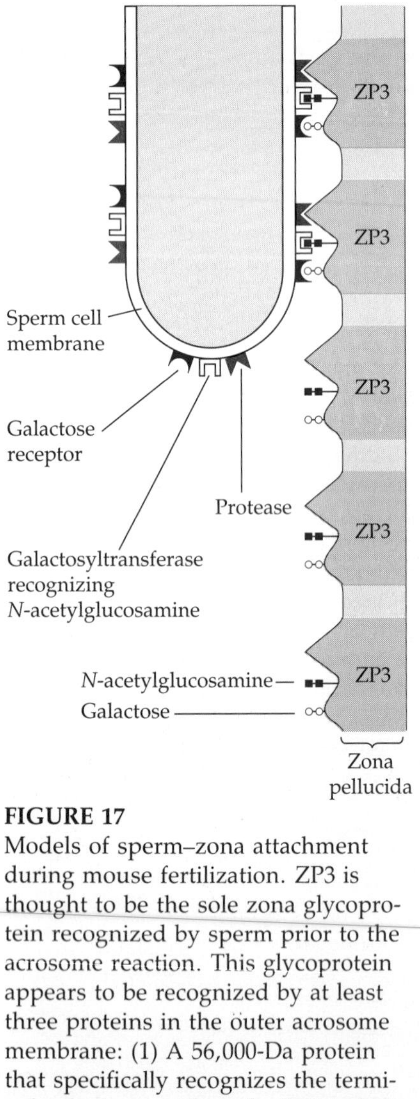

FIGURE 17
Models of sperm–zona attachment during mouse fertilization. ZP3 is thought to be the sole zona glycoprotein recognized by sperm prior to the acrosome reaction. This glycoprotein appears to be recognized by at least three proteins in the outer acrosome membrane: (1) A 56,000-Da protein that specifically recognizes the terminal galactose residue of a ZP3 carbohydrate chain. (2) A sperm cell surface galactosyltransferase that recognizes an exposed N- acetylglucosamine residue of a ZP3 carbohydrate chain. (3) A protease recognizes (and presumably doesn't digest) a peptide region of the zona glycoprotein.

SIDELIGHTS & SPECULATIONS

Induction of the mammalian acrosome reaction by ZP3

Once the capacitated sperm has bound to the zona pellucida, how does the mammalian acrosome reaction take place? The reaction appears to take place on the ZP3 glycoprotein, since modifying or destroying the peptide chain of ZP3 can eliminate the acrosome reaction even though the binding of sperm to ZP3 is unaffected (Endo et al., 1987a, Leyton and Saling, 1989a). Investigators are presently looking at the ways by which binding to ZP3 might cause changes within the sperm cell that could lead to acrosomal exocytosis. First, ZP3 may stimulate exocytosis by aggregating the ZP3 receptors in the sperm cell membrane. Leyton and Saling (1989b) incubated capacitated mouse sperm with fragments of ZP3 that were able to bind to sperm. They did not cause the acrosome reaction to occur. However, when these ZP3 fragments were artificially bound together by anti-ZP3 antibodies, the acrosome reaction did take place (Figure 18). Therefore, it appears that the clustering together of the sperm receptors for ZP3 is necessary for the acrosome reaction.

But this still does not tell us what reactions are taking place. Since membrane fusion events (including the acrosome reactions of many invertebrates) require calcium ions, scientists began to look at the possibility that the binding of ZP3 might trigger a release of calcium ions within the sperm. In the 1980s, a major pathway for the release of calcium ions had been discovered in a variety of cells. It was found that when epinephrine binds to its receptors in neuronal cell membranes, when growth factors bind to

their receptors in the membranes of peripheral tissues, or when antigens bind to their receptors on the surfaces of lymphocytes, the same reactions occur to release calcium ions from storage. The receptor was found to be coupled to a GTP-binding protein ("G-protein") that becomes activated once the receptor binds its specific ligand. The activated G-protein, in turn, activates an enzyme called phospholipase C. This enzyme splits the membrane lipid phosphatidylinositol 4,5-bisphosphate into diacylglycerol and inositol trisphosphate. Inositol trisphosphate is able to free bound calcium ions from membrane compartments. (This pathway will be described later in the chapter when we discuss egg activation.) Such a G-protein has been discovered in mammalian sperm and appears to be activated by the binding of ZP3. Moreover, if activation of the G-protein is inhibited, the sperm can bind ZP3 but cannot undergo the acrosome reaction (Endo et al., 1987b; Wilde and Kopf, 1989). These two mechanisms mentioned here (and there are others being studied) are not mutually exclusive. In lymphocytes, for instance, both the clustering of antigen receptors and the activation of G-proteins by the antigen-bound receptors are thought to be critical for the release of sequestered calcium ions in these cells.

FIGURE 18
Evidence that the crosslinking of sperm receptors for ZP3 are necessary to initiate the acrosome reaction. Fragments of ZP3 can bind to sperm but not activate the acrosome reaction. However, when these fragments are crosslinked by antibodies, the acrosome reaction takes place.

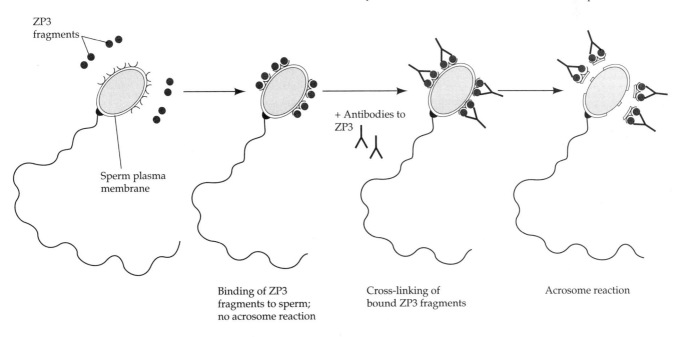

ZP3 fragments

Sperm plasma membrane

Binding of ZP3 fragments to sperm; no acrosome reaction

+ Antibodies to ZP3

Cross-linking of bound ZP3 fragments

Acrosome reaction

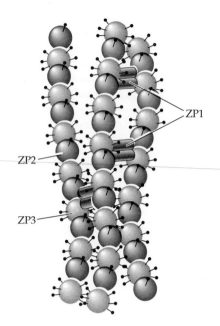

ZP1

ZP2

ZP3

FIGURE 19

Diagrammatic representation of the fibrillar structure of the mouse zona pellucida. The major strands of the zona are composed of repeating dimers of proteins ZP2 and ZP3. These strands are occasionally crosslinked together by ZP1, forming a meshlike network. (After Wassarman, 1989.)

Secondary binding of sperm to zona pellucida. During the acrosome reaction, the anterior portion of the sperm plasma membrane is shed from the sperm (Figure 10). This is where the ZP3-binding proteins are located. Yet, the sperm must still remain bound to the zona in order to lyse a path through it. In mice, it appears that the secondary binding to the zona is accomplished by proteins in the inner acrosomal membrane that bind specifically to ZP2 (Bleil et al., 1988). Whereas acrosome-intact sperm will not bind to the ZP2 glycoprotein, acrosome-reacted sperm will. Moreover, antibodies against the ZP2 protein will not prevent the binding of acrosome-intact sperm to the zona, but will inhibit the attachment of acrosome-reacted sperm. The structure of the zona consists of repeating units of ZP3 and ZP2, occasionally crosslinked by ZP1 (Figure 19). It appears that the acrosome-reacted sperm transfer their binding from ZP3 to the adjacent ZP2 molecules. After a mouse sperm has entered the egg, the egg cortical granules release their contents. One of the proteins released by these granules is a protease that specifically alters ZP2 (Moller and Wassarman, 1989). This would inhibit acrosome-reacted sperm from moving further towards the egg.

It is not known which of the mouse sperm proteins bind to the ZP2, and it is possible that one or more of the proteins thought to direct primary binding is actually involved with secondary binding. In porcine sperm, secondary zona binding appears to be mediated by PROACROSIN. Proacrosin becomes the protease ACROSIN that has long been known to be involved in digesting the zona pellucida. However, proacrosin is also a fucose-binding protein that serves to maintain the connection between acrosome-reacted sperm and the zona pellucida (Jones et al., 1988). It is possible that proacrosin binds to the zona and is then converted into the active enzyme that locally digests the zona pellucida.

SIDELIGHTS & SPECULATIONS

Contraception by antibodies

If male and female animals are injected with extracts of sperm cells, many will produce anti-sperm antibodies and become infertile. In fact, approximately 5 percent of people attending infertility clinics have antibodies in their serum or genital tract secretions that react against sperm. In 1981, Yanagimachi and his colleagues showed that the antisera produced by guinea pigs against sperm blocked the attachment of sperm to the zona pellucida. These animals had been sterilized by a vaccination.

In 1988, Primakoff and co-workers found that a particular protein on guinea pig sperm could induce this type of sterility. When this sperm protein, PH-20, was injected into male or female guinea pigs, 100 percent of them became sterile for several months. The blood sera of these sterile guinea pigs had extremely high concentrations of antibodies to PH-20. PH-20 is present in both the plasma membrane and the inner acrosomal membrane of guinea pig sperm. Its presence throughout the inner acrosomal mem-

brane of sperm that have undergone the acrosome reaction suggests that it could be important in the secondary binding of the sperm and zona. The antiserum from female guinea pigs sterilized by injections of PH-20 not only bound specifically to this protein, but they also blocked sperm–zona adhesion in vitro. The contraceptive effect lasted several months, after which time fertility was restored.

Another approach to long-term immunological contraception has been to produce antibodies against the zona pellucida. In mice, females injected with ZP3 form antibodies against this protein and are rendered infertile (Millar et al., 1989).

A human analogue of the PH-20 protein is not yet known, but certain sperm antigens display a similar pattern of localization in the sperm. Similarly, the human zona proteins and their functions are not as clearly established as those in the mouse. Nevertheless, these experiments show that the principle of immunological contraception is well founded.

(A) 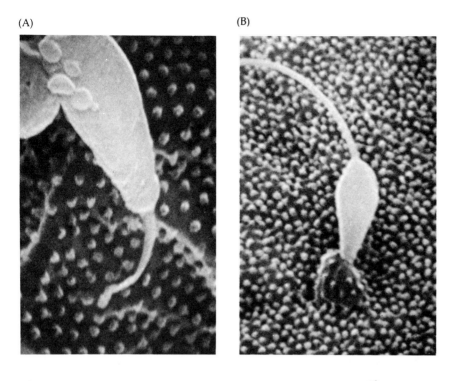 (B)

(C) 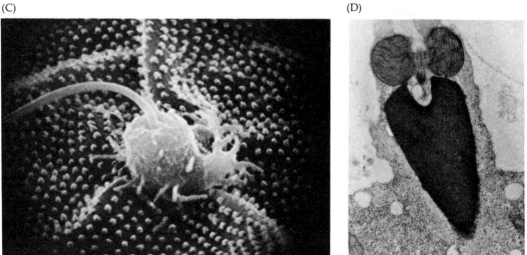 (D)

FIGURE 20
Scanning electron micrographs of the entry of sperm into sea urchin eggs. (A) Contact of sperm head with egg microvillus through the acrosomal process. (B) Formation of fertilization cone. (C) Sperm becomes internalized within the egg. (D) Transmission electron micrograph of sperm internalization through the fertilization cone. (A–C from Schatten and Mazia, 1976; photographs courtesy of G. Schatten; D courtesy of F. J. Longo.)

Gamete fusion and prevention of polyspermy

Fusion between egg and sperm cell membranes

Recognition of sperm by the vitelline envelope or zona is followed by the lysis of that portion of the envelope in the region of the sperm head (Colwin and Colwin, 1960; Epel, 1980). This lysis is followed by the fusion of the sperm cell membrane with the cell membrane of the egg.

The entry of a sperm into the sea urchin egg is illustrated in Figure 20. The egg surface is covered with small extensions called MICROVILLI; sperm–egg fusion appears to cause the polymerization of actin and the extension of several microvilli (Summers et al., 1975; Schatten and Schatten, 1980, 1983) to form the FERTILIZATION CONE. This cytoplasmic protuberance averages 7 μm in length and 2 μm in width. Homology between the egg and the sperm is again demonstrated, because the transitory fertilization cone, like the acrosomal process, appears to be extended by the polymerization of actin. The sperm and egg membranes join together, and material from the sperm cell membrane can later be found on the egg membrane (Gundersen et al., 1986). The sperm nucleus and tail pass through the cytoplasmic bridge, which is widened by the actin polymerization. Longo (1986) has suggested that the membranes for the fertilization cone in the sea urchin *Arbacia* come largely from the surrounding microvilli (which become shorter as a result). Not all sperm membrane components may be mixing with the egg membrane, however. Some material appears to be localized at the point of sperm entry (not necessarily in the membrane). Manes and Barbieri (1976, 1977) have speculated that this localization of sperm components in a restricted area provides an asymmetry to the cell that helps direct the plane of first cleavage and the movement of cytoplasmic components within the egg.

In the sea urchin, all regions of the egg are capable of fusing with sperm; in certain amphibians and many invertebrates, there are specialized regions of the membrane for sperm recognition and fusion (Vacquier, 1979). Although not much is known about the mechanism for gamete fusion, Hirao and Yanagimachi (1978) have demonstrated that membrane phospholipids are extremely important in this reaction. This finding is in accord with present models of membrane fusion, which emphasize the role of membrane lipid molecules.

Fusion is an active process, often mediated by specific "fusogenic" proteins. Proteins such as the HA protein of influenza virus and the F protein of Sendai virus are known to promote cell fusion, and it is possible that bindin is also such a protein. Glabe (1985) has shown that sea urchin bindin will cause phospholipid vesicles to fuse together and that, like the viral fusogenic proteins, bindin contains a long stretch of hydrophobic amino acids near its *N*-terminus. In abalones, the lysin that dissolves the vitelline envelope has also been found to have fusogenic activity (Hong and Vacquier, 1986), and a protein in the head of guinea pig sperm is essential for sperm–egg membrane fusion in that species (Primakoff et al., 1987). It appears, then, that one of the proteins on the sperm head is capable of stimulating the fusion of the sperm and egg cell membranes.

Prevention of polyspermy

As soon as one sperm has entered the egg, the fusibility of the egg membrane, which was so necessary to get the sperm inside the egg, becomes a dangerous liability. In sea urchins, as in most animals studied, any sperm that enters the egg can provide a haploid nucleus and a centriole to the egg. In normal MONOSPERMY, where only one sperm enters the egg, a haploid sperm nucleus and a haploid egg nucleus combine to form the diploid nucleus of the fertilized egg (ZYGOTE), thus restoring the chromosome number appropriate for the species. The centriole, coming from the sperm, will divide to form the two poles of the mitotic spindle during cleavage.

The entrance of multiple sperm—POLYSPERMY—leads to disastrous consequences in most organisms. In the sea urchin, fertilization by two sperm results in a triploid nucleus wherein each chromosome is represented three times rather than twice. Worse, it means that instead of a bipolar mitotic spindle that separates the chromosomes into two cells, the triploid chromosomes would be divided into as many as four cells. Because there is no mechanism to ensure that each of the four cells receives the proper number and type of chromosomes, the chromosomes would be apportioned unequally. Some cells would receive extra copies of certain chromosomes and other cells would lack them (Figure 21). Theodor Boveri demonstrated in 1902 that such cells either die or develop abnormally.

Species have evolved ways to prevent the union of more than two haploid nuclei. The most common way is to prevent the entry of more than one sperm into the egg. The sea urchin egg has two mechanisms to avoid polyspermy: a fast reaction, accomplished by an electrical change in the egg plasma membrane, and a slower reaction, caused by the exocytosis of the cortical granules.

The fast block to polyspermy. The egg cell membrane has to be able to fuse with the sperm membrane, but it must lose this ability immediately after the entry of the first sperm (Just, 1919). The fast block to polyspermy achieves that goal by changing the electrical potential of the egg membrane.

The cell membrane provides a selective barrier between the egg cytoplasm and the outside environment, and the ionic concentration of the egg is greatly different from that of its surroundings. This concentration difference is especially true for sodium ions (Na^+) and potassium ions (K^+). Seawater has a particularly high sodium ion concentration, whereas the egg cytoplasm has relatively little free sodium. The reverse is the case with potassium ions. This condition is maintained by the cell membrane, which steadfastly inhibits the entry of sodium into the oocyte and prevents potassium ions from leaking into the environment. If an electrode is inserted into the egg and a second electrode is placed outside the oocyte, one can measure the constant potential difference across the egg plasma membrane. This RESTING MEMBRANE POTENTIAL is generally about 70 millivolts, usually expressed as −70 mV, because the inside of the cell is negatively charged with respect to the exterior.

Within 1–3 seconds after the binding of the first sperm, the membrane potential shifts to a positive level (Longo et al., 1986). A small influx of sodium ions into the egg is permitted, thereby bringing the potential

FIGURE 21
Aberrant development in a dispermic sea urchin egg. (A) Fusion of three haploid nuclei, each containing 18 chromosomes, and the division of the two sperm centrioles to form four mitotic poles. (B) The 54 chromosomes randomly assort on the four spindles. (C) At anaphase of the first division, the duplicated chromosomes are pulled to the four poles. (D) Four cells containing different numbers and types of chromosomes are formed, thereby causing the early death of the embryo (E). (After Boveri, 1907.)

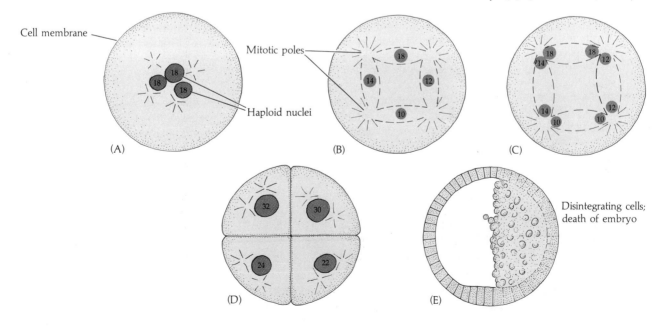

FIGURE 22
Membrane potential of sea urchin eggs before and after fertilization. Before the addition of sperm, the potential difference across the egg cell membrane is about −70 mV (the inside of the cell is more negatively charged than the environment). As a sperm enters the egg, the potential shifts in a positive direction within 0.1 second. (After Jaffe, 1980.)

difference to about +20 mV (Figure 22). Although sperm can fuse with membranes having a resting potential of −70 mV, they cannot readily fuse with membranes having a positive resting potential. It is not known how the binding or entry of a sperm signals the opening of the sodium channels, but Gould and Stephano (1987) have provided what may be an important clue to understanding this process. They isolated from *Urechis* sperm an acrosomal protein that is able to open the sodium channels of unfertilized *Urechis* eggs. Moreover, when these eggs are exposed to this protein, the rate of sodium influx and the resulting membrane-potential shift is very similar to that produced by live sperm. The opening of the sodium channels in the egg appears to be caused by the binding of the sperm to the egg.

Jaffe (1976) and her co-workers have found that polyspermy can be induced in eggs if the eggs are artificially supplied with electric current that keeps their membrane potential negative. Conversely, fertilization can be prevented entirely by artificially keeping the membrane potential of eggs positive (Jaffe, 1976). The fast block to polyspermy can also be circumvented by lowering the concentration of sodium ions in the water (Figure 23). If the sodium ions are not sufficient to cause the positive shift in membrane potential, polyspermy occurs (Gould-Somero et al., 1979; Jaffe, 1980). An electrical block to polyspermy also occurs in frogs (Cross and Elinson, 1980) but probably not in most mammals (Jaffe and Cross, 1983).

It is not known how the differences in membrane potential act on the sperm to block secondary fertilization. One hypothesis is that the egg plasma membrane contains a voltage-sensitive sperm-binding site. Another hypothesis is that the sperm carries a voltage-sensitive component in its cell membrane and that this might be a positively charged "fusion protein." The insertion of this protein into the egg cell membrane would be regulated by the electrical charge across the membrane. Iwao and Jaffe (1989) have provided evidence for the second model by showing that the

FIGURE 23
Polyspermy in eggs of the sea urchin *L. pictus* induced by fertilization in low-sodium seawater. (A) Control eggs developing in 490 mM Na$^+$. (B) Polyspermy in eggs fertilized in 120 mM Na$^+$ (choline was substituted for sodium). The eggs were photographed during the first cleavage. (C) Table showing rise of polyspermy with decrease in sodium ions. (From Jaffe, 1980; photographs courtesy of L. A. Jaffe.)

(A)　　　　　　(B)　　　　　　(C)

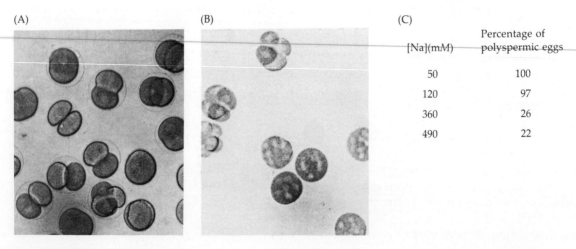

[Na](mM)	Percentage of polyspermic eggs
50	100
120	97
360	26
490	22

sperm of species having voltage-dependent fertilization contain a voltage sensor. The investigators took amphibian eggs that did not have voltage-dependent blocks to polyspermy and artificially brought their membrane potential to +40 mV. This prevented further entry of sperm from species having voltage-dependent blocks to polyspermy, but it did not impede sperm from species having no electrical block to polyspermy. It appears, then, that the fast electrical block to polyspermy is effective when the sperm has a component that is sensitive to the voltage differences.

The slow block to polyspermy. Sea urchin eggs (and many others) have a second mechanism to ensure that multiple sperm do not enter the egg cytoplasm. The fast block to polyspermy is transient, and the membrane potential of the sea urchin egg remains positive for only about a minute. This brief potential shift may not be sufficient to permanently prevent polyspermy, for Carroll and Epel (1975) have demonstrated that polyspermy can still occur if the sperm bound to the vitelline envelope are not somehow removed. This removal is accomplished by the CORTICAL GRANULE REACTION, a slower, mechanical, block to polyspermy that becomes active about 1 minute after sperm–egg attachment.

Directly beneath the sea urchin egg membrane are 15,000 cortical granules about 1 μm in diameter (Figure 6B). These vesicles contain hyaline, proteases, a peroxidase, and mucopolysaccharides. Upon sperm entry, these cortical granules fuse with the egg plasma membrane, releasing their contents into the area between the egg plasma membrane and the vitelline envelope. The proteins linking the vitelline envelope to the egg are dissolved by the released proteolytic enzymes, and the newly released mucopolysaccharides produce an osmotic gradient that causes water to rush into the space between the cell membrane and the vitelline envelope. The vitelline envelope is thus elevated, whereupon it is called the FERTILIZATION ENVELOPE. The cortical granule discharge changes this membrane, too. First, the proteases modify or strip off the bindin receptor and any sperm attached to it (Vacquier et al., 1973; Glabe and Vacquier, 1978). Second, the peroxidase hardens the fertilization envelope by cross-linking tyrosine residues on adjacent proteins (Foerder and Shapiro, 1977). As shown in Figure 24, the fertilization envelope starts to form at the site of sperm entry and continues its expansion around the egg. As this envelope forms, sperm are released. This process starts about 20 seconds after sperm attachment and is complete by the end of the first minute of fertilization. As the fertilization envelope forms, HYALIN, a protein that is stored in the cortical granules, forms a coating around the egg (Hylander and Summers, 1982). The cell extends elongated microvilli whose tips attach to the hyaline layer. This hyalin envelope provides support for the blastomeres during cleavage.

In mammals, the cortical granule reaction does not create a fertilization envelope, but the effect is the same. Released enzymes modify the zona pellucida sperm receptors such that they can no longer bind sperm (Bleil and Wassarman, 1980). This modification process is called the ZONA REACTION. Florman and Wassarman (1985) have proposed that the cortical granules of mouse eggs contain an enzyme that clips off the terminal sugar residues of ZP3, thereby releasing bound sperm from the zona and preventing the attachment of other sperm.

The mechanism for the cortical granule reaction is similar to that of the acrosome reaction. In the presence of increased intracellular free calcium ions, the cortical granule membranes fuse with the egg plasma membrane, thereby causing the exocytosis of their contents (Figure 25). Following the fusion of the cortical granules about the point of sperm

FIGURE 24

Formation of the fertilization envelope and removal of excess sperm. Sperm were added to sea urchin eggs and the suspension was fixed in formaldehyde to prevent further reactions. (A) Ten seconds after sperm addition. Sperm are seen surrounding the egg. (B,C) Twenty-five and 35 seconds after insemination, a fertilization envelope forms around the egg, starting at the point of sperm entry. (D) Fertilization envelope is complete and excess sperm are removed. (From Vacquier and Payne, 1973; photographs courtesy of V. D. Vacquier.)

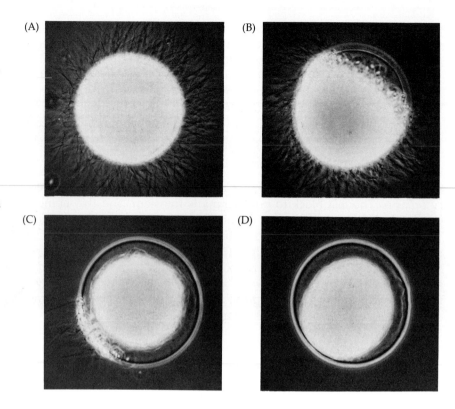

entry, a wave of cortical granule exocytosis propagates around the cortex to the opposite side of the egg.

The release of calcium from intracellular storage can be monitored visually by calcium-activated dyes such as aequorin (isolated from luminescent jellyfish). These dyes emit light when they bind free calcium ions. Eggs are injected with the dye and then fertilized. Figure 26 shows the striking wave of calcium release that propagates across the egg. Starting at the point of sperm entry, a band of light traverses the cell (Gilkey et al., 1978; Steinhardt et al., 1977). The entire release of calcium ions is complete in roughly 30 seconds in sea urchin eggs, and the free calcium ions are bound almost as soon as they are released. If two sperm enter the egg cytoplasm, calcium ion release can be seen starting at the two separate points on the cell surface (Hafner et al., 1988).

Several experiments have demonstrated that calcium ions are directly responsible for propagating the cortical reaction and that the calcium ions are stored within the egg itself. The first evidence came in 1974 when unfertilized eggs were treated with the ionophore A23187. This drug transports calcium ions across membranes, allowing them to traverse the otherwise impermeable barriers. Placing sea urchin eggs into seawater containing A23187 causes the cortical granule reaction and the elevation of the fertilization envelope. Moreover, this reaction occurs in the absence of any calcium ions in the seawater. Therefore, A23187 caused the release of calcium ions already sequestered in organelles within the egg (Chambers et al., 1974; Steinhardt and Epel, 1974). Further studies (Fulton and Whittingham, 1978; Hollinger and Schuetz, 1976) have shown that calcium ions will initiate cortical granule reactions when injected into sea urchin, mouse, and frog eggs and that procaine, an anesthetic that prevents the release of bound calcium, can inhibit the sea urchin egg cortical granule reaction.

The internal calcium ions are probably stored in the endoplasmic

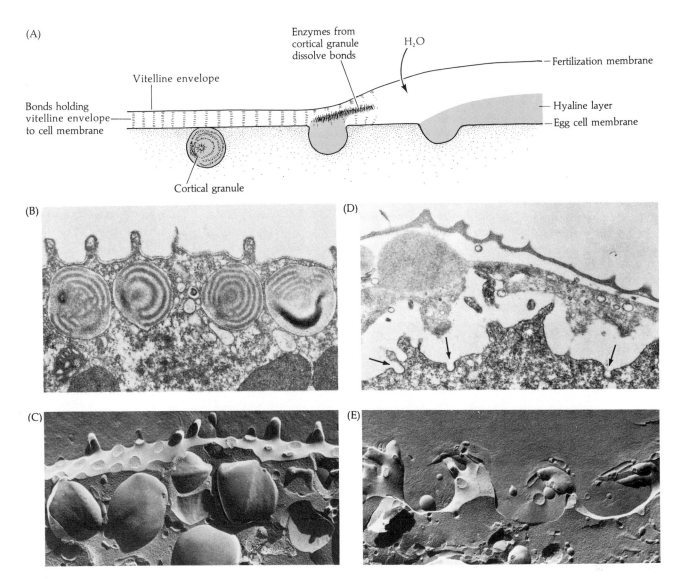

(A) Schematic diagram labels: Enzymes from cortical granule dissolve bonds; H_2O; Fertilization membrane; Vitelline envelope; Hyaline layer; Egg cell membrane; Bonds holding vitelline envelope to cell membrane; Cortical granule

FIGURE 25

Cortical granule exocytosis. (A) Schematic diagram showing the events leading to the formation of the fertilization envelope and the hyaline layer. As cortical granules undergo exocytosis, they release peptidases, which cleave the bonds joining the vitelline envelope to the cell membrane. Mucopolysaccharides released by the cortical granules form an osmotic gradient, thereby causing water to enter and swell the space between the vitelline envelope and the cell membrane. Other enzymes released from the cortical granules act to harden the fertilization envelope and to release sperm bound to it. (B,C) Transmission and scanning electron micrographs of the cortex of an unfertilized sea urchin egg. (D,E) Transmission and scanning electron micrographs of the same region of a recently fertilized egg showing the raising of the fertilization envelope and the points at which the cortical granules fused with the plasma membrane of the egg (arrows in D). (A after Austin, 1965. B–E from Chandler and Heuser, 1979; photographs courtesy of D. E. Chandler.)

reticulum of the egg. In sea urchins and frogs, whose eggs undergo a cortical granule reaction, this reticulum is pronounced in the cortex and surrounds the cortical granules (Gardiner and Grey, 1983; Luttmer and Longo, 1985; Figure 27). In the clam *Spisula*, which does not have a cortical granule reaction (and which uses external calcium ions to activate the egg),

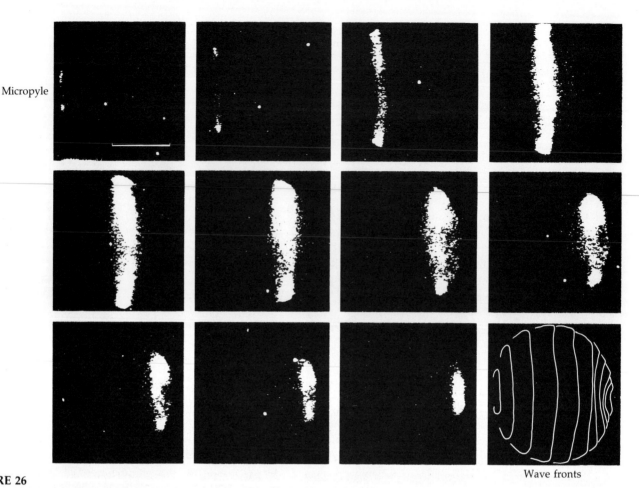

Micropyle

Wave fronts

FIGURE 26

Wave of free calcium ions propagated in the egg of the medaka fish. The egg has been injected with aequorin and oriented with its micropyle (the point of sperm entry) to the left. Photographs taken at 10-second intervals reveal the release of calcium ions in successive parts of the egg. The last frame traces the leading edges of the 11 wave fronts. (From Gilkey et al., 1978; photographs courtesy of L. F. Jaffe.)

the cortical endoplasmic reticulum is very sparse. When sea urchin eggs were centrifuged to stratify their contents, the source of calcium release was seen in the region containing the endoplasmic reticulum (Eisen and Reynolds, 1985). In the frog *Xenopus*, the cortical endoplasmic reticulum becomes tenfold more abundant during the maturation of the egg and disappears locally within a minute after the wave of exocytosis occurs in any particular region of the cortex. Jaffe (1983) likens this calcium-sequestering cortical endoplasmic reticulum to the sarcoplasmic reticulum of skeletal and cardiac muscle. Once initiated, the release of calcium is self-propagating. Free calcium is able to release sequestered calcium from its storage sites, thus causing a wave of calcium ion release and cortical granule exocytosis.

Variations in polyspermy-preventing strategies exist throughout nature. In mammals, polyspermy is minimized by the small number of sperm that reach the site of fertilization (Braden and Austin, 1954). The block to polyspermy in hamsters and mice appears to be controlled by the release of sperm binding sites on the zona pellucida (Miyazaki and Igusa, 1981; Jaffe et al., 1983). Rabbits, however, rely completely on a membrane-level block to polyspermy and nobody will argue with their success. Lastly, certain animals have defenses to polyspermy about which we know very little. In the yolky eggs of certain birds, reptiles, and salamanders, several sperm actually do enter the egg cytoplasm. In some unknown way, all but one of these sperm are induced to disintegrate in the cytoplasm after the fusion of the egg pronucleus with one of the sperm pronuclei (Ginzburg, 1985; Elinson, 1986). Whatever the mechanism, only one haploid sperm nucleus is allowed to fuse with the haploid nucleus of the egg.

FIGURE 27
Endoplasmic reticulum surrounding cortical granule in *Arbacia* egg. Endoplasmic reticulum has been stained with osmium–zinc iodide to allow visualization by transmission electron microscopy. (From Luttmer and Longo, 1985; photograph courtesy of S. Luttmer.)

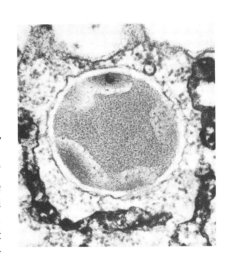

Fusion of the genetic material

In sea urchins, the sperm nucleus enters the egg perpendicularly to the egg surface. After fusion of the sperm and egg membranes, the sperm nucleus and its centriole separate from the mitochondria and the flagellum. The mitochondria and the flagellum disintegrate inside the egg, such that sperm-derived mitochondria are not found in developing or adult organisms (Dawid and Blackler, 1972; Giles et al., 1980). Thus, mitochondria are transmitted solely through the maternal parent.

The egg nucleus, once it is haploid, is called the FEMALE PRONUCLEUS. Within the egg cytoplasm, the sperm nucleus decondenses to form the MALE PRONUCLEUS. The sperm nuclear envelope vesiculates into small packets, thereby exposing the compact sperm chromatin to the egg cytoplasm. The proteins holding the sperm chromatin in its condensed, inactive, state are exchanged for similar proteins derived from the egg. This exchange permits the DECONDENSATION of the sperm chromatin (Green and Poccia, 1985). Remnants of the original sperm nuclear envelope are carried by the chromatin. Soon, new membranous vesicles aggregate along the periphery of the chromatin mass and connect with the fragments of the old envelope to produce the new membrane of the male pronucleus.

After the sperm enters the egg cytoplasm, the male pronucleus rotates 180° so that the sperm centriole is between the sperm pronucleus and the egg pronucleus. The microtubules of the centriole of the male pronucleus extend and contact the female pronucleus, and the two pronuclei migrate toward each other (Hamaguchi and Hiramoto, 1980; Bestor and Schatten, 1981). Their fusion forms the diploid ZYGOTE NUCLEUS (Figure 28). The initiation of DNA synthesis can occur either in the pronuclear stage (during migration) or after the formation of the zygote nucleus.

FIGURE 28
Nuclear events in the fertilization of the sea urchin. (A) Migration of the egg pronucleus and the sperm pronucleus in an egg of *Clypeaster japonicus*. The sperm pronucleus is surrounded by its aster of microtubules. The arrows in the lower portion of the picture indicate the time at which it was taken. (B) Fusion of pronuclei in the sea urchin egg. (A from Hamaguchi and Hiramoto, 1980; photograph courtesy of the authors. B courtesy of F. J. Longo.)

(A)

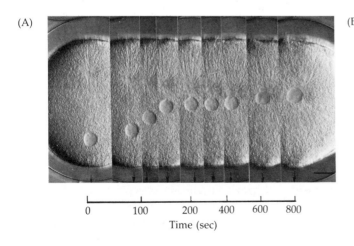

(B)

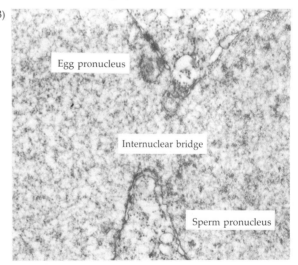

Egg pronucleus

Internuclear bridge

Sperm pronucleus

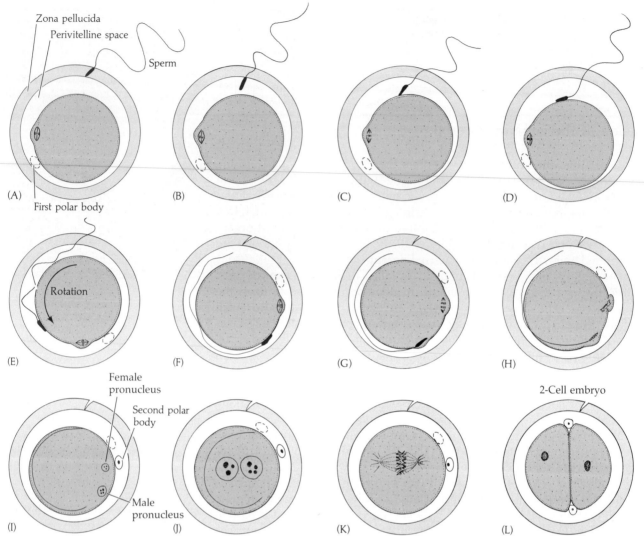

FIGURE 29

Interpretive diagram of fertilization in the hamster. (A) The sperm attaches to the zona. (B) The sperm digests its way into the perivitelline space between the zona and the egg. Upon attaching to the egg (C), it places its head parallel to the egg cell membrane (D). The firm attachment of the sperm to the egg and the whipping motion of the sperm's tail causes the egg to rotate (E) while the entire sperm enters the perivitelline space (F). The sperm and egg membranes fuse (G) and the sperm head decondenses (H). After both pronuclei swell and meet in the center of the egg (I,J), the pronuclear membranes disintegrate and the first mitosis begins (K,L). (After Yanagimachi, 1981.)

In mammals, the process of nuclear fusion takes about 12 hours, compared to less than 1 hour in the sea urchin (Figures 29 and 30). The mammalian sperm nucleus enters almost tangentially to the surface of the egg rather than approaching it perpendicularly, and it fuses with numerous microvilli. The mammalian sperm nucleus also breaks down as its chromatin decondenses and is then reconstructed by coalescing vesicles. Because the decondensation of chromatin can be induced artificially by treating the rabbit sperm nuclei with reagents that disrupt disulfide bonds, it is likely that the oocyte contains a substance that also disrupts these linkages (Calvin and Bedford, 1971; Kvist et al., 1980).

The mammalian male pronucleus enlarges while the oocyte nucleus completes its second meiotic division. Then each pronucleus migrates toward the other, replicating its DNA as it travels. Upon meeting, the two nuclear envelopes meet and break down. However, instead of producing a common zygote nucleus, the chromatin condenses into chromosomes that orient themselves upon a common mitotic spindle. Thus, a true diploid nucleus in mammals is first seen, not in the zygote, but at the 2-cell stage.

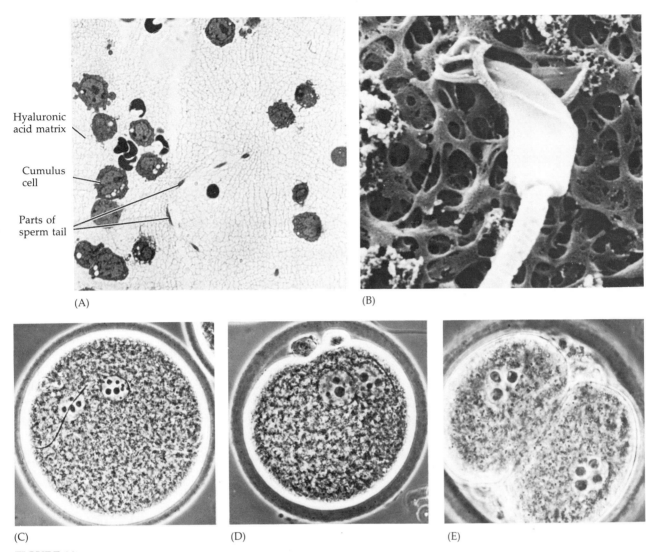

Hyaluronic
acid matrix

Cumulus
cell

Parts of
sperm tail

(A)

(B)

(C)

(D)

(E)

FIGURE 30
Fertilization in the hamster. (A) A sperm within the cumulus layer (processes
from cumulus cells are apparent). (B) Sperm head binding to the zona pellucida.
(C) Sperm entry into cell and the swelling of the sperm pronucleus. Sperm tail
can be seen within egg. (D) Apposition of sperm and egg pronuclei. (E) Two-cell
stage, showing two equally sized cells with well-defined nuclei. Debris in perivitelline space is the degenerating polar bodies. (A from Yudin et al., 1988; B from
Cherr et al., 1986; C–E from Bavister, 1980; photographs courtesy of D. Katz,
G. Cherr, and B. D. Bavister.)

SIDELIGHTS & SPECULATIONS

The nonequivalence of mammalian pronuclei

Although male and female pronuclei of mammals can be
considered genetically equivalent, studies have suggested
that they may be functionally different. Human females
sometimes develop a uterine tumor called HYDATIDIFORM

MOLE. The fetus is absent and the placental tissue is abnormally enlarged. A majority of such moles have been
shown to arise from a haploid sperm fertilizing an egg in
which the female pronucleus is absent. After entering the
egg, the sperm chromosomes duplicate themselves,
thereby restoring the diploid chromosome number. Thus,
the entire genome is derived from the sperm (Jacobs et al.,

1980; Ohama et al., 1981). Here we see a situation in which the cells survive, divide, and have a normal chromosome number, but development is abnormal. Development does not occur here when the entire genome comes from the male parent.

Evidence for the nonequivalence of mammalian pronuclei also comes from attempts to get ova to develop in the absence of sperm. This ability to develop an embryo without spermatic contribution is called PARTHENOGENESIS (Gk. "virgin birth"). The eggs of many invertebrates and some vertebrates are capable of developing normally in the absence of sperm if the egg is artificially activated. In these situations, the sperm's contribution to development seems dispensable. Mammals, however, do not exhibit parthenogenesis. Suppression of polar body formation during meiosis produces diploid mouse eggs whose inheritance is derived from the egg alone. These cells divide to form embryos with spinal cords, muscles, skeletons, and organs, including beating hearts. However, development does not continue, and by day 10–11 (halfway through the mouse's gestation), profound differences are observed between the normal and the parthenogenetic embryos (Figure 31). The parthenogenetic embryos are deteriorating and are becoming grossly disorganized (Surani et al., 1986). Development in mice cannot occur with just egg-derived chromosomes either.

The hypothesis that male and female pronuclei are different also gains support from pronuclear transplantation experiments (Surani and Barton, 1983; McGrath and Solter, 1984). Either male or female pronuclei of recently fertilized mouse eggs can be removed and added to other recently fertilized oocytes. (The two pronuclei can be distinguished because the female pronuclei is always the one beneath the polar bodies.) Thus, zygotes with two male or two female pronuclei can be constructed. Although embryonic cleavage occurs, neither of these types of eggs develops to birth, whereas some control eggs (containing one male pronucleus and one female pronucleus from different zygotes) undergoing such transplantation develop normally (Table 1). Moreover, the bimaternal or bipaternal embryos cease development at the same time as the parthenogenetic mice. Thus, though the two pronuclei are equivalent in many animals, in mammals there are important functional differences between them. These differences may reside in modifications of DNA that occur differently in the egg and sperm nuclei (Chapter 12). Both maternal and paternal pronuclei are necessary for the completion of mammalian development.

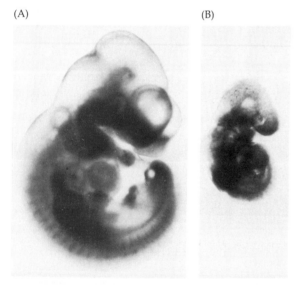

FIGURE 31
Control (A) and parthenogenetic (B; two female pronuclei) mouse embryos at 11 days' gestation. The mice were developing in the same female. In addition to being smaller and deteriorating, the parthenogenetic embryos also had much smaller placentas. (From Surani and Barton, 1983; photographs courtesy of the authors.)

TABLE 1
Pronuclear transplantation experiments

Class of reconstructed zygotes	Operation	Number of successful transplants	Number of progeny surviving
Bimaternal		339	0
Bipaternal		328	0
Control		348	18

Source: McGrath and Solter (1984).

Activation of egg metabolism

So far, then, we have discussed the mechanisms by which sperm and egg recognize each other, fuse together, and merge their respective haploid nuclei. But for fertilization to lead to development, changes must occur in the egg cytoplasm.*

*In certain salamanders, this developmental function of fertilization has been totally divorced from the genetic function. The silver salamander (*Ambystoma platineum*) is a hybrid subspecies consisting solely of females. Each female produces an egg with an unreduced chromosome number. This egg, however, cannot develop on its own. So the silver salamander mates with

The mature sea urchin egg is a metabolically sluggish cell that is reactivated by the entering sperm. This activation is merely a stimulus, however, which sets into action a preprogrammed set of metabolic events. The responses of the egg to the sperm can be divided into those "early" responses that occur within seconds of the cortical reaction and those "late" responses that take place several minutes after fertilization begins (Table 2).

Early responses

As we have seen, contact between sea urchin sperm and egg activates the two major blocks to polyspermy: the fast block, initiated by sodium influx into the cell, and the slow block, initiated by intracellular release of calcium ions. The activation of all eggs appears to depend on an increase of free calcium ions within the egg. In protostomes such as snails and worms, the calcium usually enters the egg from outside. In deuterostomes such as fish, frogs, and sea urchins, the activation is accomplished by the release of calcium ions from internal compartments, resulting in a wave of calcium ions sweeping across the egg (Jaffe, 1983).

Several experiments have shown that this release of calcium ions is essential for activating the development of the embryo. If the calcium-chelating chemical EGTA is injected into the eggs, there is no increase in the concentration of free calcium ions within the eggs and no activation of egg metabolism. Conversely, eggs can be activated artificially in the

TABLE 2
Events of sea urchin fertilization

Event	Approximate time postinsemination[a]
Sperm–egg bonding	0 seconds
Fertilization potential rise (fast block to polyspermy)	before 3 sec
Sperm–egg membrane fusion	before 30 sec
Calcium increase first detected	30 sec
Cortical vesicle exocytosis (slow block to polyspermy)	30–60 sec
Activation of NAD kinase	starts at 1 min
Increase in NADH and NADPH	starts at 1 min
Increase in O_2 consumption	starts at 1 min
Sperm entry	1–2 min
Acid efflux	1–5 min
Increase in pH (remains high)	1–5 min
Sperm chromatin decondensation	2–12 min
Sperm nucleus migration to egg center	2–12 min
Egg nucleus migration to sperm nucleus	5–10 min
Activation of protein synthesis	starts at 5–10 min
Activation of amino acid transport	starts at 5–10 min
Initiation of DNA synthesis	20–40 min
Mitosis	60–80 min
First cleavage	85–95 min

Source: Whitaker and Steinhardt (1985).
[a]Approximate times based on data from *S. purpuratus* (15–17°C), *L. pictus* (16–18°C), and *A. punctulata* (18–20°C).

the male Jefferson salamander (*A. jeffersonianum*). The sperm from the male Jefferson salamander only stimulates the egg's development (Uzzell, 1964). It does not contribute genetic material. For details of this complex mechanism of procreation, see Bogart et al., 1989.

absence of sperm by procedures that release free calcium into the oocyte. Steinhardt and Epel (1974) found that micromolar amounts of calcium ionophore A23187 elicit most of the egg responses characteristic of a normally fertilized egg. The elevation of the fertilization envelope, the rise of intracellular pH, the burst of oxygen utilization, and the increases in protein and DNA synthesis are each generated in their proper order. Moreover, this activation takes place in the total absence of calcium ions in the seawater. In most cases, development ceases before the first mitosis, because the eggs are still haploid and lack the sperm centriole needed for division.

This calcium release is responsible for activating a series of metabolic reactions (Figure 32). One of these is the activation of the enzyme NAD^+ kinase, which converts NAD^+ to $NADP^+$ (Epel et al., 1981). This change may have important consequences for the metabolism of the cell. One of these consequences involves lipid metabolism. $NADP^+$ (but not NAD^+) can be used as a coenzyme for lipid biosynthesis. Thus, the switch from NAD^+ to $NADP^+$ may be important in the construction of the many new cell membrane components required during cleavage. Another effect of this change involves oxygen consumption. A burst of oxygen reduction is seen during fertilization, and much of this "respiratory burst" is used to crosslink the fertilization membrane. The enzyme responsible for this reduction of oxygen (to hydrogen peroxide) is NADPH-dependent (Heinecke and Shapiro, 1989).

Late responses

Coupled to the increase in free intracellular calcium is a rise in intracellular pH. It is thought that these two ionic conditions (high Ca^{2+}, low H^+) act in concert to give the full spectrum of fertilization events, including protein synthesis and DNA synthesis (Winkler et al., 1980; Whitaker and Steinhardt, 1982). The rise in intracellular pH begins with a second influx of sodium ions, causing a 1:1 exchange between sodium ions from the seawater and hydrogen ions from the egg. In the sea urchin egg, nearly 75 percent of the hydrogen ions are exchanged. This loss of hydrogen ions

FIGURE 32
Model of possible interrelationships in the process of sea urchin fertilization. (After Epel, 1980, and L. A. Jaffe, personal communication).

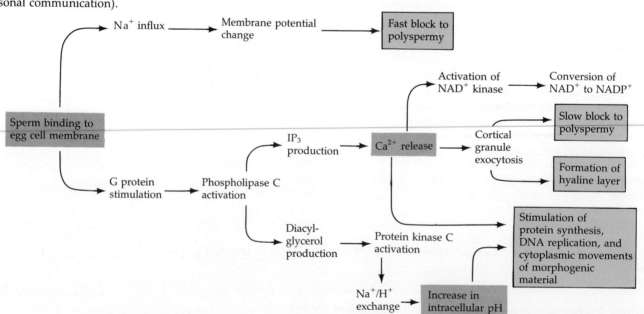

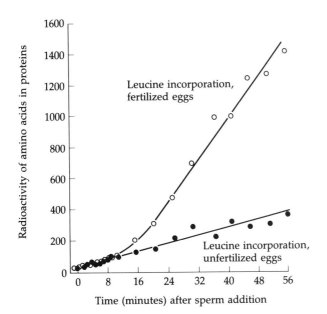

FIGURE 33
Protein synthesis in newly fertilized sea urchin eggs. [^{14}C]Leucine incorporation into the proteins of unfertilized eggs (filled circles) or fertilized eggs (open circles). (After Epel, 1967.)

causes the pH to rise from 6.8 to 7.2 and brings about enormous changes in egg physiology (Shen and Steinhardt, 1978). Although this change is believed to be caused by a calcium-mediated reaction, it has been found that any means of raising the intracellular pH can initiate many of the late fertilization responses.

The late responses of fertilization include activation of DNA synthesis and protein synthesis. The burst of protein synthesis usually occurs within several minutes after sperm entry and is not dependent on new messenger RNA synthesis (Figure 33). Rather, new protein synthesis utilizes mRNAs already present in the oocyte cytoplasm. (Much more will be said about this in Chapter 14.) Such a burst of protein synthesis can be induced artificially by placing unfertilized eggs into a solution containing ammonium ions (Winkler et al., 1980). These ions lose protons outside the cell, diffuse through the plasma membrane as NH_3, and once inside the cell, pick up protons to become NH_4^+. The resulting loss of free hydrogen ions from the cytoplasm is reflected in the increase of intracellular pH (Figure 34). Thus, bypassing the sodium–hydrogen exchange, ammonium ions are

FIGURE 34
Dual ionic control of protein synthesis during fertilization. (A) Alkalinization of egg by ammonia. Continuous recordings of intracellular pH as an egg is placed into various concentrations of NH_4Cl. (B) Protein synthesis as measured by the incorporation of radioactive valine into proteins under particular ionic conditions. The pH elevation could be inhibited by fertilizing the eggs in sodium-free seawater or elevated by ammonium ions. The increase in free calcium concentration could be blocked by EGTA or elevated by A23187. (A after Shen and Steinhardt, 1978; B after Winkler et al., 1980.)

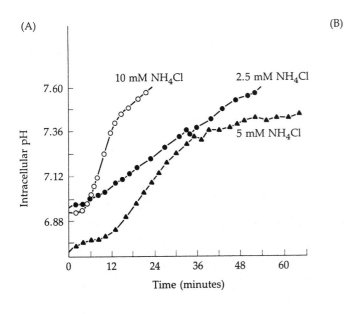

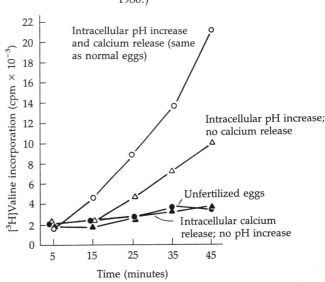

able to raise the pH of the eggs. In so doing, they activate protein synthesis and DNA synthesis.

Conversely, agents that block the rise in pH also block these late fertilization events. When newly fertilized eggs are placed into solutions containing low concentrations of sodium ions and amiloride (a drug that inhibits sodium–hydrogen exchange), protein synthesis fails to occur, the movements of the egg and sperm pronuclei are prevented, and cell division fails to take place (Dube et al., 1985).

SIDELIGHTS & SPECULATIONS

The biochemistry of egg activation

In sea urchins and vertebrates, the activation of the egg is caused by a wave of calcium ion release from internal compartments such as the endoplasmic reticulum. The rise in internal calcium is first seen 20–30 seconds after egg–sperm binding in the sea urchin *Arbacia punctulata* (Eisen and Reynolds, 1985). What happens during those 25 seconds to cause the release of sequestered calcium ions?

Recent evidence suggests that the binding of the sperm to the egg cell membrane triggers a series of reactions involving enzymes within that membrane. These enzymes synthesize a "second messenger" that can diffuse into the cytoplasm. Several studies have shown that this compound, INOSITOL 1,4,5-TRISPHOSPHATE (IP$_3$), can release se-

questered calcium ions in many cell types, including eggs (Berridge, 1985; Swann and Whitaker, 1986). Increased concentrations of intracellular IP$_3$ are seen immediately after sea urchin eggs are fertilized (Ciapa and Whitaker, 1986), and the release of calcium ions quickly follows the formation of IP$_3$.

The importance of IP$_3$ to egg activation was demonstrated by Whitaker and Irvine (1984) and Busa and his coworkers (1985). When they injected IP$_3$ directly into sea urchin and frog eggs, they activated the release of calcium ions and the cortical granule reaction. These responses were indistinguishable from the events seen during normal fertilization. Moreover, these IP$_3$-mediated effects could be thwarted by preinjecting the egg with calcium chelating agents (Turner et al., 1986), thereby confirming that IP$_3$ stimulated the release of stored calcium in the egg.

The enzyme responsible for synthesizing IP$_3$ is PHOSPHOLIPASE C. This membrane-bound enzyme splits PHOSPHATIDYLINOSITOL 4,5-BISPHOSPHATE into DIACYLGLYCEROL

FIGURE 35
Model for the initiation of fertilization ion fluxes. When sperm binds to its receptor on the egg cell membrane, the receptor activates a GTP-binding protein (G-protein) which, in turn, activates phospholipase C. Phospholipase C splits phosphatidylinositol 4,5-bisphosphate into diacylglycerol and inositol triphosphate (IP$_3$). IP$_3$ can release Ca^{2+} from the endoplasmic reticulum, while diacylglycerol can activate protein kinase C. This protein kinase stimulates the sodium/hydrogen transporter to exchange cellular hydrogen ions for extracellular sodium ions, thereby leading to the increase in pH. (A similar model, shown in Chapter 20, has been proposed for the onset of tumor growth.)

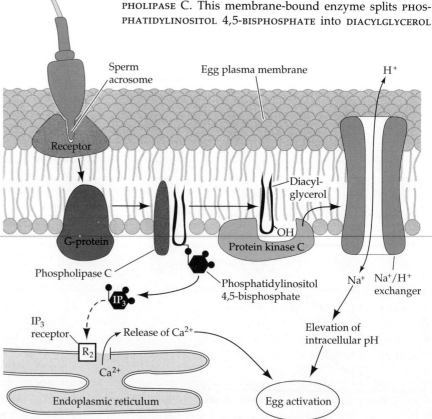

and IP$_3$ (Figure 35). The IP$_3$ released by the reaction binds to a protein on the endoplasmic reticulum that releases the stored calcium ions (Furuichi et al., 1989; Ferris et al., 1989). Meanwhile, the other product of the reaction, diacylglycerol, activates another membrane enzyme, PROTEIN KINASE C, which, in turn, may activate the Na$^+$/H$^+$ carrier that exchanges sodium ions for hydrogen ions (Swann and Whitaker, 1985; Shen and Burgart, 1986). In this way, the splitting of phosphatidylinositol 4,5-bisphosphate may bring about the activation wave of calcium and the increase in intracellular pH, both of which are necessary for egg activation and division.

But what couples the binding of the sperm and the egg to the production of IP$_3$ and diacylglycerol? Do activating molecules pass from the sperm into the egg membrane or cytoplasm, or does sperm–egg contact activate proteins already present in the egg cell membrane? The latter model is derived from studies of hormone activation. In several hormone-responsive cells, the coupling of the hormone receptor to phospholipase C is mediated through a GTP-binding protein (G-protein; see Berridge, 1985). If this is the case for eggs, G-proteins should be found in the egg plasma membrane, and their inactivation should inhibit the events of fertilization. Turner and co-workers (1986, 1987) demonstrated that this may be the case in sea urchin eggs. When they injected activators of G-proteins into eggs, the cortical granules underwent exocytosis in the absence of sperm. This activation could be inhibited if calcium chelators such as EGTA were added. It appears, then, that a G-protein is involved in regulating the release of sequestered calcium ions and the exocytosis of the cortical granules.* A model for the initial biochemical events of egg activation is depicted in Figure 35. The binding of sperm to an unidentified receptor in the egg plasma membrane changes the conformation of the receptor such that it then activates the G-protein. The G-protein, in turn, activates

phospholipase C, which cleaves phosphatidylinositol 4,5-bisphosphate into the two intracellular messengers, diacylglycerol and inositol trisphosphate. The diacylglycerol activates protein kinase C, which phosphorylates the Na$^+$/H$^+$ carrier protein, thereby enabling it to function (if stimulated by calcium ions) to elevate the intracellular pH. Because IP$_3$ releases calcium from the endoplasmic reticulum, the diffusion of IP$_3$ leads to a wave of calcium release across the egg. In this way, the binding of sperm to the egg plasma membrane can stimulate changes in the interior of the ovum.

Recent evidence also points to biochemical homologies between egg activation and the initiation of exocytosis during the acrosome reaction. When sea urchin sperm are activated by resact, their internal pH increases and calcium is freed (Shapiro et al., 1985). This appears to be accomplished by the same type of G-proteins and kinases as are found in the egg (Ramarao and Garbers, 1985; Singh et al, 1988). Similarly, as we have discussed earlier, G-proteins are seen in mouse sperm and appear to be activated by the binding of the sperm to the zona ZP3 protein (Endo et al., 1987b). It appears that the same chemistry that allows the egg to activate the sperm also enables the sperm to activate the egg.

---

*This also shows homology between the reactions necessary for exocytosis of the acrosomal vesicle (discussed earlier) and the cortical granules. In a related experiment, Kline and co-workers (1988) hypothesized that if a G-protein mediated the fertilization events by being activated by a sperm-binding receptor, then that same G-protein may be activated by a hormone if the egg contained the hormone receptor. They injected mRNA for the serotonin receptor or the acetylcholine receptor into frog eggs. These cell surface receptors were made and were seen on the egg cell membrane. In these instances, the eggs could be "fertilized" by the neural hormones serotonin and acetylcholine!

Rearrangement of egg cytoplasm

Fertilization can also initiate radical displacements of cytoplasmic materials. These rearrangements of oocyte cytoplasm are often crucial for cell differentiation later in development. As we shall see in Chapters 7 and 8, the cytoplasm of the egg frequently contains MORPHOGENIC DETERMINANTS that become segregated into specific cells during cleavage. These determinants ultimately lead to the activation or repression of specific genes and thereby confer certain properties to the cells that incorporate them. The correct spatial arrangement of these determinants is crucial for proper development.

In some species, the rearrangement of these determinants into their required orientation can be visualized because cytoplasmic pigment granules are present. One such example is the egg of the tunicate *Styela partita* (Conklin, 1905). The unfertilized egg of this animal is seen in Figure 36A. A central gray cytoplasm is enveloped by a cortical layer containing yellow lipid inclusions. During meiosis, the breakdown of the nucleus releases a clear substance that accumulates in the animal (upper) hemisphere of the egg. Within 5 minutes of sperm entry, the inner clear and cortical yellow cytoplasms migrate into the vegetal (lower) hemisphere of the egg. As the

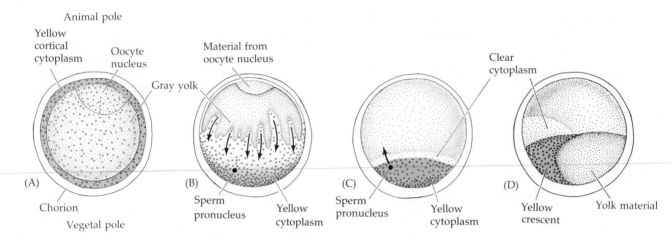

FIGURE 36

Cytoplasmic rearrangement in the egg of the tunicate *Styela partita*. (A) Before fertilization, yellow cortical cytoplasm surrounds gray yolky cytoplasm. (B) After sperm entry, the yellow cortical cytoplasm and the clear cytoplasm derived from the breakdown of the oocyte nucleus stream vegetally toward the sperm. (C) As the sperm pronucleus migrates animally toward the egg pronucleus, the yellow and clear cytoplasms move with it. (D) The final positions of the clear and yellow cytoplasms. These mark the positions where the cells give rise to the mesenchyme and muscles, respectively. (After Conklin, 1905.)

male pronucleus migrates from the vegetal pole to the equator of the cell along the future posterior side of the embryo, the lipid inclusions migrate with it. This migration forms a YELLOW CRESCENT, extending from the vegetal pole to the equator (Figure 36B–D), and brings the yellow plasm into the area where muscle cells will later form in the tunicate larva. The movement of these cytoplasmic regions is dependent upon microtubules that probably are generated by the sperm centriole (Sawada and Schatten, 1989).

Cytoplasmic movement is also seen in amphibian oocytes. In frogs, a single sperm can enter any place on the animal hemisphere; and when it does, it changes the cytoplasmic pattern of the egg. Originally, the egg is radially symmetric about the animal–vegetal axis. After sperm entry, however, the cortical (outer) cytoplasm shifts about 30° towards the point of sperm entry, relative to the inner cytoplasm (Manes and Elinson, 1980; Vincent et al., 1986). In some frogs (such as *Rana*) one region of the egg that had formerly been covered by the dark cortical cytoplasm of the animal hemisphere is now exposed (Figure 37). This underlying cytoplasm, located near the equator on the side exactly opposite the point of sperm entry, contains diffuse pigment granules and therefore appears gray. Thus, this region has been referred to as the GRAY CRESCENT (Roux,

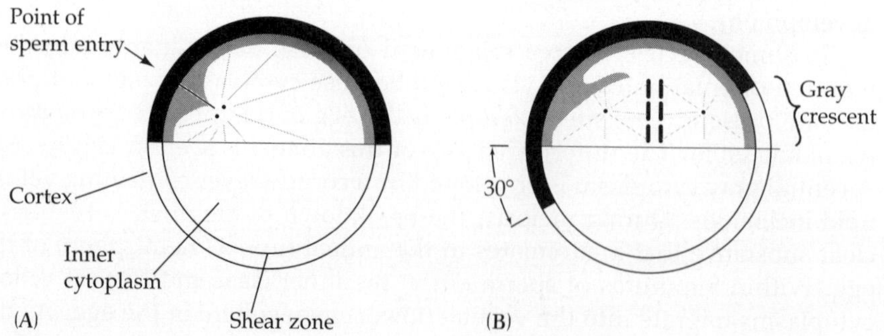

FIGURE 37

Reorganization of cytoplasm in the newly fertilized frog egg. (A) Schematic cross section of an egg midway through the first cleavage cycle. The egg has radial symmetry about its animal-vegetal axis. The sperm has entered at one side and the sperm nucleus is migrating inward. The cortex is represented like that of *Rana*, with a heavily pigmented animal hemisphere and a transparent vegetal hemisphere. (B) About 80 percent of the way into first cleavage, the cortical cytoplasm rotates 30° relative to the internal cytoplasm. This rotation will be seen to be important in that gastrulation will begin in that region opposite the point of sperm entry where the greatest displacement of cytoplasm occurs. (After Gerhart et al., 1989.)

1887; Ancel and Vintenberger, 1948). As we shall see in subsequent chapters, the gray crescent marks the region where gastrulation is initiated in amphibian embryos.

In frogs such as *Xenopus*, where no gray crescent is seen, one can still observe the rotation of the cortical cytoplasm relative to the internal, subcortical, cytoplasm. This movement was demonstrated by Vincent and his colleagues (1986). These investigators imprinted a hexagonal grid of dye (Nile Blue) onto the cytoplasm beneath the cortex while placing another type of dye (a fluorescein-bound lectin) to the egg surface. When the egg was held in one position by embedding it in gelatin, the Nile Blue dots were seen to rotate 30° relative to the fluorescent lectin spots (Figure 38). In normal, unembedded eggs, the egg surface is thought to rotate while the subcortical cytoplasm, heavy with yolk platelets, remains stabilized by gravity.

The motor for these cytoplasmic movements in amphibian eggs appears to be an array of parallel microtubules that lie between the cortical and inner cytoplasm of the vegetal hemisphere. The orientation of these microtubule tracks is parallel to the direction of cytoplasmic rotation. These microtubular tracks are first seen immediately before the rotation commences, and they disappear when rotation ceases (Elinson and Rowning, 1988). The parallel microtubules come from the egg and will form even if the egg is parthenogenetically activated, but the sperm aster appears to give the microtubules their orientation (Houliston and Elinson, 1990). Treating the egg with colchicine or ultraviolet radiation at the beginning of rotation will stop the formation of these microtubules and thereby inhibit the cytoplasmic rotations (Figure 39).

The movement of the cortical cytoplasm with respect to the inner

FIGURE 38
Photographs showing the rotation of subcortical cytoplasm relative to the cell surface cytoplasm. Recently fertilized eggs were imprinted with a hexagonal grid of Nile Blue dye (which stains the lipids in the yolk platelets). The egg was embedded in gelatin and the original positions of the dots were marked on the cell surface (circles in A). The point of sperm entry is marked with an S in (A). As the first cell cycle progressed, the dots of the subcortical cytoplasm shifted roughly 30° with respect to the outer immobilized surface of the egg (B–C). The site on the egg marking the future dorsal surface of the embryo is marked with a D in panel (C). Panel (D) shows a summary of these movements in the vegetal (lower) region of the egg. (From Vincent et al., 1986; photographs courtesy of J. C. Gerhart.)

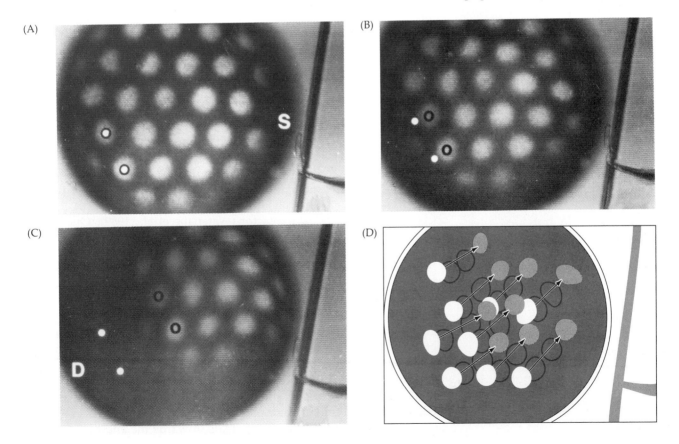

(A)

(B)

(C)

(D)

FIGURE 39

Parallel arrays of microtubules extend along the vegetal hemisphere along the future dorsal-ventral axis. (A) Parallel arrays of microtubules seen in the second half of the first cell cycle by fluorescent antibodies to tubulin. (B) Prior to cytoplasmic rotation (about midway through the first cell cycle), no array of microtubules can be seen. (C) At the end of cytoplasmic rotation, the microtubules depolymerize. (D) Egg treated with colchicine shortly after the beginning of rotation and sectioned at the same stage as (A). Rotation is blocked by such treatment. (From Elinson and Rowning, 1988; courtesy of R. Elinson).

cytoplasm causes profound movement within the inner cytoplasm. Danilchik and Denegre (1991) have labeled yolk platelets with trypan blue and watched their movement by fluorescent microscopy (the bound dye fluoresces red). During the middle part of the first cell cycle, a mass of central egg cytoplasm flows from the ventral to the future dorsal side of the embryo (see color endplate). By the end of first division, the cytoplasm of the prospective dorsal side of the embryo is distinctly different than that of the prospective vegetal side. What had been a radially symmetric embryo is now a bilaterally symmetric embryo.

As we will see in Chapters 4 and 8, these cytoplasmic movements initiate a cascade of events that determine the dorsal (back)–ventral (belly) axis of the frog. Indeed, the parallel microtubules that allow these rearrangements appear to stretch along the future dorsal–ventral axis. (Klag and Ubbels, 1975; Gerhart et al., 1983).

Preparation for cleavage

The increase in intracellular free calcium ions also sets in motion the apparatus for cell division. The mechanisms by which cleavage is initiated probably differ between species depending upon the stage of meiosis at which fertilization occurs. In many species, as we shall see in the next chapter, the rhythm of cell divisions is regulated by the synthesis and degradation of CYCLIN. Cyclin keeps the cells in metaphase, and the breakdown of cyclin enables the cells to return to interphase. In frogs, the degradation of cyclin is prevented during late oogenesis by a protein called CYTOSTATIC FACTOR (Masui et al., 1974). As will be discussed in Chapter 22, this protein is synthesized by the maturing ovum and keeps the ovum in meiotic metaphase until fertilization occurs. For the cell cycle to begin again, the cytostatic factor has to be eliminated. This occurs when the calcium ions are released from the endoplasmic reticulum. These free calcium ions activate the calcium-dependent protease CALPAIN II, which specifically digests cytostatic factor (Watanabe et al., 1989). Once the cytostatic factor is degraded, cyclin can also be degraded, and the cycles of cell division can begin anew.

Cleavage, the event that separates fertilization from embryogenesis, has a special relationship to these cytoplasmic regions. In tunicate embryos the first cleavage bisects the egg into mirror-image duplicates. From that stage on, every division on one side of that cleavage furrow has a "mirror-image" division on the opposite side. Similarly, the gray crescent is bisected by the first cleavage furrow in amphibian eggs. Thus, the position of the first cleavage is not random but tends to be specified by the point of sperm entry and the subsequent rotation of egg cytoplasm. The coordination of cleavage plane and cytoplasmic rearrangements is probably mediated through the microtubules of the sperm aster (Manes et al., 1978; Gerhart et al., 1981; Elinson, 1985). Thus, towards the end of the first cell cycle, the cytoplasm is rearranged, the pronuclei have met, DNA is replicating, and new proteins are being translated. The stage is now set for the development of a multicellular organism.

LITERATURE CITED

Afzelius, B. A. 1976. A human syndrome caused by immotile cilia. *Science* 193: 317–319.

Amos, L. A. and Klug, A. 1974. Arrangement of subunits in flagellar microtubules. *J. Cell Sci.* 14: 523–549.

Ancel, P. and Vintenberger, P. 1948. Recherches sur le determinisme de la symmetrie bilatérale dans l'oeuf des amphibiens. *Bull. Biol. Fr. Belg.* [Suppl.] 31: 1–182.

Austin, C. R. 1960. Capacitation and the release of hyaluronidase. *J. Reprod. Fert.* 1: 310–311.

Austin, C. R. 1965. *Fertilization.* Prentice-Hall, Englewood Cliffs, New Jersey.

Bavister, B. D. 1980. Recent progress in the study of early events in mammalian fertilization. *Devel. Growth Differ.* 22: 385–402.

Benau, D. A. and Storey, B. T. 1987. Characterization of the mouse sperm plasma membrane zona-binding site sensitive to trypsin inhibitors. *Biol. Reprod.* 36: 282–292.

Bentley, J. K., Shimomura, H. and Garbers, D. L. 1986. Retention of a functional resact receptor in isolated sperm plasma membranes. *Cell* 45: 281–288.

Berridge, M. J. 1985. The molecular basis of communication with the cell. *Sci. Am.* 253: 142–152.

Bestor, T. M. and Schatten, G. 1981. Antitubulin immunofluorescence microscopy of microtubules present during the pronuclear movements of sea urchin fertilization. *Dev. Biol.* 88: 80–91.

Bleil, J. D. and Wassarman, P. M. 1980. Mammalian sperm and egg interaction: Identification of a glycoprotein in mouse-egg zonae pellucidae possessing receptor activity for sperm. *Cell* 20: 873–882.

Bleil, J. D. and Wassarman, P. M. 1986. Autoradiographic visualization of the mouse egg's sperm receptor bound to sperm. *J. Cell Biol.* 102: 1363–1371.

Bleil, J. D. and Wassarman, P. M. 1988. Galactose at the nonreducing terminus of O-linked oligosaccharides of mouse egg zona pellucida glycoprotein ZP3 is essential for the glycoprotein's sperm receptor activity. *Proc. Natl. Acad. Sci. USA* 85: 6778–6782.

Bleil, J. D. and Wassarman, P. M. 1990. Identification of a ZP3-binding protein on acrosome-intact mouse sperm by photoaffinity crosslinking. *Proc. Natl. Acad. Sci. USA* 87: 5563–5567.

Bleil, J. D., Greve, J. M. and Wassarman, P. M. 1988. Identification of a secondary sperm receptor in the mouse egg zona pellucida: Role in maintenance of binding of acrosome-reacted sperm to eggs. *Dev. Biol.* 28: 376–385.

Bogart, J. P., Elinson, R. P. and Licht, L. E. 1989. Temperature and sperm incorporation in polyploid salamanders. *Science* 246: 1032–1034.

Boveri, T. 1902. On multipolar mitosis as a means of analysis of the cell nucleus. (Translated by S. Gluecksohn-Waelsch.) In B. H. Willier and J. M. Oppenheimer (eds.), *Foundations of Experimental Embryology.* Hafner, New York, 1974.

Boveri, T. 1907. Zellenstudien VI. Die Entwicklung dispermer Seeigeleier. Ein Beiträge zur Befruchtungslehre und zur Theorie des Kernes. *Jena Zeit. Naturwiss.* 43: 1–292.

Braden, A. W. H. and Austin, C. R. 1954. The number of sperm about the eggs in mammals and its significance for normal fertilization. *Aust. J. Biol. Sci.* 7: 543–551.

Busa, W. B., Ferguson, J. E., Joseph, S. K., Williamson, J. R. and Nuccitelli, R. 1985. Activation of frog (*Xenopus laevis*) eggs by inositol triphosphate. I. Characterization of Ca²⁺ release from intracellular stores. *J. Cell Biol.* 100: 677–682.

Calvin, H. I. and Bedford, J. M. 1971. Formation of disulfide bonds in the nucleus and accessory structures of mammalian spermatozoa during maturation in the epididymis. *J. Reprod. Fertil.* [Suppl.] 13: 65–75.

Carroll, E. J. and Epel, D. 1975. Isolation and biological activity of the proteases released by sea urchin eggs following fertilization. *Dev. Biol.* 44: 22–32.

Chambers, E. L., Pressman, B. C. and Rose, B. 1974. The activity of sea urchin eggs by the divalent ionophores A23187 and X-537A. *Biochem. Biophys. Res. Commun.* 60: 126–132.

Chandler, D. E. and Heuser, J. 1979. Membrane fusion during secretion: Cortical granule exocytosis in sea urchin eggs as studied by quick-freezing and freeze fracture. *J. Cell Biol.* 83: 91–108.

Cherr, G. N., Lambert, H., Meizel, S. and Katz, D. F. 1986. In vitro studies of the golden hamster sperm acrosomal reaction: Completion on zona pellucida and induction by homologous zonae pellucidae. *Dev. Biol.* 114: 119–131.

Ciapa, B. and Whitaker, M. 1986. Two phases of inositol polyphosphate and diacylglycerol production at fertilization. *FEBS Lett.* 195: 347–351.

Clermont, Y. and Leblond, C. P. 1955. Spermiogenesis of man, monkey, and other animals as shown by the "periodic acid-Schiff" technique. *Am. J. Anat.* 96: 229–253.

Colwin, A. L. and Colwin, L. H. 1963. Role of the gamete membranes in fertilization in *Saccoglossus kowalevskii* (Enteropneustra. I. The acrosome reaction and its changes in early stages of fertilization. *J. Cell Biol.* 19: 477–500.

Colwin, L. H. and Colwin, A. L. 1960. Formation of sperm entry holes in the vitelline membrane of *Hydroides hexagonis* (Annelida) and evidence of their lytic origin. *J. Biophys. Biochem. Cytol.* 7: 315–320.

Conklin, E. G. 1905. The orientation and cell-lineage of the ascidian egg. *J. Acad. Nat. Sci. Phila.* 13: 5–119.

Corselli, J. and Talbot, P. 1987. In vivo penetration of hamster oocyte-cumulus complexes using physiological numbers of sperm. *Dev. Biol.* 122: 227–242.

Cross, N. L. and Elinson, R. P. C. 1980. A fast block to polyspermy in frogs mediated by changes in the membrane potential. *Dev. Biol.* 75: 187–198.

Dan, J. C. 1967. Acrosome reaction and lysins. *In* C. B. Metz and A. Monroy (eds.), *Fertilization*, Vol. 1. Academic Press, New York, pp. 237–367.

Danilchik, M. V. and Denegre, J. M. 1991. Deep cytoplasmic rearrangements during early development in *Xenopus laevis. Development*, in press.

Davis, B. K. 1981. Timing of fertilization in mammals: Sperm cholesterol/phospholipid ratio as determinant of capacitation interval. *Proc. Natl. Acad. Sci. USA* 78: 7560–7564.

Dawid, I. B. and Blackler, A. W. 1972. Maternal and cytoplasmic inheritance of mitochondria in *Xenopus. Dev. Biol.* 29: 152–161.

De Robertis, E. D. P., Saez, F. A. and De Robertis, E. M. F. 1975. *Cell Biology*, 6th Ed., Saunders, Philadelphia.

Dube, F., Schmidt, T., Johnson, C. H. and Epel, D. 1985. The hierarchy of requirements for an elevated pH during early development of sea urchin embryos. *Cell* 40: 657–666.

Eisen, A. and Reynolds, G. T. 1985. Sources and sinks for the calcium release during fertilization of single sea urchin eggs. *J. Cell Biol.* 100: 1522–1527.

Elinson, R. P. 1985. Changes in levels of polymeric tubulin associated with activation and dorsoventral polarization of the frog egg. *Dev. Biol.* 109: 224–233.

Elinson, R. P. 1986. Fertilization in amphibians: The ancestry of the block to polyspermy. *Int. Rev. Cytol.* 101: 59–100.

Elinson, R. P. and Rowning, B. 1988. A transient array of parallel microtubules in frog eggs: Potential tracks for a cytoplasmic rotation that specifies the dorso-ventral axis. *Dev. Biol.* 128: 185–197.

Endo, Y. G., Kopf, G. S. and Schultz, R. M. 1987a. Effects of phorbol ester on mouse eggs: Dissociation of sperm receptor activity from acrosome-reaction inducing activity of the mouse zona pellucida protein, ZP3. *Dev. Biol.* 123: 574–577.

Endo, Y., Lee, M. A. and Kopf, G. S. 1987b. Evidence for the role of a guanosine nucleotide-binding regulatory protein in the zona pellucida-induced mouse sperm acrosome reaction. *Dev. Biol.* 119: 210–216.

Epel, D. 1967. Protein synthesis in sea urchin eggs: A "late" response to fertilization. *Proc. Natl. Acad. Sci. USA* 57: 899–906.

Epel, D. 1977. The program of fertilization. *Sci. Am.* 237(5): 128–138.

Epel, D. 1980. Fertilization. *Endeavour N.S.* 4: 26–31.

Epel, D., Patton, C., Wallace, R. W. and Cheung, W. Y. 1981. Calmodulin activates NAD kinase of sea urchin eggs: An early response. *Cell* 23: 543–549.

Ferris, C. D., Huganir, R. L., Supattapone, S. and Snyder, S. H. 1989. Purified inositol 1,4,5-trisphosphate receptor mediates calcium flux in reconstituted lipid vesicles. *Nature* 342: 87–89.

Florman, H. M. and Storey, B. T. 1982. Mouse gamete interactions: The zona pellucida is the site of the acrosome reaction leading to fertilization in vitro. *Dev. Biol.* 91: 121–130.

Florman, H. M. and Wassarman, P. M. 1985. O-linked oligosaccharides of mouse egg ZP3 account for its sperm receptor activity. *Cell* 41: 313–324.

Florman, H. M., Bechtol, K. B. and Wassarman, P. M. 1984. Enzymatic dissection of the functions of the mouse egg's receptor for sperm. *Dev. Biol.* 106: 243–255.

Foerder, C. A. and Shapiro, B. M. 1977. Release of ovoperoxidase from sea urchin eggs hardens fertilization membrane with tyrosine crosslinks. *Proc. Natl. Acad. Sci. USA* 74: 4214–4218.

Franklin, L. E. 1970. Fertilization and the role of the acrosomal reaction in non-mammals. *Biol. Reprod.* [Suppl.] 2: 159–176.

Fulton, B. P. and Whittingham, D. G. 1978. Activation of mammalian oocytes by intracellular injection of calcium. *Nature* 273: 149–151.

Furuichi, T., Yoshikawa, S., Miyawaki, A., Wada, K., Maeda, N. and Mikoshiba, K. 1989. Primary structure and functional expression of the inositol 1,4,5-trisphosphate-binding protein P400. *Nature* 342: 32–38.

Gardiner, D. M. and Grey, R. D. 1983. Membrane junctions in *Xenopus* eggs: Their distribution suggests a role in calcium regulation. *J. Cell Biol.* 96: 1159–1163.

Gerhart, J., Ubbels, G., Black, S., Hara, K. and Kirschner, M. 1981. A reinvestigation of the role of the grey crescent in axis formation in *Xenopus laevis*. *Nature* 292: 511–516.

Gerhart, J., Black, S., Gimlich, R. and Scharf, S. 1983. Control of polarity in the amphibian egg. *In* W. R. Jeffery and R. A. Raff (eds.), *Time, Space, and Pattern in Embryonic Development*. Alan R. Liss, New York, pp. 261–286.

Gerhart, J., Danilchik, M., Doniach, T., Roberts, S., Rowning, B. and Stewart, R. 1989. Cortical rotation of the *Xenopus* egg: Consequences for the anterioposterior pattern of embryonic dorsal development. *Development 1989* [Suppl.]: 37–51.

Giles, R. E., Blanc, H., Cann, H. M. and Wallace, D. C. 1980. Maternal inheritance of human mitochondrial DNA. *Proc. Natl. Acad. Sci. USA* 77: 6715–6719.

Gilkey, J. C., Jaffe, L. F., Ridgway, E. G. and Reynolds, G. T. 1978. A free calcium wave traverses the activating egg of the medaka, *Oryzias latipes*. *J. Cell Biol.* 76: 448–466.

Ginzburg, A. S. 1985. Phylogenetic changes in the type of fertilization. *In* J. Mlíkovsky and V. J. A. Novák (eds.), *Evolution and Morphogenesis*. Academia, Prague, pp. 459–466.

Glabe, C. G. 1985. Interaction of the sperm adhesive protein, bindin, with phospholipid vesicles. II. Bindin induces the fusion of mixed-phase vesicles that contain phosphatidylcholine and phosphatidylserine in vitro. *J. Cell Biol.* 100: 800–806.

Glabe, C. G. and Lennarz, W. J. 1979. Species-specific sperm adhesion in sea urchins: A quantitative investigation of bindin-mediated egg agglutination. *J. Cell Biol.* 83: 595–604.

Glabe, C. G. and Vacquier, V. D. 1977. Species-specific agglutination of eggs by bindin isolated from sea urchin sperm. *Nature* 267: 836–838.

Glabe, C. G. and Vacquier, V. D. 1978. Egg surface glycoprotein receptor for sea urchin sperm bindin. *Proc. Natl. Acad. Sci. USA* 75: 881–885.

Gould, M. and Stephano, J. L. 1987. Electrical response of eggs to acrosomal protein similar to those induced by sperm. *Science* 235: 1654–1656.

Gould-Somero, M., Jaffe, L. A. and Holland, L. Z. 1979. Electrically mediated fast polyspermy block in eggs of the marine worm, *Urechis caupo*. *J. Cell Biol.* 82: 426–440.

Green, G. R. and Poccia, E. L. 1985. Phosphorylation of sea urchin sperm H1 and H2B histones precedes chromatin decondensation and H1 exchange during pronuclear formation. *Dev. Biol.* 108: 235–245.

Gundersen, G. G., Medill, L. and Shapiro, B. M. 1986. Sperm surface proteins are incorporated into the egg membrane and cytoplasm after fertilization. *Dev. Biol.* 113: 207–217.

Gwatkin, R. B. L. 1976. Fertilization. *In* G. Poste and G. L. Nicolson (eds.), *The Cell Surface in Animal Embryogenesis and Development*. Elsevier North-Holland, New York, pp. 1–53.

Hafner, M., Petzelt, C., Nobiling, R., Pawley, J. B., Kramp, D. and Schatten, G. 1988. Wave of free calcium at fertilization in the sea urchin egg visualized with Fura-2. *Cell Motl. Cytoskel.* 9: 271–277.

Hamaguchi, M. S. and Hiramoto, Y. 1980. Fertilization process in the heart-urchin, *Clypeaster japonicus*, observed with a differential interference microscope. *Dev. Growth Differ.* 22: 517–530.

Hartsoeker, N. 1694. *Essai de dioptrique*. Paris.

Heinecke, J. W. and Shapiro, B. M. 1989. Respiratory oxygen burst of fertilization. *Proc. Natl. Acad. Sci. USA* 86: 1259–1263.

Hertwig, O. 1876. Beiträge zur Kenntniss der Bildung, Befruchtung, und Theilung des theirischen Eies. *Morphol. Jahr.* 1: 347–452.

Hirao, Y. and Yanagimachi, R. 1978. Effects of various enzymes on the ability of hamster egg plasma membrane to fuse with spermatozoa. *Gamete Res.* 1: 3–12.

Hollinger, T. G. and Schuetz, A. W. 1976. "Cleavage" and cortical granule breakdown in *Rana pipiens* oocytes induced by direct microinjection of calcium. *J. Cell Biol.* 71: 395–401.

Hong, K. and Vacquier, V. D. 1986. Fusion of liposomes induced by a cationic protein from the acrosomal granule of abalone spermatozoa. *Biochemistry* 25: 543–549.

Houliston, E. and Elinson, R. P. 1990. Organisation of the cytoskeleton and endoplasmic reticulum in the frog egg cortex. Abstracts of the Forty-Ninth Annual Symposium of the Society for Developmental Biology, p. 11.

Hylander, B. L. and Summers, R. G. 1982. An ultrastructural and immunocytochemical localization of hyaline in the sea urchin egg. *Dev. Biol.* 93: 368–380.

Iwao, Y. and Jaffe, L. A. 1989. Evidence that the voltage-dependent component in the fertilization porcess is contributed by the sperm. *Dev. Biol.* 134: 446–451.

Jacobs, P. A., Wilson, C. M., Sprenkle, J. A., Rosenshein, N. B. and Migeon, B. R. 1980. Mechanism of origin of complete hydatidiform moles. *Nature* 286: 714–716.

Jaffe, L. A. 1976. Fast block to polyspermy in sea urchins is electrically mediated. *Nature* 261: 68–71.

Jaffe, L. A. 1980. Electrical polyspermy block in sea urchins: Nicotine and low sodium experiments. *Dev. Growth Differ.* 22: 503–507.

Jaffe, L. A. and Cross, N. L. 1983. Electrical properties of vertebrate oocyte membranes. *Biol. Reprod.* 30: 50–54.

Jaffe, L. A., Sharp, A. P. and Wolf, D. P. 1983. Absence of an electrical polyspermy block in the mouse. *Dev. Biol.* 96: 317–323.

Jaffe, L. F. 1983. Sources of calcium in egg activation: A review and hypothesis. *Dev. Biol.* 99: 265–276.

Jones, R., Brown, C. R. and Lancaster, R. T. 1988. Carbohydrate-binding properties of boar sperm proacrosin and assessment of its role in sperm-egg recognition and adhesion during fertilization. *Development* 102: 781–792.

Just, E. E. 1919. The fertilization reaction in *Echinarachnius parma*. *Biol. Bull.* 36: 1–10.

Klag, J. J. and Ubbels, G. A. 1975. Regional morphological and cytochemical differentiation of the fertilized egg of *Discoglossus pictus* (Anura). *Differentiation* 3: 15–20.

Kline, D., Simoncini, L., Mandel, G., Maue, R., Kado, R. T. and Jaffe. L. A. 1988. Fertilization events induced by neurotransmitters after injection of mRNA into *Xenopus* eggs. *Science* 241: 464–467.

Kolliker, A. von. 1841. *Beiträge zur Kenntnis der Geschlectverhältnisse und der Samenflüssigkeit wirbelloser Thiere, nebst einem Versuch Über Wesen und die Bedeutung der sogenannten Samenthiere*. Berlin.

Kvist, U., Afzelius, B. A. and Nilsson, L. 1980. The intrinsic mechanism of chromatin decondensation and its activation in human spermatozoa. *Dev. Growth Differ.* 22: 543–554.

Leeuwenhoek, A. van. 1685. Letter to the Royal Society of London. Quoted in E. G. Ruestow, 1983, Images and ideas: Leeuwenhoek's perception of the spermatozoa. *J. Hist. Biol.* 16: 185–224.

Levine, A. E., Walsh, K. A. and Fodor, E. J. B. 1978. Evidence of an acrosin-like enzyme in sea urchin sperm. *Dev. Biol.* 63: 299–306.

Leyton, L. and Saling, P. 1989a. 95 kd sperm proteins bind ZP3 and serve as tyrosine kinase substrates in response to zona binding. *Cell* 57: 1123–1130.

Leyton, L. and Saling, P. 1989b. Evidence that aggregation of mouse sperm receptors by ZP3 triggers the acrosome reaction. *J. Cell Biol.* 108: 2163–2168.

Longo, F. J. 1986. Surface changes at fertilization: Integration of sea urchin (*Arbacia punctulata*) sperm and oocyte plasma membranes. *Dev. Biol.* 116: 143–159.

Longo, F. J., Lynn, J. W., McCulloh, D. H., Chambers, E. L. and So, F. 1986. Correlative ultrastructural and electrophysiological studies of sperm–egg interactions of the sea urchin, *Lytechinus variegatus*. *Dev. Biol.* 118: 155–166.

Lopez, L. C., Bayna, E. M., Litoff, D., Shaper, N. L., Shaper, J. H. and Shur, B. D. 1985. Receptor function of mouse sperm surface galactosyltransferase during fertilization. *J. Cell Biol.* 101: 1501–1510.

Luttmer, S. and Longo, F. J. 1985. Ultrastructural and morphometric observations of cortical endoplasmic reticulum in *Arbacia*, *Spisula*, and mouse eggs. *Dev. Growth Differ.* 27: 349–359.

Manes, M. E. and Barbieri, F. D. 1976. Symmetrization in the amphibian egg by disrupted sperm cells. *Dev. Biol.* 53: 138–141.

Manes, M. E. and Barbieri, F. D. 1977. On the possibility of sperm aster involvement in dorso-ventral polarization and pronuclear migration in the amphibian egg. *J. Embryol. Exp. Morphol.* 40: 187–197.

Manes, M.E. and Elinson, R.P. 1980. Ultraviolet light inhibits gray crescent formation in the frog egg. *Wilhelm Roux Arch. Dev. Biol.* 189: 73–76.

Manes, M. E., Elinson, R. P. and Barbieri, F. D. 1978. Formation of the amphibian gray crescent: Effects of colchicine and cytochalasin-B. *Wilhelm Roux Arch. Dev. Biol.* 185: 99–104.

Masui, Y. 1974. A cytostatic factor in amphibian oocytes: Its extraction and partial characterization. *J. Exper. Zool.* 187: 141–147.

McGrath, J. and Solter, D. 1984. Completion of mouse embryogenesis requires both the maternal and paternal genome. *Cell* 37: 179–183.

Meizel, S. 1984. The importance of hydrolytic enzymes to an exocytotic event, the mammalian sperm acrosome reaction. *Biol. Rev.* 59: 125–157.

Metz, C. B. 1978. Sperm and egg receptors involved in fertilization. *Curr. Top. Dev. Biol.* 12: 107–148.

Millar, S., Chamow, S. M., Baur, A. W., Oliver, C., Robey, F. and Dean, J. 1989. Vaccination with a synthetic zona pellucida peptide produces long-term contraception in female mice. *Science* 246: 935–938.

Miller, R. L. 1978. Site-specific agglutination and the timed release of a sperm chemoattractant by the egg of the leptomedusan, *Orthopyxis caliculata*. *J. Exp. Zool.* 205: 385–392.

Miller, R. L. 1985. Sperm chemo-orientation in the metazoa. *In* C. B. Metz, Jr. and A. Monroy (eds.), *Biology of Fertilization*, Vol. 2. Academic Press, New York, pp. 275–337.

Miyazaki, S. and Igusa, Y. 1981. Fertilization potential in golden hamster eggs consists of recurring hyperpolarizations. *Nature* 290: 702–704.

Moller, C. C. and Wassarman, P. M. 1989. Characterization of a proteinase that cleaves zona pellucida glycoprotein ZP2 following activation of mouse eggs. *Dev. Biol.* 132: 103–112.

Moy, G. W. and Vacquier, V. D. 1979. Immunoperoxidase localization of bindin during the adhesion of sperm to sea urchin eggs. *Curr. Top. Dev. Biol.* 13: 31–44.

Ogawa, K., Mohri, T. and Mohri, H. 1977. Identification of dynein as the outer arms of sea urchin sperm axonemes. *Proc. Natl. Acad. Sci. USA* 74: 5006–5010.

Ohama, K., Kajii, T., Okamoto, E., Fukada, Y., Imaizumi, K., Tsukahara, M., Kobayashi, K. and Hagiwara, K. 1981. Dispermic origin of XY hydatidiform moles. *Nature* 292: 551–552.

Poirier, G. R. and Jackson, J. 1981. Isolation and characterization of two proteinase inhibitors from the male reproductive tract of mice. *Gamete Res.* 4: 555–569.

Prevost, J. L. and Dumas, J. B. 1824. Deuxieme mémoire sur la génération. *Ann. Sci. Nat.* 2: 129–149.

Primakoff, P., Hyatt, H. and Tredick-Kline, J. 1987. Identification and purification of a sperm cell surface protein with a potential role in sperm–egg membrane fusion. *J. Cell Biol.* 104: 141–149.

Primakoff, P., Lathrop, W., Woolman, L., Cowan, A., and Myles, D. 1988. Fully effective contraception in male and female guinea pigs immunized with the sperm protein PH-20. *Nature* 335: 543–546.

Ramarao, C. S. and Garbers, D. L. 1985. Receptor-mediated regulation of guanylate cyclase activity in spermatozoa. *J. Biol. Chem.* 260: 8390–8396.

Rossignol, D. P., Earles, B. J., Decker, G. L. and Lennarz, W. J. 1984. Characterization of the sperm receptor on the surface of eggs of *Strongylocentrotus purpuratus*. *Dev. Biol.* 104: 308–321.

Roux, W. 1887. Beiträge zur Entwicklungsmechanik des Embryo. *Arch. Mikrosk. Anat.* 29: 157–212.

Saling, P. M. 1981. Involvement of trypsin-like activity in binding of mouse spermatozoa to zonae pellucidae. *Proc. Natl. Acad. Sci. USA* 78: 6231–6235.

Saling, P. M. 1989. Mammalian sperm interaction with extracellular matrices of the egg. *Oxford Reviews of Reproductive Biology* 11: 339-388.

Saling, P. M., Sowinski, J. and Storey, B. T. 1979. An ultrastructural study of epididymal mouse spermatozoa binding to zonae pellucida *in vitro*: Sequential relationship to acrosome reaction. *J. Exp. Zool.* 209: 229–238.

Saunders, J. W., Jr. 1970. *Principles and Patterns of Animal Development*. Macmillan, New York.

Sawada, T. and Schatten, G. 1989. Effects of cytoskeletal inhibitors on ooplasmic segregation and microtubule organization during fertilization and early development in the ascidian *Molgula occidentalis*. *Dev. Biol.* 132: 331–342.

Schackmann, R. W. and Shapiro, B. M. 1981. A partial sequence of ionic changes associated with the acrosome reaction of *Strongylocentrotus purpuratus*. *Dev. Biol.* 81: 145–154.

Schackmann, R. W., Eddy, E. M. and Shapiro, B. M. 1978. The acrosome reaction of *Strongylocentrotus purpuratus* sperm: Ion requirements and movements. *Dev. Biol.* 65: 483–495.

Schatten, G. and Mazia, D. 1976. The penetration of the spermatozoan through the sea urchin egg surface at fertilization: Observations from the outside on whole eggs and from the inside on isolated surfaces. *Exp. Cell Res.* 98: 325–337.

Schatten, G. and Schatten, H. 1983. The energetic egg. *The Sciences* 23(5): 28–35.

Schatten, H. and Schatten, G. 1980. Surface activity at the plasma membrane during sperm incorporation and its cytochalasin-B sensitivity: Scanning electron micrography and time-lapse video microscopy during fertilization of the sea urchin *Lytechinus variegatus*. *Dev. Biol.* 78: 435–449.

Schroeder, T. E. 1979. Surface area change at fertilization: Resorption of the mosaic membrane. *Dev. Biol.* 70: 306–326.

SeGall, G. K. and Lennarz, W. J. 1979. Chemical characterization of the component of the jelly coat from sea urchin eggs responsible for induction of the acrosome reaction. *Dev. Biol.* 71: 33–48.

Shapiro, B. M. Schackmann, R. W., Tombes, R. M. and Kazozoglan, T. 1985. Coupled ionic and enzymatic regulation of sperm behavior. *Curr. Top. Cell Regul.* 26: 97–113.

Shen, S. S. and Burgart, L. J. 1986. 1,2-Diacylglycerols mimic phorbol 12-myristate 13-acetate activation of the sea urchin egg. *J. Cell Physiol.* 127: 330–340.

Shen, S. S. and Steinhardt, R. A. 1978. Direct measurement of intracellular pH during metabolic depression of the sea urchin egg. *Nature* 272: 253–254.

Shur, B. D. and Hall, N. G. 1982a. Sperm surface galactosyltransferase activities during in vitro capacitation. *J. Cell Biol.* 95: 567–573.

Shur, B. D. and Hall, N. G. 1982b. A role for mouse sperm surface galactosyltransferase in sperm binding for the egg zona pellucida. *J. Cell Biol.* 95: 574–579.

Shur, B. D. and Neely, C. A. 1988. Plasma membrane association, purification, and partial characterization of mouse sperm b1,4-galactosyltransferase. *J. Biol. Chem.* 263: 17706–17714.

Singh, S. and several others 1988. Membrane guanylate cycles is a cell surface receptor with homology to protein kinases. *Nature* 334: 708–712.

Steinhardt, R. A. and Epel, D. 1974. Activation of sea urchin eggs by a calcium ionophore. *Proc. Natl. Acad. Sci. USA* 71: 1915–1919.

Steinhardt, R., Zucker, R. and Schatten, G. 1977. Intracellular calcium release at fertilization in the sea urchin egg. *Dev. Biol.* 58: 185–196.

Summers, R. G. and Hylander, B. L. 1975. Species-specificity of acrosome reaction and primary gamete binding in echinoids. *Exp. Cell Res.* 96: 63–68.

Summers, R. G., Hylander, B. L., Colwin, L. H. and Colwin, A. L. 1975. The functional anatomy of the echinoderm spermatozoon and its interaction with the egg at fertilization. *Am. Zool.* 15: 523–551.

Surani, M. A. H. and Barton, S. C. 1983. Development of gynocogenetic eggs in the mouse: Implications for parthenogenetic embryos. *Science* 222: 1034–1036.

Surani, M. A. H., Barton, S. C. and Norris, M. L. 1986. Nuclear transplantation in the mouse: Hereditable differences between parental genomes after activation of the embryonic genome. *Cell* 45: 127–136.

Swann, K. and Whitaker, M. 1985. Stimulation of the Na/H exchanger of sea urchin eggs by phorbol ester. *Nature* 314: 274–277.

Swann, K. and Whitaker, M. 1986. The part played by inositol trisphosphate and calcium in the propagation of the fertilization wave in sea urchin eggs. *J. Cell Biol.* 103: 2333–2342.

Talbot, P. 1985. Sperm penetration through oocyte investments in mammals. *Am. J. Anat.* 174: 331–346.

Talbot, P., DiCarlantonio, G., Zao, P., Penkala, J. and Haimo, L. T. 1985. Motile cells lacking hyaluronidase can penetrate the hamster oocyte cumulus complex. *Dev. Biol.* 108: 387–398.

Tilney, L. G., Bryan, J., Bush, D. J., Fujiwara, K., Mooseker, M. S., Murphy, D. B. and Snyder, D. H. 1973. Microtubules: Evidence for thirteen protofilaments. *J. Cell Biol.* 59: 267–275.

Tilney, L. G., Kiehart, D. P., Sardet, C. and Tilney, M. 1978. Polymerization of actin. IV. Role of Ca^{2+} and H^+ in the assembly of actin and in membrane fusion in the acrosomal reaction of echinoderm sperm. *J. Cell Biol.* 77: 536–560.

Tombes, R. M. and Shapiro, B. M. 1985. Metabolite channeling: A phosphocreatine shuttle to mediate high energy phosphate transport between sperm mitochondrion and tail. *Cell* 41: 325–334.

Turner, P. R., Jaffe, L. A. and Fein, A. 1986. Regulation of cortical vesicle exocytosis in sea urchin eggs by inositol 1,4,5-trisphosphate and GTP-binding protein. *J. Cell Biol.* 102: 70–76.

Turner, P. R., Jaffe, L. A. and Primakoff, P. 1987. A cholera toxin-sensitive G-protein stimulates exocytosis in sea urchin eggs. *Dev. Biol.* 120: 577–583.

Uzzell, T. M. 1964. Relations of the diploid and triploid species of the *Ambystoma jeffersonianum* complex. *Copeia* 1964: 257–300.

Vacquier, V. D. 1979. The interaction of sea urchin gametes during fertilization. *Am. Zool.* 19: 839–849.

Vacquier, V. D. and Moy, G. W. 1977. Isolation of bindin: The protein responsible for adhesion of sperm to sea urchin eggs. *Proc. Natl. Acad. Sci. USA* 74: 2456–2460.

Vacquier, V. D. and Payne, J. E. 1973. Methods for quantitating sea urchin sperm in egg binding. *Exp. Cell Res.* 82: 227–235.

Vacquier, V. D., Tegner, M. J. and Epel, D. 1973. Protease release from sea urchin eggs at fertilization alters the vitelline layer and aids in preventing polyspermy. *Exp. Cell Res.* 80: 111–119.

Vincent, J. P., Oster, G. F. and Gerhart, J. C. 1986. Kinematics of gray crescent formation in *Xenopus* eggs. The displacement of subcortical cytoplasm relative to the egg surface. *Dev. Biol.* 113: 484–500.

Voet, D. and Voet, J. G. 1990. *Biochemistry.* Wiley, New York.

Ward, G. E., Brokaw, C. J., Garbers, D. L. and Vacquier, V. D. 1985. Chemotaxis of *Arbacia punctulata* spermatozoa to resact, a peptide from the egg jelly layer. *J. Cell Biol.* 101: 2324–2329.

Wassarman, P. 1987. The biology and chemistry of fertilization. *Science* 235: 553–560.

Wassarman, P. M. 1989. Fertilization in mammals. *Sci. Am.* 256(6): 78–84.

Watanabe, N., Vande Woude, G. F., Ikawa, Y. and Sagata, N. 1989. Specific proteolysis of the *c-mos* proto-oncogene product by calpain on fertilization of *Xenopus* eggs. *Nature* 342: 505–517.

Whitaker, M. and Irvine, R. F. 1984. Inositol 1,4,5-triphosphate microinjection activates sea urchin eggs. *Nature* 312: 636–639.

Whitaker, M. and Steinhardt, R. 1982. Ionic regulation of egg activation. *Q. Rev. Biophys.* 15: 593–666.

Whitaker, M. J. and Steinhardt, R. 1985. Ionic signalling in the sea urchin egg at fertilization. *In* C. B. Metz and A. Monroy (eds.), *Biology of Fertilization*, Vol. 3. Academic Press, Orlando, Florida, pp. 167–221.

Wilde, M. W. and Kopf, G. S. 1989. Activation of a G-protein in mammalian sperm by an egg-associated extracellular matrix, the zona pellucida. *J. Cell Biol.* 109: 251a.

Wilson, W. L. and Oliphant, G. 1987. Isolation and biochemical characterization of the subunits of the rabbit sperm acrosome stabilizing factor. *Biol. Reprod.* 37: 159–169.

Winkler, M. M., Steinhardt, R. A., Grainger, J. L. and Minning, L. 1980. Dual ionic controls for the activation of protein synthesis at fertilization. *Nature* 287: 558–560.

Yanagimachi, R. 1981. Mechanisms of fertilization in mammals. *In* L. Mastroianni and J. D. Biggers (eds.), *Fertilization and Embryonic Development In Vitro.* Plenum, New York, pp. 81–182.

Yanagimachi, R. and Noda, Y. D. 1970. Electron microscope studies of sperm incorporation into the golden hamster egg. *Am. J. Anat.* 128: 429–462.

Yanagimachi, R., Okada, A. and Tung, K. S. K. 1981. Sperm autoantigens and fertilization: Effects of anti-guinea pig sperm autoantibodies on sperm–ovum interactions. *Biol. Reprod.* 24: 512–518.

Yudin, A., Cherr, G. and Katz, D. 1988. The structure of the cumulus matrix and zona pellucida in the golden hamster: A new view of sperm interaction with oocyte-associated extracellular matrices. *Cell Tiss. Res.* 251: 555–564.

3

Cleavage: Creating multicellularity

One must show the greatest respect towards any *thing that increases exponentially, no matter how small.*

—GARRETT HARDIN (1968)

Were I to live a thousand years I should still be as deeply moved to see a frog's egg at the moment of cleavage.

—JEAN ROSTAND (1962)

Remarkable though it is, fertilization is but the initiatory step in development. The zygote, with its new genetic potential and its new arrangement of cytoplasm, now begins the production of a multicellular organism. In all animal species known, this begins by a process called cleavage, a series of mitotic divisions whereby the enormous volume of egg cytoplasm is divided into numerous smaller nucleated cells. These cleavage-stage cells are called BLASTOMERES.

In most species (mammals being the chief exception), the rate of cell division and the placement of the blastomeres with respect to one another is completely under the control of the proteins and mRNAs stored in the oocyte by the mother. The zygotic genome, transmitted by mitosis to all the other cells, does not function in early cleavage embryos. Few, if any, mRNAs are made until relatively late in cleavage, and the embryo can divide properly even when chemicals are used to inhibit transcription. Also, in most species there is no net increase in embryonic volume during cleavage. This differs from most cases of cell proliferation where there is a period of cell growth between mitoses; a cell will expand to nearly twice its volume, then divide. This growth produces a net increase in the total volume of cells while maintaining a relatively constant ratio of nuclear volume to cytoplasmic volume. During embryonic cleavage, however, cytoplasmic volume does not increase. Rather, the enormous volume of zygote cytoplasm is divided into increasingly smaller cells. First the egg is divided in half, then quarters, then eighths, and so forth. This division of egg cytoplasm without growth is accomplished by abolishing the growth period between divisions, while the cleavage of nuclei occurs at a rate never seen again (even in tumor cells). A frog egg, for example, can divide into 37,000 cells in just 43 hours; and mitosis in cleavage-stage *Drosophila* occurs every 10 minutes for over 2 hours.

This increase in cell number can be appreciated by comparing cleavage with other states of development. Figure 1 shows the logarithm of cell number in a frog embryo plotted against the time of development (Sze, 1953). It illustrates a sharp discontinuity between cleavage and gastrulation. One consequence of this rapid division is that the ratio of cytoplasmic volume to nuclear volume gets increasingly smaller as cleavage progresses. In many types of embryos, this manifold decrease in the cytoplasmic volume to nuclear volume ratio is crucial in timing the activation of certain genes.

In the frog *Xenopus laevis*, transcription of new messages is not activated until after 12 divisions. At that time, the rate of cleavage decreases, the blastomeres become motile, and nuclear genes begin to be transcribed. It is thought that some factor in the egg is being titrated by the newly made chromatin, as the time of this transition can be changed by altering the amount of chromatin in each nucleus: the more chromatin present in the embryo, the earlier the transition. If the amount of chromatin present is double the normal amount of chromatin, the transition occurs one division earlier (Newport and Kirschner, 1982a,b). Thus, cleavage begins soon after fertilization and ends when the embryo achieves a new balance between nucleus and cytoplasm.

PATTERNS OF EMBRYONIC CLEAVAGE

Cleavage is a very well-coordinated process under genetic regulation. The pattern of embryonic cleavage particular to a species is determined by two major parameters: (1) the amount and distribution of yolk protein within the cytoplasm, and (2) those factors in the egg cytoplasm influencing the angle of the mitotic spindle and the timing of its formation.

The amount and distribution of yolk determines where cleavage can occur and the relative size of the blastomeres. When one pole of the egg is relatively yolk-free, the cellular divisions occur there at a faster rate than at the opposite pole. The yolk-rich pole is referred to as the VEGETAL POLE; the ANIMAL POLE is relatively low in yolk concentration. The zygote nucleus is frequently displaced toward the animal pole. Table 1 provides a classification of cleavage types and shows the influence of yolk on the

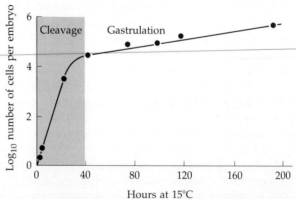

FIGURE 1
Formation of new cells during the early development of the frog *Rana pipiens*. (After Sze, 1953.)

TABLE 1
Classification of cleavage types

Cleavage pattern	Position of yolk	Cleavage symmetry	Representative animals
Holoblastic (complete cleavage)	Isolecithal (oligolecithal) (sparse, evenly distributed yolk)	Radial	Echinoderms, *Amphioxus*
		Spiral	Most molluscs, annelids, flatworms, and roundworms
		Bilateral	Ascidians
		Rotational	Mammals
	Mesolecithal (moderately telolecithal)	Radial	Amphibians
Meroblastic (incomplete cleavage)	Telolecithal (dense yolk concentrated at one end of egg)	Bilateral	Cephalopod molluscs
		Discoidal (bilateral)	Reptiles, fishes, birds
	Centrolecithal (yolk concentrated in center of egg)	Superficial	Most arthropods

cleavage symmetry and pattern. In general, yolk inhibits cleavage. In zygotes with relatively little yolk (isolecithal and mesolecithal eggs), cleavage is HOLOBLASTIC, meaning that the cleavage furrow extends through the entire egg. Zygotes containing large accumulations of yolk protein undergo MEROBLASTIC cleavage, wherein only a portion of the cytoplasm is cleaved. The cleavage furrow does not penetrate into the yolky portion of the cytoplasm. Meroblastic cleavage can be DISCOIDAL, as in birds' eggs, or SUPERFICIAL, as in insect zygotes, depending on whether the yolk deposit is located to one side (telolecithal) or in the center of the cytoplasm (centrolecithal).

Yolk is an evolutionary adaptation that enables an embryo to develop in the absence of an external food source. Animals developing without large yolk concentrations, such as sea urchins, usually form a larval stage fairly rapidly. This larval stage can feed itself, and development then continues from this free-swimming larva. Mammalian eggs, which also lack large quantities of yolk, adopt another strategy, namely, making a placenta. As we shall see, the first differentiation of the mammalian embryo sets aside the cells that will form the placenta. This organ will supply food and oxygen for the embryo during its long gestation.

At the other extreme are the eggs of fishes, reptiles, and birds. Most of their cell volumes are yolk. The yolk must be sufficient to nourish these animals, as they develop without a larval stage or placental attachment. The correlation between heavy yolk concentration and lack of larval forms is seen in certain species of frogs. Certain tropical frogs such as *Eleutherodactylus* and *Arthroleptella* lack a tadpole stage. Rather, they provision their eggs with an enormous yolk concentration (Lutz, 1947). The eggs do not need to be laid in water because the tadpole stage has been eliminated. This will be discussed further in Chapter 19.

However, the yolk is just one factor influencing a species' pattern of cleavage. There are also inherited patterns of cell division that are superimposed upon the restraints of the yolk. This can readily be seen in isolecithal eggs, in which very little yolk is present. In the absence of a large concentration of yolk, four major cleavage types can be observed: radial holoblastic cleavage, spiral holoblastic cleavage, bilateral holoblastic cleavage, and rotational holoblastic cleavage.

Radial holoblastic cleavage

The sea cucumber, *Synapta*

RADIAL HOLOBLASTIC cleavage is the simplest form of cleavage to understand. It is characteristic of echinoderms and the protochordate *Amphioxus*. The furrows in this type of cleavage are oriented parallel to and perpendicular to the animal–vegetal axis of the egg. Figure 2 shows the cleavage patterns of the sea cucumber, *Synapta digita*. After the union of the pronuclei, the axis of the first mitotic spindle is formed perpendicular to the animal–vegetal axis of the egg. Therefore, the first cleavage furrow passes directly through the animal and vegetal poles, creating two equal-sized daughter cells. This cleavage is said to be MERIDIONAL because it passes through the two poles like a meridian on a globe. The mitotic spindles of the second cleavage are at right angles to the first but are still perpendicular to the animal–vegetal axis of the egg. The cleavage furrows appear simultaneously in both blastomeres and also pass through the two poles. Thus, the first two divisions are both meridional and are perpendicular to each other. The third division is EQUATORIAL: the mitotic spindles of each blastomere are now positioned parallel to the animal–vegetal axis and the resulting cleavage furrow separates the two poles from each other, dividing the embryo into eight equal blastomeres. Each blastomere in the animal half of the embryo is now directly above a blastomere of the vegetal half.

The fourth division is again meridional, producing two tiers of 8 cells each, while the fifth division is equatorial, producing four tiers of 8 cells each. Successive divisions produce 64-, 128-, and 256-cell embryos, with meridional divisions alternating with equatorial divisions. The resulting embryo consists of blastomeres arranged in horizontal rows along a central cavity. At both poles of the embryo, blastomeres move toward each other to create a hollow sphere composed of a single cell layer. This hollow sphere is called the BLASTULA, and the central cavity is referred to as the BLASTOCOEL.

At any time during the cleavage of *Synapta*, an embryo bisected through any meridian will produce two mirror-image halves. This type of

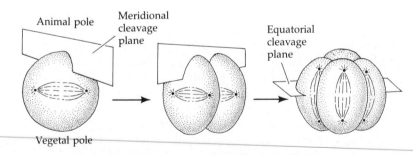

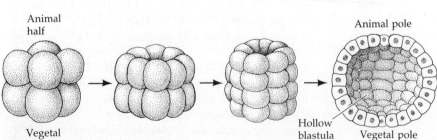

FIGURE 2
Holoblastic cleavage in the echinoderm *Synapta digita*, leading to the formation of a hollow blastula, as shown in the cutaway view in the last panel. (After Saunders, 1982.)

symmetry is characteristic of a sphere or cylinder and is called radial symmetry. Thus, *Synapta* is said to have radial holoblastic cleavage.

Sea urchins

Sea urchins also have a radial holoblastic cleavage, but with some important modifications. The first and second cleavages are very similar to those of *Synapta*; both are meridional and are perpendicular to each other. Similarly, the third cleavage is equatorial, separating the two poles from one another (Figure 3). In the fourth cleavage, however, events are very different. The four cells of the animal tier split meridionally into eight blastomeres, each having the same volume. These cells are called MESOMERES. The vegetal tier, however, undergoes an unequal equatorial cleavage to produce four large cells, the MACROMERES, and four smaller MICROMERES at the vegetal pole (Figure 4). As the 16-cell embryo cleaves, the eight mesomeres divide equatorially to produce two "animal" tiers, an_1 and an_2, one above the other. The macromeres divide meridionally, forming a tier of eight cells below an_2. The micromeres also divide, producing a small cluster beneath the larger tier. All the cleavage furrows of the sixth division are equatorial; and the seventh cleavage is meridional, producing a 128-cell blastula.

In 1939, Sven Hörstadius performed a simple experiment demonstrat-

(A)

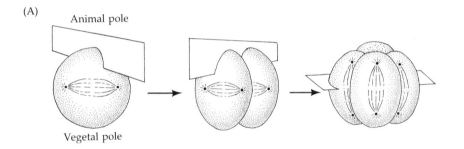

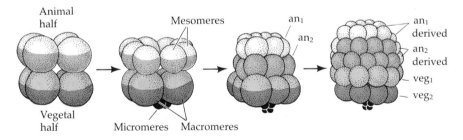

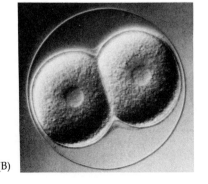

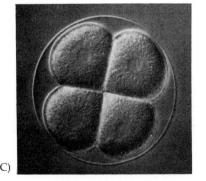

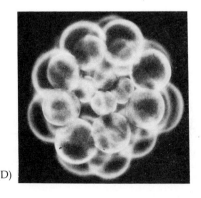

FIGURE 3
Cleavage in the sea urchin. (A) Planes of cleavage in the first three divisions and the formation of particular tiers of cells in divisions 3–6. (B–D) Photomicrographs of live embryos of the sea urchin *Lytechinus pictus*, looking down upon the animal pole. (B) 2-cell stage; (C) 4-cell stage of sea urchin development. (D) The 32-cell stage, shown without the fertilization membrane to allow visualization of the animal pole mesomeres, the central macromeres, and the vegetal micromeres, which angle into the center. (Photographs courtesy of G. Watchmaker.)

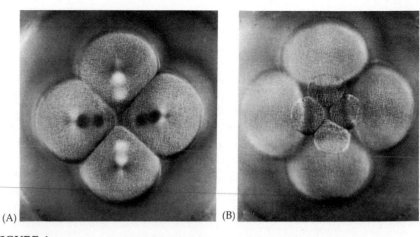

(A)　　　　　　　　　　(B)

FIGURE 4

Formation of the micromeres during the fourth division of sea urchin embryos. The vegetal poles of the embryos are viewed from below. (A) The location and orientation of the mitotic spindle at the bottom portion of the vegetal cells is shown by viewing the living embryo with polarized light. (B) The cleavage through these asymmetrically placed spindles has produced micromeres and macromeres. (From Inoué, 1982; photographs courtesy of S. Inoué.)

ing that the timing and placement of each sea urchin cleavage is independent of preexisting cleavages. He showed that if he inhibited the first one, two, or three cleavages by shaking the eggs or placing them into hypotonic seawater, the unequal (fourth) cleavage division that forms the micromeres would still occur at the appropriate time (Figure 5). Thus, Hörstadius concluded that there are three factors that determine cleavage in the 8-cell embryo: (1) there are "progressive changes in the cytoplasm which cause spindles formed after a certain time after fertilization to lie in a certain direction"; (2) there must be micromere-forming material in the vegetal cytoplasm; and (3) there must be some mechanism by which the micromere-forming material is activated at the correct time (Hörstadius, 1973).

The blastula stage of sea urchin development begins at the 128-cell stage. Here the cells form a hollow sphere surrounding a central blastocoel (Figure 6). Every cell is in contact with the proteinaceous fluid of the blastocoel and with the hyaline layer within the fertilization membrane. During this time, contacts between the cells are tightened. Dan-Sohkawa and Fujisawa (1980) have analyzed this process in starfish embryos and have shown that the closing of the hollow sphere is contemporaneous with the formation of TIGHT JUNCTIONS between the blastomeres. These junctions unite the loosely connected cells into a tissue and seal off the blastocoel from the outside environment (Figure 7). As the cells continue to divide, the cell layer expands and thins out. During the period, the blastula remains one cell-layer thick.

Two theories have been offered to explain the concomitant expansion and blastocoel formation. Dan (1960) hypothesized that the motive force of this expansion is the blastocoel itself. As the blastomeres secrete proteins into the blastocoel, the blastocoel fluid becomes syrupy. This blastocoel sap absorbs large quantities of water by osmosis, thereby swelling and putting pressure on the blastomeres to expand outward. This pressure would also align the long axis of each cell so that division would never be inward toward the blastocoel. This would create further expansion by having the population oriented in one plane only. Wolpert and Gustafson (1961) and Wolpert and Mercer (1963) have proposed that pressure from the blastocoel is not needed to get this effect. They emphasize the role of

Fertilization

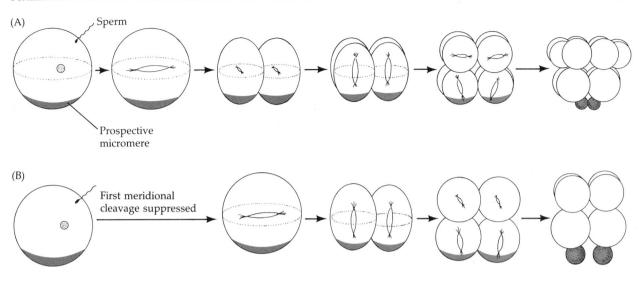

(A)
Sperm

Prospective
micromere

(B)

First meridional
cleavage suppressed

(C)

Equatorial
cleavage suppressed

(D)

Fertilized egg cut in half

FIGURE 5
Micromere formation is dependent on time of development, not on the number
of divisions. (A) Normal sequence of sea urchin cleavage, the micromeres forming at the fourth cleavage. (B,C) The first or third cleavage is suppressed by
hypotonic seawater. (D) A fertilized egg is cut in half meridionally, so that half
the normal cleavage pattern occurs. In all cases, micromeres (shaded) appear at
approximately the same time regardless of the number of cleavages. (After Hörstadius, 1939.)

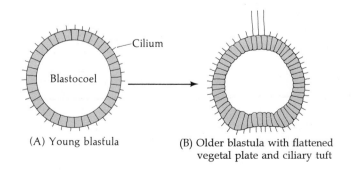

Cilium

Blastocoel

(A) Young blastula

(B) Older blastula with flattened
vegetal plate and ciliary tuft

FIGURE 6
Sea urchin blastulae. (A) Early sea urchin blastula shows a single layer of
rounded cells surrounding a large
blastocoel. (B) As division continues,
the cells of the late blastula show differences in shape as the vegetal plate
cells elongate. (After Giudice, 1973.)

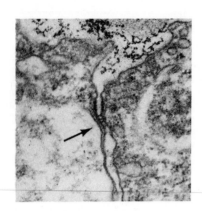

FIGURE 7
Area of contact at the junction of two cells of a 1024-cell starfish blastula. Membrane regions cross over at these junctions (arrow) to form tight seals. (From Dan-Sohkawa and Fujisawa, 1980; photograph courtesy of the authors.)

differential adhesiveness of the cells to one another and to the hyaline layer. They find that as long as the cells remain strongly attached to the hyaline layer, the cells have no alternative but to expand. This expansion creates the blastula rather than the other way around.

The blastula cells develop cilia on their outer surfaces (Figure 8), thereby causing the blastula to rotate within the fertilization envelope. Soon afterward, the cells secrete a HATCHING ENZYME that enables them to digest the fertilization membrane. The embryo is now a free-swimming HATCHED BLASTULA.

Amphibians

Cleavage in most frog and salamander embryos is radially symmetrical and holoblastic, just like echinoderm cleavage. The amphibian egg, however, contains much more yolk. This yolk, which is concentrated in the vegetal hemisphere, acts as an impediment to cleavage. Thus, the first division begins at the animal pole and slowly extends down into the vegetal region (Figure 9). In the axolotl salamander, the cleavage furrow extends through the animal hemisphere at a rate close to 1 mm/min. The cleavage furrow bisects the gray crescent and then slows down to a mere 0.02-0.03 mm/min as it approaches the vegetal pole (Hara, 1977).

Figure 10A is a scanning electron micrograph showing the first cleavage in a frog egg. One can see the folds in the cleavage furrow and the difference between the furrows in the animal and vegetal hemispheres. While the first cleavage furrow is still trying to cleave the yolky cytoplasm of the vegetal hemisphere, the second cleavage has already started near the animal pole. This cleavage is at right angles to the first one and is also meridional (Figure 10B). The third cleavage, as expected, is equatorial. However, because of the vegetally placed yolk, this cleavage furrow in amphibian eggs is much closer to the animal pole. It divides the frog embryo into four small animal blastomeres (micromeres) and four large blastomeres (macromeres) in the vegetal region (Figure 10C). This unequal holoblastic cleavage establishes two major embryonic regions: a rapidly dividing region of micromeres near the animal pole and a more slowly dividing macromere area. As cleavage progresses, the animal region becomes packed with numerous small cells while the vegetal region contains only a relatively small number of large, yolk-laden macromeres (Figure 9).

In amphibians, the embryos containing 16 to 64 cells are commonly called MORULAE (Latin for "mulberry," which they vaguely resemble). At the 128-cell stage, the blastocoel becomes apparent and the embryo is considered to be a blastula. Actually, the formation of the blastocoel has been traced back to the very first cleavage furrow. Kalt (1971) demonstrated that in the frog *Xenopus laevis*, the first cleavage furrow widens in the animal hemisphere to create a small intercellular cavity, which is sealed

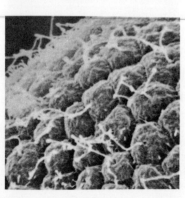

FIGURE 8
Ciliated blastula cells. Each cell develops a single cilium. (Photographs courtesy of W. J. Humphreys.)

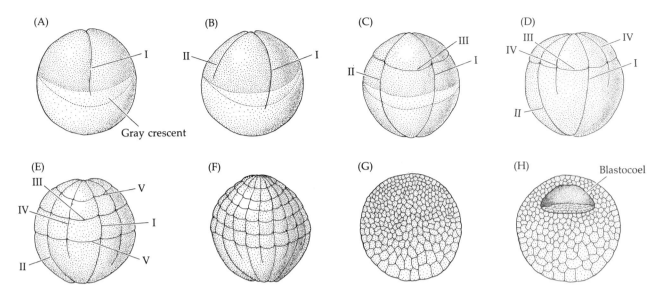

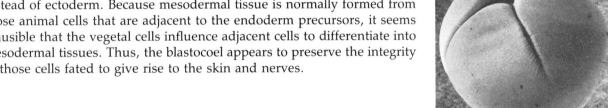

Gray crescent

Blastocoel

FIGURE 9

Cleavage of a frog egg. Cleavage furrows, designated by Roman numerals, are numbered in order of appearance. (B) The vegetal yolk impedes the cleavage such that the second division begins in the animal region of the egg before the first division has divided the vegetal cytoplasm. (C) The third division is displaced toward the animal pole. The vegetal hemisphere ultimately contains longer and fewer blastomeres than the animal half. (After Carlson, 1981.)

off from the outside by tight intercellular junctions (Figure 11). This cavity expands during subsequent cleavages to become the blastocoel.

The blastocoel probably serves two major functions in frog embryos. First it is a cavity that permits cell migration during gastrulation. It also prevents the cells beneath it from interacting prematurely with the cells above it. When Nieuwkoop (1973) took embryonic newt cells from the roof of the blastocoel and placed them next to the yolky vegetal cells from the base of the blastocoel, these animal cells produced mesoderm tissue instead of ectoderm. Because mesodermal tissue is normally formed from those animal cells that are adjacent to the endoderm precursors, it seems plausible that the vegetal cells influence adjacent cells to differentiate into mesodermal tissues. Thus, the blastocoel appears to preserve the integrity of those cells fated to give rise to the skin and nerves.

Spiral holoblastic cleavage

SPIRAL CLEAVAGE is characteristic of annelids, turbellarian flatworms, nemertean worms, and all molluscs except cephalopods. It differs from radial cleavage in numerous ways. First, the eggs do not divide in parallel or perpendicular orientations to the animal–vegetal axis of the egg. Rather,

FIGURE 10

Scanning electron micrograph of the cleavage of a frog egg. (A) First cleavage, (B) second cleavage (4 cells), and (C) fourth cleavage (16 cells), the latter showing the size discrepancy between the animal and vegetal cells after arising from the third division. (A from Beams and Kessel, 1976; photographs courtesy of the authors; B and C courtesy of L. Biedler.)

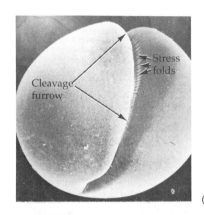

Cleavage furrow

Stress folds

(A)

(B)

(C)

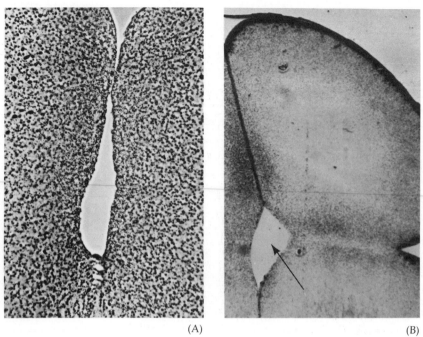

(A)　　　　　　　　　　　　　　　　(B)

FIGURE 11
Formation of the blastocoel in a frog egg. (A) First cleavage plane, showing a small cleft, which later develops into the blastocoel. (B) 8-cell embryo showing a small blastocoel (arrow) at the junction of the three cleavage planes. (From Kalt, 1971; photographs courtesy of M. R. Kalt.)

FIGURE 12
Diagram showing the arrangement of four and eight soap bubbles in a slightly concave dish. The thermodynamic arrangement maximizes contact and is very reminiscent of spirally cleaving embryos.

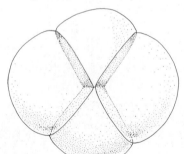

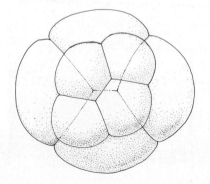

cleavage is at oblique angles, forming the "spiral" arrangement of daughter blastomeres. Second, the cells touch each other at more places than do those of radially cleaving embryos. In fact, they take the most thermodynamically stable packing orientation, much like that of adjacent soap bubbles (Figure 12). Third, spirally cleaving embryos usually undergo fewer divisions before they begin gastrulation. This makes it possible to know the fate of each individual cell of the blastula. When the fates of the individual cells from annelid, flatworm, and mollusc embryos were compared, the same cells were seen in the same places, and their general fates were identical (Wilson, 1898). The blastulae so produced have no blastocoel and are called STEREOBLASTULAE.

Figures 13 and 14 depict the cleavage of the mollusc embryo. The first two cleavages are nearly meridional, producing four large macromeres (labeled A, B, C, and D). In many species, the blastomeres are different sizes (D being the largest), a characteristic that allows them to be individually identified. In each successive cleavage, each macromere buds off a small micromere at its animal pole. Each successive quartet of micromeres is displaced to the right or to the left of its sister macromere, creating the spiral relationship characteristic of the cleavage. Looking down on the embryo from the animal pole, the upper ends of the mitotic spindle appear to alternate clockwise and counterclockwise. This causes alternate micromeres to form obliquely to the left and to the right of their macromere. At the third cleavage, the A macromere will give rise to two daughter cells, macromere 1A and micromere 1a. The B, C, and D cells behave similarly, producing the first quartet of micromeres. In most species, the micromeres are to the right of their macromeres (looking down on the animal pole), an arrangement indicating a dextral (as opposed to sinistral) spiral. At the fourth cleavage, the 1A macromere divides to form macromere 2A and micromere 2a; and micromere 1a divides to form two more micromeres,

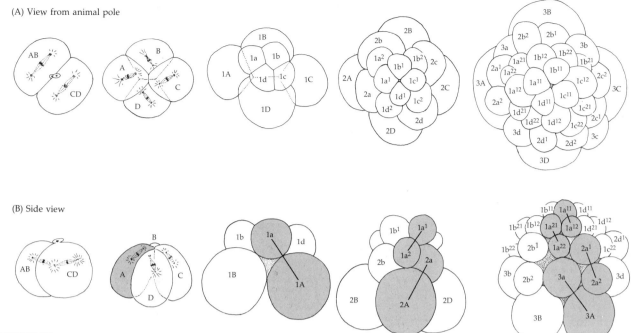

(A) View from animal pole

(B) Side view

FIGURE 13

Spiral cleavage of the mollusc *Trochus* viewed from the animal pole (A) and from one side (B). In panel (B), the cells derived from the A blastomere are shaded. The mitotic spindles, sketched in the early stages, divide the cells unequally and at an angle to the vertical and horizontal axes.

1a1 and 1a2. Further cleavage yields blastomeres 3A and 3a from the 2A macromere; and micromeres such as $1a^2$ divide to produce cells such as $1a^{21}$ and $1a^{22}$.

The orientation of the cleavage plane to the left or to the right is controlled by cytoplasmic factors within the oocyte. This was discovered by analyzing mutations of snail coiling. Some snails have their coils opening to the right of their shells, whereas other snails have their coils opening to the left. Usually, the rotation of coiling is the same for all members of a given species. Occasionally, though, mutants are found. For instance, in the species in which the coils open on the right, occasional individuals will be found with coils that open on the left. Crampton (1894) analyzed the embryos of such aberrant snails and found that their early cleavage differed from the norm. The orientation of the cells after second cleavage was different (Figure 15). This was the result of a different orientation of the mitotic apparatus in the sinistrally coiling snails. All subsequent divisions in left-coiling embryos are mirror images of those of dextrally coiling embryos. In Figure 15, one can see that the position of the 4d blastomere (which is extremely important, as its progeny will form the mesodermal organs) is different in the two types of spiraling embryos. Eventually, the two snails are formed with their bodies on different sides of the coil opening.

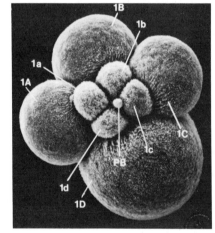

(A)

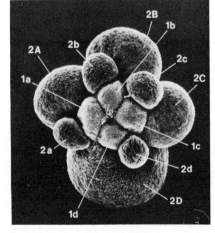

(B)

FIGURE 14

Spiral cleavage of the snail *Ilyanassa*. The D blastomere is larger than the others, allowing the identification of each cell. Cleavage is dextral. (A) 8-cell stage. (B) Mid-fourth cleavage wherein the macromeres have already divided into large and small spirally oriented cells. (From Craig and Morrill, 1986; photographs courtesy of the authors.)

(A) Left-handed coiling

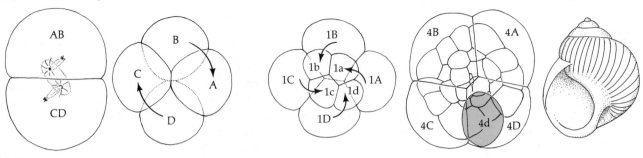

(B) Right-handed coiling

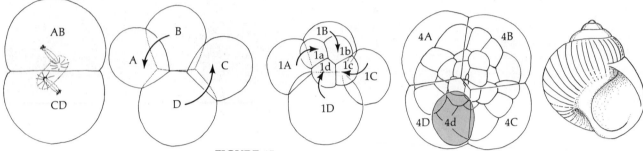

FIGURE 15

Looking down upon the animal pole of right-handed and left-handed snails. The origin of right-handed and left-handed snail coiling can be traced to the orientation of the mitotic spindle at second cleavage. The left-handed (A) and right-handed (B) snails develop as mirror images of each other. (After Morgan, 1927.)

The direction of snail shell coiling is controlled by a single pair of genes (Sturtevant, 1923; Boycott et al., 1930). In the snail *Limnaea peregra*, most individuals are dextrally coiled. Rare mutants exhibiting left-handed coiling were found and mated with wild-type snails. These matings show that there is a "right-handed" allele (*D*), which is dominant to the "left-handed" allele (*d*). However, the direction of cleavage is not determined by the genotype of the developing snail but by the genotype of the snail's mother. A *dd* female snail can only produce sinistrally coiling offspring, even when the offspring's genotype is *Dd*. Thus, a *Dd* individual will coil either left or right depending on the genome of its mother.

The matings produce a chart like this:

DD (female) × *dd* (male)   → *Dd* (all right-coiling)
DD (male)   × *dd* (female) → *Dd* (all left-coiling)
Dd   × *Dd*   → 1 *DD* : 2 *Dd* : 1*dd* (all right-coiling)

Thus, the genetic factors involved in snail coiling are brought to the embryo in the oocyte cytoplasm. It is the genotype of the ovary in which the oocyte develops that determines which orientation the cleavage will take. When Freeman and Lundelius (1982) injected a small amount of cytoplasm from dextrally coiling snails into the eggs of *dd* mothers, the resulting embryos coiled to the right. Cytoplasm from sinistrally coiling snails did not affect the right-coiling embryos. This confirmed the view that the wild-type mothers were placing a factor into their eggs that was absent or defective in the *dd* mothers.

SIDELIGHTS & SPECULATIONS

Adaptation by modifying embryonic cleavage

Evolution is caused by the hereditary alteration of embryonic development. (When we say that the contemporary one-toed horse evolved from a five-toed ancestor, we are stating that during the evolution of the horse, the pattern of embryonic bone formation changed to bring about a one-toed organism.) Sometimes we are able to identify a modification of embryogenesis that has enabled the organism to survive in an otherwise inhospitable environment. One such modification, discovered by Frank Lillie in 1898, is brought about by altering the typical pattern of spiral cleavage in the unionid family of clams.

Unlike most clams, *Unio* and its relatives live in swift-flowing streams. Streams usually create a problem for the dispersal of larvae: because the adults are sedentary, the free-swimming larvae would always be carried downstream by the current. These clams, however, have solved this problem by effecting two changes in their development. The first alters embryonic cleavage. In the typical cleavage of molluscs, either all the macromeres are equal in size or the 2D blastomere is the largest cell at that embryonic stage. However, the division of the *Unio* is such that the 2d blastomere gets the largest amount of cytoplasm (Figure 16). This cell divides to produce most of the larval structures, including a gland capable of producing a massive shell. These larvae (called GLOCHIDIA) resemble tiny bear traps, and they have sensitive hairs that cause the valves of the shell to snap shut when they are touched by the gills or fins of a wandering fish. They "hitchhike" with the fish until they are ready to drop off and metamorphose into adult clams. In this manner, they can spread upstream.

In some species, the glochidia are released from the female's brood pouch and merely wait for a fish to come wandering by. Some other species, such as *Lampsilis ven-*

FIGURE 17
Phony "fish" atop the unionid clam, *Lampsilis ventricosa*. The "fish" is actually the brood pouch and mantle of the clam. (Photograph courtesy of J. H. Welsh.)

tricola, have increased the chances of its larvae finding a fish by yet another modification of their development (Welsh, 1969). Many clams develop a thin mantle that flaps around the shell and surrounds the brood pouch (marsupium). In some unionids, the shape of the brood pouch and the undulations of the mantle mimic the shape and swimming behavior of a minnow. To make the deception all the better, they develop a black "eyespot" on one end and a flaring "tail" on the other. The "fish" seen in Figure 17 is not a fish at all, but the brood pouch and mantle of the clam beneath it. When a predatory fish is lured within range, the clam discharges the glochidia from the brood pouch.

In such a manner, the modification of existing developmental patterns have permitted unionid clams to survive in environments that would otherwise be inhospitable.

FIGURE 16
Formation of glochidium larvae by the modification of spiral cleavage. After the 8-cell embryo is formed (A), placement of the mitotic spindle causes most of the D cytoplasm to enter the 2d blastomere (B). This large 2d blastomere divides (C) to eventually give rise to the large "bear-trap" shell of the larva (D). (After Raff and Kaufman, 1983.)

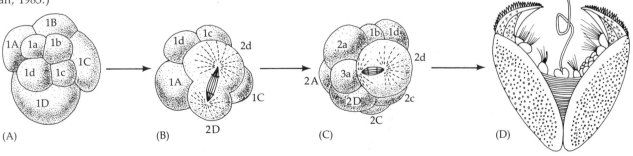

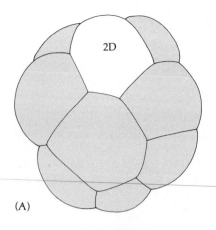

(A)

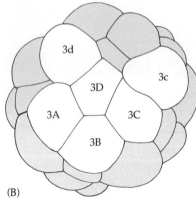

(B)

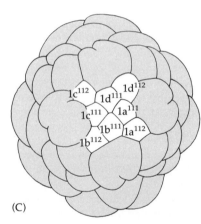

(C)

FIGURE 18

Intercellular communication during mollusc cleavage. Diagrammatic representations show the extent to which the fluorescent dye Lucifer Yellow (white area) has spread after injection into specific macromeres of *Patella vulgata*. (A) Labeling pattern 20 minutes after injection of the 2D macromere of the 16-cell embryo indicates no transfer of dye to adjacent cells. (B) Vegetal view of a 32-cell embryo after the injection of dye into the 3D macromere, showing the spread of the dye to adjacent cells (except $2d^{22}$). (C) Animal view of the same embryo about 45 minutes after injection, showing that dye has spread from the 3D macromere to the central group of animal micromeres. (After de Laat et al., 1980.)

Another exciting discovery concerning molluscan cleavage is that certain blastomeres communicate with each other. In those molluscs with equal-sized blastomeres at the 4-cell stage, the determination of which cell will give rise to the mesodermal precursor cell is accomplished between the fifth and sixth cleavages. At this time, the 3D macromere extends inward and contacts the micromeres of the animal pole. Without this contact, the 4d cell given off by the 3D macromere does not produce mesoderm (Biggelaar and Guerrier, 1979).* de Laat and co-workers (1980) demonstrated that at the time of contact (and not before) small molecules are capable of diffusing between the 3D macromere and the central micromeres. They injected the dye Lucifer Yellow (molecular weight 457) into one of the macromeres. When added before the 32-cell stage, the dye resides only in that single macromere and its progeny. However, when the dye is injected into the 3D blastomere at the beginning of the 32-cell stage, the first tier of micromeres picks up the yellow color (Figure 18). Low-molecular-weight material is transferred from one cell to the other at just that time when the micromeres alter the developmental potential of the 3D blastomere. Transmission electron microscopy shows that at this time, GAP JUNCTIONS appear on the surfaces of these cells. These modifications of the cell membrane facilitate the intercellular transfer of small molecules and ions between neighboring cells, and they are found in many tissues in both embryonic and adult animals. (The microanatomy of these junctions will be discussed in Chapter 15.)

Bilateral holoblastic cleavage

Bilateral holoblastic cleavage is found primarily in ascidians (tunicates). Figure 19 shows the cleavage pattern of a tunicate, *Styela partita*. The most striking phenomenon in this type of cleavage is that the first cleavage plane establishes the only plane of symmetry in the embryo, separating the embryo into its future right and left sides. Each successive division orients itself to this plane of symmetry, and the half-embryo formed on one side of the first cleavage is the mirror image of the half-embryo on the other side. The second cleavage is meridional, like the first division, but unlike the first division, it does not pass through the center of the egg. Rather, it creates two large anterior cells (A and D) and two smaller posterior cells (B and C). Each side now has a large and a small blastomere. At the next three divisions, differences in cell size and shape highlight the bilateral symmetry of these embryos. At the 32-cell stage, a small blastocoel is formed and gastrulation begins.

As mentioned in the last chapter, certain tunicates, such as *S. partita*,

*Don't worry about those mollusc embryos with unequal-sized blastomeres at the 4-cell stage. We shall see much of them in Chapter 8.

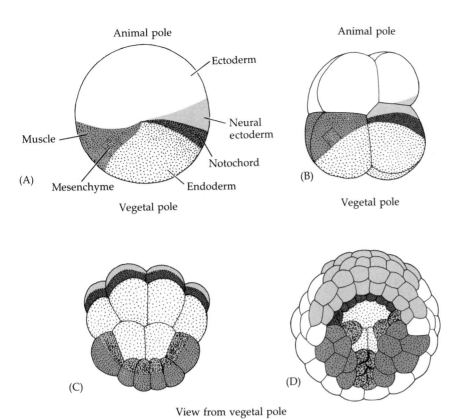

Animal pole

Ectoderm

Neural
ectoderm

Muscle

Notochord

(A)

Mesenchyme   Endoderm

Vegetal pole

Animal pole

(B)

Vegetal pole

(C)   (D)

View from vegetal pole

FIGURE 19
Bilateral symmetry in a tunicate egg.
(A) Uncleaved egg. (B) 8-cell embryo,
showing the blastomeres and the fates
of various cells. It can be viewed as
two 4-cell halves; from here on each
division on the right side of the embryo has a mirror-image division on
the left. (C,D) Views of later embryos
from the vegetal pole. (The regions of
cytoplasm destined to form particular
organs are labeled in panel A, and are
coded by color throughout the diagram.) (After Balinsky, 1981.)

contain colored cytoplasmic regions. During cleavage, these plasms become partitioned into different cells. Moreover, the type of cytoplasm the
cells receive determines their eventual fates. Cells receiving clear cytoplasm become ectoderm; those containing yellow cytoplasm give rise to
mesodermal cells; the cells that incorporate the slate-gray inclusions become endoderm; and the light gray cells become the neural tube and
notochord. These colored plasms are localized bilaterally around the plane
of symmetry, so they are bisected by the first cleavage furrow into the
right and left halves of the embryo. The second cleavage causes the prospective mesoderm to lie in the two posterior cells, while the prospective
neural and chordamesoderm are to be formed from the two anterior cells.
The third division further partitions these determinants such that the
mesoderm-forming cells are confined to the two vegetal posterior blastomeres, and the chordamesoderm cells are likewise restricted to the two
vegetal anterior cells. The fate of each cell of the early *Styela* embryo has
been followed and will be discussed in detail in Chapter 7.

Rotational holoblastic cleavage

It is not surprising that mammalian cleavage has been the most difficult
to study. Mammalian eggs are among the smallest in the animal kingdom,
making them hard to manipulate experimentally. The human zygote, for
instance, is only 100 μm in diameter, barely visible to the eye, and less
than one-thousandth the volume of a *Xenopus* egg. Also, mammalian
zygotes are not produced in numbers comparable to sea urchin or frog
embryos. Usually fewer than 10 eggs are ovulated by a female at a given
time, so it is difficult to obtain enough material for biochemical studies.
As a final hurdle, the development of mammalian embryos is accom-

FIGURE 20

Development of a human embryo from fertilization to implantation. The egg "hatches" from the zona upon reaching the uterus, and it is probable that the zona prevents the cleaving cells from sticking to the oviduct rather than traveling to the uterus. (After Tuchmann-Duplessis et al., 1972.)

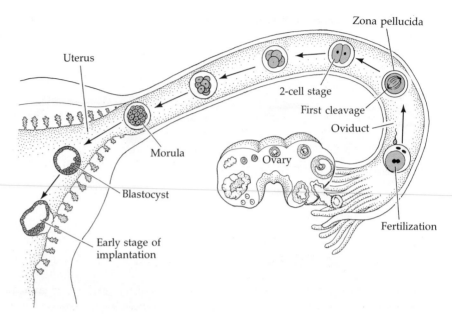

plished within another organism rather than outside of it. Only recently has it been possible to duplicate some of these internal conditions and observe development in vitro.

With all these difficulties, knowledge of mammalian cleavage was worth waiting for, as mammalian cleavage turned out to be strikingly different from most other patterns of embryonic cell division. The mammalian oocyte is released from the ovary into the oviduct (Figure 20). Fertilization occurs in the ampulla of the oviduct, a region close to the ovary. Meiosis is completed at this time, and first cleavage begins about a day later. Cleavages in mammalian eggs are among the slowest cleavages occurring in the animal kingdom—about 12 to 24 hours apart. Meanwhile, the cilia in the oviduct push the embryo toward the uterus; the first cleavages occur along this journey.

There are several features of mammalian cleavage that distinguish it from other cleavage types. The first feature is the relative slowness of the divisions. The second fundamental difference is the unique orientation of mammalian blastomeres with relation to one another. The first cleavage is a normal meridional division; however, in the second cleavage one of the two blastomeres divides meridionally and the other divides equatorially

FIGURE 21
Comparison of early cleavage of (A) echinoderms and (B) mammals. (After Gulyas, 1975.)

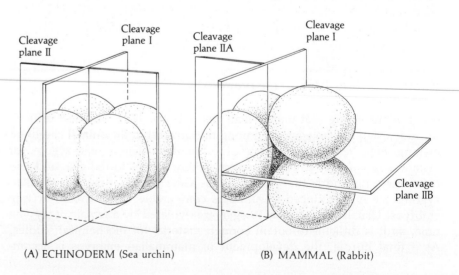

(A) ECHINODERM (Sea urchin)   (B) MAMMAL (Rabbit)

(Figure 21). This type of cleavage is called ROTATIONAL CLEAVAGE (Gulyas, 1975).

The third major difference between mammalian cleavage and that of most other embryos is the marked asynchrony of early division. Mammalian blastomeres do not all divide at the same time. Thus, mammalian embryos do not increase evenly from 2- to 4- to 8-cell stages but frequently contain odd numbers of cells. Also, unlike most all other animals, the mammalian genome is activated during early cleavage and the genome produces the proteins necessary for cleavage to occur. In the mouse and goat, the switch from maternal to zygotic control occurs at the 2-cell stage (Piko and Clegg, 1982; Prather, 1989).

Compaction

Perhaps the most crucial difference between mammalian cleavage and all other types involves the phenomenon of COMPACTION. As seen in Figure 22, mammalian blastomeres through the 8-cell stage form a loose arrangement with plenty of space between them. Following the third cleavage, however, the blastomeres undergo a spectactular change in their behavior. They suddenly huddle together, maximizing their contact with the other blastomeres and forming a compact ball of cells (Figures 22C,D and 23). This tightly packed arrangement is stabilized by tight junctions that form between the outside cells of the ball, sealing off the inside of the sphere (Figure 24). The cells within the sphere form gap junctions, thereby enabling small molecules and ions to pass between the cells.

The cells of the compacted embryo divide to produce a 16-cell morula. This morula consists of a small group of internal cells (1–2 cells) surrounded by a larger group of external cells (Barlow et al., 1972). Most of the descendants of the external cells become the TROPHOBLAST (or TROPHECTODERM) cells. This group of cells produces no embryonic structures. Rather, the cells form the tissue of the CHORION, the embryonic portion of the placenta. These are the tissues that enable the fetus to get oxygen and nourishment from the mother, that secrete hormones so that the mother's uterus will keep the fetus, and that produce regulators of the immune response so that the mother will not reject the embryo as she would an organ graft. However, they are not able to produce any cell of the embryo

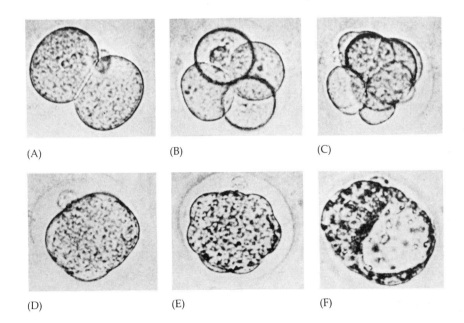

(A) (B) (C)

(D) (E) (F)

FIGURE 22
The cleavage of a single mouse embryo in vitro. (A) 2-cell stage; (B) 4-cell stage; (C) early 8-cell stage; (D) compacted 8-cell stage; (E) morula; (F) blastocyst. (From Mulnard, 1967; photographs courtesy of J. G. Mulnard.)

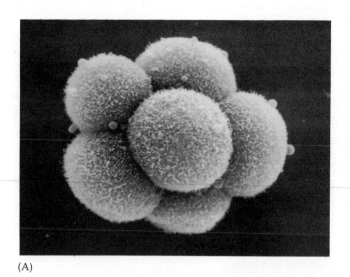
(A)

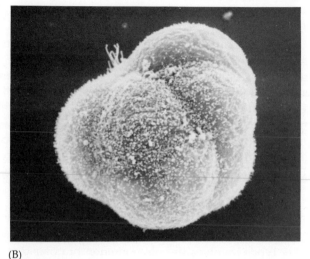
(B)

FIGURE 23
Scanning electron micrograph of uncompacted (A) and compacted (B) 8-cell mouse embryos. (Photographs courtesy of C. Ziomek.)

itself. These trophoblast cells are necessary for IMPLANTING the embryo in the uterine wall (Figure 25).

The embryo is derived from the descendants of the inner cell(s) of the 16-cell stage, supplemented by an occasional cell dividing from the trophoblast during the transition to the 32-cell stage (Pederson et al., 1986; Fleming, 1987). These cells generate the INNER CELL MASS (ICM) that will give rise to the embryo. These cells not only look different from the trophoblast cells, they also synthesize different proteins at this early developmental stage. By the 64-cell stage, the inner cell mass and the trophoblast cells have become separate cell layers, neither of which contributes cells to the other group (Dyce et al., 1987). Thus, the distinction between trophoblast and inner cell mass blastomeres represents the first differentiation event in mammalian development.

Initially the morula does not have an internal cavity. However, during a process called CAVITATION, the trophoblast cells secrete fluid into the morula to create a blastocoel. The inner cell mass is positioned on one side of the ring of trophoblast cells (Figures 22, 24, and 25). This structure is the BLASTOCYST and is another hallmark of mammalian cleavage.

FIGURE 24
Schematic diagram of compaction and the formation of the blastocyst. (A,B) 8-cell embryo; (C) morula; (D) blastocyst.

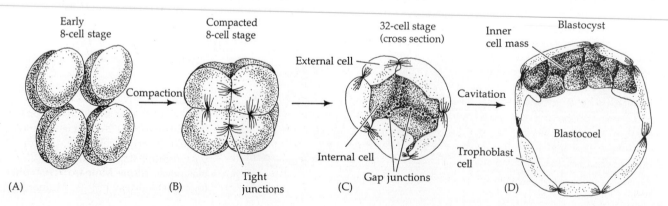

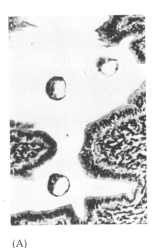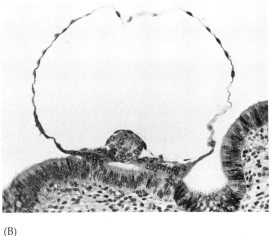

FIGURE 25
Implantation of mammalian blastocysts into the uterus. (A) Mouse blastocysts entering uterus. (B) Initial implantation of the blastocyst onto the uterus in a rhesus monkey. (A from Rugh, 1967; B courtesy of the Carnegie Institution of Washington; Chester Reather, photographer.)

(A)   (B)

SIDELIGHTS & SPECULATIONS

The cell surface and the mechanism of compaction

Compaction creates the circumstances that bring about the first differentiation in mammalian development—the separation of trophoblast from inner cell mass. How is this done? There is growing evidence that compaction is mediated by events occurring at the cell surfaces of adjacent blastomeres.

First, prior to compaction, each of the eight blastomeres undergoes extensive membrane changes known as POLARIZATION. Different components of the cell surface migrate to different regions of the cell (Ziomek and Johnson, 1980). This can be seen by tagging certain cell surface molecules with fluorescent dyes. One such tag, which recognizes a class of glycoproteins, shows that at the 4-cell stage these glycoproteins are randomly distributed throughout the membrane (Figure 26A). However, at the mid-8-cell stage, these molecules are found predominantly at the poles farthest away from the center of the aggregate (Figure 26B). This phenomenon appears to be influenced by cell–cell interactions, because it takes place on a pair of attached blastomeres that are isolated together but not on individual isolated cells.

Second, specific cell surface proteins are seen to play a role in compaction. One such molecule is UVOMORULIN,* a 120,000-Da glycoprotein found on the cell surfaces of these cell types. This protein is synthesized at the 2-cell stage and is seen to be uniformly spread throughout the cell membrane. However, as compaction occurs, uvomorulin is seen to become restricted to those sites on cell membranes that are in contact with adjacent blastomeres. Antibodies to this molecule cause the decompaction of the morula and inhibit the cell–cell attachments of teratocarci-

noma cells (Figure 27; Peyrieras et al., 1983; Johnson et al., 1986). The carbohydrate portion of this glycoprotein may be essential to its function, as TUNICAMYCIN (a drug that inhibits the glycosylation of proteins) also prevents compaction.

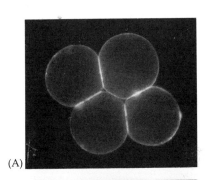

(A)

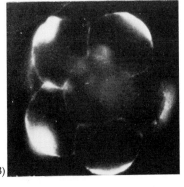

(B)

FIGURE 26
Polarization of membrane components in 8-cell stage mouse blastomeres. (A) Homogeneous nonpolar distribution of membrane components labeled with fluorescent concanavalin A at the 4-cell stage. (B) Heterogeneous, polar distribution of these components at the 8-cell stage. (A from Fleming et al., 1986; B from Levy et al., 1986. Photographs courtesy of the authors.)

*Also called E-cadherin or L-CAM, this cell surface molecule is important in several morphogenic events and will be discussed in Chapter 15.

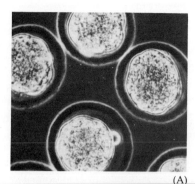

(A)

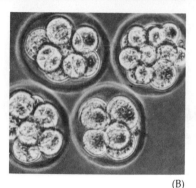

(B)

Third, recent experiments have also shown that the phosphatidylinositol pathway (discussed for sperm and egg activation in Chapter 2) may also be important for

FIGURE 27

Prevention of compaction by antiserum directed against the cell surface adhesion glycoprotein, uvomorulin. (A) Normal compaction occurring in the absence of antiserum. (B) Proliferation without compaction occurring in the presence of antibodies to uvomorulin. (Photographs courtesy of C. Ziomek.)

initiating compaction. If 4-cell mouse embryos are placed into media containing drugs that activate protein kinase C, premature compaction occurs. Similarly, diacylglycerides can transiently cause these 4-cell embryos to undergo compaction. When this occurs, the uvomorulin is seen to accumulate specifically at the junctions between the blastomeres (Winkel et al., 1990). These results suggest that the activation of the cell-surface protein kinase C may initiate compaction by shifting the localization of uvomorulin.

Fourth, the cell membrane may also be modified during compaction by cytoskeletal reorganization. Microvilli, extended by actin microfilaments, appear on adjacent cell surfaces and attach one cell to the other. These microvilli may be the sites where uvomorulin is functioning to mediate intercellular adhesion. The flattening of the blastomeres against one another may therefore be brought about by the shortening of the microvilli through actin depolymerization (Pratt et al., 1982; Sutherland and Calarco-Gillam, 1983). Thus, there is growing evidence that compaction is caused by changes in the architecture of the blastomere cell surface.

Formation of the inner cell mass

The creation of an inner cell mass distinct from the trophoblast is *the* crucial process of early mammalian development. How is a cell directed into one or the other of these paths? How is a cell informed that it is either to give rise to a portion of the adult mammal or that it is to give rise to a rather remarkable supporting tissue that will be discarded at birth? Observations of living embryos suggest that this momentous decision is merely a matter of a cell's being in the right place at the right time. Up through the 8-cell stage, there are no obvious differences in the biochemistry, morphology, or potency of any of the blastomeres. However, compaction forms inner and outer cells with vastly different properties. By labeling the various blastomeres, numerous investigators have found that the cells that happen to be on the outside will form the trophoblast whereas the cells that happen to be inside will generate the embryo (Tarkowski and Wróblewska, 1967; Sutherland et al., 1990).* Hillman and co-workers (1972) have shown that when each blastomere of a 4-cell mouse embryo is placed on the outside surface of a mass of aggregated blastomeres, the external, transplanted cells will only give rise to trophoblast tissue. Therefore, it seems that whether a cell becomes trophoblast or embryo depends on whether it was an external or an internal cell after compaction.

*The inner cells have been found to come most frequently from the first cell to divide at the 2-cell stage. This cell usually produces the first pair of blastomeres to reach the 8-cell stage, and these cells usually divide so that they are inside the loosely aggregated cluster of blastomeres (Graham and Kelly, 1977).

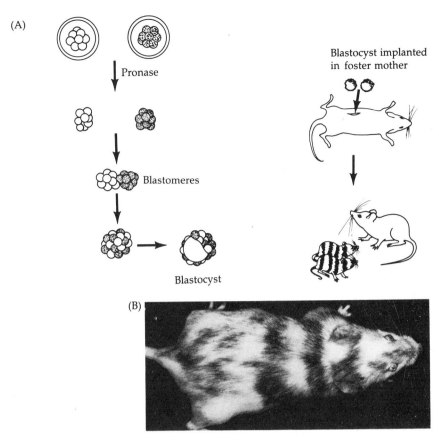

(A)

Pronase

Blastomeres

Blastocyst

Blastocyst implanted
in foster mother

(B)

FIGURE 28
Production of allophenic mice. (A) The
experimental procedures used to produce allophenic mice. Early 8-cell embryos of genetically distinct mice (here,
those with coat-color differences) are
isolated from mouse oviducts and
brought together after their zonae are
removed by proteolytic enzymes. The
cells form a composite blastocyst,
which is implanted into the uterus of a
foster mother. (B) An adult allophenic
mouse showing contributions from the
pigmented (black) and unpigmented
(white) embryos. (Photograph courtesy
of B. Mintz.)

If most of the cells of the blastocyst give rise to the trophoblast, exactly
how many cells actually form the embryo? One way to answer this question is to produce ALLOPHENIC MICE. Allophenic mice are the result of two
early-cleavage (usually 4- or 8-cell) embryos that have been aggregated
together to form a composite embryo. As shown in Figure 28, the zonae
pellucidae of two genetically different embryos are removed and the embryos brought together to form a common blastocyst. These prepared
blastocysts are implanted into the uterus of the foster mother. When they
are born, the allophenic offspring have some cells from each embryo. This
is readily seen when the aggregated blastomeres come from mouse strains
that differ in their coat colors. When blastomeres from white and black
strains are aggregated, the result is commonly a mouse with black and
white bands (Figure 28B). If there were only one cell in the blastocyst that
gave rise to the embryo, this result would not be possible; the offspring
would be either all white or all black. If two cells of the blastocyst were
responsible for producing the embryo, we would expect the allophenic
pattern to be expressed only half the time (1WW:2WB:1BB). Should there
be three embryo-producing cells, the chance that the embryo would have
an allophenic pattern would then increase to 75 percent (1WWW:3WWB:
3WBB:1BBB), and the 4-cell situation would give an 87.5 percent chance
of providing two-color mice. The experimental data of Mintz (1970) are
that 73 percent of the double embryos yield allophenic mice, thus suggesting that three blastomeres of the blastocyst produce the entire embryo.
Markert and Petters (1978) have shown that three early 8-cell embryos can
unite to form a common compacted morula (Figure 29 and color endpaper)
and that the resulting mouse can have the coat colors of the three different
strains. Therefore, while it is not certain that three is the absolute number
of blastomeres that form the embryo, we can be fairly certain that the

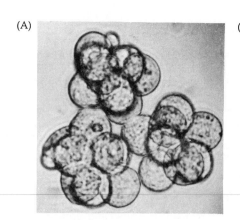
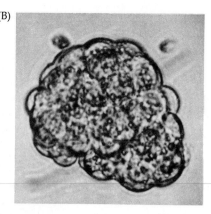
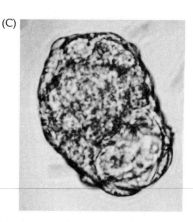

FIGURE 29

Aggregation and compaction of three 8-cell mouse embryos to form a single compacted morula. Cells from three different embryos (A) are aggregated together (B) to form a morula (C), which undergoes compaction to form a single blastocyst (D). The resulting allophenic mouse is shown in the color portfolio. (From Markert and Petters, 1978; photographs courtesy of C. Markert.)

number is not much greater and that most of the cells of the blastocyst never contribute to the adult organism.

Escape from the zona pellucida

While the embryo is moving through the oviduct en route to the uterus, the blastocyst expands. The cell membranes of the trophectoderm cells contain a sodium pump (a Na^+/K^+-ATPase) facing the blastocoel and transporting sodium ions into the central cavity. This accumulation of sodium ions draws in water osmotically, thus enlarging the blastocoel (Borland, 1977; Wiley, 1984). During this time it is essential that the zona pellucida prevent the blastocyst from adhering to the oviduct walls. When such adherence does take place in humans, it is called an "ectopic pregnancy" (or "tubal pregnancy"). This is a dangerous condition, because the implantation of the embryo into the oviduct can cause life-threatening hemorrhage. When it reaches the uterus, however, the embryo must hatch from the zona so that it can adhere to the uterine wall.

The mouse blastocyst hatches from the zona by lysing a small hole in it and squeezing through that hole as the blastocyst expands (Figure 30). A trypsin-like protease, STRYPSIN, is located on the cell membrane and lyses a hole in the fibrillar matrix of the zona (Perona and Wassarman, 1986; Yamazaki and Kato, 1989).

Once out, the blastocyst can make direct contact with the uterus. Here the trophoblast cells will secrete other proteases, such as COLLAGENASE, STROMELYSIN, and PLASMINOGEN ACTIVATOR. These protein-digesting enzymes digest the extracellular matrix of the uterine tissue, enabling the blastocyst to bury itself in the uterine wall (Strickland et al., 1976; Brenner et al., 1989).

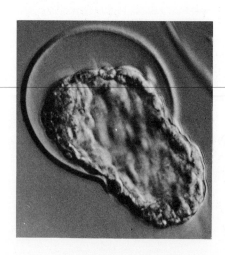

FIGURE 30

Mouse blastocyst hatching from the zona pellucida. (Photograph from Mark et al., 1985, courtesy of E. Lacy.)

SIDELIGHTS & SPECULATIONS

Twins

Human twins are classified into two major groups: MON-OZYGOTIC (one-egg; identical) twins and DIZYGOTIC (two-egg; fraternal) twins. Fraternal twins are the result of two separate fertilization events, whereas identical twins are formed from a single embryo whose cells somehow dissociated from one another. This fact implies that an isolated mammalian blastomere can give rise to an entire embryo. In 1952 Seidel supported this notion by destroying one cell of a 2-cell rabbit embryo. The resulting blastomere was able to develop into a complete adult. Even a single blastomere of an 8-cell mouse embryo can develop successfully into a complete adult (Kelly, 1977). Gardiner and Rossant (1976) have also shown that if ICM cells (but not trophoblast cells) are injected into blastocysts, they contribute to the new embryo. Thus, it is probable that identical twins are produced by the separation of early blastomeres or even by the separation of the inner cell mass into two regions within the same blastocyst.

This seems to be exactly what is happening in roughly 0.25 percent of human births. About 33 percent of identical twins have two complete and separate chorions, indicating that separation occurred before the formation of the trophoblast tissue at day 5. The remaining identical twins share a common chorion, suggesting that the split occurs within the inner cell mass after the trophoblast has formed (Figure 31A). By day 9, the human embryo has completed the construction of another extraembryonic layer, the AMNION. This tissue forms the amniotic sac (or water sac) which surrounds the embryo with amniotic fluid and protects it from desiccation and abrupt movement (Chapter 4). If the separation of the embryo were to come between the formation of the chorion on day 5 and the amnion on day 9, then the resulting embryos should have one chorion and two amnions (Figure 31B). This happens in about two-thirds of human identical twins. A small percentage of

FIGURE 31
Schematic diagram showing the timing of human monozygotic twinning with relation to extraembryonic membranes. (A) Splitting occurs before the formation of the trophectoderm, so each twin has its own chorion and amnion. (B) Splitting occurs after trophectoderm formation but before amnion formation, resulting in twins having individual amniotic sacs but sharing one chorion. (C) Splitting after amnion formation leads to twins in one amniotic sac and a single chorion. (After Langman, 1981).

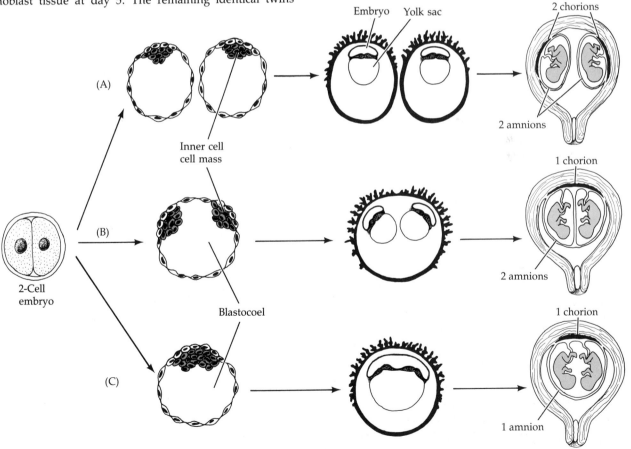

identical twins are born within a single chorion and amnion (Figure 31C). This means that the division of the embryo came after day 9. Such newborns are at risk of being conjoined ("Siamese") twins.

The ability to produce an entire embryo from cells that would have normally contributed to only a portion of it is called REGULATION (discussed in Chapter 8). Regulation is also seen in the ability of two or more early embryos to form one allophenic mouse rather than twins, triplets, or a multiheaded monster. There is even evidence (Chappelle et al., 1974; Mayr et al., 1979) that allophenic regulation can occur in humans. These allophenic individuals have two genetically different cell types (XX and XY) within the same body, each with its own set of genetically defined characteristics. The simplest explanation for the existence of such a phenomenon is that these individuals resulted from the aggregation of two embryos, one male and one female, which were developing at the same time. If this explanation is correct, then two fraternal twins fused to create a single composite individual.

Meroblastic cleavage types

As mentioned earlier in this chapter, yolk concentration plays an important role in cell cleavage. Nowhere is this more apparent than in the meroblastic cleavage types. Here, the large concentrations of yolk prohibit cleavage in all but a small portion of the egg cytoplasm. In DISCOIDAL cleavage, cell division is limited to a small disc of yolk-free cytoplasm atop a mound of yolk; in SUPERFICIAL CLEAVAGE, the centrally located yolk permits cleavage only along the peripheral rim of the egg.

Discoidal cleavage

Discoidal cleavage is characteristic of fishes, birds, and reptiles. Figure 32 shows the cleavage of the avian egg. The bulk of the oocyte is taken over by the yolk, allowing cleavage to occur only in the blastodisc at the animal pole of the egg. Because these cleavages do not extend into the yolky cytoplasm, the early cleavage-stage cells are actually continuous at their bases. After a single-layered blastoderm is formed, equatorial cleavages divide this layer into a tissue three to four cell layers thick. Between the blastoderm and the yolk is a space, called the SUBGERMINAL CAVITY. At this stage, two distinct regions of the blastodisc can be identified: the AREA PELLUCIDA, composed of the cells above the subgerminal cavity; and the AREA OPACA, consisting of the darker cells at the margin of the blastodisc and yolk.

By the time the hen has laid the egg, the blastoderm contains some 60,000 cells. Some of these cells are shed into the subgerminal cavity to form a second layer (Figure 33). Thus, soon after laying, the chick egg contains two layers of cells: the upper epiblast and the lower hypoblast. Between them lies the blastocoel.

The yolky eggs of fishes develop similarly, cell division occurring only in the animal pole blastodisc. Scanning electron micrographs of fish egg cleavage show beautifully the incomplete nature of discoidal cleavage (Figure 34).

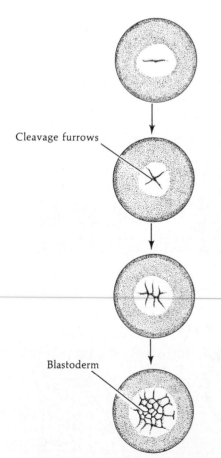

Cleavage furrows

Blastoderm

FIGURE 32
Discoidal cleavage in a chick egg, viewed from the animal pole. The cleavage furrows do not penetrate the yolk, and a blastoderm consisting of a single layer of cells is produced.

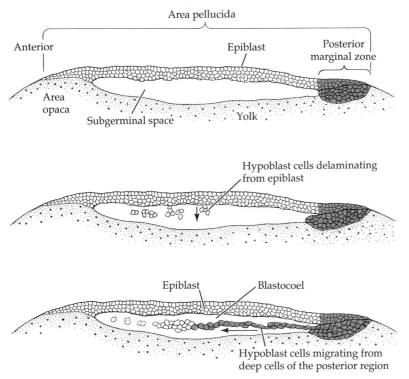

FIGURE 33
Formation of the hypoblast in the avian egg. The hypoblast cells form from a delamination of epiblast cells and from the migration of cells from the posterior margin of the embryo. The secondary hypoblast forms from the subsurface cells of the posterior marginal zone (color) whereas the superficial cells of the posterior marginal zone (gray) form the early primitive streak. (After Torrey, 1962, and Stern, 1990.)

FIGURE 34
Discoidal cleavage in a zebra fish, creating a cellular region above the dense yolk. BD in panel (A) signifies the blastodisc region. (From Beams and Kessel, 1976; photographs courtesy of the authors.)

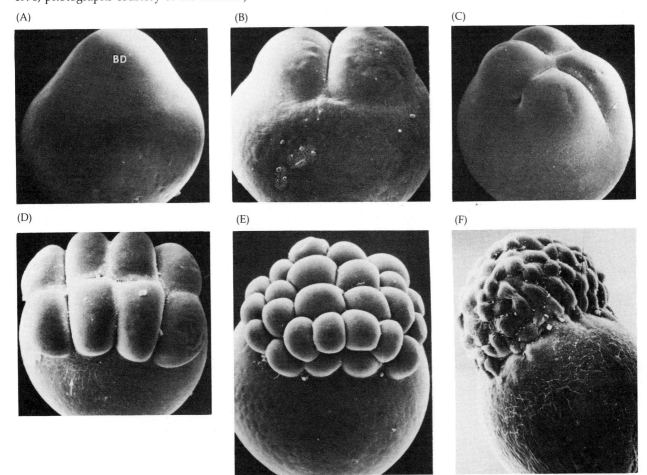

Superficial cleavage

Insect eggs undergo superficial cleavage, whereby a large mass of centrally located yolk confines cleavage to the cytoplasmic rim of the egg. One of the fascinating details of this cleavage type is that the cells do not form until after the nuclei have divided. Cleavage of an insect egg is shown in Figure 35. The zygote nucleus undergoes several mitotic divisions within the central portion of the egg. In *Drosophila*, 256 nuclei are produced by a series of nuclear divisions averaging 8 minutes each. The nuclei then migrate to the periphery of the egg, where the mitoses continue, albeit at a progressively slower rate. The embryo is now called a SYNCYTIAL BLASTODERM, meaning that all the cleavage nuclei are contained within a common cytoplasm. No cell membranes exist other than that of the egg itself. Those nuclei migrating to the posterior pole of the egg soon become enveloped by new cell membranes to form the POLE CELLS of the embryo. These cells give rise to the germ cells of the adult. Thus, one of the first events of insect development is the separation of the future germ cells from the rest of the embryo.

After the pole cells have been formed, the oocyte membrane folds inward between the nuclei, eventually partitioning off each nucleus into a single cell (Figure 36). This creates the CELLULAR BLASTODERM, with all the cells arranged in a single-layered jacket around the yolky core of the egg. In *Drosophila* this layer consists of approximately 6000 cells and is formed within 4 hours of fertilization.

Although the nuclei originally divide within a common cytoplasm,

FIGURE 35

Superficial cleavage in a *Drosophila* embryo. The numeral above each embryo corresponds to the number of minutes after deposition of the egg; the numeral at the bottom indicates the number of nuclei (energids) present. Pole cells (which will form the germ cells) are seen at the 512-nuclei stage even though the cellular blastoderm does not form until nearly 3 hours later. The times are representative, as the duration of each division cycle depends, in part, on the temperature at which the egg is incubated.

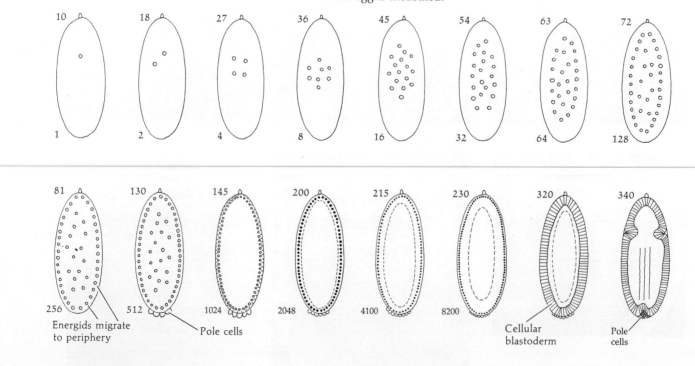

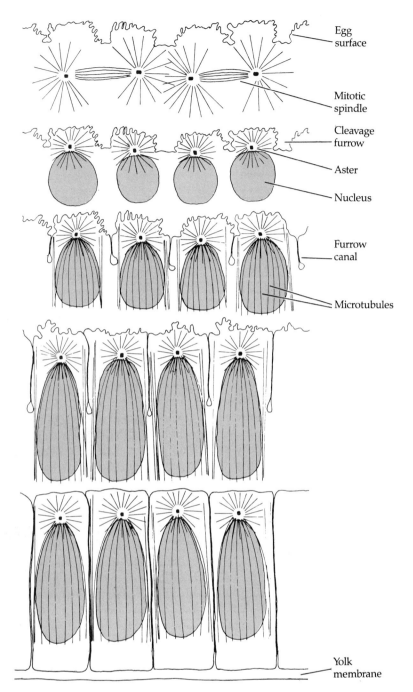

Egg
surface

Mitotic
spindle

Cleavage
furrow

Aster

Nucleus

Furrow
canal

Microtubules

Yolk
membrane

FIGURE 36
Schematic diagram of nuclear elongation and cellularization in *Drosophila* blastoderm. (After Fullilove and Jacobson, 1971.)

this does not mean that the cytoplasm is itself uniform. Karr and Alberts
(1986) have shown that each nucleus within the syncytial blastoderm is
contained within its own little territory of cytoskeletal proteins. When the
nuclei reach the periphery during the tenth cleavage cycle, each nucleus
becomes surrounded by microtubules and microfilaments. The nuclei and
their associated cytoplasmic islands are called ENERGIDS. Figure 37 shows
the nuclei and their essential microfilament and microtubule domains in
the prophase of the twelfth mitotic division.

After the nuclei reach the periphery, the time required to complete

(A)

 Prophase 12

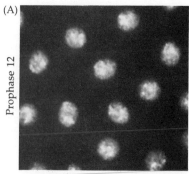

Nuclei

(B)

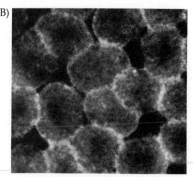

Microfilaments

(C)

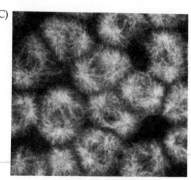

Microtubules

FIGURE 37

Localization of the cytoskeleton around nuclei in the syncytial blastoderm of *Drosophila*. *Drosophila* embryo entering mitotic prophase of its twelfth division was sectioned and triple-stained. (A) The nucleus localized by a dye that binds to DNA. (B) Microfilaments were identified using a fluorescent antibody to actin. (C) Microtubules were recognized by a fluorescent antibody to tubulin. Cytoskeletal domains can be seen surrounding each nucleus. (From Karr and Alberts, 1986; photographs courtesy of T. L. Karr.)

each of the next four divisions becomes gradually longer. While cycles 1–10 are each 8 minutes long, cycle 13, the last cycle in the syncytial blastoderm, takes 25 minutes to complete. The *Drosophila* embryo forms cells in cycle 14 (i.e., after 13 divisions), and cycle 14 is asynchronous. Some groups of cells complete this cycle in 75 minutes, whereas other groups of cells take 175 minutes to complete cycle 14 (Foe, 1989). Transcription from these nuclei (which begins around the eleventh cycle) is greatly enhanced. The slowdown of *Drosophila* nuclear division and the acceleration of RNA synthesis resembles the situations in frog and sea urchin embryos. The control of this slowdown (in *Xenopus*, sea urchin, starfish, and *Drosophila* embryos) appears to be effected by the ratio of chromatin to cytoplasm (Newport and Kirschner, 1982a; Edgar et al., 1986). Edgar and his colleagues compared the early development of wild-type *Drosophila* embryos with those of a haploid mutant. The haploid *Drosophila* embryos have half the wild-type quantity of chromatin at each cell division. Hence, a haploid embryo at the eighth cell cycle has the same amount of chromatin as a wild-type embryo has at cell cycle seven. These investigators found that whereas wild-type embryos formed their cell layer during the thirteenth cycle, the haploid embryos underwent an extra, fourteenth, division before cellularization. Moreover, the lengths of cycles 11–14 in wild-type embryos corresponded to cycles 12–15 in the haploid embryos. Thus, the haploid embryos follow a pattern similar to that of the wild-type embryos, only they lag by one cell division.

If their lagging is due to the haploid mutants' having a chromatin-to-cytoplasm ratio of half the wild-type ratio at any given cycle, then one should be able to accelerate cellularization by tying off (ligating) some of the cytoplasm such that the haploid nuclei divide in a smaller volume. When this ligation was performed, the mitotic pattern of the haploid embryos was accelerated. The final blastoderm division, signaling the end of the cleavage period, is reached when there is one nucleus for each 61 μm^3 of cytoplasm. In *Xenopus*, a similar slowdown in mitotic rate is observed after the twelfth cell division. Here, too, the divisions thereafter become asynchronous. The timing of this MIDBLASTULA TRANSITION is also due to the chromatin-to-cytoplasm volume ratio. Haploid *Xenopus* embryos undergo this transition after the thirteenth cleavage cycle, while tetraploid embryos (having twice the amount of wild-type chromatin per cell) start the transition after the eleventh division (Newport and Kirschner, 1982a,b).

In both *Drosophila* and *Xenopus*, the initiation of transcription and cell motility can be induced prematurely by artificially lengthening the cell cycle. When cycloheximide (an inhibitor of protein synthesis) delays cell division, the midblastula transition is induced early in *Xenopus*, and a burst of transcription occurs in *Drosophila* (Edgar et al., 1986; Kimelman et al., 1987).

MECHANISMS OF CLEAVAGE

Regulation of cleavage cycles

Maturation promoting factor

The cell cycle of somatic cells is functionally divided into four stages (Figure 38A). After mitosis (M) there is a prereplication gap (G_1), after which time the synthesis of DNA takes place (S). After the synthetic period, there is a premitotic gap (G_2) which is followed by mitosis. The progression of these phases is regulated by growth factors that will be detailed in Chapter 20. In early cleavage blastomeres, however, cell division can be much simpler. Early sea urchin blastomeres lack G_1, replicating their DNA during the last portion (telophase) of the previous mitosis (Hinegardner et al., 1964). *Xenopus* and *Drosophila* nuclei have eliminated both the G_1 and the G_2 phases during early cleavage. (*Xenopus* embryos add those phases to the cell cycle sometime after the twelfth cleavage. *Drosophila* adds G_2 during cycle 14 and G_1 during cycle 17.) For the first twelve divisions, *Xenopus* cells divide synchronously in a biphasic cell cycle: S to M and M to S (Figure 38B; Laskey et al., 1977; Newport and Kirschner, 1982a).

The factors regulating this biphasic cycle are located in the cytoplasm. Normal enlarging *Xenopus* oocytes are arrested in first meiotic prophase. They are unable to divide. If nuclei from dividing cells are transplanted into these oocytes, they also cease dividing. When normal oocytes are stimulated by progesterone, they resume their meiotic division and stop at the metaphase of the second meiosis. If nuclei from nondividing cells (such as neurons) are placed into the cytoplasm of progesterone-treated oocytes, they, too, initiate division (Gurdon, 1968). The cytoplasm of progesterone-stimulated oocytes will still undergo periodic cortical contractions (characteristic of division), even in the absence of nuclei or centrioles. If cloned DNA fragments are injected into such enucleated embryos, their replication comes under the control of these cycles (Hara et al., 1980; Harland and Laskey, 1980; Karsenti et al., 1984). Thus, the capacity for cell division is regulated by the cytoplasm.

Some of the factors that govern DNA synthesis and cell division have been identified. The progesterone-induced factor that allows oocyte nuclei to resume their divisions is a two subunit phosphoprotein called MATURATION PROMOTING FACTOR (MPF). MPF was first discovered as the major factor responsible for the resumption of meiotic cell divisions in the ovulated frog egg (Smith and Ecker, 1969; Masui and Markert, 1971). This same factor continues to play a role after fertilization, regulating the biphasic cell cycle of early *Xenopus* blastomeres. Gerhart and co-workers (1984) showed that MPF undergoes cyclic changes in level of activity in mitotic cells. The MPF activity of early frog blastomeres is highest during M and undetectable during S. During S phase, MPF exists in an inactive state. This cyclicity is also seen in enucleated blastomeres. Newport and Kirschner (1984) demonstrated that DNA replication (S) and mitosis (M) are driven solely by the gain and loss of MPF, even in the absence of protein synthesis. Cleaving cells can be trapped in interphase by incubating them in an inhibitor of protein synthesis. When MPF is microinjected into these cells, they enter M. Their nuclear envelope breaks down and their chromatin condenses into chromosomes. After 1 hour, MPF activity is degraded, and the chromosomes return to S phase.

The small subunit of MPF has a protein kinase activity that, when

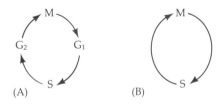

FIGURE 38
Cell cycles of somatic cells and early blastomeres. (A) Cell cycle of typical somatic cell. Mitosis (M) is followed by an "interphase" condition. This latter period is subdivided into G_1, S (synthesis), and G_2 phases. (B) Simpler biphasic cell cycle of the early amphibian blastomeres having only two states, S and M.

activated, can phosphorylate a variety of proteins. Thus, MPF may function by adding phosphate groups onto specific proteins. One such target is histone H1, which is bound to DNA. The phosphorylation of this protein may bring about chromosomal condensation. Another target is the nuclear envelope. Within 15 minutes after the addition of MPF, the three major proteins (the lamin proteins) of the nuclear envelope become hyperphosphorylated. Within the next 15 minutes, the envelope has depolymerized and is breaking apart (Arion et al., 1988; Miake-Lye and Kirschner, 1985). This purified MPF subunit has been seen to phosphorylate these nuclear envelope proteins and to bring about the disassembly of the nuclear envelope (Peter et al., 1990; Ward and Kirschner, 1990). A third target appears to be RNA polymerase (Cisek and Corden, 1989). The phosphorylation of RNA polymerase may cause the inhibition of transcription seen during mitosis.

The small subunit of MPF has been remarkably conserved through evolution and is almost identical to a mitosis-inducing phosphoprotein, p34, synthesized by the yeast *cdc2* gene (Dunphy et al, 1988; Gautier and Byers, 1988). In fact, the human gene that encodes the protein corresponding to the small subunit of *Xenopus* MPF can be inserted into the yeast genome and cause division in *cdc2*-deficient yeast mutants (Lee and Nurse, 1987). The p34 protein can exist in two forms, phosphorylated and unphosphorylated, and it is the unphosphophorylated form that has the kinase activity (Gould and Nurse, 1989).

Cyclins

How, then, is MPF regulated? Since *Xenopus* cleavage seemed to be regulated by a protein similar to that which regulates yeast cell division, it was thought that whatever regulated the yeast protein might have a counterpart in the animal embryo. One of the most important regulators of the yeast MPF-like protein is the product of the *cdc13* gene, a 56,000-Da protein called p56^{cdc13}. This gene was cloned and its sequence was found to be very similar to that of a family of eukaryotic proteins called CYCLINS (Solomon et al., 1988; Goebl and Byers, 1988). Cyclins are proteins of cleavage-stage cells that show a periodic behavior, accumulating during S phase and being degraded during mitosis (Figure 39; Evans et al., 1983; Swenson et al., 1986). The yeast p56^{cdc13} shows the same rhythmic behavior, accumulating during interphase and declining precipitously at the metaphase/anaphase border (Moreno et al., 1989). Cyclins are often encoded in the stored mRNAs of the maternal cytoplasm, and if their translation into proteins is selectively inhibited, the cell will not enter mitosis (Minshull et al., 1989). Moreover, if the degradation of cyclin is prevented, MPF remains active and the cell is frozen in metaphase (Murray et al., 1989). It is thought that cyclin activates the MPF kinase from its dormant interphase state, and that this kinase activity is crucial for initiating cell division. Thus, the control of cell division appears to be regulated by the same group of proteins in an enormous variety of organisms.

The cyclins themselves can be developmentally regulated so that they are not always active. As we mentioned in Chapter 2, the mature frog egg stops cell division by producing a protein called CYTOSTATIC FACTOR (CSF). CSF keeps the oocytes arrested in metaphase of the second meiotic division (Figure 40). It is the product of the *c-mos* gene, and it appears to act by blocking the degradation of cyclin (Karsenti et al., 1987; Sagata et al., 1989). Since cyclin is not degraded, MPF remains active, and the oocyte remains in metaphase. The release of calcium ions during fertilization inactivates CSF, allowing the degradation of cyclin and the completion of

(A)

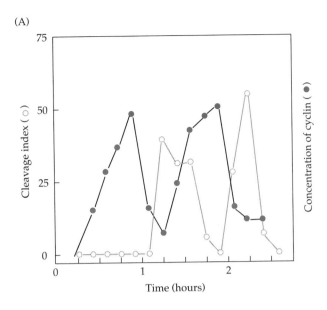

(B)

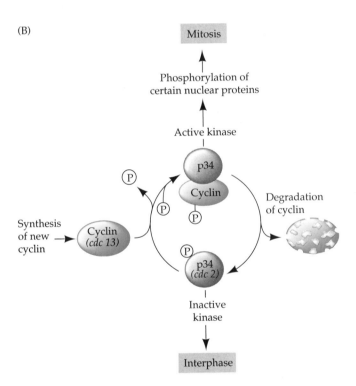

FIGURE 39
Cyclin and cell division. (A) Correlation of cyclin levels with cleavage divisions in the sea urchin embryo. Eggs were fertilized in the presence of radioactive amino acids and were analyzed every 10 minutes for the percentage of dividing cells and the presence of a particular protein that appeared to exhibit cyclicity. This protein was at its highest level during the middle of each cell cycle and was then degraded and resynthesized. Other proteins increased linearly during this same time. (B) Schematic model for the regulation of MPF activity by cyclin. During interphase, the p34 MPF kinase subunit is in its phosphorylated, inactive state. Newly synthesized cyclin complexes with this subunit, allows its phosphate to be removed, and activates its kinase activity. This kinase activity phosphorylates nuclear proteins and initiates mitosis. The cyclin is then degraded and interphase resumes. (A after Evans et al., 1983; B after Nurse, 1990.)

mitosis. After this, MPF activity is based on the cyclic rhythms of cyclin synthesis and degradation.

Cyclin is necessary for the activation of the p34 protein kinase of MPF, but it is probably not sufficient. Another developmental regulator of MPF is the product of the *string* gene in *Drosophila*. The *string* gene is homologous to the yeast gene *cdc*25 whose product activates the cdc2 protein (the small subunit of MPF). Studies by Edgar and O'Farrell (1989) show that the *string* gene is also a positive regulator of MPF and that it is needed if *Drosophila* nuclei are going to divide. During the first 13 division cycles, *string* protein is translated from mRNA stored in the oocyte. However, during the next cycle, the maternal *string* mRNA is degraded. If the nuclei

FIGURE 40

Levels of maturation promoting factor (MPF) during the early development of the frog *Xenopus laevis*. The normal maturational signal is the hormone progesterone, which stimulates the ovulation of the oocytes and the initiation of meiosis. (After Murray and Kirschner, 1989.)

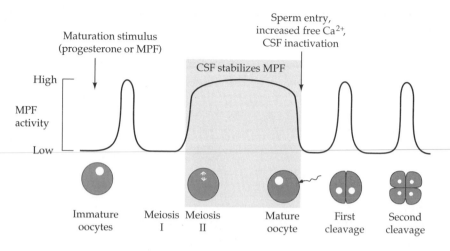

do not transcribe their own *string* mRNA, then the cells will not divide. Edgar and O'Farrell (1989) have shown that those cells that divide are synthesizing their own string protein, while those cells that are not able to join the division cycle are not making it (Figure 41). This degradation and need for resynthesis of string would explain the switch from cytoplasmic to nuclear control of division seen in cycle 14.

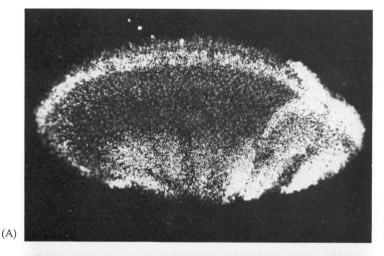

(A)

FIGURE 41

Correlation of *string* gene expression with cell division in *Drosophila* embryos. In this example (A), a late stage-14 embryo is stained with a radioactive nucleotide sequence that specifically recognizes and binds *string* mRNA. In (B), a slightly older embryo is stained with fluorescent antibodies to tubulin to show the microtubules. A comparison of the fluorescence photomicrograph and the autoradiograph obtained from the binding of the radioactive probe shows that only those cells capable of dividing synthesize *string* mRNA. (From Edgar and O'Farrell, 1989; photographs courtesy of B. A. Edgar.)

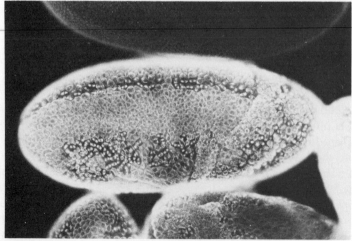

(B)

The cytoskeletal mechanisms of mitosis

Cleavage is actually the result of two coordinated processes. The first of these cyclic processes is KARYOKINESIS—the mitotic division of the nucleus. The mechanical agent of this division is the mitotic spindle, with its microtubules composed of tubulin (the same type of protein that makes the sperm flagellum). The second process is CYTOKINESIS—the division of the cell. The mechanical agent of cytokinesis is the CONTRACTILE RING of microfilaments made of actin (the same type of protein that extends the egg microvilli and the sperm acrosomal process). Table 2 presents a comparison of these systems of division. The relationship and coordination between these two systems during cleavage is diagrammed in Figure 42. Here, a sea urchin egg is shown undergoing first cleavage. The mitotic spindle and contractile ring are perpendicular to each other, and the spindle is internal to the contractile ring. The cleavage furrow eventually bisects the plane of mitosis, thereby creating two genetically equivalent blastomeres.

The actin microfilaments are found in the cortex of the egg rather than in the central cytoplasm. Under the electron microscope, the ring of microfilaments can be seen forming a distinct cortical band, 0.1 μm thick (Figure 43). This contractile ring exists only during cleavage and extends 8–10 μm into the center of the egg. It is responsible for exerting the force that splits the zygote into blastomeres, for if it is disrupted, cytokinesis stops. Schroeder (1973) has proposed a model of cleavage wherein the contractile ring splits the egg like an "intercellular purse-string," tightening about the egg as cleavage continues. This tightening of the microfilamentous ring creates the cleavage furrow.

Although karyokinesis and cytokinesis are usually coordinated, they are sometimes separated by natural or experimental conditions. Harvey (1936) found that physically enucleated sea urchin eggs could keep dividing until they formed a blastulalike structure. (Presumably the cells eventually depleted their maternal supply of mRNA and proteins and could not develop further in the absence of new transcription.) Similarly, karyokinesis can occur in the absence of cell division. We have already discussed insect eggs, in which nuclei divide several times before cell division takes place. Another way to cause this state is to treat embryos with the drug CYTOCHALASIN B. This drug inhibits the organization of microfilaments in the contractile ring. When cytochalasin B is added to seawater before the sea urchin zygote has formed its contractile ring, no cleavage furrow forms. When the drug is added after the beginning of the cleavage furrow, the existing contractile ring is disorganized and cleavage stops (Schroeder, 1972). Nuclear division, however, remains unaffected. Lillie

TABLE 2
Karyokinesis and cytokinesis

Process	Mechanical agent	Major protein composition	Location	Major disruptive drug
Karyokinesis	Mitotic spindle	Tubulin microtubules	Central cytoplasm	Colchicine, nocodazole[a]
Cytokinesis	Contractile ring	Actin microfilaments	Cortical cytoplasm	Cytochalasin B

[a]Because colchicine has been found to independently inhibit several membrane functions, including osmoregulation and the transport of ions and nucleosides, nocodazole has become the major drug used to inhibit microtubule-mediated processes (see Hardin, 1987).

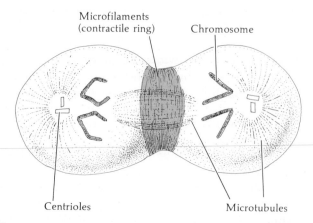

Microfilaments
(contractile ring)   Chromosome

Centrioles   Microtubules

FIGURE 42

Role of microtubules and microfilaments in cell division. In the telophase cell represented here, the chromosomes are being drawn to the centrioles by microtubules while the cytoplasm pinches in through the contraction of microfilaments.

(1902) and Whittaker (1979) have shown that even if cleavage is blocked, the nuclei continue to divide and to express their developmentally regulated products at the appropriate time.

One of the most intriguing unsolved problems of embryonic cleavage is how cytokinesis and karyokinesis are coordinated with each other. Although the detailed mechanism is not yet known, three types of evidence suggest that the microtubule asters of the mitotic apparatus dictate the location of the cleavage furrow. These asters (Figure 44) are the microtubular "rays" that extend from the poles of the mitotic spindle to the cell periphery. One indication that these asters might control the location of the cleavage furrows is that the number of furrows generated depends upon the number of asters present. Normal cleavage only occurs when a pair of asters is present. When there are no asters near the cell periphery, there is no cleavage furrow. E. B. Wilson (1901) showed that when asters are made to disappear (by adding ether to the seawater in which sea

FIGURE 43

Localization of actin microfilaments in the cleavage furrow. (A) Fluorescent labeling of actin in the first cleavage furrow (arrow) of the sea urchin *Arbacia punctulata*. (B) Cortical ring of microfilaments (arrows) around the second cleavage furrow of a zebra fish. (A from Bonder et al., 1988; B from Beams and Kessel, 1976. Photographs courtesy of the authors.)

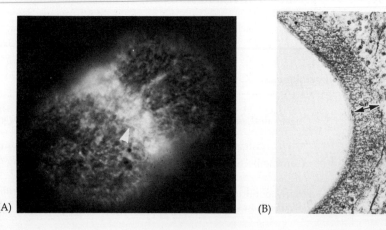

(A)   (B)

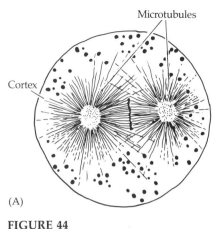

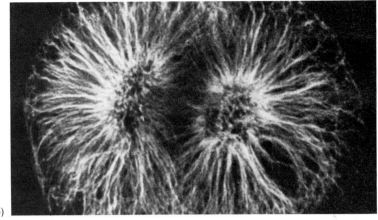

FIGURE 44
Microtubules extending from the mitotic poles through the cell to the cell cortex during first cleavage of a sea urchin egg. Views A and B are separated by nearly 100 years. (A from Wilson, 1896; B from White et al., 1987.)

urchin embryos are dividing), cleavage ceases. When the embryos are then transferred to normal seawater, the asters reappear and cleavage continues. When multiple sperm fertilize an oocyte, three or four asters can form. When there are three asters, three equidistant cleavage furrows appear, and four asters yield four cleavage furrows (see Figure 21 in Chapter 2). In more recent studies, Raff and Glover (1989) showed that if centrioles migrate into the posterior pole of the *Drosophila* embryo, they can form pole cells even in the absence of nuclei. It thus appears that asters are the *sine qua non* of cleavage.

The second type of evidence linking asters with cleavage furrow formation comes from experiments wherein the direction of cleavage is changed by placing the egg under pressure. Pflüger (1884) discovered that when a frog zygote is gently compressed between two glass plates, the direction of the first three cleavages are all perpendicular to the plane of the plates. Both Driesch and Morgan (reviewed in Morgan, 1927) made similar observations with sea urchin embryos. In both cases, the plane of the third cleavage (which is normally parallel to the equator of the oocyte) was displaced by 90°.

These and other experiments show that pressure causes the elongation of the egg in the plane of the plate and that the mitotic spindles orient themselves along the long axes of the cell. The cleavage furrows are always perpendicular to the spindle, so by changing the location of the mitotic spindle, experimenters can change the location of the cleavage furrow.

Rappaport (1961) has extended this type of experiment by displacing the mitotic spindles to the sides of the cells. In Figure 45, a glass ball has been used to displace the asters from the center of the cell toward the periphery. The cleavage furrow that results extends only as far as the ball and does not appear on the other side. Thus, a binucleate, horseshoe-shaped cell is formed. At the next division, two spindle apparatuses form between four asters, but three cleavage furrows are generated! Each arm of the horseshoe has its own mitotic spindle and cleavage furrow as expected, but a third furrow appears between the two asters at the top of the arms (Figure 45C). This demonstrates clearly that if two asters are close enough together, their interactions cause the formation of a cleavage furrow even in the absence of a mitotic spindle between them. Here we see again that cell division can occur without nuclear division as long as the asters are present.

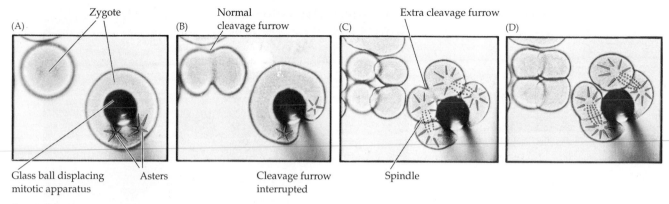

(A) Zygote

(B) Normal cleavage furrow

(C) Extra cleavage furrow

(D)

Glass ball displacing mitotic apparatus · Asters

Cleavage furrow interrupted · Spindle

FIGURE 45

Creation of a new cleavage furrow by displacement of the asters. The diagrammatic representation of asters and spindles has been superimposed on the photographs of the cells. (A,B) Glass ball displaces the spindle and interrupts the cleavage furrow. (C,D) Upon mitosis, each centrosome generates two asters. In addition to the cleavage furrow dividing the spindles, an extra cleavage furrow forms between two adjacent asters. (After Rappaport, 1961; photographs courtesy of R. Rappaport.)

The formation of new membranes

Our last general consideration of embryonic cleavage involves the formation of new cell membranes. Are these membranes newly synthesized or are they merely extensions of the oocyte cell membrane? The answer is probably that both mechanisms contribute to the internal cell membranes. In sea urchin eggs, the zygote's cell membrane is extended into numerous microvilli. During cleavage, these microvilli shrink in size, suggesting that the egg plasma membrane is being "stretched out" as the furrows are formed. However, the surface area of the egg cell membrane is not sufficient to account for all the membrane present toward the end of cleavage. Thus, new membrane is probably synthesized as well.

Evidence that new membrane components may be synthesized during the early cleavage of some species comes from studies done on amphibian embryos with pigmented cortical regions. Figure 46 shows the first cleavage furrow of a pigmented frog zygote. Whereas the original membrane has a pigmented cortical region associated with it, the new membrane is white. This new membrane also has electrical conductance properties different from those of the original membrane. This evidence, though not conclusive, suggests that in frogs the blastomere membranes may be predominantly newly synthesized, even at the first cleavage division.

One of the major questions involving this large amount of new membrane concerns how it becomes connected to older membranes during cleavage furrow formation. Byers and Armstrong (1986) have radiolabeled membrane components of newly fertilized *Xenopus* eggs and followed the

FIGURE 46

Formation of new membranes in the first cleavage of a *Xenopus* egg. (A) Old membrane has pigment granules associated with it. The new membrane appears clear, because it does not have these associated granules. (B) Autoradiograph of membrane proteins in the first cleavage furrow. Surface proteins had been radiolabeled prior to division. White spots indicate regions containing proteins found in the zygote surface before division. (A from de Laat and Bluemink, 1974. B from Byers and Armstrong, 1986. Photographs courtesy of the authors.)

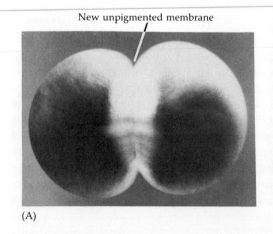

New unpigmented membrane

(A)

(B)

redistribution of these molecules by autoradiography. During the first cleavage, the membrane of the outer surface of the embryo and the membrane of the leading edge of the cleavage furrow are heavily labeled. Between them is a large region devoid of radioactive label (Figure 46B). Thus, the furrow membrane is a mosaic of different parts. The membrane at the leading end of the furrow is derived from the preexisting radiolabeled outer surface of the egg, but most of the membrane of the furrow has been derived from regions that were inaccessible to surface labeling. Byers and Armstrong speculate that the domain of heavily labeled membrane at the leading edge of the furrow contains the membrane anchors for the underlying ring of cortical microfilaments.

SUMMARY

Cleavage is the collection of processes by which the fertilized egg is converted into a multicellular structure. The specific type of cleavage depends upon the evolutionary history of the species and on the ability of the egg to provide the nutritional requirements of the embryo. Some animals need large quantities of yolk to support development to the time when feeding begins, whereas other animals quickly form a feeding larval stage and do not require large amounts of yolk in the egg cytoplasm. In mammals, a peculiar type of cleavage forms the beginnings of the placenta, which will supply nutrients to the embryo. Yet we have seen that underneath the incredible diversity of cleavage types there is an underlying unity of function and mechanism. In all cases, karyokinesis and cytokinesis must be coordinated and the egg divided into cellular regions. During cleavage, the ratio of nucleus to cytoplasm is restored, and developmentally important information is sequestered in various cell regions. In most animals, cleavage is regulated by factors present in the egg cytoplasm and only later does nuclear transcription occur. It appears that the biochemistry regulating these divisions is similar, if not identical, among all the animal phyla, and that the biochemistry of cell division may be the same throughout all eukaryotes. In the next chapter we shall see how these cleavage-stage blastomeres move about and interact with each other to lay down the framework of the body.

LITERATURE CITED

Arion, D., Meijer, L., Brizuela, L. and Beach, D. 1988. cdc2 is a component of the M phase-specific histone H1 kinase: Evidence for identity with MPF. *Cell* 55: 371–378.

Balinsky, B. I. 1981. *Introduction to Embryology*, 5th Ed. Saunders, Philadelphia.

Barlow, P., Owen, D. A. J. and Graham, C. 1972. DNA synthesis in the preimplantation mouse embryo. *J. Embr. Exp. Morphol.* 27: 432–445.

Beams, H. W. and Kessel, R. G. 1976. Cytokinesis: A comparative study of cytoplasmic division in animal cells. *Am. Sci.* 64: 279–290.

Biggelaar, J. A. M. van der and Guerrier, P. 1979. Dorsoventral polarity and mesentoblast determination as concomitant results of cellular interactions in the mollusc *Patella vulgata*. *Dev. Biol.* 68: 462–471.

Bonder, E. M., Fishkind, D. J., Henson, J. H., Cotran, N. M. and Begg, D. A. 1988. Actin in cytokinesis: Formation of the contractile apparatus. *Zool. Sci.* 5: 699-711.

Borland, R. M. 1977. Transport processes in the mammalian blastocyst. *Dev. Mammals* 1: 31–67.

Boycott, A. E., Diver, C., Garstang, S. L. and Turner, F. M. 1930. The inheritance of sinestrality in *Limnaea peregra* (Mollusca:

Pulmonata). *Philos. Trans. R. Soc. Lond.* [Series B] 219: 51–131.

Brenner, C. A., Adler, R. R., Rappolee, D. A., Pedersen, R. A. and Werb, Z. 1989. Genes for extracellular matrix-degrading metalloproteases and their inhibitor, TIMP, are expressed during early mammalian development. *Genes Dev.* 3: 848–859.

Byers, T. J. and Armstrong, P. B. 1986. Membrane protein redistribution during *Xenopus* first cleavage. *J. Cell Biol.* 102: 2176–2184.

Carlson, B. M. 1981. *Patten's Foundations of Embryology*. McGraw-Hill, New York.

Chappelle, A. de la, Schroder, J., Rantanen, P., Thomasson, B., Niemi, M., Tilikainen, A., Sanger, R. and Robson, E. E. 1974. Early fusion of two human embryos? *Ann. Human Genet.* 38: 63–75.

Cisek, L. J. and Corden, J. L. 1989. Phosphorylation of RNA polymerase by the murine homologue of the cell-cycle control protein *cdc2*. *Nature* 339: 679–684.

Craig, M. M. and Morrill, J. B. 1986. Cellular arrangements and surface topography during early development in embryos of *Ilyanassa obsoleta*. *Int. J. Invert. Reprod. Dev.* 9: 209–228.

Crampton, H. E. 1894. Reversal of cleavage in a sinistral gastropod. *Ann. N.Y. Acad. Sci.* 8: 167–170.

Dan, K. 1960. Cytoembryology of echinoderms and amphibia. *Int. Rev. Cytol.* 9: 321–367.

Dan-Sohkawa, M. and Fujisawa, H. 1980. Cell dynamics of the blastulation process in the starfish, *Asterina pectinifera*. *Dev. Biol.* 77: 328–339.

de Laat, S. W. and Bluemink, J. G. 1974. New membrane formation during cytokinesis in normal and cytochalasin B-treated eggs of *Xenopus laevis*. II. Electrophysical observations. *J. Cell Biol.* 60: 529–540.

de Laat, S. W., Tertoelen, L. G. J., Dorresteijn, A. W. C and van der Biggelaar, J. A. M. 1980. Intercellular communication patterns are involved in cell determination in early muscular development. *Nature* 287: 546–548.

Dunphy, W. G., Brizuela, L., Beach, D. and Newport, J. 1988. The *Xenopus cdc2* protein is a component of MPF, a cytoplasmic regulator of mitosis. *Cell* 54: 423–431.

Dyce, J., George, M., Goodall, H. and Fleming, T. P. 1987. Do trophectoderm and inner cell mass cells in the mouse blastocyst maintain discrete lineages? *Development* 100: 685–698.

Edgar, B. A. and O'Farrell, P. H. 1989. Genetic control of cell division patterns in the *Drosophila* embryo. *Cell* 57: 177–187.

Edgar, B. A., Kiehle, C. P. and Schubiger, G. 1986. Cell cycle control by the nucleo-cytoplasmic ratio in early *Drosophila* development. *Cell* 44: 365–372.

Evans, T., Rosenthal, E., Youngblom, J., Distel, D. and Hunt, T. 1983. Cyclin: A protein specified by maternal mRNA in sea urchin eggs that is destroyed at each cleavage division. *Cell* 33: 389–396.

Fleming, T. P. 1987. Quantitative analysis of cell allocation to trophectoderm and inner cell mass in the mouse embryo. *Dev. Biol.* 119: 520–531.

Fleming, T. P., Pickering, S. J., Qasim, F. and Maro, B. 1986. The generation of cell surface polarity in mouse 8-cell blastomeres: The role of cortical microfilaments analyzed using cytochalasin D. *J. Embryol. Exp. Morphol.* 95: 169–191.

Foe, V. 1989. Mitotic domains reveal early commitment of cells in Drosophila embryos. *Development* 107: 1–25.

Freeman, G. and Lundelius, J. W. 1982. The developmental genetics of dextrality and sinistrality in the gastropod *Limnea peregra*. *Wilhelm Roux Arch. Dev. Biol.* 191: 69–83.

Fullilove, S. L. and Jacobson, A. G. 1971. Nuclear elongation and cytokinesis in *Drosophila montana*. *Dev. Biol.* 26: 560–577.

Gardiner, R. C. and Rossant, J. 1976. Determination during embryogenesis in mammals. *Ciba Found. Symp.* 40: 5–18.

Gautier, C., Norbury, C., Lohka, M., Nurse, P. and Maller, J. 1988. Purified maturation-promoting factor contains the product of *Xenopus* homolog of the fission yeast cell cycle control gene *cdc2⁺*. *Cell* 54: 433–439.

Gerhart, J. C., Wu, M. and Kirschner, M. 1984. Cell dynamics of an M-phase-specific cytoplasmic factor in *Xenopus laevis* oocytes and eggs. *J. Cell Biol.* 98: 1247–1255.

Giudice, A. 1973. *Developmental Biology of the Sea Urchin Embryo*. Academic Press, New York.

Goebl, M. and Byers, B. 1988. Cyclin in fission yeast. *Cell* 54: 739–740.

Gould, K. and Nurse, P. 1989. Tyrosine phosphorylation of the fission yeast cdc2⁺ protein kinase regulates entry into mitosis. *Nature* 342: 39–45.

Graham, C. F. and Kelly, S. J. 1977. Interactions between embryonic cells during early development of the mouse. *In* M. Karkinen-Jaaskelainen, L. Saxén and L. Weiss (eds.), *Cell Interactions in Differentiation*. Academic Press, New York, pp. 45–57.

Gulyas, B. J. 1975. A reexamination of the cleavage patterns in eutherian mammalian eggs: Rotation of the blastomere pairs during second cleavage in the rabbit. *J. Exp. Zool.* 193: 235–248.

Gurdon, J. B. 1968. Changes in somatic nuclei inserted into growing and maturing amphibian oocytes. *J. Embryol. Exp. Morphol.* 20: 401–414.

Hara, K. 1977. The cleavage pattern of the axolotl egg studied by cinematography and cell counting. *Wilhelm Roux Arch. Entwicklungsmech. Org.* 181: 73–87.

Hara, K., Tydeman, P. and Kirschner, M. W. 1980. A cytoplasmic clock with the same period as the division cycle in *Xenopus* eggs. *Proc. Natl. Acad. Sci. USA* 77: 462–466.

Hardin, J. D. 1987. Archenteron elongation in the sea urchin embryo is a microtubule independent process. *Dev. Biol.* 121: 253–262.

Harland, R. M. and Laskey, R. A. 1980. Regulated replication of DNA microinjected into eggs of *X. laevis*. *Cell* 21: 761–771.

Harvey, E. G. 1936. Parthenogenetic merogony or cleavage without nuclei in *Arbacia punctulata*. *Biol. Bull.* 71: 101–121.

Hinegardner, R. T., Rao, B. and Feldman, D. E. 1964. The DNA synthetic period during early development of the sea urchin egg. *Exp. Cell Res.* 36: 53–61.

Hillman, N., Sherman, H. I. and Graham, C. F. 1972. The effects of spatial arrangement of cell determination during mouse development. *J. Embryol. Exp. Morphol.* 28: 263–278.

Hörstadius, S. 1939. The mechanics of sea urchin development, studied by operative methods. *Biol. Rev.* 14: 132–179.

Hörstadius, S. 1973. *Experimental Embryology of Echinoderms*. Clarendon Press, Oxford.

Inoué, S. 1982. The role of self-assembly in the generation of biologic form. *In* S. Subtelny and B. P. Green (eds.), *Developmental Order: Its Origin and Regulation*, Alan R. Liss, New York, pp. 35–76.

Johnson, M. H., Chakraborty, J., Handyside, A. H., Willison, K. and Stern, P. 1979. The effect of prolonged decompaction on the development of the preimplantation mouse embryo. *J. Embryol. Exp. Morphol.* 54: 241–261.

Johnson, M. H., Chisholm, J. C., Fleming, T. P. and Houliston, E. 1986. A role for cytoplasmic determinants in the development of the early mouse embryo. *J. Embryol. Exp. Morphol.* [Suppl.]: 97–117.

Kalt, M. R. 1971. The relationship between cleavage and blastocoel formation in *Xenopus laevis*. I. Light microscopic observations. *J. Embryol. Exp. Morphol.* 26: 37–49.

Karr, T. L. and Alberts, B. M. 1986. Organization of the cytoskeleton in early *Drosophila* embryos. *J. Cell Biol.* 102: 1494–1509.

Karsenti, E., Newport, J., Hubble, R. and Kirschner, M. 1984. The interconversion of metaphase and interphase microtubule arrays, as studied by the injection of centrosomes and nuclei into *Xenopus* eggs. *J. Cell Biol.* 98: 1730–1745.

Karsenti, E., Bravo, R. and Kirschner, M. W. 1987. Phosphorylation changes associated with early cell cycle in *Xenopus* eggs. *Dev. Biol.* 119: 442–453.

Kelly, S. J. 1977. Studies of the developmental potential of 4- and 8-cell stage mouse blastomeres. *J. Exp. Zool.* 200: 365–376.

Kimelman, D., Kirschner, M. and Scherson, T. 1987. The events of the midblastula transition in *Xenopus* are regulated by changes in the cell cycle. *Cell* 48: 399–407.

Langman, J. 1981. *Medical Embryology*, 4th Ed. Williams & Wilkins, Baltimore.

Laskey, R. A., Mills, A. D. and Morris, N. R. 1977. Assembly of SV40 chromatin in a cell-free system from *Xenopus* eggs. *Cell* 10: 237–243.

Lee, M. and Nurse, P. 1987. Complementation used to clone a human homologue of the fission yeast cell cycle control gene *cdc2⁺*. *Nature* 335: 251–254.

Levy, J. B., Johnson, M. H., Goodall, H. and Maro, B. 1986. The timing of compaction: Control of a major developmental transition in mouse early embryogenesis. *J. Embryol. Exper. Morphol.* 95: 213–237.

Lillie, F. R. 1898. Adaptation in cleavage. *In Biological Lectures of the Marine Biological Laboratory of Woods Hole.* Ginn, Boston, pp. 43–67.

Lillie, F. R. 1902. Differentiation without cleavage in the egg of the annelid *Chaetopterus pergamentaceous. Wilhelm Roux Arch. Entwicklungsmech. Org.* 14: 477–499.

Lutz, B. 1947. Trends towards non-aquatic and direct development in frogs. *Copeia* 4: 242–252.

Mark, W. H., Signorelli, K. and Lacy, E. 1985. An inserted mutation in a transgenic mouse line results in developmental arrest at day 5 of gestation. *Cold Spring Harbor Symp. Quant. Biol.* 50: 453–463.

Markert, C. L. and Petters, R. M. 1978. Manufactured hexaparental mice show that adults are derived from three embryonic cells. *Science* 202: 56–58.

Masui, Y. and Markert, C. L. 1971. Cytoplasmic control of nuclear behavior during meiotic maturation of frog oocytes. *J. Exp. Zool.* 177: 129-146.

Mayr, W. R., Pausch, V. and Schnedl, W. 1979. Human chimaera detectable only by investigation of her progeny. *Nature* 277: 210–211.

Miake-Lye, R. and Kirschner, M. W. 1985. Induction of early mitotic events in a cell-free system. *Cell* 41: 165–175.

Minshull, J., Blow, J. J. and Hunt, T. 1989. Translation of cyclin mRNA is necessary for extracts of activated *Xenopus* eggs to enter mitosis. *Cell* 56: 947–956.

Mintz, B. 1970. Clonal expression in allophenic mice. *Symp. Int. Soc. Cell Biol.* 9: 15.

Moreno, S., Hayles, J. and Nurse, P. 1989. Regulation of the p34^{cdc2} protein kinase during mitosis. *Cell* 58: 361–372.

Morgan, T. H. 1927. *Experimental Embryology.* Columbia University Press, New York.

Mulnard, J. G. 1967. Analyse microcinematographique du developpement de l'oeuf de souris du stade II au blastocyste. *Arch. Biol.* (Liege) 78: 107–138.

Murray, A. W. and Kirschner, M. W. 1989. Cyclin synthesis drives the early embryonic cell cycle. *Nature* 339: 275–280.

Murray, A. W., Solomon, M. J. and Kirschner, M. W. 1989. The role of cyclin synthesis and degradation in the control of maturation promoting factor activity. *Nature* 339: 280-286.

Newport, J. W. and Kirschner, M. W. 1982a. A major developmental transition in early *Xenopus* embryos: I. Characterization and timing of cellular changes at midblastula stage. *Cell* 30: 675–686.

Newport, J. W. and Kirschner, M. W. 1982b. A major developmental transition in early *Xenopus* embryos: II. Control of the onset of transcription. *Cell* 30: 687–696.

Newport, J. W. and Kirschner, M. W. 1984. Regulation of the cell cycle during *Xenopus laevis* development. *Cell* 37: 731–742.

Nieuwkoop, P. D. 1973. The "organization center" of the amphibian embryo: Its origin, spatial organization, and morphogenetic action. *Adv. Morphog.* 10: 1–39.

Nurse, P. 1990. Universal control mechanism regulating onset of M-phase. *Nature* 344: 503–508.

Pederson, R. A., Wu, K. and Batakier, H. 1986. Origin of the inner cell mass in mouse embryos: Cell lineage analysis by microinjection. *Dev. Biol.* 117: 581–595.

Perona, R. M. and Wassarman, P. M. 1986. Mouse blastocysts hatch *in vitro* by using a trypsin-like proteinase associated with cells of mural trophectoderm. *Dev. Biol.* 114: 42–52.

Peter, M., Nakagawa, J., Dorée, M., Labbé, J. C. and Nigg, E. A. 1990. In vitro disassembly of the nuclear lamina and M phase-specific phosphorylation of lamins by cdc2 kinase. *Cell* 61: 591–602.

Peyrieras, N., Hyafil, F., Louvard, D., Ploegh, H. L. and Jacob, F. 1983. Uvomorulin: A non-integral membrane protein of early mouse embryo. *Proc. Natl. Acad. Sci. USA* 80: 6274–6277.

Piko, L. and Clegg, K. B. 1982. Quantitative changes in total RNA, total poly(A), and ribosomes in early mouse embryos. *Dev. Biol.* 89: 362–378.

Pflüger, E. 1884. Uber die Einwirkung der Schwerkraft und anderer Bedingungen auf die Richtung der Zeiltheilung. *Arch. Ges. Physiol.* 3: 4.

Prather, R. S. 1989. Nuclear transfer in mammals and amphibia. *In* H. Schatten and G. Schatten (eds.), *The Molecular Biology of Fertilization.* Academic Press, New York, pp. 323–340.

Pratt, H. P. M., Ziomek, Z. A., Reeve, W. J. D. and Johnson, M. H. 1982. Compaction of the mouse embryo: An analysis of its components. *J. Embryol. Exp. Morphol.* 70: 113–132.

Raff, J. W. and Glover, D. M. 1989. Centrosomes, not nuclei, initiate pole cell formation in *Drosophila* embryos. *Cell* 57: 611–619.

Raff, R. A. and Kaufman, T. C. 1983. *Embryos, Genes, and Evolution: The Developmental-Genetic Basis of Evolutionary Change.* Macmillan, New York.

Rappaport, R. 1961. Experiments concerning cleavage stimulus in sand dollar eggs. *J. Exp. Zool.* 148: 81–89.

Rugh, R. 1967. *The Mouse.* Burgess, Minneapolis.

Sagata, N., Watanabe, N., Vande Woude, G. F. and Ikawa, Y. 1989. The *c-mos* proto-oncogene product is a cytostatic factor responsible for meiotic arrest in vertebrate eggs. *Nature* 342: 512–518.

Saunders, J. W., Jr. 1982. *Developmental Biology.* Macmillan, New York.

Schroeder, T. E. 1972. The contractile ring. II. Determining its brief existence, volumetric changes, and vital role in cleaving *Arbacia* eggs. *J. Cell Biol.* 53: 419–434.

Schroeder, T. E. 1973. Cell constriction: Contractile role of microfilaments in division and development. *Am. Zool.* 13: 687–696.

Seidel, F. 1952. Die Entwicklungspotenzen einer isolierten Blastomere des Zweizellenstadiums im Säugetierei. *Naturwissenschaften* 39: 355–356.

Smith, L. D. and Ecker, R. E. 1969. Role of the oocyte nucleus in the physiological maturation in *Rana pipiens. Dev. Biol.* 19: 281–309.

Solomon, M., Booher, R., Kirschner, M. and Beach, D. 1988. Cyclin in fission yeast. *Cell* 54: 738–739.

Stern, C. D. 1990. The marginal zone and its contribution to the hypoblast and primitive streak of the chick embryo. *Development* 109: 667-682.

Strickland, S., Reich, E. and Sherman, M. I. 1976. Plasminogen activator in early embryogenesis: Enzyme production by trophoblast and parietal endoderm. *Cell* 9: 231–240.

Sturtevant, M. H. 1923. Inheritance of direction of coiling in *Limnaea. Science* 58: 269–270.

Sutherland, A. E. and Calarco-Gillam, P. G. 1983. Analysis of compaction in the preimplantation mouse embryo. *Dev. Biol.* 100: 327–338.

Sutherland, A. E., Speed, T. P. and Calarco, P. G. 1990. Inner cell allocation in the mouse morula: The role of oriented division during fourth cleavage. *Dev. Biol.* 137: 13–25.

Swenson, K. L., Farrell, K. M. and Ruderman, J V. 1986. The clam embryo protein cyclin A induces entry into M phase and the resumption of meiosis in *Xenopus* oocytes. *Cell* 47: 861-870.

Sze, L. C. 1953. Changes in the amount of deoxyribonucleic acid in the development of *Rana pipiens. J. Exp. Zool.* 122: 577–601.

Tarkowski, A. K. and Wróblewska, J. 1967. Development of blastomeres of mouse eggs isolated at the 4- and 8-cell stage. *J. Embryol. Exp. Morphol.* 18: 155–180.

Torrey, T. 1962. *Morphogenesis of the Vertebrates.* Wiley, New York.

Tuchmann-Duplessis, H., David, G. and Haegel, P. 1972. *Illustrated Human Embryology,* Vol. 1. Springer-Verlag, New York.

Ward, G. E. and Kirschner, M. W. 1990. Identification of cell cycle-regulated phosphorylation cites on nuclear lamin C. *Cell* 61: 561–577.

Welsh, J. H. 1969. Mussels on the move. *Nat. Hist.* 78: 56–59.

White, J. C., Amos, W. B. and Fordham, M. 1987. An evaluation of confocal versus conventional imaging of biological structures by fluorescence light microscopy. *J. Cell Biol.* 105: 41–48.

Whittaker, J. R. 1979. Cytoplasmic determinants of tissue differentiation in the ascidian egg. *In* S. Subtelny and I. R. Konigsberg (eds.), *Determinants of Spatial Organization.* Academic Press, New York, pp. 29–51.

Wiley, L. M. 1984. Cavitation in the mouse preimplantation embryo: Na/K ATPase and the origin of nascent blastocoel fluid. *Dev. Biol.* 105: 330–342.

Wilson, E. B. 1896. *The Cell in Development and Inheritance*. Macmillan, New York.

Wilson, E. B. 1898. Cell lineage and ancestral reminiscences. *Biological Lectures of the Marine Biological Laboratory of Woods Hole*, Ginn, Boston, pp. 21–42.

Wilson, E. B. 1901. Experiments in cytology. II. Some phenomena of fertilization and cell division in etherized eggs. III. The effect on cleavage of artificial obliteration of the first cleavage furrow. *Wilhelm Roux Arch. Entwicklungsmech. Org.* 13: 353–395.

Winkel, G. K., Ferguson, J. E., Takeichi, M. and Nuccitelli, R. 1990. Activation of protein kinase C triggers premature compaction in the four-cell stage mouse embryo. *Dev. Biol.* 138: 1–15.

Wolpert, L. and Gustafson, T. 1961. Studies in the cellular basis of morphogenesis of the sea urchin embryo: Development of the skeletal pattern. *Exp. Cell Res.* 25: 280–300.

Wolpert, L. and Mercer, E. H. 1963. An electron microscopic study of the development of the blastula of the sea urchin embryo and its radial polarity. *Exp. Cell Res.* 30: 280–300.

Yamazaki, K. and Kato, Y. 1989. Sites of zona pellucida shedding by mouse embryo other than mural trophectoderm. *J. Exp. Zool.* 249: 347–349.

Ziomek, C. A. and Johnson, M. H. 1980. Cell surface interactions induce polarization of mouse 8-cell blastomeres at compaction. *Cell* 21: 935–942.

4

Gastrulation: Reorganizing the embryonic cells

*My dear fellow . . . life is infinitely stranger than anything which
the mind of man could invent. We would not dare to conceive the
things which are really mere commonplaces of existence.*

—A. CONAN DOYLE (1891)

*It is not birth, marriage, or death, but gastrulation, which is truly
the most important time in your life.*

—LEWIS WOLPERT (1986)

Gastrulation is the process of highly integrated cell and tissue migrations
whereby the cells of the blastula are dramatically rearranged. The blastula
consists of numerous cells, the positions of which were established during
cleavage. During gastrulation, these cells are given new positions and new
neighbors, and the multilayered body plan of the organism is established.
The cells that will form the endodermal and mesodermal organs are
brought inside the embryo while the cells capable of forming the skin and
nervous system are spread over the outside surface. Thus, the three germ
layers—outer ectoderm, inner endoderm, and interstitial mesoderm—are
first produced during gastrulation. In addition, the stage is set for the
interactions of these newly positioned tissues.

The movements of gastrulation involve the entire embryo, and cell
migrations in one part of the gastrulating organism must be intimately
coordinated with other movements occurring simultaneously. Although
the patterns of gastrulation vary enormously throughout the animal kingdom, there are relatively few mechanisms involved. Gastrulation usually
involves combinations of the following types of movements:

- *Epiboly.* The movement of epithelial sheets (usually of ectodermal
  cells), which spread as a unit rather than individually, to enclose the
  deeper layers of the embryo.
- *Invagination.* The infolding of a region of cells, much like the in-
  denting of a soft rubber ball when poked.

- *Involution.* The inturning or inward movement of an expanding outer layer so that it spreads over the internal surface of the remaining external cells.
- *Ingression.* The migration of individual cells from the surface layers into the interior of the embryo.
- *Delamination.* The splitting of one cellular sheet into two more or less parallel sheets.

Gastrulation affects the entire embryo, providing the embryologist with a spectacle of orchestrated cell movement second to none in animal development. It also provides the opportunity to analyze the mechanisms by which the various embryonic regions are interrelated. As we look at the phenomena of gastrulation in different types of embryos, we should keep in mind the following questions (Trinkaus, 1984):

1 *What is the unit of migration activity?* Is the migration dependent on the movement of individual cells, or are the cells part of a migrating sheet? Remarkable as it first seems, regional migrational properties may be totally controlled by cytoplasmic factors that are independent of cellularization. F. R. Lillie (1902) was able to parthenogenetically activate eggs of the annelid *Chaetopterus*

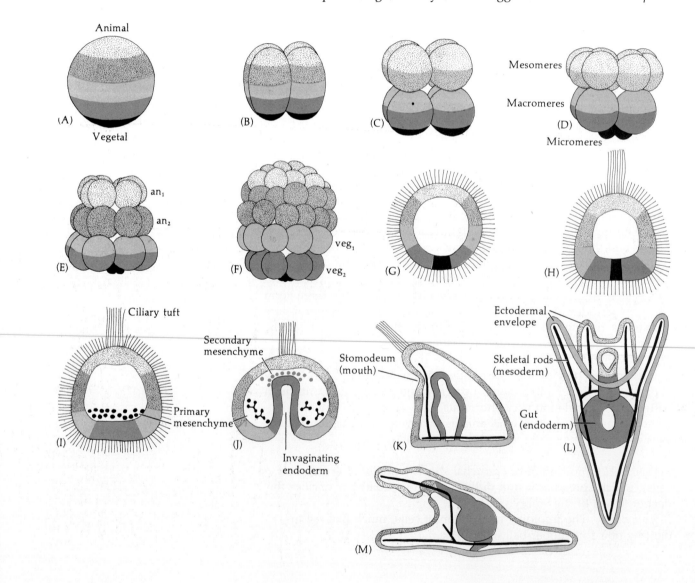

and suppress their cleavage. Many events of early development occurred even in the absence of cells. The cytoplasm of the zygote separated into defined regions, and cilia differentiated in the appropriate parts of the egg. Moreover, the outermost clear cytoplasm migrated down over the vegetal regions in a manner specifically reminiscent of the epiboly of animal hemisphere cells during normal development. This occurred at precisely the time that epiboly would have taken place during gastrulation. Thus, epiboly may be (at least in some respects) independent of the cells that form the migrating region.

2 *Is the spreading or folding of a cell sheet due to factors within the sheet or to external forces stretching or distorting it?* It is essential to know the answer to this question if we are to understand how the various cell movements of gastrulation are integrated. For instance, do involuting cells pull the epibolizing cells down toward them or are the two movements independent?

3 *Is there active spreading of the whole tissue or does the leading edge expand and drag the rest of a cell sheet passively along?*

4 *Are changes in cell shape and motility during gastrulation the consequence of changes in cell surface properties, such as adhesiveness to the substrate or to other cells?* These changes involve the relationship of the cells to other cells and to their extracellular environment.

5 *How are these changes manifest on the intracellular level?* Are rearrangements in the arrays of microtubules or microfilaments necessary for cells to migrate?

Keeping these questions in mind, we will observe the various patterns of gastrulation found in echinoderms, amphibians, birds, and mammals.*

Sea urchin gastrulation

The sea urchin blastula consists of a single layer of 1000 to 2000 cells. These cells, derived from different regions of the zygote, have different sizes and properties. Figures 1 and 2 shows the fates of the various regions of the zygote as it develops through cleavage and gastrulation to the PLUTEUS LARVA characteristic of sea urchins. Here, the fate of each cell layer can be seen through the layers' movements during gastrulation.

*Discussion of *Drosophila* gastrulation will be postponed until Chapter 18, where it occurs in the context of pattern formation.

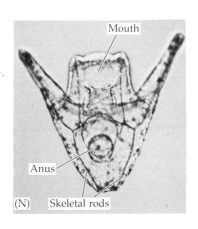

FIGURE 1
Normal sea urchin development, following the fate of the cellular layers of the blastula. (A–F) Cleavage through the 64-cell stage (2-cell stage omitted); (G) early blastula with cilia; (H) late blastula with ciliary tuft and flattened vegetal plate; (I) blastula with primary mesenchyme; (J) gastrula with secondary mesenchyme; (K) prism-stage larve; (L,M) pluteus larvae. Fates of the zygote cytoplasm can be followed through the variations of shading. (N) Photomicrograph of a live pluteus larva of the sea urchin. (A–M after Hörstadius, 1939; N courtesy of G. Watchmaker.)

FIGURE 2

Entire sequence of gastrulation in *Lytechinus variegatus*. The time shows the length of development at 25°C. (Courtesy of J. Morrill.)

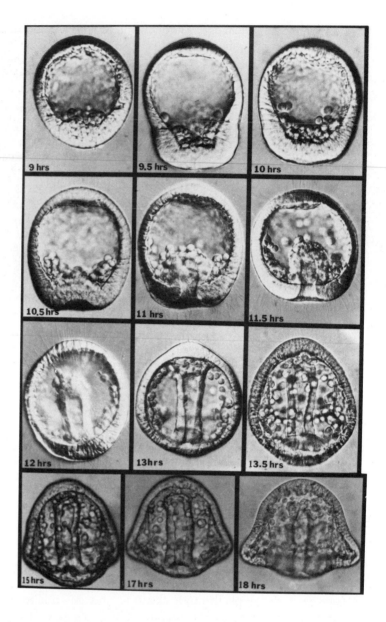

FIGURE 3

Mesenchyme blastula of the sea urchin *S. purpuratus*. (Courtesy of M. A. Harkey.)

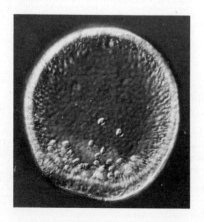

Ingression of primary mesenchyme

Approximately 24 hours after the blastula hatches from its fertilization membrane, the vegetal side of the spherical blastula begins to thicken and flatten. At the center of this flat VEGETAL PLATE, a cluster of small cells begins to change. These cells show pulsating movements on their inner surfaces, extending and contracting long, thin (30 μm × 5 μm) processes called FILOPODIA. The cells then dissociate from one another and migrate into the blastocoel. These cells are called the PRIMARY MESENCHYME and are derived from the micromeres (Figure 3).

Gustafson and Wolpert (1961) have used time-lapse films to follow the microscopic movements of these cells within the blastocoel. At first the mesenchymal cells appear to move randomly along the inner blastocoel surface, actively making and breaking filopodial connections to the wall of the blastocoel. Eventually, these cells become localized within the ventral region of the blastocoel, where their adherence is believed to be strongest. Here the primary mesenchymal cells will fuse into SYNCYTIAL CABLES, which will form the axis of the calcium carbonate spicules of the larval skeleton (Figure 4).

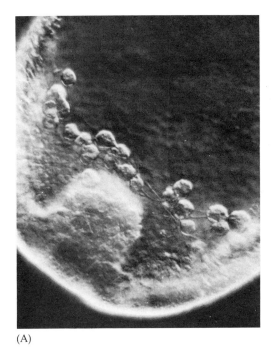

(A)

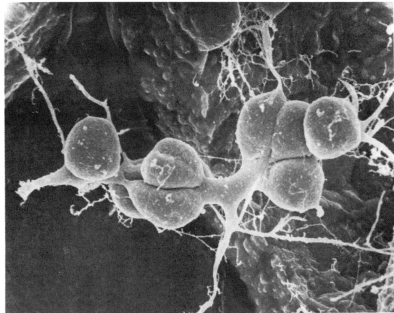

(B)

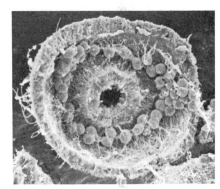

(C)

FIGURE 4
Formation of syncytial cables by mesenchymal cells of the sea urchin. (A) Primary mesenchymal cells in the early gastrula align and fuse to lay down the matrix of the calcium carbonate spicule. (B) Scanning electron micrograph of spicules formed by the fusing of primary mesenchyme cells into syncytial cables. (C) Ring of mesenchymal cells around archenteron. The animal half and the entire archenteron have been removed. (A From Harkey and Whiteley, 1980; B and C from Morrill and Santos, 1985; photographs courtesy of the authors.)

Both cytoplasmic and cell surface events are crucial to the ingression and migration of the primary mesenchymal cells. Gustafson and Wolpert (1967) proposed a model in which the ingression of the micromeres comes about through changes in their adhesion to other cells and to the extracellular matrices that surround them. Originally, all the cells of the blastula are connected on their outer surface to the HYALINE LAYER (derived from the cortical granules at fertilization) and on their inner surface by a BASAL LAMINA (an extracellular network secreted by the cells). Although it cannot be seen with light microscopy, the blastula basal lamina can be observed with the scanning electron microscope, as in Figure 5. On their lateral sides, each cell has another cell for a neighbor. In 1985, Fink and McClay confirmed Gustafson and Wolpert's speculations by measuring the adhesive strengths of the sea urchin blastomeres for the hyaline layer, for the basal lamina, and for other cells. They found that the descendants of the macromeres and mesomeres bound tightly to each other and to the hyaline layer. But they adhered only loosely to the basal lamina (Table 1). The micromeres of the blastula originally displayed a similar pattern of binding. However, the micromere pattern changed at gastrulation. Whereas the

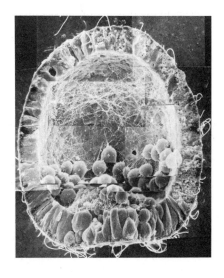

FIGURE 5
Mesenchyme blastula of *Lytechinus variegatus*. Scanning electron micrograph shows hyaline layer (outside) and extensive basal lamina within the blastocoel. (From Morrill, 1986; photograph courtesy of J. B. Morrill.)

TABLE 1

Affinities of mesenchymal and nonmesenchymal cells to cellular and extracellular components[a]

| | Dislodgement force (in dynes) | | |
Cell type	Hyaline	Gastrula cell monolayers	Basal lamina
16-cell stage micromeres	5.8×10^{-5}	6.8×10^{-5}	4.8×10^{-7}
Migratory stage mesenchymal cells	1.2×10^{-7}	1.2×10^{-7}	1.5×10^{-5}
Gastrula ectoderm and endoderm	5.0×10^{-5}	5.0×10^{-5}	5.0×10^{-7}

Source: Fink and McClay (1985).

[a]Tested cells were allowed to adhere to plates containing hyaline, extracellular basal lamina, or cell monolayers. The plates were inverted and centrifuged at various strengths to dislodge the cells. The dislodgement force is calculated from the centrifugal force needed to remove the test cells from the substrate.

FIGURE 6

Ingression of primary mesenchymal cells. (A) Interpretative diagram depicting changes in adhesive interactions in the presumptive primary mesenchymal cells (color). These cells lose their affinities for hyaline and for their neighboring blastomeres while they gain an affinity for the basal lamina. Nonmesenchymal blastomeres retain their original high affinities for hyaline and neighboring cells. (B) Scanning electron micrograph showing the ingression of the primary mesenchymal cells of *Lytechinus variegatus*. (A from Fink and McClay, 1985; B courtesy of J. B. Morrill and D. Flaherty.)

(A)

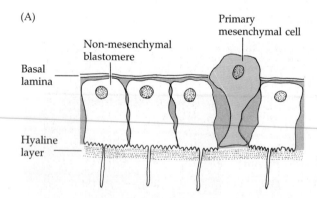

(B)

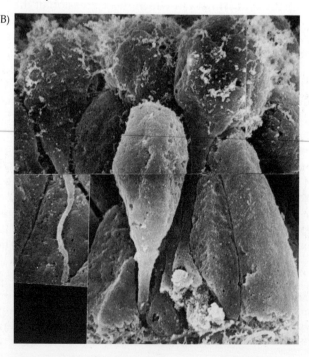

other cells retained their tight binding to the hyaline layer and to their neighbors, the primary mesenchyme precursors lost their affinity to these structures (to about 2 percent of its original value) while their affinity to the basal lamina *increased* 100-fold. This change in affinity causes the micromeres to release their attachments to the external hyaline layer and neighboring cells and, drawn by the basal lamina, migrate up into the blastocoel (Figure 6). The changes in cell affinity have been correlated with changes in cell surface molecules that occur during this time. A cell surface antigen, Meso 1, is detected at the precise time that one first observes mesenchymal movement from the blastula wall (Wessel and McClay, 1985).

As shown in Figure 5, there is a heavy concentration of extracellular lamina material around the ingressing primary mesenchymal cells (Galileo and Morrill, 1985). Moreover, once inside the blastocoel, the primary mesenchymal cells migrate along the extracellular matrix, extending their filopodia in front of them (Figure 7; Galileo and Morrill, 1985; Karp and Solursh, 1985). Two proteins appear to be important in their migration. One is FIBRONECTIN, a large (400,000-Da) glycoprotein that is a common component of basal laminae, including that of the sea urchin blastocoel (Wessel et al., 1984). Fink and McClay (1985) showed that during gastrulation the affinity of the micromeres for this particular molecule increased dramatically. The second set of molecules are sulfated glycoproteins found on the cell surface of the ingressing mesenchymal cells (Sugiyama, 1972; Heifetz and Lennarz, 1979). If the synthesis (or sulfation) of these glycoproteins is inhibited, the mesenchymal cells will enter the blastocoel but will not migrate further (Karp and Solursh, 1974; Anstrom et al., 1987; Figure 8).

(A)

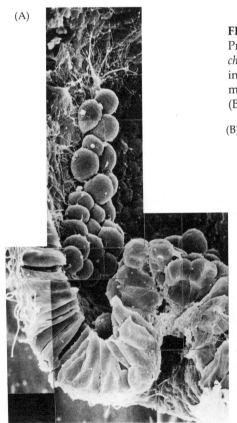

FIGURE 7
Primary mesenchymal cells migrating upon extracellular matrix of early *Lytechinus* gastrula. (A) Scanning electron micrograph of mesenchymal cells migrating upward along the lateral sides of the blastocoel. (B) Close-up of leading mesenchymal cells showing the filopodia enmeshed in the extracellular matrix. (B from Morrill and Santos, 1985; both photographs courtesy of J. B. Morrill.)

(B)

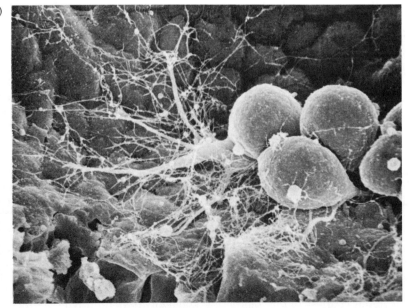

FIGURE 8

Effect of sulfate deprivation on primary mesenchyme movement in the sea urchin *Lytechinus*. (A) Normal gastrula. (B) Abnormal gastrula formed when embryos are grown in sulfate-free seawater. (From Karp and Solursh, 1974; photographs courtesy of M. Solursh.)

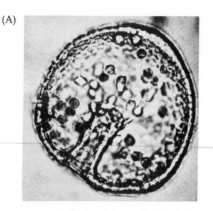 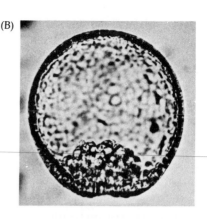

(A)   (B)

(A)

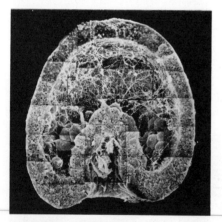

(B)

FIGURE 9

Two views of vegetal plate invagination in *Lytechinus variegatus*. (A) Scanning electron micrograph of external surface of the early gastrula showing individual cells and the invagination of the vegetal plate. (B) Montage of SEM photographs showing a cross-section through the center of the embryo. Primary mesenchymal cells can be seen migrating along the extracellular matrix. (From Morrill and Santos, 1985; photographs courtesy of J. B. Morrill.)

First stage of archenteron invagination

As the ring of primary mesenchymal cells forms in the vegetal region of the blastocoel, important changes are occurring in the cells that remain at the vegetal plate. These cells remain bound to one another and to the hyaline layer of the egg and move to fill the gaps caused by the ingression of the mesenchyme; therefore, the vegetal plate flattens further. One next sees that the vegetal plate bends inward and extends about one-quarter to one-half of the way into the blastocoel (Figure 9). Then invagination suddenly ceases. The invaginated region is called the ARCHENTERON (primitive gut), and the opening of the archenteron at the vegetal region is called the BLASTOPORE.

This first stage of vegetal plate invagination appears to be due to the forces within the vegetal plate itself and has nothing to do with pressure from the cells above it. In 1939 Moore and Burt cut echinoderm embryos into animal and vegetal halves. The vegetal halves began to invaginate even in the absence of the animal hemispheres. The invagination was only one-fourth to one-third the distance to the animal pole, about the same as in the first stage of normal invagination (Figure 10). As the folding deepened, the cut rims of the vegetal halves rolled upward and fused, forming a miniature gastrula. Ettensohn (1984) has extended this work to show that if the vegetal plate is excised hours before gastrulation begins (even before the ingression of the primary mesenchyme cells), the vegetal invagination still occurs at the appropriate time.

The hyaline layer is critical for this invagination process, since it serves as an anchoring substrate for the blastomeres that are changing their shapes (Gustafson and Wolpert, 1967). If blastulae are treated with antibodies that block the binding of these cells to hyaline layer proteins, the cells round up and no invagination occurs (Adelson and Humphreys, 1988).

Second and third stages of archenteron invagination

The invagination of the vegetal cells occurs in three discrete stages. After a brief pause, the second phase of archenteron formation begins. During this time, the archenteron extends dramatically, nearly tripling its length. In the process of extension, the wide, short gut rudiment is transformed into a long, thin tube; yet no new cells are seen to be formed. To accomplish this extension, the cells of the archenteron rearrange themselves by migrating over one another and by flattening themselves (Ettensohn, 1985; Hardin and Cheng, 1986). In at least some species of sea urchins, a third stage of archenteron elongation occurs. This last phase of archenteron

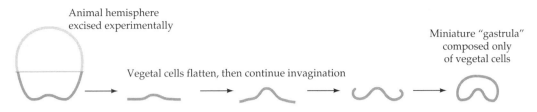

Animal hemisphere
excised experimentally

Vegetal cells flatten, then continue invagination

Miniature "gastrula"
composed only
of vegetal cells

FIGURE 10

Invagination of the vegetal plate in the absence of the animal hemisphere. Vegetal regions of the starfish *Patiria* were isolated; they succeeded in invaginating without attachment of the animal hemisphere. (After Moore and Burt, 1939.)

elongation is initiated by the tension provided by SECONDARY MESENCHYMAL CELLS, which form at the tip of the archenteron and remain there (Figure 11). Filopodia are extended from these cells through the blastocoel fluid to contact the inner surface of the blastocoel wall (Dan and Okazaki, 1956; Schroeder, 1981). The filopodia attach to the wall at the junctions between the blastoderm cells and then shorten, pulling up the archenteron. Hardin (1988) ablated the secondary mesenchyme cells with a laser, with the result that the archenteron could only elongate about two-thirds of the full length. If a few secondary mesenchyme cells were left, elongation continued, although at a slower rate than in controls. The secondary mesenchyme cells, then, play an active role in pulling the archenteron up to the blastocoel wall.

But can the secondary mesenchyme filopodia attach to any part of the blastocoel wall, or is there a specific target in the animal hemisphere that must be present for this attachment to occur? Is there a region of the blastocoel wall that is already committed to becoming the ventral side of the larva? Studies by Hardin and McClay (1990) answer these questions in the affirmative. They found that the filopodia of the secondary mesenchyme cells extend, touch the blastocoel wall at random sites, and then retract. However, when the filopodia contacted one region of the wall, they remained attached there, flattened out against this region, and pulled the archenteron towards it. When Hardin and McClay poked in the other

FIGURE 11

Midgastrula stage of the sea urchin *Lytechinus pictus*, showing filopodial extensions of secondary mesenchyme extending from the archenteron tip to the blastocoel wall. (A) Mesenchymal cells extending filopodia from the tip of the archenteron. (B) Filopodial cables connecting the blastocoel wall to the archenteron tip. The tension of the cables can be seen as they pull on the blastocoel wall at the point of attachment. (Photographs courtesy of C. Ettensohn.)

(A)   (B)

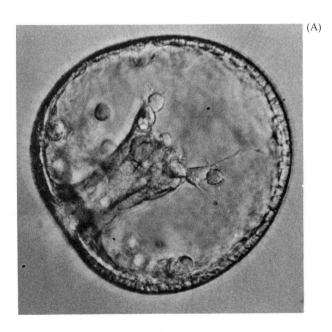

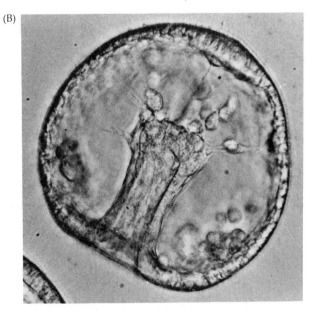

side of the blastocoel wall so that the contacts were made most readily with that region, the filopodia continued to extend and retract after touching it. Only when they found the "target" did they cease these movements. If the gastrula was constricted so that filopodia never reached the target area, the secondary mesenchyme cells continued to explore until they eventually moved off the archenteron and found the target tissue as freely migrating cells. There appears, then, to be a target region on what is to become the ventral side of the larva that is recognized by the secondary mesenchyme cells and positions the archenteron in the region where the mouth will form.

As the top of the archenteron meets the blastocoel wall in this region, the secondary mesenchymal cells disperse into the blastocoel, where they proliferate to form the mesodermal organs. Where the archenteron contacts the wall, a mouth is eventually formed. The mouth fuses with the archenteron to create a continuous digestive tube. Thus, as is characteristic for deuterostomes, the blastopore marks the position of the anus.

Amphibian gastrulation

The study of amphibian gastrulation is both one of the oldest and one of the newest areas of experimental embryology; for although amphibian gastrulation has been extensively studied for the past century, most of our theories concerning the mechanisms of these developmental movements have been revised over the past decade. The study of amphibian gastrulation has been complicated by the fact that there is no one single way that all amphibians gastrulate. Different species will employ different means toward the same goal (Smith and Malacinski, 1983; Lundmark, 1986). In recent years the most intensive investigations have focused on *Xenopus*, so we will concentrate on its mode of gastrulation.

The cellular preparation for these movements takes place during cleavage. As mentioned in Chapter 3, the amphibian embryo undergoes a MIDBLASTULA TRANSITION during which the cell cycle slows down (as a result of acquisition of G_1 and G_2 phases of the cell cycle), cell division becomes asynchronous, the cells gain the ability to move from their original positions, and the transcription of new mRNA is seen from the nucleus for the first time in the animal's life. In *Xenopus*, this transition occurs immediately after the twelfth cleavage (Newport and Kirschner, 1982).

Cell movements during amphibian gastrulation: An overview

Amphibian blastulae are faced with the same tasks as their echinoderm counterparts, namely, to bring inside those areas destined to form the endodermal organs, to surround the embryo with those cells capable of forming the ectoderm, and to place the mesodermal cells in the proper places between them. The movements whereby this is accomplished can be visualized by the technique of vital dye staining. Vogt (1929) saturated agar chips with dyes, such as Neutral Red or Nile Blue sulfate, that would stain but not damage the embryonic cells. These stained agar chips were pressed to the surface of the blastula, and some of the dye was transferred onto the contacted cells (Figure 12). The movements of each group of stained cells were followed throughout gastrulation, and the results were summarized in FATE MAPS. These maps have recently been confirmed by scanning electron microscopy and dye injection techniques (Smith and Malacinski, 1983; Lundmark, 1986).

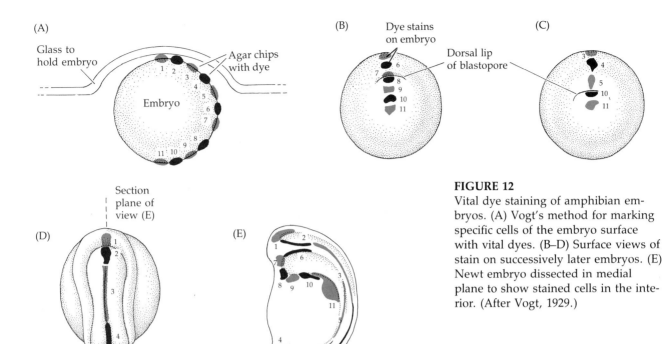

FIGURE 12
Vital dye staining of amphibian embryos. (A) Vogt's method for marking specific cells of the embryo surface with vital dyes. (B–D) Surface views of stain on successively later embryos. (E) Newt embryo dissected in medial plane to show stained cells in the interior. (After Vogt, 1929.)

Vital dye studies by Løvtrup (1975; Landstrom and Løvtrup, 1979) and by Keller (1975, 1976) have shown that cells of the *Xenopus* blastula have different fates depending on whether they are in the deep or the superficial layers of the embryo (Figure 13). In *Xenopus* (but not in *Ambystoma* or several other amphibians) the mesodermal precursors exist in the deep layer of cells, while the ectoderm and endoderm arise from the superficial layer on the surface of the embryo. The precursors for the notochord and other mesodermal tissues are located beneath the surface in the equatorial (marginal) region of the embryo. In urodeles (salamanders such as *Triturus* and *Ambystoma)* the notochord and mesoderm precursors are found in both the surface cells and the deep marginal cells (Landstrom and Løvtrup, 1979; Keller, 1976; Lundmark, 1986).

Gastrulation in frog embryos is initiated at the future dorsal side of the embryo, just below the equator in the region of the gray crescent (Figure 14). Here the local endodermal cells invaginate to form a slitlike blastopore. These cells change their shape dramatically. The main body of each cell is displaced toward the inside of the embryo while it maintains contact with the outside surface by way of a slender cytoplasmic strand (Figure 15). These BOTTLE CELLS line the initial archenteron. Thus, as in the gastrulating sea urchin, an invagination of cells initiates archenteron formation. However, unlike the sea urchin, gastrulation begins not at the

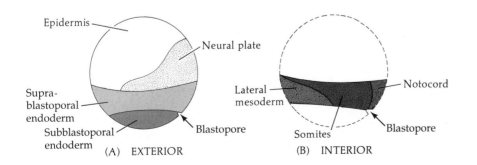

FIGURE 13
Fate map of the embryo of the frog *Xenopus laevis*. Fate map for the exterior (A) and interior (B) cells of the blastula, indicating that most of the mesodermal derivatives are formed from the interior cells. The point at which the dorsal blastopore lip forms is indicated by an arrow. (After Keller, 1976.)

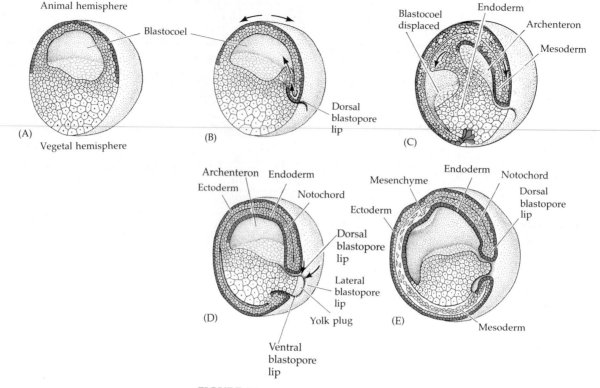

FIGURE 14

Cell movements during frog gastrulation. The sections cut through the middle of the embryo and are positioned so that the vegetal pole is toward the observer and slightly to the left. The major cell movements are indicated by arrows and the superficial animal hemisphere cells are colored so that their movements can be followed. (A) Early gastrulation. Cells move inward to form the dorsal lip of the blastopore, and mesodermal precursors involute under the roof of the blastocoel. (B,C) Midgastrulation. The archenteron forms and displaces the blastocoel, and cells migrate from the lateral and ventral blastopore lips into the embryo. The cells of the animal hemisphere migrate down towards the vegetal region, moving the blastopore to the region near the vegetal pole. Towards the end of gastrulation (D,E) the blastocoel is obliterated, the embryo becomes surrounded by ectoderm, the endoderm has been internalized, and the mesodermal cells have been positioned between the ectoderm and endoderm. (After Keller, 1986.)

most vegetal region, but in the MARGINAL ZONE near the equator of the blastula. Here the endodermal cells are not as large or as yolky as the most vegetal blastomeres.

The next phase of gastrulation involves the involution of the marginal zone cells while the animal cells undergo epiboly and converge at the blastopore. When the migrating marginal cells reach the lip of the blastopore, they turn inward and travel along the inner surface of the outer cell sheets. Thus, the cells constituting the lip of the blastopore are constantly changing. The first cells to compose the dorsal lip are the endodermal cells that invaginated to form the leading edge of the archenteron. These cells later become the pharyngeal cells of the foregut. As these first cells pass into the interior of the embryo, the blastopore lip becomes composed of involuting cells that are the precursors of the head mesoderm. The next cells involuting over the dorsal lip of the blastopore are called the CHORDAMESODERM cells. These cells will form the NOTOCHORD, a transient mesodermal "backbone" that is essential for initiating the differentiation of the nervous system (see Chapters 5 and 8).

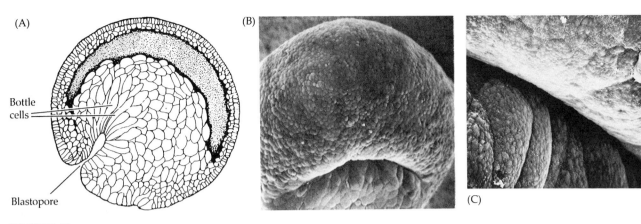

(A)

Bottle
cells

Blastopore

(B)

(C)

FIGURE 15

Structure of the blastopore lip. (A) Diagram of cells seen in a section of gastrulating amphibian embryo, showing the extension of the bottle cells from the blastopore. (B) Surface view of an early dorsal blastopore lip. The size difference between the animal and vegetal blastomeres is readily apparent. (C) Close-up of the region where the animal hemisphere cells are involuting through the blastopore lip (A after Holtfreter, 1943; scanning electron micrographs courtesy of C. Phillips.)

As the new cells enter the embryo, the blastocoel is displaced to the side opposite the dorsal blastopore lip. Meanwhile, the blastopore lip is displaced vegetally and widens as more animal hemisphere cells converge there. The widening blastopore "crescent" develops lateral lips and finally a ventral lip over which additional mesodermal and endodermal precursor cells pass. With the formation of the ventral lip, the blastopore has formed a ring around the large endodermal cells that remain exposed on the surface (Figure 16). This remaining patch of endoderm is called the YOLK PLUG; and it, too, is eventually internalized (Figure 14). At that point, all the endodermal precursors have been brought into the interior of the embryo, the ectoderm has encircled the surface, and the mesoderm has been brought between them.

Positioning of the blastopore

Having seen the general features of amphibian gastrulation, we can now look at each step in detail. Gastrulation does not exist as an independent process in the life of an animal. In fact, the preparation for gastrulation can be traced back to the literal moment of sperm–egg fusion. The unfertilized egg has a polarity along the animal–vegetal axis. The general fate of these regions can be predicted before fertilization. The surface of the animal hemisphere will become the cells of the ectoderm (skin and nerves), the vegetal hemisphere surface will form the cells of the gut and associated organs (endoderm), and the mesodermal cells will form from the internal cytoplasm around the equator. Thus, the germ layers can be mapped onto the unfertilized ovum; but this tells us nothing about which part of the egg will form the belly and which the back. The anterior–posterior (head–tail) and dorsal–ventral (back–front) axes have not yet been determined.

The dorsal–ventral and anterior–posterior axes are specified by displacements of the zygote cytoplasm during fertilization. In Chapter 2, we discussed the rotation of the cortical cytoplasm relative to the internal cytoplasm of the frog egg. The internal cytoplasm remains oriented with respect to gravity because of its dense yolk accumulation, while the cortical cytoplasm actively rotates 30° from its original position. This rotation

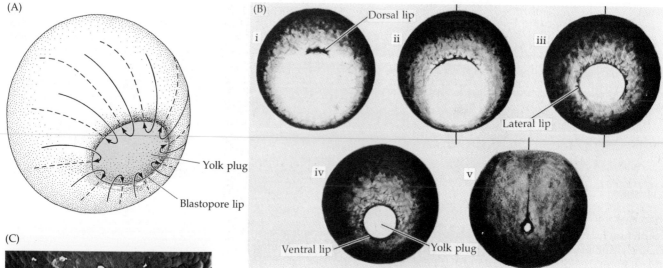

FIGURE 16

Epiboly of the ectoderm. (A) Morphogenetic movements of the cells migrating into the blastopore and then under the surface. (B) Changes in the region around the blastopore as the dorsal, lateral, and ventral lips are formed in succession. When the ventral lip completes the circle, the endoderm becomes progressively internalized. Numbers ii–v correspond to Figures 14B–E, respectively. (C) Scanning electron micrograph of *Xenopus* yolk plug and the epiboly of the ectoderm over it at the blastopore margin. (B from Balinsky, 1975; courtesy of B. I. Balinsky; C courtesy of C. Phillips.)

causes the animal–vegetal axis of the egg surface to be offset 30° relative to the animal–vegetal axis of the internal cytoplasm. In this way, a new symmetrical state is acquired. What had been a radial symmetry has become a bilateral symmetry. The inner cytoplasm moves as well, and fluorescence microscopy of early embryos has shown that the cytoplasmic patterns of presumptive dorsal cells differ from those of the presumptive ventral cells (color portfolio; Figure 17).

That region of the inner vegetal cytoplasm that is newly apposed to animal hemisphere cortical cytoplasm appears to be activated to initiate gastrulation. This usually occurs on the "gray crescent" side of the embryo, 180° opposite the side of sperm entry (Figure 18). The side where the sperm enters marks the future ventral surface of the embryo; the opposite side, where gastrulation is initiated, marks the future dorsum (back) of the embryo (Gerhart et al., 1981, 1986; Vincent et al., 1986).

While the sperm is not needed to induce these movements in the egg cytoplasm, it is important in determining the direction of this rotation. If an artificially stimulated egg is enucleated, the cortical rotation still takes place at the correct time. However, the direction of this movement is unpredictable. (In fact, in dispermic eggs, there is still only one direction of rotation.) The sperm appears to provide a spatial cue that orients the autonomous rotation of the cytoplasm, but it is the cytoplasmic rotation that is essential. If the newly fertilized egg is compressed so that the direction of cortical rotation is different from that predicted by the point of sperm entry, the blastopore lip forms in accordance with the direction of cortical rotation (Black and Vincent, 1988). Moreover, if this cortical rotation is blocked, there is no dorsal development (Vincent and Gerhart, 1987). The direction of cytoplasmic movement determines which side is to be dorsal and which side is to be ventral.

The directional bias provided by the point of sperm entry can be

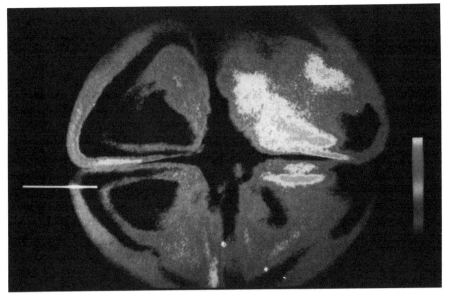

FIGURE 17
Organization of cytoplasm in the 4-cell stage *Xenopus laevis* embryo. The presumptive dorsal cells are at the left side. The oocyte has been injected with a dye that stains yolk platelets. The pattern of yolk platelets (and presumably of other cytoplasmic elements) has a different, layered, configuration on the dorsal side that is not seen on the ventral side. (Photograph courtesy of M. V. Danilchik).

overridden by mechanically redirecting the spatial relationship between the cortical and subcortical cytoplasms. This can be accomplished by rotating the egg in such a way that the internal components settle into different regions (Pflüger, 1883; Penners and Schleip, 1928; Pasteels, 1948; Gerhart et al., 1981). The fertilized egg is normally oriented so that the dense yolk platelets of the vegetal hemisphere point downward in the gravitational field. When the egg is prevented from rotating (by immersing it in a polysaccharide to dehydrate the perivitelline space between the egg and the fertilization envelope), one can turn the egg on its side, 90°, so the animal–vegetal axis is horizontal rather than vertical and the point of sperm entry faces upward (Gerhart et al., 1981; Kirschner and Gerhart, 1981; Cooke, 1986). When fertilized eggs are inclined this way for 30 minutes, starting halfway through the first cleavage cycle, almost all the embryos initiate gastrulation on the same side as sperm entry (Figure 18).

The importance of normal cytoplasmic rotation can be shown by inhibiting this displacement of material and observing the embryos that result. Treatments that cause microtubules to depolymerize (cold shock, high hydrostatic pressure, or immersion in the drug nocodazole) block cytoplasmic rotation (Manes et al., 1978). The resulting eggs cleave, but they neither gastrulate properly nor form a dorsal–ventral axis. Instead, they gastrulate symmetrically into a three-layered cylinder devoid of dorsal components such as the neural tube, mesodermal somites, and central nervous system. These embryos die when their yolk reserves are depleted (Gerhart et al., 1986). However, when eggs were inclined 90° for the 40 minutes prior to first cleavage (mimicking the normal cytoplasmic rotation), the embryos developed a normal body axis, beginning gastrulation at the side that would eventually become the dorsal surface (Sharf and Gerhart, 1983). The rearrangement of cytoplasm during the first cleavage cycle is a necessary prerequisite for normal gastrulation.

As might be expected from the preceding discussion, one might duplicate the effects of gravity by centrifuging the eggs for a brief period of time. Black and Gerhart (1985) immobilized fertilized *Xenopus* eggs in gelatin with the point of sperm entry either away from or pointing toward the center of the centrifuge rotor and then spun (10–50 × *g*) for 4 minutes. If this were done during the first 70 percent of the period before completion of first cleavage, centrifugation-mediated displacement predominated over

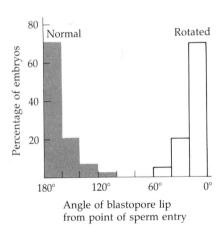

FIGURE 18
Relationship between the point of sperm entry and the dorsal blastopore lip in normal (shaded) and rotated (unshaded) frog eggs. *Xenopus* eggs were fertilized, dejellied, and placed in Ficoll to dehydrate the perivitelline space. The sperm entry point was marked with dye. Rotated eggs were inclined 90° with the sperm entry point facing upward, 50–80 minutes after fertilization. (After Gerhart et al., 1981.)

sperm-mediated displacement of cytoplasmic components. After 70 percent of the first cell cycle was completed, the cytoplasm became more rigid and resisted this displacement. Moreover, when eggs were centrifuged at $30 \times g$ for 4 minutes during the period of time 40–50 percent into the first cell cycle, two sites of gastrulation emerged, leading to conjoined twin larvae (Figure 19). Black and Gerhart hypothesize (1986) that such twinning is caused by the formation of two areas of interaction: one axis forms where the centrifugation-driven cytoplasm interacts with overlying components; the other forms after the cytoplasm, driven by gravity, returns to its original position, where it can react a second time. Twins can also be produced at normal gravity by placing the sperm entry side of the egg uppermost after removing the egg from its fertilization envelope (Gerhart et al., 1981). The ability of gravity and centrifugation to produce two axes in otherwise normal eggs and to "rescue" axis formation in eggs treated to inhibit normal cytoplasmic movements strongly suggests that the internal displacements caused by these procedures substitute for a process that is normally involved in specifying the gastrulation initiation site.

The ability to obtain two functional blastopore lips also suggests that there is nothing unique about the cortical region of the gray crescent (where gastrulation is usually initiated). The gastrulation-inducing factors appear to be created by the interactions of animal and vegetal cytoplasm, interactions that probably activate some component in the deep (as opposed to cortical) vegetal cytoplasm.

Gimlich and Gerhart (1984) performed a series of transplantation experiments that confirmed the hypothesis that the factor(s) that initiate gastrulation originally lie in the deep cytoplasm rather than in the gray crescent cortex. They demonstrated that the three most dorsal vegetal blastomeres of 64-cell *Xenopus* embryos are able to induce the formation of the dorsal lip of the blastopore and of a complete dorsal axis in UV-irradiated recipients (which otherwise would have failed to properly initiate gastrulation; Figure 20A). Moreover, these three blastomeres, which underlie the prospective dorsal lip region, can also induce a secondary invagination and axis when transplanted into the ventral side of a normal, unirradiated 64-cell embryo (Figure 20B). Thus, this small cluster of vegetal blastomeres enable their adjacent marginal cells to invaginate and form the dorsal mesodermal axis of the embryo.

It appears, then, that internal rearrangements of the cytoplasm, probably those normally oriented by the entry of the sperm, are responsible

FIGURE 19
Twin blastopores produced by rotating dejellied *Xenopus* eggs ventral side up at the time of first cleavage. (A) Two blastopores are instructed to form: the original one (opposite the point of sperm entry) and the new one created by the displacement of cytoplasmic material. (B) These eggs develop two complete axes, which form twin tadpoles, joined ventrally. (Photographs courtesy of J. Gerhart.)

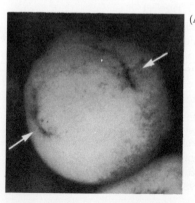

(A)

(B)

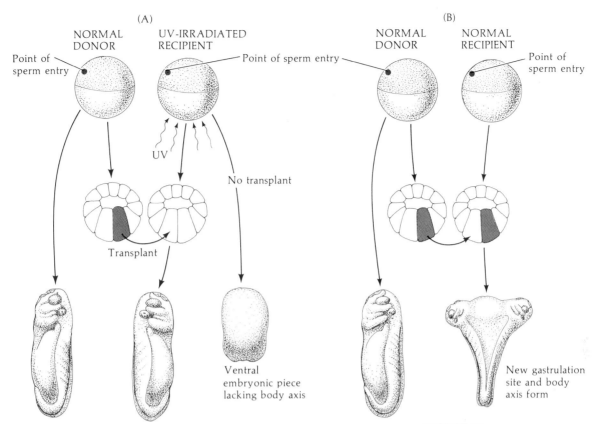

NORMAL DONOR

Point of sperm entry

UV-IRRADIATED RECIPIENT

Point of sperm entry

(B)

NORMAL DONOR

NORMAL RECIPIENT

Point of sperm entry

UV

No transplant

Transplant

Ventral embryonic piece lacking body axis

New gastrulation site and body axis form

FIGURE 20

Transplantation experiments demonstrating that the vegetal cells underlying the prospective dorsal blastopore lip regions are responsible for causing the initiation of gastrulation. (A) Rescue of irradiated embryos by transplanting the vegetal blastomeres of the most dorsal segment (color) of a 64-cell embryo into a cavity made by the removal of a similar number of vegetal cells. An irradiated zygote without this transplant fails to undergo normal gastrulation. (B) Formation of new gastrulation site and body axis by the transplantation of the most dorsal vegetal cells of a 64-cell embryo into the ventralmost vegetal region of another 64-cell embryo. (After Gimlich and Gerhart, 1984.)

for causing the asymmetric distribution of subcellular factors. This asymmetry creates in the egg a dorsal–ventral distinction that ultimately directs the position of the blastopore above a set of vegetal blastomeres opposite the point of sperm entry.

Cell movements and construction of the archenteron

Gastrulation is initiated when a group of marginal endoderm cells on the dorsal surface of the blastula sinks into the embryo. This inward movement of cells is accomplished by characteristic changes in cell–cell attachment not unlike those seen in the vegetal plate of sea urchin embryos. The inner borders of these cells lose their intercellular connections while the other border regions remain tightly bound to their neighbors. The outer portions of these cells then elongate in a direction perpendicular to the gastrula surface.

Because these cells remain in the deepest portion of the blastopore groove and appear to lead the small archenteron into the embryo, it has long been thought that the bottle cells have a major role in forming and extending the archenteron. Rhumbler (1902) suggested that the bottle cells actively migrated into the center of the gastrula as the result of lower surface tension there. They would then pull the neighboring cells into the groove as well, thus initiating the cell movement into the embryo.

In 1943–1944, the outstanding embryologist Johannes Holtfreter produced evidence that this was indeed the case. First Holtfreter isolated the early blastopore lip cells and cultured them on glass. The rounded ends of the cells attached to the glass led the movement of the remainder of the cells. Even more convincing were Holtfreter's recombination experiments in which the marginal zone cells (which would give rise to the

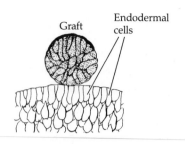

Graft   Endodermal cells

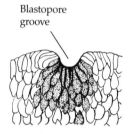

Blastopore groove

FIGURE 21

A graft of amphibian cells from the dorsal blastopore lip region sinks into a layer of endodermal cells and forms a blastopore groove. (After Holtfreter, 1944.)

dorsal blastopore lip) were combined with inner endoderm tissue. When the dorsal marginal zone cells were excised and placed on inner endoderm tissue, the blastopore cell precursors formed bottle cells and sank below the surface of the inner endoderm (Figure 21). Moreover, as they sank, they created a depression reminiscent of the early blastopore. Thus, Holtfreter claimed that the ability to invaginate into the inner endoderm was an innate property of the dorsal marginal zone cells.

More recently R. E. Keller has shown that, although the bottle cells of *Xenopus* have an intrinsic ability to invaginate into endoderm, bottle cells may play a more passive role than previously thought. Keller (1981) found that once the bottle cells are formed, they have little to do with extending the archenteron; partial or complete removal of these cells does not prevent the involution of adjacent cells into the blastopore. Gastrulation can occur without these cells. The major factor in the movement of cells into the embryo appears to be the involution of the *subsurface* marginal cells rather than the superficial ones. It appears that these subsurface, DEEP INVOLUTING MARGINAL ZONE CELLS turn inward and migrate toward the animal pole along the inside surfaces of the remaining deep cells (Figure 22) and that the superficial layer forms the lining of the archenteron merely because it is attached to the actively migrating deep cells. Even though the bottle cells do aid in forming the initial blastopore groove, this movement is dependent on their attachment to the underlying deep cells. While removal of the *bottle cells* does not affect the involution of the deep or superficial marginal zone cells into the embryo, the removal of the dorsal marginal zone *deep cells* and their replacement with animal region cells (which do not normally undergo involution) stops archenteron formation.

The peculiar bottle shape of these cells is a compromise between the intrinsic tendency of these cells to constrict apically in all directions and resistance to such constriction by the vegetal cells. The constriction of the prospective bottle cells has major consequences for further cell movements. First, the marginal zone is pulled vegetally while the vegetal cells are pushed inwards. This movement enables the ectoderm to expand vegetally and encircle the embryo. Second, the anterior mesodermal precursor cells are pushed upward toward the animal pole, where they can contact the underside of the posterior mesodermal precursors. This event positions the precursor cells near the roof of the blastocoel (Hardin and Keller, 1988).

Figure 22 attempts to integrate several studies (Keller, 1980, 1981; Keller and Schoenwolf, 1977; Hardin and Keller, 1988) into a model of *Xenopus* archenteron formation. Prior to gastrulation (Figure 22A), the prospective anterior mesoderm and posterior mesoderm comprise the deep involuting marginal zone. Gastrulation begins when the bottle cells undergo apical constriction (Figure 22B). This constriction pulls the marginal zone vegetally, pushes the anterior mesoderm toward the animal pole, and rotates the involuting marginal zone (IMZ) outward. The result (Figure 22C) is a reorientation of the vegetal end of the marginal zone (anterior mesoderm) so that it is leading the movement of cells into the blastocoel. Shortly afterward (Figure 22D), the several layers of deep IMZ cells intercalate to form one thin, deep layer. This intercalation further extends the dorsal lip vegetally. At the same time, the superficial cells spread out by dividing and flattening. Upon reaching the blastopore lip, the deep cells again change their morphology and form a mesodermal stream that migrates toward the animal pole along the inner surface of the remaining deep marginal zone cells. During the later stages of gastrulation (Figure 22E), the columnar deep cells flatten and spread out as the marginal zone continues to move ventrally (forming the enlarged blastopore about the yolk plug). The mesodermal stream continues to migrate

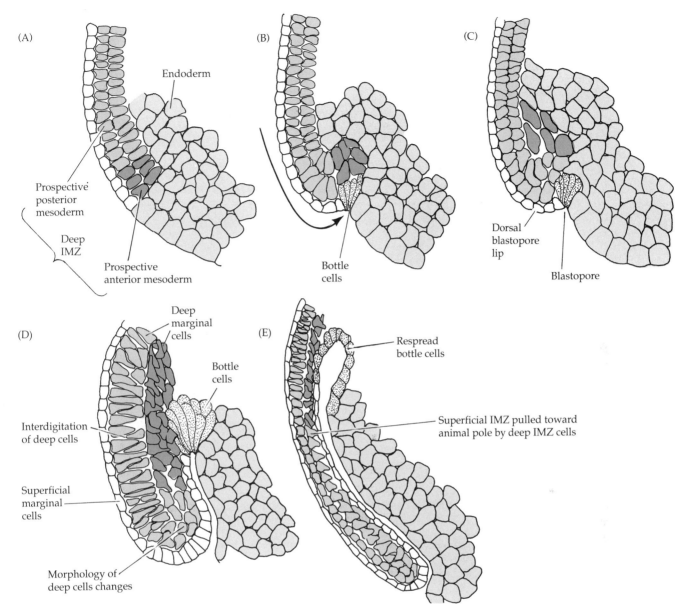

FIGURE 22
Integrative model of cell movements during early *Xenopus* gastrulation. (A) Structure of the involuting marginal zone (IMZ) prior to gastrulation. The deep IMZ consists of the prospective anterior mesoderm and the prospective posterior mesoderm. (B) Constriction of the bottle cells pushes up the prospective anterior mesoderm and rotates the IMZ outward. (C) The anterior mesodermal precursors leads the movement of the mesoderm into the blastocoel. (D) Interdigitation of the deep IMZ cells occurs. The mesoderm moves towards the animal pole, pulling along the superficial cells and the bottle cells by involution. (E) As gastrulation continues, the deep marginal cells flatten and the formerly superficial cells form the wall of the archenteron. (After Hardin and Keller, 1988.)

inward and the overlying layer of superficial cells (including the bottle cells) is passively pulled toward the animal pole, thereby forming the endodermal roof of the archenteron. Thus, although the bottle cells may be responsible for creating the initial groove, the motivating force for this involution appears to come from the deep layer of the marginal cells. Furthermore, this deep layer of cells appears to be responsible for the continued migration of cells into the embryo.

Migration of the involuting mesoderm

The migration of mesodermal precursor cells inside the embryo is part of a remarkably well coordinated series of global cellular migrations. A detailed summary of these movements can be seen in Figure 23. One of the most important aspects of amphibian gastrulation is the narrowing (*con-*

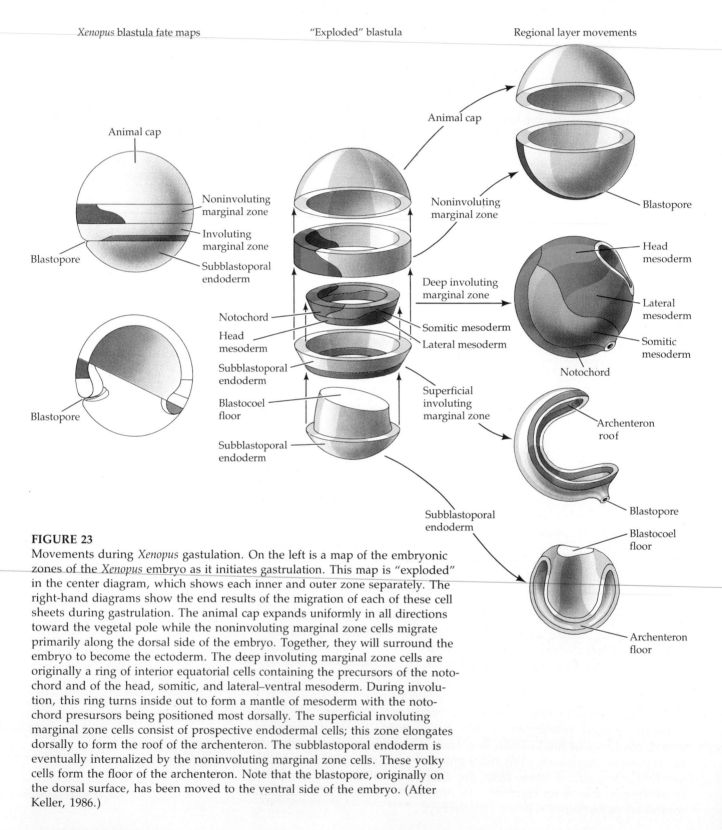

Xenopus blastula fate maps

Animal cap

Noninvoluting marginal zone

Involuting marginal zone

Blastopore

Subblastoporal endoderm

Blastopore

"Exploded" blastula

Notochord

Head mesoderm

Subblastoporal endoderm

Blastocoel floor

Subblastoporal endoderm

Animal cap

Noninvoluting marginal zone

Deep involuting marginal zone

Somitic mesoderm

Lateral mesoderm

Superficial involuting marginal zone

Subblastoporal endoderm

Regional layer movements

Blastopore

Head mesoderm

Lateral mesoderm

Somitic mesoderm

Notochord

Archenteron roof

Blastopore

Blastocoel floor

Archenteron floor

FIGURE 23

Movements during *Xenopus* gastulation. On the left is a map of the embryonic zones of the *Xenopus* embryo as it initiates gastrulation. This map is "exploded" in the center diagram, which shows each inner and outer zone separately. The right-hand diagrams show the end results of the migration of each of these cell sheets during gastrulation. The animal cap expands uniformly in all directions toward the vegetal pole while the noninvoluting marginal zone cells migrate primarily along the dorsal side of the embryo. Together, they will surround the embryo to become the ectoderm. The deep involuting marginal zone cells are originally a ring of interior equatorial cells containing the precursors of the notochord and of the head, somitic, and lateral–ventral mesoderm. During involution, this ring turns inside out to form a mantle of mesoderm with the notochord presursors being positioned most dorsally. The superficial involuting marginal zone cells consist of prospective endodermal cells; this zone elongates dorsally to form the roof of the archenteron. The subblastoporal endoderm is eventually internalized by the noninvoluting marginal zone cells. These yolky cells form the floor of the archenteron. Note that the blastopore, originally on the dorsal surface, has been moved to the ventral side of the embryo. (After Keller, 1986.)

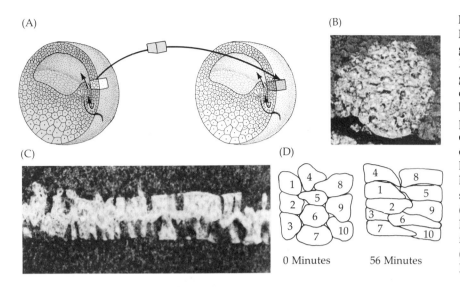

(A)

(B)

(C)

(D)

0 Minutes   56 Minutes

FIGURE 24

Rearrangement of cells during convergent extension of the mesoderm in *Xenopus* embryos. (A) The dorsal region of the IMZ$_D$ (which forms the notochord) was taken from an embryo labeled with fluorescienated dextran particles and placed into an unlabeled embryo. (B) The labeled cells (white, circular area) remained together after healing. (C) When these grafted cells helped form the notochord, they were seen to be interspersed with host cells. (D) Tracings of individual cells followed with video recorder during the formation of the notochord in vitro. (After Keller et al., 1985; Keller, 1986. Photographs courtesy of R. E. Keller.)

vergence) and lengthening (*extension*) of the INVOLUTING MARGINAL ZONE (IMZ), which is that region of cells residing immediately above the blastopore lip. This zone contains the prospective endodermal roof of the archenteron in its superficial layer (IMZ$_S$) and the prospective mesodermal cells (including those of the notochord) in its deep region (IMZ$_D$).

During gastrulation, the ANIMAL CAP and NONINVOLUTING MARGINAL ZONE CELLS expand by epiboly to cover the entire embryo. The dorsal portion of the noninvoluting marginal cells expands more rapidly than the ventral portion, thus causing the blastopore lips to move toward the ventral side. Cells from these two zones will form the ectoderm of the embryo. The deep involuting marginal zone is a ring of cells that gives rise to the notochord, head mesoderm, somitic mesoderm, and lateral–ventral mesoderm. During gastrulation, this ring involutes around the blastopore lips to form the mesodermal mantle. The head and mesodermal precursors spread across the embryo as the notochord and somitic mesodermal regions converge and extend dorsally to form the dorsal axial structures. The superficial involuting marginal zone cells also undergo CONVERGENT EXTENSION to form the endodermal tissues of the archenteron roof (Figures 23, 24). The subblastoporal endoderm is covered by the epibolizing noninvoluting marginal zone cells and forms the archenteron floor (Keller, 1986).

During the convergent extension of the deep IMZ, the cells rearrange themselves, intercalating with nearby cells to produce a thin, elongated structure from the originally broad, small band of cells (Figure 24). We have already discussed a similar phenomenon involving archenteron extension during sea urchin gastrulation. Many rows of cells have intercalated to form fewer, but longer, rows of cells. Convergent extension of the mesoderm appears to be autonomous, because the movements of cells occur even if the cells are removed from the rest of the embryo (Keller, 1986).

How are the involuting cells informed where to go once they enter the inside of the embryo? Is there a molecule on the blastocoel roof that facilitates migration? It appears that the involuting mesodermal precursors migrate toward the animal pole upon a fibronectin lattice secreted by the cells of the blastocoel roof. Shortly before gastrulation in amphibians, the presumptive ectoderm of the blastocoel roof secretes an extracellular matrix that contains fibrils of fibronectin (Figure 25; Boucaut et al., 1984; Nakatsuji et al., 1985). Like the invaginating mesenchyme of sea urchins,

FIGURE 25

Fibronectin fibrils in a salamander gastrula, visualized by fluorescent antibodies to fibronectin. Immunofluorescence reveals a fibrillar network of fibronectin on the basal surface of the ectodermal cells lining the blastocoel roof. (From Boucaut et al., 1985; photograph courtesy of J.-C. Boucaut.)

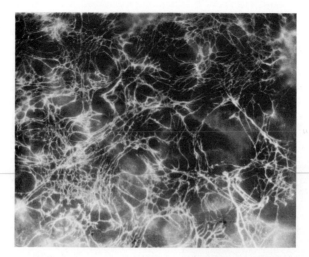

the involuting mesoderm appears to travel on these fibronectin fibers. Confirmation of this was obtained by chemically synthesizing a "phony" fibronectin that could compete with the genuine fibronectin of the extracellular matrix. Cells bind to a certain region of the fibronectin protein that contains a three-amino acid sequence (Arg-Gly-Asp). Boucaut and co-workers injected large amounts of the decapeptide containing this sequence into the blastocoels of salamander embryos shortly before gastrulation began. If fibronectin were essential for cell migration, then cells

FIGURE 26

Scanning electron micrographs of normal salamander gastrulation (A,B) and of abnormal gastrulation wherein the blastocoel was injected with the cell-binding fragment of fibronectin (C,D). A and B were injected with a control solution. A is a section during midgastrulation while B shows the yolk plug towards the end of gastrulation. C and D are the finishing stages of the arrested gastrulation where the mesodermal precursors, having bound the synthetic fibronectin, cannot recognize the normal fibronectin-lined migration route. The archenteron fails to form and the noninvoluted mesodermal precursors remain on the surface. ar, archenteron; bc, blastocoel; bl, blastopore; ec, ectoderm; en, endoderm; mes, mesoderm; yp, yolk plug. (From Boucaut et al., 1984; photographs courtesy of J. P. Thiery.)

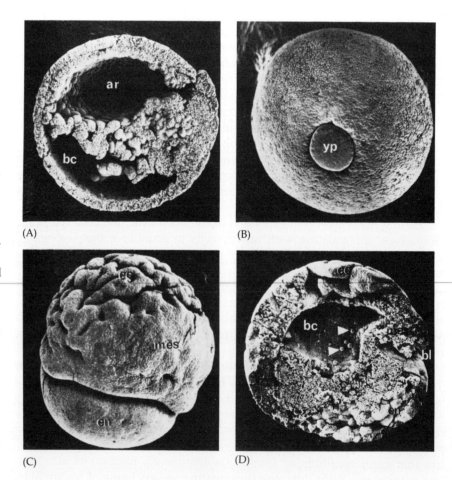

binding this soluble peptide fragment instead of the real cell-bound fibronectin should stop. Unable to find their "road," the mesodermal cells should cease their involution. This is precisely what occurred (Figure 26). No migrating cells were seen along the underside of the ectoderm. Instead, the mesodermal precursors remained outside the embryo, forming a convoluted cell mass. Other synthetic decapeptides (including other fragments of the fibronectin molecule) did not impede migration.

This fibronectin-containing extracellular matrix also appears to provide the cues for the *direction* of cell migration. Shi and colleagues (1989) excised the blastocoel roofs from early salamander gastrulae and deposited them on plastic dishes with their extracellular matrices touching the plastic (Figure 27A). The axis from the blastopore to the animal pole was marked, and after two hours the explant was removed, leaving behind its extracellular matrix. A smaller explant from the dorsal marginal zone (DMZ) is then taken from another early gastrula and is placed on the matrix with its own blastopore–animal pole axis perpendicular to that of the matrix. Do the cells of this explant migrate on the matrix and do they migrate in a particular direction? The cells are found to migrate, and this migration can be inhibited by antibodies that block the cells' ability to recognize fibronectin. Moreover, rather than migrating randomly, the DMZ cells migrate to the animal pole of the extracellular matrix that had been absorbed onto the plastic (Figure 27B).

Similar fibronectin-containing fibrils have been seen in anuran amphibian gastrulae, and Delarue and colleagues (1985) have shown that certain inviable hybrids between two species of toads die during gastrulation because they do not secrete these fibronectin fibrils. It appears, then, that the extracellular matrix of the blastocoel roof, and particularly its fibronectin component, is important in the migration of the mesodermal cells during amphibian gastrulation.

Epiboly of the ectoderm

While involution is occurring at the blastopore lips, the ectodermal precursors are expanding over the entire embryo. Keller (1980) and Keller and Schoenwolf (1977) have used scanning electron microscopy to observe the changes in both the superficial cells and the deep cells of the animal and marginal regions. The major mechanism of epiboly in *Xenopus* gastrulation appears to be an increase in cell number (through division) coupled with a concurrent integration of several deep layers into one (Figure 28). During early gastrulation, three rounds of cell division increase the volume of the deep cells in the animal hemisphere by 45 percent. At the same time, complete integration of the numerous deep cells into one layer occurs. The most superficial layer expands by cell division and flattening. The spreading of cells in the dorsal and ventral marginal zones appears

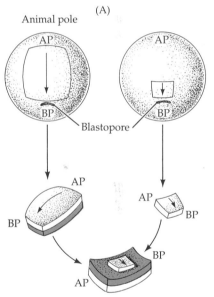

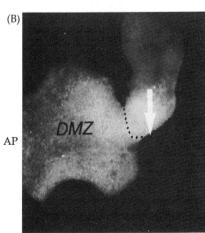

FIGURE 27
Direction of DMZ (dorsal marginal zone) cell migration depends on orientation of the extracellular matrix of the blastocoel roof. (A) Explants of blastocoel roof from blastopore to animal pole were dissected from early gastrula stage salamander embryos and placed on plastic dishes. The extracellular matrix adhered to the dish and the tissue was then removed. A smaller explant from an early gastrula, containing DMZ cells, was placed on this matrix with its own axis perpendicular to that of the matrix. (B) DMZ cells from the explant migrate towards the animal pole of the matrix. Dotted line indicates original boundary of the explant, and the white arrow represents its blastopore–animal pole axis. (From Shi et al., 1989; courtesy of the authors.)

FIGURE 28

Scanning electron micrographs of the *Xenopus* blastocoel roof showing the changes in cell shape and arrangement. Stages 8 and 9 are blastulae; stages 10 through 11.5 represent progressively later gastrulae. (From Keller, 1980; photographs courtesy of R. E. Keller.)

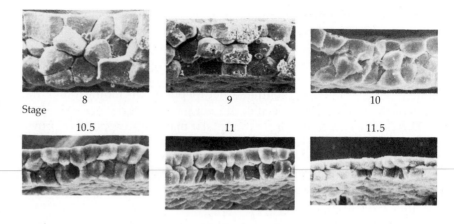

Stage   8   9   10

10.5   11   11.5

to proceed by the same mechanism, although changes in cell shape appear to play a greater role than in the animal region. The result of these expansions is the epiboly of the superficial and deep cells of the animal and marginal regions over the surface of the embryo (Keller and Danilchik, 1988). Most of the marginal region cells, as previously mentioned, involute to join the mesodermal cell stream within the embryo.

Gastrulation in *Xenopus* is the orchestration of several distinct events. The first indication of gastrulation involves the local *invagination* of endodermal bottle cells in the marginal zone at a precisely defined time and place. Next, the *involution* of marginal cells through the blastopore lip begins the formation of the archenteron. These involuting cells *migrate* along the inner surface of the blastopore roof, and the prospective chordamesoderm narrows and lengthens (*convergent extension*) in the dorsal portion of the embryo. At the same time, the ectodermal precursor cells *epibolize* vegetally by cell division and by the integration of previously independent cell layers. The result of these cell movements is the proper positioning of the three germ layers in preparation for their differentiation into the body organs. We still do not know how the cells are told to initiate and end their movements. Despite our new awareness, Keller (1986) warns: "No morphogenetic process—especially amphibian gastrulation—is satisfactorily understood at the cell or cell population level."

Gastrulation in birds

Overview of avian gastrulation

Cleavage in avian embryos creates a blastodisc above an enormous volume of yolk. This inert, underlying yolk mass imposes severe constraints on cell movements, and avian gastrulation appears at first glance to be very different from that of a sea urchin or a frog. We shall soon see, though, that there are numerous similarities between avian gastrulation and those gastrulations we have already studied. Moreover, we shall see that mammalian embryos—which do not have yolk—retain gastrulation movements very similar to those of bird and reptile embryos.

The central cells of the avian blastodisc are separated from the yolk by a subgerminal cavity and appear to be clear—hence, the center of the blastodisc is called the area pellucida. In contrast, the cells at the margin of the area pellucida appear opaque because of their contact with the yolk. They form the area opaca (Figure 29). Whereas most of the cells remain

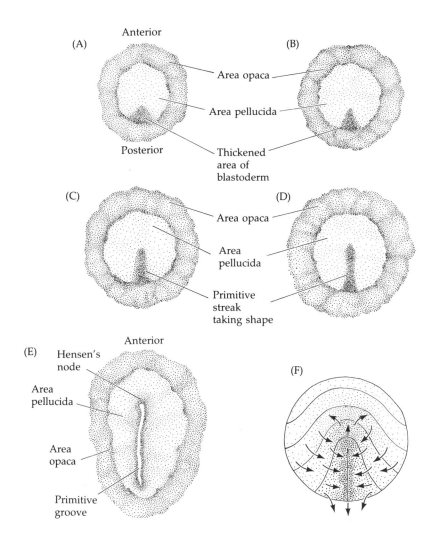

(A) Anterior
Area opaca
Area pellucida
Posterior
Thickened area of blastoderm

(B)

(C)
Area opaca
Area pellucida
Primitive streak taking shape

(D)

(E) Anterior
Hensen's node
Area pellucida
Area opaca
Primitive groove

(F)

FIGURE 29
Cell movements forming the primitive streak of the chick embryo. Dorsal view of chick blastoderm at (A) 3–4 hours, (B) 5–6 hours, (C) 7–8 hours, (D) 10–12 hours, (E) 15–16 hours. A summary of these cell movements is shown in (F). (Adapted from several sources, especially Spratt, 1946.)

at the surface, forming the EPIBLAST, certain cells migrate individually into the subgerminal cavity to form a layer called the PRIMARY HYPOBLAST (Figure 30). Shortly thereafter, a sheet of cells from the posterior margin of the blastoderm (Koller's crescent) migrate anteriorly to join the primary hypoblast, thereby forming the SECONDARY HYPOBLAST. Stern (1990) has labeled the cells of the posterior marginal zone with fluorescent dyes and has followed their movements with videomicroscopy. This enabled him to discern at least two types of cells in this posterior marginal region. The deep cells of this region form the motile cells of the secondary hypoblast, whereas the superficial cells of the region remain on the surface and

FIGURE 30
Hypoblast formation in the chick embryo. The resulting structure is a hollow blastula not unlike that of a sea urchin or a frog. (After Carlson, 1988.)

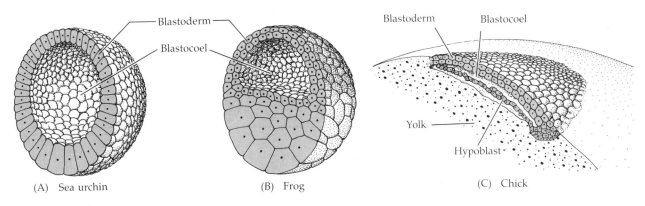

Blastoderm
Blastocoel

(A) Sea urchin

(B) Frog

Blastoderm   Blastocoel
Yolk
Hypoblast
(C) Chick

contribute to the epiblast. The two-layered blastoderm (epiblast and hypoblast) is joined together at the margin of the area opaca, and the space between the layers is a blastocoel. Thus, the structure of the avian blastodisc is not dissimilar to that of the amphibian or echinoderm blastula.

The fate map for the avian embryo is restricted to the epiblast. That is to say, the hypoblast does not contribute any cells to the developing embryo. Rather, the hypoblast cells form portions of the external membranes that nourish and protect the embryo. All three germ layers of the embryo proper (plus a considerable amount of extraembryonic membrane) are formed from the epiblastic cells. The fate map of the chick epiblast prior to gastrulation is shown in Figure 31. The presumptive notochordal cells are centrally located between the anterior presumptive ectoderm and the posterior presumptive endoderm.

The major structure characteristic of avian, reptilian, and mammalian gastrulation is the PRIMITIVE STREAK. This streak is first visible as a thickening of the cell sheet at the central posterior end of the area pellucida (Figure 29). This thickening is caused by the migration of cells from the lateral region of the posterior epiblast toward the center. As the thickening narrows, it moves anteriorly and constricts to form the definitive primitive streak. This streak extends 60–75 percent of the length of the area pellucida and marks the anterior–posterior axis of the embryo. As the cells converge to form the primitive streak, a depression forms within the streak. This depression is called the PRIMITIVE GROOVE, and it serves as a blastopore through which the migrating cells pass into the blastocoel. Thus, the primitive streak is analogous to the amphibian blastopore. At the anterior end of the primitive streak is a regional thickening of cells called the PRIMITIVE KNOT, or HENSEN'S NODE. The center of this node contains a funnel-shaped depression through which cells can pass into the blastocoel. Hensen's node is the functional equivalent of the dorsal lip of the amphibian blastopore.

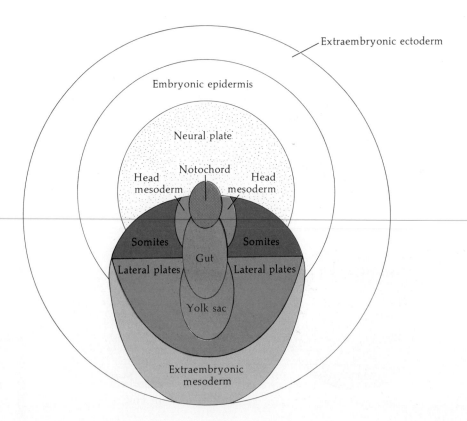

FIGURE 31
Fate map of an avian embryo immediately prior to gastrulation. The primitive streak has not formed yet, but it will eventually extend to the region of the notochord. Presumptive ectoderm is white, presumptive mesoderm is in color, and presumptive endoderm is gray. (After Balinsky, 1975.)

As soon as the primitive streak is formed, blastoderm cells begin to migrate over the lips of the primitive streak and into the blastocoel (Figure 32). Like the amphibian blastopore, the primitive streak has a continually changing cell population. Those cells migrating through Hensen's node pass down into the blastocoel and migrate anteriorly, forming head mesoderm and notochord; those cells passing through the lateral portions of the primitive streak give rise to the majority of endodermal and mesodermal tissues. Unlike the *Xenopus* mesoderm, which migrates as sheets of cells into the blastocoel, cells entering the inside of the avian embryo do so as individuals. Rather than forming a tightly organized sheet of cells, the ingressing population creates a loosely connected MESENCHYME. Moreover, there is no true archenteron formed in the avian gastrula.

As the cells enter the primitive streak, the streak elongates toward the future head region. At the same time, the secondary hypoblastic cells are continuing to migrate anteriorly from the posterior margin of the blastoderm. The elongation of the primitive streak appears to be coextensive with the anterior migration of these secondary hypoblast cells.

The first cells to migrate through the primitive streak are those des-

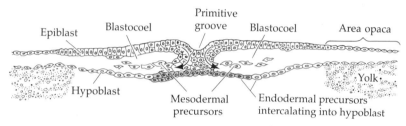

(A)   TRANSVERSE SECTION

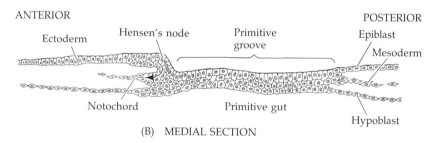

(B)   MEDIAL SECTION

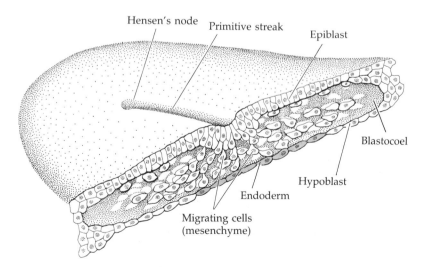

(C)   THREE-DIMENSIONAL VIEW

FIGURE 32
Migration of endodermal and mesodermal cells through the primitive streak. (A) Transverse section through a 17-hour embryo, illustrating the lateral movement of endodermal and mesodermal cells passing into the blastocoel. (B) Medial section through the same embryo, showing that those cells migrating through Hensen's node condense to form the notochord (head process). (C) Stereogram of a gastrulating chick embryo, showing the relationship of the primitive streak, the migrating cells, and the two original layers of the blastoderm. (A and B after Carlson, 1981; C after Balinsky, 1975, and Duband and Thiery, 1982.)

tined to become the foregut. This situation is again similar to that seen in amphibians. Once inside the blastocoel, these cells migrate anteriorly and eventually displace the hypoblast cells in the anterior portion of the embryo. The next cells entering the blastocoel through Hensen's node also move anteriorly, but they do not move as far ventrally as the presumptive endodermal cells. These cells remain between the endoderm and the epiblast to form the head mesoderm and the chordamesoderm (notochordal) cells. These early-ingressing cells have all moved anteriorly, pushing up the anterior midline region of the epiblast to form the HEAD PROCESS. Meanwhile, cells continue migrating inward through the primitive streak. As they enter the blastocoel, these cells separate into two streams. One stream moves deeper and joins the hypoblast along its midline, displacing the hypoblast cells to the sides. These deep-moving cells give rise to all the endodermal organs of the embryo as well as to most of the extraembryonic membranes (the hypoblast forms the rest). The second migrating stream spreads throughout the blastocoel as a loose sheet, roughly midway between the hypoblast and the epiblast. This sheet generates the mesodermal portions of the embryo and extraembryonic membranes. By 22 hours of incubation, most of the presumptive endodermal cells are in the interior of the embryo, although presumptive mesodermal cells continue to migrate inward for a longer time.

Now a second phase of gastrulation begins. While the mesodermal ingression continues, the primitive streak starts to regress, moving Hensen's node from near the center of the area pellucida to a more posterior position (Figure 33). It leaves in its wake the dorsal axis of the embryo and the head process. As the node moves further posteriorly, the remaining (posterior) portion of the notochord is laid down. Finally, the node regresses to its most posterior position, eventually forming the anal region in true deuterostome fashion. By this point, the epiblast is composed entirely of presumptive ectodermal cells.

As a consequence of this two-step gastrulation process, avian (and mammalian) embryos exhibit a distinct anterior-to-posterior gradient of developmental maturity. While cells of the posterior portions of the embryo are undergoing gastrulation, cells at the anterior end are already starting to form organs. For the next several days, the anterior end of the embryo is seen to be more advanced in its development than the posterior end.

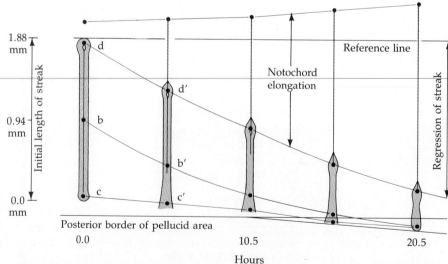

FIGURE 33
Regression of the primitive streak, leaving the notochord in its wake. Various points of the streak were followed after it achieved its maximum length. Time represents hours after achieving maximum length. (After Spratt, 1947.)

While the presumptive mesodermal and endodermal cells were moving inward, the ectodermal precursors were surrounding the yolk by epiboly. The enclosure of the yolk by the ectoderm (again reminiscent of the epiboly of amphibian ectoderm) is a Herculean task that takes the greater part of 4 days to complete and involves the continuous production of new cellular material at the expense of the yolk and the migration of the presumptive ectodermal cells along the underside of the vitelline envelope. Thus, as avian gastrulation draws to a close, the ectoderm has surrounded the yolk, the endoderm has replaced the hypoblast, and the mesoderm has positioned itself between these two regions.

Mechanisms of avian gastrulation

Role of the hypoblast. Although the hypoblast does not contribute any cells to the adult chick (Rosenquist, 1972), it is essential for its proper development. In 1932 C. H. Waddington demonstrated that this lower layer influenced the orientation of the chick embryonic axis. He separated the two cellular layers of the chick blastodisc soon after the primitive streak had formed. After bringing the two layers together again so that their original long axes were now nearly perpendicular to each other, he saw that the primitive streak bent according to the long axis of the hypoblast (Figure 34). Azar and Eyal-Giladi (1979, 1981) have extended these observations to show that the hypoblast *induces* the formation of the primitive streak. First, removal of the hypoblast stops all further development until the remaining epiblast can regenerate a new lower layer. Second, when the population of hypoblast cells derived from the posterior margin is eliminated, no primitive streak is formed. Rather, the area pellucida develops into a mass of disorganized, mesoderm-like tissue. Thus, it appears that the secondary hypoblast directs the formation and the directionality of the primitive streak.

The blastoderm of the chick embryo acts as a single integrating system to form a single embryo; for if the blastoderm is separated into parts, each having its own marginal zone, each part will form its own embryo (Spratt and Haas, 1960). Control of this field seems to reside in the posterior part of the margin whose cells contribute to the hypoblast. These posterior marginal cells not only contribute the inducing cells of the hypoblast, but also prevent other regions of the margin from inducing their own hypoblasts. Khaner and Eyal-Giladi (1989) found that if the posterior margin region is transposed to a lateral margin area (Figure 35A), the posterior tear heals and two primitive streaks emerge. Similarly, if a posterior region is reciprocally transposed with a lateral region (Figure 35B), only one primitive streak forms, and it arises from the original posterior region. However, if a posterior margin region is placed into an embryo that retains its original posterior margin (Figure 35C), only the host's original posterior margin forms the hypoblast that underlies the primitive streak. Khaner and Eyal-Giladi suggest that the marginal zone cells form a gradient of activity whose peak is at the posterior end. These posterior cells will form the hypoblast and at the same time prevent any cells with less activity from forming hypoblasts of their own.

FIGURE 34
Rotation of embryonic axis of chick embryo, caused by rotating the hypoblast with respect to the developing primitive streak at 20–24 hours incubation. The primitive streak has regressed significantly at this stage. (From Azar and Eyal-Giladi, 1979; photograph courtesy of H. Eyal-Giladi.)

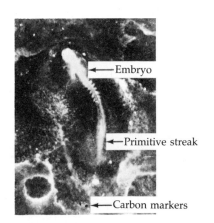

Embryo

Primitive streak

Carbon markers

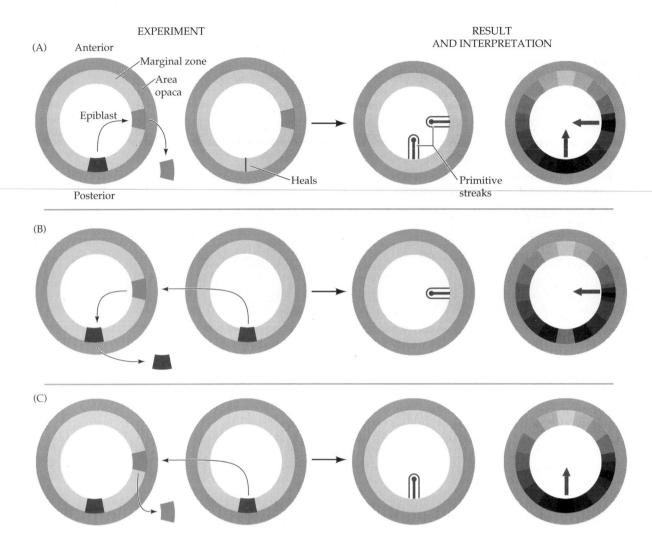

(A) Anterior

Marginal zone

Area opaca

Epiblast

Heals

Primitive streaks

Posterior

(B)

(C)

FIGURE 35

Experiments of Khaner and Eyal-Giladi demonstrating that the posterior part of the marginal zone contributes to the primitive streak-inducing cells of the hypoblast and prevents other marginal regions from creating their own hypoblasts. (After Khaner and Eyal-Giladi, 1989.)

Cell movement within the blastocoel. New evidence from the studies of Stern and Canning (1990) suggests that the epiblast is not the homogeneous, undifferentiated tissue that we have long assumed it to be. Rather, there appears to be differentiation in the epiblast cells even before primitive streak formation begins. These studies show that certain cells that are randomly scattered throughout the epiblast can be distinguished by a particular molecule (a sulfated form of glucuronic acid) on their cell surfaces. Only these cells travel towards the primitive streak to form the mesodermal and endodermal precursors, while the cells lacking this cell surface compound remain in the epiblast. Thus, movement toward the blastopore (primitive streak) is done by individual cells, not by the movement of an epithelial sheet. However, as in amphibian gastrulation, avian cells passing through the blastopore constrict their apical ends to become bottle cells (Figure 36). Once these cells are released from the primitive streak and enter the blastocoel, they flatten and enter a stream of independent, migrating cells, each advancing by the extension and contraction of filopodia. Extracellular polysaccharides may play an important role in this migration. One such complex polysaccharide is HYALURONIC ACID, a linear polymer of glucuronic acid and *N*-acetylglucosamine. This compound is made by ectodermal cells and accumulates in the blastocoel, where it coats the surfaces of the incoming cells. Fisher and Solursh (1977) have shown that when this material is digested (by injecting the enzyme hyaluronidase into the blastocoel), the mesenchymal cells clump together and fail to migrate properly. Several studies have shown that hyaluronic acid is important in keeping migrating mesenchymal cells separated from one another. Moreover, hyaluronic acid first accumulates at precisely the time the first cells enter the blastocoel. Hyaluronic acid may be able to keep the cells separated by its ability to expand in water. Once in an aqueous environment, this polymer can expand to over 1000 times its original volume. Therefore, hyaluronic acid might be an important factor in keeping the mesenchymal cells dispersed during their migration and thereby ensuring that this migration continues.

Hyaluronic acid and other polysaccharides facilitate individual cell migration, but they do not appear to direct the movement of these cells (Fisher and Solursh, 1979). Rather, cellular movement of these cells is correlated, once again, with the presence of a fibronectin meshwork in the extracellular basal lamina of the epiblast cells. This fibronectin-rich layer appears on the undersurface of the upper layer shortly before the formation of the primitive streak and disappears in the region of the streak. Within the streak, the cells separate and are then seen to migrate laterally along the fibronectin-rich basement membrane of the epiblast (Duband and Thiery, 1982). However, there is no clear evidence that this fibronectin is essential for the directed cell movements of the cells laterally away from the primitive streak. More details on the cell surface changes during gastrulation will be discussed in Chapter 15.

Epiboly of the ectoderm. During gastrulation, ectodermal precursor cells expand outward to encircle the yolk. These cells are joined to each other by tight junctions and travel as a unit rather than as individual cells. In birds, the upper surface of the area opaca is seen to adhere tightly to the lower surface of the vitelline envelope and spreads along this inner surface. The same behavior is seen in culture. New (1959) demonstrated that isolated blastoderm will spread normally on isolated vitelline envelope, and Spratt (1963) demonstrated that this spreading will not occur on other substrates. These observations suggest that the vitelline envelope is essential for the spreading of the cell sheet. Interestingly, only the marginal cells (i.e., the cells of the area opaca), attach firmly to the vitelline surface. Most of the blastoderm cells adhere loosely, if at all. These marginal cells are inherently different from the other blastoderm cells, as these cells can

FIGURE 36
Primitive streak region of a gastrulating chick embryo. Scanning electron micrograph shows epiblast cells passing into the blastocoel and extending their apical ends to become bottle cells. (From Solursh and Revel, 1978; photograph courtesy of M. Solursh.)

extend enormous (500-μm) cytoplasmic processes onto the vitelline envelope. These elongated filopodia are believed to be the locomotor apparatus of the marginal cells.

There are several lines of evidence indicating that the marginal area opaca cells are the agents of ectodermal epiboly. First, the blastoderm spreads only when the margins are expanding. If the marginal cells are removed, the epiboly of the ectoderm stops. Second, when the marginal cells are cut away from the rest of the blastoderm, they continue to expand alone. Thus, it appears that the ectodermal precursor cells are carried along by the actively migrating cells of the area opaca (Schlesinger, 1958). There also exists a specific relationship between the cell membranes of the marginal cells and the lower surface of the vitelline membrane. New (1959) showed that when the blastoderm is placed on a vitelline envelope upside down (deep layers in contact with the vitelline membrane), the edges of the blastoderm curl inward so that the marginal cells of the upper layer are once again contacting the vitelline surface (Figure 37). Lash and his co-workers (1990) expanded these results by showing that fibronectin is present on the inner surface of the vitelline envelope. Then, as in the experiment discussed earlier where Boucaut and co-workers injected the synthetic fibronectin attachment-site sequence (Arg-Gly-Ser) into the salamander blastocoel, Lash and colleagues applied the Arg-Gly-Ser sequence to the vitelline envelope as the cells were migrating on it. This treatment specifically broke the contact between the marginal cells and the vitelline envelope, caused the retraction of the marginal cell filopodia, and stopped the migration of the blastoderm. Thus, there appears to be specific recognition between the marginal cells of the area opaca and the inner surface of the vitelline envelope, a recognition essential for the epibolic migration of the ectoderm that encloses the yolk. It is probable that this attachment is mediated by cell surface receptors for the fibronectin lining the inner vitelline envelope surface.

Gastrulation in mammals

Birds and mammals are both descendants of reptilian species. Therefore, it is not surprising to find that mammalian development parallels that of reptiles and birds. What is surprising is that the gastrulation movements of reptilian and avian embryos, which evolved as an adaptation to yolky eggs, are retained even in the absence of large amounts of yolk in the mammalian embryo. The inner cell mass can be envisioned as sitting atop an imaginary ball of yolk, following instructions that seem more appropriate to its ancestors.

Instead of developing in isolation within an egg, most mammals have evolved the remarkable strategy of developing within the mother herself.

FIGURE 37
Migratory properties of chick ectodermal precursors. (A) When chick blastoderm is placed on vitelline envelope with the ectodermal precursors in contact with the vitelline surface, the marginal cells migrate and cover the vitelline envelope with ectoderm. (B) When the deep layers are placed in contact with the vitelline envelope, the blastoderm layer curls to allow the cells of the superficial layer to adhere to and migrate over the vitelline layer. The result is a closed vesicle. (After New, 1959.)

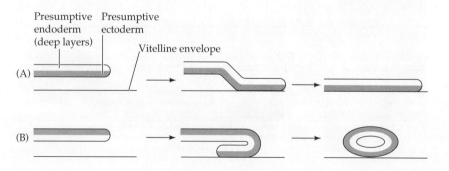

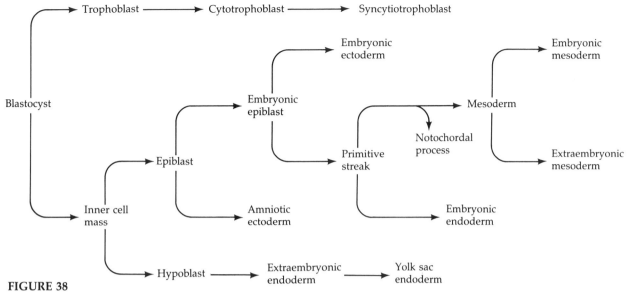

FIGURE 38

Schematic diagram showing the derivation of tissues in human and rhesus monkey embryos. (After Luckett, 1978.)

The mammalian embryo obtains nutrients directly from its mother and does not use stored yolk. This evolution has entailed a dramatic restructuring of the maternal anatomy (such as expansion of the oviduct to form the uterus) as well as the development of a fetal organ capable of absorbing maternal nutrients. This fetal organ—the PLACENTA—is derived primarily from embryonic trophoblast cells, supplemented with mesodermal cells derived from the inner cell mass.

The origins of the various early mammalian tissues are summarized in Figure 38. The first segregation of cells within the inner cell mass involves the formation of the HYPOBLAST (sometimes called the primitive endoderm) layer (Figure 39). These cells separate from the inner cell mass to line the blastocoel cavity, where they give rise to the YOLK SAC ENDODERM. As in avian embryos, these cells do not produce any part of the newborn organism. The remaining inner cell mass tissue above the hypoblast is now referred to as the EPIBLAST. The epiblast cells are split by small clefts that eventually coalesce to separate the EMBRYONIC EPIBLAST from the other epiblast cells, which form the lining of the AMNION (Figure 39C). Once the lining of the amnion is completed, it fills with a secretion called AMNIOTIC FLUID, which serves as a shock absorber to the developing embryo while preventing its desiccation.

The embryonic epiblast is believed to contain all the cells that will generate the actual embryo. It is similar to the avian epiblast. At the posterior margin of the embryonic epiblast, a localized thickening occurs, eventually producing a primitive streak through which the endodermal and mesodermal precursors migrate (Figure 40). As in avian embryos, the cells migrating between the hypoblast and epiblast layers appear to be coated with hyaluronic acid, which is first synthesized at the time of primitive streak formation (Solursh and Morriss, 1977).

While the embryonic epiblast is undergoing cell movements reminiscent of those seen in reptilian or avian gastrulation, the extraembryonic cells are making the distinctly mammalian tissues that enable the fetus to survive within the maternal uterus. Although the initial trophoblastic cells appear normal, they give rise to a population of cells wherein nuclear division occurs in the absence of cytokinesis. The initial type of cell con-

FIGURE 39

Tissue formation in the human embryo between days 7 and 12. (A,B) Human blastocyst immediately prior to gastrulation. The inner cell mass delaminates hypoblast cells that line the trophoblast, thereby forming the primitive gut and a two-layered (epiblast and hypoblast) blastodisc similar to that seen in avian embryos. The trophoblast in some mammals can be divided into the polar trophoblast, which covers the inner cell mass, and the mural trophoblast, which does not. The trophoblast divides into the cytotrophoblast, which will form the villi, and the syncytiotrophoblast, which will ingress into the uterine tissue. (C) Meanwhile, the epiblast has split into the amniotic ectoderm and the embryonic epiblast. The adult mammals forms from the cells of the embryonic epiblast. (After Gilbert, 1989.)

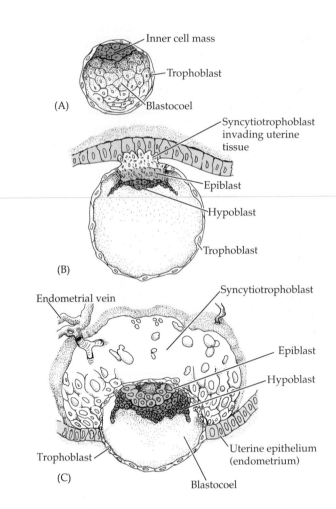

FIGURE 40

Cell movements during mammalian gastrulation. (A) Dorsal surface of the embryonic epiblast (amniotic ectoderm removed). As in chick embryos, cells migrating through Hensen's node travel anteriorly (cephalad) to form the notochord, while the remaining cells traveling through the streak migrate laterally to become the mesoderm and endoderm precursors. Dotted lines indicate internal migrations. (B) Transverse section of the embryo. (After Langman, 1981.)

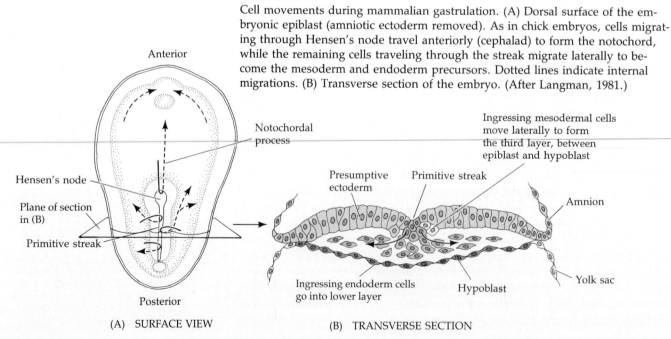

stitutes a layer called the CYTOTROPHOBLAST, whereas the multinucleated type of cell forms the SYNCYTIOTROPHOBLAST. The cytotrophoblast adheres to the uterine wall (endometrium) through a series of adhesion molecules that will be discussed in Chapter 15. The human cytotrophoblast cells also contain proteolytic enzymes that enable them to enter the uterus and remodel the uterine blood vessels so that the maternal blood bathes fetal blood vessels. This proteolytic activity ceases after the twelfth week of pregnancy (Fisher et al., 1989). The syncytiotrophoblast tissue digests its way through the uterine lining, embedding the embryo within the uterus. The uterus, in turn, sends blood vessels into this area, where they eventually contact the syncytiotrophoblast. Shortly thereafter, mesodermal tissue extends outward from the gastrulating embryo (Figure 41). Luckett (1978) has shown that this tissue had migrated through the primitive streak but became extraembryonic rather than embryonic mesoderm. This extraembryonic mesoderm joins the trophoblastic extensions and gives rise to the blood vessels that carry nutrients from the mother to the embryo. The narrow CONNECTING STALK of extraembryonic mesoderm that links the embryo to the trophoblast eventually forms the vessels of the UMBILICAL CORD. The fully developed organ, consisting of trophoblast tissue and blood vessel-containing mesoderm, is called the CHORION, and the chorion fuses with the uterine wall to create the placenta. Thus, the placenta has both a maternal portion (the uterine endometrium that is modified during pregnancy) and a fetal component, the chorion. The chorion may be very closely apposed to maternal tissues while still being readily separable (as in the CONTACT PLACENTA of the pig), or it may be so intimately integrated that the two tissues cannot be separated without damage to both the mother and the developing fetus (as in the DECIDUOUS PLACENTA of most mammals, including humans).

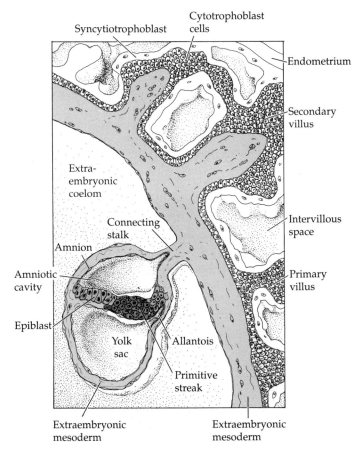

FIGURE 41

Gastrulating human embryo at the end of the third week of gestation. The amniotic cavity (between the amniotic ectoderm and the embryonic epiblast) and yolk sac (from the hypoblast) have formed, the notochord elevates the epiblast in the future anterior region of the embryo, and the trophoblast cells forming the placenta are coming into contact with the blood vessels of the uterus. The embryo is connected to the trophoblast by a connecting stalk of extraembryonic mesoderm that will shortly form the fetal blood vessels of the placenta. (After Gilbert, 1989.)

FIGURE 42

Human embryo and placenta after 40 days of gestation. The embryo lies within the amnion and its blood vessels can be seen extending into the chorionic villi. The sphere to the right of the embryo is the yolk sac. (Photograph from the Carnegie Institution of Washington; courtesy of C. F. Reather.)

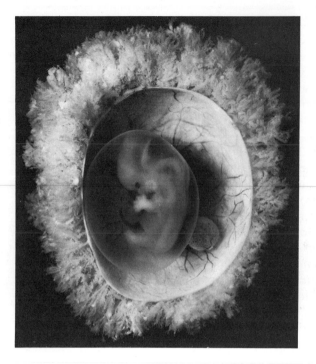

In Figure 42, one can observe the relationships between the embryonic and extraembryonic tissues of a 6-week human embryo. The embryo is seen encased in the amnion and is further shielded by the chorion. The blood vessels extending to and from the chorion are readily observable, as are the villi that project from the outer surface of the chorion. These villi contain the blood vessels and allow the chorion to have a large area exposed to the maternal blood. Thus, although fetal and maternal circulatory systems normally never merge, diffusion of soluble substances can occur through the villi (Figure 43). In this manner the mother provides the fetus with nutrients and oxygen, and the fetus sends its waste products (mainly carbon dioxide and urea) into the maternal circulation. The blood vessels of the chorionic villi form from extraembryonic mesoderm that enters the mounds of cytotrophoblast tissue called PRIMARY VILLI (Figure 44). The resulting structure, the SECONDARY VILLUS, forms during the second week of pregnancy. By the end of the third week, some of this extraembryonic mesoderm has produced blood vessels, and this TERTIARY VILLUS is able to bring nutrients and oxygen from the mother into the embryo.

The trophoblast, therefore, is necessary for adhering to and entering the uterine tissues, and the chorion acts to exchange gases and nutrients between mother and fetus. But the chorion has even further importance. It is also a endocrine organ. The syncytiotrophoblast portion of the chorion produces three hormones that are essential for mammalian development. First, it produces CHORIONIC GONADOTROPIN, a peptide hormone that is capable of causing other cells in the placenta (and in the maternal ovary) to produce PROGESTERONE. Progesterone is the steroid hormone that keeps the uterine wall thick and full of blood vessels. In primates, the ovaries can be removed after the first third of pregnancy without harm to the developing fetus because the chorion itself is able to produce the steroid needed to maintain pregnancy (Zander and Münstermann, 1956). Placental progesterone is also used by the fetal adrenal gland as a substrate for the production of biologically important corticosteroid hormones. The third hormone produced by the chorion is CHORIONIC SOMATOMAMMOTRO-

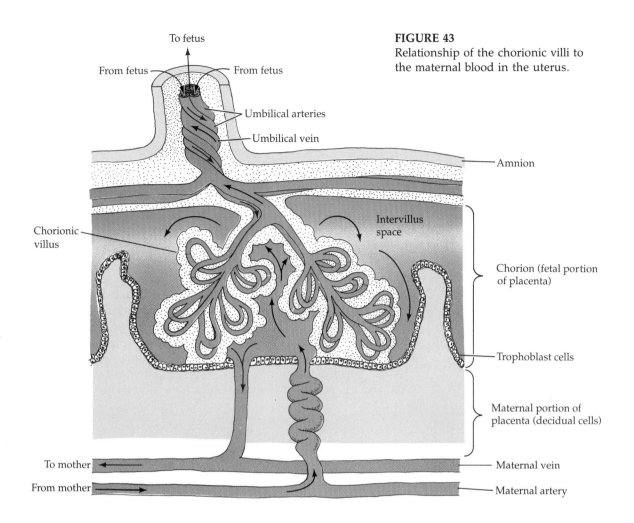

FIGURE 43
Relationship of the chorionic villi to the maternal blood in the uterus.

To fetus

From fetus   From fetus

Umbilical arteries

Umbilical vein

Amnion

Chorionic villus

Intervillus space

Chorion (fetal portion of placenta)

Trophoblast cells

Maternal portion of placenta (decidual cells)

To mother

Maternal vein

From mother

Maternal artery

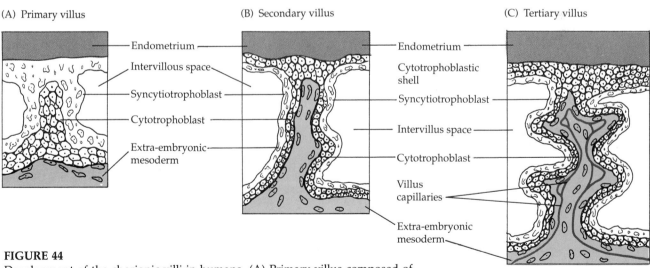

(A) Primary villus

(B) Secondary villus

(C) Tertiary villus

Endometrium

Intervillous space

Syncytiotrophoblast

Cytotrophoblast

Extra-embryonic mesoderm

Endometrium

Cytotrophoblastic shell

Syncytiotrophoblast

Intervillus space

Cytotrophoblast

Villus capillaries

Extra-embryonic mesoderm

FIGURE 44
Development of the chorionic villi in humans. (A) Primary villus composed of cytotrophoblast tissue encased in syncytiotrophoblast. (B) Secondary villus formed when underlying extraembryonic mesoderm penetrates the primary villus. Such secondary villi join adjacent villi to form the cytotrophoblastic shell that will anchor the villi to the endometrium. (C) Within the extraembryonic mesoderm, capillaries form and will connect with branches of the umbilical artery and vein. (After Gilbert, 1989.)

PIN (often called placental lactogen). This hormone is responsible for promoting maternal breast development during pregnancy, thereby enabling milk production later.

Recent studies have indicated that the chorion may have yet another function, namely, the protection of the fetus from the immune response of the mother. Each person with a normal immune system can recognize and reject foreign cells within its body. This is reflected by the rejection of skin and organ grafts from genetically different individuals. The glycoproteins responsible for this rejection are called the MAJOR HISTOCOMPATIBILITY ANTIGENS, and these are likely to differ from individual to individual. A human child expresses major histocompatibility antigens from both the mother and the father, and a mother is capable of rejecting her offspring's skin or organs because they contain paternally derived antigens. How, then, can a human fetus remain nine months within the body of its mother? Why doesn't the mother immunologically reject her fetus as she would an organ from that child? It appears that the chorion has evolved several mechanisms by which it can inhibit the immune response against the fetus (Chaouat, 1990). It can secrete soluble proteins that block the production of antibodies, and it can promote the production of certain types of lymphocytes that supress the normal immune response within the uterus. Thus, the functions of the placenta include not only physical support and nutrient exchange, but also regulation of the endocrine and immunological relationships between the mother and the fetus.

SIDELIGHTS & SPECULATIONS

Genes and mammalian gastrulation

Recent experiments have suggested that the maternal and paternal genomes have different roles during mammalian gastrulation. Mouse zygotes can be created that have only sperm-derived chromosomes or only egg-derived chromosomes (page 61). The male-derived embryos (androgenones) die with deficiencies in the embryo proper but have formed well developed chorions. Conversely, the female-derived embryos (gynogenones) die with deficiencies in their chorions, even though the actual embryo seems normal. It appears as though the sperm-derived genes are needed for proper development of the chorion, while the egg-derived genes are necessary for the normal development of the embryo itself (McGrath and Solter, 1984; Surani et al., 1984). This has been confirmed by making allophenic mice wherein blastomeres from a normal 4-cell embryo are aggregated with blastomeres from either androgenetic or gyogenetic embryos. In both cases, cells from both the normal and abnormal embryos were originally seen in all regions of the blastocyst. However, by the end of gastrulation, androgenetic cells are seen almost exclusively in the trophoblast, while gynogenetic cells are hardly ever seen in trophoblast-derived tissues (Thomson and Solter, 1989). This strongly suggests that the maternal and paternal genomes serve distinct functions during early mammalian embryogenesis.

In gastrulation, we see an incredibly well coordinated series of cell movements whereby the cleavage-stage blastomeres are rearranged and begin to interact with new neighbors. Moreover, although there are differences between the gastrulation movements of sea urchin, amphibian, avian, and mammalian embryos, certain mechanisms are common to them all. Each group has the problem of bringing the mesodermal and endodermal precursor cells inside the body while surrounding the embryo with ectodermal precursors. Given the different amounts and distributions of yolk, as well as other environmental considerations, each type of organism has evolved a way of accomplishing this goal. The stage is now set for the formation of the first organs.

5

Early vertebrate development: Neurulation and ectoderm

For the real amazement, if you wish to be amazed, is this process. You start out as a single cell derived from the coupling of a sperm and an egg; this divides in two, then four, then eight, and so on, and at a certain stage there emerges a single cell which has as all its progeny the human brain. The mere existence of such a cell should be one of the great astonishments of the earth. People ought to be walking around all day, all through their waking hours calling to each other in endless wonderment, talking of nothing except that cell.

—LEWIS THOMAS (1979)

Even more appealing than virgin forest was the jungle lying before me at that moment: the central nervous system.

—RITA LEVI-MONTALCINI (1988)

The vertebrate pattern of development

In 1828, Karl Ernst von Baer, the foremost embryologist of his day,* proclaimed, "I have two small embryos preserved in alcohol, that I forgot to label. At present I am unable to determine the genus to which they belong. They may be lizards, small birds, or even mammals." Figure 1 allows us to appreciate his quandary, and it illustrates von Baer's four general laws of embryology. From his detailed study of chick development and his comparison of those embryos to embryos of other vertebrates, von Baer derived four generalizations, illustrated here with some vertebrate examples.

1 *The general features of a large group of animals appear earlier in the embryo than do the specialized features.* All developing vertebrates (fishes, reptiles, amphibians, birds, and mammals) appear very

*K. E. von Baer discovered the notochord, the mammalian egg, and the human egg, as well as making the conceptual advances described here. His work will be discussed further in Chapter 23.

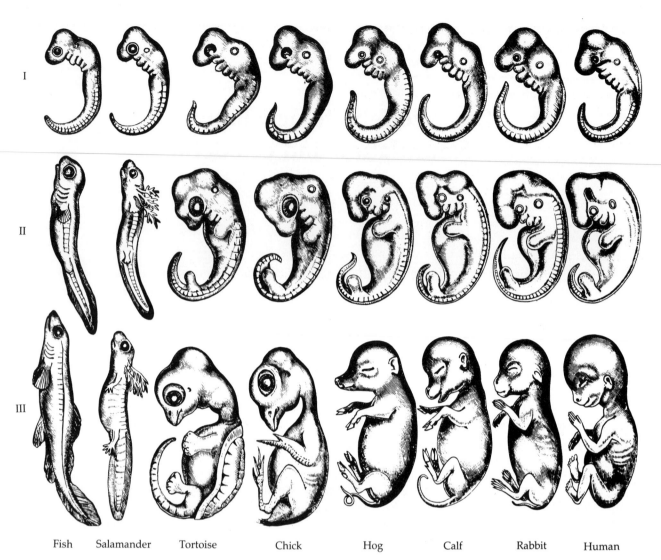

I

II

III

Fish   Salamander   Tortoise   Chick   Hog   Calf   Rabbit   Human

FIGURE 1
Illustration of von Baer's law. Early vertebrate embryos exhibit features common to the entire subphylum. As development progresses, embryos become recognizable as members of their class, their order, their family, and finally their species. (After Romanes, 1901.)

similar shortly after gastrulation. It is only later in development that the special features of class, order, and finally species emerge (Figure 1). All vertebrate embryos have gill arches, notochords, spinal cords, and pronephric kidneys.

2 *Less general characters are developed from the more general, until finally the most specialized appear.* Certain fishes use a pronephric type of kidney as adults. This type of kidney can be found in the embryos of all other types of vertebrates. However, in these more complex species, it never becomes functional but contributes to the subsequent development of more specialized types of kidneys. All vertebrates initially have the same type of skin. Only later does skin develop fish scales, reptilian scales, bird feathers, or the hair, claws, and nails of mammals. Similarly, the early development of the limb is essentially the same in all vertebrates. Only later do the differences between legs, wings, and arms become apparent.

3 *Each embryo of a given species, instead of passing through the adult stages of other animals, departs more and more from them.* The visceral clefts of embryonic birds and mammals do not resemble the gill slits of adult fishes in detail. Rather, they resemble those visceral clefts of *embryonic* fishes and other *embryonic* vertebrates.

Whereas fishes preserve and elaborate these clefts into true gill slits, mammals convert them into structures such as the eustachian tubes (between the ear and mouth).

4 *Therefore, the early embryo of a higher animal is never like [the adult of] a lower animal, but only like its early embryo.*

Von Baer saw that different groups of animals shared certain common features during early embryonic development and that these features became more and more characteristic of the species as development proceeded. Human embryos never pass through a stage equivalent to an adult fish or bird; rather, human embryos initially share characteristics in common with fish and avian embryos. Later, the mammalian and other embryos diverge, none of them passing through the stages of the other.

These conclusions were extremely important for Charles Darwin because evolutionary classification depended upon finding those similarities that demonstrated that certain diverse animals arose from a common ancestor. "Community of embryonic structure," wrote Darwin in *Origin of Species* (1859), "reveals community of descent." In fact, Darwin believed that embryology was the strongest foundation for his evolutionary theory. Von Baer, of course, was not an evolutionist when he wrote his four laws in 1828. In fact, he never became one. Von Baer believed that he had found the divine plan upon which all the organisms in a group such as the vertebrates developed. Because earlier evolutionary theories had envisioned a single, nonbranching series of transformations, von Baer's observations were often used against evolution. Darwin, however, recognized that von Baer's work supported an evolutionary hypothesis in which a common ancestor could radiate into several different types of organisms by heritable modifications of embryonic development. The reason human embryos, fish embryos, and chick embryos all had visceral clefts was because they had a common ancestor whose embryo had such visceral clefts. Moreover, the reason visceral clefts gave rise to different structures in different groups of organisms (gills in fishes, eustachian tubes in mammals) was that the ancestral plan had been modified and selected by different environments.

Von Baer also recognized that there was a common pattern to vertebrate development. The three germ layers gave rise to different organs, and their derivation was constant whether the organism was a fish, a frog, or a chick. ECTODERM formed skin and nerves; ENDODERM formed respiratory and digestive tubes; and MESODERM formed connective tissue, blood cells, heart, the urogenital system, and parts of most of the internal organs. In this chapter we shall follow the early development of ectoderm, focusing on formation of the nervous system. In the next chapter we shall follow the early development of endodermal and mesodermal organs.

Neurulation

In vertebrates, gastrulation creates an embryo having an internal endodermal layer, an intermediate mesodermal layer, and an external ectoderm. The interaction between the dorsal mesoderm and its overlying ectoderm is one of the most important interactions of all development, for it initiates ORGANOGENESIS, the creation of specific tissues and organs. In this interaction, the chordamesoderm directs the ectoderm above it to form the hollow NEURAL TUBE, which will differentiate into the brain and spinal cord. The action by which the flat layer of ectodermal cells is transformed

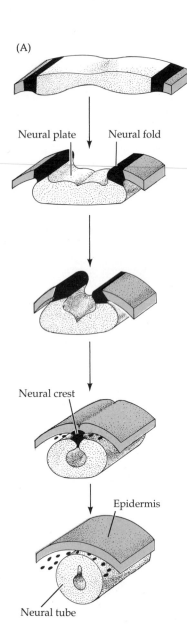

(A)

Neural plate   Neural fold

Neural crest

Epidermis

Neural tube

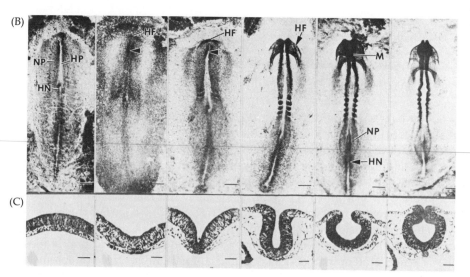

(B)

(C)

FIGURE 2

Neurulation in amphibians and amniotes. (A) Diagrammatic representation of neural tube formation. The ectodermal cells are represented either as precursors of the neural crest (black) or as precursors of the epidermis (color). The ectoderm folds in at the most dorsal point, forming an outer epidermis and an inner neural tube connected by neural crest cells. (B) Photomicrographs of neurulation in a 2-day chick embryo. (C) Neural tube formation seen in transverse cross sections of the chick embryo at the region of the future midbrain. Each photograph in C corresponds to the one above it. HF, head fold; HP, head process; HN, Hensen's node; M, midbrain; NP, neural plate. (Photomicrographs courtesy of R. Nagele.)

into a hollow tube is called NEURULATION. The events of neurulation are diagrammed in Figure 2. Here, the original ectoderm is divided into three sets of cells: (1) the internally positioned neural tube, (2) the epidermis of the skin, and (3) the neural crest cells, which migrate from the region that had connected the neural tube and epidermal tissues. An embryo undergoing such changes is called a NEURULA. The phenomenon of embryonic induction which initiates neurulation in the dorsal region of the embryo will be detailed in Chapter 8. In this chapter, we are concerned with the response of the various ectodermal tissues.

The process of neurulation in frog embryos is depicted in Figure 3. The mechanisms of neural tube formation appear to be similar in amphibians, reptiles, birds, and mammals (Gallera, 1971), so we will be considering various groups as we proceed through our survey.* The first indication that a region of ectoderm is destined to become neural tissue is a change in cell shape (Figure 4). Midline ectodermal cells become elongated, whereas those cells destined to form the epidermis become more flattened. The elongation of dorsal ectodermal cells causes these prospective neural regions to rise above the surrounding ectoderm, thus creating the NEURAL PLATE. As much as 50 percent of the ectoderm is included in this plate. Shortly thereafter, the edges of the neural plate thicken and move upward to form the NEURAL FOLDS, while a U-shaped NEURAL GROOVE appears in the center of the plate, dividing the future right and left sides of the

*Among the vertebrates, fishes generate their neural tube in a different manner. Fish neural tubes do not form from an infolding of the overlying ectoderm, but rather from a solid cord of cells, which is induced to sink into the embryo. This cord subsequently hollows out (cavitates) to form the neural tube. In mammals, only the most posterior portion of the neural tube forms by cavitation.

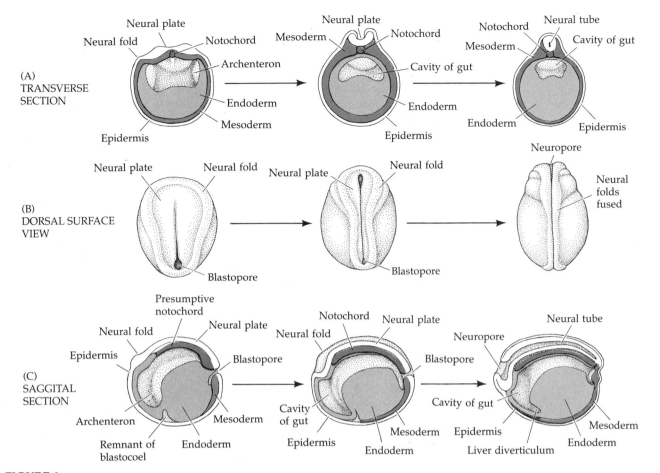

FIGURE 3

Three views of neurulation in a frog embryo, showing early (left), middle (center), and late (right) neurulae in each case. (A) Transverse section through the center of the embryo. (B) The same sequence looking down on the dorsal surface of the whole embryo. (C) Sagittal section through the median plane of the embryo. (After Balinsky, 1975.)

embryo (Figures 3 and 5). The neural folds migrate toward the midline of the embryo, eventually fusing to form the neural tube beneath the overlying ectoderm. The cells at the dorsalmost portion of the neural tube become the NEURAL CREST cells. These neural crest cells will migrate through the embryo and will give rise to several cell populations, including pigment cells and the cells of the peripheral nervous system.

The neural tube and the origins of the central nervous system

The formation of the neural tube does not occur simultaneously throughout the ectoderm. This is best seen in those vertebrates (such as birds and mammals) whose body axis is elongated prior to neurulation. Figure 6 depicts neurulation in a 24-hr chick embryo. Neurulation in the cephalic (head) region is well advanced while the caudal (tail) region of the embryo is still undergoing gastrulation. Regionalization of the neural tube also occurs as a result of changes in the shape of the tube. In the cephalic end (where the brain will form), the wall of the tube is broad and thick. Here,

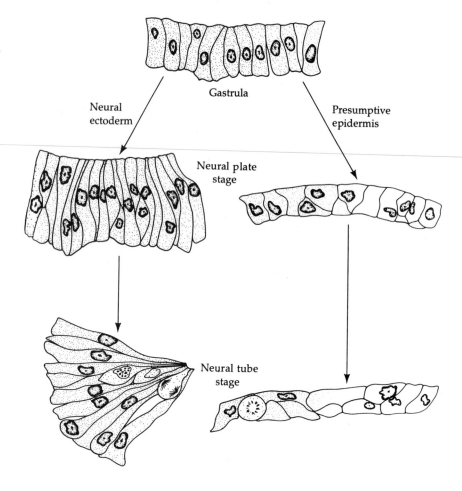

FIGURE 4

Schematic diagram of the shape changes during neurulation in the salamander. At the end of gastrulation, the cells of the ectoderm form a uniform epithelium. During neurulation, neural epithelial cells elongate to form the neural plate and then constrict at their apices to form the neural tube. Presumptive epidermal cells flatten throughout neurulation. (After Burnside, 1971.)

Gastrula

Neural ectoderm

Presumptive epidermis

Neural plate stage

Neural tube stage

FIGURE 5

Scanning electron micrograph of neural tube formation in the chick embryo. Elongated neural epithelial cells form a tube as the flattened epidermal cells are brought to the midline of the embryo. (Photograph courtesy of K. W. Tosney.)

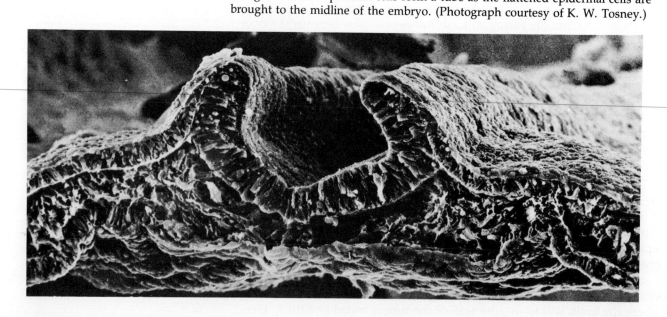

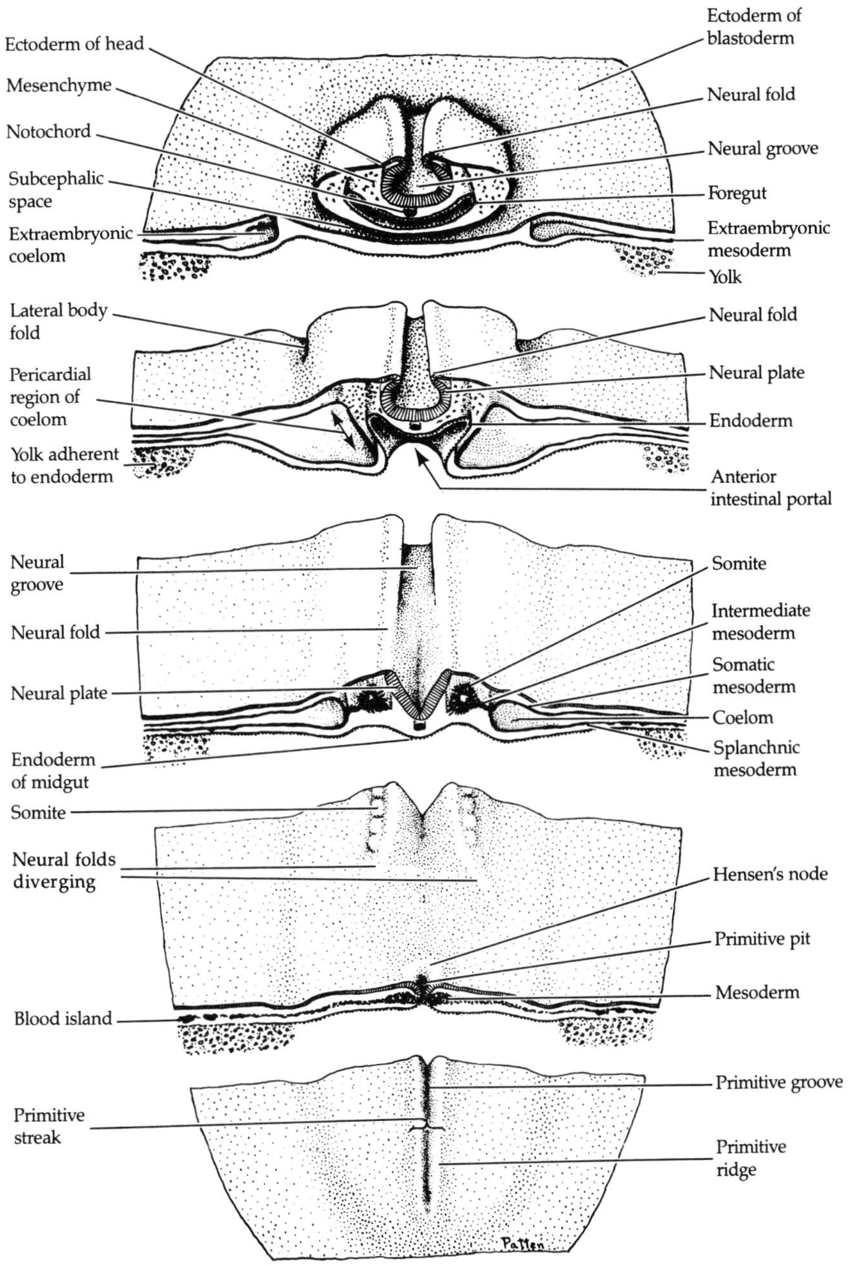
Ectoderm of head
Mesenchyme
Notochord
Subcephalic space
Extraembryonic coelom
Ectoderm of blastoderm
Neural fold
Neural groove
Foregut
Extraembryonic mesoderm
Yolk
Lateral body fold
Pericardial region of coelom
Yolk adherent to endoderm
Neural fold
Neural plate
Endoderm
Anterior intestinal portal
Neural groove
Neural fold
Neural plate
Endoderm of midgut
Somite
Intermediate mesoderm
Somatic mesoderm
Coelom
Splanchnic mesoderm
Somite
Neural folds diverging
Blood island
Hensen's node
Primitive pit
Mesoderm
Primitive streak
Primitive groove
Primitive ridge
Patten

FIGURE 6
Stereogram of a 24-hour chick embryo. Cephalic portions are finishing neurulation while the caudal portions are still undergoing gastrulation. (From Patten, 1971; after Huettner, 1949.)

a series of swellings and constrictions define the various brain compartments. Caudal to the head region, however, the neural tube remains a simple tube that tapers off toward the tail. The two open ends of the neural tube are called the ANTERIOR NEUROPORE and the POSTERIOR NEUROPORE. In mammals, these openings allow amniotic fluid to flow through the neural tube for a time (Figure 7). Failure to close the human posterior neuropore at day 27 (or its subsequent rupture shortly thereafter) results in a condition called SPINA BIFIDA, the severity of which depends upon how much of the spinal cord remains open. However, failure to close the anterior neuropore results in a lethal condition, ANENCEPHALY. Here, the forebrain remains in contact with the amniotic fluid and subsequently degenerates. Fetal forebrain development ceases, and the vault of the skull fails to form. This abnormality is not that rare in humans, occurring in

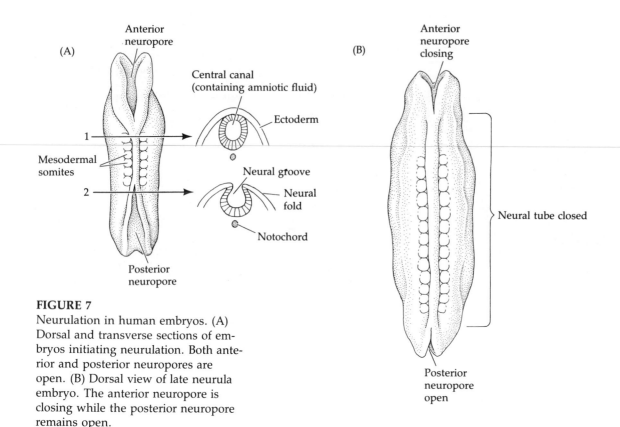

FIGURE 7
Neurulation in human embryos. (A) Dorsal and transverse sections of embryos initiating neurulation. Both anterior and posterior neuropores are open. (B) Dorsal view of late neurula embryo. The anterior neuropore is closing while the posterior neuropore remains open.

about 0.1 percent of all pregnancies. Neural tube closure defects can now be detected during pregnancy by various physical and chemical tests.

Mechanism of neural tube formation

Neural tube formation is intimately linked to changes in cell shape, and microtubules and microfilaments are both involved in these changes (Figure 8). Ectodermal cells elongate as the randomly arranged microtubules of these cells align themselves parallel to the lengthening axis. This stage of neural tube formation can be blocked by colchicine, an inhibitor of microtubule polymerization (Burnside, 1973). A second change in cell shape involves the apical constriction of cylindrical cells to form wedges. In the lateral sides of the neural tube, this change is directed by a ring of contractile microfilaments encircling the apical margins of the cells. The contraction of these microfilaments produces a "purse-string" effect, constricting the apical end of each cell. Burnside (1971) and Karfunkel (1972) have shown that when embryos are cultured in microfilament-inhibitory cytochalasin B, the neuroectodermal cells can elongate but cannot constrict to form the neural folds. The actin microfilaments are linked to myosin from which they are able to get the energy for contraction (Nagele and Lee, 1980; 1987), and they are also linked to the apical cell membrane. This membrane linkage is effected through SPECTRIN, an integral protein of the plasma membrane that can bind to microfilaments inside the cell (Sadler et al., 1986). The patterns of spectrin localization follow that of the microfilaments, suggesting that the microfilaments are actively involved in changing cell shape. Cytochalasin, however, does not effect all cells equally. The cells at the bottom of the neural groove can still constrict even in the presence of this drug, suggesting that apical constriction is not acting alone in the construction of the neural tube (Schoenwolf et al., 1988).

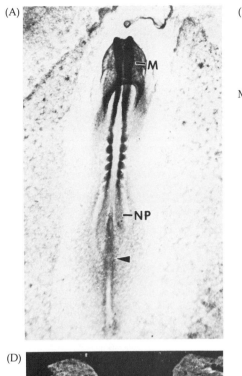

(A)

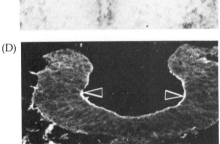

(D)

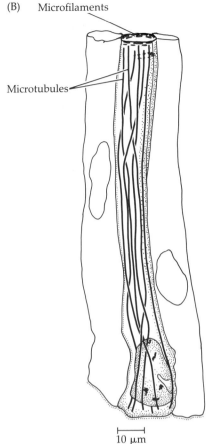

(B)

10 μm

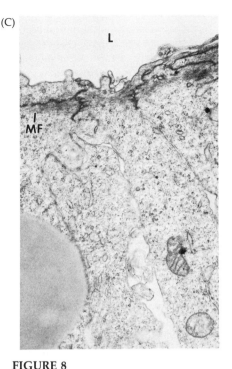

(C)

FIGURE 8
The cytoskeleton in neurulation.
(A) 2-day chick embryo showing the "head start" during neurulation. The neural folds are coming together in the region of the midbrain (M), while the neural plate (NP) lies just anterior to the regressing Hensen's node. Intermediate stages of neurulation occur between them. (B) Microtubule and microfilament orientation in neural plate cells. Microtubules align parallel to the long axis of the cell; microfilaments encircle the apex. (C) Transmission electron micrograph showing microfilaments along the apical edge of a midbrain neural tube cell. (D) Apical microfilaments seen by staining neural plate cells with fluorescent antibody to actin. The heaviest concentrations are in the regions of the steepest folding (arrows). (B from Burnside, 1971; A,C, and D from Nagele and Lee, 1987; courtesy of R. Nagele.)

Another process in neural tube formation may be the CORTICAL TRACTORING of the neural plate epithelium. Jacobson and his colleagues (1986) have proposed a hypothesis that would account for many of the specific shapes seen as the developing neural tube forms. According to this model, the cortex of the neural plate cells is in motion, beginning at the basal end and ending at the apex. The ectodermal cells are linked together by molecules that are carried in this cortical flow. Thus, if the average speed of the flow is the same in each cell, the cells remain bound to each other and no deformation of the epithelial sheet takes place (Figure 9A). However, if the speed of this "tractor" differs between two adjacent cells, the faster cell will begin to crawl basalward, out of the sheet. Since the cells are still joined together at their apices, buckling occurs. The combination of basal crawling and apical constriction would generate a movement lifting the cells above the plane of the epidermis and rolling them inward (Figure 9B). It is also possible that the shear force produced by the basal crawling might cause the apical ends of some of these cells to tear loose from the epithelium. Jacobson and co-workers speculate that this might be the origin of the neural crest cells.

Schoenwolf and Smith (1990) have summarized the above models of neural tube formation as based on three nested principles. First, all the forces needed to create a neural tube are intrinsic to the neuroepithelium itself. Second, neurulation is driven by changes in the shape of the neuroepithelial cells. Third, the forces causing these cell-shape changes are generated by the cytoskeleton. They argue, however, that these principles are inadequate to account for neural tube formation. To account for the

FIGURE 9

Cortical tractor model of neural tube formation. (A) Schematic cross section of the ectodermal epithelium showing the cortical tractors (arrows) that move the adhesive molecules linking adjacent cells. The flow begins in the basal end of each cell and ends near the apex. The adhesive molecules pile up towards the apex, forming the apical seal, which is continuously renewed. If the flow rates are equal among cells, no movements of the epithelium take place. If the rates should differ, then the faster cell crawls basally under the slower cell. (B) Interpretations of neurulation in terms of the cortical tractor model. The diagrams at the left are tracings of neurulation in newt embryos. The diagrams to the right are arrangements of cells predicted by the cortical tractor model. (After Jacobson et al., 1986.)

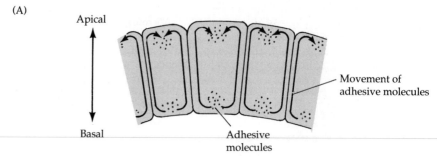

(A)

Apical

Basal

Movement of adhesive molecules

Adhesive molecules

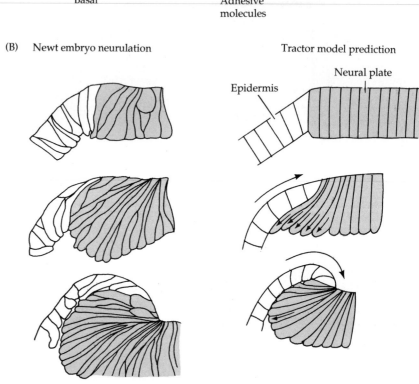

(B)   Newt embryo neurulation

Tractor model prediction

Neural plate

Epidermis

shape of the neural plate, two other principles—convergent extension and cell division—have to be considered. Schoenwolf and Alvarez (1989) have calculated that the cytoskeleton-mediated elongation of the neural plate cells can account for only 15 percent of the 50 percent reduction in neural plate width that occurs during chick neurulation. However, they have observed that the cells of the neural plate rearrange themselves similarly to the rearrangements seen during amphibian notochordal convergent extension. This would cause an increase in the length of the neural plate at the expense of its width and would account for the remaining reduction in neural plate width. Differential cell division seen in different regions of the neural plate would also contribute to the size and shape of this region. Schoenwolf also argues that differences in the cell shapes seen in different regions of the neural tube indicate that forces extrinsic to the neuroepithelium are helping to mold the neural tube. Those cells directly adjacent to the notochord and those cells at the hinges of the neural groove form bottle-like cells, whereas the other neural groove cells do not. These differences would not be expected if each cell were identical with respect to its environment. Thus, while much of neurulation can be explained solely by factors residing within the neural plate cells, some of the regional differences in the neural tube may arise from the environment in which neurulation takes place.

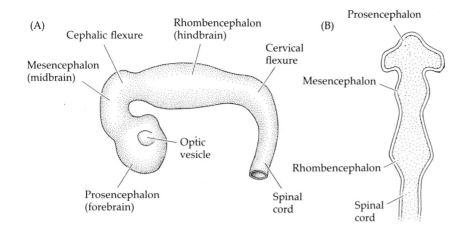

FIGURE 10
Early brain development in a 4-week human embryo. (A) Lateral view. (B) Extended diagram to illustrate the bulges of the neural tube. (After Langman, 1969.)

Differentiation of the neural tube

Formation of brain regions. The differentiation of the neural tube into the various regions of the central nervous system occurs simultaneously in three different ways. On the gross anatomical level, the neural tube and its lumen bulge and constrict to form the chambers of the brain and the spinal cord. At the tissue level, the cell populations within the wall of the neural tube rearrange themselves in various ways to form the different functional regions of the brain and the spinal cord. Finally, on the cellular level, neuroepithelial cells themselves differentiate into the numerous types of neurons and supportive (glial) cells present in the body. The early development of most vertebrate brains is similar, but because the human brain is probably the most interesting organ in the animal kingdom, we shall concentrate on the development that is supposed to make *Homo sapient*.

The early mammalian neural tube is a straight structure. However, even before the posterior portion of the tube has formed, the most anterior portion of the tube is undergoing drastic changes. In this anterior region, the neural tube balloons into three primary vesicles (Figure 10): forebrain (PROSENCEPHALON), midbrain (MESENCEPHALON), and hindbrain (RHOMBENCEPHALON). By the time the posterior end of the neural tube closes, secondary bulges—the OPTIC VESICLES—have extended laterally from each side of the developing forebrain. Moreover, the future brain has bent so that the creases demark the boundaries of the brain ventrioles. The major creases are the CEPHALIC FLEXURE and the CERVICAL FLEXURE.

The forebrain becomes subdivided into the anterior TELENCEPHALON and the more caudal DIENCEPHALON (Figure 11). The telencephalon will

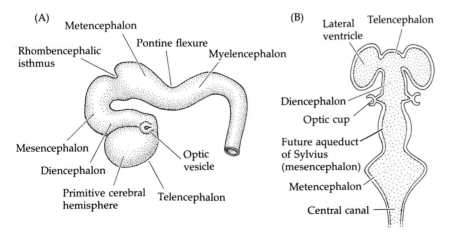

FIGURE 11
Further human brain development. Lateral view (A) and midline extended diagram (B) of a 6-week human brain showing secondary bulges of the neural tube. (After Langman, 1969.)

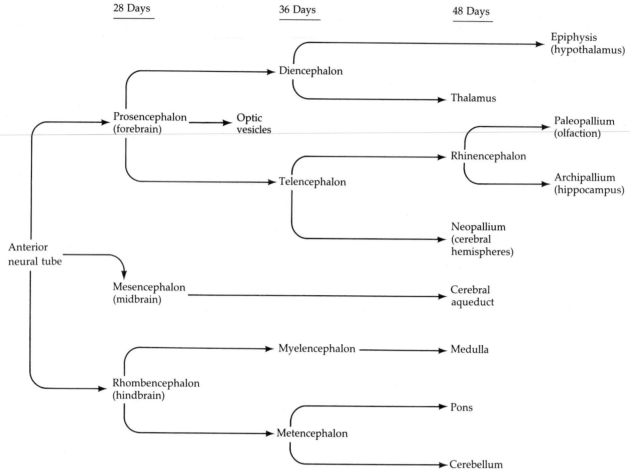

28 Days   36 Days   48 Days

Anterior neural tube

Prosencephalon (forebrain)
→ Optic vesicles
→ Diencephalon → Epiphysis (hypothalamus)
→ Diencephalon → Thalamus
→ Telencephalon → Rhinencephalon → Paleopallium (olfaction)
→ Telencephalon → Rhinencephalon → Archipallium (hippocampus)
→ Telencephalon → Neopallium (cerebral hemispheres)

Mesencephalon (midbrain) → Cerebral aqueduct

Rhombencephalon (hindbrain)
→ Myelencephalon → Medulla
→ Metencephalon → Pons
→ Metencephalon → Cerebellum

FIGURE 12
Regional specialization during early human brain development.

eventually form the CEREBRAL HEMISPHERES, and the diencephalon will form the thalamic and hypothalamic brain regions as well as the region that receives neural input from the eyes. The mesencephalon (midbrain) does not become subdivided, and its lumen eventually becomes the cerebral aqueduct. The rhombencephalon becomes subdivided into a posterior MYELENCEPHALON and a more anterior METENCEPHALON. The myelencephalon eventually becomes the MEDULLA OBLONGATA, whose neurons generate the nerves that regulate respiratory, gastrointestinal, and cardiovascular movements. The metencephalon gives rise to the CEREBELLUM, the part of the brain responsible for coordinating movements, posture, and balance. The development of specialized subdivisions in the brain is summarized in Figure 12. The hindbrain develops a segmental pattern that appears to specify the places where certain nerves originate. Periodic swellings called RHOMBOMERES divide the rhombencephalon into smaller compartments. This has been most extensively studied in the chick, where the first neurons appear in the even-numbered rhombomeres, r2, r4, and r6 (Lumsden and Keynes, 1989; Figure 13). This has great functional significance, since the neurons from r2 form the fifth (trigeminal) cranial nerve, those from r4 form the seventh (facial) cranial nerve, and the ninth (glossopharyngeal) cranial nerve exits r6.

The ballooning of the early embryonic brain is remarkable in its rate, its extent, and in its being the result primarily of an increase in cavity size, not tissue growth. In chick embryos, the brain volume expands 30-fold between days three and five of development. This rapid expansion is thought to be caused by positive fluid pressure pressing against the walls of the neural tube by the fluid within it. It might be expected that this fluid pressure might have been dissipated by the spinal cord, but this does not appear to be so. Schoenwolf and Desmond (1984; Desmond and Schoenwolf, 1986) have demonstrated that prior to the closing of the posterior neuropore, a constriction forms in the chick neural tube at the base of the brain (Figure 14). This effectively separates the presumptive brain region from the future spinal cord. (Such an occlusion also occurs in the human embryo; Desmond, 1982.) If one removes the fluid pressure in the anterior portion of such an occluded neural tube, the chick brain enlarges at a much slower rate and contains many fewer cells than are found in normal, control embryos. The occluded region of the neural tube reopens after the initial rapid enlargement of the brain ventricles.

Tissue architecture of the central nervous system. The original neural tube is composed of a germinal neuroepithelium, one cell layer thick. This is a rapidly dividing cell population. Sauer (1935) and others have shown that all of these cells are continuous from the luminal edge of the neural tube to the outside edge but that the nuclei of these cells are at different heights, thereby giving the superficial impression that the wall of the neural tube has numerous cell layers. The position of the nucleus between the luminal edge and the outer edge of the tube depends on the stage of the cell's cycle (Figure 15). DNA synthesis (S phase) occurs while the nucleus is at the outside edge of the tube, and the nucleus migrates luminally as mitosis proceeds. Mitosis occurs on the luminal side of the cell layer. During early development, 100 percent of the neural tube cells incorporate radioactive thymidine into DNA (Fujita, 1964). Shortly thereafter, certain cells stop incorporating these DNA precursors, thereby in-

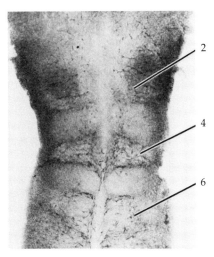

FIGURE 13
A 2-day chick-embryo hindbrain splayed to show lateral walls. Neurons were visualized with an antibody staining neurofilament proteins. Rhombomeres 2, 4, and 6 were distinguished by the high density of axons at this early developmental stage. (From Lumsden and Keynes, 1989; photograph courtesy of A. Keynes.)

(A) (B) (D)

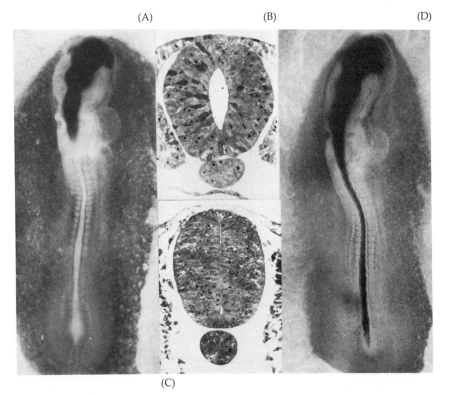

(C)

FIGURE 14
Occlusion of the neural tube to allow expansion of the future brain region. (A) Dye injected into the anterior portion of a 3-day chick neural tube will fill the brain region but does not pass into the spinal region. (B,C) Section of the chick neural tube at the base of the brain (B) before occlusion and (C) during occlusion. (D) Reopening of occlusion after initial brain enlargement allows dye to pass from brain region into spinal cord region. (Photographs courtesy of M. Desmond.)

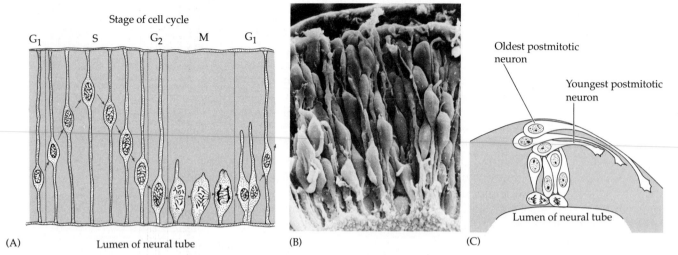

Stage of cell cycle

G₁   S   G₂   M   G₁

(A)   Lumen of neural tube

(B)   (C)   Lumen of neural tube

Oldest postmitotic neuron

Youngest postmitotic neuron

FIGURE 15

Neuroepithelial cells of the neural tube of a chick embryo. (A) Schematic section showing the position of the nucleus in a neuroepithelial cell as a function of the cell cycle. Mitotic cells are found near the center of the neural tube, adjacent to the lumen. (B) Scanning electron micrograph of a newly formed neural tube showing cells at different stages of their cell cycles. (C) After their "birthday," young neurons and neuroblasts can be seen on the periphery of the neural tube. (A After Sauer, 1935; B photomicrograph courtesy of K. Tosney; C after Walbot and Holder, 1987.)

dicating that they are no longer participating in DNA synthesis and mitosis. These are the young neuronal and glial (supporting) cells that differentiate at the periphery of the neural tube (Figure 15C; Fujita, 1966; Jacobson, 1968).

The time when its precursor last divides is called the neuron's "birthday," and different types of neuron and glial cells have their birthdays at different times. In some cases, these cells are NEUROBLASTS that will migrate away from the neural tube and differentiate into neurons. Subsequent neural differentiation is dependent upon the position these neuroblasts occupy once outside the region of dividing cells (Jacobson, 1978; Letourneau, 1977). In other cases, the nondividing cells become neurons. In regions of the central nervous system where substantial migration does not occur, the older neurons (those whose birthdays were earlier) are pushed outward by the younger postmitotic cells. Here, the outer layers of cells are older than the inner layers.

As the cells adjacent to the lumen continue to divide, the migrating cells form a second layer around the original neural tube. This layer becomes progressively thicker as more cells are added to it from the germinal neuroepithelium. This new layer is called the MANTLE ZONE (Figure 16A), and the germinal epithelium is now called the EPENDYMA. The mantle zone cells differentiate into both neurons and glia. The neurons make connections among themselves and send forth axons away from the lumen, thereby creating a cell-poor MARGINAL ZONE. Eventually, glial cells cover many of these marginal zone axons in myelin sheaths, thereby giving them a whitish appearance. Hence, the mantle zone, containing the cell bodies, is often referred to as the GRAY MATTER; and the axonal, marginal layer is often called the WHITE MATTER.

In the spinal cord and medulla, this basic three-zone pattern of ependymal, mantle, and marginal layers is retained throughout development. The gray matter (mantle) gradually becomes a butterfly-shaped structure surrounded by white matter; and both become encased in connective tissue. As the neural tube matures, a longitudinal groove—the SULCUS LIMITANS—appears to divide it into dorsal and ventral halves. The dorsal portion receives inputs from sensory neurons whereas the ventral portion is involved in effecting various motor functions.

In the brain, however, cell migration, differential growth, and selective cell death produce modifications of the three-zone pattern, especially in the cerebellum and cerebrum (Figure 16 B,C,D). Within the deeper layers

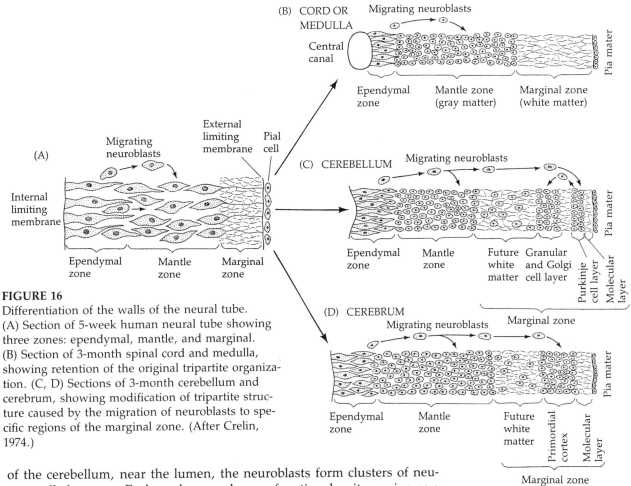

FIGURE 16
Differentiation of the walls of the neural tube.
(A) Section of 5-week human neural tube showing three zones: ependymal, mantle, and marginal.
(B) Section of 3-month spinal cord and medulla, showing retention of the original tripartite organization. (C, D) Sections of 3-month cerebellum and cerebrum, showing modification of tripartite structure caused by the migration of neuroblasts to specific regions of the marginal zone. (After Crelin, 1974.)

of the cerebellum, near the lumen, the neuroblasts form clusters of neurons called NUCLEI. Each nucleus works as a functional unit, serving as a relay station between the outer layers of the cerebellum and other parts of the brain. Other gray matter neuroblasts migrate over the outer surface of the developing cerebellum, forming a new germinal zone near the outer boundary of the neural tube. The neuroblasts formed by these germinal cells migrate back into the developing cerebellar white matter to produce a region of GRANULAR NEURONS. The original ependymal layer of the cerebellar cortex generates a wide variety of neurons and glial cells, including the distinctive and large PURKINJE NEURONS. Each Purkinje neuron has an enormous DENDRITIC APPARATUS, which spreads like a fan above a bulblike cell body (Figure 17). A typical Purkinje cell may form as many as 100,000 synapses with other neurons, more than any other neuron studied. Each Purkinje neuron also emits a slender AXON, which connects to other cells in the deep cerebellar nuclei.

The development of spatial organization, then, is critical for the proper functioning of the cerebellum. All impulses eventually regulate the activity of the Purkinje cells, which are the only output neurons of the cerebellar cortex. For this to happen, the proper cells must differentiate at the appropriate place and time. How is this accomplished?

The primary mechanism for positioning young neurons within the developing mammalian brain is glial guidance (Rakic, 1972; Hatten, 1990). Throughout the cortex, neurons are seen to ride "the glial monorail" to their respective destinations. In the cerebellum, the granule neuron precursors travel upon the long processes of the BERGMANN GLIA (Figure 18; Rakic and Sidman, 1973; Rakic, 1975). The migrating neuron is elongated, with its nucleus towards the posterior of the cell. The leading end of the

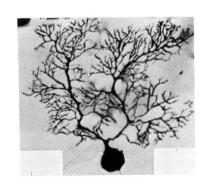

FIGURE 17
Normal mouse Purkinje neuron. (From Berry et al., 1980; photograph courtesy of M. Berry.)

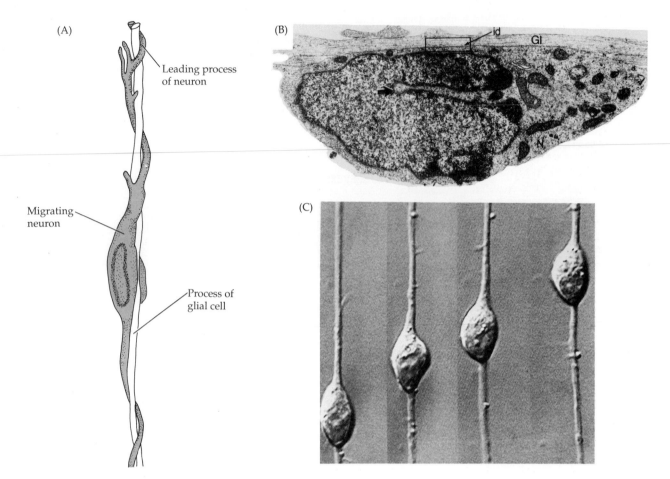

(A)

Leading process
of neuron

Migrating
neuron

Process of
glial cell

(B)

(C)

FIGURE 18
Neuronal migration upon glial cell processes. (A) Diagram of cortical neuron
migrating upon glial cell process. (B) Electron micrograph of the region where
the neuron soma adheres to the glial process. (C) Neuron migrating upon cerebellar glial process. The leading process has several filopodial extensions. It
reaches speeds up to 60 μm per hour as it progresses on the glial process. (A
after Rakic, 1975; B from Gregory et al., 1988; C from Hatten, 1990; photographs
courtesy of M. Hatten.)

neuron is a thin process that twines around the glial fiber. The neural–
glial interaction is a complex and fascinating series of events. First, when
the neuronal cell membrane makes contact with the glial membrane, the
glial cell stops proliferating and begins to differentiate and extend its glial
process. In different regions of the brain, glia generate distinct and distinctly different morphologies. Some processes are long, others short;
some are branched, others are straight. Recombination experiments have
shown that the morphologies are glial-specific and are not dependent on
the type of neuron interacting with the glial cell (Hatten, 1990). The nerve
cell then binds to the glial cell, migrates along the process, and continues
to signal the glial cell to retain its differentiated form. The neuron maintains its adhesion to the glial cell through a number of proteins, the most
important being an adhesion protein called astrotactin. If the astrotactin
on the nerve cells is masked by antibodies to that protein, the nerve cell
will fail to adhere to the glial processes (Edmondson et al., 1988).

Some insight into the mechanisms of spatial ordering has come from
the analysis of neurological mutations in mice. Over 30 mutations are
known to affect the arrangement of cerebellar neurons. Many of the cer-

ebellar mutants have been found because the phenotype of such conditions can be easily recognized. Mice with the *weaver* mutation have stunted Purkinje neurons that are often in the wrong layer. This Purkinje cell defect appears to be secondary to a genetic defect in the granule cells. This was shown by constructing chimeric mice from wild-type and *weaver* embryos (Goldowitz and Mullen, 1982). The wild-type cells could be recognized by their lower level of β-glucuronidase (which can be stained histochemically) and their different nuclear shape. In the *weaver* mouse cerebellum, there were fewer Purkinje cells and those that were present were not in their correct location in the cortex. In the chimera made with *weaver* and wild-type embryos, some of the Purkinje cells were also in the wrong region. These ectopic cells included neurons of *both* genotypes. However, all the wild-type granule cells were in their correct positions, while the granule cells of the *weaver* embryo were randomly scattered across the cerebellar cortex (Figure 19A). The defect in the *weaver* mice thus appears to involve the inability of the granule cells to migrate from the outer germinal layer to their proper cortical site.

Hatten and her colleagues (1986) have demonstrated that the granule cells of *weaver* mice fail to recognize the Bergmann glia as substrates on which to migrate. The investigators cultured different combinations of wild-type and *weaver* granule cells and glial cells (Figure 19B) and found that wild-type granule cells adhered tightly to either wild-type or *weaver* glial cells (as they would when migrating). However, granule cells from *weaver* embryos did not form close appositions to either *weaver* or wild-type glia. In 1988, Edmondson and colleagues found that the *weaver* mouse granule cells had little or no functional astrotactin on their surfaces. These data strongly suggest that the genetic defect of *weaver* involves the inability of the granule cells to recognize and migrate down the surface of the Bergmann glial cells. In wild-type mice, then, neuronal migration is often seen to result from the neuronal cells' ability to cause the differentiation of the glial cell and to then migrate upon the glial process. The mechanisms by which the movement of neurons occurs in only one direction and how the cells are informed to enter and leave the "glial monorail" is still unknown.

In the cerebrum, the three-zone arrangement is also modified. Certain neuroblasts from the mantle zone migrate upon the glia through the white matter to generate a second zone of neurons. This new mantle zone is called the NEOPALLIAL CORTEX. This cortex eventually stratifies into six

FIGURE 19
Failure of granule cell migration in the *weaver* mutant cerebellum. (A) Schematic diagram of the neurons in a chimeric mouse created by fusing a wild-type and a heterozygous *weaver* mouse embryo. The wild-type cells are in color, while the cells from the weaver embryo are shown in black. Nearly all the granule cells in abnormal positions are from the *weaver* embryo. (B) Combinations of cells cultured together from wild-type and weaver embryos. The wild-type granule cells form close appositions to either wild-type of *weaver* glial cells, but the *weaver* granule cells do not form tight associations with either type of glial cell. (A after Goldowitz and Mullen, 1982; B after Hatten et al., 1986.)

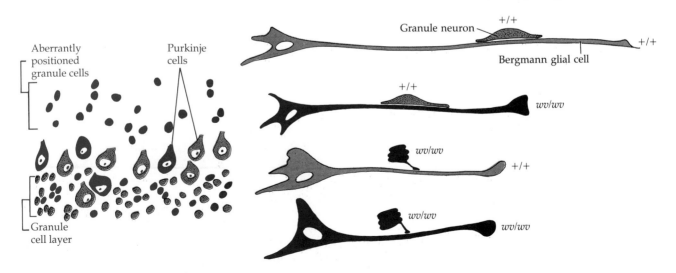

layers of cell bodies, and the adult forms of these neopallial neurons are not completed until the middle of childhood. The development of the human neopallial cortex is particularly striking. The human brain continues to develop at fetal rates even after birth (Holt et al., 1975). Based on morphological and behavioral criteria, Portmann (1941, 1945) suggested that, compared to other primates, human gestation should really be 21 months long instead of nine. However, no woman could deliver a 21-month-old infant because the head would not pass through the birth canal. Thus, humans give birth at the end of nine months rather than at the end of 21 months. Montagu (1962) and Gould (1977) have suggested that during our first year of life we are essentially extrauterine fetuses, and they speculate that much of human intelligence comes from the stimulation of the nervous system as it is forming during that first year.

Neuronal types. The human brain consists of over 10^{11} nerve cells (neurons) associated with over 10^{12} supporting glial cells. Those cells that remain as integral components of the neural tube lining become EPENDYMAL CELLS. These cells can give rise to the precursors of neurons and glial cells. It is thought that the differentiation of these precursor cells is largely determined by the environment that they enter (Rakic and Goldman, 1982), and that, at least in some cases, a given precursor cell can form both neurons and glial cells (Turner and Cepko, 1987). There is a wide variety of neuronal and glial types (as is evident from a comparison of the relatively small granule cell with the enormous Purkinje neuron). Some neurons develop only a few specialized regions where other cells can relay electrical impulses, whereas other neurons develop extensive areas for cellular interaction. The fine extensions of the cell that are used to pick up electrical impulses are called DENDRITES (Figure 20). Very few dendrites can be found on cortical neurons at birth, but one of the amazing things about the first year of life is the increase in the number of such receptive regions in the cortical neurons. During this year, each cortical neuron develops enough dendritic surface to accommodate as many as 100,000 connections with other neurons. The average cortical neuron connects with 10,000 other neural cells. This pattern of neural connections (SYN-

FIGURE 20
Diagrammatic representation of a motor neuron. Impulses received by the dendrites and the stimulated neuron can transmit electrical impulses through the axon (which may be 2–3 feet long) to its target tissue. The myelin sheath, which provides insulation for the axon, is formed by adjacent Schwann cells. (After Bloom and Fawcett, 1975.)

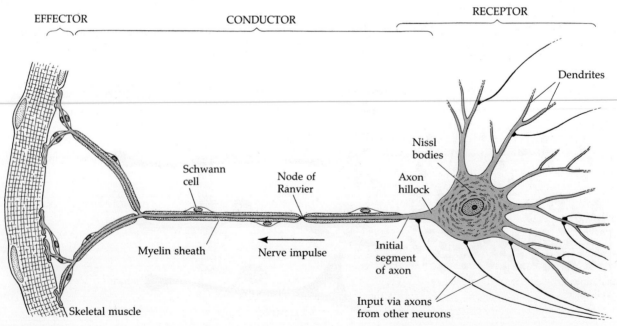

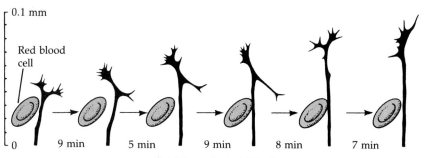

0.1 mm

Red blood cell

0

9 min   5 min   9 min   8 min   7 min

Total time elapsed: 38 min

APSES) enables the human cortex to function as the center for learning, reasoning, and memory, to develop the capacity for symbolic expression, and to produce voluntary responses to interpreted stimuli.

Another important feature of a developing neuron is its AXON. Whereas the dendrites are often numerous and do not extend far from the nerve cell body (or SOMA), axons may extend for several feet. Thus, the rhythms of the heart are controlled by nerves whose cell bodies are located as far away as the medulla oblongata. The pain receptors on one's big toe must transmit their message a long way, to the spinal cord. One of the fundamental concepts of neurobiology is that the axon is a continuous extension of the nerve cell body. At the turn of the last century, there were still numerous competing theories of axon formation. Schwann, one of the founders of the cell theory, believed that numerous neural cells linked themselves together in a chain to form an axon. Hensen (the discoverer of the embryonic node) thought that the axon formed around preexisting cytoplasmic threads between the cells. Wilhelm His (1886) and Santiago Ramón y Cajal (1890) postulated that the axon was indeed an outgrowth (albeit an extremely large one) of the nerve soma.

In 1907 Ross Harrison demonstrated the validity of the outgrowth theory in an elegant experiment that founded both the science of developmental neurobiology and the technique of tissue culture. Harrison isolated a portion of neural tube from a 3-mm frog tadpole. At this stage, shortly after the closure of the neural tube, there is no visible differentiation of axons. He placed these neuroblasts in a drop of frog lymph on a coverslip and inverted the coverslip over a depression slide so he could watch what was happening within this "hanging drop." What Harrison saw (Figure 21) was the emergence of the axons as outgrowths from the neuroblasts, elongating at about 56 μm/hr.

Such nerve outgrowth is led by the tip of the axon, which is called the GROWTH CONE (Figure 22). This cone does not proceed in a straight line but rather "feels" its way along the substrate. The growth cones move by the elongation and contraction of pointed filopodia called MICROSPIKES. These microspike filopodia contain microfilaments, which are oriented in an array parallel to the long axis of the axon. (This is similar to the situation seen in the filopodial microfilaments of secondary mesenchymal cells in echinoderms.) Treating neurons with cytochalasin B will destroy the actin microspikes and inhibit their further advance (Yamada et al., 1971; Forscher and Smith, 1988). Within the axon itself, structural support is provided by microtubules, and the axon will retract if placed in a solution of colchicine (Figure 23). Thus, the developing neuron retains the same features that we have already noted in neural tube formation, namely, elongation by microtubules and apical shape changes by microfilaments. As in most migrating cells, moreover, the exploratory filopodia of the axon

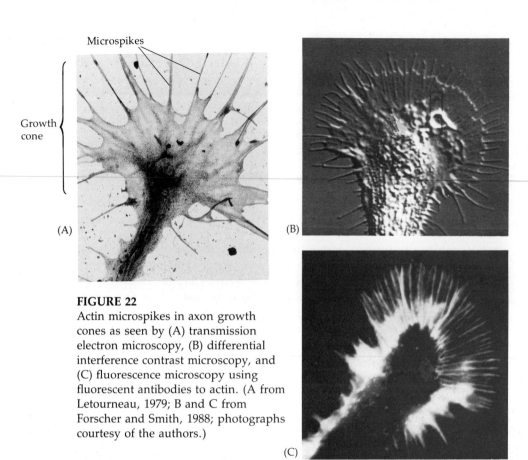

Microspikes

Growth
cone

(A)

(B)

(C)

FIGURE 22
Actin microspikes in axon growth
cones as seen by (A) transmission
electron microscopy, (B) differential
interference contrast microscopy, and
(C) fluorescence microscopy using
fluorescent antibodies to actin. (A from
Letourneau, 1979; B and C from
Forscher and Smith, 1988; photographs
courtesy of the authors.)

will attach to the substrate and exert a force that pulls the rest of the cell
forward. Axons will not grow if the growth cone fails to advance (Lamoureux et al., 1989).

Neurons transmit electrical impulses from one region to another. These
impulses usually go from the dendrites into the nerve soma, where they
are focused into the axon. To prevent dispersion of the electrical signal
and to facilitate its conduction, the axon in the central nervous system is
insulated at intervals by processes that originate from a type of glial cell
called an OLIGODENDROCYTE. An oligodendrocyte wraps itself around the
developing axon. It then produces a specialized cell membrane that is rich
in myelin basic protein and that spirals around the central axon (Figure
24). This specialized membrane is called a MYELIN SHEATH. (In the peripheral nervous system, a glial cell called the SCHWANN CELL accomplishes
this myelination.) The myelin sheath is essential for proper neural function, and demyelination of nerve fibers is associated with several severely
debilitating or lethal diseases.

The axon must also be specialized for secreting a specific neurotransmitter across the small gaps (synaptic clefts) that separate the axon of one
cell from the surface of its target cell (the soma, dendrites, or axon of a
receiving neuron or a receptor site on a peripheral organ). Some neurons
become able to synthesize and secrete acetylcholine, while other neurons
develop the enzymatic pathways for making and secreting epinephrine,
norepinephrine, octopamine, serotonin, γ-amino butyric acid, dopamine,
or some other neurotransmitter. Each neuron must activate those genes
responsible for making the enzymes that can synthesize its neurotransmitter. Thus, neuronal development involves both structural and molecular differentiation.

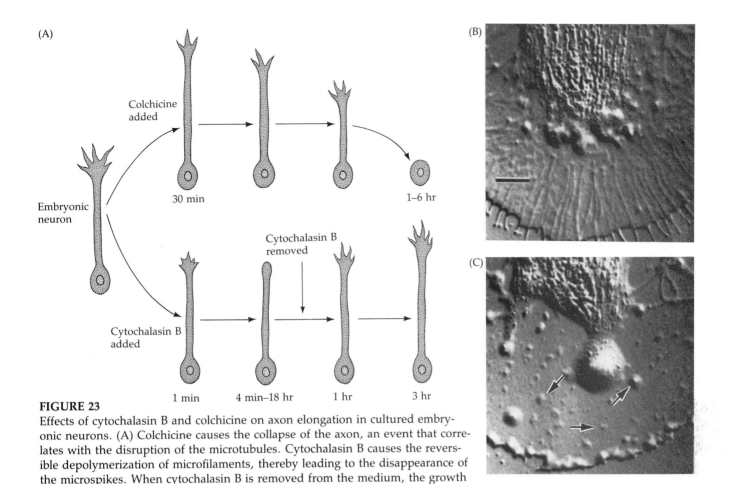

FIGURE 23

Effects of cytochalasin B and colchicine on axon elongation in cultured embryonic neurons. (A) Colchicine causes the collapse of the axon, an event that correlates with the disruption of the microtubules. Cytochalasin B causes the reversible depolymerization of microfilaments, thereby leading to the disappearance of the microspikes. When cytochalasin B is removed from the medium, the growth cone reappears. (B) Normal growth cone of an *Aplysia* neuron in culture. (C) The same neuron as B after being exposed 9 minutes to cytochalasin B. The punctate aggregates of actin (arrows) serve as foci of actin extension when the drug is removed. (B and C from Forscher and Smith, 1988; courtesy of P. Forscher.)

Development of the eye

Dynamics of optic development.   An individual gains knowledge of its environment through its sensory organs. The major sensory organs of the head develop from the interactions of the neural tube with a series of epidermal thickenings called the CRANIAL ECTODERMAL PLACODES. The most anterior placodes are the two olfactory placodes that form the ganglia for the olfactory nerves, which are responsible for the sense of smell. The auditory placodes similarly invaginate to form the inner ear labyrinth whose neurons form the acoustic ganglion, which enables us to hear. In this section, we will focus on the eye, because this organ, perhaps more than any other organ in the body, must develop with precision and perfect coordination of all its components.

The story of optic development begins at gastrulation when the involuting endoderm and mesoderm interact with the adjacent prospective head ectoderm. This interaction gives a lens-forming bias to the head ectoderm* (Saha et al., 1989). But not all parts of the head ectoderm eventually form lenses, and the lens has to be in a precise relationship with the retina. The activation of this latent lens-forming ability and the

*The inductions forming the eye will be detailed in Chapters 8 and 16.

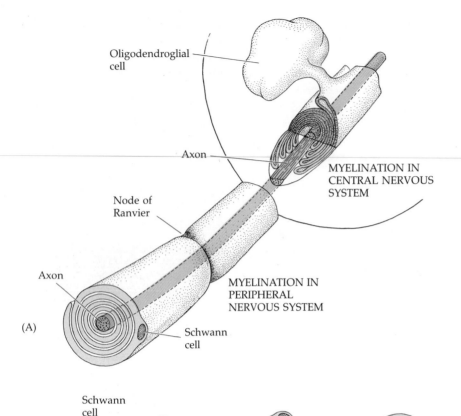

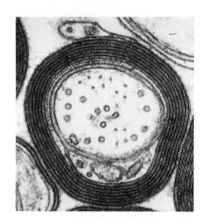

FIGURE 24
Myelination in the central and peripheral nervous systems. (A) In the peripheral nervous system, Schwann cells wrap themselves around the axon; in the central nervous system, myelination is accomplished by the processes of oligodendrocytes. (B) The mechanism of this wrapping entails the production of an enormous membrane complex. (C) Micrograph of an axon enveloped by the myelin membrane of a Schwann cell. (Photograph courtesy of C. S. Raine.)

Labels in figure: Oligodendroglial cell; Axon; MYELINATION IN CENTRAL NERVOUS SYSTEM; Node of Ranvier; Axon; MYELINATION IN PERIPHERAL NERVOUS SYSTEM; Schwann cell; (A); Schwann cell; (B); Axon

positioning of the lens in relation to the retina is accomplished by the OPTIC VESICLE. In humans, the optic vesicles begin as two bulges from the lateral walls of the 22-day embryonic diencephalon (Figure 25). These bulges continue to grow laterally from the neural tube, changing their shape to produce cup-shaped optic vesicles connected to the diencephalon by the OPTIC STALKS. Subsequently, when these vesicles contact the head ectoderm, the ectoderm thickens into the LENS PLACODES. The necessity for close contact between the optic vesicles and the surface ectoderm is seen in both experimental cases and in certain mutants. For example, in the mouse mutant *eyeless*, the optic vesicles fail to contact the surface and eye formation ceases (Webster et al., 1984).

Once formed, the lens placode reciprocates and causes changes in the optic vesicle. The vesicle invaginates to form a double-walled OPTIC CUP (Figures 25 and 26). As the invagination continues, the connection between the optic cup and the brain is reduced to a narrow slit. At the same time, the two layers of the optic cup begin to differentiate in different directions. The cells of the outer layer produce pigment and ultimately become the PIGMENTED RETINA (one of the few tissues other than the neural crest cells that can synthesize its own melanin). The cells of the inner layer proliferate rapidly and generate a variety of light-sensitive photoreceptor neurons, glia, interneurons, and ganglion cells. Collectively, these cells constitute the NEURAL RETINA. The axons from the ganglion cells of the neural retina meet at the base of the eye and travel down the OPTIC STALK. This stalk is then called the OPTIC NERVE.

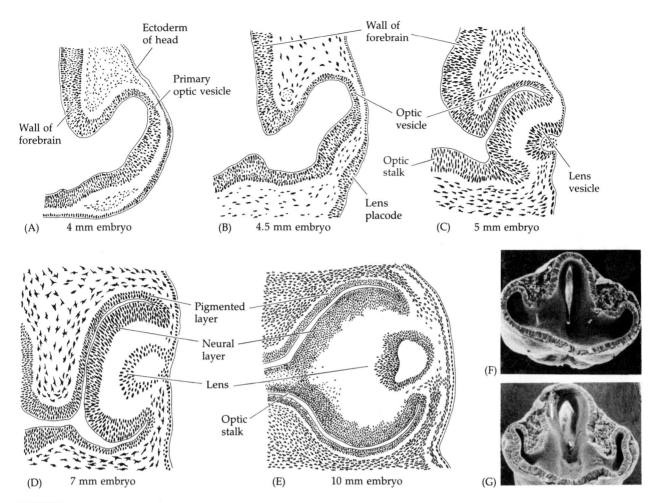

FIGURE 25
Development of the vertebrate eye. (A) Optic vesicle evaginates from the brain and contacts the overlying ectoderm. (B, C) Overlying ectoderm differentiates into lens cells as the optic vesicle folds in on itself. (D) Optic vesicle becomes the neural and pigmented retina as the lens is internalized. (E) The lens induces the overlying ectoderm to become the cornea as the optic stalk develops to carry impulses from the eye to the brain. (F) and (G) are scanning electron micrographs of chick embryos corresponding to (A) and (B). (A–E from Mann, 1964; F–G from Hilfer and Yang, 1980. Photographs courtesy of S. R. Hilfer).

Neural retina differentiation. Like the cerebral and cerebellar cortices, the neural retina develops into a layered array of different neuronal types (Figure 27). These layers include the light- and color-sensitive photoreceptor cells, the cell bodies of the ganglion cells, and the bipolar interneurons that transmit the electrical stimulus from the rods and cones to the ganglion cells. In addition, there are numerous glial cells that serve to maintain the integrity of the retina, as well as amacrine neurons, and horizontal neurons that transmit electrical impulses horizontally.

In the early stages of retinal development, cell division from a germinal layer and the migration and differential death of the resulting cells form the striated, laminar pattern of the neural retina. The formation of this

FIGURE 26
Scanning electron micrograph of the formation of the optic cup and lens placode in a chick. (From Hilfer and Yang, 1980; courtesy of S. R. Hilfer.)

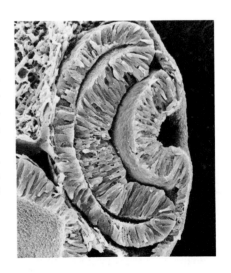

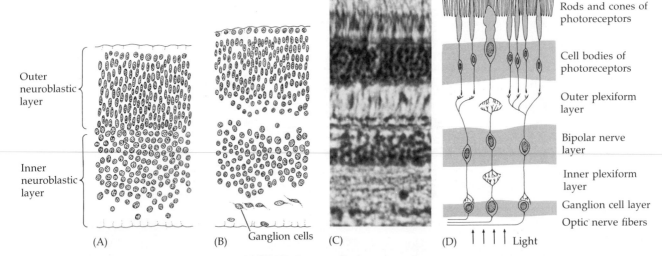

Outer
neuroblastic
layer

Inner
neuroblastic
layer

(A)

(B) Ganglion cells

(C)

(D) ↑ ↑ ↑ ↑ Light

Rods and cones of
photoreceptors

Cell bodies of
photoreceptors

Outer plexiform
layer

Bipolar nerve
layer

Inner plexiform
layer

Ganglion cell layer

Optic nerve fibers

FIGURE 27

Development of the human retina. Retinal neurons sort out into functional layers during development. (A,B) Initial separation of neuroblasts within the retina.
(C) The three layers of neurons in the adult retina and the synaptic layers between them. (D) A functional depiction of the major neuronal pathway in the
retina. Light traverses the layers until it is received by the photoreceptors. The
axons of the photoreceptors synapse with bipolar neurons that transmit the de-
polarization to the ganglion neurons. The axons of the ganglion cells join to
form the optic nerve that enters the brain. (A and B after Mann, 1964; Photograph courtesy of G. Grunwald.)

highly structured tissue is one of the most intensely studied problems of
developmental neurobiology. It has recently been shown (Turner and
Cepko, 1987) that a single retinal neuroblast precursor cell can give rise to
at least three types of neurons or to two types of neurons and a glial cell.
This analysis was performed using a most ingenious technique to label
the cells generated by one particular precursor cell. Newborn rats (whose
retinas are still developing) were injected into the back of their eyes with
a virus that can integrate into their DNA. This virus contained a β-galactosidase gene (not present in rat retina) that would be expressed in the
infected cells. A month after infecting the rat eyes, the retinas were removed and stained for the presence of β-galactosidase. Only the progeny
of the infected cells should have stained blue. Figure 28 shows one of the
stripes of cells derived from an infected precursor cell. The stain can be
seen in five rods, a bipolar neuron, and a retinal (Müller) glial cell.

FIGURE 28

Determination of the lineage of a precursor cell in the rat retina. (A) Technique whereby a virus containing a
functional β-galactosidase gene is injected into the back of the eye to infect
some of the retinal precursor cells.
After a month to 6 weeks, the eye is
removed and the retina is stained for
the presence of β-galactosidase. (B)
Stained cells forming a strip across the
neural retina, including five rods (r), a
bipolar neuron (bp), a rod terminal (t),
and a Müller glial cell (mg). The identities of these cells were confirmed by
Nomarski phase contrast microscopy.
Scale bar, 20 μm. (From Turner and
Cepko, 1987; photograph courtesy of
D. Turner.)

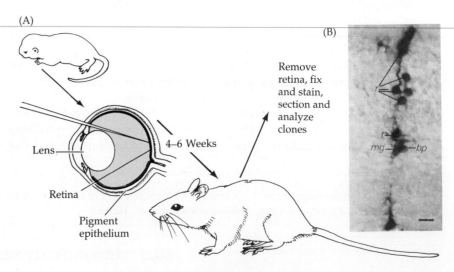

(A)

(B)

Remove
retina, fix
and stain,
section and
analyze
clones

Lens

Retina

Pigment
epithelium

4–6 Weeks

mg — bp

Of the three major neuronal types in the retina (ganglion, bipolar, and photoreceptor), the photoreceptive rods and cones are probably the last of the neurons to complete their differentiation. As they develop, the cell bodies of these outer neurons produce a bud of cytoplasm that contains several specialized organelles. These organelles elongate the bud and adjust the size and shape of the photoreactive regions (Detwiler, 1932). The cell membrane of these cells folds back upon itself to form sacs upon which the photoreceptive pigments are placed. Light induces these pigments to undergo chemical changes that ultimately result in a change of membrane potential. This change in membrane potential effects the release of neurotransmitters to a group of bipolar neurons that relay the electrical signal to the ganglion cells. The ganglion cells, whose axons bundle together to form the optic nerve, relay this information to the brain (Fesenko et al., 1985; Stryer, 1986).

SIDELIGHTS & SPECULATIONS

Why babies don't see well

Human newborns see poorly. There may be several reasons for this, but one of the most striking is the immaturity of the retinal photoreceptors. Anatomical studies by Yuodelis and Hendrickson (1986) and physical studies by Banks and Bennett (1988) have shown that the cone photoreceptors of the newborn central retina are over 7.5 μm in diameter, reaching the normal adult 2 μm width in about three years. During this time, the cone density in this region increases from 18 photoreceptors per 100 μm to 42 photoreceptors per 100 μm, and the photoreceptors develop both their outer segments (that catch the light) and their basal axonal processes. Figure 29 highlights the differences between the photoreceptors of neonatal and adult retinas. One can see that the neonatal retina has poorly differentiated photoreceptors, and those photoreceptors it does have are so wide that not many of them can fit into a given area. Banks and Bennett calculate that this enables the central region of the adult retina to absorbs light about 350 times more efficiently than the same region of the newborn retina. This low of number of photoreceptors per retinal area also causes the babies to be unable to discriminate between two points that are at a distance. This may be why a newborn responds to visual stimuli only when they are brought close to the baby's face. The development of the human retinal photoreceptor provides an excellent example of differentiation that begins early in development but which is not complete until years after birth.

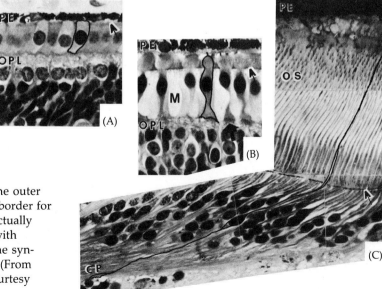

FIGURE 29
Development of the cone photoreceptors in the central region of the human retina. Stained light-microscope sections were photographed and one cone in each retina outlined for clarity. The pigment epithelium (PE), outer plexiform layer (OPL), Müller glia (M) and outer segments of the photoreceptor (OS) have been labeled. (A) Fetus of 22 weeks' gestation. (B) Newborn 5 days after birth. (C) 72-year-old. The arrow points to the outer limiting membrane, which served as the original border for the retinal neurons. The axon outlined in (C) is actually shorter than normal, thus allowing the synapse with the bipolar neuron to be shown in the picture. The synapse is formed at the cone synaptic pedicle (CP). (From Yuodelis and Hendrickson, 1986; photographs courtesy of A. Hendrickson.)

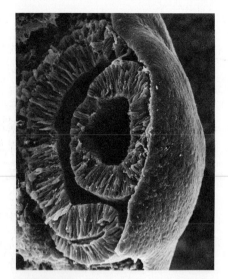

FIGURE 30
Scanning electron micrograph of the rounded and internalized lens placode (now called the optic vesicle) in the chick embryo. (Courtesy of K. W. Tosney.)

Lens and cornea differentiation. During its continued development into a lens, the lens placode rounds up and contacts the new overlying ecto-derm (Figure 30). The lens placode then induces the ectoderm to form the transparent CORNEA. Here, physical parameters play an important role in the development of the eye. Intraocular fluid pressure is necessary for the correct curvature of the cornea so that light can be focused upon the retina. The importance of such ocular pressure can be demonstrated ex-perimentally; the cornea will not develop its characteristic curve when a

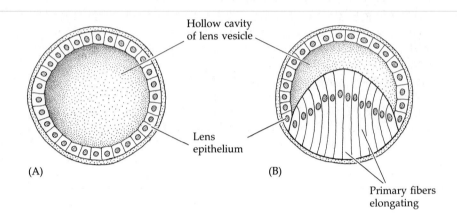

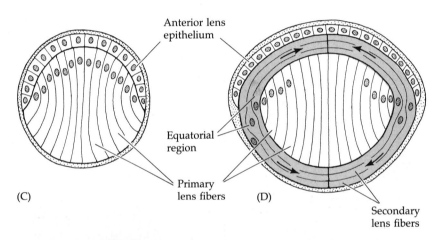

FIGURE 31
Differentiation of the lens cells (A) Lens vesicle as shown in Figure 29. (B) Elongation of the interior cells, pro-ducing lens fibers. (C) Lens filled with crystallin-synthesizing cells. (D) New lens cells derived from anterior lens epithelium. (E) As the lens grows, new fibers differentiate. (After Paton and Craig, 1974.)

small glass tube is inserted through the wall of a developing chick eye to drain away intraocular fluids (Coulombre, 1956, 1965). Intraocular pressure is sustained by a ring of scleral bones (probably derived from the neural crest), which act as inelastic restraints. Without these bones, the intraocular pressure could not be maintained, and the cornea would not be properly formed.

The differentiation of the lens tissue into a transparent membrane capable of directing light onto the retina involves changes in cell structure and shape as well as synthesis of lens-specific proteins called CRYSTALLINS. Crystallins are synthesized as cell shape changes occur, thereby causing the lens vesicle to become the definitive lens. The cells at the inner portion of the lens vesicle elongate and, under the influence of the neural retina, produce the lens fibers (Piatigorsky, 1981). As these fibers continue to grow, they synthesize crystallins, which eventually fill up the cell and cause the extrusion of the nucleus. The crystallin-synthesizing fibers continue to grow and eventually fill the space between the two layers of the lens vesicle. The anterior cells of the lens vesicle constitute a germinal epithelium, which keeps dividing. These dividing cells move toward the equator of the vesicle, and as they pass through the equatorial region, they, too, begin to elongate (Figure 31). Thus, the lens contains three regions: an anterior zone of dividing epithelial cells, an equatorial zone of cellular elongation, and a posterior and central zone of crystallin-containing fiber cells. This arrangement persists throughout the lifetime of the animal as fibers are continuously being laid down. In the adult chicken, the differentiation from an epithelial cell to a lens fiber takes two years (Papaconstantinou, 1967). The details of lens and cornea formation are discussed in Chapter 16.

Directly in front of the lens is a pigmented and muscular tissue called the IRIS. These muscles control the size of the pupil (and give an individual his or her characteristic eye color). Unlike the other muscles of the body (which are derived from the mesoderm), the iris is derived from the ectodermal layer. Specifically, the iris develops from a portion of the optic cup that is continuous with the neural retina but which does not make photoreceptors.

The neural crest and its derivatives

Although derived from the ectoderm, the neural crest has sometimes been called the fourth germ layer because of its importance. It has even been said, perhaps hyperbolically, that "the only interesting thing about vertebrates is the neural crest" (in Thorogood, 1989). The neural crest cells migrate extensively and give rise to a bewildering number of differentiated cell types including (1) the neurons and supporting glial cells of the sensory, sympathetic, and parasympathetic nervous systems; (2) the epinephrine-producing (medulla) cells of the adrenal gland; (3) the pigment-containing cells of the epidermis; and (4) skeletal and connective tissue components of the head (Table 1). The fate of the neural crest cells depends to a large degree upon where the cells migrate and settle.

Migration pathways of trunk neural crest cells

As shown in Figure 2, the neural crest is a transient structure, its cells dispersing soon after the neural tube closes. By grafting a portion of the chick neural tube and its associated crest from radioactively or genetically

TABLE 1
Major neural crest derivatives

Pigment cells	Sensory nervous system	Autonomic nervous system	Skeletal and connective tissue	Endocrine
TRUNK CREST (INCLUDING CERVICAL CREST)				

Melanocytes
Xanthophores
(erythrophores)
Iridophores
(guanophores)
in dermis,
epidermis,
and epidermal
derivatives
	Spinal ganglia
	Some contributions to
	vagal (X) root ganglia
		Sympathetic
		  Superior cervical
		  ganglion
		  Prevertebral ganglia
		  Paravertebral ganglia
		  Adrenal medulla
		Parasympathetic
		  Remak's ganglion
		  Pelvic plexus
		  Visceral and enteric
		  ganglia
		Some supportive cells
		Glia (oligodendrocytes)
		Schwann sheath cells
		Some contribution to
		meninges
			Mesenchyme of dorsal
			fin in amphibia
			Walls of aortic
			arches
			Connective tissue of
			parathyroid
				Adrenal
				medulla
				Type I cells
				of carotid
				body
				Parafollicle
				(calcitonin-
				producing)
				cells of
				thyroid

CRANIAL CREST				

Small, belated
contribution
	Trigeminal (V)
	Facial (VII) root
	Glossopharyngeal (IX)
	root (superior
	ganglia)
	Vagal (X) root (jugular
	ganglia)
	
	Supportive cells
		Parasympathetic ganglia
		  Ciliary
		  Ethmoid
		  Sphenopalatine
		  Submandibular
		  Intrinsic ganglia
		  of viscera
			Most visceral
			cartilages
			Trabeculae carneae (ant.)
			Contributes cells to
			posterior trabeculae,
			basal plate, para-
			chordal cartilages
			Odontoblasts
			Head mesenchyme
			(membrane bones)	

Sources: Weston (1970); Bockman and Kirby (1984).

marked embryos into other embryos (Weston, 1963; LeDouarin and Teillet, 1974), investigators have been able to trace the routes of neural crest cell migration (Figure 32). More recent studies have followed neural crest cell migration with fluorescent antibodies that bind almost exclusively to these migrating cells. This procedure allows the identification of individual cells within large areas. Using such antibodies, investigators in several laboratories have identified two major routes on which neural crest cells migrate in the trunk region (Rickmann et al., 1985; Bronner-Fraser, 1986; Teillet et al., 1987; Loring and Erickson, 1987; Figure 33). The first pathway is the VENTRAL PATHWAY, whose neural crest cells become the sensory and sympathetic neurons, adrenomedullary cells, and Schwann cells. This pathway extends ventrally *through* the *anterior* portion of the somite. As can be seen in Figure 34 and the color portfolio, migrating neural crest cells are able to enter the anterior section of these structures but not the posterior section. Some of these cells reach the dorsal aorta, where they give rise to the SYMPATHETIC GANGLIA. These ganglia are clusters of neurons that transmit impulses to target cells when stimulated by spinal cord neurons.*

*The parasympathetic division of the peripheral nervous system is also formed by neural crest cells migrating by this pathway, but only in the cranial and cervical regions of the embryo.

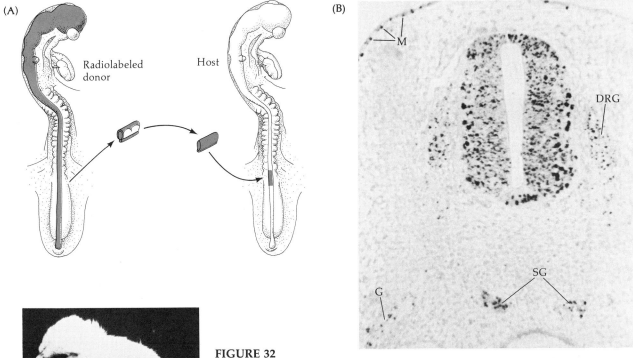

FIGURE 32

Neural crest cell migration. (A) Grafting technique for mapping neural crest cells. A piece of the dorsal axis is excised from a donor embryo; the neural tube and its associated crest are isolated and implanted into a host embryo whose neural tube and crest have been excised. When the donor crest cells are radiolabeled (with tritiated thymidine) or genetically labeled (from a different species or strain), their descendants can be traced in the host embryo as development proceeds. (B) Autoradiograph showing locations of neural crest cells that have migrated from transplanted radioactive donor neural crest to form melanoblasts (M), sympathetic ganglia (SG), dorsal root ganglia (DRG), and glial cells (G). (C) Chick resulting from the transplantation of a neural crest region from a pigmented strain of chicken into the crest region of an unpigmented strain. The crest cells that gave rise to pigment were able to migrate into the wing skin. (After Weston, 1963; photographs courtesy of J. Weston.)

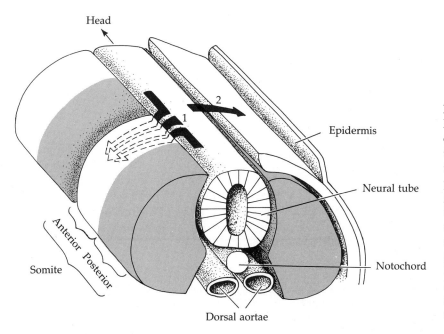

FIGURE 33

Neural crest cell migration in the trunk of the chick embryo. Path 1: Cells travel ventrally through the anterior of the somite. Those cells opposite the posterior portions of the somites migrate laterally on the neural tube until they come to an anterior region of the somite. These cells contribute to the sympathetic and parasympathetic ganglia as well as to the adrenal medullary cells and dorsal root ganglia. Path 2: Cells take a dorsolateral route beneath the ectoderm. These cells become pigment-producing melanocytes.

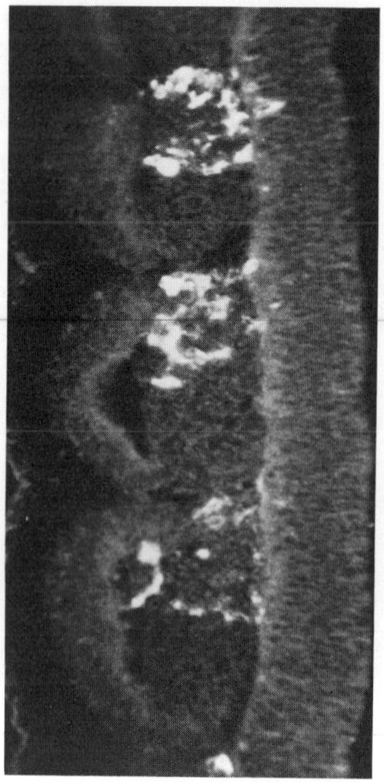

Somite / Neural tube

Anterior

Posterior

Anterior

Posterior

Anterior

Posterior

FIGURE 34

Neural crest cell migration. These fluorescence photomicrographs of longitudinal sections of a 2-day chick embryo are stained with antibody HNK-1, which selectively recognizes neural crest cells. Extensive staining is seen in the anterior, but not in the posterior, half of each somite. (From Bronner-Fraser, 1986a; photograph courtesy of M. Bronner-Fraser.)

At specific regions of the trunk, crest cells migrating along the same pathway aggregate and form the epinephrine-secreting cells of the ADRENAL MEDULLA.

Although neural crest cells emerge uniformly from the neural tube, they only enter the anterior somite. But what happens to those neural crest cells that are adjacent to a *posterior* region of the somite? The antibody recognizing neural crest cells can provide information about space, but it cannot provide information about the direction in which a specific group of cells is traveling. Teillet and co-workers (1987) combined the antibody approach with transplantation of genetically marked quail neural crest cells into chick embryos. The antibody marker recognizes and labels neural crest cells of both species; the genetic marker enables the investigators to distinguish between the quail and chick cells. These studies showed that neural crest cells opposite the posterior regions of the somites migrate anteriorly or posteriorly along the neural tube and then enter the anterior region of their own or adjacent somites. These neural crest cells join with the neural crest cells that were initially opposite the anterior portion of the somite, and they form the same structures. Thus, each dorsal root ganglion is composed of three neural crest populations: one from the neural crest opposite the anterior portion of the somite, and one from each of the adjacent neural crest regions opposite the posterior portions of the somite.

The second major route of neural crest cell migration is the DORSOLATERAL PATHWAY that leads beneath the embryonic ectoderm. The cells of this pathway differentiate into pigment cells (in birds and mammals, the MELANOCYTES). This was shown by a series of classic experiments wherein Mary Rawles (1948) and others transplanted the neural tube and crest from a pigmented strain of chicken into the neural tube of an albino chick embryo. The result (Figure 32C) was a white chicken with a specific region of colored feathers. These neural crest cells travel along the ventral surface of the ectodermal layer from the central dorsal region to the most ventral region of the skin, eventually terminating in the belly skin of the organism. Thus, the neural crest is responsible for the production of all the melanin-containing cells in the organism (with the exception of certain neural derivatives such as the pigmented retina).*

Migration pathways of cranial neural crest cells

In mammalian embryos, cranial neural crest cells migrate before the neural tube is closed (Tan and Morriss-Kay, 1985) and give rise to the facial mesenchyme cells (Johnston, 1975). The crest cells originating in the forebrain and midbrain contribute to the nasal process, palate, and mesenchyme of the first pharyngeal arch (Figure 35). This arch becomes part of the gill apparatus in fish, but in humans, it gives rise to the jawbones, the incus and malleus bones of the middle ear, and a large portion of the facial

---

*It should be noted that the production of norepinephrine and epinephrine (by sympathetic neurons and the adrenal medulla) and melanin (by melanocytes) both involve the hydroxylation of tyrosine and the subsequent oxidation of the product.

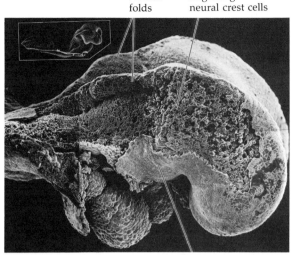

Neural folds   Migrating neural crest cells

First pharyngeal arch

FIGURE 35
Neural crest cell migration in the mammalian head. Scanning electron micrograph of a rat embryo with part of the lateral ectoderm removed from the surface. Neural crest migration can be seen over the midbrain, and the column of neural crest cells migrating into the future first pharyngeal arch is evident. (From Tan and Morriss-Kay, 1985; photograph courtesy of S.-S. Tan.)

musculature (Figure 36). The neural crest cells originating in the anterior hindbrain region generate the mesenchyme of the second pharyngeal arch, which generates the human stapes bone as well as much of the facial cartilage. The cervical neural crest cells contribute to the mesenchyme of the third, fourth, and sixth pharyngeal arches (the fifth degenerates in humans), which produce the neck bones and muscles. Indeed, the "head" is largely the product of the cranial neural crest, and the evolution of the jaws, teeth, and facial musculature arises from changes in the placement of these cells (see Chapter 23).

The mechanisms of neural crest cell migration

Thus, neural crest cells do not migrate randomly through the body; rather, they follow precise pathways. Although the nature of these pathways is still a major question, the migration of neural crest cells appears to be controlled by the substratum over which they travel. In 1941, Paul Weiss speculated that temporary linkages would form between specific molecules

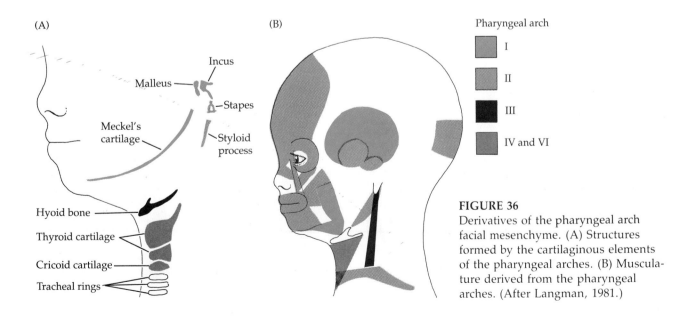

(A)

Incus
Malleus
Stapes
Meckel's cartilage
Styloid process
Hyoid bone
Thyroid cartilage
Cricoid cartilage
Tracheal rings

(B)

Pharyngeal arch

I
II
III
IV and VI

FIGURE 36
Derivatives of the pharyngeal arch facial mesenchyme. (A) Structures formed by the cartilaginous elements of the pharyngeal arches. (B) Musculature derived from the pharyngeal arches. (After Langman, 1981.)

of the cell surface and complementary molecules on the substratum. More recent evidence suggests that this is probably the case for neural crest cells. First, when the neural tube and its associated neural crest are inverted, the crest cells continue to stream out. However, they now move dorsally instead of ventrally, thereby indicating that they keep their original orientation with regard to the neural tube. Moreover, when neural crest cells or their derivatives are placed (either by transplantation or by injection) on a normal neural crest pathway in a host, they will migrate along these pathways. Other embryonic cells will not orient themselves in this manner and will stay where they are placed (Erickson et al., 1980; Bronner-Fraser and Cohen, 1980). Therefore, neural crest cells are able to recognize certain pathways in the embryo and migrate along them.

Some of the best evidence for the importance of extracellular matrices in directing neural crest cell migration comes from studies of mutant salamanders. There exists in some axolotl salamanders a mutation wherein the neural crest forms, but its cells fail to migrate along the dorsolateral pathway. This is most readily seen in the lack of pigment cells anywhere except atop the neural tube of these animals (Figure 37), and these cells eventually degenerate. When wild-type neural crests are transplanted into mutant embryos, the crest cells are unable to migrate. However, when crests from mutant embryos are transplanted into wild-type embryos, their cells migrate normally (Spieth and Keller, 1986). Thus, the defect in this mutant is in the environment that the cells encounter, not in the cells themselves. (The road is deficient, not the vehicle.) Löfberg and co-workers (1989) used this information to show that the extracellular matrix contains critical components regulating neural crest cell migration. They

FIGURE 37

Deficiency of neural crest cell migration in the *d/d* mutant of the axolotl. The larvae of wild-type axolotls (A) are characterized by pigment cells throughout the body except for the most ventral portions. In the *d/d* mutant (B), the neural crest-derived pigment cells form a stripe along the dorsal midline of the larva. Scanning electron micrographs of the embryonic neural crest shows that the neural crest cells of the wild-type embryos (C) migrate over the neural tube into the somites, while those crest cells of the mutant (D) remain atop the neural tube. (From Löfberg et al., 1989; photographs courtesy of the authors.)

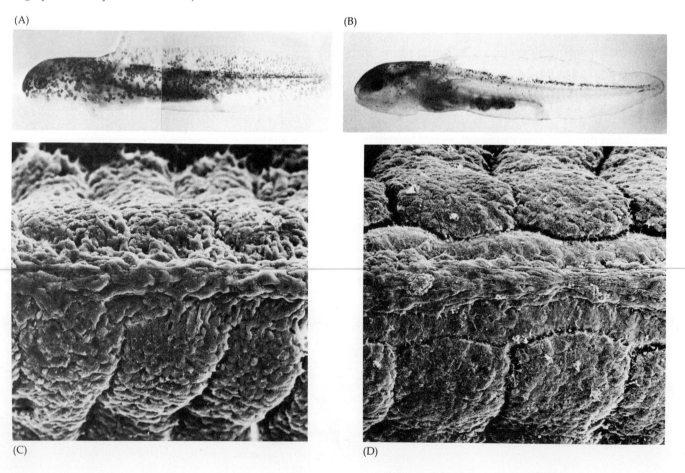

(A) (B)

(C) (D)

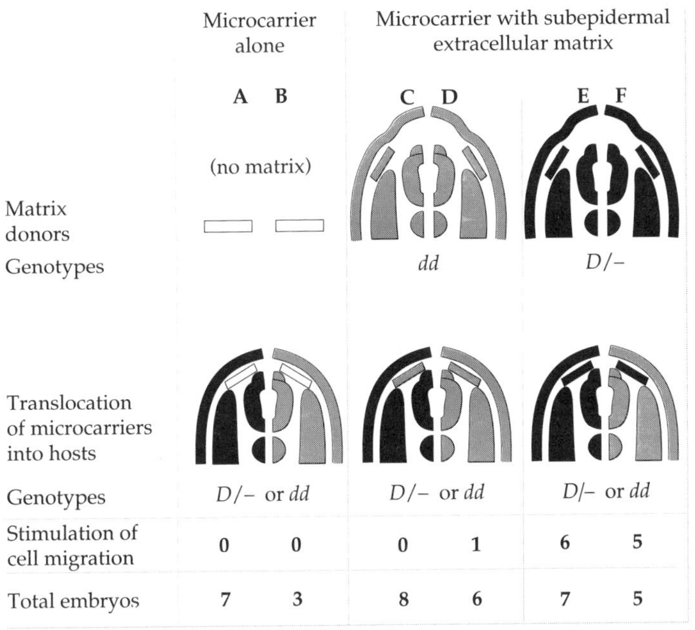

	Microcarrier alone		Microcarrier with subepidermal extracellular matrix			
	A	B	C	D	E	F
Matrix donors	(no matrix)					
Genotypes			*dd*		*D/−*	
Translocation of microcarriers into hosts						
Genotypes	*D/−* or *dd*		*D/−* or *dd*		*D/−* or *dd*	
Stimulation of cell migration	0	0	0	1	6	5
Total embryos	7	3	8	6	7	5

FIGURE 38

Importance of the extracellular matrix in neural crest cell migration. These diagrams represent the transplantation experiments wherein extracellular matrix components were carried between wild-type (dark) and mutant (color) donor and host embryos. The upper row represents the donor embryos, while the lower line shows the wild-type and mutant hosts into which the matrix-bearing microcarriers were added. The stimulation denotes the number of times that neural crest cells were seen migrating from the crest in the region of the microcarrier. (After Löfberg et al., 1989.)

absorbed, onto membrane microcarriers, extracellular matrix from the subepidermal region of the skin (through which the pigment-forming neural crest cells would migrate). They then placed these microcarriers next to the neural crests of mutant and wild-type embryos just before migration would occur. The microcarriers alone did not stimulate migration from either wild-type or mutant crests (Figure 38A,B). Microcarriers containing extracellular matrix material from mutant embryos likewise did not stimulate neural crest cell migration in either mutant or wild-type embryos (Figure 38C,D). The microcarriers containing wild-type extracellular matrix, however, were able to stimulate neural crest cell migration from both the mutant and wild-type crests (Figure 38E,F), demonstrating the importance of the extracellular matrix in neural crest cell migration.

There is much debate over which molecules direct neural crest cell migration. It is widely thought that the initiation of migration depends on the balance between the tendency of the neural crest cells to stick to each other, and the opposing tendency of the crest cells to adhere to the extracellular matrix (Perris and Bronner-Fraser, 1989). Cell migration would begin when the tendency for cell–matrix adhesion exceeded that of cell–cell binding. There is evidence that at the start of neural crest cell migration, both factors are at work: cell–cell adhesion decreases while the matrix changes to become more adhesive to these cells. Cell–cell adhesion is weakened by the disappearance of specific neural cell adhesion molecules from the crest cells' surfaces (Newgreen and Gooday, 1985; Crossin et al. 1985; Takeichi, 1988). At the same time, the extracellular surface becomes more adhesive. Perris and his co-workers (1990) have found that the extracellular matrix in the *white* axolotl (Figure 37) fails to change its components by the time the crest cells are ready to migrate.

Experiments using immunological probes have demonstrated that the extracellular matrix apposing the migrating neural crest cells is a rich mix of molecules such as fibronectin, laminin, tenascin, various collagen molecules, and proteoglycans (which we will discuss in detail in Chapter 15). The question then becomes, which of these molecules is actually crucial in directing (or merely permitting) neural crest cell migration? Experiments undertaken to address this issue have to be designed carefully, since neural crest cells may have different migration requirements in different species and even in different parts of the same embryo. One way of finding these answers is to make antibodies to the regions of the extracellular matrix molecules to which cells bind. When such antibodies are injected into the embryo and block these regions of the matrix, do they perturb neural crest cell migration? The migration of *chick cranial* neural crest cells can be severely altered when antibodies to fibronectin, fibronectin receptors, tenascin, or laminin-heparan sulfate proteoglycan are injected into the developing embryo (Poole and Thiery, 1986; Perris and Bronner-Fraser, 1989). However, these antibodies did not severely alter the migration of *chick trunk* neural crest cells.

Other extracellular matrix components besides these large proteins may also play important roles in determining the pathways of neural crest cell migration. One such factor is hyaluronic acid. It is thought that hyaluronic acid causes the formation of cell-free spaces, which neural crest cells can enter (Pratt et al., 1975; Solursh et al., 1979). This model is similar to that proposed for ingression into the avian blastocoel by mesodermal precursors, and Meier (1981) has found that hyaluronic acid does indeed accumulate in some regions where neural crest cells migrate. Hyaluronic acid may also be important as a link between the cell and the large protein molecules of the extracellular matrix. Hyaluronic acid chains can be seen between the migrating cells and the matrix, and the addition of hyaluronic acid into the embryo can alter the paths of crest cell migration (Perris and Bronner-Fraser, 1989).

In addition, neural crest cells may alter the paths over which they travel, making it more difficult for other neural crest cells to use the same path. Weston and Butler (1966) showed that when older neural crest cells were placed into younger environments, the older cells were able to migrate into all the areas normally colonized by neural crest cells. However, when younger crests were placed into older environments, most cells were restricted to the formation of dorsal root ganglia. This observation suggests that some alterations already had occurred in the environment over which these cells were to travel. Either the passage of the earlier neural crest cells modified the pathway, or further development of the somites may have obliterated the pathway. These observations also suggest that neural crest cells that are destined to form the most distal tissues (i.e, the sympathetic ganglia) are the first cells to leave the crest. This prediction was recently confirmed by Serbedzija and co-workers (1989). Despite all this work, we still do not know how the cells choose which path to take, how they know to pass through some tissues and not others, and how they know when to stop their migration.

Pluripotentiality of neural crest cells

One of the most exciting features of neural crest cells is their PLURIPOTENTIALITY. A single neural crest cell can differentiate into several different cell types depending upon its location within the embryo. For example, the parasympathetic neurons formed by the cervical neural crest cells produce acetylcholine as their neurotransmitter. They are therefore CHOLINERGIC neurons. The sympathetic neurons formed by the thoracic neural crest cells produce norepinephrine. They are ADRENERGIC neurons. But when cervical and thoracic neural crests are reciprocally transplanted, the former thoracic crest is found to produce the cholinergic neurons of the parasympathetic ganglia, and the former cervical crest forms adrenergic neurons in the sympathetic ganglia (LeDouarin et al., 1975). Thus, the thoracic crest cells are capable of developing into cholinergic neurons when they are placed into the neck, and the cervical crest cells are capable of becoming adrenergic neurons when they are placed in the trunk. Kahn and co-workers (1980) found that premigratory neural crest cells from both the thoracic and the cervical regions had the enzymes for synthesizing both acetylcholine and norepinephrine.

The pluripotentiality of some neural crest cells is such that regions of the neural crest that never produce nerves in normal embryos can be made to do so under certain conditions. Mesencephalic neural crest cells normally migrate into the eye and interact with the pigmented retina to become scleral cartilage cells (Noden, 1978). However, if this region of the neural crest is transplanted into the trunk region, it can form sensory ganglia neurons, adrenomedullary cells, glia, and Schwann cells (Schweizer et al., 1983).

Although the above studies observed populations of cells, Bronner-Fraser and Fraser (1988; 1989) have shown that *individual* neural crest cells can indeed be pluripotential. They injected fluorescent dextran beads into individual neural crest cells while the cells were still above the neural tube and then looked to see what types of cells the individual crest cells became after these cells had migrated. The progeny of a single neural crest cell could become sensory neurons, pigment cells, adrenomedullary cells, and glia (Figure 39).

This does not necessarily mean that *all* individual neural crest cells are so pluripotential. Some crest cells may be more determined than others. As the neural crest cells migrate, their potency appears to become restricted. Transplantation studies by LeDouarin and Smith (1988) suggest that many of the neural crest-derived cells that form the sensory neurons of the dorsal root ganglia are incapable of forming the autonomic neurons of the sensory ganglia, and vice versa. The mechanism for the differential success of these committed cells may involve different requirements for growth factors. The committed sensory neuron precursors seem to need a growth factor produced by the neural tube itself. If a thin, nonpermeable membrane is interposed between the neural tube and the prospective dorsal root ganglia region, the ganglia fail to form. Those neural crest cells that have continued their dorsolateral migration to the sympathetic ganglia regions survive. This effect on the dorsal root ganglia can be circumvented by coating the barrier with neural tube extract or with brain-derived neurotrophic factor, a protein derived from the neural tube (Kalcheim et al., 1987). The neural crest cells committed to form the sympathetic ganglia do not need this growth factor for survival. Rather, their differentiation

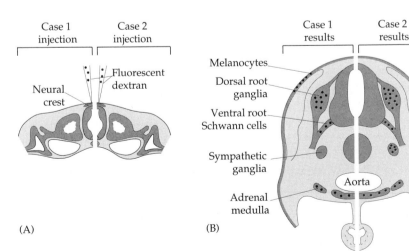

FIGURE 39

Pluripotency of neural crest cells. A single neural crest cell is injected with beads containing a highly fluorescent dextran molecule. The progeny of this cell will each receive some of these fluorescent beads. (A) Injection of beads into neural crest cell shortly before migration of the crest cells is initiated. (B) Two days later, crest-derived tissues contain labeled cells descended from the injected precursor. The figure summarizes data from two different experiments (Case 1 and Case 2). (After Lumsden, 1988.)

appears to be spurred by basic fibroblast growth factor (Kalcheim and Neufeld, 1990).

The final differentiation of the autonomic neural crest cells is also determined in large part by the environment in which these cells develop. It does not involve the selective death of those cells already committed to secreting another type of neurotransmitter (Coulombe and Bronner-Fraser, 1987). Heart cells, for example, secrete a factor (or factors) that can convert adrenergic sympathetic neurons into cholinergic neurons without changing their survival or growth (Chun and Patterson, 1977; Fukada, 1980). Identification of these factors has been difficult, because these molecules appear to be present in very small quantities, and the assay takes 2–3 weeks to complete. However, Fukada (1985) has isolated from cultured heart cells a 45,000-Da glycoprotein that induces the noradrenergic neurons of rat sympathetic ganglia to synthesize acetylcholine and form cholinergic synapses.

Thus, the fate of a neural crest cell can be directed by the hormonal milieu of the tissue environment or the substrate upon which it settles. The chick trunk neural crest cells that migrate into the region destined to become the adrenal medulla can differentiate in two directions. Such cells usually differentiate into noradrenergic sympathetic neurons. However, if these neural crest cells are given glucocorticoids like those made by the cortical cells of the adrenal gland, they differentiate into adrenal medullary cells (Anderson and Axel, 1977; Vogel and Weston, 1990). Adrenergic differentiation is also stimulated by some extracellular matrices but not others (Maxwell and Forbes, 1987). The type of matrix is seen to be important in the differentiation of the salamander neural crest cells, too. If axolotl crest cells are grown on matrices from subepidermal regions (where the pigment cells are found) they become melanocytes. However, if the same crest cells are cultured on matrices from the region of the dorsal root ganglia, they develop the neuronal phenotype (Perris et al., 1988).

From the preceding discussion, it would appear that all neural crest cells are originally identical in their potencies. This, however, is not always the case. Here, again, cranial crest cells differ from trunk crest cells, for only the cells of the cranial neural crest are able to produce the cartilage of the head. Moreover, the cranial neural crest, when transplanted into the trunk region, will participate in forming trunk cartilage that normally does not arise from neural crest components. In at least some cases, these cranial neural crest cells are instructed quite early as to what tissues they can form. Noden (1983) removed regions of chick neural crest that would normally seed the second branchial arch and replaced them with cells that would migrate into the first branchial arch. These host embryos developed two sets of lower jaw structures, since the graft-derived cells also produced a mandible. The bases for this instruction is not known, but one clue comes from a mutation in the Burmese cat (Noden and Evans, 1986). In the 1970s, a mutation occurred within this breed that gave the cats a sought-after blunt-faced phenotype. However, about 25 percent of the kittens born to parents carrying this mutation died from abnormalities of the face. Some of the facial components (such as the soft palate) were missing, while other facial components (such as the whisker pads) were duplicated. Thus it appears that this mutation can cause certain neural crest cells to produce a "wrong" type of structure. Since the prospective forebrain is abnormal before the time that the neural crest cells are dispersed, it is possible that the forebrain regions normally instruct the fates of the adjacent neural crest cells. In this mutation, the instructions are not correctly given, and the crest cells make the inappropriate structures.

The neural crest is a remarkably plastic population of cells that can give rise to an enormous variety of important vertebrate structures. The mechanisms by which these cells arrive at their respective locations and differentiate into their respective phenotypes are some of the most interesting questions in the field of developmental biology and vertebrate evolution.

The epidermis and the origin of cutaneous structures

The cells covering the embryo after neurulation form the presumptive epidermis. Originally, this tissue is one cell-layer thick, but in most vertebrates this shortly becomes a two-layered structure. The outer layer gives rise to the PERIDERM, a temporary covering that will be shed once the bottom layer differentiates to form a true epidermis. The inner layer, called the BASAL LAYER (or STRATUM GERMINATIVUM), gives rise to all the cells of the epidermis (Figure 40). The stratum germinativum first divides to give rise to another, outer population of cells that constitutes the SPINOUS LAYER. These two epidermal layers are referred to as the MALPIGHIAN

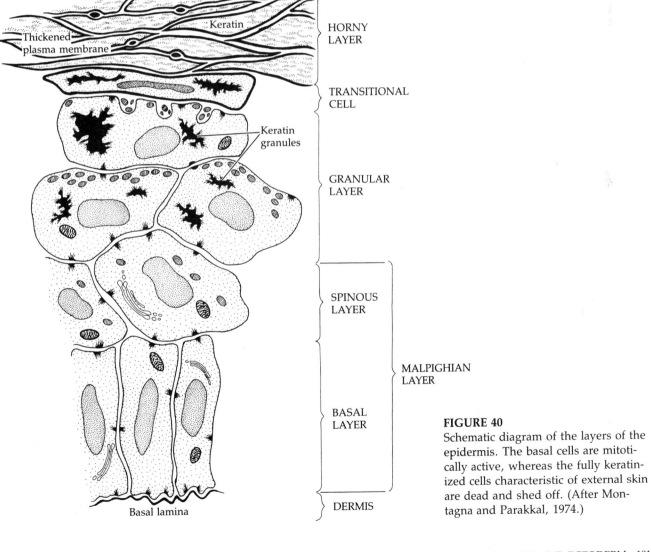

FIGURE 40
Schematic diagram of the layers of the epidermis. The basal cells are mitotically active, whereas the fully keratinized cells characteristic of external skin are dead and shed off. (After Montagna and Parakkal, 1974.)

LAYER. The cells of the Malpighian layer divide to produce the GRANULAR LAYER of the epidermis, so called because the cells are characterized by granules of the protein KERATIN. Unlike the cells remaining in the Malpighian layer, the cells of the granular layer do not divide. Rather, they begin to differentiate into skin cells (KERATINOCYTES). The keratin granules become more prominent as the cells of the granular layer age and migrate outward. Here, they form the HORNY LAYER (STRATUM CORNEUM), in which the cells have become flattened sacs of keratin protein. The nuclei have been pushed to one edge of the cell. Shortly after birth, the cells of the horny layer are shed and are replaced by new cells coming up from the granular layer. Throughout life, the dead keratinized cells of the horny layer are being shed (we humans lose about 1.5 grams each day*) and are replaced by new cells, the source of which is the mitotic cells of the Malpighian layer. The pigment cells from the neural crest also reside in the Malpighian layer, where they transfer their pigment sacs (melanosomes) to the developing keratinocyte. In adult skin, a cell born in the Malpighian layer takes roughly two weeks to reach the stratum corneum. In individuals with psoriasis, a disease characterized by the exfoliation of enormous amounts of epidermal cells, the time required is only two days (Weinstein and van Scott, 1965; Halprin, 1972). It is thought that psoriasis may result from the overexpression of transforming growth factor-α, a protein that normally stimulates epidermal cell division (Elder et al., 1989).

The epidermis alone does not make the skin. As we shall see in the next chapter, a region of mesoderm underlies the epidermis and constitutes the DERMIS of the skin. This layer consists of loose connective tissue embedded in an elastic glycoprotein matrix. The dermis and epidermis interact at specific sites to create the cutaneous appendages: hair, scales, or feathers (depending on the species), sweat glands, and apocrine glands.

The first indication that a hair follicle will form at a particular place is an aggregation of cells in the basal layer of the epidermis. This aggregation occurs at different times and different places in the embryo. The basal cells elongate and divide, sinking into the dermis and forming the hair follicle (Figure 41). At this stage, two epithelial swellings begin to grow. The cells of the upper bulge will form the SEBACEOUS GLANDS, which produce an oily secretion, SEBUM. In many mammals, including humans, the sebum mixes with the desquamated peridermal cells to form the whit-

*Most of this skin becomes "house dust" atop furniture and floors. Should you doubt this, burn some of the dust. It will smell just like singed skin.

FIGURE 41

Development of the hair follicles in fetal human skin. (A) Basal epidermal cells become columnar and bulge slightly into the dermis. (B) Epidermal cells continue to proliferate, and dermal mesenchyme cells collect at the base of the primary hair germ. (C) Elongated hair germ. The uppermost bulge develops into an apocrine sweat gland, and the central bud forms the sebaceous gland. The lowest bulb differentiates into the hair bud. (Photograph courtesy of W. Montagna.)

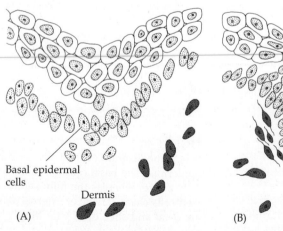

Basal epidermal cells

Dermis

(A)

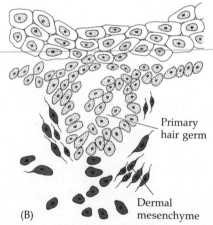

Primary hair germ

Dermal mesenchyme

(B)

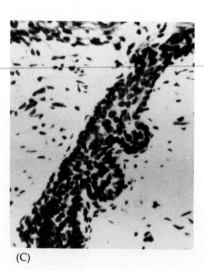

(C)

ish VERNIX CASEOSA, which surrounds the fetus at birth. The lower swelling forms the HAIR BUD, which is then split by upwelling mesenchyme. The germinal epidermis directly above this mesenchyme proliferates to form the keratinized shaft of hair.

The first hairs in the human embryo are of a thin, closely spaced type called LANUGO. This type of hair is usually shed before birth and is replaced (at least in part, by new follicles) by the short and silky VELLUS. Vellus remains on many parts of the human body usually considered hairless, such as the forehead and eyelids. In other areas of the body, vellus gives way to "terminal" hair. During a person's life, some of the follicles that produced vellus can later form terminal hair, and still later revert to vellus production. The armpits of infants, for instance, have follicles that produce vellus until adolescence. At that time, terminal shafts are generated. Conversely, in normal masculine "baldness," the scalp follicles revert back to producing unpigmented and very fine vellus (Montagna and Parakkal, 1974). The placement and pattern of hair, feathers, and scales involves the interactions of the dermis and the epidermis, and these will be discussed in more detail in Chapters 16 and 17.

The germinative cells of the epidermis also give rise to the sweat glands and, what are embryologically modified sweat glands, the mammary glands. Like the formation of the hair bulb, these glands are formed by the ingrowth of the germinative basal epidermal cells in the dermis. The epithelial–mesenchymal interactions necessary for mammary gland development will be addressed in Chapter 19.

SIDELIGHTS & SPECULATIONS

Teratology

The development of an organism is a complex orchestration of cell divisions, cell migrations, cell interactions, gene regulation, and differentiation. Any agent interfering with these processes can cause malformation in the embryo. In fact, it is estimated that about half of the total number of human conceptions do not survive to be born. Most of these embryos express their abnormality so early that they fail to implant in the uterus. Others implant but fail to establish a successful pregnancy. Thus, most abnormal embryos are spontaneously aborted before the woman even knows she is pregnant (Boué et al., 1985). Hertig and Rock (1949) showed that in 34 embryos between 1 and 17 days old, 13 showed some recognizable anomaly. Nearly 90 percent of a sample of fetuses naturally aborted before one month have developmental anomalies (Mikamo, 1970; Miller and Poland, 1970). Edmonds and co-workers (1982), using a sensitive immunological test that can detect the presence of human chorionic gonadotropin (hCG) 8 or 9 days after fertilization, monitored 112 pregnancies in normal women. Of these hCG-determined pregnancies, 67 showed no further evidence of pregnancy.

It appears, then, that many human embryos are impaired early in development and do not survive long in utero. Defects in the lungs, limbs, face, or mouth, however, would not be deleterious to the fetus (which does not depend on those organs while inside the mother), but can seriously threaten life once the baby is born. About 5 percent of all human births have a recognizable malformation, some of them benign, some very serious (McKeown, 1976).

The study of these congenital ("at birth") abnormalities is called TERATOLOGY, and agents responsible for causing these malformations are called TERATOGENS (Gk. meaning "monster-formers"). Teratogens work during certain CRITICAL PERIODS. The most critical time for any organ is when it is growing and forming its particular structures. Different organs have different critical periods, although the time from day 15 to day 60 is critical for many organs. The heart forms primarily during weeks 3 and 4, while the external genitalia are most sensitive during weeks 8 and 9. The brain and skeleton are always sensitive, from the beginning of week 3 to the end of pregnancy and beyond.

There are many types of teratogens. One class are those agents that cause gene mutations. Ionizing radiation and certain drugs can break chromosomes and alter DNA structure. For this reason, pregnant women are told to avoid unnecessary X rays, even though there is no evidence for congenital anomalies resulting from diagnostic radiation (Holmes, 1979). Congenital conditions such as achondroplastic dwarfism (an autosomal dominant leading to shortened limbs and normal torso) or Robert syndrome (an autosomal recessive disease in which the infant has severe limb reduction, cleft palate, and severe mental retardation) are examples of single-gene (Mendelianly inherited) congenital malformations. Conditions such as Down syndrome (an extra copy of chromosome 21 causing mental retardation, heart defects, and the retention of certain fetal muscle patterns) and Klinefelter syndrome (an extra X chromosome in males, causing small testes, sterility, and mild mental retardation) involve extra chromosomes.

Another class of teratogens are viruses. Gregg (1941) first documented the fact that women who had rubella (German measles) during the first third of their pregnancy had a 1 in 6 chance of giving birth to an infant with eye cataracts, heart malformations, and deafness. This was the first evidence that the mother could not fully protect the fetus from the outside environment. The earlier the rubella infection occurred during the pregnancy, the greater the risk that the embryo would be malformed. The first 5 weeks appear to be the most critical, because this is when the heart, eyes, and ears are being formed. The rubella epidemic of 1963–1965 probably resulted in about 20,000 fetal deaths and 30,000 infants with birth defects. Two other viruses, cytomegalovirus and herpes simplex virus, are also teratogenic. Cytomegalovirus infection of early embryos is nearly always fatal, but infection of later embryos can lead to blindness, deafness, cerebral palsy, and mental retardation.

Microorganisms are rarely teratogenic, but two of them can damage human embryos. *Toxoplasma gondii*, a protozoan carried by rabbits and cats (and their feces), can cross the placenta and cause brain and eye defects in the fetus. *Treponema pallidum*, the cause of syphilis, can kill early fetuses and produce congenital deafness in older ones.

The fourth class of teratogens includes drugs and environmental chemicals. Some chemicals that are naturally found in the environment can cause birth defects. Even in the pristine alpine meadows of the Rocky Mountains, teratogens are found. Here grows the skunk cabbage *Veratrum californicum*, upon which sheep can feed. If pregnant ewes eat this plant, their fetuses tend to develop severe neurological damage, including cyclopia, the fusion of two eyes in the center of the face (Figure 42). This condition also occurs in humans, pigs, and many other mammals; the affected organism dies shortly after birth (as a result of severe brain defects including the lack of a pituitary gland).

Quinine and alcohol, two substances derived from plants, can also cause congenital malformations. Quinine can cause deafness, and alcohol (more than 2–3 ounces per day) can cause physical and mental retardation in the infant. Nicotine and caffeine have not been proved to cause congenital anomalies, but women who are heavy smokers (20 cigarettes a day or more) are more likely to have infants smaller than those born to women who do not smoke. Smoking also significantly lowers the number and motility of sperm in the semen of males who smoke at least four cigarettes a day (Kulikauskas et al., 1985).

In addition, our industrial society produces hundreds of new artificial compounds that come into general use each year. Pesticides and organic mercury compounds have caused neurological and behavioral abnormalities in infants whose mothers have ingested them during pregnancy. (This was tragically seen in 1965 when a Japanese firm dumped its mercury into a lake, where it was ingested by the fish, which were eaten by pregnant women in the village of Minamata. The congenital brain damage and blindness in the resulting children became known as Minamata disease.)

In the past decade, physicians have seen the effects of the teratogen 13-*CIS*-RETINOIC ACID. Retinoic acids are analogues of vitamin A and can mimic the vitamin's effects on epithelial differentiation but are less toxic than high doses of the vitamin itself. These analogues, including 13-*cis*-retinoic acid, have been useful in treating severe cystic acne and have been available (under the name Accutane) since 1982. Because the deleterious effects resulting from the administration of large amounts of vitamin A or its analogues to various species of pregnant animals have been known since the 1950s (Cohlan, 1953; Giroud and Martinet, 1959; Kochhar et al., 1984), the drug contains a label warning that it should not be used by pregnant women. However, about 160,000 women of childbearing age (15–45 years of age) have taken this drug since it was introduced, and some of them have used it during pregnancy. Lammer and his co-workers (1985) have studied a group of women who had inadvertently exposed themselves to the retinoic acid and who had elected to remain pregnant. Of the 59 fetuses, 26 were born without any noticeable malformations, 12 aborted spontaneously, and 21 were born with obvious malformations. The malformed infants had a characteristic pattern of anomalies including absent or defective ears, absent or small jaws, cleft palate, aortic arch abnormalities, thymic deficiencies, and abnormalities of the central nervous system.

This pattern of multiple congenital anomalies is similar to that seen in rat and mouse embryos whose pregnant mothers have been given these drugs. Goulding and Pratt (1986) have placed 8-day mouse embryos in a solution containing 13-*cis*-retinoic acid at very low concentrations (2×10^{-6} M). Even at this concentration, approximately one-third of the embryos developed a very specific pattern

FIGURE 42

Head of a cyclopic lamb born of a ewe who had eaten *Veratrum californicum* early in pregnancy. The cerebral hemispheres fused, forming only one eye and no pituitary gland. (From Binns et al., 1964; photograph courtesy of J. F. James and the USDA-ARS Poisonous Plant Laboratories.)

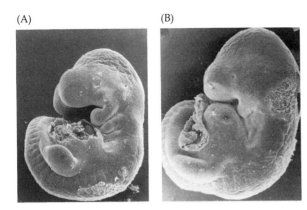

FIGURE 43

Scanning electron micrograph of mouse embryo cultured at day 8 for 48 hours in control medium (A) or in medium containing 2×10^{-6} M 13-*cis*-retinoic acid (B). The first pharyngeal arch of the treated embryo has a shortened and flattened appearance and has apparently fused with the second pharyngeal arch. (From Goulding and Pratt, 1986; photographs courtesy of the authors.)

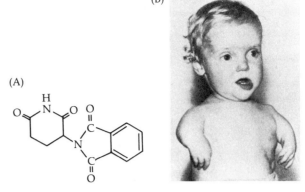

FIGURE 44

Thalidomide structure and effect. (A) Chemical structure of thalidomide. (B) Phocomelia in newborn whose mother had taken thalidomide during the first two months of pregnancy.

of anomalies, including dramatic reduction in the size of the first and second pharyngeal arches (Figure 43). The first arch eventually forms the maxilla and mandible of the jaw and two ossicles of the middle ear, while the second arch forms the third ossicle of the middle ear as well as other facial bones.

The basis for this phenomenon appears to reside in the drug's ability to inhibit neural crest cell migration from the cranial region of the neural tube. Radioactively labeled retinoic acid binds to the cranial neural crest cells and arrests both their proliferation and migration (Johnston et al., 1985; Goulding and Pratt, 1986). The binding seems to be specific to the cranial neural crest-derived cells, and the teratogenic effect of the drug is confined to a specific developmental period (days 8–10 in mice; days 20–35 in humans). Animal models of retinoic acid teratogenesis have been extremely successful at elucidating the mechanisms of teratogenesis at the cellular level. Present efforts are under way to determine the molecular effects of this drug on the cranial neural crest cells. As we will see in Chapter 17, retinoic acid is probably a substance that ordinarily directs the normal differentiation of vertebrate tissues. Its presence in large amounts can cause defects in development because these tissues are very sensitive to even small amounts of this compound and will respond to the exogenously supplied retinoic acid.

Before 1961 there was very little evidence for drug-induced malformations in humans. But in that year Lenz and McBride independently accumulated evidence that a mild sedative, THALIDOMIDE, caused an enormous increase in a previously rare syndrome of congenital anomalies. The most noticeable of these anomalies was PHOCOMELIA, a condition in which the long bones of the limbs are absent (amelia) or severely reduced (meromelia), thus causing the resulting appendage to resemble a seal flipper (Figure 44). Over 7000 affected infants were born to women who had taken this drug, and a woman need only to have taken one

tablet to have produced children with all four limbs deformed (Toms, 1962; Lenz, 1966). Other abnormalities induced by the ingestion of thalidomide included heart defects, absence of the external ears, and deformed intestines. The drug was withdrawn from the market in November 1961.

Nowack (1965) documented the PERIOD OF SUSCEPTIBILITY during which thalidomide caused these abnormalities. The drug was found to be teratogenic only during days 34 to 50 after the last menstruation (about 20 to 36 days postconception). The specificity of thalidomide action is shown in Figure 45. From day 34 to day 38, no limb abnormalities are seen. During this period, thalidomide can cause the reduction or absence of ear components. Malformations of upper extremities are seen before those of the lower limbs since the arms form slightly before the legs during development.

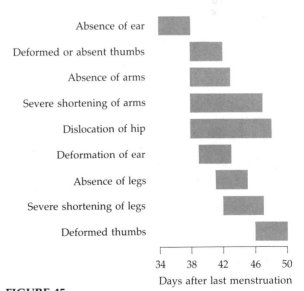

FIGURE 45

Timing of susceptibility to the teratogenic effects of thalidomide. (After Nowack, 1965.)

The thalidomide tragedy shows the limits of animal models to test the potential teratogenic effects of drugs. Different species (and strains within species) metabolize thalidomide differently. Pregnant mice and rats, the animals usually used to test such compounds, do not generate malformed pups when given thalidomide. Rabbits produce some malformed offspring, but the defects are different from those seen in affected human infants. Primates such as the marmoset appear to have a susceptibility similar to that of humans, and affected marmoset fetuses have been studied in an attempt to discover how thalidomide causes these malformations. McBride and Vardy (1983) report that the most noticeable difference seen before limb malformations concerns the size of the dorsal root ganglia and their neurons. The number of neurons in these ganglia are markedly reduced (Figure 46). These authors speculate that the neurons from these ganglia are necessary for maintaining limb development and that thalidomide works by interfering with or destroying these neurons. The molecular mechanism of this selective thalidomide toxicity is still not known, but having an animal model to study should lead to more detailed knowledge.

In terms of frequency and cost to society, the most devastating teratogen is undoubtedly ethanol. In 1973, Jones and Smith identified a syndrome of birth defects in the children of alcoholic mothers. This FETAL ALCOHOL SYN-DROME is characterized externally by small head size, an indistinct philtrum (the pair of ridges that run between the

FIGURE 46

Thalidomide effects on fetal marmosets. Top figures show phenotypes of marmoset fetuses late in gestation. Bottom figures show spinal cord cross sections at the level of the forelimbs. (A) Fetus of a control marmoset. (B) Fetus of a marmoset treated with 25 mg/kg thalidomide between days 38 and 46 of pregnancy. (From McBride and Vardy, 1983; photographs courtesy of W. G. McBride.)

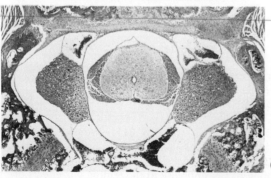

(A)

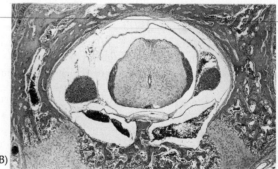

(B)

nose and mouth above the center of the upper lip), a narrow upper lip, and a low nose bridge. The brain of such a child is dramatically smaller than normal, and often shows defects in neuronal and glial migration (Clarren, 1986; Figure 47). Fetal alcohol syndrome (FAS) is the third most prevalent type of mental retardation (behind Down syndrome and spina bifida) and affects one out of every 500–750 children born in the United States (Abel and Sokol, 1987).

Children with fetal alcohol syndrome (FAS) are developmentally and mentally retarded, having a mean IQ around 68 (Streissguth and LaDue, 1987). Patients with a mean chronological age of 16.5 years were found to have the functional vocabulary of 6.5-year-olds and to have the mathematical abilities of fourth graders. Most of the adults and adolescents with FAS cannot handle money or their own lives, and they have difficulty learning from past experiences. There is great variation in the ability of mothers and fetuses to metabolize ethanol, and it is thought that 30–40 percent of the children born to alcoholic mothers who drink during pregnancy will have FAS. It is also thought that lower amounts of ethanol ingestion by the mother can lead to *fetal alcohol effect*, a less severe form of FAS, but a condition that lowers the functional and intellectual abilities of the person.

There are over 50,000 artificial chemicals presently used in our society and about 200 to 500 new materials being made each year (Johnson, 1980). The problem of screening these chemicals is of major importance, and standard protocols are expensive, long, and subject to interspecies differences in metabolism. As yet, there is no consensus on how to test a substance's teratogenicity for human embryos. Teratogenic compounds have always been with us. Certain metals and plant substances are known to cause severe developmental anomalies. However, as more and more new compounds are being produced and ingested by human populations, the fetus is placed at ever greater risk.

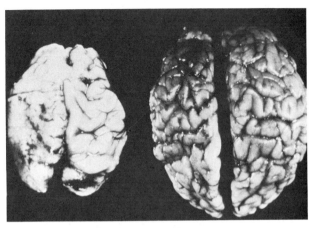

FIGURE 47
Comparison of a brain from an infant with fetal alcohol syndrome with that of a normal infant of the same age. The brain from the infant with FAS is significantly smaller, and the pattern of convolutions is obscured by glial cells that have migrated over the top of the brain. (Photograph courtesy of S. Clarren.)

SUMMARY

In this chapter we have followed the differentiation of the embryonic ectoderm into a wide variety of tissues. We have seen that the ectoderm produces three sets of cells during neurulation: (1) the neural tube, which gives rise to the neurons, glia, and ependymal cells of the central nervous system; (2) the neural crest cells, which give rise to the peripheral nervous system, pigment cells, adrenal medulla, and certain areas of head cartilage; and (3) the epidermis of the skin, which contributes to the formation of cutaneous structures such as hair, feathers, scales, and sweat and sebaceous glands, as well as forming the outer protective covering of our bodies. We also observed how the interactions of epidermal cells are involved in generating the various tissues of the eye.

In later chapters we shall discuss in more detail the induction of the neural tube, the coordinated development of the eye, and the manner in which neurons are directed to travel to specific sites, thus enabling the development of reflexes and behaviors. Meanwhile, we shall see how the endoderm and the mesoderm layers initiate the formation of their organ systems.

LITERATURE CITED

Abel, E. L. and Sokol, R. J. 1987. Incidence of fetal alcohol syndrome and economic impact of FAS-related anomalies. *Drug Alc. Depend.* 19: 51–70.

Anderson, D. J. and Axel, R. 1987. A bipotential neuroendocrine precursor whose choice of cell fate is determined by NGF and glucocorticoids. *Cell* 47: 1079–1090.

Balinsky, B. I. 1975. *Introduction to Embryology*, 4th Ed. Saunders, Philadelphia.

Banks, M. S. and Bennett, P. J. 1988. Optical and photoreceptor immaturities limit the spatial and chromatic vision of human neonates. *J. Optical Soc. A* 5: 2059–2079.

Berry, M., McConnell, P. and Sievers, J. 1980. Dendritic growth and the control of neuronal form. *Curr. Top. Dev. Biol.* 15: 67–101.

Binns, W., James, L. F. and Shupe, J. L. 1964. Toxicosis of *Veratrum californicum* in ewes and its relationship to a congenital deformity in lambs. *Ann. N.Y. Acad. Sci.* 111: 571–576.

Bloom, W. and Fawcett. D. W. 1975. *Textbook of Histology*, 10th Ed. Saunders, Philadelphia.

Bockman, D. E. and Kirby, M. L. 1984. Dependence of thymus development on derivatives of the neural crest. *Science* 223: 498–500.

Boué, A., Boué, J. and Gropp, A. 1985. Cytogenetics of pregnancy wastage. *Adv. Hum. Genet.* 14: 1–57.

Bronner-Fraser, M. 1986. Analysis of the early stages of trunk neural crest migration in avian embryos using monoclonal antibody HNK-1. *Dev. Biol.* 115: 44–55.

Bronner-Fraser, M. and Cohen, A. M. 1980. Analysis of the neural crest ventral pathway using injected tracer cells. *Dev. Biol.* 77: 130–141.

Bronner-Fraser, M. and Fraser, S. E. 1988. Cell lineage analysis reveals multipotency of some avian neural crest cells. *Nature* 335: 161–164.

Bronner-Fraser, M. and Fraser, S. 1989. Developmental potential of avian trunk neural crest cells in situ. *Neuron* 3: 755–766.

Burnside, B. 1971. Microtubules and microfilaments in newt neurulation. *Dev. Biol.* 26: 416–441.

Burnside, B. 1973. Microtubules and microfilaments in amphibian neurulation. *Am. Zool.* 13: 989–1006.

Chun, L. L. Y. and Patterson, P. H. 1977. Role of nerve growth factor in development of rat sympathetic neurons *in vitro*. Survival, growth, and differentiation of catecholamine production. *J. Cell Biol.* 75: 694–704.

Clarren, S. K. 1986. Neuropathology in the fetal alcohol syndrome. *In* J. R. West (ed.), *Alcohol and Brain Development*. Oxford University Press, New York.

Cohlan, S. Q. 1953. Excessive intake of vitamin A as a cause of congenital anomalies in the rat. *Science* 117: 535–537.

Coulombe, J. N. and Bronner-Fraser, M. 1987. Cholinergic neurones acquire adrenergic neurotransmitters when transplanted into an embryo. *Nature* 324: 569–572.

Coulombre, A. J. 1956. The role of intraocular pressure in the development of the chick eye. I. Control of eye size. *J. Exp. Zool.* 133: 211–225.

Coulombre, A. J. 1965. The eye. *In* R. DeHaan and H. Ursprung (eds.), *Organogenesis*. Holt, Rinehart & Winston, New York, pp. 217–251.

Crelin, E. S. 1974. Development of the central nervous system: A logical approach to neuroanatomy. *Ciba Clin. Symp.* 26(2): 2–32.

Crossin, K. L., Chuong, C.-M. and Edelman, G. M. 1985. Expression sequences of cell adhesion molecules. *Proc. Natl. Acad. Sci. USA* 82: 6942–6946.

Desmond, M. E. 1982. A description of the occlusion of the lumen of the spinal cord in early human embryos. *Anat. Rec.* 204: 89–93.

Desmond, M. E. and Schoenwolf, G. C. 1986. Evaluation of the roles of intrinsic and extrinsic factors in occlusion of the spinal neurocoel during rapid brain enlargement in the chick embryo. *J. Embryol. Exp. Morphol.* 97: 25–46.

Detwiler, S. R. 1932. Experimental observations upon the developing retina. *J. Comp. Neural.* 55: 473–492.

Edmonds, D. K., Lindsay, K. S., Miller, J. F., Williamson, E. and Wood, P. J. 1982. Early embryonic mortality in women. *Fertil. Steril.* 38: 447–453.

Edmondson, J. C. Liem, R. K. H., Kuster, J. C. and Hatten, M. E. 1988. Astrotactin: A novel neuronal cell surface antigen that mediates neuronal-astroglial interactions in cerebellar microcultures. *J. Cell Biol.* 106: 505–517.

Elder, J. T. and eight others. 1989. Overexpression of transforming growth factor in psoriatic epidermis. *Science* 243: 811–814.

Erickson, C. A., Tosney, K. W. and Weston, J. A. 1980. Analysis of migrating behaviour of neural crest and fibroblastic cells in embryonic tissues. *Dev. Biol.* 77: 142–156.

Fesenko, E. E., Kolenikou, S. S. and Lyubarsky, A. L. 1985. Induction by cyclic GMP of cationic conduction in plasma membranes of retinal rod outer segment. *Nature* 313: 310–313.

Forscher, P. and Smith, S. J. 1988. Actions of cytochalasins on the organization of actin filaments and microtubules in a neural growth cone. *J. Cell Biol.* 107: 1505–1516.

Fujita, S. 1964. Analysis of neuron differentiation in the central nervous system by tritiated thymidine autoradiography. *J. Comp. Neurol.* 122: 311–328.

Fujita, S. 1966. Application of light and electron microscopy to the study of the cytogenesis of the forebrain. *In* R. Hassler and H. Stephen (eds.), *Evolution of the Forebrain*. Plenum, New York, pp. 180–196.

Fukada, K. 1980. Hormonal control of neurotransmitter choice in sympathetic neuron cultures. *Nature* 287: 553–555.

Fukada, K. 1985. Purification and partial characterization of a cholinergic neuronal differentiation factor. *Proc. Natl. Acad. Sci. USA* 82: 8795–8799.

Gallera, J. 1971. Primary induction in birds. *Adv. Morphog.* 9: 149–180.

Giroud, A. and Martinet, M. 1959. Teratogenese pur hypervitaminose a chez le rat, la souris, le cobaye, et le lapin. *Arch. Fr. Pediat.* 16: 971–980.

Gizang-Ginsberg, E. and Ziff, E. B. 1990. Nerve growth factor regulates tyrosine hydroxylase gene transcription through a nucleoprotein complex that contains c-fos. *Genes and Dev.* 4: 477–491.

Goldowitz, D. and Mullen, R. J. 1982. Granule cell as a site of gene action in the *weaver* mouse cerebellum: Evidence from heterozygous mutant chimera. *J. Neurosci.* 2: 1474–1485.

Gould, S. J. 1977. *Ontogeny and Phylogeny*. Harvard University Press, Cambridge, MA.

Goulding, E. H. and Pratt, R. M. 1986. Isotretinoin teratogenecity in mouse whole embryo culture. *J. Craniofac. Genet. Dev. Biol.* 6: 99–112.

Gregg, N. M. 1941. Congenital cataract following German measles in the mother. *Trans. Opthalmol. Soc. Aust.* 3: 35.

Gregory, W. A., Edmondson, J. C., Hatten, M. E. and Mason, C. A. 1988. Cytology and neural-glial apposition of migrating cerebellar granule cells in vitro. *J. Neurosci.* 8: 1728–1738.

Halprin, K. M. 1972. Epidermal "turnover time"—a reexamination. *J. Invest. Dermatol.* 86: 14–19.

Harrison, R. G. 1907. Observations on the living developing nerve fiber. *Anat. Rec.* 1: 116–118.

Harrison, R. G. 1910. The outgrowth of a nerve fiber as a mode of protoplasmic movement. *J. Exp. Zool.* 9: 787–846.

Hatten, M. E. 1990. Riding the glial monorail: A common mechanism for glial-guided neuronal migration in different regions of the mammalian brain. *Trends Neurosci.* 13: 179–184.

Hatten, M. E., Liem, R. K. H. and Mason, C. A. 1986. *Weaver* mouse cerebellar granule neurons fail to migrate on wild-type processes *in vitro*. *J. Neurosci.* 6: 2676–2683.

Hertig, A. T. and Rock, J. 1949. A series of potentially abortive ova recovered from fertile women prior to the first missed menstrual period. *Am. J. Obstet. Gynecol.* 58: 968–993.

Hilfer, S. R. and Yang, J.-J. W. 1980. Accumulation of CPC-precipitable material at apical cell surfaces during formation of the optic cup. *Anat. Rec.* 197: 423–433.

His, W. 1886. Zur Geschichte des menschlichen Rückenmarks und der Nervenwurzeln. *Ges. d. Wissensch.* BD 13, S. 477.

Holmes, L. B. 1979. Radiation. *In* V. C. Vaughan, R. J. McKay and R. D. Behrman (eds.), *Nelson Textbook of Pediatrics*, 11th Ed. Saunders, Philadelphia.

Holt, A. B., Cheek, D. B., Mellitz, E. D. and Hill, D. E. 1975. Brain size and the relation of the primate to the non-primate. *In* D. B. Cheek (ed.), *Fetal and Postnatal Cellular Growth: Hormones and Nutrition.* Wiley, New York, pp. 23–44.

Huettner, A. F. 1949. *Fundamentals of Comparative Embryology of the Vertebrates*, 2nd Ed. Macmillan, New York.

Jacobson, A. G., Oster, G. F., Odell, G. M. and Cheng, L. Y. 1986. Neurulation and the cortical tractor model for epithelial folding. *J. Embryol. Exper. Morphol.* 96: 19–49.

Jacobson, M. 1968. Cessation of DNA synthesis in retinal ganglion cells correlated with the time of specification of their central connections. *Dev. Biol.* 17: 219–232.

Jacobson, M. 1978. *Developmental Neurobiology.* Plenum, New York.

Johnson, E. M. 1980. Screening for teratogenic potential: Are we asking the proper questions? *Teratology* 21: 259.

Johnston, M. C. 1975. The neural crest in abnormalities of the face and brain. *In* D. Bergsma (ed.), *Morphogenesis and Malformation of the Face and Brain.* Alan R. Liss, New York, pp. 1–18.

Johnston, M. C., Sulik, K. K., Webster, W. S. and Jarvis, B. L. 1985. Isotretinoin embryopathy in a mouse model: Cranial neural crest involvement. *Teratology* 31: 26A.

Jones, K. L. and Smith, D. W. 1973. Recognition of the fetal alcohol syndrome. *Lancet* 2: 999–1001.

Kahn, C. R., Coyle, J. T. and Cohen, A. M. 1980. Head and trunk neural crest *in vitro*: Autonomic neuron differentiation. *Dev. Biol.* 77: 340–348.

Kalcheim, C. R. and Neufeld, G. 1990. Expression of basic fibroblast growth factor in the nervous system of early avian embryos. *Development* 109: 203–215.

Kalcheim, C., Barde, Y.-A., Thoenen, H. and Le Douarin, N. M. 1987. *In vivo* effect of brain-derived neurotrophic factor on the survival of neural crest precursor cells of the dorsal root ganglia. *EMBO J.* 6: 2871–2873.

Karfunkel, P. 1972. The activity of microtubules and microfilaments in neurulation in the chick. *J. Exp. Zool.* 181: 289–302.

Kochhar, D. M., Penner, J. D. and Tellone, C. I. 1984. Comparative teratogenic activities of two retinoids: Effects on palate and limb development. *Terat. Carcin. Mutagen.* 4: 377–387.

Kulikauskas, V., Blaustein, A. B. and Ablin, R. J. 1985. Cigarette smoking and its possible effects on sperm. *Fertil. Steril.* 44: 526–528.

Lammer, E. J. and eleven others. 1985. Retinoic acid embryopathy. *N. Engl. J. Med.* 313: 837–841.

Lamoureux, P., Buxbaum, R. E. and Heidemann, S. R. 1989. Direct evidence that growth cones pull. *Nature* 340: 159–162.

Langman, J. 1969. *Medical Embryology*, 3rd Ed. Williams & Wilkins, Baltimore.

LeDouarin, N. and Smith, J. 1988. Development of the peripheral nervous system from the neural crest. *Annu. Rev. Cell Biol.* 4: 375–404.

LeDouarin, N. M. and Teillet, M.-A. 1974. Experimental analysis of the migration and differentiation of neuroblasts of the autonomic nervous system and of neuroectodermal mesenchyme derivatives, using a biological cell marking technique. *Dev. Biol.* 41: 162–184.

LeDouarin, N. M., Renaud, D., Teillet, M.-A. and LeDouarin, G. H. 1975. Cholinergic differentiation of presumptive adrenergic neuroblasts in interspecific chimeras after heterotopic transplantation. *Proc. Natl. Acad. Sci. USA* 72: 728–732.

LeDouarin, N. M., Cochard, P., Vincent, M., Duband, J. L., Tucker, G. C., Teillet, M.-A. and Thiery, J.-P. 1984. Nuclear, cytoplasmic, and membrane markers to follow neural crest cell migration. A comparative study. *In* R. L. Trelstad (ed.), *The Role of the Extracellular Matrix in Development.* Alan R. Liss, New York, pp. 373–398.

Lenz, W. 1962. Thalidomide and congenital abnormalities. *Lancet* 1: 45. (First reported at a 1961 symposium.)

Lenz, W. 1966. Malformations caused by drugs in pregnancy. *Am. J. Dis. Child.* 112: 99–l06.

Letourneau, P. C. 1977. Regulation of neuronal morphogenesis by cell–substratum adhesion. *Soc. Neurosci. Symp.* 2: 67–81.

Letourneau, P. C. 1979. Cell substratum adhesion of neurite growth cones, and its role in neurite elongation. *Exp. Cell Res.* 124: 127–138.

Löfberg, J., Perris, R. and Epperlin, H. H. 1989. Timing in the regulation of neural crest cell migration: Retarded maturation of regional extracellular matrix inhibits pigment cell migration in embryos of the white axolotl mutant. *Dev. Biol* 131: 168–181.

Loring, J. F. and Erickson, C. A. 1987. Neural crest cell migratory pathways in the trunk of the chick embryo. *Dev. Biol.* 121: 220–236.

Lumsden, A. 1988. Multipotent cells in the avian neural crest. *Trends Neurosci.* 12: 81–83.

Lumsden, A. and Keynes, R. 1989. Segmental patterns of neuronal development in the chick hindbrain. *Nature* 337: 424–428.

Mann, I. 1964. *The Development of the Human Eye.* Grune and Stratton, New York.

Maxwell, G. D. and Forbes, W. E. 1987. Exogenous basement-membrane-like matrix stimulates adrenergic development in avian neural crest cultures. *Development* 101: 767–776.

McBride, W. G. 1961. Thalidomide and congenital abnormalities. *Lancet* 2: 1358.

McBride, W. G. and Vardy, P. H. 1983. Pathogenesis of thalidomide teratogenesis in the marmot (*Callithrix jacchus*): Evidence suggesting a possible trophic influence of cholinergic nerves in limb morphogenesis. *Dev. Growth Differ.* 25: 361–373.

McKeown, T. 1976. Human malformations: An introduction. *Br. Med. Bull.* 32: 1–3.

Meier, S. 1981. Development of the chick embryo mesoblast: Morphogenesis of the prechordal plate and cranial segments. *Dev. Biol.* 83: 49–61.

Mikamo, K. 1970. Anatomical and chromosomal anomalies in spontaneous abortions. *Am. J. Obstet. Gynecol.* 103: 143–154.

Miller, J. R. and Poland, B. J. 1970. The value of human abortuses in the surveillance of developmental anomalies. *Can. Med. Assoc. J.* 103: 501–502.

Montagna, W. and Parakkal, P. F. 1974. The piliary apparatus. *In* W. Montagna (ed.), *The Structure and Formation of Skin.* Academic Press, New York, pp. 172–258.

Montagu, M. F. A. 1962. Time, morphology, and neoteny in the evolution of man. *In* M. F. A. Montagu (ed.), *Culture and Evolution of Man.* Oxford University Press, New York.

Nagele, R. G. and Lee, H. Y. 1980. Studies on the mechanism of neurulation in the chick: Microfilament-mediated changes in cell shape during uplifting of neural folds. *J. Exp. Zool.* 213: 391–398.

Nagele, R. G. and Lee, H. Y. 1987. Studies in the mechanism of neurulation in the chick. Morphometric analysis of the relationship between regional variations in cell shape and sites of motive force generation. *J. Exp. Biol.* 24: 197–205.

Newgreen, D. F. and Gooday, D. 1985. Control of onset of migration of neural crest cells in avian embryos: Role of Ca^{++}-dependent cell adhesions. *Cell Tiss. Res.* 239: 329–336.

Newgreen, D. F., Scheel, M. and Kaster, V. 1986. Morphogenesis of sclerotome and neural crest cells in avian embryos. *In vivo* and *in vitro* studies on the role of notochordal extracellular material. *Cell Tissue Res.* 244: 299–313.

Noden, D. M. 1978. The control of avian cephalic neural crest cytodifferentiation. I. Skeletal and connective tissue. *Dev. Biol.* 69: 296–312.

Noden, D. M. 1983. The role of the neural crest in patterning of avian cranial skeletal, connective, and muscle tissues. *Dev. Biol.* 96: 144–165.

Noden, D. M. and Evans, H. E. 1986. Inherited homeotic midfacial malformations in Burmese cats. *J. Craniofac. Gen. Dev. Biol. Supp.* 2: 249–266.

Nowack, E. 1965. Die sensible Phase bei der Thalidomide-Embryopathie. *Humangenetik* 1: 516–536.

Papaconstantinou, J. 1967. Molecular aspects of lens cell differentiation. *Science* 156: 338–346.

Paton, D. and Craig, J. A. 1974. Cataracts: Development, diagnosis, and management. *Ciba Clin. Symp.* 26(3): 2–32.

Patten, B. M. 1971. *Early Embryology of the Chick*, 5th Ed. McGraw-Hill, New York.

Perris, R. and Bronner-Fraser, M. 1989. Recent advances in defining the role of the extracellular matrix in neural crest development. *Comm. Dev. Neuro.* 1: 61–83.

Perris, R., von Boxburg, Y. and Löfberg, J. 1988. Local embryonic matrices determine region-specific phenotypes in neural crest cells. *Science* 241: 86–89.

Perris, R., Löfberg, J. Fällström, C., von Boxburg, Y., Olsson, L. and Newgreen, D. F. 1990. Structural and compositional divergencies in the extracellular matrix encountered by neural crest cells in the *white* mutant axlotl embryo. *Development* 109: 533–551.

Piatigorsky, J. 1981. Lens differentiation in vertebrates: A review of cellular and molecular features. *Differentiation* 19: 134–153.

Poole, T. J. and Thiery, J. P. 1986. *In* H. C. Slavkin (ed.), *Progress in Clinical and Biological Research*, Vol 217. Alan R. Liss, New York, pp. 235–238.

Portmann, A. 1941. Die Tragzeiten der Primaten und die Dauer der Schwangerschaft beim Menschen: Ein Problem der vergleichen Biologie. *Rev. Suisse Zool.* 48: 511–518.

Portmann, A. 1945. Die Ontogenese des Menschen als Problem der Evolutionsforschung. *Verh. Schweiz. Naturf. Ges.* 125: 44–53.

Pratt, R. M., Larsen, M. A. and Johnston, M. C. 1975. Migration of cranial neural crest cells in a cell-free, hyaluronate-rich matrix. *Dev. Biol.* 44: 298–305.

Rakic, P. 1972. Mode of cell migration to superficial layers of fetal monkey neocortex. *J. Comp. Neurol.* 145: 61–84.

Rakic, P. 1975. Cell migration and neuronal ectopias in the brain. *In* D. Bergsma (ed.), *Morphogenesis and Malformations of Face and Brain*. Birth Defects Original Article Series II (7): 95–129.

Rakic P. and Goldman, P. S. 1982. Development and modifiability of the cerebral cortex. *Neurosci. Rev.* 20: 429–611.

Rakic, P. and Sidman, R. L. 1973. *Weaver* mutant mouse cerebellum: Defective neuronal migration secondary to abnormality of Bergmann glia. *Proc. Natl. Acad. Sci. USA* 70: 240–244.

Ramón y Cajal, S. 1890. Sur l'origene et les ramifications des fibres neuveuses de la moelle embryonnaire. *Anat. Anz.* 5: 111–119.

Rawles, M. E. 1948. Origin of melanophores and their role in development of color patterns in vertebrates. *Physiol. Rev.* 28: 383–408.

Rickmann, M., Fawcett, J. W. and Keynes, R. J. 1985. The migration of neural crest cells and the growth of motor neurons through the rostral half of the chick somite. *J. Embryol. Exp. Morphol.* 90: 437–455.

Romanes, G. J. 1901. *Darwin and After Darwin*. Open Court Publishing, London.

Sadler, T. W., Burridge, K. and Yonker, J. 1986. A potential for spectrin during neurulation. *J. Embryol. Exp. Morphol.* 94: 73–82.

Saha, M., Spann, C. L. and Grainger, R. M. 1989. Embryonic lens induction: More than meets the optic vesicle. *Cell Diff. Dev.* 28: 153–172.

Sauer, F. C. 1935. Mitosis in the neural tube. *J. Comp. Neurol.* 62: 377–405.

Schoenwolf, G. C. and Alvarez, I. S. 1989. Roles of neuroepithelial cell rearrangement and division in shaping of the avian neural plate. *Development* 106: 427–439.

Schoenwolf, G. C. and Desmond, N. E. 1984. Descriptive studies of the occlusion and reopening of the spinal canal of the early chick embryo. *Anat. Rec.* 209: 251–263.

Schoenwolf, G. and Smith, J. L. 1990. Mechanisms of neurulation: Traditional viewpoint and recent advances. *Development* 109: 243–270.

Schoenwolf, G. C., Folsom, D. and Moe, A. 1988. A reexamination of the role of microfilaments in neurulation in the chick embryo. *Anat. Rec.* 220: 87–102.

Schweizer, G., Ayer-Le Lievre, C. and Le-Douarin, N. M. 1983. Restrictions in developmental capacities in the dorsal root ganglia during the course of development. *Cell Differ.* 13: 191–200.

Serbedzija, G. N., Bronner-Fraser, M. and Fraser, S. E. 1989. A vital dye analysis of the timing and pathways of avian trunk neural crest cell migration. *Development* 106: 809–816.

Soleto, C. and Changeaux, J. P. 1974. Transsynaptic degeneration "en cascade" in the cerebellar cortex of *staggerer* mutant mice. *Brain Res.* 67: 519–526.

Solursh, M., Fisher, J. and Singley, C. T. 1979. The synthesis of hyaluronic acid by ectoderm during early organogenesis in the chick embryo. *Differentiation* 14: 77–85.

Spieth, J. and Keller, R. E. 1984. Neural crest cell behavior in white and dark larvae of *Ambystoma mexicanum*: Differences in cell morphology, arrangement, and extracellular matrix as related to migration. *J. Exp. Zool.* 229: 91–107.

Streissguth, A. P. and LaDue, R. A. 1987. Fetal alcohol: Teratogenic causes of developmental disabilities. *In* S. R. Schroeder (ed.), *Toxic Substances and Mental Retardation*. American Association of Mental Deficiency, Washington, DC, pp. 1–32.

Stryer, L. 1986. Cyclic GMP cascade of vision. *Annu. Rev. Neurosci.* 9: 87–119.

Takeichi, M. 1988. The cadherins: Cell–cell adhesion molecules controlling animal morphogenesis. *Development* 102: 639–656.

Tan, S.-S. and Morriss-Kay, G. 1985. The development and distribution of the cranial neural crest in the rat embryo. *Cell Tissue Res.* 240: 403–416.

Teillet, M.-A., Kalcheim, C. and Le-Douarin, N.M. 1987. Formation of the dorsal root ganglia in the avian embryo: Segmental origin and migratory behavior of neural crest progenitor cells. *Dev. Biol.* 120: 329–347.

Thorogood, P. 1989. Review of *Developmental and Evolutionary Aspects of the Neural Crest. Trends Neurosci.* 12: 38–39.

Toms, D. A. 1962. Thalidomide and congenital abnormalities. *Lancet* 2: 400.

Turner, D. L. and Cepko, C. L. 1987. A common progenitor for neurons and glia persists in rat retina late in development. *Nature* 328: 131–136.

Vogel, K. S. and Weston, J. A. 1990. The sympathoadrenal lineage in avian embryos. II. Effects of glucocorticoids on cultured neural crest cells. *Dev. Biol.* 139: 13–23.

von Baer, K. E. 1828. *Entwicklungsgeschichte der Thiere: Beobachtung und Reflexion*. Bornträger, Konigsberg.

Walbot, V. and Holder, N. 1987. *Developmental Biology*. Random House, New York.

Webster, E. H., Silver, A. F. and Gonsalves, N. I. 1984. The extracellular matrix between the optic vesicle and the presumptive lens during lens morphogenesis in an anophthalmic strain of mice. *Dev. Biol.* 103: 142–150.

Weinstein, G. D. and van Scott, E. J. 1965. Turnover times of normal and psoriatic epidermis. *J. Invest. Dermatol.* 45: 257–262.

Weiss, P. 1941. The mechanics of nerve growth. *Third Growth Symposium. Growth* 5 [Suppl.]: 163–203.

Weston, J. 1963. A radiographic analysis of the migration and localization of trunk neural crest cells in the chick. *Dev. Biol.* 6: 274–310.

Weston, J. 1970. The migration and differentiation of neural crest cells. *Adv. Morphog.* 8: 41–114.

Weston, J. A. and Butler, S. L. 1966. Temporal factors affecting localization of neural crest cells in the chicken embryo. *Dev. Biol.* 14: 246–266.

Yamada, K. M., Spooner, B. S. and Wessells, N. K. 1971. Ultrastructure and function of growth cones and axons of cultured nerve cells. *J. Cell Biol.* 49: 614–635.

Yuodelis, C. and Hendrickson, A. 1986. A qualitative and quantitative analysis of the human fovea during development. *Vision Res.* 26: 847–855.

6

Early vertebrate development: Mesoderm and endoderm

Of physiology from top to toe I sing,
Not physiognomy alone or brain alone is worthy for the Muse,
I say the form complete is worthier far,
The Female equally with the Male I sing.

—WALT WHITMAN (1867)

Theories come and theories go. The frog remains.

—JEAN ROSTAND (1960)

In the last chapter we followed the various tissues formed by developing ectoderm. In this chapter, we shall follow the early development of the mesodermal and endodermal germ layers. Endoderm will be seen to form the lining of the digestive and respiratory tubes with their associated organs; mesoderm will be seen to generate all the organs between the ectodermal wall and the endodermal tissues.

MESODERM

The mesoderm of a neurula-stage embryo can be divided into five regions (Figure 1). The first region is the CHORDAMESODERM. This tissue forms the notochord, a transient organ whose major functions include inducing the formation of the neural tube and establishing the body axis. The second region is the DORSAL (SOMITIC) MESODERM. The term *dorsal* refers to the observation that the tissues developing from this region will be in the back of the embryo, along the spine. Located on both sides of the neural tube, this region will produce many of the connective tissues of the body: bone, muscles, cartilage, and dermis. The INTERMEDIATE MESODERM forms the urinary system and genital ducts; we will discuss this region in detail in later chapters. Further away from the notochord, the LATERAL PLATE MESODERM will give rise to the heart, blood vessels, and blood cells of the circulatory system as well as to the lining of the body cavities and all the

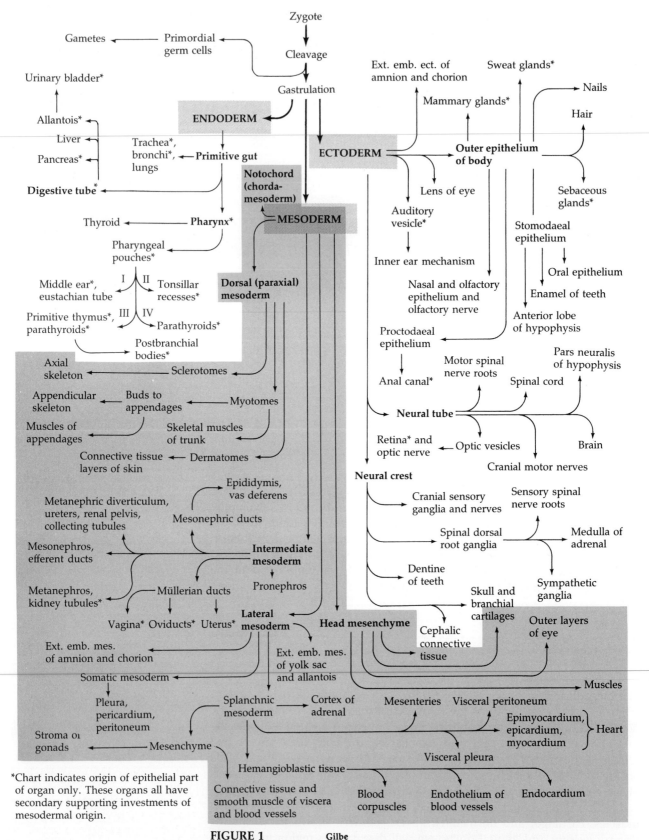

FIGURE 1   Gilbe

Chart depicting the lineage of the specialized parts of the body through the three primary germ layers. The germ cells are represented as a line of cells separate from those of the three somatic germ layers because, although the germ cell precursors are located in the presumptive endoderm or mesoderm, they are probably a unique cell type. (After Carlson, 1981.)

mesodermal components of the limbs except the muscles. It will also form a series of extraembryonic membranes that are important for transporting nutrients to the embryo. Lastly, the HEAD MESENCHYME will contribute to the connective tissues and musculature of the face.

Dorsal mesoderm: Differentiation of somites

One of the major tasks of gastrulation is to create a mesodermal layer between the endoderm and the ectoderm. As shown in Figure 2, the formation of mesodermal and endodermal organs is not subsequent to neural tube formation, but occurs synchronously. Those mesodermal cells of the chick that are not involved in notochord formation have migrated laterally to form thick bands running longitudinally along each side of the notochord and neural tube. These bands of PARAXIAL MESODERM are referred to as the SEGMENTAL PLATE (in birds) and the UNSEGMENTED MESODERM (in mammals). As the primitive streak regresses and the neural folds begin to gather at the center of the embryo, the paraxial mesoderm separates into blocks of cells called SOMITES. The first somites appear in the anterior portion of the embryo, and new somites are formed posteriorly at regular intervals, budding from the paraxial mesoderm (Figures 2D and 3). Because embryos develop at different rates when incubated at slightly different temperatures, the number of somites present is usually the best indicator as to how far development has proceeded. The total number of somites formed is characteristic of a species.

The mechanism for somite formation is not well established, but recent studies in chicks have shown that the cells of the segmental plate may be organized into whorls of cells called SOMITOMERES (Meier, 1979). The chick segmental plate is consistently seen to have 10 to 11 pairs of somitomeres that eventually become somites (Packard and Meier, 1983). Conversion

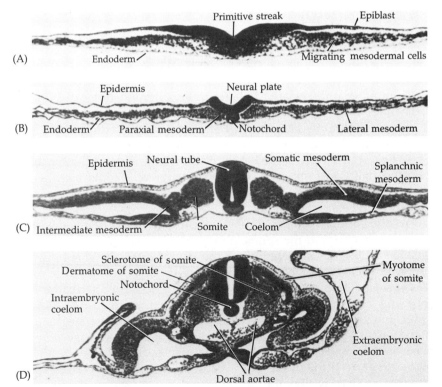

(A)
Primitive streak
Epiblast
Endoderm
Migrating mesodermal cells

(B)
Epidermis
Neural plate
Endoderm
Paraxial mesoderm
Notochord
Lateral mesoderm

(C)
Epidermis
Neural tube
Somatic mesoderm
Splanchnic mesoderm
Intermediate mesoderm
Somite
Coelom

(D)
Sclerotome of somite
Dermatome of somite
Notochord
Intraembryonic coelom
Myotome of somite
Extraembryonic coelom
Dorsal aortae

FIGURE 2
The progressive development of the chick embryo, focusing on the mesodermal component. (A) Primitive streak region, showing migrating mesodermal and endodermal precursors. (B) Formation of the notochord and paraxial mesoderm. (C,D) Differentiation of the somites, coelom, and the two aortae (which will eventually fuse). A–C are from 24-hour embryos; D is from a 48-hour embryo.

FIGURE 3

Neural tube and somites. Scanning electron micrograph showing well-formed somites and paraxial mesoderm (bottom right-hand side) that has not yet separated into distinct somites. A rounding of the paraxial mesoderm into a somitomere can be seen in the lower left side, and neural crest cells can be seen migrating ventrally from the roof of the neural tube. (Courtesy of K. W. Tosney.)

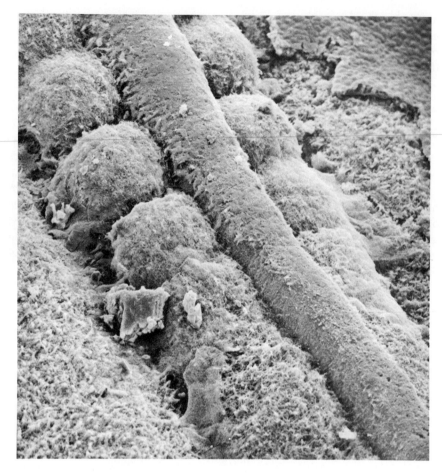

from somitomere to somite is seen as the cells of the most anterior somitomere become compacted. Lash and Yamada (1986) have speculated that fibronectin mediates the clumping of the somitomere cells. Their hypothesis is based on two types of data. First, Ostrovsky and co-workers (1984) demonstrated that the amount of fibronectin increases in the anterior region of the presumptive somite at the same time that the cells in the anterior border of this tissue start to compact. The second type of evidence comes from in vitro studies. Lash and co-workers (1984; Cheney and Lash, 1984) found that dissociated cells from the anterior portions of the chick segmental plate (which have just formed somites) readily reassociate into somite-sized clumps. Dissociated cells from the posterior portion of the segmental plate (which has not yet formed somites) do not reassociate. This suggests that the mesodermal cells become more adhesive as they gain the ability to form somites. The posterior cells could be induced to form somite-like masses by adding fibronectin to the culture medium in which they were reaggregating.

As the somite becomes a coherent entity, its cells become epithelial, and the outer cells of the somites become attached to each other by tight junctions. Like the notochord and the neural tube, the somite becomes covered with a BASAL LAMINA (Figure 4) consisting of collagen, fibronectin, laminin, and glycosaminoglycans (GAG). In fact, GAG from the neural tube and notochord appear to induce the somites to secrete their own GAG. If the GAGs are enzymatically removed from the surface of the notochord, GAG synthesis in the somites will stop until the notochord once more has elaborated GAG on its own surface (Kosher and Lash, 1975). Whether through the action of GAG or some other factor secreted by the notochord or neural tube, the *ventral* cells of the somite (those cells

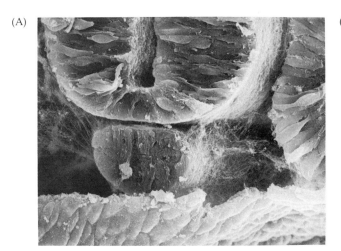

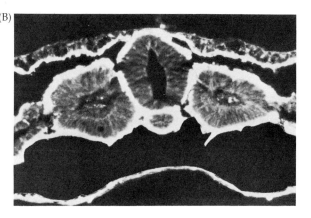

FIGURE 4
The extracellular surfaces of the somites and neighboring tissues. (A) Scanning electron micrograph of the neural tube–notochord–somite borders. The webbing of collagen and glycosaminoglycan complexes connecting these structures is important in cell migration and the intercellular reactions occurring between them. (B) Localization of fibronectin on the extracellular basal lamina of the somite and adjacent tissues in a 2-day chick embryo. The staining was accomplished by fluorescent antibodies to fibronectin. Such fibronectin-containing lamina are only seen after the somitomeres condense and become the epithelial cells of the somites. (A courtesy of K. W. Tosney; B courtesy of J. Lash.)

located farthest from the back) undergo mitosis, lose their round epithelial characteristics, and become mesenchymal cells again. The portion of the somite that gives rise to these cells is called the SCLEROTOME, and these mesenchymal cells ultimately become the CHONDROCYTES (Figures 2 and 5). Chondrocytes are responsible for secreting the special types of collagens and glycosaminoglycans (such as chondroitin sulfate) characteristic of cartilage. These particular chondrocytes will be responsible for constructing the axial skeleton (vertebrae, ribs, and so forth).

Once the cells of the sclerotome have migrated away from the somite, the remaining somitic epithelial cells (called the DERMAMYOTOME) form a bilayered, solid tube (Figures 2 and 5). The dorsal layer is called the DERMATOME, and it generates the mesenchymal connective tissue of the skin: the dermis. The inner layer of cells is called the MYOTOME, and these cells give rise to the striated muscles of both the back and the limbs (Chevallier et al., 1977).

FIGURE 5
Diagram of a transverse section through the trunk of an early 4-week (A) and late 4-week (B) human embryo. (A) The sclerotome cells are beginning to migrate away from the myotome and dermatome. (B) By the end of the fourth week, the sclerotome cells are condensing to form cartilaginous vertebrae, the dermatome is starting to form the dermis, and the myotome cells are extending ventrally down the walls of the embryo. (After Langman, 1969.)

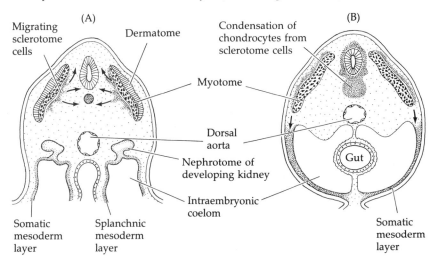

Myogenesis: Differentiation of skeletal muscle

Myogenesis is characterized by at least four stages (Quinn et al., 1990). First, a set of cells in the somite are committed to become muscle precursors, as opposed to becoming any other type of cell. Second, these cells proliferate to expand the population of potential muscle cells. Third, the muscle precursor cells differentiate into muscle cells and initiate the synthesis of muscle-specific proteins. Fourth, the muscle cells mature as innervation occurs and new forms of enzymes are made.

We will focus our attention on the differentiation of the muscle precursors into muscle cells. A skeletal muscle cell is an extremely large elongated cell that contains many nuclei. In the mid-1960s developmental biologists debated whether each of these cells (often called MYOTUBES) was derived from the fusion of several mononucleate muscle precursor cells (MYOBLASTS) or from a single myoblast that undergoes nuclear division without cytokinesis. Evidence from two independent sources demonstrates that the myotube is derived from the fusion of several mononucleate myoblasts.

One of the most significant advances in the study of muscle development came when Konigsberg (1963) demonstrated that myoblasts could differentiate into myotubes in culture. The observations of such fusion in vitro have enabled scientists to study the mechanism of this fusion and to relate it to the differentiation of the muscle cell. Cells from embryonic regions containing myoblasts are separated from each other by digesting the extracellular lamina with trypsin. These cells are then placed in collagen-coated petri dishes containing nutrients and serum (Figure 6). By 48 hours, the large majority of the cells are rapidly dividing myoblasts. Shortly thereafter, these myoblasts cease making DNA and fuse with their neighbors to produce extended myotubes (Konigsberg, 1963), which begin synthesizing muscle-specific proteins. DNA synthesis and nuclear division are not seen in the multinucleated myotubes.

The production of myotubes from myoblasts has been divided into two separate processes. The first stage involves cell migration, recognition, and alignment. When 11-day chick embryo pectoral muscle (which still contains a large amount of mononucleate myoblasts) is placed into culture, the isolated cells divide and migrate. Many of the migrating cells eventually line up in chains and then fuse to form the myotube. Nameroff and Munar (1976) have been able to separate this recognition phenomenon (the making of such chains) from the fusion event by treating the cultured cells with phospholipase C. This enzyme prohibits membrane fusion but permits recognition. Thus, in the presence of the enzyme, chains are formed but no fusion takes place. These experiments also suggest that the recognition event is responsible for stopping DNA synthesis in the myoblast. Radioactive thymidine is incorporated into the DNA of the migrating myoblasts but is not incorporated into those myoblasts within the chains (Figure 7). This observation suggests that cell–cell recognition may withdraw the myoblasts from the cell cycle.

The second stage of myogenesis is cell fusion itself. Both the recognition event and the fusion event are dependent upon calcium ions, but during fusion, the calcium ions must get into the myoblasts. Accordingly, fusion can be activated by calcium ionophores, such as A23187, which carry calcium ions across cell membranes (Shainberg et al., 1969; David et al., 1981). While the alignment phenomenon is mediated by cell membrane glycoproteins (Knudsen, 1985: Knudsen et al., 1990), the fusion events appear to be regulated by extensive reorganization of membrane components to form protein-free areas of phospholipids (Kaufman and Foster,

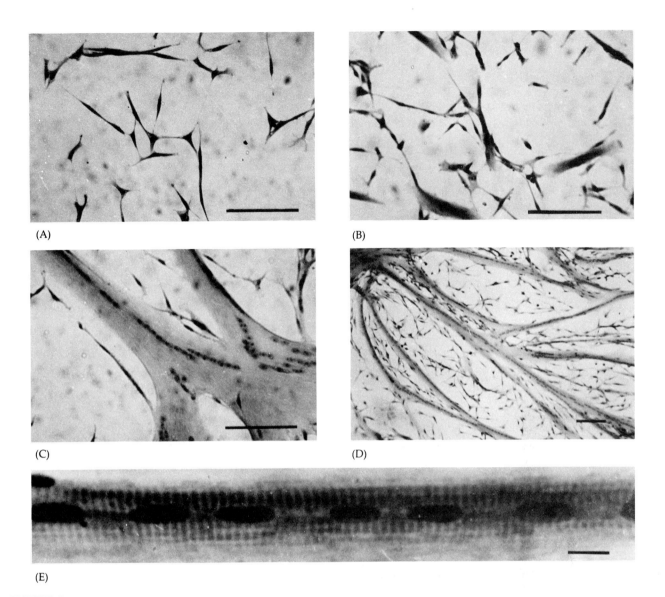

(A)

(B)

(C)

(D)

(E)

FIGURE 6
Quail myoblasts differentiating in culture. All cultures were inoculated, fixed, stained, and photographed on specific days. (A) Day 2 culture consists of unicellular myoblasts (bar represents 0.1 mm). (B) Fusion of myoblasts is first seen in day 3 culture. (C) In day 5 culture, the increase in the number of nuclei within multinucleated myotubes is apparent. (D) Lower magnification of day 5 culture, showing the extensive cell fusion. (E) Higher magnification of multinucleated myotube demonstrating the characteristic cross-striations (bar represents 0.01 mm). (From Buckley and Konigsberg, 1974; photographs courtesy of I. R. Konigsberg.)

1985; Wakelam, 1985). Kalderon and Gilula (1979) have shown that calcium-mediated membrane fusion occurs in those regions of the myoblast membranes that are enriched for phospholipids. This hypothesis would explain why phospholipase C is such a potent inhibitor of cell fusion.

Although cell–cell recognition and fusion are essential for myogenesis, species specificity is not. Myoblasts will fuse only with other myoblasts, but those myoblasts need not be of the same species. Yaffe and Feldman (1965) have shown that embryonic rat and chick myoblasts will readily fuse to form hybrid myotubes.

FIGURE 7

DNA synthesis in myoblasts. Autoradiograph of chick myoblast cells treated with phospholipase C in culture and then exposed to radioactive thymidine. Unattached myoblasts still divide and incorporate the radioactive thymidine into their DNA. The incorporated radioactive nucleotides cause the silver grains of the photographic emulsion to darken when the emulsion is developed. Lined up (but not yet fused) cells (arrows) do not incorporate the label. (From Nameroff and Munar, 1976; photograph courtesy of M. Nameroff.)

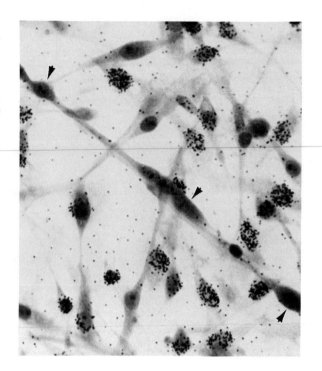

The second type of evidence for myoblast fusion came from TETRA-PARENTAL (ALLOPHENIC) MICE. These mice can be formed from the fusion of two early embryos, which regulate to produce a single mouse having two distinct cell populations (Figure 28 in Chapter 3). Beatrice Mintz and W. W. Baker (1967) fused mouse embryos that produce different types of the enzyme isocitrate dehydrogenase. This enzyme, found in all cells, is composed of two identical subunits. Thus, if myotubes are formed from one cell whose nuclei divide without cytokinesis, one would expect to find two distinct forms of the enzyme, that is, the two parental forms, in the allophenic mouse (Figure 8). On the other hand, if myotubes are formed by fusion between cells, one would expect to find muscle cells expressing not only the two parental types of enzymes (AA and BB) but also a third class composed of a subunit from each of the parental types (AB). The different forms of isocitrate dehydrogenase can be separated and identified by their mobility in an electric field. The results clearly demonstrated that although only the two parental types of enzyme were present in all the other tissues of the allophenic mice, the hybrid (AB) enzyme was present in extracts of skeletal muscle tissue. Thus, the myotubes must have been formed from the fusion of numerous myoblasts.

There is general agreement that myoblast fusion and the synthesis of muscle-specific proteins are separate events that usually occur at the same time. Once the myoblasts align and stop dividing, tropomyosin, myosin heavy chain, and the muscle-specific forms of actin and creatine kinase are produced while the myoblasts fuse. For example, in the case of creatine kinase, a dimeric protein, myoblasts only make a creatine kinase composed of B subunits (BB). At the time of fusion, the cultured myotubes begin expressing a different, muscle-specific (M) form of the enzyme, thereby creating MM dimers (Figure 9A). Myoblast fusion can be stimulated by changing the tissue culture medium from one that supports proliferation to one that lacks certain growth factors. In the limited medium, myoblasts cease to divide and will fuse within a short period of time. Devlin and Emerson (1978) have used this technique to synchronize cell fusion and look at the synthesis of myosin, actin, and tropomyosin. At first, none of

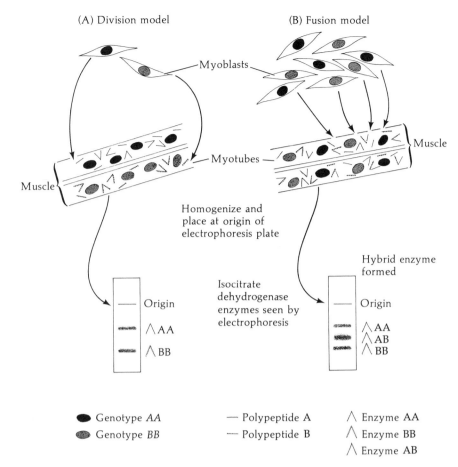

(A) Division model

(B) Fusion model

Myoblasts

Muscle

Myotubes

Muscle

Homogenize and
place at origin of
electrophoresis plate

Hybrid enzyme
formed

Isocitrate
dehydrogenase
enzymes seen by
electrophoresis

— Origin

∧ AA

∧ BB

— Origin

∧ AA
∧ AB
∧ BB

● Genotype *AA*
◍ Genotype *BB*

— Polypeptide A
--- Polypeptide B

∧ Enzyme AA
∧ Enzyme BB
∧ Enzyme AB

FIGURE 8
Diagram illustrating the two possible mechanisms of skeletal muscle formation and how to distinguish between them. Allophenic mice are made from the fusion of mouse embryos from two different strains, each strain making a different form of the enzyme isocitrate dehydrogenase. This enzyme is composed of two subunits; one strain of mouse makes AA isocitrate dehydrogenase (indicated in black) and the other makes BB (color). (A) If the enzymes are made in a single cell, or in multinucleate cells arising from the nuclear divisions within a single cell, the enzyme will be purely AA or BB. (B) If there are two different nuclei in the same cell, however, one might code for B subunits while the other might code for A, with the result that some molecules of the enzyme will be hybrid (AB). Electrophoresis can separate these three types of molecules. The presence of the AB molecule in skeletal muscle cells (but not in other cell types) confirms the fusion model. (After Mintz and Baker, 1967.)

these proteins is detectable. At the time of fusion, however, these structural proteins rapidly appear. Sutherland and Konigsberg (1983) have shown that once DNA synthesis has stopped, the muscle-specific proteins increase normally even if the fusion event is inhibited by low calcium concentrations. It appears, then, that once the cells stop dividing, they acquire the capacities to fuse and to make muscle-specific proteins.

Although there are enormous increases in the synthesis of muscle-specific proteins at the time of fusion, there seem to be differences in the mechanisms by which these muscle-specific proteins appear. Muscle-specific ("cardiac") myosin is observed for the first time soon after fusion. This is the same time that the myosin heavy-chain protein is first seen. However, the *messenger RNA* for cardiac actin is seen as soon as the myoblasts withdraw from the cell cycle, while messenger RNA for the myosin heavy chain is not synthesized until the onset of fusion (Lawrence et al., 1989; Figure 9B). Thus, the control of muscle-specific proteins is exerted at both the transcriptional and translational levels. In both cases, the fusion of the myoblasts and the accumulation of muscle-specific proteins appears to be coordinated by the cessation of DNA synthesis in the proliferating myoblasts.

Osteogenesis: Development of bones

Some of the most obvious structures derived from the somitic mesoderm are the bones. In this chapter we can only begin to outline the mechanisms of bone formation, and students wishing further details are invited to consult histology textbooks that devote entire chapters to this topic. There are two major modes of bone formation (OSTEOGENESIS), and both involve

FIGURE 9
New enzyme synthesis concomitant with myoblast fusion. (A) Decline of nonmuscle (B-type) creatine kinase subunits and concomitant replacement with muscle-specific (M-type) creatine kinase subunits as cultured embryonic chick myoblasts begin fusing. Fusion begins after 1 day of culture, and by day 2, 38 percent of the nuclei are in myotubes. (B) Schematic representation of the transcription and translation of some muscle-specific proteins near the time of cell fusion. (A after Lough and Bischoff, 1977; B after Lawrence et al., 1989.)

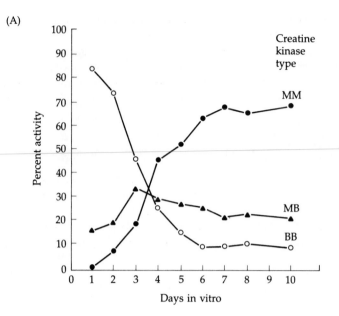

(A)

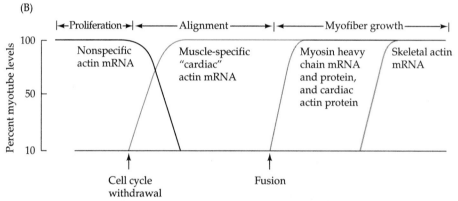

(B)

the transformation of a preexisting connective tissue into bone tissue. The conversion of primitive connective tissue into bone is called INTRAMEMBRANOUS OSSIFICATION. The replacement of cartilage by bone is called ENDOCHONDRAL OSSIFICATION.

Intramembranous ossification is the characteristic way in which the flat bones of the skull are formed. Mesenchymal cells derived from the neural crest interact with the extracellular matrix of the head epithelial cells to form bone. If the mesenchymal cells do not contact this matrix, no bone will be formed (Tyler and Hall, 1977; Hall, 1988). This was shown in vitro by Hall and colleagues (1983), who took isolated head mesenchymal cells and plated them into culture dishes. If no extracellular matrix was present on the surface of these dishes, the cells remained mesenchymal. However, if head epithelial cells had first secreted an extracellular matrix upon the surface, the cells differentiated into bone cells. The mechanism responsible for this conversion of mesenchymal cells to chondrocytes is still unknown, but recent evidence points to a particular group of molecules at the epithelial–mesenchymal junction. BONE MORPHOGENIC PROTEINS can be isolated from adult bone and injected into embryonic organs. When this is done, cartilage develops from cells within the muscle or connective tissue (Urist et al., 1984; Syftestad and Caplan, 1984). In the chick embryo, this protein has been localized to the extracellular matrix of head epithelia (Urist and Hall, 1989; Figure 10).

During intramembranous ossification, the mesenchymal cells proliferate and condense into compact nodes. Some of these cells develop into capillaries and others change their shape to become OSTEOBLASTS, cells capable of secreting the bone matrix. The secreted collagen–glycosaminoglycan matrix is able to bind calcium salts, which are brought to the region through capillaries. In this way, the matrix becomes calcified. In most cases, osteoblasts are separated from the region of calcification by a layer of the prebone (osteoid) matrix they secrete. Occasionally, though, osteoblasts become trapped in the calcified matrix and become OSTEOCYTES, bone cells. As calcification proceeds, the bony spicules radiate out from the center where ossification began (Figure 11). Furthermore, the entire region of calcified spicules becomes surrounded by compact mesenchymal cells that form the PERIOSTEUM. The cells on the inner surface of the periosteum also become osteoblasts and deposit bone matrix parallel to that of the existing spicules. In this manner, many layers of bone are formed.

Endochondral ossification involves the formation of cartilage tissue from aggregated mesenchymal cells and the subsequent replacement of this cartilage tissue by bone (Horton, 1990). The cartilage tissue is a model for the bone that follows. The bones of the vertebral column, the pelvis, and the extremities are first formed of cartilage and are later changed into bone. This remarkable process coordinates CHONDROGENESIS (cartilage production) with OSTEOGENESIS (bone growth), while these skeletal elements are simultaneously bearing a load, growing in width, and responding to local stresses. In humans, the "long bones" of the embryonic limb buds form from mesenchymal cells that form nodules in those regions that will become bone. These cells become CHONDROCYTES, and they secrete the cartilage extracellular matrix. The mesenchymal cells around them become the periosteum (Figure 12). Soon after the cartilaginous "model" is formed, the cells in the central part of the model become dramatically larger and begin secreting a different type of matrix, one that contains different types of collagen, more fibronectin, and less protease inhibitor. These cells are the HYPERTROPHIC CHONDROCYTES. Their matrix is more susceptible to invasion by blood vessel cells from the periosteum. A capillary from the periosteum then invades the center of the previously avascular cartilage shaft. As the cartilage matrix is degraded, the hypertrophic cartilage cells die, and OSTEOBLASTS (bone-forming cells) carried by the blood vessels begin to secrete bone matrix on the partially degraded cartilage. Eventually, all the cartilage is replaced by bone.

As the center of the cartilage model is converted into bone, an ossifi-

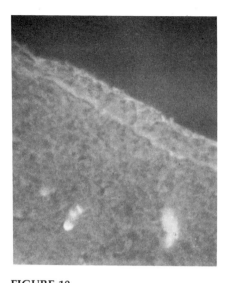

FIGURE 10
Localization of bone morphogenic protein in the future lower jaw of an embryonic chick. Fluorescent antibodies to this protein were incubated with microscopic sections of the developing chick at the time the mandible is beginning to form. The protein appears to be able to induce cartilage and then bone formation. (Photograph courtesy of M. R. Urist)

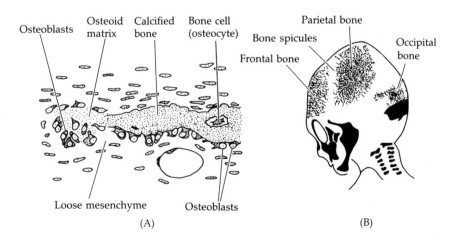

Osteoblasts Osteoid matrix Calcified bone Bone cell (osteocyte)
Parietal bone
Bone spicules
Occipital bone
Frontal bone
Loose mesenchyme Osteoblasts
(A) (B)

FIGURE 11
Schematic diagram of membranous ossification. (A) Mesenchymal cells, probably derived from the neural crest, condense to produce osteoblasts, which deposit osteoid matrix. These osteoblasts become arrayed along the calcified region of the matrix. Osteoblasts that are trapped within the bone matrix become osteocytes. (B) Spread of bone spicules from the primary ossification site in the flat skull bones of a 3-month-old human embryo. The bones shown in black are formed by endochondral ossification. (After Langman, 1969.)

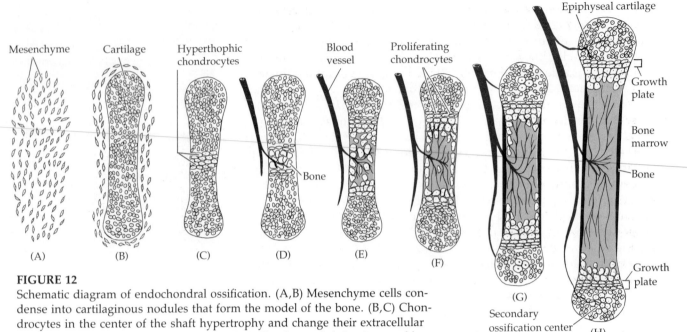

FIGURE 12
Schematic diagram of endochondral ossification. (A,B) Mesenchyme cells condense into cartilaginous nodules that form the model of the bone. (B,C) Chondrocytes in the center of the shaft hypertrophy and change their extracellular matrix, allowing blood vessels to enter. (D,E) Blood vessels bring in osteoblasts, which bind to the degenerating cartilaginous matrix and deposit bone matrix. (F–H) Formation of the epiphyseal growth plates by chondrocytes, which proliferate before undergoing hypertrophy. Secondary ossification centers also form as blood vessels enter near the tips of the bone. (After Horton, 1990.)

FIGURE 13
Proliferation of cells in epiphyseal plate in response to growth hormone. (A) Cartilaginous region in a young rat that was made growth hormone–deficient by removal of its pituitary. (B) Same region in the rat after injection of growth hormone. (I. Gersh's photographs from Bloom and Fawcett, 1975.)

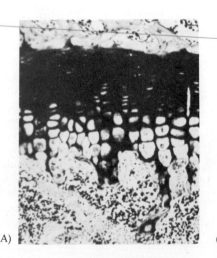

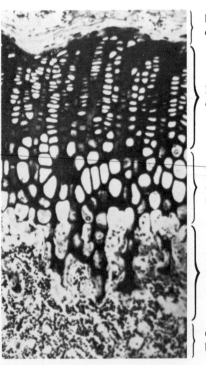

(A) (B)

cation front is formed between the newly synthesized bone and the remaining cartilage. The cartilage side of this front contains the hypertrophic cartilage that prepares the shaft for invasion by the blood vessels, and the bone side contains the osteoblast cells laying down the bone matrix. This front spreads outward in both directions from the center, as more cartilage is turned to bone. If this were all, however, there would be no growth, and our bones would just be as large as the original cartilaginous model. However, as the ossification front nears the ends of the cartilage model, the chondrocytes near the ossification front proliferate prior to undergoing hypertrophy. This pushes out the cartilaginous ends of the bone, providing a source of new cartilage. These cartilaginous regions at the end of the long bones are called EPIPHYSEAL PLATES. As this cartilage hypertrophies and the ossification front extends further outward, the remaining cartilage in the epiphyseal plate proliferates. This cartilage forms the growth area of the bone. Thus, the bone keeps growing because of the production of new cartilage cells that undergo hypertrophy, enable the blood vessels to enter, and die as the bony matrix is deposited. As long as the epiphyseal growth plates are able to produce chondrocytes, the bone continues to grow. The cells of this plate are very responsive to hormones, and their proliferation is stimulated by growth hormone and insulin-like growth factors (Figure 13; see Chapter 20). Hormones are also responsible for the cessation of growth. At the end of puberty, high levels of estrogen or testosterone cause the remaining epiphyseal plate cartilage to hypertrophy. These cartilage cells grow, die, and are replaced by bone. Without any further cartilage, growth of these bones ceases.

The replacement of chondrocytes by osteoblasts appears to be dependent on the mineralization of the extracellular matrix. In chick embryos, the source of calcium is the calcium carbonate of the eggshell, and during its development the circulatory system of the chick translocates about 120 mg of calcium from the shell to the skeleton (Tuan, 1987). When chick embryos are removed from their shells at day 3 and grown in shell-less culture (in plastic wrap) for the duration of their development, much of the calcium-deficient cartilaginous skeleton fails to mature into bony tissue (Tuan and Lynch, 1983; Figure 14).

In mammals, calcium is thought to arise from within the chondrocytes. In one hypothesis, the metabolism of the hypertrophic chondrocytes switches from aerobic metabolism to anaerobic metabolism, causing the mitochondria to cease generating ATP and to begin storing calcium. The dying chondrocyte releases this stored calcium upon the extracellular matrix. Osteoblasts bind to this calcified matrix and bone formation is initiated (Figure 15; Brighton and Hunt, 1974; Brighton, 1984). Another hypothesis stresses a different origin for the initial calcium phosphate deposits on the extracellular matrix. According to this hypothesis, the major agents of calcification are MATRIX VESICLES. These vesicles are thought to be specialized regions of the chondrocyte plasma membrane that pinch off the cell and adhere to the extracellular matrix. Once in place, they concentrate large amounts of calcium and phosphate from the surrounding lymph (Dereszewski and Howell, 1978; Wuthier et al., 1985). Indeed, these vesicles are seen in association with the earliest calcium deposits in cartilage and bone. The finding (Wuthier et al., 1985) that some of these vesicles contain calcium from the chondrocytes suggests that both models may be correct and that more than one mechanism is involved.

As new bone material is added peripherally from the internal surface of the periosteum, there is a hollowing out of the internal region to form the bone marrow cavity. This destruction of bone tissue is due to OSTEO-CLASTS, multinucleated cells that enter the bone through the blood vessels

FIGURE 14
Skeletal mineralization in 17-day chick embryos grown in shell-less (A) and normal (B) culture. The embryos were fixed and stained with Alizarin Red to stain the calcified matrix. (From Tuan and Lynch, 1983; photograph courtesy of R. Tuan.)

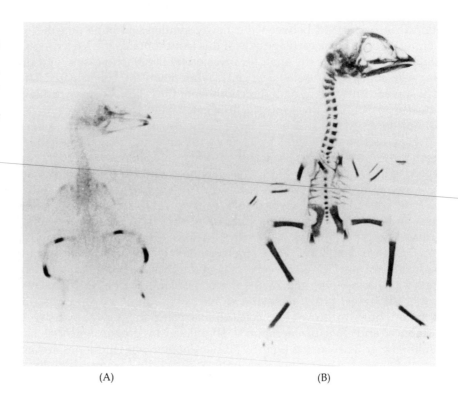

(A)   (B)

(Kahn and Simmons, 1975). Osteoclasts are probably derived from the same precursors as blood cells, and they dissolve both the inorganic and protein portions of the bone matrix (Ash et al., 1980; Blair et al., 1986). The osteoclast extends numerous cellular processes into the matrix and pumps out hydrogen ions onto the surrounding material, thereby acidifying and solubilizing it (Baron et al., 1985, 1986; Figure 16). The blood vessels also import the blood-forming cells, which will reside in the marrow for the duration of the organism's life.

FIGURE 15
Deposition of calcium by chondrocytes in the distal region of the hypertrophic zone. Calcium (stained darkly in the electron micrograph montage) is placed onto the extracellular matrix by the enlarging cells. (From Brighton and Hunt, 1974; photograph courtesy of C. T. Brighton.)

Chondrocytes

Calcium on extracellular matrix

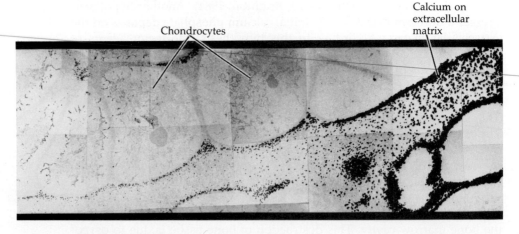

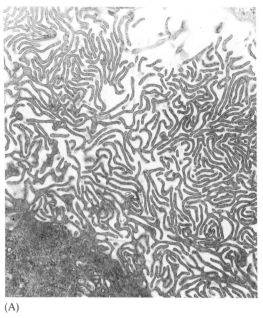

(A)

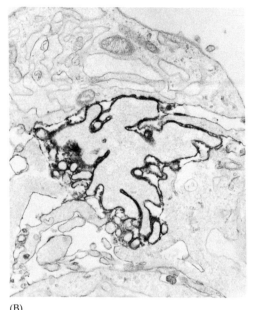

(B)

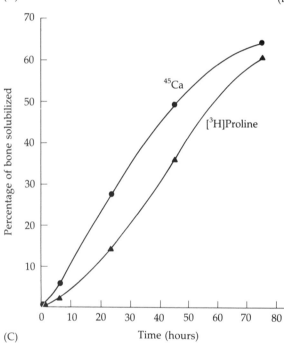

(C)

^{45}Ca

[^{3}H]Proline

Percentage of bone solubilized

Time (hours)

FIGURE 16
Osteoclast activity on the bone matrix. (A) Electron micrograph of the ruffled membrane of a chick osteoclast cultured on reconstituted bone matrix. (B) Section of ruffled membrane stained for the presence of an ATPase capable of transporting H$^+$ from the cell. The ATPase is restricted to the membrane of the cell process. (C) Solubilization of inorganic and collagenous matrix components (as measured by the release of [^{45}Ca] and [^{3}H] proline, respectively) by 10,000 osteoclasts incubated on labeled bone fragments. (A and C from Blair et al., 1986; B from Baron et al., 1986; photographs courtesy of the authors.)

Lateral plate mesoderm

Not all of the mesodermal mantle is organized into somites. Adjacent to somitic mesoderm is the INTERMEDIATE MESODERMAL REGION. This cord of mesodermal cells develops into the pronephric tubule, which is the precursor of kidney and genital ducts. The development of these organ systems will be discussed in detail in Chapters 16 and 21, respectively. Further laterally on each side we come to the LATERAL PLATE MESODERM. These plates split horizontally into the dorsal SOMATIC (or PARIETAL) MESODERM, which underlies the ectoderm, and a ventral SPLANCHNIC (or VISCERAL)

MESODERM, which overlies the endoderm (Figure 2C). Between these layers is the body cavity—the COELOM—which stretches from the future neck region to the posterior of the body. During later development, the right- and left-side coeloms fuse, and folds extend from the somatic mesoderm, dividing the coelom into separate cavities. In mammals, the coelom is subdivided into the pleural, pericardial, and peritoneal spaces, enveloping the thorax, heart, and abdomen, respectively. The mechanism for creating mesodermal somites and body linings has changed little throughout vertebrate evolution, and the development of the chick mesoderm can be compared with similar stages of frog embryos (Figure 17).

SIDELIGHTS & SPECULATIONS

Isoforms

Numerous mesodermal cell lineages are characterized by the replacement of one set of structures (cells or molecules) by another during the course of development. The molecules or cells involved in these processes are often related to each other and have been called ISOFORMS (Caplan et al., 1983). An example of a cellular isoform set is the chondrocyte and the osteoblast that replaces it. Both types of cells secrete an extensive and calcifiable extracellular matrix. Examples of molecular isoforms include the sequential appearance of embryonic hemoglobin, fetal hemoglobin, and adult hemoglobins.

During mouse development, three isoformic populations of myoblasts are formed. The first myoblasts (myoblast I), formed after the separation of the myotome from the sclerotome, undergo limited fusion to form small myotubes with only 4–6 nuclei per cell. These do not form functional muscles. The myoblast II population soon follows, and these cells fuse into large myotubes (20–100 nuclei) and supplant the myoblast I population (Bonner and Hauschka, 1974). At the same time, the major motor axons begin to enter the limb. These neurons stimulate the production of the third myoblast population (Womble and Bonner, 1980). The myotubes formed by the myoblast III cells form the muscles that characterize the newborn mouse. In the limb, one sees the formation of the characteristic M- and Z-like proteins for the first time in the myoblast III population (Jockusch and Jockusch, 1980).

Within the muscles produced by myoblast III cells, *molecular* isoforms appear and disappear. Consider the heavy (200,000-Da) myosin chains. Analyses of these proteins indicate that during fetal development, there is an embryonic myosin heavy chain. This changes at birth to a neonatal myosin heavy chain and later to the adult (fast muscle) myosin heavy chain (Whalen et al., 1979). Thus, there are two types of isoformic replacements—one on the cellular level and one on the molecular level—that are involved in muscle development.

The appearance of these isoforms can be altered by experimentation (as when nerves are prevented from reaching the limb) or by natural mutation. In humans and chicks, there are genetic mutations that cause muscular dystrophy, a disease that eventually leads to the destruction of limb muscle fibers. The muscles of human patients and chicks expressing these mutations have many features characteristic of immature muscles. This suggests a failure of the normal developmental program for muscle development. In both instances the dystrophic muscles continue to make the neonatal form of the myosin heavy chain long after it has disappeared from normal adult muscle (Bandman, 1985; Webster et al., 1988). It should be recognized, then, that the development of a tissue need not be completed as soon as it forms. There is a remodeling that occurs both intercellularly and intracellularly to generate the phenotype of the adult individual.

Formation of extraembryonic membranes

In reptiles, birds, and mammals, embryonic development has taken a new direction. Reptiles evolved a mechanism for laying eggs on dry land, thus freeing them to explore niches that were not close to ponds. To accomplish this, the embryo produced four sets of EXTRAEMBRYONIC MEMBRANES to mediate between it and the environment, and even though most mammals have evolved placentas instead of shells, the basic pattern of extraembryonic membranes remains the same. In developing reptiles, birds, and mammals, there initially is no distinction between embryonic and extraembryonic domains. However, as the body of the embryo takes shape, the

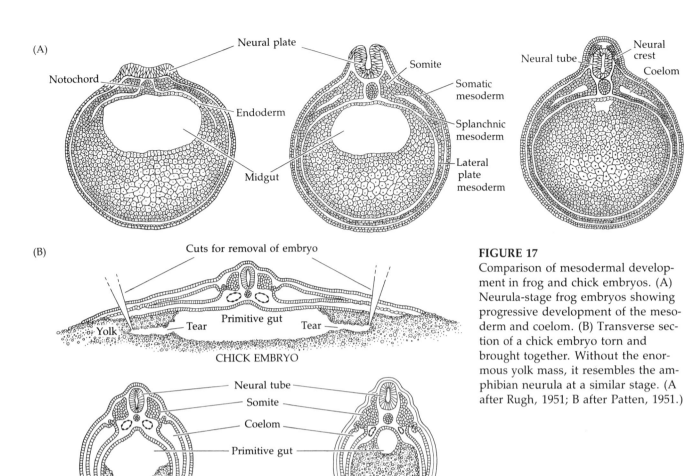

(A)

Neural plate

Notochord

Endoderm

Midgut

Somite

Somatic mesoderm

Splanchnic mesoderm

Lateral plate mesoderm

Neural tube

Neural crest

Coelom

(B)

Cuts for removal of embryo

Yolk

Tear

Primitive gut

Tear

CHICK EMBRYO

Neural tube

Somite

Coelom

Primitive gut

Yolk

CHICK EMBRYO
(removed from yolk; edges pulled together)

FROG EMBRYO

FIGURE 17
Comparison of mesodermal development in frog and chick embryos. (A) Neurula-stage frog embryos showing progressive development of the mesoderm and coelom. (B) Transverse section of a chick embryo torn and brought together. Without the enormous yolk mass, it resembles the amphibian neurula at a similar stage. (A after Rugh, 1951; B after Patten, 1951.)

border epithelia create a series of BODY FOLDS, which surround the developing embryo, thereby isolating it from the yolk and delineating which areas are to be embryonic.

The membranous folds are formed by the extension of ectodermal and endodermal epithelium underlaid with mesoderm. The combination of ectoderm and mesoderm is often referred to as the SOMATOPLEURE and forms the amnion and chorion membranes; the combination of endoderm and mesoderm—the SPLANCHNOPLEURE—forms the yolk sac and allantois. The endodermal or ectodermal tissue acts as the functioning epithelial cells; and the mesoderm generates the essential blood supply to and from this epithelium. The formation of these folds can be followed in Figure 18.

The first problem of a land-dwelling egg is desiccation. Embryonic cells would quickly dry out if they were not in an aqueous environment. This environment is supplied by the AMNION. The cells of this membrane secrete amniotic fluid; thus embryogenesis still occurs in water. This evolutionary advance is so significant and characteristic that reptiles, birds, and mammals are grouped together as the AMNIOTE VERTEBRATES.

The second problem of a land-dwelling egg is gas exchange. This exchange is provided for by the CHORION, the outermost extraembryonic membrane. In birds and reptiles, this membrane adheres to the shell, allowing the exchange of gases between the egg and the environment. In

FIGURE 18

Schematic drawings of the extraembryonic membranes of the chick. The embryo is cut longitudinally and the albumen and shell coatings are not shown. (A) 2-day embryo. (B) 3-day embryo. (C) Detailed schematic diagram of the caudal (hind) region of the chick embryo showing the formation of the allantois. (D) 5-day embryo. (E) 9-day embryo. (After Carlson, 1981.)

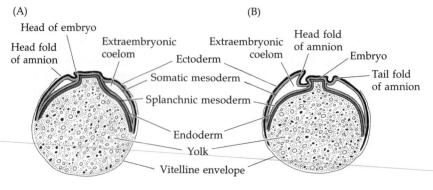

(A)

Head of embryo
Head fold of amnion
Extraembryonic coelom
Ectoderm
Somatic mesoderm
Splanchnic mesoderm
Endoderm
Yolk
Vitelline envelope

(B)

Extraembryonic coelom
Head fold of amnion
Embryo
Tail fold of amnion

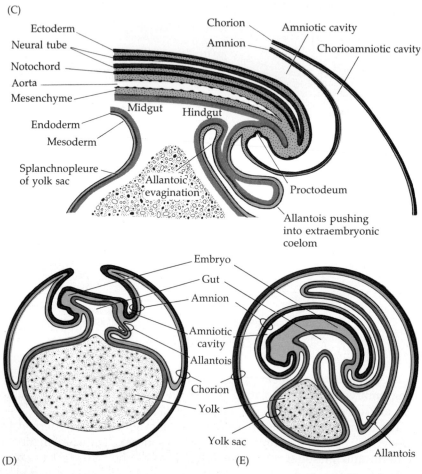

(C)

Ectoderm
Neural tube
Notochord
Aorta
Mesenchyme
Endoderm
Mesoderm
Splanchnopleure of yolk sac
Midgut
Hindgut
Allantoic evagination
Chorion
Amnion
Amniotic cavity
Chorioamniotic cavity
Proctodeum
Allantois pushing into extraembryonic coelom

Embryo
Gut
Amnion
Amniotic cavity
Allantois
Chorion
Yolk
Yolk sac
Allantois

(D)

(E)

mammals, as we have seen, the chorion has evolved into the placenta, which has many other functions besides respiration.

The ALLANTOIS acts to store urinary wastes and to mediate gas exchange. In reptiles and birds, the allantois becomes a large sac, as there is no other way to keep toxic by-products of metabolism from the developing embryo. The mesodermal layer of allantoic membrane often will reach and fuse with the mesodermal layer of the chorion to create the CHORIOALLANTOIC MEMBRANE. This extremely vascular envelope is crucial for chick development and is responsible for transporting calcium from the eggshell into the embryo for bone production. In mammals, the size of the allantois depends upon how well nitrogenous wastes can be removed by the chorionic placenta. In humans the allantois is a vestigial sac, whereas in pigs it is a large and important organ.

The YOLK SAC is the first extraembryonic membrane to be formed, as it mediates nutrition in developing birds and reptiles. It is derived from endodermal cells that grow over the yolk to enclose it. The yolk sac is connected to the midgut by an open tube, the YOLK DUCT, so that the walls of the yolk sac and the walls of the gut are continuous. The blood vessels within the mesoderm of the splanchnopleure transport nutrients from the yolk into the body, for yolk is not taken directly into the body through the yolk duct. Rather, endodermal cells digest the protein into soluble amino acids that can then be passed on to the blood vessels surrounding the yolk sac. In this way, the four extraembryonic membranes enable the embryo to develop on land.

Heart and circulatory system

The circulatory system is one of the great achievements of the lateral plate mesoderm. Consisting of a heart, blood cells, and an intricate system of blood vessels, the circulatory system provides nourishment to the developing vertebrate embryo. The circulatory system is the first functional unit in the developing embryo, and the heart is the first functional organ.

The heart

In vertebrate embryos, the heart develops in an anterior position, within the splanchnic mesoderm of the neck; only later does it move to the chest. In amphibians, the two presumptive heart-forming regions are initially found at the most anterior position of the mesodermal mantle. While the embryo is undergoing neurulation, these two regions come together in the ventral region of the embryo to form a common PERICARDIAL CAVITY. In birds and mammals, the heart also develops by fusion of paired primordia, but the fusion of these two rudiments does not occur until much later in development. In amniote vertebrates, the embryo is a flattened disc, and the lateral plate mesoderm does not completely encircle the yolk

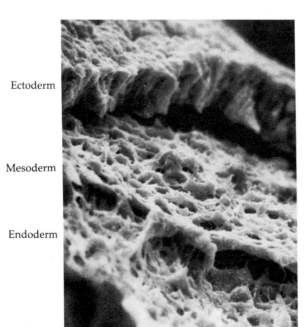

Ectoderm

Mesoderm

Endoderm

FIGURE 19
Scanning electron micrograph of the heart-forming mesoderm in the 24-hour chick embryo. The mesoderm is readily separated from the ectoderm but remains in close association with the endoderm. (From Linask and Lash, 1986; photograph courtesy of K. Linask.)

sac. The presumptive heart cells begin migrating from the lateral sides of the mesoderm (at the level of Hensen's node) when the embryo is only 18–20 hours old. At this time, the embryo is still relatively simple, consisting of only three germ layers. As the presumptive heart cells move anteriorly between the ectoderm and endoderm toward the middle of the embryo, they remain in close contact with the endodermal surface (Figure 19; Linask and Lash, 1986). When the cells reach the area where the gut has extended into the anterior region of the embryo, migration ceases. The directionality for this migration appears to be provided by the endoderm. If the cardiac region endoderm is rotated with respect to the rest of the embryo, migration of the precardiac mesoderm cells is reversed. It is thought that the endodermal component responsible for this movement is an anterior-to-posterior concentration gradient of fibronectin. Antibodies against fibronectin stop the migration while antibodies against other ex-

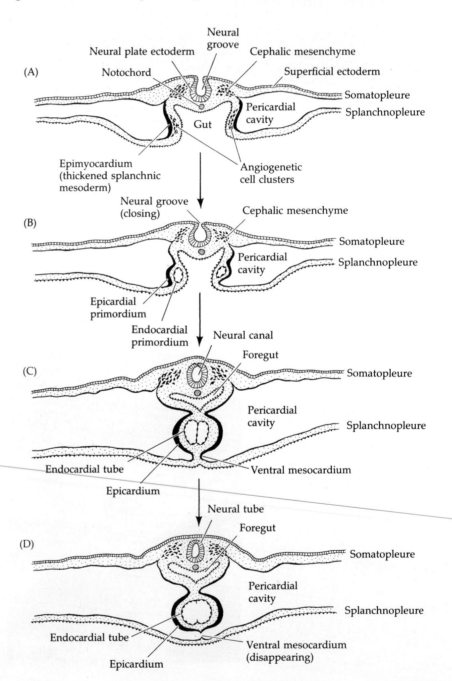

FIGURE 20

Formation of the heart. Transverse sections through the heart-forming region of the chick embryo at (A) 25 hours; (B) 26 hours; (C) 28 hours; and (D) 29 hours. (After Carlson, 1981.)

tracellular matrix components do not (Linask and Lash, 1988a,b). During this migration, the two heart-forming primordia begin to undergo significant differentiation independently of each other. The presumptive heart cells of birds and mammals form a double-walled tube consisting of an inner ENDOCARDIUM and an outer EPIMYOCARDIUM. The endocardium will form the inner lining of the heart and the outer layer will form the layer of heart muscles, which will pump for the lifetime of the organism.

As neurulation proceeds, the foregut is closed by the inward folding of the splanchnic mesoderm (Figure 20). This movement brings the two tubes together, eventually uniting the epimyocardium into a single tube. The two endocardia lie within the common chamber for a short while, but these will also fuse. At this time, the originally paired coelomic chambers unite to form the body cavity in which the heart resides. The bilateral origin of the heart can be demonstrated by surgically preventing the merger of the lateral plate mesoderm (Gräper, 1907; DeHaan, 1959). This results in a condition called CARDIA BIFIDA, wherein separate hearts form on each side of the body (Figure 21). The next step in heart formation is the fusion of the endocardial tubes to form a single pumping chamber (Figure 20C, D). This fusion occurs at about 29 hours in chick development or at 3 weeks of human gestation. The unfused posterior portions of the endocardium become the openings of the VITELLINE VEINS into the heart (Figure 22). These veins will carry nutrients from the yolk sac into the SINUS VENOSUS. The blood then passes through a valvelike flap into the atrial region of the heart. Contractions of the TRUNCUS ARTERIOSUS speed the blood into the AORTA.

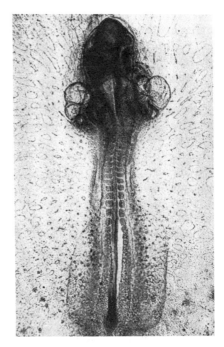

FIGURE 21
Cardia bifida in chick embryo caused by preventing the two heart primordia from fusing. (Courtesy of R. L. De-Haan.)

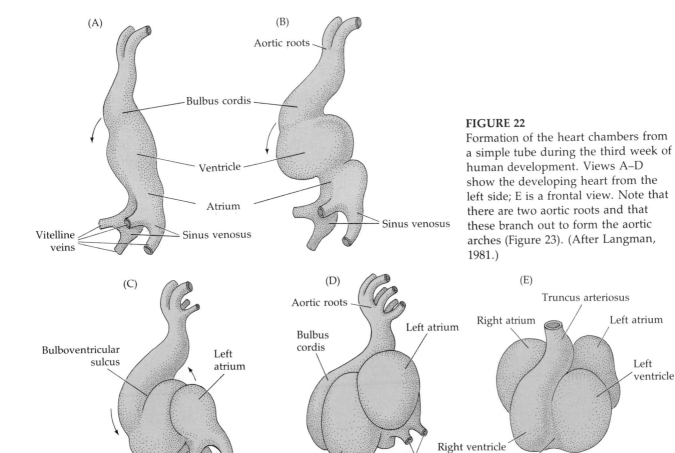

FIGURE 22
Formation of the heart chambers from a simple tube during the third week of human development. Views A–D show the developing heart from the left side; E is a frontal view. Note that there are two aortic roots and that these branch out to form the aortic arches (Figure 23). (After Langman, 1981.)

Pulsations of the heart begin while the paired primordia are still fusing. The pacemaker of this contraction is the sinus venosus. The contractions begin here and a wave of muscle contraction is then propagated up the tubular heart. In this way, the heart can pump blood even before its intricate system of valves has been completed. Heart muscle cells have their own inherent ability to contract—and isolated heart cells from 7-day rat or chick embryos will still continue to beat in petri dishes (Harary and Farley, 1963; DeHaan, 1967). In the embryo, these contractions become regulated by electrical stimuli from the medulla oblongata via the vagus nerve, and by 4 days, the electrocardiogram of a chick embryo approximates that of an adult.

Formation of blood vessels

The developing embryo has needs that are different from those of the adult organism, and its circulatory system reflects those differences. First, food is absorbed, not through the intestine, but from either the yolk or the placenta. Second, respiration is not conducted through the gills or lungs but through the chorionic or allantoic membranes. Thus, major embryonic blood vessels are constructed to serve these extraembryonic structures.

In addition to fulfilling the physiological needs of the embryo, the construction of blood vessels also is dictated by evolutionary history. The mammalian embryo will extend blood vessels to the yolk sac even though there is no yolk therein. Moreover, the blood leaving the heart loops over the foregut to form the dorsally located aorta. These six pairs of AORTIC ARCHES arch over the pharynx (Figure 23). In primitive fishes, these arches persist and enable the gills to oxygenate the blood. In the adult bird or mammal, where lungs oxygenate the blood, such a system makes little sense. Yet, in mammalian and avian embryos, all six pairs of aortic arches are formed before the system eventually becomes simplified into a single aortic arch. Thus, our embryonic condition reflects our evolutionary history, even though our physiology does not require such a structure.

The major embryonic blood vessels are those that obtain nutrients and bring them to the body and that transport gases to and from the sites of respiratory exchange. In those vertebrates with yolk, the VITELLINE (OMPHALOMESENTERIC) VEINS are formed by the aggregation of splanchnic mesodermal cells into ANGIOGENIC CLUSTERS (BLOOD ISLANDS) that line the yolk sac (Figure 24). These cells are first seen in the area opaca at the headfold stage of chick embryogenesis, when the primitive streak is at its fullest extent (Pardanaud et al., 1987). These cords soon hollow out into double-walled tubes analogous to the double tube of the heart. The inner

FIGURE 23
The aortic arches of the human embryo. (A) Originally, the truncus arteriosus pumps blood into the aorta, which branches on either side of the foregut. The six aortic arches take blood from the ventral aorta and allow it to flow into the dorsal aorta. (B) The arches begin to disintegrate or become modified; the dotted lines indicate degenerating structures. (C) Eventually, the remaining arches are modified and the adult arterial system is formed. (After Langman, 1981.)

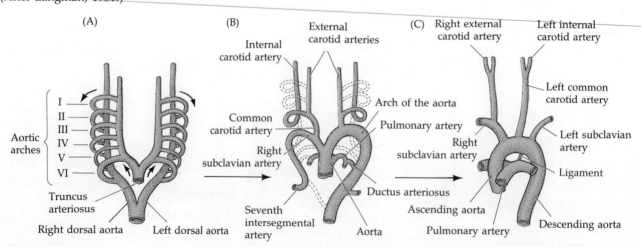

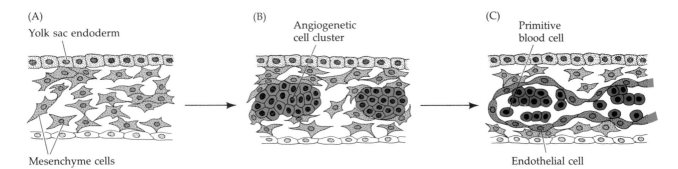

(A) Yolk sac endoderm

Mesenchyme cells

(B) Angiogenetic cell cluster

(C) Primitive blood cell

Endothelial cell

FIGURE 24
Angiogenesis. Blood vessel formation is first seen in the wall of the yolk sac where undifferentiated mesenchyme (A) condenses to form angiogenetic cell clusters (B). The centers of these clusters form the blood cells, and the outside of the clusters develop into blood vessel endothelial cells (C). (After Langman, 1981.)

wall becomes the flat ENDOTHELIAL CELL lining of the vessel and the outer cells become the smooth muscle. Between these layers is a basal lamina containing a type of collagen specific for blood vessels. It is thought (Murphy and Carlson, 1978; Kubota et al., 1988) that this basal lamina initiates the differentiation of the cell types in the vessel. The central cells of the blood islands differentiate into the embryonic blood cells. As the blood islands grow, they eventually merge to form the capillary network draining into the two vitelline veins, which bring the food and blood cells to the newly formed heart. The creation of blood vessels *de novo* from the mesoderm is called VASCULOGENESIS (Pardanaud et al., 1989). The embryonic circulatory system to and from the embryo and yolk sac is shown in Figure 25.

The capillary networks of the circulatory system—those regions where the gas and nutrient exchange actually takes place in each tissue—can arise in two different ways (Pardanaud et al., 1989). In some organs, vasculogenesis continues. Capillary networks arise independently within the tissues themselves (Auerbach et al., 1985). In such cases, the capillaries do not arise as smaller and smaller extensions of the major blood vessels

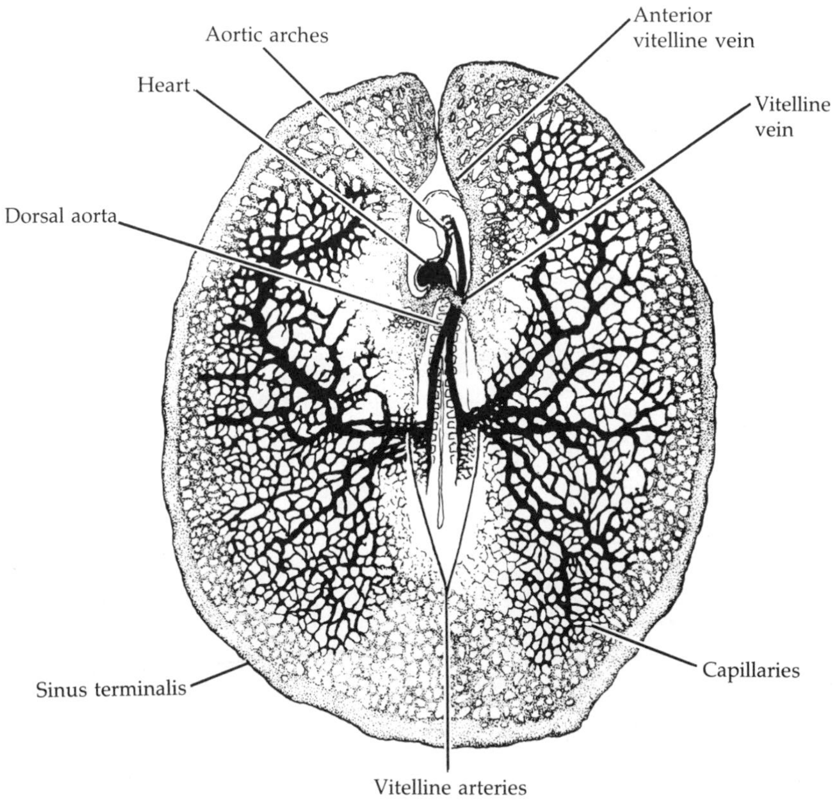

Aortic arches

Heart

Dorsal aorta

Sinus terminalis

Anterior vitelline vein

Vitelline vein

Capillaries

Vitelline arteries

FIGURE 25
Circulatory system of a 44-hour chick embryo. This ventral view shows arteries in black; the veins are stippled. The sinus terminalis is the outer limit of the circulatory system and is the site of blood cell generation. (From Carlson, 1981.)

growing from the heart. Rather, the mesoderm of each of these organs contains cells, called ANGIOBLASTS, which arrange themselves into capillary vessels. These organ-specific capillary networks eventually become linked to the extensions of the major blood vessels. In other organs (notably the limb buds, kidney, and brain), existing blood vessels are seen to proliferate into organ-forming cells (Wilson, 1983). This type of blood vessel formation where new vessels emerge from the proliferation of preexisting blood vessels is called ANGIOGENESIS. The organ-forming regions are thought to secrete ANGIOGENESIS FACTORS that promote the mitosis and migration of endothelial cells into that area. In the 3-day chick brain, a factor has been isolated that promotes the migration into the brain of endothelial cells from existing blood vessels on the brain surface (Risau, 1986).

In mammalian embryos, food and oxygen are obtained from the placenta. Thus, although the mammalian embryo has vessels analogous to the vitelline veins, the main supply of food and oxygen comes from the UMBILICAL VEIN, which unites the embryo with the placenta (Figure 26). This vein, which takes the oxygenated and food-laden blood back into the embryo, is derived from what would be the right vitelline vein in birds. The UMBILICAL ARTERY, carrying wastes to the placenta, is derived from what would have become the allantoic artery of the chick. It extends from the caudal portion of the aorta and proceeds along the allantois and then out to the placenta.

After entering the embryonic mammalian heart, the blood is pumped into a series of aortic arches, which encircle the pharynx to bring the blood dorsal. In mammals, the left member of the fourth pair of aortic arches is the only one surviving to reach the aorta. The right member of this pair has become the root of the subclavian artery. The third aortic arches have been modified to form the common carotid arteries, which supply blood to the brain and head. The sixth arch is modified to form the pulmonary artery; and the first, second, and fifth arches degenerate. The aorta and pulmonary artery, therefore, have a common opening to the heart for much of their development. Eventually, partitions form within the truncus arteriosus to create two different vessels (Figure 27). Only when the first breath of the newborn animal indicates that the lungs are ready to handle the oxygenation of the blood does the heart become modified to pump blood separately to the pulmonary artery.

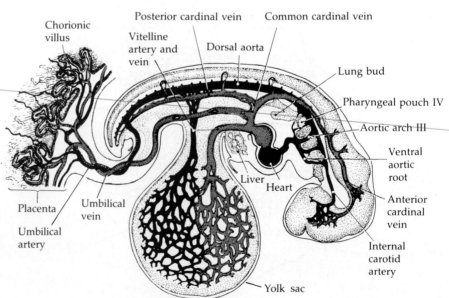

FIGURE 26
Circulatory system of a 4-week human embryo. Although at this stage all the major blood vessels are paired left and right, only the right vessels are shown. Arteries are shown in color. (From Carlson, 1981.)

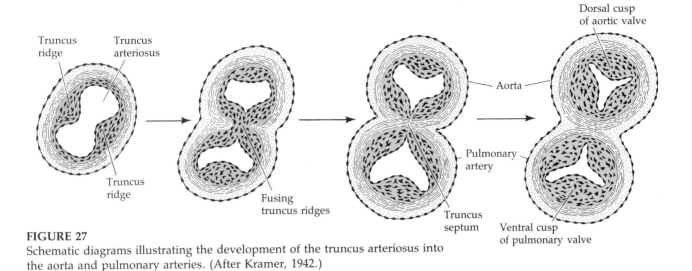

FIGURE 27
Schematic diagrams illustrating the development of the truncus arteriosus into the aorta and pulmonary arteries. (After Kramer, 1942.)

SIDELIGHTS & SPECULATIONS

Redirecting blood flow in the newborn mammal

Although the developing fetus shares with the adult the need to get oxygen and nutrients to its tissues, the physiology of the mammalian fetus differs drastically from that of the adult. Chief among these differences is the lack of functional lungs and intestines. All oxygen and nutrients must come from the placenta. This raises two questions. First, how does the fetus obtain oxygen from maternal blood; and second, how is blood circulation redirected to the lungs once the umbilical cord is cut and breathing is made necessary?

The solution to the fetus's problems in getting oxygen from its mother's blood involves the development of a FETAL HEMOGLOBIN. The hemoglobin in fetal red blood cells differs slightly from that in adult corpuscles. It has a slightly different amino acid sequence. Two of the four peptides of the fetal and adult hemoglobin chains are identical—the alpha (α) chains—but the two beta (β) chains in adult hemoglobin have been replaced by two gamma (γ) chains in the fetus (Figure 28A. The molecular

basis for this switch in globins will be discussed in Chapters 11 and 12.) Normal β chains will bind the natural regulator diphosphoglycerate, which assists in the unloading of oxygen. The γ chain isoforms do not bind diphosphoglycerate as well, and therefore they have a higher affinity for oxygen. In the low-oxygen environment of the placenta, oxy-

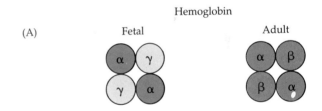

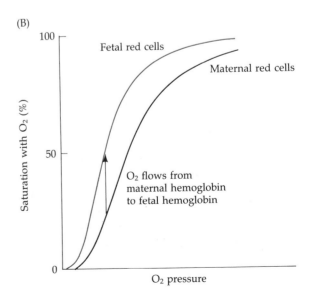

FIGURE 28
Transfer of oxygen from the mother to the fetus in human embryos. Adult and fetal hemoglobin molecules differ in their protein subunits (A). The fetal γ chain binds diphosphoglycerol less avidly than does the adult β chain. Consequently, fetal hemoglobin can bind oxygen more efficiently than can adult hemoglobin. (B) In the placenta, there is a net flow of oxygen from the mother's blood (which gives oxygen up to the tissue at the lower oxygen pressure) to the fetal blood, which is still picking it up. (B after Stryer, 1981.)

gen is released from the adult hemoglobin. In this same environment, fetal hemoglobin will not give away oxygen but will bind it. This small difference in oxygen affinity mediates the transfer of oxygen from the mother to the fetus (Figure 28B). Within the fetus, the myoglobin of the fetal muscles has an even higher affinity for oxygen, so oxygen molecules will pass from fetal hemoglobin for storage and use in the fetal muscles. Fetal hemoglobin is not deleterious to the newborn, and in humans, the replacement of fetal hemoglobin-containing blood cells by adult hemoglobin-containing blood cells is not completed until about six months after birth.

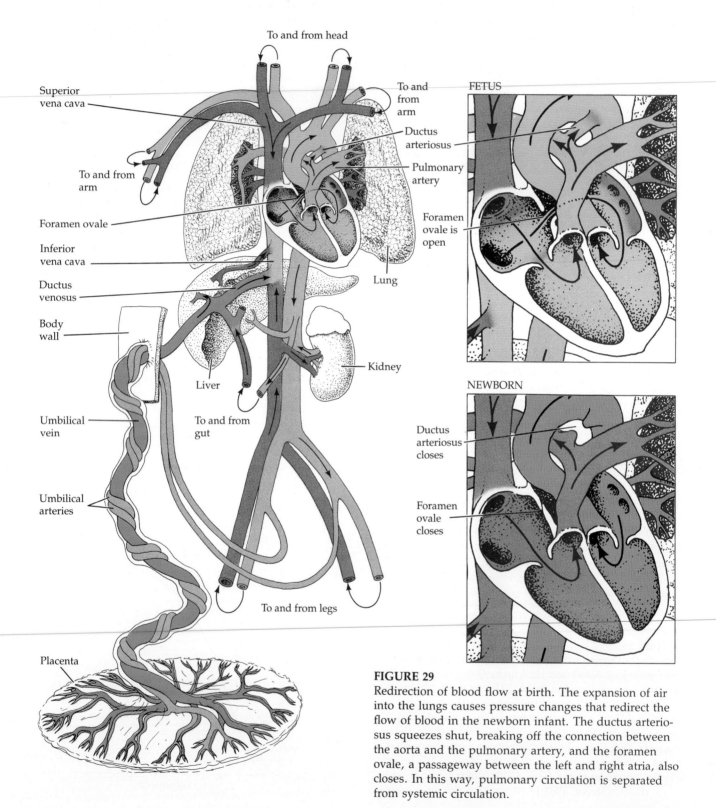

FIGURE 29
Redirection of blood flow at birth. The expansion of air into the lungs causes pressure changes that redirect the flow of blood in the newborn infant. The ductus arteriosus squeezes shut, breaking off the connection between the aorta and the pulmonary artery, and the foramen ovale, a passageway between the left and right atria, also closes. In this way, pulmonary circulation is separated from systemic circulation.

But once the fetus is not getting its oxygen from the mother, how does it restructure its circulation to get oxygen from its own lungs? During fetal development, an opening—the DUCTUS ARTERIOSUS—diverts the passage of blood from the pulmonary artery into the aorta (and thus to the placenta). Because blood does not return from the pulmonary vein in the fetus, the developing mammal has to have some other way of getting blood into the left ventricle to be pumped. This is accomplished by the FORAMEN OVALE, a hole in the septum separating the right and left

atria. Blood can enter the right atrium, pass through the foramen to the left atrium, and then enter the left ventricle (Figure 29). When the first breath is drawn, the oxygen in the blood causes the muscles surrounding the ductus arteriosus to close the opening. As the blood pressure in the left side of the heart increases, it causes a flap over the foramen ovale to close, thereby separating the pulmonary and systemic circulation. Thus, when breathing begins, the respiratory circulation is shunted from the placenta to the lungs.

SIDELIGHTS & SPECULATIONS

Tumor-induced angiogenesis

Judah Folkman (1974) has estimated that as many as 350 billion mitoses occur in each human every day. With each cell division comes the potential that the resulting cells will be malignant. Yet very few tumors actually do develop in any individual. Folkman has suggested that cells capable of forming tumors develop at a certain frequency but that a large majority are never able to form observable tumors. The reason is that solid tumor, like any other rapidly dividing tissue, needs oxygen and nutrients to survive. Without a blood supply, potential tumors either die or remain dormant. Such "microtumors" remain as a stable cell population wherein dying cells are replaced by new cells.

The critical point at which this node of tumorous cells becomes a rapidly growing tumor occurs when the pocket of cells becomes vascularized. Such exponential tumor growth is seen in Figure 30. The microtumor can expand to 16,000 times its original volume in 2 weeks after vascularization. Without the blood supply, no growth is seen (Folkman, 1974; Ausprunk and Folkman, 1977). To accomplish this vascularization, the original microtumor elaborates substances called tumor angiogenesis factors (TAF). One such TAF, called ANGIOGENIN, has been isolated from a human tumor cell population and has been shown to be a single-chain protein of around 14,400 Da (Fett et al., 1985). Tumor angiogenesis factors stimulate mitosis in endothelial cells and the cells' formation into blood vessels in the direction of the tumor. This phenomenon can be demonstrated by implanting a piece of tumor tissue within the layers of a rabbit or mouse cornea. The cornea itself is not vascularized, but it is surrounded by a vascular border, or LIMBUS. Tumor tissue induces the formation of blood vessels to come toward the tumor (Figure 31). Most other adult tissues will not induce these vessels to form. (The exceptions are antigen-stimulated lymphocytes and macrophages, which secrete angiogenesis factors that are probably involved in wound repair.) As little as 3.5 pmol of purified angiogenin can induce extensive capillary formation in the rabbit cornea. Once the blood vessels enter the tumor, the tumor cells undergo explosive growth, eventually bursting the eye.

Folkman and co-workers (1983) have shown that certain natural substances inhibit tumor-induced angiogenesis. Heparin, a complex polysaccharide found in the matrices of numerous connective tissues, is a potent inhibitor of angiogenesis, especially if administered in the presence of the steroid hormone cortisone. When mice with established tumors received injections of heparin and cortisone, their tumors markedly regressed. The same combination

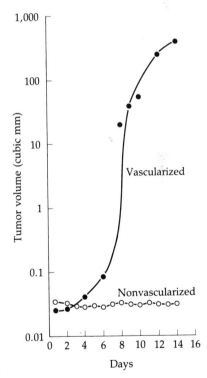

FIGURE 30
Exponential growth of a tumor after it receives blood vessels compared with that of a nonvascular tumor. Tumor spheroids were placed in the anterior chamber of rabbit eyes; some tumors became vascularized (by capillaries growing from the iris) and expanded rapidly, while others did not. (After Folkman, 1974.)

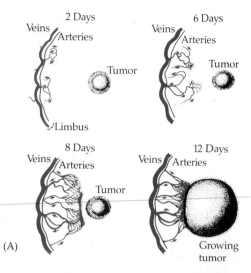

the collagenous matrix surrounding the new capillaries. This is thought to be done by the secretion of a proteolytic enzyme called PLASMINOGEN ACTIVATOR. This enzyme is important in the implantation of the blastocyst into the uterus (Chapter 3). Now, tumor cells will use this same enzyme to lyse a hole in the extracellular matrix surrounding the blood vessels. The synthesis of plasminogen activator has been correlated with the metastatic ability of tumors, and antibodies to this enzyme have been found to inhibit human tumor metastasis (Ossowski and Reich, 1983). Once on the capillary endothelial cells, the metastatic tumor cell can squeeze between them and enter the bloodstream (Kramer and Nicolson, 1979). Figure 32 shows a metastatic melanoma cell squeezing its way between capillary endothelial cells.

Metastasis requires the ability to travel from one site to another. Like embryonic cells, many tumor cells need a surface upon which to migrate, and some of these cells travel along the extracellular matrix of the blood vessel endothelial cells (Nicolson et al., 1981). This migration, like that of many embryonic cells, appears to be mediated by

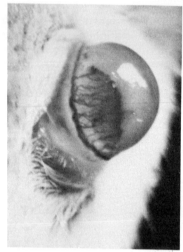

FIGURE 31

New blood vessel growth to the site of a tumor in the cornea of an albino mouse. (A) Sequence of events leading to the vascularization of a mammary adenocarcinoma tumor on days 2, 6, 8 and 12. The veins and arteries of the limbus surrounding the cornea both provide vessels. (B) Photograph of living cornea of an albino mouse, with new vessels being constructed to enter the tumor graft. (A from Muthukkaruppan and Auerbach, 1979; B, photograph courtesy of R. Auerbach.)

of heparin and cortisone also inhibited normal embryonic angiogenesis in chick embryos. A naturally occurring angiogenesis inhibitor is also produced by cartilage. This 27,600-Da protein keeps cartilage avascular (hence its white, rather than pink, appearance) and is able to inhibit angiogenesis in vivo (Moses et al., 1990).

In addition to supplying the tumor with nutrients and oxygen, the new blood vessels also provide a route that enables tumor cells to migrate to new places, thereby forming secondary tumors. This process is called METASTASIS. Indeed, most victims of cancer do not die from the original tumor but from the numerous secondary tumors spread by metastasis. To enter a blood vessel, a tumor cell has to lyse

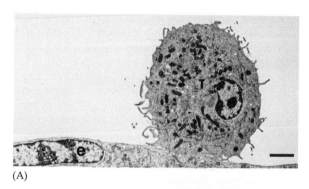

(A)

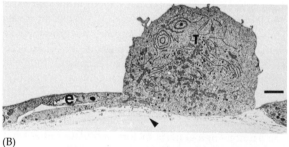

(B)

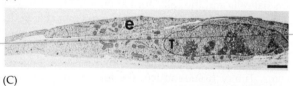

(C)

FIGURE 32

The attachment (A), invasion (B), and migration (C) of melanoma tumor cells (T) under a vascular endothelial cell (e). The melanoma cells were placed on the endothelial cells in culture, and photographs were taken at intervals of 0.5, 1, and 3 hours. (From Kramer and Nicolson, 1979; photographs courtesy of G. Nicolson.)

components of the extracellular basal lamina. When mice are injected with a certain strain of melanoma tumor cells, the cells migrate specifically to the lung, where they form secondary tumors (Figure 33). When the cells are inhibited from binding fibronectin or laminin, over 90 percent of the cells fail to reach the lungs (Humphries et al., 1986).

In their rapid division and their secretion of plasminogen activator and angiogenesis factor and in their migration patterns, tumor cells act like normal embryonic cells. There are numerous cases in which tumor cells represent adult cells that have reverted to an embryonic stage of their existence. This finding suggests new approaches to cancer therapy, including the administration of agents that prevent angiogenesis or that actually promote the "differentiation" of the "embryonic" tumor cell back into a normal, "adult" cell (Sachs, 1978; Jimenez and Yunis, 1987).

FIGURE 33

Metastasis of melanoma cells. The ability of particular melanoma cells to metastasize to the lung is inhibited by the cell-binding region of fibronectin and laminin. Melanoma cells were injected into the tail vein of mice. Fourteen days later, the animals were killed with ether and the lung metastases were counted. Each point is the average of eight mice. (From Humphries et al., 1986; photographs courtesy of K. M. Yamada.)

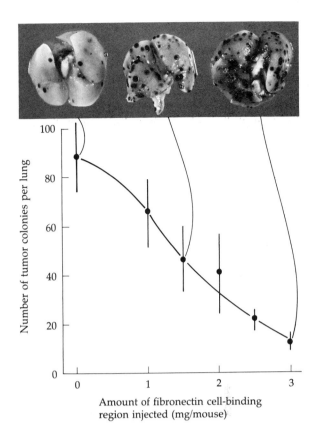

Development of blood cells

Pluripotential stem cells and hematopoietic microenvironments

Most vertebrate tissues are composed of differentiated cells that no longer divide. Myoblasts, for instance, constitute a rapidly dividing cell population until they develop into myotubes. Some tissues, such as epidermis, intestinal epithelium, and blood cells, however, retain an "embryonic" cell population within themselves; consequently, their cellular composition is always changing, even in adult animals. This phenomenon is most evident in the case of mammalian red blood cells. Each red cell lacks a nucleus and has a life span of only 120 days in circulation. A normal person will lose and replace 3×10^{11} red blood cells every day (Hay, 1966). The continuous formation of new red blood cells (as well as all the other types of blood cells) is accomplished in the bone marrow by the HEMATOPOIETIC ("blood-forming") STEM CELLS.

Stem cells are an intriguing and little-understood phenomenon. Yet our lives depend upon them. A stem cell is a cell capable of extensive proliferation, generating more stem cells (self-renewal) as well as more differentiated progeny (Siminovitch et al., 1963). Thus, a single stem cell can generate a clone containing millions of differentiated cells as well as a few stem cells. Stem cells thereby enable the continued proliferation of tissue precursors over a long period of time. Mammalian hematopoietic stem cells migrate to the bone marrow, where they will remain for the duration of the animal's life. Similarly, there are stem cells for such con-

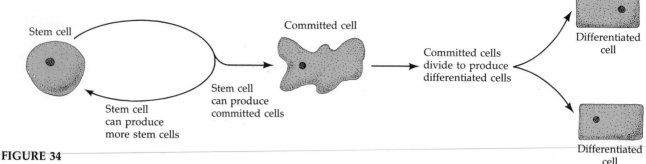

FIGURE 34
Model of the dynamics of stem cell proliferation and differentiation.

tinually renewed tissues as epidermis and sperm. Some stem cells, such as that for skeletal muscle, probably exist during fetal development (Quinn et al., 1985). The notion of a stem cell is depicted in Figure 34.

In mammals and birds, there appears to be a common PLURIPOTENTIAL HEMATOPOIETIC STEM CELL, which can give rise to red blood cells (erythrocytes), white blood cells (granulocytes, macrophages, and platelets), and immunocompetent cells (lymphocytes). The existence of such stem cells was shown by Till and McCulloch (1961), who injected bone marrow cells into lethally irradiated mice of the same genetic strain as the marrow donors. (The irradiation kills the hematopoietic cells of the host, enabling one to see the new colonies from the donor mouse.) Some of these donor cells produced discrete nodules on the spleens of the host animals (Figure 35). Microscopic studies showed these nodules to be composed of erythrocyte, granulocyte, and platelet precursors. Thus, a single cell from the bone marrow was capable of forming many of the different blood cell types. The cell responsible was called the CFU-S, the *colony forming unit* of the *spleen*. Further studies used chromosomal markers to prove that the different types of cells within the colony were formed from the same CFU-S. Here, marrow cells were irradiated so that very few survived. Many of those that did survive had abnormal chromosomes, which could be detected microscopically. When such irradiated CFU-S were injected into a mouse whose own blood-forming stem cells had been destroyed, each cell of a spleen colony, be it granulocyte or erythrocyte precursor,

FIGURE 35
Isolated blood-forming colonies. When bone marrow containing hematopoietic stem cells is injected into an irradiated mouse, discrete colonies of blood cells are seen on the surface of the spleen of that mouse. The spleen on the right has more of these colonies than does the spleen on the left. (From Till, 1981; photograph courtesy of J. E. Till.)

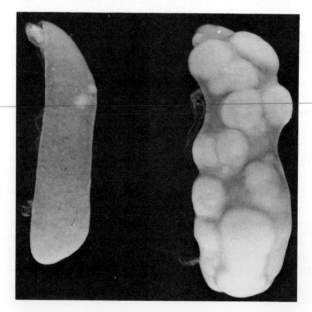

had the same chromosomal anomaly (Becker et al., 1963). An important part of the stem cell concept is the requirement that the stem cell be able to form more stem cells in addition to its differentiated cell types. This has indeed been found to be the case. When spleen colonies derived from a single CFU-S are resuspended and injected into other mice, many spleen colonies emerge (Jurśšková and Tkadleček, 1965; Humphries et al., 1979). Thus, we see that a single marrow cell can form numerous different cell types and can also undergo self-renewal; in other words, the CFU-S is a pluripotential hematopoietic stem cell.

The preceding data indicate that although the CFU-S can generate many of the blood cell types, it is not capable of generating lymphocytes. This conclusion is supported by the experiments of Abramson and her colleagues (1977), who have shown that CFU-S and lymphocytes are both derived from a different type of stem cell. When they injected irradiated bone marrow cells into mice having a hereditary deficiency of blood-forming cells, they found the same chromosomal abnormalities in both the spleen colonies and the circulating lymphocytes. This work has been confirmed by studies in which marrow cells are injected with certain viruses that become incorporated into cellular DNA at various random places. The same virally derived genes are seen in the same region of the genome in lymphocytes and in blood cells (Keller et al., 1985; Lemischka et al., 1986). In 1988, Spangrude and colleagues used immunological procedures to isolate a small population of cells (0.05 percent of the total bone marrow cells of mice) that appear to be precursor stem cells of both blood cells and lymphocytes. Any one of these cells is able to form colonies of lymphocytes and blood cells. In fact, the injection of only 30 of these cells into lethally irradiated mice enables half the mice to survive. In these survivors, the millions of lymphocytes and blood cells are all derived from only 30 donor cells.

Figure 36 is a model that summarizes several studies. The first pluripotential hematopoietic stem cell is the CFU-M,L (myeloid and lymphoid colony forming unit). This cell then gives rise to the CFU-S (blood cells) and the CFU-L (lymphocytes). The CFU-S and the CFU-L also are pluripotent stem cells because their progeny can differentiate into numerous cell types. The immediate progeny of the CFU-S, however, are *committed* stem cells. Each can produce only one type of cell in addition to renewing itself. The BFU-E (burst forming unit-erythroid), for instance, is formed from the CFU-S, and it can form only one cell type in addition to itself. This new cell is the CFU-E (colony forming unit-erythroid), which is capable of responding to the hormone ERYTHROPOIETIN to produce the first recognizable differentiated erythrocyte, the PROERYTHROBLAST. Erythropoietin is a glycoprotein that rapidly induces the synthesis of the mRNA for globin (Krantz and Goldwasser, 1965). It is produced predominantly in the kidney, and its synthesis is responsive to environmental conditions. If the level of blood oxygen falls, erythropoietin production is increased, an event leading to the production of more red blood cells. As a red blood cell matures, it becomes an ERYTHROBLAST, synthesizing enormous amounts of hemoglobin. Eventually, the mammalian erythroblast expels its nucleus, becoming a RETICULOCYTE. Reticulocytes can no longer synthesize globin mRNA but can still translate existing messages into globin. The final stage of differentiation is the ERYTHROCYTE stage. Here, no division, RNA synthesis, or protein synthesis takes place. The cell leaves the bone marrow to undertake its role of delivering oxygen to the bodily tissues. The CFU-L cells are thought to develop in a similar fashion, namely, the pluripotential stem cell generates other stem cells that are each committed to a certain type of differentiation. Thus, there are com-

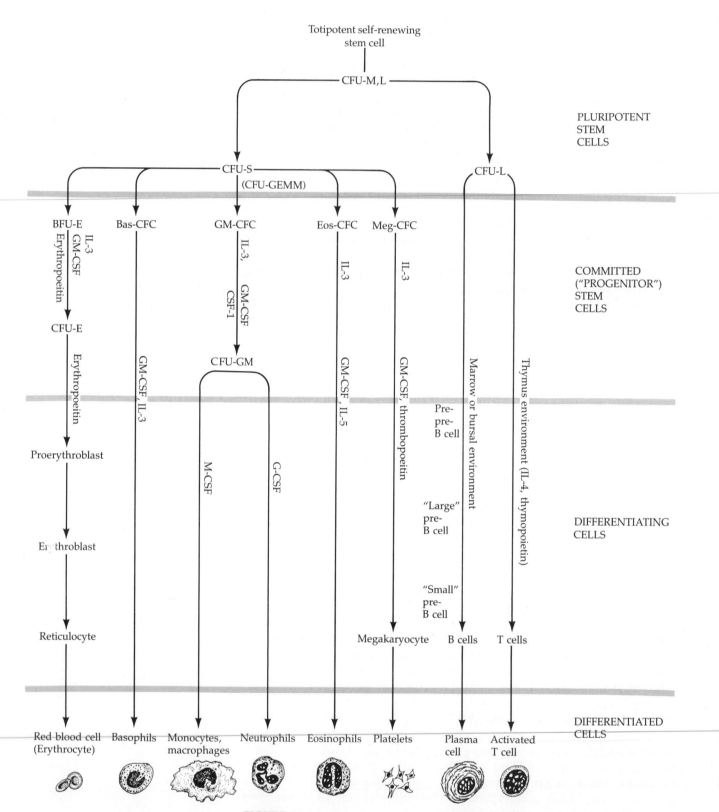

FIGURE 36

Model for the origin of mammalian blood and lymphoid cells. The growth factors are as listed in Table 1 except for BCGF and BCDF (B-cell growth and differentiation factors, respectively), which are discussed in Chapter 16. (After Quesenberry and Levitt, 1979; Whetton and Dexter, 1986; Muller-Sieburg et al., 1986; Clark and Kamen, 1987; and Foltz, 1989.)

mitted stem cells for platelets and committed stem cells for granulocytes and macrophages.

The molecules affecting the differentiation of the blood cells are a class of diffusible proteins called HEMATOPOIETIC GROWTH FACTORS, one of which is erythropoietin. Some of these factors stimulate the division and maturation of the more primitive stem cells, thus increasing the numbers of all blood cell types. Other factors are specific for certain cell lineages only. A cell's ability to respond to these factors is dependent upon the presence of receptors for these factors on its surface. The number of these receptors is quite low. There are only about 700 receptors for erythropoietin on a CFU-E, and most other progenitor cells have similar low numbers of growth factor receptors. (The exception is the receptor for CSF-1, which can number up to 73,000 per cell on certain progenitor cells.) Table 1 shows some of the known hematopoietic growth factors and their cellular targets.

At least some of these factors are made by the stromal cells (fibroblasts and other connective tissue elements) of the bone marrow itself. In the stroma of the spleen, the stem cells are predominantly committed to erythroid development. In the bone marrow, granulocyte development predominates. That short-range interactions yield these results was shown by Wolf and Trentin (1968). These investigators placed plugs of bone marrow into the spleen and then injected stem cells. Those colonies in the spleen were predominantly erythroid, whereas those forming in the marrow plugs were predominantly granulocytic. In fact, those colonies that straddled the borders were predominantly erythroid in the spleen and granulocytic in the marrow. The regions of determination are referred to as HEMATOPOIETIC INDUCTIVE MICROENVIRONMENTS (HIM).

Cell-free culture medium in which stromal cells have been growing can support the growth of lymphocyte and white blood cell precursors whose proliferation is dependent upon certain growth factors (Hunt et al., 1987; Whitlock et al., 1987). Moreover, these factors may act by binding to the extracellular matrix in specific areas of the bone marrow. Gordon and co-workers (1987) have shown that GM-CSF and multilineage growth factor IL-3 both bind to the heparan sulfate glycosaminoglycan of bone marrow stroma (Gordon et al., 1987; Roberts et al., 1988). Moreover, they remain active when bound. In this way, the growth factors may be compartmentalized, stimulating stem cells in one area to differentiate into one cell type while allowing the same type of stem cells in another area to differentiate into another cell type.

Sites of hematopoiesis

In most mammals, the major source of blood cells in the adult is the bone marrow. (In some species, like the mouse, the spleen also participates.) However, numerous experiments have shown that the hematopoietic stem cells do not originate in either the marrow or the spleen; rather, they migrate there from other regions of the embryo. In fact, the mammalian hematopoietic stem cell probably comes from *outside* the embryo proper —that is, from the yolk sac. When Moore and his co-workers removed the yolk sacs from presomite mouse embryos, no hematopoietic development took place in the embryo. Moreover, in cultures of the isolated yolk sac, erythropoiesis and CFU-S production were seen. Moore and Metcalfe (1970) demonstrated that maximum CFU-S activity in the yolk sac occurred at 10 days. In contrast, the embryonic liver begins to show the presence of stems cells only after 11 days. Moore and Metcalfe pos-

TABLE 1
Major hematopoietic growth factors

Growth factor	Species	Responsive cell types
Interleukin-3	Mouse	CFU-S, BFU-E GM-CFC (granulocyte, macrophage colony forming cell) Eos-CFC (eosinophil colony forming cell) Meg-CFC (megakaryocyte colony forming cell) Erythroid precursor cells Mast cells
Hemopoietin-1	Mouse, human	Little effect alone, but enhances ability of interleukin-3 to stimulate CFU-S
Granulocyte CSF	Human	Neutrophil precursors
Colony stimulating factor-1 (CSF-1)	Mouse, human	Some GM-CFC cells Macrophage precursors Mature macrophages
Granulocyte-macrophage	Mouse	GM-CFC
Colony stimulating factor (GM-CSF)		Eos-CFC Meg-CFC
Erythropoietin	Mouse, human	CFU-E
Interleukin-2	Mouse	T cells, B cells
Interleukin-4	Mouse	T cells, B cells
Pluripoietin	Human	CFU-S GM-CFC BFU-E
Bursin	Chick	Pre-B cells
Thymopoietin	Chick, mouse	Pre-T cells

Sources: Audhya et al. (1986); Clark and Kamen (1987); Palachios et al. (1987); Stanley et al. (1986); Whetton and Dexter (1986).

tulated that the stem cells migrate from the yolk sac to the embryonic liver and then to the spleen and bone marrow. Embryonic yolk sac cells have also been seen to carry out lymphocytic functions before any stem cell activity is present within the embryo proper (Hofman and Globerson, 1973; Dahl et al., 1980). Thus, the original mammalian pluripotential stem cell is thought to arise within the yolk sac and migrate first to the liver and then to the adult hematopoietic organs (Figure 37).

Recent evidence supports the notion that the adult stem cells of the marrow arise from the descendants of these embryonic stem cells. Wong and co-workers (1986) have cultured stem cells from the mouse yolk sac

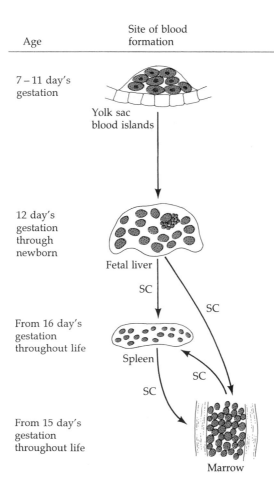

Age	Site of blood formation

7 – 11 day's gestation

Yolk sac blood islands

12 day's gestation through newborn

Fetal liver

SC

SC

From 16 day's gestation throughout life

Spleen

SC

SC

From 15 day's gestation throughout life

Marrow

FIGURE 37
The successive sites of blood formation in the embryonic mouse. Arrows indicate potential migratory paths of pluripotential hemopoietic stem cells (SC). The comparative size of the red blood cells produced in each site is shown at the right. (After Russell, 1979.)

and found that some of them gave rise to erythrocytes that expressed adult, rather than embryonic, hemoglobin. Conversely, Peschle and colleagues (1985), studying the blood cell formation in human embryos, found that the cells in 6-week human liver, just like the cells of the 5-week yolk sac, synthesize the embryonic form of hemoglobin (Figure 38).

Rossant and co-workers, however, have suggested that this migration of stem cells might not be as simple as previously envisioned. Using chimeric mice to distinguish the identity of the clones of cells, Rossant's

FIGURE 38
Early human blood cells. (A) Human erythroblasts from the human yolk sac 5 weeks after conception. (B) Section of 6-week human liver showing large red blood cells (megaloblasts). (From Peschle et al., 1985; photograph courtesy of C. Peschle.)

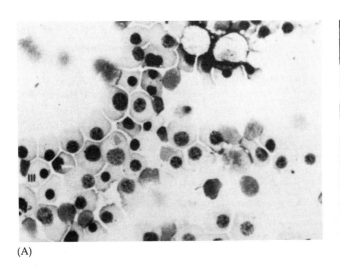

(A)

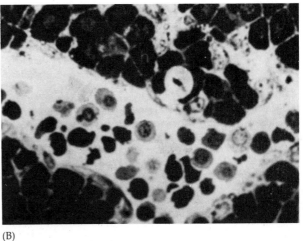

(B)

(A)

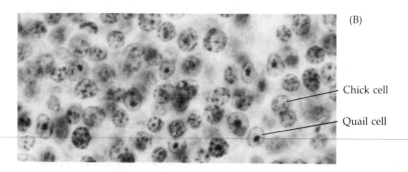

(B)

Chick cell

Quail cell

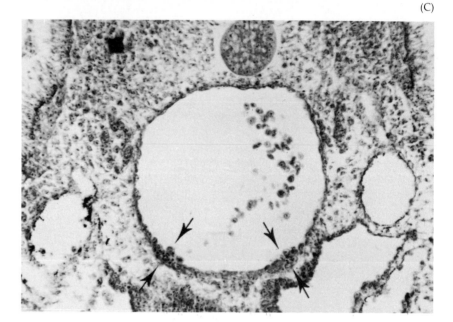

(C)

FIGURE 39

Blood cell mapping by chick–quail chimeras. (A) Photograph of a "yolk sac chimera," wherein the blastoderm of a quail was transplanted onto the yolk sac of a chick. (B) Photograph of chick and quail cells in the thymus of a chimeric animal showing the difference in the nuclear staining. The lymphoid cells are all chick whereas the structural cells of the thymus are of quail origin. (C) Section through the aorta of a 3-day chick embryo showing the cells (arrows) that give rise to the hemopoietic stem cells. If cells from this region are taken from quail embryos and placed into chick embryos, the chick embryos have quail blood. (From Martin et al., 1978; Dieterlen-Lièvre and Martin, 1981; photographs courtesy of F. Dieterlen-Lièvre.)

group (1986) showed that most, if not all, of the hematopoietic clones in the postnatal liver produced only *one* type of lineage. That is, a given clone was either a myeloid clone, an erythroid clone, or a lymphocyte clone. This result indicates either that the stem cells that migrate from the yolk sac to the liver are committed stem cells and not pluripotent ones, or that in the 17 days during which the stem cells have been in the liver, one cell lineage has predominated. Further studies will have to distinguish between these two equally exciting possibilities.

A similar situation was believed to occur in avian embryos, but birds appear to have two separate sources of hematopoietic stem cells. The first is located, as in mammalian embryos, in the yolk sac. These cells, however, appear to be a transient population. The major source of stem cells in birds seems to be local blood islands, which form within the embryo proper. This was discovered in a series of elegant experiments by Dieterlen-Lièvre, who grafted the blastoderm of chickens onto the yolk of the Japanese quail (Figure 39). Chick cells can be distinguished readily from quail cells because the quail cell nucleus stains much more darkly, thus providing a permanent marker for distinguishing the two cell types. Using these "yolk sac chimeras," Dieterlen-Lièvre and Martin (1981) showed that the yolk sac stem cells do not contribute cells to the adult animal but that the true stem cells are formed within nodes of mesoderm that line the mesentery and the major blood vessels. In the 4-day chick embryo, the aortic wall appears to be the most important source of new blood cells, and it has been found to contain numerous hematopoietic stem cells (Cormier and Dieterlen-Lièvre, 1988). In mammals, the yolk sac is still thought to be the source of all hematopoietic stem cells; but studies have suggested that these yolk sac blood precursors may themselves be derived from stem cells lining the abdominal cavity (Kubai and Auerbach, 1983).

ENDODERM

Pharynx

The function of embryonic endoderm is to construct the linings of two tubes within the body. The first tube, extending throughout the length of the body, is the digestive tube. Buds from this tube form the liver, gall-bladder, and pancreas. The second tube, the respiratory tube, forms as an outgrowth of the digestive tube, and it eventually bifurcates into two lungs. The digestive and respiratory tubes share a common chamber in the anterior region of the embryo, and this region is called the PHARYNX. Epithelial outpockets of the pharynx give rise to the tonsils, thyroid, thymus, and parathyroid glands.

The respiratory and digestive tubes are both derived from the primitive gut (Figure 40). As the endoderm pinches in toward the center of the embryo, the foregut and hindgut regions are formed. At first, the oral end is blocked by a region of ectoderm called the ORAL PLATE, or STOMODEUM. Eventually (about 22 days in human embryos), the stomodeum breaks, thereby creating the oral opening of the digestive tube. The opening itself is lined by ectodermal cells. This arrangement creates an interesting situation, because the oral plate ectoderm is in contact with the brain ectoderm, which has curved around toward the ventral portion of the embryo. The two ectodermal regions mutually interact with each other. The roof of the oral region forms Rathke's pouch and becomes the glandular part of the pituitary gland. The neural tissue on the floor of the diencephalon gives rise to the infundibular process, which becomes the neural portion of the pituitary. Thus, the pituitary gland has a dual origin; this dual nature of the pituitary is reflected in its adult functions.

The endodermal portion of the digestive and respiratory tubes begins in the pharynx. Here, the mammalian embryo produces four pairs of PHARYNGEAL POUCHES. The furrows between these pouches are sometimes called the GILL CLEFTS, because they resemble those structures found in fish embryos (Figure 41). Instead of generating gills, however, mammalian pharyngeal pouches have been modified for the terrestrial environment. The first pair of pharyngeal pouches becomes the auditory cavities of the middle ear and the associated eustachian tubes. The second pair of pouches gives rise to the walls of the tonsils. The thymus is derived from the third pair of pharyngeal pouches. This gland will direct the differentiation of T-lymphocytes during later stages of development. One pair of parathyroid glands is also derived from the third pair of pharyngeal pouches, and the other pair is derived from the fourth. In addition to these paired pouches, a small, central diverticulum is formed between the second pharyngeal pouches on the floor of the pharynx. This pocket of endoderm and mesenchyme will bud off from the pharynx and migrate down the neck to become the thyroid gland.

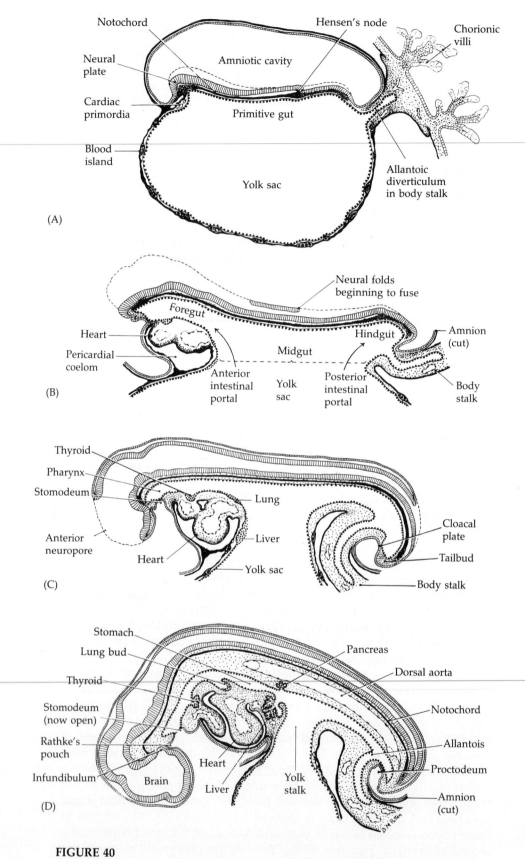

FIGURE 40
Formation of the human digestive system, depicted at about (A) 16 days, (B) 18 days, (C) 22 days, and (D) 28 days. (From Carlson, 1981.)

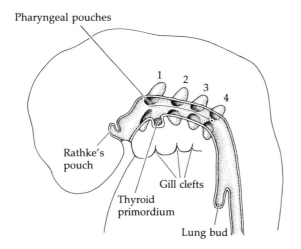

FIGURE 41
Pharyngeal pouches and gill clefts of a
5-week human embryo. (After Carlson, 1981.)

The digestive tube and its derivatives

Posterior to the pharynx, the digestive tube constricts to form the esophagus, which is followed in sequence by the stomach, small intestine, and large intestine. The endodermal cells generate only the lining of the digestive tube and its glands, for mesodermal mesenchyme cells will surround this tube to provide the muscles for peristalsis.

Figure 42 shows that the stomach develops as a dilated region close to the pharynx. More caudally, the intestines develop, and the connection between the intestine and yolk sac is eventually severed. At the caudal end of the intestine, a depression forms where the endoderm meets the

FIGURE 42
Digestive system in a 6-week human embryo. The stomach region of the digestive tube has begun to dilate, and the pancreas is represented by two buds, which will eventually fuse. (After Langman, 1981.)

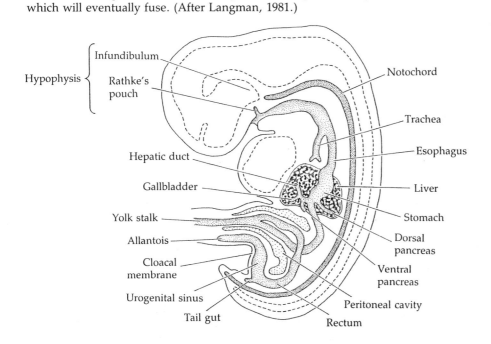

overlying ectoderm. Here, a thin CLOACAL MEMBRANE separates the two tissues. It eventually ruptures, forming the opening that will become the anus.

Liver, pancreas, and gallbladder

Endoderm also forms the lining of three accessory organs that develop immediately caudal to the stomach. The HEPATIC DIVERTICULUM is the tube of endoderm that extends out from the foregut into the surrounding mesenchyme. The mesenchyme induces the endoderm to proliferate, to branch, and to form the glandular epithelium of the liver. A portion of the hepatic diverticulum (that region closest to the digestive tube) continues to function as the drainage duct of the liver, and a branch from this duct produces the gallbladder (Figure 43).

The pancreas develops from the fusion of distinct dorsal and ventral diverticula. Both of these primordia arise from the endoderm immediately caudal to the stomach, and as they grow, they come closer together and eventually fuse. In humans, only the ventral duct survives to carry digestive enzymes into the intestine. In other species (such as the dog), both the dorsal and ventral ducts empty into the intestine.

The respiratory tube

The lungs also are a derivative of the digestive tube, even though they serve no role in digestion. In the center of the pharyngeal floor, between the fourth pair of pharyngeal pouches, the LARYNGOTRACHEAL GROOVE extends ventrally (Figure 44). This groove then bifurcates into the two branches that form the pair of bronchi and lungs. The laryngotracheal endoderm becomes the lining of the trachea, the two bronchi, and the air

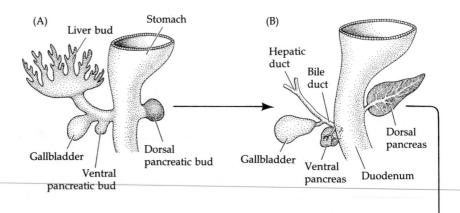

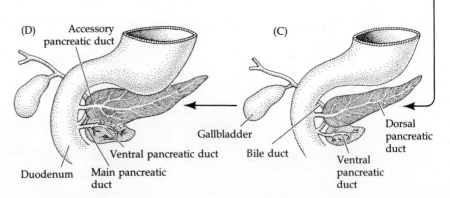

FIGURE 43
Pancreatic development in humans. At 30 days (A), the ventral pancreatic bud is close to the liver primordium, but by 35 days (B) it begins migrating posteriorly and comes into contact with the dorsal pancreatic bud during the sixth week of development (C). In most individuals, the dorsal pancreatic bud loses its duct into the duodenum. However, in about 10 percent of the population, the dual duct system persists (D). (After Langman, 1969.)

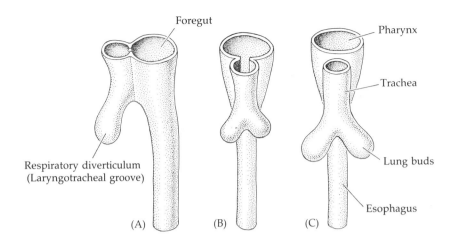

FIGURE 44

Partitioning of the foregut into the esophagus and respiratory diverticulum during the third and fourth weeks of human gestation. (A) Lateral view, end of week 3. (B,C) Ventral views, week 4. (After Langman, 1969.)

sacs (alveoli) of the lungs. As we will see in a later chapter, the branching of this endodermal tube depends upon interactions with the different types of mesodermal cells in its path.

The lungs are an evolutionary novelty, and they are among the last of the mammalian organs to fully differentiate. The lungs must be able to draw in oxygen at the baby's first breath. To accomplish this, alveolar cells secrete a SURFACTANT into the fluid bathing the lungs. This surfactant, consisting of phospholipids such as sphingomyelin and lecithin, is secreted very late in gestation, and it usually reaches physiologically useful levels around week 34 of human gestation. These compounds enable the alveolar cells to touch one another without sticking together. Thus, infants born prematurely often have difficulty breathing and have to be placed in respirators until their surfactant-producing cells mature.

SIDELIGHTS & SPECULATIONS

Teratocarcinoma

A mammalian germ cell or an early blastomere contains within it all the information needed for subsequent development. What would happen if such a cell became malignant? One type of germ cell tumor is called a TERATOCARCINOMA. This type of tumor can arise spontaneously within the testis and used to be the most deadly cancer of young males. Teratocarcinomas can also be produced experimentally by merely implanting a mammalian blastocyst in a site other than the uterus (for example, inside the connective tissue capsule that surrounds a kidney). Whether spontaneous or experimentally produced, a teratocarcinoma contains an undifferentiated stem cell population that has biochemical and developmental properties remarkably similar to those cells of the inner cell mass (Graham, 1977). Moreover, these stem cells (called EMBRYONAL CARCINOMA cells) not only divide but can also differentiate into a wide variety of tissues including gut and respiratory epithelia, muscle, nerve, cartilage, and bone (Figure 45). These differentiated cells no longer divide and are therefore not malignant. Such tumors can give rise to most of the tissue types in the body. Thus, the embryonal carcinoma cells mimic early mammalian development, but the tumor they form is characterized by random, haphazard development.

In some cases, however, one can get the embryonal carcinoma cells to differentiate in a particular manner. When embryonal carcinoma cells are placed in suspension culture and treated with retinoic acid, they form EMBRYOID BODIES that consist of a core of embryonal carcinoma cells enveloped by a rind of what appears to be endodermal tissue (Figure 46). This enveloping cell layer has many of the properties of the visceral embryonic endoderm that lines the yolk sac, and it synthesizes α-fetoprotein, a protein made only by this layer (Adamson et al., 1977; Becker and Grabel, 1990).

Although these tumors are probably derived from germ cells, the biochemistry, morphology, and cell surface behavior of embryonal carcinoma cells resemble those of inner cell mass blastomeres. In 1981, Stewart and Mintz formed a mouse from cells derived in part from a teratocarcinoma stem cell! Stem cells that had arisen in a teratocarcinoma of an agouti (yellow-tipped) strain of mice were cultured for several cell generations and were seen to maintain the characteristic chromosome complement of the parental mouse. Individual stem cells of this type were

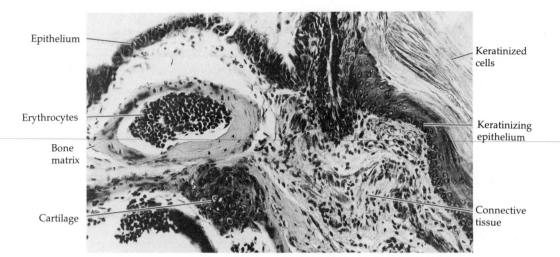

Epithelium

Erythrocytes

Bone matrix

Cartilage

Keratinized cells

Keratinizing epithelium

Connective tissue

FIGURE 45
Photomicrograph of a section through a teratocarcinoma, showing numerous differentiated cell types. (From Gardner, 1982; photograph from C. Graham, courtesy of R. L. Gardner.)

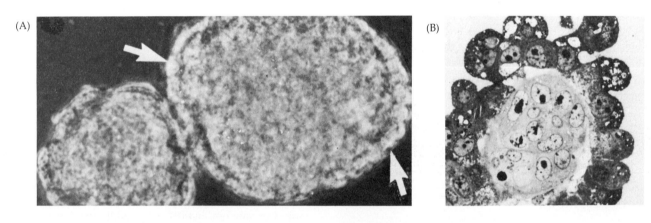

(A)

(B)

FIGURE 46
Embryoid bodies. (A) Phase-contrast photomicrograph of living embryoid bodies in culture. Arrows point to regions where the two different (inner and outer) cell types can be seen. (B) Section of such an embryoid body stained with hematoxylin and eosin. The core stem cells are seen to be surrounded by a cell type very reminiscent of early embryonic endoderm. (Courtesy of G. Martin.)

injected into the blastocysts of black mice (Figure 47). The blastocysts were then transferred to the uterus of a foster mother and live mice were born. Some of these mice had coats of two colors, indicating that the tumor cell had integrated itself into the embryo. Moreover, when mated to a mouse carrying an appropriate marker, the chimeric mouse was able to generate mice having some of the phenotypes of the tumor "parent." The malignant embryonal carcinoma cell had produced many, if not all, types of normal somatic cells and even had produced normal, functional germ cells! When mice having a tumor cell for one

parent were mated together, the resultant litter contained mice that were homozygous for a large number of genes from the tumor cell (Figure 48).

Because the cultivation of early mammalian embryos is a difficult task, researchers are looking at embryonal carcinoma cells as a unique way to analyze early mammalian development. Moreover, the ability to mutate embryonal carcinoma cells in culture and inject known mutant cells into blastocysts also provides the possibility of creating mouse mutants of nearly any gene.

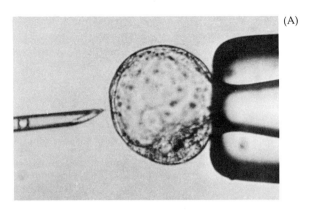

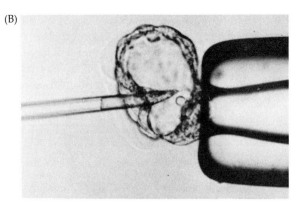

(A)   (B)

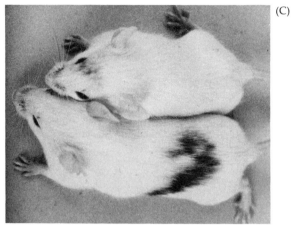

(C)

FIGURE 47
Insertion of embryonal carcinoma cell into a mouse blastocyst. (A) Teratocarcinoma stem cell in micropipette approaches blastocyst held fast by suction pipette. (B) Injection of the stem cell onto the inner cell mass. (C) Mice produced when a teratocarcinoma stem cell from a black strain of mice integrated into the inner cell mass of a white strain of mice. Black fur can be seen on heads and backs of mice, indicating that at least some of the pigment cells were of tumor origin. (A and B courtesy of K. Illmensee. C from Papaioannou, 1979; photograph courtesy of V. E. Papaioannou.)

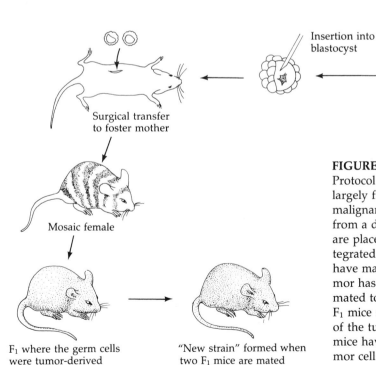

Insertion into blastocyst

Surgical transfer to foster mother

Isolate stem cell line

Malignant teratocarcinoma

Mosaic female

F₁ where the germ cells were tumor-derived

"New strain" formed when two F₁ mice are mated

FIGURE 48
Protocol for breeding mice whose genes are derived largely from tumor cells. Stem cells are isolated from a malignant mouse carcinoma and inserted into blastocysts from a different strain of mouse. The chimeric blastocysts are placed into a foster mother. If the tumor cells are integrated into the blastocyst, the mouse that develops will have many of its cells derived from the tumor. If the tumor has given rise to germ cells, the mosaic mice can be mated to normal mice to produce an F_1 generation. The F_1 mice should be heterozygous for all the chromosomes of the tumor cell. Matings between F_1 mice produce F_2 mice having some homozygous genes derived from tumor cells. (After Stewart and Mintz, 1981.)

Here we will conclude our survey of the major features of animal development. We now attend to the *mechanisms* that enable this development to take place. In Part II we focus upon the ways in which cells differentiate, and in Part III we will be concerned with how cells interact with one another to form tissues and organs.

LITERATURE CITED

Abramson, S., Miller, R. G. and Phillips, R. A. 1977. The identification in adult bone marrow of pluripotent and restricted stem cells of the myeloid and lymphoid systems. *J. Exp. Med.* 145: 1567–1579.

Adamson, E., Evans, M. J. and Magrane, G. G. 1977. Biochemical markers of the progress of differentiation in cloned teratocarcinoma cell lines. *Eur. J. Biochem.* 79: 607–615.

Ash, P. J., Loutit, J. F. and Townsend, K. M. S. 1980. Osteoclasts derived from haematopoietic stem cells. *Nature* 283: 669–670.

Audhya, T., Kroon, D., Heavner, G., Viamontes, G. and Goldstein, G. 1986. Tripeptide structure of bursin, a selective B cell-differentiating hormone of the bursa of Fabricius. *Science* 231: 987–999.

Auerbach, R., Alby, L., Morrissey, L., Tu, M. and Joseph, J. 1985. Expression of organ-specific antigens on capillary endothelial cells. *Microvasc. Res.* 29: 401–411.

Ausprunk, D. H. and Folkman, J. 1977. Migration and proliferation of endothelial cells in preformed and newly formed blood vessels during tumor angiogenesis. *Microvasc. Res.* 14: 53–65.

Bandman, E. 1985. Continued expression of neonatal myosin heavy chain in adult dystrophic skeletal muscle. *Science* 127: 780–782.

Baron, R., Neff, L., Louvard, D. and Courtoy, P. J. 1985. Cell mediated extracellular acidification and bone resorption: Evidence for a low pH in resorbing lacuna and localization of a 100-kD lysosomal membrane protein at the osteoclast ruffled border. *J. Cell Biol.* 101: 2210–2222.

Baron, R., Neff, L., Roy, C., Boisvert, A. and Caplan, M. 1986. Evidence for a high and specific concentration of (Na$^+$,K$^+$) ATPase in the plasma membrane of the osteoclast. *Cell* 46: 311–320.

Becker, A. J., McCulloch, E. A. and Till, J. E. 1963. Cytological demonstration of the clonal nature of spleen cells derived from transplanted mouse marrow cells. *Nature* 197: 452–454.

Becker, S. and Grabel, L. 1990. The use of whole-mount in situ hybridization to study spatial differentiation in F9 embryoid bodies. *Abstracts of Forty-Ninth Annual Symposium of the Society for Developmental Biology*, p. 43.

Blair, H. C., Kahn, A. J., Crouch, E. C., Jeffrey, J. J. and Teitelbaum, S. L. 1986. Isolated osteoclasts resorb the organic and inorganic components of bone. *J. Cell Biol.* 102: 1164–1172.

Bloom, W. and Fawcett, D. W. 1975. *Textbook of Histology*, 10th Ed. Saunders, Philadelphia.

Bonner, P. H. and Hauschka, S. D. 1974. Clonal analysis of vertebrate myogenesis. I. Early developmental events in the chick limb. *Dev. Biol.* 37: 317–328.

Brighton, C. T. 1984. The growth plate. *Orthoped. Clin. N. A.* 15: 571–594.

Brighton, C. T. and Hunt, R. M. 1974. Mitochondrial calcium and its role in calcification. *Clin. Orthop.* 100: 406–416.

Buckley, P. A. and Konigsberg, I. R. 1974. Myogenic fusion and the postmitotic gap. *Dev. Biol.* 37: 193–212.

Caplan, A. I., Fiszman, M. Y. and Eppenberger, H. M. 1983. Molecules and cell isoforms during development. *Science* 221: 923–927.

Carlson, B. M. 1981. *Patten's Foundations of Embryology*. McGraw-Hill, New York.

Cheney, C. M. and Lash, J. W. 1984. An increase in cell–cell adhesion in the chick segmental plate results in meristic pattern. *J. Embryol. Exp. Morphol.* 79: 1–10.

Chevallier, A., Kieny, M., Mauger, A. and Sengel, P. 1977. Developmental fate of the somitic mesoderm in the chick embryo. *In* D. A. Ede, J. R. Hinchliffe and M. Balls (eds.), *Vertebrate Limb and Somite Morphogenesis*. Cambridge University Press, Cambridge, pp. 421–432.

Clark, S. C. and Kamen, R. 1987. The human hematopoietic colony stimulating factors. *Science* 236: 1229–1237.

Cormier, F. and Dieterlen-Lièvre, F. 1988. The wall of the chick aorta harbours M-CFC, G-CFC, GM-CFC and BFU-E. *Development* 102: 279–285.

Dahl, C. A., Kahan, B. W. and Auerbach, R. 1980. Studies with mouse embryonic yolk sac cells: An approach to understanding the patterns of ontogeny of cell-mediated immune functions. *In* J. D. Horton (ed.), *Development and Differentiation of Vertebrate Lymphocytes*, Elsevier North-Holland, Amsterdam, pp. 241–253.

David, J. D., See, W. M. and Higginbotham, C. A. 1981. Fusion of chick embryo skeletal myoblasts: Role of calcium influx preceding membrane union. *Dev. Biol.* 82: 297–307.

DeHaan, R. 1959. *Cardia bifida* and the development of pacemaker function in the early chicken heart. *Dev. Biol.* 1: 586–602.

DeHaan, R. L. 1967. Regulation of spontaneous activity and growth of embryonic chick heart cells in tissue culture. *Dev. Biol.* 16: 216–249.

Dereszewski, G. and Howell, D. S. 1978. The role of matrix vesicles in calcification. *Trends Biochem. Sci.* 4: 151–153.

Devlin, R. B. and Emerson, C. P., Jr. 1978. Coordinate regulation of contractile protein synthesis during myoblast differentiation. *Cell* 13: 599–611.

Dieterlen-Lièvre, F. and Martin, C. 1981. Diffuse intraembryonic hemopoiesis in normal and chimeric avian development. *Dev. Biol.* 88: 180–191.

Fett, J. W., Strydom, D. J., Lubb, R. R., Alderman, E. M., Bethune, J. L., Riordan, J. F. and Vallee, B. L. 1985. Isolation and characterization of angiogenin, an angiogenic protein from human carcinoma cells. *Biochem.* 24: 5480–5486.

Folkman, J. 1974. Tumor angiogenesis. *Adv. Cancer Res.* 19: 331–358.

Folkman, J., Langer, R., Linhardt, R. J., Haudenschild, C. and Taylor, S. 1983. Angiogenesis inhibition and tumor regression caused by heparin or heparin fragment in the presence of cortisone. *Science* 221: 719–725.

Foltz, C. 1989. *Cytokine Pathways in the Immune System*. Genzyme, Boston.

Gardner, R. L. 1982. Manipulation of development. *In* C. R. Austin and R. V. Short (eds.), *Embryonic and Fetal Development*, Cambridge University Press, Cambridge, pp. 159–180.

Gordon, M. Y., Riley, G. P., Watt, S. M. and Greaves, M. F. 1987. Compartmentalization of a haematopoietic growth factor (GM-CSF) by glycosaminoglycans in the bone marrow microenvironment. *Nature* 326: 403–405.

Graham, C. F. 1977. Teratocarcinoma cells and normal mouse embryogenesis. *In* M. I. Sherman (ed.), *Concepts in Mammalian Embryogenesis*. MIT Press, Cambridge, MA, pp. 315–394.

Gräper, L. 1907. Untersuchungen über die Herzbildung der Vögel. *Wilhelm Roux Arch. Entwicklungsmech. Org.* 24: 375–410.

Hall, B. K. 1988. The embryonic development of bone. *Am. Sci.* 76: 174–181.

Hall, B. K., van Exan, R. J. and Brunt, S. L. 1983. Retention of epithelial basal lamina allows isolated mandibular mesenchyme to form bone. *J. Craniofac. Gen. Dev. Biol.* 3: 253-267.

Harary, I. and Farley, B. 1963. *In vitro* studies on single beating rat heart cells. II. Intercellular communication. *Exp. Cell Res.* 29: 466–474.

Hay, E. 1966. *Regeneration*. Holt, Rinehart & Winston, New York.

Hofman, F. and Globerson, A. 1973. Graftversus-host response induced in vitro by mouse yolk sac cells. *Eur. J. Immunol.* 3: 179–181.

Horton, W. A. 1990. The biology of bone growth. *Growth Genet. Horm.* 6 (2): 1–3.

Humphries, M. J., Oldern, K. and Yamada, K. M. 1986. A synthetic peptide from fibronectin inhibits experimental metastasis in murine melanoma cells. *Science* 233: 467–470.

Humphries, R. K., Jacky, P. B., Dill, F. J., Eaves, A. C. and Eaves, C. J. 1979. CFUs in individual erythroid colonies derived in vitro from adult mature mouse marrow. *Nature* 279: 718–720.

Hunt, P., Robertson, D., Weiss, D., Rennick, D., Lee, F. and Witte, O. N. 1987. A single bone marrow stromal cell type supports the in vitro growth of early lymphoid and myeloid cells. *Cell* 48: 997–1007.

Jimenez, J. J. and Yunis, A. A. 1987. Tumor cell rejection through terminal cell differentiation. *Science* 238: 1278–1280.

Jockusch, M. and Jockusch, B. M. 1980. Structural organization of the Z-line protein, α-actinin, in developing skeletal muscle cells. *Dev. Biol.* 75: 231–238.

Jursšková, V. and Tkadleček, L. 1965. Character of primary and secondary colonies of haematopoiesis in the spleen of irradiated mice. *Nature* 206: 951–952.

Kahn, A. J. and Simmons, D. J. 1975. Investigation of cell lineage in bone using a chimaera of chick and quail embryonic tissue. *Nature* 258: 325–327.

Kalderon, N. and Gilula, N. B. 1979. Membrane events involved in myoblast fusion. *J. Cell Biol.* 81: 411–425.

Kaufman, S. J. and Foster, R. F. 1985. Remodeling of the myoblast membrane accompanies development. *Dev. Biol.* 110: 1–14.

Keller, G., Paige, C., Gilboa, E. and Wagner, E. F. 1985. Expression of a foreign gene in myeloid and lymphoid cells derived from multipotent hematopoietic precursors. *Nature* 318: 149–154.

Knudsen, K. A. 1985. The calcium-dependent myoblast adhesion that precedes cell fusion is mediated by glycoproteins. *J. Cell Biol.* 101: 891–897.

Knudsen, K. A., McElwee, S. A. and Myers, L. 1990. A role for the neural cell adhesion molecule, N-CAM, in myoblast interaction during myogenesis. *Dev. Biol.* 138: 159–168.

Konigsberg, I. R. 1963. Clonal analysis of myogenesis. *Science* 140: 1273–1284.

Kosher, R. A. and Lash, J. W. 1975. Notochord stimulation of in vitro somite chondrogenesis before and after enzymatic removal of preinotochordal materials. *Dev. Biol.* 42: 362–378.

Kramer, R. H. and Nicolson, G. L. 1979. Interaction of tumor cells with vascular endothelial cell monolayers: A model for metastatic invasion. *Proc. Natl. Acad. Sci. USA* 76: 5704–5708.

Kramer, T. C. 1942. The partitioning of the truncus and conus and the formation of the membranous portion of the intraventricular septum in the human heart. *Am. J. Anat.* 71: 343–370.

Krantz, S. B. and Goldwasser, E. 1965. On the mechanism of erythropoietin induced differentiation. II. The effect on RNA synthesis. *Biochim. Biophys. Acta* 103: 325–332.

Kubai, L. and Auerbach, R. 1983. A new source of embryonic lymphocytes in the mouse. *Nature* 301: 154–156.

Kubota, Y., Kleinman, H. K., Martin, G. R. and Lawley, T. J. 1988. Role of laminin and basement membrane in the morphological differentiation of human endothelial cells into capillary-like structures. *J. Cell Biol.* 107: 1589–1598.

Langman, J. 1969. *Medical Embryology*, 3rd Ed. Williams & Wilkins, Baltimore.

Langman, J. 1981. *Medical Embryology*, 4th Ed. Williams & Wilkins, Baltimore.

Lash, J. W. and Yamada, K. M. 1986. The adhesion recognition signal of fibronectin: A possible trigger mechanism for compaction during somitogenesis. *In* R. Bellairs, D. H. Ede and J. W. Lash (eds.), *Somites in Developing Embryos*. Plenum, New York, pp. 201–208.

Lash, J. W., Seitz, A. W., Cheney, C. M. and Ostrovsky, D. 1984. On the role of fibronectin during the compaction stage of somitogenesis in the chick embryo. *J. Exp. Zool.* 232: 197–206.

Lawrence, J. B., Taneja, K. and Singer, R. H. 1989. Temporal resolution and sequential gene expression of muscle-specific genes revealed by in situ hybridization. *Dev. Biol.* 133: 235–246.

Lemischka, I. R., Raulet, D. H. and Mulligan, R. C. 1986. Developmental potential and dynamic behavior of hematopoietic stem cells. *Cell* 45: 917–927.

Linask, K. K. and Lash, J. W. 1986. Precardiac cell migration: Fibronectin localization at mesoderm–endoderm interface during directional movement. *Dev. Biol.* 114: 87–101.

Linask, K. K. and Lash, J. W. 1988a. A role for fibronectin in the migration of avian precardiac cells I. Dose-dependent effects of fibronectin antibody. *Dev. Biol.* 129: 315–323.

Linask, K. K. and Lash, J. W. 1988b. A role for fibronectin in the migration of avian precardiac cells II. Rotation of the heart-forming region during different stages and their effects. *Dev. Biol.* 129: 324–329.

Lough, J. and Bischoff, R. 1977. Differentiation of creatine phosphokinase during myogenesis: Quantitative fractionation of isozymes. *Dev. Biol.* 57: 330–344.

Martin, C., Beaupain, D. and Dieterlen-Lièvre, F. 1978. Developmental relationships between vitelline and intraembryonic haemopoiesis studied in avian yolk sac chimeras. *Cell Differ.* 7: 115–130.

Meier, S. 1979. Development of the chick mesoblast: Formation of the embryonic axis and the establishment of the metameric pattern. *Dev. Biol.* 73: 25–45.

Mintz, B. and Baker, W. W. 1967. Normal mammalian muscle differentiation and gene control of isocitrate dehydrogenase synthesis. *Proc. Natl. Acad. Sci. USA* 58: 592–598.

Moore, M. A. S. and Metcalfe, D. 1970. Ontogeny of the haemopoietic system: Yolk sac origin of in vivo and in vitro colony forming cells in the developing mouse embryo. *Br. J. Haematol.* 18: 279–296.

Moses, M. A., Sudhalter, J. and Langer, R. 1990. Identification of an inhibitor of neovascularization from cartilage. *Science* 248: 1408–1410.

Muller-Sieburg, C. E., Whitlock, C. A. and Weissman, I. L. 1986. Isolation of two early B lymphocyte progenitors from mouse marrow: A committed pre-pre-B cell and a clonogenic Thy-1[lo] hematopoietic stem cell. *Cell* 44: 653–662.

Murphy, M. E. and Carlson, E. C. 1978. Ultrastructural study of developing extracellular matrix in vitelline blood vessels of the early chick embryo. *Am. J. Anat.* 151: 345–375.

Muthukkaruppan, V. R. and Auerbach, R. 1979. Angiogenesis in the mouse cornea. *Science* 205: 1416–1418.

Nameroff, M. and Munar, E. 1976. Inhibition of cellular differentiation by phospholipase C. II. Separation of fusion and recognition among myogenic cells. *Dev. Biol.* 49: 288–293.

Nicolson, G. L., Irimura, T., Gonzalez, R. and Ruoslahti, E. 1981. The role of fibronectin in adhesion of metastatic melanoma cells to endothelial cells and their basal lamina. *Exp. Cell Res.* 135: 461–465.

Ossowski, L. and Reich, E. 1983. Antibodies to plasminogen activator inhibit tumor metastasis. *Cell* 35: 611–619.

Ostrovsky, D., Cheney, C. M., Seitz, A. W. and Lash, J. W. 1984. Fibronectin distribution during somitogenesis in the chick embryo. *Cell Differ.* 13: 217–223.

Packard, D. S. and Meier, S. 1983. An experimental study of somitomeric organization of the avian vegetal plate. *Dev. Biol.* 97: 191–202.

Palachios, R., Sideras, P. and von Boehmer, H. 1987. Recombinant interleukin 4/BSF-1 promotes growth and differentiation of intrathymic T cell precursors from fetal mice in vitro. *EMBO J.* 6: 91–95.

Papaioannou, V. 1979. Interactions between mouse embryos and teratocarcinomas. *In* N. LeDouarin (ed.), *Cell Lineage, Stem Cells, and Cell Determination*. Elsevier North-Holland, New York, pp. 141–155.

Pardanaud, L., Altmann, C., Kitos, P., Dieterlen-Lièvre, F. and Buck, C. 1987. Vasculogenesis in the early quail blastodisc as studied with a monoclonal antibody recognizing endothelial cells. *Development* 100: 339–349.

Pardanaud, L., Yassine, F. and Dieterlen-Lièvre, F. 1989. Relationship between vasculogenesis, angiogenesis, and hemopoiesis during avian ontogeny. *Development* 105: 473–485.

Patten, B. M. 1951. *Early Embryology of the Chick*, 4th Ed. McGraw-Hill, New York.

Peschle, C. and 12 others. 1985. Haemoglobin switching in human embryos: Asynchrony of zeta to alpha and epsilon to gamma switches in primitive and definitive erythropoietic lineage. *Nature* 313: 235–238.

Quesenberry, P. and Levitt, L. 1979. Hematopoietic stem cells. *N. Engl. J. Med.* 301: 755–760.

Quinn, L. S., Holtzer, H. and Nameroff, M. 1985. Generation of chick skeletal muscle cells in groups of 16 from stem cells. *Nature* 313: 692–694.

Quinn, L. S., Ong, L. D. and Roeder, R. A. 1990. Paracrine control of myoblast proliferation and differentiation by fibroblasts. *Dev. Biol.* 140: 8–19.

Risau, W. 1986. Developing brain produces an angiogenesis factor. *Proc. Natl. Acad. Sci. USA* 83: 3855–3859.

Roberts, R., Gallagher, J., Spooncer, E., Allen, T. D., Bloomfield, F. and Dexter, T. M. 1988. Heparan sulphate-bound growth factors: A mechanism for stromal cell mediated haemopoiesis. *Nature* 332: 376–378.

Rossant, J., Vijh, K. M., Grossi, C. E. and Cooper, M. D. 1986. Clonal origin of haematopoietic colonies in the postnatal mouse liver. *Nature* 319: 507–511.

Rugh, R. 1951. *The Frog: Its Reproduction and Development*. Blakiston, Philadelphia.

Russell, E. S. 1979. Hereditary anemias of the mouse: A review for geneticists. *Adv. Genet.* 20: 357–459.

Sachs, L. 1978. Control of normal cell differentiation and the phenotypic reversion of malignancy in myeloid leukemias. *Nature* 274: 535–539.

Shainberg, A., Yagil, G. and Yaffe, D. 1969. Control of myogenesis in vitro by Ca^{++} concentration in nutritional medium. *Exp. Cell Res.* 58: 163–167.

Siminovitch, L., McCulloch, E. A. and Till, J. E. 1963. The distribution of colony-forming cells among spleen colonies. *J. Cell. Comp. Physiol.* 62: 327–336.

Spangrude, G. J., Heimfeld, S. and Weissman, I. 1988. Purification and characterization of mouse hematopoietic stem cells. *Science* 241: 58–62.

Stanley, E. R., Bartocci, A., Patinkin, D., Rosendal, M. and Bradley, T. R. 1986. Regulation of very primitive, multipotent hemopoietic cells by hemopoietin-1. *Cell* 45: 667–674.

Stewart, T. A. and Mintz, B. 1981. Successive generations of mice produced from an established culture line of euploid teratocarcinoma cells. *Proc. Natl. Acad. Sci. USA* 78: 6314–6318.

Stryer, L. 1981. *Biochemistry*, 2nd Ed., Freeman, San Francisco.

Sutherland, W. M. and Konigsberg, I. R. 1983. CPK accumulation in fusion-blocked quail myocytes. *Dev. Biol.* 99: 287–297.

Syftestad, G. T. and Caplan, A. I. 1984. A fraction from extracts of demineralized adult bone stimulates conversion of mesenchymal cells into chondrocytes. *Dev. Biol.* 104: 348–356.

Till, J. E. 1981. Cellular diversity in the blood-forming system. *Am. Sci.* 69: 522–527.

Till, J. E. and McCulloch, E. A. 1961. A direct measurement of the radiation sensitivity of normal mouse bone marrow cells. *Rad. Res.* 14: 213–222.

Tuan, R. 1987. Mechanisms and regulation of calcium transport by the chick embryonic chorioallantoic membrane. *J. Exp. Zool.* [Suppl.] 1: 1–13.

Tuan, R. S. and Lynch, M. H. 1983. Effect of experimentally induced calcium deficiency on the developmental expression of collagen types in chick embryonic skeleton. *Dev. Biol.* 100: 374–386.

Tyler, M. S. and Hall, B. K. 1977. Epithelial influence on skeletogenesis in the mandible of the embryonic chick. *Anat. Rec.* 206: 61-70.

Urist, M. R. and eight others. 1984. Purification of bovine bone morphogenetic protein by hydroxyapatite chromatography. *Proc. Natl. Acad. Sci. USA* 81: 371–375.

Wakelam, M. J. O. 1985. The fusion of myoblasts. *Biochem. J.* 228: 1–12.

Webster, C., Silberstein, L., Hays, A. P. and Blau, H. M. 1988. Fast muscle fibers are preferentially affected in Duchenne muscular dystrophy. *Cell* 52: 503–513.

Whalen, R. G., Schwartz, K., Bouveret, P., Sell, S. M. and Gros, F. 1979. Contractile protein isozymes in muscle development: Identification of an embryonic form of myosin heavy chain. *Proc. Natl. Acad. Sci. USA* 76: 5197–5201.

Whetton, A. D. and Dexter, T. M. 1986. Hemopoietic growth factors. *Trends Biochem. Sci.* 11: 207–211.

Whitlock, C. A., Tidmarsh, G. F., Muller-Sieburg, C. and Weissman, I. L. 1987. Bone marrow stromal cell lines with lymphopoietic activity express high levels of a pre-B neoplasia-associated molecule. *Cell* 48: 1009–1021.

Wilson, D. 1983. The origin of the endothelium in the developing marginal vein of the chick wing bud. *Cell Differ.* 13: 63–67.

Wolf, N. S. and Trentin, J. J. 1968. Hemopoietic colony studies. V. Effect of hemopoietic organ stroma on differentiation of pluripotent stem cells. *J. Exp. Med.* 127: 205–214.

Womble, M. D. and Bonner, P. H. 1980. Developmental fate of a distinct class of chick myoblasts after transplantation of cloned cells into quail embryos. *J. Embryol. Exp. Morphol.* 58: 119–130.

Wong, P. M. C., Chung, S. W., Chui, D. H. K. and Eaves, J. 1986. Properties of the earliest clonogenic hemopoietic precursors to appear in the developing murine yolk sac. *Proc. Natl. Acad. Sci. USA* 83: 3851–3854.

Wuthier, R.-E., Chin, J. E., Hale, J. E., Register, T. C., Hale, L. V. and Ishikawa, Y. 1985. Isolation and characterization of calcium-accumulating matrix vesicles from chondrocytes of chicken epiphyseal growth plate cartilage in primary culture. *J. Biol. Chem.* 260: 15972–15979.

Yaffe, D. and Feldman, M. 1965. The formation of hybrid multinucleated muscle fibres from myoblasts of different genetic origin. *Dev. Biol.* 11: 300–317.

II

Mechanisms of cellular differentiation

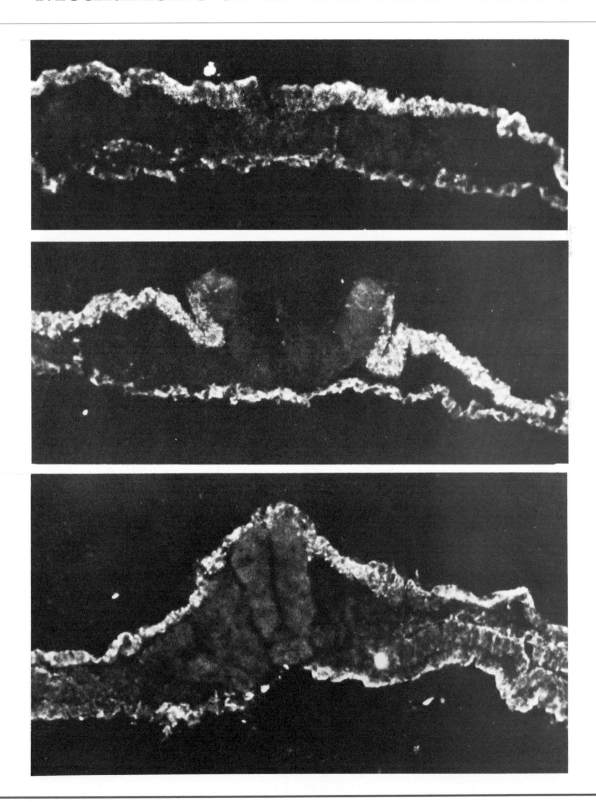

7

Determination by cytoplasmic specification

I hold it probable that in the germ cells there exist fine internal differences which predetermine the subsequent transformation to a determinant substance; not differences which are mere potencies present in the germ cells, but actual material differences so fine that we have not as yet been able to demonstrate them.

—R. VIRCHOW (1858)

Studying the period of cleavage we approach the source whence emerge the progressively branched streams of differentiation that end finally in almost quiet pools, the individual cells of the complex adult organism.

—E. E. JUST (1939)

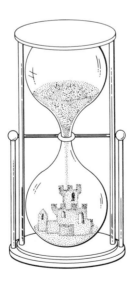

Each metazoan organism is a complex assortment of specialized cell types. For example, red and white blood cells differ not only from each other but also from the heart cells that propel them through the body. They also differ from the outstretched neurons, which conduct neural impulses from the brain to the heart, and from the glandular cells, which secrete hormones into the blood. Table 1 presents a very incomplete list of specialized cell types, their characteristic products, and their functions.

The development of specialized cell types from the single fertilized egg is called DIFFERENTIATION. This overt change in cellular biochemistry and function is preceded by a process called DETERMINATION, wherein the fate of the cell becomes committed. We know of two major ways by which this determination takes place. The first involves the CYTOPLASMIC SEGREGATION of determinative molecules during embryonic cleavage, wherein the cleavage planes separate qualitatively different regions of the zygote cytoplasm into different daughter cells. The second mode of determination is EMBRYONIC INDUCTION, which involves the interaction of cells or tissues to restrict the fates of one or both of the participants. As we shall see, both mechanisms are used in the development of any organism. This chapter will focus on those experiments demonstrating cytoplasmic segregation; the following chapter will cover determination by embryonic induction.

TABLE 1
Some differentiated cell types and their major products

Cell type	Differentiated cell product	Specialized function
Keratinocyte (skin cell)	Keratin	Protection against abrasion, desiccation
Erythrocyte (red blood cell)	Hemoglobin	Transport of oxygen
Lens cell	Crystallins	Transmission of light
B lymphocyte	Immunoglobulins	Antibody synthesis
T lymphocyte	Cell surface antigens (lymphokines)	Destruction of foreign cells; regulation of immune response
Melanocyte	Melanin	Pigment production
Pancreatic islet cells	Insulin	Regulation of carbohydrate metabolism
Leydig cell (♂)	Testosterone	Male sexual characteristics
Chondrocyte (cartilage cell)	Chondroitin sulfate; type II collagen	Tendons and ligaments
Osteoblast (bone-forming cell)	Bone matrix	Skeletal support
Myocyte (muscle cell)	Muscle actin and myosin	Contraction
Hepatocyte (liver cell)	Serum albumin; numerous enzymes	Production of serum proteins and numerous enzymatic functions
Neurons	Neurotransmitters (acetylcholine, epinephrine, etc.)	Transmission of electrical impulses
Tubule cell (♀) of hen oviduct	Ovalbumin	Egg white proteins for nutrition and protection of embryo
Follicle cell (♀) of insect oviduct	Chorion proteins	Eggshell proteins for protection of embryo

Preformation and epigenesis

Any explanation of the differentiation of the various bodily cells from the fertilized egg has to explain (1) the constant morphology of each species (i.e., that chickens beget only chickens, not crocodiles), and (2) the diversity among the bodily parts of each organism. Indeed, one of the major characteristics of development is that each species reproduces a characteristic developmental pattern. Development involves the expression of the inherited properties of the species.

In the seventeenth century, a union of development and inheritance was achieved in the hypothesis of PREFORMATIONISM. According to this view, all the organs of the adult were prefigured in miniature within the sperm or (more usually) the ovum. Organisms were not seen to be "developed," but were "unrolled." This hypothesis had the backing of both science and philosophy (Gould, 1977; Roe, 1981). First, because all organs were prefigured, embryonic development merely required the growth of existing structures, not the formation of new ones. No extra mysterious force was needed for embryonic development. Second, just as the adult organism was prefigured in the germ cells, another generation already existed in a prefigured state within the germ cells of the first prefigured generation. This corollary, called EMBOITMENT (encapsulation), assured that the species would always remain constant. Although certain microscopists claimed to see fully formed human miniatures within the sperm or egg, the major proponents of this hypothesis—Albrecht von Haller and Charles Bonnet—knew that organ systems develop at different rates and that embryonic structures need not be in the same place as those in the newborn.

The preformationists had no cell theory to provide a lower limit to the size of their preformed organisms, nor did they view mankind's tenure on Earth as potentially immortal. Rather, said Bonnet (1764), "Nature works as small as it wishes," and the human species existed in that finite time spanning Creation and Resurrection. This was in accord with the best science of its time, conforming to the French mathematician-philosopher Rene Descartes's principle of the infinite divisibility of a mechanical nature initiated, but not interfered with, by God.

Preformation was a conservative theory, emphasizing the lack of change between generations. Its principal failure was its inability to account for the variations known by the limited genetic evidence of the time. It was known, for instance, that matings between white and black parents produced children of intermediate skin color, an impossibility if inheritance and development were solely through either the sperm or the egg. In more controlled experiments, the German botanist Joseph Kölreuter (1766) had produced hybrid tobacco plants having the characteristics of both species. Moreover, by mating the hybrid to either the male or female parent, Kölreuter was able to "revert" the hybrid back to one or the other parental type after several generations. Thus, inheritance seemed to arise from a mixture of parental components. In addition, preformationism could not explain the generation of "monstrosities" and distinct deviations such as hexadactylism (six fingers per hand) when both parents were normal.

There developed, then, an alternative hypothesis: EPIGENESIS. According to this hypothesis, each adult organism develops anew from an undifferentiated condition. This view of development, having philosophical roots as far back as Aristotle, was revived by a German embryologist working in St. Petersburg, Kaspar Friedrich Wolff. By carefully observing the development of chick embryos, Wolff demonstrated that the embryonic parts developed from tissues having no counterpart in the adult organism. The heart and blood vessels (which, according to preformationism, had to be present from the beginning in order to ensure embryonic growth) could be seen to develop anew in each embryo. Similarly, the intestinal tube was seen to arise by the folding of an originally flat tissue. This latter observation was explicitly detailed by Wolff, who proclaimed (1767), "When the formation of the intestine in this manner has been duly weighed, almost no doubt can remain, I believe, of the truth of epigenesis." However, to create an organism anew each generation, Wolff had to postulate an unknown force, the *vis essentialis* ("essential force"), which, acting like gravity or magnetism, would organize embryonic development.

Preformationism best explained the continuity between generations, whereas epigenesis best explained variation and the direct observations of organ formation. A reconciliation of sorts was attempted by the German philosopher Immanuel Kant (1724–1804) and his colleague, biologist Johann Friedrich Blumenbach (1752–1840). Attempting to construct a scientific theory of racial descent, Blumenbach postulated a mechanical goal-directed force called the *Bildungstrieb* ("development force"). Such a force, he said, was not theoretical but could be shown to exist by experimentation. A *Hydra*, when cut, will regenerate its amputated parts from the rearrangement of existing elements. Some purposive organizing force could be observed in operation, and this force was a property of the organism itself. This *Bildungstrieb* was thought to be inherited through the germ cells. Thus, development could proceed epigenetically through a predetermined force inherent in the matter of the embryo (Cassirer, 1950; Lenoir, 1980). Moreover, such a force was susceptible to change; the left-handed variant of snail coiling was used as an example of such modifications in the organizing force. In this hypothesis, where epigenetic development is directed by preformed instructions, we are not far from the view held by some modern biologists that "the complete description of the organism is already written in the egg" (Brenner, 1979). However, until the rediscovery of Mendel's work at the beginning of the twentieth century, there was no consistent genetic theory in which to place such ideas of inherited variation, and each scientist was free to speculate on the mechanisms by which developmental patterns were inherited.

The French teratologists

The attempts to find a hypothesis that would explain species constancy and epigenetic development led to the creation of modern embryology. The searches for such a hypothesis were undertaken in two different intellectual traditions. One, centered in France, stressed the anatomic aspects of teratogenesis. It sought to discover the embryological errors that caused some infants to be born with developmental abnormalities. The second search for such a hypothesis was centered in Germany and focused on the physiology of developmental processes. Both research traditions began manipulating embryos to see how the developing organism would respond to these perturbations (Churchill, 1973; Fischer and Smith, 1984).

The French teratological experiments began in the 1820s with the studies of Etienne Geoffrey Saint-Hilaire and his son, Isadore. These investigators attempted to show that anomalous births were the products of disrupted fetal development rather than preformed aberrations. They sought to artificially produce developmental anomalies by altering the incubation conditions of chick egg development. Though largely unsuccessful in these attempts (their crude techniques either allowed normal development to continue or killed the embryos), they set the stage for Dareste's more refined analysis in 1877. Dareste performed thousands of experiments and traced developmental anomalies in chicks back to early stages of their development.

But the chick embryo was a poor choice of organism for studying the earliest stages of embryogenesis. If one wanted to examine whether perturbations at the earliest stages of development affected adult structures, one would have to use another organism. In 1886, Laurent Chabry began studying teratogenesis in the more readily accessible tunicate embryo. This was a fortunate choice, because these embryos develop rapidly into larvae with relatively few cell types. Chabry set out to produce specific malformations by lancing specific blastomeres of the cleaving tunicate embryo. He discovered that each blastomere was responsible for producing a particular set of larval tissues. In the absence of those cells, the larva would lack just those structures normally formed by those cells. Moreover, he observed that when particular cells were isolated from the rest of the embryo, they formed their characteristic structure apart from the context of the other cells. Thus, each of the tunicate cells appeared to be developing autonomously. This ability of each cell to develop autonomously from the other embryonic cells is often referred to as MOSAIC DEVELOPMENT, because the embryo seems to be a mosaic of self-differentiating parts.

Cytoplasmic specification in tunicate embryos

More recent studies have shown that the tunicate embryo does indeed approximate a "mosaic of self-differentiated parts" constructed from information stored in the oocyte cytoplasm. As the embryo divides, different cells incorporate different regions of cytoplasm. These different cytoplasmic regions are thought to contain MORPHOGENIC DETERMINANTS that control the commitment of the cell to a particular cell type. Studies of tunicate cell determination have been helped considerably by the eggs of certain species that segregate their cytoplasm into a series of colored regions immediately after fertilization. (This can be seen in the series of photographs in the color portfolio.)

In 1905, E. G. Conklin described how these colored plasms became apportioned into various blastomeres. The first cleavage separates the egg into right and left mirror images. From then on, each cell division on one side parallels a similar division on the other. By following the fate of *each* blastomere of the tunicate *Styela partita*, Conklin came to the astonishing conclusion that each of the colored regions of cytoplasm delineated a specific embryonic fate (Figure 1). The yellow cytoplasmic crescent gave rise to the muscle cells; the gray equatorial crescent produced the notochord and the neural tube; the clear animal cytoplasm became the larval epidermis; and the yolky gray vegetal region gave rise to the larval gut.

G. Reverberi and A. Minganti (1946) analyzed tunicate determination in a series of isolation experiments, and they, too, observed the self-differentiation of each isolated blastomere and the remaining embryo. The results of one of these experiments is shown in Figure 2. When the 8-cell embryo is separated into its four doublets (the right and left sides being equivalent), mosaic determination is the rule. The animal posterior pair of blastomeres gives rise to the ectoderm; the vegetal posterior pair produces endoderm, mesenchyme, and muscle tissue, just as expected from the fate

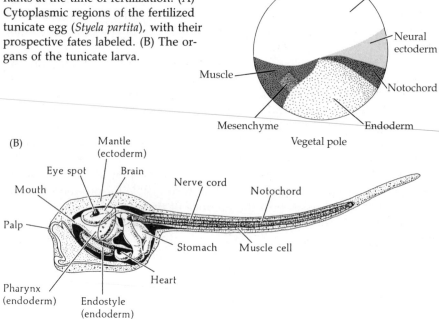

FIGURE 1
Segregation of cytoplasmic determinants at the time of fertilization. (A) Cytoplasmic regions of the fertilized tunicate egg (*Styela partita*), with their prospective fates labeled. (B) The organs of the tunicate larva.

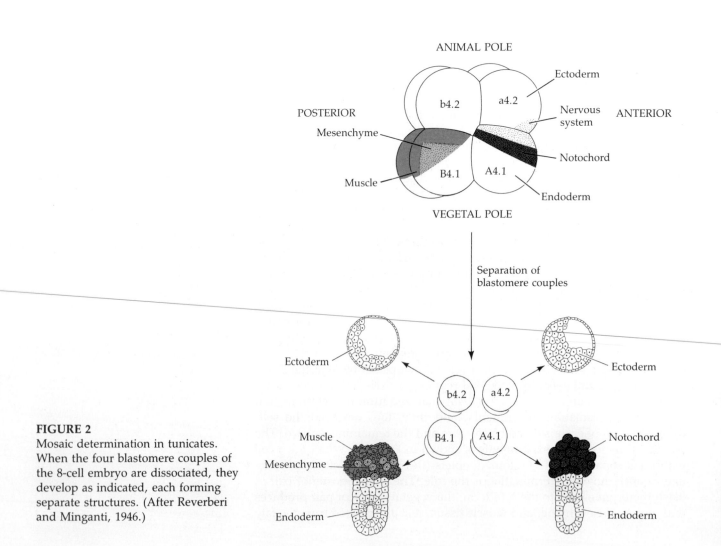

FIGURE 2
Mosaic determination in tunicates. When the four blastomere couples of the 8-cell embryo are dissociated, they develop as indicated, each forming separate structures. (After Reverberi and Minganti, 1946.)

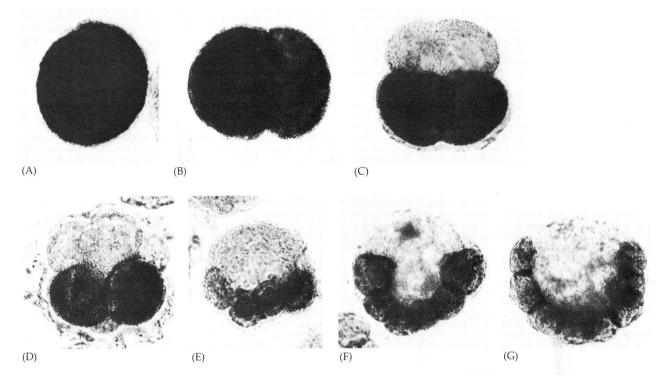

(A)　　　　　　(B)　　　　　　(C)

(D)　　　　　　(E)　　　　　　(F)　　　　　　(G)

FIGURE 5
Acetylcholinesterase localization in tunicate embryos whose cytokinesis has been arrested at various times by cytochalasin B. Localization of enzyme when cytokinesis is blocked at (A) 1-cell; (B) 2-cell; (C) 4-cell; (D) 8-cell; (E) 16-cell; (F) 32-cell; (G) 64-cell. (From Whittaker, 1973a; photographs courtesy of J. R. Whittaker.)

activating (or inactivating) specific genes. The determination of the blastomeres and the activation of certain genes are controlled by the spatial localization of morphogenic determinants within the egg cytoplasm.

The nature of tunicate morphogenic determinants

The nature of the cytoplasmic determinants remains disputed. Studies using inhibitors of transcription and translation suggest that the tunicate embryo uses two types of cytoplasmic factors. The first type is responsible for activating the transcription of certain genes in those cells that have incorporated it. Factors of this type appear to activate the acetylcholinesterase gene. When tunicate embryos are grown in the presence of the transcription inhibitor actinomycin D, no acetylcholinesterase activity emerges, suggesting that the genes for acetylcholinesterase production fail

FIGURE 6
Microsurgery on tunicate eggs enabling some yellow crescent cytoplasm (color) to enter animal cells. Pressing B4.1 blastomeres with a glass needle causes the regression of the cleavage furrow. The furrow can re-form at a more vegetal location where the cells are cut with the needle. This furrow will also separate the pairs of blastomeres in such a way that the animal pole blastomeres receive yellow crescent cytoplasm. The gray area represents vegetal cell cytoplasm. (After Whittaker, 1982.)

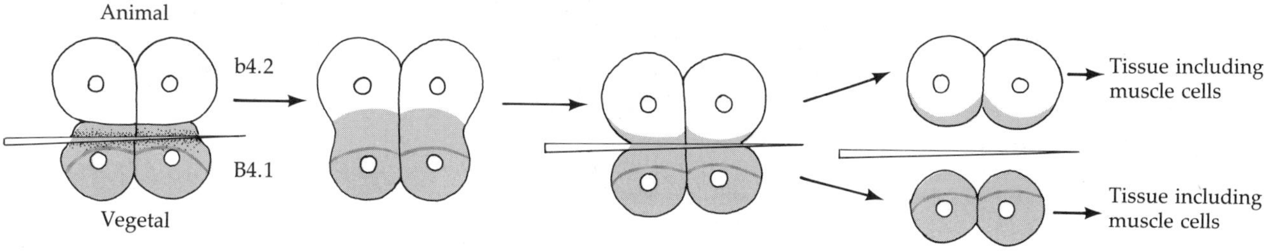

(A)

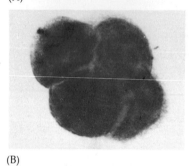

(B)

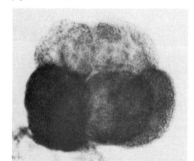

(C)

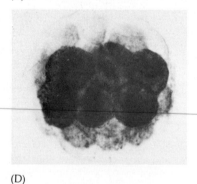

(D)

FIGURE 7
Alkaline phosphatase localization in cytochalasin B-arrested *Ciona* embryos. (A) 2-cell; (B) 4-cell; (C) 8-cell; (D) 16-cell. Staining reactions 14–16 hours after fertilization reveal the localization of alkaline phosphatase determinants. (From Whittaker, 1977; photographs courtesy of J. R. Whittaker.)

to be activated. Moreover, when RNA is purified from *Ciona* embryos of various stages and injected into frog oocytes, the oocytes will translate the newly inserted messages. By this technique, Meedel and Whittaker (1983, 1984) found that there is no detectable mRNA for acetylcholinesterase until the gastrula stage, when it appears only in the descendants of the B4.1 blastomeres. Therefore, new RNA synthesis has to occur for the production of this enzyme. Crowther and Whittaker (1984) have demonstrated that most of the cell-specific markers of tunicate development are sensitive to transcription inhibitors, so morphogenic determinants may act generally to activate specific transcription.

The same type of inhibitor experiments also indicate that certain determinants can be stored as messenger RNAs in particular regions of the egg. This situation is seen when one looks at the synthesis of intestinal alkaline phosphatase in tunicate embryos. This enzyme is synthesized by the larval endoderm and becomes segregated into those cells that form the gut (Figure 7). Actinomycin D has no effect on the synthesis of this enzyme, although inhibitors of translation prevent its expression (Whittaker, 1977). Therefore, mRNA for intestinal alkaline phosphatase must already exist in the oocyte cytoplasm. Somehow, the mRNA itself must be sequestered into the appropriate blastomere, and recent studies suggest that the cytoskeleton may be responsible (see Sidelights & Speculations).

Jeffery and co-workers (1986) have also shown that muscle-specific actin is stored as an mRNA. They find, however, that alkaline phosphatase is not produced in enucleated fragments of fertilized *Styela* eggs. If alkaline phosphatase production were entirely independent of the nucleus, then it should have been expressed. This discrepancy brings up some important points about what conclusions can be drawn from what data. Whittaker's laboratory measures levels of enzyme activity in embryos grown in inhibitors. This protocol assumes that the inhibitors enter each cell in the same way, affect each gene equally, and have no major effect on development other than specifically inhibiting transcription or translation. Jeffery's laboratory measures alkaline phosphatase activity in enucleated fragments; these investigators assume that the mRNA for this enzyme (if it exists) is not localized in a compartment of the cell that remains bound to the female pronucleus and that its translation is independent of other factors that may be transcriptionally controlled. Neither laboratory can directly measure the events of cell determination. Rather, they use a measure of cell differentiation to assay for an earlier determined state. The analysis of determination will have to wait for the identification of specific proteins or mRNAs that commit a cell to a certain fate (as opposed to those specific proteins of mRNAs that are expressed as a result of this commitment).

These intracellular morphogens have proven remarkably difficult to isolate by standard biochemical procedures. They are probably labile compounds present at low concentrations, and developmental biologists have only a limited number of embryos to work with (unlike microbial biologists, who can start off with kilograms of *E. coli*). Two alternatives to biochemical isolation have recently emerged. One is a genetic approach that seeks to identify mutant genes responsible for producing nonfunctional morphogens. The wild-type alleles for such morphogens could then be cloned. This approach can only work in an organism such as *Drosophila*, whose genome is well characterized and in which mutants can be readily screened. Work in this area has begun and will be discussed later in the chapter.

The second alternative approach is immunological, wherein antibodies are made that specifically bind to a cytoplasmic molecule. This method has been used by Nishikata and co-workers (1987) in their attempts to purify the muscle-forming determinant of crescent region. They injected yellow crescent cytoplasm into mice and let the mice make antibodies to this foreign material. Some of the antibodies bound specifically to the yellow crescent region of the egg and were able to inhibit the emergence of the muscle-specific acetylcholinesterase. Investigations are currently underway to identify the molecules bound by these antibodies and to characterize their activity during development.

SIDELIGHTS & SPECULATIONS

Intracellular localization and movements of morphogenic determinants

Whereas the various inclusions of the fertilized egg (including yolk, pigment, and soluble proteins) can easily be displaced by centrifugation, such displacement does not usually affect embryogenesis (reviewed in Morgan, 1927). It appears, then, that either the determinants are too small to be moved by centrifugation or they are somehow anchored within the egg. The lack of diffusion manifest in the cytoplasmic localization of these determinants weighs against the first possibility. Most likely, the determinants are attached to insoluble material, probably the cytoskeletal framework of the cell. This infrastructure of filaments and tubules is particularly prominent in the oocyte cortex, but it extends throughout the cell. Cervera et al. (1981) have reported that most cellular RNA in cultured cells is associated with the cytoskeletal framework. Thus, the cytoskeleton might be a means of specifically localizing cytoplasmic determinants.

The cytoskeleton can be isolated by extracting cells with nonionic detergents such as Triton X-100. The detergent solubilizes lipids, tRNA, and monoribosomes. The remaining cytoskeleton contains microtubules, microfilaments, intermediate filaments, and roughly 200 proteins, including one that is capable of binding the 5′ cap of mRNAs (Zumbe et al., 1982; Moon et al., 1983). In the tunicates *Styela* and *Boltenia*, the muscle-forming yellow crescent is characterized by an actin-containing cytoskeletal domain. This domain is originally coextensive with the unfertilized egg. After fertilization, however, the actin microfilaments contract and become segregated into those blastomeres fated to form muscle cells, taking with them the yellow pigment granules and a set of mRNAs (Jeffery and Meier, 1983; Jeffery, 1984). Figure 8 shows that the cytoskeletal framework contains the yellow pigment granules and that these granules are given their intracellular localization by the movements of the oocyte cytoplasm during fertilization. The proteins of this yellow crescent region differ significantly from the proteins of the rest of the egg, whereas the mRNAs in this region are not seen to be specific to the yellow crescent (Jeffery, 1985). The cytoskeleton may be the anchor for localizing the morphogenic factors determining embryonic cell fate.

The localization of specific proteins or mRNAs to specific regions of the egg is not limited to "mosaic" embryos.

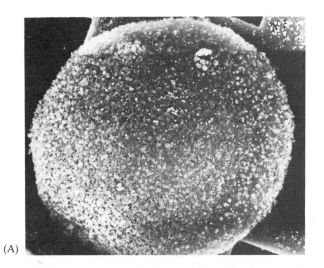

(A)

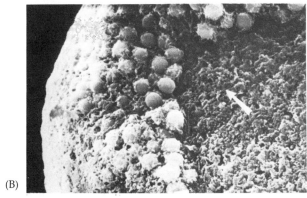

(B)

FIGURE 8

Scanning electron micrographs of tunicate eggs undergoing the segregation of morphogenic determinants. The eggs have been extracted with Triton X-100 detergent to solubilize the membrane and to allow the observation of cytoskeletal components. (A) Unfertilized *Styela* egg; yellow pigment granules and the cytoskeleton can be seen around the entire surface. (B) A newly fertilized *Boltenia* zygote segregating its yellow cytoplasm. The region containing the yellow pigment granules is elevated and consists of a plasma membrane lamina and a deeper network of filaments. The presumed direction of the pigment migration is shown by the arrow. (From Jeffery and Meier, 1983; photographs courtesy of S. Meier.)

Weeks and co-workers (1985) have shown that the animal and vegetal caps of frog eggs also contain some unique messages. These studies began with the isolation of mRNA from growing oocytes (Figure 9). Using techniques that will be detailed in Chapter 10, the investigators isolated and purified the mRNAs from different regions of the frog egg. Whereas most of the messages in the animal and vegetal caps of the cell were identical, a small percentage of animal cap DNA was not represented in the vegetal cytoplasm, and certain vegetal cap mRNA was not found in the animal cap cytoplasm. The localization of one of these unique vegetal mRNAs was confirmed when the developing frog eggs were incubated with radioactive RNA complementary to this vegetal message. (This cRNA was made by transcribing RNA from the other strand of that gene's DNA.)

FIGURE 9

Localization of a messenger RNA to the vegetal cap of *Xenopus* oocytes. Eggs at various stages in their development were incubated with a radioactive RNA complementary to the Vg1 message. The presence and location of the Vg1 message could be seen by the presence of radioactivity within any area of the egg (since the radioactive emissions would be detected by the photographic emulsion). Early in their oocyte development (A), the Vg1 mRNA is distributed throughout the egg. As egg maturation continues, however, the Vg1 message is progressively localized in the vegetal cap. The dark area near the center is the germinal vesicle of the oocyte. (From Melton, 1987; photographs courtesy of D. Melton.)

The presence of radioactivity could be detected by a photographic emulsion. Wherever the complementary RNA found this message, it would bind and remain. The radioactivity would cause a spot to occur on the photographic emulsion when the photograph was developed. Figure 9 shows that in the early oocytes, this vegetal mRNA (called Vg1) could be found throughout the oocyte. However, as oocyte development continued, it became localized to the vegetal end of the egg (Melton, 1987). The mechanism for the translocation of Vg1 to the vegetal region involves both the microfilaments and the microtubules of the cytoskeleton.

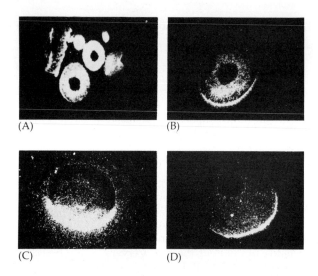

(A)   (B)

(C)   (D)

Cytoplasmic localization in mollusc embryos

The "mosaic" type of differentiation is widespread throughout the animal kingdom, especially in protostomal organisms such as ctenophores (comb jellies), annelids, nematodes, and molluscs, all of which initiate gastrulation at the future anterior end after only a few cell divisions. Molluscs provide some of the most impressive examples of "mosaic" development and of the phenomenon of CYTOPLASMIC LOCALIZATION, wherein the morphogenic determinants are found in a specific region of the oocyte.

E. B. Wilson, the outstanding American embryologist at the turn of the century, isolated early blastomeres from embryos of the mollusc *Patella coerulea* and compared their development with that of the same cells left within other embryos. Figure 10 shows one set of results published by Wilson in 1904. Not only did the isolated blastomeres follow their normal developmental fates (in this case, to produce the ciliated trochoblast cells), but they also completed the normal number of cell divisions at precisely the same time as those cells remaining within the embryo. Their cleavages were in the correct orientation, and the derived cells became ciliated at the appropriate time. Wilson (1904) concluded from these experiments that these cells possess within themselves all the factors that determine the form and rhythm of cleavage and the characteristic and complex differentiation that they undergo, wholly independent of their relation to the remainder of the embryo.

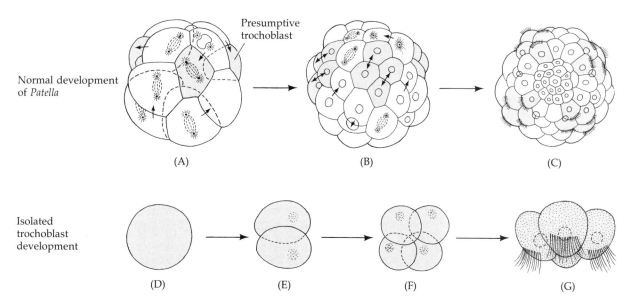

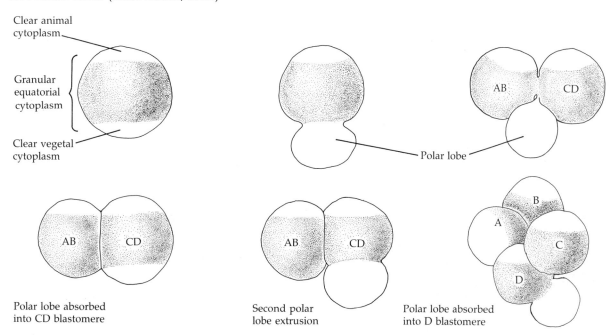

The polar lobe

In his next experiment, Wilson was able to demonstrate that such development was predicated on the segregation of specific morphogenic determinants into specific blastomeres. Certain spirally cleaving embryos (mostly in the mollusc and annelid phyla) extrude a bulb of cytoplasm immediately before first cleavage (Figure 11). This protrusion is called the POLAR LOBE. In certain species of snails, the region uniting the polar lobe to the rest of the egg becomes a fine tube. The first cleavage splits the zygote asymmetrically, so the polar lobe is connected only to the CD blastomere. In several species, nearly one-third of the total cytoplasmic volume is present in these anucleate lobes, giving them the appearance of another cell. This three-lobed structure is often referred to as the TRE-

FIGURE 10

Differentiation of trochoblast cells in the normal embryo of the mollusc *Patella* (A–C) and of trochoblast cells isolated and cultured in vitro (D–G). (A) 16-cell stage seen from the side; the presumptive trochoblast cells are shaded. (B) 48-cell stage. (C) Ciliated larval stage, seen from the animal pole. Cilia are seen on trochoblast cells. (D) Isolated trochoblast cell. (E,F) Results of first and second divisions in culture. (G) Ciliated product of F; even in isolated culture, cells become ciliated at the correct time. (After Wilson, 1904.)

FIGURE 11

Cleavage in the mollusc *Dentalium*. Extrusion and reincorporation of the polar lobe occurs twice. (From Wilson, 1904.)

Clear animal cytoplasm

Granular equatorial cytoplasm

Clear vegetal cytoplasm

Polar lobe

AB   CD

Polar lobe absorbed into CD blastomere

Second polar lobe extrusion

Polar lobe absorbed into D blastomere

FOIL-STAGE EMBRYO (Figure 12). The CD blastomere then absorbs the polar lobe material, but extrudes it again prior to second cleavage (Figure 11). After this division, the polar lobe is attached only to the D blastomere, which absorbs its material. Thereafter, no polar lobe is formed.

Wilson showed that if one removes the polar lobe at the trefoil stage, the remaining cells divide normally. However, instead of producing a normal trochophore (snail) larva, they produced an incomplete larva, wholly lacking its mesodermal organs—muscles, mouth, shell gland,* and foot. Moreover, Wilson demonstrated that the same type of abnormal embryo can be produced by removing the D blastomere from the 4-cell embryo. Wilson concluded that the polar lobe cytoplasm contained the mesodermal determinants and that these determinants give the D blastomere its mesoderm-forming capacity. Wilson also showed that the localization of the mesodermal determinants was established shortly after fertilization, thereby demonstrating that a specific cytoplasmic region of the egg, destined for inclusion into the D blastomere, contained whatever "factors" were necessary for the special cleavage rhythms of the D blastomere and for the differentiation of the mesoderm.

The morphogenic determinants sequestered within the polar lobe are probably located in the cytoskeleton or cortex and not in the diffusible cytoplasm of the embryo. A. C. Clement (1968) centrifuged the trefoil-stage embryos of the snail *Ilyanassa obsoleta*, causing the fluid part of the cytoplasm to flow back into the CD blastomere. However, when he then cut off the polar lobe, the resulting embryo *still* lacked its mesodermal derivatives. Van den Biggelaar obtained similar results when he removed the cytoplasm from the polar lobe with a micropipette. Cytoplasm from other regions of the cell flowed into the polar lobe, replacing the portion that he had removed. The subsequent development of these embryos was normal. In addition, when he added the soluble polar lobe cytoplasm to the B blastomere, duplications of structures were not seen (Verdonk and Cather, 1983). Therefore, the diffusible part of the cytoplasm does not contain these morphogenic determinants. They probably reside in the nonfluid cortical cytoplasm or on the cytoskeleton.

Clement also analyzed the further development of the D blastomere in order to observe the further appropriation of these determinants. The development of the D blastomere is illustrated in Figure 13. This macromere, having received the contents of the polar lobe, is larger than the

*The shell gland is an ectodermal organ formed through induction by mesodermal cells. Without the mesoderm, no cells are present to induce the competent ectoderm. Again, we see some limited induction within a mosaic embryo.

FIGURE 12
Polar lobes of molluscs. (A) Scanning electron micrograph of the extending polar lobe in the uncleaved egg of *Buccinum undatum*. The surface ridges are confined to the polar lobe region. (B) Section through the first cleavage, or trefoil stage, embryo of *Dentalium*. Arrow points to the large polar lobe. (Photographs courtesy of M. R. Dohmen.)

(A)

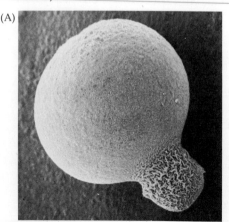

(B)

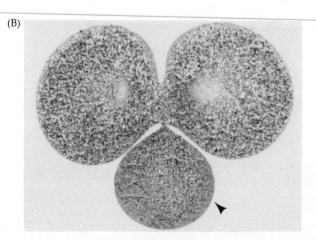

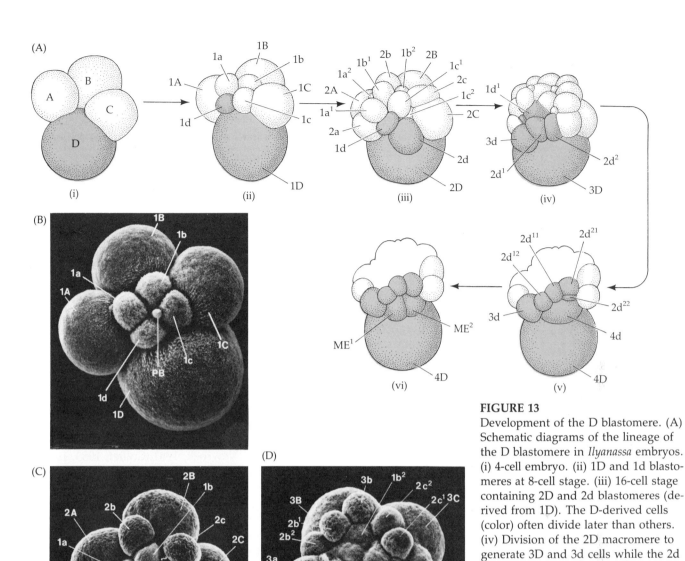

FIGURE 13
Development of the D blastomere. (A) Schematic diagrams of the lineage of the D blastomere in *Ilyanassa* embryos. (i) 4-cell embryo. (ii) 1D and 1d blastomeres at 8-cell stage. (iii) 16-cell stage containing 2D and 2d blastomeres (derived from 1D). The D-derived cells (color) often divide later than others. (iv) Division of the 2D macromere to generate 3D and 3d cells while the 2d cell divides into $2d^1$ and $2d^2$. (v) 64-cell stage. The 3D blastomere produces the 4D and 4d cells. (vi) The 4d blastomere divides symmetrically to produce the two mesentoblasts ME^1 and ME^2. (B) 8-cell embryo. (C) 12-cell embryo (2a–1d have not divided yet). (D) 32-cell embryo. (A after Clement, 1962; photographs from Craig and Morrill, 1986; courtesy of the authors.)

other three. When one removes the D blastomere or its first or second macromere derivatives (1D or 2D), one obtains an incomplete larva, lacking heart, intestine, velum (the ciliated border of the larva), shell gland, eyes, and foot. When one removes the 3D blastomere (*after* the division of the 2D cell to form the 3d blastomere), one obtains an almost normal embryo, having eyes, feet, velum, and some shell gland, but no heart or intestine (Figure 14). Therefore, some of the morphogenic determinants originally present in the D blastomere were apportioned to the 3d cell. After the 4d cell is given off (by the division of the 3D blastomere), removal of the D derivative (the 4D cell) produces no qualitative difference in development. In fact, all the essential determinants for heart and intestine formation are now in the 4d blastomere and removal of *that* cell results in a heartless

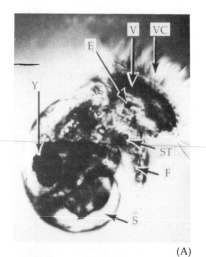

(A)

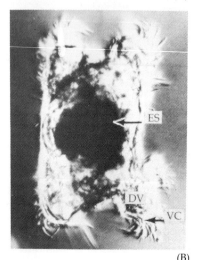

(B)

FIGURE 14
Importance of polar lobe in the development of *Ilyanassa*. (A) Normal veliger larva. (B) Aberrant larva, typical of those produced when the polar lobe of the D blastomere is removed. E, eye; F, foot; S, shell; ST, statocyst (balancing organ); V, velum; VC, velar cilia; Y, residual yolk; ES, everted stomodeum; DV, disorganized velum. (From Newrock and Raff, 1975; photographs courtesy of K. Newrock.)

and gutless larva (Clement, 1986). The 4d blastomere is responsible for forming (at its next division) the two MESENTOBLASTS, the cells that give rise to both the mesodermal (heart) and endodermal (intestine) organs.

The material in the polar lobe is also responsible for organizing the dorsal–ventral (back–belly) polarity of the embryo. When polar lobe material is allowed to pass into the AB blastomere as well as into the CD cell, twin larvae are formed that are joined at their ventral surfaces (Guerrier et al., 1978; Henry and Martindale, 1987).

Thus, experiments have demonstrated that the nondiffusible polar lobe cytoplasm is extremely important for normal mollusc development because:

1 It contains the determinants for the proper rhythm and cleavage orientation of the D blastomere.
2 It contains certain determinants (those entering the 4d blastomere and hence leading to the mesentoblasts) for mesodermal and intestinal differentiation.
3 It is responsible for permitting the inductive interactions (through the material entering the 3d blastomere) leading to the formation of the shell gland and eye.
4 It contains determinants needed for specifying the dorsal–ventral axis of the embryo.

Although the polar lobe is clearly important for normal snail development, we still do not know the mechanisms of its effects. There appear to be no major differences in mRNA or protein synthesis between lobed and lobeless embryos (Brandhorst and Newrock, 1981; Collier, 1983, 1984). One possible clue has been provided by Atkinson (1987), who has observed differentiated cells of the velum, digestive system, and shell gland within the lobeless embryo. Lobeless embryos can produce these cells, but they appear unable to organize them into functional tissues and organs. Tissues of the digestive tract can be found, but they are not connected, and myocytes are scattered around the lobeless larva but are not organized into a functional muscle tissue. Thus, the developmental functions of the polar lobe may be very complex.

Determination in the nematode *Caenorhabditis elegans*

The ability to analyze development requires appropriate organisms. Sea urchins have long been a favorite organism of embryologists because their gametes are readily obtainable in large numbers, their eggs and embryos are transparent, and fertilization and development can occur under laboratory conditions. But sea urchins are difficult to rear in the laboratory for more than one generation, making their genetics difficult to study. Geneticists, on the other hand (at least those working with multicellular eukaryotes), favor *Drosophila*. The rapid life cycle, readiness to breed, and the polytene chromosomes of the fly larva (which allow gene localization) make this animal superbly suited for hereditary analysis. But *Drosophila* development is very complex and difficult to study. A research program spearheaded by Sidney Brenner (1974) was set up to identify an organism wherein it might be possible to identify each gene involved in development as well as to trace the lineage of every single cell. Such an organism is *Caenorhabditis elegans*, a small (1 mm), free-living soil nematode (Figure 15). It has a rapid (about 16 hours) period of embryogenesis, which it can

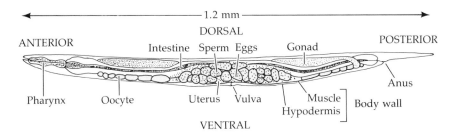

FIGURE 15

Caenorhabditis elegans. Diagram of an adult hermaphrodite. Early in its development, sperm are formed. These sperm are stored during later stages so that a mature egg passes through the sperm on its way to the vulva. In this manner, the hermaphrodite unites its own sperm with its own eggs. (After Sulston and Horvitz, 1977.)

accomplish in petri dishes, and relatively few cell types. Moreover, it is HERMAPHRODITIC, each individual containing both eggs and sperm. These roundworms can reproduce either by cross-fertilization or by self-fertilization. The body of a hermaphroditic *C. elegans* contains exactly 959 somatic cells, whose entire lineage has been mapped through its transparent cuticle (Figure 16; Sulston and Horvitz, 1977; Kimble and Hirsh, 1979). Furthermore, unlike vertebrate cell lineages, the cell lineage of *C. elegans* is almost entirely invariant from one individual to the next. There is little room for randomness (Sulston et al., 1983). (This is a consequence of the spatial ordering of cytoplasmic segregation.) *Caenorhabditis* also has a small number of genes for a multicellular organism, about 3000. This is about 35 times fewer than mammals.

The localization of cytoplasmic substances has been elegantly demonstrated in this species. One such material is the set of GERM-LINE GRANULES (P-GRANULES), which are redistributed in the zygote shortly after fertilization and become restricted to those cells capable of forming gametes (Figure 17). Using fluorescent antibodies to a component of the P-granules, Strome and Wood (1983) discovered that during the pronuclear migration in the zygote, the randomly scattered P-granules become localized in the posterior end of the zygote (toward the site of sperm entry), so that they only enter the blastomere (P1) formed from the posterior cytoplasm (Figure 18 and color portfolio). Following cleavage, the P-

FIGURE 16

Entire cell lineage chart for *C. elegans.* Each vertical line represents a cell; each horizontal line represents a cell division. (From Sulston et al., 1983.)

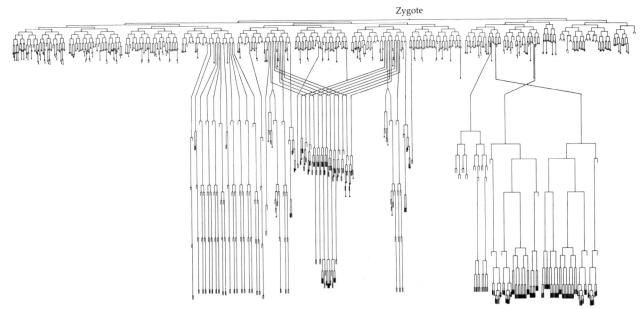

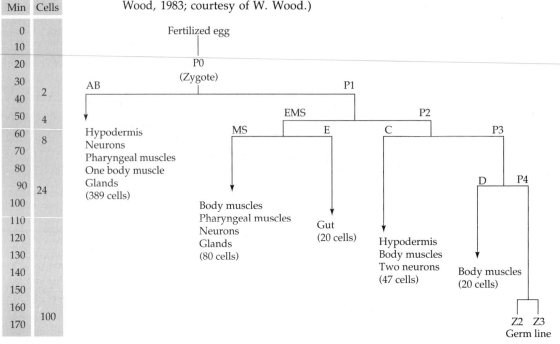

Min	Cells
0	
10	
20	
30	2
40	
50	4
60	8
70	
80	
90	24
100	
110	
120	
130	
140	
150	
160	
170	100

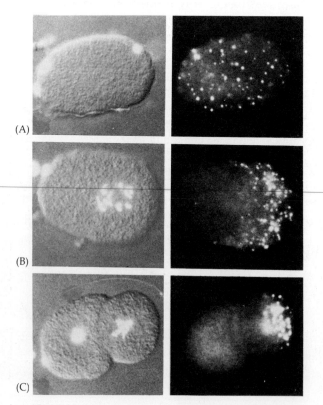

FIGURE 18

Asymmetric localization of P-granules during fertilization and first cleavage. Left-hand figures are stained to show DNA; right-hand figures are the same cells stained with fluorescent antibodies to P-granule protein. (A) A zygote prior to pronuclear migration shows random dispersal of P-granules. (B) As pronuclei come together, granules become localized to the posterior periphery of the zygote. (C) Two-cell embryo in which P1 is entering mitotic prophase; the P-granules are now positioned at the posterior periphery to be transmitted to the P2 cell. (From Strome and Wood, 1983; photographs courtesy of S. Strome.)

granules disperse throughout the P1 blastomere until the start of mitosis, when they once again migrate to the posterior end of the cell. Here they become apportioned to the P2 blastomere. Eventually, the P-granules will reside in the P4 cell, whose progeny become the sperm and eggs of the adult.

The movement of the P-granules requires microfilaments but can occur in the absence of microtubules. Treating the zygotes with cytochalasin D (a microfilament inhibitor) prevents the segregation of these granules to the posterior of the cell, whereas demecolcine (a colchicine-like microtubule inhibitor) fails to stop this movement (Strome and Wood, 1983). Once within the posterior region of the zygote, P-granules remain there even if microfilaments are then disrupted (Hill and Strome, 1987, 1990). The mechanisms for the movement and anchoring of these cytoplasmic granules remain unknown.

The developmental pattern of most *C. elegans* cells is typically mosaic, their fates being determined by internal, cytoplasmic, factors rather than by interactions with neighboring cells. The determinants for the gut lineage become localized during first cleavage to the P1 cell, which later gives rise to the gut precursor cell, E (Figure 19; Laufer et al., 1980; Edgar and McGhee, 1986). As in tunicates, however, there is some room for cell interactions. Kimble (1981) has used focused laser beams to selectively kill

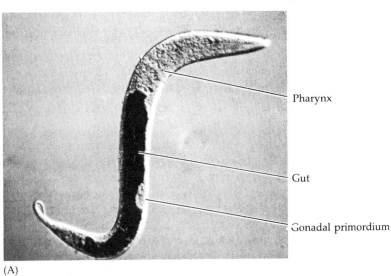

Pharynx

Gut

Gonadal primordium

(A)

FIGURE 19
Gut determination in *C. elegans*. (A) Intestinal cells of newly hatched larva stained for gut-specific esterase. (B) "Half-embryo" from AB (non-gut-forming) cell incubated in cytochalasin and stained for gut-specific esterase. No staining is seen. (C) "Half-embryo" from the P1 blastomere, showing presence of gut-specific esterase. (From Edgar and McGhee, 1986; photographs courtesy of L. Edgar.)

(B)

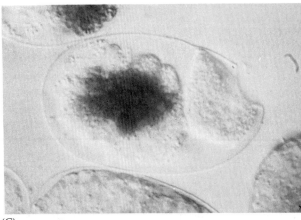

(C)

FIGURE 20
Induction in *C. elegans*. Schematic representation of experiments demonstrating that the anchor cell is necessary for vulva development. (After Alberts et al., 1983.)

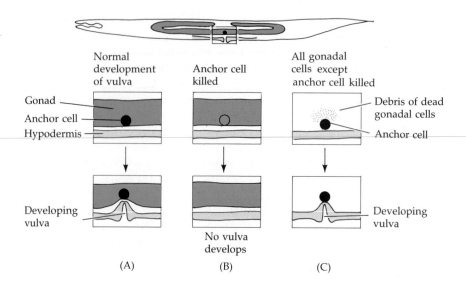

specific cells whose progeny would have formed the vulva (the passage through which the eggs are laid). One of the cells in this region becomes the ANCHOR CELL, which connects the overlying gonad to the vulva. If this cell is destroyed, the hypodermal (skin) cells will not divide to form the vulva. However, as long as the anchor cell is near the hypodermis, the vulva forms, indicating that the anchor cell induces vulva formation (Figure 20). Moreover, the six vulval precursor cells influenced by the anchor cell form an EQUIVALENCE GROUP. During normal development, the three central cells divide to form the vulva while the three outer cells divide once to form six hypodermal cells. (If the anchor cell is destroyed, all the six cells of the equivalence group divide once to form a dozen hypodermal cells.) If the three central cells are destroyed, the three outer cells, which normally form hypodermal cells, will generate vulval cells instead.

This is not the only example of cellular interactions determining the fate of *C. elegans* cells. Priess and Thomson (1987) discovered another example of cellular interaction that involves the first blastomeres of the *C. elegans* embryo. As can be seen in Figure 17, both the anterior (AB) and posterior (P1) blastomeres of the 2-cell embryo usually generate pharyngeal muscle cells. However, if the posterior cell is removed, the isolated AB cell does not generate muscle cells. By separating the muscle-forming progeny of the AB cell from the adjacent progeny of the P1 cell (the EMS blastomere), Priess and Thomson showed that the cell contact is needed between the 4- and 28-cell stages.

Genetic studies have identified one of the components of this reaction. If the maternal product of the *glp-1* gene is absent from the embryo, the P1 lineages are unaffected, and they generate the normal amount of muscle cells. However, the AB blastomere no longer produces its pharyngeal muscle cells (Preiss et al., 1987). The development of the AB blastomere in the *glp-1*-deficient embryos closely resembles that of the development of the AB blastomere that had been physically separated from P1. First, like the physically isolated AB cell, it produces neurons and hypodermal cells, but not muscles. Second, the critical period of the *glp-1* gene product appears to be between the 4- and 28-cell stages.* This was shown by using temperature-sensitive mutants. Temperature-sensitive mutants are important for developmental studies because they enable investigators to find

*The *glp-1* gene is also active in regulating postembryonic cell–cell interactions. It will be used later by the distal tip cell of the gonad to control the number of germ cells entering meiosis (Chapters 16 and 22; Austin and Kimble, 1987).

the time when a gene product is functional. The mutant protein is functional at one temperature but not at another. By shifting the temperature at various stages of development, one can determine the stage when the gene product must be active (Figure 21). Thus, as we have seen in tunicates and snails, some regulation and induction still occur in organisms whose early embryonic development is characterized by an invariant determinism.

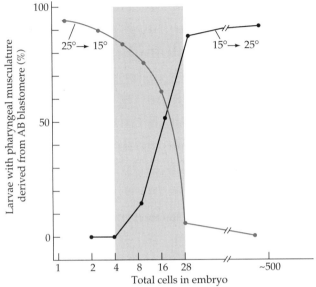

FIGURE 21
Temperature-shift experiment to determine at what stage the maternal *glp-1* gene product is active. The *glp-1* protein in this mutant functions at 25°C, but does not function at 15°C. By shifting the temperature at different embryonic stages, it was found that the *glp-1* protein was needed between the 4- and 28-cell stage. (After Priess et al., 1987.)

SIDELIGHTS & SPECULATIONS

The genetics of polarity in Drosophila

In 1936, embryologist E. E. Just criticized those geneticists who sought to explain development by looking at specific mutations affecting eye color, bristle number, and wing shape. He said that he wasn't interested in the development of the bristles on a fly's back; rather, he wanted to know how the fly embryo made the back itself. Fifty years later, embryologists and geneticists are finally answering that question.

There is evidence for three sets of cytoplasmic determinants in dipteran eggs. One set comprises determinants of germ cell formation and will be detailed later in this chapter. Another set of cytoplasmic determinants controls dorsal–ventral polarity, distinguishing the back of the larval fly from its belly. The third set of morphogenic factors establishes the anterior–posterior (head–tail) polarity of the developing insect.

Evidence that a particular gene encodes a morphogenic factor: The bicoid *gene*

The morphogen responsible for specifying the anterior portion of the *Drosophila* embryo is BICOID PROTEIN. The mRNA encoding this protein is localized in the anterior portion of the oocyte during oogenesis. Soon after fertilization, this message is translated into bicoid protein, which forms a gradient with the highest concentration at the anterior of the embryo. The details of the bicoid protein gradient and how it functions to turn on those genes responsible for the production of head and thorax structures in the fly will be detailed in Chapter 18. Female flies homozygous for the lack of *bicoid* genes give rise to embryos without any head structures. Instead, the trunks of these embryos are made of abdominal tissue and both its ends are telsons (tails). These embryos do not survive.

This genetic evidence and the correlations made be-

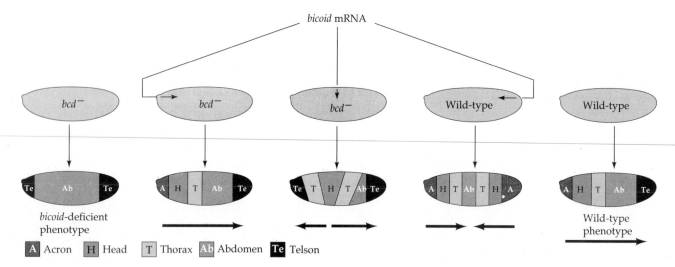

bicoid mRNA

bcd⁻ · bcd⁻ · bcd⁻ · Wild-type · Wild-type

bicoid-deficient phenotype

Wild-type phenotype

| A | Acron | H | Head | T | Thorax | Ab | Abdomen | Te | Telson |

FIGURE 22

Schematic representation of the experiments demonstrating that the *bicoid* gene encodes the morphogen responsible for head structures in *Drosophila*. The phenotypes of the *bicoid*-deficient and wild-type embryos are shown at the sides. When *bicoid*-deficient embryos are injected with *bicoid* mRNA, the point of injection forms the head structures. When the posterior pole of early cleavage wild-type embryos are injected with *bicoid* mRNA, head structures form at both poles. (After Driever et al., 1990.)

tween the presence of bicoid protein and the activation of genes involved in head formation supported the hypothesis that bicoid protein was the head morphogen in *Drosophila*. Confirmation that bicoid protein alone is the morphogen responsible for head formation came from the experiments of Driever and colleagues (1990), who injected purified *bicoid* mRNA into early-cleavage embryos (Figure 22). When injected into the anterior of *bicoid*-deficient embryos (whose mother lacked *bicoid* genes), the *bicoid* mRNA rescued the embryos and caused them to have normal anterior–posterior polarity. Moreover, any location in the embryos where the *bicoid* messages were injected became the head. If *bicoid* mRNA was injected into the center of the embryo, that middle region became the head and the regions on either side of it became thorax structures. If *bicoid* mRNA was placed in the *posterior* pole of a wild-type embryo (with its own endogenous *bicoid* message in its anterior pole), two heads emerged, one at either end. Therefore, the *bicoid* gene is now thought to encode the anterior morphogen of the *Drosophila* embryo.

Dorsal protein: Morphogen for dorsal–ventral polarity

Translocation of dorsal protein. The situation for specification of dorsal–ventral polarity of *Drosophila* is more complicated. The dorsal–ventral axis is specified by nearly 20 products placed into the oocyte by the maternal genome during egg development. Anderson and Nüsslein-Volhard (1984) have isolated ten maternal effect genes, each of whose absence is associated with the lack of ventral struc-

tures. In some cases, the maternal product is probably a protein, while in other cases the female fly places a specific mRNA into the egg.

The actual protein that distinguishes dorsum from ventrum is the protein product of the *dorsal* gene. The mRNA from the mother's *dorsal* genes is put into the egg by the mother fly's ovarian cells. However, dorsal protein is not synthesized from its maternal message until about 90 minutes after fertilization. When this protein is translated, it is found *throughout* the embryo, not just on the ventral or dorsal side. How, then, can this protein act as a morphogen if it is located everywhere the embryo? In 1989, the surprising answer was found (Rushlow et al., 1989; Steward, 1989; Roth et al., 1989). While dorsal protein can be found throughout the syncytial blastoderm of the early *Drosophila* embryo, it is transported into cell nuclei only in the *ventral* part of the embryo (Figure 23A,B). Here, dorsal protein binds to certain nuclear genes to activate or suppress their transcription. If it does not enter the nucleus, the ventralizing genes of the nucleus (probably *snail* and *twisted*) are not activated, the dorsalizing genes of the nucleus (probably *decapentaplegic* and *zerknüllt*) are not repressed, and the region of the embryo develops into dorsalized cells. This hypothesis that the dorsal–ventral axis of *Drosophila* is specified by the selective transport of the dorsal morphogen protein into the nucleus is strengthened by the analysis of mutations having an entirely dorsalized or an entirely ventralized phenotype (Figure 23C,D). In those mutants wherein all the cells are dorsalized (as is evident by their dorsal cuticle), dorsal protein does not enter the nucleus in any cell. Conversely, in those mutants wherein all cells have a ventral phenotype, dorsal protein is found in every cell nucleus.

Modification of dorsal protein. The next question is, What converts a protein that cannot enter the nucleus into a protein that can enter the nucleus? There are several maternal mutations affecting dorso-ventral polarity that may be making proteins that modify the structure of dorsal protein. One such mutation affecting the dorsal–ventral axis is called *snake*. Females homozygous for this mutation produce eggs that develop into completely dorsalized em-

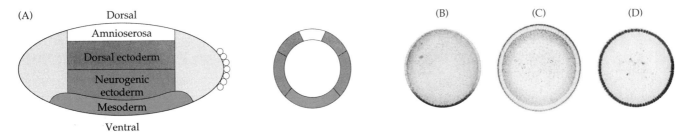

(A) Dorsal | Ventral | Lateral view | Transverse section
(B) (C) (D)

FIGURE 23

Inclusion of dorsal protein into the nuclei of ventral, but not lateral or dorsal, nuclei. (A) Fate map of the mid-*Drosophila* embryo. The most ventral part becomes the mesoderm, the next higher portion becomes the neurogenic ectoderm. The epidermal ectoderm takes up most of the dorsal hemisphere. The dorsalmost region becomes the amnioserosa, an extraembryonic layer that surrounds the embryo. (B–D) Cross sections of embryos stained with antibody to show the presence of dorsal protein. In all cases, the dark stain represents the dorsal protein. (B) Cross section of wild-type embryo showing dorsal protein in the ventralmost nuclei. (C) Cross section of a dorsalized mutant, stained as in B, showing no localization of dorsal protein in any nucleus. (D) Cross section of a ventralized mutant, showing that dorsal protein has entered the nucleus of every cell. (A after Rushlow et al., 1989; B–D from Roth et al., 1989; photographs courtesy of the authors.)

sequence. It is possible (but not yet proven) that the protein products of *snake* (and other "dorsalizing" genes) act to modify dorsal protein so that it can enter the nucleus and generate ventral cell types.

The asymmetric signal from the follicle. Even if dorsal protein is the effective morphogen that is modified so that it can enter the nucleus, we still need to know how it comes to be modified only in the ventral region and not in the dorsal region of the embryo. One of the clues comes from the protein Toll.

Toll protein, the product of the *Toll* gene, is also a maternal product placed into the egg. The recessive mutation *Toll* has a similar dorsalized phenotype, and injections of mRNA from wild-type eggs will restore the dorsal–

bryos (as dorsal protein failed to enter any nucleus). Such embryos lack the structures and cuticle typical of the larval belly. These eggs from *snake* females can be "rescued" by the injection of mRNA from the cytoplasm of wild-type eggs (Figure 24). Here, then, is a situation in which a maternal gene is encoding an mRNA necessary for the dorsal–ventral specification. In the absence of this product, all the cells develop dorsally but when the mRNA is supplied to deficient embryos, their normal development is restored.

The wild-type *snake* allele has been cloned (DeLotto and Spierer, 1986) and found to encode an amino acid sequence similar to that of trypsin and clotting-factor proteases. Such enzymes are often involved in activating inactive precursors by cleaving off an inhibitory amino acid

FIGURE 24

Rescue of larva by injection of wild-type mRNA into eggs destined to have the *snake* phenotype. (A) Deformed larva consisting entirely of dorsal cells. Larvae like these developed from eggs of a female homozygous for the *snake* allele. (B) Wild-type appearance of such larvae developing from *snake* eggs that received injections of mRNA from wild-type eggs. (From Anderson and Nüsslein-Volhard, 1984; photographs courtesy of C. Nüsslein-Volhard.)

(A)
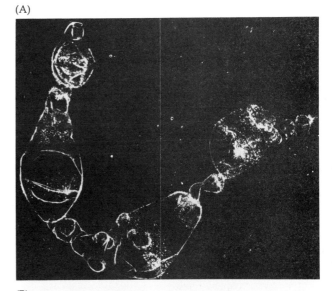

(B)
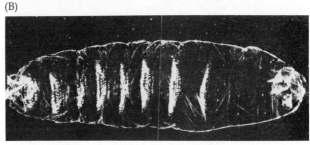

FIGURE 25

Relationship between the point of mRNA injection and the position of the dorsal–ventral axis when *Toll⁻* embryos receive wild-type mRNA. (A) Normal dorsal–ventral pattern with ventral furrow and associated structures. (B) Cells of *Toll⁻* embryos differentiate into dorsal epidermis only. (C–E) The position of injected wild-type cytoplasm (black dot) defines the placement of the dorsal–ventral axis. (Note, however, that it never extends all the way along the anterior–posterior axis.) (After Anderson et al., 1985.)

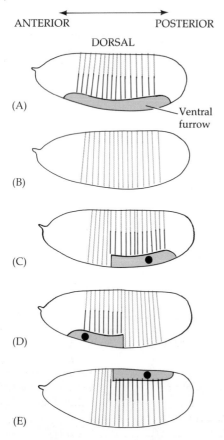

ventral polarity of eggs laid by *Toll⁻/Toll⁻* mothers. But unlike the case of *snake* or the other nine maternal genes, the site of the injection is important. Whichever part of the egg is injected becomes the ventral region of the rescued embryo (Anderson et al., 1985). This suggests that the *Toll⁻/Toll⁻* eggs lack a dorsal–ventral axis (whereas in *snake*, the ventral region occurs in its normal place). It appears, then, that wild-type *Toll* product is involved in the establishment of the dorsal–ventral axis (Figure 25).

Another clue into how dorsal protein can be modified only in one side of the embryo comes from studies on the *torpedo* mutation (Schüpbach, 1987). Maternal deficiency of this gene causes the ventralization of the embryo. Moreover, the *torpedo* gene is active in the ovarian follicle cells, not in the embryo. This was discovered by making germ line chimeras. Schüpbach transplanted germ cell precursors from wild-type embryos to embryos whose mothers carried the *torpedo* mutation. Conversely, she transplanted those germ cells from the *torpedo* embryos to wild-type embryos (Figure 26). The results were surprising in that the wild-type eggs produced ventralized embryos when these eggs had developed within *torpedo* mutant follicles. The *torpedo* mutant eggs were able to produce normal embryos if the eggs developed within a wild-type ovary. Thus, the wild-type *torpedo* gene is needed in the follicle cells, not in the egg itself.

It has been hypothesized (Schüpbach, 1990) that the initial signal for dorsal–ventral polarity comes from the interactions between the oocyte and the follicle cells surrounding it. These interactions may cause the *torpedo* gene

FIGURE 26

Germ line chimeras made by interchanging pole cells (germ cell precursors) between wild-type embryos and embryos from mothers homozygous for the *torpedo* gene. These transplants produce wild-type females whose embryos come from eggs from the mutant mothers, and *torpedo*-deficient embryos with wild-type eggs. The eggs from the *torpedo*-deficient mothers produced normal embryos if the eggs developed in the wild-type ovary, while the wild-type eggs produced ventralized embryos if the eggs developed in the mutant ovary.

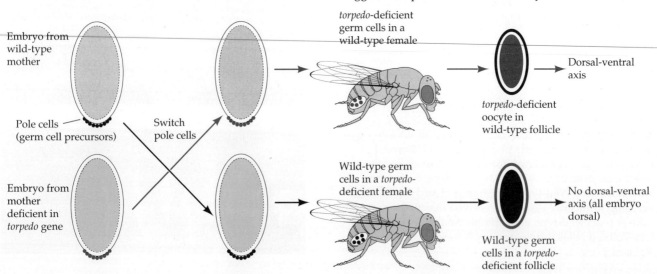

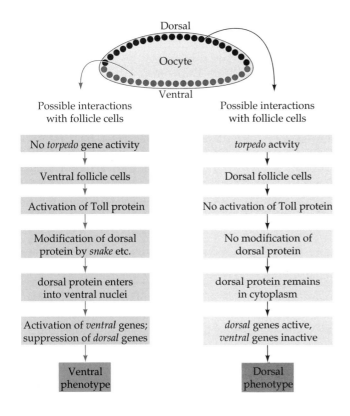

Dorsal

Oocyte

Ventral

Possible interactions with follicle cells	Possible interactions with follicle cells
No *torpedo* gene activity	*torpedo* actvity
Ventral follicle cells	Dorsal follicle cells
Activation of Toll protein	No activation of Toll protein
Modification of dorsal protein by *snake* etc.	No modification of dorsal protein
dorsal protein enters into ventral nuclei	dorsal protein remains in cytoplasm
Activation of *ventral* genes; suppression of *dorsal* genes	*dorsal* genes active, *ventral* genes inactive
Ventral phenotype	Dorsal phenotype

FIGURE 27

Possible sequence of morphogenetic events leading to the specification of the dorsal–ventral axis in the *Drosophila* egg. Interactions between the oocyte and the follicle cells cause the activation of the *torpedo* gene only in those follicle cells on the dorsal side. The phenotype of the dorsal follicle cells includes the production of an inhibitor of the Toll protein. In the absence of Toll function, the dorsal protein is not modified and it remains in the cytoplasm. These cells become the dorsal surface of the embryo. On the ventral side, Toll protein is active, and this initiates a cascade of protein activity that modifies dorsal protein, allowing it to enter the nuclei on the ventral side of the embryo. These cells become the ventral portion of the embryo.

to be active in forming the dorsal follicle cells. (Indeed, the follicle cells of *torpedo* mutants are also ventralized.) Since loss-of-function mutants of *torpedo* have ventralized embryos, the dorsal follicle cells may be producing a factor that inhibits the activation of Toll protein. Toll protein, which appears to be a receptor on the cell surface of the embryo, could respond to signals from the ventral follicle cells and locally activate the cascade of proteins that modifies dorsal protein. Thus, the dorsal protein would be modified in the portion of the embryo closest to the ventral follicle cells and would enter only those nuclei closest to these ventral follicle cells. The dorsal protein on the side

closest to the dorsal follicle cells would remain unmodified and would not enter into the nuclei (Figure 27). This hypothesis is not yet proven, and many points remain unclear. The nature of the interaction between oocyte and follicle cell is largely unknown, and so are the modifications that must be made to enable dorsal protein to enter the nucleus. Even so, we are now able to frame testable hypotheses that can be confirmed or invalidated by experimentation.

Research into the genetics of *Drosophila* pattern formation may well become the golden spike of developmental genetics. For decades, there have been two major gaps in attempts to integrate genetics and development. The first was our lack of knowledge concerning the molecules involved in specifying the primary embryonic axes. How is a back distinguished from a belly? The second gap in our knowledge concerned the interaction of the cytoplasmic determinants in the egg with the nuclei of the cells that incorporated them. The genes responsible for these processes are now known and some of the gene products have been identified. Within a short time we should be able to explain how cytoplasmic determinants specify the fate of embryonic cells.

Cytoplasmic localization of germ cell determinants

Cytoplasmically localized determinants are found throughout the animal kingdom. The most frequently observed determinants are those responsible for the determination of germ cell precursors, that is, those cells that give rise to gametes. Even in many embryos in which other aspects of early development are regulative, those cells containing a certain region of egg cytoplasm are destined to become germ cell precursors.

Germ cell determination in nematodes

Theodor Boveri (1862–1915) was the first person to look at an organism's chromosomes throughout its development. In so doing, he discovered a fascinating feature in the development of the roundworm *Parascaris aequorum* (formerly *Ascaris megalocephala*). This nematode has only two chromosomes per haploid cell, thus allowing detailed observations of the individual chromosomes. The cleavage plane of the first embryonic division is unusual in that it is equatorial, separating the animal from the vegetal half of the zygote (Figure 28A). More bizarre, however, is the behavior of the chromosomes in the subsequent division of these first two blastomeres. The ends of the chromosomes in the animal-derived blastomere fragment into dozens of pieces just before cleavage of this blastomere. This phenomenon is called CHROMOSOME DIMINUTION, because only a portion of the original chromosome survives. Numerous genes are lost in these cells when the chromosome fragments are not included in the newly formed nuclei (Tobler et al., 1972). Meanwhile, in the vegetal blastomere, the chromosomes remain normal. During the second division, the animal cell splits meridionally while the vegetal cell again divides equatorially. Both vegetally-derived cells have normal chromosomes. However, the chromosomes of the more animally located of these two vegetal blastomeres fragment before third division. Thus, at the 4-cell stage, only one cell—the most vegetal—contains a full set of genes. At successive cleavages, somatic nuclei are given off from this line, until at the 16-cell stage, there are only two cells that have undiminished chromosomes. One of these two blastomeres gives rise to the germ cells; the other blastomere eventually undergoes chromosome diminution and forms somatic cells. The chromosomes are kept intact only in those cells destined to form the germ line. If this were not the case, the genetic information would degenerate from one generation to the next. The cells that have undergone chromosome diminution generate the somatic cells.

Boveri has been called the last of the great "observers" of embryology and the first of the great experimenters. (Both Spemann and Wilson dedicated their major books to him.) Not content with observing the retention of the full chromosome complement solely by the germ cell precursors, he set out to test whether a specific region of cytoplasm protects the nuclei within it from diminution. If so, any nucleus happening to reside in this region should be protected. Boveri (1910) tested this by centrifuging *Parascaris* eggs shortly before their first cleavage. This treatment shifted the

FIGURE 28

Distribution of germ plasm (color) during cleavage of normal (A) and centrifuged (B) zygotes of *Parascaris*. (A) The germ plasm is normally conserved in the most vegetal blastomere, as shown by the lack of chromosomal diminution in that particular cell. Thus, at the 4-cell stage, the embryo has one stem cell for its gametes. (B) When the first cleavage is displaced 90° by centrifugation, both resulting cells have vegetal germ plasm and neither cell undergoes chromosome diminution. After the second cleavage, these two cells give rise to germinal stem cells. (After Waddington, 1966.)

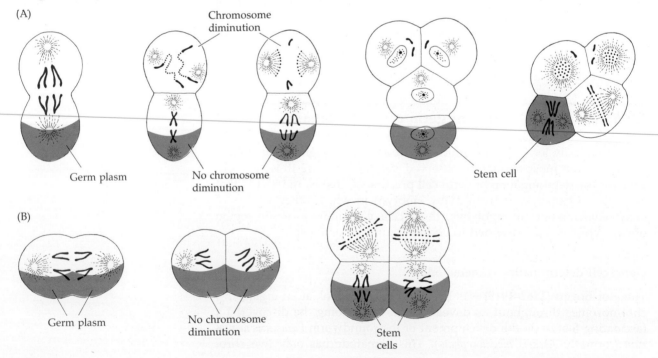

(A)

Chromosome diminution

Germ plasm

No chromosome diminution

Stem cell

(B)

Germ plasm

No chromosome diminution

Stem cells

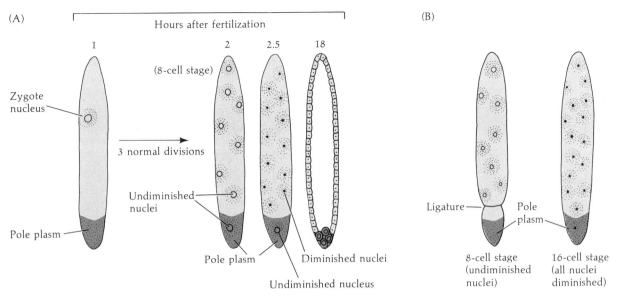

FIGURE 29

Segregation of the germ line and chromosome elimination in *Wachtiella*. (A) Chromosome elimination depends on location. The first three divisions are normal. Thereafter all nuclei except a single nucleus at the posterior pole lose 32 of their 40 chromosomes. (B) When a ligature is tied around the 8-cell zygote near its posterior end so that no nucleus enters the pole region, all nuclei undergo elimination at the 16-cell stage, even if the connection to the pole plasm is re-established. (After Geyer-Duszynska, 1959.)

orientation of the mitotic spindle. When the spindle forms perpendicular to its normal orientation, both resulting blastomeres should contain some of the vegetal cytoplasm (Figure 28B). Indeed, Boveri found that after the first division neither nucleus underwent chromosomal diminution. However, the next division was equatorial along the animal–vegetal axis. Here the resulting animal blastomeres both underwent diminution whereas the two vegetal cells did not. Boveri concluded that the vegetal cytoplasm contained a factor (or factors) that protected nuclei from chromosomal diminution and determined them to be germ cells.

Germ cell determination in insects

Certain insect eggs also contain a germ plasm that appears to act very similarly to the one observed in *Parascaris*. In the midge *Wachtiella persicariae*, most nuclei lose 32 of their original 40 chromosomes! However, two undiminished nuclei are found at the posterior pole of the egg and do not divide for a period of time (Figure 29). These two nuclei eventually give rise to the germ cells.* When nuclei are prevented by a ligature from migrating into the posterior pole region, every nucleus undergoes diminution, and the resulting midge is sterile. When the ligature is loosened and a diminished nucleus enters the posterior pole, germ cells are made but never differentiate into functional gametes (Geyer-Duszynska, 1959). Kunz and co-workers (1970) have shown that the eliminated chromatin contains genes that are active during germ cell production.

*This and the *Parascaris* example sound like perfect evidence for Weismann's hypothesis of the segregation of nuclear determinants (to be discussed in Chapter 8). These cases of chromosome diminution and elimination are exceptions to the general rule (Chapters 9 and 10) that the nuclei of differentiated cells retain unused genes; there is no evidence that different somatic cells in *Wachtiella* or *Parascaris* retain different parts of the genome.

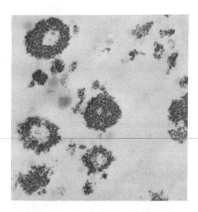

FIGURE 30
Electron micrograph of polar granules from particulate fraction from *Drosophila* pole cells. (Courtesy of A. P. Mahowald.)

FIGURE 31
Scanning electron micrograph of the pole cells of a *Drosophila* embryo just prior to completion of cleavage. The pole cells can be seen at the upper end of this picture. (Courtesy of A. P. Mahowald.)

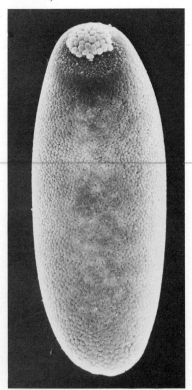

The germinal cytoplasm of insects is different from that of any other cytoplasm in the egg. Hegner (1911) found that when he removed or destroyed this region of beetle eggs before pole cell formation had occurred the resulting embryos had no germ cells and were sterile. This posterior POLE PLASM is conveniently marked with POLAR GRANULES (Figure 30). Although their role in germ cell determination is not known, their constant association with the pole plasm and the POLE CELLS derived from it makes them a convenient marker for this region. This region of pole cells is easily identifiable under a scanning electron microscope (Figure 31).

Recent work on the pole cell cytoplasm has focused primarily on *Drosophila* embryos. For the first two hours after fertilization, the *Drosophila* embryo develops as a SYNCYTIUM. The nuclei divide without any corresponding cellular division and eventually form a BLASTODERM LAYER that contains about 3500 nuclei. Each nucleus is eventually enclosed in a cellular membrane, so the blastoderm layer becomes the CELLULAR BLASTODERM. Transplantation experiments (Zalokar, 1971; Illmensee, 1968, 1972) have shown that all syncytial nuclei are equivalent and totipotent. The cells of the cellular blastoderm, however, are precisely determined. A similar conclusion was reached by Schubiger and Wood (1977), who ligated *Drosophila* eggs during various stages of their development. *Drosophila* nuclei do not undergo diminution, and any syncytial-stage nucleus can give rise to germ cells or to somatic cells. Their determination depends upon the region of the egg into which they migrate. One of these regions is the posterior pole plasm. Again, this pole plasm appears to contain the morphogenic determinants for germ cell production.

Geigy (1931) showed that irradiating the pole plasm with ultraviolet light produced sterile flies; Okada and co-workers (1974) extended this line of experimentation by showing that the addition of pole plasm from unirradiated donor embryos could cure the sterility of irradiated eggs (Figure 32). No other part of the cytoplasm could accomplish this reversal of sterility.

The autonomy of this cytoplasmic region and its ability to determine any *Drosophila* nucleus was shown in 1974 by the ingenious experiments of Karl Illmensee and Anthony Mahowald. In these experiments (outlined in Figure 33), an incredibly small amount (5–100 picoliters) of anucleate pole plasm was transferred from wild-type *Drosophila* eggs into the *anterior* pole of genetically marked eggs, prior to their cellularization. These donor eggs carried the chromosomal mutations *multiple wing hair* and *ebony*. After the cellular blastoderm had formed, the cells in the anterior pole of the recipient embryo resembled normal posterior pole cells, having incorporated the polar granules and developed a typical pole cell morphology. To test whether or not these cells had become functioning germ cell precursors, Mahowald and Illmensee transplanted these modified anterior cells into the posterior region of cleaving embryos containing their own genetically marked pole cells.* These new host embryos were marked by different mutations (recessives *yellow*, *white*, and *singed*). When these embryos developed, they all became flies bearing the mutations *yellow*, *white*, and *singed*. These flies were mated to other flies carrying the same mutations. In most cases (88/92), these matings produced individuals identical to both parents. However, in four cases, wild-type progeny emerged, indicating that some of the germ cells in these flies derived from the transplanted cells; the germ plasm from one embryo was able to cause the

*The reason for doing this was so that these cells might be incorporated into the developing gonads. There is evidence from other organisms that the pole plasm also contains determinants for the proper *migration* of germ cells (Züst and Dixon, 1977; Ikenishi and Kotani, 1979).

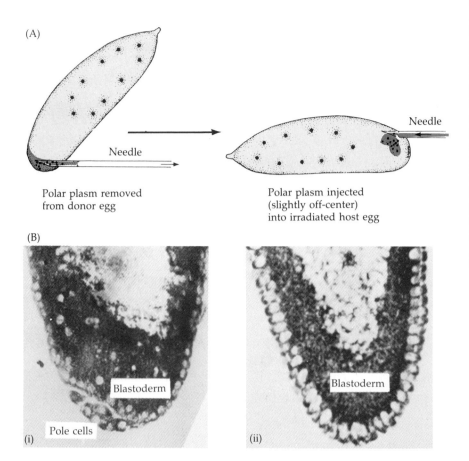

(A)

Needle

Polar plasm removed
from donor egg

Needle

Polar plasm injected
(slightly off-center)
into irradiated host egg

(B)

(i) Blastoderm / Pole cells

(ii) Blastoderm

(iii) Blastoderm / Pole cells

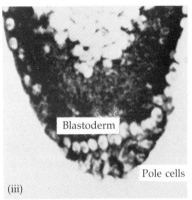

FIGURE 32
Ability of pole plasm to correct radiation-induced sterility. (A) Technique of pole plasm transplantation from unirradiated donor to irradiated host. (B) Longitudinal sections of the posterior portion of the *Drosophila* embryo fixed at the completion of cleavage. (i) Normal embryo with complete blastoderm and pole cells. (ii) Embryo irradiated during early cleavage. Blastoderm has formed, but pole cells are absent. (iii) Embryo irradiated during early cleavage but subsequently injected with pole plasm from normal embryos. Pole cells and blastoderm are both seen. (From Okada et al., 1974.)

anterior nuclei of another embryo to develop into functioning germ cells! This technique has also proven useful in showing the time of the germ cell determinant's localization. Illmensee and his colleagues (1976) found that this determinant is able to function before fertilization and becomes localized in the developing oocyte at about the same time that the yolk reaches the posterior end of the egg.

Nature has also provided confirmation of the importance of both pole plasm and the polar granules. Female *Drosophila* homozygous for the *grandchildless* mutation produce normal but sterile offspring [GG (male) × gg (female) → Gg (sterile)]. Mahowald and colleagues (1979) have shown that when such females are mated with normal males, the nuclei of the resulting embryos never migrate into the pole plasm of the egg. No pole cells are formed, and the resulting adults have no primordial germ cells with which to produce gametes. Another maternal-effect mutation—*agametic*—causes the absence of germ cells in about half the gonads of offspring derived from homozygous female flies. Here, the normal number of pole cells form, but the polar granules degenerate shortly after fertilization (Engstrom et al., 1982). Transplantation experiments demonstrate that the defect is in the polar cytoplasm and not in the ovarian environment. Thus, we now have fairly strong evidence that the polar granules are directly concerned with germ cell determination.

In *Drosophila*, these polar granules have been isolated (Waring et al., 1978) and appear to be composed of both protein and RNA. The protein (there only seems to be one major protein in the granules) has a molecular weight of 95,000; it also has basic properties. It is synthesized anew during oogenesis, so the polar granules are not inherited directly through the egg

FIGURE 33
Ability of germ plasm to determine the fate of cells that contain it. Pole plasm from wild-type *Drosophila* eggs is transplanted into the anterior (but not the posterior) pole of genetically marked (mutant) embryos. Subsequently, the cells in the anterior pole resemble the normal germ cell precursors seen at the posterior pole. These anterior pole cells (color) are then transplanted into host posterior regions (marked with different mutations) so that their descendants can migrate to the gonads. When these flies are mated to other flies with the second series of mutations (which did not have transplanted cells), some of the progeny are wild-type, indicating that the germ cells of these progeny come from a nonparental strain of fly. (After Illmensee and Mahowald, 1974.)

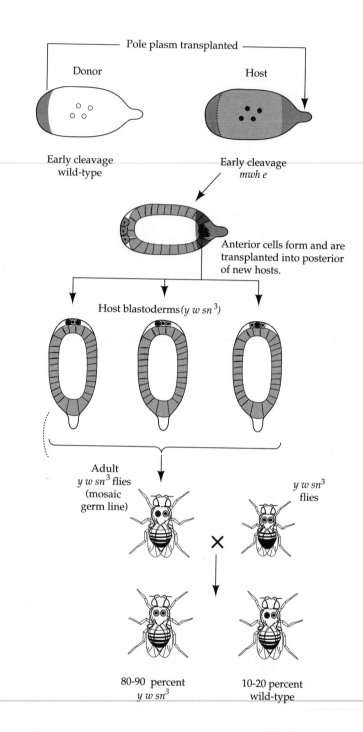

cytoplasm. The RNA component is present in the oocyte polar granules but is no longer seen once the pole cells are formed (Mahowald, 1971a,b). It is probable that this RNA component is extremely important in germ cell formation. Okada and Kobayashi (1987) injected RNA from *Drosophila* eggs and early cleavage embryos into embryos whose pole cells were sterilized by UV irradiation. The resulting embryos formed pole cells and developed into flies that contained germ cells.

Polar granules change during development. Before fertilization, the dense, membranous granules cluster around the mitochondria and then disperse before the nuclei reach the pole. The pole cells eventually incorporate these granules. When the pole cells migrate toward the gonadal ridge, the granules further decondense into filaments called NUAGE. This

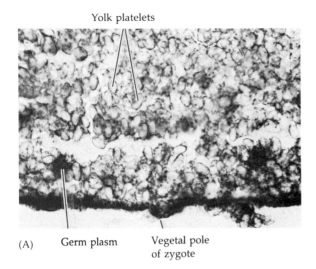

Yolk platelets

(A)   Germ plasm   Vegetal pole
of zygote

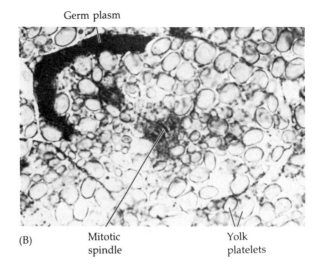

Germ plasm

(B)   Mitotic   Yolk
spindle   platelets

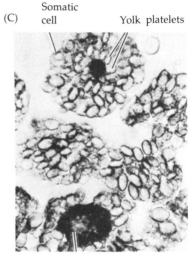

(C)   Somatic cell   Yolk platelets

Germ cell

FIGURE 34

Germ plasm of frog embryos. (A) Germ plasm (dark regions) near the ventral pole of a newly fertilized zygote. (B) Germ plasm-containing cell in endodermal region of blastula in mitotic anaphase. Note the germ plasm entering into only one of the two yolk-laden daughter cells. (C) Primordial germ cell and somatic cells near the floor of the blastocoel in early gastrula. (Photographs courtesy of A. Blackler.)

material clusters about the nuclear envelope. This behavior is seen throughout the animal kingdom (Eddy, 1975) and is probably important in gametogenesis.

Germ cell determination in amphibians

Cytoplasmic localization of germ cell determinants has also been observed in vertebrate embryos. Bounoure (1934) showed that the vegetal region of fertilized frog eggs contains a material with staining properties similar to that of *Drosophila* pole plasm (Figure 34). He was able to trace this cortical cytoplasm into the few cells in the presumptive endoderm that would normally migrate into the genital ridge. Blackler (1962) showed that these cells are the primordial germ cell precursors. He removed the endodermal region from the neurula of one frog and inserted it into the neurula endoderm of another (Figure 35). The donor frogs were genetically distinguishable because their cells contained only one nucleolus instead of the usual two. The host frogs were wild-type (i.e., two nucleoli per nucleus). After the operation, the donor frogs were sterile, indicating that the neurula could not regulate for the production of germ cells and that the correct region had been excised. The host frogs, however, were fertile. Moreover, when these frogs were mated to normal adults, the offspring were a mixture of one-nucleolus and two-nucleoli frogs. Therefore, some of the gametes came from the endodermal region of the *donor* embryo. (If none had come from the donor, one would get 100 percent two-nucleoli frogs. If every germ cell were derived from the donor, one would expect a 1:1 ratio, as is shown in Figure 35. In several experiments, this latter figure was actually obtained.)

With his associates, Bounoure (1939) showed that when ultraviolet light is applied to the vegetal poles of fertilized frog eggs, the resulting adults are sterile but otherwise normal. In 1966, L. Dennis Smith was able to extend these studies by irradiating various regions of the frog embryo

FIGURE 35

Demonstration of primordial germ cells in early tadpole endoderm. A block of ventral tissue from a strain of frog with one nucleolus per cell is transplanted into the analogous region in a host tadpole with two nucleoli per cell. The donor tadpole metamorphoses into a sterile adult, while the host tadpole develops into a frog with both host and donor gametes. This fact can be demonstrated by crossing that frog with a normal (two-nucleoli) adult; the results of such a cross are shown in the flow chart. Some of the offspring have only one nucleolus per cell, indicating that they received one nucleolus from one parent (the two-nucleoli frog) and no nucleolus from the other (the two-nucleoli frog with some one-nucleolus germ cells). (After Blackler, 1966.)

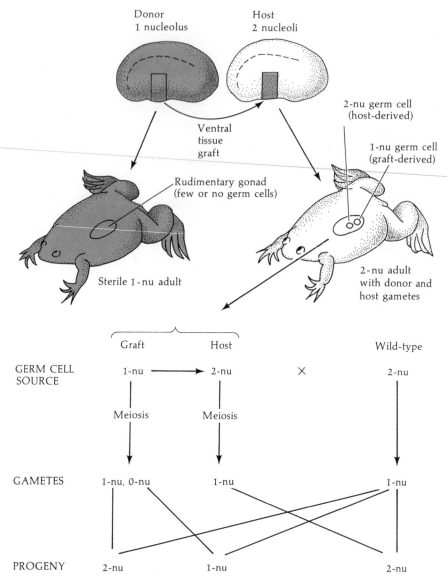

with ultraviolet light. He found that ultraviolet irradiation of the animal hemisphere did not affect normal development, whereas irradiation of the vegetal pole produced tadpoles that lacked germ cells. No large primordial germ cells were seen to enter the genital ridge. Smith showed the importance of this region of cytoplasm by transferring samples of cytoplasm from normal zygotes into the ventral region of irradiated zygotes. When the animal region cytoplasm was added to the irradiated ventral regions, there was no effect, and the resulting tadpoles lacked the large germ cells. When vegetal pole cytoplasm was added, however, primordial germ cells were ultimately seen in the genital ridge. More recent studies (Züst and Dixon, 1977) have shown that primordial germ cells do arrive at the genital ridge of tadpoles whose germ plasm is irradiated at the 2- or 4-cell stage. However, there are significantly fewer of these cells; they are only about one-tenth the size of normal primordial germ cells, have aberrantly shaped nuclei, and enter the genital ridge much later than their counterparts from unirradiated eggs. So, like the *Drosophila* pole plasm, the cytoplasm from the ventral region of frog zygotes contains a determinant for germ cell formation and migration that is sensitive to ultraviolet irradiation and that can be transferred through the fluid portion of the cytoplasm.

SUMMARY

We have evidence in certain organisms that the determination of a cell's fate is due to the portion of egg cytoplasm that it acquires during cleavage. Such a cell differentiates independently of other cells, and organisms that utilize such a mechanism tend toward a mosaic, or determinate, type of development. Embryos exhibiting this form of development include those of molluscs, tunicates, and nematodes. The placement of morphogens within the egg cytoplasm, their redistribution during egg development and fertilization, and the patterns of cell cleavage are important for determining which cell acquires which fate. Each of these phenomena is a function of the egg. Although most of the development of these organisms follows the mosaic pattern, some interactive determination is also seen. In tunicates, the nervous system and some of the muscles are formed by inductive interactions between blastomeres, and snails and nematodes also have certain organs formed in this interactive manner. In the next chapter, we will look at certain organisms in which inductive interactions constitute the primary mechanism of determining cell fate.

LITERATURE CITED

Alberts, B., Bray, D., Lewis, J., Raff, M., Roberts, K. and Watson, J. D. 1983. *Molecular Biology of the Cell*. Garland, New York.

Anderson, K., Bokla, L. and Nüsslein-Volhard, C. 1985. Establishment of dorsal–ventral polarity in the *Drosophila* embryo: The induction of polarity by the *Toll* gene product. *Cell* 42: 791–798.

Anderson, K. V. and Nüsslein-Volhard, C. 1984. Information for the dorsal–ventral pattern of the *Drosophila* embryo is stored as maternal mRNA. *Nature* 311: 223–227.

Atkinson, J. W. 1987. An atlas of light micrographs of normal and lobeless larvae of the marine gastropod *Ilyanassa obsoleta*. *Int. J. Invert. Reprod. Dev.* 9: 169–178.

Austin, J. and Kimble, J. 1987. glp-1 is required in the germ line for regulation of the decision between mitosis and meiosis in C. *elegans*. *Cell* 51: 589–600.

Blackler, A. W. 1962. Transfer of primordial germ cells between two subspecies of *Xenopus laevis*. *J. Embryol. Exp. Morphol.* 10: 641–651.

Blackler, A. W. 1966. The role of a "germinal plasm" in the formation of primordial germ cells in *Rana pipiens*. *Dev. Biol.* 14: 330–347.

Bonnet, C. 1764. *Contemplation de la Nature*. Marc-Michel Ray, Amsterdam.

Bounoure, L. 1934. Recherches sur la lignée germinale chez la grenouille rousse aux premiers stades du développement. *Ann. Sci. Natl. 10e wer.* 17: 67–248.

Bounoure, L. 1939. *L'origine des Cellules Reproductries et le Probleme de la Lignée Germinale*. Gauthier-Villars, Paris.

Boveri, T. 1910. Über die Teilung centrifugierter Eier von *Ascaris megalocephala*. *Wilhelm Roux Arch. Entwicklungsmech. Org.* 30: 101–125.

Brandhorst, B. P. and Newrock, K. M. 1981. Post-transcriptional regulation of protein synthesis in *Ilyanassa* embryos and isolated polar lobes. *Dev. Biol.* 83: 250–254.

Brenner, S. 1974. The genetics of *Caenorhabditis elegans*. *Genetics* 77: 71–94.

Brenner, S. 1979. Cited in H. F. Judson, *The Eighth Day of Creation*. Simon and Schuster, New York, p. 219.

Cassirer, E. 1950. Developmental mechanics and the problem of cause in biology. *In* E. Cassirer (ed.), *The Problem of Knowledge*. Yale University Press, New Haven.

Cervera, M., Dreyfuss, G. and Penman, S. 1981. Messenger RNA is translated when associated with the cytoskeletal framework in normal and VSV-infected HeLa cells. *Cell* 23: 113–120.

Chabry, L. M. 1887. Contribution a l'embryologie normale tératologique des ascidies simples. *J. Anat. Phys. Norm. Pathol.* 23: 167–321.

Churchill, F. B. 1973. Chabry, Roux, and the experimental method in nineteenth century embryology. *In* R. N. Giere and R. S. Westfall (eds.), *Foundations of Scientific Method: The Nineteenth Century*. Indiana University Press, Bloomington, pp. 161–205.

Clement, A. C. 1962. Development of *Ilyanassa* following removal of the D macromere at successive cleavage stages. *J. Exp. Zool.* 149: 193–215.

Clement, A. C. 1968. Development of the vegetal half of the *Ilyanassa* egg after removal of most of the yolk by centrifugal force, compared with the development of animal halves of similar visible composition. *Dev. Biol.* 17: 165–186.

Clement, A. C. 1986. The embryonic value of the micromeres in *Ilyanassa obsoleta*, as determined by deletion experiments. III. The third quartet cells and the mesentoblast cell, 4d. *Int. J. Invert. Reprod. Dev.* 9: 155–168.

Collier, J. R. 1983. The biochemistry of molluscan development. *In* N. H. Verdonk, J. A. M. van den Biggelaar and A. S. Tompa (eds.), *The Mollusca*, Vol. 3. Academic, New York, pp. 215–252.

Collier, J. R. 1984. Protein synthesis in normal and lobeless gastrulae of *Ilyanassa obsoleta*. *Biol. Bull.* 167: 371–377.

Conklin, E. G. 1905a. The organization and cell lineage of the ascidian egg. *J. Acad. Nat. Sci. Phila.* 13: 1–119.

Conklin, E. G. 1905b. Organ-forming substances in the eggs of ascidians. *Biol. Bull.* 8: 205–230.

Conklin, E. G. 1905c. Mosaic development in ascidian eggs. *J. Exp. Zool.* 2: 145–223.

Craig, M. M. and Morrill, J. B. 1986. Cellular arrangements and surface topology during early development in embryos of *Ilyanassa obsoleta*. *Int. J. Invert. Reprod. Dev.* 9: 209–228.

Crowther, R. J. and Whittaker, J. R. 1984. Differentiation of histotypic ultrastructural features in cells of cleavage-arrested early ascidian embryos. *Wilhelm Roux Arch. Dev. Biol.* 194: 87–98.

Dareste, C. 1877. *Recherches sur la Production Artificielle des Monstruosités ou Essais de Teratogenie Experimentale*. Reinwald, Paris.

DeLotto, R. and Spierer, P. 1986. A gene required for the specification of dorsal–ventral pattern in *Drosophila* appears to encode a serine protease. *Nature* 323: 688–692.

Deno, T., Nishida, H. and Satoh, N. 1984. Autonomous muscle cell differentiation in partial ascidian embryos according to the newly verified cell lineages. *Dev. Biol.* 104: 322–328.

Driever, W., Siegel, V. and Nüsslein-Volhard, C. 1990. Autonomous determination of anterior structures in the early *Drosophila* embryo by the *bicoid* morphogen. *Development* 109: 811–820.

Eddy, E. M. 1975. Germ plasm and the differentiation of the germ line. *Int. Rev. Cytol.* 43: 229–280.

Edgar, L. G. and McGhee, J. D. 1986. Embryonic expression of a gut-specific esterase in *Caenorhabditis elegans*. *Dev. Biol.* 114: 109–118.

Engstrom, L., Caulton, J. H., Underwood, E. M. and Mahowald, A. P. 1982. Developmental lesions in the agametic mutant of *Drosophila melanogaster*. *Dev. Biol.* 91: 163–170.

Fischer, J.-L. and Smith, J. 1984. French embryology and the "mechanics of development" from 1887 to 1910: L. Chabry, Y. Delage and E. Bataillon. *Hist. Phil. Life Sci.* 6: 25–39.

Geigy, R. 1931. Action de l'ultra-violet sur le pôle germinal dans l'oeuf de *Drosophila melanogaster* (castration et mutabilité). *Rev. Suisse Zool.* 38: 187–288.

Geyer-Duszynska, I. 1959. Experimental research on chromosome diminution in *Cecidomiidae* (Diptera). *J. Exp. Zool.* 141: 381–441.

Gould, S. J. 1977. *Ontogeny and Phylogeny*. Belknap Press, Cambridge, MA.

Guerrier, P., van den Biggelaar, J. A. M., Dongen, C. A. M. , and Verdonk, N. H. 1978. Significance of the polar lobe for the determination of dorsoventral polarity in *Dentalium vulgare* (da Costa). *Dev. Biol.* 63: 233–242.

Hegner, R. W. 1911. Experiments with chrysomelid beetles. III. The effects of killing parts of the eggs of *Leptinotarsa decemlineata*. *Biol. Bull.* 20: 237–251.

Henry, J. J. and Martindale, M. Q. 1987. The organizing role of the D quadrant as revealed through the phenomenon of twinning in the polychaete *Chaetopterus variopedatus*. *Wilhelm Roux Arch. Dev. Biol.* 196: 449–510.

Hill, D. P. and Strome, S. 1987. An analysis of the role of microfilaments in the establishment and maintainance of asymmetry in *Caenorhabditis elegans* zygotes. *Dev. Biol.* 125: 75–84.

Hill, D. P. and Strome, S. 1990. Brief cytochalasin-induced disruption of microfilaments during a critical interval in 1-cell *C. elegans* embryos alters the positioning of developmental instructions to the 2-cell embryo. *Development* 108: 159–172.

Ikenishi, K. and Kotani, M. 1979. Ultraviolet effects on presumptive primordial germ cells (pPGCs) in *Xenopus laevis* after the cleavage stage. *Dev. Biol.* 69: 237–246.

Illmensee, K. 1968. Transplantation of embryonic nuclei into unfertilized eggs of *Drosophila melanogaster*. *Nature* 219: 1268–1269.

Illmensee, K. 1972. Developmental potencies of nuclei from cleavage, preblastoderm, and syncytial blastoderm transplanted into unfertilized eggs of *Drosophila melanogaster*. *Wilhelm Roux Arch. Entwicklungsmech. Org.* 170: 267–298.

Illmensee, K. and Mahowald, A. P. 1974. Transplantation of posterior polar plasm in *Drosophila*. Induction of germ cells at the anterior pole of the egg. *Proc. Natl. Acad. Sci. USA* 71: 1016–1020.

Illmensee, K., Mahowald, A. P. and Loomis, M. R. 1976. The ontogeny of germ plasm during oogenesis in *Drosophila*. *Dev. Biol.* 49: 40–65.

Jeffery, W. R. 1984. Spatial distribution of messenger RNA in the cytoskeletal framework of ascidian eggs. *Dev. Biol.* 103: 482–492.

Jeffery, W. R. 1985. Identification of proteins and mRNAs in isolated yellow crescents of ascidian eggs. *J. Embryol. Exp. Morphol.* 89: 275–287.

Jeffery, W. R., Bates, W. R., Beach, R. L. and Tomlinson, C. R. 1986. Is maternal mRNA a determinant of tissue-specific proteins in ascidian embryos? *J. Embryol. Exp. Morphol.* 97 [Suppl.] 1–14.

Jeffery, W. R. and Meier, S. 1983. A yellow crescent cytoskeletal domain in ascidian eggs and its role in early development. *Dev. Biol.* 96: 125–143.

Just, E. E. 1936. Quoted in Harrison, R. G. 1937. Embryology and its relations. *Science* 85: 369–374.

Kimble, J. 1981. Lineage alterations after ablation of cells on the somatic gonad of *Caenorhabditis elegans*. *Dev. Biol.* 87: 286–300.

Kimble, J. and Hirsch, D. 1979. The postembryonic cell lineages of the hermaphrodite and male gonads in *Caenorhabditis elegans*. *Dev. Biol.* 70: 396–417.

Kölreuter, J. G. 1761–1766. *Vorläufige Narchricht von einigen das Geschlecht der Pflanzen beteffenden Versuchen und Beobachtungen, nebst Fortsetzugen 1, 2 und 3.* Leipzig.

Kunz, W., Trepte, H. H. and Bier, K. 1970. On the function of the germ line chromosomes in the oogenesis of *Wachtiella persiariae* (Cecidomyiidae). *Chromosoma* 30: 180–192.

Laufer, J. S., Bazzicalupo, P. and Wood, W. B. 1980. Segregation of developmental potential in early embryos of *Caenorhabditis elegans*. *Cell* 19: 569–577.

Lenoir, T. 1980. Kant, Blumenbach, and vital materialism in German biology. *Isis* 71: 77–108.

Mahowald, A. P. 1971a. Polar granules of *Drosophila*. III. The continuity of polar granules during the life cycle of *Drosophila*. *J. Exp. Zool.* 176: 329–343.

Mahowald, A. P. 1971b. Polar granules of *Drosophila*. IV. Cytochemical studies showing loss of RNA from polar granules during early stages of embryogenesis. *J. Exp. Zool.* 176: 329–343.

Mahowald, A. P., Caulton, J. H. and Gehring, W. J. 1979. Ultrastructural studies of oocytes and embryos derived from female flies carrying the *grandchildless* mutation in *Drosophila subobscura*. *Dev. Biol.* 69: 118–132.

Meedel, T. H. and Whittaker, J. R. 1983. Development of transcriptionally active mRNA for larval muscle acetylcholinesterase during ascidian embryogenesis. *Proc. Natl. Acad. Sci. USA* 80: 4761–4765.

Meedel, T. H. and Whittaker, J. R. 1984. Lineage segregation and developmental autonomy in expression of functional muscle acetylcholinesterase in the ascidian embryo. *Dev. Biol.* 105: 479–487.

Meedel, T. H., Crowthier, R. J., and Whittaker, J. R. 1987. Determinative properties of muscle lineages in ascidian embryos. *Development* 100: 245–260.

Melton, D. 1987. Translocation of a localized maternal mRNA to the vegetal pole of *Xenopus* oocytes. *Nature* 328: 80–82.

Moon, R. T., Nicosia, R. F., Olsen, C., Hille, M. and Jeffery, W. R. 1983. The cytoskeletal framework of sea urchin eggs and embryos: Developmental changes in the association of messenger RNA. *Dev. Biol.* 95: 447–458.

Morgan, T. H. 1927. *Experimental Embryology.* Columbia University Press, New York.

Newrock, K. M. and Raff, R. A. 1975. Polar lobe specific regulation of translation in embryos of *Ilyanassa obsoleta*. *Dev. Biol.* 42: 242–261.

Nishida, H. 1987. Cell lineage analysis in ascidian embryos by intracellular injection of a tracer enzyme. III. Up to the tissue restricted stage. *Dev. Biol.* 121: 526–541.

Nishikata, T., Mita-Miyazawa, I., Deno, T. and Satoh, N. 1987. Monoclonal antibodies against components of the myoplasm of the ascidian *Ciona intestinalis* partially block the development of muscle-specific acetylcholinesterase. *Development* 100: 577–586.

Okada, M., Kleinman, I. A. and Schneiderman, H. A. 1974. Restoration of fertility in sterilized *Drosophila* eggs by transplantation of polar cytoplasm. *Dev. Biol.* 37: 43–54.

Okada, M. and Kobayashi, S. 1987. Maternal messenger RNA as a determinant of pole cell formation in *Drosophila* embryos. *Dev. Growth Diff.* 29: 185–192.

Ortolani, G. 1959. Richerche sulla induzione del sistema nervoso nelle larve delle Ascidie. *Boll. Zool.* 26: 341–348.

Priess, R. A. and Thomson, J. N. 1987. Cellular interactions in early *C. elegans* embryos. *Cell* 48: 241–250.

Priess, R. A., Schnabel, H. and Schnabel, R. 1987. The *glp-1* locus and cellular interactions in early *C. elegans* embryos. *Cell* 51: 601–611.

Reverberi, G. and Minganti, A. 1946. Fenomeni di evocazione nello sviluppo dell'uovo di Ascidie. Risultati dell'indagine spermentale sull'ouvo di *Ascidiella aspersa* e di *Ascidia malaca* allo stadio di 8 blastomeri. *Pubbl. Staz. Zool. Napoli* 20: 199–252. (Quoted in Reverberi, 1971, p. 537.

Roe, S. 1981. *Matter, Life, and Generation: Eighteenth-Century Embryology and the Haller–Wolff Debate*. Cambridge University Press, Cambridge.

Roth, S., Stein, D., Nüsslein-Volhard, C. 1989. A gradient of nuclear localization of the *dorsal* protein determines dorsoventral pattern in the *Drosophila* embryo. *Cell* 59: 1189–1202.

Rushlow, C. A., Han, K., Manley, J. L. and Levine, M. 1989. The graded distribution of the *dorsal* morphogen is initiated by selective nuclear transport in *Drosophila*. *Cell* 59: 1165–1177.

Satoh, N. 1979. On the "clock" mechanism determining the time of tissue- specific development during ascidian embryogenesis. I. Acetylcholinesterase development in cleavage-arrested embryos. *J. Embryol. Exp. Morphol.* 54: 131–139.

Schubiger, G. and Wood, W. J. 1977. Determination during early embryogenesis in *Drosophila melanogaster*. *Am. Zool.* 17: 565–576.

Schüpbach, T. 1987. Germ-line and soma cooperate during oogenesis to establish the dorsoventral pattern of egg sell and embryo in *Drosophila melanogaster*. *Cell* 49: 699–707.

Schüpbach, T. 1990. Oocyte/follicle cell interactions in *Drosophila* patterning. Forty-Ninth Annual Symposium of the Society of Developmental Biology. Alan R. Liss, New York.

Smith, L. D. 1966. The role of a "germinal plasm" in the formation of primordial germ cells in *Rana pipiens*. *Dev. Biol.* 14: 330–347.

Steward, R. 1989. Relocalization of the dorsal protein from the cytoplasm to the nucleus correlates with its function. *Cell* 59: 1179–1188.

Strome, S. and Wood, W. B. 1983. Generation of asymmetry and segregation of germ-like granules in early *Caenorhabditis elegans* embryos. *Cell* 35: 15–25.

Sulston, J. and Horvitz, H. R. 1977. Postembryonic cell lineages of the nematode *Caenorhabditis elegans*. *Dev. Biol.* 56: 110–156.

Sulston, J. E., Schierenberg, J., White, J. and Thomson, N. 1983. The embryonic cell lineage of the nematode *Caenorhabditis elegans*. *Dev. Biol.* 100: 64–119.

Tobler, H., Smith, K. D. and Ursprung, H. 1972. Molecular aspects of chromatin elimination in *Ascaris lumbricoides*. *Dev. Biol.* 27: 190–203.

Tung, T. C., Wu, S. C., Yel, Y. F., Li, K. S. and Hsu, M. C. 1977. Cell differentiation in ascidians studied by nuclear transplantation. *Scientia Sinica* 20: 222–233.

Verdonk, N. H. and Cather, J. N. 1983. Morphogenic determination and differentiation. In N. H. Verdonk, J. A. M. van den Biggelaar and A. S. Tompa (eds.), *The Mollusca*. Academic Press, New York, pp. 215–252.

Waddington, C. H. 1966. *Principles of Development and Differentiation*. Macmillan, New York.

Waring, G. L., Allis, C. D. and Mahowald, A. P. 1978. Isolation of polar granules and the identification of polar granule–specific protein. *Dev. Biol.* 66: 197–206.

Weeks, D. L., Rebagliati, M. R., Harvey, R. P. and Melton, D. A. 1985. Localized maternal mRNAs in *Xenopus laevis* eggs. *Cold Spring Harbor Symp. Quant. Biol.* 50: 21–30.

Whittaker, J. R. 1973a. Segregation during ascidian embryogenesis of egg cytoplasmic information for tissue–specific enzyme development. *Proc. Natl. Acad. Sci. USA* 70: 2096–2100.

Whittaker, J. R. 1973b. Evidence for the localization of an RNA template for endodermal alkaline phosphatase in an ascidian egg. *Biol. Bull.* 145: 459–460.

Whittaker, J. R. 1977. Segregation during cleavage of a factor determining endodermal alkaline phosphatase development in ascidian embryos. *J. Exp. Zool.* 202: 139–153.

Whittaker, J. R. 1979. Cytoplasmic determinants of tissue differentiation in the ascidian egg. *In* S. Subtelny and I. R. Konigsberg (eds.), *Determinants of Spatial Organization*. Academic Press, New York, pp. 29–51.

Whittaker, J. R. 1982. Muscle cell lineage can change the developmental expression in epidermal lineage cells of ascidian embryos. *Dev. Biol.* 93: 463–470.

Whittaker, J. R. 1987. Cell lineages and determinants of cell fate in development. *Am. Zool.* 27: 607–622.

Whittaker, J. R., Ortolani, G. and Farinella-Ferruzza, N. 1977. Autonomy of acetylcholinesterase differentiation in muscle lineage cells in ascidian embryos. *Dev. Biol.* 55: 196–200.

Wilson, E. B. 1904. Experimental studies on germinal localization. I. The germ regions in the egg of *Dentalium*. II. Experiments on the cleavage- mosaic in *Patella* and *Dentalium*. *J. Exp. Zool.* 1: 1–72.

Wolff, K. F. 1767. De formatione intestinorum praecipue. *Novi Commentarii Academine Scientarum Imperialis Petropolitanae*.

Zalokar, M. 1971. Transplantation of nuclei in *Drosophila melanogaster*. *Proc. Natl. Acad. Sci. USA* 68: 1539–1541.

Zumbe, A., Stahli, C. and Trachsel, H. 1982. Association of a M_r 50,000 cap-binding protein with the cytoskeleton in baby hamster kidney cells. *Proc. Natl. Acad. Sci. USA* 79: 2927–2931.

Züst, B. and Dixon, K. E. 1977. Events in the germ cell lineage after entry of the primordial germ cells into the genital ridges in normal and UV-irradiated *Xenopus laevis*. *J. Embryol. Exp. Morphol.* 41: 33–46.

8

Determination by progressive cell–cell interactions

Actually, there begins in each of the three [germ] layers a particular development, and each one strives to achieve its goal; only each is not yet sufficiently independent to produce that for which it is destined. Each one still needs the help of its companions; and therefore, all three, until each reaches a specific level, work mutually together although destined for different ends.

—CHRISTIAN PANDER (1817)

We are standing and walking with parts of our body which could have been used for thinking had they developed in another part of the embryo.

—HANS SPEMANN (1943)

In the last chapter we saw that determination can be caused by the partitioning of specific cytoplasmic substances into the blastomeres during embryonic cleavage. We also saw that this mode of determination was not sufficient to account for all the determinative events of embryogenesis. In some instances, interaction between two blastomeres was required for a particular type of tissue to be formed. In this chapter, we will see that this interactive mode of determination is extremely important, especially in deuterostomes such as sea urchins and vertebrates. The discovery of this type of determination has its roots in the failure of the mosaic theories of development formulated in Germany at the turn of the last century.

August Weismann: The germ plasm theory

In 1883, August Weismann began proposing a theory that integrated such diverse biological phenomena as heredity, development, regeneration, sexual reproduction, and evolution by natural selection. Weismann was a brilliant chemist and physician who turned his attention to the problems of embryonic development and metamorphosis. In 1863, when a severe eye disorder curtailed his microscopy, he used his time to ponder the ways by which the germ cells produce their differentiated progeny. This quest freed Weismann from a strict Darwinian type of explanation (i.e., the use of embryology to support theories of evolutionary descent) and set embryology on a new, physiological course that emphasized experimental manipulation over comparative observation.

The cytological work of the 1870s had given Weismann important new information concerning sexual generation. Hermann Fol and Richard Hertwig had independently observed the union of egg and sperm and their pronuclei. Researchers such as van Beneden and Strasburger had demonstrated that each somatic nucleus contained a defined number of chromosomes depending on the species and that this number was halved during germ cell maturation and restored during fertilization. Using these scant data, Weismann proposed a mechanical model of cellular differentiation, the GERM PLASM THEORY.

First, the sperm and egg were posited to provide equal chromosomal contributions, both quantitatively and qualitatively, to the new organism. Second, chromosomes were postulated to carry the inherited potentials of this new organism and were the basis for the continuity between generations. (Embryologists were thinking in these terms some 15 years before the rediscovery of Mendel's work. Weismann [1892, 1893] also speculated that these nuclear determinants of inheritance functioned by elaborating substances that became active in the cytoplasm!) However, not all the determinants on the chromosomes were thought to enter every cell of the embryo; instead of dividing equally, the chromosomes were hypothesized to divide in such a way that different nuclear determinants entered different cells. Weismann, a former military physician, likened these determinants to specialized army brigades. Whereas the fertilized egg would carry the full complement of all determinants, certain cells would retain the "blood-forming" brigades while other cells would retain the "muscle-forming" determinants. Only in the nuclei of those cells destined to become gametes (the germ cells) were all types of determinants retained. The nuclei of all other cells had only a fraction of the original determinant types.

Weismann's hypothesis proposed the continuity of the germ plasm and the diversity of the somatic lines. Differentiation was due to the "segregation of the nuclear determinants" into the various cell types and was accomplished by "the architecture of the germ plasm," that is, the germ cell nucleus. The chromosomes, while appearing equal in all cells, would be unequal in their qualities. The germ cell line was totally *independent* of the somatic cells. Hence, there could be no inheritance of characteristics acquired by the somatic cells. Weismann obtained support for this model by cutting off the tails of newborn mice for nineteen generations. The mice of each succeeding generation had tails of normal length, indicating that the germ line was insulated from insults to the somatic tissue.*

Weismann's germ plasm theory is depicted in Figure 1. It emphasizes the continuity and immortality of the germ line contrasted with the temporary nature of the adult organism, showing, as physiologist Michael

*At this time, the major alternative view was that of pangenesis. This theory, advocated as a "provisional hypothesis" by Charles Darwin, posited that each somatic cell contained particles ("pangenes") that migrated back into the sex cells to provide for the transmission of that cell's characteristics. According to this theory, Weismann should have obtained mice with shorter tails. More recently, Thomas Jukes, commenting on these results, quoted Hamlet's intuition that "there's a divinity that shapes our ends, rough-hew them how we will." It should be noted that the independence of the germ line from the somatic line is not absolute in all organisms. In sponges, flatworms, hydroids, and colonial tunicates, germ cells can develop from somatic tissues, and genetic changes made to those somatic tissues can be inherited (Berrill and Liu, 1948; Buss, 1987).

FIGURE 1

Weismann's theory of inheritance. The germ cell gives rise to the differentiating somatic cells of the body (indicated in color), as well as to new germ cells. (From Wilson, 1896.)

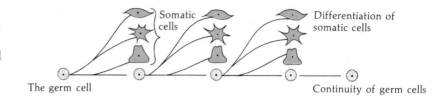

Somatic cells

Differentiation of somatic cells

The germ cell

Continuity of germ cells

Foster noted, that "the animal body is in reality a vehicle for ova." E. B. Wilson, who claimed that his remarkable textbook *The Cell in Development and Inheritance* (1896) grew out of Weismann's hypothesis, also saw the implications of Weismann's scheme:

> The death of the individual involves no breach of continuity in the series of cell divisions by which the life of the race flows onwards. The individual dies, it is true, but the germ-cells live on, carrying with them, as it were, the traditions of the race from which they have sprung, and handing them on to their descendants.

Wilhelm Roux: Mosaic development

Weismann had intuited that chromosomes were the bearers of the inherited information for development. More importantly, though, he had proposed a hypothesis that could be tested immediately. Weismann had claimed that when the first cleavage division separated the future right half of the embryo from the future left half, there would be a separation of "right" determinants from "left" determinants in the resulting blastomeres. This assertion was tested by Wilhelm Roux, a young German embryologist. Roux, too, had been an ardent Darwinian (his first major work in embryology proposed a competition for survival between embryonic cells), but, like Weismann, he felt that development needed to be studied analytically. In 1888, Roux published the results of a series of experiments in which he took 2- and 4-cell frog embryos and destroyed some of the cells of each embryo with a hot needle. Weismann's hypothesis would predict the formation of right or left half-embryos. Roux obtained half-morulae, just as Weismann had predicted (Figure 2). These developed into half-neurulae having a complete right or left side, with one medullary fold, one ear pit, and so on. He therefore concluded that the frog embryo was a mosaic of self-differentiating parts and that it was likely that each cell received a specific set of determinants and differentiated accordingly. With this series of experiments, Roux inaugurated his program of devel-

FIGURE 2

Roux's attempt to show mosaic development. Destroying one cell of a 2-cell frog embryo results in the development of only one-half of the embryo.

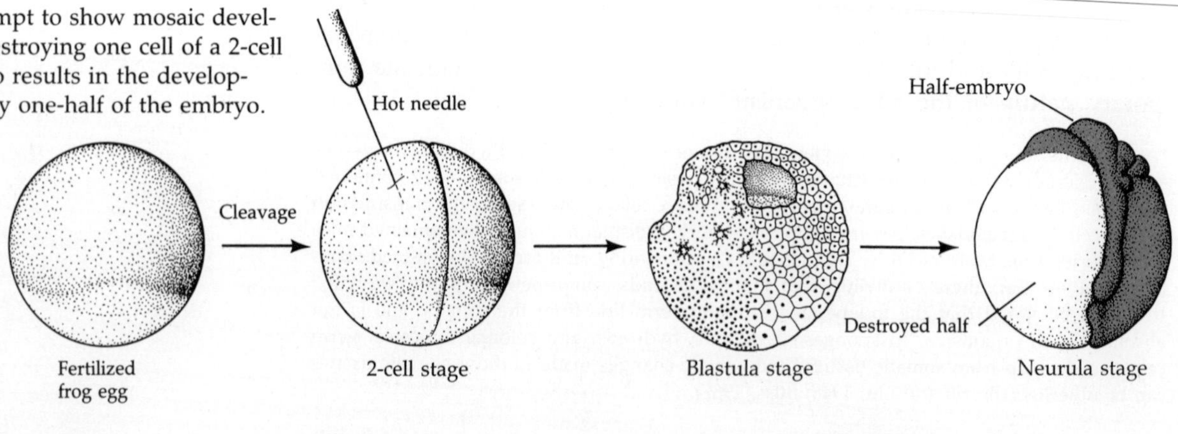

Hot needle

Half-embryo

Cleavage

Destroyed half

Fertilized frog egg

2-cell stage

Blastula stage

Neurula stage

opmental mechanics (*Entwicklungsmechanik*), the physiological approach to embryology. No longer, insisted Roux, would embryology merely be the servant of evolutionary studies. Rather, embryology would assume its role as an independent experimental science.

Hans Driesch: Regulative development

Nobody appreciated this experimental approach to embryology more than another former student of Haeckel's, Hans Driesch. Driesch's goal was to explain development in terms of the laws of physics and mathematics. His initial investigations were similar to those of Roux. Roux's experiments were, technically, *defect* studies that answered the question of how the remaining blastomeres of an embryo would develop when a subset of them was destroyed. Driesch (1892) sought to extend this research by performing *isolation* experiments. Sea urchin blastomeres were separated from each other by vigorous shaking (or, later, by placing them in calcium-free seawater). To Driesch's surprise, each of the blastomeres from a 2-cell embryo developed into a complete larva. Similarly, when Driesch separated the blastomeres from 4- and 8-cell embryos, some of the cells produced entire pluteus larvae (Figure 3). Here was a result drastically different from that predicted by Weismann or Roux. Rather than self-differentiating into its future embryonic part, each blastomere could regulate its development so as to produce a complete organism. This phenomenon is called REGULATIVE DEVELOPMENT.

Regulative development was also demonstrated in another experiment by Driesch. In sea urchin eggs, the first two cleavage planes are meridional, passing through the animal and vegetal poles, whereas the third division is equatorial, dividing the embryo into four upper and four lower cells (see Figure 3 in Chapter 3). Driesch (1893) changed the direction of the third cleavage by gently compressing the early embryos between two glass plates, thus causing the third division to be meridional like the preceding two cleavages. After he released the pressure, the fourth division was equatorial. This procedure reshuffled the nuclei, causing a nucleus that normally would be in the region destined to form endoderm to

FIGURE 3
Driesch's demonstration of regulative development. (A) A normal pluteus larva. (B) Smaller, but normal, plutei that each developed from one blastomere of a dissected 4-cell embryo. (All larvae are drawn to the same scale.) Note that the four larvae derived in this way are not identical, despite their ability to generate all the necessary cell types. (After Hörstadius and Wolsky, 1936.)

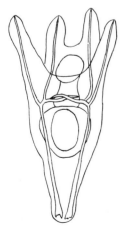

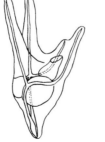 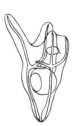 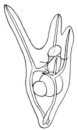 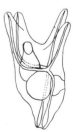

(A) Normal pluteus larva

(B) Plutei developed from single cells of 4-cell embryo

now be in the presumptive ectoderm region. Some nuclei that would normally have produced dorsal structures were now found in the ventral cells (Figure 4). If the segregation of nuclear determinants had occurred (as had been proposed by Weismann and Roux), the resulting embryo should be strangely disordered. However, Driesch obtained normal larvae from these embryos.

The consequences of these experiments were momentous both for embryology and for Driesch personally. First, Driesch had demonstrated that the "prospective potency" of an isolated blastomere (those cell types it was possible for it to form) was greater than its "prospective fate" (the cell types it would normally give rise to over the unaltered course of its development). According to Weismann and Roux, the prospective potency and the prospective fate of a blastomere should be identical. Second, Driesch concluded that the sea urchin embryo was a "harmonious equipotential system," because all these potentially independent parts functioned together to form a single organism. Third, the fate of a nucleus depended solely on its location in the embryo. Driesch (1894) hypothesized a series of events wherein development proceeded by the interactions of the nucleus and cytoplasm:

> Insofar as it contains a nucleus, every cell, during ontogenesis, carries the totality of all primordia; insofar as it contains a specific cytoplasmic cell body, it is specifically enabled by this to respond to specific effects only. . . . When nuclear material is activated, then, under its guidance, the cytoplasm of its cell that had first influenced the nucleus is in turn changed, and thus the basis is established for a new elementary process, which itself is not only the result but also a cause.

This strikingly modern concept of nuclear–cytoplasmic interaction and nuclear equivalence eventually caused Driesch to abandon science. He could no longer envision the embryo as a physical machine, because it could be subdivided into parts that were each capable of reforming the entire organism. In other words, Driesch had come to believe that development could not be explained by physical forces. He was driven to invoke a vital force, *entelechy* ("internal goal-directed force"), to explain how development proceeds. Essentially, the embryo was imbued with an internal psyche and wisdom to accomplish its goals despite the obstacles embryologists placed in its path. Unable to explain his results in the terms of the physics of his day, Driesch renounced the study of developmental phys-

FIGURE 4

Driesch's pressure-plate experiment for altering the distribution of nuclei. (A) Normal cleavage from 8- to 16-cell sea urchin embryos seen from the animal pole (upper sequence) and from the side (lower sequence). (B) Abnormal cleavage planes formed under pressure, as seen from the animal pole and from the side. (After Huxley and deBeer, 1934.)

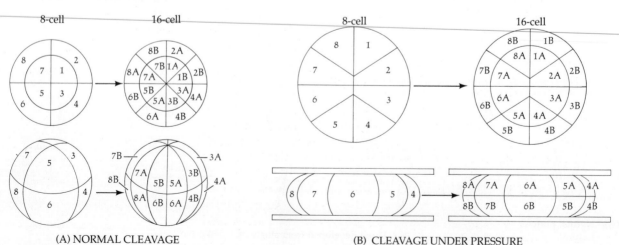

(A) NORMAL CLEAVAGE   (B) CLEAVAGE UNDER PRESSURE

TABLE 1
Experimental procedures and results of Roux and Driesch

Investigator	Organism	Type of experiment	Conclusion	Interpretation concerning potency and fate
Roux (1888)	Frogs (*Rana fusca* and *Rana esculenta*)	Defect	Mosaic (self-differentiating) development	Prospective potency equals prospective fate
Driesch (1892)	Sea urchin (*Echinus microtuberculatus*)	Isolation	Regulative development	Prospective potency is greater than prospective fate
Driesch (1893)	Sea urchin (*Echinus* and *Paracentrotus*)	Recombination	Regulative development	Prospective potency is greater than prospective fate

iology and became a philosophy professor, proclaiming vitalism until his death in 1941.

The differences between Roux's experiments and those of Driesch are summarized in Table 1. The difference between isolation and defect experiments and the importance of the interactions provided by the destroyed blastomeres were highlighted in 1910 when J. F. McClendon showed that isolated frog blastomeres behaved just like separated sea urchin cells. Therefore, the mosaic-like development of frog blastomeres in Roux's study was an artifact of the defect experiment. Something in or on the dead blastomere still informed the live cells that it existed. We have also seen that early mammalian blastomeres have a regulative type of development. As we discussed in Chapter 3, each isolated blastomere of a mouse inner cell mass is capable of generating an entire fertile mouse. The ability of two or more early mouse or rat embryos to fuse into one normal embryo (Figure 29 in Chapter 3) and the phenomenon of identical twins also attest to the regulative ability of mammalian blastomeres. Therefore, even though Weismann and Roux pioneered the study of developmental physiology, their proposition that differentiation is caused by the segregation of nuclear determinants was shortly shown to be incorrect.

Sven Hörstadius: Potency and oocyte gradients

But Driesch was not 100 percent correct either. As we saw in the last chapter, there are numerous animals that do develop largely as a mosaic of self-differentiating parts. More importantly, though, even the sea urchin embryo is not a collection of completely equipotential cells. In a series of experiments conducted from 1928 to 1935, Swedish biologist Sven Hörstadius separated various layers of early sea urchin embryos with fine glass needles and observed their subsequent development (Hörstadius, 1928, 1939). When the 8-cell embryo was divided meridionally through the animal and vegetal poles, both halves produced pluteus larvae, just as Driesch had foretold. But when embryos at the same stage were split equatorially (separating animal and vegetal poles), neither part developed into a complete larva (Figure 5). Rather, the animal half became a hollow ball of ciliated epidermal cells (called a DAUERBLASTULA), and the vegetal half developed into a slightly abnormal embryo with an expanded gut.

FIGURE 5

Early asymmetry in the sea urchin embryo. (A) When the four animal blastomeres are separated from the four vegetal blastomeres and each half is allowed to develop, the animal cells form a ciliated dauerblastula and the vegetal cells form a larva with an expanded gut. (B) When the 8-cell embryo is split so that each half contains animal and vegetal cells, small, normal-appearing larvae develop.

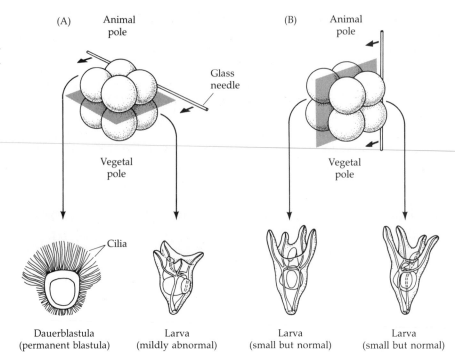

(A) Animal pole

Glass needle

Vegetal pole

(B) Animal pole

Vegetal pole

Cilia

Dauerblastula
(permanent blastula)

Larva
(mildly abnormal)

Larva
(small but normal)

Larva
(small but normal)

Hörstadius was able to duplicate these results by cutting unfertilized sea urchin eggs in half and fertilizing the halves separately. In sea urchins, egg fragments (MEROGONES) can divide and develop even if they have only a haploid nucleus. If a sperm enters the half that lacks the haploid egg nucleus, the merogone will still develop (Figure 6). When the egg had been split meridionally, normal embryos formed from both halves of the egg. However, when the oocyte had been equatorially cut, fertilization produced either the ciliated animal ball or the expanded vegetal gut. Therefore, even in sea urchin embryos, there appears to be some degree

FIGURE 6

Asymmetry in the sea urchin egg. (A) When Hörstadius split the sea uchin egg meridionally so that both merogones contained animal and vegetal cytoplasm, small, normal-appearing plutei developed. (B) When the sea urchin egg was divided into animal and vegetal halves (merogones) and the halves were allowed to be fertilized by sperm, the animal half developed into a ciliated dauerblastula and the vegetal half produced a pluteus with an expanded gut.

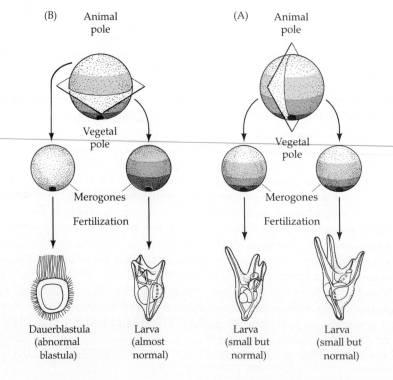

(B) Animal pole

Vegetal pole

Merogones

Fertilization

(A) Animal pole

Vegetal pole

Merogones

Fertilization

Dauerblastula
(abnormal blastula)

Larva
(almost normal)

Larva
(small but normal)

Larva
(small but normal)

of mosaicism, at least along the animal–vegetal axis. Recent studies have shown a similar degree of mosaicism within the frog embryo (as will be discussed later).

These observations led Hörstadius to perform some of the most exciting experiments in the history of embryology. First, Hörstadius (1935) traced the normal development of each of the six tiers of cells of the 64-cell sea urchin embryo. As shown in Figure 7A, the animal cells and the first vegetal layer normally produce ectoderm; the second vegetal layer gives rise to endoderm and some larval mesoderm; and the micromeres generate the mesodermal skeleton.

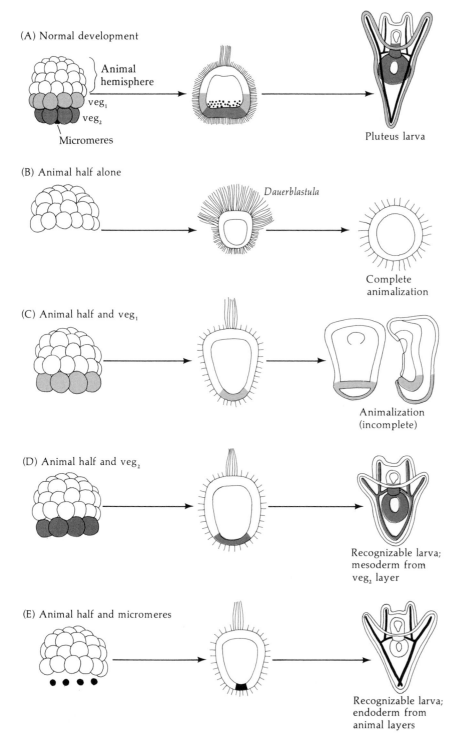

(A) Normal development

Animal hemisphere

veg_1

veg_2

Micromeres

Pluteus larva

(B) Animal half alone

Dauerblastula

Complete animalization

(C) Animal half and veg_1

Animalization (incomplete)

(D) Animal half and veg_2

Recognizable larva; mesoderm from veg_2 layer

(E) Animal half and micromeres

Recognizable larva; endoderm from animal layers

FIGURE 7
Hörstadius's demonstration of a vegetalizing gradient. (A) Fate of each cell layer of the 64-cell sea urchin embryos through blastula to pluteus stage. The different cell layers are marked as in Figure 1 of Chapter 4. (B) Fate of the isolated animal half. (C) Recombination of animal half plus the veg_1 tier of cells. (D) Recombination of animal half plus the veg_2 tier of cells. (E) Recombination of animal half plus the micromeres. (After Hörstadius, 1939.)

Next, Hörstadius removed the fertilization membrane from the 64-cell embryos, separated the tiers with fine glass needles, and recombined them in various ways. The isolated animal hemisphere, alone, became a dauerblastula of ciliated ectoderm cells (Figure 7B). Such an embryo was said to be "animalized." When Hörstadius recombined an isolated animal hemisphere with the veg_1 tier (Figure 7C), the resulting larva was less animalized. Ciliary development was suppressed and a portion of the gut was formed. However, when the animal hemisphere was combined with the veg_2 tier (Figure 7D), a normal-looking pluteus larva developed. In this combination, the veg_2 cells, which normally form only the archenteron and its derivatives, now formed skeletal structures as well. Similarly, when the animal half was recombined with just the micromeres (Figure 7E), a small, normal-looking pluteus was formed, but in this case the endoderm was completely derived from the animal cells. Here, the gut was formed by cells that would normally have created the ciliated ectoderm. These experiments showed that the animal cells had the genetic potential to be gut cells even at the 64-cell stage. More importantly, it also suggested that the ability to suppress "animalization" was distributed as a GRADIENT. The micromeres were stronger "vegetalizers" than the veg_2 layer; but the veg_2 layer, in turn, was stronger than the veg_1 tier.

Hörstadius's next series of experiments suggested the existence of an animalizing factor that also occurred as a gradient. Micromeres were combined with each of the tiers of the 64-cell embryo. An isolated an_1 tier (Figure 8A) would produce an ectodermal dauerblastula. As additional micromeres were added, a more complete embryo developed. The combination of just the an_1 tier with four micromeres resulted in the production of a normal pluteus. The an_2 tier formed a normal pluteus with only two micromeres (Figure 8B); and four micromeres added to the an_2 tier caused

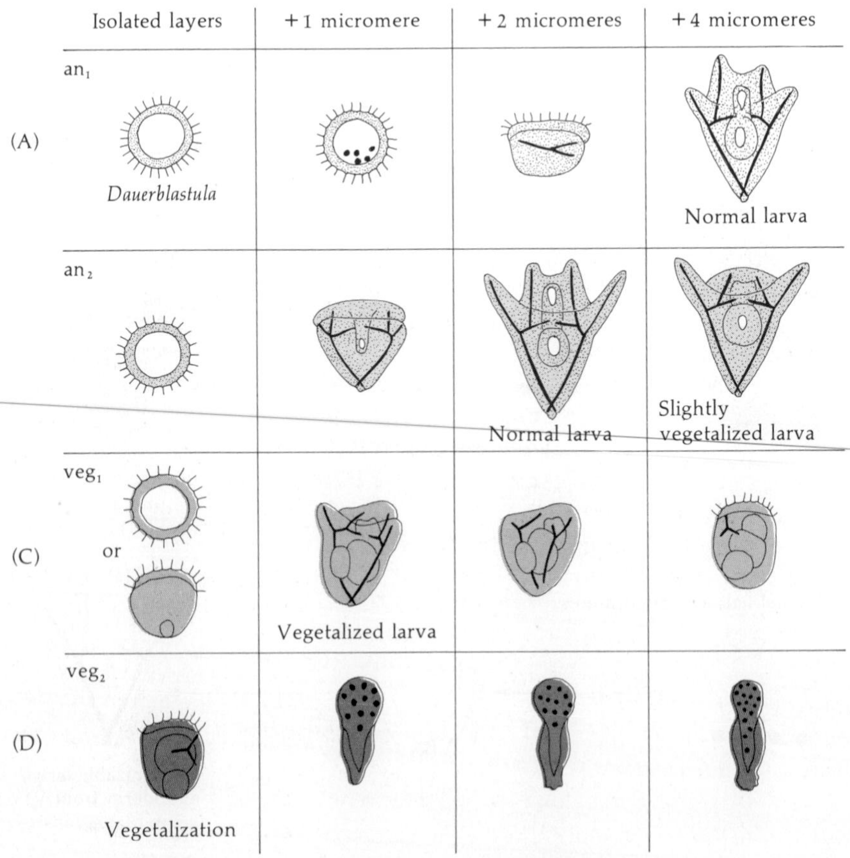

FIGURE 8

Hörstadius's demonstration of an animalizing gradient. Each tier of a 32-cell sea urchin embryo is isolated and then recombined with 0, 1, 2, or 4 micromeres. It takes four micromeres to form a normal pluteus larva from cells derived from the an_1 tier (A), but only two are needed to produce a pluteus from the an_2 tier. The veg_1 layer produces a vegetalized larva with only one micromere (C), and the veg_2 tier lacks enough animalizing properties to form a normal pluteus (D). (After Hörstadius, 1935.)

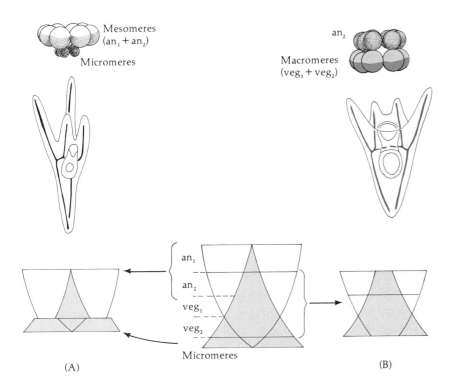

Mesomeres ($an_1 + an_2$)

Micromeres

an_2

Macromeres ($veg_1 + veg_2$)

an_1
an_2
veg_1
veg_2
Micromeres

(A)

(B)

an abnormally expanded endoderm. The veg_1 layer would also produce a dauerblastula when alone. However, even one micromere caused a severe expansion of gut tissue (Figure 8C) and the isolated veg_2 tier showed this tendency to "vegetalize" without the aid of any micromeres (Figure 8D). Thus, there appeared to be a gradient of "animalization" proceeding in strength from an_1 to the micromeres.

The most logical explanation of these results is that of two opposed gradients: a vegetalizing gradient with its maximal activity at the vegetal pole, and an animalizing gradient with its maximal activity at the animal region. Exactly such a system of dual gradients had been proposed by Hörstadius's adviser, J. Runnström, in 1929. Because the relative ratios of substances would be important for development (rather than their absolute amounts), the recombination of the two poles would reestablish all the intermediate positions. Similarly, the intermediate cells would still have a maximum and minimum for both gradients and could thus give rise to an entire pluteus as well (Figure 9). This dual-gradient model has been extremely useful in explaining other recombination experiments. For example, when micromeres are transplanted from the vegetal pole to a region near the center of a 32-cell embryo, the implanted micromeres invaginate into the host blastocoel, thereby causing the formation of a well developed secondary archenteron (Figure 10). When the micromeres are added to the animal pole, a very small secondary gut forms. So far neither the vegetal factor nor the animal factor has been isolated. Protein denaturers (heavy metals, NaSCN, Evans Blue dye) appear to diminish the vegetalizing gradient and thus animalize the embryo. Respiratory inhibitors (CO, KCN, NaN_3, Li) appear to vegetalize the embryo. Thus, von Ubisch (1919) was able to get a normal pluteus from an isolated animal hemisphere grown in dilute lithium chloride. The lithium appeared to weaken the animal portion of the gradient within the animal cells, thereby enabling balanced development of this hemisphere.

We have, then, a model for regulative development based on relative concentration gradients within the oocyte. When animal cells are com-

FIGURE 10
(A) Micromeres (black) are transplanted from the vegetal pole to a region between the animal and vegetal halves of a 32-cell embryo. (B) The micromeres invaginate into the blastocoel. (C) A secondary archenteron forms and eventually (D) a pluteus larva that has two archenterons is formed. (E) These separate archenterons later fuse into a large gut. (After Hörstadius, 1935.)

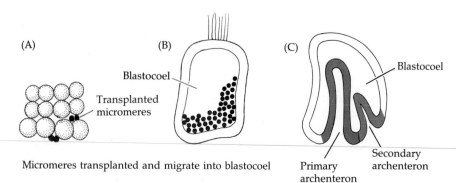

Micromeres transplanted and migrate into blastocoel

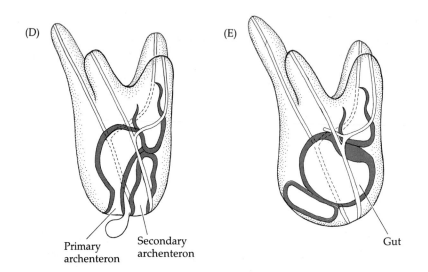

bined with micromeres only, a normal pluteus is formed, with its gut cells coming from animal cells that would normally have formed the ectoderm. Thus, the potency is still greater than the fate. But the animal cells alone would have given rise only to an animalized blastula, so the ability of one cell to form the entire larva is gone. In other words, potency is restricted. Any cell with a particular ratio of animal and vegetal substances will produce a certain type of cell. Thus, animal pole cells, with a high animal factor : vegetal factor ratio, would normally produce ectodermal tissue. When recombined with a strong source of the vegetal factor, however, certain animal cells will produce gut tissue. When the embryo is separated so that each half has a complete animal and vegetal gradient (i.e., a meridional separation along the animal–vegetal axis), a complete larva forms. The blastomeres will stay regulative as long as they are not completely formed from animal or vegetal cytoplasm. It is not surprising, then, that after the 32-cell stage, most of the individual blastomeres can no longer give rise to complete larvae (Morgan, 1895), and the micromeres of the 16-cell stage are already incapable of doing so (Hagstrom and Lonning, 1965; Okazaki, 1975). Even in an embryo that undergoes regulative development, there comes a time when the potencies of its cells are restricted.

Forming an integrated organism

Recent studies have confirmed and extended many of Hörstadius's observations. More importantly, they have drawn attention to a previously overlooked area of developmental biology: the importance of negative inductive interactions. Driesch referred to the embryo as an "equipotential harmonious system" because each of the composite cells had surrendered most of its potential in order to form part of a single complete organism. Each cell could have become a complete animal on its own, yet didn't. What made the cells cooperate instead of becoming autonomous entities? In the case of snails and tunicates, the answer was simple. The maternal cytoplasm does not permit each cell to become autonomous; each cell can only develop into a portion of the embryo. In sea urchins and other embryos that show regulation, the answer is more complex.

In these instances, there appears to be negative induction—induction that restricts the potential of the embryonic cells. Jon Henry and colleagues in Rudolf Raff's laboratory (1989) have shown that if one isolates pairs of cells from the animal cap of a 16-cell sea urchin embryo, these cells can give rise to both ectodermal and mesodermal components. However, their capacity to form mesoderm is severely restricted if they are aggregated with other animal cap pairs. Thus, the presence of neighbor cells, even of the same kind, restricts the potencies of both partners. Ettensohn and McClay (1988) have shown that potency is also restricted when a cell is combined with its neighbors along the animal–vegetal axis (Figure 11). They have shown that when primary mesenchymal cells are removed from the mesenchyme blastula of sea urchin embryos, a group of secondary mesenchyme cells replaces them. These secondary mesenchyme cells come from the vegetal plate adjacent to the region where the primary mesenchyme cells formed. They have shown that the presence of normal primary mesenchyme cells inhibits the surrounding cells from forming primary mesenchyme. If they are removed, these macromeres will form the primary mesenchyme cells. Thus, there is negative induction occurring here. Just as in normal induction, where one cell type induces its neighbor to produce a new cell type, here we see that one cell can restrict the potency of its neighbor.

This restriction in potency was followed on the molecular level by Hurley and co-workers (1989), who dissociated 16-cell sea urchin embryos into individual blastomeres. The loss of cell contacts between the isolated blastomeres caused dramatic alterations in the synthesis of messenger RNAs from these cells. Some mRNA, such as that for mesenchymal actin, was expressed by a far greater than normal number of the dissociated cells. Other messages that become common in gastrulating embryos, however, were hardly synthesized at all in the dissociated cells. Thus, the contact between the blastomeres of the cleaving sea urchin blastula restrict each other's potency so that they construct a single embryo.

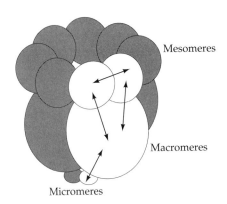

FIGURE 11
Inductive interactions that restrict potency in the early cleavage stages of the sea urchin embryo. Double-headed arrows illustrate the mutually restrictive interactions between adjacent cells. (After Henry et al., 1989.)

SIDELIGHTS & SPECULATIONS

Gradients of activated regulatory factors

Recent studies of sea urchin development conducted in the laboratory of Eric Davidson have suggested a mechanism that would produce the results predicted by Hörstadius's gradients (Davidson, 1989). By injecting fluorescently labeled beads into each of the cells of the 8-cell stage embryo, Cameron and colleagues (1987) were able to follow the fate of each of these blastomeres (Figure 12). They discovered that the cell fates in a normal sea urchin embryo were invariant; that is, the same cells gave rise to the same tissues in every embryo. Moreover, Davidson's laboratory

found that the blastula can be divided into five territories, each territory being the progeny of a small group of cells. As seen in Figure 12, the five territories are the prospective oral ectoderm (the ciliated bands along the mouth), the prospective aboral ectoderm (outer surface of the larva), the prospective vegetal plate (gut endoderm and secondary mesenchyme), the prospective skeletal mesenchyme, and the small micromeres. The cells whose progeny comprise each territory are called FOUNDER CELLS. A founder cell is the first cell to arise within a lineage, all of whose progeny contribute to a single territory. For instance, Figure 12 shows that there are 11 founder cells for the aboral ecto-

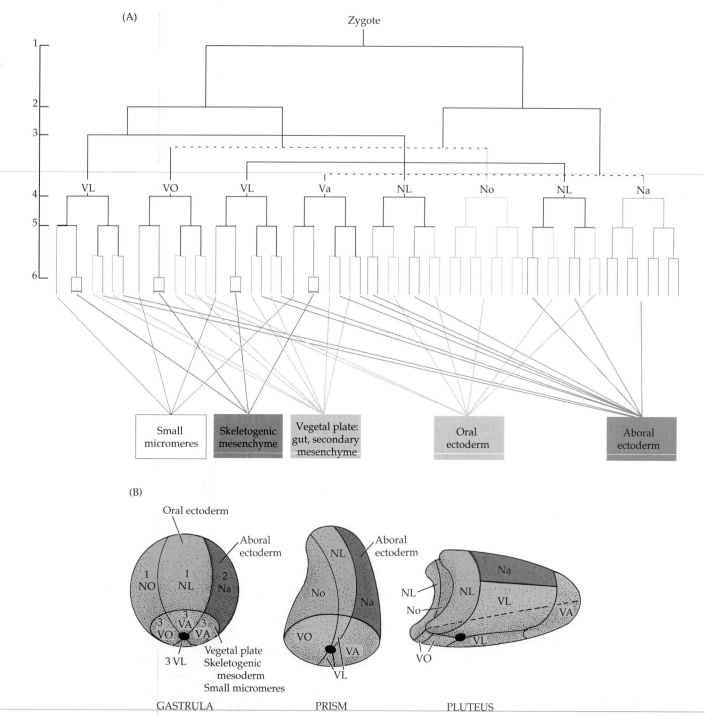

FIGURE 12

Cell fate in the sea urchin embryo. (A) Diagrammatic representation of the first six cleavages of the sea urchin embryo. The sequence of division cycles and cell fates are the same for most urchins that form pluteus larvae (Okaizaki, 1975; Davidson, 1989). Cleavages are numbered, and the eight cells of the third cleavage are labeled as animal (N) or vegetal (V); these cells form the lateral (L), oral (O), or aboral (A) surfaces. (B) Representation of the fates of the descendants of the first eight blastomeres mapped onto the gastrula, prism, and pluteus stages of *Paracentrotus lividus*. (After Cameron et al., 1987, and Davidson, 1989.)

derm territory and 4 founder cells for the skeletogenic mesenchyme territory. Note also that there is a territory that was unknown to embryologists until recently. This is the "small micromere" territory. This group of cells is formed along with the micromeres that become the skeletogenic mesenchyme. However, they do not ingress immediately into the blastocoel during gastrulation. Eventually they contribute to the coelomic pouches.

This leads to a paradox: in each embryo of the species, the cell lineage is invariant. Yet Driesch, Hörstadius, and generations of college students have shown that the sea urchin embryo has the ability to regulate its development. How are both observations possible? Davidson has proposed a model that seeks to integrate the observations of lineage invariance, regulative development, and the findings that within each territory certain genes are activated and certain genes are suppressed. The model is based on the invariant cleavage planes in the embryo and on the observation that the animal–vegetal axis is already polarized before fertilization. According to the model proposed by Davidson (1989), there exist within the egg cytoplasm four major types of gene regulatory factors. These are molecules, probably proteins, that bind to certain sequences of DNA and activate the genes adjacent to them. (More will be said about this in Chapter 12, but the regulatory proteins that activate bacterial operons provide a similar situation.) These factors are present in varying degrees throughout the egg, but in the zygote, probably the skeletogenic mesenchyme factor is the only one that is active (Figure 13).

During normal development, cleavage separates the cells. Each neighboring cell would interact with the membranes of its adjacent cells to activate enzymes that would alter the neighboring cell's cytoplasm. This alteration in cytoplasm would activate the specific DNA-binding factors in this region. By the end of blastulation, the five regions would be specified. At this point, the territories would constitute both the precursor cells for certain regions of the pluteus larva and the regions of specific gene activation. For example, the cytoskeletal actin gene, which is expressed only in the aboral ectoderm cells, cannot be activated anywhere else in the embryo. Similarly, if one injects the gene for the skeletal matrix protein anywhere else in the embryo except the region of skeletogenic mesenchyme, it will not be activated (Hough-Evans et al., 1987; Sucov et

al., 1988). Davidson's model is that regulatory factors that exist throughout the embryo are activated by interactions between neighboring cells. These regulatory factors determine which genes can be expressed within that territory of cells, thereby determining cell fate. However, if the embryo is perturbed, the neighbors differ. New fates will ensue as the new vegetalmost cells activate their skeletogenic mesenchyme determinants and start activation of their neighbors. Similarly, lithium is proposed to function by activating those factors that convert the vegetalmost cells into cells having the micromere's inducing functions. Thus, the developmental fates of the sea urchin embryo may be determined by the activities of cell surfaces in activating specific regulatory factors needed for gene expression.

FIGURE 13
Davidson's model of cell fate determination in sea urchins. In the zygote, the animal–vegetal axis is specified. Along this axis are inactive DNA-binding regulatory factors, represented by open circles. The filled circles indicate that the factor is active (unless attached to a specific inhibitor). In the zygote, only the skeletogenic mesenchyme factors are active. During cleavage, the cell surfaces of the skeletogenic micromeres induce the vegetal cells above them to activate their vegetal plate factors only. The cell membranes of the vegetal plate cells then induce the cells above them to activate the ectoderm factors. The oral–aboral axis appears to be specified by the end of the second cleavage. This axis might form a gradient of some substance that releases the inhibition of aboral ectoderm regulatory factor. In so doing, the future oral ectoderm would be distinguished from the future aboral ectoderm. (After Davidson, 1989.)

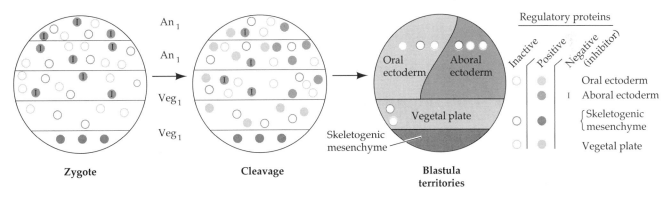

Hans Spemann: Progressive determination of embryonic cells

In the previous section, we saw evidence for regulative development. We noted that the two major aspects of regulation—first, that an isolated blastomere has a potency greater than its normal embryonic fate, and second, that rearranged blastomeres develop according to their new locations—hold true during the early stages of sea urchin cleavage. Eventually, however, the blastomeres become committed to certain fates. Hörstadius was able to relate this restriction in potency to the cytoplasmic constitution of the egg, as blastomeres could regulate only as long as they had sufficient material from both the animal and vegetal parts of the egg. In 1918, Hans Spemann of the University of Freiburg discovered that a similar situation existed in the salamander egg. The experiments by which he and his colleagues analyzed this phenomenon over the next 20 years form the basis for much of our knowledge of embryonic physiology and won Spemann a Nobel prize in 1935.

Spemann, like Roux and Driesch, sought to test Weismann's hypothesis, and, by an ingenious method, he demonstrated that early newt blastomeres had identical nuclei, each capable of producing an entire larva. Shortly after fertilizing a newt egg, Spemann took a baby's hair and lassoed the zygote in the plane of the first cleavage. He then partially constricted the egg, causing all the nuclear divisions to remain on one side of the constriction. Eventually, often as late as the 16-cell stage, a nucleus would escape across the constriction into the non-nucleated side. Cleavage then began on this side, too, whereupon the lasso was tightened until the two halves were completely separated. Twin larvae developed, one slightly older than the other (Figure 14). Spemann concluded from this that early amphibian nuclei were genetically identical and each was capable of giving rise to an entire organism. In this respect, amphibian blastomeres were similar to those of the sea urchin.

However, when Spemann performed a similar experiment with the constriction still longitudinal but perpendicular to the plane of the first cleavage (separating the future dorsal and ventral regions rather than right and left sides), he obtained a different result altogether. The nuclei con-

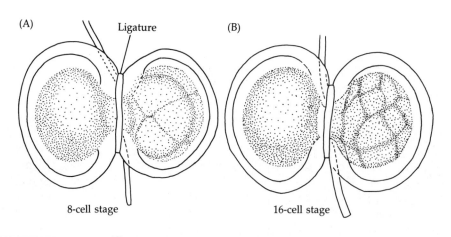

(A) Ligature (B)

8-cell stage 16-cell stage

(C)

140 days

FIGURE 14
Spemann's demonstration of nuclear equivalence in newt cleavage. (A) When the fertilized egg of the newt *Triturus taeniatus* was constricted by a ligature, the nucleus was restricted to one-half of the embryo. The cleavage on that side of the embryo reached the 8-cell stage while the other side remained undivided. (B) At the 16-cell stage, a single nucleus entered the as-yet undivided half, and the ligature was constricted to complete the separation of the two halves. (C) After 140 days, each side had developed into a normal embryo. (After Spemann, 1938.)

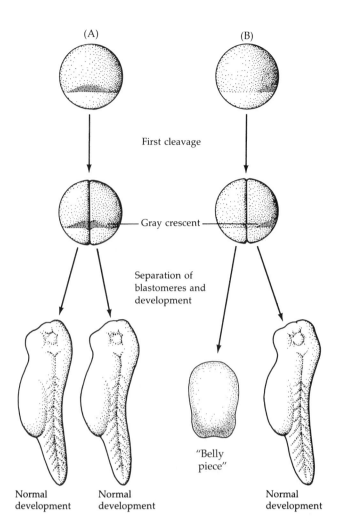

(A)   (B)

First cleavage

Gray crescent

Separation of
blastomeres and
development

"Belly
piece"

Normal   Normal   Normal
development   development   development

FIGURE 15
Asymmetry in an amphibian egg. (A) When the plane of first cleavage divides the egg into two blastomeres that each get one-half of the gray crescent, each experimentally separated cell develops into a normal embryo. (B) When only one of the two blastomeres receives the entire gray crescent, it alone forms a normal embryo. The other piece lacks dorsal structures and remains a mass of unorganized tissues. (After Spemann, 1938.)

tinued to divide on both sides of the constriction, but only one side would give rise to a normal larva. The other side would produce an unorganized tissue mass, which Spemann called the Bauchstück—the belly piece. This tissue mass was a ball of epidermal cells (ectoderm) containing blood and mesenchyme (mesoderm) and gut cells (endoderm) but no dorsal structures such as nervous system, notochord, or somites (Figure 15).

Why should these two experiments give different results? Could it be that when the egg was divided perpendicular to the first cleavage plane, some cytoplasmic substance was not equally distributed to the two halves? Fortunately, the salamander egg was a good place to look for answers. As we have seen in Chapters 2 and 4, there are dramatic movements in the cortical cytoplasm following the fertilization of amphibian eggs, and in some amphibians these movements expose a gray, crescent-shaped area of cytoplasm in the region directly opposite the point of sperm entry. Moreover, the first cleavage plane normally bisects this region equally into the two blastomeres. If these cells are then separated, two complete larvae develop. However, should this cleavage plane be aberrant (either in the rare natural event or in an experiment in which an investigator constricts a hair-loop lasso perpendicular to the normal cleavage plane), the gray crescent material passes into only one of the two blastomeres. Spemann found that when these two blastomeres are separated, only the blastomere containing the gray crescent developed normally.

It appears, then, that something in the gray crescent region is essential for proper embryonic development. But how does it function? What role does it play in normal development? The most important clue came from the fate map of this area of the egg, for it showed that the gray crescent region gave rise to the cells that initiate gastrulation. These cells form the dorsal lip of the blastopore. As was shown in Chapter 4, the cells of the dorsal blastopore lip are somehow "programmed" to invaginate into the blastula, thus initiating gastrulation and the formation of the archenteron. Because all future amphibian development depends on the interaction of cells rearranged during gastrulation, Spemann speculated that the importance of gray crescent material lay in its ability to initiate gastrulation and that crucial developmental changes occurred during gastrulation.

In 1918, Spemann demonstrated that enormous changes in cell potency did indeed take place during gastrulation. He found that the cells of the *early* gastrula were uncommitted with respect to their eventual differentiation, but that the fates of the *late* gastrula cells were fixed. Spemann exchanged tissues between the early gastrulae of two differently pigmented species of newts (Figure 16). When a region of prospective epidermal cells was transplanted to an area where the neural plate formed, the transplanted cells gave rise to neural tissue. When prospective neural plate cells were transplanted to the region fated to become belly skin, these cells became epidermal (Table 2). Thus, early newt gastrula cells were not yet committed to a specific type of differentiation. Their prospective potencies were still greater than their prospective fates. These cells were said to exhibit DEPENDENT DEVELOPMENT because their ultimate fates depended on their location in the embryo. However, when the same heteroplastic (interspecies) transplantation experiments were performed on *late* gastrulae, Spemann obtained completely different results. Rather than regulating their differentiation in accordance with their new location, the transplanted cells exhibited INDEPENDENT (or AUTONOMOUS) DEVELOPMENT. Their prospective fate was fixed, and the cells developed independently of their new embryonic location. Specifically, prospective neural cells now developed into brain tissue even when placed in the region of prospective epidermis, and prospective epidermis formed skin even in the region of the prospective neural tube. Within the time separating early

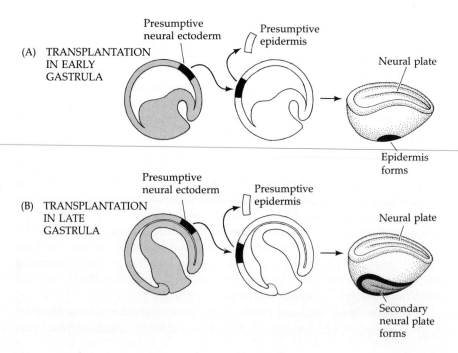

FIGURE 16
Determination of ectoderm during newt gastrulation. Presumptive neural ectoderm from one newt embryo is transplanted into a region in another embryo that normally becomes epidermis. (A) When the transfer is done in the early gastrula, the presumptive neural tissue develops into epidermis and only one neural plate is seen. (B) When the same experiment is performed on late gastrula tissues, the presumptive neural cells form neural tissue, thereby causing two neural regions to form on the host. (After Saxén and Toivonen, 1962.)

TABLE 2
Results of tissue transplantation during early and late gastrula stages in the newt

Donor region	Host region	Differentiation of donor tissue	Conclusion
EARLY GASTRULA			
Prospective neurons	Prospective epidermis	Epidermis	Dependent development
Prospective epidermis	Prospective neurons	Neurons	Dependent development
LATE GASTRULA			
Prospective neurons	Prospective epidermis	Neurons	Independent development (determined)
Prospective epidermis	Prospective neurons	Epidermis	Independent development (determined)

and late gastrulation, the potencies of these groups of cells had become restricted to their eventual paths of differentiation. These cells are now said to be DETERMINED.

Determination refers to the cell's commitment to eventually differentiate into a specific cell type and into no other tissue. This term should not be confused with SPECIFICATION, which refers to the fixing of positional information such as the embryonic axes. The axes can be *specified* early in development, as they are in frogs, but this does not *determine* cell fate. The epidermis, for instance, exists on the dorsal, ventral, anterior, posterior, left, and right sides of the embryo. Therefore, in the late gastrula, those animal pole cells on the same side as the blastopore lip are now committed (determined) to produce neural tissue in whatever location they might be placed (including a petri dish). They can no longer regulate their differentiation into other cell types. It should be noted that the criteria for determination are completely operational. There are no obvious changes occurring in the cells and no overt differentiation can yet be seen. The molecular basis of determination remains one of the major unsolved puzzles of development.

Hans Spemann and Hilde Mangold: Primary embryonic induction

The most spectacular transplantation experiments were published by Hans Spemann and Hilde Mangold in 1924. They showed that when tissues are placed in new locations, the dorsal lip of the blastopore is the only self-differentiating region in the early gastrula. When dorsal blastopore lip tissue from an early gastrula was transplanted into the ventral ectoderm of another gastrula, it not only continued to be blastopore lip, but it also initiated gastrulation and embryogenesis in the surrounding tissue. In these experiments, Spemann and Mangold used differently pigmented embryos from two species of newt, the darkly pigmented *Triturus taeniatus* and the nonpigmented (clear) *Triturus cristatus*. When Spemann and Mangold prepared these heteroplastic transplants, they were able to readily identify host and donor tissues on the basis of color. The dorsal blastopore lips (the dorsal marginal zone tissue) of early *T. cristatus* gastrulae were removed and implanted into the regions of early *T. taeniatus* gastrulae fated to become ventral epidermis (Figure 17). Unlike the other early gastrula tissues, which developed according to their new location, the donor blastopore lip did not become belly skin. Rather, it invaginated just as it would normally have done (showing self-determination) and disappeared beneath the vegetal cells. The light-colored donor tissue then continued to self-differentiate into the chordamesoderm and other mesodermal structures that constituted the original fate of that blastopore tissue. As the axis was formed, host cells began to participate in the production of the new embryo, becoming organs that normally they never would have formed. Thus, a somite could be seen containing both colorless (donor) and pigmented (host) tissue. What was even more spectacular, the dorsal blastopore lip cells were able to interact with the host tissues to form a complete neural plate. Eventually, a secondary embryo formed, face-to-face with its host (Figure 17). These technically difficult experiments have recently been repeated using a nuclear marker, and the results of Spemann and Mangold were confirmed (Gimlich and Cook, 1983; Recanzone and Harris, 1985).*

*Hilde Proescholdt Mangold died in a tragic accident when her gasoline heater exploded. At the time she was 26 years old, and her paper was just being published. The experiment, which formed the basis for her doctoral thesis, is one of the very few doctoral theses in biology that have been directly concerned with the awarding of a Nobel prize. For more information about this remarkable individual and her times, see Hamburger, 1984.

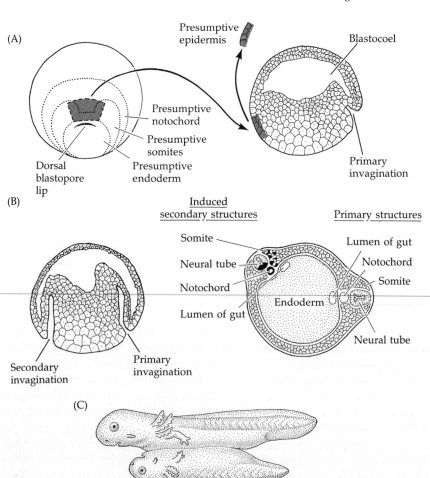

FIGURE 17
Self-differentiation of the dorsal blastopore lip tissue. Dorsal blastopore lip from early gastrula (A) is transplanted into another early gastrula in the region that normally becomes ventral epidermis. (B) Tissue invaginates and forms a second archenteron and then a second embryonic axis. Both donor and host tissues are seen in this new neural tube, notochord, and in these somites. (C) Eventually, a second embryo forms that is joined to host.

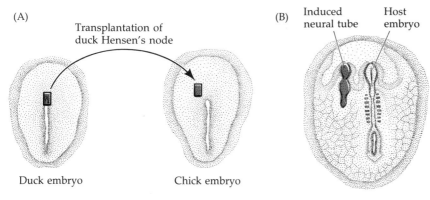

FIGURE 18
Induction of a new embryonic axis by Hensen's node. (A) Hensen's node tissue is removed from a duck embryo and implanted in a host chick embryo. (B) An accessory neural tube is induced at the graft site. (After Waddington, 1933.)

The process by which one embryonic region interacts with a second region to cause the latter tissue to differentiate in a specific direction is called INDUCTION. Because there are numerous inductions during embryonic development, this key induction wherein the blastopore lip cells induce the dorsal axis and the neural tube is traditionally called PRIMARY EMBRYONIC INDUCTION.* Spemann referred to the dorsal blastopore lip cells as the ORGANIZER because (a) they induced the host's ventral tissues to change their fates to form a neural tube and dorsal mesodermal tissue, and (b) they organized these host and donor tissues into a secondary embryo with clear anterior–posterior and dorsal–ventral axes. It is now known (thanks largely to Spemann and his students) that the interaction of the chordamesoderm and ectoderm is not sufficient to "organize" the entire embryo. Rather, it initiates a series of inductive events. We also know that the dorsal blastopore lip is active in organizing secondary embryos in *Amphioxus*, cyclostomes, and a variety of amphibians. The anterior portion of the primitive streak (that is, Hensen's node, the region initiating gastrulation in birds and mammals) is similarly effective in organizing secondary embryos in these classes of vertebrates (Waddington, 1933; Figure 18).

Regional specificity of induction

One of the most fascinating phenomena in primary embryonic induction is the regional specificity of the neural structures that are produced. Forebrain (archencephalic), hindbrain (deuterencephalic), and spinocaudal regions of the neural tube must all be properly organized in an anterior-to-posterior direction. Thus, the chordamesoderm tissue of the archenteron roof induces not only the neural tube but also the specific regions of the neural tube. This region-specific induction was shown by Otto Mangold (1933) in a series of experiments wherein various regions of the *Triturus* (newt) archenteron roof were transplanted into early gastrula embryos (Figure 19). After removing the superadjacent neural plate, four successive sections of the archenteron roof were excised from embryos that had just completed gastrulation and were placed into the blastocoels of early gastrulae. The most anterior portion of the archenteron roof induced balancers and portions of the oral apparatus (Figure 19A); the next most anterior section induced the formation of various head structures, including nose, eyes, suckers, and otic (hearing organ) vesicles (Figure 19B). The third

*This classic term has been the source of confusion because the induction of the neural tube by the notochord is no longer considered to be the first inductive process in the embryo. We will shortly discuss inductive events that precede this "primary" induction.

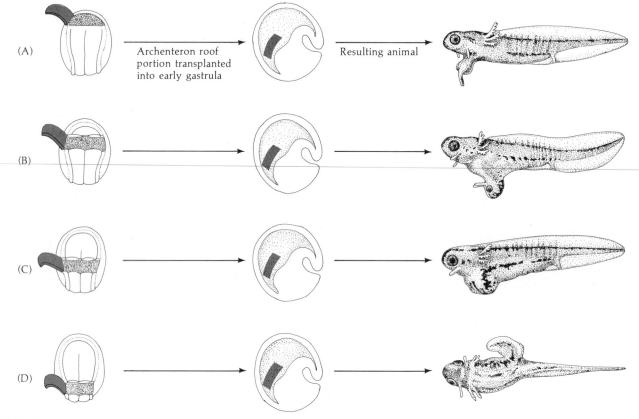

FIGURE 19

Regional specificity of induction can be demonstrated by implanting different regions (color) of the archenteron roof into early *Triturus* gastrulae. The resulting animals have secondary parts. (A) Head with balancers; (B) head with balancers, eyes, and forebrain; (C) posterior part of head, deuterencephalon, and otic vesicles; and (D) trunk–tail segment. (After Mangold, 1933.)

section induced the otic vesicles without any other head organs (Figure 19C); and the most posterior segment induced the formation of dorsal trunk and tail mesoderm (Figure 19D). (The induction of dorsal mesoderm—rather than the dorsal ectoderm of the nervous system—by the posterior end of the notochord was confirmed by Bijtel [1931] and Spofford [1945], who showed that the posterior fifth of the neural plate gives rise to tail somites and the posterior portions of the pronephric kidney duct.)

To further study this phenomenon, Holtfreter (1936) made a blastopore "sandwich," enclosing the dorsal blastopore lip between slices of undifferentiated ectoderm. The blastopore lips from the *early* gastrula stages induced the most archencephalic (anterior) structures, whereas the blastopore lips of *later* embryos caused the differentiation of the more posterior neural elements. There are two general models to explain this. One model holds that the early and late blastopore lips differ in the *quantity* of some substance. Large quantities of that substance produce either anterior or posterior neural structures, while small quantities produce the other. Midbrain regions would have an intermediate quantity of this substance. The second model holds that the early and late blastopore lips produce *qualitatively* different substances. Regional specification would be caused by the interaction of two substances secreted by the cells of the chordamesoderm. A high concentration of one would give forebrain development, while a high concentration of the other would elicit the formation of spinal cord and trunk structures. Mixtures of the two substances would elicit the midbrain and hindbrain regions.

Evidence for the latter model came from studies involving artificial tissue-specific inducers. Guinea pig bone marrow, for instance, was found to induce mesodermal structures only. Pellets of guinea pig liver, however, could only cause the induction of forebrain structures. Toivonen and Saxén (1955) implanted these inducers together within the blastocoel of the same early gastrula. Whereas the liver would only have induced forebrain and the bone marrow would only have induced mesoderm, the two together induced all the normal forebrain, hindbrain, spinal cord, and trunk mesoderm. Thus, primary neural induction may also be due to a double gradient (Figure 20).

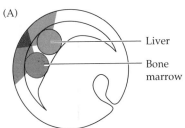

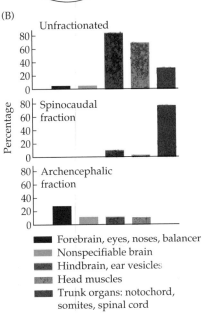

FIGURE 20
Evidence for a double gradient of inducers in the archenteron roof. (A) Simultaneous implantation of an archencephalic inducer (guinea pig liver) and a mesodermal inducer (guinea pig bone marrow pellets) into young newt gastrulae produces all of the dorsal regions of the brain and somites. (B) Separation of a midbrain (deuterencephalic)-inducing extract of chick embryos into an archencephalic-inducing fraction and a spinocaudal-inducing fraction. (After Toivonen and Saxén, 1955, and Tiedmann et al., 1963.)

The mechanisms of primary embryonic induction

Despite the enormous amount of research that has been performed on amphibian embryos, we still do not know the basic mechanisms of primary embryonic induction. One reason for this confusion may be that different groups of amphibians could use different means for arriving at the same outcome. The results from experiments on urodeles (such as the salamanders studied by Spemann, Holtfreter, and Mangold) may not be pertinent to the ways that frogs (such as *Xenopus*) form their dorsal axis and neural tube. Different assays used in different laboratories also cannot be directly compared. In the past five years, numerous laboratories have focused their efforts on explaining embryonic induction in one amphibian—*Xenopus laevis*. While there is little agreement as to the precise mechanisms by which this dorsal information is formed and transmitted, a consensus has emerged concerning the general outline of primary embryonic induction in this organism.

The first act of induction takes place at fertilization when the cytoplasm of the radially symmetric egg rotates relative to the cortex, thereby causing an asymmetry that specifies the dorsal–ventral axis. This mixing of cytoplasmic regions appears to activate dorsalizing determinants in the vegetal cells that form at the side opposite the point of sperm entry. In the second act, the descendants of these vegetal cells induce the cells above them to become Spemann's organizer. In other words, the endodermal precursors that received these dorsalizing determinants induce the cells above them to become presumptive chordamesoderm. The other vegetal cells induce the marginal cells above them to become the lateral and ventral mesoderm. Thus, there is an induction before "primary induction." In the third act, the dorsal ectoderm is converted into neural tissue. There are two scenarios for this act. Some investigators, following the results of Mangold, favor the view that the chordamesoderm acts both to neuralize the ectoderm and to assign the regional specificities to that neural tube. Other investigators contend that the neurulation of the ectoderm is determined by signals from the dorsal blastopore lip that travel through the ectoderm itself, while a second set of signals from the chordamesoderm are responsible for the regional specificity of this induction.

The specification of dorsoventral polarity at fertilization

As we saw in Chapters 2 and 4, dorsoventral specification is accomplished by the rotation of the inner cytoplasm of the egg relative to the cortex. This rotation creates a region of vegetal cytoplasm where the dorsal information (what Wakahara [1989] calls the "anterodorsal structure-forming activity") has been activated. If this rotation is inhibited by UV light, the embryo will not form dorsoanterior structures (Vincent and Gerhart, 1987). Furthermore, Render and Elinson (1986) and Wakahara (1989) cut eggs into fragments before and after this rotation. If the egg was cut before rotation, both sides developed dorsoanterior structures—head, noto-chord, neural tube. If the cut was made after the rotation had occurred, one fragment developed a head, heart, and some dorsal mesodermal structures, while the other half developed essentially into a *Bauchstück*, having little or no dorsal mesoderm and no nervous system. It appears, then, that this cytoplasmic rotation moves the dorsal activating determinants to the future dorsal side of the egg.

Localization of dorsoanterior determinants in the vegetal cells during cleavage

The cytoplasmic rotation that occurs during fertilization presumably creates an area of cytoplasm that is enriched with dorsoanterior-forming determinants (Figure 21). When the 4-cell embryo is separated into the two dorsal and the two ventral halves, each half behaves differently. If allowed to develop, the dorsal halves will, as expected from Spemann's experiment, form dorsoanterior structures. In fact, the dorsoanterior structures are larger than in control embryos. The ventral halves develop into embryos with marked dorsoanterior deficiencies (Kageura and Yamana, 1983; Cooke and Webber, 1985). These embryos are similar to those formed by cutting the eggs after rotation. By the 8-cell stage, the vegetal dorsal pair of blastomeres contains nearly all the dorsal-forming activity of the frog embryo (see Sidelights & Speculations). Removing this pair of blastomeres stops the formation of dorsoanterior structures, while the substitution of one of these vegetal dorsal blastomeres into vegetal ventral region creates twins (Gimlich and Gerhart, 1984; Kageura and Yamana, 1986).

The location of the dorsoanterior structure-forming agents in the 32- and 64-cell embryos has been delineated by transplantaton. Gimlich (1985, 1986) showed that the prospective chordamesoderm cells of the 32-cell embryo (i.e., the dorsal marginal zone cells; C1 and D1 in Figure 21), but not their ventral or lateral equatorial counterparts, were also able to induce the formation of the body axis. These dorsal marginal zone (DMZ) transplants also induced neighboring host cells to form the body axis, including the dorsoanterior somites and central neural cells (Figure 22). In Chapter 4, we discussed the dramatic transplantation experiments of Gimlich and Gerhart (1984). When the dorsalmost vegetal cell of the 64-cell *Xenopus* blastula (a descendant of the D1 cell in Figure 21) was transplanted into a 64-cell *Xenopus* embryo whose axis would not form due to its having been

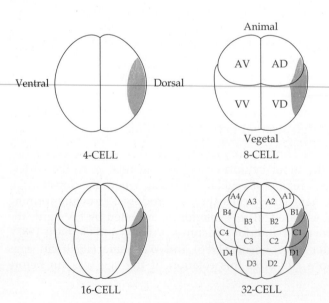

FIGURE 21
Hypothetical boundaries for the location of "dorsal morphogenic activities" in the 4–32 cell stage *Xenopus* embryo. The region is thought to remain stable throughout these cleavages. There is general agreement that dorsal information is contained in the C1 and D1 blastomeres of the 32-cell embryo. There is a possibility that it can extend into B1 at that time. (After Elinson and Kao, 1989).

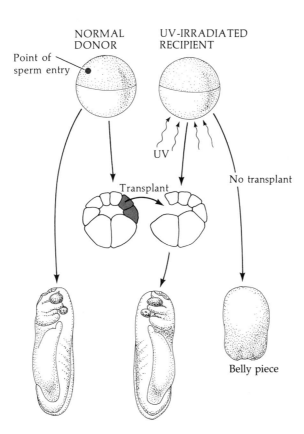

Point of
sperm entry

UV

Transplant

No transplant

Belly piece

FIGURE 22

Rescue of irradiated embryos by the transplantation of dorsal marginal cells of normal 32-cell embryos. Without such a rescue, irradiated embryos fail to gastrulate, forming "belly pieces." If the dorsal marginal zone cells of a normal embryo are transplanted into the irradiated embryo, they will induce the formation of mesoderm and the body axis, thereby producing a normal tadpole. (After Gimlich, 1986.)

irradiated at fertilization, the transplanted cell induced the formation of the dorsoanterior structures. The vegetal cell would go on to form endodermal structures, while the mesoderm it induced to form was derived from host tissue. The experiment diagrammed in Figure 20 of Chapter 4 can be seen as analogous to that of Spemann and Mangold. But, whereas Spemann and Mangold obtained two body axes by transplanting DMZ cells, Gimlich and Gerhart obtained the same phenomenon by transplanting the dorsalmost *vegetal* cells—the cells beneath the DMZ.

The conclusion that dorsoanterior information was contained within the D1 (dorsalmost vegetal) blastomere was further demonstrated by recombination experiments (Figure 23). Dale and Slack (1987) recombined single vegetal blastomeres from a 32-cell *Xenopus* embryo with the uppermost animal tier of a fluorescently labeled embryo of the same stage. The dorsalmost vegetal cell, as expected, induced the animal pole cells to become dorsal mesoderm. The remaining vegetal cells generally induced these animal cells to produce either intermediate or ventral mesodermal tissues. Thus, dorsal vegetal cells can induce animal cells to become dorsal mesodermal tissue.

At about the 32-cell stage, DMZ cells begin aquiring the ability to become dorsal mesoderm on their own (Gimlich, 1986). When Takasaki and Konishi (1989) transplanted the DMZ blastomeres of *Xenopus laevis* (B1 and C1) into the ventral marginal zone of *Xenopus borealis*, the dorsal blastomeres produced descendants that differentiated into chordamesoderm and somite tissues (Figure 24). These cells induced the host tissues to form a secondary neural tube with a head and tail. None of these embryos had complete anterior structures (the cement gland and nose), indicating that not all the information from the D1 blastomeres is present in these cells. Malacinski and his colleagues (1980) noted a similar pattern in the axolotl salamander embryo, where dorsal-inducing activity is first seen at the 64-cell stage, in the dorsalmost vegetal blastomeres.

FIGURE 23

Regional specificity of mesoderm induction shown by recombining cells of 32-cell *Xenopus* embryos. The animal pole cells of 32-cell embryos were combined with individual vegetal blastomeres. The animal pole cells were labeled with fluorescent beads to identify their descendants. The inductions resulting from these recombinations are summarized at right. (After Dale and Slack, 1987.)

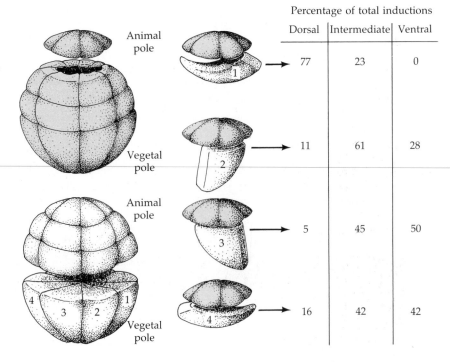

Percentage of total inductions

	Dorsal	Intermediate	Ventral
1	77	23	0
2	11	61	28
3	5	45	50
4	16	42	42

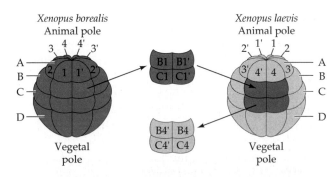

FIGURE 24

Interspecific transplantation of dorsal marginal (equatorial) blastomeres from 32-cell stage *Xenopus borealis* into cavity made by removing ventral marginal cells from *X. laevis*. Chart depicts the structures within the secondary embryo. Colored areas represent donor cells; gray areas represent host cells, striped areas indicate that both donor and host cells comprised the structure; white areas indicate that no structure of this type was formed in the secondary embryo. (After Takasaki and Konishi, 1989.)

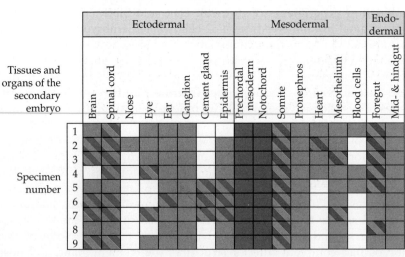

SIDELIGHTS & SPECULATIONS

The mosaic qualities of frog eggs

Using techniques that permit the survival of isolated *Xenopus* blastomeres throughout the larval period, Kageura and Yamana (1984; Yamana and Kageura, 1987) have shown that the 8-cell *Xenopus* embryo is more mosaic than had previously been thought.

First, they confirmed Spemann's observations that both halves of an 8-cell embryo split along the plane of the first cleavage give rise to normal embryos. Since both lateral halves can form complete embryos, it would appear that all the cells needed for complete embryogenesis were found in each lateral half. Each half consists of an animal dorsal

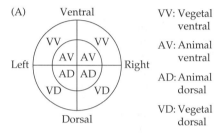

(A) Ventral

VV: Vegetal ventral
AV: Animal ventral
AD: Animal dorsal
VD: Vegetal dorsal

Left — Right

Dorsal

(B) Deletion of four cells

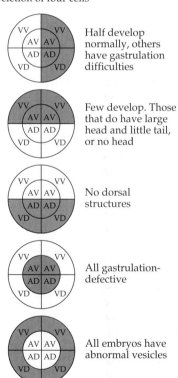

Half develop normally, others have gastrulation difficulties

Few develop. Those that do have large head and little tail, or no head

No dorsal structures

All gastrulation-defective

All embryos have abnormal vesicles

cell, an animal ventral cell, a vegetal dorsal cell, and a vegetal ventral cell. Are each of these cells necessary for development, or can one be excised without injury to the larva? To answer this, one more cell was deleted from the lateral half-embryos (Figure 25). The removal of the vegetal dorsal cell from the developing lateral half caused the formation of embryos with little or no head structures. The removal of the vegetal ventral cell from the developing half led to the formation of larvae with large heads and little tail. When either of the two animal pole cells was removed from the half-embryos, the resulting blastula failed to gastrulate properly, leaving the mesoderm and endoderm exposed on the surface. This was probably due to the absence of epidermal precursor cells. It appears, then, that in order to regulate an entire larva, an embryo needs at least the four types of cells composing the right or left sides of the 8-cell embryo.

FIGURE 25
Analysis of the isolation experiments. (A) Key to the positions of the four sets of blastomeres in the 8-cell *Xenopus* embryo, looking down on the animal pole. (B) Results of experiments wherein half the embryo was removed. (C) Results of experiments wherein the lateral half of 8-cell embryos developed in the absence of one further blastomere. (After Kageura and Yamana, 1984.)

(C) Deletion of five cells

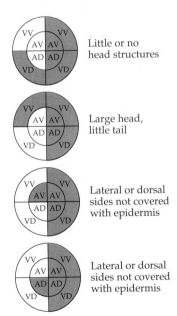

Little or no head structures

Large head, little tail

Lateral or dorsal sides not covered with epidermis

Lateral or dorsal sides not covered with epidermis

FIGURE 26
Summary of the experiments of Nieuwkoop and those of Nakamura and Takasaki, showing mesodermal induction by vegetal endoderm. Isolated animal cap cells become a mass of ciliated epidermis, isolated vegetal cells generate gutlike tissue, and isolated equatorial (marginal) cells become mesoderm. If the animal cap cells are combined with vegetal hemisphere cells, many of the cells of the animal cap generate mesodermal tissue.

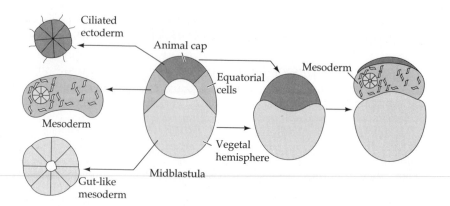

Induction of mesodermal specificity by the endoderm

Nieuwkoop (1969, 1973, 1977) demonstrated the importance of the vegetal endoderm in inducing the mesoderm. He removed the equatorial cells from the blastula and showed that neither the animal nor the vegetal caps produced mesodermal tissue. However, when the two caps were recombined, the animal cap cells were induced to form mesodermal structures such as notochord, muscles, kidney cells, and blood cells. Moreover, the polarity of this induction (whether the region of animal cells formed notochord or muscles, and so on) depended on the dorsal–ventral polarity of the endodermal cap. In a related series of experiments, Nakamura and Takasaki (1970) showed that explants from the equatorial region of the midblastula (the piece removed in Nieuwkoop's experiments) were able to form mesoderm in culture. Explants from the same regions of 32- and 64-cell embryos failed to form mesoderm, becoming ciliated ectodermal cells. These experiments, summarized in Figure 26, suggested (1) that prospective mesoderm cells were not determined immediately but achieved their prospective fates by the progressive interaction of animal and vegetal cells, and (2) that the equatorial precursors already had their mesodermal fate determined before gastrulation.

This has been confirmed (in general, if not in the specifics)* by the transplant experiments discussed earlier. The frequency with which the DMZ transplants can produce the complete axis in irradiated host embryos starts off very low and increases with age. Conversely, the mesoderm-inducing activity of the dorsal *vegetal* cells declines over this period (Gimlich, 1985; Boterenbrood and Nieuwkoop, 1973). It appears, then, that the cells of the dorsal vegetal quadrant in some way activate (or produce) factors within the DMZ cells, and that the complete "inductive mesoderm" phenotype develops gradually as the vegetal cells influence the marginal cells above them.

The ventral and lateral vegetal cells also have roles in specifying the mesoderm. Whereas the dorsalmost vegetal cells specify the axial mesoderm components (notochord and somites), the remaining vegetal cells specify the intermediate (muscle, mesenchyme) and ventral (mesenchyme, blood, pronephric kidney) types of mesoderm (Figure 23). While the dorsalmost vegetal cell induced the animal pole cells to become dorsal me-

*It is difficult to compare explant experiments and transplant experiments, since they are asking the cells to perform different tasks. Also, the third cleavage of the frog egg is not uniformly in the same place in all frog embryos. Due to yolk concentrations, it is sometimes lower or higher (see Elinson and Kao, 1989). Thus, in some instances, the B1 cell may have mesodermal properties and in other instances it may not.

soderm, the remaining vegetal cells induced these same cells to become either intermediate or ventral mesodermal tissues. Although this experiment demonstrated two distinct inductive processes (one from the dorsalmost vegetal cells to induce the dorsal marginal cells and one from the other vegetal cells to induce the intermediate and ventral mesoderm), it did not explain how ventral mesoderm is distinguished from intermediate mesoderm.

Dale and Slack (1987) provide evidence that can be explained by a third inductive signal, coming from the dorsal maginal cells, that "dorsalizes" the marginal cells adjacent to them. When the ventral marginal cells are isolated, they will give rise primarily to ventral mesodermal tissues. However, if they are cultured adjacent to dorsal marginal cells, they generate intermediate mesodermal tissue. Thus, there is evidence for a three-step specification of the mesoderm (Figure 27): (1) the induction of the organizer activity by the dorsalmost vegetal cells, (2) the induction of the ventral and lateral mesoderm by the other vegetal cells, and (3) the dorsalization of the ventral marginal cells adjacent to the dorsal marginal cells to produce the intermediate mesoderm.

Induction of the mesoderm influences gene activity

According to the above model, there are only two types of tissues specified in the blastula—the ectoderm and the endoderm. The mesoderm is specified by the induction of animal cap ectodermal cells by the endoderm. This has been borne out by recent studies wherein a cell's RNA can be identified. (Such procedures will be detailed in Chapter 10.) Sargent and

FIGURE 27
Inductive interactions during early *Xenopus* development. During oogenesis, the animal–vegetal axis arises. Fertilization causes cytoplasmic rearrangements that subdivide the vegetal region into dorsovegetal (DV) and ventrovegetal (VV) areas. During cleavage, mesodermal induction occurs such that the DV region induces the organizer activity (O) in the dorsal marginal cells above it, while the VV induces the cells above to become mesodermal (M). The polarity of this mesoderm (M1, M2, M3, M4) is a reflection of the underlying polarity of the ventral cells. During gastrulation, the ventral and lateral mesoderm go to the sides of the gastrula (not shown), while the dorsal mesoderm expands and induces the polarity in the ectodermal cells above it. This causes these ectodermal cells to become the different regions of the neural tube (N1, N2, N3, N4). The polarity of the endoderm is thus transferred to the neural tissue. The uninduced ectoderm (A) becomes epidermis (E). (After Smith et al., 1985.)

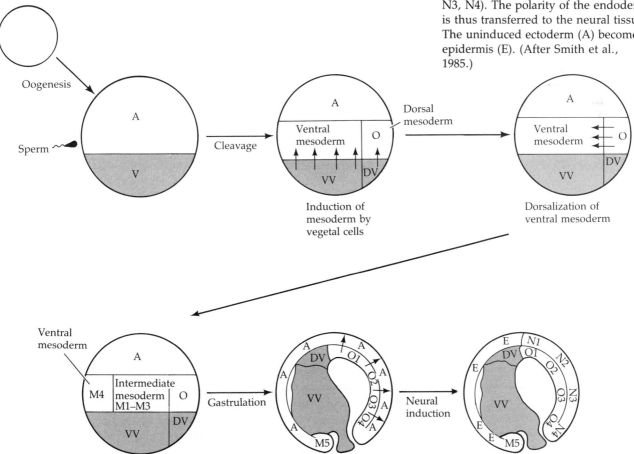

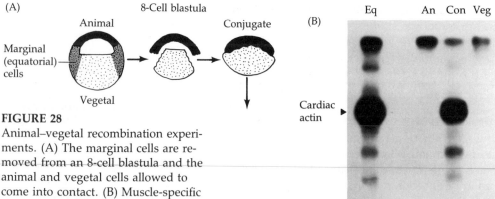

(A)

Marginal (equatorial) cells

Animal

8-Cell blastula

Vegetal

Conjugate

(B)

Eq   An Con Veg

Cardiac actin

FIGURE 28

Animal–vegetal recombination experiments. (A) The marginal cells are removed from an 8-cell blastula and the animal and vegetal cells allowed to come into contact. (B) Muscle-specific actin was found in the marginal (equatorial) region and in the recombined animal and vegetal cells, but not in vegetal or animal cells alone. In addition, a portion of the muscle-specific actin gene was made radioactive and was allowed to hybridize with its message, if the message was present. A nuclease was then added that would destroy any single-stranded DNA but leave DNA–RNA hybrids. The resulting DNA–RNA hybrids were subjected to electrophoresis on polyacrylamide gels and detected by autoradiography. The radioactive gene fragment was protected from digestion only when muscle-specific actin mRNA was present. (After Gurdon et al., 1985b; photograph courtesy of J. Gurdon.)

his co-workers (1986) have dissociated early blastula (32- to 128-cell stage) *Xenopus* embryos into their component cells by removing the fertilization envelope and placing the embryos into medium lacking calcium and magnesium ions. They then asked whether these dispersed cells could make mRNA specific to a particular germ layer after an appropriate amount of time had elapsed. They found that an endoderm-specific gene (for a particular gut protein) and an ectoderm-specific gene (for a particular cytoskeletal protein) were activated to make new mRNA even when the cells were dispersed. This finding suggests that these cells are regulated autonomously by cytoplasmic factors within themselves. The mesoderm-specific gene (for α-actin), however, is not activated when the cells are separated. When embryos were dissociated as late as the 128-cell stage, no α-actin mRNA can be detected in the progeny of the dissociated cells. But this gene can be activated if the dispersed cells are clumped together and thus allowed to interact with one another. It appears, then, that mesoderm-specific gene expression requires cell contact between at least two cell types.

Gurdon and his co-workers (1985) have also confirmed Nieuwkoop's observations. If Nieuwkoop's morphological studies are correct, α-actin gene activity should be induced in animal hemisphere cells when such cells (which would normally give rise to ectodermal tissues) are placed directly upon vegetal cells. Gurdon and his co-workers dissected away the marginal cells of midblastula embryos and recombined the animal pole cells with the vegetal cell mass (Figure 28). They found that the animal pole cells began transcribing mesoderm-specific α-actin message. The vegetal cells had induced a mesoderm-specific gene to be turned on in the presumptive ectodermal tissue. At the same time, these interactions appear to cause the loss of ectoderm-specific gene activity (Sargent et al., 1986).

Search for the mesoderm inducers

As illustrated in Figure 27, the mesoderm is thought to be induced by three (sets of) factors. Two factors (from the dorsal vegetal cells and from the ventral vegetal cells) induce the formation of a ring of mesoderm containing an organizer region (i.e., those cells above the dorsal vegetal cells). The organizer then sends out another factor, which induces the regionally specific structures of the mesoderm (such as dorsally located notochord and somites and peripherally located blood cells). Although the identity of these agents remains unknown, recent studies have shown that they may be identical or very similar to PEPTIDE GROWTH FACTORS, which are known to regulate mammalian cell proliferation (and which will be discussed in Chapter 20). The factor elaborated by the ventral vegetal cells may be similar to FIBROBLAST GROWTH FACTOR (FGF). Slack and his co-workers (1987) have shown that when the animal hemisphere of mid-blastula (1024–2048 cells) *Xenopus* embryos is exposed to low concentrations of mammalian FGF, it forms mesodermal tissue such as blood cells, muscles, and mesenchyme. Kimelman and Kirschner (1987) showed that when *Xenopus* animal pole cells are incubated with FGF, these cells are induced to transcribe the mesoderm-specific α-actin genes. Moreover, like FGF, the natural mesodermal inducer binds specifically to the polysaccharide heparin. The *Xenopus* FGF-like protein has been found in *Xenopus* oocytes and appears to be very similar to the mammalian FGF (Kimelman et al., 1988; Slack and Isaacs, 1989). It thus appears that a protein very similar to fibroblast growth factor is responsible for the induction of ventral mesoderm in frog embryos.

The factor responsible for generating the dorsal mesoderm (notochord and axial muscles) may be a 24,000-Da protein analogous to the mammalian hormone ACTIVIN (Smith et al., 1990). The first clue that such a protein might be an inducer of dorsal mesoderm came with the discovery that a clone of cultured *Xenopus* cells, the XTC cell line, secreted a protein that that caused animal hemisphere cells from midblastula *Xenopus* embryos to become notochord and muscle (but not blood, and rarely kidney) mesodermal tissue (Smith, 1987; Smith et al., 1988). The XTC medium appears to act in two ways: it blocks epidermal differentiation, and it stimulates mesodermal differentiation. First, it prevents the animal cap cells from differentiating into epidermal cells; individual animal cap cells exposed to XTC medium will not differentiate into the epidermal cells that they would otherwise become. However, for the XTC medium to cause these cells to differentiate along the mesodermal pathway, the animal cap cells have to be aggregated together. Individual animal cap cells will not become mesodermal if exposed to XTC medium (Symes et al., 1988; Gurdon, 1988).

The next clue was that the XTC mesoderm-inducing factor appeared to be in the family of proteins resembling TRANSFORMING GROWTH FACTOR-β (TGF-β). First, one of these proteins, TGF-β2, was found to induce α-actin gene expression in animal cap cells, albeit at lower levels than the XTC factor (Figure 29). Second, antibodies against TGF-β2 blocked 80 percent of the mesodermal induction of animal cap cells by XTC medium (Rosa et al., 1988). Third, Roberts and her colleagues (1991) found that TGF-β3 was 10 times more active than TGF-β2 in inducing the α-actin mRNA in animal pole cells. However, in other assays, the XTC mesoderm-inducing factor could not substitute for TGF-β, suggesting that the two molecules were related but were not the same. When Smith and co-workers (1990; van dan Eijnden-van Raaij et al., 1990) purified the XTC mesoderm-inducing factor, it was found to induce mesoderm formation in animal caps at concentrations as low as 0.2 ng/ml. TGF-β3, however, could induce mesoderm formation only at concentrations of approximately 1 ng/ml. The amino acid sequence of the purified mesoderm inducing factor showed that it was a member of the TGF-β family of proteins, but not TGF-β or the related protein, the product of the Vg1 mRNA. Rather, it resembled ACTIVIN A, a protein that is involved in releasing follicle-stimulating hormone from the pituitary and in stimulating red blood cell development. The functional equivalence between the *Xenopus* mesoderm-inducing factor was shown in reciprocal assays. The XTC protein could substitute for activin A in stimulating the differentiation of erythroid precursors into erythrocytes. Conversely, purified activin A was able to induce notochord, muscle, and also neural differentiation in animal cap tissue.

FIGURE 29

Efficacy of TGF-β2 in inducing mesoderm-specific (cardiac) actin in animal cap blastomeres. Increasing amounts of TGF-β2 produce increasing amounts of cardiac actin message in *Xenopus* animal cap tissue. Similar amounts of the related protein TGF-β1 do not induce the specific actin. Bar graphs show the level of cardiac actin message induced by vegetal halves, XTC medium, and XTC medium after treatment with antibodies to TGF-β2. (After Rosa et al., 1988.)

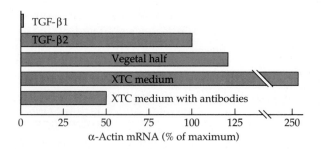

Experiments by Asashima's laboratory (1990) and by D. A. Melton's laboratory (Sokol et al., 1990) confirmed that activin A could induce dorsal mesoderm in animal caps explants. The paper by Sokol and colleagues is especially exciting, since in their hands activin A induced the animal cap cells to produce dorsal mesoderm and to generate an embryo-like structure with a head (including eyes and a brain) at one end and a tail at the other.

But is activin A the *natural* inducer of dorsal mesoderm? Is it found in the early embryo? Thomsen and co-workers (1990) cloned the two genes for *Xenopus* activin—*activin A* and *activin B*. Using these genes to find the *activin* mRNAs (in the procedure outlined in Figure 28), they first detected *activin A* mRNA in the late gastrula, but were able to see activin B message in the *Xenopus* blastula. This meant that activin B (a less common form of activin) was a better candidate for the dorsal mesoderm inducer than activin A. Thomsen and co-workers then demonstrated that activin B would also cause isolated animal caps to become embryo-like "bodies." Activin B appears to be able to induce the animal cap cells to produce the different types of mesoderm in a polarized manner, and this mesoderm appears to be able to induce neural tube structures in the cells above it. If this is so, activin B should act like the dorsalmost vegetal blastomeres. Thomsen and colleagues tested this by injecting *activin B* mRNA into a *ventral* equatorial blastomere of a 32-cell *Xenopus* embryo. The result (shown in the color portfolio) was the formation of a secondary embryonic axis!

Activin B can also induce axial structures in the chick epiblast, and *activin B* mRNA has been identified in the hypoblast at the time when axial mesoderm is being induced (Mitrani et al., 1990). It seems, then, that activin B is an excellent candidate for the dorsal mesoderm inducer.

SIDELIGHTS & SPECULATIONS

Regional specificity, mesoderm-inducing factors, and the Xhox3 gene

Although activin may be responsible for *inducing* the mesoderm, it may not be sufficient to cause the *regionalization* of the mesoderm. Even the axes formed by activin B are not complete secondary embryos. Green and Smith (1990) have recently shown that at a given dosage of activin A, both notochord and muscles are formed. Thus there is no simple correlation between activin concentration and type of mesoderm produced. Rather, regional specification of the dorsal–ventral and anterior–posterior axes may result from the interaction of a gradient of activin with other peptide factors, such as FGF, TGF-β, and Vg1. Recent studies demonstrate that these peptide growth factors may play im-

portant roles in the regional specification of the mesoderm and the resulting neural tube.

Ruiz i Altaba and Melton (1989a,b,c) looked for a gene that might be responsible for the anterior–posterior polarity of the embryo. In *Drosophila*, there are a series of genes that are concerned with the specification of the body axes, and many of these genes are characterized by a 180-base pair stretch of DNA called the homeobox. This homeobox produces a 60-amino acid region that binds to specific regions of *Drosophila* DNA. Employing techniques that will be discussed in Chapter 11, Ruiz i Altaba and Melton used a homeobox region from a *Drosophila* gene to find a homeobox-containing gene in frog embryos. This gene was called *Xhox3* (*Xenopus homeobox-3*).

Next, when these investigators isolated the messenger

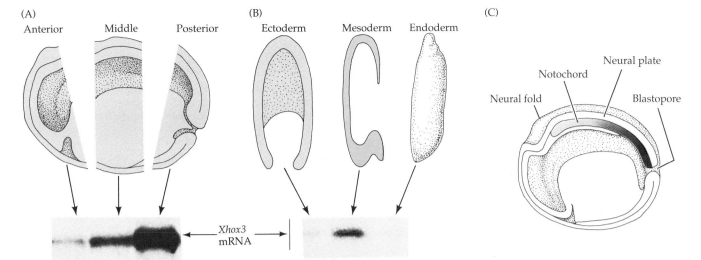

(A) Anterior — Middle — Posterior (B) Ectoderm — Mesoderm — Endoderm (C) Neural fold — Notochord — Neural plate — Blastopore

FIGURE 30

Gradient of *Xhox3* mRNA in late gastrula mesoderm. The *Xhox3* mRNA is recognized by a radioactive segment of the *Xhox3* DNA that binds specifically to its complement. (A) Gradient of *Xhox3* mRNA wherein there is little *Xhox3* message in the anterior region of the gastrula, somewhat more in the middle of the gastrula, and abundant *Xhox3* mRNA in the posterior region of the gastrula. (B) Dissection into ectoderm, endoderm, and mesoderm shows that most of the *Xhox3* mRNA is located in the mesoderm. (C) Interpretive drawing of a gradient of *Xhox3* mRNA in the notochord. (Photographs courtesy of D. A. Melton.)

RNA from a series of developing *Xenopus* embryos, they found that the *Xhox3* mRNA was seen shortly after midblastula transition. It peaked during the late gastrula-early neurula stages and disappeared shortly thereafter. Moreover, a series of dissection experiments (Figure 30) showed that *Xhox3* mRNA was found only in the mesoderm and that it formed a gradient whose posterior end had from 5–10 times more *Xhox3* message than the anterior end.

Since the regional specificity of the neural ectoderm is thought to reflect underlying differences in the mesoderm, the *Xhox3* gene product looked like an excellent candidate for the molecule that would specify positional information. First, it had its peak of activity during late gastrulation and early neurulation (when the regional specificity of the mesoderm is known to be fixed). Second, its activity was localized primarily, if not exclusively, in the mesoderm; and third, there was a difference in concentration between the anterior and posterior portions of the mesoderm.

Other experiments provided further evidence that *Xhox3* was important in specifying the anterior–posterior axis of *Xenopus*. First, two correlations were made. In lithium-treated embryos, the entire mesodermal mantle acts as a dorsoanterior organizer (Kao and Elinson, 1988).* Such

*This teratogenic effect appears to be caused by the ability of lithium ions to block the enzymes that allow the recycling of phosphatidylinositol. This blockage causes a depletion of the secondary messengers inositol trisphosphate and diacylglycerol. The addition of inositol to the lithium-treated embryos can circumvent the block, which suggests that the activation of the phosphatidylinositol breakdown pathway (discussed in Chapter 2) is involved in the induction of the dorsal mesoderm (Busa and Gimlich, 1989).

embryos develop enormous notochords that give rise to multiple eyes and cement glands. If the *Xhox3* gene were involved in specifying the anterior–posterior axis, such lithium-treated embryos should have a notochord almost devoid of *Xhox3* mRNA. Indeed, that was found to be the case. Conversely, the mesoderm of ultraviolet-treated embryos that form only posterior structures were found to have large amounts of *Xhox3* message in their mesoderm.

Another prediction of this hypothesis is that if anterior mesodermal cells could be made to express the *Xhox3* product, they would differentiate into posterior mesoderm and cause their overlying ectoderm to become posterior, rather than anterior, neural tube. Ruiz i Altaba and Melton therefore injected individual blastomeres with large concentrations of *Xhox3* mRNA. The animal pole of the cleaving embryo gives rise to cells that are predominantly in the anterior part of the larva. When *Xhox3* mRNA was injected

FIGURE 31

Anterior defects caused by injection of *Xhox3* mRNA into the animal poles of cleaving *Xenopus* eggs. If injected with a control mRNA (for β-globin protein), the embryo develops into a normal tadpole (top). If injected with the same amount of *Xhox3* mRNA, embryos develop into larvae with reduced or absent anterior structures. (From Ruiz i Altaba and Melton, 1989b; photographs courtesy of D. A. Melton.)

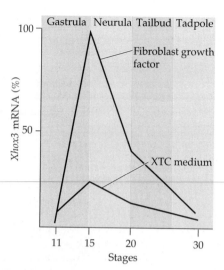

FIGURE 32

Different growth factors induce different amounts of *Xhox3* mRNA. Sets of 50 isolated animal caps collected at different stages of development were treated with basic fibroblast growth factor or XTC medium at levels that induced a similar amount of mesoderm-specific actin. The mRNA was collected and the *Xhox3* message was quantitated. FGF induced 5–10 times the amount of *Xhox3* mRNA induced by an equivalent amount of XTC factor. (After Ruiz i Altaba and Melton, 1989c.)

into the animal portion of the cleaving egg, the gradient that had existed was abolished. Instead, all regions of these embryos contained high amounts of *Xhox3* message. Of these injected embryos, 70 percent of them gave rise to larvae with defects in the anterior portion of their bodies (Figure 31). Injecting control mRNA into these blastomeres did not cause these defects. It is possible, then, that the level of *Xhox3* expression controls the specificity of the anterior–posterior axis.

How is this graded pattern of *Xhox3* expression established? Here we come back to the mesoderm-inducing factors. *Xhox3* can be induced in animal cap cells if the animal cap is combined with a vegetal cap. The *Xhox3* message appears at the same time as the message for mesoderm-specific actin. Moreover, *Xhox3* mRNA can be expressed in the animal cap tissue by incubating the animal cap in XTC medium or basic fibroblast growth factor (Ruiz i Altaba and Melton, 1989c). Nonrelated peptide growth factors do not induce this expression, and the induction of *Xhox3* mRNA is again seen concomitant with that of mesoderm-specific actin message. Thus, the same growth factors that induce the animal cap cells to become mesoderm also induce the expression of the *Xhox3* gene. (*Xhox3* does not induce the animal cells to become mesoderm, as the injection of *Xhox3* message into animal cap cells does not cause it to turn on their genes for mesoderm-specific α-actin.)

When animal caps were taken at different stages of development and treated with amounts of growth factors capable of inducing actin gene expression, a striking pattern emerged. The FGF needed to induce this mesodermal differentiation caused a large rise in the level of *Xhox3* mRNA in the animal cap. The XTC medium needed to

FIGURE 33

Peptide growth factors induce mesoderm with different anterior–posterior specificity. (A–C). Animal caps are isolated from blastulae and are treated in FGF, XTC medium, or a control saline solution. The caps are then implanted into the blastocoels of early gastrulae. A majority of the XTC-treated caps induce heads, a majority of the FGF-treated caps induce trunks and tails, and the control caps induce no extra structures. (D) Endogenous dorsal mesoderm from untreated *Xenopus* gastrulae also induce head structures. (From Ruiz i Altaba and Melton, 1989c; photographs courtesy of D. A. Melton.)

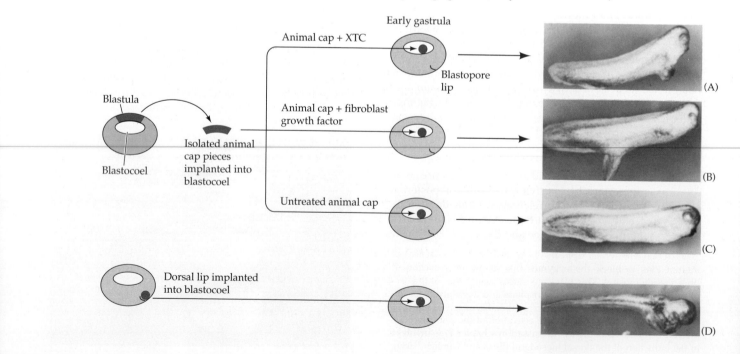

produce the same level of mesoderm induction caused only a small rise in the level of *Xhox3* message (Figure 32). Thus, XTC medium that contains the activin-like factor is correlated with inducing *dorsoanterior* mesoderm and evoking a *small* increase in *Xhox3* mRNA, whereas FGF is correlated with the induction of *ventroposterior* mesoderm, and is seen to cause *large* increases in the level of *Xhox3* mRNA. This agrees with the finding that *Xhox3* message is graded throughout the mesoderm, with its highest concentration around the posterior region by the blastopore.

But was the XTC-induced mesoderm really dorsal mesoderm and could it actually induce anterior structures? To answer this, Ruiz i Altaba and Melton (1989c) repeated Mangold's 1933 experiments with a slight variation. They isolated the animal cap of a blastula and incubated the cap in either FGF, XTC medium, or in a balanced salt solution containing neither of these growth factors. They then implanted the animal caps into the blastocoels of early *Xenopus* gastrulae (Figure 33). A majority of those animal caps treated with high levels of XTC medium produced head structures. Those animal caps incubated in FGF never produced head structures but would often produce tails. The untreated animal caps did not induce any extra structures in the larva. It appears, then, that both morphological criteria and functional criteria show that XTC medium induces dorsal anterior mesoderm, while FGF induces ventroposterior mesoderm. In both cases, this induction can be correlated with the levels of *Xhox3* mRNA induced by the respective growth factors.

These studies open up new questions concerning where the endogenous growth factors are found and how they become regionally and temporally expressed. Moreover, these studies also indicate that some other factor besides *Xhox3* must be acting to effect regional specificity. This fact can be deduced from the observation that the anterior neural structures are induced by mesoderm containing little if any *Xhox3*, and from the observation that the anterior cells induced by mesoderm loaded with *Xhox3* message did not become posterior cells, but rather disoriented and undifferentiated anterior cells.

Proposed mechanisms of neural induction

The nonspecificity of neural induction

We now come to the stage during which the neural ectoderm is determined (i.e., what Spemann thought of as primary induction). Although Spemann thought that searching for an organizer molecule was folly, his students pressed forward. They discovered that identifying the active inducer was not going to be easy. The problem was a lack of specificity; it seemed a huge variety of things could induce a neural plate. These compounds included turpentine, formaldehyde, and methylene blue dye, as well as dead archenteron and an assortment of adult tissues from several phyla. On the basis of such eclectic induction, Holtfreter (1948) suggested that the real inducer lay within the ectoderm itself and that it was released by mild cytolysis. Anything causing such sublethal damage would suffice, even a mildly alkaline or hypertonic medium. Barth and Barth (1969) have proposed a modification of this view, according to which the various compounds all act by releasing bound sodium ions within the ectodermal cells. If there is a masked inducer that is released by these nonspecific means (as well as by natural induction), it has not yet been found. John and co-workers (1984) provide evidence for a masked neuralizing factor within the riboprotein particles of ectoderm cells, but this factor has not yet been purified.

One possible reason nobody has been able to isolate *the* naturally occurring neural inducing factor is that there may be two simultaneous factors at work. If this were the case, no single molecule isolated from the chordamesoderm would be the inducer. There is now evidence (Davids et al., 1987; Davids, 1988; Otte et al., 1988, 1989) that there are two reactions needed for neural induction. The first is the activation of protein kinase C on the ectoderm cell surfaces, and the second is an increase in the concentration of cyclic AMP within the ectodermal cells. These studies have shown that if one and not the other of these events occurs, no neural tissue is formed. However, if one artificially activates protein kinase C *and* adenyl cyclase on the ectodermal cell membranes, neural tissue is formed. In this model, neural induction is accomplished by two reactions, and each reaction might be initiated by a different molecule.

Mechanisms of neural induction in *Xenopus*

Spemann considered two alternatives to explain the induction of a secondary neural tube by the dorsal blastopore lip. First, the organizer might be sending out its inducing signals tangentially through the plane of the ectoderm. Second, the organizer might induce the chordamesoderm, which then induces the overlying ectoderm to become neural tissue. Originally Spemann favored the first model, in which the dorsal lip of the blastopore directly induced the ectoderm (Spemann, 1918, 1927; Hamburger, 1988). However, the experiments by his students Mangold and Holtfreter supported the second model, at least in salamanders. Mangold, as discussed above, showed that the dorsal mesoderm could induce regionally specific neural tube structures, and Holtfreter (1933) prevented the interaction between the chordamesoderm and the dorsal ectoderm by removing the fertilization envelope and placing the embryos in a hypertonic solution. This caused the mesoderm to exogastrulate instead of involuting (Figure 34). In such embryos, where the dorsal mesoderm is separate from the dorsal ectoderm, Holtfreter saw no neural tube formation and concluded that the chordamesoderm was necessary to induce neural tube formation.

During the past five years, however, studies on *Xenopus laevis* have shown that the signal produced by the dorsal blastopore lip that goes through the *ectoderm* may be at least as important as the chordamesoderm signal. It had long been assumed that all early gastrula ectoderm had equal ability to be induced. Many investigators in the field now think that the dorsal ectoderm is already biased to become neural plate before the dorsal mesoderm even reaches it. Sharpe and co-workers (1987) showed that the *dorsal* ectoderm of early gastrulae (before any contact with underlying mesoderm has occurred) is much more readily induced than *ventral* ectoderm. As their markers for neural determination, they used two mRNAs for proteins found in neural, but not epidermal, tissues. Both these mRNAs are induced in the prospective neural region soon after neural

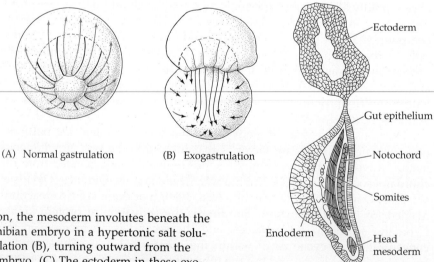

(A) Normal gastrulation (B) Exogastrulation

Ectoderm

Gut epithelium

Notochord

Somites

Endoderm

Head mesoderm

(C) Differentiation in exogastrulation

FIGURE 34
Exogastrulation. (A) In normal gastrulation, the mesoderm involutes beneath the ectoderm. However, by placing the amphibian embryo in a hypertonic salt solution, the mesoderm undergoes exogastrulation (B), turning outward from the ectoderm instead of involuting into the embryo. (C) The ectoderm in these exogastrulae does not form neural tissue. (After Holtfreter and Hamburger, 1955.)

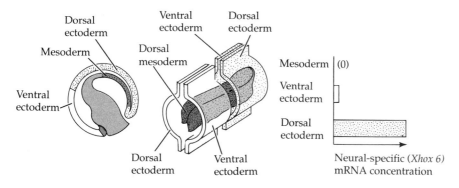

FIGURE 35
Neural gene expression in induction experiments using dorsal or ventral ectoderm. Ectoderm from the ventral and dorsal third of the embryo was wrapped around a piece of chordamesoderm. The fragments were then separated and tested for the presence of neural mRNAs. The mesoderm had none, the ventral ectoderm had little, and the dorsal ectoderm had high concentrations. (After Sharpe et al., 1987.)

plate formation begins. Sharpe and colleagues found that when *dorsal* ectoderm was combined with chordamesoderm, the neural mRNAs were induced in the ectoderm. However, when the same experiment was performed using *ventral* ectoderm, these mRNAs were not seen (Figure 35). This suggested that before the neurula stage, the dorsal ectoderm was already different than the ventral ectoderm.

Carey Phillips and his co-workers (London et al., 1988; Savage and Phillips, 1989) have provided evidence that these differences between prospective epidermis and prospective neural plate are present at the 8-cell stage. At that time, the animal pole blastomeres on the future dorsal side of the embryo lack a protein—Epi 1—that characterizes the animal pole cells on the future ventral side. During early gastrulation, the area of cells lacking Epi 1 is extended by signals transmitted from the dorsal blastopore lip through the ectoderm. This signal establishes a sharp boundary at what will become the lateral and anterior edges of the neural plate. Thus, they found that before the chordamesoderm involuted, the presumptive epidermal cells already expressed Epi 1, and the presumptive neural plate cells already lacked it. They also showed that the dorsal blastopore lip was capable of inducing these differences. When ventral ectoderm (prospective epidermis) was isolated from early gastrula cells, it expressed this Epi 1 protein (Figure 36). However, if this ventral ectoderm was placed in contact with dorsal blastopore lip from an early-to-mid gastrula, the prospective epidermis lost the protein. The loss of this protein is normally observed only on those ectoderm cells that will form the neural plate. Thus, a signal from the dorsal blastopore lip appears to bias the dorsal ectoderm to become neural cells rather than epidermis.

In addition to these studies, which show that the ventral ectoderm differs from the dorsal ectoderm prior to chordamesoderm contact, other investigations suggest that chordamesoderm is not needed to get neural tissue to differentiate. These studies are far more controversial. First, Dixon and Kintner (1989) showed that if the *Xenopus* ectoderm is removed from the involuting chordamesoderm before the mesoderm has made contact with the ectoderm, the ectoderm still synthesizes its neural cell adhesion molecule, N-CAM.* Second, when Phillips and Doniach (1990) repeated Mangold's experiments using *Xenopus* embryos, they did not get Mangold's results. They found that neither anterior nor posterior chordamesoderm of a late *Xenopus* gastrula could induce a neural tube (of any region) when placed into the blastocoel cavity of an early *Xenopus* gastrula. Neither was the N-CAM protein induced. However, when dorsal blastopore lip cells were placed into these blastocoels, they were found to be very effective at inducing secondary axes. These blastopore lips were also

*N-CAM is a glycoprotein that aids in holding the neural tube cells together. It is used here as a marker of neural differentiation that is more sensitive than observing the morphology of cells under a microscope.

FIGURE 36

Expression of the protein Epi 1 in the *Xenopus* ectoderm. Expression of Epi 1 was monitored with antibodies to Epi 1. (A) Isolated ventral ectoderm expresses Epi 1. (B) Ventral ectoderm in contact with the dorsal blastopore lip loses Epi 1 expression. The cells of the neural plate are devoid of Epi 1 expression, as seen in a neurula observed in cross section (C) or looking down at the dorsal surface of the neurula (D). (A and B from Savage and Phillips, 1989; C and D from London et al., 1988; photographs courtesy of C. Phillips.)

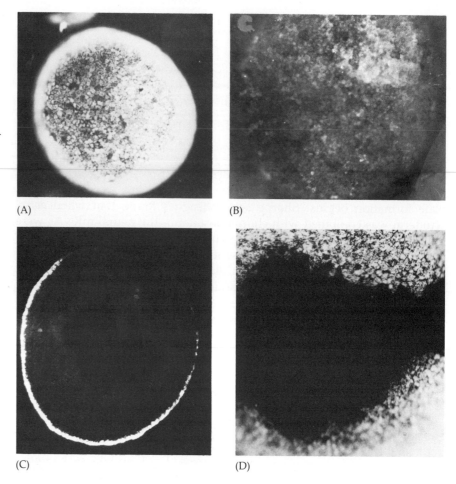

(A) (B)

(C) (D)

able to induce neural structures (and N-CAM) when combined laterally with isolated ventral ectoderm.

Phillips and Doniach believe that two sets of blastopore lip cells are necessary for neural induction. The cells that are at the blastopore lip in its earliest stage are able to send the signal to bias the ectoderm towards neural induction, but they are not able to induce neural proteins such as N-CAM in adjacent ectoderm. However, shortly afterwards the cells that comprise the dorsal blastopore lip are very effective at inducing N-CAM in the ectoderm. These N-CAM-inducing cells are those at the border between the involuting marginal zone and the noninvoluting marginal zone at the beginning of gastrulation. When isolated from the gastrula, these cells are able to induce N-CAM in adjacent ventral ectoderm.

It is possible that there are several signals involved in neural induction. The first set of signals biases the dorsal ectoderm toward becoming neural plate. One signal is present as early as the 8-cell stage; the other signal may be secreted by the early dorsal blastopore lip and travels through the ectoderm. The second set of signals may come from both the later dorsal lip (through the ectoderm) and from the chordamesoderm beneath the ectoderm. These latter signals may interact with each other. Dixon and Kintner (1989) provided evidence that both these signals are needed for complete neural determination. They found that a low level of two neural-specific mRNAs can be synthesized from animal cap ectoderm when it is placed in contact with the dorsal blastopore lip. Similarly, a very low level of these neural-specific messages are made when the ectoderm is placed in contact with involuting chordamesoderm. However, when the ectoderm is placed in such a manner that both types of contact occur, the ectodermal cells transcribe large amounts of these neural messages (Figure 37).

Phillips and Doniach propose that the blastopore lip signal sent through the ectoderm determines the dorsal ectoderm to be neural, and that the signal coming through the chordamesoderm determines the regional specificity of the neural structure produced. This conclusion is based on their findings that the chordamesoderm can influence the neural differentiation produced by the blastopore lip. When chordamesoderm is incubated next to ventral ectoderm, no neural structures arise in the ventral ectoderm. When ectoderm is incubated with blastopore lip tissue, the ectoderm is induced to form a tubelike structure that is wider than the spinal cord but lacks any convolutions or eye structures. However, when the ventral ectoderm is incubated with blastopore lip tissue and *anterior* chordamesoderm, many of the ectodermal explants will form N-CAM-expressing tubes with anterior-like structures, including eyes. When incubated with dorsal blastopore lip and posterior chordamesoderm, the ectoderm forms a compact N-CAM-expressing tube reminiscent of the spinal cord (Figure 38). However, the results of Dixon and Kintner and of Phillips and Doniach have been disputed by Sharpe and Gurdon (1990). Using different neural-specific mRNAs, they find that, although the dorsal ectoderm is biased for neural induction, the inducing signal comes solely from the chordamesoderm.

At the moment, we have very little data as to what these signals are or how they interact. It is possible that the ectodermal signals (whatever they are) act by the protein kinase C and cAMP pathways to determine the neural plate, while the Xhox 3 gradient acts to pattern its regional specificities. It is obvious that although our understanding of induction has increased enormously within the past five years, we still have much more work to do. Surveying the field in 1927, Spemann remarked,

> What has been achieved is but the first step; we still stand in the presence of riddles, but not without hope of solving them. And riddles with the hope of solution—what more can a scientist desire?

The challenge still remains.

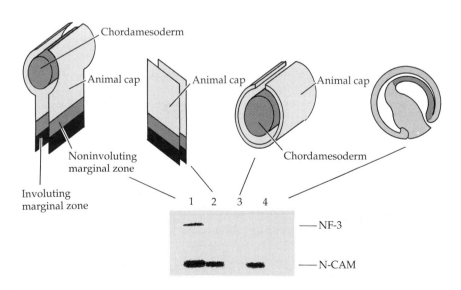

FIGURE 37
Evidence that the dorsal blastopore lip sends two signals for neural development. The mRNAs for two neural proteins, N-CAM (neural cell adhesion molecule) and NF-3 (a neurofilament protein) were assayed. In lane 1, the dorsal blastopore lip (black) was in contact with the presumptive (animal cap) ectoderm, as was the chordamesoderm (color). In lane 2, only the dorsal lip was in contact with the laterally adjacent ectoderm. In lane 3, only the chordamesoderm was in contact with the ectoderm from beneath it. Lane 4 represents the mRNA from control embryos. In all cases, two specimens were used. (After Dixon and Kintner, 1989.)

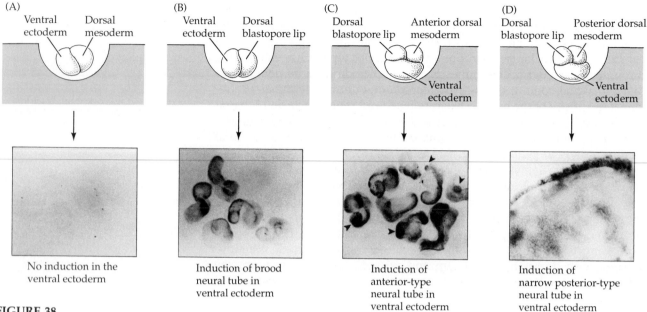

(A)	(B)	(C)	(D)
Ventral ectoderm / Dorsal mesoderm	Ventral ectoderm / Dorsal blastopore lip	Dorsal blastopore lip / Anterior dorsal mesoderm / Ventral ectoderm	Dorsal blastopore lip / Posterior dorsal mesoderm / Ventral ectoderm
No induction in the ventral ectoderm	Induction of brood neural tube in ventral ectoderm	Induction of anterior-type neural tube in ventral ectoderm	Induction of narrow posterior-type neural tube in ventral ectoderm

FIGURE 38

Modulation of neural induction by the chordamesoderm. (A) Grafting chorda-mesoderm to ventral ectoderm does not induce neural structures in the ectoderm. No N-CAM stain or neural tube is seen. (B) Grafting dorsal blastopore lip to ventral ectoderm induces in the ectoderm a wide tubular structure that stains positively for N-CAM. (C) Grafting dorsal blastopore lip and anterior chordamesoderm to ventral ectoderm induces an "anterior-type" neural tube in the ectoderm. These tubes stain positively for N-CAM and often show eyes (arrowheads). (D) Grafting dorsal blastopore lip and posterior chordamesoderm to ventral ectoderm induces a "posterior-type" neural tube in the ectoderm. The photograph is a close-up of the narrow N-CAM-positive tube. (After Phillips and Doniach, 1990; photographs courtesy of C. Phillips).

Competence and secondary induction

The primary inductive interactions, although complex, cannot construct the entire embryo. However, the formation of the neural tube, dorsal mesoderm, pharyngeal endoderm, and other tissues creates the conditions for a cascade of inductive events. Interactions by which one tissue interacts with another to specifically direct its fate are called SECONDARY INDUCTIONS.

Any system of embryonic induction has at least two components: a tissue capable of producing the inducing stimulus, and a tissue capable of receiving and responding to it. So far we have been looking at the specificity of production; now we must look at the specificity of responding cells. This ability to respond in a specific manner to a given stimulus is called COMPETENCE. We have already seen that in the early gastrula, an implanted blastopore lip can induce a new neural plate and embryonic axis just about anywhere in the embryo where it can meet ectoderm. However, with increasing embryonic age the ectoderm loses this ability to respond, and the implantation of a dorsal blastopore lip beneath the prospective epidermis of a neurula stage embryo will not cause it to form a new neural plate. It has lost its competence to respond to the new blastopore lip.

Although the late neurula ectoderm is no longer competent to respond to the blastopore lip, it has become competent to respond to new inducers. This competence can be localized to particular areas. During gastrulation and early neurulation, the *head* ectoderm (but not the *trunk* ectoderm) becomes competent to form the lens, nose, and ear placodes. This competence is acquired by its being acted upon by the neural plate region (see Figure 8 in Chapter 16; Henry and Grainger, 1990). Thus, the head region of the neurula is now competent to respond to contact from the optic vesicle (derived from the forebrain) to become lens.

Moreover, once a tissue has been induced, it can serve to induce other tissues. This has been seen in the formation of the motor neurons of the spinal cord. Jessell and his colleagues (1990; Placzek et al., 1990) have shown that once the neural tube is induced, it can still respond to inductive signals from the notochord. Those neural tube cells adjacent to the notochord (the ventralmost midline cells) become the floor plate cells.* If notochord fragments are taken from one embryo and transplanted to the lateral sides of a host neural tube, the host neural tube will form another set of floor plate cells at the sides of the neural tube. If a piece of notochord is removed from an embryo, the neural tube adjacent to the deleted region has no floor plate cells (Figure 39). These floor plate cells, once induced, serve to induce the formation of motor neurons on either side of them. We have seen, then, that the induction of the ectoderm to form neural cells also initiates a cascade wherein some of those induced cells begin to induce other cells. In one case, the induction of the lens is initiated by the anterior neural plate cells, and in the other case, the ventral floor plate cells induce the formation of motor neurons.

SUMMARY

In this and the preceding chapter, we have discussed the mechanisms of embryonic cell determination. We have seen that there are two general modes by which determination occurs. The first involves the partitioning of morphogenic determinants into specific cells and the subsequent determination of those cells contingent upon which region of zygote cytoplasm they enclose. Each of these cells will differentiate autonomously, irrespective of their surrounding cells. Organisms where this type of determination is prevalent tend toward a mosaic type of development.

The second mode of determination involves the interactions of cells somewhat later in development. Here, cells develop according to their positions within the embryo. Embryos whose cells are determined primarily in this interactive manner tend toward a regulative type of development. In this chapter, for instance, we considered a pathway in which endodermal cells induce the cells above them to become mesoderm. The dorsal mesoderm (notochord and dorsal blastopore lip) induce the neural determination and regional specification of the dorsalmost ectoderm cells. The notochord continues to induce the formation of the floor plate cells

*The floor plate cells are important for directing the migration of the commisural axons to the ventral midline of the embryo (Chapter 15).

FIGURE 39
Cascade of inductions initiated by notochord on the newly formed neural tube. (A) Two cell types in the newly formed neural tube. Those cells closest to the notochord become the ventral floor plate cells. Motor neurons emerge on the ventrolateral sides. (B) If a second notochord is transplanted adjacent to the neural tube, it induces a new set of floor plate cells and two new sets of motor neurons. (C) If the ventral floor plate cells are transplanted adjacent to the neural tube, new sets of motor neurons differentiate. (D) Representation of the inductive interactions between these cells. (After Jessel et al., 1990.)

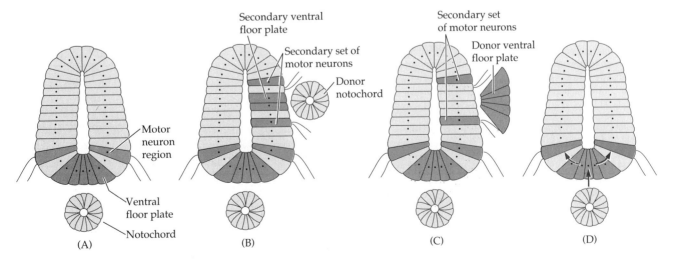

in the neural tube, and the floor plate cells induce the formation of motor neurons. By progressive cell–cell interactions, we have proceeded from the vegetal cells of the early frog blastula to the formation of specific types of neurons in the larval neural tube. Moreover, the pathways can diverge at each point. The notochord also induces cartilage formation in the somites, and the neural tube helps induce the formation of lenses.

It is important to realize that both mechanisms are used during the development of any particular organism and that there is a continuum from mosaic to regulative development. Mosaic animals such as tunicates and snails use regulative interactions to form certain tissues, and regulative animals such as frogs have regions of cytoplasm that contain morphogenic determinants such as polar granules. It is also important to note that both mosaic and regulative modes of determination are traceable to the heterogeneously distributed materials in the zygote cytoplasm.

LITERATURE CITED

Asashima, M., Nikano, H., Shimada, K., Kinoshita, K., Ishii, K., Shibai, H. and Veno, N. 1990. Mesodermal induction in early amphibian embryos by activin A (erythroid differentiation factor). *Wilhelm Roux Arch. Dev. Biol.* 198: 330–335.

Barth, L. G. and Barth, L. J. 1969. The sodium dependence of embryonic induction. *Dev. Biol.* 20: 236–262.

Berrill, N. J. and Liu, C. K. 1948. Germplasm, Weismann, and hydrozoa. *Q. Rev. Biol.* 23: 124–132.

Bïjtel, J. H. 1931. Über die Entwicklung der Schwanzes bei Amphibien. *Wilhelm Roux Arch. Entwicklungsmech. Org.* 125: 448–486.

Boterenbrood, E. C. and Nieuwkoop, P. D. 1973. The formation of the mesoderm in urodelean amphibians. V. Its regional induction by the endoderm. *Wilhelm Roux Arch. Entwicklungsmech. Org.* 173: 319–332.

Busa, W. B. and Gimlich, R. L. 1989. Lithium-induced teratogenesis in frog embryos is prevented by a polyphosphoinositide cycle intermediate or a diacylglycerol analog. *Dev. Biol.* 132: 315–324.

Buss, L. 1987. *The Evolution of Individuality.* Princeton University Press, Princeton, NJ.

Cameron, R. A., Hough-Evans, B., Britten, R. J. and Davidson, E. H. 1987. Lineage and fate of each blastomere of the eight-cell sea urchin embryo. *Genes Dev.* 1: 75–85.

Cooke, J. and Webber, J. A. 1985. Dynamics of the control of body pattern in the development of *Xenopus laevis.* I. Timing and pattern in the development of dorsoanterior and posterior blastomeres, isolated at the 4-cell stage. *J. Embryol. Exp. Morphol.* 88: 85–99.

Czihak, G. 1971. Echinoids. *In* G. Reverberi (ed.), *Experimental Embryology of Marine and Freshwater Invertebrates.* Elsevier North-Holland, Amsterdam, pp. 363–506.

Dale, L. and Slack, J. M. W. 1987. Regional specificity within the mesoderm of early embryos of *Xenopus laevis. Development* 100: 279–295.

Davids, M. 1988. Protein kinases in amphibian ectoderm induced for neural differentiation. *Wilhelm Roux Arch. Dev. Biol.* 197: 339–344.

Davids, M., Loppnow, B., Tiedemann, H. and Tiedemann, H. 1987. Neural differentiation of amphibian gastrula ectoderm exposed to phorbol ester. *Wilhelm Roux Arch. Dev. Biol.* 196: 137–140.

Davidson, E. H. 1989. Lineage-specific gene expression and the regulative capacities of the sea urchin embryo: a proposed mechanism. *Development* 105: 421–446.

Dixon, J. E. and Kintner, C. R. 1989. Cellular contacts required for neural induction in *Xenopus laevis* embryos: evidence for two signals. *Development* 106: 749–757.

Driesch, H. 1892. The potency of the first two cleavage cells in echinoderm development. Experimental production of partial and double formations. *In* B. H. Willier and J. M. Oppenheimer (eds.), *Foundations of Experimental Embryology.* Hafner, New York.

Driesch, H. 1893. Zur Verlagerung der Blastomeren des Echinideneies. *Anat. Anz.* 8: 348–357.

Driesch, H. 1894. *Analytische Theorie de organischen Entwicklung.* W. Engelmann, Leipzig.

Elinson, R. P. and Kao, K. R. 1989. The location of dorsal information in frog early development. *Dev. Growth Diff.* 31: 423–430.

Ettensohn, C. A. and McClay, D. R. 1988. Cell lineage conversion in the sea urchin embryo. *Dev. Biol.* 125: 396–409.

Gimlich, R. L. 1985. Cytoplasmic localization and chordamesoderm induction in the frog embryo. *J. Embryol. Exp. Morphol.* 89 [Suppl.] 89–111.

Gimlich, R. L. 1986. Acquisition of developmental autonomy in the equatorial region of the *Xenopus* embryo. *Dev. Biol.* 116: 340–352.

Gimlich, R. L. and Cook, J. 1983. Cell lineage and the induction of nervous sys-

tems in amphibian development. *Nature* 306: 471–473.

Gimlich, R. L. and Gerhart, J. C. 1984. Early cellular interactions promote embryonic axis formation in *Xenopus laevis. Dev. Biol.* 104: 117–130.

Green, J. B. A. and Smith, J. C. 1990. Graded changes in dose of a *Xenopus* activin A homologue elicit stepwise transitions in embryonic cell fate. *Nature* 347: 391–394.

Gurdon, J. B. 1988. A community effect in animal development. *Nature* 336: 772–774.

Gurdon, J. B., Fairman, S., Mohun, T. J. and Brennan, S. 1985. The activation of muscle-specific actin genes in *Xenopus* development by an induction between animal and vegetal cells of a blastula. *Cell* 41: 913–922.

Hagstrom, B. E. and Lonning, S. 1965. Studies in cleavage and development of isolated sea urchin blastomeres. *Sarsia* 18: 1–9.

Hamburger, V. 1984. Hilde Mangold, codiscoverer of the organizer. *J. Hist. Biol.* 17: 1–11.

Hamburger, V. 1988. *The Heritage of Experimental Embryology: Hans Spemann and the Organizer.* Oxford University Press, Oxford.

Henry, J. J. and Grainger, R. M. 1990. Early tissue interactions leading to embryonic lens formation in *Xenopus laevis. Dev. Biol.* 141: 149–163.

Henry, J. J., Amemiya, S., Wray, G. A. and Raff, R. A. 1989. Early inductive interactions are involved in restricting cell fates of mesomeres in sea urchin embryos. *Dev. Biol.* 136: 140–153.

Holtfreter, J. 1933. Die totale exogastrulation, eine Selbstablosung des ektoderms vom entomesoderm. *Wilhelm Roux Arch. Entwicklungsmech. Org.* 129: 669–793.

Holtfreter, J. 1936. Regional induktionen in Xenoplastisch zusammengesetzten explantaten. *Wilhelm Roux Arch. Entwicklungsmech. Org.* 134: 466–561.

Holtfreter, J. 1948. Concepts on the mechanism of embryonic induction and its relation to parthenogenesis and malignancy. *Symp. Soc. Exp. Biol.* 2: 17–48.

Holtfreter, J. and Hamburger, V. 1955. Amphibians. *In* B. H. Willier, P. A. Weiss and V. Hamburger (eds.), *Analysis of Development.* Saunders, Philadelphia, pp. 230–295.

Hörstadius, S. 1928. Über die Determination des Keimes bei Echinodermen. *Acta Zool.* 9: 1–191.

Hörstadius, S. 1935. Über die Determination im Verlaufe der Eiasche bei Seeigeln. *Pubbl. Stn. Zool. Napoli* 14: 251–479.

Hörstadius, S. 1939. The mechanics of sea urchin development studied by operative methods. *Biol. Rev.* 14: 132–179.

Hörstadius, S. and Wolsky, A. 1936. Studien über die Determination der Bilateralsymmetrie des jungen Seeigelkeimes. *Wilhelm Roux Arch. Entwicklungsmech. Org.* 135: 69–113.

Hough-Evans, B. R., Franks, R. R., Cameron, R. A., Britten, R. J. and Davidson, E. H. 1987. Correct cell type-specific expression of a fusion gene injected into sea urchin eggs. *Dev. Biol.* 121: 576–579.

Hurley, D. L., Angerer, L. M. and Angerer, R. C. 1989. Altered expression of spatially regulated embryonic genes in the progeny of separated sea urchin blastomeres. *Development* 106: 567–579.

Huxley, J. S. and deBeer, G. R. 1934. *Elements of Experimental Embryology.* Cambridge University Press, Cambridge.

Jessell, T., Placzek, M., Tessier-Lavigne, M., Wagner, M., Yamada, T. and Dodd, J. 1990. Control of axial polarity and cell differentiation in the developing central nervous system. Lecture at Forty-ninth Annual Meeting, Society for Developmental Biology, Georgetown University, Washington, D.C.

John, M., Born, J., Tiedemann, He. and Tiedemann, Hi. 1984. Activation of a neuralizing factor in amphibian ectoderm. *Wilhelm Roux Arch. Dev. Biol.* 193: 13–18.

Jukes, T. 1966. *Molecules and Evolution.* Columbia University Press, New York.

Kageura, H. and Yamana, K. 1983. Pattern regulation in isolated halves and blastomeres of early *Xenopus laevis. J. Embryol. Exp. Morphol.* 74: 221–234.

Kageura, H. and Yamana, J. 1984. Pattern regulation in defect embryos of *Xenopus laevis. Dev. Biol.* 101: 410–415.

Kageura, H. and Yamana, J. 1986. Pattern formation in 8-cell composite embryos of *Xenopus laevis. J. Embryol. Exp. Morphol.* 91: 79–100.

Kao, K. R. and Elinson, R. P. 1988. The entire mesodermal mantle behaves as Spemann's organizer in dorsoanterior enhanced *Xenopus laevis* embryos. *Dev. Biol.* 127: 64–77.

Kimelman, D. and Kirschner, M. 1987. Synergistic induction of mesoderm by FGF and TGF-β and the identification of an mRNA coding for FGF in the early *Xenopus* embryo. *Cell* 51: 869–877.

Kimelman, D., Abraham, J. A., Haaparanta, T., Palisi, T. M. and Kirschner, M. W. 1988. The presence of fibroblast growth factor in the frog egg: Its role as a natural mesoderm inducer. *Science* 242: 1053–1056.

London, C., Akers, R. and Phillips, C. 1988. Expression of Epi 1, an epidermis-specific marker in *Xenopus laevis* embryos, is specified prior to gastrulation. *Dev. Biol.* 129: 380–389.

Malacinski, G. M., Chung, H. M. and Asashima, M. 1980. The association of primary embryonic organizer activity with the future dorsal side of amphibian eggs and early embryos. *Dev. Biol.* 77: 449–462.

Mangold, O. 1933. Über die Induktionsfahigkeit der verschiedenen Bezirke der Neurula von Urodelen. *Naturwissenschaften* 21: 761–766.

McClendon, J. F. 1910. The development of isolated blastomeres of the frog's egg. *Am. J. Anat.* 10: 425–430.

Mitrani, E., Ziv, P. T., Thomsen, G., Shimoni, Y., Melton, D. A. and Bril, A. 1990. Activin can induce the formation of axial structures and is expressed in the hypoblast of the chick. *Cell* 63: 495–501.

Morgan, T. H. 1895. Studies on the "partial" larvae of *Sphaerechinus. Wilhelm Roux Arch. Entwicklungsmech. Org.* 2: 81–126.

Nakamura, O. and Takasaki, H. 1970. Further studies on the differentiation capacity of the dorsal marginal zone in the morula of *Triturus pyrrhogaster. Proc. Japan Acad.* 46: 700–705.

Nieuwkoop, P. D. 1969. The formation of the mesoderm in urodele amphibians. I. Induction by the endoderm. *Wilhelm Roux Arch. Entwicklungsmech. Org.* 162: 341–373.

Nieuwkoop, P. D. 1973. The "organisation center" of the amphibian embryo: Its origin, spatial organisation and morphogenetic action. *Adv. Mophog.* 10: 1–39.

Nieuwkoop, P. D. 1977. Origin and establishment of embryonic polar axes in amphibian development. *Curr. Top. Dev. Biol.* 11: 115–132.

Otte, A. P., Koster, C. H., Snoek, G. T. and Durston, A. J. 1988. Protein kinase C mediates neural induction in *Xenopus laevis. Nature* 334: 818–620.

Otte, A. P., van Run, P., Heideveld, M., van Driel, R. and Durston, A. J. 1989. Neural induction is mediated by cross-talk between the protein kinase C and cyclic AMP pathways. *Cell* 58: 641–648.

Okazaki, K. 1975. Spicule formation by isolated micromeres of the sea urchin embryo. *Am. Zool.* 15: 567–581.

Phillips, C. and Doniach, T. 1990. Lecture at Forty-ninth Annual Meeting, Society for Developmental Biology, Georgetown University, Washington, D.C.

Placzek, M., Tessier-Lavigne, M., Yamada, T., Jessell, T. and Dodd, J. 1990. Mesodermal control of neural cell identity: Floor plate induction by the notochord. *Science* 250: 985–988.

Recanzone, G. and Harris, W. A. 1985. Demonstration of neural induction using nuclear markers in *Xenopus. Wilhelm Roux Arch. Dev. Biol.* 194: 344–354.

Render, J. and Elinson, R. P. 1986. Axis determination in polyspermic *Xenopus* eggs. *Dev. Biol.* 115: 425–433.

Roberts, A. B. and 9 others. 1991. Mesoderm induction in *Xenopus laevis* distinguishes between the various TGF-β isoforms. *Growth Fact.* (in press).

Rosa, F., Roberts, A. B., Danielpour, D., Dart, L. L., Sporn, M. B. and Dawid, I. B. 1988. Mesoderm induction in amphibians: The role of TGF-β2-like factors. *Science* 239: 783–785.

Roux, W. 1888. Contributions to the developmental mechanics of the embryo. On the artificial production of half-embryos by destruction of one of the first two blastomeres and the later development (postgeneration) of the missing half of the body. *In* B. H. Willier and J. M. Oppenheimer (eds.), *Foundations of Experimental Embryology.* Hafner, New York, pp. 2–37.

Ruiz i Altaba, A. and Melton, D. A. 1989a. Bimodal and graded expression of the *Xenopus* homeobox gene *Xhox3* during embryonic development. *Development* 106: 173–183.

Ruiz i Altaba, A. and Melton, D.A. 1989b. Involvement of the *Xenopus* homeobox gene *Xhox3* in pattern formation along the antero-posterior axis. *Cell* 57: 317–326.

Ruiz i Altaba, A. and Melton, D. A. 1989c. Interaction between peptide growth factors and homeobox genes in the establishment of antero-posterior polarity in frog embryos. *Nature* 341: 33–38.

Runnström, J. 1929. Über Selbstdifferenzierun und Induktion bei dem Seeigelkeim. *Wilhelm Roux Arch. Entwicklungsmech. Org.* 117: 123–145.

Sargent, T. D., Jamrich, M. and Dawid, I. 1986. Cell interactions and the control of gene activity during early development of *Xenopus laevis. Dev. Biol.* 114: 238–246.

Savage, R. and Phillips, C. 1989. Signals from the dorsal blastopore lip region during gastrulation bias the ectoderm toward a nonepidermal pathway of differentiation in *Xenopus laevis. Dev. Biol.* 133: 157–168.

Saxén, L. and Toivonen, S. 1962. *Embryonic Induction.* Prentice-Hall, Englewood Cliffs, NJ.

Sharpe, C. R. and Gurdon, J. B. 1990. The induction of anterior and posterior neural genes in *Xenopus laevis. Development* 109: 765–774.

Sharpe, C. R., Fritz, A., De Robertis, E. M. and Gurdon, J. B. 1987. A homeobox-containing marker of posterior neural differentiation shows the importance of predetermination in neural induction. *Cell* 50: 749–758.

Slack, J. W. M. and Isaacs, H. V. 1989. The presence of basic fibroblastic growth factor in the early *Xenopus* embryo. *Development* 105: 147–153.

Slack, J. M. W., Darlington, B. G., Heath, J. K. and Godsave, S. F. 1987. Mesoderm induction in early *Xenopus* embryos by heparin–binding growth factor. *Nature* 326: 197–200.

Smith, J. C. 1987. A mesoderm-inducing factor is produced from a *Xenopus* cell line. *Development* 99: 3–14.

Smith, J. C., Dale, L. and Slack, J. M. W. 1985. Cell lineage labels and region-specific markers in the analysis of inductive interactions. *J. Embryol. Exp. Morphol.* 89 [Suppl.]: 317–331.

Smith, J. C., Yaqoob, M. and Symes, K. 1988. Purification, partial characterization, and biological effects of the XTC mesoderm-inducing factor. *Development* 103: 591–600.

Smith, J. C., Price, B. M. J., van Nimmen, K. and Huylebroeck, D. 1990. Identification of a potent *Xenopus* mesoderm-inducing factor as a homologue of activin A. *Nature* 345: 729–731.

Sokol, S., Wong, G. and Melton, D. A. 1990. A mouse macrophage factor induces head structures and organizes a body axis in *Xenopus*. *Science* 249: 561–564.

Spemann, H. 1918. Über die Determination der ersten Organanlagen des Amphibienembryo. *Wilhelm Roux Arch. Entwicklungsmech. Org.* 43: 448–555.

Spemann, H. 1927. Neue arbieten über organization in der tierischen entwicklung. *Naturwissenschaften* 15: 946–951.

Spemann, H. 1938. *Embryonic Development and Induction.* Yale University Press, New Haven.

Spemann, H. and Mangold, H. 1924. Induction of embryonic primordia by implantation of organizers from a different species. *In* B. H. Willier and J. M. Oppenheimer (eds.), *Foundations of Experimental Embryology.* Hafner, New York, pp. 144–184.

Spofford, W. R. 1945. Observations on the posterior part of the neural plate in *Amblystoma*. *J. Exp. Zool.* 99: 35–52.

Sucov, H. M., Hough-Evans, B. R., Franks, R. R., Britten, R. J. and Davidson, E. H. 1988. A regulatory domain that directs lineage-specific expression of a skeletal matrix protein in the sea urchin embryo. *Genes Dev.* 2: 1238–1250.

Symes, K., Yaqoob, M. and Smith, J. C. 1988. Mesoderm induction in *Xenopus laevis*: responding cells must be in contact for mesoderm formation but suppression of epidermal differentiation can occur in single cells. *Development* 104: 609-618.

Takasaki, H. and Konishi, H. 1989. Dorsal blastomeres in the equatorial region of the 32-cell *Xenopus* embryo autonomously produce progeny committed to the organizer. *Dev. Growth, Diff.* 31: 147–156.

Thomsen, G., Woolf, T., Whitman, M., Sokol, S., Vaughan, J., Vale, W. and Melton, D. A. 1990. Activins are expressed early in *Xenopus* embryogenesis and can induce axial and mesoderm anterior structures. *Cell* 63: 485–493.

Tiedmann, H., Becker, U. and Tiedmann, H. 1963. Chromatographic separation of a hindbrain-inducing substance into mesodermal- and neural-inducing subfractions. *Biochem. Biophys. Acta* 74: 557–560.

Toivonen, S. and Saxén, L. 1955. The simultaneous inducing action of liver and bone marrow of the guinea pig in implantation and explantation experiments with embryos of *Triturus*. *Exp. Cell Res.* [Suppl.] 3: 346–357.

van dan Eijnden-van-Raaij, A. J. M., van Zoelent, E. J. J., van Nimmen, K., Koster, C. H., Snoek, G. T., Durston, A. J. and Huylebroeck, D. 1990. Activin-like factor from a *Xenopus* laevis cell line responsible for mesoderm induction. *Nature* 345: 732–734.

Vincent, J.-P. and Gerhart, J. C. 1987. Subcortical rotation in *Xenopus* eggs: An early step in embryonic axis specification. *Dev. Biol.* 123: 526–539.

von Ubisch, L. 1919. Über die Determination der larvalen Organe und der Imaginalanlage bei Seeigeln. *Wilhelm Roux Arch. Entwicklungsmech. Org.* 117: 80–122.

Waddington, C. H. 1933. Induction by the primitive streak and its derivatives in the chick. *J. Exp. Biol.* 10: 38–46.

Waddington, C. H. 1940. *Organisers and Genes.* Cambridge University Press. Cambridge.

Wakahara, M. 1989. Specification and establishment of dorsal–ventral polarity in eggs and embryos of *Xenopus laevis*. *Dev. Growth Diff.* 31: 197–207.

Weismann, A. 1892. *Essays on Heredity and Kindred Biological Problems.* Translated by E. B. Poulton, S. Schoenland and A. E. Shipley. Clarendon, Oxford.

Weismann, A. 1893. *The Germ-Plasm: A Theory of Heredity.* Translated by W. Newton Parker and H. Ronnfeld. Walter Scott Ltd., London.

Wilson, E. B. 1896. *The Cell in Development and Inheritance.* Macmillan, New York.

Yamana, K. and Kageura, H. 1987. Reexamination of the "regulative development" of amphibian embryos. *Cell Differ.* 20: 3–10.

9

Genomic equivalence and differential gene expression: Embryological investigations

Heredity is effected by the transmission of a nuclear preformation which in the course of development finds expression in a process of cytoplasmic epigenesis.

—E. B. WILSON (1925)

Two cells are differentiated with respect from one another if, while they harbor the same genome, the pattern of proteins they synthesize is different.

—F. JACOB AND J. MONOD (1963)

Developmental genetics is the study of how the inherited potential of the fertilized egg is expressed during the life of an organism. When we observe the developing embryo, it becomes apparent that different cell types are expressing different genes. Hemoglobin, for instance, characterizes the red blood cells, whereas crystallin is found only in the lens cells of our eyes. The cells of the neural retina are capable of transmitting electrical impulses across large distances, whereas the adjacent cells of the pigmented retina are darkened by melanin granules and lack electrical conductivity. Yet each of these cell types arose from the mitotic divisions of the same fertilized egg. Each should contain the same nuclear information. Development, then, involves the differential expression of specific genes at specific places and times. The problem of developmental genetics then becomes: How is the genetic information regulated so that cells become different?

The central hypothesis of developmental genetics has been that cell differentiation occurs in the absence of genetic alteration. Within any organism, each somatic cell is thought to contain an identical complement of genes. Each different cell type would then utilize different genes from the common inherited repertoire. The basis for this hypothesis of DIFFERENTIAL GENE EXPRESSION comes from both genetics and embryology. In this chapter we will look at those studies seeking to determine whether or not the genome has undergone irreversible changes during development.

Genomic equivalence

Certain large, nondividing cells of larval flies such as *Drosophila* and *Chironomus* contain POLYTENE CHROMOSOMES. Such chromosomes undergo DNA replication in the absence of mitosis and therefore contain 512, 1024, or even more parallel DNA double helices instead of just one (Figures 1 and 2). These chromosomes never undergo mitosis and are visible under the light microscope, where they exhibit characteristic banding patterns. In *Drosophila*, approximately 5150 individual bands have been counted in the haploid genome. In some tissues, thick bands are seen that, upon stretching, resolve into two or more thinner bands. Several genetic studies (see Judd and Young, 1973) have suggested a correlation between the number of these bands, called CHROMOMERES, and the number of genes in the fly (Swanson et al., 1981). Beermann (1952) demonstrated that these chromosomes and their banding patterns were constant throughout the larval organism (Figure 3) and that no loss of any chromosomal region could be seen when different cell types were compared. When it became possible to study the individual chromosomes of vertebrate nuclei, Tjio and Puck (1958) also found chromosome constancy throughout the different tissues of the adult organism. As we will see later, various studies have demonstrated that DNA extracted from several different somatic tissues of the same individual is identical in each cell type.

The second support for the hypothesis of genomic equivalence came from embryology. Driesch and Spemann clearly demonstrated (Chapter 8) that the nuclei of early blastomeres of sea urchins and salamanders were TOTIPOTENT—capable of generating every type of differentiated cell. In both sea urchin and salamander embryos, a cell that would normally have produced only a small fraction of the embryo was found to be capable of generating the entire organism. The nuclei of such cells would have had

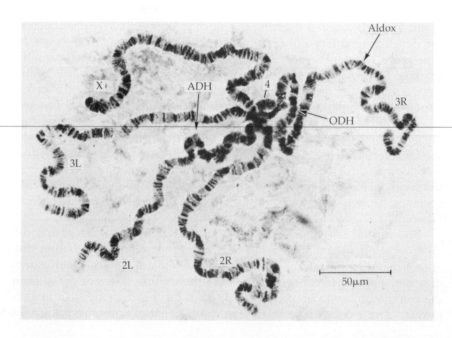

FIGURE 1
Polytene chromosomes from the salivary gland cells of *Drosophila melanogaster*. The four chromosomes are connected at their centromere regions, forming a dense chromocenter. The structural genes for alcohol dehydrogenase (ADH), aldehyde oxidase (Aldox), and octanol dehydrogenase (ODH) have been mapped to the assigned positions on these chromosomes. (From Ursprung et al., 1968; photograph courtesy of H. Ursprung.)

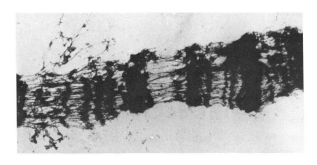

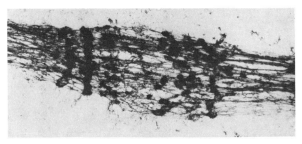

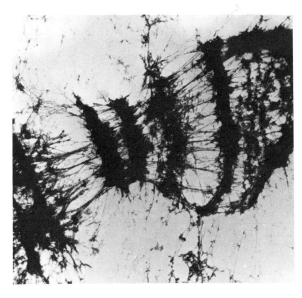

FIGURE 2
Electron microscopic views of the band (dark) and interband (light) regions of *Drosophila* polytene chromosomes. The bands are highly condensed in comparison to the interband chromatin. The figures represent different degrees of stretching so that the fine structure of the bands may be seen. (From Burkholder, 1976; photographs courtesy of G. D. Burkholder.)

to retain those genes for the products of all the other cell types. Likewise, Spemann showed that early salamander gastrula cells could change their prospective fates when transplanted into another area of the embryo. Could these embryological observations be extrapolated to cells that had already made a commitment to differentiate in a certain direction? Does a cell type that has become differentiated or determined still retain other potencies? Two lines of evidence lead to the conclusion that this is indeed the case.

Transdetermination

Upon hatching, a *Drosophila* larva has two distinct cell populations. About 10,000 cells form the larval tissue. Most of these cells have polytene chromosomes, and they grow by expanding to about 150 times their original volume. In addition, about 1000 nonpolytene diploid cells occur in clusters

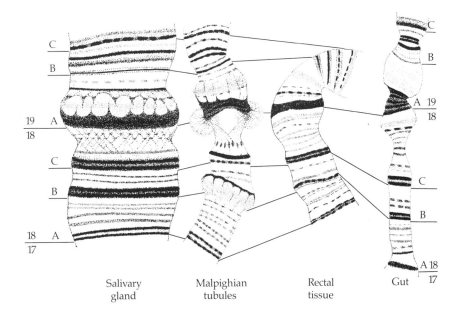

Salivary gland

Malpighian tubules

Rectal tissue

Gut

FIGURE 3
A region of the polytene chromosome set of the midge *Chironomus tentans*. Note the constancy of band number in different tissues. (After Beermann, 1952.)

FIGURE 4

The locations and developmental fates of the imaginal discs in *Drosophila melanogaster*. (After Fristrom et al., 1969.)

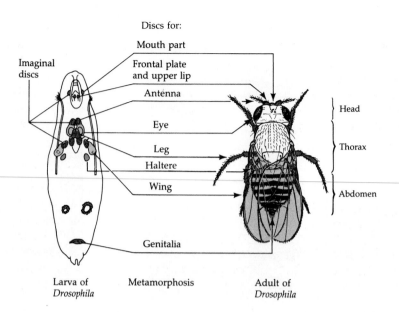

Discs for:

Mouth part

Frontal plate and upper lip

Imaginal discs

Antenna

Eye

Leg

Haltere

Wing

Genitalia

Head

Thorax

Abdomen

Larva of *Drosophila*   Metamorphosis   Adult of *Drosophila*

FIGURE 5

Testing the potency of imaginal discs. Discs can be excised and placed in adult flies, where they divide. If they are removed from the adults after various periods of incubation and are transplanted into normal larvae, they will become adult structures after metamorphosis. (After Markert and Ursprung, 1971.)

throughout the larva. These clusters of undifferentiated cells are called IMAGINAL DISCS (from the Latin *imago*, meaning "adult"), and they divide throughout the larval growth period. During metamorphosis, the hormone HYDROXYECDYSONE triggers enormous changes throughout the organism (Chapter 18). The larval cells degenerate while the imaginal disc cells are signaled to differentiate into the organs of the adult fly. Figure 4 shows the location of the *Drosophila* imaginal discs and the structures into which they develop.

Larval imaginal disc cells are determined. For example, an eye disc can be removed from one larva and implanted into the abdomen of a second larva. After metamorphosis, the fly developing from the second larva will have an extra eye in its abdomen. If only a portion of the eye

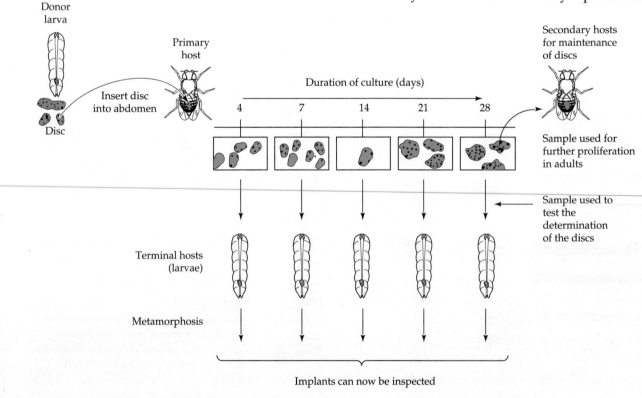

Donor larva

Insert disc into abdomen

Disc

Primary host

Duration of culture (days)

4   7   14   21   28

Secondary hosts for maintenance of discs

Sample used for further proliferation in adults

Sample used to test the determination of the discs

Terminal hosts (larvae)

Metamorphosis

Implants can now be inspected

disc is transplanted, only a portion of the eye will develop. Transplantation of a disc or a disc fragment into a *larva*, then, provides an excellent system for testing what structure that disc or disc segment is determined to produce at metamorphosis. When discs are transplanted into *adult* flies, however, no differentiation takes place. Rather, the imaginal disc cells continue to proliferate. These proliferating cells can be continually cultured by transplanting them from adult fly to adult fly. At the same time, they can also be tested for their state of determination by removing pieces of the growing discs and placing them back into metamorphosing larvae (Figure 5).

Ernst Hadorn and his co-workers used these discs to demonstrate that a cell can change its developmental commitment (Hadorn, 1968). Usually fragments of a disc determined to give rise to antennae would continue to produce antennal structures every time they were tested, even after several serial transplantations in adult flies. However, an occasional disc would surprise the scientists. Instead of monotonously producing antennal structures, portions of the antenna disc would form parts of the leg, mouth, or wing. This is called TRANSDETERMINATION. Instead of developing the "proper" organ, the imaginal cells develop into some other part of the adult fly. For example, a disc normally determined to develop into an antenna can yield structures appropriate for the fly leg (Figure 6). Moreover, like the original state of determination, the transdetermined state is relatively stable and is inherited by the disc cells over many generations of cell division.

Transdetermination happens more frequently after several passages through adult flies and occurs preferentially in certain directions (Figure

FIGURE 6

Transdetermination between antenna and leg structures. (A) Transdetermination in an antennal disc transplanted as shown in Figure 5. In addition to normal antennal structures (AIII, the third antennal segment; Ar, aristae), there are leg structures such as tarsal bristles (Ta) and their associated bracts (b). (B) Head of an adult fly carrying the mutation *Antennapedia*. In this mutant, the antennae are almost completely transformed into normal legs. Such mutations, where one structure is transformed into another, are called homoeotic mutations (Chapter 18). Although the mechanism of transdetermination in transplanted discs probably differs from that of homeotic mutations, they both demonstrate the change in fate of the representative discs. (A from Gehring, 1969; courtesy of W. J. Gehring. Photograph in B courtesy of J. Haynie.)

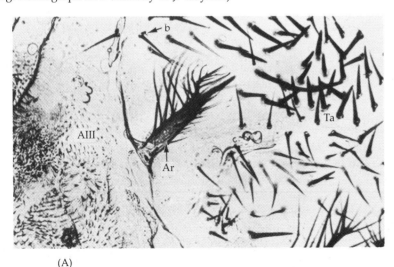

(A)

(B)

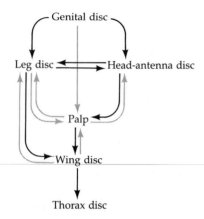

FIGURE 7
Routes of transdetermination in imaginal discs. The black arrows represent frequently observed changes; the gray arrows show rarer events.

7). A genital disc, for instance, can generate leg structures, but leg discs have not been seen to differentiate into parts of the genitalia. Although the cause of this directionality remains unsolved, it is clear that determined cells can give rise to cell types other than those dictated by their normal fate. Imaginal disc cells, therefore, have retained the genes for the specific products of some other differentiated cell types.

Metaplasia

The study of salamander eye regeneration has demonstrated that even adult *differentiated* cells can retain their potential to produce other cell types. In the salamander, removal of the neural retina promotes its regeneration from the pigmented retina, and a new lens can be formed from the cells of the dorsal iris. This latter type of regeneration (called Wolffian regeneration after its first observer) has been intensively studied by Tuneo Yamada and his colleagues (Yamada, 1966; Dumont and Yamada, 1972). They find that after removal of a lens, a series of events leads to the production of a new lens by the iris (Figure 8).

1 The nuclei of the iris cells change their shape.
2 The dorsal iris cells begin to produce enormous amounts of ribosomes.
3 The DNA of these cells begins to replicate, and cell divisions soon follow.
4 The cells dedifferentiate. They throw out their melanosomes (the differentiated products that give the eye its characteristic color), and these melanosomes are ingested by macrophages that have entered the wounded site.
5 Dorsal iris cells continue to divide, forming dedifferentiated tissue in the region of the removed lens.
6 The dedifferentiated iris cells now start synthesizing the differentiated products of lens cells, the crystallin proteins. These proteins are made in the same order as in normal lens development.
7 Once a new lens has formed, the cells on the dorsal side of the iris cease mitosis.

These events are not the normal route by which the lens is formed. As you will recall, in embryogenesis the lens develops from a layer of epithelial cells induced by the underlying neural ectoderm. The formation of the lens by the differentiated cells of the iris represents METAPLASIA, the transformation of one differentiated cell type into another. Therefore, the evidence of genetics and developmental biology points to the hypothesis of differential gene expression from genetically identical nuclei.

Amphibian cloning: Restriction of cell potency

The ultimate test of whether or not the nucleus from a differentiated cell has undergone any irreversible functional restriction would be to have that nucleus generate every other type of differentiated cell in the body. In 1938, Hans Spemann suggested a "somewhat fantastical" experiment for determining whether various cellular genomes were indeed identical. One would have to implant a nucleus from some differentiated cell into a host egg whose own nucleus had been removed. If each nucleus were identical to the zygote nucleus, this new nucleus would be capable of

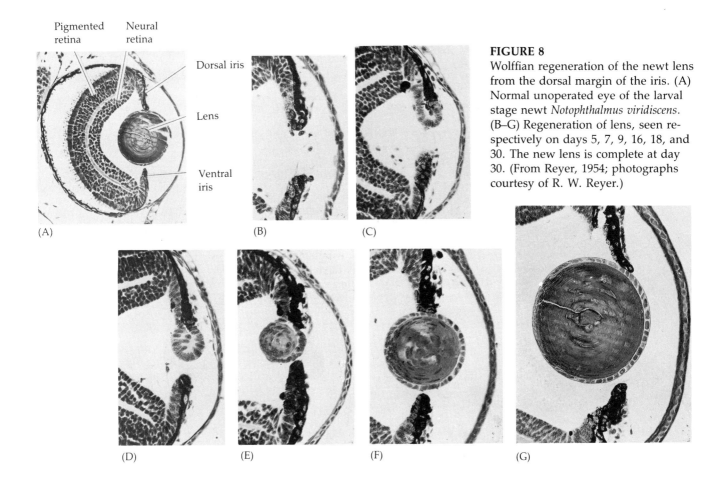

Pigmented retina

Neural retina

Dorsal iris

Lens

Ventral iris

(A)

(B)

(C)

(D)

(E)

(F)

(G)

FIGURE 8

Wolffian regeneration of the newt lens from the dorsal margin of the iris. (A) Normal unoperated eye of the larval stage newt *Notophthalmus viridiscens*. (B–G) Regeneration of lens, seen respectively on days 5, 7, 9, 16, 18, and 30. The new lens is complete at day 30. (From Reyer, 1954; photographs courtesy of R. W. Reyer.)

directing the entire development of the organism. Before such an experiment could be done, however, three techniques had to be perfected: (1) a method for enucleating host eggs without destroying them; (2) a method for isolating intact donor nuclei; and (3) a method for transferring such nuclei into the egg without damaging either the nucleus or the oocyte.

These techniques were developed by Robert Briggs and Thomas King. First, they combined the enucleation of the egg with its parthenogenetic activation. When an oocyte from the leopard frog (*Rana pipiens*) is pricked with a clean glass needle, the egg undergoes all the cytological and biochemical changes associated with fertilization. The cortical granules burst, internal cytoplasmic rearrangements of fertilization occur, and meiosis occurs near the animal pole of the cell. This meiotic spindle can easily be located as the pigment granules at the animal pole move away from it, and puncturing the oocyte at this site will cause the spindle and its chromosomes to flow outside the egg (Figure 9). The host egg is now considered to be both activated (in that fertilization reactions necessary to initiate development have been completed) and enucleated. The transfer of nuclei into the eggs is accomplished by disrupting donor cells and transferring a released nucleus into the oocyte through a micropipette. Some cytoplasm accompanies the nucleus to its new home, but the ratio of donor to recipient cytoplasm is only $1:10^5$, and the donor cytoplasm does not seem to affect the outcome of the experiments.

In 1952, Briggs and King demonstrated that blastula cell nuclei could direct the development of complete tadpoles when transferred into the oocyte cytoplasm. Spemann had shown that blastula cells alone were not

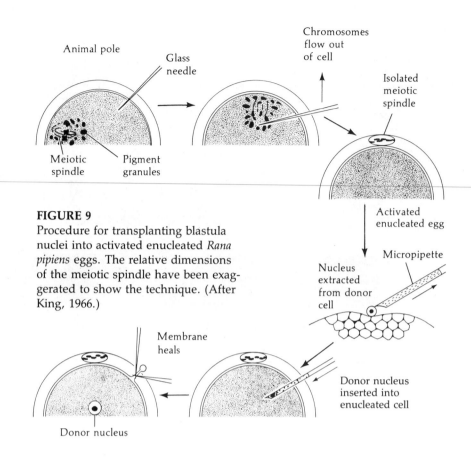

Animal pole

Glass
needle

Chromosomes
flow out
of cell

Isolated
meiotic
spindle

Meiotic
spindle

Pigment
granules

FIGURE 9

Procedure for transplanting blastula
nuclei into activated enucleated *Rana
pipiens* eggs. The relative dimensions
of the meiotic spindle have been exag-
gerated to show the technique. (After
King, 1966.)

Activated
enucleated egg

Micropipette

Nucleus
extracted
from donor
cell

Membrane
heals

Donor nucleus
inserted into
enucleated cell

Donor nucleus

FIGURE 10

This handsome specimen of *Rana pipiens* was derived from the transplantation of a blastula nucleus into an activated enucleated egg. (Photograph
courtesy of M. DiBerardino and
N. Hoffner.)

yet determined, so their nuclei were already known to be totipotent. If
the nuclear transfer system worked, such blastula nuclei might be able to
promote complete development. They did. Sixty percent of all transferred
nuclei were able to direct the development of the oocytes into functional,
swimming tadpoles, and all these tadpoles were diploid (a result suggesting that the nucleus was from the donor cells). Thus, the nuclear transfer
system worked and could be used to study nuclear potency (Figure 10).

What happens when nuclei from more advanced stages are transferred
into activated enucleated oocytes? The results of King and Briggs (1956)
are outlined in Figure 11. Whereas most blastula nuclei could produce
entire tadpoles, there was a dramatic decrease in the ability of nuclei from
later stages to direct development to the tadpole stage. When nuclei from
the *somatic cells* of tailbud-stage tadpoles were used as donors, no nucleus
was able to direct normal development. However, *germ cell* nuclei from
tailbud-stage tadpoles (which eventually will give rise to a complete organism after fertilization) were capable of directing normal development
in 40 percent of the blastulae that developed (Smith, 1956). Thus, somatic
cells appear to lose their ability to direct complete development as they
become determined and differentiated.

The decrease in nuclear potency was shown to be stable and tissue-
specific. Endoderm nuclei from late gastrula embryos were amplified by
serial transplantation (Figure 12). Here, one nucleus is transferred into an
enucleated oocyte and allowed to produce thousands of identical blastula
nuclei. These blastula nuclei can then be transferred into more enucleated
oocytes. In this way, more copies of the original nucleus are prepared and
the potential of that nucleus can be evaluated. This technique is called
NUCLEAR CLONING. In the transfer prior to serial transplantation, there was
wide variation in the stages to which individual nuclei could direct devel-

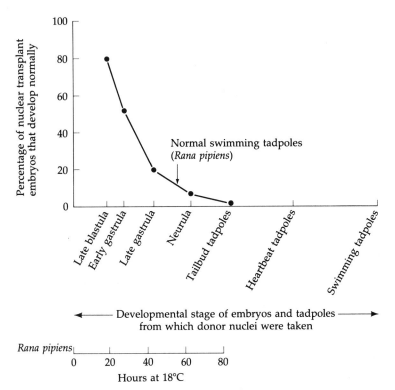

FIGURE 11
Graph of successful nuclear transplants as a function of the developmental age of the nucleus. The abscissa represents the stage at which the donor nucleus (from *Rana pipiens*) is isolated and inserted into the activated enucleated oocyte. The ordinate shows the percentage of those transplants capable of producing blastulae that could then direct development to the swimming tadpole stage. (After McKinnell, 1978.)

opment. Some nuclei could direct development all the way through the swimming tadpole stage, while other nuclei would abort development at gastrulation. Although King and Briggs found this "normal" variation *among* their clones of endoderm nuclei, they found little variation *within* these clones. Each clone had a "characteristic" phenotype, and often the stage of arrest was similar for all the embryos produced from a single, cloned endoderm nucleus. This was the case through several clonal generations. Moreover, when aberrant larvae were produced, they were all

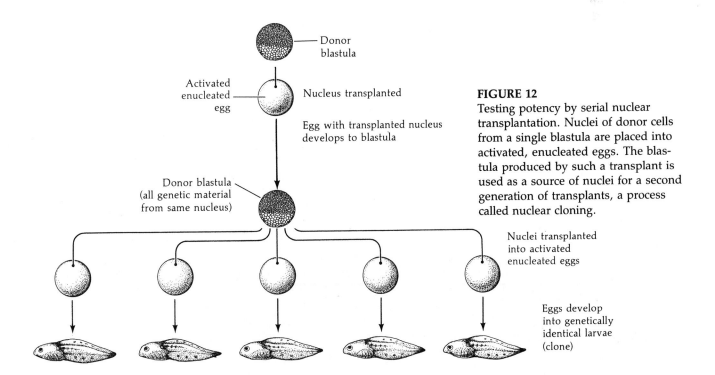

FIGURE 12
Testing potency by serial nuclear transplantation. Nuclei of donor cells from a single blastula are placed into activated, enucleated eggs. The blastula produced by such a transplant is used as a source of nuclei for a second generation of transplants, a process called nuclear cloning.

aberrant in the same way. They had endodermal structures (notably, gut) but were lacking several mesodermal or ectodermal derivatives. The endoderm nuclei appeared to be good at forming endoderm but were restricted in their ability to form ectoderm or mesoderm. DiBerardino and King (1967) found a similar loss of potency in nuclei from ectodermal cells. Here the aberrant tadpoles had excellent neural differentiation but lacked endodermal structures. Thus, the progressive restriction of nuclear potency during development appears to be the general rule. It is possible that some differentiated cell nuclei differ from others.

Amphibian cloning: Exceptions to restriction

There are other explanations for the limited potency of differentiated cell nuclei. When transferring a nucleus of a differentiated cell into oocyte cytoplasm, one is asking the nucleus to revert back to physiological conditions it is not used to. The cleavage nuclei of frogs divide at a rapid rate, whereas some differentiated cell nuclei divide rarely if at all. Failure to replicate DNA rapidly can lead to chromosomal breakage, and such chromosomal abnormalities have been seen in many cells of the cloned tadpoles. John Gurdon and his colleagues, using slightly different methods of nuclear transplantation, have obtained results suggesting that many of the nuclei of differentiated cells remain totipotent.

A major difference between the experiments of Gurdon and those of Briggs and King concerns the organism being studied. Gurdon isolated nuclei from *Xenopus laevis*, the South African clawed frog. *Xenopus* (Figure 13) is a much more primitive frog than *Rana*, lacking the eyelids, tympani (ears), and even the tongue so characteristic of more recently evolved species of frogs. *Xenopus* also has different developmental properties. Unlike the leopard frog, adult *Xenopus* can regenerate lost limbs, and early

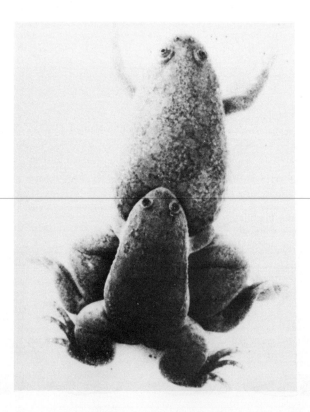

FIGURE 13
Mating pair of *Xenopus laevis*. The smaller male grasps the female and fertilizes the newly shed eggs externally. (From Deuchar, 1975.)

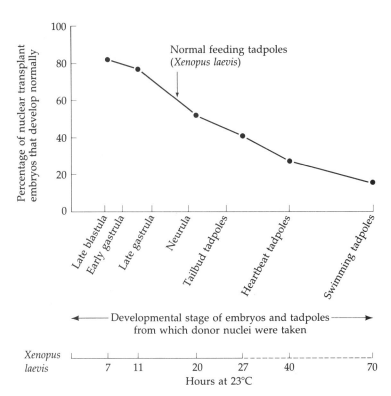

FIGURE 14
Graph of successful transplants in *Xenopus laevis*. Compared with the graph in Figure 11, *Xenopus* nuclei taken from later stages are generally more successful than nuclei taken from later stages of *Rana pipiens*. (After McKinnell, 1978.)

development in *Xenopus* is about three times as rapid as early development in *Rana pipiens*. Specifically, *R. pipiens* takes 80 hours to reach tailbud-stage tadpoles; *Xenopus* accomplishes the same amount of development in only 26 hours. Thus, nuclei from *Xenopus* tailbud endoderm cells are as young as early gastrula nuclei in *Rana* (McKinnell, 1978).

Gurdon, too, found a progressive loss of potency with increased development (Figure 14). The exceptions to this rule, however, proved very interesting. Gurdon had transferred the intestinal endoderm of feeding *Xenopus* tadpoles to activated enucleated eggs. These donor nuclei contained a genetic marker (one nucleolus per cell instead of the usual two) that could be used to distinguish them from host nuclei. Out of 726 nuclei transferred, only 10 promoted development to the feeding tadpole stage. Serial transplantation (which here involved placing an intestinal nucleus into an egg and, when the egg had become a blastula, transferring the nuclei of the blastula cells into several more eggs) increased the yield to 7 percent (Gurdon, 1962). In some instances nuclei from intestinal epithelial cells were capable of generating all the cell lineages—neurons, blood cells, nerves, and so forth—of a living tadpole. Moreover, seven of these tadpoles (from two original nuclei) metamorphosed into fertile adult frogs (Gurdon and Uehlinger, 1966). These nuclei were totipotent (Figure 15).

King and his colleagues, however, criticized these experiments, pointing out that (1) not enough precautions were taken to make certain that primordial germ cells—which migrate through and often stay in the gut —were not used as sources of nuclei, and (2) the intestinal epithelial cell of such a young tadpole may not qualify as a truly differentiated cell type. Such cells of feeding tadpoles still contain yolk platelets (DiBerardino and King, 1967; McKinnell, 1978; Briggs, 1979).

To answer these criticisms, Gurdon and his colleagues cultured epithelial cells from adult frog foot webbing. These cells were shown to be differentiated, because each of them contained keratin, the characteristic protein of adult skin cells. When nuclei from these cells were transferred

FIGURE 15
Procedure used to obtain mature frogs from the intestinal nuclei of *Xenopus* tadpoles. The wild-type egg (2-nu) is irradiated to destroy the maternal chromosomes, and an intestinal nucleus from a 1-nu tadpole is inserted. In some cases there is no division; in some cases, the embryo is arrested in development; but in other cases, an entire, new frog is formed and has a 1-nu genotype. (After Gurdon, 1968, 1977.)

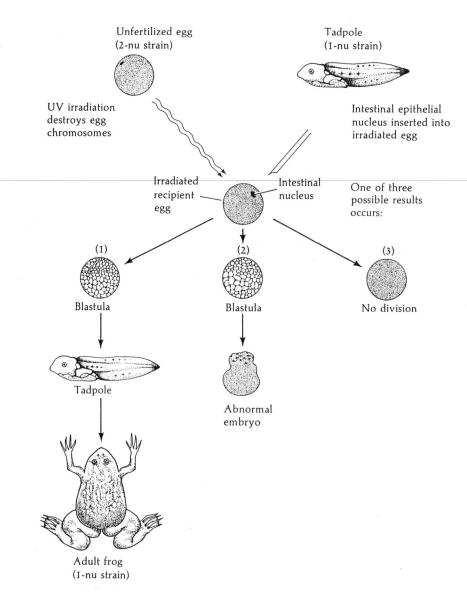

Unfertilized egg
(2-nu strain)

Tadpole
(1-nu strain)

UV irradiation destroys egg chromosomes

Intestinal epithelial nucleus inserted into irradiated egg

Irradiated recipient egg

Intestinal nucleus

One of three possible results occurs:

(1) (2) (3)

Blastula

Blastula

No division

Tadpole

Abnormal embryo

Adult frog
(1-nu strain)

into activated, enucleated *Xenopus* oocytes, none of the first generation transfers progressed further than neurulation. By serial transplantation, however, numerous tadpoles were generated, but these tadpoles all died prior to feeding (Gurdon et al., 1975). A similar developmental arrest was reported by Wabl and colleagues (1975), who attempted to produce entire frogs from lymphocyte nuclei. In a recent review of this field, DiBerardino (1987) concluded that "to date, no nucleus of a documented specialized cell nor of an adult cell has yet been shown to be totipotent."

One can look at these amphibian cloning experiments in two ways. First, one can recognize a general restriction of potency concomitant with development. This restriction is genetically determined and characteristic of the donor nucleus type. Second, one can readily see that the differentiated cell genome is remarkably potent in its ability to produce all the cell types of the amphibian tadpole. A single nucleus derived from a frog red blood cell (which neither replicates nor synthesizes mRNA) can undergo over 100 divisions after being transplanted into an activated oocyte and still retain the ability to generate swimming tadpoles (Orr et al., 1986; DiBerardino, 1989). In other words, even if there is debate over the *totipotency* of such nuclei, there is little doubt that they are extremely *pluripotent*. Certainly many unused genes in skin or blood cells can be reactivated to produce the nerves, stomach, or heart of a swimming tadpole.

Cloning

One can see (as we have also seen while discussing the work of Roux and Driesch) that the details of experimental technique can greatly influence the answers one obtains to a given question. By slight modifications of the cloning procedure, even late-stage *Rana pipiens* nuclei can develop into normal larvae. Sally Hennen has shown that the developmental success of donor nuclei can be increased by treating such nuclei with spermine and by cooling the egg to give the nucleus time to adapt to the egg cytoplasm (Hennen, 1970). By so treating *Rana pipiens* tailbud-stage endoderm nuclei, she obtained from 60 percent of those nuclei normal larvae that could support development through the blastula stage (compared with 0 percent in the untreated controls). The experiments on cultured skin cell transplants also used spermine but did not result in complete development to an adult frog from a differentiated cell nucleus.

Cloning human beings from previously differentiated cells seems to be the goal of several newspaper writers and novelists. (Novelists tend to clone politically important persons such as Hitler or Kennedy; periodical writers extend this technique to athletes, movie actors, and rock stars.) It should be obvious from the preceding discussion that cloning a fully developed individual from differentiated cells is a formidable undertaking, and the results are not conclusive. Even in amphibians, the nuclei of differentiated adult cells have not been able to generate adult animals when they are replaced into activated enucleated eggs.

Even if adult frogs could be generated from such differentiated cell nuclei, this ability could not be extrapolated to human cells. Besides the ethical questions, there are more difficulties associated with working with the human organism. Whereas the amphibian female produces hundreds of eggs simultaneously, human females produce only a few ripe ova each month. Moreover, there may be significant differences in the ability of oocyte cytoplasm to be directed from a nucleus of a more advanced stage cell.

Nuclear transplantation has been accomplished in mice by removing the two pronuclei of the zygote and then replacing them with pronuclei from other zygotes (McGrath and Solter, 1983). This is done by the procedure shown in Figure 16. One-cell embryos are first incubated in cytochalasin and colchicine to relax the microfilaments and microtubules of the cytoskeleton. While the embryo is held in place by a suction pipette, the enucleation pipette penetrates the zona pellucida. The enucleation pipette does not pierce the cell membrane. Rather, it is positioned adjacent to the pronuclei, and the region containing the pronuclei is sucked into the pipette (A). The enucleation pipette is then withdrawn, and the bud of pronuclei-containing cytoplasm separates from the egg. This cytoplasm is contained in a plasma membrane (B). The pipette containing the membrane-bounded pronuclei is moved to a drop containing inactivated Sendai virus, whose viral coat can cause membrane fusion. After aspirating some Sendai

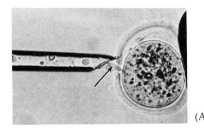
(A)

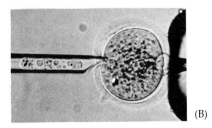
(B)

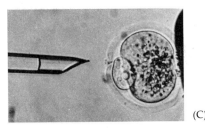
(C)

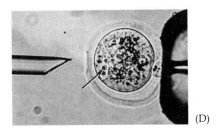
(D)

FIGURE 16
Procedure for transferring nuclei into an activated enucleated mammalian egg. Single-cell embryo incubated in colcemid and cytochalasin is held on a suction pipette. The enucleation pipette pierces the zona pellucida and sucks up adjacent cell membrane and the area of the cell containing the pronuclei. The enucleation pipette is then withdrawn (A) and the pronuclei-containing cytoplasm is removed from the egg. The cell membrane is not broken and the continuity of the membrane-bounded cytoplasm is indicated by the arrow. (B) The cell membrane forms a vesicle around the pronuclear cytoplasm within the enucleation pipette. (C) This vesicle is mixed with Sendai virus and is inserted into the perivitelline space between the zona pellucida and another enucleated egg. (D) The Sendai virus mediates the fusion of the enucleated egg and the membrane-bounded pronuclei, allowing the pronuclei (arrow) to enter the cell. The polar body, seen beside the suction pipette, does not enter into the fusion reaction. (From McGrath and Solter, 1983; courtesy of the authors.)

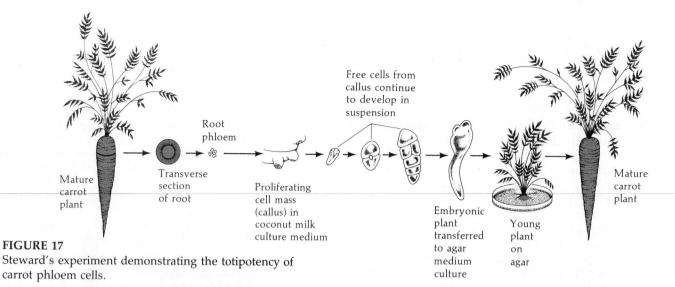

FIGURE 17
Steward's experiment demonstrating the totipotency of carrot phloem cells.

Labels in figure:
Mature carrot plant — Transverse section of root — Root phloem — Proliferating cell mass (callus) in coconut milk culture medium — Free cells from callus continue to develop in suspension — Embryonic plant transferred to agar medium culture — Young plant on agar — Mature carrot plant

virus, the pipette is moved to a zygote whose pronuclei have been similarly removed. The zona is penetrated, and the membrane-bounded pronuclei are deposited within the perivitelline space, between the zona and the egg cell membrane (C). The embryo is then incubated at 37°C, until the membranes fuse (D). In this way, the two donor pronuclei are found to reside within the host cytoplasm. When these embryos are cultured for 5 days, they form a blastocyst that can be implanted into the uterus of a pseudopregnant female adult. The resulting mice display the phenotype of the donor nucleus.

Whereas over 90 percent of enucleated mouse zygotes receiving pronuclei from other zygotes develop successfully to the blastocyst stage, not a single embryo (out of 81) developed even this far when nuclei from 4-cell embryos were transferred into the enucleated zygotes. Similarly, nuclei from 8-cell embryos and the inner cell mass would not support preimplantation development (McGrath and Solter, 1984). Unlike sea urchins and amphibians, the nuclei of early mouse blastomeres (known to be totipotent) do not support full development. These experiments probably fail because blastomere nuclei cannot function properly in zygote cytoplasm. Thus, the cloning of Elvis Presley from fully differentiated cells is not something we should count on. However, not all mammalian blastomeres are the same, and mammalian species differ greatly in the time of gene activation and implantation into the uterus. Using modifications of the above technigue, Willadsen (1986) has produced full-term lambs from the transplanted nuclei of 8-cell-stage blastomeres, and Prather and co-workers (1987) have reported the production of two calves from the implantation of early cleavage bovine nuclei into enucleated eggs. Although mouse blastomere nuclei could not direct all of development, sheep and bovine nuclei appear to be able to do so. If confirmed, this finding may have important agricultural consequences.

The only cases where the nuclei of differentiated cells of adult organisms are readily seen to be capable of directing the development of another adult organism occurs in

plants. This ability has been demonstrated dramatically in cells from carrots and tobacco. In 1958, F. C. Steward and his colleagues established a procedure by which the differentiated tissue of carrot roots could give rise to an entire new plant (Figure 17). Small pieces of the phloem tissue were isolated from the carrot and rotated in large flasks containing coconut milk. This fluid (which is actually the liquid endosperm of the coconut seed) contains the factors and nutrients necessary for plant growth and the hormones required for plant differentiation. Under these conditions, the tissues proliferate and form a disorganized mass of tissue called a CALLUS. Continued rotation causes the shearing away of individual cells from the callus and into suspension. These individual cells give rise to rootlike nodules of cells that continue to grow as long as they remain in suspension. When these nodules are placed into a medium solidified with agar, the rest of the plant is able to develop, ultimately forming a complete, fertile carrot plant (Steward et al., 1964; Steward, 1970).

Although the process from a single cell to a flowering plant cannot be observed under the rotating culture conditions, Vasil and Hildebrandt (1965) have been able to follow these events by isolating single tobacco cells and watching their development either directly or by time-lapse cinematography. Like the carrot cells, the tobacco cells were able to give rise to a plantlet that eventually could flower and seed.

The tobacco cells provide us with an example of nuclear totipotency. All the genes needed to produce an entire plant exist within the nucleus of a differentiated cell. But plants and animals develop differently; the vegetative propagation of plants by cuttings (i.e., portions of the plant that, when nourished, regenerate the missing parts) is common agricultural practice. Moreover, in contrast to animals (in which germ cells are set off as a distinctive cell lineage early in development), plants normally derive their gametes from somatic cells. It is not overly surprising, then, that a single plant cell can differentiate into other cell types and form a genetically identical clone (from the Greek *klon*, meaning "twig").

So far we can conclude that differential gene loss is not the cause of differentiation. The nuclei of differentiated cells contain most, if not all, of the genes of the zygote, and these genes can be expressed under the appropriate conditions. The process of differentiation, then, involves the selective expression of different portions of a common genetic repertoire. With the advent of molecular biology and gene cloning techniques, the problems of genome constancy and differential gene expression have been reinvestigated. The next chapter extends this discussion into contemporary molecular biology.

LITERATURE CITED

Beermann, W. 1952. Chromomeren Konstanz und spezifische modifikationen der Chromosomenstruktur in der Entwicklung und Organdifferenzierung von *Chironomus tentans*. *Chromosoma* 5: 139–198.

Briggs, R. 1979. Genetics of cell type determination. *Int. Rev. Cytol.* [Suppl.] 9: 107–127.

Briggs, R. and King, T. J. 1952. Transplantation of living nuclei from blastula cells into enucleated frogs' eggs. *Proc. Natl. Acad. Sci. USA* 38: 455–463.

Burkholder, G. D. 1976. Whole mount electron microscopy of polytene chromosome from *Drosophila melanogaster*. *Can. J. Genet. Cytol.* 18: 67–77.

Deuchar, E. M. 1975. *Xenopus: The South African Clawed Frog.* Wiley-Interscience, New York.

DiBerardino, M. A. 1987. Genomic potential of differentiated cells analyzed by nuclear transplantation. *Am. Zool.* 27: 623–644.

DiBerardino, M. A. 1989. Genomic activation in differentiated somatic cells. *In* M. A. DiBerardino and L. D. Etkin (eds.), *Developmental Biology: A Comprehensive Synthesis.* Plenum, New York, pp. 175–198.

DiBerardino, M. A. and King, T. J. 1967. Development and cellular differentiation of neural nuclear transplants of known karyotypes. *Dev. Biol.* 15: 102–128.

Dumont, J. N. and Yamada, T. 1972. Dedifferentiation of iris epithelial cells. *Dev. Biol.* 29: 385–401.

Fristrom, J. W., Raikow, R., Petri, W. and Stewart, D. 1969. In vitro evagination and RNA synthesis in imaginal discs of *Drosophila melanogaster*. *In* E. W. Hanly (ed.), *Problems in Biology: RNA in Development.* University of Utah Press, Salt Lake City, pp. 381–402.

Gehring, W. J. 1969. Problems of cell differentiation in *Drosophila*. *In* E. W. Hanley (ed.), *Problems in Biology: RNA in Development.* University of Utah Press, Salt Lake City, pp. 231–244.

Gurdon, J. B. 1962. The developmental capacity of nuclei taken from intestinal epithelial cells of feeding tadpoles. *J. Embryol. Exp. Morphol.* 10: 622–640.

Gurdon, J. B. 1968. Transplanted nuclei and cell differentiation. *Sci. Am.* 219(6): 24–35.

Gurdon, J. B. 1977. Egg cytoplasm and gene control in development. *Proc. R. Soc. Lond.* [B] 198: 211–247.

Gurdon, J. B. and Uehlinger, V. 1966. "Fertile" intestinal nuclei. *Nature* 210: 1240–1241.

Gurdon, J. B., Laskey, R. A. and Reeves, O. R. 1975. The developmental capacity of nuclei transplanted from keratinized cells of adult frogs. *J. Embryol. Exp. Morphol.* 34: 93–112.

Hadorn, E. 1968. Transdetermination in cells. *Sci. Am.* 219(5): 110–120.

Hennen, S. 1970. Influence of spermine and reduced temperature on the ability of transplanted nuclei to promote normal development in eggs of *Rana pipiens*. *Proc. Natl. Acad. Sci. USA* 66: 630–637.

Judd, B. H. and Young, M. W. 1973. An examination of the one cistron–one chromomere concept. *Cold Spring Harbor Symp. Quant. Biol.* 38: 573–579.

King, T. J. 1966. Nuclear transplantation in amphibia. *Meth. Cell Physiol.* 2: 1–36.

King, T. J. and Briggs, R. 1956. Serial transplantation of embryonic nuclei. *Cold Spring Harbor Symp. Quant. Biol.* 21: 271–289.

Markert, C. L. and Ursprung, H. 1971. *Developmental Genetics.* Prentice-Hall, Englewood Cliffs, NJ.

McGrath, J. and Solter, D. 1983. Nuclear transplantation in the mouse embryo by microsurgery and cell fusion. *Science* 220: 1300–1302.

McGrath, J. and Solter, D. 1984. Inability of mouse blastomere nuclei transferred to enucleated zygotes to support development in vitro. *Science* 226: 1317–1319.

McKinnell, R. G. 1978. *Cloning: Nuclear Transplantation in Amphibia.* University of Minnesota Press, Minneapolis.

Orr, N. H., DiBerardino, M. A. and McKinnell, R. G. 1986. The genome of frog erythrocytes displays centuplicate replications. *Proc. Natl. Acad. Sci. USA* 83: 1369–1373.

Prather, R. S., Barnes, F. L., Sims, M. M., Robl, J. M., Eyestone, W. H. and First, N. L. 1987. Nuclear transplantation in the bovine embryo: Assessment of donor nuclei and recipient oocyte. *Biol. Reprod.* 37: 859–866.

Reyer, R. W. 1954. Regeneration in the lens in the amphibian eye. *Q. Rev. Biol.* 29: 1–46.

Smith, L. D. 1956. Transplantation of the nuclei of primordial germ cells into enucleated eggs of *Rana pipiens*. *Proc. Natl. Acad. Sci. USA* 54: 101–107.

Steward, F. C. 1970. From cultured cells to whole plants: The induction and control of their growth and morphogenesis. *Proc. R. Soc. Lond.* [B] 175: 1–30.

Steward, F. C., Mapes, M. O. and Smith, J. 1958. Growth and organized development of cultured cells. I. Growth and division of freely suspended cells. *Am. J. Bot.* 45: 693–703.

Steward, F. C., Mapes, M. O., Kent, A. E. and Holsten, R. D. 1964. Growth and development of cultured plant cells. *Science* 143: 20–27.

Swanson, C. P., Merz, T. and Young, W. J. 1981. *Cytogenetics: The Chromosome in Division, Inheritance, and Evolution,* 2nd Ed. Prentice-Hall, Englewood Cliffs, NJ.

Tjio, J. H. and Puck, T. T. 1958. The somatic chromosomes of man. *Proc. Natl. Acad. Sci. USA* 44: 1229–1237.

Ursprung, H., Smith, K. D., Sofer, W. H. and Sullivan, D. T. 1968. Assay systems for the study of gene function. *Science* 160: 1075–1081.

Vasil, V. and Hildebrandt, A. C. 1965. Differentiation of tobacco plants from single isolated cells in microcultures. *Science* 150: 889–892.

Wabl, M. R., Brun, R. B. and DuPasquier, L. 1975. Lymphocytes of the toad *Xenopus laevis* have the gene set for promoting tadpole development. *Science* 190: 1310–1312.

Willadsen, S. M. 1986. Nuclear transplantation in sheep embryos. *Nature* 320: 63–65.

Yamada, T. 1966. Control of tissue specificity: The pattern of cellular synthetic activities in tissue transformation. *Am. Zool.* 6: 21–31.

10

Genomic equivalence and differential gene expression: Molecular investigations

Seek simplicity and distrust it.

—ALFRED NORTH WHITEHEAD (1919)

For it is not cell nuclei, not even individual chromosomes, but certain parts of certain chromosomes from certain cells that must be isolated and collected in enormous quantities for analysis; that would be the precondition for placing the chemist in such a position as would allow him to analyse [the hereditary material] more minutely than the morphologists.

—THEODOR BOVERI (1904)

Embryologists have asked: Is the set of genes found in each cell type of an organism the same as those that were found in the zygote? Molecular biologists frame the same question: Is the DNA of each cell the same despite the different proteins made by each cell type? The question remains the same, but the techniques used to answer it have become more sophisticated. Instead of analyzing the generation of organs or embryos, we can now look at individual sequences of DNA and monitor whether or not these genes are present in a cell and whether or not they are actively transcribing RNA. In order to further explore this question, then, we must become familiar with the techniques of molecular biology.

Molecular biology techniques

Nucleic acid hybridization

Our modern understanding of eukaryotic genes and their RNA products is derived largely from experiments involving NUCLEIC ACID HYBRIDIZATION. This technique involves annealing single-stranded pieces of RNA and DNA to allow complementary strands to form double-stranded hybrids. For example, if DNA is cut into small pieces and each piece dissociated into two single strands (i.e., DENATURED), each strand should find and reunite with its complementary partner, given sufficient time. The conditions of this renaturation must be such that all specific binding between complementary strands is maintained while nonspecific matchings are dissociated. This is usually accomplished by varying the temperature or the ionic conditions in the solution in which such renaturation is taking place (Wetmur and Davidson, 1968).

Similarly, RNA synthesized from a particular region of DNA would be expected to bind to the strand from which it was transcribed. Thus, RNA is expected to hybridize specifically with the gene that codes for it. In order to measure this hybridization, one of the nucleic acid strands is usually radiolabeled. One such experiment will illustrate how this technique can be used. Unlabeled DNA from a frog liver can be denatured (its strands separated in alkali) and immobilized on nitrocellulose filter paper. Radioactive RNA can simultaneously be prepared by feeding radioactive RNA precursors either to another frog or to frog cells growing in culture. One can then isolate ribosomes and extract from them radioactive 28 S ribosomal RNA (rRNA), which is added to the DNA-containing filters. We would expect the radioactive RNA to bind only to the DNA pieces containing the genes for the 28 S ribosomal RNA. The amount of binding can be quantified by measuring the radioactivity of these filters after washing away the unbound RNA.

The procedure and results for such an experiment are shown in Figure 1. Wallace and Birnstiel (1966) isolated DNA from normal tadpoles with two nucleoli per cell, from homozygous mutant tadpoles that lacked nucleoli altogether, and from heterozygous tadpoles with only one nucleolus per cell. The isolated DNA from each group was denatured. Fifty micrograms of DNA were then placed on each of dozens of filter papers; one group of filters contained denatured DNA from the wild-type tadpoles, a second group of filters had DNA from the mutant tadpoles, and a third group of filters had DNA from the heterozygous tadpoles. Then various amounts of radioactive 28 S ribosomal RNA were hybridized with the three types of DNA on the filters. Some filters received small amounts of the radioactive rRNA, other filters received more. Wallace and Birnstiel found that the radioactive rRNA bound well to the normal tadpole DNA, eventually saturating all the available DNA. There was no specific binding to the DNA from the tadpoles lacking nucleoli, and the DNA from the heterozygous tadpoles bound only one-half as much ribosomal RNA as their normal counterparts. Thus, it was shown that the genes for 28 S ribosomal RNA are transcribed from the nucleolar region of the tadpole genome.

One technical problem associated with nucleic acid hybridization is the difficulty of getting enough radioactivity into the RNA molecule. This problem can be circumvented by isolating the RNA and making a complementary DNA (cDNA) copy in the presence of radioactive precursors. This can be done in a test tube containing the RNA, a short stretch of DNA (called a primer), radioactive DNA precursors, and the viral enzyme REVERSE TRANSCRIPTASE. This enzyme is capable of making DNA from an RNA template (Figure 2). Because the DNA is synthesized in vitro, one need not worry about the dilution of the radioactive precursor. Furthermore, such a cDNA can hybridize with both the gene that produced the RNA (albeit the other strand) and the RNA itself. It is therefore extremely useful in detecting small amounts of specific RNAs.

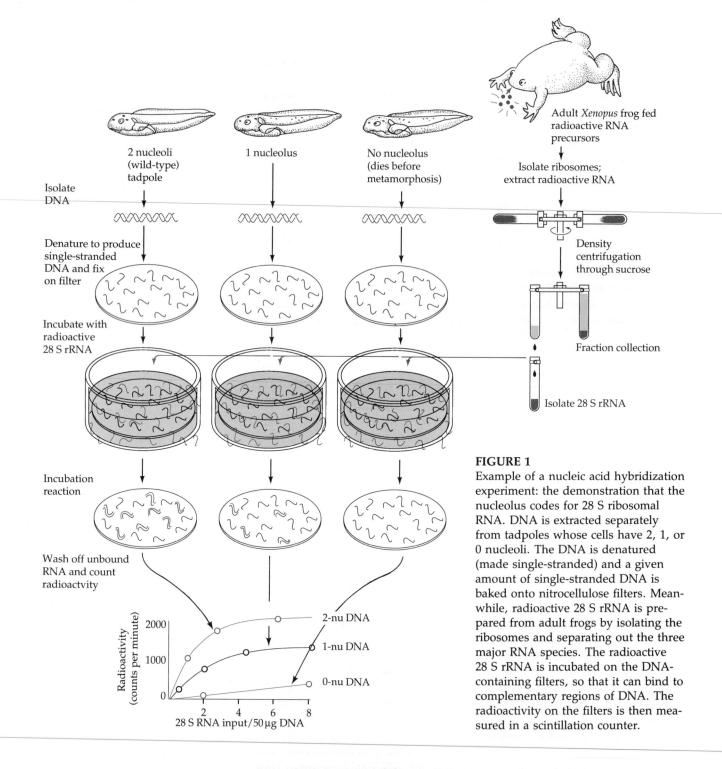

FIGURE 1

Example of a nucleic acid hybridization experiment: the demonstration that the nucleolus codes for 28 S ribosomal RNA. DNA is extracted separately from tadpoles whose cells have 2, 1, or 0 nucleoli. The DNA is denatured (made single-stranded) and a given amount of single-stranded DNA is baked onto nitrocellulose filters. Meanwhile, radioactive 28 S rRNA is prepared from adult frogs by isolating the ribosomes and separating out the three major RNA species. The radioactive 28 S rRNA is incubated on the DNA-containing filters, so that it can bind to complementary regions of DNA. The radioactivity on the filters is then measured in a scintillation counter.

Gene cloning: From entire genomes

More recently, nucleic acid hybridization techniques have enabled developmental biologists to isolate individual genes through a technique called gene cloning. If one wanted to physically isolate a single human gene, for instance, the problem would be enormous. The human genome has enough DNA to code for approximately 2×10^5 genes of about 15,000 base pairs each. Ordinary biochemical procedures would not allow a single gene to be resolved from the rest. Gene cloning, however, allows the purification of specific regions of DNA as well as their amplification into an enormous number of copies.

The first step in this process involves cutting the nuclear DNA into various discrete pieces. This is done by incubating the DNA with a RE-STRICTION ENDONUCLEASE (more commonly called a "restriction enzyme"). These endonucleases are usually bacterial enzymes that recognize specific sequences of DNA (see Table 1) and cleave the DNA at these sites (Nathans and Smith, 1975). For example, when human DNA is incubated with the enzyme BamHI (from *Bacillus amyloliquifaciens* strain H), the DNA is cleaved at every site where the sequence GGATCC occurs. The products are variously sized pieces of DNA, all ending with G on one end and GATCC on the other (Figure 3).

The next step is to incorporate these DNA fragments into CLONING VECTORS. These vectors are often circular DNA molecules that replicate in bacterial cells independently of the bacterial chromosome. Usually, either drug-resistant plasmids or specially modified viruses (which are especially useful for cloning large DNA fragments) are used. Such a plasmid can be constructed so as to have only one BamHI-sensitive site. This plasmid can be opened by incubating it with that restriction enzyme. After being opened, it can be mixed with fragmented DNA. In numerous cases, the cut DNA pieces will become incorporated into these vectors (because their ends are complementary to the vector's open ends), and the pieces can be joined covalently by placing them in a solution containing the enzyme DNA LIGASE. This whole process yields bacterial plasmids that each contain a single piece of human DNA. These are called RECOMBINANT PLAS-MIDS, or usually RECOMBINANT DNA (Cohen et al., 1973; Blattner et al., 1978).

The plasmid illustrated in Figure 3 is pBR322, a cloning vector often used by molecular biologists. It has two drug resistance genes, tc^R, which confers resistance to tetracycline, and Ap^R, which makes the bacterium immune to ampicillin. These drug resistance genes, so baneful to physi-

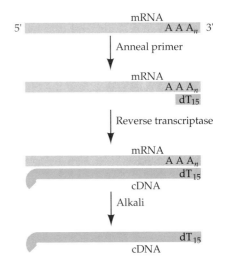

FIGURE 2

Method for preparing complementary DNA (cDNA). Most messenger RNA (mRNA) contains a long stretch of adenosine residues (AAA_n) at the 3' end of the message (to be discussed in Chapter 11); therefore, investigators anneal a "primer" consisting of 15 deoxythymidine residues (dT_{15}) to the 3' end of the message. Reverse transcriptase then transcribes a complementary DNA strand, starting at the dT_{15} primer. The cDNA can be isolated by raising the pH of the solution, thereby denaturing the double-stranded hybrid and cleaving the RNA.

TABLE 1
Commonly used restriction enzymes

Enzyme	Derivation	Recognition and cleavage site[a]
EcoRI	*Escherichia coli*	G▾A A T T C C T T A A⌃G
BamHI	*Bacillus amyloliquifaciens*	G▾G A T C C C C T A G⌃G
HindIII	*Haemophilus influenzae*	A▾A G C T T T T C G A⌃A
SalI	*Streptomyces albus*	G▾T C G A C C A G C T⌃G
SmaI	*Serratia marcescens*	C C C▾G G G G G G⌃C C C
HhaI	*Haemophilus haemolyticus*	G C G▾C C⌃G C G
HaeIII	*Haemophilus aegyptius*	G G▾C C C C⌃G G
AluI	*Arthrobacter luteus*	A G▾C T T C⌃G A

[a] All restriction enzyme recognition sites have a center of symmetry. The double-stranded sequence read in one direction is identical to the sequence read backwards in the other direction.

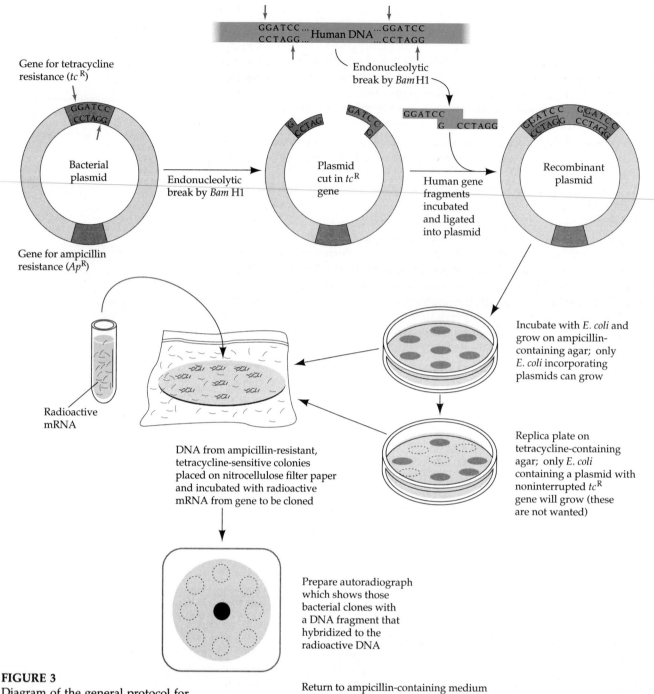

Gene for tetracycline resistance (tc^R)

GGATCC
CCTAGG

Bacterial plasmid

Gene for ampicillin resistance (Ap^R)

Endonucleolytic break by *Bam* H1

Human DNA

GGATCC... ...GGATCC
CCTAGG... ...CCTAGG

Endonucleolytic break by *Bam* H1

GGATCC
CCTAGG

Plasmid cut in tc^R gene

Human gene fragments incubated and ligated into plasmid

GGATCC
CCTAGG

G CCTAGG

GGATCC GGATCC
CCTAGG CCTAGG

Recombinant plasmid

Incubate with *E. coli* and grow on ampicillin-containing agar; only *E. coli* incorporating plasmids can grow

Replica plate on tetracycline-containing agar; only *E. coli* containing a plasmid with noninterrupted tc^R gene will grow (these are not wanted)

Radioactive mRNA

DNA from ampicillin-resistant, tetracycline-sensitive colonies placed on nitrocellulose filter paper and incubated with radioactive mRNA from gene to be cloned

Prepare autoradiograph which shows those bacterial clones with a DNA fragment that hybridized to the radioactive DNA

FIGURE 3
Diagram of the general protocol for cloning DNA, using as an example the insertion of a human DNA sequence into a plasmid with one *Bam*HI-sensitive site.

Return to ampicillin-containing medium and grow the clone(s) that contain the gene of interest

cians, become invaluable to gene cloners, for the one *Bam*HI site in this plasmid lies in the middle of the tc^R gene. The putative recombinant plasmids made in the above manner are then incubated with wild-type *E. coli* (sensitive to both drugs) under conditions that enable the bacteria to take in plasmids. But not every bacterium incorporates a plasmid and not every plasmid has incorporated a foreign gene (since it is possible for the "sticky ends" of the restriction enzyme site to renature with themselves). To screen for those bacteria that have incorporated plasmids, the treated *E. coli* are grown on agar containing ampicillin. Only those bacteria that have incorporated plasmids survive. Portions of each surviving colony are then placed on agar containing tetracycline. If the bacteria grow, the tc^R

gene must be intact, and the plasmid has merely re-formed itself by sticking its own ends together. But if the bacteria grown on ampicillin do *not* survive on tetracycline, then the tetracycline resistance gene is no longer functional, implying the tc^R gene has been interrupted by a restriction enzyme fragment from the foreign genome. Those colonies on the ampicillin-containing medium that did not grow on tetracycline-containing medium are saved.

These colonies are then screened for the presence of the particular gene. Cells from each of the *E. coli* colonies that contains recombinant plasmids are placed on a nitrocellulose filter. When these cells are lysed, their DNA binds to the filter. Next, the DNA strands are separated by heating, and the filter is incubated in a solution containing the radioactive RNA (or its cDNA copy) of the gene one wishes to clone. If a plasmid contains that gene, its DNA should be on the paper and only that DNA should be able to bind the radioactive RNA or cDNA. Therefore, only those areas will be radioactive. The radioactivity of these regions is detected by autoradiography. Sensitive X-ray film is placed over the treated paper. The high-energy electrons emitted by the radioactive RNA sensitize the silver grains in the film, causing them to turn dark when the film is developed. Eventually, a black spot is produced over each colony containing the recombinant plasmid carrying that particular gene (Figure 3). This colony is then isolated and grown to produce billions of bacteria, each containing hundreds of identical recombinant plasmids.

The recombinant plasmids can be separated from the *E. coli* chromosome by centrifugation, and incubating the plasmid DNA with *Bam*HI releases the human DNA fragment that contains the gene. This fragment can then be separated from the plasmid DNA, so the investigator has micrograms of purified DNA sequences containing a specific gene.

Although this procedure sounds very logical and easy, the numbers of colonies that have to be screened is often astronomical. The number of random fragments that must be cloned in order to obtain the gene you want gets larger with the increasing complexity of the organism's genome.* To be 99 percent certain of cloning the desired gene from *E. coli*, one needs to screen some 1500 colonies. For mammals, this number increases to 800,000; rather than growing the bacteria in standard petri dishes, the colonies are often grown on cafeteria trays (Blattner et al., 1978; Slightom et al., 1980).

Gene cloning: Complementary DNA libraries

It is sometimes more useful to clone only the protein-encoding exons of the gene. Here, one uses an alternative cloning technique that starts with cytoplasmic RNA rather than with genomic DNA. As shown in Figure 2, one can make DNA strands complementary to mRNA strands. By taking the procedure a step further (with the aid of DNA polymerase), one can make this single-stranded cDNA into double-stranded cDNA. These strands of DNA can be inserted into plasmids by adding the appropriate "ends" onto them. Appending a GATCC/G fragment onto the blunt ends of such a DNA piece creates an artificial *Bam*HI restriction cut and enables the piece to insert itself into a virus or plasmid cut with that enzyme (Figure 4A).

Such collections of clones derived from mRNAs are often called LIBRARIES. Thus, one can have a mouse liver library representing all the genes active in making liver proteins. One can also have a *Xenopus* vegetal oocyte library representing messages present only in a particular part of

---

*Complexity is the term referring to the number of different types of genes within a nucleus.

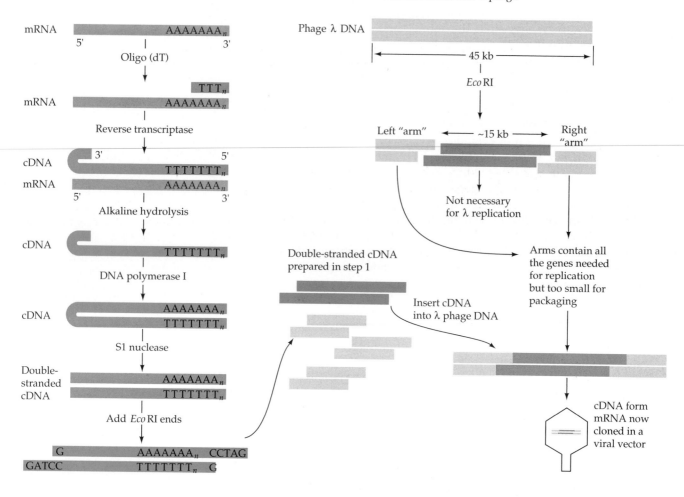

(A) Preparation of clonable cDNA

(B) Insertion of double-stranded cDNA into λ phage

the cell. Genes cloned in this manner are very important because they lack introns. When they are added to bacterial cells, these genes can be transcribed and then translated into the proteins they encode.

Libraries have been extremely useful in studying development. As an example, we will follow the steps used by Wessel and co-workers (1989) to detect differences in the RNAs in different parts of the gastrulating sea urchin embryo. To find these endoderm-specific messenger RNAs in the sea urchins, Wessel and co-workers prepared a cDNA library from gastrulating embryos. The messenger RNA of these samples (most of the RNA of eukaryotic cells is ribosomal) was isolated by running the samples through oligo-dT beads that capture the poly(A) tails of the messages. Then the mRNA population was converted into a cDNA population by using reverse transcriptase (Figure 2). By using *E. coli* polymerase I, the single-stranded cDNA was then made double-stranded. Next, commercially available *Eco*RI "ends" were ligated onto the double-stranded cDNAs. This made them clonable into vectors that were cut with *Eco*RI restriction enzyme. This DNA was then mixed with the arms of a genetically modified λ phage (Figure 4B). This phage is so constructed that if DNA is inserted into the phage, it interrupts a gene encoding for a particular enzyme. When grown on certain media (Figure 4C), the phages that have incorporated the DNA produce white plaques, while the phages that do not incorporate any new DNA synthesize the enzyme, metabolize media components, and produce blue plaques. In this way, approximately 4 million recombinant phages were generated, each containing some cDNA representing a mRNA molecule.

(C) Preparing library of phage clones

(D) Screening of phage clone library

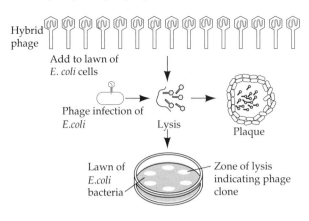

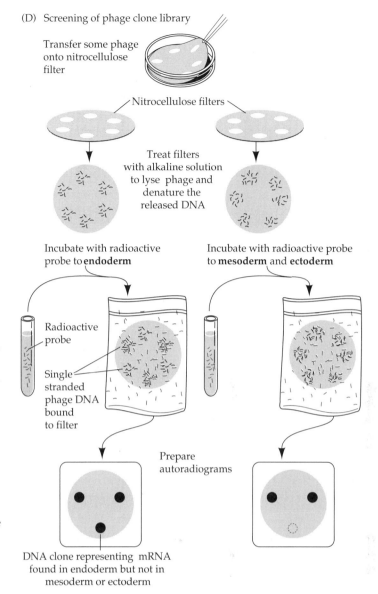

FIGURE 4

Diagram of protocol used to make cDNA libraries. (A) Messenger RNA is isolated, made into cDNA, and this cDNA is made double-stranded. (B) By adding restriction fragment ends, the cDNA "genes" can be inserted into specially modified bacteriophages. (C) Phages containing the recombinant DNA will lyse *E. coli*, forming plaques. Biochemical techniques can distinguish plaques of recombinant phages from those that lack the inserted gene. (D) The plaques are transferred to nitrocellulose paper and treated with alkali to lyse the phages and denature the DNA in place. These filters are then incubated in radioactive probes (usually cDNA) from a tissue. In the case of the differential cDNA library screening discussed in the text, the same phage library was screened with radioactive probes from two different tissues. This allowed the researchers to look for an mRNA that would be found in one type of tissue but not in the other.

The next steps involved screening these recombinant phages. Which ones might represent mRNAs found in endoderm and not in the other cell layers? Wessel and his colleagues isolated the mRNA populations from mesoderm, ectoderm, and endoderm. They then made labeled cDNAs from each of these mRNA populations using radioactive precursors. They now had three collections of radioactive cDNA molecules, each representing the mRNA population from one of the three germ layers.

The recombinant phages representing the mRNAs of the gastrulating sea urchin embryo were grown and their DNA was placed onto nitrocellulose paper (Figure 4D). The DNA was then placed in alkaline solutions to make them single-stranded. Each nitrocellulose paper had dozens of such phages on it, and each recombinant DNA was placed onto two nitrocellulose papers. One of these papers was incubated with radioactive cDNA made from the total mRNA of the *endoderm*; the other paper was incubated with radioactive probe to *mesoderm and ectoderm*. The filters were washed to remove any unhybridized radioactive cDNA, dried, and exposed on X-ray film. If an mRNA was present in the endoderm but not in either the ectoderm or mesoderm, the recombinant DNA made from that message should bind radioactive cDNA from the endoderm but should not find an mRNA anywhere else. As a result, that spot of recombinant DNA from the endoderm should be radioactive (since it bound radioactive cDNA from the endoderm) but the same clone should not be radioactive when exposed to ectodermal or mesodermal mRNA. This was found to be the case. One recombinant phage in particular was found to bind radioactive cDNA made from endodermal mRNA, but not from mesodermal or ectodermal mRNAs. Hence, it represented an mRNA found in the endoderm and not anywhere else. The phage containing this gene can now be grown in large quantities and characterized (see Sidelights & Speculations).

SIDELIGHTS & SPECULATIONS

Once the gene is cloned . . .

. . . there's a great deal more to be done. In subsequent chapters, it will become obvious that cloned genes are revolutionizing our understanding of development. Some of the techniques for using this recombinant DNA are outlined below.

Sequencing

Sequence data can tell us the structure of the encoded protein and can identify regulatory DNA sequences that certain genes have in common. (For example, as we will discuss in the next chapter, all genes regulated by the presence of the hormone progesterone contain nearly the

FIGURE 5
Schematic for the dideoxy method of DNA sequencing.

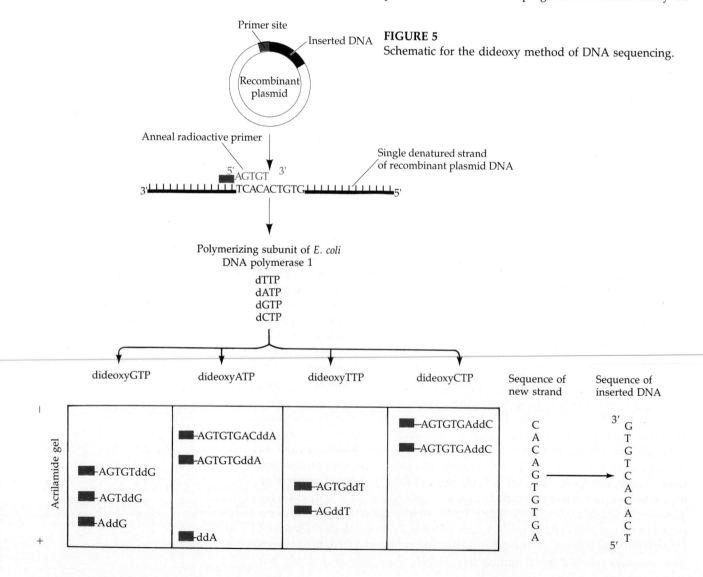

same DNA sequence near the start of the protein-coding region of the gene.) The simplicity of the Sanger "dideoxy" sequencing technique (Sanger et al., 1977) has made it a standard procedure in many molecular biology laboratories. One starts with the phage or plasmid carrying the cloned gene and isolates a single strand of the circular DNA (Figure 5). One then anneals a radioactive *primer* of DNA (about 20 base pairs) complementary to the phage DNA immediately 5' to that of the cloned gene. (Because these sequences are known, oligonucleotide primers can be readily synthesized or purchased commercially.) The primer has a free 3' end to which more nucleotides can be added. One places the primed DNA and all four deoxyribonucleoside triphosphates into four test tubes. Each of the test tubes contains the polymerizing subunit of DNA polymerase and a different *di*deoxynucleoside triphosphate: one tube contains dideoxy G, one tube contains dideoxy A, and so forth. The structure of the deoxy- and the dideoxynucleotides is shown in Figure 6. Whereas a deoxyribonucleotide has no hydroxyl (OH) group on the 2' carbon of its sugar, a dideoxyribonucleotide lacks hydroxyl groups on both the 2' and 3' carbons. So, even though a dideoxyribonucleotide can be bound to a growing chain of DNA by DNA polymerase, it stops the chain's growth because, lacking a 3' hydroxyl group, no new nucleotide can bind to it. Thus, when the DNA polymerase is synthesizing DNA from the

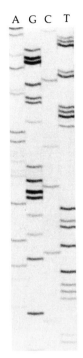

FIGURE 7

Sample of DNA sequenced by the dideoxy method. (Courtesy of G. Guild.)

primer, it will be complementary to the cloned gene. In the tube with dideoxy A, however, every time the polymerase puts an A into the growing chain, there is a chance that the dideoxy A will be placed there instead of the deoxy A. If this happens, the chain stops. Similarly, in the tube with dideoxy G, the chain has the potential to stop every time a G is inserted. (This has been likened to a Greek folk dance in which some small percentage of the potential dancers have an arm in a cast.) Because there are millions of chains being made in each tube, each tube will contain a population of chains, some stopped at the first possible site, some at the last, and some at sites in between. The tube with dideoxy A, for instance, will contain chains of different discrete lengths, each ending at an A residue.

The resulting radioactive DNA fragments are separated by ELECTROPHORESIS. The mixtures of fragments are placed into slots on one side of a gel and an electric current is then passed through the gel. The negatively charged DNA fragments migrate towards the positive pole, the smaller fragments moving faster than the larger ones.* Fragments with the fewest nucleotides run faster than those with more (Figures 5 and 7). By reading up the "ladders," one obtains the DNA sequence complementary to the cloned gene.

One can compare the DNA sequence one obtains with other sequences (there are now computer data banks to do this, which saves much graduate student eyestrain). For

---

*Given the same charge-to-mass ratio, smaller fragments obtain a faster velocity than larger ones when propelled by the same energy. This is a function of the kinetic energy formula ($E = mv^2/2$); the velocity becomes inversely proportional to the square root of the mass.

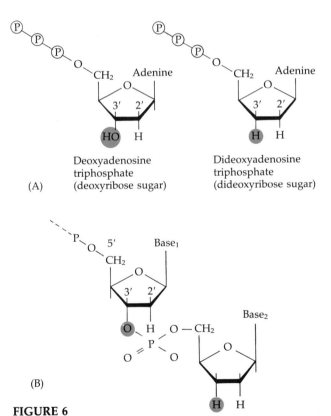

(A)

Deoxyadenosine triphosphate (deoxyribose sugar)

Dideoxyadenosine triphosphate (dideoxyribose sugar)

(B)

FIGURE 6

Comparison of deoxy- and dideoxynucleotides. (A) Structures of the two types of nucleotides. The difference is highlighted in color. (B) The 3' end of a chain that has been terminated by incorporation of dideoxynucleotide, because it has no free 3' hydroxyl group for further DNA polymerization.

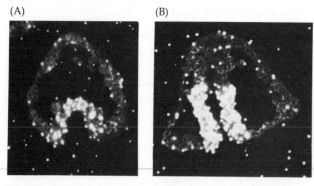

(A)   (B)

FIGURE 8

In situ hybridization showing the localization of an endoderm-specific mRNA. The radioactive cDNA used as a probe was prepared from the cloned gene made from endoderm-specific mRNA (Figure 4). This radioactive cDNA binds to mRNA in the endoderm of early gastrula-stage sea urchins (A) and in the mid- and hindgut endoderm of late gastrulating sea urchins (B). (From Wessel et al., 1989; photograph courtesy of G. Wessel.)

instance, the cDNA isolated by Wessel and co-workers from the sea urchin endoderm (in the experiment mentioned above) turned out have no homology to any known DNA sequence, indicating that it is a previously unknown gene.

In situ hybridization

Once one has a gene, one can determine where it is active by the process of IN SITU HYBRIDIZATION. This technique, developed by Mary Lou Pardue and Joseph Gall (1970), allows one to visualize the positions of specific nucleic acids within cells and tissues. When one adds radioactive cDNA to appropriately fixed cells on a microscope slide, the radioactive cDNA will only bind if the complementary mRNA is present. After the unbound cDNA is washed off, the slides are covered with a transparent photographic emulsion and are stored in the dark. The radioactivity from the bound cDNA sensitizes the silver grains of the emulsion, causing silver to precipitate when the slide is dipped into photographic developer. These spots appear black when viewed directly or white when viewed under darkfield illumination. These spots will be directly above the places where the radioactive cDNA had bound. Thus, one can visualize those cells (or even regions of cells) that have accumulated a specific type of messenger RNA. Figure 8 shows an in situ hybridization using the cDNA found to be specific for endodermal cells. As one can see, it finds mRNAs only in the endoderm of the early sea urchin gastrula. However, as gastrulation continues, it becomes localized even more precisely—to the hindgut-midgut region of the endodermal tube.

Transferring genes

In many cases, one wants to take a gene, modify it, and then place it back into a eukaryotic organism. This has been made possible by TRANSFECTION, a technique whereby

DNA fragments are incorporated into the chromosomal DNA of a cell. The cloned DNA is mixed with a selectable marker gene (such as the *thymidine kinase* gene of *Herpes simplex* virus) and some large DNA from some other species. This mixture is incubated with calcium phosphate and forms DNA-containing crystals that are phagocytosed by the cultured cells. If the cultured cells are deficient in thymidine kinase, the culture can be grown under conditions that allow only those cells that have taken up the DNA–calcium phosphate crystals (and therefore express the *thymidine kinase* gene) to survive (Perucho et al.,1980; Robins et al., 1981). Figure 9 shows an experiment wherein human globin genes were mixed with the *Herpes thymidine kinase* gene and large stretches of hamster DNA. The colonies selected for the presence of thymidine kinase showed multiple human globin genes interspersed with hamster DNA

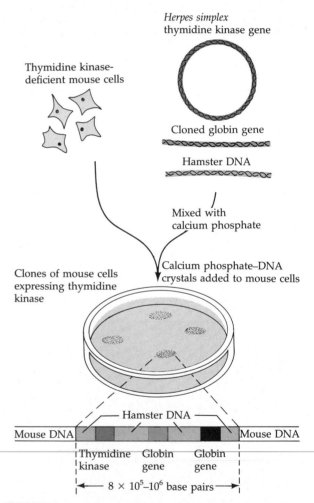

FIGURE 9

Transfection of globin gene DNA from one species into the genome of another. Cloned human globin genes, a marker gene (for thymidine kinase) enabling selection, and long stretches of another foreign genome are co-precipitated as crystals in calcium phosphate. The crystals containing the DNA are taken up by the cells and, under the proper intracellular conditions, the DNA is integrated into the chromosomes of the host cells. (After Watson et al., 1983.)

within a mouse chromosome (Watson et al., 1983). Such techniques have recently been used to transfer genes into embryonal carcinoma stem cells of mice. These stem cells can integrate into mouse embryos, subsequently generating portions of the adult mouse tissues (see Chapter 6). When such adult mice mate, these new genes can be transferred into every cell of their progeny (Gossler et al., 1986).

The calcium phosphate procedure is not very efficient, so other techniques have been devised. One is to microinject the DNA directly into the nucleus (or pronucleus) of a cell (Capecchi, 1980; Figure 10). With sophisticated equipment and a great deal of practice, one can microinject about 1000 cells an hour (on a good day) and have as much as 50 percent of the injected DNA integrate into the cells' chromosomes. This technique has been used to add human β-globin genes to mouse zygotes deficient in β-globin. The resulting mice had a functioning human β-globin gene, which cured their genetic disease (Constantini et al., 1986). Another valuable technique is ELECTROPORATION. Here, cells and DNA are mixed together in the presence of a high-voltage pulse that allows the DNA to enter the cells. This technique has been used to push the gene for hypoxanthine phosphoribosyl transferase (HPRT) into embryonal carcinoma cells that were deficient in this enzyme (Thompson et al., 1989). Not only did these cells synthesize HPRT, but when they were injected into mouse blastocoels, some of them became part of the embryo and contributed to the germ line (see Chapter 6). These chimeric mice were mated to mice that lacked this gene, so that the offspring containing HPRT obtained this allele from the electroporesed DNA.

A recent innovation has been the use of RETROVIRAL VECTORS. Retroviruses are RNA-containing viruses. Within a host cell, they make a DNA copy of themselves (using their own reverse transcriptase); the copy becomes double-stranded and integrates into a host chromosome. This integration is accomplished as a result of two identical sequences (long terminal repeats) at the ends of the DNA.

Retroviral vectors are made by removing the viral packaging genes (needed for the exit of viruses from the cell) from the center of a mouse retrovirus. This extraction creates a vacant site where other genes can be placed. By using the appropriate restriction enzymes, researchers can remove genes from a cloned phage or plasmid and reinsert the gene into the retroviral vectors. A part of this vacant site can also be occupied by a selectable marker (such as *Herpes simplex* virus thymidine kinase). These retroviral vectors infect mouse cells with an efficiency approaching 100 percent. The mice having stable genes derived from other individuals (by microinjection, virus infection, or the incorporation of modified embryonal carcinoma stem cells) are called TRANSGENIC mice. For example, when mouse hematopoietic stem cells were cultured in the presence of a retroviral vector containing a human β-globin gene, the stem cells incorporated the human gene into their genome. These stem cells were injected back into irradiated mice and repopulated their circulatory systems with red blood cells containing, in part, human hemoglobin (Dzierzak et al., 1988). One can now change the genome of mammalian blood cells and return them to the organism. This holds out the prospect for various types of human gene therapy as well as the rapid genetic alteration of livestock.

In *Drosophila*, new genes can be carried into a fly via P-ELEMENTS. These DNA sequences are mobile regions of DNA that can integrate like viruses into any region of the *Drosophila* genome. Moreover, they can be isolated, and cloned genes can be inserted into the center of the P-element. When the recombined P-element is injected into the *Drosophila* oocyte, it can integrate into the DNA and provide the embryo with this new gene (Spradling and Rubin, 1982).

DNA and RNA blotting

DNA BLOTTING (also called SOUTHERN BLOTTING) is a technique for determining whether a cell contains a certain gene in its genome (Southern, 1975). This knowledge can be useful in determining the gene's function. Lang and co-workers (1985) used a retroviral vector to insert the gene for granulocyte–macrophage colony stimulating factor (GM-CSF) into cells that do not usually synthesize the factor but are able to respond to it. Certain cells treated with the recombinant vector began to grow out of control, forming malignant tumors when placed back into mice. Did these cells each get an extra copy of *GM-CSF*? Is this extra gene transcribing *GM-CSF* message? The first question was answered using a DNA blot.

Lang and co-workers isolated the DNA from cells of these malignant colonies and treated it with a restriction enzyme. The fragments were then separated on an electrophoresis gel. They then denatured the DNA fragments into single strands by placing the gel in an alkaline solution. After returning the gel to a neutral pH, they placed it on wet filter paper atop a plastic support (Figure 11). Nitrocellulose paper (capable of binding single-stranded DNA) was placed directly over the gel and covered with multiple layers of dry paper towels. The filter paper beneath the gel extended into a trough of high ionic strength buffer. The buffer traveled through the gel up through the nitrocellu-

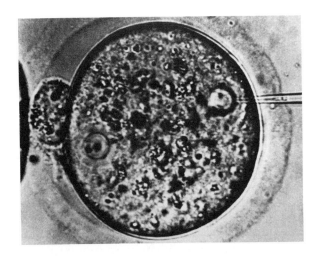

FIGURE 10
Injection of DNA (from cloned genes) into a nucleus (in this case, a pronucleus of a mouse egg). (From Wagner et al., 1981; photograph courtesy of T. E. Wagner.)

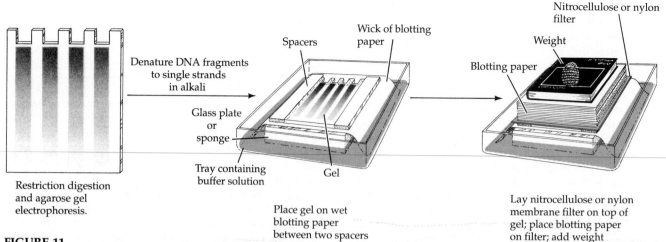

Denature DNA fragments
to single strands
in alkali

Spacers

Wick of blotting
paper

Nitrocellulose or nylon
filter

Weight

Blotting paper

Glass plate
or
sponge

Tray containing
buffer solution

Gel

Restriction digestion
and agarose gel
electrophoresis.

Place gel on wet
blotting paper
between two spacers

Lay nitrocellulose or nylon
membrane filter on top of
gel; place blotting paper
on filter; add weight

FIGURE 11

Southern blotting apparatus. After electrophoresing fragments of DNA into a gel, the DNA is denatured into single strands. The gel is then placed on a support on top of
a filter paper saturated with high-ionic strength buffer.
Nitrocellulose paper or a nylon filter is placed on the gel,
and towels are placed atop the filter. The transfer buffer

makes its way through the gel, nitrocellulose paper, and
towels by capillary action, taking the DNA with it. The
single-stranded DNA is stopped by the nitrocellulose paper. The positions of the DNA in the paper directly reflect the positions of the DNA fragments in the gel.

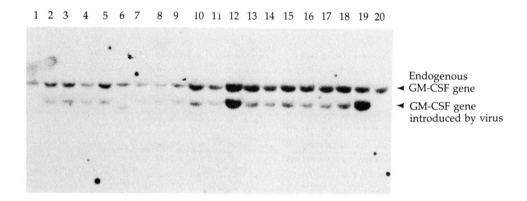

1 2 3 4 5 6 7 8 9 10 11 12 13 14 15 16 17 18 19 20

Endogenous
◄ GM-CSF gene

◄ GM-CSF gene
introduced by virus

FIGURE 12

Integration of *GM-CSF*-containing retrovirus into the
mouse cell genome. Southern blot of the genomes of 20
clones. The DNA of each clone was isolated and treated
with restriction enzyme *Sac*I, separated on a gel, blotted,
and hybridized with a radioactive cDNA to the *GM-CSF*
gene. Lanes 2–19 contain DNA from malignant cells infected with the retrovirus containing the *GM-CSF* gene.

Lane 1 contains DNA from a nonmalignant clone infected
with a similar virus lacking the *GM-CSF* gene; lane 20
contains DNA from an uninfected mouse embryo. The
upper bands represent the *GM-CSF* gene from the normal
mouse genome. The lower bands represent the *GM-CSF*
gene introduced by the virus. (From Lang et al., 1985;
photograph courtesy of R. A. Lang.)

lose filter and into the towels. The DNA, however, was
stopped by the nitrocellulose filter. Thus, the DNA was
transferred from the gel to the nitrocellulose paper. After
baking it onto the filter (otherwise it would have come off),
the DNA fragments were incubated with radioactive cDNA
to the gene of interest (mRNA can also be used). An autoradiograph of the filter paper showed where the radioactive DNA had found its match. The results from Lang and
co-workers (Figure 12) show that in addition to two normal
mouse genes for GM-CSF, which are on a restriction fragment 7800 base pairs long, the malignant cells had a virally
derived *GM-CSF* gene on a restriction fragment (known to

be virally derived on the basis of the viral DNA sequence)
5500 base pairs long. Each of the malignant clones of cells
showed this pattern.

The expression of these genes can be demonstrated by
an RNA BLOT (often called a NORTHERN BLOT). Whereas
Southern blots transfer *DNA* from a gel to paper, Northern
blots (not named after their inventor) transfer *RNA* from
gel to paper in the same way. Northern blots are useful in
determining whether a specific mRNA has accumulated in
the cytoplasm of a particular cell. In the study by Lang and
colleagues, cytoplasmic RNA was isolated from malignant
cells infected with the retrovirus containing the *GM-CSF*

gene and from nonmalignant cells that had been infected with a similar retrovirus lacking the *GM-CSF* gene. The RNAs from each clone of cells were separated by size on an electrophoresis gel and transferred to nitrocellulose paper. After the paper was incubated with cDNA to the *GM-CSF* gene, all the malignant clones were found to be transcribing the *GM-CSF* gene. The other cells were not (Figure 13). Lang and his colleagues concluded that the newly introduced *GM-CSF* gene, controlled by the virus, was being transcribed in these cells. Moreover, these cells, which grow in the presence of GM-CSF, had become capable of synthesizing their own GM-CSF. In this manner, they stimulated themselves to continuously divide, thereby producing tumors. (This confirmed a hypothesis of Todaro and co-workers, who had predicted such a phenomenon in 1977. These experiments on the induction of cancer will be discussed in much greater detail in Chapter 20.)

One important use of the Northern blot is determining the time at which a particular gene is being expressed. An investigator can extract RNA from embryos at various stages of their development and electrophorese the RNAs side-by-side on a gel. After transferring the separated RNA (by blotting) to a piece of nitrocellulose paper, the RNA-containing paper is incubated in a solution containing a radioactive single-stranded DNA fragment from a particular gene. This DNA only sticks to regions where the complementary RNA was located. Thus, if the mRNA for that gene is present at a certain embryonic stage, the radioactive DNA binds to it and can be detected by autoradiography. Autoradiographs of this type are called DEVELOPMENTAL Northern blots. A developmental Northern blot is shown for the expression of the above-mentioned endoderm-specific gene during sea urchin development (Figure 14A). One can see that mRNA for this protein is first synthesized

FIGURE 14
Northern blots for an endoderm-specific gene of the sea urchin *Lytechinus variegatus*. (A) Developmental Northern blot showing stage-specific accumulation of the mRNA from this gene. Total mRNA (10 µg per stage) was electrophoresed into an agarose gel. The gel was blotted onto treated paper, and the mRNAs were bound to the paper. The paper was then incubated with radioactive cDNA from the endoderm-specific clone. This mRNA was found to be synthesized during the mesenchyme blastula stage and increased throughout development. (B) Northern blot at the prism stage showing that the mRNA is present in the endoderm but not in the ectoderm or mesoderm. Total RNA from the endoderm was electrophoresed (lane 2) next to total mRNA from the rest of the sea urchin (lane 1). Binding of the radioactive cDNA detected only mRNA in the endoderm. (From Wessel et al., 1989; photographs courtesy of G. Wessel.)

during the mesenchyme blastula stage and is continually made during the rest of development. The Northern blot shown in Figure 14B shows that the accumulation of this mRNA in the prism stage is restricted solely to the endoderm (Wessel et al., 1989).

Site-specific mutagenesis and nick translation

One of the most powerful applications of recombinant DNA technology has been the ability to mutate a particular nucleotide of a cloned gene and then determine whether or not that change affected the function of that gene. The specific mutation is accomplished by oligonucleotide-directed SITE-SPECIFIC MUTAGENESIS. First, the sequence of the DNA is determined, and an oligonucleotide is artificially synthesized that is very similar to the normal sequence, except for one (or more) mismatches. This 12- to 15-base oligonucleotide is then mixed with the single strand of a recombinant plasmid containing the cDNA version of the wild-type gene. The oligonucleotide will anneal with the plasmid even though one of its base pairs does not match. The mismatched oligonucleotide can then serve as a primer to replicate the remainder of the plasmid and cloned DNA by *E. coli* DNA polymerase I. The ends of the replicated plasmids are then covalently joined by DNA ligase and the plasmids introduced into *E. coli*. Half of the replicating plasmids will restore the wild-type sequence of the *E. coli* host, whereas half will create a mutant DNA with a single base substitution (Figure 15). (Using additional genetic tricks, more mutant than wild-type clones are made.) Mutant and wild-type clones can be distinguished by their efficiency in hybridizing exact copies of the specified region (Gillam et al., 1980; Smith, 1985). The

1 2 3 4 5 6 7 8 9 10 11

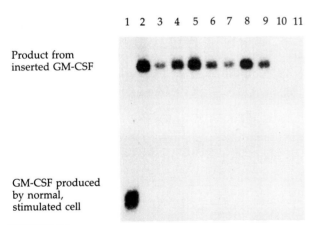

Product from inserted GM-CSF

GM-CSF produced by normal, stimulated cell

FIGURE 13
Detection of virus-specific mRNAs encoding GM-CSF. Northern blots of the mRNAs from eight clones of cells infected with the retrovirus carrying the mouse *GM-CSF* gene (lanes 2–9) and two clones infected with a similar virus that lacked the *GM-CSF* DNA (lanes 10 and 11). The virally coded transcript has more DNA between the initiation site and is therefore larger than the normal mouse mRNA for GM-CSF (lane 1), so it remains near the top of the gel. (From Lang et al., 1985; photograph courtesy of R. A. Lang.)

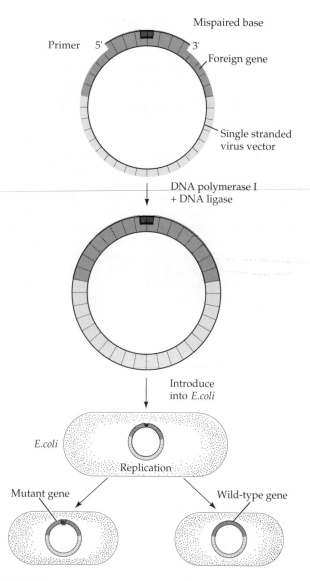

FIGURE 15

Oligonucleotide-directed site-specific mutagenesis. Oligonucleotide containing a one-base substitution is annealed to the complementary strand of a cloned DNA. The remainder of the cloned strand can be synthesized by *E. coli* DNA polymerase I, using the oligonucleotide as a primer. The circle is covalently closed by ligase. After reintroducing the plasmid into *E. coli*, the molecule replicates wild-type and mutant genes. (After Watson et al., 1983.)

The cloned gene is isolated from the plasmid and treated with a low concentration of DNase I. This enzyme nicks the DNA at random places, creating free 3' hydroxyl termini to which DNA polymerase I can bind. One of the enzymatic peculiarities of *E. coli* DNA polymerase I is that it will add nucleotides at one end of while deleting them "ahead" of the nick (Figure 16). The result is the "movement" of the nick ("nick translation") along the DNA (Kelly et al., 1970; Rigby et al., 1977). After the original bases are replaced by radiolabeled bases, the DNA fragments can have over 10^8 cpm (counts per minute) per microgram of DNA.

The polymerase chain reaction

The POLYMERASE CHAIN REACTION (PCR) is a method of in vitro cloning that can generate large quantities of a specific DNA fragment from a small amount of different genes (Saiki et al., 1985). The standard methods of cloning use living microorganisms to amplify recombinant DNA. PCR, however, can amplify a single DNA molecule several millionfold in a few hours, and it does it in the test tube. This technique has been extremely useful in cases where there is very little nucleic acid to study. Preimplantation mouse

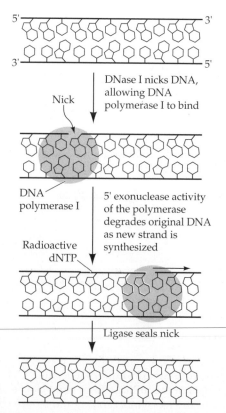

FIGURE 16

Nick translation. A nick is introduced into double-stranded DNA by DNase I. As *E. coli* DNA polymerase I repairs the nick, it excises the bases ahead of it. If radioactive nucleoside triphosphates are present, they will be used to repair these holes. The result is a strand of DNA that is very heavily radiolabeled.

plasmid containing the mutated gene can then be placed back into the genome of an animal to see if the gene still functions, or alternatively, the mutant gene can be isolated from these plasmids, translated into a mutant protein, and assayed to see if it still functions even though specific amino acids have been changed.

NICK TRANSLATION also uses DNA polymerase, but in a different manner. With this technique, investigators can replace one strand of a cloned gene with radiolabeled nucleotides, thereby generating highly radioactive probes for Southern blots, Northern blots, and in situ hybridization.

embryos, for instance, have very little mRNA, and one cannot obtain millions of such embryos to study. After converting the mRNA into cDNA, though, the PCR generates the equivalent of millions of embryos' worth of cDNA. This has allowed the identification of specific growth factor mRNAs in the preimplantation mouse embryo (Rappolee et al., 1988).

The principle of PCR is illustrated in Figure 17. A single-stranded DNA fragment is flanked by two oligonucleotide primers that hybridize to opposite strands of the target sequence. (If one is trying to isolate the gene or mRNA for a specific protein of known sequence, these flanking regions can be prepared by synthesizing oligonucleotides that would encode the amino end of the protein and complementary oligonucleotides to those that would encode the carboxyl end of the protein.) The 3' ends of these primers face each other, so that replication is through the target DNA. Once the first primer has hybridized, the DNA polymerase can synthesize a new strand. However, this is not normal *E. coli* DNA polymerase. Rather, it is DNA polymerase from the bacteria *Thermus aquaticus*. This bacteria lives in hot springs (such as those in Yellowstone National Park) where the temperature reaches nearly 90°C. Its DNA polymerase can withstand temperatures near boiling. PCR takes advantage of this evolutionary adaptation. Once the second strand is made, it is heat-denatured from its complement. The second primer is added, and now both strands can synthesize new DNA. Repeated cycles of denaturation and synthesis will amplify this region of DNA in geometric fashion. After 20 such rounds, that specific region has been amplified 2^{20} (a little more than a million) times.

FIGURE 17
Protocol for initiating the polymerase chain reaction. DNA is denatured, and two sets of primers are hybridized to its opposite ends. (If one is looking for the gene encoding a specific sequence, as in this diagram, specific primers can be made.) Using thermostable DNA polymerase from *T. aquaticus*, each DNA strand synthesizes its complement. These strands are then denatured and the primers are hybridized to them, starting the cycle again. In this way, the number of new strands having the sequence between the two primers increases exponentially. (The other fragments increase at a much slower rate and become minor background contaminants.)

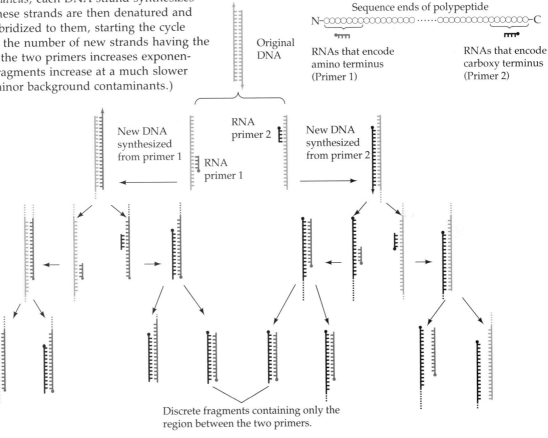

Differential gene expression

With the advent of such sophisticated biochemical techniques, the hypothesis of differential gene expression became testable on the molecular level. The hypothesis can be divided into three testable postulates.

1 Every cell nucleus contains the complete genome established in the fertilized egg. In molecular terms, the DNAs of all differentiated cells are identical.

2 The unused genes in differentiated cells are not destroyed or mutated, and they retain the potential for being expressed.

3 Only a small percentage of the genome is being expressed in each cell, and a portion of the RNA synthesized is specific for that cell type.

Genomic equivalence

In the previous two chapters, we have already analyzed some of the genetic and embryological evidence for genomic equivalence. The ability of nuclei derived from blood cells or larval intestinal cells to generate a complete frog tadpole is impressive testimony that unused genes are still present in these nuclei. Similarly, the ability of a salamander iris to form the differentiated structures of the lens shows that the lens genes were still present, although unused, in the iris. Evidence that each cell of an individual contains the same genes as any other cell has also been obtained from several studies involving nucleic acid hybridization.

The first large-scale molecular analysis to test the identity of DNA throughout the organism was done by Brian McCarthy and B. H. Hoyer in 1964. They found that single-stranded DNA derived from any mouse cell type was equally effective in inhibiting the hybridization of single-stranded radioactive mouse DNA to embryonic mouse genes. This strongly suggested that the DNAs of all cell types had the same number and types of sequences.

Evidence for the presence of specific genes in differentiated cells that do not synthesize the product of these genes comes from in situ hybridization. Polytene chromosomes can be denatured in such a way that the strands of the numerous DNA helices are separated while the chromosomes are still on a microscope slide. Radioactive RNA or cDNA is added to the denatured DNA and then washed off the slide after an appropriate incubation time. During that incubation, the RNA is able to bind to the regions of DNA that encoded it. After washing, the slides are covered with a transparent photographic emulsion. The radioactivity from any bound RNA sensitizes the silver grains of the emulsion, causing a black dot to form over the chromosomal band when the emulsion is developed. Therefore, an RNA for a specific product should be able to reveal the band from which it was synthesized.

Figure 18 shows an autoradiograph localizing the genes for a specific

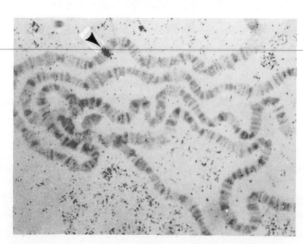

FIGURE 18
In situ hybridization of yolk protein cDNA to the polytene chromosomes of the larval *Drosophila* salivary gland. The dark grains (arrow) show where radioactive yolk protein cDNA (made from yolk protein mRNA) bound to the chromosomes. (From Barnett et al., 1980; photograph courtesy of P. C. Wensink.)

yolk protein of *Drosophila*. DNA from *Drosophila* was cloned and screened with partially purified yolk-protein mRNA. Of those clones binding the yolk-protein mRNA, several were found to contain DNA capable of directing the synthesis of a yolk protein. These clones were grown and the yolk-protein genes isolated. A radioactive probe was made from one of these clones and hybridized with the preparations of polytene salivary gland chromosomes. As illustrated in the figure, this cDNA binds to one specific band. However, larval salivary glands do not produce this protein. The only cells normally synthesizing yolk proteins are the ovaries and fat body cells of the adult female fly. Thus, a gene that is active solely in the adult fat body and ovarian cells was shown to be present in larval salivary gland chromosomes.

Gene stability

It has also been possible to show that unused genes of differentiated cells can be activated under certain conditions and that they can produce proteins specific for other cell types. In mammals, the clearest evidence for the reactivation of unused genes comes from the laboratory of Mary Weiss (Peterson and Weiss, 1972; Brown and Weiss, 1975), who fused together differentiated cells of various kinds. Cell fusion can occur naturally (as in muscle development) or can be mediated by agents such as inactivated Sendai virus (mouse measles) or polyethylene glycol. Such fusion creates a situation in which two nuclei reside in a common cytoplasm (Figure 19). Under the appropriate conditions, the two nuclei of a fused cell will enter mitosis at the same time and eventually form a hybrid nucleus containing the chromosomes of both parental cell types. In most instances, when two different types of cells are fused together, the resulting hybrid lacks the differentiated traits that characterize the parental cells. By fusing rat liver tumor cells to mouse fibroblasts, Weiss was able to isolate hybrids having two sets of liver chromosomes to one fibroblast set. These cells retained their ability to make rat liver-specific proteins such as albumin, aldolase,

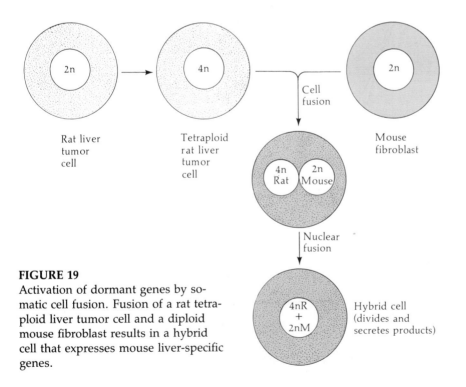

FIGURE 19
Activation of dormant genes by somatic cell fusion. Fusion of a rat tetraploid liver tumor cell and a diploid mouse fibroblast results in a hybrid cell that expresses mouse liver-specific genes.

and tyrosine aminotransferase (TAT). More surprising, they also synthe-sized *mouse* albumin, mouse aldolase, and mouse TAT—three liver pro-teins that no fibroblast ever synthesizes. The mouse fibroblast had retained liver-specific genes in a form that was capable of expression under certain circumstances. This situation fits a general rule in animal development: no irreversible genetic changes occur during cellular differentiation.

Exceptions to genome equivalence: Changes in lymphocyte genes

The techniques of molecular biology have also uncovered an intriguing exception to the rule stated above: the differentiation of lymphocytes. According to the general rule, unused genes are present and potentially functional in differentiated cells. The genetic repertoire is the same in all these tissues. The genome of each lymphocyte cell, however, can be different from any of the other cells in the body, including other lympho-cyte cells. One of these types of lymphocytes is the B CELL, the cell that synthesizes proteins called antibodies. The genome of each B cell is rear-ranged so that each B cell is able to make one type of antibody.

Antibodies are produced when a foreign substance—the ANTIGEN—comes into contact with B cells, which reside in the lymph nodes and spleen. Even before contact with an antigen, each of the resting B cells makes antibody molecules. However, they do not secrete them. Rather, the antibody molecules are inserted into their cell membranes. Each B lymphocyte makes an antibody that recognizes one and only one antigenic shape. Therefore, an antibody recognizing the protein shell of polio virus would not be expected to recognize cholera toxin, *E. coli* membranes, or influenza virus. Once the membrane-bound antibody binds a specific an-tigen molecule, however, the B cell divides numerous times and differ-entiates into an antibody-*secreting* PLASMA CELL (Figures 20 and 21). (The mechanism by which this differentiation occurs will be detailed in Chapter 16.) Only those B cells having the ability to bind the specific antigen are stimulated to multiply and secrete antibodies. According to this model,

FIGURE 20

Clonal selection model of antibody for-mation. B lymphocytes obtain their cell-surface specificity before contact with antigen. Only one type of anti-body is in the cell membrane. B lym-phocytes whose membrane-bound anti-bodies have bound an antigen proliferate to produce clones of plasma cells that secrete antibody with the same specificity as the cell-surface an-tibody that bound the antigen. (After Edelman, 1970.)

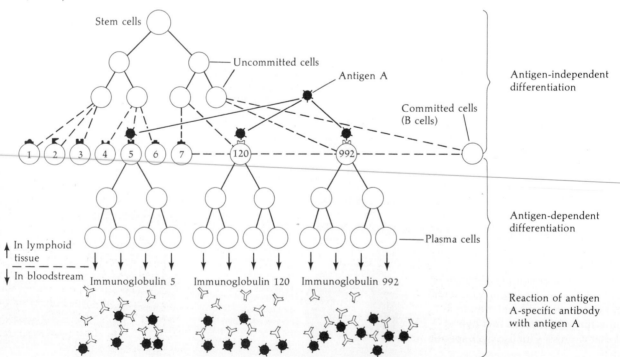

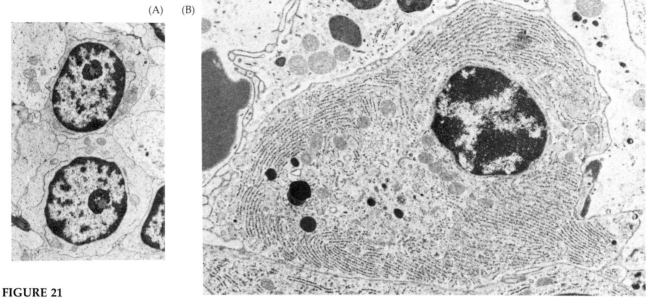

(A)   (B)

FIGURE 21

(A) B lymphocyte and (B) plasma cell. Note the expansion of the rough endo-plasmic reticulum as the B lymphocyte becomes the antibody-secreting plasma cell. (Photographs courtesy of L. Weiss.)

called the CLONAL SELECTION THEORY (Burnett, 1959), each B cell acquires its specificity *before* contact with the antigen. That is, out of the millions of possible antibodies it can make, each B lymphocyte "chooses" one of them and places antibodies of that specificity on its cell surface.

The mechanism of this selection of antibody specificity involves the creation of new genes during B lymphocyte differentiation. The antibody protein on the cell surface consists of two pairs of polypeptide subunits (Figure 22). There are two identical heavy chains and two identical light chains. The chains are linked together by disulfide bonds. The specificity

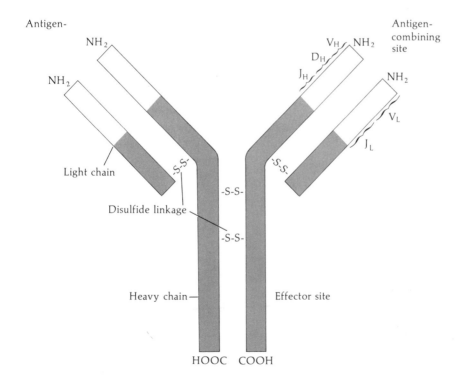

FIGURE 22

Structure of a typical immunoglobulin (antibody) protein. Two identical heavy chains are connected by disulfide linkages. The antigen-combining site is composed of the variable regions (white) of the heavy and light chains, whereas the effector site of the antibody (which controls whether it agglutinates antigens, binds to macrophages, or enters into mucous secretions) is determined by the amino acid sequence of the heavy chain constant (color) region.

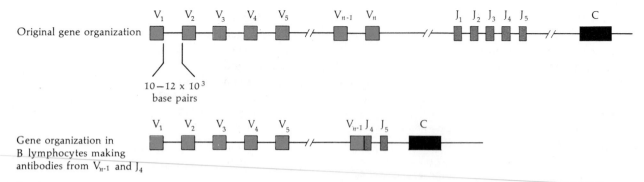

Original gene organization

V_1 V_2 V_3 V_4 V_5 V_{n-1} V_n J_1 J_2 J_3 J_4 J_5 C

$10-12 \times 10^3$
base pairs

Gene organization in
B lymphocytes making
antibodies from V_{n-1} and J_4

V_1 V_2 V_3 V_4 V_5 V_{n-1} J_4 J_5 C

FIGURE 23

Rearrangement of the light-chain genes during B lymphocyte development. Prior to antigenic stimulation, one of the 300 or more V gene segments combines with one of the five J gene segments and moves closer to the constant (C) gene segment.

of the antibody molecule (i.e., whether it will bind to a polio virus, an *E. coli* cell, or some other molecule) is determined by the amino acid sequence of the VARIABLE REGION. This region is made up of the amino-terminal ends of one heavy chain and one light chain. The variable regions of the antibody molecules are attached to CONSTANT REGIONS, which give the antibody its effector properties. For example, the heavy chain constant region of cell-surface immunoglobulin molecules anchors the protein into the cell membrane.

Creation of antibody light chain genes. Antibody heavy and light chain genes are organized in segments. Light chain genes contain three segments (Figure 23). The first gene segment codes for the light chain variable (V) region. It contains about 300 different sequences, all of which code for the first 97 amino acids of the light chain of an antibody. The second segment, the J region, consists of four or five possible DNA sequences for the last 15–17 residues of the variable portion of the light chain of an antibody. The third segment is the constant (C) region of the light chain. In B lymphocyte development, one of the 300 V regions and one of the five J regions combine to form the variable portion of the antibody gene. This is done by moving a V region sequence to a J region sequence, a rearrangement that eliminates the intervening DNA.

This gene rearrangement was first shown by Nobumichi Hozumi and Susumu Tonegawa. Hozumi and Tonegawa (1976) isolated DNA from a mouse embryo and from a light chain-secreting B cell tumor.* They digested these two DNAs separately with the restriction enzyme *Bam*HI, which cleaved the DNA wherever it located the sequence GGATCC. The result was a series of DNA fragments, the size of each fragment determined by the length of the DNA molecule between two cleavage sites.

These DNA fragments were placed at one end of a gelatinous slab and an electric current was passed through the gel (Figure 24). As the DNA migrated toward the positive electrode, the smaller fragments moved

FIGURE 24

Electrophoretic separation of DNA fragments. DNA cut by restriction enzyme *Bam*HI is placed in a slot in a gel. In an electric current, the smaller pieces move faster than the larger ones. After separation of the fragments (on the basis of size), the gel is sliced, and the DNA fragments eluted for hybridization.

*Tumors were used because they produce an enormous amount of one specific immunoglobulin (and of the mRNA for that immunoglobulin).

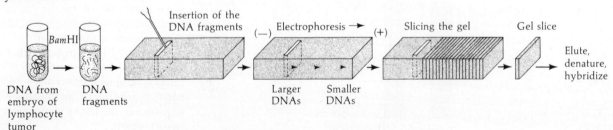

Insertion of the DNA fragments

*Bam*HI

DNA from embryo of lymphocyte tumor

DNA fragments

(−) Electrophoresis → (+)

Larger DNAs   Smaller DNAs

Slicing the gel

Gel slice

Elute, denature, hybridize

faster than the larger ones, effectively separating by size. The gel (containing the DNA spread throughout it) was cut into several pieces, each slice containing pieces of DNA of a certain size.* The DNA from each slice of the gel was eluted and denatured. Part of this DNA was hybridized with radioactive RNA that coded for the entire light chain and had been isolated from the original B cell tumor. The other part was hybridized with radioactive RNA coding only for the C region of the light chain (the 3′ half of a longer RNA message). The DNA from the embryo bound the light chain mRNA in two slices. The DNA of the first slice had a molecular weight (MW) of about 6 million, and the molecular weight of the DNA in the second band was 3.9 million. When the mouse embryo DNA was reacted with the light chain C region mRNA, only the 6 million MW DNA bound the RNA. Thus, in the mouse embryo, the C region was coded within DNA fragments having a molecular weight of 6 million (between *Bam*HI sites), while the V region was coded within a region of 3.9 million (Figure 25).

The lymphocyte tumor DNA, however, gave a very different result. The only lymphocyte DNA that bound the light chain mRNA had a molecular weight of 2.4 million. Moreover, it bound the light chain C region mRNA segment as well. Both the C and the V regions were found to be coded on the same fragment of DNA! The simplest explanation (and one confirmed by numerous other laboratories and methods; see Brack et al., 1978; Bernard et al., 1978; Seidman et al., 1979) was that two gene fragments, one coding for the light chain C region and one coding for a specific light chain V region, had fused together *to form a new gene*. A new gene had been created during the development of the lymphocyte. Their model for such gene synthesis is presented in Figure 26.

Creation of antibody heavy chain genes. The heavy chain genes of antibodies contain even more segments than the light chain genes. Heavy chain gene segments include a V region (200 different sequences for the first 97 amino acids), a D region (10–15 different sequences of 3–14 amino acids) and a J region (four sequences for the last 15–17 amino acids of the V region). The next region is the C region. The heavy chain variable region is formed by adjoining one V sequence and one D sequence to one J sequence (Figure 27 A,B). This VDJ variable region sequence is now adjacent to the first C region of the heavy chain—the C region specifically for antibodies that can be inserted into the plasma membrane. Thus, an antibody molecule is formed from two genes created during the antigen-independent stage of B lymphocyte development. This antibody is placed in the cell membrane as an antigen receptor.

Class switching in the heavy chain gene. Upon stimulation by antigen, the B cell divides and differentiates into an antibody-secreting PLASMA CELL. At first, the antibodies produced by these cells contain the same C region as before. However, as antibody synthesis continues, the C region can change. The first C region is called a mu constant (Cμ) region. The second C region—that of the secreted antibody—may remain μ (albeit with a modification that allows secretion), but it can also be a gamma (γ), an epsilon (ε), or an alpha (α) constant region. Thus, a single heavy chain variable region can be seen first on a μ constant region and later on, say, a γ constant region. This phenomenon is called CLASS SWITCHING. (The heavy chain CLASS of an antibody determines its mode of operation: μ and

---

*The size of the DNA in each piece was determined by running DNAs of known sizes in an adjacent column.

(A) EXPERIMENT PROTOCOL

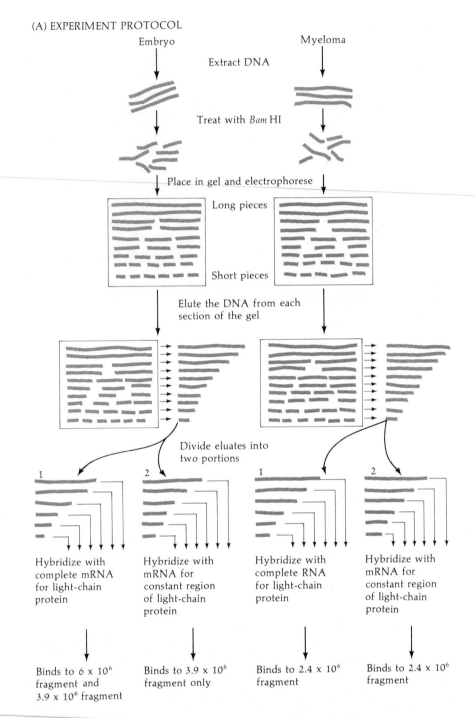

FIGURE 25

Protocol and results of the Hozumi and Tonegawa experiment. DNA from mouse embryo cells and B cell tumors (myelomas) were separately digested in BamHI, separated by electrophoresis and eluted from the gels. After being denatured, each eluted DNA sample was hybridized to either mRNA coding for the variable and constant regions of the immunoglobulin light chain (whole) or a fragmented mRNA encoding only the constant region of that light chain protein (3' half). For the embryonic DNA, the variable and constant regions of the light chain protein were found on two different pieces of DNA (the constant region being on a piece having a molecular weight (MW) of 3.9×10^6, the variable region being on a DNA fragment of 6×10^6). In the lymphocyte tumor, the constant and variable regions are together on a single DNA fragment having a molecular weight of 2.4×10^6. (After Hozumi and Tonegawa, 1976.)

(B) RESULTS

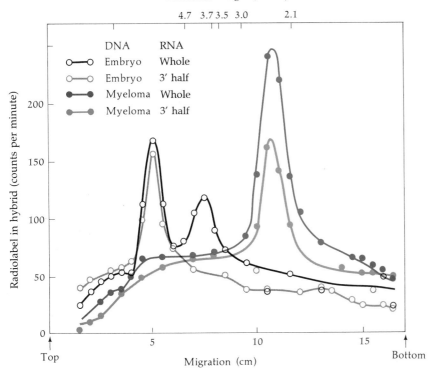

Molecular weight (x 10⁻⁶)

4.7 3.7 3.5 3.0   2.1

DNA   RNA
○—○ Embryo   Whole
○—○ Embryo   3′ half
●—● Myeloma Whole
●—● Myeloma 3′ half

Radiolabel in hybrid (counts per minute)

200

150

100

50

0

Top   5   10   15   Bottom

Migration (cm)

γ chains promote lysis, agglutination, or macrophage digestion of an antigen; ε chains induce inflammatory responses; and α chains allow the antibodies to be secreted into mucus, tears, sweat, and milk.) Class switching is accomplished by taking the entire variable region gene segment (VDJ) and translocating it from in front of the μ constant region and placing it adjacent to the γ, ε, or α constant region (Figure 27C,D). This process results in the deletion of the μ constant region gene segment from the genome (Davis et al., 1980a,b; Cory et al., 1980; Rabbitts et al., 1980; Yaoita and Honjo, 1980). The deleted part of the chromosome forms a circle containing the μ constant region and whatever other regions were between the VDJ sequence and the new constant region (von Schwedler et al., 1990). As this circle lacks a centrosome, it is presumably lost upon cell division and is degraded in the cytoplasm.

Thus, the genome of each plasma cell clone can be markedly different from that of any other cell. First, it has created a variable region gene sequence by bringing together different DNA segments. In all other organs, these DNA segments are separated; in B lymphocytes and plasma

FIGURE 26

Model of the changes in DNA between embryonic cells and the B lymphocyte, according to the data of Hozumi and Tonegawa (1976).

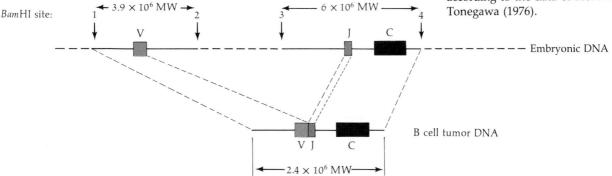

*Bam*HI site:

← 3.9 × 10⁶ MW →   ←——— 6 × 10⁶ MW ———→
1   2   3   4

V   J   C

Embryonic DNA

V J   C   B cell tumor DNA

←——— 2.4 × 10⁶ MW ———→

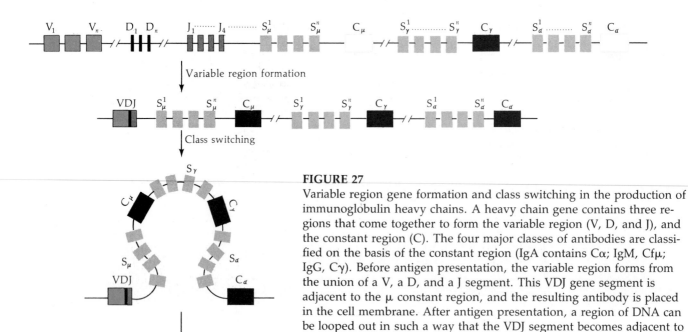

FIGURE 27

Variable region gene formation and class switching in the production of immunoglobulin heavy chains. A heavy chain gene contains three regions that come together to form the variable region (V, D, and J), and the constant region (C). The four major classes of antibodies are classified on the basis of the constant region (IgA contains $C\alpha$; IgM, $Cf\mu$; IgG, $C\gamma$). Before antigen presentation, the variable region forms from the union of a V, a D, and a J segment. This VDJ gene segment is adjacent to the μ constant region, and the resulting antibody is placed in the cell membrane. After antigen presentation, a region of DNA can be looped out in such a way that the VDJ segment becomes adjacent to another constant region (in this case, the α constant region, which enables the antibodies to enter mucous secretions). This class switching is mediated by a series of switch (S) sequences adjacent to each of the constant regions. (After Davis et al., 1980a,b.)

cells, they have been brought together. Second, in many plasma cells, a portion of the genome (namely, the DNA of the μ heavy chain constant region) has been eliminated from the nucleus. A specific part of the genetic repertoire has been lost in the development of the plasma cell; new genes are created while other genes are destroyed.

What can we conclude from the preceding discussion? Present evidence suggests that the nuclei of differentiated cells do indeed retain most of their specific genetic information in a form that can be expressed under appropriate conditions. It is clear, however, that some loss of genetic material accompanies at least one form of cellular differentiation. As yet, we have no way of knowing how many other cases of irreversible genetic change there may be during animal development, but our present knowledge of lymphocyte gene rearrangements suggests that irreversible genetic loss is a result, not a cause, of cellular differentiation.

SIDELIGHTS & SPECULATIONS

Gene alterations

Plasma cells have a genome deficient in certain DNA sequences that are needed for differentiation. Another cell of the immune system, the T lymphocyte, also deletes a portion of its genome in the construction of its antigen receptor (Fujimoto and Yamagishi, 1987). The enzymes responsible for mediating the DNA recombination events appear to be the same in the B cell and T cell lineages. Called RECOMBINASES (Schatz et al., 1989; Oettinger et al., 1990), these two proteins recognize the signal regions of DNA immediately upstream from the recombinable DNA. Moreover,

the genes for these enzymes are active only in the pre-B cells and pre-T cells wherein the genes are being recombined. These recombinase genes are not active in mature B and T cells or in any nonlymphocyte cell.

But this is only one form of genomic instability. Other examples are known (Borst and Greaves, 1987). The parasite *Trypanosoma brucei*, which causes sleeping sickness, alters its genome to produce different cell-surface glycoproteins (thereby escaping the host's immune system). A similar mechanism appears to be responsible for the alteration of mating types in yeast (Hoeijmakers et al., 1980; Haber et al., 1980; Strathern et al., 1979). Here, each haploid cell contains both mating-type genes. The determination of mating type depends upon which of them is in a particular position on the chromosome. The positions are switched from one division to the next. Thus, in one generation, the *a* mating-type gene is in the "on" locus, whereas in the next generation, the α gene is in that position while the *a* gene waits nearby.

In each of these three cases, the genes involved are responsible for cell-surface recognition phenomena. Such phenomena are found throughout vertebrate development (Part III), so it would not be too surprising if an irreversible genomic change were responsible for the loss of totipotency.

Another type of gene alteration comes from the insertion of TRANSPOSABLE ELEMENTS (or TRANSPOSONS) into a gene. Transposons are migratory pieces of DNA that can integrate throughout the genome. When such a sequence interrupts a structural gene, the gene is inactivated. The best-known examples are in bacteria, maize, and *Drosophila*. In maize, a transposon inserted in or near the gene for kernel pigment will produce a colorless kernel (McClintock, 1952; Peterson, 1980). However, upon the transposon's removing itself from the gene, pigment synthesis is restored. The result is a variegated phenotype (Figure 28). In *Drosophila*, the *white-apricot* mutation is caused by the insertion of a transposable element into the DNA of the *white* gene (Green, 1980; Gehring and Paro, 1980). So far, such changes in the positions of transposons are not known to direct any developmental processes. Transposons can insert into genomic DNA much as retroviruses do, and they can be used like retroviruses to bring new genetic material into *Drosophila* cells (Spradling and Rubin, 1982).

It is clear, then, that the genome is a dynamic entity and not a perfectly stable structure. Although the changeability of the nucleus came as a surprise to many biologists, it was predicted by the founder of the gene theory, Thomas Hunt Morgan. Morgan, writing in 1927, acknowledged that

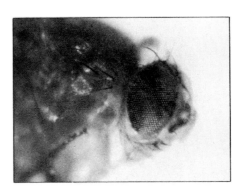

FIGURE 28
The mosaic eye of this fruit fly was created by the spontaneous removal of a DNA transposon that had interrupted a gene that encoded a protein necessary for the production of wild-type eye pigment. The white areas represent cells whose genes have retained the transposons, while the wild-type (dark) regions contain those cells that have lost the transposon. (Photograph courtesy of G. M. Rubin.)

"the most common genetic assumption is that the genes remain the same throughout this time [of development]." He reasoned that

> [T]he basic constitution of the gene remains always the same, the postulated addition or changes in the gene being of the same order as those that take place in the protoplasm. If the latter can change in differentiation in a new environment without losing its fundamental properties, why not the genes also? This question is clearly beyond the range of present evidence, but as a possibility, it need not be rejected. The answer, for or against such an assumption, will have to wait until evidence can be obtained by experimental investigation.

Differential RNA synthesis

The third postulate—that only a small portion of the genome is active in making tissue-specific products—has also been tested in numerous organisms. Cytoplasmic RNA from insects and vertebrates hybridizes with less than 10 percent of the possible DNA sites available, but evidence shows that this RNA contains tissue-specific sequences. Becker (1959), studying *Drosophila*, and Beermann (1952), studying the larval gall midge *Chironomus*, found that there were regions of the chromosomes that were "puffed out." These puffs appeared in different places on the chromosomes in different tissues, and their appearances changed with the development of these cells (Figure 29). Furthermore, certain puffs could be induced or inhibited by certain physiological changes caused by heat or hormones (Clever, 1966; Ashburner, 1972; Ashburner and Berondes, 1978).

These puffs are now known to represent the spinning out of the polytene chromosome into a more loosely compacted arrangement (Figure 30); Beermann also presented the first evidence that these puffs were sites of active messenger RNA synthesis (Beermann, 1961). He found two in-

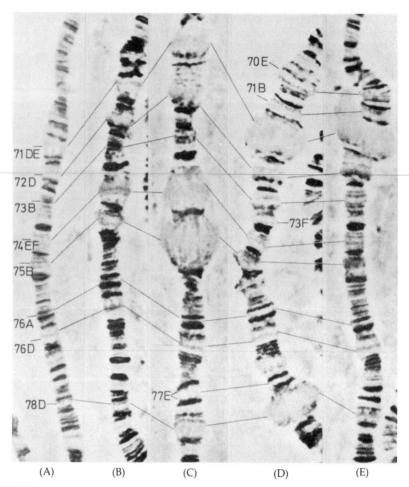

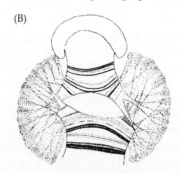

FIGURE 29
Puffing sequence of a portion of chromosome 3 in the larval *Drosophila melanogaster* salivary gland. (A,B) 110-hour larva; (C) 115-hour larva; (D,E) prepupal stage (4 hours apart). Note the puffing and regression of bands 74EF and 75B. Other bands (71DE, 78D) puff later, but most do not puff at all during this time. (Courtesy of M. Ashburner.)

FIGURE 30
Proximal end of chromosome 4 from the salivary gland of *Chironomus pallidivitatus*, showing the enormous puff, BR2. (A) Phase-contrast photomicrograph of stained preparation showing extended puff on the polytene chromosome. (B) Diagrammatic representation of the BR2 region undergoing puffing. (A from Grossbach, 1973; photograph courtesy of U. Grossbach. B after Beermann, 1963.)

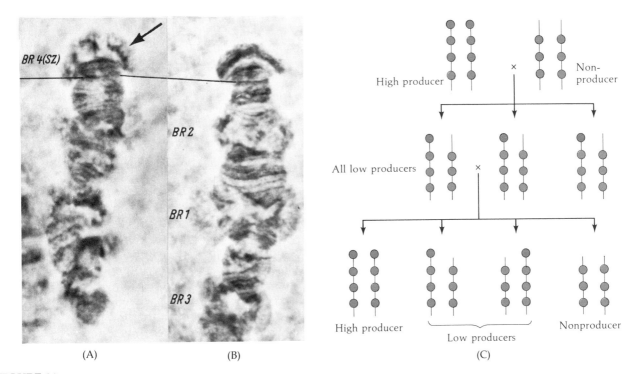

(A)   (B)   (C)

FIGURE 31

Correlation of puffing patterns with specialized functions in salivary gland cells of *Chironomus pallidivitatus*. (A) Chromosome 4 from a salivary gland cell producing a granular secretion and showing an additional Balbiani ring [4(SZ)]. (B) Chromosome 4 from a clear salivary cell, showing only Balbiani rings 1, 2, and 3 (BR1, BR2, BR3). (C) Genetic evidence that the synthesis of a major salivary protein is dependent on the formation of BR4(SZ) puffs. Larvae with high amounts of granular secretions have salivary gland cells with BR4(SZ) puffs on both chromosomes 4 (color), whereas those larvae withouts these secretions have no such puffs. Intermediate producers have only one chromosome 4 with a puffed BR4(SZ) region in each salivary cell making the secretion. (B from Beermann, 1961; photographs courtesy of W. Beermann.)

terbreeding species of *Chironomus* that differed: one produced a major salivary protein, whereas the other did not (Figure 31). The producers had a large puff (Balbiani ring) at a certain band in the polytene chromosomes in cells of the larval salivary gland; that puff was absent in the nonproducers. When producer was mated with nonproducer, the hybrid larva produced an intermediate amount of salivary protein. When two hybrid flies were mated, the ability to produce salivary protein segregated in proper Mendelian fashion (1 high producer:2 intermediate producers:1 nonproducer). Moreover, whereas high producers were found to have two puffs (one on each homologous chromosome), the intermediate producers had only one puff. The non-producers lacked any puffs at these sites. Beermann concluded that the genetic information needed for the synthesis of this salivary protein was present in this distal chromosome band and that the synthesis of that product depended upon that band becoming a puff region.

Further proof that chromosomal puffs make messenger RNA comes from studies of Balbiani ring 2 puffs in *Chironomus tentans*. Because of its exceptionally large size, Balbiani ring 2 (BR2) can be isolated by microdissection. Its products can be analyzed by electrophoresis and autoradiography (Lambert and Daneholt, 1975). Figure 32 shows the isolation of

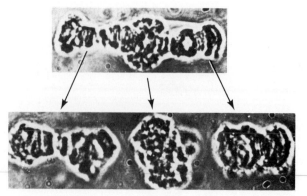

FIGURE 32

Isolation of BR2 region of *Chironomus tentans* by micromanipulation. The intact chromosome 4 can be divided into three regions, one containing BR2. (From Lambert and Daneholt, 1975; photograph courtesy of B. Lambert.)

BR 2

BR2 from chromosome 4 of *C. tentans*. The nuclear sap and cytoplasm of these salivary gland cells can also be collected to monitor the fate of the Balbiani ring RNA. Balbiani ring 2 transcription was demonstrated by incubating isolated salivary glands with radioactive RNA precursors. Radioactive RNA could then be extracted from the BR2 portion of the dissected chromosome (Lambert, 1972). This RNA was found to sediment at 75 S, thereby indicating an exceptionally large RNA of about 50,000 bases. This 75 S RNA hybridized specifically to the BR2 region of the chromosome, thereby showing that the puffed DNA—and no other locus—had been actively transcribing it (Figure 33). Unlike most RNAs, the 75 S BR2 RNA did not decrease its size noticeably when it passed through the nucleus into the cytoplasm. This characteristic made it especially valuable because, if it were coding for a protein, the large number of ribosomes binding to it would form a complex with an extraordinarily high molecular weight. Therefore, salivary glands were again incubated with radioactive RNA precursors and the cytoplasmic polysome complexes were sedimented through a sucrose gradient. Some of the polysomes had very high sedimentation values, due to the binding of the mRNA with a large numbers of ribosomes. The mRNA from these large polysomes was extracted (by dispersing the ribosomes with EDTA) and 75 S RNA was obtained. This RNA hybridized specifically to the BR2 region of the chromosome (Wieslander and Daneholt, 1977). Thus, an RNA transcribed from a specific band of DNA that puffs in the larval salivary gland is seen later

FIGURE 33

Transcription from the BR2 region of chromosome 4 of *Chironomus tentans* salivary gland cells. (A) Toluidine blue-stained chromosome preparation. Arrows show the points at which the chromosome was cut for in vitro RNA synthesis. (B) In situ autoradiograph after BR2 RNA was hybridized to the chromosome preparation. (From Lambert, 1972; photographs courtesy of B. Lambert.)

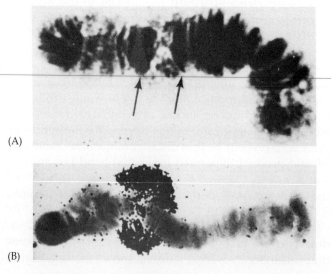

(A)

(B)

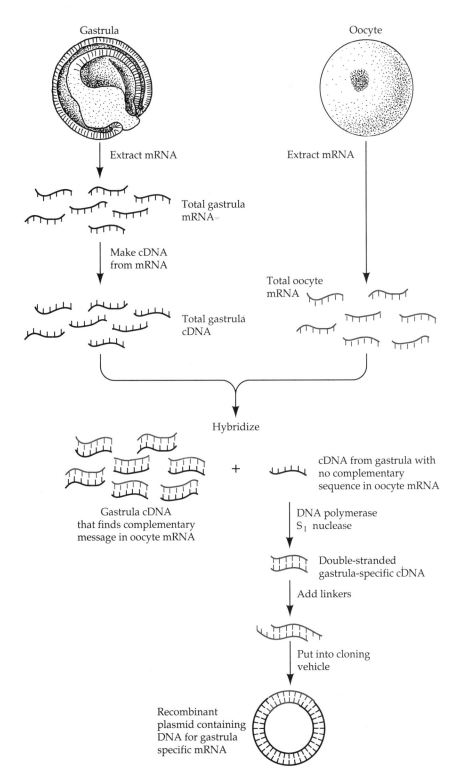

Gastrula

Oocyte

Extract mRNA

Extract mRNA

Total gastrula
mRNA

Make cDNA
from mRNA

Total oocyte
mRNA

Total gastrula
cDNA

Hybridize

Gastrula cDNA
that finds complementary
message in oocyte mRNA

+

cDNA from gastrula with
no complementary
sequence in oocyte mRNA

DNA polymerase
S_1 nuclease

Double-stranded
gastrula-specific cDNA

Add linkers

Put into cloning
vehicle

Recombinant
plasmid containing
DNA for gastrula
specific mRNA

FIGURE 34
Subtraction cloning of differentially expressed gastrula genes in *Xenopus laevis*. cDNA was made to messages isolated from gastrulae and then hybridized to oocyte mRNA. Those gastrula cDNAs that did not find complementary sequences in the oocyte mRNAs were the products of genes active in the gastrula but not in the oocyte. These genes were cloned by making the cDNA double-stranded and adding linkers so that they could enter into cloning vehicles.

to be making proteins on cytoplasmic ribosomes. Therefore, the puffs on salivary chromosomes are actively making messenger RNA.

The ability to correlate puffs with active gene transcription provided some of the first evidence for tissue-specific gene expression. For example, the salivary glands of the last larval stage of *Drosophila* are responsible for secreting the glue that will adhere the pupal case to the solid substrate. Genetic analysis has located the genes for several glue proteins. These genes puff out (a) only during the last half of the last larval stage, and (b) only in the salivary gland (Korge, 1975). Northern blot analysis of the mRNA for these proteins has confirmed that they are expressed in both a tissue-specific and temporally specific manner.

Temporally specific gene activity. Differential gene activity occurs both in time and space. Cloned genes have also been valuable for determining the times at which specific genes are being transcribed. Sargent and Dawid (1983) have isolated from *Xenopus* gastrulae the mRNA that was not present in the egg. To do this, they extracted the gastrula mRNA and made cDNA copies of these messages. They mixed these gastrula cDNAs with large quantities of oocyte mRNA. If the oocyte mRNA hybridized with gastrula complementary DNA, that meant that the cDNA was derived from an mRNA present in both oocyte and gastrula stages. These double-stranded hybrid molecules were removed by filtration, thereby leaving a population of gastrula-specific cDNAs. These cDNAs were then made double-stranded (by DNA polymerase) and inserted into cloning vehicles. This technique is called SUBTRACTION CLONING. It generates a stage-specific library of clones whose mRNA is found in some stages but not others (Figure 34).

Sargent and Dawid then took embryos from the one-cell stage through the tailbud tadpole stage and separately isolated their RNAs. The RNAs were spotted onto nitrocellulose filters so that each filter had RNAs from all stages. After the RNAs were baked onto the filter, single-stranded DNA derived from a particular clone was made radioactive and incubated with the filters. If a gene was being transcribed at a given stage, the radioactive cDNA from that gene would find its complement in the mRNAs from that stage on the filter. The nonbound cDNA was then washed off, and the binding of radioactive cDNA observed by autoradiography. This technique is called DOT BLOTTING. Figure 35 shows the time courses of gene expression for 17 genes that are active at various stages of gastrulation. None of

FIGURE 35
Developmental "dot blots," showing the times when 17 *Xenopus* genes were transcriptionally active. Specific mRNA accumulation in the cytoplasm is scored by baking total mRNA from gene embryonic stages onto nitrocellulose paper and incubating that strip of paper with radiolabeled DNA derived from a gastrula-specific cDNA clone. By placing these strips together, one gets a time course for the activity of the specific genes. The r5 line represents a ribosomal RNA control that should always be present. (From Jamrich et al., 1985; photograph courtesy of I. Dawid and M. Sargent.)

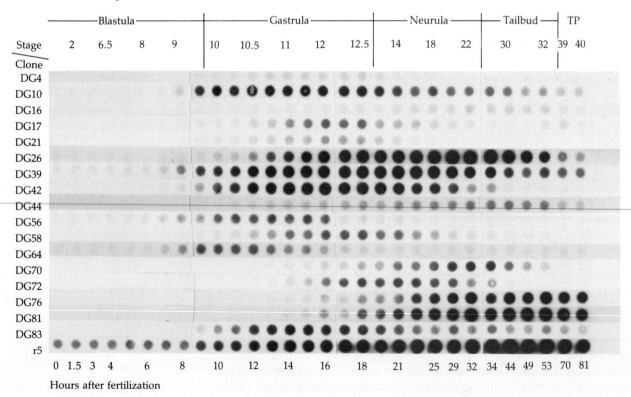

them is expressed earlier than the midblastula transition at 7 hours. Some genes (DG64, DG39) are expressed immediately afterward, whereas other genes (such as DG72 or DG81) begin to be transcribed at midgastrula, about 7 hours later. Some genes (DG76, DG81) appear to "stay on" once activated, while the activity of other genes (DG56 or DG21) is far more transient.

Whereas dot blotting can show the *temporal* specificity of gene expression, the *spatial*, tissue-specific character of gene expression can be demonstrated by in situ hybridization. An example of this was shown in Figure 8, in which a radioactive probe identified an accumulation of a particular mRNA solely in the sea urchin larval endoderm and not in the mesodermal or ectodermal cells. Thus, we can use in situ hybridization to visualize the spatial patterns of differential RNA synthesis or accumulation.

SUMMARY

We can conclude from the studies presented in this chapter that

1 Most differentiated cell types retain unused genes in a form that allows their expression under appropriate conditions.
2 There exist certain specific cases, such as the plasma cell, where the loss of genetic material is associated with cellular differentiation. (Yet even here, the loss of specific DNA sequences is the result of, and not the cause of, differentiation.)
3 There exist mRNAs whose synthesis is developmentally regulated; that is to say, these mRNAs are produced only by certain cells at certain stages of development. (This does not imply that *all* mRNAs are so regulated or that *all* development is regulated by differential gene transcription.)

We have, then, the paradigm of differential gene expression: differentiated cells contain the same genes but regulate their expression so that different cells make different proteins. The remaining chapters of this section will investigate the mechanisms by which differential gene expression takes place in space and time.

LITERATURE CITED

Ashburner, M. 1972. Patterns of puffing activity in the salivary glands of *Drosophila*. VI. Induction by ecdysone in salivary glands of *D. melanogaster* cultured in vitro. *Chromosoma* 38: 255–281.

Ashburner, M. and Berondes, H. D. 1978. Puffing of polytene chromosomes. In *The Genetics and Biology of Drosophila*, Vol. 2B. Academic Press, New York, pp. 316–395.

Barnett, T., Pachl, C., Gergen, J. P. and Wensink, P. C. 1980. The isolation and characterization of *Drosophila* yolk protein genes. *Cell* 21: 729–738.

Becker, H. J. 1959. Die Puffs der Speicheldrüsenchromosomen von *Drosophila melanogaster*. I. Beobachtungen zum Verhalten des Puffmusters im Normalstamm und bei zwei Mutanten, giant- und lethal-giant larvae. *Chromosoma* 10: 654–678.

Beermann, W. 1952. Chromomerenkonstanz und spezifische Modifikationen der Chromosomenstruktur in der Entwicklung und Organdifferenzierung von *Chironomus tentans*. *Chromosoma* 5: 139–198.

Beermann, W. 1961. Ein Balbiani-ring als Locus einer Speicheldrüsen-Mutation. *Chromosoma* 12: 1–25.

Beermann, W. 1963. Cytological aspects of information transfer in cellular differentiation. *Am. Zool.* 3: 23–28.

Bernard, O., Hozumi, N. and Tonegawa, S. 1978. Sequences of mouse immunoglobulin light chain genes before and after somatic change. *Cell* 15: 1133–1144.

Blattner, F. R., Blechl, A. E., Denniston-Thompson, K., Faber, H. E., Richards, J. E., Slightom, J. L., Tucker, P. W. and Smithies, O. 1978. Cloning human fetal γ-globin and mouse α-type globin DNA: Preparation and screening of shotgun collections. *Science* 202: 1279–1283.

Borst, P. and Greaves, D. R. 1987. Programmed gene rearrangements altering gene expression. *Science* 235: 658–667.

Brack, C., Hirama, M., Lenhard-Schuller, R. and Tonegawa, S. 1978. A complete immunoglobulin gene is created by somatic recombination. *Cell* 15: 1–14.

Brown, J. E. and Weiss, M. C. 1975. Activation of production of mouse liver enzymes in rat hepatoma–mouse lymphoid hybrids. *Cell* 6: 481–493.

Burnett, F. M. 1959. *The Clonal Selection Theory of Immunity.* Vanderbilt University Press, Nashville, TN.

Capecchi, M. R. 1980. High efficiency transformation by direct microinjection of DNA into cultured mammalian cells. *Cell* 22: 479–488.

Clever, U. 1966. Induction and repression of a puff in *Chironomus tentans. Dev. Biol.* 14: 421–438.

Cohen, S. N., Chang, A. C. Y., Boyer, H. W. and Helling, R. B. 1973. Construction of biologically functional bacterial plasmids in vitro. *Proc. Natl. Acad. Sci. USA* 70: 3240–3244.

Constantini, F., Chada, K. and Magran, J. 1986. Correction of murine β-thalassemia by gene transfer into the germ line. *Science* 233: 1192–1194.

Cory, S., Jackson, J. and Adams, J. M. 1980. Deletions in the constant region locus can account for switches in immunoglobulin heavy chain expression. *Nature* 285: 450–456.

Davis, M. M., Calame, K., Early, P. W., Livant, D. L., Joho, R., Weissman, I. L. and Hood, L. 1980a. An immunoglobulin heavy chain gene is formed by at least two recombinational events. *Nature* 283: 733–739.

Davis, M. M., Kim, S. K. and Hood, L. 1980b. Immunoglobulin class switching: Developmentally regulated DNA rearrangements during differentiation. *Cell* 22: 1–2.

Dzierzak, E. A., Papayannopoulou, T., and Mulligan, R. C. 1988. Lineage-specific expression of a human β-globin gene in murine bone marrow transplant recipients reconstituted with retrovirus-transduced stem cells. *Nature* 331: 35–41.

Edelman, G. M. 1970. The structure and function of antibodies. *Sci. Am.* 223(2): 34–42.

Fujimoto, S. and Yamagishi, H. 1987. Isolation of an excision product of T-cell receptor α-chain gene rearrangements. *Nature* 327: 242–243.

Gehring, W. J. and Paro, R. 1980. Isolation of a hybrid plasmid with homologous sequences to a transposing element of *Drosophila melanogaster. Cell* 10: 897–904.

Gillam, S., Astell, C. R. and Smith, M. 1980. Site specific mutagenesis using oligodeoxyribonucleotides: Isolation of a phenotypically silent φX174 mutant, with a specific nucleotide deletion, at a very high frequency. *Gene* 12: 129–137.

Gossler, A., Doetschman, T., Korn, R., Serfling, E. and Kemler, R. 1986. Transgenesis by means of blastocyst-derived stem cell lines. *Proc. Natl. Acad. Sci. USA* 83: 9065–9069.

Green, M. M. 1980. Transposable elements in *Drosophila* and other diptera. *Annu. Rev. Genet.* 14: 109–120.

Grossbach, U. 1973. Chromosome puffs and gene expressions in polytene cells. *Cold Spring Harbor Symp. Quant. Biol.* 38: 619–627.

Haber, J. E., Rogers, D. T. and McCusker, J. H. 1980. Homothallic conversions of yeast mating types occur by intrachromosomal recombination. *Cell* 22: 277–289.

Hoeijmakers, J. H. J., Frasch, A. C. C., Bernards, A., Borst, P. and Cross, G. A. M. 1980. Novel expression-linked copies of the genes for variant surface antigens in trypanosomes. *Nature* 284: 78–80.

Hozumi, N. and Tonegawa, S. 1976. Evidence for somatic rearrangement of immunoglobulin genes coding for variable and constant regions. *Proc. Natl. Acad. Sci. USA* 73: 3628–3632.

Jamrich, J., Sargent, T. D. and Dawid, I. 1985. Altered morphogenesis and its effects on gene activity in *Xenopus laevis* embryos. *Cold Spring Harbor Symp. Quant. Biol.* 50: 31–34.

Kelly, R. G., Cozarelli, N., Deutscher, M. P., Lehman, I. R. and Kornberg, A. 1970. Enzymatic synthesis of deoxyribonucleic acid. *J. Biol. Chem.* 245: 39.

Korge, G. 1975. Chromosome puff activity and protein synthesis in larval salivary glands of *Drosophila melanogaster. Proc. Natl. Acad. Sci. USA* 72: 4550–4554.

Lambert, B. 1972. Repeated DNA sequences in a Balbiani ring. *J. Mol. Biol.* 72: 65–75.

Lambert, B. and Daneholt, B. 1975. Microanalysis of RNA from defined cellular components. *Meth. Cell Biol.* 10: 17–47.

Lang, R. A., Metcalf, D., Gough, N. M., Dunn, A. R. and Gonda, T. J. 1985. Expression of a hemopoietic growth factor cDNA in a factor-dependent cell line results in autonomous growth and tumorigenicity. *Cell* 43: 531–542.

McCarthy, B. J. and Hoyer, B. H. 1964. Identity of DNA and diversity of messenger RNA molecules in normal mouse tissues. *Proc. Natl. Acad. Sci. USA* 52: 915–922.

McClintock, B. 1952. Chromosome organization and genic expression. *Cold Spring Harbor Symp. Quant. Biol.* 16: 13–47.

Morgan, T. H. 1927. *Experimental Embryology.* Columbia University Press, New York.

Nathans, D. and Smith, H. O. 1975. Restriction endonucleases in the analysis and restructuring of DNA molecules. *Annu. Rev. Biochem.* 44: 273–293.

Oettinger, M. A., Schatz, D. G., Gorka, C. and Baltimore, D. 1990. *RAG-1* and *RAG-2*, adjacent genes that synergistically activate V(D)J recombination. *Science* 248: 1517–1522.

Pardue, M. L. and Gall, J. G. 1970. Chromosomal localization of mouse satellite DNA. *Science* 168: 1356–1358.

Perucho, M., Hanahan, D. and Wigler, M. 1980. Genetic and physical linkage of exogenous sequences in transformed cells. *Cell* 22: 309–317.

Peterson, J. A. and Weiss, M. C. 1972. Expression of differentiated functions in hepatoma cell hybrids: Induction of mouse albumin production in rat hepatoma–mouse fibroblast hybrids. *Proc. Natl. Acad. Sci. USA* 69: 571–575.

Peterson, R. A. 1980. Instability among the components of a regulatory element transposon in maize. *Cold Spring Harbor Symp. Quant. Biol.* 45: 447–455.

Rabbitts, T. H., Forster, A., Dunnick, W. and Bentley, D. L. 1980. The role of gene deletion in the immunoglobulin heavy chain switch. *Nature* 283: 351–356.

Rappolee, D. A., Brenner, C. A., Schultz, R., Mark, D., and Werb, Z. 1988. Developmental expression of *PDGF, TGF-α* and *TGF-β* genes in preimplantation mouse embryos. *Science* 241: 1823–1825.

Rigby, P. W. J., Dieckmann, M., Rhodes, C. and Berg, P. 1977. Labelling deoxyribonucleic acid to high specific activity by in vitro nick-translation with DNA polymerase I. *J. Mol Biol.* 113: 237–251.

Robins, D. M., Ripley, S., Henderson, A. S. and Axel, R. 1981. Transforming DNA integrates into the host chromosome. *Cell* 23: 29–39.

Saiki, R. K., Scharf, S., Faloona, F., Mullis, K. B., Horn, G. T., Erlich, H. A., and Arnheim, N. 1985. Enzymatic amplification of β-globin genomic sequences and restriction site analysis for diagnosis of sickle cell anemia. *Science* 230: 1350–1354.

Sanger, F., Nicklen, S. and Coulson, A. R. 1977. DNA sequencing with chain-terminating inhibitors. *Proc. Natl. Acad. Sci. USA* 74: 5463–5467.

Sargent, T. D. and Dawid, I. 1983. Differential gene expression in the gastrula of *Xenopus laevis. Science* 222: 135–139.

Schatz, D. G., Oettinger, M. A. and Baltimore, D. 1989. The V(D)J recombination activating gene, *RAG-1. Cell* 59: 1035–1048.

Seidman, J. G., Max, E. E. and Leder, P. 1979. A κ-immunoglobulin gene is formed by site-specific recombination without further somatic mutation. *Nature* 280: 370–375.

Slightom, J. L., Blechl, A. E. and Smithies, O. 1980. Human fetal γ^G and γ^A globin genes: Complete nucleotide sequences suggest that DNA can be exchanged between these duplicated genes. *Cell* 21: 627–638.

Smith, M. 1985. In vitro mutagenesis. *Annu. Rev. Genet.* 19: 423–462.

Southern, E. M. 1975. Detection of specific sequences among DNA fragments separated by gel electrophoresis. *J. Mol. Biol.* 98: 503–517.

Spradling, A. C. and Rubin, G. M. 1982. Transposition of cloned P elements into *Drosophila* germ line chromosomes. *Science* 218: 341–347.

Strathern, J. N., Newton, C. S., Herskowitz, I. and Hicks, J. B. 1979. Isolation of a circular derivative of yeast chromosome. III. Implications for the mechanism of mating type interconversion. *Cell* 18: 309–319.

Thompson, S., Clarke, A. R., Pow, A. M., Hooper, M. L. and Melton, D. W. 1989. Germ line transmission and expression of a corrected HPRT gene produced by gene targeting in embryonic stem cells. *Cell* 56: 313–321.

Todaro, G. J., DeLarco, J. E., Nissley, S. P. and Rechler, M. M. 1977. MSA and EGF receptors on sarcoma virus-transformed cells and human fibrosarcoma cells in culture. *Nature* 267: 526–528.

von Schwedler, U., Jäck, H.-M. and Wabl, M. 1990. Circular DNA is a product of the immunoglobulin class switch rearrangement. *Nature* 345: 452–456.

Wagner, T. E., Hoppe, P., Jollick, J. D., Scholl, D. R., Hodinka, R. L. and Gault, J. B. 1981. Microinjection of rabbit β-globin gene into zygotes and its subsequent expression in adult mice and offspring. *Proc. Natl. Acad. Sci. USA* 78: 6376–6380.

Wallace, H. and Birnstiel, M. O. 1966. Ribosomal cistrons and the nucleolar organizer. *Biochim. Biophys. Acta* 14: 296–310.

Watson, J. D., Tooze, J. and Kurtz, D. T. 1983. *Recombinant DNA: A Short Course.* Freeman, New York.

Weislander, L. and Daneholt, B. 1977. Demonstration of Balbiani ring RNA sequence in polysomes. *J. Cell Biol.* 73: 260–264.

Wessel, G. M., Goldberg, L., Lennarz, W. J. and Klein, W. H. 1989. Gastrulation in the sea urchin is accompanied by the accumulation of an endoderm-specific mRNA. *Dev. Biol.* 136: 526–538.

Wetmur, J. G. and Davidson, N. 1968. Kinetics of renaturation of DNA. *J. Mol. Biol.* 31: 349–370.

Yaoita, Y. and Honjo, T. 1980. Deletion of immunoglobulin heavy chain genes from expressed allelic chromosome. *Nature* 286: 850–853.

11

Transcriptional regulation of gene expression: Transcriptional changes during development

While my companion contemplated with a serious and satisfied spirit the magnificent appearances of things, I delighted in investigating their causes. . . . Curiosity, earnest research to learn the hidden laws of nature, gladness akin to rapture, as they unfolded to me, are among the earliest sensations I can remember.

—MARY WOLLSTONECRAFT SHELLEY (1817)

Hence, we cannot categorically deny that perhaps we may be able to grind genes in a mortar and cook them in a beaker after all.

—H. J. MULLER (1922)

A field of science is defined by its problems. Until the 1920s, embryology and genetics were one science, and many people whom we now consider pioneering geneticists were originally trained in embryology. However, during the 1920s, geneticists began defining their problems as those concerning the *transmission* of inherited potentials, whereas the embryologists saw their problems as concerning the *expression* of these inherited traits. The two sciences separated and developed their own rules of evidence, sets of problems, journals, and vocabularies.

Many biologists saw this division as unnatural, but no theory served to bridge the two sciences. F. R. Lillie, whose work had revolutionized the study of fertilization (a place where genetics and embryology must have something to do with each other!), lamented that a re-synthesis of the two disciplines could not occur until we could solve the paradox of how nuclei with identical genes could generate different types of cells. "The dilemma at which we have arrived appears to be unsolvable at present" (Lillie, 1927).

But was it? Hans Driesch, in 1894, had proposed an interactive hypothesis involving nuclei (which he thought contained the inherited potentials) and the cytoplasm. According to this hypothesis (quoted in Chapter 8), the cytoplasm activated the nucleus, which would then form new materials for the cytoplasm. This new cytoplasm would react with the nucleus in new ways, and the cycle would continue.

The following summer, Driesch collaborated on a series of experiments with T. H. Morgan, a young American embryologist who had come to Europe to work with him. In 1934, after receiving fame for his discoveries in genetics, Morgan updated Driesch's ideas. Instead of the prevailing view that every gene had to be active in every cell, Morgan speculated that in different types of cells, different "batteries of genes" would be activated by the cytoplasm. During cleavage, the egg divided in such a way that different nuclei were placed in different regions of the egg cytoplasm. "The initial differences in the protoplasmic regions may be supposed to affect the activity of the genes. The genes will then in turn affect the protoplasm, which will start a new series of reciprocal reactions. In this way we can picture to ourselves the gradual elaboration and differentiation of the various regions of the embryo."

In the 1960s, evidence accumulated for this hypothesis, and gene activation came to be viewed largely as a process of selective gene transcription. Moreover, the analysis of differential gene expression in bacteria appeared to confirm the view that different batteries of genes are transcriptionally activated as cells differentiate. Today we are aware that there are numerous ways to regulate gene expression and that controls have evolved at the levels of transcription, RNA processing, translation, and protein modification. In this chapter, we will introduce certain genes whose activities are controlled by differential transcription. These include genes active in most cells (such as the ribosomal genes) and those genes whose products characterize specific cell types only (such as the ovalbumin and globin genes). In the following chapter, we will focus on the mechanisms of this transcriptional control to see how certain regions of the chromatin are activated while others remain repressed.

Heterochromatin

Chromatin is grossly divided into two major categories, depending on its state of condensation in interphase nuclei. Most of the chromosomal material decondenses during interphase and loses the dark-staining properties that caused it to be seen as individual chromosomes. This loosely coiled material is called EUCHROMATIN. Some chromosomal regions, however, do not decondense. Rather, they retain their tight coiling and their heavy-staining properties throughout interphase. This material is called HETEROCHROMATIN. Heterochromatin is found throughout the animal and plant kingdoms, and it has several properties, besides staining, that distinguish it from euchromatin. First, heterochromatin is relatively, if not completely, inactive in RNA synthesis; second, heterochromatin is the last DNA to replicate at each cell division; and third, heterochromatin suppresses crossing-over between chromatids during meiosis. If a gene is usually in a euchromatic region of chromatin, it can be inactivated by making that region heterochomatic.

There are two types of heterochromatin: constitutive and facultative. CONSTITUTIVE HETEROCHROMATIN is always seen at the same position on both members of a homologous pair of chromosomes. Usually it is found at the centromeres, and it is often composed of highly repetitive DNA sequences. FACULTATIVE HETEROCHROMATIN is formed by the condensation of chromatin at specific stages of the organism's life cycle and is usually present in only one of the two members of a homologous chromosome pair. In other words, sometime during development, specific regions of DNA are rendered transcriptionally inactive by the condensation of their

FIGURE 1

Heterochromatization of the paternally derived chromosomes in male mealy bugs. The male mealy bug becomes functionally haploid and only the euchromatic chromosomes are packaged in the sperm. These chromosomes become heterochromatic if the resulting organism is male.

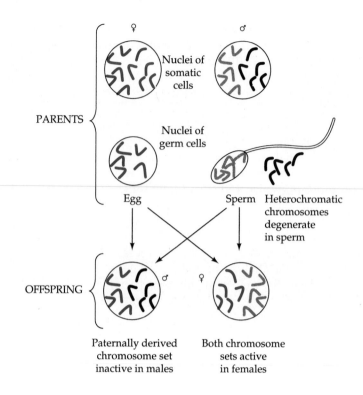

DNA into heterochromatin. Facultative heterochromatin has been found to be a widespread mechanism of gene regulation, and we will discuss its features in insects and mammals.

Paternal heterochromatin in mealy bugs

One of the most impressive feats of facultative heterochromatization occurs in the male mealy bug *Planococcus citri* (Figure 1). Females of this species have no facultative heterochromatin. In males, however, the entire paternally derived haploid chromosome set becomes heterochromatic. Even though this chromosome set was euchromatic in the father, it becomes heterochromatic when transmitted to the son. The male *Planococcus*, then, has a euchromatic haploid chromosome set derived from the mother and a heterochromatic haploid set derived from the father. In male meiosis, the heterochromatic chromosomes disintegrate, leaving only the euchromatic ones to be packaged into sperm. If the offspring is to be male, these chromosomes will condense.

A result of this arrangement is that all the gene expression in male coccids comes from maternally derived chromosomes (except for the few tissues in the males where the heterochromatization is reversed). This genetic inertness of the paternally derived chromosome set in males was elegantly demonstrated by Brown and Nelson-Rees (1961), who irradiated male and female coccids and mated them to unirradiated animals. When females were irradiated and mated, there were many deaths in the male and female offspring produced. This result showed that X-rays could induce dominant lethal mutations. However, when the male parents were irradiated, there were few deaths among the male progeny but many deaths among the daughters; the heterochromatization of the irradiated chromosomes in male offspring prevented the expression of dominant lethal mutations.

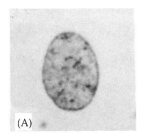 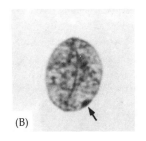 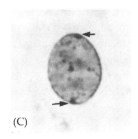

FIGURE 2
Nuclei of human oral epithelial cells stained with Cresyl Violet. (A) Cell from a normal XY male, showing no Barr body. (B) Cell from a normal XX female, showing a single Barr body (arrow). (C) Cell from a female with three X chromosomes. Two Barr bodies can be seen and only one X chromosome is active per cell. (From Moore, 1977.)

Mammalian X chromosome inactivation

In mammals, facultative heterochromatin is seen in the phenomenon of X chromosome inactivation. In 1949, Barr and Bertram discovered a deeply staining nuclear body residing on the nuclear envelope of female cat neurons. This BARR BODY was later demonstrated in numerous female mammals, including humans, and was shown to be an inactive X chromosome (Figure 2). Further research established that each somatic cell in female mammals contained an inactive X chromosome. This means that even though female cells have two X chromosomes and male cells have only one, only one X chromosome in the female cells can be transcriptionally active. This equalization is called DOSAGE COMPENSATION. Dosage compensation must occur early in development. Using a mutated X chromosome that would not inactivate, Tagaki and Abe (1990) showed that the expression of two X chromosomes per cell in mouse embryos leads to ectodermal cell death and the absence of mesoderm formation, eventually causing embryonic death at day 10 of gestation.

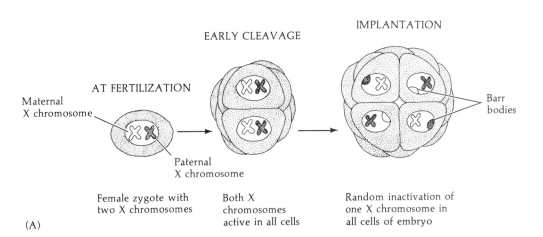

FIGURE 3
X chromosome inactivation in mammals. (A) Schematic diagram illustrating random X chromosome inactivation. The inactivation is believed to occur at about the time of implantation. (B) A female mouse heterozygous for the X-linked coat color gene *dappled*. Distinctly different pigmented regions are seen. (Photograph courtesy of M. F. Lyon.)

One of the earliest analyses of X chromosome inactivation was performed by Mary Lyon (1961), who observed coat color patterns in mice. If a mouse is heterozygous for an autosomal gene controlling hair pigmentation, then the mouse resembles one of the two parents or has a color intermediate between the two. In any case, the mouse will be a single color. But if a female mouse is heterozygous for a pigmentation gene on the X chromosome, a different result is seen (Figure 3). Patches of one parental color alternate with patches of the other parental color. Lyon proposed the following hypothesis to account for these results:

1 In the early development of female mammals, both X chromosomes are active.
2 As development proceeds, one X chromosome is turned off in each cell.
3 This inactivation is random. In some cells, the paternally derived X chromosome is inactivated; in other cells, the maternally derived X chromosome is shut off.
4 This process is irreversible. Once an X chromosome has been inactivated, the same X chromosome is inactivated in all that cell's progeny. (The areas of pigment in these mice are large patches, not a "salt and pepper" pattern.) Thus, all tissues in female mammals are mosaics of two cell types.

Some of the most impressive evidence for this model comes from biochemical studies on clones of human cells. In humans, there is a genetic disease—Lesch-Nyhan syndrome—that is characterized by the lack of the X-linked enzyme hypoxanthine phosphoribosyltransferase (HPRT). This disease is transmitted through the X chromosome—that is, males who have this mutation in their one X chromosome suffer (and die) from the disease. In females, however, the presence of the mutant HPRT gene can be masked by the other X chromosome, which carries the wild-type allele. A woman who has sons with this disease is said to be a carrier, as she has a mutant HPRT gene on one chromosome and a wild-type HPRT gene on the other X chromosome. If the Lyon hypothesis is correct, each cell

FIGURE 4
Retention of X chromosome inactivation. Approximately 30 cells from a woman heterozygous for HPRT deficiency were placed into a petri dish and allowed to grow. The cells were visualized by autoradiography after incubation in a medium containing radioactive hypoxanthine. Cells with HPRT incorporate the radiolabeled compound into their RNA and darken the photographic emulsion placed over them. The clones of cells without HPRT appear lighter because their cells cannot incorporate the radioactive compound. (From Migeon, 1971; photograph courtesy of B. Migeon.)

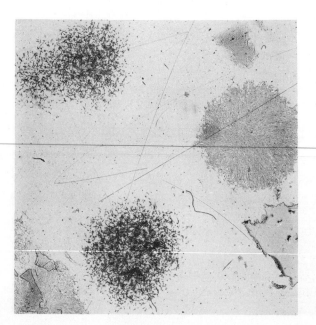

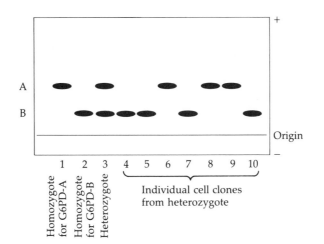

FIGURE 5
Two populations of cells in human females. Electrophoresis of skin cells from women heterozygous for G6PD indicates that both chromosomes are synthesized, but in different cells. When heterozygous skin cells are cultured (line 3), both types of enzymes are present. However, in each clone of skin cells (lines 4–10), only one form of the enzyme is seen.

from such a woman should be making either the active or the inactive HPRT, depending on which X chromosome is active. Barbara Migeon (1971) tested this prediction by taking individual skin cells from a woman heterozygous for the HPRT gene and placing them in culture. Each of these cells divided to form a clone of cells. When Migeon stained the clones for the presence of wild-type HPRT, approximately half of the clones had the enzyme and the other half did not (Figure 4).

Another X chromosome gene codes for glucose 6-phosphate dehydrogenase (G6PD). There are two common electrophoretic variants of this enzyme: G6PD-A and G6PD-B. Males can be either A or B, but females can be A, B, or AB with regard to their G6PD phenotype. Skin taken from heterozygous women show both G6PD variants (Figure 5). However, when individual cells from these heterozygotes are cloned, the isolated clones express either one of the two possible variants. No clone expressed both (Davidson et al., 1963).

The Lyon hypothesis of X chromosome inactivation provides an excellent account of differential gene inactivation at the level of transcription. Some interesting exceptions to the general rules further show its importance. First, the hypothesis holds true only for *somatic* cells. In female *germ* cells, the inactive X chromosome is reactivated shortly before the cells enter meiosis (Kratzer and Chapman, 1981; Gartler et al., 1980). Thus, in mature oocytes, both X chromosomes are active. This is shown in Figure 6. Glucose 6-phosphate dehydrogenase is a dimeric enzyme, so each somatic cell has enzymes constructed of either two A subunits or two B subunits. In a heterozygous female, some cells will have G6PD composed of two A subunits, whereas other cells will have G6PD composed of two B subunits (depending on which X chromosome is active in any particular cell). If *both* X chromosomes are functional in the same cell, however, one expects to find some G6PD composed of one A subunit and one B subunit. This is precisely what is seen when oocytes are tested (Gartler et al., 1973; Migeon and Jelalian, 1977). An AB *heterodimer* is seen, evidence that both X chromosomes have been transcriptionally active in the same cell.

The second class of exceptions concerns the randomness of X chromosome inactivation. Randomness is certainly the rule, but in some instances preferential inactivation of the paternal X chromosome is seen. In the trophoblast tissues of female mice (but not of humans), expression of the maternal X predominates—to the extent of an almost complete absence of paternal X expression. In marsupials, the paternally derived X chro-

FIGURE 6
Both X chromosomes are active in mammalian oocytes. Electrophoresis of cells from the ovary (lanes 1 and 2) and lung (lane 3) of a 14-week human fetus heterozygous for G6PD. The lung cells express the A (AA) and B (BB) forms of the enzyme, whereas the ovary cells also contain a heterodimer (AB) of G6PD. The heavy A and B bands of the ovary reflect the fact that the ovarian cells themselves express only the A and B enzymes; the heterodimer is expressed only in the oocytes. (Photograph courtesy of B. Migeon.)

mosome is also preferentially inactivated throughout the embryo (Sharman, 1971; Cooper et al., 1971). We still do not know why this happens, but it may ultimately provide an important clue to understanding the mechanics of X chromosome inactivation.

Third, X chromosome inactivation does not extend throughout the entire length of the X chromosome. Almost all known X chromosome genes map to the long arm of the X chromosome. However, on the short arm of the X chromosome, there are several genes (such as that for steroid sulfatase) that do not undergo this dosage-related inactivation (Mohandas et al., 1980; Brown and Willard, 1990). Thus, heterochromatization may not extend entirely throughout the X chromosome.*

The fourth exception really ends up being the proof of the rule. There are some male mammals with coat color patterns that would not be expected unless the animals exhibit X chromosome inactivation. Male calico and tortoiseshell cats are among these examples. These spotted coat patterns are normally seen in females and are thought to result from the Lyon effect. But rare males exhibit these coat patterns as well. How can this be? It turns out that these cats are XXY. The Y chromosome makes them male (see Chapter 21), but one X chromosome undergoes inactivation, just as in females, so there is only one active X per cell (Centerwall and Benirschke, 1973). Thus, these cats have cells with a Barr body and random X chromosome inactivation. It is clear, then, that one mechanism for transcription-level control of gene regulation is to make a large number of genes heterochromatic and, thus, transcriptionally inert.

*Brown and co-workers (1991a,b) have suggested that one of these "uninactivated" genes may be a candidate for producing the signal that inactivates the human X chromosome. This gene, called *Xist*, is necessary for X chromosome inactivation and synthesizes its transcript only when it is on the inactivated X.

SIDELIGHTS & SPECULATIONS

The timing of X chromosome inactivation

The inactivation of one of the two mammalian X chromosomes occurs at different times in different tissues. Within the embryo, inactivation takes place in all cells of the epiblast prior to the differentiation of tissues (Nesbitt, 1974; Gartler and Riggs, 1983). Such X inactivation also is seen in teratocarcinoma cells. As mentioned in Chapter 6, these cells resemble the early inner cell mass blastomeres in that they are totipotent and can differentiate into other cell types. McBurney and Strutt (1980) showed that both X chromosomes are active in these teratocarcinoma stem cells, but as they differentiate, one of the X chromosomes in each cell becomes late replicating and transcriptionally inert.

In the extraembryonic tissues, however, X chromosome inactivation can occur nonrandomly and at different times. The first evidence for X chromosome inactivation in the mouse is seen in the trophectoderm, wherein the paternally derived X chromosome is specifically inactivated (Tagaki, 1974; West et al., 1977). This pattern resembles the situation in female marsupials, wherein the paternally derived X chromosome is inactivated throughout the embryo (Cooper et al., 1971; Samollow et al., 1987). Subsequent X

chromosome inactivations in mice occur in the hypoblast tissue and finally in the embryonic epiblast (Monk and Harper, 1979). Figure 7 shows the cycle of X chromosome inactivation in the mouse.

Just because X chromosome inactivation happens this way in the mouse placenta does not mean it happens this way in the placenta of all mammals.* In human chorionic villi, some cells contain two active X chromosomes, and those X chromosomes that are inactivated can be reactivated (Migeon et al., 1985, 1986). Moreover, X chromosome inactivation in the human placenta appears to be random; either the paternally or maternally derived X chromosome can be turned off.

In the next chapter, we will discuss the molecular mechanism for maintaining the transcriptional inactivity on one of the two X chromosomes.

*Even Leonardo da Vinci was fooled by differences between placentas. In his famous drawing of the human fetus within its placenta, the placenta depicted is that of a cow, not of a human. As Renfree (1982) has noted, this drawing serves as a warning to those who would make uncritical extrapolations from one mammalian species to another.

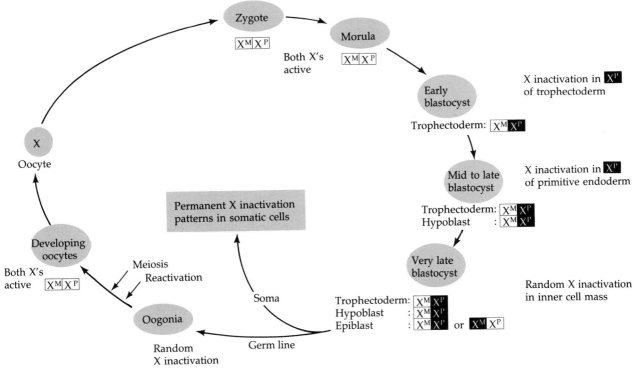

FIGURE 7

The cycle of X chromosome inactivation in the mouse. X^M and X^P are the maternal and paternal X chromosomes, respectively. The inactive chromosomes are represented in black boxes. The primordial germ cells have a condensed X chromosome (thought to be inactive) before they migrate into the gonads, and both X chromosomes appear to be active shortly after the germ cells arrive there (Witschi, 1957; Kratzer and Chapman, 1981). (After Gartler and Riggs, 1983.)

Amplified genes

Amplification of ribosomal RNA genes

One strategy for transcribing enormous amounts of a particular RNA is to make many more copies of the specific gene. This strategy is used only rarely, and it is mainly seen in the amplification of insect and amphibian ribosomal RNA genes and in the chorion (eggshell) genes of certain insects such as *Drosophila*. In many respects, the ribosome can be considered as a differentiated cell product of the oocyte. This is especially so in amphibians, where the oocyte contains around 200,000 times the number of ribosomes found in most somatic cells. For one cell, that number of ribosomes is truly spectacular. In fact, in those zygotes lacking the nucleolar organizing regions that make ribosomal RNA, development can still continue on the stored ribosomes up until the feeding tadpole stage (at which time the organism finally dies from the lack of ribosomes needed for growth).

Synthesis of ribosomal RNA in amphibian oocytes occurs during the prolonged prophase of the first meiotic division. Specifically, rRNA synthesis takes place during the months-long diplotene stage, after pairing and replication of the homologous chromosomes.

The ribosomal RNA genes of the frog *Xenopus laevis* are among the best studied; they are diagrammed in Figure 8. There is a 5' leader sequence followed by the coding sequence for 18 S rRNA. This is followed by a transcribed spacer, and the gene for the 5.8 S rRNA gene, another transcribed spacer, and then the gene for the 28 S rRNA. There are no introns within these genes, and the entire unit is transcribed into a 40 S rRNA precursor, which is then processed to form the three rRNAs. (The 5 S rRNA is made from another region of the genome, and these 5 S genes are present in thousands of copies in the genome and are not amplified.)*

Mechanism of ribosomal gene amplification

There are two homologous regions of large-ribosomal-RNA genes per diploid frog cell. Each region contains roughly 450 copies of the unit described in the preceding section and the copies are separated within the region by nontranscribed spacers. All of the ribosomal RNA gene units on a chromosome are identical, but the nontranscribed spacers vary in length from unit to unit, even on the same chromosome. We would expect, then, that the diplotene oocyte (in which the DNA has already replicated) would contain 4×450, or roughly 1800, genes coding for rRNA. But apparently this is not enough! Rather than a few thousand copies, one finds nearly a million rRNA genes. One sees not four nucleoli, but thousands (Figure 9). These extra nucleoli reside within the nucleus but are not attached to any chromosome, as nucleoli usually are.

By comparing the hybridization of radioactive 28 S ribosomal RNA to oocyte and somatic cell DNA, it was determined that the rRNA genes

Repeated genes denotes inherited multiplicity wherein a given sequence is present numerous times in the genome of every cell. The 5 S ribosomal RNA genes, represented thousands of times per genome, constitutes a set of repeated genes. *Amplified* genes refers to those DNA sequences that are reiterated in specific cells only. Their multiplicity is not inherited. The 28 S, 18 S, and 5.8 S rRNA genes are repeated (about 900 times per cell) *and* amplified (in the oocyte).

FIGURE 8

Organization of the genes for 18 S, 5.8 S, and 28 S ribosomal RNA in amphibians. Each transcription unit (TU) is flanked by nontranscribed spacer regions (NTS). The transcription unit contains a transcribed leader sequence (TL), the 18 S rRNA gene, a transcribed spacer (TS), the 5.8 S rRNA gene, another transcribed spacer, and the gene for 28 S rRNA. Beneath the diagram is an electron micrograph of tandemly arranged rRNA genes actively transcribing in the newt oocyte. The RNA gets progressively larger as it is transcribed, and dozens of transcripts are made simultaneously. (After Watson, 1976; photograph courtesy of O. L. Miller, Jr.)

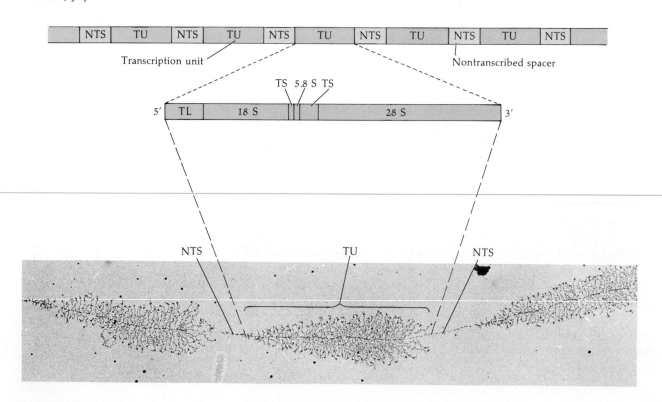

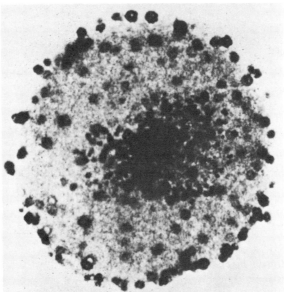

FIGURE 9
Nucleus isolated from an oocyte of
Xenopus laevis. The deeply stained
spots represent the extrachromosomal
nucleoli. (From Brown and Dawid,
1968; courtesy of D. D. Brown.)

were amplified 1500 times over their normal amount. If each nucleolus
contained 450 gene copies, we would estimate that there are 6.8×10^5
genes for the large-ribosomal-RNA precursors in the amphibian oocyte.
This amplification can also be seen on cesium chloride gradients. When
DNA is fragmented and then centrifuged in such gradients, the fragments
sediment according to their density. The DNA of *Xenopus laevis* somatic
cells has an average buoyant density of 1.699 grams per cubic centimeter.
But the density pattern for the sheared DNA from oocytes is slightly
different (Figure 10). A second, or "satellite," peak is seen; this DNA is
derived from the amplified rRNA genes. The base composition of this
DNA gives it a higher density (1.729 g/cm^3) than the majority of the
Xenopus DNA. The average DNA of *Xenopus laevis* is 40 percent GC,
whereas the ribosomal genes are 67 percent GC. Thus, the ribosomal genes
are denser than most other genes. In the somatic tissues, the ribosomal
genes make up only 0.06 percent of the genome and do not form a
separate, detectable satellite peak. However, when these genes are amplified 1500 times, they make a band of dense DNA easily distinguished
from the main band. This DNA will hybridize to ribosomal DNA but not
to any other genes. Clearly, we are observing the specific amplification of
a gene, an amplification allowing the transcription of enormous amounts
of RNA (Brown and Dawid, 1968).

The next question is, How are these genes amplified? Although we
do not know the exact details of the mechanism, we do have some clues.
We know, for instance, that the extrachromosomal nucleoli are made
during the pachytene stage of meiotic prophase, just before diplotene.
Indeed, there is a "pachytene cap," which represents a large mass of
extrachromosomal DNA. Thus, the extra genes are made during the pachytene stage of meiosis and are transcribed during the diplotene stage.
When the oocyte nucleus disintegrates during the first meiotic division,
the extra nucleoli are thrown into the cytoplasm and are destroyed.

One of the most interesting clues to the mechanism of amplification
is that in any given extrachromosomal nucleolus, the nontranscribed
spacer regions are all the same length (Wellauer et al., 1976). In the
chromosomal gene clusters of the same cells, however, there is marked
heterogeneity of spacer lengths; so the extrachromosomal nucleoli are not

FIGURE 10
Isolation of extrachromosomal (amplified) nucleoli from *Xenopus laevis*. DNA
from somatic cell nuclei and oocyte
nuclei were isolated; each was mixed
with a standard amount of dAT (for
aligning the results) and centrifuged in
cesium chloride to equilibrium. In both
cases, the major band of DNA had a
density of 1.699 g/cm^3. In the oocyte,
a "satellite" band of greater density is
also seen. (After Brown and Dawid,
1968.)

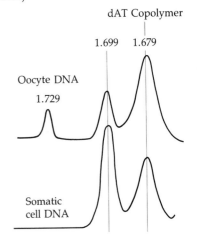

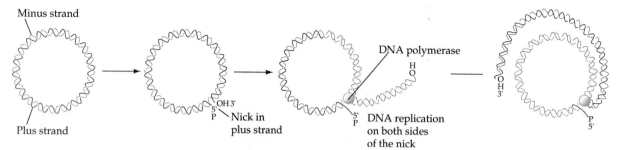

Minus strand

Plus strand

OH 3'
5' P
Nick in plus strand

DNA polymerase
H O H N

5' P
DNA replication on both sides of the nick

OH 3'

P 5'

FIGURE 11

The rolling circle model of replication by which many identical units could be synthesized. The circle would include one ribosomal RNA transcription unit and one nontranscribed spacer. By this method, extrachromosomal nucleoli with the same length NTS between their genes could be formed.

merely copies of the entire chromosomal nucleolus. One model that could explain the generation of hundreds of ribosomal gene units separated by identical nontranscribed spacers is the rolling circle model. This replicative mechanism is used by many viruses. In the oocyte, a single ribosomal gene unit would detach in some way from the chromosome and form a circle. The circle would be nicked (one strand broken) and DNA synthesis would begin in both directions from this site. In such a way, identical copies of the ribosomal genes and their spacers could be formed (Figure 11). Electron micrographs of amplified genes (Hourcade et al., 1973; Rochaix et al., 1974) give support to this model. They reveal closed circles typical of the extrachromosomal nucleoli, as well as "lariat" structures believed to be the rolling circle intermediates (Figure 12).

Visualization of transcription from amplified genes

The amplified ribosomal RNA genes of amphibian oocytes were the first genes to be isolated and purified. They were also the first genes whose transcription could actually be seen by electron microscopy. Knowing that these genes would be intensely active in transcribing rRNA during the diplotene stage of meiosis, Miller and Beatty (1969) used low ionic strength buffers to disperse the chromatic strands for electron microscopy. The circles of DNA unwound to reveal the configuration seen in Figure 8.

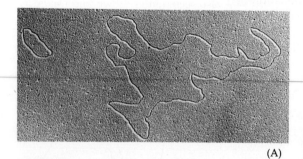

(A)

FIGURE 12

Isolated extrachromosomal rRNA in *Xenopus laevis* oocytes. Electron micrographs show circles (A) and "lariats" (B) suggestive of the rolling circle mechanism of replication. (From Rochaix et al., 1974; courtesy of A. Bird.)

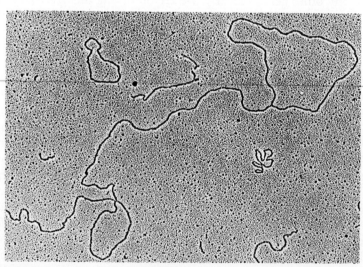

(B)

There is an axial core of DNA upon which rRNA is synthesized. Numerous rRNA chains are transcribed from any one unit, and several such units are seen in this picture. In each unit, transcription begins at the small end of the "Christmas tree" and continues until the large-ribosomal-RNA precursor is formed. At each junction of the DNA with an RNA transcript, there is a molecule of RNA polymerase. The length of the RNA in these large transcripts corresponds to about 7200 bases, which is roughly the size estimated for the ribosomal precursor. Between the transcribing units of DNA are the "nontranscribed spacers." Electron micrographs can truly give us an accurate picture of eukaryotic gene transcription in action.

SIDELIGHTS & SPECULATIONS

RNA polymerase I and ribosomal gene transcription

Transcription requires the interaction of RNA polymerase with DNA. But not just any piece of DNA will suffice. (If it did, transcription could begin anywhere and the cell would be full of truncated mRNAs.) The region of DNA specialized for binding RNA polymerase is called the PROMOTER. In eukaryotic cells, there are three different types of RNA polymerases, each having particular functions and properties (Rutter et al., 1976). RNA POLYMERASE I is found in the nucleolar region of the nucleus and is responsible for transcribing large rRNAs; RNA POLYMERASE II transcribes messenger RNA precursors; and RNA POLYMERASE III transcribes small RNAs such as transfer RNA, 5 S ribosomal RNA, and other small DNA sequences. These polymerases have different sensitivities to ionic conditions and drugs; Table 1 lists their properties.

None of the eukaryotic RNA polymerases can bind directly to DNA. Rather, there are families of DNA-binding proteins that first bind to DNA and, once bound, interact with the RNA polymerase to initiate RNA synthesis. RNA polymerase I is unique in that it recognizes only one type of promoter sequence, that of the large-rRNA gene. Whereas RNA polymerases II and III have to recognize a large variety of different genes, this is not the case with RNA polymerase I. Bell and collaborators (1990) have identified the proteins that bind human and mouse RNA polymerase I to the rRNA promoters and found subtle, but important, differences in them. These factors were identified by passing nuclear extracts through columns containing the promoter regions of the rRNA genes bound to insoluble beads. Most of the nuclear proteins pass through the column, whereas those that bind to rRNA promoters remain. These can be isolated after the other proteins have been washed away. Using this proceedure, two types of proteins, UBF and SL1, were isolated. Human and mouse UBF were found to have the same molecular weights and DNA footprinting showed that they bound to the same region of DNA. The second auxillary protein, SL1, was also similar in mice and humans.

The UBF and SL1 proteins form a tight complex on the

TABLE 1
General classification of eukaryotic RNA polymerases

Property	RNA polymerase I	RNA polymerase II	RNA polymerase III
Cellular location	Nucleolar	Nucleoplasmic	Nucleoplasmic
Proportion of cellular activity (%)	50–70	20–40	10
Major transcription product	Large rRNA precursors	mRNA precursors	tRNA and 5 S rRNA
α-Amanitin concentration (μg/ml) needed for 50% inhibition:			
Mammalian polymerases	>400	0.025	20
Insect polymerases	—	0.03	>1000
Yeast polymerases	600	1.0	>2000

Source: Lewin (1980).

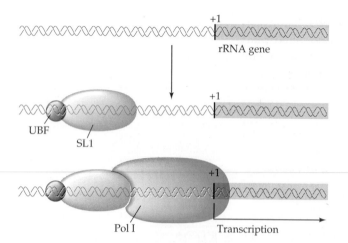

FIGURE 13
Model for the transcription of the large ribosomal genes by RNA polymerase I. Transcription factors UBF and SL1 bind to the promoter region just upstream from the site where ribosomal RNA transcription is initiated (+1). The binding of these two factors is much more stable than the binding of either one alone. RNA polymerase I binds to SL1 and is stabilized on the DNA. Transcription then begins. (After Bell et al., 1990.)

rRNA gene, and the binding of the complex to the promoter is much stronger than the binding of either of the proteins separately. Figure 13 shows the relationships between these proteins as inferred from binding studies and DNase protection experiments. As can be seen, the RNA polymerase is stabilized on the gene by its binding to SL1. The DNA binding of SL1 to the DNA, moreover, is stabilized by its binding with UBF. In this way, RNA polymerase is able to bind to the proper places in the genome and efficiently initiate the transcription of the large ribosomal RNA gene.

Drosophila chorion genes

The chorion genes of *Drosophila melanogaster* are also known to be amplified during the synthesis of the chorion proteins of the egg. These proteins are synthesized by the ovarian follicle cells. Prior to chorion gene expression, the entire genome of the follicle cells undergoes extra rounds of DNA synthesis to achieve 16 times the haploid DNA content. After these replications, the chorion protein genes are selectively replicated about tenfold more. This amplification does not happen in tissues other than the ovarian follicle cell (Spradling and Mahowald, 1980). Spradling (1981) showed that this amplification is due to additional rounds of DNA synthesis only in specific regions of the genome. These regions of DNA are characterized by several branches, each containing the amplified chorion genes (Figures 14 and 15). Here, amplification occurs but the genes remain attached to the chromosomes.

Selective gene transcription

Most genes do not utilize selective gene amplification as a mechanism for large-scale gene expression. In developing red blood cells, where hemoglobin constitutes 98 percent of the protein being synthesized, no selective amplification of the globin genes is seen. In the silk gland of caterpillars

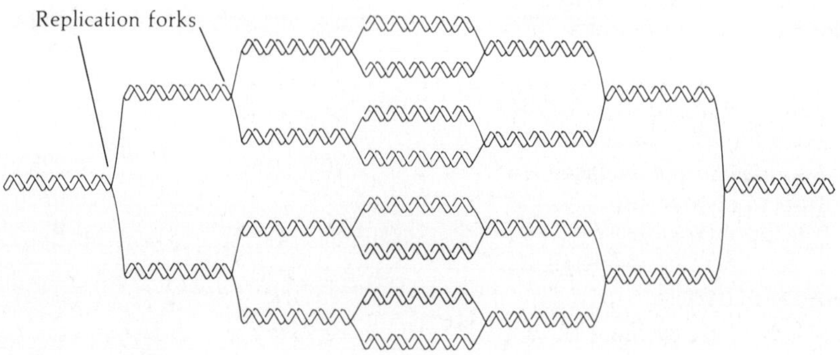

FIGURE 14
Amplification of chorion genes within the chromosome. Diagram of local amplification by multiple rounds of DNA synthesis in a region already containing multiple copies of the chorion genes.

producing tremendous amounts of the silk protein fibroin, the entire genome is amplified by polyteny. Even so, there is no selective amplification of the fibroin genes. Thus, we expect that certain genes can be selectively transcribed.

Chromosomal puffs and lampbrush chromosomes

The nature of polytene chromosomes and their puffs was discussed in Chapter 10. It should be realized that the data presented there illustrates the transcription-level control of specific genes. When ecdysone is added to salivary glands, certain puffs are produced and others regress. The puffing is mediated by the binding of ecdysone to specific places on the chromosomes. This can be shown by cross-linking ecdysone to the chromatin so that it stays where it has been bound. Then rabbit antibodies recognizing ecdysone are added and the unbound antibodies washed away. Fluorescein-labeled goat antibody against rabbit immunoglobulins is added last, and the unbound goat antibodies are washed away. Ultimately, the fluorescent tag should be located wherever a goat antibody has bound to a rabbit antibody; and rabbit antibodies should bind only to ecdysone. In this manner, a fluorescent label should appear at every site on the chromosome where ecdysone has been bound (Gronemeyer and Pongs, 1980). The result (Figure 16) shows that almost all the ecdysone-sensitive puff sites bind ecdysone.

Ecdysone-sensitive puffs occurring during the late stages of the third instar larva (as it prepares to form the pupa) can be grossly divided into three categories. There are those puffs that ecdysone causes to regress; there are those puffs that ecdysone induces rapidly; and there are those puffs that are first seen several hours after ecdysone stimulation. The early puffs are stimulated to form within 1 hour after the addition of ecdysone. Moreover, they appear to synthesize products that are necessary for the induction of the later puffs. If protein synthesis inhibitors are added to cultured *Drosophila* salivary glands soon after the early puffs have formed, ecdysone will not stimulate the formation of the later puffs. Similarly, if one of the early puffs is eliminated by mutation, the later puffs are not

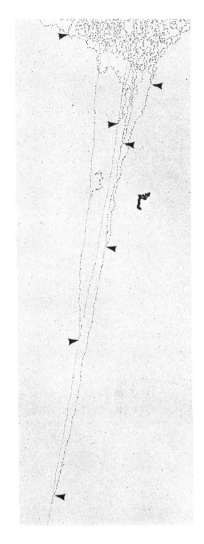

FIGURE 15
Electron micrograph of chorion gene amplification in the follicle cells of *Drosophila*. Arrows point to the sites of DNA replication. (From Osheim and Miller, 1983; photograph courtesy of O. L. Miller, Jr.)

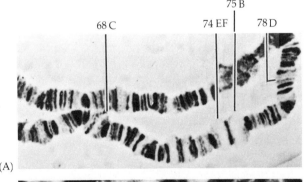

(A)

(B)

FIGURE 16
Localization of ecdysone on the polytene chromosomes of *Drosophila*. After ecdysone is linked in situ on the chromosomes, it can be recognized by fluorescent antibodies. (A) Phase-contrast micrograph of a segment of salivary gland chromosomes. (B) Same segment seen under ultraviolet light, showing immunofluorescence at ecdysone-sensitive areas. (Courtesy of O. Pongs.)

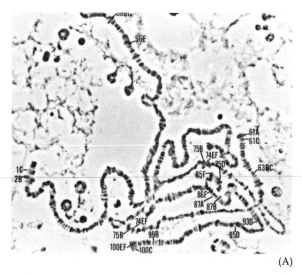

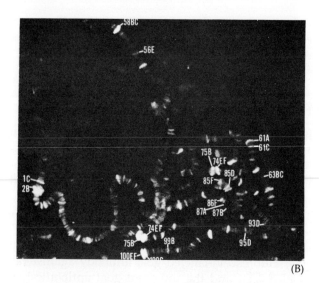

(A)

(B)

FIGURE 17

Immunofluorescent localization of a protein that is released upon DNase I digestion of *Drosophila* chromatin. When antibodies are made against that protein and are bound to the polytene chromosomes, they can be recognized by other, fluorescent antibodies. (A) Phase-contrast micrograph. (B) Same region seen under ultraviolet light, showing the labeling of ecdysone-sensitive sites. (Figures 16 and 17 can be compared with Figure 29 in Chapter 10, which shows the normal puffing sequence.) (From Mayfield et al., 1978; photographs courtesy of S. Elgin.)

formed. Presumably, the later puffs need both ecdysone and the early puff products (Ashburner, 1974).

It is worth noting that ecdysone not only stimulates the puffing of certain regions but also causes the regression of certain existing larval puffs. One of these latter puffs is present at position 68C of the left arm of chromosome 3. This puff site contains the gene for Sgs3, a "glue protein" responsible for attaching the pupal case to a solid surface (Korge, 1975). Only in the salivary gland does this region of the chromosome puff out at the last instar. The addition of ecdysone to cultured salivary glands causes the rapid regression of this puff and the cessation of transcription from this gene (Crowley and Meyerowitz, 1984). Thus, the glue can be synthesized when it is needed to adhere the larva to a substrate, but afterward the gene is shut off, while metamorphosis occurs.

As we will discuss more fully in the next chapter, transcriptional activity is associated with changes in the chromatin, and these changes can be mediated by the binding of nonhistone chromatin proteins. Digestion of *Drosophila* chromatin with the enzyme DNase I releases a 63,000-Da protein that will bind to all the ecdysone-sensitive loci of third instar larvae (Figure 17). This protein is thought to be responsible for establishing or maintaining the chromatin structure necessary for active genes (Mayfield et al., 1978).

SIDELIGHTS & SPECULATIONS

Activation of specific genes during insect metamorphosis

Some of the most dramatic evidence for the activation of genes during development comes from insect metamorphosis. As shown in Figure 29 of the previous chapter, the onset of metamorphosis brings about the activation of numerous genes, as is seen by the new puffs that emerge from the chromosomes. During the last day of larval life, a series of puffs forms in a precise sequence. Moreover, this exact pattern of puffing could be induced in fly nuclei by injecting earlier stage fly larvae with ecdysone (the hor-

mone that stimulates molting and metamorphosis) or by adding ecdysone into the culture medium where the cells had been growing in vitro (Figure 18).

In *Drosophila*, this sequence of gene activation has been studied in detail. A small set of genes (about six) puff out within minutes of ecdysone exposure. These genes are activated directly by ecdysone and are independent of protein synthesis. A much larger set of genes (over 100) reacts to ecdysone only after several hours. These "late-acting" genes require protein synthesis to be active (Ashburner, 1974). The early puffs normally remain active about 4 hours before regressing. However, if protein synthesis is inhib-

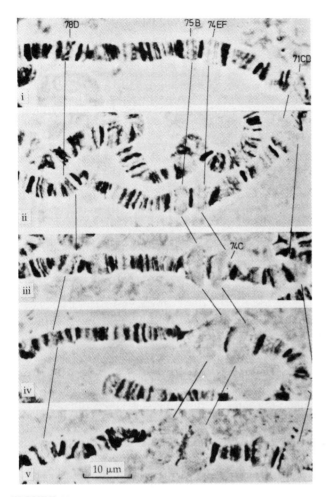

FIGURE 18
Ecdysone-induced puffs in cultured salivary gland cells of
D. melanogaster. The chromosomal region here is the
same as that shown in Figure 29 of Chapter 10. Puffing
cycle induced by ecdysone. (i) Unincubated control. (ii–v)
Ecdysone-stimulated chromosomes at 25 minutes, 1 hour,
2 hours, and 4 hours, respectively. (Photographs courtesy
of M. Ashburner.)

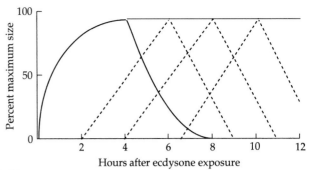

FIGURE 19
Summary of the responses of certain *Drosophila* genes to
ecdysone. *Drosophila* cells were cultured and then exposed to ecdysone. The normal response of the early puff
genes is shown by the solid black curve. The colored line
indicates the response of the early puff genes when they
were exposed to ecdysone in the continuing presence of a
protein synthesis inhibitor. When protein synthesis is inhibited, the late puff genes (dashed lines) are not activated. (After Ashburner, 1990.)

FIGURE 20
Model for the control of gene activity by ecdysone. Ecdysone would bind to its receptor protein. This complex
would bind to the early puff genes and activate them and
would bind to the late puff genes and repress them. The
protein products of the early genes would activate the
repressed late genes and would inhibit their own transcription. (After Ashburner, 1974.)

ited, the regression of the puff does not take place (Figure
19). This suggests that the protein products of the early
puffs are directly or indirectly responsible for turning themselves off. Ashburner hypothesized that the early genes
are responsible both for the activation of the late genes and
for shutting themselves down. His model for this is shown
in Figure 20. In 1990, evidence for this model was obtained
on the molecular level. Three genes that were involved
with the model were cloned. The first was the gene for the
ecdysone receptor protein. The protein encoded by this
gene was found to bind specifically to the early puff loci,
as predicted by Ashburner's model (Koelle et al., quoted
in Thummel et al., 1990). The other two cloned genes were
each responsible for an early puff: the E74 gene that produces the 74EF puff, and the E75 gene that produces the
75B puff (Segraves and Hogness, 1990; Thummel et al.,
1990). Both E74 and E75 appear within minutes of ecdysone
addition, normally regress after 4 hours, and are dependent
upon ecdysone. If ecdysone is withdrawn, their puffs regress immediately.

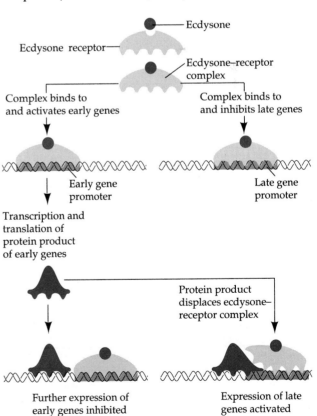

Thummel and co-workers used Northern blots to show that these genes were activated during times of high ecdysone levels, especially at the onset of metamorphosis. They also showed that before ecdysone treatment, cultured cells had no detectable mRNA for either of these proteins, but that their mRNA concentrations rose rapidly when ecdysone was added to the medium. The protein products of these genes share characteristics of known DNA-binding proteins, and the protein encoded by E75 can bind to the E74 gene. This would be a mechanism for turning off the early gene set. It is still unknown whether the protein products of these early puff genes bind to the late puff genes. We also do not yet know the function of these late puff genes. The early puffs appear to be active in all cells of the larva, but the later puffs are thought to be tissue-specific. When we know the function and regulation of these genes, we will be in a much better position to understand how hormones control such striking developmental changes as those of insect metamorphosis.

Puffs represent the decondensation of DNA in specific areas of larval insect chromosomes. Because the chromosomes are polytene, the thousands of strands produce the characteristic puff appearance. An analogous unwinding of DNA from nonpolytene chromosomes is seen in amphibian oocytes. These structures are called LAMPBRUSH CHROMOSOMES. During the diplotene stage of meiosis, the compacted amphibian chromosomes stretch out large loops of DNA (color portfolio), retracting them after this stage is completed.

The notion that this unfolding of the chromosomes enables certain genes to be expressed can be demonstrated by in situ hybridization. Oocyte chromosomes can be prepared, denatured, and incubated with radioactive RNA that encodes a specific protein. After the unbound RNA is washed away, autoradiography visualizes the precise location of the gene. Figure 21 shows diplotene chromosome I of the newt *Triturus cristatus* after incubation with radioactive histone mRNA. It is obvious that a histone gene (or set of histone genes) is located on one of these loops of the lampbrush chromosome (Old et al., 1977). Electron micrographs of gene transcripts from lampbrush chromosomes look very much like the transcripts seen on the amplified ribosomal RNA genes. The same "Christmas tree" polarity is seen, but usually (the exception being gene families like the histones) only one transcription unit is observed for each loop (Hill and MacGregor, 1980).

Ovalbumin synthesis

The enormous egg of the laying hen is protected by numerous extra layers, which are secreted upon it during its passage through the oviduct (Figure 22). The ovum is gathered by the INFUNDIBULUM of the left oviduct (the right oviduct having atrophied in most birds) and is fertilized shortly thereafter. The egg then passes sequentially into the MAGNUM, where ovalbumin and the other egg-white proteins are secreted; into the ISTHMUS, where the shell membranes are put on; and finally into the UTERUS, where water, salts, and the calcium carbonate shell are added.

The major product of the tubular cells of the magnum is ovalbumin, and in laying hens ovalbumin constitutes more than 50 percent of the tubular cell protein. The production of ovalbumin depends on the presence of the sex hormone estrogen. Estrogen enters the cell and is picked up by a receptor protein, which brings it into the nucleus. The estrogen-receptor complex travels through the nuclear pores and binds to the chromatin, where it can stimulate transcription.

The estrogen dependence of ovalbumin production is readily seen when young female chicks are injected with estrogen (Palmiter and Schimke, 1973). These injections cause the tubular glands of the magnum

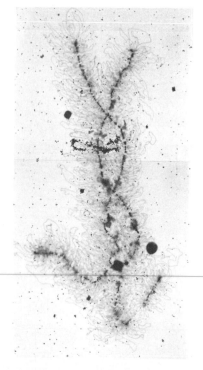

FIGURE 21
Localization of histone genes on a lampbrush chromosome. Genes have been visualized by in situ hybridization and autoradiography. (From Old et al., 1977; courtesy of H. G. Callan.)

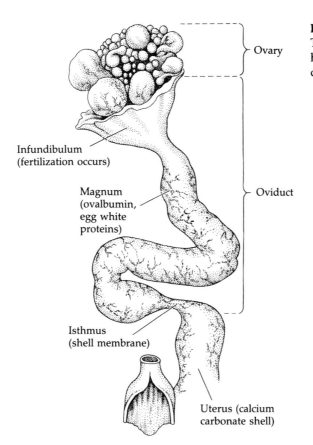

FIGURE 22
The reproductive system of a mature hen. Numerous secretions coat the egg during its passage within the oviduct.

Ovary

Oviduct

Infundibulum (fertilization occurs)

Magnum (ovalbumin, egg white proteins)

Isthmus (shell membrane)

Uterus (calcium carbonate shell)

to differentiate and to synthesize the major egg white proteins: ovalbumin, conalbumin, ovomucoid, and lysozyme. During 2 weeks of estrogen treatment, ovalbumin levels rise from undetectable amounts to greater than 50 percent of all newly synthesized cellular proteins (Figure 23). Withdrawal of the estrogen causes a decline in ovalbumin production even though the cells retain their differentiated state. After 2 weeks of estrogen withdrawal, no ovalbumin can be detected.

This experiment demonstrates only that the synthesis of ovalbumin is

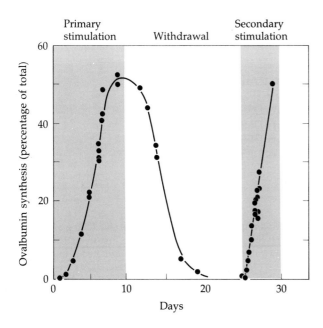

FIGURE 23
The effect of estrogen on ovalbumin synthesis in the oviduct of 4-day-old chicks. Chicks were given daily injections of estrogen (primary stimulation); after 10 days the injections were stopped. Two weeks after this withdrawal, the injections were resumed (secondary stimulation). (After Palmiter and Schimke, 1973.)

dependent upon the presence of the hormone. It does not tell us at what level this hormone works. Conceivably, it could either increase transcription, or stabilize the ovalbumin protein or its message, or even aid in the transport of the ovalbumin message through the nuclear membrane. The next step was to find a correlation between estrogen and the presence of ovalbumin mRNA (Harris et al., 1975; McKnight and Palmiter, 1979). To do this, a complementary DNA was made from isolated ovalbumin message. The ovalbumin message was obtained by passing the cytoplasmic mRNA from oviduct cells over a column containing oligo(dT)-cellulose beads (Figure 24). These beads bind mRNA because the deoxythymidine chains of the beads bind the poly(A) tails of the mRNA. All the other nucleic acids pass through. The mRNA is then eluted from the column and the different mRNAs separated by electrophoresis. The ovalbumin message can be readily isolated (as it is the most prevalent message in the cell), and it is then translated in vitro to make certain that it codes for ovalbumin. The ovalbumin message is then used as a template for reverse transcriptase, and a cDNA probe is made. This probe enables one to "fish out" any complementary sequence, and it can recognize a single copy of the ovalbumin. The data in Table 2A show that the estrogen treatment is responsible for inducing the ovalbumin message. Even more important, other experiments using the cDNA probe have shown that estrogen stimulates the appearance of ovalbumin-coding sequences in the *nucleus* (Roop et al., 1978; Table 2B). Thus, the production of ovalbumin by the tubule cells of the magnum is primarily regulated at the level of gene transcription.

TABLE 2
Transcriptional regulation of ovalbumin

A. Induction of cytoplasmic ovalbumin messenger RNA during primary and secondary stimulation with estrogen

Hormonal state of oviduct[a]	Molecules $mRNA_{ov}$ per tubular gland cell[b]
UNSTIMULATED	None detectable
+4 days DES	20,000
+9 days DES	44,000
+18 days DES	48,000
WITHDRAWN FOR 12 DAYS	0-4
+0.5 hour DES	9
+1.0 hour DES	50
+4.0 hours DES	2,300
+8.0 hours DES	5,100
+29.0 hours DES	17,000

B. Induction of nuclear RNA transcripts containing ovalbumin sequences

Tissue	RNA molecules containing ovalbumin-coding sequences per tubule gland cell nucleus[c]
ESTROGEN-STIMULATED	
OVIDUCT	3,075
ESTROGEN-WITHDRAWN	
OVIDUCT	2
+2 hours DES	6
+4 hours DES	71
+8 hours DES	460
+16 hours DES	995
+48 hours DES	3,049

Source (A): Harris et al. (1975).
Source (B): Roop et al. (1978).
[a]Ten-day-old White Leghorn chicks received daily subcutaneous injections of diethylstilbestrol (DES, a synthetic estrogen; 2.5 mg in oil) and were killed at the indicated times. For experiments involving secondary stimulation with estrogen, the chicks were first treated with DES for 10 days, followed by 11 days of withdrawal from hormone. On day 12 of withdrawal, chicks were given one subcutaneous injection of 2.5 mg of DES and oviducts were collected at the indicated time intervals.
[b] ^{3}H-labeled cDNA was made to ovalbumin message and was hybridized to total oviduct RNA extracted from chicks treated as indicated. The number of molecules of ovalbumin mRNA was calculated.
[c]Determined by hybridization to $cDNA_{ov}$.

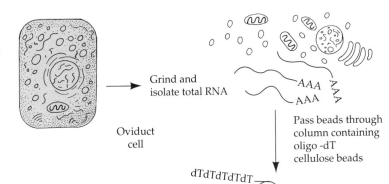

FIGURE 24
Diagram of protocol used to isolate ovalbumin
mRNA to make cDNA probe. (See text for details.)

Oviduct
cell

Grind and
isolate total RNA

Pass beads through
column containing
oligo -dT
cellulose beads

mRNA sticks by
poly(A) tail

Elute and subject
to electrophoresis

Globin gene transcription

One of the best documented cases for the transcriptional regulation of
differentiation is the activation of the globin gene in developing red blood
cells. In chick and human embryos, changes in hemoglobin synthesis can
be observed during development. In chick erythroblasts, the earliest transcription of hemoglobin sequences can be monitored. When the posterior
region of the area opaca of a 20- to 23-hour chick embryo is isolated, blood
islands are observed to contain the precursors of the red blood cells.
Complementary DNA probes to globin messages show that these precursor cells are not yet transcribing globin genes. However, after two more
cell divisions (35 hours of development), the cells (now called erythroblasts) are rapidly synthesizing hemoglobin. Somehow during this time
the globin genes are turned on.

Another type of transcription-level regulation occurs later in development. In many species, including chicks and humans, the embryonic
or fetal hemoglobin differs from that found in adult red blood cells. A
schematic diagram of human hemoglobin types and the genes that code
for them is shown in Figure 25. Human EMBRYONIC HEMOGLOBIN consists
largely of two zeta (ζ) globin chains, two epsilon (ϵ) globin chains, and
four molecules of heme. During the second month of human gestation,
ζ- and ϵ-globin synthesis abruptly ceases, while alpha (α) and gamma

FIGURE 25
Sequential gene activation in hemoglobin synthesis during development.

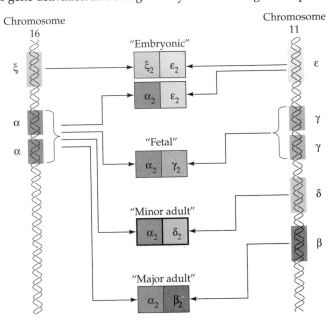

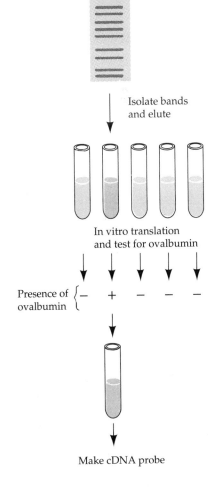

Isolate bands
and elute

In vitro translation
and test for ovalbumin

Presence of
ovalbumin $\left\{ \begin{array}{ccccc} - & + & - & - & - \end{array} \right.$

Make cDNA probe

FIGURE 26

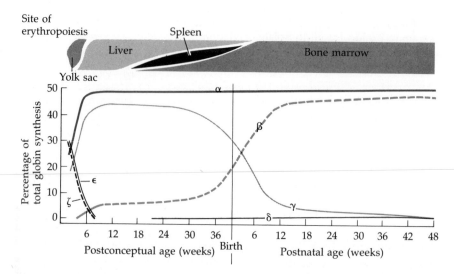

(γ) globin synthesis increase (Figure 26). The association of two γ-globin chains with two α-globin chains produces FETAL HEMOGLOBIN. At 3 months' gestation, the beta (β) globin and delta (δ) globin genes begin to be active, and their products slowly increase while γ-globin gradually declines. This switchover is greatly accelerated after birth, and fetal hemoglobin is replaced by adult hemoglobin, $\alpha_2\beta_2$. The normal adult hemoglobin profile is 97 percent $\alpha_2\beta_2$, 2–3 percent $\alpha_2\delta_2$, and 1 percent $\alpha_2\gamma_2$.

In humans, the ζ- and α-globin genes are on chromosome 16, but the ε-, γ-, δ-, and β-globin genes are linked together, in order of appearance, on chromosome 11. It appears, then, that there is a mechanism that directs the sequential switching of the chromosome 11 genes from embryonic, to fetal, to adult globins. The mechanism is so far unknown, but one clue comes from certain individuals who, though perfectly healthy, never shut off their fetal hemoglobin synthesis. In such cases, the γ-globin genes are still active in the adult and the β-globin genes are not yet turned on. The possible molecular mechanisms for how this switch takes place will be discussed in the next chapter.

A transcriptional regulatory protein: The control of 5 S ribosomal RNA genes

Central promoter element

The 5 S rRNA gene is transcriptionally regulated throughout development (Korn, 1982). In each haploid genome, there are 20,000 "oocyte" 5 S genes (that are active only in the oocyte) and 400 "somatic" 5 S genes (that are functional in every cell). They are all transcribed during early oogenesis. During late oogenesis, all the 5 S genes are turned off until the midblastula transition takes place. At midblastula, both oocyte and somatic 5 S genes are expressed, but by the end of the blastula stage, only the somatic 5 S genes are active.

The RNA polymerase controlling the production of 5 S rRNA genes is RNA polymerase III, and it has a unique promoter. Whereas the promoters for RNA polymerases I and II are in the 5′ flanking regions of the gene (as they are in bacteria), the promoters for RNA polymerase III are in the center of the gene. This was discovered by making deletions in various areas in and around the *Xenopus* 5 S rRNA genes and seeing if the genes would still be transcribed when mixed with ribonucleotide triphosphates, RNA polymerase III, and nuclear extracts. Accurate transcription still occurred even after the 5′ and 3′ flanking regions from the *Xenopus* 5 S rRNA gene were enzymatically removed (Figure 27). Only when the nucleotides from positions 50 to 83 were removed did transcription cease (Sakonju et al., 1980; Bogenhagen et al., 1980). Moreover, there appears to be a conserved region within the genes transcribed by RNA polymerase III. The sequence AGCAGGGT is found between positions 55 and 62 of the *Xenopus* 5 S rRNA gene, and similar sequences (within two base pairs) are found within *Bombyx mori* (moth) tyrosine tRNA and adenovirus *VA-I* genes, all of which are transcribed by RNA polymerase III.

Transcriptional regulation by TFIIIA

As we discussed in the case of RNA polymerase I, the binding of eukaryotic RNA polymerases to their promoters is mediated by special proteins called transcription factors. When isolated 5 S ribosomal RNA genes of *Xenopus laevis* are incubated with RNA polymerase III, accurate transcription does not take place. However, faithful transcription does occur when 5 S rRNA *chromatin* is used, or when the 5 S rRNA genes are incubated in a nuclear extract. There is a 38,500-Da protein that binds to the control region of the 5 S rRNA gene and serves to direct the RNA polymerase III to bind (Ng et al., 1979; Engelke et al., 1980). Without this protein, the 5 S rRNA gene is inactive.

This protein, called TFIIIA (the first *Transcription Factor for RNA polymerase III*) is specific for the 5 S rRNA gene (it does not bind to the promoters of other RNA polymerase III-dependent genes) and is itself developmentally regulated. There are 10^{12} TFIIIA molecules per cell in the oocyte, but only 10^3 TFIIIA molecules per somatic cell (Ginsberg et al., 1984; Scotto et al., 1989). The binding of TFIIIA to the internal promoter of the 5 S rRNA genes is very weak. It is greatly stabilized by the nonspecific factor TFIIIC (Lassar et al., 1983; Pieler et al., 1987). This tightly bound complex, however, does not begin initiation until another factor, TFIIIB, has bound. TFIIIB does not bind to the gene, but to the TFIIIA–TFIIIC complex (Wolffe and Brown, 1988). Once RNA polymerase III binds to this complex, transcription begins.

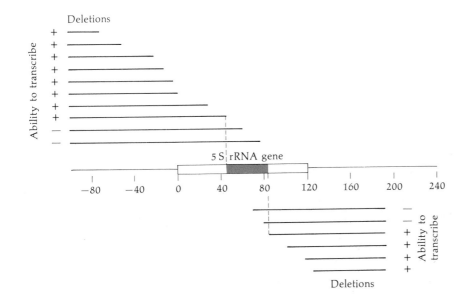

FIGURE 27

Deletion map of a 5 S rRNA gene of *Xenopus laevis*. The horizontal lines represent the deleted portions of DNA in plasmids containing the 5 S rRNA gene. A plus mark adjacent to a line indicates that the remaining plasmid DNA still supports accurate transcription of 5 S rRNA. A negative sign indicates failure to be transcribed accurately. A control region within the gene, roughly 30 base pairs in length, is defined by these deletions. (After Brown, 1981.)

It appears, then, that the type and amount of 5 S rRNA depends on the level of TFIIIA in the nucleus. But how is that control effected? Brown and Schlissel (1985) injected cloned oocyte and somatic 5 S rRNA genes into oocytes and blastula nuclei and showed that TFIIIA has a much higher (over 200-fold) affinity for the control region of the somatic 5 S rRNA genes than it does for the oocyte 5 S rRNA genes. Moreover, when purified TFIIIA was injected into late blastula-stage embryos (where only the somatic 5 S rRNA genes are active), the oocyte 5 S genes become activated as well. These observations indicated that at low concentrations of TFIIIA, the only genes binding TFIIIA are of the somatic type, but at higher levels the oocyte 5 S rRNA genes also become activated.

The basis for this difference in the ability of the 5 S genes to be activated by TFIIIA was found to reside in three nucleotides that differed between the oocyte and somatic 5 S genes. While these three base substitutions were responsible for changing the stability of the 5 S gene–TFIIIA complex, they did not directly alter TFIIIA–promoter binding. Rather, the changes in nucleotides caused the TFIIIA protein to become displaced from the promoter by TFIIIC (Wolffe and Brown, 1988). The somatic 5 S rRNA gene appears to form a more stable transcription initiation complex than does the oocyte 5 S rRNA gene and thereby yields more efficient transcription.

TFIIIA binding and function

The binding sites for TFIIIA on the 5 S rRNA gene have been established by mutation analysis, DNase I footprinting, guanosine methylation, and phosphate blockage (Sakonju and Brown, 1982; Pieler et al., 1985). Mutation analysis has established three regions needed for proper transcription: the nucleotides at positions 50–61, 70–71, and 80–89. The sequence of nucleotides 50–61 is found in all other genes transcribed by RNA polymerase III. In this same region, the somatic 5 S gene differs from the oocyte 5 S gene; and this difference is thought to give the two genes their different affinities for TFIIIA. DNase I "footprinting" is done by binding the protein (in this case, TFIIIA) to DNA made radioactive at one end (as is done for sequencing) and then digesting the DNA with this relatively nonspecific DNase. Each strand should be broken only once (on average). The resulting DNA is placed in a gel and separated by electrophoresis, as if it were being sequenced. One obtains the expected oligonucleotide "ladders" seen in sequencing gels, where each oligonucleotide is one base shorter than the one above it. But in those regions where the protein has protected the DNA from the DNase I digestion, the corresponding oligonucleotides are absent. Such a pattern for TFIIIA binding to the *Xenopus* oocyte 5 S RNA gene is shown in Figure 28. Three binding sites are readily observed.

By modifying guanosine residues (by methylation) or phosphate groups (by ethylation) one can see whether this pattern of protection changes. Methylating certain groups of guanosines (70 and 71; 80–89) and ethylating certain groups of phosphates (70 and 71; 80–87) inhibits TFIIIA binding and prevents this protein from protecting these regions of DNA.

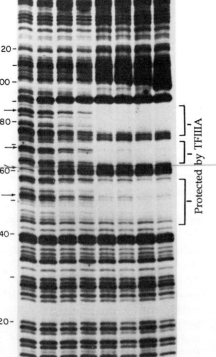

FIGURE 28

Autoradiogram of radioactive 5 S DNA incubated with increasing concentrations of TFIIIA and subsequently cleaved with DNase I. The fragments generated by this technique were separated by electrophoresis and autoradiographed. The areas where the DNA was protected show up as blank areas. (From Sakonju and Brown, 1982; photograph courtesy of D. D. Brown.)

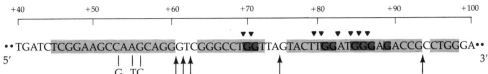

```
    +40        +50        +60        +70        +80        +90       +100
     |          |          |          |          |          |          |
                                      ▾▾           ▾▾  ▾  ▾▾▾
•• TGATCTCGGAAGCCAAGCAGGGTCGGGCCTGGTTAGTACTTGGATGGGAGACCGCCTGGGA ••
5′              | | |     ↑↑↑            ↑               ↑          3′
                G TC
```

These data are summarized in Figure 29 and show the three binding sites for TFIIIA on the 5 S RNA gene.

Whereas there are three major regions in the 5 S promoter that bind TFIIIA, the TFIIIA protein has nine DNA binding sites. Each of these "DNA binding fingers" is a globular domain whose central amino acids tend to be basic. These domains are linked together in a tandem array and are each stabilized by a centrally located zinc ion (Figure 30). Miller and co-workers (1985) suggest that this arrangement enables the TFIIIA to "hold on" to the gene as the RNA polymerase moves along. Some of the binding regions will maintain the hold on the DNA as the others relinquish it.

The rate-limiting function required for the transcription of *Xenopus* 5 S rRNA genes is the concentration of the TFIIIA protein. But what regulates the transcription of the TFIIIA gene? We know that this gene is developmentally regulated and that the amount of 5 S rRNA is a reflection of the abundance of TFIIIA. The TFIIIA gene, then, is controlled by other transcription factors that bind to its own promoters (Scotto et al., 1989). It is probable that levels of the proteins that bind to the TFIIIA gene promoters activate the transcription of TFIIIA just as levels of TFIIIA regulate the expression of the 5 S rRNA gene. Eventually, we come back to the inequalities of regulator molecules in the oocyte cytoplasm that can differently regulate the identical genes of each nucleus.

FIGURE 29

Summary of the interactions between TFIIIA and the internal control region of the *Xenopus* somatic 5 S RNA gene. Guanosine residues that, when methylated, interfere with TFIIIA binding are shown in color. The DNA phosphate groups that make contact with TFIIIA are marked with solid triangles, and sequences protected by TFIIIA against DNase I digestion (arrows) are in a gray box. Bases that differ in the oocyte-type 5 S rRNA gene are indicated beneath the DNA strand. (After Sakonju and Brown, 1982.)

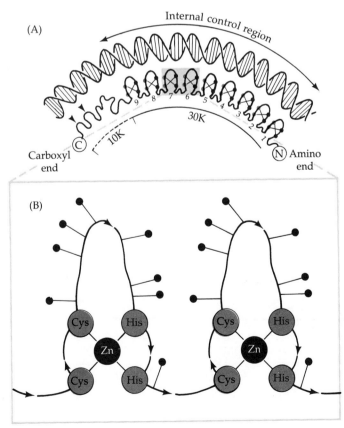

FIGURE 30

Structure of transcriptional regulator protein TFIIIA in relation to the internal control region DNA helix. (A) Orientation of the protein in relation to the DNA. The nine "fingers" binding to the 5 S promoter are looped out. (B) Close-up of the DNA-binding fingers stabilized by zinc; basic residues (black circles) touch the DNA. (After Miller et al., 1985.)

Transcriptional control of determination: The lineage switch genes

So far in our discussion of transcriptional control, we have looked at genes whose products are the result of differentiation rather than its cause. The production of globin, for instance, is but the last stage in the differentiation of a red blood cell. Are there cases where transcription is able to cause the cell to become a blood cell rather than a bone cell? Can genes acting early in development direct the cell's fate? In 1940, C. H. Waddington, one of the foremost theorists in developmental biology, predicted the existence of "switch" genes. These genes, he said, would act like the binary switches in a train yard that cause the trains to follow one path and not the other (Figure 31). Such genes would be responsible for somehow activating a battery of genes for one developmental pathway rather than for an alternative pathway.

Recently, a group of genes has been discovered that have the properties of such "switch genes." The cells wherein these genes are expressed have the potential to become either of two cell types, and the presence or absence of the gene product determines which lineage the cell generates. In these studies, one sees the convergence of embryology and molecular genetics.

The *lin-12* gene of *Caenorhabditis*

The *lin-12* locus of the roundworm *Caenorhabditis elegans* controls two alternative cell fates (Greenwald et al., 1983). In the wild-type embryo, there exist two neighboring cells, Z1.ppp and Z4.aaa, that interact with each other (Kimble, 1981). One or the other of them will give rise to a uterine anchor cell (ac) while the other cell generates the ventral uterine precursor cell (vu). In recessive mutants of *lin-12*, the gene is not transcribed. Both cells become anchor cells. In dominant mutants, where the level of *lin-12* transcription is too high, both cells become ventral uterine precursors (Table 3). The *lin-12* gene appears to control this "binary" switch between two alternative developmental pathways. One of the most interesting aspects of *lin-12* is that it affects the bipotentiality of several sets of cells in different parts of the body. As shown in the table, the ventral and

FIGURE 31
Photograph of a railroad switching yard presented by Joseph Needham (1936) to illustrate the hypothesis of his friend and collaborator, C. H. Waddington. The tracks represent different developmental pathways that a boxcar could follow. The binary branches can be readily observed.

TABLE 3
Cell fates in *lin-12* mutants

Mutant	Z1.ppp	Z4.aaa	M.v.(l/r)pa	M.d.(l/r)pa
		Cells		
lin-12(+)	Ventral uterine precursor cell or anchor cell	Ventral uterine precursor cell or anchor cell	Sex mesoblast	Coelomocyte
lin-12(0)	Anchor cell	Anchor cell	Coelomocyte	Coelomocyte
lin-12(d)	Ventral uterine precursor cell	Ventral uterine precursor cell	Sex mesoblast	Sex mesoblast

Source: Greenwald (1987).

dorsal m(l/r)pa cells become the sex mesoblast and the coelomocytes, respectively. However, in the dominant *lin-12* mutants, both become sex mesoblasts, whereas in the recessive *lin-12* mutants, both become coelomocytes.

The *Notch* gene of *Drosophila*

The *Notch* mutation of *Drosophila* also channels a bipotential cell into one of two alternative paths. Here, the choice is between becoming a skin (hypodermal) cell or a neuroblast. Soon after gastrulation, a region of about 1800 ectodermal cells lies along the ventral midline of the *Drosophila* embryo. These cells have the potential to form the ventral nerve cord of the insect, and about one-quarter of these cells become neuroblasts whereas the rest become the precursors of the hypodermis. The cells that give rise to neuroblasts are intermingled with those cells destined to give rise to hypodermal precursors. Thus, each ectodermal cell in the nerve-forming regions of the fly embryo can give rise to either hypodermal or neural precursor cells (Hartenstein and Campos-Ortega, 1984). In the absence of *Notch* gene transcription in the embryo, the cells develop into neural precursors rather than into a mixture of hypodermal and neural precursor cells (Figure 32; Lehmann et al., 1983; Artavanis-Tsakonis et al.,

FIGURE 32
Representation of the effect of the *Notch* mutation. In wild-type embryos, the neurogenic ectodermal cells generate both the neuroblasts and the skin (hypodermal) cells. In *Notch*-deficient embryos, however, all the neurogenic ectoderm generates neuroblasts.

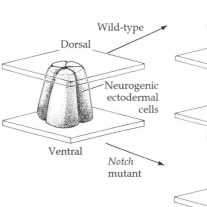
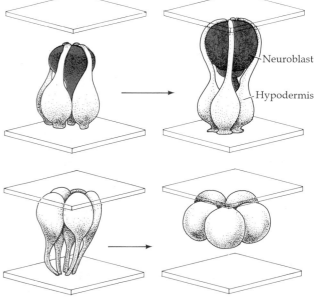

1983). These embryos die, having a gross excess of neural cells at the expense of the ventral and head hypodermis (Poulson, 1937; Hoppe and Greenspan, 1986). The *Notch* gene has been cloned (Kidd et al., 1983; Yedvobnick et al., 1985) and found to be transcribed during the early half of embryogenesis (and later in the early pupal stage). Both Notch and lin-12 proteins share remarkable sequence homologies to one another. The are both transmembrane proteins that may act as receptors for signals from adjacent cells (Yochem et al., 1988).*

The *MyoD1* gene of vertebrates

Another important switch gene is *MyoD1*. When DNA isolated from myoblast cells is transfected into certain strains of adipose cells, the adipose cells are turned into muscle cells. DNA isolated from fibroblasts or other cell types cannot accomplish this conversion (Lassar et al., 1986). By subtractive hybridization (see Chapter 10), a myoblast-specific cDNA clone was found that could also effect this change in differentiated phenotype. The gene encoding this cDNA was called *Myoblast Determination 1*, or more commonly, *MyoD1* (Davis et al., 1987). *MyoD1* is expressed only in muscle cells, and it appears to be a "master" switch gene in that it can convert other cell types into muscles if this gene is active in them. If the *MyoD1* gene is cloned into a viral vector so that it is under the control of a constitutively active viral promoter (hence, always "on"), it can be transfected into numerous cell types. When this happens, the pigment cell, nerve cell, fat cell, fibroblast, or liver cell is converted into a muscle cell (Figure 33; Weintraub et al., 1989). Thus, *MyoD1* appears to be sufficient to activate the muscle-specific genes that make up the muscle phenotype.

In *Xenopus laevis*, the *MyoD1* gene is activated by induction (Figure 34). In situ hybridization of *Xenopus* gastrulae with radioactive *MyoD1* cDNA shows that *MyoD1* message has accumulated specifically in those areas destined to become mesoderm. By the neurula stage, *MyoD1* mRNA expression is seen in the somites and other regions of skeletal muscle, such as limb primordia (Hopwood et al., 1989). Similar *MyoD1* expression patterns are also seen in the mouse (Sassoon et al., 1989).

MyoD1 encodes a nuclear DNA-binding protein that can bind to regions of the DNA adjacent to muscle-specific genes and activate these genes. For instance, the MyoD1 protein appears to directly activate the muscle-specific creatine phosphokinase gene by binding to the DNA im-

---

*In Chapter 16, we will discuss the roles of the lin-12 and Notch proteins in receiving inductive signals from neighboring cells.

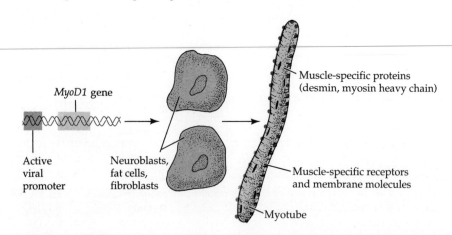

FIGURE 33
Summary of several experiments wherein the *MyoD1* gene is activated by a viral promoter and transfected into nonmuscle cells. MyoD1 protein binds to DNA and appears to override the original regulators of the cell phenotype and convert the cells into myocytes.

MyoD1 gene

Active viral promoter

Neuroblasts, fat cells, fibroblasts

Muscle-specific proteins (desmin, myosin heavy chain)

Muscle-specific receptors and membrane molecules

Myotube

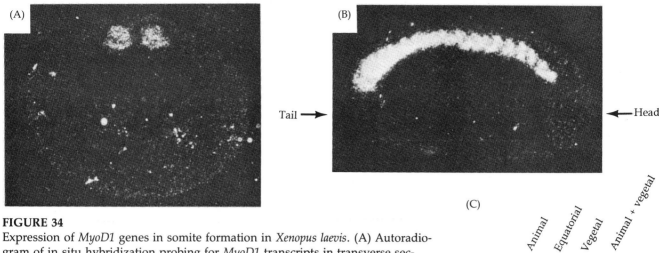

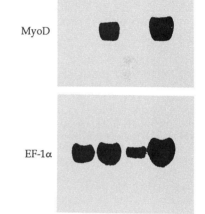

FIGURE 34

Expression of *MyoD1* genes in somite formation in *Xenopus laevis*. (A) Autoradiogram of in situ hybridization probing for *MyoD1* transcripts in transverse sections of *Xenopus* early neurula. *MyoD1* message accumulation is seen in the two somite regions. (B) Sagittal section of a late neurula stage *Xenopus* embryo prepared similarly to (A). The accumulation of *MyoD1* mRNA is seen throughout the somites. It is already receding in the head and will eventually recede in all somites after each has formed. (C) *MyoD1* is found in the pre-mesodermal equatorial cells of the *Xenopus* blastula and can be induced by combining animal cells with vegetal cells. Northern blot was incubated with radioactive probes to *MyoD1* mRNA and (as an internal control, since it is found in all cells) the mRNA for translation elongation factor 1α. (From Hopwood et al., 1989; photographs courtesy of J. B. Gurdon).

mediately upstream from it (Lassar et al., 1989). Similarly, there are two MyoD1 binding sites on the DNA adjacent to the α subunit gene of the chicken muscle acetylcholine receptor (Piette et al., 1990). It also directly activates itself. Once the *MyoD1* gene is activated, its protein product binds to the DNA immediately upstream of the *MyoD1* gene and keeps this gene from being turned off (Thayer et al., 1989). In other cases, the effects of *MyoD1* may be indirect. Not all genes involved in producing the muscle phenotype may be directly activated by MyoD1 protein. *MyoD1* probably acts indirectly by turning on other regulatory genes that then activate the structural muscle-specific genes.

One of these other regulatory genes may be MYOGENIN. The myogenin gene was found by subtractive hybridization in a search for muscle-specific RNAs expressed during the terminal differentiation of myotubes. Transfection of active myogenin genes into fibroblasts and other cells can also convert these cells into muscles. Myogenin message is at very low levels in myoblasts, peaks during myotube formation, and then declines in differentiated myotubes (Wright et al., 1989). *MyoD1* expression leads to *myogenin* expression, and the transfection of myogenin cDNA reciprocates by causing *MyoD1* expression. Thus, there is a reciprocal positive feedback loop such that when either *myogenin* or *MyoD1* is activated, the other gene will likewise be activated (Thayer et al., 1989). In the mouse, myogenin mRNA accumulates in the somites two days before myoD1 transcripts, which suggests that these two muscle-specific regulatory factors may play different roles in muscle determination and differentiation (Sassoon et al., 1989). The relationship between *MyoD1*, *myogenin*, and some of the other newly discovered muscle determination genes has yet to be delineated.

SUMMARY

Differential gene expression can be effected at the level of transcription in several ways. As in heterochromatization, large portions of chromosomes can be rendered genetically inert; or, as in the case of globin or ovalbumin genes, a specific gene can be activated in a particular type of cell at a particular time. In some cases where large amounts of normally synthesized gene product are required (as in the ribosomal genes of the amphibian oocyte), genes can be amplified (40 S RNA) and positive transcriptional factors can be synthesized (TFIIIA for 5 S rRNA). Transcriptional regulation may also direct the fate of a cell. The transcription of these "lineage switch genes" is essential for the establishment of cellular identity. In this chapter, we have shown that certain genes are differentially controlled at the level of transcription. But how is this regulation carried out? The molecular basis for differential gene transcription will be the subject of Chapter 12.

LITERATURE CITED

Artavanis-Tsakonis, S., Muskavitch, M. A. T. and Yedvobnick, Y. 1983. Molecular cloning of *Notch*, a locus affecting neurogenesis in *Drosophila melanogaster*. *Proc. Natl. Acad. Sci. USA* 80: 1977–1981.

Ashburner, M. 1974. Sequential gene activation by ecdysone in polytene chromosomes of *Drosophila melanogaster*. II. Effects of inhibitors of protein synthesis. *Dev. Biol.* 39: 141–157.

Ashburner, M. 1990. Puffs, genes, and hormones revisited. *Cell* 61: 1–3.

Barr, M. L. and Bertram, E. G. 1949. A morphological distinction between neurones of the male and female, and the behavior of the nucleolar satellite during accelerated nucleoprotein synthesis. *Nature* 163: 676.

Bell, S. P., Jantzen, H.-M. and Tjian, R. 1990. Assembly of alternative multiprotein complexes directs rRNA promoter selectivity. *Genes Dev.* 4: 943–954.

Bogenhagen, D. F., Sakonju, S. and Brown, D. D. 1980. A control region in the center of the 5 S RNA gene directs specific initiation of transcription. II. The 3′ border of the region. *Cell* 19: 27–35.

Brown, C. J. and Willard, H. F. 1990. Localization of a gene that escapes inactivation to the X chromosome proximal short arm: Implications for X inactivation. *Am. J. Hum. Genet.* 46: 273–279.

Brown, C. and 6 others. 1991a. A gene from the region of the human X inactivation center is expressed exclusively from the inactive X chromosome. *Nature* 349: 38–44.

Brown, C. and 9 others. 1991b. Localization of the X chromosome inactivation center on the human X chromosome. *Nature* 349: 82–84.

Brown, D. D. 1981. Gene expression in eukaryotes. *Science* 211: 667–674.

Brown, D. D. and Dawid, I. B. 1968. Specific gene amplification in oocytes. *Science* 160: 272–280.

Brown, D. D. and Schlissel, M. S. 1985. A positive transcription factor controls the differential expression of two 5 S RNA genes. *Cell* 42: 759–767.

Brown, S. W. and Nelson-Rees, W. A. 1961. Radiation analysis of a lecanoid genetic system. *Genetics* 46: 983–1007.

Centerwall, W. R. and Benirschke, K. 1973. Male tortoiseshell and calico (T-C) cats. *J. Hered.* 64: 272–278.

Cooper, D. W., Vandeberg, J. L., Sharmen, G. B. and Poole, W. E. 1971. Phosphoglycerate kinase polymorphism in kangaroos provides further evidence for paternal X inactivation. *Nature New Biol.* 230: 155–157.

Crowley, T. E. and Meyerowitz, E. M. 1984. Steroid regulation of RNAs transcribed from the *Drosophila* 68C polytene chromosome puff. *Dev. Biol.* 102: 110–124.

Davidson, R. G., Nitowsky, H. M. and Childs, B. 1963. Demonstration of two populations of cells in the human female heterozygous for glucose-6-phosphate dehydrogenase variants. *Proc. Natl. Acad. Sci. USA* 50: 481–485.

Davis, R. L., Weintraub, H. and Lassar, A. B. 1987. Expression of a single transfected cDNA converts fibroblasts into myoblasts. *Cell* 51: 987–1000.

Driesch, H. 1894. *Analytische Theorie der Organischen Entwickelung*. Wm. Engelmann, Leipzig. Quoted in J. Oppenheimer 1967), *Essays in the History of Embryology and Biology*. MIT Press, Cambridge, MA, p. 76.

Engelke, D. R., Ng, S.-Y., Shastry, B. S. and Roeder, R. G. 1980. Specific interaction of a purified transcription factor with an internal control region of 5 S RNA genes. *Cell* 19: 717–728.

Gartler, S. M. and Riggs, A. D. 1983. Mammalian X-chromosome inactivation. *Annu. Rev. Genet.* 17: 155–190.

Gartler, S. M., Liskay, R. M. and Grant, N. 1973. Two functional X chromosomes in human fetal oocytes. *Exp. Cell Res.* 82: 464–466.

Gartler, S. M., Rivest M. and Cole, R. E. 1980. Cytological evidence for an inactive X chromosome in murine oogonia. *Cytogenet. Cell Genet.* 28: 203–207.

Ginsberg, A. M., King, D. O. and Roeder, R. G. 1984. *Xenopus* 5 S gene transcription factor, TFIIIA: Characterization of a cDNA clone and measurement of RNA levels throughout development. *Cell* 39: 479–489.

Greenwald, I. S. 1987. The *lin-12* locus of *Caenorhabditis elegans*. *BioEssays* 6: 70–73.

Greenwald, I. S., Sternberg, P. W. and Horvitz, H. R. 1983. The *lin–12* locus specifies cell fates in *Caenorhabditis elegans*. *Cell* 34: 435–444.

Gronemeyer, H. and Pongs, O. 1980. Localization of ecdysterone on polytene chromosomes of *Drosophila melanogaster*. *Proc. Natl. Acad. Sci. USA* 77: 2108–2112.

Harris, S. E., Rosen, J. M., Meens, A. R. and O'Malley, B. W. 1975. Use of specific probe for ovalbumin messenger RNA to quantitate estrogen-induced transcripts. *Biochemistry* 14: 2072–2081.

Hartenstein, V. and Campos-Ortega, J. A. 1984. Early neurogenesis in wild-type *Drosophila melanogaster*. *Wilhelm Roux Arch. Dev. Biol.* 193: 308–325.

Hill, R. S. and MacGregor, H. C. 1980. The development of lampbrush chromosome-type transcription in the early diplotene oocytes of *Xenopus laevis*: An electron microscope analysis. *J. Cell Sci.* 44: 87–101.

Hoppe, P. E. and Greenspan, R. J. 1986. Local function of the *Notch* gene for embryonic ectodermal pathway choice in *Drosophila*. *Cell* 46: 773–783.

Hopwood, N. D., Pluck, A. and Gurdon, J. B. 1989. MyoD expression in the forming somites is an early response to mesoderm induction in Xenopus embryos. *EMBO J.* 8: 3409–3417.

Hourcade, D., Dressler, D. and Wolfson, J. 1973. The amplification of ribosomal RNA genes involves a rolling circle intermediate. *Proc. Natl. Acad. Sci. USA* 70: 2926–2930.

Karlsson, S. and Nienhaus, A. W. 1985. Developmental regulation of human globin genes. *Annu. Rev. Biochem.* 54: 1071–1108.

Kidd, S., Lockett, T. J. and Young, M. W. 1983. The *Notch* locus of *Drosophila melanogaster*. *Cell* 34: 421–433.

Kimble, J. 1981. Alterations in cell lineage following laser ablation of cells in the somatic gonad of *Caenorhabditis elegans*. *Dev. Biol.* 87: 286–300.

Korge, G. 1975. Chromosome puff activity and protein synthesis in larval salivary glands of *Drosophila melanogaster*. *Proc. Natl. Acad. Sci. USA* 72: 4550–4554.

Korn, L. 1982. Transcription of *Xenopus* 5 S ribosomal RNA genes. *Nature* 295: 101–105.

Kratzer, P. G. and Chapman, V. M. 1981. X-chromosome reactivation in oocytes of *Mus caroli*. *Proc. Natl. Acad. Sci. USA* 78: 3093–3097.

Lassar, A. B., Martin, P. L. and Roeder, R. G. 1983. Transcription of class III genes: Formation of preinitiation complexes. *Science* 222: 740–748.

Lassar, A. B., Paterson, B. M. and Weintraub, H. 1986. Transfection of a DNA locus that mediates the conversion of 10T1/2 fibroblasts into myoblasts. *Cell* 47: 649–656.

Lassar, A. B., Buskin, J. N., Lockshon, D., Davis, R. L., Apone, S., Hauschka, S. D. and Weintraub. H. 1989. MyoD is a sequence-specific DNA binding protein requiring a region of *myc* homology to bind to the muscle creatine kinase enhancer. *Cell* 58: 823–831.

Lehmann, R., Jimenez, F., Dietrich, U. and Campos-Ortega, J. A. 1983. On the phenotype and development of mutants of early neurogenesis in *Drosophila melanogaster*. *Wilhelm Roux Arch. Dev. Biol.* 192: 62–74.

Lewin, B. 1980. *Gene Expression 2*. Wiley-Interscience, New York.

Lillie, F. R. 1927. The gene and the ontologic process. *Science* 66: 361–368.

Lyon, M. F. 1961. Gene action in the X-chromosome of the mouse (*Mus musculus* L.). *Nature*: 190: 372–373.

Mayfield, J. E., Serunian, L. A., Silver, L. M. and Elgin, S. C. R. 1978. A protein released by DNase I digestion of *Drosophila* nuclei is preferentially associated with puffs. *Cell* 14: 539–544.

McBurney, M. W. and Strutt, B. J. 1980. Genetic activity of X chromosomes in pluripotent female teratocarcinoma cells and their differentiated progeny. *Cell* 21: 357–364.

McKnight, G. S. and Palmiter, R. D. 1979.

Transcriptional regulation of the ovalbumin and conalbumin genes by steroid hormones in chick oviduct. *J. Biol. Chem.* 254: 9050–9058.

Migeon, B. R. 1971. Studies of skin fibroblasts from ten families with HGPRT deficiency, with reference to X-chromosomal inactivation. *Am. J. Hum. Genet.* 23: 199–209.

Migeon, B. R. and Jelalian, K. 1977. Evidence for two active X chromosomes in germ cells of female before meiotic entry. *Nature* 269: 242–243.

Migeon, B. R., Wolf, S. F., Axelman, J., Kaslow, D. C. and Schmidt, M. 1985. Incomplete X chromosome dosage compensation in chorionic villi of human placenta. *Proc. Natl. Acad. Sci. USA* 82: 3390–3394.

Migeon, B. R., Schmidt, M., Axelman, J. and Cullen, C. R. 1986. Complete reactivation of X chromosomes from human chorionic villi with a switch to early DNA replication. *Proc. Natl. Acad. Sci. USA* 83: 2182–2186.

Miller, D. L., Jr. and Beatty, B. R. 1969. Visualization of nucleolar genes. *Science* 164: 955–957.

Miller, J., McLachlan, A. D. and Klug, A. 1985. Repetitive zinc-binding domains in the protein transcription factor IIIA from *Xenopus* oocytes. *EMBO J.* 4: 1609–1614.

Mohandas, T., Sparkes, R. S., Hellkuhl, B., Brzeschik, K. H. and Shapiro, L. J. 1980. Expression of an X-linked gene from an inactive human X chromosome in mouse–human hybrid cells: Further evidence for the non-inactivation of the steroid sulfatase locus in man. *Proc. Natl. Acad. Sci. USA* 77: 6759–6763.

Monk, M. and Harper, M. I. 1979. Sequential X chromosome inactivation coupled with cellular differentiation in early mouse embryos. *Nature* 281: 311–313.

Moore, K. L. 1977. *The Developing Human*. Saunders, Philadelphia.

Morgan, T. H. 1934. *Embryology and Genetics*. Columbia University Press, New York, pp. 9–10.

Needham, J. 1936. *Order and Life*. Yale University Press, New Haven.

Nesbitt, M. N. 1974. Chimeras vs. X inactivation mosaics: Significance of differences in pigment distribution. *Dev. Biol.* 38: 202–207.

Ng, S.-Y., Parker, C. S. and Roeder, R. G. 1979. Transcription of cloned *Xenopus laevis* RNA polymerase III in reconstituted systems. *Proc. Natl. Acad. Sci. USA* 76: 136–140.

Old, R. W., Callan, H. G. and Gross, K. W. 1977. Localization of histone gene transcripts in newt lampbrush chromosomes by *in situ* hybridization. *J. Cell Sci.* 27: 57–80.

Osheim, Y. N. and Miller, O. L. 1983. Novel amplification and transcriptional activity of chorion genes in *Drosophila melanogaster* follicle cells. *Cell* 33: 543–653.

Palmiter, R. D. and Schimke, R. T. 1973. Regulation of protein synthesis in chick oviduct. III. Mechanism of ovalbumin "superinduction" by actinomycin D. *J. Biol. Chem.* 248: 1502–1512.

Pieler, T., Oei, S.-L., Hamm, J., Engelke, U. and Erdmann, V. A. 1985. Functional domains of the *Xenopus laevis* 5 S gene promoter. *EMBO J.* 4: 3751–3756.

Pieler, T., Hamm, J. and Roeder, R. G. 1987. The 5 S gene internal control region is composed of three distinct sequence elements, organized as two functional domains with variable spacing. *Cell* 48: 91–100.

Piette, J., Bessereau, J.-L., Huchet, M. and Changeaux, J.-P. 1990. Two adjacent MyoD1-binding sites regulate expression of the acetylcholine receptor α-subunit gene. *Nature* 345: 353–355.

Poulson, D. F. 1937. Chromosomal deficiencies and the embryonic development of *Drosophila melanogaster*. *Proc. Natl. Acad. Sci. USA* 23: 133–137.

Renfree, M. B. 1982. Implantation and placentation. *In* C. R. Austin and R. V. Short (eds.), *Embryonic and Fetal Development*. Cambridge University Press, Cambridge, pp. 26–69.

Rochaix, J. D., Bird, A. and Bakken, A. 1974. Ribosomal RNA gene amplification by rolling circles. *J. Mol. Biol.* 87: 473–488.

Roop, D. R., Nordstrom, J. L., Tsai, S. Y., Tsai, M.-J. and O'Malley, B. W. 1978. Transcription of structural and intervening sequences in the ovalbumin gene and identification of potential ovalbumin mRNA precursors. *Cell* 15: 671–685.

Rutter, W., Jr., Valenzuela, P., Bell, G. E., Holland, M., Hager, G. L., Degennaro, L. J. and Bishop, R. J. 1976. The role of DNA-dependent RNA polymerase in transcriptive specificity. *In* E. M. Bradbury and K. Javeherian (eds.), *The Organization and Expression of the Eukaryotic Genome*. Academic Press, New York, pp. 279–293.

Sakonju, S. and Brown, D. D. 1982. Contact points between a positive transcription factor and the *Xenopus* 5 S RNA gene. *Cell* 31: 395–405.

Sakonju, S., Bogenhagen, D. F. and Brown, D. D. 1980. A control region in the center of the 5 S RNA gene directs specific initiation of transcription. I. The 5′ border of the region. *Cell* 19: 13–25.

Samollow, P. B., Ford, A. L. and VandeBerg, J. L. 1987. X-linked gene expression in the Virginia opossum: Differences between the paternally derived *Gpd* and *Pgk-A* loci. *Genetics* 115: 185–195.

Sassoon, D., Lyons, G., Wright, W. E., Lin, V., Lassar, A., Weintraub, H. and Buckingham, M. 1989. Expression of two myogenic regulatory factors myogenin and MyoD1 during mouse embryogenesis. *Nature* 341: 303–307.

Scotto, K. W., Kaulen, H. and Roeder, R. G. 1989. Positive and negative regulation of the gene for transcription factor TFIIIA in *Xenopus laevis* oocytes. *Genes Dev.* 3: 651–662.

Segraves, W. A. and Hogness, D. S. 1990. The *E75* ecdysone-inducible gene responsible for the 75B early puff in Drosophila encodes two new members of the steroid receptor superfamily. *Genes and Dev.* 4: 204–209.

Sharman, G. B. 1971. Late DNA replication in the paternally derived X chromosome of female kangaroos. *Nature* 230: 231–232.

Spradling, A. C. 1981. The organization and amplification of two chromosomal domains containing *Drosophila* chorion genes. *Cell* 27: 193–201.

Spradling, A. C. and Mahowald, A. P. 1980. Amplification of genes for chorion proteins during oogenesis in *Drosophila melanogaster*. *Proc. Natl. Acad. Sci. USA* 77: 1096–1100.

Tagaki, N. 1974. Differentiation of X chromosomes in early female mouse embryos. *Exp. Cell Res.* 86: 127–135.

Tagaki, N. and Abe, K. 1990. Detrimental effects of two active X chromosomes on early mouse development. *Development* 109: 189–201.

Thayer, M. J., Tapscott, S. J., Davis, R. L., Wright, W. E., Lassar, A. B. and Weintraub, H. 1989. Positive autoregulation of the myogenic determination gene *MyoD1*. *Cell* 58: 241–248.

Thummel, C. S., Burtis, K. C. and Hogness, D. S. 1990. Spatial and temporal patterns of *E74* transcription during *Drosophila* development. Cell 61: 101–111.

Waddington, C. H. 1940. *Organisers and Genes*. Cambridge University Press, Cambridge.

Watson, J. D. 1976. *Molecular Biology of the Gene*, 3rd Ed. Benjamin Cummings, Menlo Park, CA.

Weintraub, H., Tapscott, S. J., Davis, R. L., Thayer, M. J., Adam, M. A., Lassar, A. B. and Miller, D. 1989. Activation of muscle-specific genes in pigment, nerve, fat, liver, and fibroblast cell lines by forced expression of MyoD. *Proc. Nat. Acad. Sci. USA* 86: 5434–5438.

Wellauer, P. K., Dawid, I. B., Brown, D. D. and Reeder, R. H. 1976. The arrangement of length heterogeneity in repeating units of amplified and chromosomal ribosomal DNA from *Xenopus laevis*. *J. Mol. Biol.* 105: 487–506.

West, J. D., Frels, W. I., Chapman, V. M. and Papaioannou, V. E. 1977. Preferential expression of the maternally derived X chromosome in the mouse yolk sac. *Cell* 12: 873–882.

Witschi, E. 1957. Sex chromatin and sex differentiation in human embryos. *Science* 126: 1288–1290.

Wolffe, A. P. and Brown, D. D. 1988. Developmental regulation of two 5 S ribosomal RNA genes. *Science* 241: 1626–1632.

Wright, W. E., Sassoon, D. A. and Lin, V. K. 1989. Myogenin, a factor regulating myogenesis, has a domain homologous to MyoD. *Cell* 56: 607–617.

Yedvobnick, B., Muskavitch, M. A. T., Wharton, K. A., Halpern, M. E., Paul, E., Grimwade, B. G. and Artavanis-Tsakonas, S. 1985. Molecular genetics of *Drosophila* neurogenesis. *Cold Spring Harbor Symp. Quant. Biol.* 50: 841–854.

Yochem, J., Weston, K. and Greenwald, I. 1988. The *Caenorhabditis elegans 1in-12* gene encodes a transmembrane protein with overall similarity to *Drosophila Notch*. *Nature* 335: 547–550.

12

Transcriptional regulation of gene expression: The mechanisms of differential gene transcription

But whatever the immediate operations of the genes turn out to be, they most certainly belong to the category of developmental processes and thus belong to the province of embryology. This central problem of fundamental biology at the present time is being attacked from many sides, both by physiologists and biochemists and by geneticists; but it is essentially an embryological problem.

—C. H. WADDINGTON (1956)

We have entered the cell, the mansion of our birth, and have started the inventory of our acquired wealth.

—ALBERT CLAUDE (1974)

EUKARYOTIC PROTEIN-ENCODING GENES

Exons and introns

One of the fundamental mechanisms of differential gene expression is the transcription of different genes in different cell types. This differential gene transcription entails the orchestration of numerous factors that bind to specific regions of the DNA, causing the DNA to be bound or modified in some cells and not in others. To begin discussing the mechanisms by which this cell-specific modification can take place, we must first discuss the structure of the eukaryotic gene. The first thing that becomes apparent is that most eukaryotic genes are not like most prokaryotic genes, as the eukaryotic genes are not colinear with their peptide products. Rather, the 5' and 3' ends of eukaryotic messenger RNA come from noncontiguous regions on the chromosome. Between the regions of DNA coding for the proteins (EXONS) are intervening sequences (INTRONS) that have nothing whatever to do with the amino acid sequence of the protein.* The structure of the human β-globin gene is shown in Figure 1. This gene consists of the following elements:

1 A PROMOTER REGION responsible for the binding of RNA polymerase and for the subsequent initiation of transcription. This promoter region of the human β-globin gene has (as will be shown later) three distinct units and extends from 95 to 26 base pairs before the transcription initiation site (i.e., from −95 to −26).

2 The sequence ACATTTG, where *transcription* is initiated. This is often called the CAP SEQUENCE, because it represents the 5′ end of the RNA, which will receive a "cap" of modified nucleotides soon after it is transcribed (as we shall see later). The specific cap sequence varies between genes.

3 The ATG CODON for the initiation of *translation*. This codon is located 50 base pairs after the initiation point of transcription. The intervening sequence of 50 nucleotide pairs between the initiation points of transcription and translation is called the LEADER SEQUENCE.

4 The first exon containing 90 base pairs coding for amino acids 1–30 of human β-globin.

5 An intron containing 130 base pairs with no coding sequences for hemoglobin.

6 An exon containing 222 base pairs coding for amino acids 31–104.

7 A large intron—850 base pairs—having nothing to do with the globin protein structure.

8 An exon containing 126 base pairs coding for amino acids 105–146.

9 A TRANSLATION TERMINATION CODON, TAA.

10 A 3′ UNTRANSLATED REGION, which, although transcribed, is not translated into protein. This region includes the sequence AATAAAA, which is needed to place a "tail" of some 200–300 adenylate residues on the RNA transcript. This poly(A) tail is inserted into the RNA about 20 bases downstream of the AAUAA motif of the RNA. However, transcription continues beyond the AATAAA site for about 1000 nucleotides before being terminated. Within this 3′ transcribed but untranslated sequence (about 600–900 base pairs from the AATAAA site) is an enhancer sequence that is necessary for the temporal and tissue-specific expression of the β-globin gene in adult red blood cell precursors (Trudel and Constantini, 1987).

The original nuclear RNA transcript for such a gene contains the capping sequence, the leader sequence, the exons, the introns, and the 3′ nontranslated region (Figure 2). In addition, both its ends become modified. A CAP consisting of methylated guanosine is placed on the 5′ end of the RNA in opposite polarity to the RNA itself. Thus, whereas all the bases

*The term exon has taken on two overlapping meanings. In the original sense, it is anatomically defined as a nucleotide sequence whose RNA "exits" the nucleus. It has taken on the functional definition of referring to a protein-encoding nucleotide sequence. For discussion here, we will use the latter definition, and we will define leader sequences and 3′ untranslated sequences apart from the exons. Certain eukaryotic genes (such as the genes for histones) lack intervening sequences, and any hypothesis concerning intron function must take these exceptions into account.

(A)

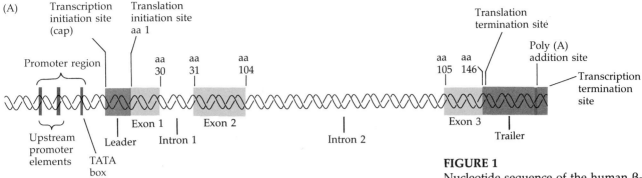

Transcription
initiation site
(cap)

Translation
initiation site
aa 1

Promoter region

aa
30
aa
31
aa
104

Translation
termination site

Poly (A)
addition site

aa
105
aa
146

Transcription
termination
site

Upstream
promoter
elements

TATA
box

Leader

Exon 1

Intron 1

Exon 2

Intron 2

Exon 3

Trailer

FIGURE 1

Nucleotide sequence of the human β-globin gene. (A) Schematic representation of the locations of the promoter region, transcription initiation (cap) site, leader sequence, exons and introns of the β-globin gene. Exons are in black, and the numbers flanking them indicate the amino acid positions they code for in β-globin. (B) The nucleotide sequence of the β-globin gene, shown from the 5′ to the 3′ end of the RNA. The promoter sequences are boxed, as are the translation initiation and termination codes ATG and TAA. The large capital letters correspond to exons, and the amino acids for which they code are abbreviated above them. The small capital letters are the bases of the intervening sequences. The codons represented by capital letters after the translation terminator are in the globin mRNA but are not translated into proteins. Within this group is the sequence thought to be needed for polyadenylation. A G in the first intron (arrow) is mutated to an A in one form of β+-thalassemia. (Sequence from Lawn et al., 1980.)

(B)

```
ccctgtggagccacaccctagggttggccaatctactcccaggagcagggagggcaggagccagggctgggcataaaa

gtcagggcagagccatctattgcttACATTTGCTTCTGACACAACTGTGTTCACTAGCAACCTCAAACAGACACCATG
                         ValHisLeuThrProGluGluLysSerAlaValThrAlaLeuTrpGlyLysValAsnValAspGluValGlyGlyGlu
                         GTGCACCTGACTCCTGAGGAGAAGTCTGCCGTTACTGCCCTGTGGGGCAAGGTGAACGTGGATGAAGTTGGTGGTGAG

AlaLeuGlyArg
GCCCTGGGCAGGttggtatcaaggttacaagacaggtttaaggagaccaatagaaactgggcatgtggagacagagaag

actcttgggtttctgataggcactgactctctctgcctattggtctattttcccacccttaggCTGCTGGTGGTCTAC
                                                           ↓                    LeuLeuValValTyr
ProTrpThrGlnArgPhePheGluSerPheGlyAspLeuSerThrProAspAlaValMetGlyAsnProLysValLys
CCTTGGACCCAGAGGTTCTTTGAGTCCTTTGGGGATCTGTCCACTCCTGATGCTGTTATGGGCAACCCTAAGGTGAAG

AlaHisGlyLysLysValLeuGlyAlaPheSerAspGlyLeuAlaHisLeuAspAsnLeuLysGlyThrPheAlaThr
GCTCATGGCAAGAAAGTGCTCGGTGCCTTTAGTGATGGCCTGGCTCACCTGGACAACCTCAAGGGCACCTTTGCCACA

LeuSerGluLeuHisCysAspLysLeuHisValAspProGluAsnPheArg
CTGAGTGAGCTGCACTGTGACAAGCTGCACGTGGATCCTGAGAACTTCAGGGTGAGTCTATGGGACCCTTGATGTTTT

CTTTCCCCTTCTTTTCTATGGTTAAGTTCATGTCATAGGAAGGGGAGAAGTAACAGGGTACAGTTTAGAATGGGAAAC

AGACGAATGATTGCATCAGTGTGGAAGTCTCAGGATCGTTTTAGTTTCTTTTATTTGCTGTTCATAACAATTGTTTTC

TTTTGTTTAATTCTTGCTTTCTTTTTTTTTCTTCTCCGCAATTTTTACTATTATACTTAATGCCTTAACATTGTGTAT

AACAAAAGGAAATATCTCTGAGATACATTAAGTAACTTAAAAAAAAAACTTTACACAGTCTGCCTAGTACATTACTATT

TGGAATATATGTGTGCTTATTTGCATATTCATAATCTCCCTACTTTATTTTCTTTTATTTTTAATTGATACATAATCA

TTATACATATTTATGGGTTAAAGTGTAATGTTTTAATATGTGTACACATATTGACCAAATCAGGGTAATTTTGCATT

TGTAATTTTAAAAAATGCTTTCTTCTTTTAATATACTTTTTTGTTTATCTTATTTCTAATACTTTCCCTAATCTCTTT

CTTTCAGGGCAATAATGATACAATGTATCATGCCTCTTTGCACCATTCTAAAGAATAACAGTGATAATTTCTGGGTTA

AGGCAATAGCAATATTTCTGCATATAAATATTTCTGCATATAAATTGTAACTGATGTAAGAGGTTTCATATTGCTAA

TAGCAGCTACAATCCAGCTACCATTCTGCTTTTATTTTATGGTTGGGATAAGGCTGGATTATTCTGAGTCCAAGCTAG
                                                                          LeuLeuGlyAsnValLeuValCysValLeuAla
GCCCTTTTGCTAATCATGTTCATACCTCTTATCTTCCTCCCACAGCTCCTGGGCAACGTGCTGGTCTGTGTGCTGGCC

HisHisPheGlyLysGluPheThrProProValGlnAlaAlaTyrGlnLysValValAlaGlyValAlaAsnAlaLeu
CATCACTTTGGCAAAGAATTCACCCCACCAGTGCAGGCTGCCTATCAGAAAGTGGTGGCTGGTGTGGCTAATGCCCTG

AlaHisLysTyrHis
GCCCACAAGTATCACTAAGCTCGCTTTCTTGCTGTCCAATTTCTATTAAAGGTTCCTTTGTTCCCTAAGTCCAACTAC

TAAACTGGGGGATATTATGAAGGGCCTTGAGCATCTGGATTCTGCCTAATAAAAAACATTTATTTTCATTGCaatgat

gtatttaaattatttctgaatattttactaaaaagggaatgtgggaggtcagtgcatttaaaacataaagaaatgatg

agctgttcaaaccttgggaaaatacactatatcttaaactccatgaaagaaggtgaggctgcaaccagctaatgcaca

ttggcaacagcccctgatgcctatgccttattcatccctcagaaaaggattcttgtagaggcttgatttgcaggttaa

agttttgctatgctgtattttacattacttattgttttagctgtcctcatgaatgtcttttcactacccatttgctta

tcctgcatctctctcagccttgact ...
```

in the message precursor are linked 5′ to 3′, the cap structure is linked 5′ to 5′. This means that there is no free 5′ phosphate group on the nuclear RNA (Figure 3). Messenger RNA molecules are likewise "capped," although it is not certain whether the mRNA cap is the original one it received in the nucleus. The 5′ cap is also known to be necessary for the binding of mRNA to the ribosome (Shatkin, 1976).

The 3' terminus is usually modified in the nucleus by having roughly 200 adenylate residues added on as a tail. These adenylic acid residues are put together enzymatically and are added to the transcript. They are not part of the gene sequence. Both the 5' and 3' modifications may protect the RNA from exonucleases (Sheiness and Darnell, 1973; Gedamu and Dixon, 1978), thereby stabilizing the message and its precursor. These poly(A) tails get progressively shorter as the mRNA ages. Newly synthesized globin message in mouse and rabbit cells has about 150 adenylate residues, whereas older messages have 100, 60 or 40 extra adenylates (Merkel et al., 1975; Nokin et al., 1976). Marbaix and co-workers (1975) have shown that globin message lacking poly(A) is degraded rapidly when injected into *Xenopus* oocytes. Globin mRNA with its poly(A) tails lasts over 20 hours.

Introns are very important in that they can contain numerous regulatory sequences that can effect the rate of DNA transcription or its exit from the nucleus. This can be shown in individuals with certain types of β⁺-THALASSEMIA, a disease characterized by low levels of β-globin.* These

*In this and the next chapter, the β-thalassemias will be used to illustrate defective gene regulation. People with β⁺-thalassemia synthesize lower than normal amounts of β-globin. People with β°-thalassemia do not synthesize any β-globin. As we shall see, β-thalassemias are often caused by mutations in promoters or enhancers, or in splice-producing sequences within introns.

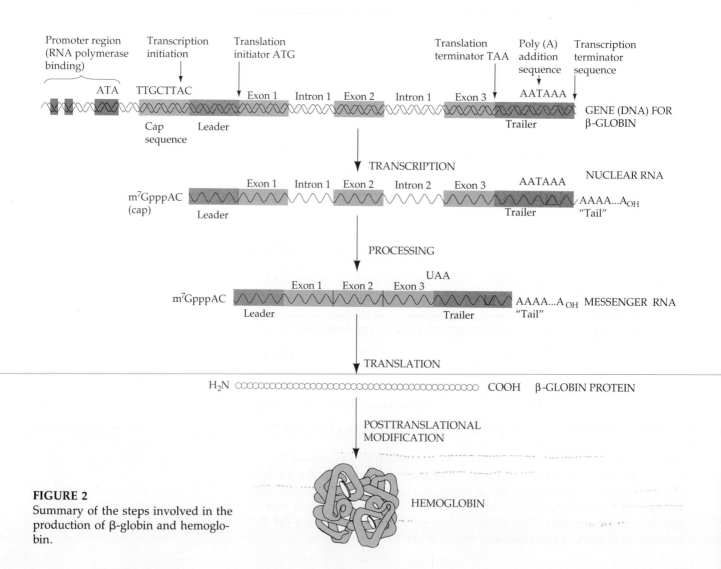

FIGURE 2
Summary of the steps involved in the production of β-globin and hemoglobin.

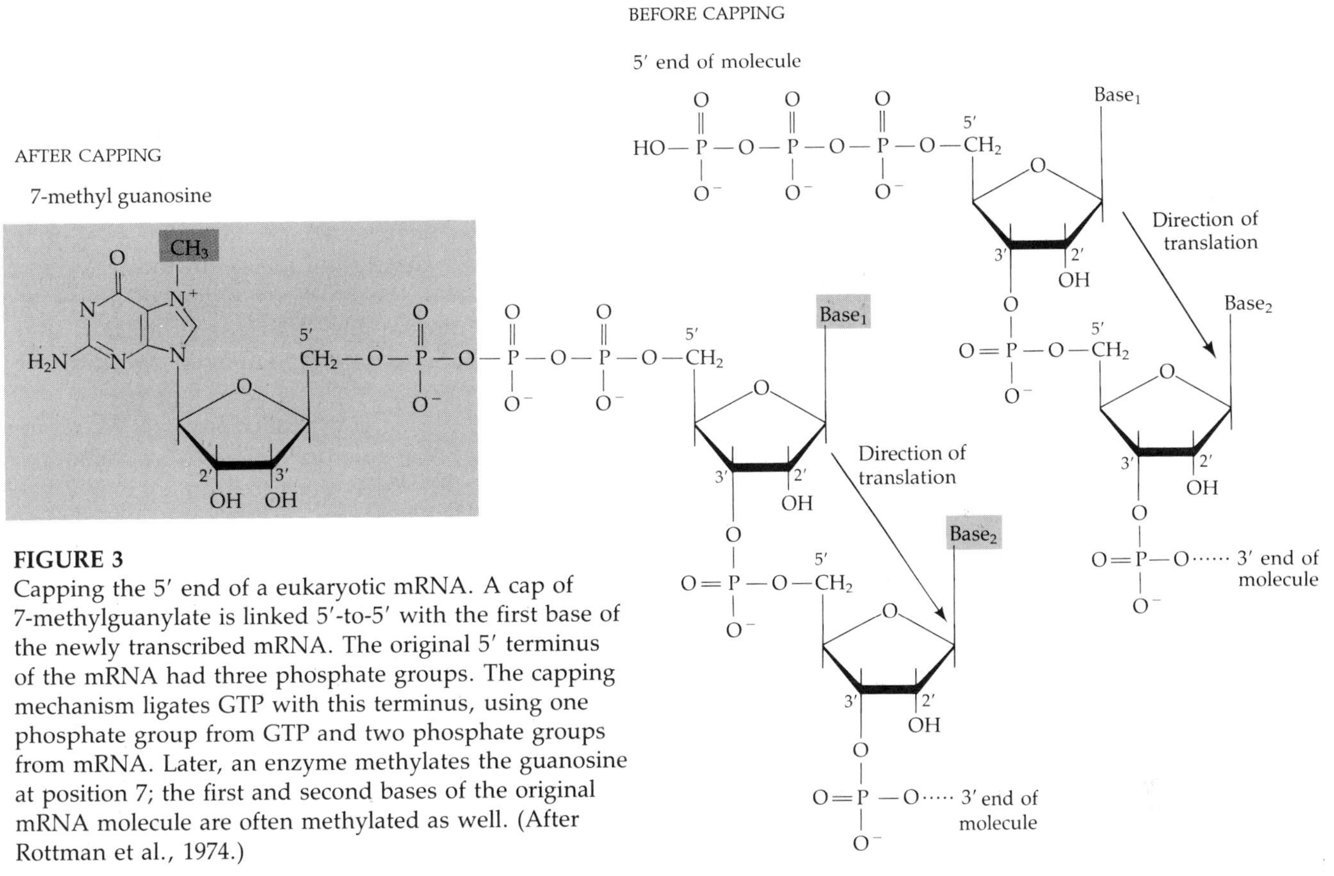

5′ end of molecule

AFTER CAPPING

7-methyl guanosine

FIGURE 3

Capping the 5′ end of a eukaryotic mRNA. A cap of 7-methylguanylate is linked 5′-to-5′ with the first base of the newly transcribed mRNA. The original 5′ terminus of the mRNA had three phosphate groups. The capping mechanism ligates GTP with this terminus, using one phosphate group from GTP and two phosphate groups from mRNA. Later, an enzyme methylates the guanosine at position 7; the first and second bases of the original mRNA molecule are often methylated as well. (After Rottman et al., 1974.)

people do not lack the genes for β-globin, and transcription of the RNA precursor from these genes appears to be normal. Moreover, the mRNA from such patients gets translated properly into the amino acid sequence expected for human β-globin. The defect is in the *amount* of β-globin message entering the cytoplasm from the nucleus, and the disease appears to be at the level of RNA processing. Maquat and colleagues (1980) showed this to be the case by a "pulse-chase" experiment. Developing red blood cells were isolated from patients' bone marrow and given a 12-minute "pulse" of radioactive nucleotides. After 12 minutes, transcription was stopped with actinomycin D. At various intervals, RNA was harvested from the cells, subjected to electrophoresis to separate RNA molecules by size, and then hybridized to β-globin cDNA. In normal cells, the large β-globin mRNA precursor was seen to become β-globin mRNA within a few minutes. However, in the β$^+$-thalassemic cells, there was a buildup of the intermediates. By 30 minutes very few message-size RNAs existed, and the intermediates were probably digested within the nucleus so that very few got into the cytoplasm (Maquat et al., 1980; Kantor et al., 1980). Thus, β$^+$-thalassemia appears to be a defect in the processing of the β-globin message precursor. When the β-globin genes from several of these individuals were sequenced, they were found to contain mutations in the first *intron*. The mutation in one of these genes is shown in Figure 1 (Spritz et al., 1981). The structure of the intron appears to be important for the processing and transport of RNA from the nucleus to the cytoplasm. The mechanism of such splicing and its importance for differential gene expression will be discussed in Chapter 13.

Promoter structure and function

In order to transcribe DNA into RNA, certain sequences must be present in the DNA and certain TRANSCRIPTION FACTORS must be present in the nucleus. These transcription factors must be able to find the DNA sequences despite the fact that most of the DNA is packaged in nucleosomes. It is currently thought that rendering a gene competent to transcribe mRNA involves (a) the binding of transcription factors to the DNA and (b) the exclusion of nucleosomes from the promoter region of the gene. The interactions between specific transcription factors and the DNA they bind cause the phenomenon of tissue-specific and temporally specific gene transcription.

There are two types of regulatory elements needed to effect transcription at the proper sites. The first set of regulatory elements are called CIS-REGULATORS. These represent specific DNA sequences on a given chromosome. *Cis*-regulators only act on adjacent genes. The second set are called TRANS-REGULATORS. These are soluble molecules (including proteins and RNAs) that are made by one gene and interact with genes on the same or different chromosomes. If one recalls gene induction in the *lac* operon in *E. coli*, one will remember that a repressor gene makes a repressor protein that interacts with the operator sequence of the *lac* operon genes. In this case, the operator DNA is a *cis*-regulatory element because it controls only the *lac* operon on its own chromosome. (A mutant operator sequence on the other chromosome may or may not bind the repressor protein.) The repressor protein, however, is a *trans*-regulator, because it can be made by one chromosome and bind to the *cis*-regulatory operator on another chromosome (Figure 4).

In eukaryotic genes that encode messenger RNA, two types of *cis*-regulatory DNA sequences have been discovered, promoters and enhancers. PROMOTERS are typically located immediately "upstream" from the site where transcription begins and are about 100 base pairs long. The promoter site is required for the binding of RNA polymerase II and the accurate initiation of transcription. As mentioned in the previous chapter, eukaryotic RNA polymerases require additional protein factors to bind to the promoter. The ENHANCER activates the utilization of the promoter, controlling the efficiency and rate of transcription from that particular promoter. Enhancers can activate only *cis*-linked promoters (i.e., promoters on the same chromosome), but they can do so at great distances (some as great as 50 kilobases away from the promoter). Moreover, enhancers do not need to be at the 5' ("upstream") side of the gene. They can be at the 3' end, in the introns, or even on the complementary DNA strand (Maniatis et al., 1987). Enhancers also work by binding specific proteins.

Promoters of genes that transcribe relatively large amounts of mRNA have similar structures. They have an ATA sequence (sometimes called the TATA box or the GOLDBERG-HOGNESS BOX) about 30 base pairs upstream from the site where transcription begins, and one or more UPSTREAM PROMOTER ELEMENTS further upstream. The upstream promoter element is often a variation of the sequence CAAT, but other upstream promoter elements have also been found (Grosschedl and Birnstiel, 1980; McKnight and Tjian, 1986) (Figure 5).

The promoter region of the β-globin gene was first analyzed by experiments testing the specific transcription from cloned DNA. Cloned genes can be accurately transcribed when they are placed into the nuclei of frog oocytes or fibroblasts or when they are incubated with purified RNA polymerase in the presence of nucleotides and cell extracts (Wasylyk

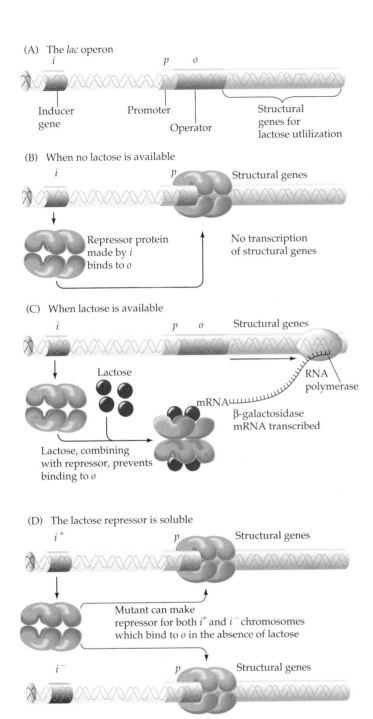

(A) The *lac* operon

Inducer gene

Promoter

Operator

Structural genes for lactose utilization

(B) When no lactose is available

Repressor protein made by *i* binds to *o*

No transcription of structural genes

(C) When lactose is available

Structural genes

Lactose

RNA polymerase

mRNA

β-galactosidase mRNA transcribed

Lactose, combining with repressor, prevents binding to *o*

(D) The lactose repressor is soluble

Structural genes

Mutant can make repressor for both i^+ and i^- chromosomes which bind to *o* in the absence of lactose

Structural genes

FIGURE 4

Differential gene regulation in *E. coli*, showing *cis*- and *trans*-regulatory elements. In its wild-type, inducible state (A–C), no β-galactosidase RNA is transcribed unless lactose is present. When no lactose is available (B), a repressor protein (R), made by gene *i*, binds to the operator site (*o*), inhibiting transcription by RNA polymerase from the promoter (*p*). When lactose is present (C), it combines with the repressor protein, so the repressor cannot bind to the DNA and transcription ensues. (D) The soluble nature of this repressor is shown in studies on mutant *E. coli*. When haploid bacteria cells with an i^- gene are made partially diploid with the wild-type *i* gene (i^+), wild-type repressor is manufactured and is able to make the original β-galactosidase gene inducible. The repressor protein is a *trans*-regulatory element. The promoter and operator sequences are *cis*-regulatory elements.

et al., 1980). After the transcription of a gene is confirmed, one uses restriction enzymes to make specific deletions in the gene or in the regions surrounding it. One can then see whether such a modified gene will still be accurately transcribed. These studies (Grosveld et al., 1982; Dierks et al., 1983) showed that the first 109 base pairs preceding the cap site were sufficient for the correct initiation of β-globin gene transcription by RNA polymerase II.

Myers and co-workers (1986) then refined this analysis by cloning the region of a mouse globin gene from 106 base pairs upstream from the start of transcription (−106) through the first 475 base pairs (+475) of the first exon. These clones were subjected to in vitro mutagenesis. In this way, 130 different single base substitutions were introduced into the promoter

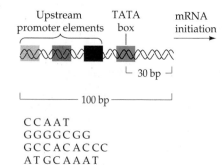

CCAAT
GGGCGG
GCCACACCC
ATGCAAT

FIGURE 5

Typical promoter region for a protein-coding eukaryotic gene. The gene diagrammed here contains a TATA box and three upstream promoter elements. Examples of such upstream elements are shown beneath the diagram. (After Maniatis et al., 1987.)

region of the globin gene. These cloned genes were placed into plasmids containing an enhancer from a gene normally expressed in all tissues. These recombinant plasmids were then introduced by transfection into cultured cells that do not usually produce globin. Would the cells transcribe a truncated globin message (475 bases) from these clones? Figure 6 shows the results. In most cases, mutating a base in the 5′ flanking region did not affect the efficiency of globin gene transcription. But there were three clusters of nucleotides wherein mutations drastically reduced transcription. One cluster was in the Goldberg-Hogness TATA box, another was in the CAAT upstream promoter element, and a third was in the CACCC region, about 95 to 87 base pairs upstream from the cap site.

The CAAT and TATA boxes have been found to be critical elements in numerous eukaryotic promoters (Efstratiadis et al., 1980) (Table 1), but the CACCC sequence is hardly ever seen except in the β-globin gene promoters in several species. In humans, this sequence appears to be critical. A naturally occurring mutation in this sequence causes a total loss of β-globin gene transcription (Orkin and Kazazian, 1984). Two mutations, at positions −78 and −79, have actually increased transcription to three times the wild-type level. It is thought that these changes facilitate the interaction of the promoter with the *trans*-regulatory proteins.

Trans-regulatory proteins binding to promoters

Formation of the transcription initiation complex. Before discussing tissue-specific transcription, we must first discuss the general mechanism of gene transcription. There are three types of RNA polymerases in most eukaryotic cells, and the enzyme responsible for the transcription of messenger RNAs is RNA POLYMERASE II. Purified RNA polymerase II, however, cannot recognize the promoter sequence in the absence of other factors in the nucleus. Without these factors, transcription is initiated randomly. Thus, accurate transcription will not occur if cloned genes are incubated with purified RNA polymerase II and nucleotide triphosphates. But if nuclear extracts are added, accurate transcription commences. What are these factors that allow transcription to be initiated? Recent studies (Buratowski et al., 1989; Sopta et al., 1989) have shown that five nuclear proteins are necessary for the proper initiation of transcription by RNA polymerase II (Figure 7). In the first step, TFIID, perhaps in association with TFIIA, recognizes and binds to the TATA box. This was shown by DNase protection experiments wherein TFIID was shown specifically to protect the TATA region. However, this complex did not appear to be

FIGURE 6

The effect of specific point mutations in the mouse β-globin promoter on the ability of that promoter to initiate transcription. Each line represents the transcription level of a mutant promoter relative to the transcription level of a wild-type globin promoter assayed concurrently. The black dots represent nucleotides for which no mutation had been generated. Below the histogram is a diagram showing the position of the TATA box and the two upstream promoter elements of the mouse β-globin gene. (From Myers et al., 1986.)

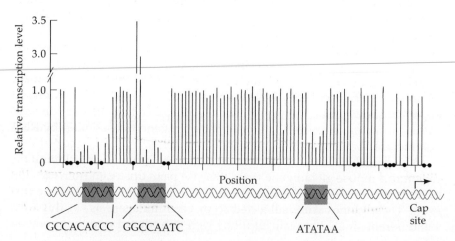

TABLE 1
Basic promoter elements common to several genes

Gene	CAAT region[a]		TATA region[a]	
Histone				
Sea urchin H2A	GGACAATTG	(−85)	TATAAAA	(−34)
Sea urchin H2B	GACCAATGA	(−92)	TATAAAA	(−26)
Sea urchin H3	GACCAATCA	(−75)	TATAAAT	(−30)
Drosophila H2A	AGTCAATTC		TATAAAT	
Drosophila H3	CGTCAAATG		TATAAGT	
Globin				
Mouse α	AGCCAATGA	(−88)	CATATAA	(−29)
Human α2	AGCCAATGA	(−70)	CATAAAC	(−28)
Collagen				
Chick α2 type 1	GCCCATTGC	(−78)	TATAAAT	
Insulin				
Human	GGCCAGGCG	(−73)	TATAAAG	(−29)
Rat I	GGCCAAACG	(−78)	TATAAAG	(−30)
Ovalbumin	GGTCAAACT	(−74)	TATATAT	(−31)
Conalbumin	GGACAAACA	(−81)	TATAAAA	(−30)
Silk fibroin	GTACAAATA	(−93)	TATAAAA	(−29)

Source: Efstratiadis et al. (1980); Vogeli et al. (1981).
[a]Numbers in parentheses correspond to the position upstream from the point where transcription is initiated.

stable without the addition of TFIIA. Next, TFIIB binds to the complex. DNase protection experiments suggest that this protein may bind to only one strand of DNA and extend past the transcription initiation site. RNA polymerase will not bind to the promoter without TFIIB and TFIID being present. It is possible that TFIIB acts as a bridge between TFIID and RNA polymerase II. The next steps involve the formation of the first phosphodiester bond of RNA. TFIIE, a DNA-dependent ATPase, appears to be essential for generating the energy for transcription to occur (Bunick et al., 1982; Sawadogo and Roeder, 1984). TFIIF (also called the RAP3/74 protein) also has enzymatic activity and is needed to unwind the DNA helix.

General and specific promoter-binding proteins. There are other *trans*-regulatory proteins that bind to the promoters of many genes. These proteins must interact directly or indirectly with the transcription initiation apparatus on the TATA box in order to regulate transcription. The nuclear protein CTF binds to the CAAT upstream promoter element and is needed to obtain transcription from several CAAT-containing promoters. The amino terminus of this protein is sufficient to recognize the CAAT element but needs the proline-rich carboxyl end to activate transcription (Mermod et al., 1989). The Sp1 protein binds specifically to the GGGCGG through a zinc-finger region on its carboxyl end but regulates transcriptional activity via its amino terminus (Dynan and Tjian, 1985; Kadonaga et al., 1988). It is probable that these factors are found in all cells, and so cannot regulate differential gene expression. However, they may be involved with the interaction between the promoter region and the enhancer region in ways that do result in the differential transcription of particular genes in particular cells.

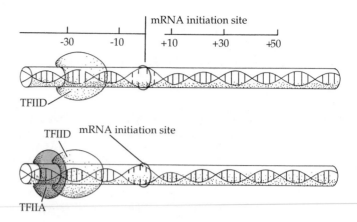

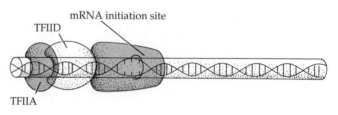

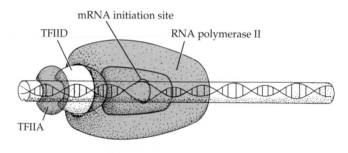

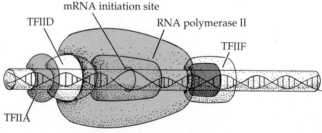

FIGURE 7

The formation of the active eukaryotic initiation complex. The diagrams represent the complexes formed on the TATA box by four transcription factors and RNA polymerase II. (After Buratowski et al., 1989.)

In some cells, the same *cis*-regulatory element can be bound by more than one *trans*-regulatory factor. For example, the CAAT element can be recognized by different proteins in different cells. In some cases, these interactions stimulate transcription, as when the liver-specific nuclear protein C/EBP binds to CAAT elements to activate the transcription of liver-specific genes (Friedman et al., 1989). One of the liver-specific genes activated by the C/EBP protein is blood clotting factor IX. Mutations in this clotting factor gene cause hemophilia B. In some patients, the cause of this disease has been traced to mutations in the C/EBP promoter region that prevent the binding of C/EBP to this gene (Crossley and Brownlee, 1990). The sperm-specific H2B histone genes are also regulated by proteins

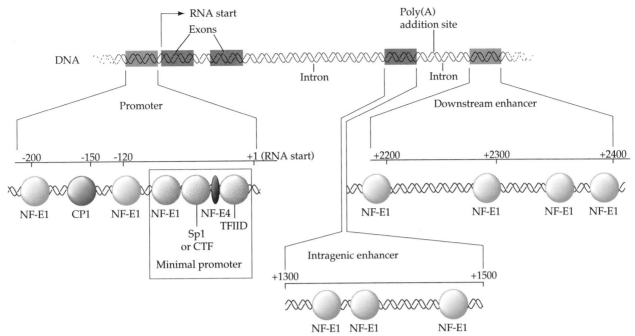

FIGURE 8

Schematic summary of the regulatory regions flanking the human β-globin gene. The promoter region contains a "minimal promoter" plus three sites used in erythroid-specific gene expression. The enhancer 3′ to the gene is important in the temporal expression of the β-globin gene. NF-E1 and NF-E4 are erythroid-specific promoter-binding factors. (After deBoer et al., 1988.)

that bind to the promoter CAAT element. These genes are turned off in most sea urchin cells by a repressor protein that binds to the same CAAT site and displaces the CAAT-binding protein that would otherwise activate transcription (Barberis et al., 1987).

In any developmentally regulated promoter, there appears to be both ubiquitous and tissue-specific promoter elements. In the β-globin gene, for instance, the CAAT element is probably recognized by a ubiquitous *trans*-regulatory factor, while the CAC element is recognized by an erythroid-specific protein (Figure 8; Mantovani, 1988). Moreover, at 120 and 200 base pairs upstream from the point of transcription initiation, there are two promoter sites that bind the erythroid-specific protein NF-E1. Between them, at −150, is a region of DNA that binds the ubiquitous nuclear protein CP1. DeBoer and his colleagues (1988) have shown that β-globin synthesis is inducible in cultured blood cells only if the −150 site is present along with one or the other NF-E1-binding sites. No one site can direct transcription alone. Thus, there appears to be cooperation between the specific and the ubiquitous *trans*-regulatory proteins.

Enhancer structure and function

In addition to the promoters, another type of *cis*-regulatory region can provide a mechanism for differential gene transcription. These DNA sequences are called ENHANCERS. Although they were first discovered in viruses, they are now known to exist in all eukaryotic cells. One of the first cellular enhancers found was seen to control the cell specificity of immunoglobulin gene transcription. Gillies and his co-workers (1983) transfected a cloned immunoglobulin heavy chain gene into cultured B

FIGURE 9

Tissue specificity of enhancer element effect. The immunoglobulin heavy chain gene was isolated and cloned from an IgG-producing myeloma cell line. In these cells, the VDJ segments had been translocated (as discussed in Chapter 10) to a region near the Cγ locus. Some of these clones were kept intact, while in others various regions of the intron between the VDJ and the Cγ exons were excised by restriction enzymes. DNA from the resulting clones was transfected into myeloma cells that had lost their own immunoglobulin genes. The accumulated mRNA from the transfected cells was isolated and separated by electrophoresis on a polyacrylamide gel, along with messenger RNA from the untransformed mouse fibroblast and from the original myeloma cell. The RNA was blotted onto nitrocellulose paper and hybridized with a radioactive restriction enzyme fragment of the Cγ region. If the Cγ region was being transcribed from these cloned genes, the radioactive probe should bind to it. The probe detected Cγ message only from those clones that contained a certain region of the intron (indicated by the dot). When transfected into mouse fibroblast cells, however, even the entire cloned gene would not transcribe Cγ mRNA. (Protocol and data of Gillies et al., 1983.)

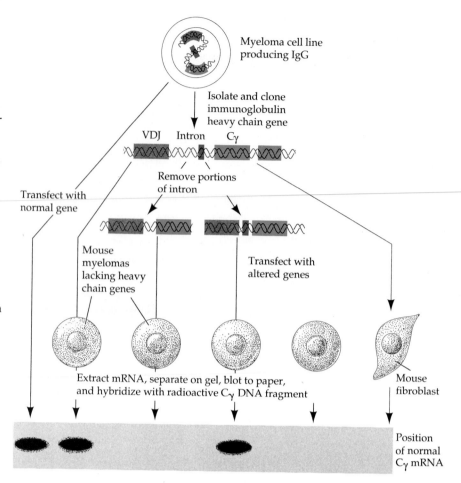

lymphocyte tumor cells that had lost the ability to make their own heavy chain. These transfected myeloma cells were then able to synthesize the heavy chain encoded by the incorporated gene. However, if Gillies and colleagues added the same gene, but without a small region of the intron between the variable and the constant regions, they observed very little transcription of the gene. There was a region within the intron necessary for transcription (Figure 9).

This requirement for the enhancer sequence would explain why one does not see random transcription from all the promoter sequences in the variable-region genes, for the promoter must be brought near the enhancer in order to function. This happens during gene rearrangement. During class switching, the enhancer region remains by the VDJ piece (Figure 10). Enhancers also explain tissue-specific transcription: the cloned immunoglobulin genes are *not* transcribed when inserted into the nuclei of cells other than B lymphocytes (Gillies et al., 1983; Banerji et al., 1983). Moreover, when the enhancer region of the immunoglobulin heavy chain is inserted into a cloned gene for β-globin, it stimulates the transcription of that hemoglobin gene over 100-fold when it is inserted into a myeloma cell. Thus, we see a specific region of DNA that stimulates transcription of nearby genes in a cell-specific fashion.

Temporally specific enhancers

Enhancer sequences can regulate the timing of gene expression, the cell specificity of gene expression, or both. One of the most striking temporal

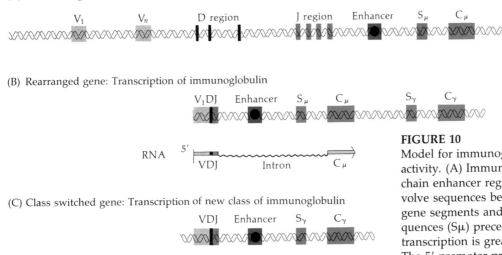

(A) Germ–line gene: No transcription

V_1 V_n D region J region Enhancer S_μ C_μ S_γ C_γ

(B) Rearranged gene: Transcription of immunoglobulin

V_1DJ Enhancer S_μ C_μ S_γ C_γ

RNA 5′ VDJ Intron C_μ

(C) Class switched gene: Transcription of new class of immunoglobulin

VDJ Enhancer S_γ C_γ

RNA 5′ VDJ Intron C_γ

FIGURE 10
Model for immunoglobulin enhancer activity. (A) Immunoglobulin heavy chain enhancer region appears to involve sequences between the J region gene segments and the switch sequences ($S\mu$) preceding $C\mu$. If deleted, transcription is greatly diminished. The 5′ promoter precedes each of the V region gene segments and is originally very distant from the enhancer. (B) The VDJ gene rearrangement brings one promoter near the enhancer and allows transcription to take place. (C) During class switching, the enhancer stays with the VDJ segments as they are placed near a new constant region ($C\gamma$).

switches in gene activity is the midblastula transition (MBT) seen in frog embryos. Krieg and Melton (1987) showed that certain genes contained an enhancer that was specifically activated at this time. As shown in Figure 35 of Chapter 10, genes can be cloned that are activated at this transitional stage. Krieg and Melton fused the first 500 base pairs of the 5′ flanking region of one of these genes onto the *Xenopus* β-globin gene. The β-globin gene is usually not transcribed when injected into fertilized *Xenopus* eggs until late in development. If the 5′ upstream sequences of the MBT-activated gene contained such a temporal enhancer, one would expect to see globin message starting at the time of the midblastula transition. Northern blots showed that this is indeed the case. No globin transcripts were seen until the time of the midblastula transition, at which time new globin messages could be detected. Further deletions have located the midblastula enhancer in a 74-base-pair sequence about 700 base pairs upstream of the normal gene promoter.

The human β-globin gene appears to have its own temporally specific enhancer located between 600 and 900 base pairs downstream from the AATAAA site. If a DNA fragment containing this region is placed near a human γ-globin gene, this fetal gene can be activated at the time the β-globin gene is usually expressed (Trudel and Constantini, 1987). As shown on the right-hand side of Figure 8, this enhancer site contains four binding regions for the erythroid-specific NF-E1 transcription factor (deBoer et al., 1988).

Tissue-specific enhancers

β-Globin enhancers. In addition to the 3′ enhancer that appears to regulate the *temporal* expression of the β-globin gene, there are also two other enhancers that regulate the *tissue-specificity* of β-globin gene expression. One enhancer is actually located within the third exon of the β-globin gene itself (Behringer et al., 1987) and allows the transcription of β-globin genes solely in erythroid cells. The second enhancer is actually a "master" enhancer controlling the entire β-globin gene family (ε-, γ-, β-, and δ-globins) on human chromosome 11 (Grosveld et al., 1987). Deletion of this

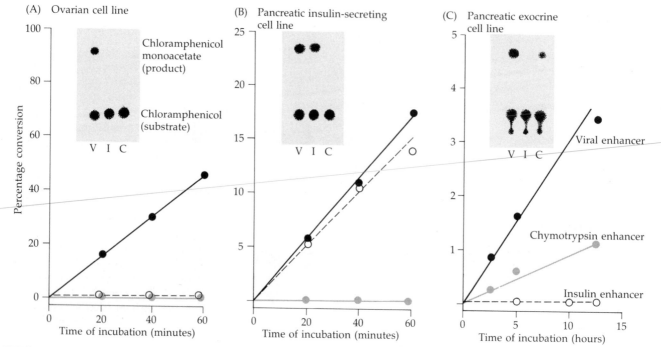

FIGURE 11

Tissue specificity of pancreatic gene enhancers. The 5′ flanking regions of the insulin gene (I) and the chymotrypsin gene (C) were separately inserted next to the gene for bacterial chloramphenicol acetyltransferase (CAT). As a positive control, the enhancer from Rous sarcoma virus (V), which appears to operate in all cell types, was also placed by the CAT gene. The three clones were transfected into three types of cells: (A) ovarian cells (which synthesize neither chymotrypsin nor insulin; (B) endocrine pancreatic cells (which normally synthesize insulin); and (C) exocrine pancreatic cells (which normally synthesize chymotrypsin). The CAT activity is assayed on the cell lysates. The inserts show typical autoradiograms of the CAT assay wherein radioactive chloramphenicol (the substrate of the CAT reaction) can be separated from chloramphenicol monoacetate (the product of the CAT reaction) by chromatography. (After Walker et al., 1983.)

enhancer* region causes the silencing of all these genes. Conversely, if this enhancer is placed adjacent to genes that are not usually expressed in red blood cells (such as the T-cell-specific *thy-1* gene) and then transfected into erythroid precursor cells, these new proteins become expressed in red blood cells. This effect is specific for red blood cell precursors, since only they would have the appropriate *trans*-regulatory factors to bind to this region (Blom van Assendelft et al., 1989). It is possible that the "master" enhancer functions to "open up" this entire region to *trans*-regulating factors.

Pancreatic enhancers. Enhancers appear to regulate the temporal and tissue-specific expression of all differentially regulated genes. In some cases, genes active in adjacent cell types are seen to have different enhancers. In the pancreas, for instance, the exocrine protein genes (chymotrypsin, amylase, and trypsin) have enhancers different from that of the endocrine protein insulin. Both these enhancers lie in the 5′ flanking sequences of their respective genes. Walker and colleagues (1983) placed these flanking regions onto the gene for bacterial chloramphenicol acetyltransferase (CAT), a gene whose enzyme product is not found in mammalian cells. CAT activity is easy to assay in mammalian cells and is used as a "reporter gene" to tell investigators whether a particular enhancer is functioning. They then transfected these hybrid genes into (a) ovary cells, (b) an insulin-secreting cell line, and (c) an exocrine cell line, and measured the activity of the marker enzyme in each of these cells. As shown in Figure 11, neither enhancer sequence caused the enzyme to be made in the ovarian cells. In the insulin-secreting cell, however, the 5′ flanking region of the insulin gene enabled the chloramphenicol acetyltransferase gene to be expressed, but the 5′ flanking region of the chymotrypsin gene did not. Conversely, when the clones were placed into the exocrine pancreatic cell line, the chymotrypsin 5′ flanking sequence allowed chloram-

*This major enhancer is also called the locus activating region (LAR) or the dominant control region (DCR). This far-upsteam site may restructure the chromatin in this region to make it more accessible to *trans*-regulatory factors.

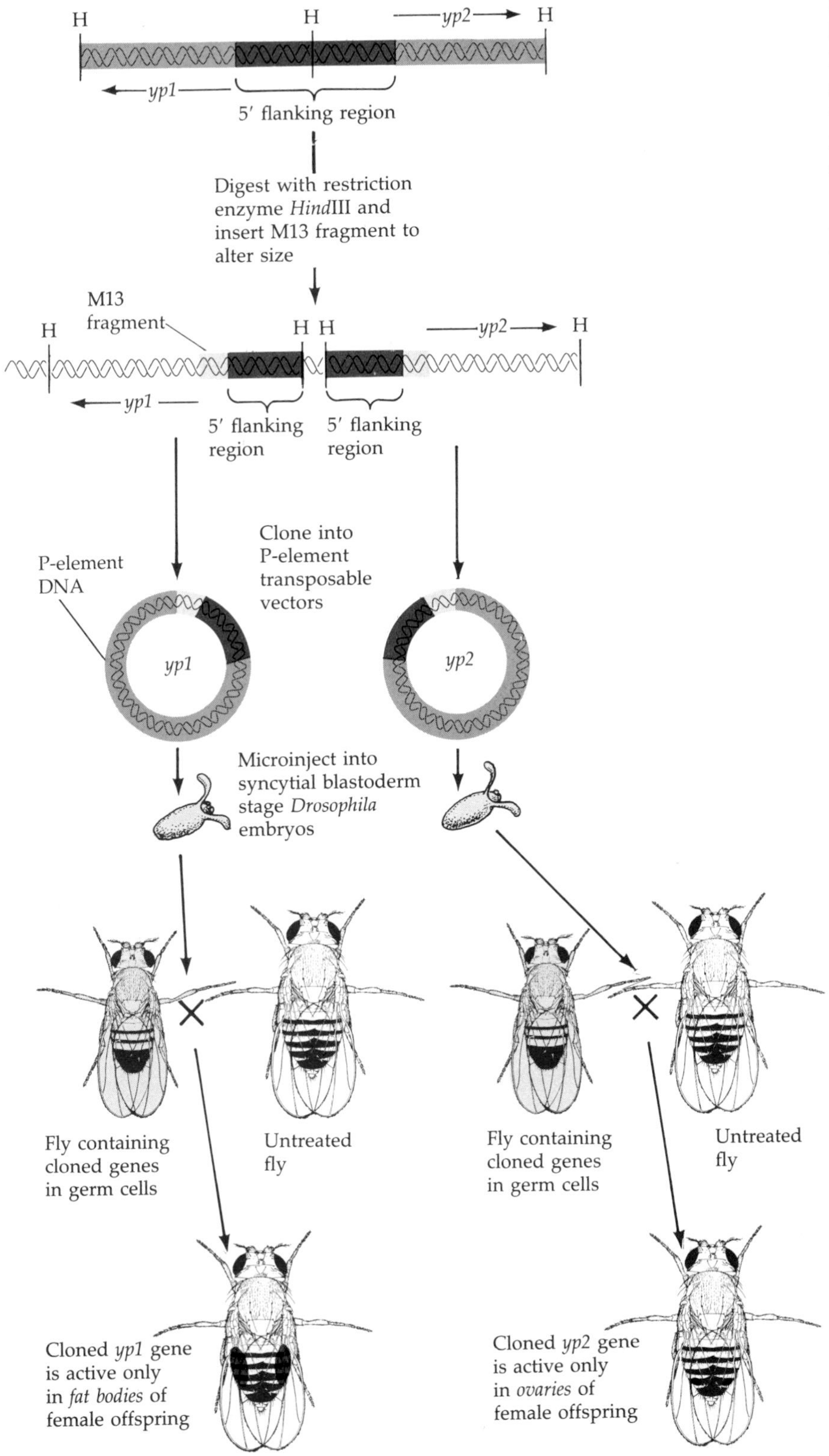

FIGURE 12

Protocol for defining two separate enhancer elements controlling tissue specificity of yolk protein transcription. Yolk protein genes and their 5' flanking elements are separated from each other by restriction enzymes (*Hind*II, represented as H). The genes are then modified by the addition of extra DNA into an exon and cloned into a transposable P-element. The recombinant P-element is microinjected into a syncytial blastoderm embryo, where it becomes incorporated into some of the nuclei. The resulting fly can contain the P-element in its germ line, and it is mated to an untreated fly. The offspring that received the P-element genes can be identified by their eye color (because of a genetic marker in the P-element) and the organs can be tested for the presence of the *yp1* and *yp2* transcripts.

phenicol acetyltransferase expression, while the insulin enhancer did not. Thus, the expression of genes in exocrine and endocrine cells of the pancreas appears to be controlled by different enhancers.

Yolk protein enhancers. It is not unusual to find the same protein being synthesized in more than one specialized cell type. This is the case for the

Drosophila yolk protein, which is made in adult female flies. Two of these yolk protein genes, *yp1* and *yp2*, are adjacent to each other but are transcribed in opposite directions (Figure 12), and both genes are transcribed in the ovary and in fat bodies. These genes can be separated from one another by a restriction enzyme that cleaves them in their common 5′ flanking region (Garabedian et al., 1985). Each of these genes can then be cloned separately into a transposable P-element vector that can insert into the genome of *Drosophila* eggs.

Before transforming the fly DNA with these clones, however, an extra piece of plasmid DNA was inserted into the cloned *yp* genes so that they could be distinguished from the endogenous gene products. Garabedian and co-workers found that, like the endogenous messages, the transcripts from these clones were sex-specific (they occur only in females) and stage-specific (found only in adults). However, the *yp2* mRNA was found only in the ovary and the *yp1* mRNA was found only in the fat bodies. It appears that the original yolk protein genes had had two enhancers at their 5′ ends, one allowing transcription in the ovary, one allowing transcription in the fat bodies (Figure 13). Thus, both genes were normally transcribed in both cell types. The restriction enzyme, however, cut between these two sites, giving *yp1* the fat body-specific enhancer and *yp2* the ovary-specific enhancer. Garabedian and co-workers (1986) then found that any gene placed next to the fat body enhancer could be transcribed specifically in the fat body, and any gene placed adjacent to the ovary enhancer could be injected into a fly embryo and expressed specifically in the ovaries.

Pituitary enhancer: The role of Pit-1 (GHF1) Just as in promoters, there are cell-specific *trans*-regulatory factors that bind to enhancers. One well characterized enhancer-binding protein is Pit-1 (also called GHF1), a DNA-binding protein only found in anterior pituitary cells. DNA footprinting experiments have shown that Pit-1 binds to two sites in the upstream promoter element of the human growth hormone gene. When cloned growth hormone genes are placed in nuclear extracts of nonpituitary cells, these genes are not transcribed, whereas transcription will take place if the growth hormone genes are placed into nuclear extracts of anterior pituitary cells. Moreover, the addition of Pit-1 to the nonpituitary nuclear extract will enable it to transcribe the growth hormone gene (Bodner and Karin, 1987). It appears, then, that the specific expression of the growth hormone gene in the anterior pituitary is mediated by a tissue-specific promoter binding protein. Conversely, when foreign genes are cloned next

FIGURE 13
Location of yolk protein enhancer elements. (A) Expression of the bacterial β-galactosidase gene in *Drosophila* ovaries when fused to region −44 to −350 upstream of the *yp2* gene. Those cells expressing β-galactosidase stain darkly. (B) Summary of data on yolk protein enhancers in *Drosophila*. Recent evidence suggests a third element, located between −100 and +1 of the *yp2* gene, that elevates the transcription of both genes. (Photograph courtesy of P. Wensink.)

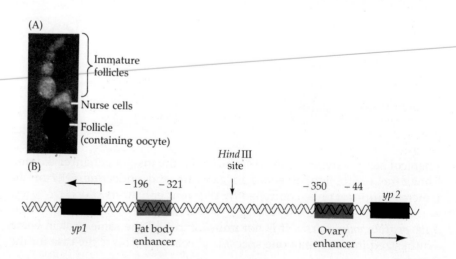

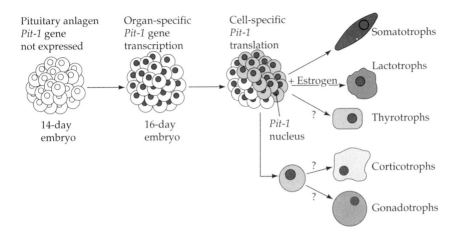

Pituitary anlagen
Pit-1 gene
not expressed

Organ-specific
Pit-1 gene
transcription

Cell-specific
Pit-1
translation

Somatotrophs

Lactotrophs

+ Estrogen

Thyrotrophs

Pit-1
nucleus

Corticotrophs

Gonadotrophs

14-day
embryo

16-day
embryo

FIGURE 14
Determination of cell type by combinations of *trans*-regulatory proteins. The mRNA for the *Pit-1* transcription factor is transcribed in all cell types destined to reside in the anterior pituitary. However, Pit-1 protein is only translated in thyrotroph, somatotroph, and lactotroph cells. Somatotrophs synthesize growth hormone upon the induction of Pit-1. Lactotrophs synthesize some prolactin concomitant with *Pit-1* translation but produce significant prolactin only when costimulated with estrogen (and its *trans*-regulatory receptor protein). The mechanism distinguishing thryroid-stimulating hormone transcription is not yet known. (After Simmons et al., 1990.)

to the Pit-1 binding sequences and are incorporated into the mouse genome, these genes show pituitary-specific transcription (Behringer et al., 1988).

The Pit-1 enhancer-binding protein is itself developmentally regulated (Simmons et al., 1990; Dollé et al., 1990). Transcription of the mouse *Pit-1* gene is detectable within a day following the histological appearance of Rathke's pouch, the anlage of the anterior pituitary, but translation of this mRNA into protein does not occur for another 2–3 days. Growth hormone mRNA is first detected when the *Pit-1* gene is translated. Interestingly, in two types of anterior pituitary cells (the corticotrophs and gonadotrophs which synthesize adrenocorcicotropic hormone [ACTH] and gonadotropins, respectively), *Pit-1* mRNA is made but remains untranslated (Figure 14). Only in the pituitary cells of somatotrophs (growth hormone-producing), lactotrophs (prolactin producing), and thyrotrophs (which synthesize thyroid stimulating hormone) is the *Pit-1* message translated into the DNA-binding nuclear protein. The growth hormone gene appears to respond directly to the presence of Pit-1. The prolactin gene, which activates milk production during pregnancy, is maximally stimulated by Pit-1 in the presence of estrogen. The estrogen is needed to activate another region of the enhancer. In this way, combinations of enhancer-binding factors can fine tune organ-specific transcription into cell-type specific transcription. The Pit-1 enhancer-binding protein not only mediates the transcription of the differentiated products of these cells, but it is necessary for the formation of the anterior pituitary cells in Rathke's pouch. The *dwarf* mutation in mice is caused by a mutation in Pit-1 protein. These mice lack the thyrotroph, lactotroph, and somatotroph cells of the anterior pituitary (Li et al., 1990).

Hormone-responsive elements

During development it is not uncommon to see COORDINATE REGULATION of gene expression. This phenomenon occurs when one type of cell expresses several cell-specific proteins at the same time or when a hormone induces the expression of several genes in different cell types. Accumulating evidence indicates that, at least in some instances, coordinate gene regulation is brought about when different structural genes are linked to the same or similar enhancers. In the pancreas, the enhancers for ten exocrine proteins share a 20-base pair consensus sequence, suggesting that these similar sequences play a role in activating these genes in the exocrine cells of the pancreas (Boulet et al., 1986).

Steroid hormones are known to increase the transcription of several specific genes. The hormone binds to a hormone-specific receptor protein, converting that receptor into a conformation that is able to enter the nucleus and bind particular DNA sequences (Miesfeld et al., 1986; Green and Chambon, 1988). The DNA sequences capable of binding nuclear hormone receptors are called HORMONE-RESPONSIVE ELEMENTS and they can be either in enhancers or in promoters. One set of steroids includes the glucocorticoid hormones (cortisone, hydrocortisone, and the synthetic hormone dexamethasone), which regulate protein metabolism and reduce tissue inflammation. They appear to act in all cell types. Several genes, including mammalian metallothionein genes and certain viral tumor genes, are activated by glucocorticoids, so it was reasonable to hypothesize that the glucocorticoid responsive element for the hormone-bound receptor lay near those genes. One way of determining the glucocorticoid responsive element of a gene is to perform a binding competition assay (Figure 15). One first fixes double-stranded calf thymus DNA (which is presumed to have many glucocorticoid responsive elements) to cellulose filters and adds the radioactive glucocorticoid-receptor protein complex. The complex is allowed to bind to the DNA, and the bound radioactivity is counted (Figure 15A). Next, the same assay is run with an excess of some other DNA in solution (Figure 15B). If this competitor DNA has the glucocorticoid responsive element, it will compete for the radioactive complex and less radioactivity will be bound to the calf thymus DNA (Pfahl, 1982). By using various restriction enzyme-derived fragments of DNA, Karin and co-workers (1984) found that the sequence capable of being bound by the glucocorticoid responsive element was TGGTACAAATGTTCT. These glucocorticoid binding sequences were shown to act as enhancers: when the glucocorticoid response element was ligated onto genes that are not usually hormone dependent, those genes gained the responsiveness to the glucocorticoid (Figure 16; Chandler et al., 1983).

The steroid responsive elements are extremely similar to one another and are recognized by closely related proteins. These steroid receptor

FIGURE 15
The protocol for determining the site of the glucocorticoid enhancer. See text for details.

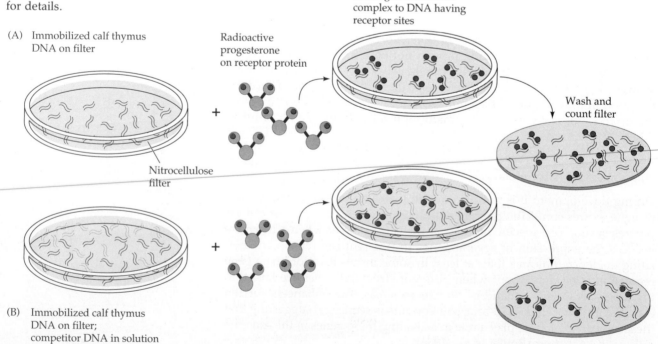

(A) Immobilized calf thymus DNA on filter

Radioactive progesterone on receptor protein

Binding of radioactive complex to DNA having receptor sites

Wash and count filter

Nitrocellulose filter

(B) Immobilized calf thymus DNA on filter; competitor DNA in solution

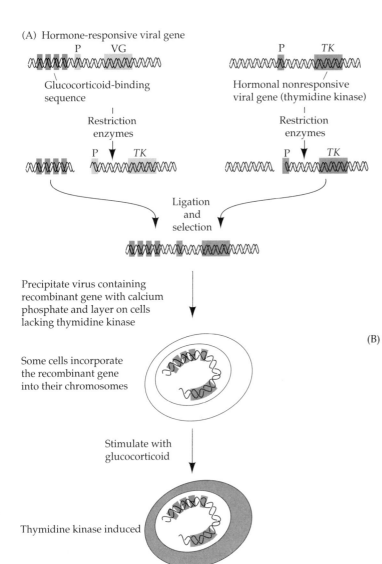

(A) Hormone-responsive viral gene

Glucocorticoid-binding sequence

Hormonal nonresponsive viral gene (thymidine kinase)

Restriction enzymes

Restriction enzymes

Ligation and selection

Precipitate virus containing recombinant gene with calcium phosphate and layer on cells lacking thymidine kinase

Some cells incorporate the recombinant gene into their chromosomes

Stimulate with glucocorticoid

Thymidine kinase induced

FIGURE 16

Test for the glucocorticoid enhancer sequence. (A) A recombinant virus containing the glucocorticoid-responsive enhancer of the mouse mammary tumor virus and the *thymidine kinase* gene of *Herpes simplex* virus can integrate into the genome of a cell lacking *thymidine kinase* gene. Upon treatment with glucocorticoids, the recombinant gene transcribes viral thymidine kinase. P, promoter region; *TK, thymidine kinase* gene; VG, viral gene of mouse mammary tumor virus. (B) Electron micrograph of glucocorticoid enhancer elements, showing hormone receptor bound to that region of the gene. (A modified from Chandler, 1983; B from Payvar et al., 1983; photograph courtesy of K. R. Yamamoto.)

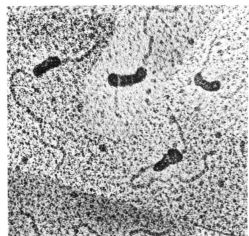

(B)

proteins each contain three functional domains: (1) a HORMONE-BINDING DOMAIN, (2) a DNA-BINDING DOMAIN that recognizes the hormone-responsive element, and (3) a TRANSACTIVATION DOMAIN, which is involved in mediating the signal to initiate transcription. As shown in Figure 17, these functions can overlap. This is especially true in activating transcription, where all the domains appear to have some role (Beato, 1989). For transcriptional activation to occur, the receptor has to enter the nucleus and dimerize with a similar hormone-binding protein. The binding of hormone to the first domain appears to be necessary both for dimerization and for the ability of the DNA binding region to recognize the hormone-responsive element (Kumar et al., 1987).

The binding of the receptor protein to the hormone-responsive enhancer element is accomplished by a zinc finger region in the DNA binding domain (Green et al., 1988). When chimeric proteins are made wherein the zinc finger domain of the estrogen receptor substitutes for the same region of a glucocorticoid receptor, the protein recognizes the DNA containing estrogen-responsive elements and causes the gene to be responsive to glucocorticoids. The critical amino acids appear to reside at the "knuckle" of the zinc finger (Danielsen et al., 1989; Umesono and Evans, 1989). Changing as few as two amino acids at the knuckle of the zinc

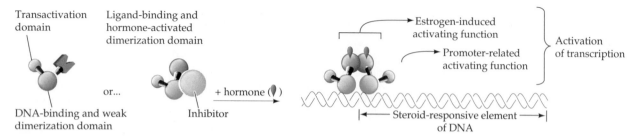

Transactivation domain

Ligand-binding and hormone-activated dimerization domain

DNA-binding and weak dimerization domain

or...

Inhibitor

+ hormone (🔻)

Estrogen-induced activating function

Promoter-related activating function

Activation of transcription

Steroid-responsive element of DNA

FIGURE 17

Generalized structure of a steroid hormone binding protein. The functions ascribed to each fraction have been determined by analyzing the effects of mutations in each of these regions and by making chimeric protein molecules having regions derived from different receptor proteins. A similar structure is seen in the retinoic acid receptor and the thyroid hormone receptor. (After Beato, 1989.)

FIGURE 18

Cooperativity between the glucocorticoid responsive element and the CACCC box. Plasmids were constructed containing a glucocorticoid responsive element (colored box) and (A) a wild-type CACCC box, (B) a mutated CACCC box, or (C) an unrelated DNA sequence. These plasmids were separately transfected into cultured human cells and their inducibility was tested by adding the glucocorticoid dexamethasone to the cells' medium. The result shows that, while the presence of the glucocorticoid responsive element was sufficient for some inducibility, the combination of glucocorticoid responsive element and a functional CACCC box gave three times as much stimulation. (After Schüle et al., 1988.)

finger region will change the specificity of the binding protein. Thus, while these DNA-binding domains of the hormone receptor proteins are very similar, they can distinguish subtle differences in the enhancer sequences. For example, the (palindromic) sequence 5′–GGTCACTGTGACC–3′ is a strong estrogen-responsive enhancer element that will bind the estrogen-containing receptor protein. Two symmetrical mutations in this sequence, making it 5′–GGACACTGTGTCC–3′, will convert this DNA into a glucocorticoid-responsive enhancer (Martinez et al., 1987; Klock et al., 1987). Given the similarities among hormone receptor proteins and the similarities among hormone-responsive elements, it is very likely that each steroid hormone mediates its transcriptional activation by the same general mechanism.

But what is this mechanism? The way in which the DNA-bound hormone receptor stimulates the initiation of transcription is largely unknown. It is generally thought that these receptors are, in essence, hormone-responsive enhancer-binding proteins, and, like other enhancer-binding proteins, the hormone receptor proteins interact with other *trans*-regulatory proteins on the DNA. This interaction is seen to occur when the glucocorticoid receptor binds to DNA near a CCAAT box (which binds CTF) or near a CACCC box (which binds SP1). First, removal of a CACCC element from the promoter of the tryptophan oxygenase gene abolishes its ability to be induced by glucocorticoids (Danesch et al., 1987). Second, placing the marker gene CAT on a plasmid containing the glucocorticoid responsive element and wild-type and mutant CACCC elements shows that both wild-type elements are needed for maximal glucocorticoid enhancement (Figure 18; Schüle et al., 1988; Strähle et al., 1988). Thus, the hormone receptor protein interacts synergistically with other *trans*-regulatory proteins to stimulate the transcription of hormonally responsive genes.

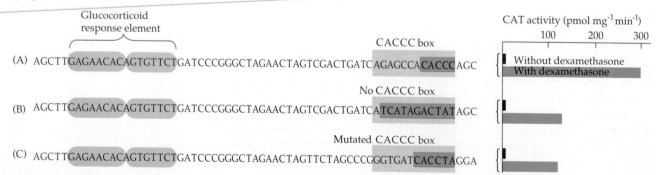

Glucocorticoid response element

CAT activity (pmol mg⁻¹ min⁻¹)

CACCC box

(A) AGCTTGAGAACACAGTGTTCTGATCCCGGGCTAGAACTAGTCGACTGATCAGAGCCACACCCAGC

Without dexamethasone
With dexamethasone

No CACCC box

(B) AGCTTGAGAACACAGTGTTCTGATCCCGGGCTAGAACTAGTCGACTGATCATCATAGACTATAGC

Mutated CACCC box

(C) AGCTTGAGAACACAGTGTTCTGATCCCGGGCTAGAACTAGTTCTAGCCCGGGTGATCACCTAGGA

Transcription of immunoglobulin light chain genes

One of the best studied cases of transcriptional gene regulation has been that of mouse immunoglobulin genes. Within the past five years, investigators have identified the *cis-* and *trans-*regulatory elements necessary for the B cell-specific transcription of these genes.

Deletion mapping of cloned genes has shown that the promoter of the immunoglobulin light chain contains several critical regions for transcription. One of these upstream promoter elements is called an "octa" box (Bergman et al., 1984; Parslow et al., 1984) because of its eight base pairs, ATTTGCAT. This octa sequence has been found in every light chain immunoglobulin promoter studied. The heavy chain immunoglobulin gene promoters have an "inverted"

octa box, ATGCAAAT (Figure 19). When the octa box is placed upstream from a globin gene in a myeloma cell, the globin gene transcription increased 11–18 times. This increase is seen only in lymphoid cells and has not been observed in fibroblasts (Wirth et al., 1987).

The enhancer sequence of the light chain immunoglobulin genes is located within the first intron between the VJ sequence and the constant region sequence (Queen and Baltimore, 1983; Bergman et al., 1984). When this se-

FIGURE 19

Conserved octa box in promoters of immunoglobulin genes. The heavy chain octa box is the inverted form of the light chain octamer (reading T for A, G for C, and so on in the opposite direction). The number on the right refers to distance from the end of the octa box to the start of the first exon. (From Parslow et al., 1984.)

MOUSE LIGHT CHAIN GENES

		Octa box		Distance from ATG
	TGCCTAGACTGTATCTTGCG	ATTTGCAT	ATTACATTTTCAGTAACCACAA	107
K2	GCTGTGCCTACCCTTTGCTG	ATTTGCAT	GTACCCAAAGCATAGCTTACTG	100
M1738	ATCCTAACTGCTTCTTAATG	ATTTGCAT	ATCCTCACTACATCGCCTTGGG	91
M41	ATCCTAACTGCTTCTTAATA	ATTTGCAT	ACCCTCACTGCATCGCCTTGGG	92
M167	----CAGCACTGACCAATGG	ATTTGCAT	AATGCTCCCTAGGGTCCACTTC	106
T1	GCAATAACTGGTTCCCAATG	ATTTGCAT	GCTCTCACTTCACTGCCTTGGG	97
T2	AGCAACATGAAGACAGTATG	ATTTGCAT	AAGTTTTTCTTTCTTCTAATGT	109
K21C	AAACAGTACATACTCCGCTG	ATTTGCAT	ATGAAATAATTTTATAACAGCC	93
M11	ACTTCCTTATTTGATGACTC	CTTTGCAT	AGATCCCTAGAGGCCAGCACAG	73
λ$_I$	TAAACCTGTAAATGAAAGTA	ATTTGCAT	TACTAGCCCAGCCCAGCCCATA	106
λ$_{II}$	TAAACCTGTAAATGAAAGTA	ATTTGCAT	TACTAGCCCAGCCCAGCCCATA	105

HUMAN LIGHT CHAIN GENES

h101	GCCTGCCCCATCCCCTGCTC	ATTTGCAT	GTTCCCAGAGCACAACCTCCTG	96
h122	GCCTGCCCCATCCCCTGCTG	ATTTGCCT	GTTCCTAGAGCACAGCCCCCTG	102
h100	TCATTCTTGCATCTGTTGAA	ATTTTCAT	TTTCAAAAAAACACAGCCAACT	96

HEAVY CHAIN GENES

V$_H$167	TAATGATATAGCAGAAAGAC	ATGCAAAT	TAGGCCACCCTCATCACATGAA	118
V$_H$105	GTAATGCACTGCTCATGAAT	ATGCAAAT	CACCTGGGTCTATGGCAGTAAA	114
V$_H$111	GTAATGCACTGCTCATGAAT	ATGCAAAT	CACGCAAGTCTTTGGCAGTAAA	108
V$_H$104	GAAGTACCCTGCTCATGAAT	ATGAAAAT	TACCCAAGTCTATGGTAGTAAA	108
V$_H$108	AAAGTCCCCTGCTCATGAAT	ATGCAAAT	TACCGTTCTCTATGTTGGTTAA	109
V$_H$101	TCTCTCAGGAACCTCCCCCA	ATGCAAAG	CAGCCCTCAGGCAGAGGATAAA	85

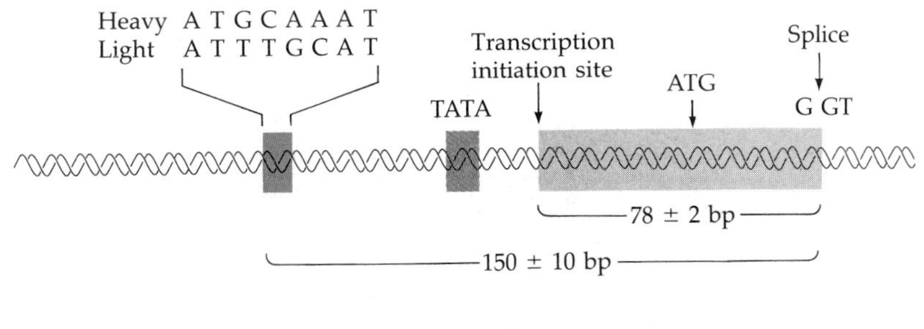

Heavy ATGCAAAT
Light ATTTGCAT

TATA

Transcription initiation site

ATG

Splice
G GT

78 ± 2 bp

150 ± 10 bp

quence is translocated onto cloned globin genes, globin gene transcription can also occur specifically in lymphocytes (Picard and Schaffner, 1984).

These enhancers and promoters are brought into proximity during the gene rearrangements occurring during lymphocyte maturation. Mather and Perry (1982) have shown that nuclear transcripts from rearranged light chain genes (having the promoter and enhancer in proximity) are over 16,000 times more abundant than transcripts from nonrearranged genes. Moreover, the light chain immunoglobulin enhancer stimulates its own promoters 20 times more efficiently than it stimulates the promoters of SV40 virus or the metallothionein gene. This suggests a specific synergistic interaction between these two elements (Garcia et al., 1986). Rearrangement itself is not sufficient for this gene activation, since a rearranged immunoglobulin gene will not actively transcribe when placed in a fibroblast or liver cell. There have to be cell-specific *trans*-regulatory factors.

Two promoter-binding factors were identified by Staudt and co-workers in 1986. They incubated a small piece of DNA containing the octa box in nuclear extracts from various cells. They then ran the resulting products on a gel. If the nuclear extract did not have a protein that bound to this DNA, the small DNA fragment should migrate rapidly through the gel. If a protein *did* bind to this DNA, however, the migration should be impeded. This mobility shift assay (Figure 20) demonstrated that every nucleus had at least one factor capable of binding to the DNA fragment. However, the B-cell lineage (pre-B cells, B cells, and plasma cells) contained, in addition, another factor capable of binding specifically to the octa box DNA. These binding proteins were isolated, and the lymphocyte-specific protein was termed Oct-2 (NF-A2).

Similar assays were used to find a nuclear protein restricted to B-cell lineage cells that binds specifically to the enhancer sequences of light chain immunoglobulin genes (Sen and Baltimore, 1986a; Atchinson and Perry, 1987). One such protein, NF-κB, was found only in mature B cells and plasma cells, and it recognized the sequence 5'–

GGGACTTTCC–3' in the κ light chain enhancer. These cells are the only cells known to synthesize immunoglobulins.* Active NF-κB was found only in mature B cells and activated T cells, and it was not found in any other cell type (including pre-B cells and unactivated T cells). Moreover, when pre-B cells are induced to become B cells, active NF-κB protein appears.

This brings the problem of differential gene activity to another level: what controls the synthesis of the B-cell specific *trans*-regulatory factors such as Oct-2 or NF-κB? (That is, to explain how immunoglobulins are produced in a cell-specific manner, we now have to explain how these nuclear proteins are generated in a cell-specific manner.) It appears that the NF-κB protein is actually present in numerous cell types, but is bound by a 65,000-Da protein called IκB (Inhibitor of kappa). In its bound state, NF-κB cannot bind to DNA. Thus, NF-κB can only recognize the kappa-chain enhancer region in those cells that do not make IκB, i.e., in mature B cells. (Sen and Baltimore, 1986b; Baeuerle and Baltimore, 1988). Needless to say, the next step is to find what regulates IκB and how this gene is activated in all cells but mature B cells.

In addition to the positive regulation of immunoglobulin gene transcription seen in B cells, there is also negative regulation of immunoglobulin production in non-B cells. There appear to be several sites flanking the immunoglobulin genes that are bound by proteins that inhibit the genes' transcription (Calame, 1989). These DNA sites are called SILENCERS. Proteins capable of binding to the silencer regions of immunoglobulin genes have been seen in non-B cells. Thus, transcription may be stimulated or inhibited

*The pre-B cell can make the heavy chain but not the light chain of the immunoglobulin protein. One might wonder what the nonspecific octa box-binding protein is doing in non-B cells. It appears that histone H2B gene has an octa box in its enhancer and that the binding of the nonspecific octa box-binding protein confers on it a temporal specificity, thereby causing the gene to be transcribed during the S phase of the cell cycle (Fletcher et al., 1987). Why the Oct-1 protein does not activate the immunoglobulin genes in every cell is still not known.

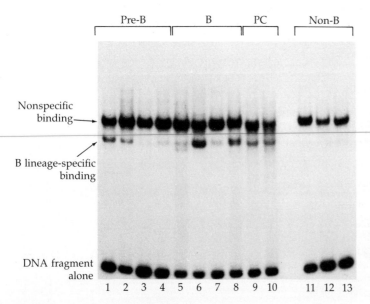

FIGURE 20
Electrophoretic mobility shift assay. Nuclear extracts from B-cell lineage cells (pre-B, B, and plasma cell) and non-B cells (cervical, fibroblast, and red blood cell precursor cell lines) were mixed with a small segment of DNA containing the octamer. After incubation, the mixtures were electrophoretically separated on a gel, blotted to nitrocellulose paper, and hybridized with radioactive DNA complementary to the octamer sequence. Without any binding, the octamer-containing fragment rapidly moves to the bottom of the gel. All nuclei contain a protein that nonspecifically binds to the octamer and greatly impedes migration. The nuclei from B-cell lineage cells, however, also contain another protein that inhibits migration by binding to the octamer sequence. (From Staudt et al., 1986; photograph courtesy of D. Baltimore.)

by *trans*-regulatory proteins depending upon a cell's developmental history.

The analysis of cell-specific transcription of immunoglobulin light chain genes has moved, then, to a level beyond looking at the terminal differentiated cell product. We know that the transcription of this immunoglobulin gene depends upon the prior activity of two nuclear proteins, Oct-2 and NF-κB, whose activity is seen only in the B cell lineage. It remains now to determine how these two genes become differentially expressed in the developing lymphocyte.

Analysis of the immunoglobulin enhancer has explained other biological problems as well. As we will discuss in Chapter 20, lymphocyte tumors are produced when a normal growth factor gene is translocated into the region of an immunoglobulin enhancer. This proximity to the immunoglobulin enhancer causes the growth factor gene to be transcribed at the level and times of antibodies. Another problem involves the replication of the AIDS virus in T cells. When HIV infects a T cell, it induces the formation of active NF-κB (Sen and Baltimore, 1986b; Lenardo and Baltimore, 1989). The production of active NF-κB stimulates the T cell genes that have enhancer sites for this protein. These genes include the gene for T cell growth factor and the gene for the T cell growth factor receptor. Thus, the T cell is stimulated to proliferate. At the same time, HIV also has NF-κB enhancer elements that enable it to transcribe its products rapidly, too. It is obvious that NF-κB plays a very important role in normal and altered lymphocyte development.

Cooperation between promoter and enhancer elements

As mentioned above, the immunoglobulin enhancer works best with its own promoter. This has been seen in several other genes as well. One of the most interesting promoter-enhancer interactions is postulated to occur in developing erythroid cells. In chick erythroid cells, the transcription factor NF-E4 is found only in definitive red blood cells. This protein binds to a DNA sequence in the β-globin gene promoter. It also binds to two *trans*-regulatory *proteins* that are bound to the β-globin gene *enhancer*. Without this factor, the β-globin gene does not form a stable structure. But once NF-E4 binds the promoter and enhancer regions together, then the gene is ready for transcription (Figure 21; Gallarda et al., 1989). The contact between enhancer and promoter may be critical for the transcriptional activation of developmentally regulated genes.

This interaction between promoter and enhancer may be the mechanism for the switch from embryonic to adult globin genes during development. Choi and Engel (1988) found that in the absence of the β-globin gene, ε-globin genes were continually active, even in the adult chick. Furthermore, the replacement of the ε-globin promoter by the β-globin promoter resulted in the expression of ε-globin only in the adult stages. These results suggested that the β-globin promoter contained a region that was critical for stage-specific regulation. The data also suggested that there was competition between the β and ε promoters for the same enhancer. Analyzing the proteins that bind to the enhancer and promoters, Gallarda and colleagues (1989) have proposed that in *embryonic* red blood cells, the interaction between the ε-globin promoter and the β-globin enhancer is stabilized by ε-globin promoter-binding factors that can interact with the β-globin enhancer-binding proteins. The β-globin promoter lacks such binding proteins and thus cannot interact with the enhancer. Therefore, the ε-globin gene is transcribed. In *immature* red blood cells, the developing erythrocyte transcribes two proteins that stabilize the interactions between the β-globin promoter and its enhancer. As a result, the enhancer is thermodynamically more stable when joined with the β-gene promoter than with the ε-gene promoter. Transcription of ε-globin ceases. However, the transcription of β-globin does not begin at high levels until NF-E4 is produced in the *mature* red blood cell. This protein completes the stabilization of the enhancer with the β-globin gene pro-

FIGURE 21

Model for promoter-enhancer interaction in regulating the transcription of the chick β-globin gene. The globin gene in the primitive red blood cell precursor has three major 5' promoter elements that can interact with four DNA elements in the 3' enhancer. Some of these sites are bound by *trans*-regulatory factors while others are not. As erythroid development continues, new *trans*-factors are synthesized that bind to the promoter and enhancer and that lead to protein–protein interactions between the enhancer and promoter regions. In the definitive red blood cell stage (where hemoglobin synthesis is greatly increased), a new *trans*-regulatory factor, NF-E4, stabilizes the promoter-enhancer interaction and permits efficient globin synthesis. (After Gallarda et al., 1989.)

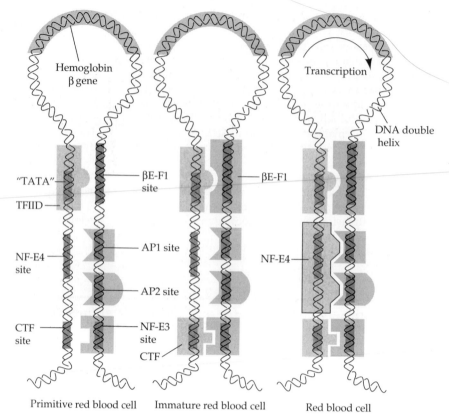

moter. A similar situation may exist in the switch from fetal to adult human hemoglobin. Competition for the major β-globin region enhancer has been seen between the human γ- and β-globin genes (Enver et al., 1990), and the switch between γ- and β-globins is prevented if a person has a mutation in the promoter region that prevents the binding of an erythrocyte-specific β-globin promoter-binding factor (Martin et al., 1989).

SIDELIGHTS & SPECULATIONS

Families of trans-regulatory factors

The *trans*-regulatory factors that bind to DNA can be grouped together in families based on similarities in structure. We have already discussed one of these families, the steroid hormone receptors. These proteins share a common framework structure in their respective DNA binding sites, and slight differences in the amino acids at the binding site can determine whether the protein binds to a glucocorticoid responsive element or an estrogen responsive element on the DNA.

Another important family of *trans*-regulatory factors are the HOMEODOMAIN PROTEINS. This homeodomain was first seen in proteins that were regulators of segment identity in *Drosophila*. Mutations of these proteins caused one segment to be transformed into another (a transformation known as "homeosis" that will be discussed in detail in Chapter 18). These transcription factors are believed to be

important in the differentiation of the central nervous system in flies and vertebrates. In mice, for instance, the expression of homeodomain proteins often coincides with rhombomere boundaries (Wilkinson et al., 1989; Keynes and Lumsden, 1990). As is represented in Figure 22, the homeodomain protein Hox 2.9 is expressed only in rhombomere r4. Homeodomain proteins Hox 2.6, 2.7, and 2.8 are found throughout the spinal cord but stop at a rhombomere border. It is thought that these homeodomain proteins activate batteries of genes that specify the particular properties of that segment. In *Drosophila*, the presence of certain homeodomain-containing proteins is necessary for the determination of specific neurons. Without these transcription factors, the fates of these neuronal cells are altered (Doe et al., 1988).

Several of the homeodomain proteins of *Drosophila melanogaster* have been cloned, sequenced, and tested for their ability to regulate transcription. Table 2 shows nine hom-

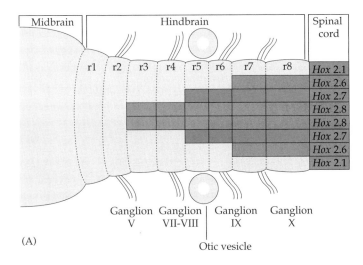

(A)

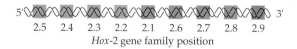

(B)

FIGURE 22

Expression patterns of the Hox-2 cluster of homeodomain proteins in the 9.5-day embryonic mouse hindbrain. (A) Diagram of the hindbrain rhombomeres (r1–8) showing the positions of the otic vesicle and the cranial ganglia. To the right are represented the domains of homeodomain gene expression. (B) Order of the *Hox-2* gene family on mouse chromosome 11. (After Wilkinson et al., 1989.)

eodomain-containing *Drosophila* proteins and the DNA sequences they recognize. The recognition of specific promoters by homeodomain-containing proteins has been shown to be essential for *Drosophila* development. The bicoid protein, for instance, binds to the promoters on the *hunchback* gene. This activates the *hunchback* gene, which is necessary for *Drosophila* head and thorax formation (Driever and Nüsslein-Volhard, 1989; Struhl et al., 1989). (If this gene is absent or inactivated, the mutant larvae lack these structures, hence the name of the gene.) As in the steroid hormone receptors, small changes in the amino acid composition of the DNA binding site can change the DNA sequence recognized by the protein. Treisman and her colleagues (1989) found that a single amino acid substitutions can change the DNA binding specificity of paired protein to that of bicoid or fushi tarazu proteins.

The POU domain

Some transcription factors have both a homeodomain and a second DNA-binding region. This region that comprises the homeodomain and the second DNA-binding region is called the POU domain (pronounced "pow"). The initials stand for the first letters of the four proteins first seen to have such domains: Pit-1 (the pituitary-specific factor that activates growth hormone and prolactin genes), Oct-1 (the ubiquitous protein that recognizes the octa box), Oct-2 (the B cell-specific protein that recognizes the octa box and activates immunoglobulin genes), and unc-86 (a nematode gene product involved in determining neuronal cell fates). The location of the POU domains within these proteins is shown in Figure 23. The homeodomain of Pit-1 recognizes the sequence ATATTCAT, while the homeodomain of Oct-2 recognizes the similar sequence ATTTGCAT. If the DNA element recognized by Pit-1 is mutated in two places, it becomes an Oct-2 binding site, and the prolactin gene becomes expressed in B lymphocytes (Elsholtz et al., 1990).

TABLE 2

Major homeodomain proteins of *Drosophila melanogaster* and their DNA binding sites

Protein	DNA-binding site(s)
Abdominal B	TAATTTGCAT
	TCAATTAAAT
Antennapedia	TAATAATAATAATAA
bicoid	TCCTAATCCC
engrailed	TCAATTAAAT
even-skipped	TCAATTAAAT
	TAATAATAATAATAA
	TCAGCACCG
fushi tarazu	TCAATTAAAT
	TAATAATAATAATAA
paired	TCAATTAAAT
Ultrabithorax	TAATAATAATAATAA
zerknült	TCAATTAAAT

FIGURE 23

Position of the POU domain in the first four POU proteins. Between the POU-specific domain and the homeodomain is a nonconserved region. (After Herr et al., 1988.)

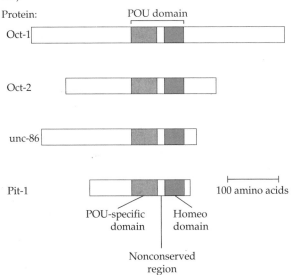

Thus, a two-base change in the POU-factor binding site switches pituitary-specific transcription to lymphoid-specific transcription.

Searching for new POU domain genes, He and co-workers (1989) synthesized all the possible oligomers that could encode the conserved amino acids common to the four known POU domain proteins. These were used as primers for the polymerase chain reaction, and cDNA to rat and human brain mRNAs were used as templates. The resulting amplification yielded four new POU genes that are expressed in the brain. Figure 24 shows the expression of two of these POU proteins. The mechanisms by which these POU and homeodomain proteins may specify the types of neurons produced in the different segments of the mammalian and insect nervous systems is currently an extremely active area of investigation.

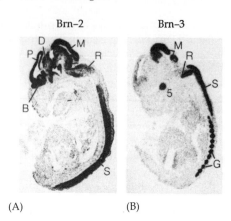

Brn–2   Brn–3

(A)   (B)

Helix–loop–helix motifs

Another prominent arrangement seen in enhancer and promoter DNA binding proteins is the "helix-loop-helix." The muscle-specific transcription factors myoD1 and myogenin contain this motif, as do several *Drosophila* proteins that determine the cells of *Drosophila*'s peripheral nervous system: the products of the *daughterless, hairy,* and *extramacrochaetae* genes. These proteins bind to DNA through a region of basic amino acids that precedes the first loop. The helices contain hydrophobic amino acids at every third or fourth position so that the helix presents a surface of hydrophobic residues to the environment. This enables the protein to pair by hydrophobic interaction (the "leucine zipper") with the same (or related) protein that also displays such a surface (Jones, 1990).

FIGURE 24
Patterns of POU domain expression in the 16d embryonic rat brain. Microscopic sections were incubated with radioactive cDNA to the messages for Brn-2 and Brn-3 POU-domain proteins. (A) Protein Brn-2 is expressed at high levels in the brain stem (B), pallium (P), diencephalon (D), mesencephalon (M), and spinal cord (S). (B) Protein Brn-3 shows expression in parts of the mesencephalon, the rhombencephalon (R), and anterior spinal cord, as well as in the retina (5) and sensory ganglia (G). (From He et al., 1989: photographs courtesy of M. G. Rosenfeld.)

DNA methylation

Correlations between gene activity and lack of methylation

It is often assumed that a gene contains exactly the same nucleotides when it is active as when it is inactive. A β-globin gene in a red blood cell precursor should have the same nucleotides as the β-globin in a fibroblast of the same animal. However, in 1948, R. D. Hotchkiss discovered a "fifth base" in DNA, 5-methylcytosine. This base is made enzymatically after DNA is replicated, and about 5 percent of the cytosines in mammalian DNA are converted to 5-methylcytosine. This conversion can only occur when the cytidine residue is followed by a guanosine (CpG). Recent studies have shown that the degree to which the cytosines of a gene are methylated may control the gene's transcription. In other words, DNA METHYLATION might change the structure of the gene and, in so doing, regulate its activity.

The first evidence that DNA methylation helps regulate gene activity comes from studies showing a correlation between gene activity and non-methylation (hypomethylation), especially in the promoter region of the gene. In developing human and chick red blood cells, the DNA involved in globin synthesis is completely (or nearly completely) unmethylated, whereas the same genes are highly methylated in cells that do not produce globin (Figure 25A). Fetal liver cells that produce hemoglobin in early development have unmethylated genes for fetal hemoglobin. These genes

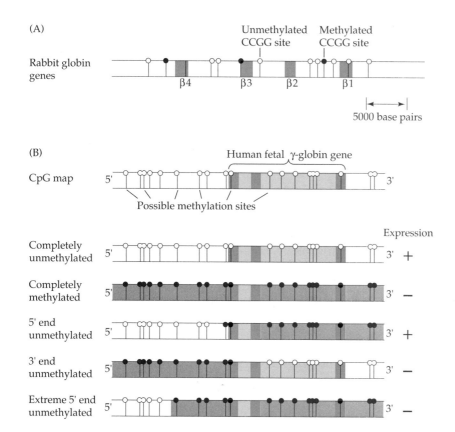

(A)

Rabbit globin genes

Unmethylated CCGG site   Methylated CCGG site

β4   β3   β2   β1

5000 base pairs

(B)

CpG map

Human fetal γ-globin gene

5'   3'

Possible methylation sites

Expression

Completely unmethylated   5'   3'   +

Completely methylated   5'   3'   −

5' end unmethylated   5'   3'   +

3' end unmethylated   5'   3'   −

Extreme 5' end unmethylated   5'   3'   −

FIGURE 25
Summary of methylation in globin genes. (A) Methylation in rabbit β-globin genes (β1, β2, β3, β4). Each gene of this cluster is represented by a colored box. The positions of CCGG sequences are indicated by stemmed circles. The dark circles represent sites that are totally methylated only in cells not transcribing globin genes. (B) Transcription from a cloned human fetal γ-globin gene is controlled by methylation. The top diagram represents a map of the potential methylation sites in and around the cloned gene. Exons are deeply colored, introns are more lightly colored. Possible methylation sites (CpG) are indicated as open dots. Beneath this map are diagrams of the methylation pattern of five clones used in this study. Open circles represent nonmethylated sites. Black circles in dark area represent regions where the DNA has been methylated. The data to the right indicate whether the globin genes were transcribed. (A after Shen and Maniatis, 1980; B after Busslinger et al., 1983.)

become methylated in the adult tissue (van der Ploeg and Flavell, 1980; Groudine and Weintraub, 1981).

Organ-specific methylation patterns are also seen in the chick ovalbumin gene; the gene is unmethylated in the oviduct cells but is methylated in other chick tissues (Mandel and Chambon, 1979). Demethylation accompanies class switching in immunoglobulin synthesis (Rogers and Wall, 1981) and correlates with the ability of murine lymphocytes to produce the metal-binding protein metallothionein I (Compere and Palmiter, 1981). Thus, the absence of DNA methylation correlates well with the tissue-specific expression of certain genes.

Another correlation was made with constitutive genes (those genes that are active in all somatic cells). While these genes often lack the TATA and CAAT promoter elements associated with tissue-specific genes, they usually have "CpG ISLANDS" at their 5' ends (Bird et al., 1985; Melton et al., 1986; Wolf and Migeon, 1985). However, if these CpG islands are methylated, as in the case of X-chromosome inactivation, the gene becomes inactive (Wolf et al., 1984; Lock et al., 1987).

The second type of evidence for DNA methylation as a regulatory process comes from experiments in which the expression of cloned genes is altered by introducing or removing methyl groups from their cytosine residues. One such experiment by Busslinger and co-workers (1983) provided further evidence for the notion that hypomethylation is critical for human β-globin gene transcription. When these investigators cloned globin genes and added them to cells by coprecipitation with calcium phosphate, the cells ingested the DNA and, in many cases, incorporated the DNA into the nucleus. In such cases, the cloned globin gene was transcribed. By protecting certain regions of the cloned globin genes from methylation before adding them to the cells, it is possible to create clones in which the globin genes have identical sequences but different methylation patterns (Figure 25B). A completely unmethylated gene is transcribed, whereas a completely methylated gene (methyl groups on every appropriate C residue) is not transcribed. Using partially methylated clones, Busslinger and co-workers showed that methylation in the 5' region of the globin gene (nucleotides −760 to +100) prevents transcription. Thus, methylation in the 5' end of a gene appears to play a direct role in the regulation of gene expression.

A third type of evidence comes from experiments in which otherwise inactive genes are activated by demethylating them while they are still inside the nucleus. A correlation has been made between the activation of silent genes and the loss of methyl groups from their cytosines. In Chapter 10, we saw that inactive liver-specific genes could be activated in mouse fibroblasts when these fibroblasts were fused to liver cells. This activation is accompanied by the demethylation of these liver-specific genes, especially at their 5' regions (Sellem et al., 1985). Once the 5' regions of these genes are demethylated, they resemble the liver-specific genes of the original liver cell and actively transcribe RNA.

Another experimental mechanism for demethylating genes is to treat cells with the drug 5-AZACYTIDINE. This drug is thought to inhibit methyltransferase, thereby causing the demethylation of DNA that would otherwise be methylated (Razin et al., 1984). When 5-azacytidine is given to colonies of mouse fibroblastic C3H10T1/2 cells, these cells are often converted into myocytes, adipocytes, or chondrocytes (Taylor and Jones, 1982; Konieczny and Emerson, 1984). These converted phenotypes are stable, suggesting that the methylation pattern is propagated through numerous rounds of cell division. Lassar and co-workers (1986) showed that in those cells that became muscle, the drug treatment demethylated to *myoD1* gene, allowing it to be transcribed into the protein that could transform the fibroblasts into myocytes.

Mechanisms for transcriptional inactivation by DNA methylation

There are several ways in which the methylation of DNA can interfere with transcriptional activity. First, methylation may inhibit the binding of *trans*-regulatory factors to the DNA. The enhancers of cAMP-responsive genes contain a consensus sequence, TGACGTCA, that binds a *trans*-regulatory factor for the activation of the gene. Transcription from such genes is dependent on the binding of this factor, and the factor does not bind to the consensus sequence if the central C in that sequence is methylated. In this way, the methylation of that base stops transcription from cAMP-responsive genes (Iguchi-Ariga and Schaffner, 1989).

A second mechanism by which DNA methylation can stop transcription involves the placement of nucleosomes along the DNA. While unmethylated DNA assumes a structure that is relatively sensitive to DNase I (and is presumably accessible by *trans*-regulatory proteins), fully methylated stretches of DNA form structures that are resistant to DNase I and that are transcriptionally inactive (Keshet et al., 1986; Tazi and Bird, 1990).

A third mechanism involves the *maintenance*, rather than the creation, of an inactive state. This mechanism may be responsible for the correlation between hypomethylation and globin gene activity. Enver and co-workers (1988) fused human fetal erythroid cells to a mouse tumor cell line and observed the methylation patterns in different clones as the γ-globin genes were switched off and the β-globin genes were activated. Before the switch, the globin genes were unmethylated, but they became methylated after the switch had occurred. However, during the time of the switch, clones were found that had turned off the γ-globin gene. These γ-globin genes still hypomethylated. This suggests that *trans*-acting regulatory factors may be necessary for initiating the inactive state of the genes, but that methylation is used to maintain that state.

This use of methylation to lock in the transcriptionally inactive state is seen in the inactivation of the mammalian X chromosome. The first evidence for some *"cis"* difference in DNAs came when Liskay and Evans (1980) transfected the X-linked gene for hypoxanthine phosphoribosyl-transferase (HPRT) into HPRT⁻ mouse cells. When they used a clone of cells in which the gene for HPRT was on the *inactive* X chromosome, the DNA was unable to cause the HPRT⁻ host cell to express the enzyme. When the DNA was taken from a clone of cells having the HPRT gene on its *active* X chromosome, the transformed cells expressed the enzyme. A year later, DNA methylation was implicated in this regulation, when Mohandas and his co-workers found that 5-azacytidine could locally reactivate genes on the inactive X chromosome. Further investigations, using probes that could recognize methylated and unmethylated DNA fragments, showed that the 5′ CpG islands of rat genes on the active X chromosome are hypomethylated, whereas the same CpG islands on the inactive X show extensive methylation (Wolf et al., 1982, 1984; Keith et al., 1986). When the inactive X chromosome was reactivated (either naturally, as in chorionic villi, or experimentally with 5-azacytidine), these clusters became hypomethylated (Wolf et al., 1984). Although the initiators of X-chromosome inactivation (the molecules responsible for determining which X chromosome is inactivated in each cell) have not yet been found, it is very likely that DNA methylation maintains this inactive state and propagates it from one cell generation to another (Kaslow and Migeon, 1987).

If this demethylation hypothesis is correct, the pattern of methylated and demethylated cytosines must be stably inherited over many cell divisions. This is indeed the case. A specific methylation pattern introduced into a cell is stable for over 100 cell generations (Wigler et al., 1981; Stein et al., 1982). Moreover, Stein and co-workers have suggested a mechanism for the propagation of this pattern. When they transfected a cell with hemimethylated DNA (i.e., DNA in which only one strand is methylated), the CG methylations were found on all the clonal progeny of that cell after many cellular generations. They postulate that a CG-specific methylating enzyme uses methylated cytidines as templates to methylate the cytidine on the opposite strand (Figure 26). This would also explain why methylated cytidines are always followed by guanosines.

DNA methylation and gametic imprinting

In Chapter 2, we saw that the genomes of the mammalian sperm and egg were not identical. Zygotes did not develop properly if their nuclei were derived from two egg pronuclei or from two sperm pronuclei. There are also rare conditions in mice and humans where a mutant gene causes severe or lethal defects if inherited from one parent, but does not produce these deficiencies if inherited from the other parent (McGrath and Solter, 1984; Nicholls et al., 1989). It is now generally thought that the differences between the gamete pronuclei involve differences in their DNA methylation patterns. The primordial germ cell nuclei of both male and female mammals are strikingly hypomethylated (Monk et al., 1987; Driscoll and Migeon, 1990), but both the sperm and egg genes undergo extensive methylation as the gametes mature. It appears that during germ cell formation, previous information is erased, and then, during meiosis, new

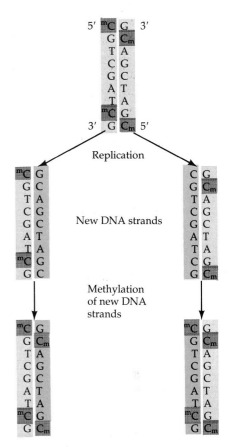

FIGURE 26

Model for the propagation of methylation patterns. When DNA replicates, only one of the two strands (the "old" strand) retains the original methylation pattern. The other strand (the "new" strand) is unmethylated. A CG-specific methylating enzyme would be able to bind to CG pairs where a C residue was methylated and then methylate the C residue on the complementary strand. (After Browder, 1984.)

information is introduced onto the genome. The pattern of methylation on a given gene can differ between egg and sperm, and these methylation differences are seen in the chromosomes of embryonic cells (Sanford et al., 1987; Reik et al., 1987; Sapienza et al., 1987). Thus, methylation differences between sperm and egg genes may specify whether a gene came from the father or mother. This maternal and paternal IMPRINTING adds additional information to the inherited genomes, information that may regulate spatial and temporal gene activity and chromosome behavior. Swain and co-workers (1987) have constructed a strain of transgenic mice in which a tumor-producing gene (to be discussed in Chapter 20) was inserted into the mouse genome. If this gene was inherited from the male parent, it was transcribed specifically in the heart and in no other tissue. If this gene was inherited from the female parent, it was not expressed at all. The pattern of expression correlated with the degree of methylation. This gene is methylated during egg maturation and is demethylated during sperm formation. Such gamete-specific methylation differences provide a plausible explanation for the failure of parthenogenetic mammals to develop and for the need for both male and female pronuclei in the zygote.

SIDELIGHTS & SPECULATIONS

Probing methylation sites

When DNA methylation occurs, it is almost always on the cytidine of a CpG pair. The distribution of these methylated and unmethylated CG doublets is determined by cutting the DNA with two restriction enzymes, HpaII and MspI (McGhee and Ginder, 1979). Both these enzymes cut at the same site, CCGG; but HpaII will not cut DNA if the central C is methylated, whereas MspI will cut whether the sequence is methylated or not. Therefore, DNA from a particular cell type can be digested separately with HpaII and MspI and the cleaved DNA fragments separated in a gel by electrophoresis, blotted onto a nitrocellulose filter, and hybridized with a radioactive gene-specific probe. Differences in band patterns in the autoradiograms of the MspI-cleaved and HpaII-cleaved fragments can then be ascribed to methylation differences.

Figure 27 shows the results of an experiment in which DNA from sperm was isolated and treated with HpaII or MspI. The probe was a radioactive DNA from the second exon of the β-globin gene. The autoradiogram of fragments from the MspI digestion shows that this probe binds to DNA fragments that have 1400 base pairs between CCGG sites. The HpaII digestion autoradiogram shows that in the sperm these sites are methylated and that this DNA sequence now resides in a 25,000-base-pair stretch of DNA where all CCGG sites are methylated (Groudine and Conklin, 1985).

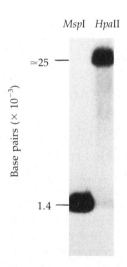

FIGURE 27
Detection of methylation sites on DNA. DNA was isolated from chicken sperm and digested with either MspI (lane 1) or HpaII (lane 2). The fragments were separated by electrophoresis, blotted onto paper, and hybridized to a radioactive DNA probe from the second exon of the β-globin gene. This probe bound to a fragment 1400 bases long in the MspI digest, but to a fragment about 25,000 bases long in the HpaII digest. (From Groudine and Conklin, 1985; photograph courtesy of M. Groudine.)

Methylation is obviously an important mechanism of gene regulation, but it is by no means used by every gene. It appears to be widely used in vertebrates, but even here, some genes are able to be transcribed while

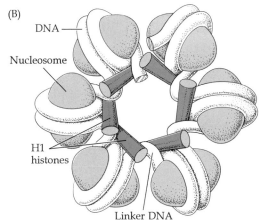

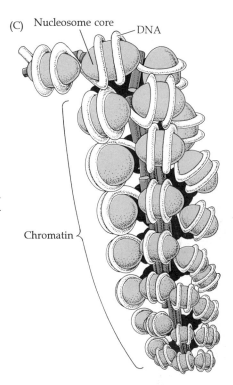

FIGURE 28

Nucleosome and chromatin structure. (A) Model of nucleosome structure. About 140 base pairs of DNA encircle the histone octamer (containing two copies each of histones H2A, H2B, H3, and H4), and about 60 base pairs of DNA link the nucleosomes together. (B) Model for the arrangement of nucleosomes into the highly compacted, solenoidal chromatin structure. (C) Model of nucleosome structure as seen by X-ray crystallography at a resolution of 0.33 nm. The central biconcave protein unit (white) is the H3–H4 tetramer. Each of the dark ovoids flanking the tetramer is an H2A–H2B dimer. This model is derived from chick erythrocyte chromatin. (A from Burlingame et al., 1985.)

fully methylated (Gerber-Huber et al., 1983). Moreover, DNA methylation does not appear to occur in *Drosophila*, where tissue-specific gene transcription is well documented.

EUKARYOTIC CHROMATIN

Nucleosomes

Up to this point, we have limited our discussion of messenger RNA transcription to the structure of the gene itself. But genes do not exist in an uncovered state within the nucleus, readily accessible to RNA polymerase or any enhancer or promoter binding protein. Rather, eukaryotic chromosomes contain as much protein (by weight) as nucleic acid, and this DNA–protein complex is called CHROMATIN. The most abundant of the chromatin proteins are the HISTONES. These five proteins—H1, H2A, H2B, H3, and H4—are highly basic polypeptides containing a large proportion of the positively charged amino acids lysine and arginine. Another histone, H5, is found in a few cell types (notably avian and amphibian erythrocytes) in which the DNA is extremely tightly packed. In these cells, H5 replaces H1.

The five major histones are present throughout the animal, plant, and protist kingdoms and have changed very little during the course of evolution. Such conservation usually implies that the proteins must be playing an extremely important role in all the cells studied; for histones, the major function is the packaging of DNA into specific coiled structures called nucleosomes. The nucleosome is the basic unit of chromatin structure; it is composed of histone octamer (two molecules each of H2A, H2B, H3, and H4) wrapped about with two loops of approximately 140 base pairs of DNA (Kornberg and Thomas, 1974; Figure 28). Between each nucleosome are another 60 or so base pairs of DNA; these "linker" nucleotides can be covered by H1. This arrangement is suggested by the cleavage pattern of chromatin when it is subjected to small amounts of a deoxyribonuclease (DNase) derived from a certain bacterium (*Micrococcus*). This treatment cleaves the DNA at several sites, and the resulting chromatin fragments can be separated from each other by centrifugation through a sucrose gradient (Figure 29). When the DNA from the treated chromatin is extracted and separated on gels, one obtains a clearly defined pattern. Instead of showing random-sized pieces of DNA, the DNA appears to be cleaved into pieces containing multiples of 200 base pairs (Hewish and Burgoyne, 1973; Noll, 1974). When these fragments of DNase-treated chromatin were observed under the electron microscope, the fragment containing 200 base pairs of DNA was seen as a spherical body with a small tail. The 400-base-pair sequence is found in two connected bodies, and the 600-base-pair sequence has three connected spherical units (Finch et al., 1975). Thus, it was concluded that the basic chromatin subunit contains about 200 base pairs of DNA and that micrococcal DNase preferentially cleaves the DNA between nucleosomes.

Chromatin can thus be visualized as a string of nucleosome beads linked by 10–100 base pairs of DNA. While classical geneticists have lik-

FIGURE 29

Characterization of nucleosome multimers. (A) Separation of discrete groups of particles formed by treating rat liver chromatin with micrococcal DNase is achieved by sucrose density centrifugation. (B) Electron micrographs of the particles in four sucrose gradient peaks, showing nucleosome monomers, dimers, trimers, and tetramers. (C) The particles of chromatin in each of the peaks were isolated, and the DNA extracted from the chromatin and analyzed by electrophoresis. Discrete jumps of 200 base pairs are seen. The right-hand side of the gel shows DNA extracted after DNase digestion and before sedimentation on sucrose. (B and C from Finch et al., 1975.)

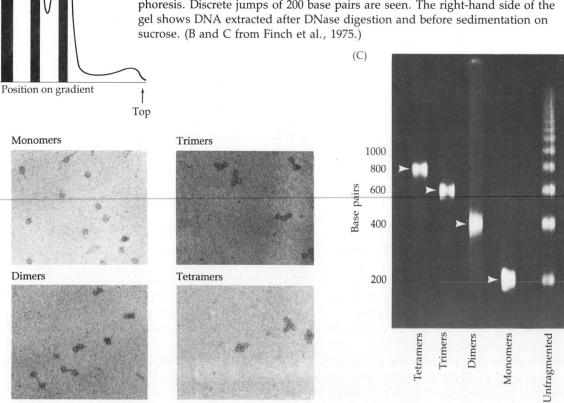

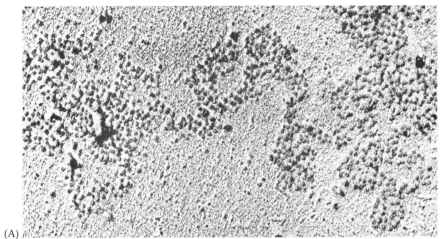

(A)

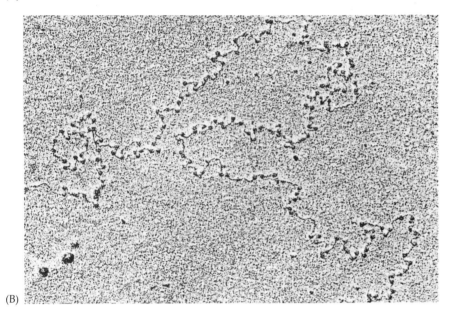

(B)

FIGURE 30

Role of H1 in compaction of chromatin. (A) Chicken liver chromatin observed in the electron microscope. The beads represent the nucleosomes. (B) The same chromatin after the removal of histone H1 by salt elution. The chromatin has become far less compact. The bar indicates 0.25 μm. (From Oudet et al., 1975; photographs courtesy of P. Chambon.)

ened genes to "beads on a string," molecular geneticists liken genes to "string on the beads."

Histone H1 is not present on the nucleosomes but can be bound to the linker DNA between them, depending on the physiological state of the cell. These nucleosomes are packed into tight structures themselves, and H1 appears to be involved in the winding of nucleosomes about each other, especially during preparation for cell division (Figure 30; Weintraub, 1984). This H1-dependent conformation of nucleosomes inhibits the transcription of genes in somatic cells by packing adjacent nucleosomes together into tight arrays that prohibit the access of transcription factors and RNA polymerases (Thoma et al., 1979). Workman and Roeder (1987) found that when DNA containing the promoter was incorporated into nucleosomes, these genes were unable to be transcribed. However, when the TATA box-binding protein TFIID was added to the genes before or during nucleosome formation, the resulting chromatin was transcriptionally active. The cell-type-specific activation of the genes can thereby be seen as having two components. First, the chromatin region must unfold so that the gene and its promoter are accessible to transcription factors and RNA polymerase. Second, these transcription factors must be present in the nucleus to bind to the DNA so exposed.

Activating repressed chromatin

Accessibility to *trans*-regulatory factors

It is incredible that DNA can become accessible to *trans*-regulatory factors at all. There is enough DNA in a single human body to extend the diameter of the solar system (Crick, 1966), and this enormous length must be packed up tightly into the nuclei of our cells. Yet our developmental library can be accessed specifically in each cell type. Solution hybridization experiments suggest that there is a minimum of 10,000 tissue-specific genes in the genome of most vertebrates; so it is not surprising that in any given cell type most of these genes will be repressed. It is generally thought, then, that the "default" condition of chromatin is a repressed state and that tissue-specific genes become activated by locally interrupting the repressive factors (Weintraub, 1985). As mentioned above, the major mechanism of general gene repression is probably the compaction of DNA into

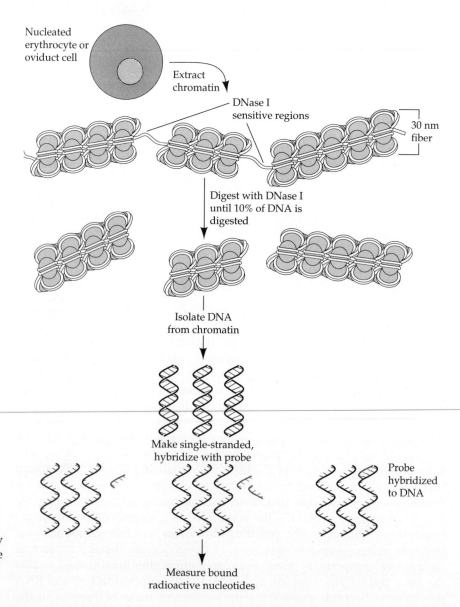

FIGURE 31
Protocol for determining the specificity of DNase I digestion of chromatin. See Table 3 for results of digestion experiment.

TABLE 3
Binding studies with DNase I-treated chromatin

Source of DNase I-treated chromatin	Radioactive cDNA probe	Percentage of maximum binding of radioactive cDNA to DNA extracted from treated chromatin
Chick red blood cell DNA (no DNase treatment)	Globin cDNA	94
Chick brain cell chromatin	Globin cDNA	90–100
Chick fibroblast chromatin	Globin cDNA	90–100
Chick red blood cell chromatin	Globin cDNA	25
Chick red blood cell chromatin	Ovalbumin cDNA	90–100

Source: Weintraub and Groudine (1976).

clusters of nucleosomes. As the DNA gets wrapped into these tight configurations, the *trans*-regulatory agents capable of initiating transcription are not able to find the DNA sequences with which they can bind (Schlissel and Brown, 1984).

There is evidence that transcribed genes in any particular cell are more accessible to the RNA polymerase and *trans*-regulatory factors than are the nontranscribed genes. (In other words, the repressive system has been locally abrogated in these regions.) The major evidence for this comes from studies concerning the DNase susceptibility of specific genes in different cells. The accessibility of a gene to nuclear proteins can be detected by treating the chromatin of a tissue with small amounts of DNase I (a pancreatic type of DNase different from that of *Micrococcus*). The DNA of the treated chromatin is extracted and mixed with radioactive cDNA for a particular gene (Figure 31). If the cDNA finds sequences to bind to, then the gene has been protected by chromatin proteins from digestion—it was not accessible to the DNase, and it probably would not be accessible to RNA polymerase either. However, if the cDNA probe does not find sequences to bind to, then the gene has been exposed to the DNase, and probably would be accessible to RNA polymerase and *trans*-regulatory factors.

The DNase I sensitivity of a given gene was found to be dependent upon the cell type in which it resides (Table 3; Weintraub and Groudine, 1976). When chromatin from developing chick *red blood cells* was treated with DNase I and its DNA extracted and mixed with radioactive globin cDNA, the globin cDNA found little with which to bind. The globin genes in the chromatin had been digested by the small amounts of DNase I. However, treating *brain cell* chromatin with the same amounts of DNase I did not destroy the globin genes. Therefore, the globin gene was accessible to outside enzymes in developing red blood cell chromatin but not in brain cell chromatin.

Similarly, the ovalbumin gene is susceptible to DNase I digestion in oviduct chromatin but not in red blood cell chromatin. When the chromatin is treated with DNase I and the DNA is extracted, ovalbumin cDNA is able to find sequences in the erythrocyte preparation but not in the DNA from the treated oviduct chromatin. Here, then, we have clear correlation of differential gene regulation and chromatin structure.

DNase-hypersensitive sites

The DNase I-sensitive sites include large stretches of DNA that contain the genes active in the particular cell. These regions usually contain nucleosomes, but they are not packed together. It is thought that as RNA polymerase passes through a region of the gene containing a nucleosome, the nucleosome temporarily loosens so that the polymerase can pass through. The histones appear to remain attached to the DNA at their amino and carboxyl termini, while the central region of the histones lose their attachment to the DNA, thereby accommodating the movement of the transcriptional apparatus on the gene (Lorch et al., 1988; Nacheva et al., 1989). Figure 32 shows transcription occurring on DNA wrapped around nucleosomes.

In addition to these DNase I-sensitive sites, there exist the *DNase I-hypersensitive* sites. These sites, identified on Southern blots by small radioactive DNA fragments, are destroyed by extremely minute amounts of DNase I, indicating that they are extremely accessible to outside molecules. This accessibility appears to arise from the nearly total absence of nucleosomes in this region of DNA (Elgin, 1988). The DNase I-hypersensitive sites mark regions of the chromatin, such as active promoters and enhancers, where DNA-binding proteins are bound. DNase I-hypersensitive regions are therefore associated with tissue-specific developmentally regulated genes (Elgin, 1981; Conklin and Groudine, 1984). For example, globin genes in red blood cells and their immediate precursors contain DNase I-hypersensitive sites, but globin genes in other cells do not (Stalder et al., 1980). In the red blood cell precursors from fetal liver, DNase I-hypersensitive sites occur within 200 base pairs 5' to the cap site of the β-, γ-, and δ-globin genes. When the chromatin is extracted from adult bone marrow cells, only the δ- and β- (adult) globin genes have DNase I hypersensitivity (Groudine et al., 1983). The 5' flanking region of the chick vitellogenin gene contains several hypersensitive sites in the chromatin from the liver of laying hens; but these sites are not present in male liver chromatin, embryonic liver chromatin, or brain or lymphocyte chromatin (Burch and Weintraub, 1983). When a DNase I-hypersensitive site was deleted from the 5' region of a developmentally regulated salivary protein gene in *Drosophila*, no transcripts were detectable, and no RNA polymerase

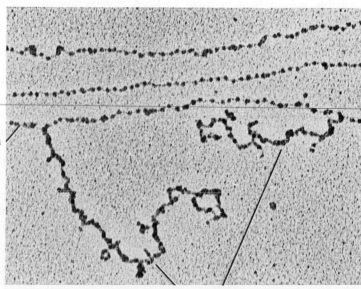

FIGURE 32

Transcription of chromatin from a region of *Oncopeltus* DNA containing nucleosomes. Two RNA transcripts can be seen. (Photograph courtesy of V. Foe.)

Chromatin

RNA transcripts

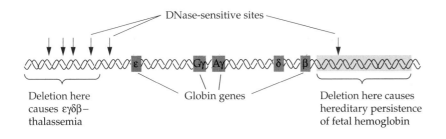

DNase-sensitive sites

Deletion here
causes εγδβ–
thalassemia

Globin genes

Deletion here causes
hereditary persistence
of fetal hemoglobin

FIGURE 33
Diagram of the human β-globin family of genes on chromosome 11. The erythroid-specific hypersensitive region is located 6–22 kilobases upstream of the ε-globin gene. The five DNase I-hypersensitive sites within this region are marked by arrows. A sixth DNase I-hypersensitive site downstream from the β globin gene is also marked. Its function is presently unknown, but it is within a region whose presence is required for normal globin gene switching during development. (After Ryan et al., 1989.)

molecules bound to the gene's promoter (Shermoen and Beckendorf, 1982; Steiner et al., 1984).

These hypersensitive sites often reside within or adjacent to sites that have enhancer functions. Zaret and Yamamoto (1984), studying the glucocorticoid-responsive enhancer of the mouse mammary tumor virus, demonstrated that before the addition of the hormone to cells containing the virus, this enhancer sequence showed no special DNase I sensitivity. After administering the hormone, a discrete DNase I-hypersensitive site developed in this region. The formation of the hypersensitive site coincided with the initiation of viral gene transcription; when the hormone is withdrawn, both the hypersensitive site and viral gene transcription disappear. Zaret and Yamamoto speculate that the interaction between the glucocorticoid receptor complex and the enhancer DNA alters the chromatin configuration to facilitate transcription from the nearby promoter.

The gene for serum albumin is transcribed in mouse liver at rates over 1000 times greater than in any other mouse tissue. In liver only, there are seven DNase I-hypersensitive sites that map to the promoter and enhancer regions. Using a gel retardation assay, Liu and co-workers (1988) found that liver nuclei contain *trans*-regulatory factors that bind specifically to the DNA sequences within these regions. In some instances, the binding of a factor to one hypersensitive site can cause the formation of new hypersensitive sites. This is seen to occur in the chick oviduct. A binding site for progesterone has been mapped within an ovalbumin gene hypersensitive region in oviduct cells (Mulvihill et al., 1982), and the administration of estrogen is able to induce the formation of four DNase I-sensitive sites in the promoter region of that same gene (Kaye et al., 1986). Upon withdrawing the estrogen, three of these sites disappear. The promoters and enhancers of numerous tissue-specific genes are seen to lie within DNase I-hypersensitive sites that allow for the binding of constitutive and tissue-specific transcription factors.

In the chromosome containing the human β-globin gene family, there are DNase-hypersensitive regions in several areas containing promoters and enhancers. Moreover, some of these regions become more sensitive to DNase during the switch from fetal to adult hemoglobin (Groudine et al., 1983). There is also a region of DNA that is exceptionally sensitive to DNase I that is found 6–22 kilobases upstream of the ε-globin gene (Figure 33). This is the "master enhancer" region described earlier in this chapter, and it contains five sites that are hypersensitive only in erythroid precursor cells and appears necessary for activating high levels of erythroid-specific transcription from these genes. Ryan and co-workers (1989) constructed transgenic mice containing the human β-globin gene and its immediate promoters and enhancers. These transgenic mice made only small amounts of human β-globin (less that 0.3 percent of the total cell β-globin). However, when they added the "master enhancer" DNase I-hypersensitive region, human β-globin accounted for over half the total globin in these mice. This result explains the medical observations (Tuan et al.,

1987) that certain patients who lacked this upstream DNase I-hypersensitive region had deficiencies of γ-, δ-, and β-globins, even though their genes for these proteins were intact and the globin genes on the other chromosome functioned normally.

DNase hypersensitivity and methylation pattern

The relationship between DNase-hypersensitive sites, enhancer elements, and methylation sites is difficult to unravel. In chicks, the major yolk protein gene, vitellogenin II, is under the regulation of estrogen and is only synthesized in the livers of egg-laying hens. The chromatin of the liver already has four DNase I-hypersensitive sites in this gene before hormone stimulation. That is, the sites exist before transcription occurs at this locus (Burch and Weintraub, 1983). These hypersensitive sites (marked by the black arrows in Figure 34D) are in the middle and 3' flanking regions of the genes in liver chromatin and are absent from the gene when it is in the chromatin of erythrocytes, fibroblasts, or neurons. Upon estrogenic stimulation, three other DNase I-hypersensitivity sites emerge. These are at the 5' flanking region of the gene and are associated with enhancer elements that bind the estrogen receptor (Burch and Weintraub, 1983). In addition, a CG site in this region that is usually methylated in embryos and roosters becomes unmethylated by the addition of estrogen (depicted by a circle in Figure 34). Upon estrogen withdrawal, one of the

FIGURE 34
Correlation of transcription with DNase I hypersensitivity and hypomethylation in the enhancer sites of the chick vitellogenin II gene. (A) The DNase I hypersensitivity assay is performed by digesting liver chromatin 0, 3, 4.5, and 8 hours after estrogen treatment. The left-hand lane of each time sample represents undigested chromatin. At 8 hours, the presence of newly cleaved fragments indicates that DNase cleavage is occurring in this region. The radioactive probe recognizes DNA in the 5' end of the gene. C, B1, and B2 correspond to sites on the gene shown in (D). (B) The methylation assay is performed on the liver DNA at various times after hormone treatment. DNA was digested with *Hpa*II. Until 8 hours, the probe recognized the DNA in a large *Hpa*II fragment. By 8 hours, a second fragment is seen, indicating that a previously methylated CCGG has become unmethylated in this region. (C) Dot blot of cytoplasmic mRNA probed with cDNA to vitellogenin. Transcription is first seen at around 8 hours. (D) Summary of results. Before hormone addition, there are three DNase I-hypersensitive sites (black arrows) in the liver chromatin and none on the chromatin of other cells. Addition of hormone causes three new DNase I-hypersensitive sites (colored arrows) to form on the 5' end of the gene in the liver, along with the hypomethylation of this region (filled-in circle). In nonhepatic cells, neither hypomethylation nor DNase I-hypersensitivity was seen. Withdrawal of the hormone leads to the reduction of one of these hormone-specific hypersensitive sites in liver chromatin and a lack of transcriptional activity. (From Burch and Weintraub, 1983; photograph courtesy of the authors.)

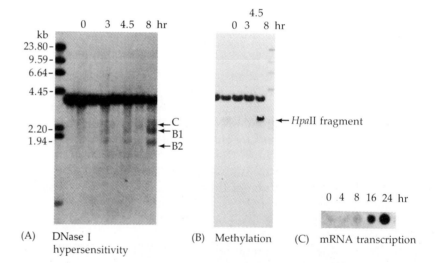

(A) DNase I hypersensitivity   (B) Methylation   (C) mRNA transcription

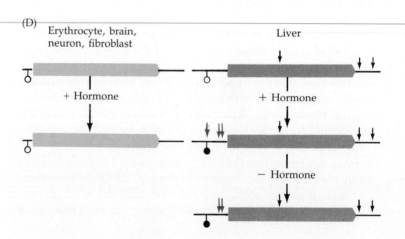

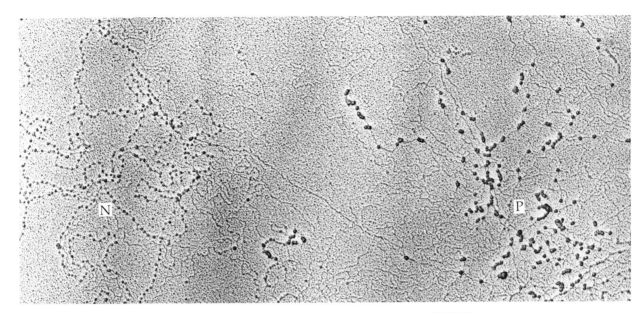

N

P

FIGURE 35

Active and inactive regions of the *Xenopus* genome. Isolated *Xenopus* chromatin was treated with RNase to remove nascent RNA strands and was fixed for electron microscopy. Two regions of chromatin are seen: in region P (right), large protein molecules (believed to be RNA polymerase) are present; and, in region N (left), nucleosomes without polymerase are seen. (From Labhart and Koller, 1982; photograph courtesy of T. Koller.)

5′ hypersensitivity sites is abolished, although the other sites (and the demethylation pattern) remain. Neither the demethylation nor the creation of hypersensitive sites is seen in the vitellogenin gene in other cell types.

Keshet and co-workers (1986) have suggested that demethylation is necessary to create DNase I-hypersensitive sites. When unmethylated globin genes were transfected into mouse fibroblast nuclei, they were found to have DNase-sensitive sites (regardless of the transcriptional ability of the gene). When the same genes were methylated at every CpG site, the DNase-sensitive regions failed to form, presumably because their methylation caused them to be packaged into an inaccessible form.

Regulation of gene accessibility

How can these DNase I-hypersensitive regions originate if the DNA is uniformly bound up in nucleosomes? How can factors recognize a specific stretch of "inaccessible" chromatin? Two answers have been considered. One is that there are nucleosome-free regions of chromatin, and the other is that there are factors that bind to nucleosomes causing them to relax their hold on the DNA.

The absence of nucleosomes would ensure that all the DNA in that region was accessible to RNA polymerase and that transcription could be efficiently accomplished. In the milkweed bug *Oncopeltus fasciatus*, the portion of the genome containing the ribosomal RNA genes is almost totally devoid of nucleosomes, and the DNA in this region is very actively transcribed (Foe et al., 1976). Electron microscopy has confirmed this finding in *Oncopeltus* and in some other species (Figure 35; Labhart and Koller, 1982).

The transcription of mouse globin genes may also involve disruptions of nucleosome spacing. In cells that are not in the erythroid lineage, nucleosomes appear to be regularly spaced throughout the globin genes. In red blood cell precursors, however, this normal distribution is interrupted so that the 5′ end of the gene (from −200 to +500) appears to be devoid of nucleosomes. As the cells are induced to synthesize hemoglobin, further nucleosome disruption may also occur (Cohen and Sheffery, 1985; Benezra et al., 1986).

Other studies also indicate that most (if not all) of the DNase I-hypersensitive regions are devoid of nucleosomes. Rose and Garrard (1984) have shown that the immunoglobulin light chain genes become hypomethylated and DNase I-sensitive as they proceed from pre-B cells to B cells. Moreover, at the same time, the repeating nucleosome pattern changes, causing large regions of the immunoglobulin gene to be relatively free of nucleosomes. The nucleosomes that do remain appear to be devoid of histone H1 and to contain instead HMG 14 and 17 proteins (see below).

It is possible that the way in which methyl groups interact with the histones enables nucleosomes to form only on methylated DNA and not on unmethylated DNA (Keshet et al., 1986). This restriction would result in regions of DNase hypersensitivity. Once these regions were established, it would become easier for other *trans*-regulatory elements to find these nucleosome-free regions. This arrangement would explain why the chick liver vitellogenin gene contains a set of DNase-hypersensitive sites before hormone addition. The binding of the *trans*-regulatory hormone receptor would then create the new hypersensitive sites associated with active transcription.

The formation of an active region of chromatin need not demand a total absence of nucleosomes. Many actively transcribing genes are located within regions of normally spaced nucleosomes. However, nucleosomes can be modified at certain specific loci. One of the most important of these modifications appears to involve the binding of two peptides called HIGH MOBILITY GROUP (HMG) PROTEINS 14 and 17.

When chromatin is digested by DNase I, there is a preferential release of the two nonhistone chromatin proteins, HMG 14 and HMG 17. These low-molecular-weight proteins can be extracted from chromatin with 0.35 M NaCl. They appear to confer DNase I sensitivity (as opposed to hypersensitivity) on genes when they bind to the nucleosomes in their vicinity. When HMG 14 and 17 proteins are removed from the chromatin, the differential susceptibility of specific genes to DNase I treatment is simultaneously removed. Thus, when the HMG proteins are extracted from red blood cell chromatin, the globin genes are no longer specifically susceptible to DNase-I digestion (Figure 36). However, when these proteins are added back to such depleted chromatin, the differential sensitivity of the globin genes is restored (Weisbrod and Weintraub, 1979; Weisbrod et al., 1980; Gazit et al., 1980). The ability of these proteins to restore sensitivities is still based on the chromatin. HMG 14 and HMG 17 isolated from the brain will restore the DNase I-sensitivity of globin genes in red blood cells but not in brain cells. Thus, two nonhistone chromatin proteins appear necessary (but not in themselves sufficient) to allow the transcription of specific genes in different tissues.

Nucleosomes can also be displaced by *trans*-regulatory factors. Nucleosomes routinely form on the glucocorticoid response element of certain genes. When the mouse mammary tumor virus integrates into a cellular genome, six nucleosomes are seen to form in the promoter region. When these cells are exposed to glucocorticoids, the glucocorticoid-bound receptor binds to one of the DNA on one of these nucleosomes and initiates a process whereby the nucleosome is displaced. Once the nucleosome is gone, other *trans*-regulatory promoter-binding factors, TFIID, and RNA polymerase soon bind to this region of DNA and initiate transcription. Moreover, if the nucleosome is stabilized (by the acetylation of its histones), this displacement fails to occur (Bresnick et al., 1990). It is possible, then, that some *trans*-regulatory proteins may be active in finding their DNA sequence on the nucleosome and actively displacing the nucleosome, while other *trans*-regulatory factors can find their DNA sequence only after the nucleosome has been removed.

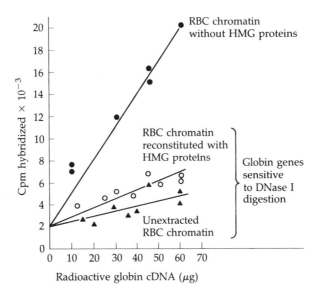

FIGURE 36
Reconstruction of DNase I sensitivity of the globin gene. The HMG proteins were removed from chick red blood cell (RBC) chromatin. One fraction was maintained without HMG proteins and the other fraction had them restored. Sufficient DNase was added to both fractions and to unextracted RBC chromatin to digest 10 percent of the DNA. Radioactive globin cDNA found complementary sequences much more readily when the chromatin lacked the HMG proteins. (Data from Weisbrod and Weintraub, 1979.)

Association of active DNA with the nuclear matrix

Attachment of active chromatin to a nuclear matrix

Within the nucleus, the replicative enzymes must somehow find their sites for initiating DNA synthesis; the transcriptional factors and polymerases must find their promoters and enhancers; the RNA processing factors must find their RNA splicing sites; and the messenger RNA must efficiently find the pores through which to exit the nucleus. This is a great deal to ask of molecules in solution. We would have to expect the various factors involved in transcription to be floating around in the nuclear sap, bumping randomly into DNA. The RNA so formed would then be spliced and bump around the nucleus until it found a nuclear pore through which to leave.

An alternative model suggests that RNA is transcribed on a solid substrate wherein all the enzymes needed for transcription, processing, and transport are collected together. There are precedents for thinking in such terms. The electron transport chain of the mitochondria is such an ordered aggregate, and the DNA synthetic enzymes of bacteria have long been known to reside on the inner surface of the cell membrane. What one has to ask, then, are the following questions: Is there a nuclear reticulum wherein such enzymes might be found? If such a network exists, are the transcriptionally active genes located there? If such a network exists, are the RNA synthesizing enzymes found there?

A nuclear matrix can be isolated by dissolving nuclei in lipid detergents and solubilizing most of the DNA with DNases (Berezney and Coffey, 1977; Capco et al., 1982). Transmission electron microscopy of such complexes shows a meshwork of proteins that extends throughout the nucleus and connects to the cytoskeleton at the nuclear envelope (Figure 37). This matrix is seen in all eukaryotic nuclei thus far examined (Wilson, 1895; Nelson et al., 1986).

When one isolates such a matrix, DNase has removed about 98 percent of the DNA. Is the DNA that is still bound to this matrix (and presumably protected from the DNase by being so tightly associated with it) enriched for actively transcribing genes? There is evidence that this is so for at least

FIGURE 37
Transmission electron micrograph (×47,000) of a portion of nuclear matrix (left) and surrounding cytoplasm. Cytoskeletal filaments are clearly visible. The nuclei of mouse fibroblast cells were isolated, extracted with detergent to remove lipids, and further treated with DNase I. (From Capco et al., 1982; photograph courtesy of S. Penman.)

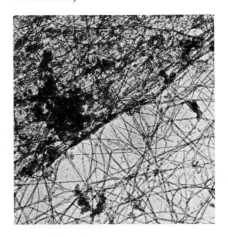

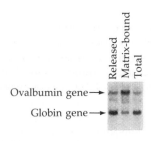
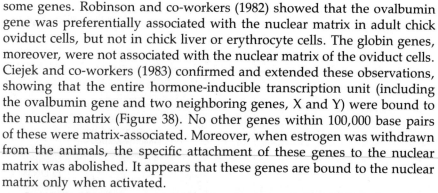

Ovalbumin gene →
Globin gene →

FIGURE 38

Southern blot showing DNA fragments protected by the chick oviduct nuclear matrix and those released from nuclei when treated with *Eco*RI. The DNA from the matrix-bound and released fractions were collected and put on an electrophoresis gel that would separate out the components by their sizes. The resulting gel was Southern blotted and incubated with radioactive cDNA for ovalbumin and β-globin. The globin gene was not found to be associated with the matrix of these cells, while the ovalbumin gene was retained by the matrix. (From Ciejek et al., 1983; photograph courtesy of B. W. O'Malley.)

some genes. Robinson and co-workers (1982) showed that the ovalbumin gene was preferentially associated with the nuclear matrix in adult chick oviduct cells, but not in chick liver or erythrocyte cells. The globin genes, moreover, were not associated with the nuclear matrix of the oviduct cells. Ciejek and co-workers (1983) confirmed and extended these observations, showing that the entire hormone-inducible transcription unit (including the ovalbumin gene and two neighboring genes, X and Y) were bound to the nuclear matrix (Figure 38). No other genes within 100,000 base pairs of these were matrix-associated. Moreover, when estrogen was withdrawn from the animals, the specific attachment of these genes to the nuclear matrix was abolished. It appears that these genes are bound to the nuclear matrix only when activated.

In 1985, Hutchinson and Weintraub showed that DNase I-sensitive sites were not found uniformly throughout the nucleus. They treated nuclei with DNase I and then performed a nick-translation reaction on these intact nuclei. The labeled DNA, then, should represent just the actively transcribing (i.e., DNase I-sensitive) genes. The results of such treatment (Figure 39) showed the DNase I-sensitive DNA existed at the periphery of the nuclei and along channels or fibers that connect to the nuclear envelope. It is possible, then, that active genes are specifically associated with the nuclear envelope or matrix.

The third type of evidence for nuclear matrix participation in transcription involves the finding that most (some report 95 percent) of the newly synthesized RNA appears to be attached to the nuclear matrix (Miller et al., 1978; Herman et al., 1978; Mariman et al., 1982; van Eekelen and van Venrooij, 1981). This binding appears to be mediated by a particular nuclear matrix protein. Given that active genes, RNA polymerase, and nascent transcripts appear to be bound to a nuclear matrix, Jackson and Cook (1985) have proposed that transcription does not occur by a mobile polymerase traveling down the length of a gene. Rather, they envision the RNA polymerase as tethered to the nuclear matrix, with the DNA traveling through it.

There is also some evidence that active DNA may be attached to the nuclear matrix by AT-rich DNA sequences called MATRIX-ASSOCIATED REGIONS (or SCAFFOLD-ASSOCIATED REGIONS) (Gasser and Laemmli, 1986). Most of these *cis*-regulatory elements map near or within enhancers or promoters. The importance of these regions was shown by Stief and co-workers (1989), who identified two matrix-associated regions of the chick lysozyme gene. In this case, the matrix-associated regions were not in the enhancer and so could be separated from it. When they fused the chick lysozyme enhancer and promoter to the reporter gene, CAT, and transfected the clone into lysozyme-producing cells, it did not produce much CAT protein. They then made a similar gene containing the promoter, enhancer, and CAT sequences, and flanked by the two matrix-associated regions. When this clone was transfected into the lysozyme-producing cells, CAT synthesis was enormously enhanced (Figure 40).

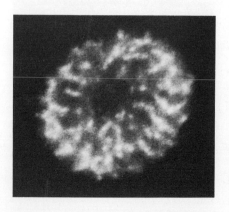

FIGURE 39

Presence of active chromatin along the nuclear periphery and channels. Erythrocyte nuclei were treated with DNase I, which is thought to nick actively transcribing chromatin regions. This nick was "healed" by nick translation within the nucleus in the presence of nucleotides whose presence could be detected by fluorescence. The labeled nucleotides were found at the periphery of the nucleus and along structures leading inward from the nuclear envelope. (From Hutchinson and Weintraub, 1985; photograph courtesy of N. Hutchinson.)

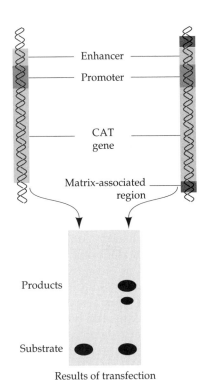

Products

Substrate

Results of transfection

FIGURE 40
Importance of matrix-associated regions in transcription. When clones consisting of lysozyme promoter, enhancer, and the CAT gene are transfected into a lysozyme-secreting cell line, very little CAT protein is made, as assayed by CAT enzyme activity. However, if the two matrix-associated regions are also included in the cloned gene, much more CAT protein can be found in these cells. (After Stief et al., 1989.)

Topoisomerases and gene transcription

In several studies, matrix-associated regions have been seen to contain or be adjacent to a DNA sequence that is recognized by an enzyme—TOPO-ISOMERASE II—that might be essential for transcription. Recent studies have suggested that the unwinding of the DNA helix is very important in facilitating transcription. Transcriptionally active chromatin has to be twisted to enable the strands to unwind (Ryoji and Worcel, 1984), and this twisting is accomplished by supercoiling the DNA helix (Figure 41). When supercoiled circular plasmids are injected into the nuclei of *Xenopus* oocytes, they are transcribed. When the same plasmids are linearized (by opening them with a restriction enzyme), they are not transcribed when injected into the cells (Harland et al., 1983; Pruitt and Reeder, 1984). Villeponteau and co-workers (1984) have shown that DNase I-sensitive sites in active genes are formed only when the genes are under torsional strain. The enzyme responsible for twisting the DNA and allowing the strands to separate is topoisomerase II. Using antibodies to this protein, Berrios and colleagues (1985) demonstrated that topoisomerase II is located on the nuclear matrix–nuclear envelope complex.

Several studies (Cockerill and Garrard, 1986; Adachi et al., 1989; Scheuermann and Chen, 1989) have shown that the DNA sequence bound by topoisomerase is often adjacent to a matrix-associated region of DNA.

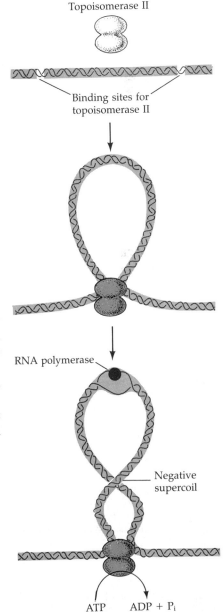

FIGURE 41
Supercoiling of DNA during transcription. Topoisomerase II joins together two regions of DNA and introduces supercoiling by transiently breaking and recombining the DNA strands. As a result of the strain, a portion of the double helix separates into two strands, enabling RNA ploymerase (and presumably other *trans*-regulatory factors) to initiate transcription. The topoisomerase binding sites have been found on matrix-bound DNA (Cockerill and Garrard, 1986). (After Darnell et al., 1986.)

FIGURE 42

One of four regions of the heavy chain immunoglobulin gene enhancer protected by the NF-μNR protein. NF-μNR was added to the enhancer region DNA and the DNA was digested with DNase. Only those sequences covered by NF-μNR would be preserved. The protected region (gray) includes a matrix-associated region sequence and a topoisomerase II binding site (colored). (After Scheuermann and Chen, 1989.)

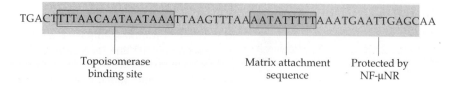

TGACT**TTTAACAATAATAAA**TTAAGTTTAA**AATATTTTT**AAATGAATTGAGCAA

Topoisomerase binding site   Matrix attachment sequence   Protected by NF-μNR

This suggests that the anchoring of the DNA to the matrix may be necessary to prevent free rotation of the DNA so as to permit the matrix-bound topoisomerases to twist the chromatin. (Cockerill and Garrard, 1986).

If binding to the nuclear matrix is essential for RNA transcription, it is possible that negative regulatory proteins such as silencers may inhibit this association. This possibility was suggested by Scheuermann and Chen (1989), who isolated a protein that inhibits the transcription of the heavy chain immunoglobulin gene. This protein, NF-μNR, is expressed in early stages of B cell development, but is absent from the mature B cells that transcribe large amounts of immunoglobulin genes. It is also present in non-B cells. This protein binds to four places flanking the heavy chain enhancer. The sequences it protects include two matrix-associated region consensus sequences and two topoisomerase II consensus sequences (Figure 42). It is possible, then, that when NF-μNR is present in nuclei, it would bind to the enhancer flanking regions and prevent the association of the heavy chain immunoglobulin gene with the nuclear matrix and topoisomerase. When this protein was not present, these associations would occur and transcription of this gene would result.

Z-DNA and hairpin loops

One effect of supercoiling may be to cause DNA to assume alternative conformations differing from the standard Watson-Crick B-form double helix. The supercoiling of DNA can bring about the formation of either "left-handed" Z-DNA or hairpin loops. Hairpin loops are postulated to form when, for thermodynamic reasons, base pairing can exist between two complementary regions of the same strand more readily than between strands. Figure 43 shows possible intrastrand base pairing at the 5′ end of the chick 22 collagen gene. Larsen and Weintraub (1982) have speculated that these distortions of B-DNA induced by supercoiling might provide recognition sites for *trans*-regulatory proteins and that, once bound, these proteins would inhibit nucleosome formation in that region of DNA.

Another conformational change that might be brought about by torsion is Z-DNA (Johnston and Rich, 1985; Kohwi-Shigematsu and Kohwi, 1985). Most nuclear DNA is the right-handed, double-helical B-DNA described by Watson and Crick. However, Alexander Rich and his colleagues discovered that certain CG-rich sequences, such as CGCGCG, preferentially form *left-handed* double helices (Wang et al., 1979). The physical properties of this left-handed structure differ from those of standard DNA. The left-handed helix has 12 base pairs per turn (rather than 10) and forms a zigzag conformation (hence the name Z-DNA). Z-DNA also lacks the major groove that allows the binding of certain molecules to B-DNA (Figure 44).

It had previously been thought that the transition from B-DNA to Z-DNA could only occur under high salt conditions. However, Behe and Felsenfeld (1981) demonstrated that Z-DNA will form naturally under physiological salt concentrations if the cytidine residues of an alternating GC polymer are methylated. Thus, methylation of cytidine may enable B-

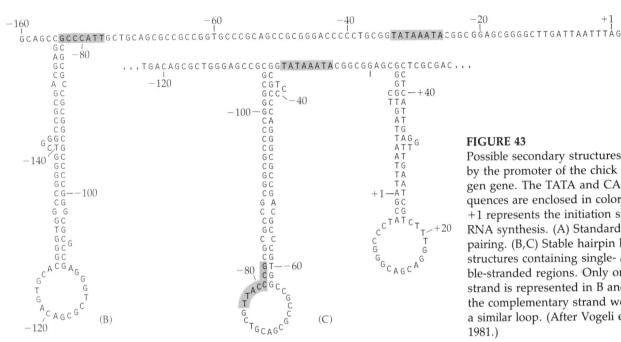

FIGURE 43
Possible secondary structures formed by the promoter of the chick α2 collagen gene. The TATA and CAAT sequences are enclosed in colored boxes; +1 represents the initiation site of RNA synthesis. (A) Standard base pairing. (B,C) Stable hairpin loop structures containing single- and double-stranded regions. Only one DNA strand is represented in B and C, but the complementary strand would form a similar loop. (After Vogeli et al., 1981.)

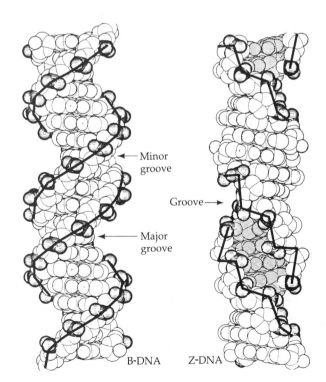

FIGURE 44
Comparison of right-handed (B) and left-handed (Z) DNA, showing the differences in the two helical structures. (From Wang et al., 1979; courtesy of A. Rich.)

DNA to enter the Z configuration. Moreover, whereas B-DNA promotes nucleosome formation from histones, Z-DNA actively prevents nucleosome formation (Nickol et al., 1982). Histones will indeed bind to Z-DNA, but they will not associate into the nucleosome octamer.

SUMMARY

At the moment, we have in our hands the pieces of one or more puzzles. We are not certain that we have all the pieces; neither do we know how many puzzles there are, for it is likely there are several ways to activate transcription. Researchers over the past five years have discovered that the formation of "active genes" is not enough for gene transcription. One also needs "active chromatin." The relationships between alternative gene structures (methylated versus unmethylated) and alternative chromatin structures (HMG-modified nucleosomes versus nonmodified nucleosomes, nucleosome-free regions versus nucleosome-containing regions) are presently being elucidated. While the roles of H1 histone in the general suppression of transcription and of *trans*-regulatory proteins in activating specific local regions appear to be key elements in gene regulation, it remains to be seen how the chromatin is informed as to what structure to take.

There are probably some reciprocal interactions occurring between the DNA and the chromatin. Through these interactions the DNA may tell the chromatin to form in a certain manner and this change in chromatin structure may enable the DNA to subsequently change its form in a particular way. One model might be the following: when the DNA is replicated, nucleosomes are removed. At this time, methylation patterns can be changed or retained, and certain *trans*-acting factors could bind to the DNA and prevent nucleosomes from forming there. The nucleosomes that reform would be positioned by these patterns of methylation and protein binding. The binding of *trans*-regulatory factors would also inhibit H1 condensation. Alternatively, nucleosomes in the hypomethylated regions might undergo a conformational change so that they can bind HMG 14 or HMG 17, or some of the *trans*-regulatory factors may be able to find their DNA elements on the nucleosomes and bring about the removal of the nucleosome from this region. The exposed DNA may then be able to bind further *trans*-regulatory factors, or possibly might bind to the topoisomerases on the nuclear matrix. The binding of TFIID and RNA polymerase would then initiate transcription. It is probable that there is no single way that every gene acts and that what works in one case might not work for all. As Albert Claude remarked, we have just begun the inventory of our acquired wealth.

LITERATURE CITED

Adachi, Y., Käs, E. and Laemmli, U. 1989. Preferential, cooperative binding of DNA topoisomerase II to scaffold-associated regions. *EMBO J.* 8: 3997–4006.

Atchinson, M. L. and Perry, R. P. 1987. The role of κ-enhancer and its binding factor NF-κB in the developmental regulation of κ gene transcription. *Cell* 48: 121–128.

Baeuerle, P. A. and Baltimore, D. 1988. IκB: A specific inhibitor of the NF-κB transcription factor. *Science* 242: 540–545.

Banerji, J., Olson, L. and Schaffner, W. 1983. A lymphocyte-specific cellular enhancer is located downstream of the joining region in immunoglobulin heavy chain genes. *Cell* 33: 729–740.

Barberis, A., Superti-Furga, G. and Busslinger, M. 1987. Mutually exclusive interaction of the CAAT-binding factor and of a displacement protein with overlapping sequences of a histone gene promoter. *Cell* 50: 347–359.

Beato, M. 1989. Gene regulation by steroid hormones. *Cell* 56: 335–344.

Behe, M. and Felsenfeld, G. 1981. Effects of methylation on a synthetic polynucleotide. The B–Z transition in poly(dG-m⁵dC)·poly-(dG-m⁵dC). *Proc. Natl. Acad. Sci. USA* 78: 1619–1623.

Behringer, R., Hammer, R., Brinster, R., Palmiter, R. and Townes, T. 1987. Two 3′ sequences direct adult erythroid-specific expression of human β-globin genes in transgenic mice. *Proc. Natl. Acad. Sci. USA* 84: 7056–7060.

Behringer, R. R., Mathews, L. S., Palmiter, R. D. and Brinster, R. L. 1988. Dwarf mice produced by genetic ablation of growth-hormone expressing cells. *Genes Dev.* 2: 453–461.

Benezra, R., Cantor, C. R. and Axel, R. 1986. Nucleosomes are phased along the mouse β-major globin gene in erythroid and non-erythroid cells. *Cell* 44: 697–704.

Berezney, R. and Coffey, D. S. 1977. Nuclear matrix: Isolation and characterization of a framework structure from rat liver nuclei. *J. Cell Biol.* 73: 616–637.

Bergman, Y., Rice, D., Grosschedl, R. and Baltimore, D. 1984. Two regulatory elements for immunoglobulin κ light chain gene expression. *Proc. Natl. Acad. Sci. USA* 81: 7041–7045.

Berrios, M., Osheroff, N. and Fisher, P. A. 1985. In situ localization of DNA topoisomerase II, a major polypeptide component of the *Drosophila* nuclear matrix fraction. *Proc. Natl. Acad. Sci. USA* 82: 4142–4146.

Biggin, M. D. and Tjian, R. 1989. Transcription factors and the control of *Drosophila* development. *Trends Genet.* 5: 377–383.

Bird, A. P., Taggart, M. H., Frommer, M., Miller, O. J. and MacLeod, D. 1985. A fraction of the mouse genome is derived from islands of non-methylated CpG-rich DNA. *Cell* 40: 91–99.

Blom van Assendelft, G., Hanscombe, O., Grosveld, F. and Greaves, D. R. 1989. The β-globin dominant control region activates homologous and heterologous promoters in a tissue-specific manner. *Cell* 56: 969–977.

Bodner, M. and Karin, M. 1987. A pituitary-specific *trans*-acting factor can stimulate transcription from the growth hormone promoter in extracts of non-expressing cells. *Cell* 50: 267–275.

Boulet, A. M., Erwin, C. R. and Rutter, W. J. 1986. Cell-specific enhancers in the rat exocrine pancreas. *Proc. Natl. Acad. Sci. USA* 83: 3599–3603.

Bresnick, E. H., John, S., Berard, D. S., LeFebvre, P. and Hager, G. L. 1990. Glucocorticoid receptor-dependent disruption of a specific nucleosome on the mouse mammary tumor virus promoter is prevented by sodium butyrate. *Proc. Natl. Acad. Sci. USA* 87: 3977–3981.

Browder, L. W. 1984. *Developmental Biology*, 2nd Edition. Saunders, Philadelphia.

Bunick, D., Zandomeni, R., Ackerman, S. and Weinmann, R. 1982. Mechanism of RNA polymerase II-specific initiation of transcription *in vitro*: ATP requirement and uncapped runoff transcripts. *Cell* 29: 877–886.

Buratowski, S., Hahn, S., Guarente, L. and Sharp, P. A. 1989. Five initiation complexes in transcription initiation by RNA polymerase II. *Cell* 56: 549–561.

Burch, J. B. and Weintraub, H. 1983. Temporal order of chromatin structural changes associated with activation of the major chicken vitellogenin gene. *Cell* 33: 65–76.

Burlingame, R. W., Love, W. E., Wang, B.-C., Hamlin, R., Xuong, N.-H. and Moudrianakis, E. N. 1985. Crystallographic structure of the octameric histone core of the nucleosome at a resolultion of 3.3 Å. *Science* 228: 246–253.

Busslinger, M., Hurst, J. and Flavell, R. A. 1983. DNA methylation and the regulation of globin gene expression. *Cell* 34: 197–206.

Calame, K. L. 1989. Immunoglobulin gene transcription: Molecular mechanisms. *Trends Genet.* 5: 395–399.

Capco, D. G., Wan, K. M. and Penman, S. 1982. The nuclear matrix: Three-dimensional architecture and protein composition. *Cell* 29: 847–858.

Chandler, V. L., Maier, B. A. and Yamamoto, K.R. 1983. DNA sequences bound specifically by glucocorticoid receptor in vitro render a heterologous promoter hormone responsive in vivo. *Cell* 33: 489–499.

Choi, O.-R. and Engel, J. D. 1988. Developmental regulation of β-globin gene switching. *Cell* 55: 17–26.

Ciejek, E. M., Tsai, M.-J. and O'Malley, B. W. 1983. Actively transcribed genes are associated with the nuclear matrix. *Nature* 306: 607-609.

Cockerill, P. N. and Garard, W. T. 1986. Chromosome loop anchorage of the κ immunoglobulin gene occurs next to the enhancer in a region containing topoisomerase II sites. *Cell* 44: 273–282.

Cohen, R. B. and Sheffery, M. 1985. Nucleosome disruption precedes transcription and is largely limited to the transcribed domain of globin genes in murine erythroleukemia cells. *J. Mol. Biol.* 182: 109–129.

Compere, S. J. and Palmiter, R. D. 1981. DNA methylation controls the inducibility of the mouse metallothionein-I gene in lymphoid cells. *Cell* 25: 233–240.

Conklin, K.F. and Groudine, M. 1984. Chromatin structure and gene expression. *In* A. Razin, H. Cedar and A. D. Riggs (eds.), *DNA Methylation*. Springer-Verlag, New York, pp. 293–351.

Crick, F. H. C. 1966. *Of Molecules and Men*. University of Washington Press, Seattle; p. 58.

Crossley, M. and Brownlee, G. G. 1990. Disruption of a C/EBP binding site in the factor IX promoter is associated with haemophilia B. *Nature* 345: 444–446.

Danesch, U., Gloss, B., Schmid, W., Schötz, G. and Renkawitz, R. 1987. Glucocorticoid induction of the rat tryptophan oxygenase gene is mediated by two widely separated glucocorticoid-responsive elements. *EMBO J.* 6: 625–630.

Danielsen, M. Hinck, L. and Ringold, G. M. 1989. Two amino acids within the knuckle of the first zinc finger specify DNA response element activation by the glucocorticoid receptor. *Cell* 57: 1131–1138.

Darnell, J., Lodish, H. and Baltimore, D. 1986. *Molecular Cell Biology*. Scientific American Books, New York.

deBoer, E., Antoniou, M., Mignotte, V. and Grosveld, F. 1988. The human β-globin promoter: nuclear factors and erythroid specific induction of transcription. *EMBO J.* 7: 4203–4212.

Dierks, P., van Ooyen, A., Chochran, M. D., Dobkin, C., Reiser, J. and Weissman, C. 1983. Three regions upstream from the cap site are required for efficient and accurate transcription of the rabbit β-globin gene in mouse 3T6 cells. *Cell* 32: 695–706.

Doe, C. Q., Smouse, D. and Goodman, C. S. 1988. Control of neuronal fate by the Drosophila segmentation gene *even-skipped*. *Nature* 333: 376–378.

Dollé, P., Castrillo, J.-L., Theill, L. E., Deerinck, T., Ellisman, M. and Karin, M. 1990. Expression of GHF-1 protein in mouse pituitaries correlates both temporally and spatially with the onset of growth hormone gene activity. *Cell* 60: 809–820.

Driever, W. and Nüsslein-Volhard, C. 1989. The *bicoid* protein is a positive regulator of *hunchback* transcription in the early Drosophila embryo. *Nature* 337: 138–143.

Driscoll, D. J. and Migeon, B. R. 1990. Evidence that human single-copy genes which are methylated in male meiotic germ cells are unmethylated in female meiotic germ cells and fetal germ cells of both sexes. In press.

Dynan, W. S. and Tjian, R. 1985. Control of eukaryotic messenger RNA synthesis by sequence-specific DNA-binding proteins. *Nature* 316: 774–778.

Efstratiadis, A. and fourteen others. 1980. The structure and evolution of the human β-globin gene family. *Cell* 21: 653–668.

Elgin, S. 1981. DNase I hypersensitive sites of chromatin. *Cell* 27: 413–415.

Elgin, S. C. R. 1988. The formation and function of DNase I hypersensitive sites in the process of gene activation. *J. Biol. Chem.* 263: 19259–19262.

Elsholtz, H. P., Albert, V. R., Treacy, M. N. and Rosenfeld, M. G. 1990. A two-base change in a POU factor-binding site switches pituitary specific to lymphoid-specific gene expression. *Genes Dev.* 4: 43–51.

Enver, T., Zhang, J.-W., Papayannopoulou, T. and Stamatoyannopoulos, G. 1988. DNA methylation: A secondary event in globin gene switching? *Genes Dev.* 2: 698–706.

Enver, T., Raich, N., Ebens, A. J., Papayannopoulou, T., Costantini, F. and Stamatoyannopoulos, G. 1990. Developmental regulation of human fetal-to-adult globin gene switching in transgenic mice. *Nature* 344: 309–312.

Finch, J. T., Noll, M. and Kornberg, R. D. 1975. Electron microscopy of defined lengths of chromatin. *Proc. Natl. Acad. Sci. USA* 72: 3320–3322.

Fletcher, C., Heintz, N. and Roeder, R. 1987. Purification and characterization of OTF-1, a transcription factor regulating cell cycle expression of a human histone H2b gene. *Cell* 51: 773–781.

Foe, V. E., Wilkinson, L. E. and Laird, C. D. 1976. Comparative organization of active transcription units in *Oncopeltus fasciatus*. *Cell* 9: 131–146.

Friedman, A. D., Landschulz, W. H. and McKnight, S. L. 1989. CCAAT/enhancer binding protein activates the promoter of the serum albumin gene in cultured hepatoma cells. *Genes Dev.* 3: 1314–1322.

Gallarda, J. L., Foley, K. P., Yang, Z. and Engel, J. D. 1989. The β-globin stage selector element factor is erythroid-specific promoter/enhancer binding protein NF-E4. *Genes Dev.* 3: 1845–1859.

Garabedian, M. J., Hung, M.-C. and Wensink, P. C. 1985. Independent control elements that determine yolk protein gene expression in alternative *Drosophila* tissues. *Proc. Natl. Acad. Sci. USA* 82: 1396–1400.

Garabedian, M. J., Shepherd, B. M. and Wensink, P. C. 1986. A tissue- specific transcription enhancer from the *Drosophila* yolk protein 1 gene. *Cell* 45: 859–867.

Garcia, J. V., Bich-Thuy, L. T., Stafford, J. and Queen, C. 1986. Synergism between immunoglobulin enhancers and promoters. *Nature* 322: 383–385.

Gasser, S. M. and Laemmli, U. K. 1986. Cohabitation of scaffold binding regions with upstream/enhancer elements of three developmentally regulated genes in *D. melanogaster*. *Cell* 46: 521–530.

Gazit, B., Panet, A. and Cedar, H. 1980. Reconstitution of deoxyribonuclease I-sensitive structure on active genes. *Proc. Natl. Acad. Sci. USA* 77: 1787–1790.

Gedamu, L. and Dixon, G. H. 1978. Effect of enzymatic decapping on protamine messenger RNA translation in wheat-germ S-30. *Biochem. Biophys. Res. Commun.* 85: 114–124.

Gerber-Huber, S., May, F. E. B., Westley, B. R., Felber, B. K., Hosbach, H. A., Andres, A.-C. and Ryffel, G. U. 1983. In contrast to other *Xenopus* genes the estrogen-induced vitellogenin genes are expressed when totally methylated. *Cell* 33: 43–51.

Gillies, S. D., Morrison, S. L., Oi, V. T. and Tonegawa, S. 1983. A tissue-specific transcription enhancer element is located in the major intron of a rearranged immunoglobulin heavy chain gene. *Cell* 33: 717–728.

Green, S. and Chambon, P. 1988. Nuclear receptors enhance our understanding of transciptional regulation. *Trends Genet.* 4: 309–314.

Green, S., Kumar, V., Thenlaz, I., Wahli, W. and Chambon, P. 1988. The *N*-terminal DNA binding zinc finger of the oestrogen and glucocorticoid receptors determines target gene specificity. *EMBO J.* 7: 3037–3044.

Grosschedl, R. and Birnstiel, M. L. 1980. Spacer DNA upstream from the TATAATA sequence are essential for promotion of

H2A histone gene transcription in vivo. *Proc. Natl. Acad. Sci. USA* 77: 7102–7106.

Grosveld, F. , de Boer, E., Shewmaker, C. K. and Flavell, R. A. 1982. DNA sequences necessary for the transcription of the rabbit β-globin gene in vivo. *Nature* 295: 120–126.

Grosveld, F., Blom van Assendelft, G., Greaves, D. R. and Kollias, G. 1987. Position-dependent, high-level expression of the human β-globin gene in transgenic mice. *Cell* 51: 975–985.

Groudine, M. and Conklin, K. F. 1985. Chromatin structure and de novo methylation of sperm DNA: Implications for activation of the paternal genome. *Science* 228: 1061–1068.

Groudine, M. and Weintraub, H. 1981. Activation of globin genes during chick development. *Cell* 24: 393–401.

Groudine, M., Kohwi-Shigematsu, T., Gelinas, R., Stamatoyannopoulos, G. and Papayannopoulo, T. 1983. Human fetal to adult hemoglobin switching: Changes in chromatin structure of the β-globin gene locus. *Proc. Natl. Acad. Sci. USA* 80: 7551–7555.

Harland, R. M., Weintraub, H. and McKnight, S. L. 1983. Transcription of DNA injected into *Xenopus* oocytes is influenced by template topology. *Nature* 302: 38–43.

He, X., Treacy, M. N., Simmons, D. M., Ingraham, H. A., Swanson, L. W. and Rosenfeld, M. G. 1989. Expression of a large family of POU-domain regulatory genes in mammalian brain development. *Nature* 340: 35–42.

Herman, R., Weymouth, L. and Penman, S. 1978. Heterogeneous nuclear RNA–protein fibers in chromatin depleted nuclei. *J. Cell Biol.* 78: 663–674.

Herr, W. and several others. 1988. The POU domain: A large conserved region in the mammalian *pit-1, oct-1, oct-2*, and *Caenorhabditis elegans unc-86* gene products. *Genes Dev.* 2: 1513–1516.

Hewish, D. R. and Burgoyne, L. A. 1973. Chromatin substructure. The digestion of chromatin DNA at regularly spread sites by nuclear DNase. *Biochem. Biophys. Res. Commun.* 52: 504–510.

Hotchkiss, R. D. 1948. The quantitative separation of purines, pyrimidines, and nucleosides by paper chromatography. *J. Biol. Chem.* 175: 315–332.

Hutchinson, N. and Weintraub, H. 1985. Localization of DNase I-sensitive sequences to specific regions of interphase nuclei. *Cell* 43: 471–482.

Iguchi-Ariga, S. M. M. and Schaffner, W. 1989. CpG methylation of the cAMP-responsive enhancer/promoter sequence TGACGTCA abolishes specific factor binding as well as transcriptional activation. *Genes Dev.* 3: 612–619.

Jackson, D. A. and Cook, P. R. 1985. Transcription occurs at a nucleoskeleton. *EMBO J.* 4: 919–925.

Jackson, D. A., McCready, S. J. and Cook,

P. R. 1981. RNA is synthesized at the nuclear cage. *Nature* 292: 552–555.

Johnston, B. H. and Rich, A. 1985. Chemical probes of DNA conformation: Detection of Z-DNA at nucleotide resolution. *Cell* 42: 713–724.

Jones, N. 1990. Transcriptional regulation by dimerization: Two sides to an incestuous relationship. *Cell* 61: 9 -11.

Kadonaga, J. T., Courey, A. J., Ladika, J. and Tjian, R. 1988. Distinct regions of Sp1 modulate DNA binding and transcriptional activation. *Science* 242: 1566–1570.

Kantor, J. A., Turner, P. H. and Nienhuis, A. W. 1980. β Thalassemia: Mutations which affect processing of the β-globin mRNA precursor. *Cell* 21: 149–157.

Karin, M., Haslinger, A., Holtgreve, H., Richards, R. I., Krautner, P., Westphal, H. M. and Beato, M. 1984. Characterization of DNA sequences through which cadmium and glucocorticoid hormones induce human metallothionein-II gene. *Nature* 308: 513–519.

Kaslow, D. C. and Migeon, B. R. 1987. DNA methylation stabilizes X chromosome inactivation in eutherians but not in marsupials: Evidence for multistep maintenance of mammalian X dosage compensation. *Proc. Natl. Acad. Sci. USA* 84: 6210–6214.

Kaye, J. S., Pratt-Kaye, S., Bellard, M., Dretzen, G., Bellard, F. and Chambon, P. 1986. Steroid hormone dependence of four DNase I-hypersensitive regions located within the 7000-bp 5′ flanking segment of the ovalbumin gene. *EMBO J.* 5: 277–285.

Keith, D. H., Singersam, J. and Riggs, A. D. 1986. Active X-chromosome DNA is unmethylated at eight CCGG sites clustered in a guanine-plus-cytosine-rich island at the 5′ end of the gene for phosphoglycerate kinase. *Mol. Cell Biol.* 6: 4122–4125.

Keshet, I., Lieman-Hurwitz, J. and Cedar, H. 1986. DNA methylation affects the formation of active chromatin. *Cell* 44: 535–543.

Keynes, R. and Lumsden, A. 1990. Segmentation and the origin of regional diversity in the vertebrate central nervous system. *Neuron* 2: 1–9.

Klock, G., Strahle, U. and Schutz, G. 1987. Oestrogen and glucocorticoid responsive elements are closely related but distinct. *Nature* 329: 734–736.

Kohwi-Shigematsu, T. and Kohwi, Y. 1985. Poly(dG)–Poly(dC) sequences, under torsional stress, induce an altered DNA conformation upon neighboring DNA sequences. *Cell* 43: 199–206.

Konieczny, S. F. and Emerson, C. P., Jr. 1984. 5-Azacytidine induction of stable mesodermal stem cell lineages from 10T1/2 cells: Evidence for regulatory genes controlling determination. *Cell* 38: 791–800.

Kornberg, R. D. and Thomas, J.D. 1974. Chromatin structure: Oligomers of histones. *Science* 184: 865–868.

Krieg, P. A. and Melton, D. A. 1987. An enhancer responsible for activating transcription at the mid-blastula transition in *Xenopus* development. *Proc. Natl. Acad. Sci. USA* 84: 2331–2335.

Kumar, V., Green, S., Stack, G., Berry, M., Jin, J.-R. and Chambon, P. 1987. Functional domains of the human estrogen receptor. *Cell* 51: 941–951.

Labhart, P. and Koller, T. 1982. Structure of the active nucleolar chromatin of *Xenopus laevis* oocytes. *Cell* 28: 279–292.

Larsen, A. and Weintraub, H. 1982. An altered DNA conformation detected by S1 nuclease occurs at specific regions in active chick globin chromatin. *Cell* 29: 609–622.

Lassar, A. B., Paterson, B. M. and Weintraub, H. 1986. Transfection of a DNA locus that mediates the conversion of 10T1/2 fibroblasts into myoblasts. *Cell* 47: 649–656.

Lawn, R. M., Efstratiadis, A., O'Connell, C. and Maniatis, T. 1980. The nucleotide sequence of the human β-globin gene. *Cell* 21: 647–651.

Lenardo, M. J. and Baltimore, D. 1989. NF-κB: A pleiotropic mediator of inducible and tissue-specific gene control. *Cell* 58: 227–229.

Li, S., Crenshaw, E. B. III, Rawson, E. J., Simmons, D. M., Swanson, L. W. and Rosenfeld, M. G. 1990. *Dwarf* locus mutants lacking three pituitary cell types result from mutations in the POU-domain gene *pit-1*. *Nature* 347: 528–533.

Liskay, R. M. and Evans, R. 1980. Inactive X chromosome DNA does not function in DNA-mediated cell transformation for the hypoxanthine phosphoribosyltransferase gene. *Proc. Natl. Acad. Sci. USA* 77: 4895–4898.

Liu, J-K., Bergman, Y., and Zaret, K. S. 1988. The mouse albumin promoter and a distal upstream site are simultaneously DNase I hypersensitive in liver chromatin and bind similar liver-abundant factors *in vitro*. *Genes Dev.* 2: 528–541.

Lock, L. F., Takagi, N. and Martin, G. R. 1987. Methylation of the HPRT gene on the inactive X chromosome occurs after chromosome inactivation. *Cell* 48: 39–46.

Lorch, Y., LaPointe, J. W. and Kornberg, R. D. 1988. On the displacement of histones from DNA by transcription. *Cell* 55: 743–744.

Mandel, J. L. and Chambon, P. 1979. DNA methylation differences: Organ-specific variations in methylation pattern within and around ovalbumin and other chick genes. *Nucleic Acid Res.* 7: 2081–2103.

Maniatis, T., Goodbourn, S. and Fischer, J. A. 1987. Regulation of inducible and tissue-specific gene expression. *Science* 236: 1237–1245.

Mantovani, R. and several others. 1988. An erythroid-specific nuclear factor binding to the proximal CACCC box of the β-globin gene promoter. *Nucl. Acid Res.* 16: 4299–4313.

Maquat, L. E., Kinniburgh, A. J., Beach, L. R., Honig, G. R., Lazerson, J., Ershler, W. B. and Ross, J. 1980. Processing of human β-globin mRNA precursor to mRNA is defective in three patients with β⁺ thalassemias. *Proc. Natl. Acad. Sci. USA* 77: 4287–4291.

Marbaix, G., Huez, G., Burny, A., Cleuter, Y., Hubert, E., Lecleroq, M., Chantrenne, H., Soreq, H., Nudel, U. and Littauer, U.Z. 1975. Absence of polyadenylate segment in globin messenger RNA accelerates its degradation in *Xenopus* oocytes. *Proc. Natl. Acad. Sci. USA* 72: 3065–3067.

Mariman, E. C. M., van Eekelen, C. A. G., Reinders, R. J., Berns, A. J. M. and van Venrooij, W. J. 1982. Adenoviral heterogeneous nuclear RNA is associated with the host nuclear matrix during splicing. *J. Mol. Biol.* 154: 103–119.

Martin, D. I. K., Tsai, S.-F. and Orkin, S. H. 1989. Increased γ-globin expression in a nondeletion HPFH mediated by an erythroid-specific DNA-binding factor. *Nature* 338: 435–437.

Martinez, E., Givel, F. and Wahl, W. 1987. An estrogen-responsive element as an inducible enhancer: DNA sequence requirements and conversion to a glucocorticoid-responsive element. *EMBO J.* 6: 3719–3727.

Mather, E. L. and Perry, R. P. 1982. Transcriptional regulation of the immunoglobulin-V genes. *Nucleic Acid Res.* 9: 6855–6867.

McGhee, J. D. and Ginder, G. D. 1979. Specific DNA methylation sites in the vicinity of the chick β-globin genes. *Nature* 280: 419–420.

McGrath, J. and Solter, D. 1984. Maternal *T^hp* lethality in the mouse is a nuclear, not a cytoplasmic, defect. *Nature* 308: 550–551.

McKnight, S. and Tjian, R. 1986. Transcriptional selectivity of viral genes in mammalian cells. *Cell* 46: 795–805.

Melton, D. W., McEwan, C., McKie, A. B. and Reid, A. M. 1986. Expression of the mouse HPRT gene: Deletion analysis of the promoter region of an X-chromosome linked housekeeping gene. *Cell* 44: 319–328.

Merkel, C. G., Kwan, S.-P. and Lingrel, J. B. 1975. Size of the poly(A) region of newly synthesized globin mRNA. *J. Biol. Chem.* 250: 3725–3728.

Mermod, N., O'Neill, E. A., Kelly, T. J. and Tjian, R. 1989. The proline-rich transcriptional activator of CTF/NF-1 is distinct from the replication and DNA binding domain. *Cell* 58: 741–753.

Miesfeld, R. and seven others. 1986. Genetic complementation of a glucocorticoid receptor deficiency by a cloned receptor cDNA. *Cell* 46: 389–399.

Miller, T. E., Huang, C.-Y. and Pogo, A. O. 1978. Rat liver nuclear skeleton and ribonucleoprotein complexes containing HnRNA. *J. Cell Biol.* 76: 675–691.

Mohandas, T., Sparkes, R. S. and Shapiro, L. J. 1981. Reactivation of an inactive human X chromosome: Evidence for X inactivation by DNA methylation. *Science* 211: 393–396.

Monk, M., Boubelik, M. and Lehnert, S. 1987. Temporal and regional changes in DNA methylation in the embryonic, extraembryonic, and germ cell lineages during mouse embryo development. *Development* 99: 371–382.

Mulvihill, E. R., Lepennec, J. P. and Chambon, P. 1982. Chick oviduct progesterone receptor: Location of specific regions of high-affinity binding in cloned DNA fragments of hormone-responsive genes. *Cell* 28: 621–632.

Myers, R. M., Tilly, K. and Maniatis, T. 1986. Fine structure genetic analysis of a β-globin promoter. *Science* 232: 613–618.

Nacheva, G. A., Gushin, D. Y., Preobrazhenskaya, O. V., Karpov, V. L., Ebralidse, K. K. and Mirzabekov, A. D. 1989. Change in the pattern of histone binding to DNA upon transcriptional activation. *Cell* 58: 27–36.

Nelson, W. G., Pienta, K. J., Barrack, E. R. and Coffey, D. S. 1986. The role of the nuclear matrix in the organization and function of DNA. *Annu. Rev. Biophys. Bioph. Chem.* 15: 457–475.

Nicholls, R. D., Kroll, J. H. M., Butler, M. G., Karma, S. and Lalande, M. 1989. Genetic imprinting suggested by maternal heterodisomy in non-deletion Prader-Willi syndrome. *Nature* 342: 281–285.

Nickol, J., Behe, M. N. and Felsenfeld, G. 1982. Effect of the B–Z transformation in poly(dG-m⁵dC)·poly(dG-m⁵dC) on nucleosome formation. *Proc. Natl. Acad. Sci. USA* 79: 1771–1775.

Nokin, P., Burny, A., Huez, G. and Marbaix, G. 1976. Globin mRNA from anaemic rabbit spleen. Size of its polyadenylate segment. *Eur. J. Biochem.* 68: 431–436.

Noll, M. 1974. Subunit structure of chromatin. *Nature* 251: 249–251.

Orkin, S. and Kazazian, H. H. 1984. The mutation and polymorphism of the human β-globin gene and its surrounding DNA. *Annu. Rev. Genet.* 18: 131–171.

Oudet, P., Gross-Bellard, M. and Chambon, P. 1975. Electron microscope and biochemical evidence that chromatin structure is a repeating unit. *Cell* 4: 281–300.

Parslow, T. G., Blair, D. L., Murphy, W. J. and Granner, D. K. 1984. Structure of the 5'ends of immunoglobulin genes: A novel conserved sequence. *Proc. Natl. Acad. Sci. USA* 81: 2650–2654.

Payvar, F. and seven others. 1983. Sequence-specific binding of glucocorticoid receptor to MTV DNA at sites within and upstream of the transcribed region. *Cell* 35: 381–392.

Pfahl, M. 1982. Specific binding of the glucocorticoid-receptor complex to the mouse mammary tumor proviral promoter region. *Cell* 31: 475–482.

Picard, D. and Schaffner, W. 1984. A lymphocyte-specific enhancer in the mouse immunoglobulin κ gene. *Nature* 307: 80–82.

Pruitt, S. C. and Reeder, R. H. 1984. Effect of topological constraint on transcription of ribosomal DNA in *Xenopus* oocytes. *J. Mol. Biol.* 174: 121–139.

Queen, C. and Baltimore, D. 1983. Immunoglobulin gene transcription is activated by downstream sequence elements. *Cell* 33: 741–748.

Razin, A., Cedar, H. and Riggs, A. 1984. *DNA Methylation: Biochemistry and Biological Significance.* Springer-Verlag, New York.

Reik, W., Collick, A., Norris, M. L., Barton, S. C. and Surani, M. A. 1987. Genomic imprinting determines methylation of paternal alleles in transgenic mice. *Nature* 328: 248–250.

Robinson, S. I., Nelkin, B. D. and Vogelstein, B. 1982. The ovalbumin gene is associated with the nuclear matrix of chicken oviduct cells. *Cell* 28: 99–106.

Rogers, J. and Wall, R. 1981. Immunoglobulin heavy-chain genes: Demethylation accompanies class switching. *Proc. Natl. Acad. Sci. USA* 78: 7497–7501.

Rose, S. and Garrard, W. T. 1984. Differentiation-dependent chromatin alterations precede and accompany transcription of immunoglobulin light chain genes. *J. Biol. Chem.* 259: 8534–8544.

Rottman, F. A., Shatkin, A. J. and Perry, R. P. 1974. Sequences containing methylated nucleotides at the 5′ termini of messenger RNAs: Possible implications for processing. *Cell* 3: 197–199.

Ryan, T. M., Behringer, R. B., Martin, N. C., Townes, T. M., Palmiter, R. D. and Brinster, R. L. 1989. A single erythroid-specific DNase I super-hypersensitivity site activates high levels of human β-globin gene expression in transgenic mice. *Genes and Dev.* 3: 314–323.

Ryoji, M. and Worcel, A. 1984. Chromatin assembly in *Xenopus* oocytes: In vitro studies. *Cell* 37: 21–32.

Sanford, J. P., Clark, H. J., Chapman, V. M. and Rossant, J. 1987. Differences in DNA methylation during oogenesis and spermatogenesis and their persistence during early embryogenesis in the mouse. *Genes Dev.* 1: 1039–1046.

Sapienza, C., Peterson, A. C., Rossant, J. and Balling, R. 1987. Degree of methylation of transgenes is dependent on gamete of origin. *Nature* 328: 251–254.

Sawadogo, M. and Roeder, R. G. 1984. Energy requirement for specific transcription by the human RNA polymerase II system. *J. Biol. Chem.* 259: 5321–5326.

Scheuermann, R. H. and Chen, U. 1989. A developmental-specific factor binds to suppressor sites flanking the immunoglobulin heavy-chain enhancer. *Genes Dev.* 3: 1255–1266.

Schlissel, M. S. and Brown, D. D. 1984. The transcriptional regulation of *Xenopus* 5 S RNA genes in chromatin: The roles of active stable transcription complex and histone H1. *Cell* 37: 903–913.

Schüle, R., Muller, M., Kaltschmidt, C. and Renkawitz, R. 1988. Many transcription factors interact synergistically with steroid receptors. *Science* 242: 1418–1420.

Sellem, C. H., Weiss, M. C. and Cassio, D. 1985. Activation of a silent gene is accompanied by its demethylation. *J. Mol. Biol.* 181: 363–371.

Sen, R. and Baltimore, D. 1986a. Multiple nuclear factors interact with the immunoglobulin enhancer sequences. *Cell* 46: 705–716.

Sen, R. and Baltimore, D. 1986b. Inducibility of κ immunoglobulin enhancer-binding protein NF-κB by a post-translational mechanism. *Cell* 47: 921–928.

Shatkin, A. J. 1976. Capping of eucaryotic mRNAs. *Cell* 9: 645–653.

Sheiness, D. and Darnell, J. E. 1973. Polyadenylic segment in mRNA becomes shorter with age. *Nature New Biol.* 241: 265–268.

Shen, C. K. J. and Maniatis, T. 1980. Tissue-specific DNA methylation in a cluster of rabbit β-like globin genes. *Proc. Natl. Acad. Sci. USA* 77: 6634–6638.

Shermoen, A. W. and Beckendorf, S. K. 1982. A complex of DNAase I-hypersensitive sites near the *Drosophila* glue protein gene, Sgs-4. *Cell* 29: 601–607.

Simmons, D. M., Voss, J. W., Ingraham, H. A., Holloway, J. M., Broide, R. S., Rosenfeld, M. G. and Swanson, L. W. 1990. Pituitary cell phenotypes involve cell-specific Pit-1 mRNA translation and synergistic interactions with other classes of transcription factors. *Genes. Dev.* 4: 695–711.

Smale, S. T. and Baltimore, D. 1989. The "initiator" as a transcription control element. *Cell* 57: 103–113.

Sopta, M., Burton, Z. F. and Greenblatt. J. 1989. Structure and associated DNA-helicase activity of a general transcription initiation factor that binds to RNA polymerase II. *Nature* 341: 410–414.

Spritz, R. A. and nine others. 1981. Base substitution in an intervening sequence of a β⁺-thalassemic human globin gene. *Proc. Natl. Acad. Sci. USA* 78: 2455–2459.

Stalder, J., Larsen, A., Engel, J. D., Dolan, M., Groudine, M. and Weintraub, H. 1980. Tissue-specific DNA cleavages in the globin chromatin domain introduced by DNase I. *Cell* 20: 451–460.

Staudt, L. M., Singh, H., Sen, R., Wirth, T., Sharp, P. A. and Baltimore, D. 1986. A lymphoid-specific protein binding to the octamer motif of immunoglobulin genes. *Nature* 323: 640–643.

Stein, R., Gruenbaum, Y., Pollack, Y., Razin, A. and Cedar, H. 1982. Clonal inheritance of the pattern of methylation in mouse cells. *Proc. Natl. Acad. Sci. USA* 79: 61–65.

Steiner, E. K., Eissenbert, J. C. and Elgin, S. C. R. 1984. A cytological approach to the ordering of events in gene activation using the Sgs-4 locus of *Drosophila melanogaster*. *J. Cell Biol.* 99: 233–238.

Stief, A., Winter, D. M., Strätling, W. H. and Sippel. A. E. 1989. A nuclear DNA attachment element mediates elevated and position-dependent gene activity. *Nature* 341: 343–345.

Strähle, U., Schmid, W. and Schütz, G. 1988. Synergistic action of the glucocorticoid receptor with transcription factors. *EMBO J.* 7: 3389–3395.

Struhl, G., Struhl, K. and Macdonald, P. M. 1989. The gradient morphogen *bicoid* is a concentration-dependent transcriptional activator. *Cell* 57: 1259–1273.

Swain, J. L., Stewart, T. A. and Leder, P. 1987. Parental legacy determines methylation and expression of an autosomal transgene: A molecular mechanism for parental imprinting. *Cell* 50: 719–727.

Taylor, S. M. and Jones, P. A. 1982. Changes in phenotypic expression in embryonic and adult cells treated with 5-azacytidine. *J. Cell Physiol.* 111: 187–194.

Tazi, J. and Bird, A. P. 1990. Alternative chromatin structure at CpG islands. *Cell* 60: 909–920.

Thoma, F., Koller, T. and Klug, A. 1979. Involvement of histone H1 in the organization of the nucleosome and of the salt-dependent superstructures of chromatin. *J. Cell Biol.* 83: 403–427.

Treisman, J., Gönczy, P., Vashishta, M., Harris, E. and Desplan, C. 1989. A single amino acid can determine the DNA binding specificity of homeodomain proteins. *Cell* 59: 553–562.

Trudel, M. and Constantini, F. 1987. A 3′ enhancer contributes to the stage-specific expression of the human β-globin gene. *Genes Dev.* 1: 954–961.

Tuan, D., Abeliovich, A., Lee-Oldham, M. and Lee, D. 1987. Identification of regulatory elements in human β-like globin genes. *In* G. Stamatoyannopoulos and A. W. Nienhuis (eds.), *Developmental Control of Globin Gene Expression*. Alan R. Liss, New York, pp. 211–220.

Umesono, K. and Evans, R. M. 1989. Determinants of target gene specificity for steroid/thyroid hormone receptors. *Cell* 57: 1139–1146.

van der Ploeg, L. H. T. and Flavell, R. D. 1980. DNA methylation in the human γ–δ–β globin locus in erythroid and non-erythroid cells. *Cell* 19: 947–958.

van Eekelen, C. A. G. and van Venrooij, W. J. 1981. HnRNA and its attachment to a nuclear protein matrix. *J. Cell Biol.* 88: 554–563.

Villeponteau, B., Lundell, M. and Martinson, H. 1984. Torsional stress promotes the DNase hypersensitivity of active genes. *Cell* 39: 469–478.

Vogeli, G., Ohkubo, H., Sobel, M. E., Yamada, Y., Pastan, I. and de Crombrugghe, B. 1981. Structure of the promoter for chicken type I collagen gene. *Proc. Natl. Acad. Sci. USA* 78: 5334–5338.

Walker, M. D., Edlund, T., Boulet, A. M. and Rutter, W. J. 1983. Cell-specific expression controlled by the 5' flanking region of the insulin and chymotrypsin genes. *Nature* 306: 557–561.

Wang, A. H. J., Quigley, G. J., Kolpack, F. J., Crawford, J. L., vanBoom, J. H., van der Marel, G. and Rich, A. 1979. Molecular structure of a left-handed double helical DNA fragment at atomic resolution. *Nature* 282: 680–686.

Wasylyk, B., Kedinger, C., Corden, J., Brison, D. and Chambon, P. 1980. Specific in vitro initiation of transcription on conalbumin and ovalbumin genes and comparison with adenovirus 2 early and late genes. *Nature* 285: 367–373.

Weintraub, H. 1984. Histone H1-dependent chromatin superstructures and the suppression of gene activity. *Cell* 38: 17–27.

Weintraub, H. 1985. Assembly and propagation of repressed and derepressed chromosomal states. *Cell* 42: 705–711.

Weintraub, H. and Groudine, M. 1976. Chromosomal subunits in active genes have an altered configuration. *Science* 193: 848–856.

Weisbrod, S. and Weintraub, H. 1979. Isolation of a subclass of nuclear proteins responsible for conferring a DNase I-sensitive structure on globin chromatin. *Proc. Natl. Acad. Sci. USA* 76: 630–634.

Weisbrod, S., Groudine, M. and Weintraub, H. 1980. Interaction of HMG 14 and 17 with actively transcribing genes. *Cell* 19: 289–301.

Wigler, M., Levy, D. and Perucho, M. 1981. The somatic replication of DNA methylation. *Cell* 24: 33–40.

Wilkinson, D. G., Bhatt, S., Cook, M., Boncinelli, E. and Krumlauf, R. 1989. Segmental expression of *Hox*-2 homeobox-containing genes in the developing mouse hindbrain. *Nature* 341: 405–409.

Wilson, E. B. 1895. *An Atlas of the Fertilization and Karyogenesis of the Ovum.* Macmillan, New York.

Wirth, T., Staudt, L. and Baltimore, D. 1987. An octamer oligonucleotide upstream of a TATA motif is sufficient for lymphoid-specific promoter activity. *Nature* 329: 174–178.

Wolf, S. F. and Migeon, B. R. 1984. Clusters of CpG dinucleotides implicated by nuclease sensitivity as control elements of housekeeping genes. *Nature* 314: 467–469.

Wolf, S. F., Jolly, D. J., Lunnen, K., Friedmann, T., and Migeon, B. R. 1982. Methylation of the hypoxanthine phosphoribosyltransferase locus on the human X chromosome: Implications for X-chromosome inactivation. *Proc. Natl. Acad. Sci. USA* 81: 2806–2810.

Wolf, S. F., Dintgis, S., Toniolo, D., Persico, G., Lunnen, K. D., Axelman, J. and Migeon, B. R. 1984. Complete concordance between glucose-6-phosphate dehydrogenase activity and hypomethylation of 3' CpG clusters: Implication for X chromosome dosage compensation. *Nucleic Acid Res.* 12: 9333–9348.

Workman, J. L. and Roeder, R, G. 1987. Binding of transcription factor TFIID to the major late promoter during in vitro nucleosome assembly potentiates subsequent initiation by RNA polymerase II. *Cell* 51: 613–622.

Zaret, K. S. and Yamamoto, K. R. 1984. Reversible and persistent changes in chromatin structure accompany activation of a glucocorticoid-dependent enhancer element. *Cell* 38: 29–38.

13

Control of development by RNA processing

It may be in the interpretation and analysis of differentiation that the new concepts derived from the study of microorganisms will prove of the greatest value . . . Eventually, however, differentiation will have to be studied in differentiated cells.

—J. MONOD AND F. JACOB (1961)

Between the conception
And the creation . . .
Between the potency
And the existence
Between the essence
And the descent
Falls the Shadow.

—T. S. ELIOT (1936)

The essence of differentiation is the production of different subsets of proteins in different types of cells. In bacteria, differential gene expression can be effected at the levels of transcription, translation, and protein degradation. In eukaryotes, however, another possible level of regulation exists, namely, control at the level of RNA processing. This chapter will present some of the recent evidence suggesting that this type of regulation is crucial for development, and that different cells can process the same transcribed RNA in different ways, thus creating different populations of cytoplasmic messenger RNAs from the same set of nuclear transcripts.

During the 1960s, developmental biologists had a strong bias toward explaining differentiation in terms of transcriptional regulation of gene expression. First, research on polytene chromosomes and their products indicated that transcriptional regulation did occur during insect development. Second, DNA–RNA hybridization studies showed that whereas the DNA of different cell types was identical, their RNAs were different (McCarthy and Hoyer, 1964). Third, gene expression could be studied more readily in bacteria than in eukaryotic cells, and this research demonstrated the importance of transcriptional regulation in producing different physiological states in *Escherichia coli*. The *lac* operon of *E. coli*, for instance, was able to transcribe mRNA only in the presence of a specific inducer molecule (in this case, lactose). This seemed a powerful analogue for differential gene expression in eukaryotic cells.

Heterogeneous nuclear RNA

It soon became apparent, however, that messenger RNA was not the primary transcription product of eukaryotic cells. Rather, there was a NUCLEAR RNA (nRNA) that was much larger than messenger RNA and that had a much shorter half-life than the cytoplasmic messages. Because they also varied widely in size, these nuclear RNA molecules were called HETEROGENEOUS NUCLEAR RNA (hnRNA). The hnRNA molecules were approximately $5–19 \times 10^3$ nucleotides long, compared with the 2×10^3 bases usually found in mRNA, and they decayed with a half-life of minutes instead of hours (for review, see Lewin, 1980). In sea urchin embryos, the half-life of mRNA is about 5 hours whereas the half-life of nuclear RNA is about 20 minutes. The relationship between this heterogeneous nuclear RNA and the cytoplasmic mRNA was disputed until 1977, at which time several laboratories found mRNA sequences within the large nuclear RNA molecules.* Bastos and Aviv (1977) isolated mouse globin message and made complementary DNA (cDNA) from it by using reverse transcriptase (Figure 1). They then conjugated this cDNA to cellulose beads. After hemoglobin-synthesizing cells were grown with radioactive uridine, the radiolabeled RNA was extracted and passed over this column. Any RNA sequence coding for globin should bind to the column. No other RNA sequence should be "fished out" in such a manner. The RNA bound to the column was extracted from the column by washing it with low ionic solutions, which broke the bonds holding the complementary molecules together. When cells were labeled for 5 minutes, most of the radioactive RNA binding to the column had a sedimentation value of 27 S and a size of approximately 5000 nucleotides (nearly eight times larger than the cytoplasmic globin message).

However, when the labeling was stopped and the RNA extracted 5 minutes later, the globin sequences were found in a smaller 15 S piece. Finally, the globin-coding RNA was found in a 10 S (600-nucleotide) sequence characteristic of the cytoplasmic message. Thus, heterogeneous nuclear RNA appears to contain the precursors of cytoplasmic globin mRNA. When Hames and Perry (1977) made cDNA to the *entire* message population of cultured mouse cells, they obtained a similar result: the cDNA probe could bind to enormously large nuclear RNA sequences. Therefore, eukaryotic DNA transcribes large nuclear RNAs, which serve as the precursors to the smaller cytoplasmic messages.

Complexity of nuclear and messenger RNAs

The next question involved the number of different types of RNA sequences found in the nucleus and cytoplasm. Does the nucleus contain a greater diversity of sequences than the cytoplasm, or are they about the same? The answer to this question will tell us whether more of the genome is transcribed into nRNA than the amount we can measure from mRNA. The diversity of nucleotide sequences is called COMPLEXITY. To measure complexity, a fixed amount of radioactive RNA is added to an excess

*Although the terms *nuclear RNA* (nRNA) and *heterogeneous nuclear RNA* (hnRNA) are often used interchangeably, they do have slightly different meanings. hnRNA usually refers to message precursors (sometimes called pre-mRNA), whereas nRNA refers to the total unfractionated RNA extracted from the nucleus. We will use nRNA, rather than hnRNA, throughout.

FIGURE 1
Protocol for isolating nuclear RNA (nRNA) containing globin-coding sequences.

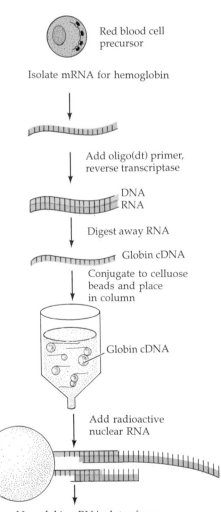

Red blood cell precursor

Isolate mRNA for hemoglobin

Add oligo(dt) primer, reverse transcriptase

DNA
RNA

Digest away RNA

Globin cDNA

Conjugate to celluose beads and place in column

Globin cDNA

Add radioactive nuclear RNA

Non-globin nRNA elutes from column; Add low ionic buffer

Radioactive RNA eluted containing globin nRNA sequences

FIGURE 2

Reassociation of nucleic acids from various sources. The curve represents the normal kinetics expected when each gene is present in the same frequency. Because the hybridization reaction demands that two complementary sequences come together, the rate of hybridization can be described as $dC/dt = -kC^2$, where C is the concentration of single-stranded sequences present at time t, and k is the association constant. Note that the ordinate is plotted with 0 percent at the top and that the C_0t axis is represented logarithmically. Also note that the DNA of *E. coli* is about 100 times as many sequences as the DNA of the T4 bacteriophage has. (After Britten and Kohne, 1968.)

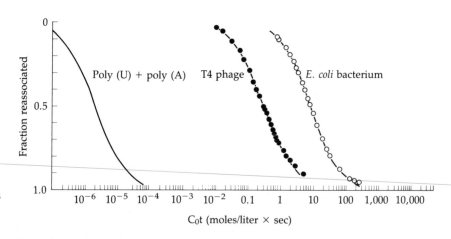

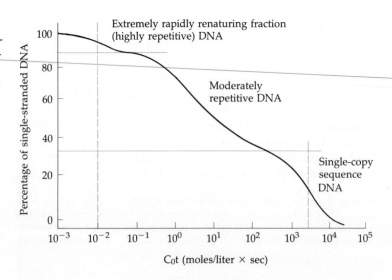

FIGURE 3

Reassociation of DNA from a typical eukaryotic organism. DNA is sheared to a small size—about 500 base pairs —and denatured by alkali into single strands. The pH is adjusted and the DNA allowed to renature. The curve shows a rapidly renaturing fraction that represents highly repetitive DNA sequences, a fraction of moderately repetitive DNA, and a fraction of single-copy DNA sequences. The highly repetitive sequences can be seen here to be represented about a million times more frequently than the single-copy genes. Highly repetitive DNA is usually localized to the centromeres and serves a structural function. The single-copy DNA includes genes such as those for enzymes and structural proteins. The moderately repetitive DNA is thought to contain those regulatory sequences found in the introns and 5' and 3' flanking regions. (After Hood et al., 1975.)

amount of denatured DNA and allowed to renature to it. The more complex the RNA, the slower its rate of renaturation to DNA. In other words, if there were only one type of RNA present (complexity = 1), this RNA sequence would be distributed throughout the reaction mixture and would find its DNA homologues rather quickly. However, if that same small amount of RNA contained a thousand sequences (complexity = 1000), each individual sequence would be a thousandfold more dilute and would have a small chance of finding its homologue in the same amount of time. The double-stranded DNA–RNA hybrid can be readily separated from the unhybridized RNA, and the percentage of the radioactive RNA that has found its complement can be calculated.

The analysis of complexity is often graphed on C_0t curves. C_0 refers to the original concentration of labeled nucleic acid (in moles per liter) and t refers to the time (in seconds). The extent of reassociation should be a function of both concentration and time. Thus, the percentage of renatured RNA is plotted against the C_0t value (Figure 2). This type of analysis has been performed with DNA as well as with RNA and has revealed that the genome contains some DNA sequences represented millions of times (the highly repetitive fraction), some sequences represented thousands of times (the moderately repetitive fraction), and some DNA (such as those genes coding for enzymes) that are essentially represented once per haploid genome (Figure 3) (Britten and Kohne, 1968).

By this method the nRNA complexity was determined to be much

TABLE 1
Approximate kinetic characteristics and relative complexity of messenger and
nuclear RNA populations in various species

| Organism | Proportion of nRNA converted to poly(A)⁺-containing mRNA | |
	% of original mass	% of original complexity
Drosophila	14–20	28–40
Aedes (mosquito)	~3	~13
Sea urchin	4–7	17–29
Cultured human carcinoma (HeLa) cells	3–6	15–30

Source: Lewin (1980).

greater than that of mRNA from the same cells. First, certain nRNA
sequences associated with DNA very rapidly. This indicated that they
contained RNA transcribed from repetitive DNA sequences. In the sea
urchin blastula, cytoplasmic mRNA contains no sequences that hybridize
to repetitive DNA, whereas approximately 25 percent of the nRNA sheared
to obtain 1100-nucleotide pieces can associate with DNA at low C_0t values.
Because the average length of sea urchin repetitive sequences in nRNA is
approximately 300 bases, it is estimated that 70 percent of the nRNA
molecules contain at least one repetitive sequence (Smith et al., 1974).
However, if the RNA is reacted with purified nonrepetitive DNA (so that
the repetitive sequences do not complicate matters), the complexity of
nRNA is usually found to be 4 to 20 times greater than that of mRNA
from the same organ or cell type (Table 1). From such studies, it is estimated that in the sea urchin embryo, only 10–20 percent of nRNA complexity gets into the cytoplasmic mRNA population. In rat liver, only 11
percent of the nRNA sequence types become message, and in cultured
insect cells, the figure is as low as 5 percent. We can conclude that the
nuclear transcripts contain (1) a small population of nRNA, which is the
precursor to the mRNA of the cell, and (2) a large population of nRNA,
which turns over rapidly within the nucleus. The population of nRNA
molecules contains numerous sequences different from those that become
processed into mRNA.

Control of early development by nRNA processing

According to the previous section, the genome is transcribing many more
sequences than those that become mRNA. This conclusion suggests that
much of the control of differentiation may take place *after* transcription by
the differential processing of specific mRNA precursors. Studies on sea
urchins and rats confirm this.

In 1977, Kleene and Humphreys obtained single-copy sea urchin DNA
by denaturing blastula DNA and isolating whatever did not reassociate at
a C_0t value of 200. Nuclear RNA from blastula cells was found to bind to
15 percent of this DNA. Similarly, pluteus-stage nRNA, even when present
in great excess, also bound to only 15 percent of this DNA. Because only
one strand of DNA is complementary to RNA, 15 percent of DNA hybridized corresponds to 30 percent of the genome. Therefore, about 30 percent
of the genome is actively transcribing blastula nRNA and about 30 percent

FIGURE 4

Hybridization of nuclear RNA with single-copy [³H]DNA. Radioactive single-copy DNA was mixed with either blastula RNA, pluteus RNA, or a mixture of blastula and pluteus RNAs. The mixtures were incubated to allow all complementary sequences to pair. In all three cases, about 15 percent of the DNA hybridized to the RNA. (After Kleene and Humphreys, 1977.)

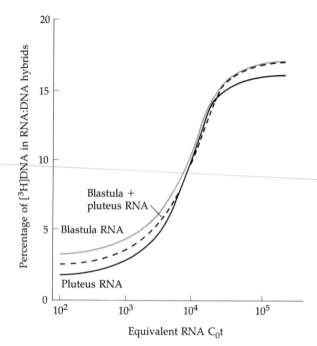

of the genome is actively transcribing pluteus nRNA. Are these two sets of DNA sequences the same or are they different? This question was approached by mixing blastula and pluteus nuclear RNAs and adding them to the denatured single-copy DNA. If the sequences were totally different, one would expect 30 percent of the DNA to be bound (i.e., 60 percent of the genome would be coding for the combined set of blastula and pluteus messages). If they were identical, one would expect 15 percent of the DNA to be bound. The result is shown in Figure 4. The mixture bound to only 15 percent of the DNA. The nRNA sequences of the blastula and pluteus bound to the same DNA. Within experimental error, the nRNA was identical in blastula and pluteus cells.

To confirm this unexpected observation, Kleene and Humphreys (1977, 1985) isolated radioactive single-copy DNA and mixed it with pluteus or blastula RNA at a high C_0t value (so that all DNA sequences could bind the RNA if their complementary RNA sequences were present). They then isolated those DNA fragments that would not bind to blastula nRNA (blastula-null DNA) and those DNA fragments that would not bind to the pluteus nRNA (pluteus-null DNA). When blastula-null DNA was denatured and mixed to new samples of *pluteus* nRNA and when pluteus-null DNA was mixed with *blastula* nRNA, no hybridization above background levels was observed (Figure 5). Therefore, the DNA sequences transcribed during the blastula stage are the same as DNA sequences transcribed during the pluteus stage.

Some of the most impressive evidence for processing-level control of development, however, came from the laboratory of Eric Davidson and Roy Britten. They first discovered that in the sea urchin *Strongylocentrotus purpuratus*, the complexity of the mRNAs becomes progressively less as development proceeds (Galau et al., 1976). Oocyte mRNA binds to approximately 4.5 percent of the genome and has a complexity (measured by C_0t) of about 37,000 kilobases. Given an average mRNA length of 2000 bases, this represents about 18,500 different types of messenger RNAs. Blastula mRNA binds to only 3.1 percent of the genome and represents about 13,000 types of mRNA, and gastrula mRNA binds to only 2.0 percent of the DNA and has 8000 species of messages. By the time tissues have

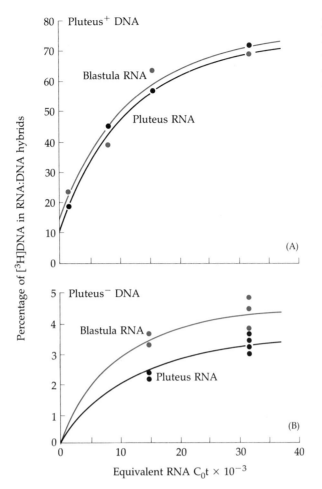

FIGURE 5
Hybridization of "pluteus-plus" DNA and "pluteus-null" DNA by pluteus and blastula RNA. (A) Both pluteus and blastula RNA bound "pluteus-plus" DNA to the same extent. (B) Neither RNA bound significantly to the "pluteus-null" DNA. (After Kleene and Humphreys, 1977.)

differentiated, they each contain about 6000 mRNA types, which hybridize to only 0.75 percent of the genome (Figure 6). Thus, the amount of the genome expressed in cytoplasmic mRNA is progressively restricted.

They next asked whether this restriction was due to changes in genomic transcription. In other words, is the nRNA similarly restricted? To

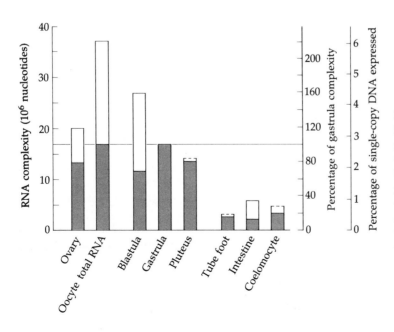

FIGURE 6
Sets of structural genes expressed as messenger RNAs in various sea urchin tissues. The shaded portion of each bar represents the amount of mRNA in each tissue that is identical to gastrula mRNA. The unshaded portion of each bar represents the amount of mRNA that is not shared in common with gastrula messages. Complexity is indicated in three ways on the ordinate; the horizontal line indicates 100 percent of gastrula mRNA complexity. (After Galau et al., 1976.)

FIGURE 7

Specificity of sea urchin blastula mDNA. Hybridization of blastula mDNA (cDNA to blastula mRNA) with blastula mRNA and intestinal cytoplasmic RNA. (After Wold et al., 1978.)

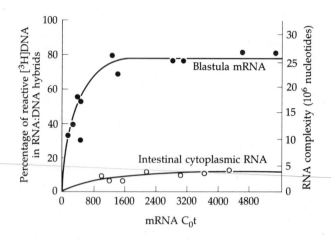

answer this question, Wold and her colleagues (1978) isolated blastula mRNA and hybridized it to denatured radioactive DNA. The resulting hybrids were eluted, denatured, and the DNA isolated from the RNA bound to it. This produced a DNA probe that would recognize blastula mRNA sequences. This DNA was highly specific for blastula mRNA sequences (Figure 7): about 78 percent of this "blastula cDNA" could bind to blastula messages whereas less than 10 percent of it could bind to adult intestinal mRNA.

They could then use this "blastula mDNA" to find any blastula mRNA sequences in the RNA of various differentiated tissues. The results are shown in Figure 8 and Table 2. Blastula mRNA sequences were found in the RNA of all three adult tissues tested. About 80 percent of the blastula message DNA hybridized to gastrula, intestine, and coelomocyte nRNAs. Thus, only 10 percent of a probe capable of recognizing blastula *message* sequences will react with intestinal *messenger* RNA, but about 80 percent of it will react with intestinal *nuclear* RNA. Because the control reaction (blastula mDNA with blastula mRNA) also gave a value of 75–80 percent, it appears that *all* of the blastula message sequences are present as nRNA

FIGURE 8

Identity (within experimental limits) of nuclear RNA from differentiated sea urchin tissues. (A) Hybridization of blastula mDNA to *nuclear* RNAs of intestine, coelomocytes, and gastrulae. The hybridization mixtures were incubated at a C_0t high enough to allow all complementary sequences to pair. All blastula sequences were found in each nuclear RNA population—although not in their cytoplasm (Figure 5). (B) Diagram of model based on differential RNA processing. In both cell types, the same RNAs (a, b, c, d, e) are transcribed, but in one cell type, sequences c, d, and e are processed into the cytoplasm, whereas in another cell type sequences a, b, and c are processed into the cytoplasm. (A from Wold et al., 1978.)

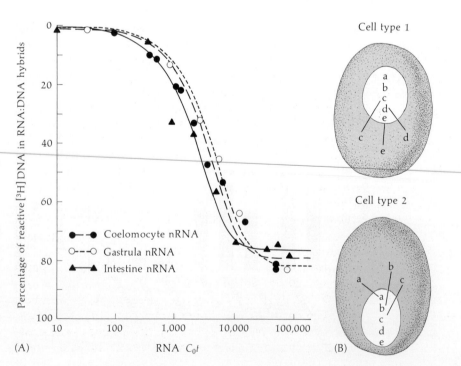

TABLE 2
Intertissue comparisons of structural gene sequences in messenger RNA and nuclear RNA

Reference tracer complementary to	Normalized reaction with parent mRNA		Normalized reaction with other mRNA		Normalized reaction with nRNA	
	mRNA	%	mRNA	%	nRNA	%
SEA URCHIN						
Blastula mRNA (single copy DNA)	Blastula	100	Intestine	12	Intestine	97
			Coelomocyte	13	Coelomocyte	101
MOUSE						
Brain mRNA (total cDNA)	Brain	100	Kidney	78	Kidney	102
Brain mRNA (cDNA representing rare messages)	Brain	100	Kidney	56	Kidney	100

Source: Davidson and Britten (1979).

transcripts in all the nuclei of intestine, coelomocyte, and gastrula cells, despite their absence from the cytoplasm. The nucleus is transcribing blastula-specific sequences even in differentiated coelomocytes and intestine cells. It appears, then, that the control of sea urchin gene expression occurs predominantly at the level of RNA processing. Moreover, similar experiments on rats and mice suggest the near identity of nuclear RNAs from brain, liver, and kidneys (Chikaraishi et al., 1978).

Evidence for unprocessed message precursors in the nucleus

Because evidence from sea urchins shows that the population of nRNAs in the various differentiated cells are identical (within the limits of resolution), it would be expected that probes for specific mRNAs of one cell type should find them in nuclei of nonexpressive cells. By using cDNA made from globin message, investigators have found globin mRNA precursors in the total cellular RNA from mouse liver, brain, and cultured cell lines (Humphries et al., 1976) and in the nRNA of uninduced Friend erythroleukemia cells (Gottesfeld and Partington, 1977) and the *Xenopus* oocyte (Perlman et al., 1977). Low-level transcription of ovalbumin genes has been detected in chicken liver, spleen, brain, and heart (Axel et al., 1976; Tsai et al., 1979). This value approximates one molecule of ovalbumin coding sequence in every two cell nuclei; even RNAs that are known to be regulated transcriptionally can be found in very low amounts in the nuclei of inappropriate cell types.

Some of the best evidence for specific processing-level control of RNA transcripts comes from experiments on viruses in mouse cells. Here, certain cell types can splice nRNA to create certain messages and other cells cannot. Polyoma virus and simian vacuolating virus 40 (SV40) infect cells and integrate into host cell chromosomes. Here, they produce their own RNA, which makes virus-specific products. Like the RNA of their host cell, polyoma and SV40 transcripts must first be spliced to delete introns. Only after the viral RNA is processed can stable viral messages accumulate in the cytoplasm. (The messages of certain other viruses that replicate in the cytoplasm do not need to be spliced.) Undifferentiated teratocarcinoma stem cells (Chapter 6) will not support SV40 or polyoma virus infection or gene expression. However, if the stem cells are allowed to differentiate into somatic cell types, they suddenly become susceptible to infection by those two viruses. The basis for this initial nonpermissiveness appears to be the failure of RNA processing. Undifferentiated teratocarcinoma cells will not splice the SV40 message precursors, whereas the differentiated cells will. In the stem cells, large message precursors accumulate and are eventually degraded. Only the differentiated cell types are able to process this nuclear RNA to a cytoplasmic message (Segal et al., 1979). The viruses that do not need RNA processing for gene expression (Sindbis and vaccinia) are perfectly capable of growing in both the teratocarcinoma stem cell and its differentiated derivatives. The observations that SV40 RNA splicing occurs only in differentiated mouse cells (Topp et al., 1977) and teratocarcinoma derivatives suggest that cell development may be dependent on the expression of the splicing enzymes themselves. The agents responsible for processing SV40 RNA undoubtedly have more appropriate roles to play in the cell, and they may be responsible for the processing of several message precursors necessary for cell differentiation.

Creating alternative proteins from the same gene: Differential RNA processing in the immune system

Differential RNA processing has been seen to be extremely important in the formation of several specific proteins. The best-studied examples concern the generation of different immunoglobulin (antibody) proteins during the differentiation of the B lymphocyte. The expression of antibody-coding genes during development of lymphocytes and their responses to antigen involve numerous events of selection and rearrangement of gene segments. Much of this selection and rearrangement occurs at the DNA level, as we saw in Chapters 10 and 12. But several aspects of the regulation of antibody gene expression take place at the RNA processing level.

During the development of a B lymphocyte, before it becomes an antibody-secreting cell it produces samples of the type of antibody it will be capable of secreting and inserts them into the plasma membrane. Here they function as antigen receptors. Depending on the stage of development and whether or not the cell has been exposed to the type of antigen with which its antibodies will react, the surface immunoglobulins may be of either the IgM or the IgD class. Initially they are of the IgM class; after the first antigen stimulation, both IgM and IgD may be made simultaneously; and later only IgD may be on the cell surface. But the "variable" (antigen binding) portion of the immunoglobulin remains the same throughout the life history of the specific lymphocyte and its progeny cells. It has now been shown (Maki et al., 1981) that switching between IgM and IgD occurs at the RNA processing level. A single RNA transcript is synthesized, and it contains the exons coding for the variable (antigen-binding) region plus the mu (IgM) and the delta (IgD) constant regions, separated by various introns. Whether the mRNA that reaches the cytoplasm will code for IgM or IgD depends on which of two alternative splicing mechanisms are used. If the delta (δ) exons are spliced out and discarded, IgM will be made. If the mu (μ) exons are removed by splicing, however, IgD will be made. During transient periods when both splicing mechanisms are being utilized, both classes of immunoglobulins are produced (Figure 9).

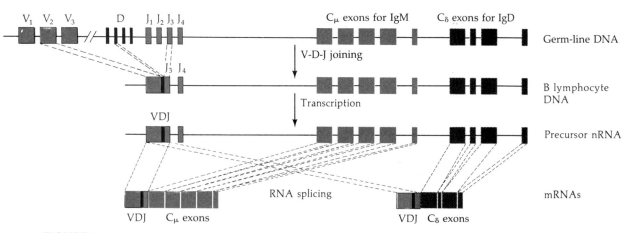

FIGURE 9
Proposed model for the expression of IgM and IgD in B lymphocytes. The heavy chain gene is made during lymphocyte development by the translocation of V, D, and J region sequences to a position adjacent to the constant region sequences. The transcript of this gene includes both the μ (IgM) and δ (IgD) constant regions along with the newly constructed variable region. Alternative pathways of processing can delete μ constant region (Cμ) exons or δ constant region (Cδ) exons. (After Liu et al., 1980.)

The decision to make IgM or IgD is not the only decision mediated by differential RNA processing. The placement of these proteins—whether they are to be inserted into the plasma membrane or secreted—is also determined in this manner.

IgD is usually confined to the cell membrane, and it is thought that serum IgD may be a result merely of accidental leakage or cell death. IgM, however, can exist in two distinct forms, membrane-bound (mIgM) and secreted (sIgM). Membrane-bound IgM acts as an antigen receptor, whereas secreted IgM is the first type of antibody found in the blood when an organism is exposed to an antigen for the first time. Several investigators have found that sIgM and mIgM differ from each other at their carboxyl ends. The molecules are identical except that the membrane-bound form contains a hydrophobic "tail" that keeps it inserted in the membrane (Figure 10). The mRNAs for membrane and secreted forms of IgM also differ. Although both of them are found to be transcribed from the same Cμ gene, they are processed differently (Rogers et al., 1980; Early et al., 1980; Alt et al., 1980). The secreted IgM contains regions encoded by the *VDJ* genes and exons $C\mu_1$, $C\mu_2$, $C\mu_3$, and $C\mu_4$. It also contains a terminal portion that allows it to be secreted. The membrane-bound IgM contains the same arrangement except that instead of the "secretion" terminal portion, it has added a portion encoded by two more exons, $C\mu_5$ and $C\mu_6$, which gives it a hydrophobic tail that can integrate into the lymphocyte membrane. The decision as to whether the molecule is to become sIgM or mIgM is determined by RNA splicing (Figure 11).

There is also a third step in antibody formation where differential RNA processing plays an important role. Each B lymphocyte makes only one type of antibody molecule, which is encoded by only one of the two homologous chromosomes having such genes. This phenomenon is called ALLELIC EXCLUSION. What happens to the other chromosome-bearing antibody genes? Perry and co-workers (1980) have found nuclear RNAs corresponding to unrearranged or inappropriately rearranged antibody genes, a finding suggesting that the "unexpressed" antibody genes are being transcribed, but are not being processed into mRNA.

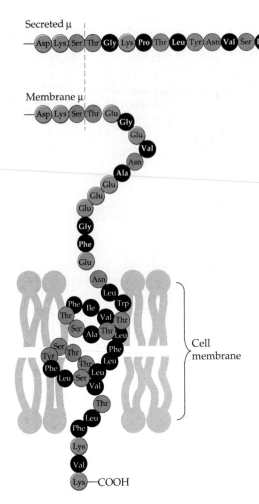

FIGURE 10
The carboxyl terminal amino acids of membrane (m) and secreted (s) forms of IgM. (A) Carboxyl ends of the μ_s and μ_m chains. (B) Possible configuration of the membrane-bound chain; this configuration anchors the molecule in the cell membrane. Hydrophobic residues are shown in black, charged residues are in color, and relatively uncharged polar residues are shown in gray. (After Rogers et al., 1980.)

FIGURE 11
Proposed alternative splicing patterns for secreted μ and membrane μ mRNA. P includes the leader sequence; V encodes the variable (VDJ) region; the possible constant region exons of the μ (IgM) constant region follow downstream. The spliced RNAs (introns) are shown as dashed lines. (After Early et al., 1980.)

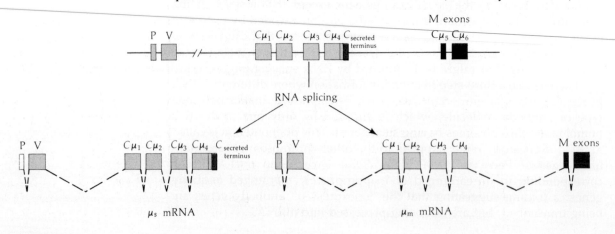

Differential RNA processing:
Generating new proteins in different times and places

Although it has been studied most intensely in immunoglobulin gene expression, differential RNA processing has been found to control the alternative forms of expression of over 50 proteins (Breitbart et al., 1987). In some instances, RNA splicing can generate different proteins in different cells. The processing of a particular mRNA precursor in certain thyroid cells will yield the message for the hormone calcitonin. The same mRNA precursor in nerve cells, however, will be processed into the message for a neuropeptide, CGRP (Figure 12; Amara et al., 1982; Crenshaw et al., 1987). This phenomenon is interesting in its own right and also because another alternatively spliced transcript generates substance P in neurons and substance K in the same thyroid cells that make calcitonin. Crenshaw and co-workers speculate that there may be neuron-specific splice-regulating factors that could function on a variety of alternately splicable precursors. This would enable the generation of a tissue-specific array of proteins.

In a similar way, the five different human fibronectin proteins are thought to be generated from the one type of fibronectin gene. The diverse (and in some instances, organ-specific) forms of fibronectin come from different mRNAs generated by splicing together different exons of the fibronectin mRNA precursor (Tamkun et al., 1984; Hynes, 1987). Some forms of fibronectin are found on the pathways over which embryonic cells migrate whereas other forms are not, suggesting that the alternative splicing of the fibronectin gene may be important for different embryonic functions (ffrench-Constant and Hynes, 1989). Alternative splicing of mRNA precursors is also seen to generate different forms of the neural cell adhesion molecule (NCAM) in different tissues (Cunningham et al., 1987; Thompson et al., 1989). The soluble serum protein that binds growth hormone and the membrane-bound growth hormone receptor are encoded by the same gene. The nuclear transcript is alternatively spliced to make these two proteins (Baumbach et al., 1989). Thus, alternative RNA splicing can generate a family of proteins from a single gene.

If the same message precursor is processed differently in different cell types, then one would expect the mRNA produced to depend upon the processing factors present in those cells. Breitbart and Nadal-Ginard (1987) have shown that the RNA for troponin T, a muscle-specific protein responsible for conferring calcium-ion sensitivity to actinomyosin, is generated in the muscle cell nucleus by splicing together certain particular exons. The ability to correctly process the troponin T precursor RNA appears to be directed by a muscle-specific splicing factor. Nonmuscle cells do not put the correct exons together, and neither do mononucleate myoblast cells. Only the multinucleated myotube cells have a factor that can properly process the precursor nRNA for this muscle-specific protein.

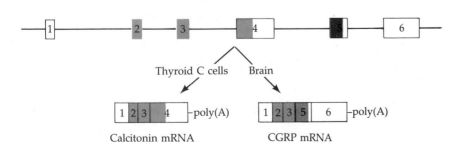

Thyroid C cells   Brain

Calcitonin mRNA   CGRP mRNA

FIGURE 12
Alternative splicing of the calcitonin/ CGRP gene. In thyroid C cells, exons 1, 2, 3, and 4 are utilized to make the message for calcitonin. In brain cells, exons 1, 2, 3, 5, and 6 are used to synthesize the message for the CGRP neuropeptide. (From Crenshaw et al., 1987.)

The biochemistry of RNA processing

How are the message precursors spliced? How might different messages be generated from the same message precursor? Studies on the biochemistry of RNA processing are just beginning to bring light to this relatively unknown area. Two of the most important discoveries in the past few years have been the identification of a branched RNA intermediate and the identification of catalytic ribonucleoprotein particles within the nucleus.

Splice sites

Comparison of the DNA sequences of many different genes revealed certain similarities at the intron–exon junctions. The base sequence of the intron usually begins with GT, and the base sequence of the exon usually ends with AG (Table 3). In fact, the "consensus sequence" for 5' ends of introns is AG/GT (i.e., most exons end in AG and the intron begins with GT) and the 3' sequence of most introns is TTNCAG (where *N* can be any nucleotide). These sites appear to be very important for intron processing. In the disease β-thalassemia, there is a genetic deficiency of β-globin production. The mutations, however, are not in the exons coding for the globin protein; rather, they are in the *introns* or regulatory regions. In one of these thalassemia mutations, the sequence TTGGTCT in the first intron is mutated to TTAGTCT. However, the usual 3' splicing site for this intron is TTAG/GCT. Thus, the mutation makes an artificial splicing site, and in so doing creates the possibility for erroneous splicing (Spritz et al., 1981). In fact, the thalassemic β-globin precursor splices at the wrong junction over 90 percent of the time (Busslinger et al., 1981).

Branch sites

In addition to the splicing junctions, there is one other sequence necessary for proper cleavage of the introns. This is the BRANCH SITE. This site is located within the intron, generally about 20–50 nucleotides upstream from the 3' splice site and has the consensus sequence UAUAAC. In most cases, U can be replaced by C, and A can be replaced by G. The penultimate (underlined) A residue, however, is invariant. This A residue is critical, since it will create a

TABLE 3
Nucleotide sequences at the junctions between coding regions and intervening sequences[a]

Organism	Gene and intron	Coding sequence	Intervening sequence	Coding sequence
Chicken	Ovalbumin 1	... TCAAAAG	GTAGGC ... TGCTCTAG	ACAACTC ...
	Ovalbumin 2	... AAATAAG	GTGAGC ... AATTACAG	GTTGTTC ...
	Ovalbumin 3	... AGCTCAG	GTACAG ... GTATTCAG	TGTGGCA ...
	Ovalbumin 4	... CCTGCCA	GTAAGT ... CTTTACAG	GAATACT ...
	Ovalbumin 5	... ACAAATG	GTAAGT ... TCTTAAAG	GAATTAT ...
	Ovalbumin 6	... GACTGAG	GTATAT ... TGCTCTAG	CAAGAAA ...
	Ovalbumin 7	... TGAGCAG	GTATAT ... CCTTGCAG	CTTGAGA ...
Mouse	β^maj globin 1	... TGGGCAG	GTGAGC ... CTTTTTAG	GCTGCTG ...
Rabbit	β globin 1	... TGGGCAG	GTGAGC ... CTTTTTAG	GCTGCTG ...
Mouse	β^maj globin 2	... CTTCAGG	GTGAGT ... TCCCACAG	CTCCTG ...
Rabbit	β globin 2	... CTTCAGG	GTGAGT ... TCCTACAG	CTCCTG ...
Mouse	α globin 2	... CTTCAAG	GTATGC ...	
	V_γII immunoglobulin 1	... TGCTCAG	GTCAGC ... GTTTGCAG	GAGCCA ...
	C_γ1 immunoglobulin 2	... TGTACAG	GTAAGT ... ATCCTTAG	TCCAGA ...
Rat	Insulin I & II–1	... CAAGCAG	GTACTC ... TCTTCCAG	GTCATTG ...
	Insulin II–2	... CCACAAG	GTAAGC ... CCTGGCAG	TGCACA ...
Bombyx mori	Silk fibroin	... TCTGCAG	GTGAGT ... TGTTTCAG	TATGTCG ...
Consensus sequence		AG	GT TTNCAG	

Source: Lewin (1980).

[a]The DNA sequences homologous to the RNA are presented here with the junctions between exons and introns aligned. The exact point of splicing cannot be assigned because the last bases in the first coding region are often repeated in the right end of the intron. At the bottom of the table, the consensus sequence common to most exon/intron boundaries (AG/GT) is identified, as is the consensus sequence for most intron/exon junctions (TT*N*CAG).

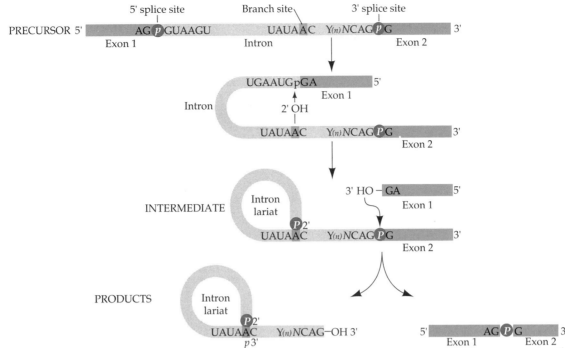

FIGURE 13

Postulated mechanism for splicing an mRNA precursor. The top figure represents an intron of an mRNA precursor and its two flanking exons. The boundary consensus sequences and the branch point consensus sequence are indicated. In splicing, the 2′ OH of the branch site forms a covalent bond to the phosphate adjacent to one of the nucleotides of the 5′ splice site. This creates a "lariat" structure and a free exon 1. The terminal OH group of this exon can then replace the intron at the 5′ end of the second exon. (Modified from Sharp, 1987.)

phosphodiester "branch" with the phosphate of the 5′ splice site (Figure 13). It does this by using its free 2′ hydroxyl group to combine with the phosphate of the GMP at the 5′ splicing site. In effect the GMP phosphate, instead of holding on to the first exon, releases it and hooks on to the 2′ hydroxyl group of the branch site AMP. This maneuver releases the first exon and makes a "lariat" structure as an intermediate.

The free 3′ OH terminus of the first exon then binds to the phosphate at the 3′ splice junction downstream from this lariat, thereby substituting itself for the intron. The result is the linkage of the two exons with the AG group coming from the first exon and the G coming from the second (Sharp, 1987).

This lariat-shaped intermediate with its 2′–5′ phospho-

diester bond was discovered by using a cloned fragment of the human β-globin gene containing the first two exons and the surrounded intron (Ruskin et al., 1984). The RNA transcribed from such a clone was placed into an extract of human cell nuclei capable of splicing this "precursor" at the appropriate sites (Krainer et al., 1984). This in vitro reaction takes about an hour to complete, and the intermediates can be isolated on gels (because the lariat structure impedes their migration during electrophoresis). Chemical analysis of these intermediates confirmed that the branch site A had two phosphodiester linkages, one from its 2′ OH group and one from its 3′ OH group. Shortly thereafter, Zeitlin and Efstratiadis (1984) identified similar intermediates for β-globin transcripts being processed in vivo in fetal rabbits.

FIGURE 14

Binding of U1 snRNA to 5′ splice region of an intron. Consensus sequence of the 5′ exon/intron boundary is complementary to the 5′ end of the U1 snRNA. The (m) indicates a methyl group and the asterisk marks a post-transcriptional modification of uridine. (After Padgett et al., 1985.)

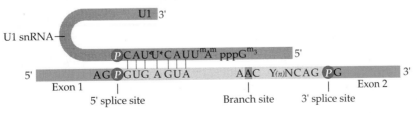

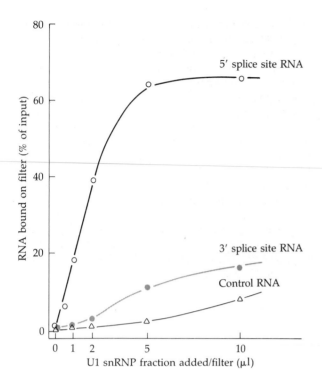

FIGURE 15
Specific binding of the U1 snRNP to purified RNA fragments. Little specific binding is seen when the U1 snRNP is combined with bulk RNA or with the 3' splicing junction. When combined with fragments containing the 5' splicing junction, this snRNP binds to the RNA. (From Tatei et al., 1984.)

teins. These small ribonucleoprotein particles are called snRNPs ("snirps"). There are five major snRNAs—U1, U2, U4, U5, and U6—each present between 2×10^5 and 10^6 copies per nucleus, and each containing from 56 to 217 nucleotides. In addition, numerous other snRNAs are present in smaller amounts. Most snRNPs contain one snRNA molecule complexed with approximately ten proteins. The U4 and U6 snRNAs are together in one particle called the U4/U6 snRNP (Maniatis and Reed, 1987).

The first suggestion that snRNPs were involved in splicing came from the observation that the 5' terminal region of U1 snRNA was complementary to the 5' splice site of the introns (Figure 14; Lerner et al., 1980; Rogers and Wall, 1980). More recent experiments (Mount et al., 1983; Tatei, 1984) have shown that purified U1 snRNP binds specifically to the 5' splicing site sequences in the RNA precursors (Figure 15). The U2 snRNP binds specifically to the branch site by base pairing (Wu and Manley, 1989; Zhuang and Weiner, 1989). If U2 snRNP is mixed with mRNA precursors and then RNase is added, the only RNA protected from digestion (since it is covered by the U2 snRNP) is the branch site and the nucleotides surrounding it (Black et al., 1985). The U5 snRNP binds to the 3' splice site, since this snRNP is able to protect the 3' splicing site from RNase digestion (Chabot et al., 1985). This interaction is thought to be mediated by the recognition of the 3' splice sequence by the protein rather than by the RNA portion of the snRNP (Tazi et al., 1986).

Splicing occurs, however, only after snRNPs interact

The roles of small nuclear ribonucleoprotein particles

In metazoan nuclei, these splicing reactions do not happen spontaneously.* Rather, they are directed by catalytic particles composed of small nuclear RNAs (snRNAs) and pro-

*Remarkably enough, such spontaneous splicing does occur in removing the introns of yeast mitochondrial genes and the introns of the ribosomal RNA in the protist *Tetrahymena* (see Sharp, 1987).

FIGURE 16
Formation of spliceosomes in mammalian cell nuclei. (A) The proposed splicing pathway in mammalian cells, wherein a cascade of snRNPs forms a spliceosome on the intron of the pre-mRNA. (B) Isolation of spliceosomes bound to fragment of human β-globin pre-mRNA. (A after Mattaj and Hamm, 1989; B from Reed et al., 1988; photograph courtesy of the authors.)

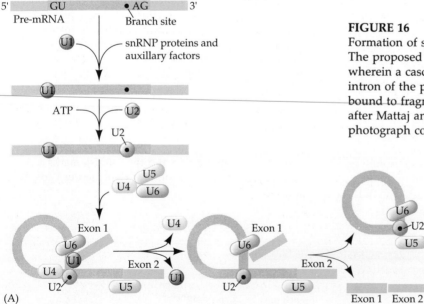

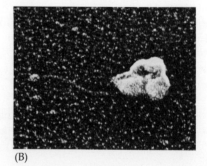

with one another to form a SPLICEOSOME on the message precursor. Such particles have been isolated from live cells and from in vitro assemblages of snRNPs and pre-mRNAs (Figure 16). The splicosome assembly is probably initiated by the binding of U1 snRNP to the 5′ end splice junction of the intron. Although the details may differ between phyla, it is generally thought that the U2 snRNP binds next, selecting the branch site (Nelson and Green, 1989). The U4/U6 snRNP binds to the U5 snRNP and this particle binds to the assembly of U1 and U2 on the pre-mRNA (Pikielny et al., 1989). After the U4 (but not the U5 or U6 snRNPs) leaves this complex, the spliceosome is ready to cleave the pre-mRNA. The spliceosome, like the ribosome, is involved in bringing different RNAs into proximity with one another and synthesizing new covalent bonds. The observation that certain snRNAs and their associated proteins can change during development has heightened speculations that tissue-specific or temporally specific alternative RNA splicing may be directed by these different snRNPs (McAllister et al., 1988; Mattaj and Hamm, 1989).

It should be remembered that spliceosomes do not meet "naked" mRNA precursors. These pre-mRNAs are themselves bound by highly conserved basic proteins, and these proteins fold the RNA into specific structures (Skoglund et al., 1983, 1986). The globin nRNA–protein complex, for instance, is folded in such a way that certain regions of the message precursor are more exposed than others (Patton and Chae, 1985; Eperon et al., 1988). Thus, the folding of RNA by these proteins may allow certain splice sites to be recognized more efficiently than other possible sites (a property that becomes important when introns extend for 10,000 or more nucleotides). Some of these nRNA-binding proteins may also be important for splicing. A class of these proteins (C-proteins) is found in the spliceosome, and antibodies against these proteins inhibit cleavage and branch formation (Choi et al., 1986).

Alternative RNA processing and sex determination

Even if differential RNA processing does occur to generate some of the proteins characteristic of differentiated cell types, it does not necessarily mean that alternative RNA splicing is used for earlier events that determine cell fate. Recent studies, however, demonstrate that differential RNA processing plays a major role in determining the sexual phenotype of *Drosophila* (Baker et al., 1987; Boggs et al., 1987).

As we will see in Chapter 21, the development of the sexual phenotype in *Drosophila* is mediated by a series of genes that convert the X-chromosome:autosome ratio into either a male or a female cell. When the X:autosome ratio is 1 (i.e., when there are two X chromosomes per diploid cell), the embryo develops into a female fly. When the ratio is 0.5 (i.e., when the fly is XY with only one X chromosome per haploid cell) the embryo develops into a male (Figure 17).

One of the key genes of this pathway is *transformer*. This gene is necessary for the production of females, and its loss results in male flies, irrespective of the chromosomal ratio. Throughout the larval period, the *transformer* gene actively synthesizes a transcript that is processed into a general messenger RNA (found in both females and males) or into a female-specific mRNA (Figure 18A). Only females contain the alternatively spliced message, and it may be the only functional message made by this gene. The general mRNA found in both males and females contains a stop codon (UGA) early in the second exon. The small protein produced by

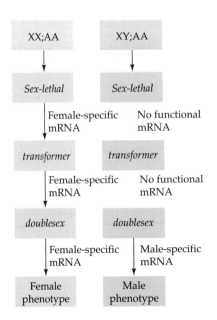

FIGURE 17

Sex determination in *Drosophila*. This simplified scheme shows that the X:autosome ratio is monitored by the *Sex-lethal* gene. If this gene is active, the *transformer* genes are active, and *transformer* processes a functional female-specific message. In the presence of the female-specific transformer product, the *doublesex* gene transcript is processed in a female-specific fashion, leading to the production of the female phenotype. If the *transformer* gene does not make a female-specific product (that is, if the *Sex-lethal* gene is not activated), the *doublesex* transcript is spliced in the male-specific manner, leading to the realization of the male phenotype.

FIGURE 18

Sex-specific transcripts during *Drosophila* development. (A) Female-specific and general transcription products generated from the *transformer* gene by alternative RNA splicing. The arrows point to the UGA stop codon, which is present in the general transcript exons but not included (because it is in an intron) in the female-specific transcript from the same gene. The boxes represent exons, the thin lines introns. (B) Alternative RNA splicing in the *doublesex* gene. During pupation, the *doublesex* gene will synthesize male-specific or female-specific gene products depending on whether the *tra* and *tra-2* genes are active. (A after Boggs et al., 1987; B after Baker et al., 1987.)

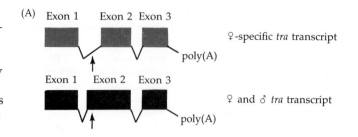

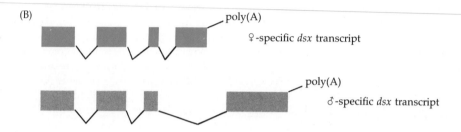

this mRNA is not functional, and the general nonspecific transcript has no bearing on sex determination (Belote et al., 1989). However, in the female-specific message, this UGA codon is in an intron which gets spliced out during mRNA formation and does not interfere with the translation of the message. In other words, the female transcript is the only functional transcript of this gene. In fact, when the cDNA of this female-specific transcript is incorporated into the genomes of XY flies, these flies become female. The protein encoded by the female-specific mRNA appears to be an arginine-rich peptide about 196 amino acids long (Boggs et al., 1987).

What makes the *transformer* gene process a female-specific transcript in XX cells and not in XY cells? It appears that sex-specific alternative splicing of the *transformer* gene involves competition between two possible 3' (acceptor) splice sites in the intron. Sosnowski and her colleagues (1989) have provided evidence that this competition is altered by the presence or absence of a functional *Sex-lethal* gene. *Sex-lethal* is one of the first genes on the sex phenotype pathway, and it acts prior to *transformer*. If the X:autosome ratio is 1, the first gene to be activated appears to be *Sex-lethal*.[*] This gene does not appear to be active in XY embryos or larvae. When the *Sex-lethal* gene is active, the *transformer* gene makes both the general and the female-specific transcript and the fly becomes female. If the *Sex-lethal* gene is absent or mutated, the *transformer* gene makes only the nonfunctional general transcript, and the fly becomes male. It appears that the product of the *Sex-lethal* gene is controlling which of the 3' splice sites are being used.

There are two major ways that the *Sex-lethal* gene could control which splice site is used to make the intron (Figure 19). One way is to block the use of the general acceptor site so that the alternative, female-specific, acceptor site is used. The other model is to activate the female-specific acceptor site in a positive manner. Sosnowski and colleagues have provided evidence for the first model. Using in vitro mutagenesis, they have constructed *transformer* genes where the general 3' acceptor site has been rendered nonfunctional by mutation. When these genes are inserted by P-elements into the genomes of XY embryos, these embryos begin making the female-specific transcript and become female flies. Inoue and co-workers (1990) have shown that the protein product of the *Sex-lethal* gene binds

*As we will see in Chapter 21, the maintenance of *Sex-lethal* gene activity is also a matter of differential RNA splicing, but we will save that discussion until then.

directly to the *transformer* pre-mRNA at or near the non-specific acceptor site and that this binding is sufficient for the generation of female-specific mRNAs. It appears, then, that if the general 3' acceptor splice site of *transformer* is blocked (either by mutation or by the product of the *Sex-lethal* gene), the alternative, female-specific, acceptor site is used. The result is a female fly.

In the pathway of *Drosophila* sex determination, the *transformer* gene is thought to control the expression of a pivotal gene called *doublesex* (*dsx*). This gene is needed for the production of either sexual phenotype, and mutations of *doublesex* can reverse the expected sexual phenotype, causing XX embryos to become males or XY embryos to become females. During pupation, *doublesex* makes a transcript that can be processed in two alternative ways. It can generate a female-specific mRNA or a male-specific mRNA (Figure 18B) (Nagoshi et al., 1988). In females and males, the first three exons are the same. However, the fourth exons are different. The male-specific RNA deletes a large section of the precursor RNA that includes the female-specific exon.

Baker and his co-workers speculate that the sex-determining genes acting prior to *doublesex* (such as *transformer*) act to make a splicing factor that enables the *doublesex* message precursor to be processed into the female-specific mRNA. (In this case, the product of *transformer* would be a splicing regulator similar to that of *Sex-lethal*.) If the *transformer* genes are absent (as in mutations) or not activated (as in XY flies; see Chapter 21), the *doublesex* transcript is processed in the male-specific manner, and male flies are generated. Research into *Drosophila* sex determination shows that differential RNA processing plays enormously important roles throughout development.

Determining the 3' end of the RNA

Cis-acting elements

The 3' end of most eukaryotic mRNAs is created by two sequential chemical reactions. First, pre-mRNA is cleaved at a specific site to leave a 3'-hydroxyl group. After this 3' end is generated, about 200 adenylate molecules—poly(A)—are added to that end. This 3' end cleavage and polyadenylation are done prior to splicing, so the splicing reactions are performed on polyadenylated mRNA precursors. The 3' processing reactions appear to be regulated by both *cis*- and *trans*-acting factors. The first *cis*-acting element is the AAUAAA sequence. The primary RNA transcript contains a 3' "trailer" sequence that extends far beyond the point at which the translation terminates. Within this trailer is the sequence AAUAAA, which is essential for the cleavage of the message 10–30 bases downstream from this site (Proudfoot and Brownlee, 1976). Mutations of this sequence prevent the 3' end formation of mRNA (Wickens and Stephenson, 1984; Orkin et al., 1985). If the AATAAA site of human β-globin is mutated to AACAAA, the 3' terminus of the new RNA is located 900 nucleotides downstream from the normal poly(A) addition site, within 15 nucleotides of the first AAUAAA in the 3' flanking region of the transcript. The poly(A) tail appears to be added in a biphasic manner (Sheets and Wickens, 1989). The addition of the first ten adenylate residues is dependent upon the AAUAAA site. The addition of the remaining residues, however, does not depend on AAUAAA, but on the presence of those ten adenylate residues that start the poly(A) tail.

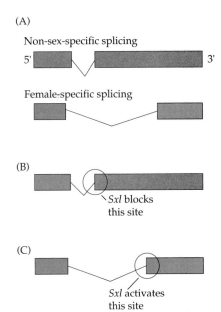

(A)

Non-sex-specific splicing

5'   3'

Female-specific splicing

(B)

Sxl blocks this site

(C)

Sxl activates this site

FIGURE 19
Two major ways in which the product of the *Sex-lethal* gene might cause the female-specific 3' intron splice site of *transformer-1* mRNA to be utilized. (A) The two alternative splicing modes of the *transformer* gene: the processing of an inactive transcript in both males and females, and the active transcript processed only in females. (B) Block of the general splice site. (C) Activation of the female-specific splice site. (After Belote et al., 1989.)

Another *cis*-acting element is a GU- or U-rich sequence located further downstream (3') from the AAUAAA region. This sequence appears to be critical for the efficient cleavage of the nRNA at the 3' processing site (McDevitt et al., 1984; Christofori and Keller, 1988).

Trans-acting factors

The *cis*-acting elements are thought to enable the binding of specific *trans*-acting proteins that will perform the tasks of cleaving the message precursor and polyadenylating its 3' end. The fractionation of proteins from human cancer cell nuclei has resulted in the isolation of five nuclear proteins that are necessary, and perhaps sufficient, to cleave and polyadenylate the mRNA precursor (Christofori and Keller, 1988; Takagaki et al.,1989). The first factor is a 290,000-Da protein, SPECIFICITY FACTOR, which can recognize the *cis*-sequences where cleavage and polyadenylation can occur (Figure 20). The second nuclear protein is the POLY(A) POLYMERASE enzyme that synthesizes the polyadenylate tail. Two other proteins, CLEAVAGE FACTORS I AND II, are the enzymes that actually cleave the pre-mRNA, but they cannot synthesize the polyA tail nor can they bind specifically to the AAUAAA signal sequence. When these four factors are mixed together, they can cut and polyadenylate pre-mRNAs in vitro. A fifth nuclear protein, CLEAVAGE STIMULATORY FACTOR, can significantly increase the efficiency of this reaction.

It should be remembered that the mechanisms by which the RNA precursors are cleaved and the poly(A) tail attached have yet to be explained. Moreover, as we saw earlier, there are sometimes alternative sites of poly(A) addition, and it is not known what controls these alternative poly(A) site selections. In the case of IgM production, there appears to be competition between the polyadenylation and the processing of the same exon. Polyadenylation of the $C\mu_4$ exon would yield the secretory form of IgM, while the processing of that exon would enable the membrane-bound form of IgM to be made. Peterson and Perry (1989) have shown that the relative efficiencies of these sites differ such that the synthesis of membrane IgM is favored over secreted IgM. It is not known what occurs during plasma cell differentiation to change this condition so that the secreted form of IgM can be synthesized.

There are some instances where transcription is terminated prematurely, and the resulting transcript does not receive its poly(A) tail. This deficiency causes the truncated message to be quickly degraded. At least two proteins involved in regulating cell proliferation appear to be regulated in this manner. In nongrowing cells, transcription of these genes is initi-

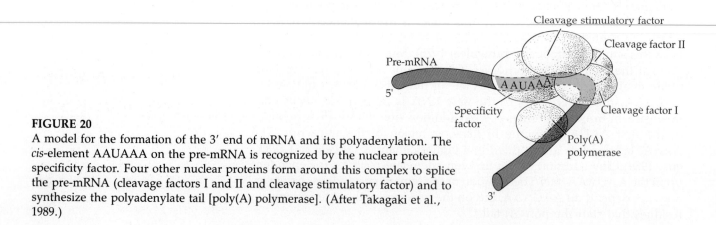

FIGURE 20

A model for the formation of the 3' end of mRNA and its polyadenylation. The *cis*-element AAUAAA on the pre-mRNA is recognized by the nuclear protein specificity factor. Four other nuclear proteins form around this complex to splice the pre-mRNA (cleavage factors I and II and cleavage stimulatory factor) and to synthesize the polyadenylate tail [poly(A) polymerase]. (After Takagaki et al., 1989.)

ated, but the transcript is not completed. Only in growing cells is the complete RNA made (Bentley and Groudine, 1987; Bender et al., 1987). In at least one of these cases, mutations near the 3' end of the first exon (where the premature termination can occur) cause the completion of transcription on cells that normally would prematurely terminate the transcript (Cesarman et al., 1987). In such cases, the mitosis-promoting protein is made, and the cells become malignant.

In some instances, the 3' processing of message precursors can occur without polyadenylation. This is routinely seen in the processing of the histone transcripts which contain neither introns nor poly(A) tails. Here, the participation of U7 snRNP has been demonstrated in the relatively simple cleavage reaction (Strub and Birnstiel, 1987; Mowry and Steitz, 1987).

Transport out of the nucleus

One of the largest mysteries of RNA processing is how it is connected with the transport of the message out of the nucleus. It has long been assumed that mRNA exits through the nuclear pores, and recent evidence supports this view. By microinjecting gold-labeled RNA or protein molecules into the nucleus or cytoplasm of *Xenopus* oocytes, Dworetzky and Feldherr (1988) have shown that RNA exits the nucleus through these pores and that protein can enter the nucleus through them as well (Figure 21). Featherstone and her co-workers (1988) have found that a monoclonal antibody directed against nuclear pore proteins is able to stop the transport of RNAs out of the nucleus while having no effect on transcription or RNA processing. This does not seem to be the result of the antibody nonspecifically occluding the pores, since the diffusion of small proteins into the nucleus remained unaffected.

FIGURE 21

Nuclear pores and RNA transport. (A) Nuclear pores from *Xenopus laevis* oocytes. The ring of the pore is formed by eight units. Inside this ring is a central "plug" that may be RNA exiting the nucleus or a regulatory element of the pore itself. (B) When RNA coated with gold particles is injected into the *Xenopus* oocyte, its export from the nucleus (N) to the cytoplasm (C) can be monitored by electron microscopy. (Other polymers complexed with gold are not exported.) This figure shows gold-coated tRNA molecules being transported through the channels of the nuclear pores. (A from Unwin and Mulligan, 1982; B from Dworetzky and Feldherr, 1988; photographs courtesy of the authors.)

(A)

(B)

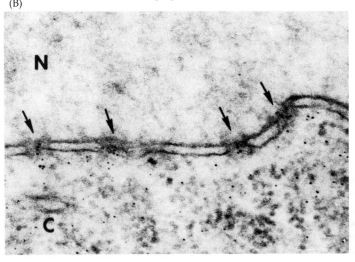

The processing of RNA out of the nucleus and into the cytoplasm seems too efficient to be accomplished by the simple diffusion of mRNA molecules bouncing against the nuclear envelope. One possible alternative (mentioned in Chapter 12) holds that RNA is transcribed and processed on the nuclear matrix, and that the mRNA is delivered to the pores on this solid substrate. If this were the case, then it is probable that the nuclear matrix would also contain snRNPs and other processing proteins. Zeitlin and co-workers (1987) have found unspliced and partially spliced message precursors for rabbit β-globin on the nuclear matrix, and Ciejek and co-workers (1982) have found partially spliced ovalbumin message precursors solely on the matrix of chick oviduct nuclei. Research on viral mRNAs also provides evidence for the active translocation of pre-mRNAs to the nuclear envelope. Adenovirus message is transcribed on the nuclear matrix and undergoes translocation to the nuclear envelope before entering the cytoplasm. A particular viral protein is necessary for this translocation to occur, and if the viral gene encoding this protein is mutated, viral mRNAs accumulate in the nucleus without reaching the nuclear envelope (Leppard and Shenk, 1989).

If this model is correct, then it might be possible for the nuclear pores to regulate development by permitting the exit of some messages but not others (Blobel, 1985). In other words, nuclear pores may differ among cell types. Evidence for this has come from studies of the muscle-specific expression of an acetylcholine receptor protein. Berman and his colleagues (1990) found that the pre-mRNA encoding this protein was specifically located at the periphery of muscle-cell nuclei (Figure 22). Interestingly, the pre-mRNA for α-actin, another muscle-specific protein, did not appear to collect at the nuclear envelope. It thus appears that nuclei can treat different pre-mRNAs in different fashions, indicating another step in the differential regulation of protein expression.

The mechanism by which mRNAs get into the cytoplasm is not well characterized. The cytoplasmic side of the nuclear envelope is often seen to be studded with ribosomes. Could the ribosomes grab the message as it comes out of the pores and pull it from the nucleus as it translates it? While we do not know the answer to this, there is at least one piece of evidence in its favor. In a certain type of human β-thalassemia, there is a mutation in codon 39 such that it now codes for the termination of translation. Therefore, it is understandable that no β-globin protein is made from this gene. But this does not explain why there should be a deficiency of β-globin *mRNA* in the cytoplasm. Studies by Humphries and co-workers (1984) show that when this message is made in vitro from a cloned gene, it is not less stable than wild-type β-globin mRNA, and Orkin and Kazazian (1984) speculate that translation may be needed on the cytoplasmic side of the nuclear envelope in order to get the globin message out. When

FIGURE 22

Localization of an acetylcholine receptor pre-mRNA to the nuclear envelope in myotubes. Cultured myotubes were hybridized in situ with a radioactive probe to the first intron in this pre-mRNA. (A) Autoradiograph of the myotube showing the accumulation of acetylcholine receptor pre-mRNA around the nuclei. (B) The same slide as A, but stained to show the location of the nuclei. (From Berman et al., 1990; photographs courtesy of S. Berman.)

(A)

(B)

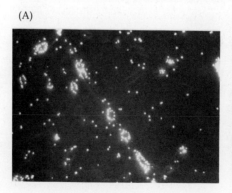
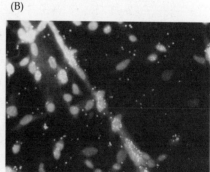

this translation ceases (because of the termination codon), so would the transport of message out of the nucleus.

There is much that needs to be known before one can determine the points at which developmental regulation of RNA splicing takes place. It might take place at the binding of snRNPs, the formation of the spliceosome, the generation of the 3' end, or the transport of RNA into the cytoplasm. At the moment, we are unable to point to an intranuclear mechanism for selecting certain specific message precursors to be processed into tissue-specific cytoplasmic mRNA.

SUMMARY

RNA processing must be viewed as a major means of regulating differential gene expression. Evidence indicates that it determines the cell-specific mRNA population in developing sea urchins and that it may do likewise in mice and rats. DNA and RNA sequencing have shown that differential RNA processing can create different proteins in different cell types (calcitonin or CGRP; IgM or IgD) or at different times within the same cell lineage (membrane-bound and secreted IgM; myosins). The regulation of differential processing is also thought to be responsible for determining the sexual phenotype of *Drosophila*. The mechanism underlying such differential RNA processing may give us insights into the very core of cell differentiation and embryonic determination.

LITERATURE CITED

Alt, F. W., Bothwell, A. L. M., Knapp, M. N., Siden, E., Mather, E., Koshland, M. and Baltimore, D. 1980. Synthesis of secreted and membrane-bound immunoglobulin μ heavy chains is directed by mRNAs that differ at their 3' ends. *Cell* 20: 293–301.

Amara, S. G., Jonas, V., Rosenfeld, M. G., Ong, E. S. and Evans, R. M. 1982. Alternative RNA processing in calcitonin gene expression generates mRNAs encoding different polypeptide products. *Nature* 298: 240–244.

Axel, R., Feigleson, P. and Schutz, G. 1976. Analysis of the complexity and diversity of mRNA from chicken liver and oviduct. *Cell* 7: 247–254.

Baker, B., Nagoshi, R. N. and Burtin, K. C. 1987. Molecular genetic aspects of sex determination in *Drosophila*. *BioEssays* 6: 66–70.

Bastos, R. N. and Aviv, H. 1977. Globin RNA precursor molecules: Biosynthesis and processing erythroid cells. *Cell* 11: 641–650.

Baumbach, W. R., Horner, D. L. and Logan, J. S. 1989. The growth hormone-binding protein in rat serum is an alternatively spliced form of the rat growth hormone receptor. *Genes Dev.* 3: 1199–1205.

Belote, J. M., McKeown, M., Boggs, R. T., Ohkawa, R. and Sosnowski, B. A. 1989. Molecular genetics of *transformer*, a genetic switch controlling sexual differentiation in *Drosophila*. *Dev. Genet.* 10: 143–154.

Bender, T.P., Thompson, C.B. and Kuehl, W.M. 1987. Differential expression of c-*myb* mRNA in murine B lymphomas by a block to transcription elongation. *Science* 237: 1473–1476.

Bentley, D. L. and Groudine, M. 1987. A block to elongation is largely responsible for decreased transcription of c-*myc* in differentiated HL60 cells. *Nature* 321: 702–706.

Berman, S. A., Bursztajn, S., Bowen, B. and Gilbert, W. 1990. Localization of an acetylcholine receptor intron to the nuclear membrane. *Science* 247: 212–214.

Black, D. L., Charbot, B. and Steitz, J. 1985. U2 as well as U1 small nucleus ribonucleoproteins are involved in pre-mRNA splicing. *Cell* 42: 737–750.

Blobel, G. 1985. Gene gating: A hypothesis. *Proc. Natl. Acad. Sci. USA* 82: 8527–8529.

Boggs, R. T., Gregor, P., Idriss, S., Belote, J. M. and McKeown, M. 1987. Regulation of sexual differentiation in *D. melanogaster* via alternative splicing of RNA from the *transformer* gene. *Cell* 50: 739–747.

Breitbart, R. E. and Nadal-Ginard, B. 1987. Developmentally induced, muscle-specific factors control the differential splicing of alternative and constitutive troponin T exons. *Cell* 49: 793–803.

Breitbart, R. E., Andreadis, A. and Nadal-Ginard, B. 1987. Alternative splicing: A ubiquitous mechanism for the generation of multiple protein isoforms from single genes. *Annu. Rev. Biochem.* 56: 467–495.

Britten, R. J. and Kohne, D. E. 1968. Repeated sequences in DNA. *Science* 161: 529–540.

Busslinger, M., Moschonas, N. and Flavell, R. A. 1981. β⁺ thalassemia: Aberrant splicing results from a single point mutation in an intron. *Cell* 27: 289–298.

Cesarman, E., Dalla-Favera, R., Bentley, D. and Groudine, M. 1987. Mutations in the first exon are associated with altered transcription in c-*myc* in Burkitt lymphoma. *Science* 238: 1272–1275.

Chabot, B., Black, D. L., LeMaster, D. M. and Steitz, J. A. 1985. The 3' splice site of pre-messenger RNA is recognized by a small ribonucleoprotein. *Science* 230: 1344–1349.

Chikaraishi, D. M., Deeb, S. S. and Sueoka, N. 1978. Sequence complexity of nuclear RNAs in adult rat tissues. *Cell* 13: 111–120.

Choi, Y.-D. and Dreyfuss, G. 1984. Isolation of the heterogeneous nuclear RNA–ribonucleoprotein complex (hnRNP): A unique supramolecular assembly. *Proc. Natl. Acad. Sci. USA* 81: 7471–7475.

Choi, Y.-D., Grabowski, P. J., Sharp, P. A. and Dreyfuss, G. 1986. Heterogenous nuclear riboproteins: Role in RNA splicing. *Science* 231: 1534–1539.

Christofori, G. and Keller, W. 1988. 3′ cleavage and polyadenylation of mRNA precursors in vitro requires a poly(A) polymerase, a cleavage factor, and a snRNP. *Cell* 54: 875–889.

Ciejek, E. M., Nordstrom, J. L., Tsai, M. J. and O'Malley, B. W. 1982. Ribonucleic acid precursors are associated with the chick oviduct nuclear matrix. *Biochemistry* 21: 4945–4953.

Crenshaw, E. B., III, Russo, A. F., Swanson, L. W. and Rosenfeld, M. G. 1987. Neuron-specific alternative RNA processing in transgenic mice expressing a metallothionein–calcitonin fusion gene. *Cell* 49: 389–398.

Cunningham, B. A., Hemperly, J. J., Murray, B. A., Prediger, A. E., Brackenbury, R. and Edelman, G. M. 1987. Neural cell adhesion molecule: Structure, immunoglobulin-like domains, cell surface modulation, and alternative RNA splicing. *Science* 236: 799–806.

Davidson, E. H. and Britten, R. J. 1979. Regulation of gene expression: Possible role of repetitive sequences. *Science* 204: 1052–1059.

Dworetzky, S. and Feldherr, C. 1988. Translocation of RNA-coated gold particles through the nuclear pores of oocytes. *J. Cell Biol.* 106: 575–584.

Early, P., Rogers, J., Davis, M., Dalame, K., Bond, M., Wall, R. and Hood, L. 1980. Two mRNAs can be produced from a single immunoglobulin gene by alternative RNA processing pathways. *Cell* 20: 313–319.

Eperon, L. P., Graham, I. R., Griffiths, A. D. and Eperon, I. C. 1988. Effects of RNA secondary structure on alternative splicing of pre-mRNA: Is folding limited to a region behind the transcribing RNA polymerase? *Cell* 54: 393–401.

Featherstone, C., Darby, M. K., and Gerace, L. 1988. A monoclonal antibody against the nuclear pore complex inhibits nucleocytoplasmic transport of protein and RNA *in vivo*. *J. Cell Biol.* 107: 1289–1297.

ffrench-Constant, C. and Hynes, R. O. 1989. Alternative splicing of fibronectin is temporally and spatially regulated in the chicken embryo. *Development* 106: 375–388.

Galau, G. A., Klein, W. H., Davis, M. M., Wold, B. J., Britten, R. J. and Davidson, E. H. 1976. Structural gene sets active in embryos and adult tissues of the sea urchin. *Cell* 7: 487–505.

Gottesfeld, J. M. and Partington, G. A. 1977. Distribution of messenger RNA coding sequences in fractionated chromatin. *Cell* 12: 953–962.

Hames, B. D. and Perry, R. P. 1977. Homology relationship between the messenger RNA and heterogeneous nuclear RNA of mouse L cells. A DNA excess hybridization study. *J. Mol. Biol.* 109: 437–453.

Hood, L. E., Wilson, J. H. and Wood, W. B. 1975. *The Molecular Biology of Eukaryotic Cells.* Benjamin, New York.

Humphries, R. K., Ley, T. J., Anagnou, N. P., Baur, A. W. and Nienhuis, A. W.

1984. β°-39 thalassemia gene: A premature termination codon causes β-mRNA deficiency without affecting cytoplasmic β-mRNA stability. *Blood* 64: 23–32.

Humphries, S., Windass, J. and Williamson, J. 1976. Mouse globin gene expression in erythroid and non-erythroid tissue. *Cell* 7: 267–277.

Hynes, R. O. 1987. Fibronectins: A family of complex and versatile adhesive glycoproteins derived from a single gene. *Harvey Lect.* 81: 133–152.

Inoue, K., Hoshijima, K., Sakamoto, H. and Shimura, Y. 1990. Binding of the *Drosophila Sex-lethal* gene product to the alternative splice site of *transformer* primary transcript. *Nature* 344: 461–463.

Kleene, K. C. and Humphreys, T. 1977. Similarity of hnRNA sequences in blastula and pluteus stage sea urchin embryos. *Cell* 12: 143–155.

Kleene, K. C. and Humphreys, T. 1985. Transcription of similar sets of rare maternal RNAs and rare adult RNAs in sea urchin blastulae and adult coelomocytes. *J. Embryol. Exp. Morphol.* 85: 131–149.

Krainer, A. R., Maniatis, T., Ruskin, B. and Green, M. R. 1984. Normal and mutant human β-globin pre-messenger RNAs are faithfully and efficiently spliced in vitro. *Cell* 36: 993–1005.

Leppard, K. N. and Shenk, T. 1989. The adenovirus E1B 55kd protein influences mRNA transport via an intranuclear effect on RNA metabolism. *EMBO J.* 8: 2329–2336.

Lerner, M. R., Boyle, J. A., Mount, S. M., Wolin, S. L. and Steitz, J. A. 1980. Are snRNPs involved in splicing? *Nature* 283: 220–224.

Lewin, B. 1980. *Gene Expression 2*, 2nd Ed. Wiley-Interscience, New York, pp. 694–760.

Liu, C.-P., Tucker, P. W., Mushinski, F. and Blattner, F. R. 1980. Mapping of heavy chain genes for mouse immunoglobulins M and G. *Science* 209: 1348–1353.

Maki, R., Roeder, W., Traunecker, A., Sidman, C., Wabl, M., Rasjhke, W. and Tonegawa, S. 1981. The role of DNA rearrangement and alternate mRNA processing in the expression of immunoglobulin delta genes. *Cell* 24: 353–365.

Maniatis, T. and Reed, R. 1987. The role of small nuclear ribonucleoprotein particles in pre-mRNA splicing. *Nature* 325: 673–678.

Mattaj, I. W. and Hamm, J. 1989. Regulated splicing in early development and stage-specific U snRNPs. *Development* 105: 183–189.

McAllister, G., Amara, S. G. and Lerner, M. R. 1988. Tissue-specific expression and cDNA cloning of small nuclear ribonucleoprotein-associated polypeptide N. *Proc. Nat. Acad. Sci. USA* 85: 5296–5300.

McCarthy, B. J. and Hoyer, B. H. 1964. Identity of DNA and diversity of RNA in normal mouse tissues. *Proc. Natl. Acad. Sci. USA* 52: 915–922.

McDevitt, M. A., Imperiale, M. J., Ali, H. and Nevins, J. R. 1984. Requirement of a downstream sequence for the generation of poly(A) addition site. *Cell* 37: 329–338.

Mount, S. M., Pettersson, I., Hinterberger, M., Karmas, A. and Steitz, J. 1983. The U1 small nuclear RNA–protein complex selectively binds a 5′ splice site in vitro. *Cell* 33: 509–518.

Mowry, K. L. and Steitz, J. A. 1987. Identification of the human H7 snRNP as one of several factors involved in the 3′ end maturation of histone premessenger RNAs. *Science* 238: 1682–1687.

Nagoshi, R. N., McKeown, M., Burtis, K. C., Belote, J. M. and Baker, B. S. 1988. The control of alternative splicing at genes regulating sexual differentiation in *D. melanogaster*. *Cell* 53: 229–236.

Nelson, K. K. and Green, M .R. 1989. Mammalian U2 snRNP has a sequence-specific RNA-binding activity. *Genes Devel.* 3: 1562–1571.

Orkin, S. H. and Kazazian, H. H. 1984. The mutation and polymorphism of the human β-globin gene and its surrounding DNA. *Annu. Rev. Genet.* 18: 131–171.

Orkin, S. H., Cheng, T.-C., Antonarakis, S. E. and Kazazian, H. H., Jr. 1985. Thalassemia due to a mutation in the cleavage–polyadenylation signal of the human β-globin gene. *EMBO J.* 4: 453–456.

Padgett, R. A., Grabowski, P. J., Konarska, M. M. and Sharp, P. A. 1985. Splicing messenger RNA precursors: Branch sites and lariat RNAs. *Trends Biochem. Sci.* 10: 154–157.

Patton, J. R. and Chae, C.-B. 1985. Specific regions of the intervening sequences of β-globin RNA are resistant to nuclease in 50 S heterogeneous nuclear RNA–protein complexes. *Proc. Natl. Acad. Sci. USA* 82: 8414–8418.

Perlman, S. M., Ford, P. J. and Rosbash, M. M. 1977. Presence of tadpole and adult globin RNA sequences in oocytes of *Xenopus laevis*. *Proc. Natl. Acad. Sci. USA* 74: 3835–3839.

Perry, R. P., Kelley, D. E., Coleclough, C., Seidman, J. G., Leder, P., Tonegawa, S., Matthyssens, G. and Weigart, M. 1980. Transcription of mouse κ chain genes: Implications for allelic exclusion. *Proc. Natl. Acad. Sci. USA* 77: 1937–1941.

Peterson, M. L. and Perry, R. 1989. The regulated production of μ_m and μ_s mRNA is dependent upon the relative efficiencies of the fs poly(A) site usage and the $C\mu_4$-to-M1 splice. *Mol. Cell Biol.* 9: 726–738.

Pikielny, C. W., Bindereif, A. and Green, M. R. 1989. In vitro reconstitution of sn-RNPs: a reconstituted U4/U6 snRNP participates in splicing complex formation. *Genes Dev.* 3: 479–487.

Proudfoot, N. J. and Brownlee, G. G. 1976. 3′ Non-coding region sequences in eukaryotic messenger RNA. *Nature* 263: 211–214.

Reed, R., Griffith, J. and Maniatis, T. 1988. Purification and visualization of native spliceosomes. *Cell* 53: 949–961.

Rogers, J. and Wall, R. 1980. A mechanism for RNA splicing. *Proc. Natl. Acad. Sci. USA* 77: 1877–1879.

Rogers, J., Early, P., Carter, C., Calame, K., Bond, M., Hood, L. and Wall, R. 1980. Two mRNAs with different 3′ ends encode membrane-bound and secreted forms of immunoglobulin μ chain. *Cell* 20: 303–312.

Ruskin, B., Krainer, A. R., Maniatis, T. and Green, M. R. 1984. Excision of an intact intron as a novel lariat structure during pre-mRNA splicing in vitro. *Cell* 38: 317–331.

Segal, S., Levine, A. J. and Khoury, G. 1979. Evidence for non-spliced SV40 RNA in undifferentiated murine teratocarcinoma stem cells. *Nature* 280: 335–338.

Sharp, P. A. 1987. Splicing of messenger RNA precursors. *Science* 235: 766–771.

Sheets, M. D. and Wickens, M. 1989. Two phases in the addition of a poly(A) tail. *Genes Dev.* 3: 1401–1412.

Skoglund, U., Andersson, K., Bjorkroth, B., Lamb, M. M. and Daneholt, B. 1983. Visualization of the formation and transport of a specific hnRNP particle. *Cell* 34: 847–855.

Skoglund, U., Andersson, K., Strandberg, B. and Dancholt, B. 1986. Three-dimensional structure of a specific pre-messenger RNP particle established by electron microscope tomography. *Nature* 319: 560–564.

Smith, M. J., Hough, B. R., Chamberlain, M. E. and Davidson, E. H. 1974. Repetitive and non-repetitive sequence in sea urchin heterogeneous nuclear RNA. *J. Mol. Biol.* 85: 103–126.

Sosnowski, B. A., Belote, J. M. and McKeown, M. 1989. Sex-specific alternative splicing of RNA from the *transformer* gene results from sequence-specific splice site blockage. *Cell* 58: 449–459.

Spritz, R.A. and nine others. 1981. Base substitution in an intervening sequence of a β⁺-thalassemic human globin gene. *Proc. Natl. Acad. Sci. USA* 78: 2455–2459.

Strub, K. and Birnstiel, M. L. 1986. Genetic complementation in the *Xenopus* oocyte: Co-expression of sea urchin histone and U7 RNAs restores 3′ processing of H3 pre-mRNA in the oocyte. *EMBO J.* 5: 1675–1682.

Takagaki, Y., Ryner, L. C. and Manley, J. L. 1989. Four factors required for 3′-end cleavage of pre-mRNAs. *Genes Dev.* 3: 1711–1724.

Tamkun, J. W., Schwartzbauer, J. E. and Hynes, R. O. 1984. A single rat fibronectin gene generates three different mRNAs by alternative splicing of a complex exon. *Proc. Natl. Acad. Sci. USA* 81: 5140–5144.

Tatei, K., Takemura, K., Mayeda, A., Fujiwara, Y., Tanaka, H., Ishimura, A. and Oshima, Y. 1984. U1 RNA–protein complex preferentially binds to both 5′ and 3′ splice junction sequences in RNA or single-stranded DNA. *Proc. Natl. Acad. Sci. USA* 81: 6281–6285.

Tazi, J., Alibert, C., Temsamani, J., Reveillaud, I., Cathala, G., Brunel, C. and Jeanteur, P. 1986. A protein that specifically recognizes the 3′ splice site of mammalian pre-mRNA introns is associated with a small ribonucleoprotein. *Cell* 47: 755–766.

Thompson, J. and 7 others. 1989. Alternative splicing of the neural cell adhesion molecule gene generates variant extracellular domain structure in skeletal muscle and brain. *Genes Dev.* 3: 348–357.

Topp, W., Hall, J. D., Rifkin, D., Levine, A. J. and Pollack, R. 1977. The characterization of SV40-transformed cell lines derived from mouse teratocarcinoma: Growth properties and differentiated characteristics. *J. Cell Physiol.* 93: 269–276.

Tsai, S. Y., Tsai, M.-J., Lin, C.-T. and O'Malley, B. W. 1979. Effect of estrogen on ovalbumin gene expression in differentiated nontarget tissues. *Biochemistry* 18: 5726–5731.

Unwin, P. N. T. and Mulligan, R. A. 1982. A large particle is associated with the perimeter of the nuclear pore complex. *J. Cell Biol.* 93: 63–75.

Wickens, M. and Stephenson, P. 1984. Role of the conserved AAUAAA sequence: Four AAUAAA point mutants prevent 3′ end formation. *Science* 226: 1045–1051.

Wold, B. J., Klein, W. H., Hough-Evans, B. R., Britten, R. J. and Davidson, E. H. 1978. Sea urchin embryo mRNA sequences expressed in nuclear RNA of adult tissues. *Cell* 14: 941–950.

Wu, J. and Manley, J. L. 1989. Mammalian pre-mRNA branch site selection by U2 snRNP involves base pairing. *Genes Dev.* 3: 1553–1561.

Zeitlin, S. and Efstratiadis, A. 1984. In vivo splicing of the rabbit β-globin pre-mRNA. *Cell* 39: 589–602.

Zeitlin, S., Parent, A., Silverstein, S. and Efstratiadis, A. 1987. Pre-mRNA splicing and the nuclear matrix. *Mol. Cell. Biol.* 7: 111–120.

Zhuang, Y. and Weiner, A. M. 1989. A compensatory base change in human U2 snRNA can suppress a branch site mutation. *Genes Dev.* 3: 1545–1552.

14

Translational and posttranslational regulation of developmental processes

We must not conceal from ourselves the fact that the causal investigation of organisms is one of the most difficult, if not the most difficult problem which the human intellect has attempted to solve, and that this investigation, like every causal science, can never reach completeness, since every new cause ascertained only gives rise to fresh questions concerning the cause of this cause.

—WILHELM ROUX (1894)

There is no rest for the messenger til the message is delivered.

—JOSEPH CONRAD (1920)

TRANSLATIONAL REGULATION OF DEVELOPMENT

After a messenger RNA has been transcribed, processed, and exported from the nucleus, it still needs to be translated in order to form the protein encoded in the genome. In this chapter we shall see that regulation at the level of translation is an extremely important mechanism in the control of gene expression. In such cases, the message is already present but may or may not be translated, depending on certain cellular conditions. Thus, translational control of gene expression can be used when a burst of protein synthesis is needed immediately (as we shall find to be the case in the newly fertilized egg), or it can be used as a fine-tuning mechanism to ensure that a very precise amount of protein is made from the available supply of messages (as is the case in hemoglobin synthesis). We shall also see that there are several ways to effect translational control and that different cells have evolved different means to do so.

Mechanism of eukaryotic translation

Translation is the process by which the information contained in the nucleotide sequence of mRNA instructs the synthesis of a particular polypeptide. This process, outlined in Figure 1, has been divided into three phases—initiation, elongation, and termination—and it is regulated by soluble proteins called (appropriately) initiation factors, elongation factors,

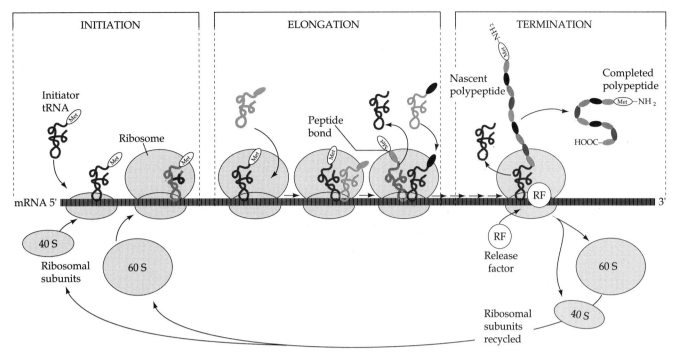

FIGURE 1

Schematic representation of the events of eukaryotic translation. The initiation steps bring together the 40 S and 60 S ribosomal subunits, mRNA, and the initiator tRNA, which is complexed to the amino acid methionine (Met). During elongation, amino acids are brought to the polysome and peptide bonds are formed between the amino acids. The sequence of amino acids in the growing protein is directed by the sequence of nucleic acid codons in the mRNA. After the last peptide bond of the protein has been made, one of the codons UAG, UGA, or UAA signals the termination of translation. The ribosomal subunits and message can be reutilized.

and termination factors (Safer, 1989; Hershey, 1989). Initiation consists of the reactions wherein the first aminoacyl-transfer RNA and the mRNA are bound to the ribosome. The only transfer RNA (tRNA) capable of initiating translation is a special initiator tRNA (tRNA$_i$), which carries the amino acid methionine. As shown in Figure 2, the first reactions involve the formation of an INITIATION COMPLEX consisting of methionyl-initiator tRNA bound to a 40 S ("small") ribosomal subunit. This reaction is catalyzed by the active form of EUKARYOTIC INITIATION FACTOR 2 (eIF2-GTP), which binds the initiator Met-tRNA to the 40 S ribosomal subunit. Note that this binding occurs in the absence of mRNA. The mRNA is added next. A CAP-BINDING PROTEIN (eIF4F) binds to the 7-methylguanosine cap at the 5' end of the message, and eukaryotic initiation factors 4A and 4B attach to or near the cap binding protein. Without the cap, the binding of mRNA to the ribosomal subunit is not often completed (Shatkin, 1976, 1985). The 40 S ribosomal subunit then travels down the message until it reaches an AUG codon in the proper context. Kozak (1986) has shown that just any AUG will not do. For the 40 S ribosomal subunit to stop and initiate translation, the nucleotides around AUG are also important. By mutating cloned genes and assaying the translation of their RNAs, Kozak found the "optimum" sequence to be ACC<u>AUG</u>G. Mutations in the flanking nucleotides could reduce translation 20-fold. The importance of the flanking sequence also has been seen in vivo. Morle and co-workers (1985) have reported a patient whose α-thalassemia (deficiency of the α-globin subunit) was due to a change in this sequence from ACCAUGG to CCCAUGG. The binding of the 40 S subunit to the AUG of the message positions the initiator tRNA over the AUG codon. Only after the mRNA has been properly positioned on the small ribosomal subunit can the 60 S ("large") ribosomal subunit bind. This binding completes the initiation reaction. During this process, the GTP on eIF2 is hydrolyzed to GDP. In order for eIF2 to pick up a new initiator tRNA, it is regenerated into eIF-GTP by eIF2B.

Elongation involves the sequential binding of aminoacyl-tRNAs to the ribosome and the formation of peptide bonds between the amino acids as

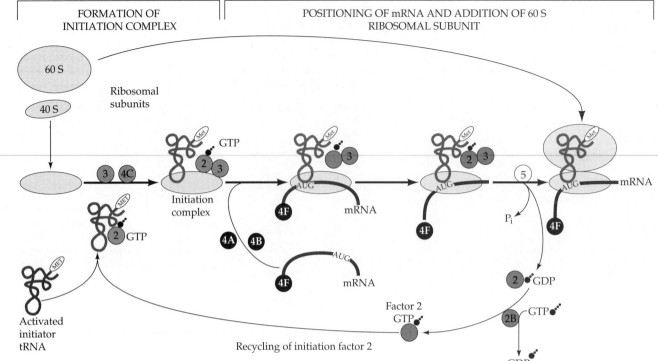

FIGURE 2

Initiation phase of eukaryotic translation. All the initiation factors are represented as circles. The first complex is made by the union of the 40 S ribosomal subunit with initiator-tRNA carrying methionine. The initiator-tRNA has been complexed with the active (GTP) form of initiation factor 2 (shown here in color). After this complex has formed, the mRNA is positioned with the aid of cap-binding protein (eIF4F) and other eIF4 subunits. Once the mRNA is in place, eIF5 mediates the joining of the 60 S ribosomal subunit and the release of the previous initiation factors. The eIF2, now in its inactive (GDP) form, is reactivated by eIF2B. (After Hershey, 1989.)

they sequentially relinquish their tRNA carriers (Figure 1). As amino acids are joined together, the ribosome travels down the message, thereby exposing new codons for tRNA binding. This allows another ribosome to initiate on the 5′ end of the message and begin its traveling. Thus, any mRNA usually will have several ribosomes attached to it. This structure is then called a polyribosome—or, more commonly, a POLYSOME (Figure 3). The termination of protein synthesis takes place when one of the mRNA codons UAG, UAA, or UGA is exposed on the ribosome. These nucleotide triplets (called termination codons) are not recognized by tRNAs and hence do not code for any amino acids. Rather, they are recognized by release factors, which hydrolyze the peptide from the last tRNA, freeing it from the ribosome. The ribosome separates into its two subunits, and the cycle of translation begins anew.

Translational control of coordinated protein synthesis: Hemoglobin production

One of the major problems in genetic regulation is the coordinated production of several products from different regions of the genome. When a developing red blood cell synthesizes hemoglobin, it must ensure that the α-globin chains, β-globin chains, and heme molecules are in a 2:2:4 ratio (Figure 4). Any major deviation from this ratio results in severely debilitating diseases.

Recent evidence has shown that the heme molecule regulates the proportional synthesis of the hemoglobin components. It accomplishes this feat in two ways. First, excess heme (i.e., heme that has not been bound to protein such as globin) will shut off its own synthesis (Karibian and London, 1965). It does this by inactivating δ-aminolevulinate synthase (DALA synthetase), the first enzyme in the pathway for the production

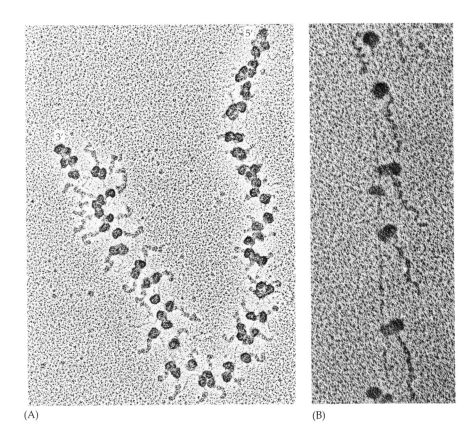

(A)

(B)

FIGURE 3
Individual polysome transcribing the giant messenger RNA from the BR2 puff of *Chironomus tentans*. (A) Electron micrograph of a polysome containing 74 ribosomes. The nascent proteins can be seen extending from the ribosomes and growing as the ribosomes move from the 5′ end of the message to the 3′ end. Near the 3′ end are ribosomes from which the protein has detached. (B) Higher magnification of such a polysome; the polysome has been stretched during the preparation of the specimen. The relationship of the mRNA to the ribosomal subunits and the nascent polypeptide can be seen. (From Francke et al., 1982; photographs courtesy of J. E. Edstrom.)

of heme (Figure 5). Thus, when there is more heme present than there are molecules to bind it, no more heme will be produced. Second, excess heme is seen to stimulate the production of globin proteins (Gribble and Schwartz, 1965; Zucker and Schulman, 1968). When heme (as its oxidized form, hemin) is added to a cell-free translation system that includes all the factors needed to translate mRNAs (Table 1), the synthesis of globin

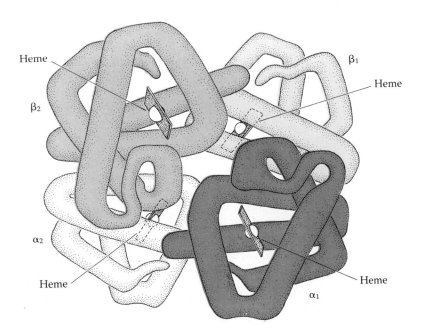

FIGURE 4
The structure of adult human hemoglobin, which has four polypeptide chains (two α, two β) and four heme molecules. (After Dickerson and Geis, 1983.)

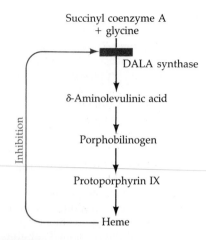

FIGURE 5

Feedback regulation of heme synthesis. (After Harris, 1975.)

is greatly enhanced (Figure 6A). Therefore, if there is no globin to bind the heme, the excess heme will shut off its own synthesis and stimulate the production of more globin!

Several laboratories have investigated how a molecule as small as heme could regulate protein synthesis. In 1972, Adamson and colleagues demonstrated that the stimulatory effect of heme on globin synthesis could be mimicked by adding to the translation system those proteins that were loosely associated with the ribosomes. Because such solutions are rich in translation initiation factors, each factor was tested separately. It was found that eukaryotic initiation factor 2 (eIF2) restored protein synthesis to heme-deficient lysates in the translation system (Figure 6B). This initiation factor is responsible for combining with the initiator tRNA and complexing it to the 40 S ribosomal subunit.

What, then, was the relationship between heme and eIF2? To answer this, London and his co-workers (Levin et al., 1976; Ranu et al., 1976) added heme-deficient lysates to heme-supplemented translation systems. They found that a portion of the heme-deficient lysate could actually depress the synthesis of globin in the translation system to which it was added. This finding indicated that an inhibitor was present. Furthermore, this inhibiting fraction had enzymatic activity. It contained a kinase capable of phosphorylating eIF2.

Phosphorylated eIF2 eventually halts translation. Normally, once the ribosomal subunits come together, eIF2 is released as a complex with GDP (Raychaudhury et al., 1985). For eIF2 to be used again in initiation, it must complex with eIF2B (recycling factor). This eIF2B acts to exchange GTP for GDP (see Figure 2), and the resulting eIF2-GTP complex is able to enter another round of initiation. However, if the alpha subunit of the eIF2 is phosphorylated, the eIF2B recycling factor binds but cannot let go (Thomas et al., 1985; Gross et al., 1985). Eventually, all the eIF2B (whose concentration is 10- to 20-fold lower than that of eIF2) gets bound on these complexes, and translation stops. Addition of eIF2B to heme-deficient lysates will restore protein synthesis to the levels of heme-supplemented systems (Grace et al., 1984).

To summarize, heme appears to regulate globin synthesis in the following manner (Figure 7).

1 In the absence of heme, a specific protein kinase will phosphorylate initiation factor 2.

TABLE 1

Components of the in vitro translation system containing rabbit reticulocyte lysate

Component	Concentration (in 100 μl)	Component	Concentration (in 100 μl)
Reticulocyte lysate (1:1)	50 μl	KCl	76 mM
Tris-HCl (pH 7.6) buffer	10 mM	Proportional amino acid mixture	6–170 μM
ATP	1 mM	[^{14}C]Leucine	0.8 μCi
GTP	0.2 mM	"Cold" leucine	26 μM
Creatine phosphate	5 mM	Hemin	10–30 μM
Creatine phosphokinase	10 μg	H$_2$O to bring the total reaction	
Magnesium acetate	2 mM	to 100 μl volume	

Source: London et al. (1976).

TABLE 3
Tennant's data on the control of early gastrulation events in hybrid sea urchin embryos

Species of sea urchin	Archenteron invagination (hours)	Mesenchyme formation (hours)	Site of origin of primary mesenchyme cells
Cidaris (♀)	20–33	23–26 (follows invagination)	Archenteron tip
Lytechinus (♂)	9	8 (precedes invagination)	Archenteron base and sides
Hybrid (*Cidaris* ♀ × *Lytechinus* ♂)	20	24 (follows invagination)	Archenteron base and sides

Source: Davidson (1976).

been accomplished by physical, chemical, and genetic means. Physical enucleation of the oocyte was accomplished by E. B. Harvey (1940). She placed unfertilized sea urchin eggs into a sucrose solution having the same density as the eggs themselves. When such eggs were centrifuged, their contents stratified, and the egg eventually split into two spheres (Figure 10A). The lighter half contained the nucleus, whereas the heavier half was enucleated. Harvey then parthenogenetically activated the enucleated halves by placing them into hypotonic seawater. These halves ("parthenogenetic merogones") cleaved, formed abnormal blastulae (that lacked a blastocoel), and successfully hatched (Figure 10B). No further development was seen. Similar studies involving frog eggs have also shown that cleavage can even occur in the total absence of chromosomes (provided the egg is injected with centrioles to compensate for those normally provided by the sperm). Here, too, an abnormal blastula is created. Physical enucleation studies suggested that certain eggs were capable of development through the midblastula stage, even in the absence of a nucleus.

Stored messenger RNA

Evidence that the oocyte controlled early development by storing messenger RNAs was first obtained by Brachet and co-workers (1963) and by Denny and Tyler (1964). The results of these investigations demonstrated

FIGURE 10
Cleavage in the absence of a nucleus. (A) Physical enucleation of *Arbacia* oocyte by centrifugation in a sucrose solution. (B) Development of enucleated portion of oocyte into parthenogenetic merogones having no nuclei. Although the cleavage is often abnormal, some enucleated oocytes develop into blastulae and hatch abnormal larvae even in the absence of products from a nucleus. (After Harvey, 1940.)

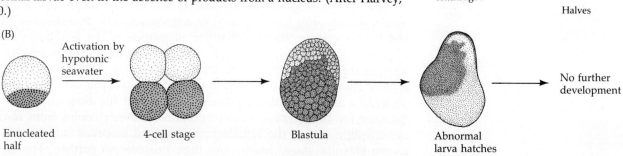

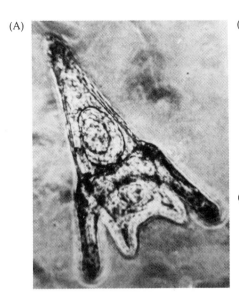

(A)

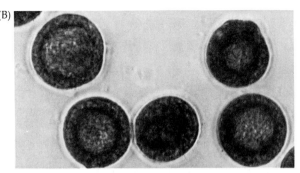

(B)

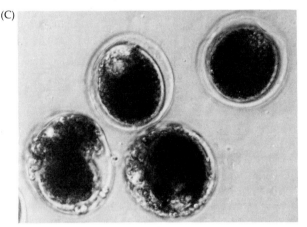

(C)

FIGURE 11
Effect of protein synthesis inhibitors on the development of the sea urchin *Lytechinus*. (A) Control pluteus larva. (B) Development arrested at blastula stage when actinomycin D prevents transcription of new RNAs. (C) Development arrested at a single-cell stage when emetine or cycloheximide prevents translation of new proteins.

that enucleated and activated sea urchin half-eggs contained RNA and that they could synthesize proteins at a rate comparable to that of normally fertilized eggs. Because there could be no transcription from these cells, stored mRNA was implicated. Craig and Piatigorsky (1971) demonstrated that this increase in protein synthesis could not have come from the transcription of mitochondrial DNA in the oocyte. It seemed that the oocyte had stored messenger RNAs that were not translated until after fertilization.

A gentler means of enucleation was made possible when it was discovered that the drug actinomycin D inhibited RNA synthesis, and transcription could be eliminated by placing the newly fertilized egg in a solution of this drug. Gross and Cousineau (1964) found that when sea urchin eggs were treated with enough actinomycin D to shut down 94 percent of their RNA synthesis, the embryos still became blastulae (Figure 11B). That this amount of development was dependent on stored messages and not on preformed proteins was shown by treating the fertilized eggs with emetine or cycloheximide. These drugs inhibit translation, and embryos fertilized in the presence of these inhibitors did not develop at all (Figure 11C).

Several investigators demonstrated that a burst of protein synthesis occurs shortly after fertilization. The actinomycin D experiments of Gross and his co-workers showed that the magnitude of the fertilization burst of protein synthesis in chemically enucleated embryos was exactly the same as that in controls (Figure 12). By 6–10 hours, however, a decline in protein synthesis was observed in the treated embryos, and they did not undergo a second burst of protein synthesis at the blastula stage. The message for this second burst of protein synthesis comes from nuclear transcription. Thus, the actinomycin D-treated embryos can develop up to the blastula stage, hatch, and then proceed no further. The stored

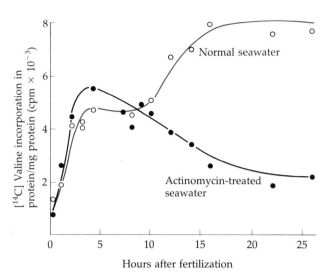

FIGURE 12
Protein synthesis in embryos of the sea urchin *Arbacia punctulata* fertilized in the presence or absence of actinomycin D. For the first few hours, there is no significant difference in the rate of new protein synthesis. A burst of new protein synthesis beginning in the midblastula stage (colored curve) represents the translation of newly transcribed messages and therefore is not seen in the embryos growing in actinomycin D. (After Gross and Cousineau, 1964.)

mRNAs of the oocyte are sufficient to take development only through the hatched blastula stage.

Studies in amphibians have given similar results, although the stimulus for the increased rate of protein synthesis appears to be the ovulation, rather than the fertilization, of the egg. In activated enucleated frog eggs, the proteins synthesized shortly after fertilization are not only synthesized in the proper amounts, they are also the same type of protein that is normally made (Smith and Ecker, 1965; Ecker and Smith, 1971). Here, too, it appears that the oocyte cytoplasm is "preprogrammed" to carry out early developmental processes even in the absence of a nucleus. Eventually, transcription is initiated in the nuclei of the embryo; yet even this transcription of the embryonic nuclei may be activated by the maternal factors in the oocyte. This concept is supported by the investigations of the *o* mutant of the axolotl. This is a maternal effect mutation wherein homozygous females produce eggs that are successfully fertilized and are totally normal until the late cleavage and early blastula stages (Briggs and Cassens, 1966). At midblastula, eggs shed by an *o/o* female are seen to have slower mitoses. These eggs will form a dorsal blastopore lip but will always arrest in gastrulation. Malacinski (1971) and Carroll (1974) have shown that in blastulae from wild-type females, new RNA and proteins are starting to be synthesized. However, the blastulae from eggs of *o/o* mothers do not undergo this burst of protein synthesis and have a pattern of proteins identical to that produced by enucleated zygotes (Figure 13). Briggs and Cassens (1966) demonstrated that the embryos from *o/o* mothers lacked a factor that activated the nuclear genome during blastula stages. In the absence of such a factor, the only development that occurs is that which can be supported by the stored oocyte mRNA. In amphibians, there is enough stored oocyte material to enable the embryo to enter gastrulation. Without new RNA synthesis, however, no further development can occur.

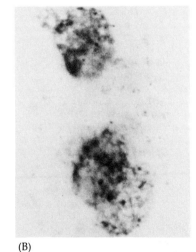

FIGURE 13

Incorporation of [³H]uridine into RNA of wild-type and mutant embryos. Blastula-stage embryos were incubated in the radioactive RNA precursor for 3 hours, washed, fixed, stained, and observed by autoradiography. (A) Normal embryo cells showing intense radioactivity indicating RNA synthesis. (B) Embryo from an *o/o* female. Stain is present, but no significant labeling is seen, indicating that little or no transcription has occurred. (From Carroll, 1974.)

Physical, chemical, and genetic enucleations of the egg make it clear that stored messages do indeed exist in the cytoplasm of the oocyte and that the products of these messages support the early development of the embryo.

Characterization of maternal messages

Eric Davidson and his colleagues have estimated the complexity of the oocyte mRNA in a manner similar to their analysis of nRNA complexity (Chapter 13). RNA (in large excess) was hybridized to denatured DNA, and the half C_0t value of the hybridization was found to be proportional to the amount of different RNA sequences present. By this analysis, they estimated that each oocyte (in numerous phyla) had enough different nucleotide sequences to account for roughly 1600 copies each of 20,000–50,000 RNA types (Galau et al., 1976; Hough-Evans et al., 1977). This is the greatest message complexity of any known cell type, and it reflects the enormous developmental potential of the oocyte. However, only a few of these messages have been characterized. The use of protein synthesis inhibitors has indicated that the proteins involved in blastomere adhesion are encoded by maternal mRNAs (Shiokawa et al., 1983), and the use of cell-free translation systems and cDNA probes has enabled the identification of histone mRNA (Skoultchi and Gross, 1973) and tubulin mRNA (Raff et al., 1972) in the oocyte cytoplasm. The products of these messages are of obvious importance for the formation of chromatin and the mitotic spindle during cleavage. Biochemical assays have permitted the identification of the messages for ribonucleotide reductase (an enzyme essential for the formation of deoxyribose nucleotides from stored ribonucleotides) and for the hatching enzyme (an enzyme that permits the blastulae to digest their fertilization membranes) (Raff, 1980). In clam oocytes, ribonucleotide reductase is stored in a remarkable manner. The large subunit is stored as a protein in the oocyte cytoplasm. The small subunit is stored as an untranslated maternal message. Only after fertilization can the newly synthesized small subunit combine with the preformed larger subunit to generate the functional enzyme (Standart et al., 1986).

The utilization of recombinant DNA techniques and the polymerase chain reaction procedure have enabled investigators to isolate stored mRNA for actin and other proteins. This has been especially useful in determining the types of stored mRNAs in mammalian oocytes. If older biochemical techniques had been used, it would take thousands of these small oocytes to get enough material to analyze. The polymerase chain reaction has enabled investigators to discover that some of the stored mRNAs in mammalian eggs are used in the process of implanting the embryo into the uterus. The mouse egg, for instance, contains the mRNAs for proteases that are secreted by trophoblast cells in order to digest the extracellular matrix of the uterus. This enables the embryo to implant therein (Brenner et al., 1989). The mouse egg also contains maternal messages encoding growth factors that are probably secreted to stimulate uterine cell proliferation (Rappollee et al., 1988). Thus, in the mammal, some of the maternal mRNAs may not be so much for the development of the embryo itself as they are to make proteins that modify the mother's uterine environment.

Interestingly enough, some stored mRNAs are not uniformly distributed throughout the oocyte. We have already seen that certain RNA sequences show cytoplasmic localization in snail and tunicate eggs, but such localizations have also been observed in sea urchin eggs and frog eggs (Rodgers and Gross, 1978). Rebagliati and co-workers (1985) have

shown that, whereas most of the maternal messages are found uniformly throughout the unfertilized *Xenopus* egg, some stored mRNAs are localized in the animal or vegetal poles of the cytoplasm. They extracted poly(A)-containing RNA from oocytes and used reverse transcriptase and DNA polymerase to convert the RNAs into a population of double-stranded DNAs. These double-stranded DNAs were then inserted into cloning vectors and grown separately in *E. coli*. About two million clones were derived in this manner. The DNA from these clones (the *Xenopus* oocyte library) was then transferred to two pieces of filter paper and denatured under conditions yielding single-stranded DNA. The investigators then cut the animal or vegetal poles from the egg and extracted the poly(A)-containing RNA from these regions. Radioactive cDNAs were made from the RNAs and one group of the DNA-containing filters was incubated in the cDNAs from the animal messages while the other group was incubated in the cDNAs from the vegetal messages. When the binding of radioactive cDNAs was measured, most clones bound equal amounts of cDNA from the animal and vegetal poles, indicating that those messages were equally distributed. However, about 1.2 percent of the clones only bound cDNA made from animal pole messages, and about 0.2 percent of the clones only bound to the cDNA derived from vegetal pole mRNA.

The DNA of the clones specific for animal or vegetal messages could then be used to identify the localized mRNAs. RNA was extracted from whole eggs or their animal and vegetal poles and run on gels. The electrophoretically separated RNAs were blotted onto nitrocellulose paper (Northern blots) and probed with radioactive DNA from each of the region-specific clones. Two of these results are shown in Figure 14. Thus, the cytoplasmic localization of preformed oocyte messages can be seen in both mosaic and regulative eggs.

Mechanisms for translational control of oocyte messages

There are at present at least five hypotheses for the regulation of oocyte mRNA translation. Three of them involve the availability of messenger RNAs, whereas the other two involve the efficiency of mRNA translation. Although these hypotheses may be seen as competing with one another, it is probable that most species use more than one of these mechanisms to regulate oocyte mRNA translation.

Masked messages. This hypothesis contends that the oocyte messages are physically masked by proteins so that the mRNA cannot attach to ribosomes. Messenger RNA is never found devoid of proteins. However, the type of protein associated with the RNA can vary. In 1966, Spirin

FIGURE 14

Demonstration of localized messages in the animal and vegetal poles of the *Xenopus* oocyte. RNA was obtained from the entire egg (T), the animal cap (A), or the vegetal cap (V) and electrophoretically separated on gels. The RNA was transferred to paper by the Northern blot procedure, and the paper was incubated with radioactive DNA from clones derived from cDNA complementary to oocyte messages. The radioactive DNA from clone *An*2 hybridizes to a message present in the animal pole but not in the vegetal pole. The reverse distribution is seen for the messages hybridizing to the DNA from clone *Vg*1. (The *Vg*1 RNA is the same message referred to in Chapter 8 as encoding a growth factor-like substance that may be important in mesodermal determination.) (From Rebagliati et al., 1985; photographs courtesy of D. Melton.)

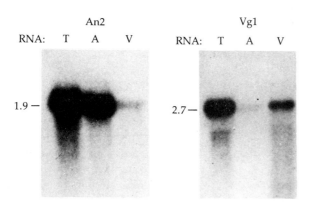

An2
RNA: T   A   V

1.9 —

Vg1
RNA: T   A   V

2.7 —

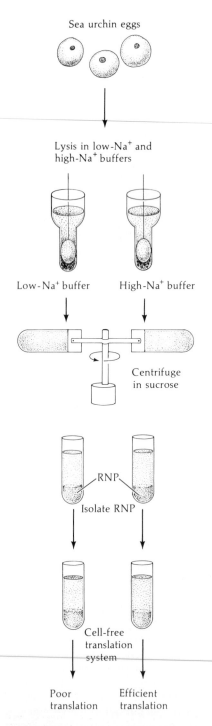

Sea urchin eggs

Lysis in low-Na⁺ and high-Na⁺ buffers

Low-Na⁺ buffer   High-Na⁺ buffer

Centrifuge in sucrose

RNP

Isolate RNP

Cell-free translation system

Poor translation   Efficient translation

FIGURE 15

Demonstration of the presence of mRNA in RNP. Protein-bound RNA cannot be translated in low-sodium buffer, but it can be translated into proteins in buffer containing high concentrations of sodium ions, which cause the mRNA to dissociate from the RNP.

proposed that the mRNA of the oocyte was stored in INFORMOSOMES, ribonucleoprotein complexes wherein the mRNA was masked. The masked messages would be unable to bind to the ribosomes and thus would not be translated. At fertilization, the proteins masking the message would be released (possibly because of the ionic changes occurring during fertilization) and the message would be free to initiate translation.

Support for this hypothesis followed shortly. In 1968, Infante and Nemer found that the unfertilized sea urchin egg contained ribonucleoprotein (RNP) particles that sedimented more slowly than ribosomes, and Gross and co-workers (1973) found that these particles contained various messages.

A series of experiments in Rudolf Raff's laboratory demonstrated that the mRNA of unfertilized eggs is indeed stored in a nontranslatable form that is susceptible to ionic changes such as those occurring during fertilization. Nonribosomal RNP particles can be isolated from mature sea urchin oocytes and are found to contain RNA with poly(A) tails (Jenkins et al., 1978). Thus, RNP particles seem to contain messenger RNA. These RNP particles were isolated in two different ionic solutions (Figure 15). Some were isolated in "oocyte" buffer containing 0.35 mM K⁺ and 5 mM Mg²⁺, an ionic condition approximating that of unfertilized eggs. Other particles were isolated into 0.35 mM Na⁺, an ionic concentration that should remove the proteins bound noncovalently to the mRNA. The mRNA-containing particles prepared in the "oocyte" buffer were very poorly translated in a cell-free system. However, the RNA extracted into the sodium-containing buffer were able to support protein synthesis almost as well as isolated egg mRNA. It thus appears that the influx of sodium during fertilization may destabilize the RNP particle, thereby allowing its mRNA to be translated (Raff, 1980).

One of the difficulties in purifying mRNPs is that it is extremely hard to be certain that the mRNP one has isolated is identical to the mRNP that is in the unfertilized egg. It is difficult to ascertain whether the purified product has not picked up or lost proteins along the way. However, certain oocyte-specific proteins have recently been implicated in the masking of stored messages. Amphibian oocytes contain mRNA binding proteins different from those of somatic cells, and some of these oocyte-specific mRNA binding proteins appear to recognize specific sequences (Audet et al., 1987). These differences may have important functional significance. When globin mRNA is injected into *Xenopus* oocytes, it is readily translated. However, if this message is first mixed with proteins isolated from *Xenopus* oocyte RNP particles, very little translation is observed. When other RNA-binding proteins were isolated from somatic tissues and mixed with globin mRNA, they did not affect the translatability of this injected message (Richter and Smith, 1984). Moreover, when globin mRNA is injected into oocytes, it acquires a different population of proteins than messages that are not recruited into polysomes (Swiderski and Richter, 1988). Thus, the oocyte-specific proteins of RNP particles are implicated in the translational regulation of stored maternal messages.

It is known that whereas the unfertilized sea urchin egg has less than 1 percent of its ribosomes incorporated into polysomes, the cleavage-stage blastomeres have nearly 20 percent of their ribosomes in such structures (Figure 16). Since new mRNA is not being made at that time, it appears that preexisting messages are being recruited into polysomes. It has been confirmed by experiments showing that the mRNA from the oocyte RNP becomes incorporated into embryonic polysomes. Young and Raff (1979) showed that radioactively labeled RNA that was originally in the sea urchin oocyte RNPs later became associated with blastomere polysomes.

A similar phenomenon is observed in the early development of the surf clam. Here, Rosenthal and his co-workers (1980) found that when protein-free mRNAs from oocytes and embryos were placed in cell-free translation systems, they both coded for the same proteins. In other words, the oocytes and embryos contained identical sets of mRNA. However, different subsets of messages are on the polysomes in oocytes and embryos. Messages for three prominent embryo-specific proteins (A, B, and C) are seen as untranslated RNPs in the oocyte cytoplasm, whereas they are seen as polysomal mRNA in embryonic blastomeres. Conversely, messages for three characteristic oocyte proteins (X, Y, and Z) are found on the polysomes of oocytes, but not on those of embryos. The recruitment of mRNA can be seen from the untranslated RNP to the translationally active polysomes.

Message recruitment is also seen in *Drosophila* embryos. Mermod and co-workers (1980) isolated poly(A)-containing RNA from oocyte RNP, oocyte polysomes, and embryo polysomes. This mRNA was then translated in a cell-free system containing radioactive amino acids. They found that the mRNA from *Drosophila* oocyte mRNPs coded for proteins different from those encoded by the mRNAs of the oocyte polysomes (Figure 17A,B). However, the same proteins that were encoded in the mRNA of the stored mRNPs are synthesized from the mRNA isolated from *embryo* polysomes (Figure 17C). Thus, mRNA formerly stored in the oocyte RNP particles has been recruited into the embryo polysomes. In this manner, gene expression is controlled by the initiation of translation.

Uncapped messages. The 5' and 3' ends of messenger RNA are necessary for efficient translation. We have already seen how differences in the length of the 3' poly(A) tail can effect differential RNA translation in *Spisula* oocytes. Certain moths use a mechanism for translational control involving changes at the 5' cap (Kastern et al., 1982). In order to be efficiently

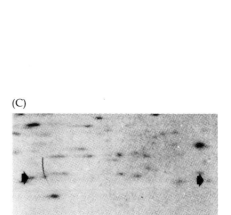

FIGURE 16
Increase in the percentage of ribosomes incorporated into polysomes during early sea urchin development. (After Humphreys, 1971.)

FIGURE 17
Evidence for the recruitment of RNA from oocyte RNP into embryo polysomes. Poly(A)-containing RNA was isolated from oocyte RNP (A), oocyte polysomes (B), and embryo polysomes (C). The RNA was translated in vitro in the presence of radioactive amino acids and then passed over a column of DNA to obtain the subset of proteins capable of binding to DNA. These proteins were eluted from the column and subjected first to iontophoresis in one dimension (to separate the proteins by their electric charge) and then to electrophoresis in the second dimension (to separate the proteins by their mass). The autoradiographs of these gels show that three proteins seen in the two-dimensional gels were synthesized from the oocyte RNP and were later synthesized from embryonic polysomal messages (arrows in A and C). These three proteins were not made from the RNAs extracted from oocyte polysomes (arrows in B). (From Mermod et al., 1980; courtesy of J. Mermod.)

(A)

(B)

(C)

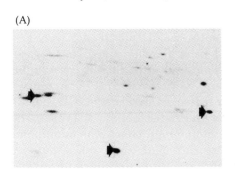 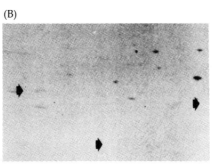

70 Minutes

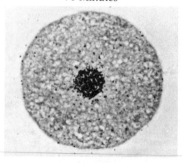

80 Minutes

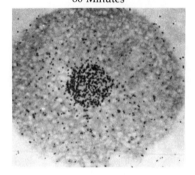

80 Minutes

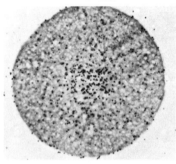

90 Minutes

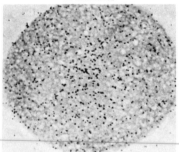

FIGURE 18
Sequestering of the sea urchin oocyte histone messages. A cDNA probe recognizing histone message is hybridized to sea urchin eggs fixed at various times after fertilization. Autoradiography shows the message to be sequestered in the maternal pronucleus until its breakdown 80–90 minutes after sperm entry. (From DeLeon et al., 1983; photographs courtesy of L. and R. Angerer.)

translated, almost all eukaryotic messages need a 7-methylguanosine "cap" on their 5' ends (Shatkin, 1976). The stored messages of the tobacco hornworm moth oocyte have a nonmethylated cap. The guanosine is present, but the methyl group has not been added to it. Such messages are not translated into proteins in a cell-free system. However, at fertilization, there is a burst of methylation in these oocytes and the caps are completed. The mRNAs with the completed caps are then able to bind to the ribosomes and initiate translation. Whether by masking the message or not completing its 5' end modifications, the initiation of translation is inhibited. The messages can remain untranslated until the appropriate signal is given.

Sequestered messages. Some studies have disputed the evidence that oocyte RNPs are untranslatable. Rather, it is possible that the protein synthetic apparatus is compartmentalized, thereby preventing the mRNA (within the RNP) from getting close to the ribosomes (Moon et al., 1982). The histone mRNAs of sea urchin oocytes seem to be regulated by this type of restriction. The histone messages of the oocyte are not found in the cytoplasm. Rather, they are localized in the large pronucleus of the unfertilized egg. It is only when the pronucleus breaks down, at the end of fertilization, that the histone mRNA gets into the cytoplasm (Figure 18; DeLeon et al., 1983). This may not be the case for other messages. Less than 0.1 percent of the total messenger RNA of the unfertilized egg is found in pronuclei (Angerer and Angerer, 1981), and those RNPs containing the messages for actin and tubulin are predominantly cytoplasmic (Showman et al., 1982). The observation that some maternal messages as well as individual ribosomes are bound to the cytoskeleton (Moon et al., 1983) suggests that the cytoskeleton may also act to separate mRNAs from ribosomes.

Changes in translational efficiency. In the preceding models of translational regulation, it is assumed that the translational apparatus is capable of efficiently translating any message but that mRNA and the ribosomes are kept apart by physical or chemical means. This need not be the case. The initial low pH of the oocyte is itself able to impede protein synthesis. As discussed in Chapter 2, there is an enormous release of hydrogen ions during sea urchin fertilization, resulting in an elevation of cytoplasmic pH from 6.9 to 7.4. Winkler and Steinhardt (1981) prepared an in vitro cell-free translation system from unfertilized sea urchin eggs. The pH of the resulting suspension was then retained or altered by dialysis, and protein synthesis was measured by the incorporation of radioactive valine into proteins. No exogenous mRNA was added. Figure 19 shows the results of one such experiment. Increasing the pH from oocyte levels (pH 6.9) to zygote levels (pH 7.4) occasioned a burst of protein synthesis mimicking that seen during fertilization.

Hille and her colleagues (1985; Danilchik et al., 1986) have suggested that the change in pH activates the translational apparatus of the egg. Ribosomes and initiation factors derived from unfertilized eggs were less active in translation than ribosomes derived from fertilized eggs. Moreover, the injection of exogenous globin message into unfertilized eggs did not increase the amount of protein being synthesized. The globin mRNA was translated at the expense of other messages, suggesting that there is a limiting amount of some portion of the translational apparatus. The limiting factor is probably a translational initiation factor. The addition of eIF-2B (the GTP-binding recycling factor) or eIF-4F (cap-binding protein) to a lysate prepared from unfertilized eggs increased the translational efficiency of that lysate (Colin et al., 1987; Lopo et al., 1988). It is probable that the alkalinization of egg cytoplasm serves both to unmask the mRNA and to activate initiation factors. Support for this notion also comes from Winkler and his colleagues (1985), who have seen a threefold increase in the binding of mRNA to ribosomes once the pH has been raised. Moreover, they do not see the binding of unfertilized egg mRNPs to these ribosomes, suggesting that the sea urchin egg regulates translation by changing the availability of messages as well as by activating translational initiation factors. The egg has evolved numerous ways of regulating the translation of its stored mRNA, and species are able to use several of these mechanisms at once.

Maternal mRNA and embryonic cleavage

Cleavage in sea urchins requires continued protein synthesis from stored maternal mRNAs. If sea urchins are fertilized in the presence of an inhibitor of translation, no cleavage results, even though the two pronuclei fuse. No mitotic spindle forms, the chromosomes do not condense, and the nuclear membrane does not break down (Wagenaar and Mazia, 1978). Thus, it appears that some of the proteins specified by maternal messages are involved in cell division during cleavage.

This hypothesis has gained support from the discovery (Evans et al., 1983) of a class of proteins called CYCLINS. These proteins, discussed in Chapter 3, are encoded by maternal mRNAs, and they are not detectable in the unfertilized egg. However, their synthesis increases dramatically after fertilization. What is striking about cyclins is that they are destroyed upon cell division and have to be resynthesized anew from the stored messages after the completion of each cleavage. Cyclin synthesis is seen to decline as the embryo nears the end of the blastula stage. Although originally seen in sea urchins, cyclins have also been detected in other species. Proteins A and B in the surf clam (mentioned earlier) are cyclins, and their levels correlate with the cell cycle.

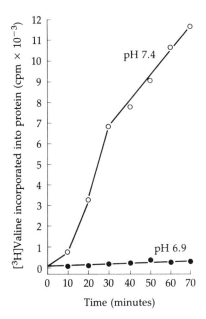

FIGURE 19
Evidence for the inefficiency of protein synthesis at prefertilization pH levels. The in vitro translation system made from unfertilized eggs is kept at pH 6.9 or dialyzed to pH 7.4. Endogenous messages are translated much more efficiently at the postfertilization pH. (After Winkler and Steinhardt, 1981.)

SIDELIGHTS & SPECULATIONS

The activation of the embryonic genome

The timing of developmental events differs enormously among animal species. At 24 hours after fertilization, *Drosophila* larvae have hatched and are busily eating; amphibian embryos are either at late gastrula or early neurula stages; and the sea urchin embryo is a late blastula or early gastrula with hundreds of cells. Mammalian embryos take their time. At 24 hours of development, a mouse zygote has divided once, and a human egg is still about six hours to its first cleavage. As one might expect from such variations, the embryonic genome is activated at different stages in different species (Table 4). *Xenopus* embryos show a stark transition after the twelfth cleavage (Newport and Kirschner, 1982). Before this time, nuclear transcription in normal embryos is undetectable, but after their midblastula transition, transcription proceeds at a rapid rate (Figure 20). In sea urchins, there seems to be no time when the embryonic nucleus is not functioning. Low levels of transcription (including new histone messages) can be seen from pronuclei even before they fuse (Poccia et al., 1985). These newly

transcribed messages join the larger pool of maternal mRNA. The chromatin of the first four cleavages is made primarily with histones stored in the oocyte cytoplasm and with histones synthesized from maternal messages. From the 16-cell stage onward, however, most histones are synthesized from messages transcribed from embryonic cell nuclei (Goustin and Wilt, 1981). This pattern is in marked contrast to that of *Xenopus* embryos, where a large pool of maternally stored histone protein and a large supply of stored oocyte histone message are utilized by thousands of cells.

In *Drosophila*, nuclear transcription is first seen after the tenth nuclear division. This is the first cycle with a G2 phase, and the G2 phase increases in length from 10 minutes after the tenth cycle to 60 minutes following the fourteenth. In the G2 phase of the fourteenth cycle, the genome is transcribing at the highest level of activity seen during embryogenesis (Anderson and Lengyel, 1979; Weir and Kornberg, 1985). Edgar and Schubiger (1986) have shown that *Drosophila* nuclei become competent to transcribe at cycle 10 but that most genes need a longer G2 phase to

TABLE 4
Activation of embryonic genomes and duration of functional maternal mRNA

Organism	Time of first observable transcription from nucleus	Time of major new nuclear transcription	Longevity of functional maternal message
Mammal (*Mus musculus*)	Late 1-cell stage (11–17 hr)	8- to 16-cell early morula (day 3)	4-cell cleavage stage (day 2–4)
Amphibian (*Xenopus laevis*)	Early cleavage (≤32-cell) midblastula (3 hr)	12th cleavage (4000-cell) midblastula (7 hr)	Neurula stage (15–30 hr)
Echinoderm (*S. purpuratus* and other sea urchins)	Zygote (pronuclear stage) (≤0.5 hr)	Midblastula (~128 cells) (11 hr)	Late blastula (15 hr)
Insect (*Drosophila melanogaster*)	Syncytial blastoderm after 10th nuclear division (2.5 hr)	Cellular blastoderm after 14th nuclear division (3.5 hr)	Mid-organogenesis (~15 hr)

Sources: Wilt (1964); Poccia et al. (1985); Weir and Kornberg (1985); Clegg and Piko (1983); Gilbert and Solter (1985); Woodland and Ballentine (1980); Davidson (1986); Edgar and Schubiger (1986); Shiokawa et al. (1989). Times indicate incubation at an appropriate temperature.

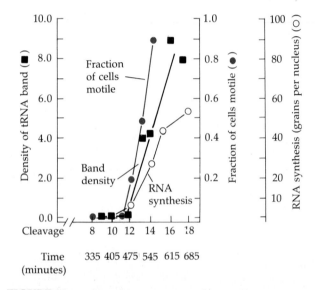

FIGURE 20
Activation of transcriptions and cell motility in *Xenopus* after the twelfth cell division. Transcription was scored autoradiographically by the number of reduced silver grains over the nuclei of embryos immersed in radioactive uridine and by the activation of a cloned tRNA gene whose radioactive product could be assayed by measuring the density of the band of the autoradiographed gel. Motility was scored by determining the fraction of cells showing pseudopodia or blebbing in video recordings. (After Davidson, 1986.)

become activated. The high transcriptional activity of cycle 14 embryos can be induced prematurely by artificially extending the G2 period of younger embryos with cycloheximide. This activation can be accomplished with embryos as young as the tenth cycle but not earlier. It appears, then, that most genes become capable of activation during cycle 10, but that they do not initiate their transcription until cycle 14.

The mouse, unlike the other organisms discussed above, has a distinct time when maternal message is used, followed by a separate time when these RNAs are replaced by nuclear transcripts. The maternal mRNAs last about two days—roughly the same amount of time as those of the other phyla—and then, during the second day, the maternal messages are rapidly degraded (Clegg and Piko, 1983; Paynton et al., 1988). As the gene products encoded by the maternal messages decay, they are replaced by new proteins made from mRNA that is newly transcribed from the nucleus. In most instances, the sperm-derived chromosomes are probably activated simultaneously with the egg-derived chromosomes (Gilbert and Solter, 1985).

In all animal species observed, there is a period of time when the phenomena of early development are controlled by messages and proteins stored in the oocyte cytoplasm. In most species (mammals being the exception), the nuclear genome is activated long before the maternal messages are degraded, so both sets of mRNAs are being translated simultaneously. Eventually, as the maternal messages are degraded on day 1 or 2, the transcripts from the embryonic genome become more important.

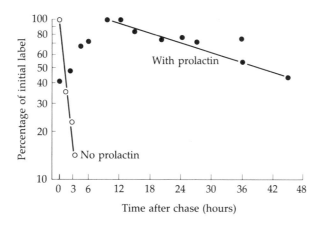

FIGURE 21
Degradation of casein mRNA in the presence and absence of prolactin. Cultured mammary cells were given radioactive RNA precursors (pulse) and, after a given time, were washed and fed nonradioactive precursors (chase). The casein RNA synthesized during the pulse time was then isolated and counted. In the absence of prolactin, the newly synthesized casein mRNA decayed rapidly, with a half-life of 1.1 hours. When the same experiment was done in medium containing prolactin, the half-life was extended to 28.5 hours. (After Guyette et al., 1979.)

Hormonal stabilization of specific messenger RNAs

Differentiated gene products are often synthesized in response to hormonal induction. In some of these cases, hormones do not increase the transcription of certain messages but act at the level of translation. One such case involves the synthesis of casein by lactating mammals. Casein is the major phosphoprotein of milk and is therefore a differentiated product of the mammary gland. As will be discussed more fully in Chapter 19, the mammary gland is prepared by the sequential actions of several hormones. Prolactin, however, is the hormone responsible for lactation— that is, actual milk production. Prolactin augments the transcription of casein messages only about twofold; its major effect appears to be the stabilization of casein mRNA (Guyette et al., 1979). Prolactin somehow increases the longevity of the casein message so it exists 25 times longer than most other messages in the cell (Figure 21). Consequently, each casein mRNA can be used for more rounds of translation. In this way, a greater than normal number of casein molecules can be synthesized from each casein message. Table 5 summarizes these data and shows that other hormones also increase the stability of specific messenger RNAs.

TABLE 5
Stabilization of specific messenger RNAs by hormones

mRNA	Cell or tissue	Regulatory effector	Half-life + Effector	Half-life − Effector
Vitellogenin	*Xenopus* liver	Estrogen	500 hr	16 hr
Albumin	*Xenopus* liver	Estrogen	10 hr	3 hr
Vitellogenin	Avian liver	Estrogen	22 hr	~2.5 hr
Apo VLDL II	Avian liver	Estrogen	26 hr	3 hr
Casein	Rat mammary gland	Prolactin	92 hr	5 hr
Growth hormone	Rat pituitary cell cultures	Dexamethasone and thyroxine	20 hr	2 hr
Insulin	Rat pancreatic islet cells	Glucose	77 hr	29 hr
Ovalbumin	Hen oviduct	Estrogen, progesterone	~24 hr	2–5 hr

Source: Shapiro et al. (1987).

The widespread use of translational regulation

Translational control is certainly a major mechanism of developmental regulation. We have seen its importance in the burst of protein synthesis following fertilization in many species, in balancing the synthesis of the components of hemoglobin, and in the hormone-induced synthesis of casein. There are many more cases of translational regulation throughout the animal kingdom. The reactivation of dehydrated brine shrimp embryos is also regulated at the level of translation. These embryos (often sold as "sea monkeys") begin development and can then lie dormant for long periods of time. During this dormancy, no protein synthesis occurs. It has been found that an inhibitor of protein synthesis exists in these embryos and that upon reactivation, a 20-fold increase in the level of functional eIF2 occurs (Filipowicz et al., 1975). Moreover, Sierra et al. (1977) have shown that this inhibitor may be a protein kinase. Thus, the same mechanism that regulates hemoglobin synthesis may also be used by brine shrimp embryos.

The coordinate and opposite regulation of the two major mammalian iron-binding proteins, ferritin and the transferrin receptor, has recently been elucidated (see Klausner and Harford, 1989). Both the ferritin and the transferrin receptor mRNAs contain regions that bind an iron-responsive protein (IRE-BP). The ferritin message has this sequence on its leader sequence (5' to the protein-encoding region), while the transferrin receptor message contains two of these sequences on its 3' trailer portion. When cellular iron is in low supply, the iron-binding protein cannot bind iron and is in a conformation that binds to these mRNAs. When it binds to the leader sequence of the ferritin message, it blocks its translation, thus preventing the synthesis of this iron-storage protein. Simultaneously, the binding protein binds to the 3' end of the transferrin receptor message, thereby stabilizing it against degradation and enabling more transferrin receptors to be produced. The transferrin receptors act to bring more iron into the cell (Figure 22).

One of the most unexpected mechanisms of translational control has recently been seen in the regulation of the apolipoprotein-B proteins. Apo-B proteins are components of serum lipid-carrier proteins and are thought to play a major role in the genesis of atherosclerosis. Apo-B48 (48,000 Da) is synthesized in the intestines and becomes part of the chylomicron complexes necessary for the absorption and transport of dietary cholesterol and triglycerides. Apo-B100 (100,000 Da) is made in the liver and is the major protein component of the very low, low, and intermediate density lipid carrier proteins. The apo-B100 and apo-B48 proteins are transcribed from the same gene, and no differential RNA processing is seen to occur to generate different mRNAs for these two proteins. Analysis of apo-B cDNAs indicates that the apo-B message in the *liver* encodes the entire apo-B100 peptide. The *intestinal* message, however, differs from the liver message by only one base. A C-to-U transition has occurred, changing a normal glutamine codon (CAA) into a terminator codon (UAA) at codon 2153. This difference results in the formation of the shorter apo-B48 protein in the intestine (Powell et al., 1987; Chen et al., 1987). This is an instance where a specific base change is made in an existing RNA, thereby changing the message. Up to this time, it had been an axiom of molecular biology that the nucleotide sequence of a message, once transcribed, could not be altered. Such RNA EDITING is an exceptionally rare event and has only been seen elsewhere in certain mitochondrial messages.

We will also see (Chapter 22) that translational control is important for

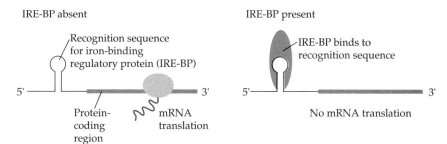

FERRITIN mRNA

IRE-BP absent

Recognition sequence
for iron-binding
regulatory protein (IRE-BP)

5'———————————3'

Protein-
coding
region

mRNA
translation

IRE-BP present

IRE-BP binds to
recognition sequence

5'———————————3'

No mRNA translation

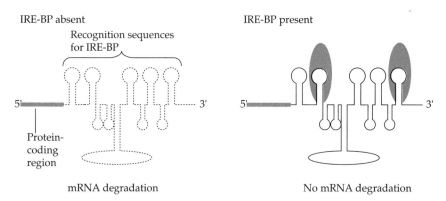

TRANSFERRIN RECEPTOR mRNA

IRE-BP absent

Recognition sequences
for IRE-BP

5'———————————3'

Protein-
coding
region

mRNA degradation

IRE-BP present

5'———————————3'

No mRNA degradation

FIGURE 22
Coordinate and opposite translational regulation of ferritin and the transferrin receptor. Both messages contain regions that are recognized by an iron-binding regulatory protein (IRE-BP). In the absence of intracellular iron, this protein binds to these messages, inhibiting the translation of ferritin mRNA and stabilizing the mRNA for the transferrin receptor. (After Klausner and Harford, 1989.)

regulating which mRNAs get translated into protein during the maturation of vertebrate eggs and sperm. Translational control, then, is an important and widely used mechanism for regulating gene expression in development.

POSTTRANSLATIONAL REGULATION OF GENE EXPRESSION

Once a protein is made, it becomes part of a larger level of organization. It may become part of the structural framework of the cell or it may become involved in one of the myriad enzymatic pathways for the synthesis or breakdown of cellular metabolites. In either case, the individual protein is now part of a complex "ecosystem" that integrates it into a relationship with numerous other proteins. Thus, we do not abandon the regulation of gene expression after synthesizing a protein, for several changes can still take place. First, some newly synthesized proteins are inactive without further modifications. These modifications can involve the cleaving away of certain inhibitory sections of the protein or may involve the binding of a small compound to enhance its activity. Second, some proteins may be selectively inactivated. In some cases, inactivation involves the degradation of the protein itself; in other cases, inactivation may be brought about by the binding of an inhibitory ligand. Third, some proteins must be "addressed" to their specific intracellular destinations. The cell is not merely a sack of enzymes; and proteins will often be sequestered in certain regions such as membranes, lysosomes, or mitochondria. Fourth, some proteins need to assemble with other proteins to form a functional unit.

The hemoglobin protein, the microtubule, and the ribosome are all examples of numerous proteins joining together to form a functional unit. Therefore, the expression of genetic information can still be influenced at the posttranslational level.

Activation of proteins by posttranslational mechanisms

Many newly synthesized polypeptides are inactive unless certain amino acid residues are cleaved away. This situation is especially evident in the case of peptide hormones. The precursor of insulin, for instance, is a long polypeptide with three disulfide bridges. The hormone is active only after the middle and the beginning of the peptide have been cleaved away. The two remaining pieces, held together by disulfide bonds, constitute the active molecule (Figure 23). In the cases of the mollusc egg-laying prohormone or the mammalian pro-ACTH-endorphin, several small peptide hormones are made as one long precursor (Figure 24). This one polypeptide can be cleaved to form adrenocorticotropic hormone (ACTH), γ-lipotropin (γ-LPH), and β-endorphin. Further processing of these hormones depends on the specific cell type in which they are found. In the anterior pituitary cells, ACTH is an end product that can be secreted to stimulate the production of steroids by the adrenal cortex. However, in the cells of the intermediate lobe of the pituitary, ACTH is cleaved to form α-melanocyte stimulating hormone (α-MSH; Herbert et al., 1981; Fisher et al., 1988).

Some proteins can only work after they have been modified by the covalent attachment of smaller molecules. Phosphorylase kinase, for example, is inactive if it is not itself phosphorylated, and collagen, one of the major structural proteins of the body, cannot function unless its prolines and certain lysines are hydroxylated. The cell cycle is a cascade of

FIGURE 23

Amino acid sequences of rat insulin II (colored circles) and its precursors. The insulin molecule is inactive until the presequence and the connecting middle sequence have been cleaved away. The amino acid sequence of the first 24 residues of preproinsulin was determined by sequencing the gene for rat preproinsulin II (Chan et al., 1979) and deducing the protein sequence.

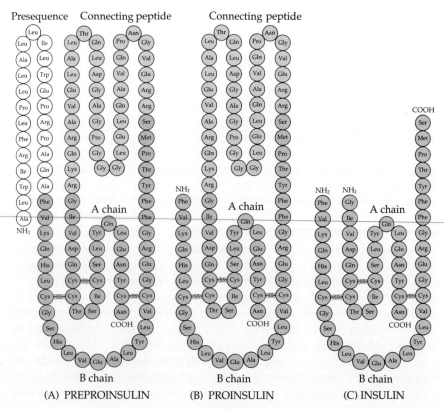

(A) PREPROINSULIN

(B) PROINSULIN

(C) INSULIN

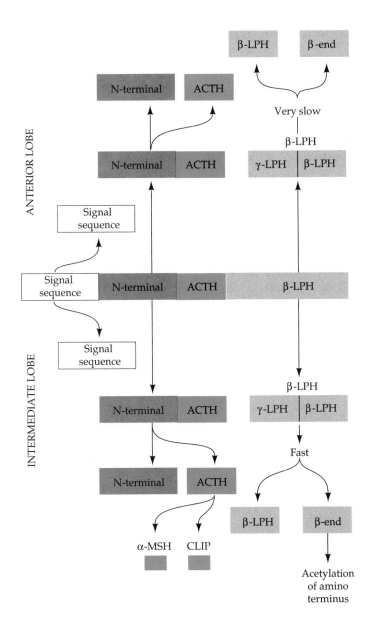

FIGURE 24
Protein processing of pro-ACTH-endorphin (center) in the intermediate and anterior lobes of the rat pituitary gland. ACTH, adrenocorticotropic hormone; β-end, β-endorphin; γ-LPH, γ-lipotropin; α-MSH, α-melanocyte stimulating hormone. (After Herbert et al., 1981.)

protein phosphorylations and dephosphorylations (Dunphy and Newport, 1989; Cyert and Thorner, 1989). The phosphorylation of *Xenopus* maturation promoting factor (MPF) is high during interphase but absent during division. Activation of this protein requires that it be dephosphorylated (Figure 25). This dephosphorylation is blocked by p13, the protein that normally regulates MPF. Thus, p13 locks the cell into interphase by preventing the activation of MPF. Once the MPF is activated, moreover, it expresses a kinase activity that phosphorylates histone H1 and other nuclear components. Once it is phosphorylated, however, it is catalytically active. Phosphorylation can also inactivate a protein, as we have seen in our discussion of eIF2.

Such posttranslational modification of histones may be important in regulating transcription itself. In the slime mold *Physarum polycephalum*, there are no cellular boundaries and all the nuclei are synchronously regulated within a common cytoplasm. This makes it an excellent organism in which to study the biochemistry of the cell cycle. Here, histone H4 can exist in several forms, depending on its state of acetylation. Highly acetylated (2–4 acetyl groups) H4 is associated with transcriptional activity

FIGURE 25
Diagram of some of the phosphorylations and dephosphorylations that regulate the activity of *Xenopus* MPF to create the cell cycle. (After Dunphy and Newport, 1989.)

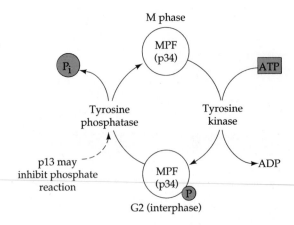

(Chahal et al., 1980). Conversely, the phosphorylation of histone H1 appears to activate its chromatic condensing activity, thereby curtailing transcription. In *Physarum*, H4 acetylation and H1 phosphorylation are inversely expressed. Thus, the gross transcriptional activity of chromatin may be regulated by the phosphorylation or acetylation of histone proteins.

Differential protein inactivation

In some instances, proteins are dangerous to retain within the cell. For instance, those proteins concerned with the timing of cell division have to be degraded rapidly if DNA synthesis is to be coordinated with cell growth. Rogers and co-workers (1986) have shown that these rapidly degraded proteins (those that have a half-life of less than 2 hours) have one or more regions rich in the amino acids proline, glutamate, serine, and threonine. Longer-lived proteins rarely have such clusters, or, if they do, these regions may be hidden deeply within the protein configuration. However, the single most important determinant of protein longevity appears to be its amino-terminal amino acid. By mutating a cloned gene for β-galactosidase, Bachmair and colleagues (1986) changed the first codon (AUG) into codons specifying fifteen other amino acids. When yeast cells were transformed with these sixteen plasmids carrying the wild-type gene or one of the mutated alleles, the differences in protein longevity were dramatic (Table 6). Whereas normal β-galactosidase (with methionine as its amino terminus) has a half-life of over 20 hours within the cell, other amino-terminal amino acids could reduce this value to less than 3 minutes.

This correlation between protein longevity and its amino-terminal amino acid was also seen to hold for naturally occurring proteins. Moreover, this correlation fits other observations that the amino termini of proteins are posttranslationally modified when cells die in vivo (Soffer, 1980; Shyne-Athwal et al., 1986). Here, destabilizing amino acids (Arg, Lys, Leu, Phe, and Tyr) are transferred from aminoacylated tRNA to the amino terminus of the proteins. Bachmair and colleagues (1986, 1989) have proposed that a specific factor binds to other cellular proteins to target them for degradation and that the probability of this factor binding to any given protein depends upon that protein's amino terminus. Once bound, this factor would signal the binding of UBIQUITIN to that protein. The conjugation of ubiquitin (a small peptide that forms covalent bonds with a lysine residue of the targeted protein) marks the protein for degradation.

TABLE 6
Correlation of protein stability with the identity of its amino acid terminal residue

Residue X in X-β-galactosidase	Half-life of X-β-galactosidase
Met Ser Ala Thr Val Gly	>20 hr
Ile Glu	~30 min
Tyr Gln	~10 min
Phe Leu Asp Lys	~3 min
Arg	~2 min
Pro	~7 min

Source: Bachmair et al. (1986).

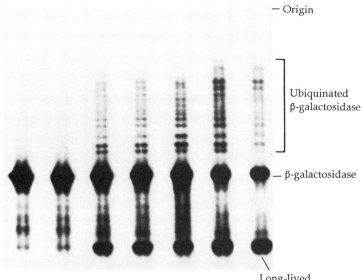

N-terminus: Met Val Ile Tyr Gln Leu Arg

— Origin

Ubiquinated
β-galactosidase

— β-galactosidase

Long-lived
degradation product

FIGURE 26

Degradation of β-galactosidase bearing different amino-terminal amino acids. Plasmids carrying the different genes (each plasmid carrying one gene for β-galactosidase, but each specifying a different amino terminus) were inserted into yeast cells, which were then labeled for 5 minutes with radioactive methionine. The cells were lysed, and the β-galactosidase was isolated by immunoprecipitating it with antibodies directed to the enzyme. The proteins recognized by the antibodies (and therefore precipitated) were separated from the antibodies and nonprecipitated proteins, run on a gel, and autoradiographed. Those β-galactosidase proteins beginning with methionine (Met) or valine (Val) showed very little degradation. Those starting with isoleucine (Ile), tyrosine (Tyr), or glutamine (Gln) showed substantial degradation (accumulation of molecules at the low end of the gel) along with the binding of several ubiquitin peptides (the ladders in the upper part of the gel showing the intact protein to be heavier). The β-galactosidases starting with leucine (Leu) or arginine (Arg) showed even more degradation. (From Bachmair et al., 1986; photograph courtesy of A. Bachmair.)

Figure 26 shows that those β-galactosidase proteins targeted for degradation are complexed with as many as 15 ubiquitin molecules.

Cells differ in the proteins they degrade. One such example is lactate dehydrogenase (LDH), which catalyzes the reversible interconversion of pyruvate and lactate. LDH is coded by two different genes: one that encodes LDH-A subunits and one that encodes LDH-B subunits. These genes are on different chromosomes. The functional LDH enzyme is a tetramer, meaning that it is composed of four subunits. These subunits can be of either the A or the B variety. If both genes are expressed, one expects to find five different forms of LDH: A4, A3B1, A2B2, A1B3, and B4. Alternative forms of an enzyme are called ISOZYMES; these LDH isozymes have different catalytic efficiencies under different cellular conditions. If the two LDH genes are being expressed equally and recombination is random, one would expect the ratio of isozyme types to be 1(A4) : 4(A3B1) : 6(A2B2) : 4(A1B3) : 1(B4). These isozymes have different electrical charges, because the B subunit is more negatively charged than the A subunit. Therefore, the five isozymes can be separated by electrophoresis and stained for catalytic activity (Figure 27). When equal quantities of the

FIGURE 27
Electrophoretic gel showing the different amounts of the five lactate dehydrogenase (LDH) isozymes found in various tissues of the rat. (From Markert and Ursprung, 1971.)

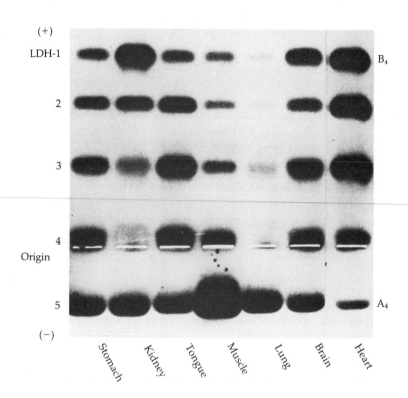

subunits are combined artificially, one indeed finds the 1:4:6:4:1 ratio. However, when one looks at individual organs, large deviations from this ratio are found. Muscle LDH is predominantly made of A subunits, whereas testes, heart, and kidney isozymes are composed mostly of B subunits (Markert, 1968).

Using anti-LDH-antibodies to precipitate any newly synthesized A subunits, Fritz and co-workers (1969) discovered that the rate of synthesis of LDH-A does not vary greatly from tissue to tissue. However, the rate of its degradation varies enormously. The degradation of LDH-A is 22 times greater in heart muscle than in skeletal muscle. So the control of LDH isozymes is not at the level of LDH transcription or translation, but at the level of enzyme degradation.

Subcellular localization of proteins by posttranslational modifications: Targeting proteins for membranes and lysosomes

After they are synthesized, proteins have to be properly positioned. Thus, a protein may ultimately be located in the soluble cytoplasm, the mitochondria, the lysosomes, the endoplasmic reticulum, or the nucleus; it may even be secreted from the cell.

Those proteins destined for lysosomes, endoplasmic reticulum, or secretion are passed into the lumen of the rough endoplasmic reticulum while they are still being translated (Figure 28). These proteins (including collagen and the peptide hormones discussed earlier) contain a SIGNAL SEQUENCE of approximately 30–50 amino acids, which is recognized by a SIGNAL RECEPTOR COMPLEX that floats in the cytoplasm. When this complex (which contains six different peptide chains and a small RNA molecule) binds to the growing peptide of a cytoplasmic ribosome, translation stops.

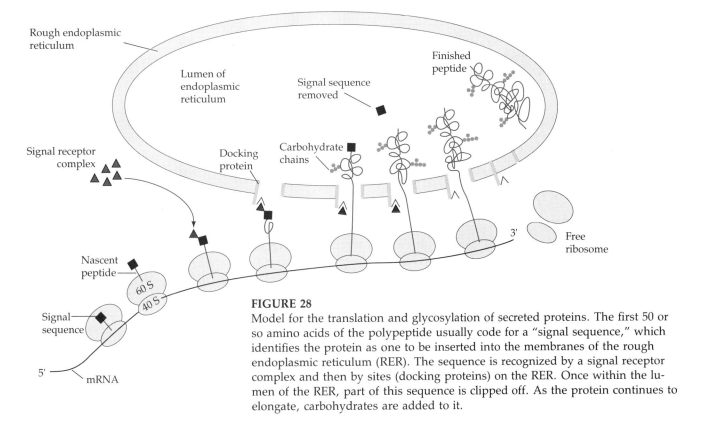

FIGURE 28

Model for the translation and glycosylation of secreted proteins. The first 50 or so amino acids of the polypeptide usually code for a "signal sequence," which identifies the protein as one to be inserted into the membranes of the rough endoplasmic reticulum (RER). The sequence is recognized by a signal receptor complex and then by sites (docking proteins) on the RER. Once within the lumen of the RER, part of this sequence is clipped off. As the protein continues to elongate, carbohydrates are added to it.

The complex then joins to a DOCKING PROTEIN on the endoplasmic reticulum and translation resumes. The growing polypeptides pass through the reticular membrane, and, once inside the lumen, this signal sequence is removed and the remaining protein can be modified (Blobel and Dobberstein, 1975; Walter and Blobel, 1983; Weidmann et al., 1987). If the protein (or protein precursor) contains a hydrophobic region, it is possible that the protein will become part of the endoplasmic reticulum and later become part of the cell membrane. One of the unsolved problems in this field concerns how proteins become localized in particular regions of the cell membrane, for numerous cell types have plasma membranes where specific proteins are found in one place in the membrane only. We saw such a phenomenon occur in the polarity of mammalian blastomeres before compaction. Recent data (Mostov et al., 1987) suggest that there are particular amino acid sequences that are able to localize membrane proteins to the apical or basal regions of the cell.

Some peptides in the endoplasmic reticulum become modified by the covalent addition of mannose 6-phosphate to the protein. This sugar acts as an address label that packages the glycoprotein for lysosomes. Mannose 6-phosphate is recognized by receptors at specific areas of the endoplasmic reticulum that concentrate those enzymes having this address label. This region of the reticulum then buds off to form the lysosome (Sly et al., 1981). When binding does not occur, these proteins are secreted from the cell. Evidence for this comes from a lethal genetic syndrome, I-cell disease, wherein patients are unable to put the mannose 6-phosphate address label on the proteins. The lysosomes of these individuals lack the appropriate enzymes, which have been mistakenly secreted into the blood.

Some proteins are recognized by specific binding proteins, which localize them in certain regions of the cell. In mouse liver, β-glucuronidase is found both in the lysosomes and on the endoplasmic reticulum. Although the mannose 6-phosphate label can get this enzyme into the lysosomes, β-glucuronidase can be bound to the liver endoplasmic reticulum by the protein EGASYN (Swank and Paigen, 1973), which specifically recognizes this protein. In mutant mice lacking egasyn, no β-glucuronidase is found on the liver endoplasmic reticulum (Lusis et al., 1976). Normally, the localization of β-glucuronidase is different in various cell types. Tissues such as liver and kidney are rich in egasyn and have β-glucuronidase on their endoplasmic reticula. Brain and spleen cells, however, appear to be entirely deficient in egasyn, and these have all their β-glucuronidase in their lysosomes (Lusis and Paigen, 1977). So the intracellular localization of proteins is posttranslationally regulated and may differ from tissue to tissue.

Targeting proteins for the nucleus and mitochondria

The accumulation of a particular protein in the nucleus involves both the *selective entry* of the protein into the nucleus and the *selective retention* of those proteins that do enter (Dingwall et al., 1982; Kalderon et al., 1984). These properties appear to be governed by specific amino acid sequences in the protein. In 1984, Kalderon and co-workers identified a seven amino acid sequence in a nuclear protein; this sequence could alter the location of a cytoplasmic protein. The nuclear protein was the T antigen of the SV40 virus. It had been known that when a mutation in the gene for this protein changed the lysine at position 128 into other amino acids, the protein remained in the cytoplasm. This finding suggested that the amino acid sequence around position 128 was a nuclear localization signal. This hypothesis was tested directly by constructing a recombinant plasmid that contained as one unit almost all of the gene for the cytoplasmic enzyme pyruvate kinase and the 75-base-pair sequence coding for amino acids 114–138 of the SV40 T antigen. These recombinant plasmids were microinjected into cells in tissue culture. In some cases, the fused gene was incorporated into the chromosomes, transcribed, and translated into a protein that consisted of pyruvate kinase attached to 25 amino acids of the T antigen protein. When these cells were stained with fluorescent antibodies to pyruvate kinase, the protein was found in the nucleus (Figure 29A). Reducing the size of the T antigen gene sequence caused different proteins to be made. Figure 29 shows that the ability to localize in the nucleus was conferred by a very basic sequence of seven amino acids: Pro–Lys–Lys–Lys–Arg–Lys–Val. This does not mean that every nuclear protein uses this particular sequence for its localization or that every nucleus recognizes this amino acid signal. Other nuclear proteins appear to have different nuclear recognition signals (Hall et al., 1984; Davey et al., 1985).

The actual import of proteins into the nucleus is an active process that can be divided into two signal-dependent steps (Newmeyer and Forbes, 1988; Richardson et al., 1988). In the first step, proteins with the nuclear translocation signal bind to the pores of the nuclear envelope. The nuclear pores probably recognize the nuclear translocation signal through pore-specific glycoproteins. If these glycoproteins are absent or functionally blocked, this binding does not occur (Finlay and Forbes, 1990). The binding of nucleoproteins to the pore can occur in the absence of ATP, and if ATP hydrolysis is inhibited, these proteins accumulate at the pores without being moved into the nucleus. The second step requires ATP and involves the translocation of the proteins from the pores into the nucleus.

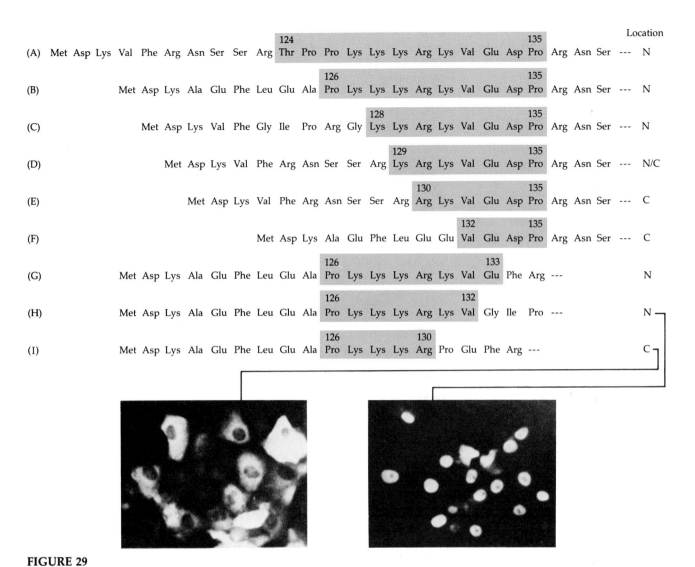

FIGURE 29

Identification of amino acid sequence that targets protein to the nucleus. The 25 amino acid residues flanking lysine-128 of the SV40 T antigen (A) directed pyruvate kinase into the nucleus. The colored sequence represents the closest residues. (B–I) Variation of this sequence, identifying the residues necessary for the accumulation of the protein in the nucleus. The photographs show the intracellular localization of (H) and (I). (From Kalderon et al., 1984; photographs courtesy of D. Kalderon.)

Targeting a protein to the mitochondria is much more complicated, since the mitochondrion itself contains four separate compartments. A protein can be directed to the outer or inner mitochondrial membranes, the space between the membranes, or the inner mitochondrial matrix (see Figure 30). Unlike the targeting sequences directing proteins to the nucleus, most mitochondrial proteins are directed by amino acid sequences that are readily separable from the mature protein (analogous to the signal sequences for localizing proteins into the endoplasmic reticulum). The identities of these targeting sequences have been obtained by techniques similar to those used to discover the nuclear targeting sequences (van Loon et al., 1986; Hurt and van Loon, 1986). The major intracellular targeting signal for mitochondrial localization is a relatively short sequence having basic amino acids periodically arranged (the exact length and exact

FIGURE 30

Model for the targeting of proteins to specific regions within the mitochondria. All mitochondrial proteins derived from nuclear genes contain a matrix-targeting signal sequence (open circles). If transport is to be halted before reaching the matrix, a second, hydrophobic, signal (black circles) enables the protein (colored circles) to be retained by the membrane. Whether or not it stays in the membrane or goes into the intermembrane space depends on whether the protein is cleaved from this signal sequence. The entry of the precursors at regions where the two membranes come close to each other is suggested by Schleyer and Neupert (1985). The model is after that of Hurt and van Loon (1986) and van Loon and Schatz (1987).

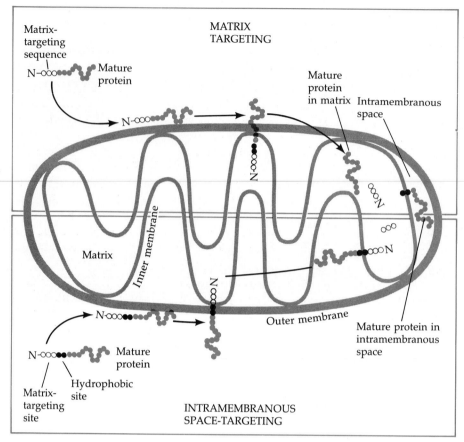

amino acid sequence varies between mitochondrial proteins). If this sequence is in a protein, it will not only be directed to the mitochondria, but it will be translocated across both membranes into the mitochondrial matrix. If a hydrophobic region lies between the mature protein and the matrix-targeting sequence, it is thought that this blocks the transfer of the protein into the matrix. This causes the protein to be part of the inner membrane. If the mature protein is readily cleaved from this hydrophobic region, it is found in the intramembranous space (van Loon and Schatz, 1987).

In addition to these "*cis*" signals found on the proteins themselves, translocation into the mitochondria is also facilitated by "*trans*" factors in the cytoplasm and the mitochondrial membranes. One of these factors appears to be a 19,000-Da protein that is embedded in the outer mitochondrial membrane. Called MOM19, this protein is probably the binding site that recognizes the *cis*-sequences of the incoming polypeptides. Antibodies against MOM19 inhibit the binding and transport of numerous polypeptide precursors into the mitochondria (Söllner et al., 1989). Some of the other *trans* factors include the ubiquitous group of proteins called "heat shock proteins" or "stress proteins." When cells or organisms are subjected to heat shock or stress, genes for these proteins are selectively transcribed. The reasons for the synthesis of these proteins (and their conservation throughout the animal kingdom) remained obscure until two laboratories demonstrated that the heat shock proteins facilitated the transport of proteins from the cytosol into the mitochondria and endoplasmic reticulum (Deshaies et al., 1988; Chirico et al., 1988).

Therefore, once a protein is synthesized, it does not just float about in the cytosol. Rather, it can be directed to enter into specific compartments of the cell.

Supramolecular assembly

The last form of posttranslational regulation that we will consider involves the assembly of these newly synthesized proteins into a functional unit. We have already discussed several proteins that can spontaneously assemble themselves into functional units. Hemoglobin and lactate dehydrogenase, for example, both contain four polypeptide chains, which are noncovalently joined to form the functional protein. On a slightly higher level, the proteins tubulin and actin both exist in polymerized and nonpolymerized forms. We have seen that the polymerization of actin from globular monomers into microfilaments is essential for the extension of the acrosomal process during fertilization. Similarly, tubulin is assembled into the microtubules of the mitotic apparatus, only to be converted into monomers as mitosis finishes. These tubulin units can then be reassembled into new microtubules, which are important in establishing cell shape.

The growth of a microtubule is regulated by the hydrolysis of GTP molecules that are covalently associated with the tubulin protein. Newly polymerized tubulin is bound to GTP, and it forms stable associations with other tubulin molecules. As tubulin proteins age (or when cellular remodeling occurs), GTP is hydrolyzed to GDP. The tubulin bound to GDP does not form as stable an attachment to its neighboring tubulin molecules. This reduced stability is thought to bring about the depolymerization of the microtubule (Figure 31). As we shall see in the next chapter, the differential growth of microtubules can have profound effects on cellular and embryonic morphogenesis.

The ability to form fibers is found in a relatively small subset of cellular proteins, and disastrous consequences can occur if other proteins gain this ability by mutation. Such a phenomenon occurs in sickle cell anemia. The one-amino acid substitution (glutamine to valine in the sixth amino acid position in β-globin) causes the formation of a hydrophobic pocket that allows the hemoglobin tetramers to interact when deoxygenated. Such fibers stretch the cell, thereby giving it its characteristic sickle shape, and decrease its elasticity (Figure 32).

Several proteins form assemblies with nucleic acids. Histones, for instance, are useless unless assembled with DNA to form the nucleosome particle. The eukaryotic 80 S ribosome consists of approximately 70 proteins and 4 different ribosomal RNAs. The proteins and nucleic acids have to assemble together to be functional, and it is remarkable that this assembly occurs quickly and efficiently within all cells. It should not be a total surprise, then, to discover that whole infectious viruses can be obtained by incubating the component proteins and nucleic acids together. The proteins that appear to "mask" the messenger RNA in oocytes are also examples of proteins that can form specific complexes with nucleic acids. The TFIIIA protein that complexes with the 5 S rRNA in the *Xenopus* 5 S

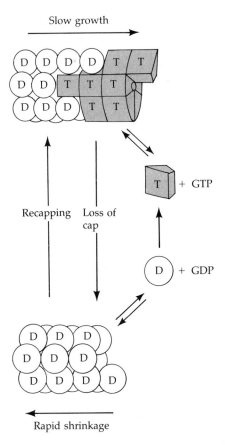

FIGURE 31
Model for the growth and depolymerization of microtubules. Growing microtubules add GTP-containing tubulin subunits (represented as T) that form close associations with other tubulin subunits. As subunits age, the bound GTP of tubulin is hydrolyzed to GDP. The GDP-bearing tubulin (represented as D) is more readily dissociated from the microtubules. When hydrolysis surpasses assembly, the GTP-tubulin "cap" disappears, and the microtubule shrinks rapidly. (After Kirschner and Mitchison, 1986.)

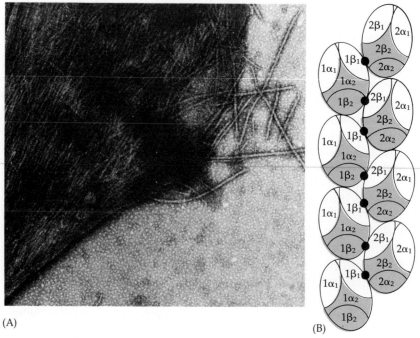

(A)

(B)

FIGURE 32

Fiber formation in sickle cell hemoglobin. (A) Electron micrograph of deoxygenated sickle cell hemoglobin fibers spilling out of an osmotically ruptured erythrocyte. (B) Model of fiber formation in which the mutated β-globin is able to adhere with another region of the β-globin peptide of another hemoglobin tetramer. (A from Wellems and Josephs, 1979; photograph courtesy of the authors. B after Dickerson and Geis, 1983.)

rRNA gene also complexes with the 5 S rRNA in the oocyte to form a stable 7 S particle (Pelham and Brown, 1980). Thus, spontaneous assembly is another important form of posttranslational control.

Collagen: An epitome of posttranslational regulation

By looking at the synthesis of collagen, one of the most important structural proteins of the body, we can demonstrate the importance and variety of posttranslational control mechanisms (for reviews, see Prockop et al., 1979; Davidson and Berg, 1981).

First, the collagen mRNA sequence is translated into a procollagen peptide. The amino terminal of the procollagen contains a signal sequence that enables it to enter the lumen of the rough endoplasmic reticulum. This signal peptide is cleaved from the remainder of the collagen molecule. Within the endoplasmic reticulum, individual collagen molecules (also called tropocollagen) encounter three hydroxylating enzymes. Two of these enzymes hydroxylate proline residues to 3-hydroxyproline and 4-hydroxyproline; the third converts lysines to hydroxylysyl residues. Only certain proline and lysine residues are in the correct positions to be recognized by these enzymes (Figure 33).

As the tropocollagen chains are being hydroxylated, sugar residues can be added to the hydroxylysines. The first enzyme—galactosyltransferase— adds a galactose to the hydroxylysine; the second enzyme — glucosyltransferase—adds a glucose unit to the hydroxylysylgalactose. The region near the carboxyl end is modified by glucosamine and mannose.

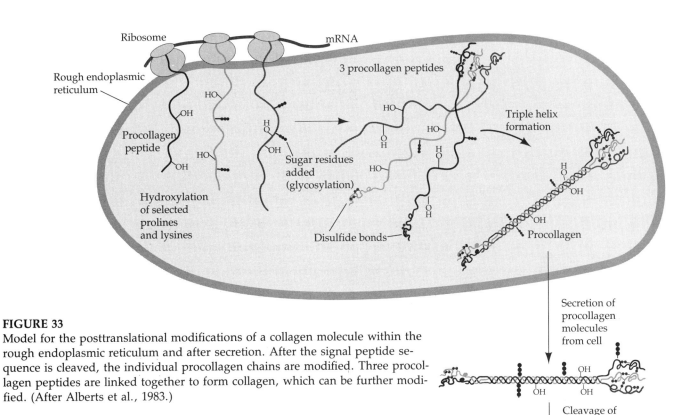

FIGURE 33

Model for the posttranslational modifications of a collagen molecule within the rough endoplasmic reticulum and after secretion. After the signal peptide sequence is cleaved, the individual procollagen chains are modified. Three procollagen peptides are linked together to form collagen, which can be further modified. (After Alberts et al., 1983.)

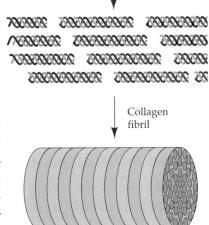

The next step involves the formation of intrachain disulfide bonds and the organization of three tropocollagen polypeptides into a collagen triple helix. Disulfide bonds are formed at the carboxyl ends between adjacent collagen molecules, establishing the preconditions for forming a triple helix. If the prolines have been properly hydroxylated, the three chains fold over one another to produce a large triple helical region bounded on either side by globular areas.

The collagen molecule is secreted in this form to the outside of the cell. Even here, processing continues. First, two proteolytic enzymes remove the globular domains at the amino and carboxyl ends. This trimming creates a pure triple helical fragment. Next, these fragments spontaneously assemble into fibrils. These fibrils, however, lack the necessary tensile strength to function as structural proteins until the various triple helices are joined together covalently by further crosslinking. This bonding is accomplished by enzymatically modifying the lysyl of hydroxylysyl residues and linking them together. Thus, there are an impressive number of essential steps by which a protein is modified after translation. (As we shall see in the next two chapters, collagen is not only an extremely important structural protein, but it also is thought to be responsible for certain embryonic interactions.)

We have now looked at the various levels regulating gene expression: transcription, RNA processing, translation, and posttranslational modification. Each type of regulation has been found to be important in controlling gene expression during development. We are getting closer to answering the mystery of differentiation, and the new techniques of gene cloning (as Boveri predicted in 1904) should enable us to understand differentiation on a chemical level. However, the question of differentiation of individual cells is not the only problem in developmental biology. We still need to understand how the cells of the body interact to develop at the correct time and at the correct place. In Part III, we analyze *cellular interactions* in development that lead to the formation of tissues and organs.

LITERATURE CITED

Adamson, S. D., Yau, P. M. P., Herbert, E. and Zucker, W. V. 1972. Involvement of hemin, a stimulatory fraction from ribosomes, and a protein synthesis inhibitor in the regulation of hemoglobin synthesis. *J. Mol. Biol.* 63: 247–264.

Alberts, B., Bray, D., Lewis, J., Raff, M., Roberts, L. and Watson, J. D. 1983. *Molecular Biology of the Cell*. Garland, New York.

Anderson, K. W. and Lengyel, J. A. 1979. Rates of synthesis of major classes of RNA in *Drosophila* embryos. *Dev. Biol.* 70: 217–231.

Angerer, L. M. and Angerer, R. C. 1981. Detection of poly A$^+$ RNA in sea urchin eggs and embryos by quantitative *in situ* hybridization. *Nucleic Acids Res.* 9: 2819–2840.

Audet, R. G., Goodchild, J. and Richter, J. D. 1987. Eukaryotic initiation factor 4A stimulates translation in microinjected *Xenopus* oocytes. *Dev. Biol.* 121: 58–68.

Bachmair, A. and Varshausky, A. 1989. The degradation signal in a short-lived protein. *Cell* 56: 1019–1032.

Bachmair, A., Finley, D. and Varshausky, A. 1986. In vivo half-life of a protein is a function of its amino-terminal residue. *Science* 234: 179–186.

Blobel, G. and Dobberstein, B. 1975. Transfer of proteins across membranes. I. Presence of proteolytically processed and unprocessed nascent immunoglobulin light chains on membrane bound ribosomes of murine myeloma. *J. Cell Biol.* 67: 835–851.

Brachet, J., Decroly, M., Ficq, A. and Quertier, J. 1963. Ribonucleic acid metabolism in fertilized and unfertilized sea urchin eggs. *Biochim. Biophys. Acta* 72: 660–662.

Brenner, C. A., Adler, R. R., Rappolee, D. A., Pedersen, R. A. and Werb, Z. 1989. Genes for extracellular matrix-degrading metalloproteases and their inhibitor, TIMP, are expressed during early mammalian development. *Genes Dev.* 3: 848–859.

Briggs, R. and Cassens, G. 1966. Accumulation in the oocyte nucleus of a gene product essential for embryonic development beyond gastrulation. *Proc. Natl. Acad. Sci. USA* 55: 1103–1109.

Carroll, C. R. 1974. Comparative study of the early embryonic cytology and nucleic acid synthesis of *Ambystoma mexicanum* normal and *o* mutant embryos. *J. Exp. Zool.* 187: 409–422.

Chahal, S. S., Matthews, H. R. and Bradbury, E. M. 1980. Acetylation of histone H4 and its role in chromatin structure and function. *Nature* 287: 76–79.

Chan, S. J., Noyes, B. E., Agarwal, K. L. and Steiner, D. F. 1979. Construction and selection of recombinant plasmids containing full-length complementary DNAs corresponding to rat insulins I and II. *Proc. Natl. Acad. Sci. USA* 76: 5036–5040.

Chen, S.-H. and 12 others. 1987. Apolipoprotein B48 is the product of a messenger RNA with an organ-specific in-frame stop codon. *Science* 238: 363–366.

Chirico, W. J., Waters, M. G. and Blobel, G. 1988. 70K heat shock related proteins stimulate protein translocation into microsomes. *Nature* 332: 805–810.

Clegg, K. B. and Piko, L. 1983. Poly(A) length, cytoplasmic adenylation, and synthesis of poly(A)$^+$ RNA in early mouse embryos. *Dev. Biol.* 95: 331–341.

Clemens, M. J., Henshaw, E. C., Rahaminoff, H. and London, I. M. 1974. Met-tRNA$_{fmet}$ binding to 40 S ribosomal units: A site for the regulation of initiation of protein synthesis by hemin. *Proc. Natl. Acad. Sci. USA* 71: 2946–2950.

Colin, A. M., Brown, B. D., Dholakia, J. N., Woodley, C. L., Wahba, A. J. and Hille, M. B. 1987. Evidence for simultaneous derepression of messenger RNA and the guanine nucleotide exchange factor in fertilized sea urchin eggs. *Dev. Biol.* 123: 354–363.

Craig, S. P. and Piatigorsky, J. 1971. Protein synthesis and development in the absence of cytoplasmic RNA synthesis in non-nucleate egg fragments and embryos of sea urchins: Effect of ethidium bromide. *Dev. Biol.* 24: 213–232.

Cyert, M. S. and Thorner, J. 1989. Putting it on and taking it off: Phosphoprotein phosphatase involvement in cell cycle regulation. *Cell* 57: 891–893.

Danilchik, M. V., Yablonka-Reuveniz, Z., Moon, R. T., Reed, S. K. and Hille, M. B. 1986. Separate ribosomal pools in sea urchin embryos: Ammonia activates a movement between pools. *Biochemistry* 25: 3696–3702.

Davey, J., Dimmock, N. J. and Colman, A. 1985. Identification of the sequence responsible for the nuclear accumulation of the influenza virus nucleoprotein in *Xenopus* oocytes. *Cell* 40: 667–675.

Davidson, E. H. 1976. *Gene Activity in Early Development*, 2nd Ed. Academic Press, New York.

Davidson, E. H. 1986. *Gene Activity in Early Development*, 3rd Ed. Academic Press, New York.

Davidson, J. M. and Berg, R. H. 1981. Posttranslational events in collagen biosynthesis. *In* A. R. Hand and C. Oliver (eds.), *Methods in Cell Biology*, Vol. 23. Academic Press, New York, pp. 119–136.

DeLeon, C. V., Cox, K. H., Angerer, L. M. and Angerer, R. C. 1983. Most early variant histone mRNA is contained in the pronucleus of sea urchin eggs. *Dev. Biol.* 100: 197–206.

Denny, P. C. and Tyler, A. 1964. Activation of protein synthesis in non-nucleate fragments of sea urchin eggs. *Biochem. Biophys. Res. Commun.* 145: 245–249.

Deshaies, R. J., Koch, B. D., Werner-Washburne, M., Craig, E. A., and Schekman, R. 1988. A subfamily of stress proteins facilitates translocation of secretory and mitochondrial precursor polypeptides. *Nature* 332: 800–805.

Dickerson, R. E. and Geis, I. 1983. *Hemoglobin*. Benjamin/Cummings, Menlo Park, CA.

Dingwall, C., Sharnick, S. V. and Laskey, R. A. 1982. A polypeptide domain that specifies migration of nucleoplasmin into the nucleus. *Cell* 30: 449–458.

Driesch, H. 1898. Über rein-mütterliche Charaktere und Bastardlarven von Echiniden. *Wilhelm Roux Arch. Entwicklungsmech. Org.* 7: 65–102.

Dunphy, W. G. and Newport, J. W. 1989. Fission yeast p13 blocks mitotic activation of tyrosine dephosphorylation of the *Xenopus cdc2* protein kinase. *Cell* 58: 181–191.

Dworkin, M. B. and Dworkin-Rastl, E. 1985. Changes in the RNA titers and polyadenylation during oogenesis and oocyte maturation in *Xenopus laevis*. *Dev. Biol.* 112: 451–457.

Ecker, R. E. and Smith, L. D. 1971. The nature and fate of *Rana pipiens* proteins synthesized during maturation and early cleavage. *Dev. Biol.* 24: 559–576.

Edgar, B. A. and Schubiger, G. 1986. Parameters controlling transcriptional activation during early *Drosophila* development. *Cell* 44: 871–877.

Evans, T., Rosenthal, E., Youngblom, J., Distel, D. and Hunt, T. 1983. Cyclin: A protein specified by maternal mRNA in sea urchin eggs that is destroyed at each cleavage division. *Cell* 33: 389–396.

Fagard, R. and London, I. M. 1981. Relationship between phosphorylation and activity of heme-regulated eukaryotic initiation factor 2 kinase. *Proc. Natl. Acad. Sci. USA* 78: 866–870.

Filipowicz, W., Sierra, J. J. and Ochoa, S. 1975. Polypeptide chain initiation in eukaryotes: Initiation factor MP in *Artemia salina* embryos. *Proc. Natl. Acad. Sci. USA* 72: 3947–3951.

Finlay, D. R. and Forbes, D. J. 1990. Reconstitution of biochemically altered nuclear pores: Transport can be eliminated and restored. *Cell* 60: 17–29.

Fisher, J. M., Sossin, W., Newcomb, R. and Scheller, R. H. 1988. Multiple neuropeptides derived from a common precursor are differentially packaged and transported. *Cell* 54: 813–822.

Francke, C., Edstrom, J. E., McDowell, A. W. and Miller, O. L. 1982. Microscopic visualization of a discrete class of giant translation units in salivary gland cells of *Chironomus tentans*. *EMBO J.* 1: 59–62.

Fritz, P. I., Vesell, E., White, E. L. and Pruit, K. M. 1969. Roles of synthesis and degradation in determining tissue concentration of LDH-5. *Proc. Natl. Acad. Sci. USA* 62: 558–565.

Galau, G., Kelin, W. H., Davis, M. M., Wold, B., Britten, R. J. and Davidson, E. H. 1976. Structural gene sets active in embryos and adult tissues of the sea urchin. *Cell* 7: 487–505.

Gilbert, S. F. and Solter, D. 1985. Onset of paternal and maternal *Gpi-2* expression in preimplantation mouse embryos. *Dev. Biol.* 109: 515–517.

Goustin, A. S. and Wilt, F. H. 1981. Protein synthesis, polyribosomes, and peptide elongation in early development of *Strongylocentrotus purpuratus*. *Dev. Biol.* 82: 32–40.

Grace, M., Bagchi, M., Ahmad, F., Yeager, T., Olson, C., Chakravarty, I., Nasron, N., Banerjee, A. and Gupta, N. K. 1984. Protein synthesis in rabbit reticulocytes: Characteristics of the protein factor RF that reverses inhibition of protein synthesis in heme-deficient reticulocyte lysates. *Proc. Natl. Acad. Sci. USA* 79: 6517–6521.

Gribble, T. J. and Schwartz, H. C. 1965. Effect of protoporphyrin on hemoglobin synthesis. *Biochim. Biophys. Acta* 103: 333–338.

Gross, K. W., Jacobs-Lorena, M., Baglioni, G. and Gross, P. R. 1973. Cell-free translation of maternal messenger RNA from sea urchin eggs. *Proc. Natl. Acad. Sci. USA* 70: 2614–2618.

Gross, M., Redman, R. and Kaplansky, D. A. 1985. Evidence that the primary effects of phosphorylation of eukaryotic initiation factor 2α in rabbit reticulocyte lysate is inhibition of the release of eukaryotic initiation factor 2-GDP from 60 S ribosomal subunits. *J. Biol. Chem.* 260: 9491–9500.

Gross, P. R. and Cousineau, G. H. 1964. Macromolecular synthesis and the influence of actinomycin D on early development. *Exp. Cell Res.* 33: 368–395.

Guyette, W. A., Matusik, R. J. and Rosen, J. M. 1979. Prolactin-mediated transcriptional and post-transcriptional control of casein gene expression. *Cell* 17: 1013–1023.

Hall, M. N., Hereford, L. and Herskowitz, I. 1984. Targeting of *E. coli* β-galactosidase to the nucleus in yeast. *Cell* 36: 1057–1065.

Harris, H. 1975. *Principles of Human Biochemical Genetics*. Elsevier North-Holland, New York.

Harvey, E. B. 1940. A comparison of the development of nucleate and non-nucleate eggs of *Arbacia punctulata*. *Biol. Bull.* 79: 166–187.

Herbert, E., Phillips, M. and Budarf, M. 1981. Glycosylation steps involved in processing of pro-corticotropin-endorphin in mouse pituitary tumor cells. *Methods Cell Biol.* 23: 101–118.

Hershey, J. W. B. 1989. Protein phosphorylation controls translation rates. *J. Biol. Chem.* 264: 20823–20826.

Hille, M. B., Danilchik, M. V., Colin, A. M. and Moon, R. T. 1985. Translational control in echinoid eggs and early embryos. *In* R. H. Sawyer and R. M. Showman (eds.), *The Cellular and Molecular Biology of Invertebrate Development*. University of South Carolina Press, pp. 91–124.

Holt, J. T., Gopal, T. V., Moulton, A. D. and Nienhuis, A. W. 1986. Inducible production of c-*fos* antisense RNA inhibits 3T3 cell proliferation. *Proc. Natl. Acad. Sci. USA* 83: 4794–4798.

Hough-Evans, B. R., Wold, B. J., Ernst, S. G., Britten, R. J. and Davidson, E. H. 1977. Appearance and persistence of maternal RNA sequences in sea urchin development. *Dev. Biol.* 260: 258–277.

Humphreys, T. 1971. Measurements of messenger RNA entering polysomes upon fertilization in sea urchins. *Dev. Biol.* 26: 201–208.

Hurt, E. C. and van Loon, A. P. G. M. 1986. How proteins find mitochondria and intramitochondrial compartments. *Trends Biochem. Sci.* 11: 204–207.

Infante, A. and Nemer, M. 1968. Heterogeneous RNP particles in the cytoplasm of sea urchin embryos. *J. Mol. Biol.* 32: 543–565.

Jenkins, N. A., Kaumeyer, J. R., Young, E. M. and Raff, R. A. 1978. A test for masked message: The template activity of messenger ribonucleoprotein particles isolated from sea urchin eggs. *Dev. Biol.* 63: 279–298.

Kabat, D. and Chappell, M. R. 1977. Competition between globin messenger ribonucleic acids for a discriminating initiation factor. *J. Biol. Chem.* 252: 2684–2690.

Kalderon, D., Roberts, B. L., Richardson, W. D. and Smith, A. W. 1984. A short amino acid sequence able to specify nuclear location. *Cell* 39: 499–509.

Karibian, D. and London, I. M. 1965. Control of heme synthesis by feedback inhibition. *Biochem. Biophys. Res. Commun.* 18: 243–249.

Kastern, W. H., Swindlehurst, M., Aaron, C., Hooper, J. and Berry, S. J. 1982. Control of mRNA translation in oocytes and developing embryos of giant moths. I. Functions of the 5' terminal "cap" in the tobacco hornworm *Manduca sexta*. *Dev. Biol.* 89: 437–449.

Kirschner, M. and Mitchison, T. 1986. Beyond self-assembly: From microtubules to morphogenesis. *Cell* 45: 329–342.

Klausner, R. D. and Harford, J. B. 1989. *Cis-trans* models for post-transcriptional gene regulation. *Science* 246: 870–872.

Kozak, M. 1986. Point mutations define a sequence flanking the AUG initiator codon that modulates translation by eukaryotic ribosomes. *Cell* 44: 283–292.

Levin, D., Ranu, R., Ernst, V. and London, I. M. 1976. Regulation of protein synthesis in reticulocyte lysates: Phosphorylation of methionyl-tRNAf binding factor by protein kinase activity of translational inhibitor isolated from heme-deficient lysates. *Proc. Natl. Acad. Sci. USA* 73: 3112–3116.

Littauer, V. Z. and Soreq, H. 1982. The regulatory function of poly(A) and adjacent 3' sequences in translated RNA. *Prog. Nucleic Acid Res. Mol. Biol.* 27: 53–83.

Lodish, H. F. 1971. Alpha and beta globin messenger ribonucleic acid. Different amounts and rates of translation. *J. Biol. Chem.* 246: 7131–7138.

London, I. M., Clemens, M. J., Ranu, R. S., Levin, D. H., Cherbas, L. F. and Ernst, V. 1976. The role of hemin in the regulation of protein synthesis in erythroid cells. *Fed. Proc.* 35: 2218–2222.

Lopo, A. C., MacMillan, S. and Hershey, J. W. B. 1988. Translational control in early sea urchin embryogenesis: Initiation factor eIF4F stimulates protein synthesis in lysates from unfertilized eggs of *S. purpuratus*. *Biochemistry* 27: 351–357.

Lusis, A. J. and Paigen, K. 1977. Relationship between levels of membrane-bound glucuronidase and the associated protein egasyn in mouse tissues. *J. Cell Biol.* 731: 728–735.

Lusis, A. J., Tomina, S. and Paigen, K. 1976. Isolation, characterization, and radioimmunoassay of murine egasyn, a protein stabilizing glucuronidase membrane binding. *J. Biol. Chem.* 251: 7753–7760.

Malacinski, G. M. 1971. Genetic control of qualitative changes in protein synthesis during early amphibian (Mexican axolotl) embryogenesis. *Dev. Biol.* 26: 442–451.

Markert, C. L. 1968. The molecular basis for isozymes. *Ann. N.Y. Acad. Sci.* 151: 14–40.

Markert, C. L. and Ursprung, H. 1971. *Developmental Genetics*. Prentice-Hall, Englewood Cliffs, NJ.

Meijlink, F., Curran, T., Miller, A. D. and Verma, I. M. 1985. Removal of a 67-base pair sequence in the non-coding region of protooncogene *fos* converts it to a transforming gene. *Proc. Natl. Acad. Sci. USA* 82: 4987–4991.

Mermod, J. J., Schatz, G. and Croppa, M. 1980. Specific control of messenger translation in *Drosophila* oocytes and embryos. *Dev. Biol.* 75: 177–186.

Moon, R. T., Danilchik, M. V. and Hille, M. 1982. An assessment of the masked messenger hypothesis: Sea urchin egg messenger ribonucleoprotein complexes are efficient templates for in vitro protein synthesis. *Dev. Biol.* 93: 389–403.

Moon, R. T., Nicosia, R. F., Olsen, C., Hille, M. B. and Jeffery, W. R. 1983. The cytoskeletal framework of sea urchin eggs and embryos: Developmental changes in the association of messenger RNA. *Dev. Biol.* 95: 447–458.

Morle, F., Lopez, B., Henni, T. and Godet, J. 1985. α-Thalassaemia associated with the deletion of two nucleotides at positions −2 and −3 preceding the AUG codon. *EMBO J.* 4: 1245–1250.

Mostov, K. E., Breitfeld, P. and Harris, J. M. 1987. An anchor-minus form of polymeric immunoglobulin receptor is secreted predominantly apically in Madin-Darby canine kidney cells. *J. Cell Biol.* 105: 2031–2036.

Newmeyer, D. D. and Forbes, D. J. 1988. Nuclear import can be separated into distinct steps in vitro: Nuclear pore binding and translocation. *Cell* 52: 641–653.

Newport, J. and Kirschner, M. 1982. A major developmental transition in early *Xenopus* embryos. II. Control of the onset of transcription. *Cell* 30: 687–696.

Palatnik, C. M., Wilkins, C. and Jacobson, A. 1984. Translational control during early *Dictyostelium* development: Possible involvement of poly(A) sequences. *Cell* 36: 1017–1025.

Pavlakis, G. N., Lockard, R. E., Vamvakopolous, N., Rieser, L., Rajbhandary, U. L. and Vournakis, J. N. 1980. Secondary structure of mouse and rabbit α- and β-globin mRNAs: Differential accessibility of initiator AUG codons towards nucleases. *Cell* 19: 91–102.

Paynton, B. V., Rempel, R. and Bachvarova, R. 1988. Changes in states of adenylation and time course of degradation of maternal mRNAs during oocyte maturation and early embryonic development in the mouse. *Dev. Biol.* 129: 304–314.

Pelham, H. R. B. and Brown, D. D. 1980. A specific transcription factor that can bind either the 5 S RNA gene or 5 S RNA. *Proc. Natl. Acad. Sci. USA* 77: 4170–4174.

Pelletier, J. and Sonenberg, N. 1985. Insertional mutagenesis to increase secondary structure within the 5′ noncoding region of a eukaryotic mRNA reduces translational efficiency. *Cell* 40: 515–526.

Poccia, D., Wolff, R., Kragh, S. and Williamson, P. 1985. RNA synthesis in male pronuclei of the sea urchin. *Biochim. Biophys. Acta* 824: 349–356.

Powell, L. M., Wallis, S. C., Pease, R. J., Edwards, Y. H., Knott, T. J. and Scott, J. 1987. A novel form of tissue-specific RNA processing produces apolipoprotein-B48 in intestine. *Cell* 50: 831–840.

Prockop, D. J., Kivirikko, K. I., Tuderman, L. and Guzman, N. A. 1979. The biosynthesis of collagen and its disorders. *N. Engl. J. Med.* 301: 13–23.

Raff, R. A. 1980. Masked messenger RNA and the regulation of protein synthesis in eggs and embryos. *In* D. M. Prescott and L. Goldstein (eds.), *Cell Biology: A Comprehensive Treatise*, Vol. 4. Academic Press, New York, pp. 107–136.

Raff, R. A., Colot, H. V., Solvig, S. E. and Gross, P. R. 1972. Oogenetic origin of messenger RNA for embryonic synthesis of microtubule proteins. *Nature* 235: 211–214.

Ranu, R. S., Levin, D. H., Delaunay, J., Ernst, U. and London, I. M. 1976. Regulation of protein synthesis in rabbit reticulocyte lysates: Characteristics of inhibition of protein synthesis by a translational inhibitor from heme-deficient lysates and its relationship to the initiation factor which binds Met-tRNA$_f$. *Proc. Natl. Acad. Sci. USA* 73: 2720–2726.

Rappollee, D., Brenner, C. A., Schultz, R., Mark, D., and Werb, Z. 1988. Developmental expression of PDGF, TGF-α, and TGF-β genes in preimplantation mouse embryos. *Science* 241: 1823–1825.

Ray, B. K. and 8 others. 1983. Role of mRNA competition in regulating translation: Further characterization of mRNA discriminatory initiation factors. *Proc. Natl. Acad. Sci. USA* 80: 663–667.

Raychaudhury, P., Chaudhuri, A. and Maitra, U. 1985. Formation and release of eukaryotic initiation factor 2 GDP complex during eukaryotic ribosomal polypeptide chain initiation complex formation. *J. Biol. Chem.* 260: 2140–2145.

Rebagliati, M. R., Weeks, D. L., Harvey, R. P. and Melton, D. A. 1985. Identification and cloning of localized maternal RNAs from *Xenopus* eggs. *Cell* 42: 769–777.

Restifo, L. L. and Guild, G. M. 1986. Poly(A) shortening of coregulated transcripts in *Drosophila*. *Dev. Biol.* 115: 507–510.

Richardson, W. D., Mills, A. D., Dilworth, S. M., Laskey, R. A. and Dingwall, C. 1988. Nuclear protein migration involves two steps: rapid binding at the nuclear envelope followed by slower translocation through nuclear pores. *Cell* 52: 655–664.

Richter, J. D. and Smith, L. D. 1984. Reversible inhibition of translation by *Xenopus* oocyte-specific proteins. *Nature* 309: 378–380.

Rodgers, W. H. and Gross, P. R. 1978. Inhomogeneous distribution of egg RNA sequences in the early embryo. *Cell* 14: 279–288.

Rogers, S., Wells, R. and Rechsteiner, M. 1986. Amino acid sequences common to rapidly degraded proteins: The PEST hypothesis. *Science* 234: 364–368.

Rosenthal, E. T. and Ruderman, J. V. 1987. Widespread change in the translation and adenylation of maternal messenger RNA following fertilization of *Spisula* oocytes. *Dev. Biol.* 121: 237–246.

Rosenthal, E., Hunt, T. and Ruderman, J. V. 1980. Selective translation of mRNA controls the pattern of protein synthesis during early development of the surf clam, *Spisula solidissima*. *Cell* 20: 487–494.

Safer, B. 1989. Nomenclature of initiation, elongation and termination factors for translation in eukaryotes. *Eur. J. Biochem.* 186: 1–3.

Sarkar, G., Edery, I., Gallo, R. and Sonenberg, N. 1984. Preferential stimulation of rabbit α-globin mRNA translation by a cap-binding protein complex. *Biochim. Biophys. Acta* 783: 122–129.

Schleyer, M. and Neupert, W. 1985. Transport of proteins into mitochondria: Translocational intermediates spanning contact sites between outer and inner membranes. *Cell* 43: 339–350.

Shapiro, D. J., Blume, J. E. and Nielsen, D. A. 1987. Regulation of messenger RNA stability in eukaryotic cells. *BioEssays* 6: 221–226.

Shatkin, A. J. 1976. Capping of eukaryotic mRNAs. *Cell* 9: 645–653.

Shatkin, A. J. 1985. mRNA cap binding proteins: Essential factors for initiating translation. *Cell* 40: 223–224.

Shaw, G. and Kamen, R. 1986. A conserved AU sequence from the 3′ untranslated region of GM-CSF mRNA mediates selective mRNA degradation. *Cell* 46: 659–667.

Shiokawa, K., Tashiro, K., Oka, T. and Yamana, K. 1983. Contribution of maternal mRNA for maintenance of Ca^{2+}-dependent reaggregating activity in dissociated cells of *Xenopus laevis* embryos. *Cell Diff.* 13: 247–255.

Shiokawa, K., Misumi, Y., Nakamura, N., Yamana, K. and Oh-uchida, M. 1989. Changes in the patterns of RNA synthesis in early embryogenesis of *Xenopus laevis*. *Cell Diff. Devel.* 28: 17–26.

Showman, R. M., Wells, D. E., Anstrom, J., Hursh, D. A. and Raff, R. A. 1982. Message-specific sequestration of maternal histone mRNA in the sea urchin egg. *Proc. Natl. Acad. Sci. USA* 79: 5944–5947.

Shyne-Athwal, S., Riccio, R. P., Chakraborty, G. and Ingolia, N. A. 1986. Protein modification by amino acid addition is increased in crushed sciatic but not optic nerves. *Science* 231: 603–607.

Sierra, J. M., de Haro, C., Datta, A. and Ochoa, S. 1977. Translational control by protein kinases in *Artemia salina* and wheat germ. *Proc. Natl. Acad. Sci. USA* 74: 4356–4359.

Skoultchi, A. and Gross, P. R. 1973. Maternal histone messenger RNA: Detection by molecular hybridization. *Proc. Natl. Acad. Sci. USA* 70: 2840–2844.

Sly, W. S., Fischer, H. D., Gonzalez-Noriega, A., Grubb, J. H. and Natowicz, M. 1981. Role of the 6-phosphomannosyl-enzyme receptor in intracellular transport and adsorptive pinocytosis of lysosomal enzymes. *Methods Cell Biol.* 23: 191–214.

Smith, L. D. and Ecker, R. C. 1965. Protein synthesis in enucleated eggs of *Rana pipiens*. *Science* 150: 777–779.

Soffer, R. L. 1980. Biochemistry and biology of aminoacyl-transfer RNA-protein transferases. *In* D. Soll, J. N. Abelson and P. R. Schimmel (eds.), *Transfer RNA: Biological Aspects*. Cold Spring Harbor Laboratory, Cold Spring Harbor, NY, pp. 493–505.

Söllner, T., Griffiths, G., Pfaller, R., Pfanner, N. and Neupert, W. 1989. MOM19, an import receptor for mitochondrial precursor proteins. *Cell* 59: 1061–1070.

Spirin, A. S. 1966. On "masked" forms of messenger RNA in early embryogenesis and in other differentiating systems. *Curr. Top. Dev. Biol.* 1: 1–38.

Standart, N., Hunt, T. and Ruderman, J.V. 1986. Differential accumulation of ribonucleotide reductase subunits in clam oocytes: The large subunit is stored as a polypeptide, the small subunit as untranslated mRNA. *J. Cell Biol.* 103: 2129–2136.

Swank, R. T. and Paigen, K. 1973. Biochemical and genetic evidence for a macromolecular β-glucuronidase complex in microsomal membranes. *J. Mol. Biol.* 77: 371–389.

Swiderski, R. E. and Richter, J. D. 1988. Photocrosslinking of proteins to maternal mRNA in *Xenopus* oocytes. *Dev. Biol.* 128: 349–358.

Tennant, D. H. 1914. The early influence of spermatozoa upon the characters of echinoid larva. *Carnegie Inst. Wash. Publ.* 182: 127–138.

Thomas, N. S. B., Matts, R. L., Levin, D. H. and London, I. M. 1985. The 60S ribosomal subunit as a carrier of eukaryotic initiation factor 2 and the site of reversing factor activity during protein synthesis. *J. Biol. Chem.* 260: 9860–9866.

van Loon, A. P. G. M. and Schatz, G. 1987. Transport of protons into the mitochondrial intramembranous space: The "sorting" domain of cytochrome c_1 presequence is a stop-transfer sequence specific for the mitochondrial inner membrane. *EMBO J.* 6: 2441–2448.

van Loon, A. P. G. M., Brandli, A. W. and Schatz, G. 1986. The presequences of two imported mitochondrial proteins contain information for intracellular and intramitochondrial sorting. *Cell* 44: 801–812.

Wagenaar, E. B. and Mazia, D. 1978. The effect of emetine on the first cleavage division of the sea urchin, *Strongylocentrotus purpuratus*. In E. R. Dirksen, D. M. Prescott and L. F. Fox (eds.), *Cell Reproduction: In Honor of Daniel Mazia*. Academic Press, New York, pp. 539–545.

Walter, P. and Blobel, G. 1983. Subcellular distribution of signal recognition particle and 7SL-RNA determined with polypeptide-specific antibodies and complementary DNA probe. *J. Cell Biol.* 97: 1693–1699.

Weidmann, M., Kurzchalia, T. V., Hartmann, E. and Rapoport, T. A. 1987. A signal sequence receptor in the endoplasmic reticulum membrane. *Nature* 328: 830–833.

Weir, M. P. and Kornberg, T. 1985. Patterns of *engrailed* and *fushi tarazu* transcripts reveal novel intermediate stages of *Drosophila* segmentation. *Nature* 318: 433–439.

Wellems, T. E. and Josephs, R. 1979. Crystallization of deoxyhemoglobin S by fiber alignment and fusion. *J. Mol. Biol.* 135: 651–674.

Wilson, T. and Treisman, R. 1988. Removal of poly(A) and consequent degradation of c-*fos* mRNA facilitated by 3' AU-rich sequences. *Nature* 336: 396–399.

Wilt, F. H. 1964. Ribonucleic acid synthesis during sea urchin embryogenesis. *Dev. Biol.* 9: 299–313.

Winkler, M. M. and Steinhardt, R. A. 1981. Activation of protein synthesis in a sea urchin cell-free system. *Dev. Biol.* 84: 432–439.

Winkler, M. M., Nelson, E. M., Lashbrook, C. and Hershey, J. W. B. 1985. Multiple levels of regulation of protein synthesis at fertilization in sea urchin eggs. *Dev. Biol.* 107: 290–300.

Woodland, H. R. and Ballantine, J. E. M. 1980. Paternal gene expression in developing hybrid embryos of *Xenopus laevis* and *Xenopus borealis*. *J. Embryol. Exp. Morphol.* 60: 359–372.

Young, E. M. and Raff, R. A. 1979. Messenger ribonucleoprotein particles in developing sea urchin embryos. *Dev. Biol.* 72: 24–40.

Zucker, W. V. and Schulman, H. M. 1968. Stimulation of globin-chain initiation by hemin in the reticulocyte cell-free system. *Proc. Natl. Acad. Sci. USA* 59: 582–589.

III

Cell interactions in development

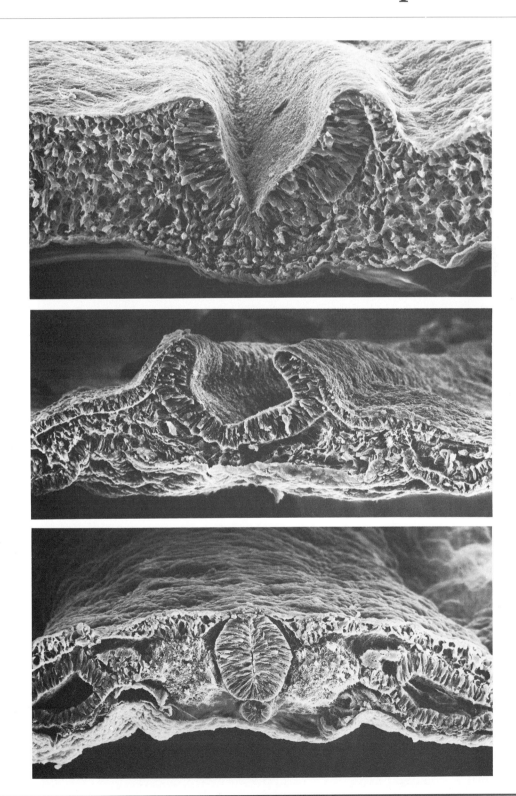

15

The spatial ordering of cells: The roles of the cell surface

But nature is not atomized. Its patterning is inherent and primary, and the order underlying beauty is demonstrably there; what is more, the human mind can perceive it only because it is itself part and parcel of that order.

—PAUL WEISS (1960)

I am fearfully and wonderfully made.

—PSALM 139 (CA. 500 B.C.)

A body is not merely a collection of randomly distributed cell types, for development involves not only the *differentiation* of cells but also their *morphogenesis* into multicellular arrangements such as tissues and organs. When one observes the detailed anatomy of a tissue such as the retina, one sees an intricate and precise arrangement of many different types of cells. In the next seven chapters we will discuss the ways by which the cells of the developing embryo change to create the functional organs of the body. There are four major questions that form the framework for discussions on morphogenesis.

- *How are tissues formed from cells?* The renal epithelium, for instance, forms a tight tube of cells with a thick basement membrane that is used for filtering the blood. How are these cells stuck together?
- *How are organs constructed from tissues?* The epithelial cells of the renal tubule must form a tube with the mesenchyme outside it. A sheet of cells would not be an appropriate structure for urine flow and resorption; nor would such tubules serve any function if they were full of connective tissue. Moreover, the filtering units of the kidney must somehow be connected to the ureter, through which the urine passes to the bladder. All these connections must be precisely ordered.
- *How do migrating cells reach their destinations and how do organs form in particular locations?* Some of the cells that produce the kidney origi-

nate in the anterior of the embryo and must migrate caudally. Similarly, the germ cells of vertebrates do not originally reside in the gonads. How are cells instructed to travel along certain routes and to stop when they reach a particular destination?

• *How do organs and their cells grow, and how is their growth coordinated throughout development?* The kidneys of college students do not grow at the rate at which they grew when the students were newborns. Blood cell precursors, however, divide at essentially fetal rates throughout one's life. What controls these differences in growth rate?

All these questions concern aspects of cell behavior. Indeed, it is generally thought that morphogenesis is brought about through a limited repertoire of cellular processes: (1) the direction and number of cell divisions; (2) cell shape changes; (3) cell migration; (4) cell growth; (5) cell death; and (6) changes in the composition of the cell membrane and extracellular matrix. We have mentioned some of these processes earlier, especially in the context of gastrulation and neurulation—two of the major morphogenetic events of the early embryo.

There appear to be two major ways by which cells communicate with one another to effect morphogenesis. The first is through diffusible substances that are made by one type of cell and that change the behavior of other types of cells. These substances include hormones, growth factors, and morphogens (and will be detailed in subsequent chapters). The second method involves interactions between the surfaces of adjacent cells. During the period of organogenesis, individual cells and groups of cells change their relative positions and become associated with other cell types. Organ and tissue formation involve the selective interactions between neighboring cells to create the proper arrangement of the different cell types.

Therefore, cells must have the ability to selectively recognize other cells, adhering to some and migrating over others. It is generally accepted that the molecular events mediating the selective recognition of cells and their formation into tissues and organs occur at the cell surface. Whereas the dominant paradigm of developmental genetics is differential gene expression, the dominant paradigm of morphogenesis involves *differential cell affinities*. In this chapter, we will analyze the ways by which adjacent cell surfaces can interact during development such that cells become localized in their respective sites within tissues and organs.

Differential cell affinity

Stationary cultures

The modern analysis of morphogenesis begins with the experiments of Townes and Holtfreter in 1955. Taking advantage of the discovery that amphibian tissues become dissociated into single cells when placed in alkaline solutions, they prepared single-cell suspensions from each of the three amphibian germ layers and from neurula tissues. Two or more of these single-cell suspensions could be combined in various ways, and when the pH was normalized, the cells adhered to each other, forming aggregates on agar-coated petri dishes. By using the embryos from species having cells of different sizes and colors, Townes and Holtfreter were able to follow the behavior of the recombined cells (Figure 1).

The results of their experiments were striking. First, they found that

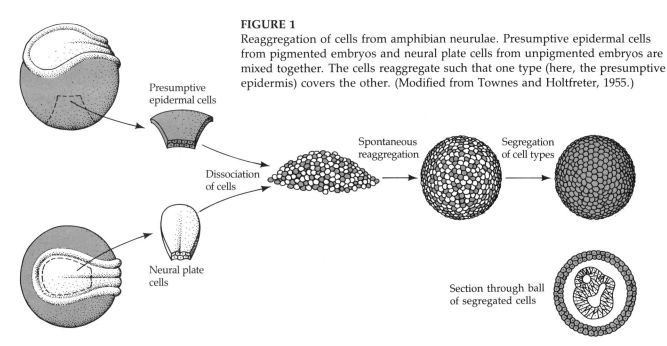

FIGURE 1

Reaggregation of cells from amphibian neurulae. Presumptive epidermal cells from pigmented embryos and neural plate cells from unpigmented embryos are mixed together. The cells reaggregate such that one type (here, the presumptive epidermis) covers the other. (Modified from Townes and Holtfreter, 1955.)

Presumptive
epidermal cells

Dissociation
of cells

Spontaneous
reaggregation

Segregation
of cell types

Neural plate
cells

Section through ball
of segregated cells

the reaggregated cells became spatially segregated. That is, instead of remaining mixed, each cell type SORTED OUT into its own region. Thus, when epidermal and mesodermal cells were brought together to form a mixed aggregate, the epidermal cells were found at the periphery of the aggregate and the mesodermal cells were found inside. In no case did the recombined cells remain randomly mixed. In most cases, one tissue type completely enveloped the other.

Second, they found that the final positions of the reaggregated cells reflected their embryonic positions. We have already seen that the mesoderm migrates centrally to the epidermis, adhering to the inner epidermal surface (Figure 2A). The mesoderm also migrates centrally with respect to endoderm (Figure 2B). However, when the three germ layer cells are mixed together (Figure 2C), the endoderm separates from the ectoderm and mesoderm and is then enveloped by them. In its final configuration, the ectoderm is on the periphery, the endoderm is internal, and the mesoderm lies in the region between them. Holtfreter interpreted this in terms of SELECTIVE AFFINITY. The inner surface of the ectoderm has a positive affinity for the mesodermal cells, while it has a negative affinity for the endoderm. The mesoderm has positive affinities for both the ectodermal and endodermal cells. The mimicry of normal embryonic structure by cell aggregates is also seen in the recombination of epidermis and neural plate cells (Figure 2D). The presumptive epidermal cells migrate to the periphery as before; the neural plate cells migrate inward, forming a structure reminiscent of the neural tube. When axial mesoderm cells are added to the suspension of presumptive epidermal and neural plate cells, the cell segregation results in an external epidermal layer, a centrally located neural tissue, and a layer of mesodermal somites and mesenchyme between them (Figure 2E). Somehow, the cells are able to sort out into their proper embryological positions. Such preferential affinities have also been noted by Boucaut (1974), who injected individual cells from specific germ layers back into blastocoels of amphibian embryos. He found that these cells migrate to their appropriate germ layer. Endodermal cells find positions in the host endoderm, whereas ectodermal cells are found only in host ectoderm. Thus, selective affinity appears to be important for imparting positional information to embryonic cells.

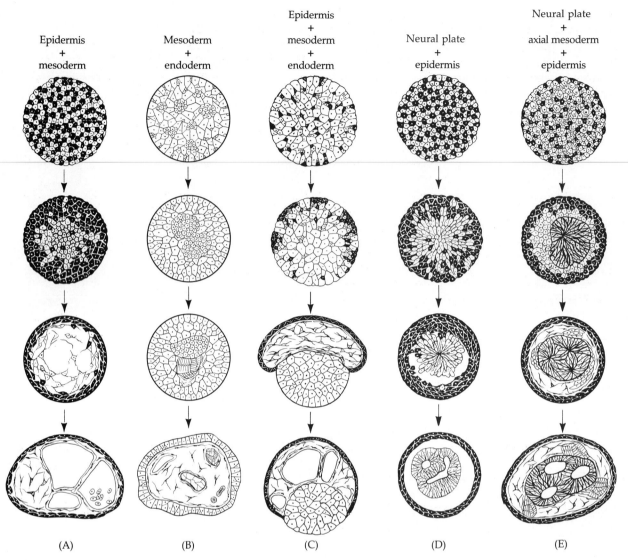

Epidermis
+
mesoderm

Mesoderm
+
endoderm

Epidermis
+
mesoderm
+
endoderm

Neural plate
+
epidermis

Neural plate
+
axial mesoderm
+
epidermis

(A)   (B)   (C)   (D)   (E)

FIGURE 2

Sorting out and reorganization of embryonic spatial relationships in aggregates of embryonic amphibian cells. (Modified from Townes and Holtfreter, 1955.)

The third conclusion of Holtfreter and his colleagues was that selective affinities change during development. This should be expected, because embryologic cells do not retain a single stable relationship with other cells. For development to occur, cells must interact differently with other cell populations at specific times. Such changes in cellular affinity were dramatically confirmed by Trinkaus (1963), who showed a clear correlation between adhesive changes in vitro and changes in embryonic cell behavior. When the discoidal embryos of teleost fishes initiate gastrulation, the cells at the margin of the blastodisc flatten and begin to migrate over the egg cytoplasmic layer. Trinkaus found that when he isolated these cells from the *blastula*, they did not adhere to an artificial substrate, whereas the same cells isolated from *gastrulating* embryos would stick to the surface and migrate. Moreover, when the blastula cells were kept in culture for a duration similar to the time required for their entry into gastrulation, the cells began to flatten and adhere to the artificial substrate, just as they would have done in the embryo.

One also sees such changes in cell affinity during sea urchin gastrulation. In Chapter 4, for instance, the experiments of Fink and McClay (1985) showed the changes in affinity in the precursors of the primary mesenchymal cells during early gastrulation. At the onset of gastrulation, these cells lose their affinity for neighboring cells and for the hyaline layer while simultaneously acquiring affinity for the fibronectin fibers lining the blastocoel. Later in gastrulation, these cells acquire the ability to migrate toward a particular portion of the blastocoel wall. Moreover, these cells, which had been "antisocial" toward each other since their ingression into the blastocoel, must now adhere together to form a mesodermal ring around the invaginating endoderm. These changes in adhesion are temporally specific and are also specific to the primary mesenchyme cells (McClay and Ettensohn, 1987). Such changes in cell affinity are extremely important in the processes of morphogenesis.

Rotary cultures

The reconstruction of aggregates from older embryos of birds and mammals was accomplished and extended by two techniques pioneered by A. A. Moscona. The first technique was the use of trypsin to dissociate the cells from one another (Moscona, 1952). This enzyme cleaves the proteins at the cell surface and between cells, thereby breaking many of the bonds that join the cells together. The second technique was that of rotary aggregation (Moscona, 1961). Instead of permitting the cells to settle on a culture dish (to which they had little or no affinity), Moscona gently swirled the mixed cell suspension in a flask. Here, cells are thrown together in a vortex and will aggregate only if the strength of their mutual adhesion is greater than the shearing force produced by the swirling medium. Within these initially random aggregates, cells move around and sort out according to their cell type. In so doing, they mimic the structure of the original tissue. Figure 3 shows the "reconstruction" of skin tissue from a 15-day embryonic mouse. The skin cells are separated by proteolytic enzymes and are then aggregated in a rotary culture. The epidermal cells migrate to the periphery and the dermal cells migrate toward the center. By 72 hours, the epidermis has been reconstituted, a keratin layer has formed, and hair follicles are seen in the dermal region. The aggregation procedure has enabled the individual cells to reconstruct the tissue. Similarly, suspensions of isolated embryonic kidney cells reaggregate to form tubules, and retinal cells come together to reconstitute the structure of the neural retina. Such reconstruction of complex tissues from single cells is called HISTOTYPIC AGGREGATION. Thus, the cells of the embryonic organs maintain the morphogenetic information that enables them to reform the tissue- and organ-specific structures.

In some species, the entire organism can be reconstructed by aggregating a suspension of separated cells. The blastomeres of the sea urchin blastula can reaggregate to form blastula-like structures that can then develop into normal pluteus larvae (Figure 4) (Giudice, 1962; Spiegel and Spiegel, 1975). All three cell types (micromeres, macromeres, and mesomeres) must be present for this to happen, which indicates that the cells did not revert to a dedifferentiated state and then redifferentiate according to their new positions (Spiegel and Spiegel, 1975).

FIGURE 3

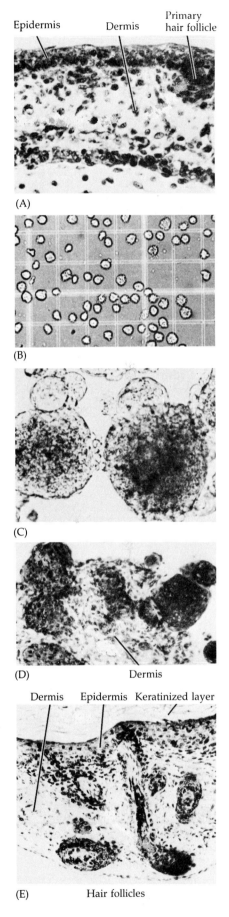

Reconstruction of skin from a suspension of skin cells from a 15-day embryonic mouse. (A) Section through the embryonic skin showing epidermis, dermis, and primary hair follicle. (B) Suspension of single skin cells from both the dermis and epidermis. (C) Aggregates after 24 hours. (D) Section through an aggregate, showing migration of epidermal cells to the periphery. (E) Further differentiation of aggregates (72 hours), showing reconstituted epidermis and dermis, complete with hair follicles and keratinized layer. (From Monroy and Moscona, 1979; courtesy of A. Moscona.)

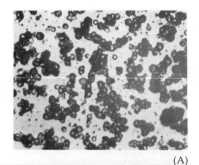

(A)

(B)

FIGURE 4
Reaggregation of blastomeres dissociated from the 16-cell embryo of the sea urchin *Arbacia punctulata*. (A) Dissociated cells. (B) Reconstructed pluteus larva after 25 hours' aggregation. (From Spiegel and Spiegel, 1975; courtesy of the authors.)

Cells, then, do not sort randomly, but can actively move to create tissue organization. What forces direct cell movement during morphogenesis?

Modes of cellular migration

A motile cell (like a randomly motile person) would not be expected to move far from its origin over any period of time if the environment did not provide some cues for it to move to one place rather than to another. What mechanisms permit cells to migrate to specific areas within the embryo? Long-range migrations can be caused by chemotaxis, haptotaxis, galvanotaxis, and contact guidance. Shorter-range interactions (such as those responsible for the mutual interactions of cells during organ formation) can be caused by contact inhibition and by thermodynamic interactions at the cell surfaces.

Chemotaxis

CHEMOTAXIS is defined as cellular locomotion directed in response to a concentration gradient of a chemical factor in solution (Harris, 1954; Armstrong, 1985). Cells would sense the chemical and migrate toward higher concentrations of this substance until they reached the source secreting it. Even though chemotaxis is easily conceptualized, it is difficult to prove. First, it is not sufficient to measure the accumulation of cells at a particular place, since they may simply be trapped there. Second, a soluble factor may merely activate cell motility, and if there is only one path possible, the outcome will mimic that of chemotaxis (Zigmond, 1978; Trinkaus,

FIGURE 5
Chemotaxis of lymphocyte precursors to diffusible compound from the thymus. Cells were placed on a coverslip between two wells of culture medium (A). The media mix between the wells, forming a gradient and the migration of cells is recorded by a video camera. Video camera tracings of cell movement are shown (B) when both wells contained control medium and (C) when one of the wells contained control (unconditioned) medium and the other contained medium in which embryonic quail thymic epithelium had been grown. (After Champion et al., 1986.)

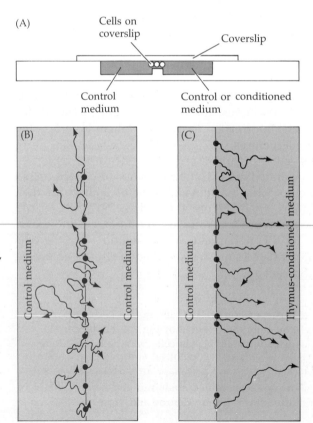

1985). We have already encountered two examples of chemotaxis in development. Sea urchin sperm appear to behave chemotactically to low-molecular-weight substances (such as resact) in the egg jelly, and the angiogenesis factors secreted by tumors can cause the migration of capillary endothelial cells toward the tumor.

Does chemotaxis act during the morphogenesis of the embryo? There is at least one case wherein the migration of cells seems to be caused by a specific chemotactic molecule. The migration of lymphocyte precursors from the bone marrow into the embryonic thymus (wherein they generate the T cells of the immune system) is caused by soluble factors secreted by the thymus cells. Champion and co-workers (1986) placed quail bone marrow cells into a 1-mm region connecting two wells (Figure 5A). When normal medium was placed in both wells, there was no net migration of the bone marrow cells toward either well (Figure 5B). However, when one of the wells contained medium in which embryonic thymus epithelial cells had been cultured, the bone marrow cells migrated specifically toward that well (Figure 5C). The chemotactic compound secreted by the thymus cells has been identified as β_2-microglobulin, an 11,000-Da peptide that is also known to stabilize certain proteins in the cell membrane (Dargemont et al., 1989). Although this compound has not yet been tested in vivo, it is probable that the migration of lymphoid precursors into the developing thymus is directed by a gradient of a small peptide secreted by the embryonic thymic epithelial cells.

Haptotaxis

Gradients do not have to be in solution. An adhesive molecule could be present in increasing amounts along an extracellular matrix. A cell that is constantly making and breaking adhesions with such a molecule would move from a region of low concentration to an area where that adhesive molecule is more highly concentrated. Such a phenomenon is called HAPTOTAXIS (Carter, 1967; Curtis, 1969). Harris (1973) showed that cells in culture will migrate up a concentration gradient when such a substance is added to the plastic of their tissue culture dishes; but the conditions are very stringent. If the substrate is not sticky enough, cells will wander from the path; if the substrate is too sticky, the cells will become stuck in place.

Poole and Steinberg (1982) have evidence that the migration of the pronephric duct cells in salamanders is regulated by haptotaxis. The pronephric duct rudiment is important for kidney and gonad development (Chapter 16). The duct rudiment segregates from the dorsal mesoderm as a solid cord of cells, five to six somites in length, along the ventrolateral border of the segmenting somites. This group of cells is first seen as an ovoid mass near the head of the embryo, but as the embryo develops, this pronephric rudiment extends along the ventrolateral border of the somites, eventually reaching the cloaca (where the urine will be excreted). The duct grows by the migration of the original population of cells. Division is not seen, the individual cells do not appear to elongate, and dye placed on the cells at the posterior end of the pronephric rudiment in an early embryo is found near the cloaca in later stages (Figure 6). Thus, the cells of the pronephric rudiment appear to migrate upon a specific path from one place to another along the surface of the embryo.

The cues guiding this migration appear to be polarized (i.e., localized as a gradient) on the surface of the embryo. The migration cannot be due to chemotaxis, because the removal of the cloacal region does not stop migration, nor does placing the cloacal region in a different position in the embryo redirect the path of the pronephric rudiment. The migration

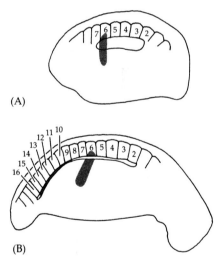

(A)

(B)

FIGURE 6
Elongation of the pronephric duct by cell migration. (A) A portion of the embryo is stained with a vital dye (color) across the region of the newly formed pronephric duct rudiment. (B) Several hours later, the dye mark in the pronephric duct has extended into the caudal region of the embryo, away from its original source. (After Poole and Steinberg, 1982.)

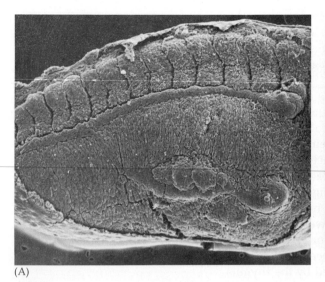

(A)

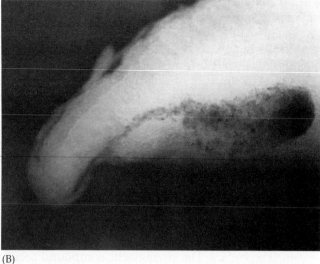

(B)

FIGURE 7

Grafted pronephric duct rudiment (bottom) extends a duct dorsocaudally across the host's flank mesoderm to fuse with the host's pronephric duct. (A) Scanning electron micrograph showing relationship between pronephric duct and somites. (B) Wild-type pronephric duct grafted to albino host to show migration is only directed dorsally (towards somites) and caudally (towards cloaca). (A from Poole and Steinberg, 1982; B from Zackson and Steinberg, 1986; photographs courtesy of M. S. Steinberg.)

FIGURE 8

Galvanotaxis in culture. *Xenopus* neurons extend axons parallel to voltage gradient of 170 mV/mm. The anode (+) and cathode (−) are labeled at the sides of the photograph. (From Hinkle et al., 1981; courtesy of K. R. Robinson.)

also cannot be due to the contours of the embryo, and no electrical gradient is observed (see below). When a pronephric rudiment is transplanted from one embryo into the ventral region of another, the secondary duct always travels dorsally across the flank mesoderm to meet the original one and then migrates caudally beneath the somites toward the cloaca (Figure 7). The transplanted ducts never migrate ventrally or toward the head, regardless of the orientation in which the graft is inserted (Zackson and Steinberg, 1987). Recent data (Zackson and Steinberg, 1988, 1989) suggest that there is a gradient of the enzyme alkaline phosphatase across the mesodermal surface of the salamander embryo and that this gradient exists ventral-to-dorsal and anterior-to-posterior. Removal of this gradient by alkaline phosphatase inhibitors inhibits the migration of the pronephric duct. It is possible, then, that gradients of molecules exist along cell surfaces that orient cells in their migrations through the embryo.

Galvanotaxis

There is one other possible source of polar gradients in the embryo: electrically charged ions. Potential differences between cells and their environment can be critical in development (as in fertilization). Do voltage differences exist between embryonic regions, and could they be important in morphogenesis?

In 1920, Harrison's laboratory reported that growing nerve fibers could align themselves along electrical lines of force (Ingvar, 1920). However, these experiments could not be repeated in the same laboratory 14 years later (Weiss, 1934). Thereafter, the idea that electrical currents could influence cell migration (GALVANOTAXIS) fell out of favor until Jaffe and Nuccitelli invented a probe that could detect extremely small electrical currents in living organisms (Jaffe, 1981; Nuccitelli, 1984). These minute electric fields, measuring between 10 and 100 mV/mm, are sufficient to alter the direction of nerve growth or to accelerate it in the direction of the negative pole (Figure 8; Hinkle et al., 1981; Jaffe and Poo, 1979). This cathodal effect can be seen in electrical gradients as shallow as 7 mV/mm. It is possible that the electrical current allows an influx of Ca^{2+} into a particular region of the growth cone, thereby causing cytoskeletal assembly and mobility in that particular direction (Cooper and Schliwa, 1985). Embryonic chick fibroblasts also migrate toward the negative pole when cultured in a small, steady electrical field (Nuccitelli and Erickson, 1983). Larger electric currents can be detected in early chick embryos (Jaffe and Stern, 1979), in the regeneration blastema of several amphibian limbs (Borgens, 1982), and in the *Cecropia* moth ovary (Woodruff and Telfer, 1974). The current in the moth ovary is probably important in selectively transporting (by a kind of electrophoresis) material from follicle cells into the oocyte (Chapter 21). However, the role of such currents in directing axonal migration and cell patterning in vivo is still uncertain.

Contact guidance

So far, we have limited our discussion to the chemical or ionic factors that could affect directed cell migration. However, physical factors may also play important roles. The cell is influenced by the physical terrain over which it passes and can be led to one place rather than another by this CONTACT GUIDANCE. If physical barriers allow only one channel to be open, motile cells will have little choice but to migrate through it.

Weiss (1934) showed that growing nerve fibers follow the stress contours of a plasma clot rather than move out of shallow grooves. Moreover, by altering the lines of stress, nerve fibers could be made to run parallel to one another (Figure 9). Weiss noted (1955) that if the substratum were stressed in certain ways, the channels that were created acted as guides for cell migration. At that time, however, he could not envision what might cause these stresses. In 1980 Harris showed that fibroblasts can dramatically alter the substrate on which they are cultured. When they are cultured on silicone rubber or collagen, they deform their substrate, generating stress folds practically identical to the ones seen by Weiss (Figure 10). The traction generated by clusters of fibroblasts can gather in

FIGURE 9

Contact guidance of axons by the stress fibers of a plasma clot. Along the random meshes, the axons travel randomly along the clot fibers and branch frequently. Where the fibrous matrix consists of parallel grooves, the axons are made to run parallel to one another. (From Weiss, 1955.)

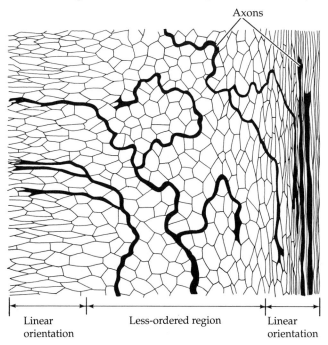

Linear orientation — Less-ordered region — Linear orientation

FIGURE 10

Stress fibers made by a single embryonic chick fibroblast crawling upon a sheet of silicone rubber. The cell is approximately 80 mm wide. (From Harris, 1984; photograph courtesy of A. K. Harris.)

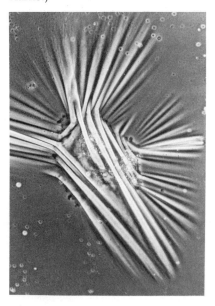

collagen from its environment to alter the shape of the substrate. If muscle cells are added to such a substrate, the randomly scattered muscle cells become aligned into well-formed functional units (Stopak and Harris, 1982). Contact guidance is also hypothesized to account for the migration of mesenchymal cells inside the fish fin. The collagen-containing fibrils at the base of the developing pectoral fin in bony fish embryos have been shown to provide a substrate that favors the movement of mesenchymal cells from the trunk into the fin region (Wood and Thorogood, 1987).

Contact inhibition of movement

In addition to the above-mentioned mechanisms of migration that act at relatively long distances, local phenomena can also provide directional cues for cell migration. Cells move by extending a thin process called a LAMELLIPODIUM. When the lamellipodium of one migrating cell contacts the surface of another cell, the lamellipodium is paralyzed and disappears. This process is called CONTACT INHIBITION of movement (Figure 11). A new lamellipodium is then formed elsewhere on the cell, thereby taking the cell away from its neighbor (Abercrombie and Ambrose, 1958). The net result is the migration of motile cells away from central masses. (This phenomenon is seen primarily in mesenchymal cells and does not occur when the cells are joined to each other in epithelial sheets, except when a free margin is exposed.)

Contact inhibition may provide the stimulus for the migration of the neural crest cells away from the original crest (Rosavio et al., 1983), and it has been seen to operate in the healing of wounds. When epidermal cells are removed from *Xenopus* tadpole tails, the underlying basal cells exposed on both sides of the wound area form lamellipodia within 5–10 seconds. These lamellipodia rapidly travel across the basement membrane to cover the wound area. Upon touching each other, the opposing lamellipodia cross slightly, adhere to each other, and stop their forward motion (Radice, 1980).

Thermodynamic model of cell interactions

While the above models explain several important features of morphogenesis, they have not been able to explain the sorting out phenomena noted by Townes and Holtfreter. In 1964 Malcolm Steinberg proposed a model that explained the directions of cell sorting on thermodynamic principles. Using cells derived from trypsinized embryonic tissues, Steinberg showed that certain cell types always migrate centrally when combined with some cell types, but migrate peripherally when combined with

FIGURE 11
Contact inhibition. The lamellipodium of one fibroblast approaches (A) and contacts (B) the membrane of another fibroblast. While retaining these contacts, the cell that had the ruffled membrane passes beneath (underlaps) the other cell (C) and then moves away from it (D). (From Erickson et al., 1980; photographs courtesy of C. A. Erickson.)

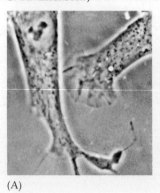

(A)

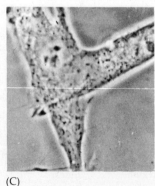

(B)

(C)

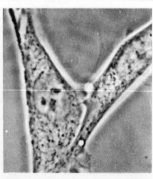

(D)

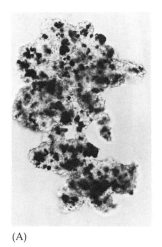

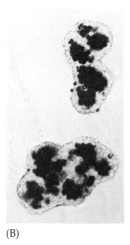

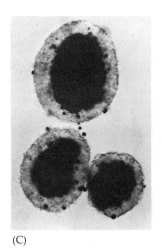

(A)   (B)   (C)

FIGURE 12
Aggregates formed by mixing 7-day old chick embryo neural retina (unpigmented) cells with pigmented retina (dark) cells. (A) 5 hours after the single-cell suspensions were mixed, aggregates of randomly distributed cells are seen. (B) At 19 hours, the pigmented retina cells are no longer seen on the periphery. (C) By two days, a great majority of the pigmented retina cells are located in a central internal mass, surrounded by the neural retina cells. (The scattered pigmented cells are probably dead cells.) (From Armstrong, 1989; photographs courtesy of P. B. Armstrong.)

others. Figure 12 illustrates the interactions between cultures of pigmented retina cells and neural retina cells. When single cell suspensions of these two cell types are mixed together, they form aggregates of randomly arranged cells. However, after several hours, the pigmented retina cells are no longer seen on the periphery of these aggregates, and by two days, two distinct layers are seen; the pigmented retina lies internal to the neural retina cells. The same type of interactions can also be seen when spherical aggregates of tissues are placed in contact with each other. One of the tissues will envelop the other. The final topography is independent of the starting positions (Figure 13).

Moreover, such interactions form a hierarchy (Steinberg, 1970). If the final position of one cell type A is internal to a second cell type B, and the final position of B is internal to a third cell type C, then the final position of A will always be internal to C. Thus, to use the above example, pigmented retina cells migrate internally to neural retina cells, and heart cells migrate centrally to pigmented retina. Therefore, heart cells migrate internally to neural retina cells. This observation led Steinberg to propose that the mixed cells interact to form an aggregate with the smallest interfacial free energy (Figure 14). In other words, the cells rearrange themselves into the most thermodynamically stable pattern. If cell types A and B have different strengths of adhesion and if the strength of A–A connections is greater than the strength of A–B or B–B connections, sorting will occur, with the A cells becoming central. If the strength of A–A connections is less than or equal to the strength of A–B connections, then the aggregate will remain as a random mix of cells. Last, if the strength of A–A connections is far greater than the strength of A–B connections—in other words, A and B cells show essentially no adhesivity toward one another—then A cells and B cells will form separate aggregates.

All that is needed for sorting out to occur is that cells differ in the strengths of their adhesion. In the simplest form of this model, all cells could have the same type of "glue" distributed on the cell surface. The amount of this cell surface product or the cellular architecture that allows the substance to be differentially concentrated could cause a different

FIGURE 13
Spreading of one cell type over another. The final position of aggregates composed of two tissue types is independent of their initial position. An identical final condition is formed whether the tissues are made into single cell suspensions and then reaggregated or the two tissues are kept intact and brought into contact. (After Armstrong, 1989.)

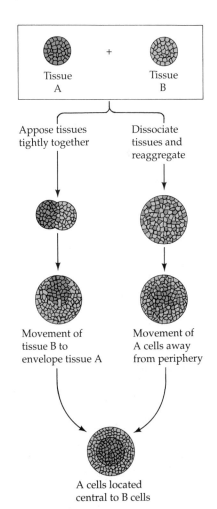

Tissue A + Tissue B

Appose tissues tightly together

Dissociate tissues and reaggregate

Movement of tissue B to envelope tissue A

Movement of A cells away from periphery

A cells located central to B cells

FIGURE 14

Sorting out as a process tending toward the maximum thermodynamic stability. (A) Sorting out occurs when the average strength of adhesions between different types of cells (w_{ab}) is less than the average of homotypic (a–a or b–b) adhesive strengths (w_{aa}, w_{bb}). The more adhesive type of cells becomes centrally located. (B) If the strength of the a–b adhesions is greater than or equal to the average of the homotypic adhesions, no sorting will occur, because the system has already achieved a thermodynamic equilibrium, and the mixture of cell types will be random. (C) If the a–b bonds are much weaker than the average of the homotypic adhesions, complete separation will ensue. (This is characteristic of oil and water, for example.)

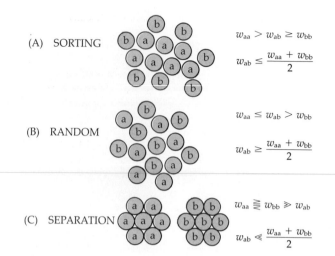

(A) SORTING

$$w_{aa} > w_{ab} \geq w_{bb}$$

$$w_{ab} \leq \frac{w_{aa} + w_{bb}}{2}$$

(B) RANDOM

$$w_{aa} \leq w_{ab} > w_{bb}$$

$$w_{ab} \geq \frac{w_{aa} + w_{bb}}{2}$$

(C) SEPARATION

$$w_{aa} \gtreqless w_{bb} \gg w_{ab}$$

$$w_{ab} \ll \frac{w_{aa} + w_{bb}}{2}$$

amount of stable contacts to be made between cell types. Alternatively, the thermodynamic differences could be caused by several types of adhesion molecules. This thermodynamic model is called the DIFFERENTIAL ADHESION HYPOTHESIS. In this hypothesis, the early embryo can be viewed as existing in an equilibrium state until some change in gene activity changes the cell surface molecules. The movements that occur are those that seek to restore the cells to a new equilibrium configuration.

In vivo evidence for the thermodynamic model

Until recently it was extremely difficult to design an experiment to test this model of cell migration in vivo; but evidence for this hypothesis is emerging from studies of limb regeneration in salamanders. As will be detailed in Chapter 17, regenerating salamander limbs have certain remarkable attributes. When one amputates a forelimb at the *upper arm*, the remaining stump forms a dedifferentiated mass of cells at its tip (the regeneration blastema), which divides and differentiates to form a new limb. The new limb tissue starts at the amputation site, in this case forming the remainder of the limb from the upper arm downward. When the forelimb is amputated at the *wrist*, a similar regeneration blastema forms. However, it does not start making upper arm tissue, elbow, and ulna bones. Rather, it "knows" its location and regenerates only the wrist and digits.

How is this "positional memory" stored? Nardi and Stocum (1983) demonstrated that when two salamander limb blastemas from the same level of origin are placed together, they fuse, but that neither tissue surrounds the other (Figure 15). However, when the blastemas were from different levels, the more proximal (closer to the body) blastema surrounded the more distal one. It appears, then, that the adhesive properties of these cells form a gradient along the proximodistal axis, being greatest at the wrist and lowest at the upper arm.

Crawford and Stocum (1988) were able to relate this in vitro sorting of cells to processes in the living, regenerating limb. Blastemas from the wrist, elbow, or upper arm were grafted into the blastema–stump junction of a hindlimb regenerating from the midthigh. The forelimb blastemas migrated distally to the corresponding level of the host hindlimb and then regenerated a new structure (Figure 16). The *upper arm* blastema immediately regenerated a complete arm from the midthigh level. The *elbow*

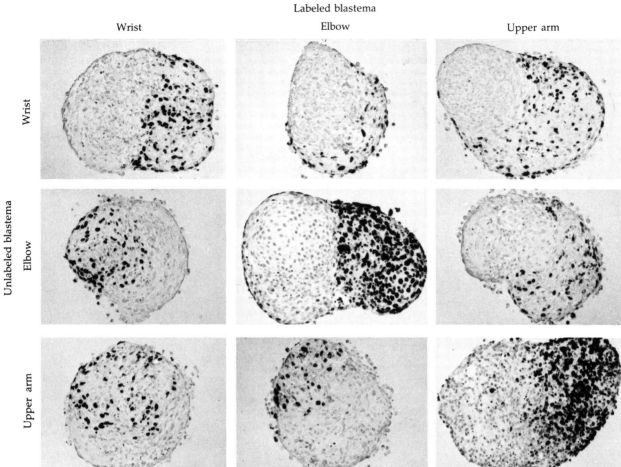

Labeled blastema

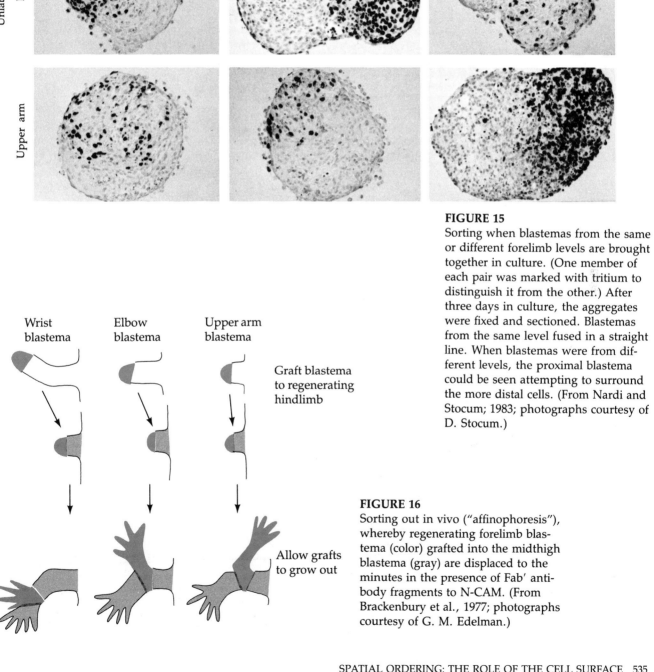

FIGURE 15
Sorting when blastemas from the same or different forelimb levels are brought together in culture. (One member of each pair was marked with tritium to distinguish it from the other.) After three days in culture, the aggregates were fixed and sectioned. Blastemas from the same level fused in a straight line. When blastemas were from different levels, the proximal blastema could be seen attempting to surround the more distal cells. (From Nardi and Stocum; 1983; photographs courtesy of D. Stocum.)

FIGURE 16
Sorting out in vivo ("affinophoresis"), whereby regenerating forelimb blastema (color) grafted into the midthigh blastema (gray) are displaced to the minutes in the presence of Fab' antibody fragments to N-CAM. (From Brackenbury et al., 1977; photographs courtesy of G. M. Edelman.)

blastema moved to the level of the knee and then formed the remainder of the arm from this point onward; and the *wrist* blastema was displaced to the end of the regenerating hindlimb, where it formed a wrist adjacent to the tarsus of the foot. These data suggest that the hierarchies of cellular sorting out seen in vitro reflect differences that are used by the body in constructing new organs in vivo.

The changing structure of the cell surface

Structure of the cell membrane

The formation of tissues and organs is mediated by events occurring at the cell surfaces of adjacent cells. The cell surface includes the cell plasma membrane, the molecules directly beneath the membrane and associated with it, and the molecules found in the extracellular spaces. Eukaryotic cells are surrounded by a complex molecular border called the PLASMA (or cell) MEMBRANE. Our present understanding of this membrane structure is summarized by the FLUID MOSAIC MODEL illustrated in Figure 17 (Singer and Nicolson, 1972). First, there are two layers of phospholipids, the polar ends of which are oriented toward the aqueous solutions on either side of the phospholipid sea. Some of the proteins traverse the membrane from one side to the other, whereas other proteins extend only part way through the phospholipid bilayer. The distribution of proteins gives rise to the "mosaic" nature of the membrane. The ability of these proteins to move laterally within the phospholipid matrix is evidence of its "fluid" nature. Most membranes also have substantial amounts of carbohydrates. These complex sugars are attached to lipids and proteins on the outside of the membrane.

Some of the best evidence for the fluid nature of the membrane is provided by somatic cell fusions. Such an experiment was performed by Frye and Edidin (1970) with human and mouse cells. Antibodies were prepared to human cell proteins by injecting these cells into rabbits, and antibodies specific to mouse cell proteins were similarly prepared. The antibody against mouse proteins was complexed with a fluorescein dye, which glows green under ultraviolet light, and the antibody against human proteins was complexed with a rhodamine dye, which glows red. Thus, the proteins on the two cells could be labeled: human proteins became red, mouse proteins became green. When these cells were fused together and marked with the labeled antibodies, the mouse and human proteins were first seen on separate halves of the composite cell (Figure 18). However, when the fused cells were treated with labeled antibodies an hour later, the proteins were completely intermixed. Because protein synthesis inhibitors did not block this mixing, the synthesis of new membrane components was ruled out. Thus, the proteins are able to move within a fluid matrix. The fluidity of the lipid bilayer also makes possible the insertion of new lipid and protein components into a preexisting membrane.

The diffusion of proteins in the membrane is an example of passive diffusion and requires no energy from the cell. However, specific proteins may be forced to move in a certain way. When antibodies are made to a specific membrane protein and added in excess to cells containing it, that protein is seen to be redistributed into patches and then into "caps" (Figure 19). This capping is dependent on energy and on the microtubule and microfilament systems of the cell (Edelman, 1976; Nicolson, 1976). Such

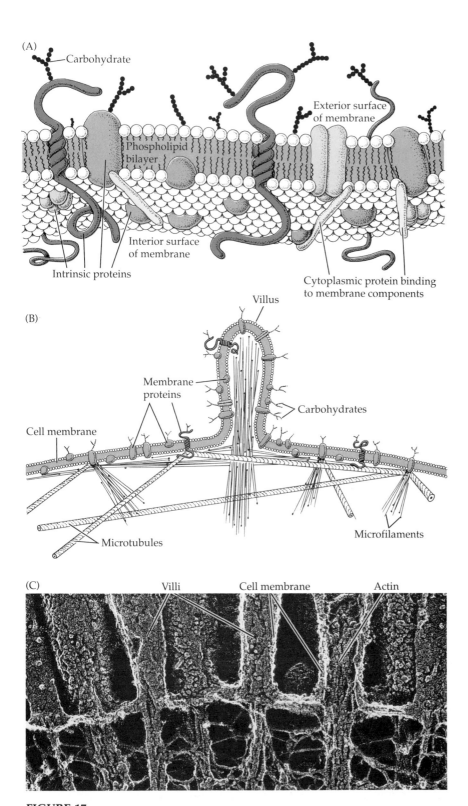

(A) Carbohydrate

Exterior surface
of membrane

Phospholipid
bilayer

Interior surface
of membrane

Intrinsic proteins

Cytoplasmic protein binding
to membrane components

(B) Villus

Membrane
proteins

Carbohydrates

Cell membrane

Microtubules

Microfilaments

(C) Villi   Cell membrane   Actin

FIGURE 17
The structure of the cell surface. (A) Fluid mosaic model of the cell membrane.
(B) Underlying cytoskeleton. Contractile fibers consisting of actin microfilaments
and tubulin microtubules are linked to each other and to the cell membrane.
Membrane glycoproteins may be linked by extracellular proteins as well as by
internal fibers. (C) Quick-frozen, deep-etched scanning electron micrograph
showing the surface of an intestinal cell. Actin extends into the villi and is linked
within the cell by thick and thin filaments directly beneath the cell membrane.
(After Hirokawa et al., 1983; photograph courtesy of N. Hirokawa.)

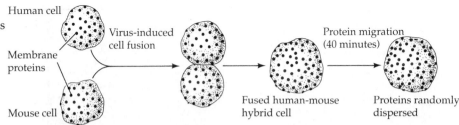

FIGURE 18
Lateral mobility of membrane proteins is shown in cell fusion experiment.

Human cell

Virus-induced cell fusion

Membrane proteins

Mouse cell

Fused human-mouse hybrid cell

Protein migration (40 minutes)

Proteins randomly dispersed

capping also occurs naturally, placing specific molecules in a particular region of the membrane. This process is seen in the polarization of mammalian blastomeres prior to compaction (Chapter 3).

Actin microfilaments are seen to be attached to the proteins of the cell membrane (Figures 17B,C; 20) and are important for cell locomotion. Here, the actin microfilaments associate with myosin, α-actinin, and tropomyosin to create a contractile network not unlike that found in muscle. This contractile network is thought to be anchored to the cell membrane through a protein called TALIN, which connects the actin microfilaments to a cell membrane protein, INTEGRIN. In most vertebrate cells, this network provides the basis for cell extension and migration.

The cell membrane also contains proteins that are capable of interacting with the outside environment. Certain proteins have their active sites pointing outward, toward other cells. The most familiar of these types of proteins are the TRANSPORT PROTEINS, which facilitate the movement of ions and nutrients into the cell, and the RECEPTOR PROTEINS, which bind hormones and drugs. These proteins—either directly, or indirectly through other proteins—also serve to transmit information into the cytoplasm and affect a cell's behavior. In addition, there are three classes of cell membrane molecules (usually proteins) that are particularly involved in making specific interactions with other cells (Edelman and Thiery, 1985):

- *Cell adhesion molecules.* These proteins are involved in cell–cell adhesion. They can unite cells into epithelial sheets and condense mesenchymal cells.
- *Substrate adhesion molecules.* These molecules are involved in binding cells to their extracellular matrices. They include components of the extracellular matrix and the cell surface receptors for these molecules. Substrate adhesion molecules permit the movement of mesenchymal cells and allow the spatial separation of epithelial sheets.
- *Cell junctional molecules.* These molecules provide communication pathways between the cytoplasm of adjacent cells and provide permeability barriers and mechanical strength to epithelial sheets.

The local patterns of expression of these cell surface molecules are thought to provide a major link between the one-dimensional genetic code and the three-dimensional organism. By modulating the appearance of these molecules, the genetic potentials of the genome can become manifest in the mechanical processes of morphogenesis.

(A)

(B)

(C)

FIGURE 19
Specific movement of cell membrane proteins. Attachment of antibodies to specific cell surface molecules that are uniformly distributed in the cell membrane (A) causes them to cluster into discrete patches (B), which migrate to a single pole (C). In these photographs, the cell surface molecules (themselves antibody molecules) have been visualized by tagging them with a fluorescent dye. (Photograph courtesy of G. M. Edelman.)

(A)

(B)

FIGURE 20
Connection of internal microfilaments to the cell membrane. Sagittal (A) and cross-sectional (B) electron micrographs of intestinal villi show actin microfilaments attached to the plasma membrane at its uppermost surface and its sides. The attachment is mediated by accessory proteins such as α-actinin. (Compare with Figure 7C.) (From Mooseker and Tilney, 1975; photographs courtesy of M. Mooseker.)

Cell membrane changes with development

The expression of different membrane components changes in time and space. Different cell types display different cell surface components and these components change as the cell develops. Such tissue-specific membrane components are often recognized by antisera and are therefore called DIFFERENTIATION ANTIGENS (Boyse and Old, 1969).

Specific differentiation antigens can now be identified by MONOCLONAL ANTIBODIES (Figure 21). These antibodies are usually made by injecting foreign cells into mice (or mouse cells of one strain into mice of another strain). The mouse B lymphocytes will begin producing antibodies against each foreign component on these cells, with each B lymphocyte producing an antibody with only one such specificity. These lymphocytes are then "immortalized" by fusing them with B lymphocyte tumor cells (myelomas) that has been mutated so that (1) they can no longer synthesize their own antibodies and (2) they lack the purine salvage enzyme hypoxanthine-phosphoribosyltransferase (HPRT). This latter alteration means that the myeloma cells can only make purine nucleotides de novo and cannot utilize purines from the culture medium. After fusion, the cells are then grown in a medium containing aminopterin. This drug inhibits the de novo purine synthetic pathway. Thus, unfused myeloma cells die of purine starvation. They cannot make purine nucleotides using the HPRT-mediated salvage pathway, and the aminopterin blocks the de novo pathway as well. Normal B lymphocytes do not divide in culture, so they, too, die. The fused product of the B lymphocyte and the myeloma cell—a HYBRIDOMA—proliferates, having the purine salvage enzyme from the B lymphocyte and the growth properties from the tumor. Moreover, each of

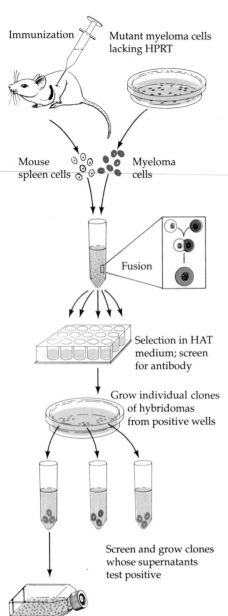

Immunization / Mutant myeloma cells lacking HPRT

Mouse spleen cells / Myeloma cells

Fusion

Selection in HAT medium; screen for antibody

Grow individual clones of hybridomas from positive wells

Screen and grow clones whose supernatants test positive

FIGURE 21

Protocol for making monoclonal antibodies. Spleen cells from an immunized mouse are fused with mutated myeloma cells lacking the enzyme HGPRT. Cells are grown in a medium containing hypoxanthine, aminopterin, and thymidine (HAT). Unfused myeloma cells cannot grow in this medium because aminopterin blocks the only way they have of making purine nucleosides. B cells die in this medium even though they contain an enzyme (HGPRT) that would allow them to utilize the hypoxanthine placed in the medium. The fused cells (hybridomas) grow and divide. The wells in which the hybridomas grow are screened for the presence of the effective antibody, and the cells from positive wells are plated at densities low enough to allow individual cells to give rise to discrete clones. These clones are isolated and screened for the effective antibody. Such an antibody is monoclonal. The hybridomas producing this antibody can be grown and frozen. (After Yelton and Scharff, 1980.)

these hybridomas secretes the specific antibody of the B lymphocyte. The medium in which hybridomas are growing is then tested for antibody that binds to the original population of cells. Such antibody, having a single B lymphocyte as its original source, is called a monoclonal antibody. Monoclonal antibodies can be produced in enormous amounts and can recognize antigens (proteins, lipids, or carbohydrates) that are only weakly expressed (Köhler and Milstein, 1975).

Monoclonal antibodies directed against specific types of cells have uncovered numerous differentiation antigens appearing at different times and places during development. Figure 22 shows different cell surface molecules at different *spatial* layers of the newly hatched chick neural retina. Each of the monoclonal antibodies recognizes a different molecule

FIGURE 22

Cell surface specificity in the chick neural retina. (A) Phase-contrast photograph of a section through newly hatched chick neural retina. (B) Retinal section stained with a fluorescent monoclonal antibody that recognizes retinal (but not other neuronal) cells. (C) Retinal section stained with fluorescent monoclonal antibody that recognizes neuronal processes but not cell bodies in the retina. (D) Retinal section stained with fluorescent monoclonal antibody that recognizes antigens on a subset of nerve cell processes in the outer and inner synaptic layer. (Photograph courtesy of G. Grunwald.)

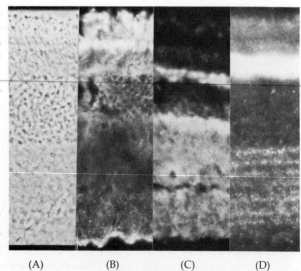

Outer segments of photoreceptors
Somas of photoreceptors (outer nuclear layer)
Outer synaptic layer
Inner nuclear layer (interneuron soma)
Inner synaptic layer
Ganglion cell soma
Ganglion cell axons

(A)   (B)   (C)   (D)

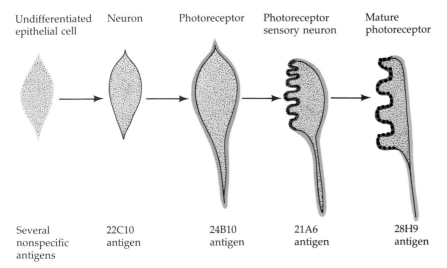

Undifferentiated epithelial cell	Neuron	Photoreceptor	Photoreceptor sensory neuron	Mature photoreceptor
Several nonspecific antigens	22C10 antigen	24B10 antigen	21A6 antigen	28H9 antigen

FIGURE 23
Temporal changes in the cell membrane correlated with the morphogenesis of *Drosophila* retinal photoreceptor cells. As differentiation proceeds, different antigens become expressed on the cell membrane. (After Venkatesh et al., 1985.)

in the cell membrane. As is evident from this composite photograph, the membranes of all the cells of the neural retina are not the same. In fact, regions of the same cell membrane can differ; for example, the membranes of the axons and the membranes of the nerve soma contain some different molecules. Figure 23 shows *temporal* changes in the cell membrane of a single *Drosophila* epithelial cell as it develops into a retinal photoreceptor. Monoclonal antibodies were obtained after mice had been injected with homogenates of *Drosophila* head tissue, and a panel of antibodies was tested on the cells of the larval eye imaginal disc that was differentiating into eye structures. As soon as the undifferentiated epithelial cells of the disc show neuronal properties, they express the 22C10 antigen. This antigen is also found in other neuronal cell types. Shortly thereafter, though, the cell begins to express another cell membrane molecule, the 24B10 antigen. This molecule is only seen in those neurons destined to become photoreceptors. At subseqent stages (about 80 hours later) the 21A6 antigen becomes expressed on certain regions of the maturing photoreceptors, and another antigen, 28H9, is characteristic of the terminally differentiated retinal photoreceptor (Zipursky et al., 1984).

Thus, cell membranes of different cell types contain different molecules, and these molecules can change during the maturation of the particular cell.

SIDELIGHTS & SPECULATIONS

From protein to gene

Since differentiation antigens are often proteins whose expression is regulated in time and space, and since these changes often correlate with specific morphological transitions (as shown in Figure 23), it would be interesting to know how their genes are regulated. For example, an understanding of how the 24B10 protein becomes expressed might provide clues to the genetic mechanisms of neuronal diversity. How does one accomplish this "reverse genetics," going from protein to gene?

First, one binds the monoclonal antibody to resinous beads and passes retinal homogenates through a column

of such material (Figure 24). (This is called an immunoaffinity column.) The antibody binds only to the antigen it originally recognized, and the bead-bound protein is then eluted from the beads (by ionic solutions) and is run on an electrophoresis gel to separate it from any possible contaminants. The region of the gel containing this protein is cut from the gel, and the protein is eluted from the gel matrix and partially sequenced. Radioactive oligonucleotides are then synthesized that would bind to a DNA sequence capable of encoding such a protein. In the case of 24B10, these were used to screen a library of recombinant DNA clones containing regions of the *Drosophila* genome. The *Drosophila* DNA of a positive clone was sequenced to see if

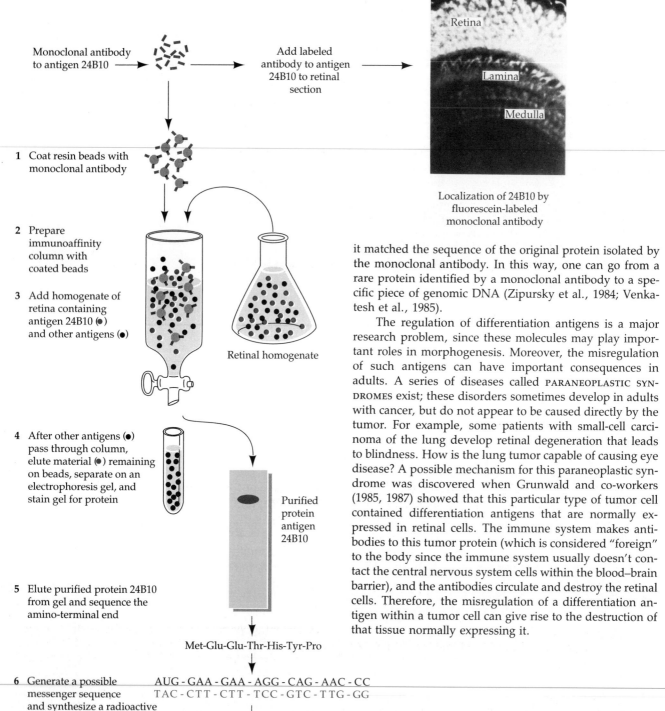

Monoclonal antibody to antigen 24B10 → → Add labeled antibody to antigen 24B10 to retinal section →

Retina

Lamina

Medulla

Localization of 24B10 by fluorescein-labeled monoclonal antibody

1 Coat resin beads with monoclonal antibody

2 Prepare immunoaffinity column with coated beads

3 Add homogenate of retina containing antigen 24B10 (●) and other antigens (●)

Retinal homogenate

4 After other antigens (●) pass through column, elute material (●) remaining on beads, separate on an electrophoresis gel, and stain gel for protein

Purified protein antigen 24B10

5 Elute purified protein 24B10 from gel and sequence the amino-terminal end

Met-Glu-Glu-Thr-His-Tyr-Pro

6 Generate a possible messenger sequence and synthesize a radioactive complementary sequence

AUG - GAA - GAA - AGG - CAG - AAC - CC
TAC - CTT - CTT - TCC - GTC - TTG - GG

7 Use this probe to screen the phage library of *Drosophila* genome; sequence positive clone

TCC ATG TTC GAT CGC GAG ATG GAG GAG ACG CAT TAC CCG CCC TGC ACC TAC AAC GTG ATG TGC
Ser | Met | Phe | Asp | Arg | Glu | Met | Glu | Glu | Thr | His | Tyr | Pro | Pro | Cys | Thr | Tyr | Asn | Val | Met | Cys

Expected sequence

8 Isolate and characterize gene

it matched the sequence of the original protein isolated by the monoclonal antibody. In this way, one can go from a rare protein identified by a monoclonal antibody to a specific piece of genomic DNA (Zipursky et al., 1984; Venkatesh et al., 1985).

The regulation of differentiation antigens is a major research problem, since these molecules may play important roles in morphogenesis. Moreover, the misregulation of such antigens can have important consequences in adults. A series of diseases called PARANEOPLASTIC SYNDROMES exist; these disorders sometimes develop in adults with cancer, but do not appear to be caused directly by the tumor. For example, some patients with small-cell carcinoma of the lung develop retinal degeneration that leads to blindness. How is the lung tumor capable of causing eye disease? A possible mechanism for this paraneoplastic syndrome was discovered when Grunwald and co-workers (1985, 1987) showed that this particular type of tumor cell contained differentiation antigens that are normally expressed in retinal cells. The immune system makes antibodies to this tumor protein (which is considered "foreign" to the body since the immune system usually doesn't contact the central nervous system cells within the blood–brain barrier), and the antibodies circulate and destroy the retinal cells. Therefore, the misregulation of a differentiation antigen within a tumor cell can give rise to the destruction of that tissue normally expressing it.

FIGURE 24
Protocol for finding the gene that encodes a protein identified by a monoclonal antibody. (After Venkatesh et al., 1985; photograph courtesy of S. Benzer.)

Cell adhesion molecules

Identifying cell adhesion molecules and their roles in development

The sorting out studies of Holtfreter's and Steinberg's laboratories did not identify the molecules involved in these differential cell adhesions. Roth (1968; Roth et al., 1971) demonstrated that different cell types displayed selective cell adhesion independent of cell sorting. He modified the rotary aggregation assay by incubating ^{3}H-labeled cartilage cells and ^{14}C-labeled hepatocytes in a rotating solution containing small aggregates of unlabeled cartilage cells. By measuring the ^{14}C- and ^{3}H-labeled cells in these aggregates, he demonstrated that the cartilage aggregates specifically picked up cartilage cells. Similar experiments extended this finding to liver and muscle cells as well (Figure 25). These studies demonstrated that different cell types could use different adhesion molecules.

The next task is to identify those molecules mediating cell adhesion and to discover how they accomplish this feat. Two major approaches have lead to the identification of cell adhesion molecules. The first approach uses antibodies to identify molecules involved in cell adhesion, whereas the second approach uses the observation that calcium is often necessary for cell adhesion to find calcium-binding cell adhesion molecules.

Antibodies were first used to identify cell adhesion molecules in *Dictyostelium* (see Chapter 1). Gerisch and colleagues (Beug et al., 1970) prepared antibodies against *Dictyostelium* and then chemically split the antibodies so that only their monovalent antigen-binding regions—the so-called Fab' fragments—remained (Figure 26A). (The divalent antibodies had to be split, for if they remained divalent, they might artificially join the cells together, and the effect could not be measured.) This led to the discovery of the 80-kDa glycoprotein that mediates cell–cell adhesion during slime mold aggregation. This approach was used to study embryonic cell adhesion by Edelman and his colleagues (Brackenbury et al., 1977) and led to the isolation of NEURAL CELL ADHESION MOLECULE (N-CAM). In this study, chick neural retina cells were injected into rabbits, and rabbit antibodies were made to the neural retina cell surface. Once the antibodies

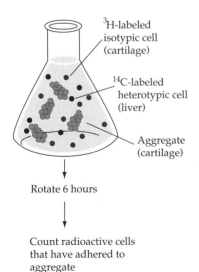

Rotate 6 hours

Count radioactive cells that have adhered to aggregate

³H-labeled isotypic cell (cartilage)

¹⁴C-labeled heterotypic cell (liver)

Aggregate (cartilage)

FIGURE 25

Specificity of cell–cell attachment. Collecting aggregates, each consisting of one type of cells, are placed in a rotating culture containing single cells of both the same (isotypic) and different (heterotypic) types. The single isotypic and heterotypic cells were previously labeled with different radioactive isotopes. After 6 hours, the aggregates are collected, washed, and counted for both isotypic and heterotypic cells that adhered to the aggregate as shown in the table. (Data from Roth, 1968.)

Count of radioactive cells that have adhered to aggregate

Aggregate type	Labeled single cells in suspension*		
	Cartilage	Liver	Pectoral muscle
Cartilage	100	6	48
Liver	10	100	0
Pectoral muscle	38	49	100

*Percentage of mean number of cells collected by isotopic aggregates.

(A)

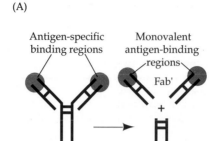

Antigen-specific
binding regions

Monovalent
antigen-binding
regions
Fab'

Antibody

(B)

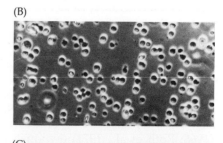

(C)

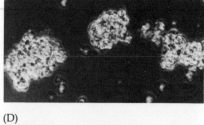

(D)

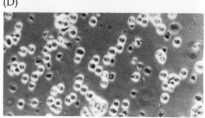

FIGURE 26
Aggregation of neural retinal cells
from 10-day chick embryos. (A) Creation of antibody fragments (Fab'). (B)
Single cells prior to aggregation. (C)
Aggregates produced after 30 minutes
of rotation. (D) Inhibition of aggregation when retinal cells were rotated 30
minutes in the presence of Fab' antibody fragments to N-CAM. (From
Brackenbury et al., 1977; photographs
courtesy of G. M. Edelman.)

were made and found to bind to neural retina cells (and to some other
cells as well), the antibodies were split into their Fab' fragments. When
these Fab' fragments were added to the culture media, they were able to
inhibit the aggregation of the chick neural retina cells (Figure 26B). Moreover, when isolated neural retina membranes were added to this mixture,
they bound the Fab' fragments, thus enabling the retina cells to aggregate.
By extracting the components of these membranes, Thiery and his colleagues (1977) found that the Fab' fragments bound to a 140,000-Da protein. When this protein was placed into the solution of aggregating retina
cells, it alone inhibited the effect of the Fab' fragments. It was concluded
that the 140,000-Da protein, N-CAM, is a cell surface component with a
very important role in neuronal aggregation.

The second approach to identifying cell adhesion molecules was used
by Takeichi and colleagues who found a family of proteins for which
calcium ions were critical, both for their adhesive functions and for the
protection they provided these proteins against trypsin. These calcium-
dependent cell adhesion glycoproteins are called CADHERINS (Takeichi,
1987).

Cadherins

Cell adhesion molecules (CAMs) can be divided into three groups: the
cadherins, whose cell adhesive properties are dependent on calcium ions;
the *immunoglobulin superfamily CAMs*, whose cell-binding domains resemble those of antibody molecules; and the *saccharide CAMs*, in which the
CAM protein recognizes a carbohydrate residue on an adjacent cell. Table
1 lists some of the recently discovered general cell adhesion molecules.

Cadherins are calcium-dependent CAMs that appear to be crucial to
the spatial segregation of cells and to the organization of animal form.
Three major cadherin classes have been identified. E-CADHERIN (homolo-

TABLE 1
General classification of major cell adhesion molecules (CAMs)

Class	CAM
Cadherins (calcium-dependent)	N-cadherin (a.k.a. A-CAM) P-cadherin E-cadherin (a.k.a. L-CAM, uvomorulin)
Immunoglobulin superfamily CAMs	N-CAM Ng-CAM (a.k.a. L1, NILE) Neurofascin TAG-1 LFA-1 CD4 glycoprotein (HIV receptor)
Saccharide-mediated CAMs	Bindin Mouse sperm galactosyltransferase Neural retina N-acetyl- galactosyltransferase Neurite laminin–heparan sulfate receptor Proacrosin Mouse compaction galactosyltransferase gp90^{MEL} (lymphocyte homing receptor)

gous to chick L-CAM) is expressed on all early mammalian embryonic cells, even at the one-cell stage. Later, this molecule is restricted to epithelial tissues of embryos and adults. P-CADHERIN appears to be expressed primarily on the extraembryonic cells and the uterine wall (Nose and Takeichi, 1986). It is possible that P-cadherin facilitates the connection of the trophoblast with the uterus since P-cadherin on the uterine cells is seen to contact P-cadherin on the trophoblastic cells of mouse embryos (Kadokawa et al., 1989). N-CADHERIN is first seen on the mesodermal cells in the gastrulating embryo as they lose their E-cadherin expression (Hatta and Takeichi, 1986). Migrating neural crest cells show neither N- or E-cadherins, but re-express N-cadherin when they reaggregate. N-CAM and N-cadherin are often found on the same cells, but in different regions. On the neuroepithelial cells of the neural tube, N-CAM is concentrated near the outer margin whereas N-cadherin staining is more intense at the luminal surface (Hatta et al., 1987; Rutishauser and Jessell, 1988). The appearance or disappearance of N-cadherin often occurs at sites of active cell rearrangement—such as somite dispersal, during which the sclerotome and dermatome cells lose their N-cadherin as they start to leave the epithelial structure of the somite (Hatta et al., 1987). Cells expressing N-cadherin readily sort out from N-cadherin-negative cells in vitro, and Fab' antibodies against cadherins will convert a three-dimensional histotypic aggregate of cells into a monolayer (Takeichi et al., 1979). When E-cadherin genes are transfected and expressed in cultured mouse fibroblasts (which usually do not express this protein), E-cadherin is seen on their cell surfaces and the fibroblasts become tightly connected to each other (Nagafuchi et al., 1987).

Differential cadherin expression can also explain the homotypic sorting out data presented earlier. As we discussed earlier, Roth and co-workers had shown that liver cells tended to collect liver cells and that retinal cells collected other retinal cells. Takeichi (1987) demonstrated that the retinal cells express N-cadherin while the liver cells express E-cadherin, and that the sorting out would be expected because of this difference in cadherin expression. He also suggested that the observations of Townes and Holtfreter may likewise be explained by differential cadherin expression. Support for this came from studies wherein different cadherin genes were transfected into mouse fibroblasts that do not usually express either cadherin type. Fibroblasts expressing E-cadherin adhered to other E-cadherinbearing fibroblasts, while P-cadherin fibroblasts stuck to other fibroblasts expressing P-cadherin. Moreover, when embryonic lung tissue was dissociated and allowed to recombine in the presence of untreated or Ecadherin-bearing fibroblasts, the E-cadherin-expressing fibroblasts became integrated into the epithelial lung tubules (which express E-cadherin), while the untreated fibroblasts became associated with the mesenchymal cells (that do not express any cadherin) (Nose et al., 1988).

All the above-mentioned experiments have been performed on cells in culture. Recently, however, in vivo studies have been published by Fujimoro and colleagues (1990). They injected mRNA for chicken N-cadherin into one of the two first-cleavage blastomeres of the *Xenopus* embryo. Those neurulae that expressed extra N-cadherin were often characterized by clumps of cells and thickened tissue layers. In some animals wherein both the epidermis and the neural tube expressed this extra N-cadherin, the neural tube would not separate from the epidermis. Thus, the cadherins are probably playing major roles in the assortment of cells into tissues.

The cadherin molecules have three major domains. There is an extracellular domain that mediates adhesion, a transmembrane domain that spans the cell membrane, and a cytoplasmic domain. Site-specificic mutagenesis on cloned E-cadherin genes demonstrated that this cytoplasmic domain is critical for cadherin function. This region is seen to link cadherin to the cytoskeleton (Nagafuchi and Takeichi, 1989). This suggests a mechanism by which extracellular binding could influence cell shape and behavior.

Immunoglobulin superfamily CAMs

Many of the cell adhesion molecules that do not need calcium ions for their function have a very similar structure (Figure 27). This structure, with its extracellular globular domains held in place by disulfide bonds, resembles that of the immunoglobulin molecule, and it is considered likely that the immunoglobulins are derived from this group of cell adhesion

FIGURE 27
Three members of the "immunoglobulin superfamily." The IgM molecule has two heavy chains, each with five domains, and two light chains with two domains. N-CAM is a single polypeptide chain with five domains. Its anchor into the membrane can either be a transmembrane amino acid sequence in the protein or a lipid. Ng-CAM is a transmembrane protein with six globular domains. The insect cell adhesion molecules fasciclin II and neuroglian resemble N-CAM and Ng-CAM, respectively.

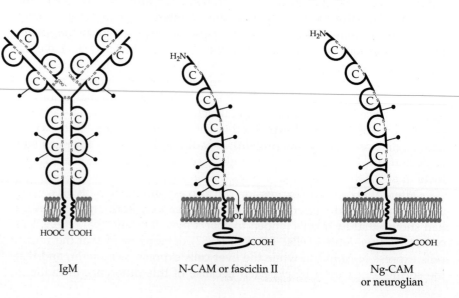

IgM   N-CAM or fasciclin II   Ng-CAM or neuroglian

molecules (Williams and Barclay, 1988; Lander, 1989). Thus, these glycoproteins are called the "immunoglobulin superfamily."*

The best characterized of all these cell adhesion molecules is the neural cell adhesion molecule, N-CAM. Although the N-CAM protein is encoded by one gene per haploid genome, it can occur in three sizes, depending on differential RNA splicing. The predominant forms are 180,000 Da and 140,000 Da in mass and contain three domains: an extracellular domain, a membrane-spanning domain, and a cytoplasmic domain. (The 120,000-Da form does not span the membrane and appears to consist of an extracellular domain covalently bound to a membrane lipid.) The 180,000-Da form may be able to bind to the neural–glial cell adhesion molecule, Ng-CAM, whereas the other forms cannot (Edelman, 1986).

N-CAM molecules on adjacent cells are thought to interact with each other to construct the bonds between the cells. N-CAM-containing neural retina cells can bind to lipid vescicles that contain N-CAM in their bilayer. If anti-N-CAM Fab' fragments are placed onto retinal cells, only the N-CAM molecules should be coated. These treated retinal cells are then unable to bind lipid vesicles containing only N-CAM in their membranes (Rutishauser et al., 1982). Electron microscopy confirms that N-CAM molecules can interact with each other to form complex multimeric structures (Hall and Rutishauser, 1987). Thus, N-CAM can mediate the adhesion of cells only if the cells all contain N-CAM on their surfaces. Each region of cell–cell contact involves many interactions among N-CAM molecules.

N-CAM molecules may be given different functions by modifying them with sialic acid residues. Sialic acid is a complex (10-carbon) saccharide derived from the conjugation of D-mannose with pyruvic acid, and it carries a negative charge. Highly glycosylated N-CAM is about 30 percent sialic acid by weight. Less glycosylated N-CAM consists of about 10 percent sialic acid. The difference between the highly and less sialylated molecules appears to be the length of the sialic acid polymers (Hoffman and Edelman, 1983). The sialic acid molecules are covalently linked to the N-CAM protein at three sites on the middle of the molecule, within the extracellular domain (Cunningham et al., 1983).

Sialic acid can severely reduce the adhesiveness of one cell toward another. Cells with a relatively low amount of sialic acid on their N-CAM aggregate four times more readily than those with the high levels of sialic acid (Hoffman and Edelman, 1983). As the embryo gets older, most of the N-CAM proteins progress from the high sialic acid to the low sialic acid forms (Rieger et al., 1985; Chuong and Edelman, 1985a,b). In this manner, N-CAM may play a role in both stimulating two adjacent cells to form regions of contact or in inhibiting such contact. If the neighboring cells both express the low sialic acid form of N-CAM, cell adhesion is promoted. However, if the cells contain the highly sialated forms of N-CAM, cell adhesion will be inhibited (Figure 28; Rutishauser et al., 1988).

The distribution of cell adhesion molecules

The spatiotemporal distribution of CAMs is seen using fluorescent antibodies. N-CAM and L-CAM originally are expressed on all cells of the chick embryo at the beginning of gastrulation. However, when the neural plate is induced, a dramatic change takes place. The neural plate cells show N-CAM expression, while the surrounding ectodermal cells show only L-CAM (Figure 29; Edelman et al., 1983). Subsequent development

*The designation "superfamily" is often given because the different classes of immunoglobulin molecules themselves constitute a "family." These other members of the superfamily have structures resembling immunoglobulins, but are not exactly "close" family.

FIGURE 28

Two mechanisms by which N-CAM could regulate cell–cell adhesions. (A) Expression of low sialic acid N-CAM molecules on adjacent cell surfaces leads to the formation of numerous bonds between N-CAM molecules that promote junction formation from sub-units (represented as elongated ovals) on the adjacent cell membranes. (B) Highly sialated forms of N-CAM can inhibit cell–cell interactions that would otherwise have resulted from other adhesive molecules (represented as balls and sockets) on the adjacent cell membranes. When the polysialic acid residues are removed, the two cells can adhere. (After Rutishauser et al., 1988.)

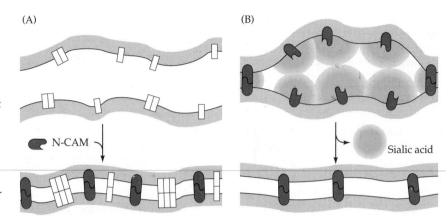

shows the modulation of these molecules during induction or tissue reorganization. Neural crest cells, for instance, originally contain N-CAM on their cell surfaces. When they begin their migrations, they lose N-CAM expression, and they regain N-CAM as the crest cells aggregate into ganglia (Crossin et al., 1985; Edelman, 1985). After neural induction occurs and L-CAM expression diminishes from neural derivatives, N-CAM appears on the neural tube cells and another adhesion molecule—Ng-CAM—becomes expressed on postmitotic neurons. Ng-CAM is strongly expressed on axons whose migration is being directed on the surfaces of glial cells (Thiery et al., 1985; Daniloff et al., 1986).

Differential CAM expression is critical at boundaries between two groups of cells. At such places, the body segregates different cells into different regions. Notochord cells do not enter into the neural tube, nor do dermal cells trespass into the epidermis. Such segregation may be accomplished by the adjacent populations having different CAMs. As we will see in the next chapter, feathers are induced when mesodermally derived mesenchymal cells condense together to form a ball of cells directly beneath the epidermis of the chicken skin. The ectodermal cells are linked together by L-CAM, while the CAM-negative mesenchymal cells collect

FIGURE 29

Change in CAM distribution in the neurulating chick embryo. Embryos were stained with fluorescent antibodies to (A) N-CAM and (B) L-CAM. N-CAM is present in large amounts in the chordamesoderm and neural plate cells, but is not seen appreciably in the epidermal ectoderm. L-CAM, in contrast, is seen primarily in the epidermal ectoderm and not in the neural plate cells. Before neurulation, both N-CAM and L-CAM were on each of these cells. (C) Fate map of the chick embryo with the areas that will express different CAMs superimposed on it. Smooth muscle (Sm) and blood-forming regions (Ha) have no known CAM associated with them. (From Edelman, 1985; photographs courtesy of G. M. Edelman.)

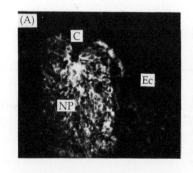

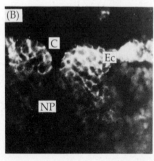

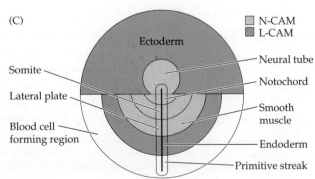

FIGURE 30

Distribution of different cell adhesion molecules at tissue boundaries. As the mesodermal cells collect to induce the feather bud in the ectoderm, the newly aggregated mesenchymal cells express N-CAM (A), while the ectodermal cells express L-CAM (B) in their cell membranes. (From Chuong and Edelman, 1985a; photographs courtesy of G. M. Edelman.)

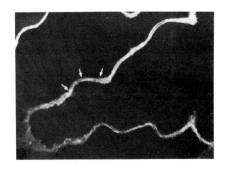

(A)

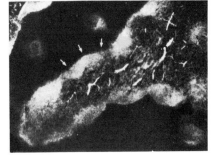

(B)

to form an aggregate that expresses N-CAM (Figure 30). Throughout the developing feather, different groups of cells become separated from one another as a result of their ability to express N-CAM, L-CAM, or both proteins (Chuong and Edelman, 1985a,b).

In addition to helping generate and maintain boundaries between different cell types, CAMs also appear to play roles in the binding of one cell type to another. This role is especially important in the connections made from an axon to its target organ, and recent studies have shown that N-CAM is needed for the proper attachment of axons to specific target tissue. Tosney and her co-workers (1986) have shown that muscle cells begin to express N-CAM in their cell membranes around the time when axons that bear N-CAM are first seen near them. However, once synapses between the nerve and muscle are established, the only regions of the muscle that retain N-CAM are those at the neuromuscular junctions (Covault and Sanes, 1986). When antibodies against N-CAM are placed into specific regions of the developing embryo, the connections between the neuron and its target tissue can be dramatically distorted. Fraser and his colleagues (1988) have shown that antibodies to N-CAM distort the recognition of axons from the frog retina to their specific targets in the optic tectum of the brain. Landmesser and her colleagues (1988) have similarly demonstrated that the innervation of embryonic chick skeletal muscle is affected by anti-N-CAM antibodies. In both cases, the effects can be explained by the ability of the antibodies to interfere with N-CAM's roles in tethering axons together and in maintaining connections between the nerve axons and their target tissues.

While some CAMs are used in many places in the embryo, other CAMs have a much more restricted role. Such cell adhesion molecules are thought to be expressed at very specific times and places during development, and this has made them more difficult to identify and characterize. Some of these CAMs are responsible for enabling axons traveling in the same direction to bind together. This binding is called FASCICULATION, and the resulting bundle of axons produces the nerve. (The optic nerve, for instance, is a fasciculated bundle of axons originating in the ganglion cells of the neural retina.) As we will see in Chapter 17, an insect axon can change its direction when it reaches another axon and travel along that axon's surface. This phenomenon is thought to happen only when the two axons contain the same adhesion molecule in their cell membranes. It appears that an insect axon contains many of these cell adhesion molecules on its cell surface and that the surface molecules differ at different sites along the axon (Bastiani et al., 1985, 1987). These surface compounds have been called FASCICLINS (Figure 31) and some of them are also in the immunoglobulin superfamily (Harrelson and Goodman, 1988).

Saccharide-mediated cell adhesion

In the mid-1940s Weiss and Tyler independently proposed that molecules at adjacent cell surfaces interacted with each other in a lock-and-key fashion to provide the spatial information of development. Interactions between specific complementary molecules, they said, would account for

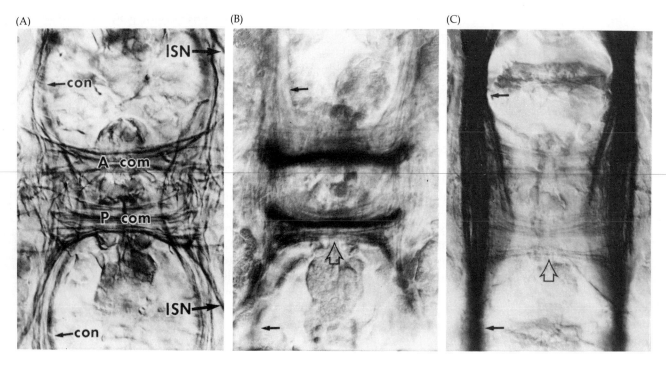

FIGURE 31

Fasciclin expression in the developing grasshopper nervous system. (A) Scaffold of fasciculated axons in a grasshopper embryo as seen by Nomarski microscopy. A com and P com are the anterior and poterior commissures whose axons traverse the segment; ISN is the intersegmental neuron, and con is a connective neuron. (B,C) Embryonic nervous system as in (A), but stained with monoclonal antibodies made to the cell surface fasciclin glycoproteins. The antibody on (B) recognizes a subset of axons in the anterior and posterior commissures, while the antibody in (C) binds to a membrane glycoprotein of most longitudinal axon fascicles. The arrows show the same locations on (B) and (C). Note that the antibody stains only a portion of each axon. (From Bastiani et al., 1987; photographs courtesy of C. Goodman.)

the relationships formed by moving cells during organ formation. In 1970 Saul Roseman postulated that the molecules responsible for these interactions were a subset of enzymes called GLYCOSYLTRANSFERASES. These membrane-bound enzymes are routinely found in the endoplasmic reticulum and Golgi vesicles, where they are responsible for adding sugar residues onto peptides to make glycoproteins. There are numerous glycosyltransferases, each one specific for a given sugar and some showing substrate specificity as well. Thus, a galactosyltransferase is an enzyme capable of transferring galactose from an activated donor molecule (UDP-galactose) to an acceptor. There may be several galactosyltransferases with affinities for different acceptor molecules.

The presence of cell surface glycosyltransferases can be revealed in several ways. One method is to add the labeled sugar in its activated form. The labeled sugar is then transferred to the cell surface and becomes

FIGURE 32

Autoradiograph of a 10-somite chick embryo incubated with UDP-[^{3}H]galactose, or sugar donor. Insoluble radioactivity (black grains) indicates that this radioactive sugar was transferred by the surfaces of migrating cells to other cells or the matrices around them. (From Shur, 1977a; photograph courtesy of B. D. Shur.)

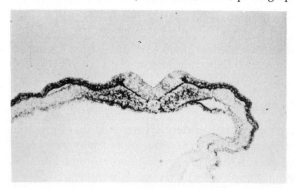

insoluble. Its presence can be detected by autoradiography (Figure 32; Roth et al., 1971b; Shur, 1977a). When neural retina cells are incubated with radioactive UDP-galactosamine, over 90 percent of the labeled galactosamine is transferred onto their cell surfaces (Balsamo et al., 1986). Another technique employs a fluorescently labeled antibody made against glycosyltransferase. Using this method, the neural retina galactosyltransferase is first seen throughout the embryonic neural retina, but later becomes restricted to the outer layers of the retina (Figure 33).

Figure 34 shows two possible reactions that might be carried out by cell surface glycosyltransferases. In one instance, binding is accomplished between one cell and another cell by means of the mutual recognition of glycosyltransferases and sugar residues (acceptors). This reaction constitutes adhesion, and was seen in Chapter 2 when we discussed the interactions between sperm cell surface galactosyltransferase and the carbohydrate components of the zona pellucida (which is the extracellular matrix of the mammalian egg). In addition to adhesion reactions, the binding of cell to substrate can be made transitory (as in the recognition of a substrate by migrating cells) by the presence of the activated sugar. Such adhesion and catalysis could occur between two cells or between the cell surface glycosyltransferase and the substratum, such as the basal lamina over which the cell migrates.

Glycosyltransferases may also be involved in those cell movements characteristic of gastrulation and morphogenesis. During these times, the cells migrate upon carbohydrate substrates, especially glycosaminoglycans (Toole, 1976). Shur (1977a,b) demonstrated that the migrating cells of the chick neurula had intense cell surface glycosyltransferase activity (Figure 32). The migrating primitive streak cells and neural crest cells incorporated several sugars from sugar nucleotides onto their cell surfaces (shown by autoradiography). Shur has suggested that the migrating cells use their glycosyltransferases as receptors for the substrate oligosaccharides. Adhesion would be accomplished by the specific lock-and-key interactions between the substrate and the glycosyltransferase. As the cells move forward, these bonds would be broken by catalysis involving the addition of a new carbohydrate residue (Figure 35). Such glycosyltransferase proteins have been seen on the lamellopodia of migrating mesenchyme cells and

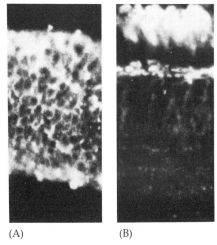

(A)   (B)

FIGURE 33
(A) Monoclonal antibody to a cell surface glycosyltransferase (capable of transferring N-acetylgalactosamine phosphate) binds to the cell membranes of a 10-day chick neural retina. (B) At birth, the glycosyltransferase is restricted to the outer layers. (From Balsamo et al., 1986a; photographs courtesy of G. Grunwald.)

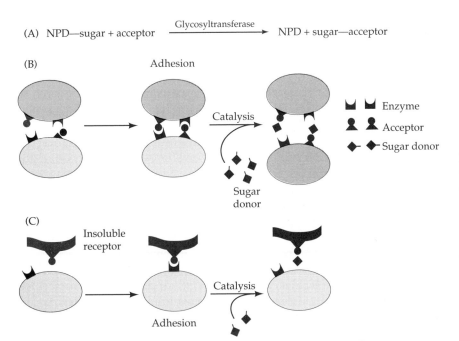

FIGURE 34
Cell surface interactions through glycosyltransferases. (A) The standard glycosyltransferase reaction, in which a sugar is transferred from a nucleotide carrier to an acceptor. (B,C) Intercellular reactions that may be mediated by cell surface glycosyltransferases (see text). If the activated sugar were absent, adhesion would result (as is thought to occur during fertilization). (Modified from Pierce et al., 1980.)

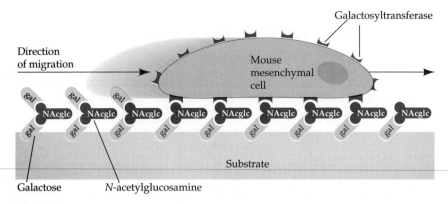

Direction of migration

Galactosyltransferase

Mouse mesenchymal cell

NAcglc NAcglc NAcglc NAcglc NAcglc NAcglc NAcglc NAcglc NAcglc NAcglc

gal gal gal gal gal gal gal gal gal gal gal gal gal gal gal gal gal gal gal gal

Substrate

Galactose   N-acetylglucosamine

FIGURE 35

Model for migration of embryonic mouse mesenchymal cells. Cell migrates on substrate containing exposed galactose-N-acetylglucosamine residues. Cells recognize and adhere to that substrate by a cell surface galactosyltransferase. Cells move by catalyzing the addition of galactose to those exposed residues, thus dissociating the cell from its substrate and allowing the cell surface enzymes to fill new, ungalactosylated sites on the matrix. (From Shur, 1982.)

on the growth cones of elongating axons (Eckstein and Shur, 1989; Begovac and Shur, 1990). This galactosyltransferase recognizes a specific oligosaccharide on the laminin of the basal lamina. If this enzyme is removed or its specificity changed, the neurons do not extend processes on laminin, and the mesenchymal cells fail to migrate.

Two predictions from this model are that (1) the addition of activated sugars should perturb normal development and (2) that migrating cells should leave new sugar groups on the substrates over which they migrate. The addition of large concentrations of sugar nucleotides to developing embryos does interfere with normal development (Shur, 1977a,b; Shur et al., 1979). Although the direct cause of these abnormalities has not been ascertained, the simplest explanation is that the sugar nucleotides interfere with the cell surface glycosyltransferases. Turley and Roth (1979) showed that when cells prelabeled with radioactive galactose were allowed to migrate on surfaces containing glycosaminoglycans, the cells put the sugar groups onto the substrate. Using different glycosaminoglycans, it was found that the more labeling the cells accomplished, the less migration occurred. The extent of migration, then, may depend on how many exposed carbohydrate groups are on the substrate. A large number of sites may promote strong adhesions, thereby slowing migration. Runyan and his co-workers (1986) have shown that certain migrating cells (such as neural crest cells) can use their glycosyltransferases to bind to glycoprotein components in natural basal laminae.

Glycosyltransferases may promote adhesion as well as migration. As we have seen in Chapter 2, the glycosyltransferase on the mouse sperm cell membrane recognizes the lactosamine sequences on the egg zona pellucida. During embryonic cleavage, glycosyltransferase–lactosamine binding contributes to the adhesion between the early blastomeres (Bayna et al., 1986) and may mediate part of the adhesion between uterine cells (Dutt et al., 1987).

The cell membrane, then, has several mechanisms by which it can form attachments with other cell membranes. It can use calcium-independent cell adhesion molecules, calcium-dependent cell adhesion molecules, restricted cell adhesion molecules, and glycosyltransferases. This does not exhaust its repertoire. As was briefly mentioned above, the cell can also bind specifically to particular components of the extracellular matrix. It is to these components that we now turn our attention.

Morphogenesis by cell–substrate interactions

The extracellular matrix

Cells also synthesize macromolecules that are secreted into extracellular spaces and mediate the relationships among the cells and their environment. These compounds are called SUBSTRATE ADHESION MOLECULES and they form the EXTRACELLULAR MATRIX. Such matrices may play several roles in development. In some instances, they serve to separate two adjacent groups of cells and prevent any interactions. In other cases, the extracellular matrix may serve as the substrate upon which cells migrate, or it may even induce the differentiation of certain cell types. One type of matrix is shown in Figure 36. Here, a sheet of epithelial cells is adjacent to a layer of loose mesenchymal tissue. The epithelial cells have formed a tight extracellular layer called the BASAL LAMINA; the mesenchymal cells secrete a loose RETICULAR LAMINA. Together, these layers constitute the BASEMENT MEMBRANE of the epithelial cell sheet. There are three major components of most extracellular matrices: collagen, proteoglycans, and matrix glycoproteins.

COLLAGEN is a family of glycoproteins containing a large percentage of glycine and proline residues. As we have seen earlier, many of the proline and lysine groups have been posttranslationally modified to hydroxyproline and hydroxylysine. Because collagen is the major structural support of almost every animal organ, it constitutes nearly half the total body protein. Most of this collagen is type I—found in the extracellular matrices of skin, tendons, and bones. This type makes up 90 percent of the collagen in the body. In addition, there are numerous other types of collagen that serve special functions (Table 2). Type II collagen is most evident as the secretion of cartilage cells, but it is also found in the notochord and in the vitreous body of the eye. Type III collagen is most evident in blood vessels, and types IV and V are found in various basal laminae produced by epithelial cells (Vuorio, 1986). In subsequent chapters, we will discuss the morphogenetic roles that collagen plays in causing the branching of organ rudiments and the differentiation of corneal tissue.

PROTEOGLYCANS are specific types of glycoproteins wherein (1) the weight of the carbohydrate residues far exceeds that of the protein and (2) the carbohydrates are linear chains composed of repeating disacchar-

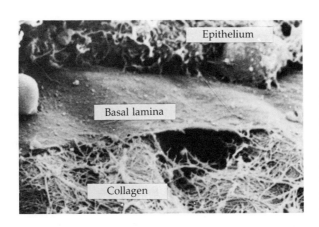

FIGURE 36
Location and formation of extracellular matrices in the chick embryo. Extracellular matrix at the junction of the epithelial cells (above) and mesenchymal cells (below). The epithelial cells synthesize a reticular lamina made primarily of collagen. (Photograph courtesy of R. L. Trelsted.)

TABLE 2
Structural and functional heterogeneity of collagen

Collagen type	Characteristics and function	Localization
I	Low carbohydrate content Thick, tightly packed, 67-nm banded fibrils, giving high tensile strength; most abundant collagen	Skin, bone, tendons
II	High carbohydrate content Small 67-nm banded fibrils to form compressible elastic matrix	Cartilage, vitreous body
III	High carbohydrate content Triple-helical, interchain disulfide bonds Small 67-nm banded fibrils to provide support and elasticity	Skin, blood vessels, internal organs (not bone or tendon)
IV	"Chicken wire"-like fibrils act as a cell attachment site and as a selective barrier. Binds to laminin No fibrils	Basement membranes
V	Abundant collagen Flexible 67-nm banded fibrils Function unknown	Pericellular space, bone, dermis
VI	Microfibrils with 105- to 110-nm periodicity; no fibrils Anchors nerves, blood vessels to tissues	Blood vessels, kidneys, skin, placenta, muscle, liver
VII	Long chain collagen Involved in anchoring fibrils	Choriamniotic membranes
VIII	Three collagenous domains linked in tandem by two noncollagenous domains; no fibrils Function unknown	Endothelial cells, neural crest-derived extracellular matrices
IX	Contains three collagenous, four noncollagenous domains and attached glycosaminoglycans Function unknown; short chains	Intersections of type II collagen in cartilage
X	Short chain collagen Functions in the mineralization of collagen	Hypertrophic (growth zone) cartilage
XI	Three chains Function unknown	Cartilage
XII	Discovered by cDNA cloning Function unknown	Cartilage
XIII	Discovered by cDNA cloning Function unknown	Cartilage, bone, epidermis, skeletal muscle

Sources: Vuorio (1986); Hostikka (1990).

ides. Usually one of the sugars of the disaccharide has an amino group, so the repeating unit is called a GLYCOSAMINOGLYCAN. Table 3 lists the common glycosaminoglycans, and the basic proteoglycan structure is illustrated in Figure 37. The interconnection of protein and carbohydrate forms a weblike matrix, and in many motile cell types, the proteoglycan surrounds the cells and is thought to mediate against their coming together (Figure 38). The consistency of the extracellular matrix depends upon the ratio of collagen to proteoglycan. Cartilage, which has a high percentage of proteoglycans, is soft, whereas tendons, which are predominantly collagen fibers, are tough. In basal laminae, proteoglycans predominate, forming a molecular sieve in addition to providing structural support.

One such proteoglycan, SYNDECAN, is a 33,000-Da protein that carries chondroitin sulfate and heparan sulfate glycosaminoglycans. It is found in the cell membrane and is especially prominent on the basal surfaces of

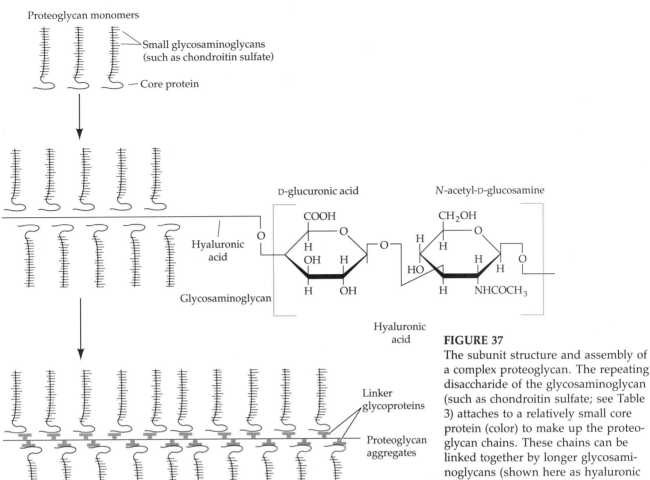

Proteoglycan monomers

Small glycosaminoglycans
(such as chondroitin sulfate)

Core protein

D-glucuronic acid

N-acetyl-D-glucosamine

COOH

CH₂OH

Hyaluronic
acid

Glycosaminoglycan

Hyaluronic
acid

Linker
glycoproteins

Proteoglycan
aggregates

FIGURE 37
The subunit structure and assembly of a complex proteoglycan. The repeating disaccharide of the glycosaminoglycan (such as chondroitin sulfate; see Table 3) attaches to a relatively small core protein (color) to make up the proteoglycan chains. These chains can be linked together by longer glycosaminoglycans (shown here as hyaluronic acid, and magnified in the right-hand side of the drawing) to produce complex networks. Linker glycoproteins stabilize these latter associations. (Modified from Cheney and Lash, 1981.)

epithelial cells where they contact the basal lamina. Syndecan, while anchored to the cell membrane, can attach to fibronectin and several types of collagen, and it appears to be necessary for establishing and/or maintaining epithelial cell sheets. If its synthesis is selectively inhibited in epithelial cells (by transfecting anti-sense mRNA into them), these cells lose their adhesivity to the basal lamina and become mesenchymal in appearance. Moreover, if fibroblasts are transfected with syndecan DNA attached to an active promoter, these cells will condense into tight aggregates. As we will see in Chapter 16, syndecan may be very important in

TABLE 3
Repeating disaccharide units of the most common glycosaminoglycans of matrix proteoglycans

GAG	Repeating disaccharide unit	Distribution
Hyaluronic acid	Glucuronic acid-N-acetylglucosamine	Connective tissues, bone, vitreous body
Chondroitin sulfate	Glucuronic acid-N-acetylglucosamine sulfate	Cartilage, cornea, arteries
Dermatan sulfate	[Glucuronic or Iduronic acid]-N-acetylgalactosamine sulfate	Skin, heart, blood vessels
Keratan sulfate	Galactose-N-acetylglucosamine sulfate	Cartilage, cornea
Heparan sulfate	[Glucuronic or Iduronic acid]-N-acetylgalactosamine sulfate	Lung, arteries, cell surfaces

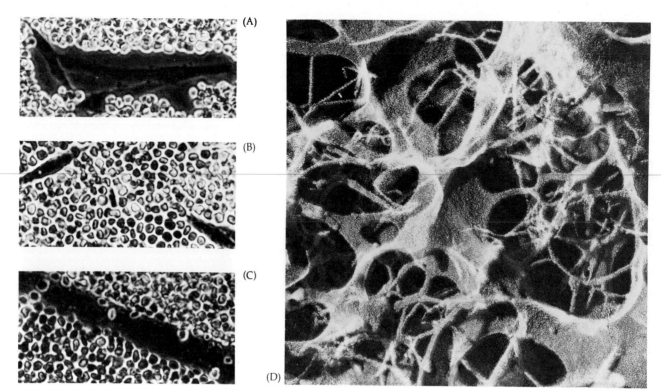

FIGURE 38

The proteoglycan coat surrounding mobile cells. (A) Hyaluronidate coat surrounds chick myoblasts. Myoblasts in culture exclude small particles (in this case fixed red blood cells) for a significant distance from the cell border. (B) When the myoblasts are treated with hyaluronidase (which dissolves hyaluronic acid), this extracellular coat vanishes. (C) The coat also vanishes as the myoblasts cease dividing and join together as they differentiate. (D) Electron micrograph of hyaluronidate in aqueous solution shows a branching fibrillar network. (A–C from Orkin et al., 1985, courtesy of B. Toole; D from Hadler et al., 1982, courtesy of N. M. Hadler.)

development, since it is found to be synthesized in kidney, tooth, and limb bud mesenchyme cells when they are induced to condense (Vainio et al., 1989; Thesleff et al., 1989; Bernfield and Sanderson, 1990).

The same glycosaminoglycans may have different roles in different tissues. Heparan sulfate proteoglycans, for instance, are universal components of basement membranes and can act in the nervous system to stimulate the proliferation of Schwann cells. Axons of the dorsal root ganglia have heparan sulfate on some of their cell surface proteins and the removal of this proteoglycan prevents the associated Schwann cells from proliferating around them (Ratner et al., 1985). In the developing limb, heparan sulfate (but not chondroitin sulfate) helps induce the cartilaginous phenotype and the growth of the cartilage nodules (San Antonio et al., 1987).

Extracellular glycoproteins

Extracellular matrices also contain a variety of other specialized molecules, such as fibronectin, laminin, and entactin. These large glycoproteins probably are responsible for organizing the collagen, proteoglycan, and cells into an ordered structure. FIBRONECTIN is a very large (460,000-Da) glycoprotein dimer synthesized by fibroblasts, chondrocytes, endothelial cells, macrophages, and certain epithelial cells, such as hepatocytes and amniocytes. One of the functions of fibronectin is to serve as a general adhesive molecule linking cells to various substrates such as collagen and proteoglycans. Fibronectin also organizes the extracellular matrix by having several distinct binding sites, whose interaction with the appropriate molecules results in the proper alignment of cells with their extracellular matrix (Figure 39).

As we have seen in earlier chapters, fibronectin has also been seen to play an important role in cell migration, as the road over which certain migrating cell types travel are paved with this protein. Mesodermal cell

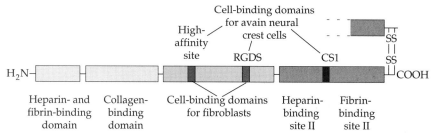

FIGURE 39

Structure and binding domains of fibronectin. The rectangles represent protease-resistant domains. The fibroblast cell binding domain consists of two units, the RGDS site and the high affinity site, both of which are essential for cell binding. Avian neural crest cells have another site which is necessary for their motility on a fibronectin substrate. Other regions of fibronectin enable it to bind to collagen, heparin,* and other molecules of the extracellular matrix. (After Dufour et al., 1988.)

migration during gastrulation is seen on the fibronectin surfaces of many species, and the movement of these cells ceases when fibronectin is locally removed. One group of chick mesodermal cells, the precardiac cells, migrates upon fibronectin to travel from the lateral sides of the embryo to the midline. If chick embryos are injected with antibodies to fibronectin or with the soluble cell-binding amino acids of fibronectin, the precardiac cells fail to migrate to the midline and two separate hearts develop. Fluorescent antibodies to fibronectin have demonstrated a gradient of fibronectin in the path of migration between the endoderm and the mesoderm. If this region is cut and rotated, the heart cells follow the gradient to new positions away from the midline (Linask and Lash, 1988a,b). Thus, fibronectin appears to play a major role in the migration of the precardiac cells to the embryonic midline. Other cell types, such as the germ cell precursors of frog embryos, also travel over cells that secrete fibronectin onto their surfaces (Heasman et al., 1981).

The cell-binding sites on fibronectin are located about three-quarters down the length of the peptides from their amino termini. By testing progressively shorter fragments of fibronectin that retained cell-binding activity, Pierschbacher and Ruoslahti (1984) showed that a simple tetrapeptide—arginine-glycine-aspartate-serine (RGDS, in the unit-letter abbreviations)—is capable of permitting cells to attach to the fibronectin substrate. Substitutions in the first three amino acids drastically reduced cell binding (Yamada and Kennedy, 1985), indicating that the RGD tripeptide was the important unit for the cell's recognition of fibronectin. Moreover, when the RGD sequence is injected into embryos, the fibronectin-dependent migrations cease, because the cells bind to the synthetic peptide (Boucaut et al., 1984; Linask and Lash, 1988a).

Some cells can recognize other regions of fibronectin in addition to the RGD site. A second cell-binding site, called the "high-affinity site," is located about 300 amino acids towards the amino-terminal end of the protein (Figure 39). The high-affinity site and the RGD site are both needed for the adhesion of fibroblasts to the extracellular matrix (Obara et al., 1988). Avian neural crest cells bind to both the RGD site and the high-affinity site, but they also can bind to another portion of the fibronectin molecule, the CS1 site. This latter site is able to promote the attachment

*"Heparin" is a portion of the heparin proteoglycan secreted by mast cells and basophils. "Heparan" and "heparan sulfate" are names given to similar glycosaminoglycans found in the extracellular matrix or on the cell surface. It is presumed that the binding sites for heparin are also those of heparan sulfate (Bernfield and Sanderson, 1990).

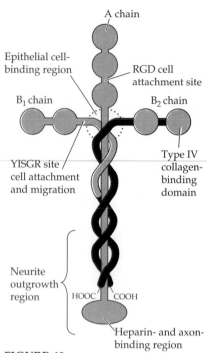

FIGURE 40
Structure of laminin and proposed binding regions.

A chain

Epithelial cell-binding region

RGD cell attachment site

B₁ chain

B₂ chain

YISGR site cell attachment and migration

Type IV collagen-binding domain

Neurite outgrowth region

HOOC   COOH

Heparin- and axon-binding region

FIGURE 41
Inhibition of cell adhesion by tenascin. Fibronectin and tenascin were both plated onto a tissue culture dish. Fibroblasts were added to the dish and allowed to adhere and migrate. The result shows that fibronectin was a much preferred substrate over tissue culture plastic, whereas cells did not adhere to or migrate well upon tenascin. (Photograph courtesy of R. Chiquet.)

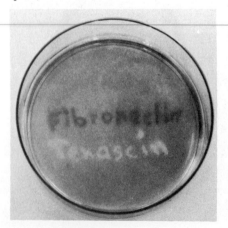

of neural crest cells to the matrix, but the RGD, high-affinity site, and the CS1 site must all be present for migration to occur (Dufour et al., 1988). The CS1 site does not exist in all fibronectin molecules since it is in an exon that can be spliced out in certain types of fibronectin. Thus, not all types of fibronectin will promote the migration of neural crest cells.

In addition to its function in cell migration, fibronectin appears to be a regulator of other developmental functions. For example, mesenchymal cells and myoblasts have substantial amounts of cell surface fibronectin, but this external protein diminishes as the cells differentiate into chondrocytes and myotubes, respectively (Lewis et al., 1978; Chen, 1977; Furcht et al., 1978). The removal of fibronectin from the myoblast cell surface is seen to be necessary for the fusion of these cells into myotubes (Podleski et al., 1979).

LAMININ is a major component of basal laminae. It is made of three peptide chains, and, like fibronectin, it can bind to collagen, glycosaminoglycans, and cells (Figure 40). The collagen bound by laminin is type IV (specific for basal lamina), and the cell-binding region of laminin chiefly recognizes epithelial cells and neurons. The adhesion of epithelial cells to laminin (which they sit upon and utilize) is much greater than the affinity of mesenchymal cells for fibronectin (which they have to bind to and release if they are to migrate). Like fibronectin, laminin is seen to play roles in assembling extracellular matrices, promoting cell adhesion and growth, changing cell shape, and permitting cell migrations (Hakamori, 1984).

Not all the large extracellular glycoproteins are substrates for cell adhesion. TENASCIN (also called cytotactin) resembles fibronectin for about half the length of the molecule, and it is found transiently in several extracellular matrices during embryonic development. However, different cells react in different ways to tenascin. Some cells stick to it, while other cells round up and detach from tenascin (Figure 41; Spring et al., 1989). Different relative amounts of fibronectin and tenascin may be able to generate substrates of various degrees of adhesiveness. In addition, tenascin appears to increase the synthesis and secretion of proteases from the cells that are on it (Werb et al., 1990). Both these characteristics may be important in generating paths for cell migration and in remodeling the extracellular matrix during development (Tan et al., 1987; Bronner-Fraser, 1988; Wehrle and Chiquet, 1990).

Integrins: Cellular receptors for fibronectin and laminin

The ability of a cell to bind these glycoproteins depends on its expressing a cell membrane receptor for the "cell binding site" of these large molecules. The main fibronectin receptors have been purified by using monoclonal antibodies that block the attachment of cells to fibronectin, by isolating those membrane proteins that bind to the RGD sequence (Leptin, 1986), and by isolating proteins bound by antibodies capable of inhibiting cell–fibronectin adhesion (Knudsen et al., 1985; Chen et al., 1985).

The fibronectin receptor complex was found not only to bind fibronectin on the *outside* of the cell, but also to bind cytoskeletal proteins on the *inside* of the cell. Thus the fibronectin receptor complex appears to span the cell membrane and unite two types of matrices. On the inside of the cell, it serves as the anchorage site for the actin microfilaments that move the cell; meanwhile, on the outside of the cell, it binds to the fibronectin of the extracellular matrix (Figure 42). Horwitz and co-workers (1986; Tamkun et al., 1986) have called this family of receptor proteins INTEGRINS because they integrate the extracellular and intracellular scaf-

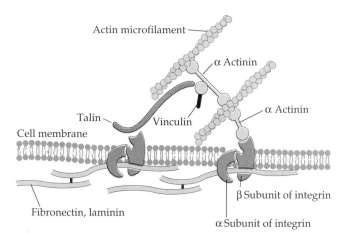

FIGURE 42
Speculative diagram relating the binding of cytoskeleton to the extracellular matrix through the integrin molecule. (After Burridge et al., 1990.)

folds, allowing them to work together. Integrin proteins have been found to span the cell membrane of numerous cell types and to bind to RGD sequences of several adhesive proteins in extracellular matrices, including vitronectin (found in the basal lamina of the eye), fibronectin, and laminin (Ruoslahti and Pierschbacher, 1987). This dual binding probably enables the cell to move by contracting the actin microfilaments against the fixed extracellular matrix (Figure 43).

Integrins contain one α subunit and one β subunit, and the binding specificity of the integrin depends upon its combination of subunits and upon certain conditions in the cellular environment (Hemler et al., 1987, Hemler, 1990). As shown in Table 4, the different binary combinations of α and β subunits enable the integrin to bind to particular extracellular molecules. However, an integrin can often bind more than one molecule, and its specificity can differ according to the cellular environment. The $\alpha2\beta1$ integrin, for example, has the potential to bind both collagen and laminin. On some cells it does indeed bind to both molecules, but in some cells this integrin binds only to laminin, and, in other cells, it binds only to collagen. In some instances, ionic conditions can cause this integrin to favor one substrate over another. Often one sees different integrins in the same organ. The cells of the kidney tubule originate from an aggregate of mesenchymal cells. However, as they develop, cells that give rise to Bowman's capsule epithelium express the $\alpha3\beta1$ integrin while all tubule cells express the $\alpha6\beta1$ integrin. The binding of the A-chain of laminin by the $\alpha6\beta1$ integrin is probably essential for forming the polarized epithelium of the kidney tubules (Korhonen et al., 1990; Sorokin et al., 1990).

FIGURE 43
Cell migration. (A) Phase-contrast micrograph of a fibroblast cell in motion. The thin, ruffled edge (lamellipodium) leads the way, attaching to the matrix and pulling the rest of the cell to it. (B) Indirect immunofluorescence staining of the actin microfilaments of a cell extending a ruffled membrane. The actin fibers radiate from the ordered cytoskeletal lattice into the ruffled membrane of the lamellipodium. (A courtesy of N. Wessels; B from Lazarides, 1976, courtesy of E. Lazarides.)

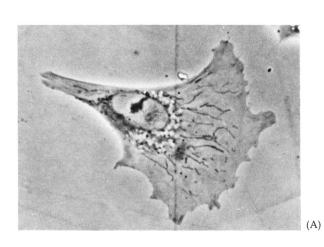

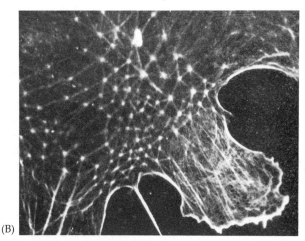

(A)   (B)

TABLE 4
Integrin subunit combinations and ligands[a]

Major β subunits[b] and integrins	α Subunits										
	α1	α2	α3	α4	α5	α6	αV	αIIb	α2L	α2M	αX
β1 (VLA INTEGRINS)											
Collagen	▓	▓	▓								
Laminin	▓	▓	▓			▓					
Fibronectin			▓	▓	▓		▓				
Vitronectin							▓				
β2 (LEUCAN INTEGRINS)											
I-CAM									▓	▓	
Fibronogen										▓	
Clotting factor X										▓	
C3B										▓	▓
β3 (CYTOADHESION INTEGRINS)											
Vitronectin							▓	▓			
Fibrinogen							▓	▓			
von Willibrand factor							▓	▓			
Bone sialo-protein							▓				
Fibronectin							▓	▓			
β4 (EPITHELIAL INTEGRINS)											
Component of desmosomes						▓					
Basal lamina of epithelia and Schwann cells						▓					

[a]The empty spaces in the grid indicate combinations of subunits that have not been observed.
[b]Only the four major β subunits are shown here. At least eight β subunits are now known to exist.

The importance of integrins for morphogenesis is dramatically illustrated during *Drosophila* embryogenesis. Like vertebrate integrins, *Drosophila* integrins are composed of two subunits (α and β) that span the cell membrane. In the two known *Drosophila* integrins, the β subunits are identical, but the α subunits differ. These two integrins often work together to affect tissue and cell adhesion during development. In the development of the *Drosophila* wing, two epithelial sheets are brought together. The PS1 integrin is found on the basal surface of the presumptive *dorsal* wing epithelium, while the PS2 integrin is on the upper surface of the presumptive *ventral* wing epithelium. During metamorphosis, these two epithelia meet and adhere to form the two-layered wing blade. Mutations in the integrins cause the wing to have regions where the two wing epithelia come apart, as evidenced by bubbles between the two blades (Wilcox et al., 1989; Brower and Jaffe, 1989). Some of the mutants of *Drosophila* integrins are lethal, since integrin is needed to attach muscles to the epidermis and the gut wall. In the lethal mutation *lethal (1) myospheroid*, there is a deficiency in the genes encoding the β subunit of the *Drosophila* integrins. In the absence of this subunit, neither integrin is formed. The somatic muscles are contracted into spheres that lack attachments to the body wall or gut (Leptin et al., 1989).

While both integrin subunits are essential for binding extracellular matrix molecules, it is the cytoplasmic region of the β subunit that appears to play a critical role in the transmembrane connection to the internal cytoskeleton. Cells with mutant β1 subunits that lack parts of this region fail to bind the cytoskeleton, even though they can bind to fibronectin by their extracellular domains (Solowska et al., 1989; Hayashi et al., 1990). The binding of the cytoplasmic domain to talin and fibulin has also been shown biochemically (Horwitz et al., 1986; Argraves et al., 1989). As depicted in Figure 42, there is an assembly of proteins on either side of the integrin molecule. On the outer surface is the complex assortment of extracellular matrix proteins. On the inner surface are the cytoskeletal components, especially those involved with locomotion. The principle binding of the actin filaments is through talin or α-actinin (Burridge, 1991).

During the migration of cells, the contractile actin microfilaments must attach to the substrate in order to pull the cell. Integrin is seen to provide this attachment. However, unlike the case of epithelial cells, which adhere tightly to a basement membrane, the attachment of migrating cells to the basal lamina must be continually made and broken during cell migration. Beckerle and her colleagues (1987) have demonstrated the existence of a calcium-dependent protease that cleaves talin and that is specifically localized to the sites where integrin binds to the substratum. It is possible that this protease could cleave the bridge between the fibronectin receptor and the cytoskeleton. Stickel and Wang (1988) have shown that when synthetic peptides break the contact between integrin and fibronectin, the α-actinin and vinculin are released from the adhesion regions.

Integrins are not the only molecules capable of binding to laminin and fibronectin. While the integrin receptor binds to an RGD sequence in the A-chain of laminin, two other laminin receptor proteins bind to a different sequence (YIGSR) in the B1-chain (Graf et al., 1987; Yow et al., 1988). As mentioned earlier, another protein recognizes carbohydrate groups on the laminin molecule as well. The receptors have different affinities for laminin, and these might be important in their function (Horwitz et al., 1985). The α3β1 integrin of fibroblasts, for example, has a relatively low affinity for laminin ($K_d = 10^{-6}$ M), whereas the affinity for laminin of a laminin receptor from epithelial cells is much higher ($K_d = 2 \times 10^{-9} M$). The receptor that is used may be important in allowing the cells to use laminin either as a basement membrane (in which case the affinity of the receptor would be high) or as a substrate for migration (in which lower affinity receptors would be used).

Differential adhesion resulting from multiple adhesion systems

Although we have been discussing these adhesion systems as separate units, it is probable that the morphogenetic processes of cell–cell interaction are brought about by combinations of cell adhesion molecules. The adhesion of axons to cultures of muscle cells can be inhibited only by adding a combination of antibodies directed at N-CAM, N-cadherin, and integrin (Chuong et al., 1987). Similarly, Bixby and colleagues (1987, 1988) found that the migration of granule cell neurons upon the glial cells of the cerebellum is regulated by specific combinations of Ng-CAM, N-cadherin, and integrin acting at different layers of the cerebellum. The initial attachment of the mouse blastocyst to the uterine wall also appears to be mediated through several adhesion systems. First, the trophoblastic cells have receptors for the collagen and heparan sulfate proteoglycans of the uterine endometrium, and interference with this binding can impede implantation

(Carson et al., 1988; Farach et al., 1987). Second, Dutt and co-workers (1987) have shown that the trophoblast can adhere to uterine cells through cell surface galactosyltransferases. The substrate for this saccharide CAM, lactosaminoglycan, is present on the uterine cells when they are induced by estrogen. Third, Kadokawa and co-workers (1989) have shown P- and E-cadherin on both the trophoblast and the uterine tissue at the site of implantation.

Importance of the extracellular matrix to differentiation

In many cells there is a correlation between morphogenesis and gene expression. Both these processes may be coordinated by the extracellular matrix. When rat Sertoli cells—the major structural cells of the testis—are cultured on plastic petri dishes, they lose their differentiated morphology, forming a flat monolayer of cells. However, when they are cultured atop a basement membrane (containing laminin, type IV collagen, heparan sulfate, nidogen, and entactin), they form columnar monolayers identical to normal Sertoli cells in the testis (Figure 44). These Sertoli cells contain a well-formed cytoskeleton and basally located junctions similar to the cells in vivo, and the synthesis of androgen-binding protein and transferrin is much higher in the cells growing on the basal lamina (Hadley et al., 1985).

Matrix-induced differentiation has also been seen in endothelial cells, wherein the laminin of the extracellular matrix causes the endothelial cells to differentiate their polar cell type and form tubular structures (Grant et

(A)

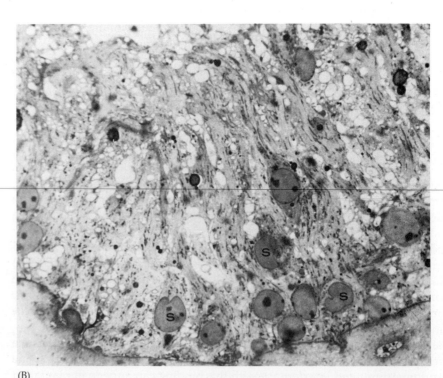

FIGURE 44
Basal lamina and cell differentiation. Light micrographs of rat Sertoli cells grown for two weeks (A) on tissue culture plastic dishes, and (B) on tissue culture plastic dishes coated with basal lamina. The photographs are taken at the same magnification, 1200×. (From Hadley, et al., 1985; photographs courtesy of M. Dym.)

(B)

al., 1989). In sea urchins, certain genes are selectively activated when the cells contact a collagen-containing basal lamina (Wessel et al., 1989). This type of matrix-induced differentiation has been seen in several other instances and may be driven by the fixing of cell surface receptors on one side of the cell. As we will see in the next chapter, the change in the placement of these cell membrane molecules may have profound effects within the cell.

Junctional modifications in morphogenesis

The extracellular matrix can be used to keep cells apart or to bring cells together. When cells are connected by the extracellular matrix, usually a basal lamina is formed on only one side. This arrangement allows the adjacent cells to interact through the lateral sides of their membranes. There are three major types of cell membrane specializations between cells: the tight junction, the desmosome, and the gap junction.

TIGHT JUNCTIONS are the most intimate of cellular interactions (Figure 45A), and they are used to separate the extracellular space on one side of an epithelial sheet from the extracellular space on the other side. This ability to form distinct compartments was seen earlier, when we noted the tight junctions forming in the external cells of the mammalian blastocyst. The tight junction (sometimes called the occluding zonule) can separate the plasma membrane into functional regions. In kidney and intestinal epithelia, for instance, there is an asymmetric distribution of integral membrane proteins. The digestive hydrolases and phosphatases are on the apical surface at the lumen, while the sodium and potassium pumps are located in the basal cell membrane. If the tight junctions are dismantled by lowering the calcium ion concentration of these cultured cells, the apical and basal membrane proteins can mix (Ziomek et al., 1980; Palade, 1985).

The DESMOSOME is often seen when cells are joined together to form an impermeable tissue. In the cytoplasms of both adjacent cells, there are local thickenings of fibrillar protein. The space between the cells is 200–350 nm wide and is filled with a glycoprotein cement (Figure 45B). These junctions are also very hard to separate.

GAP JUNCTIONS often can be seen when adjacent cells are 20 to 40 nm apart. Through this space, fine connections are seen (Figure 45C). These connections do not serve to separate compartments; rather, they serve as communication channels between the adjacent cells. Cells so linked are said to be "coupled," and small molecules (<1500 mw) and ions can freely pass from one cell to another. In most embryos, at least some of the early blastomeres are connected by gap junctions, thereby enabling ions and small soluble molecules to pass readily between them. We have previously seen that the inner cell mass of mammalian embryos is coupled together by gap junctions, and that during the development of certain molluscs, gap junctions form at the precise time that information needs to be transferred from one blastomere to another.

The importance of gap junctions in development has been demonstrated in amphibian and mammalian embryos (Warner et al., 1984). When antibodies to gap junction proteins were microinjected into one specific cell of an 8-cell *Xenopus* blastula, the progeny of that cell, which are usually coupled through gap junctions, could no longer pass ions or small molecules from cell to cell. Moreover, the tadpoles that resulted from such treated blastulae showed defects specifically relating to the developmental fate of the injected cell (Figure 46). The progeny of such a cell did not die,

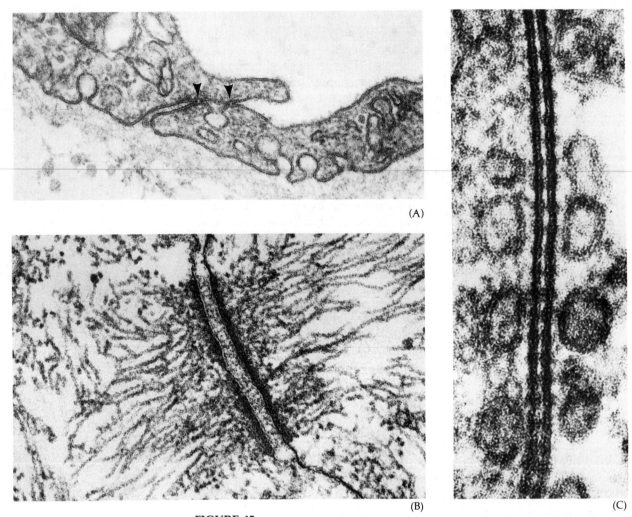

(A)

(B)

(C)

FIGURE 45
Membrane specialization at the junctions of two cells. (A) Tight junctions. Two adjacent endothelial cells of a capillary come together and the membranes fuse (arrows) to form a tight seal. (B) Desmosomes. Electron micrograph of desmosomes on two adjacent cells from newt epidermis. Protein thickening is seen on both cells, and a glycoprotein matrix fills the intercellular space. (C) Gap junctions. Electron micrograph of an area of gap junction between two apposed cells. The beaded appearance is caused by rows of gap junction proteins, which form tunnels between the two cells. (A courtesy of J. E. Rash; B from Kelly, 1966, courtesy of D. E. Kelly; C from Peracchia and Dulhunty, 1976, courtesy of C. Peracchia.)

FIGURE 46
Developmental effects of gap junctions. Section through *Xenopus* tadpole in which one of the blastomeres at the 8-cell stage had been injected with (A) a control antibody or (B) antibody against gap junction protein. The side formed by the injected blastomere lacks its eye and has abnormal brain morphology. (From Warner et al., 1984; photographs courtesy of A. E. Warner.)

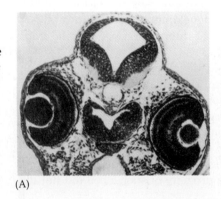

(A)

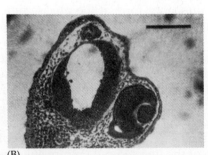

(B)

but they were unable to undergo their normal development (Warner et al., 1984). In the uncompacted and compacted mouse embryo, the eight blastomeres are connected to one another by gap junctions. If the junctions of uncompacted embryos are blocked by these antibodies, the blastomeres continue to divide, but compaction fails to occur (Lo and Gilula, 1979; Lee et al., 1987). Gap junctions appear to be essential for the communication of developmentally important information between cells.

In 1782 the French essayist Denis Diderot posed the question of morphogenesis in the fevered dream of a noted physicist. This character could imagine that the body was formed from myriads of "tiny sensitive bodies" that collected together to form an aggregate, but he could not envision how this aggregate could become an animal. Recent studies have shown that this ordering is due to the molecules on the surfaces of these cells. In subsequent chapters we will look more closely at some of these morphogenetic interactions.

LITERATURE CITED

Abercrombie, M. and Ambrose, E. J. 1958. Interference microscope studies of cell contacts in tissue culture. *Exp. Cell Res.* 15: 332–345.

Argraves, W. S., Dickerson, K., Burgess, W. H. and Ruoslahti, E. 1989. Fibulin, a novel protein that interacts with the fibronectin receptor β subunit cytoplasmic domain. *Cell* 58: 623–629.

Armstrong, P. B. 1985. The control of cell motility during embryogenesis. *Cancer Metas. Rev.* 4: 59–80.

Armstrong, P. B. 1989. Cell sorting out: The self-assembly of tissues *in vitro*. *CRC Critical Reviews in Biochemistry and Molecular Biology* 24: 119–149.

Balsamo, J., Pratt, R. S. and Lilien, J. 1986. Chick neural retina *N*-acetylgalactosaminyltransferase/acceptor complex: Catalysis involves transfer of *N*-acetylgalactosamine phosphate to endogenous acceptors. *Biochemistry* 25: 5402–5407.

Bastiani, M. J., Doe, C. Q., Helfand, S. L. and Goodman, C. S. 1985. Neuronal specificity and growth cone guidance in the grasshopper and *Drosophila* embryos. *Trends Neurosci.* 8: 257–266.

Bastiani, M. J., Harrelson, A. L., Snow, P. M. and Goodman, C. S. 1987. Expression of fasciclin I and II glycoproteins on subsets of axon pathways during neuronal development in the grasshopper. *Cell* 48: 745–755.

Bayna, E. M., Runyan, R. B., Scully, N. F., Lopez, L. C., Reichner, J. and Shur, B. D. 1986. Cell surface receptors during mammalian fertilization and development. *Mol. Cell. Biochem.* 72: 141–151.

Beckerle, M. C., Burrdige, K., DeMartino, G. N. and Croall, D. E. 1987. Colocalization of calcium-dependent protease II and one of its substrates and sites of adhesion. *Cell* 51: 569–577.

Begovac, P. C. and Shur, B. D. 1990. Cell surface galactosyltransferase mediates the initiation of neurite outgrowth from PC12 cells on laminin. *J. Cell Biol.* 110: 461–470.

Bernfield, M. and Sanderson, D. 1990. Syndecan, a morphogenetically regulated cell surface proteoglycan that binds extracellular matrix and growth factors. *Phil. Trans. Roy. Soc. London* [A]. 327: 171–186.

Beug, H., Gerisch, G., Kempff, S., Riedel, V. and Cremer, G. 1970. Specific inhibition of cell contact formation in *Dictyostelium* by univalent antibodies. *Exp. Cell Res.* 63: 147–158.

Bixby, J., Pratt, R., Lilien, J. and Reichardt, L. F. 1987. Neurite outgrowth on muscle cell surfaces involves extracellular matrix receptors as well as Ca^{2+}-dependent and independent cell adhesion molecules. *Proc. Natl. Acad. Sci. USA* 84: 2555–2559.

Bixby, J. L., Lilien, J. and Reichardt, L. F. 1988. Identification of the major proteins that promote neuronal process outgrowth on Schwann cells *in vitro*. *J. Cell Biol.* 107: 353–361.

Borgens, R. B. 1982. What is the role of naturally produced electric current in vertebrate regeneration and healing? *Int. Rev. Cytol.* 76: 245–300.

Boucaut, J. C. 1974. Étude autoradiographique de la distribution de cellules embryonnaires isolées, transplantées dans le blastocèle chez *Pleurodeles waltii* Michah (Amphibien, Urodele). *Ann. Embryol. Morphol.* 7: 7–50.

Boucaut, J. C., Darribère, T., Poole, T. J., Aoyama, H., Yamada, K. M. and Thiery, J.-P. 1984. Biologically active synthetic peptides as probes of embryonic development: A competitive peptide inhibitor of fibronectin function inhibits gastrulation in amphibian embryos and neural crest cell migration in avian embryos. *J. Cell Biol.* 99: 1822–1830.

Boyse, E. A. and Old, L. J. 1969. Some aspects of normal and abnormal cell surface genetics. *Annu. Rev. Genet.* 3: 269–289.

Brackenbury, R., Thiery, J.-P., Rutishauser, U. and Edelman, G. M. 1977. Adhesion among neural cells of the chick embryo. I. Immunological assay for molecules involved in cell–cell binding. *J. Biol. Chem.* 252: 6835–6840.

Bronner-Fraser, M. 1988. Distribution of tenascin during cranial neural crest development in the chick. *J. Neurosci. Res.* 21: 135–147.

Brower, D. L. and Jaffe, S. M. 1989. Requirement for integrins during *Drosophila* wing development. *Nature* 342: 285–287.

Burridge, K., Nuckolls, G., Otey, C., Pavalko, F., Simon, K. and Turner, C. 1990. Actin–membrane interactions in focal adhesions. *Cell Diff. Dev.* In press.

Carson, D. D., Tang, J.-P. and Gay, S. 1988. Collagens support embryo attachment and outgrowth *in vitro*: Effects of the Arg-Gly-Asp sequence. *Dev. Biol.* 127: 368–375.

Carter, S. B. 1967. Haptotaxis and the mechanism of cell motility. *Nature* 213: 256–260.

Champion, S., Imhof, B. A., Savagnier, P. and Thiery, J.-P. 1986. The embryonic thymus produces chemotactic peptides involved in the homing of hemopoietic precursors. *Cell* 44: 781–790.

Chen, L. B. 1977. Alteration in cell surface LETS protein during myogenesis. *Cell* 10: 393–400.

Chen, W. T., Hasegawa, E., Hasegawa, T., Weinstock, C. and Yamada, K. M. 1985. Development of cell-surface linkage complexes in cultured fibroblasts. *J. Cell Biol.* 100: 1103–1114.

Cheney, C. M. and Lash, J. W. 1981. Diversification within embryonic chick somites: Differential response to notochord. *Dev. Biol.* 81: 288–298.

Chuong, C.-M. and Edelman, G. M. 1985a. Expression of cell adhesion molecules in embryonic induction: I. Morphogenesis of nestling feathers. *J. Cell Biol.* 101: 1009–1026.

Chuong, C.-M. and Edelman, G. M. 1985b. Expression of cell adhesion molecule in embryonic induction: II. Morphogenesis of adult feathers. *J. Cell Biol.* 101: 1027–1043.

Chuong, C.-M., Crossin, K. L. and Edelman, G. M. 1987. Sequential expression and differential function of multiple adhesion molecules during the formation of cerebellar cortical layers. *J. Cell Biol.* 104: 331–342.

Cooper, M. S. and Schliwa, M. 1985. Electrical and ionic controls of tissue cell locomotion in DC electric fields. *J. Neurosci. Res.* 13: 223–244.

Covault, J. and Sanes, J. R. 1986. Distribution of N-CAM in synaptic and extrasynaptic portions of developing and adult skeletal muscle. *J. Cell Biol.* 102: 716–730.

Crawford, K. and Stocum, D. L. 1988. Retinoic acid coordinately proximalizes regenerate pattern and blastema differential affinity in axolotl limbs. *Development* 102: 687–698.

Crossin, K. L., Chuong, C.-M. and Edelman, G. M. 1985. Expression sequences of cell adhesion molecules. *Proc. Natl. Acad. Sci. USA* 82: 6942–6946.

Cunningham, B. A., Hoffman, S., Rutishauser, U., Hemperly, J. J. and Edelman, G. M. 1983. Molecular topography of the neural cell adhesion molecule N-CAM: Surface orientation and location of sialic acid-rich and binding regions. *Proc. Natl. Acad. Sci. USA* 80: 3116–3120.

Curtis, A. S. G. 1969. The measurement of cell adhesiveness by an absolute method. *J. Embryol. Exp. Morphol.* 22: 305–325.

Daniloff, J. K., Chuong, C.-M., Levi, G. and Edelman, G. M. 1986. Differential distribution of cell adhesion molecules during histogenesis of the chick nervous system. *J. Neurosci.* 6: 739–758.

Dargemont, C. and several others. 1989. Thymotaxin, a chemotactic protein, is identical to β_2-microglobulin. *Science* 246: 803–806.

Diderot, D. 1782. *D'Alembert's Dream.* Reprinted in J. Barzun and R. H. Bowen (eds.), *Rameau's Nephew and Other Works* (1956). Doubleday, Garden City, NY, p. 114.

Dufour, S., Duband, J.-L., Humphries, M. J., Obara, M., Yamada, K. M. and Thiery, J. P. 1988. Attachment, spreading and locomotion of avian neural crest cells are mediated by multiple adhesion sites on fibronectin molecules. *EMBO J.* 7: 2661–2671.

Dutt, A., Tang, T.-P. and Carson, D. D. 1987. Lactosaminoglycans are involved in uterine epithelial cell adhesion in vitro. *Dev. Biol.* 119: 27–37.

Eckstein, D. J. and Shur, B. D. 1989. Laminin induces the stable expression of surface glycosyltransferases on lamellipodia of migrating cells. *J. Cell Biol.* 108: 2507–2517.

Edelman, G. M. 1976. Surface modulation in cell recognition and growth. *Science* 192: 218–226.

Edelman, G. M. 1985. Expression of cell adhesion molecules during embryogenesis and regeneration. *Exp. Cell Res.* 161: 1–16.

Edelman, G. M. 1986. Cell adhesion molecules in the regulation of animal form and pattern. *Annu. Rev. Cell Biol.* 2: 81–116.

Edelman, G. M. and Thiery, J.-P. 1985. *The Cell in Contact: Adhesions and Junctions as Morphogenic Determinants.* Wiley, New York.

Edelman, G. M., Gallin, W. J., Delouvée, A., Cunningham, B. A. and Thiery, J.-P. 1983. Early epochal maps of two different cell adhesion molecules. *Proc. Natl. Acad. Sci. USA* 80: 4384–4388.

Erickson, C. A., Tosney, K. W. and Weston, J. A. 1980. Analysis of migratory behavior of neural crest and fibroblastic cells in embryonic tissues. *Dev. Biol.* 77: 142–156.

Farach, M. C., Tang, J. P., Decker, G. L. and Carson, D. D. 1987. Heparin heparan sulfate is involved in attachment and spreading of mouse embryos *in vitro. Dev. Biol.* 123: 401–410.

Fink, R. and McClay, D. R. 1985. Three cell recognition changes accompany the ingression of sea urchin primary mesenchyme cells. *Dev. Biol.* 107: 66–74.

Fraser, S., E., Carhart, M. S., Murray, B. A., Chuong, C.-M. and Edelman, G. E. 1988. Alterations in the *Xenopus* retinotectal projection by antibodies to *Xenopus* N-CAM. *Dev. Biol.* 129: 217–230.

Frye, L. D. and Edidin, M. J. 1970. The rapid intermixing of cell surface antigens after the formation of mouse–human heterokaryons. *J. Cell Sci.* 7: 319–335.

Fujimori, T., Miyatani, S. and Takeichi, M. 1990. Ectopic expression of N-cadherin perturbs histogenesis in *Xenopus* embryos. *Development* 110: 97–104.

Furcht, L. T., Mosher, D. F. and Wendelschafer-Crabb, G. 1978. Immunocytochemical localization of fibronectin (LETS protein) on the surface of L6 myoblasts: Light and electron microscope studies. *Cell* 13: 263–271.

Giudice, G. 1962. Restitution of whole larvae from disaggregated cells of sea urchin embryos. *Dev. Biol.* 5: 402–411.

Graf, J., Ogle, R. C., Robey, F. A., Sasaki, M., Martin, G. R., Yamada, Y. and Kleinman, H. K. 1987. A pentapeptide from the laminin B1 chain mediates cell adhesion and binds to the Mr-67000 laminin receptor. *Biochemistry* 26: 6896–6900.

Grant, D. S., Tashiro, K.-I., Bartolome, S.-R., Yamada, Y., Martin, G. R. and Kleinman, H., K. 1989. Two different laminin domains mediate the differentiation of human endothelial cells into capillary-like structures in vitro. *Cell* 58: 933–943.

Grunwald, G. B., Simmonds, M. A., Klein, R. and Kornguth, S. E. 1985. Autoimmune basis for visual paraneoplastic syndrome in patients with small-cell lung carcinoma. *Lancet* 1985(II): 658–661.

Grunwald, G. B., Kornguth, S. E., Towfighi, J., Sassani, J., Simmonds, M. A., Housman, C. M. and Papadopoulos, N. 1987. Autoimmune basis for visual paraneoplastic syndrome in patients with small cell lung carcinoma. *Cancer* 60: 780–786.

Hadler, N. M., Dourmash, R. R., Nermut, M. V. and Williams, L. D. 1982. Ultrastructure of a hyaluronic acid matrix. *Proc. Natl. Acad. Sci. USA* 79: 307–309.

Hadley, M. A., Byers, S. W., Suárez-Quian, C. A., Kleinman, H. K. and Dym, M. 1985. Extracellular matrix regulates Sertoli cell differentiation, testicular card formation, and germ cell development in vitro. *J. Cell Biol.* 101: 1511–1522.

Hakamori, S., Fukuda, M., Sekiguchi, K. and Carter, W. G. 1984. Fibronectin, laminin, and other extracellular glycoproteins. *In* K. A. Picz and A. H. Reddi (eds.), *Extracellular Matrix Biochemistry.* Elsevier, New York, pp. 229–275.

Hall, A. K. and Rutishauser, U. 1987. Visualization of neural cell adhesion molecule by electron microscopy. *J. Cell Biol.* 104: 1579–1586.

Harrelson, A. L. and Goodman, C. S. 1988. Growth cone guidance in insects: Fasciclin-II is a member of the immunoglobulin superfamily. *Science* 242: 700–708.

Harris, A. K. 1973. Behavior of cultured cells on substrate of various adhesiveness. *Exp. Cell Res.* 77: 285–297.

Harris, A. K. 1980. Silicone rubber substrate: A new wrinkle in the study of cell locomotion. *Science* 208: 176–179.

Harris, A. K. 1984. Cell traction and the generation of anatomical structure. *In* W. Jäger and J. D. Murray (eds.), *Modeling of Patterns in Time and Space.* Springer-Verlag, Berlin, p. 103–122.

Harris, H. 1954. Role of chemotaxis in inflammation. *Physiol. Rev.* 34: 529–562.

Hatta, K. and Takeichi, M. 1986. Expression of N-cadherin adhesion molecules associated with early morphogenetic events in chick development. *Nature* 320: 447–449.

Hatta, K. Takagi, S., Fujisawa, H. and Takeichi, M. 1987. Spatial and temporal expression pattern of N-cadherin cell adhesion molecules correlated with morphogenetic processes of chicken embryos. *Dev. Biol.* 120: 215–227.

Hayashi, Y., Haimovich, B., Reszka, A., Boettiger, D. and Horwitz, A. 1990. Expression and function of chicken integrin b1 subunit and its cytoplasmic domain mutants in mouse NIH3T3 cells. *J. Cell Biol.* 110: 175–184.

Heasman, J., Hines, R. D., Swan, A. P., Thomas, V. and Wylie, C. C. 1981. Primordial germ cells of *Xenopus* embryos: The role of fibronectin in their adhesion during migration. *Cell* 27: 437–447.

Hemler, M. E. 1990. VLA proteins in the integrin family: Structures, functions, and their role on leukocytes. *Annu. Rev. Immunol.* 8: 365–400.

Hemler, M. E., Huang, C. and Schwartz, L. 1987. The VLA protein family. Characterization of five distinct cell surface heterodimers each with a common 130,000 molecular weight β subunit. *J. Biol. Chem.* 262: 3300–3309.

Hinkle, L., McCaig, C. D. and Robinson, K. R. 1981. The direction of growth of differentiating neurons and myoblasts from frog embryos in an applied electric field. *J. Physiol.* 314: 121–135.

Hirokawa, N., Cheney, R. E. and Willard, M. 1983. Localization of a protein of the fodrin-spectrin-TW260/240 family on the mouse intestinal brush border. *Cell* 32: 953–965.

Hoffman, S. B. and Edelman, G. M. 1983. Kinetics of homophilic binding by E and A forms of the neural cell adhesion molecule. *Proc. Natl. Acad. Sci. USA* 80: 5762–5766.

Horwitz, A., Duggan, K., Greggs, R., Decker, C. and Buck, C. 1985. The cell substrate attachment (CSAT) antigen has properties of a receptor for laminin and fibronectin. *J. Cell Biol.* 101: 2134–2144.

Horwitz, A., Duggan, K., Buck, C., Beckerle, M. C. and Burridge, K. 1986. Interaction of plasma membrane fibronectin receptor with talin-α actinin transmembrane linkage. *Nature* 320: 531–533.

Hostikka, S. L. 1990. Human type IV collagen. *Acta Univ. Oulu.*

Ingvar, S. 1920. Reactions of cells to galvanic current in tissue cultures. *Proc. Am. Soc. Exp. Biol. Med.* 17: 198–199.

Jaffe, L. F. 1981. The role of ionic current in establishing developmental pattern. *Philos. Trans. R. Soc. Lond. B* 295: 553–566.

Jaffe, L. F. and Poo, M.-M. 1979. Neurites grow faster toward the cathode than the anode in a steady field. *J. Exp. Zool.* 209: 115–128.

Jaffe, L. F. and Stern, C. D. 1979. Strong electrical currents leave the primitive streak of chick embryos. *Science* 206: 569–571.

Kadokawa, Y., Fuketa, I., Nose, A., Takeichi, M. and Nakatsuji, N. 1989. Expression of E- and P-cadherin in mouse embryos and uteri during the periimplantation period. *Dev. Growth Diff.* 31: 23–30.

Kelly, D. E. 1966. Fin structure of desmosomes, hemidesmosomes, and an adepidermal globular layer in developing newt epidermis. *J. Cell Biol.* 28: 51–72.

Knudsen, K., Horwitz, A. F. and Buck, C. 1985. A monoclonal antibody identifies a glycoprotein complex involved in cell-substratum adhesion. *Exp. Cell Res.* 157: 218–226.

Köhler, G. and Milstein, C. 1975. Continuous cultures of fused cells secreting antibody of predefined specificity. *Nature* 256: 495–497.

Korhonen, M., Ylänne, J., Laitinen, L. and Virtanen, I. 1990. The α_1–α_6 subunits of integrins are characteristically expressed in distinct segments of developing and adult human nephron. *J. Cell Biol.* 111: 1245–1254.

Lander, A. D. 1989. Understanding the molecules of neural cell contacts: Emerging patterns of structure and function. *Trends Neurosci.* 12: 189–195.

Landmesser, L., Dahm, L., Schultz, K. and Rutishauser, U. 1988. Distinct roles for adhesion molecules during innervation of embryonic chick muscle. *Dev. Biol.* 130: 645–670.

Lazarides, E. 1976. Actin, α-actinin, and tropomyosin interaction in the structural organization of actin filaments in nonmuscle cells. *J. Cell Biol.* 68: 202–219.

Lee, S., Gilula, N. B. and Warner, A. E. 1987. Gap junctional communication and compaction during preimplantation stages of mouse development. *Cell* 51: 851–860.

Leptin, M. 1986. The fibronectin receptor family. *Nature* 321: 728.

Leptin, M., Bogaert, T., Lehmann, R. and Wilcox, M. 1989. The function of PS integrins during *Drosophila* embryogenesis. *Cell* 56: 401–408.

Lewis, C. A., Pratt, R. M., Pennypacker, J. P. and Hassell, J. R. 1978. Inhibition of limb chondrogenesis in vitro by vitamin A: Alteration in cell surface characteristics. *Dev. Biol.* 64: 31–47.

Linask, K. L. and Lash, J. W. 1988a. A role for fibronectin in the migration of avian precardiac cells. I. Dose-dependent effects of fibronectin antibody. *Dev. Biol.* 129: 315–323.

Linask, K. L. and Lash, J. W. 1988b. A role for fibronectin in the migration of avian precardiac cells II. Rotation of the heart-forming region during different stages and its effects. *Dev. Biol.* 129: 324–329.

Lo, C. and Gilula, N. B. 1979. Gap junctional communication in the preimplantation mouse embryo. *Cell* 18: 399–409.

McClay, D. R. and Ettensohn, C. A. 1987. Cell recognition during sea urchin gastrulation. *In* W. F. Loomis (ed.), *Genetic Regulation of Development*. Alan R. Liss, New York, pp. 111–128.

Monroy, A. and Moscona, A. A. 1979. *Introductory Concepts in Developmental Biology*. University of Chicago Press, Chicago.

Mooseker, M. S. and Tilney, L. G. 1975. Organization of an actin filament–membrane complex: Filament polarity and membrane attachment in the microvilli of intestinal epithelial cells. *J. Cell Biol.* 67: 725–743.

Moscona, A. A. 1952. Cell suspension from organ rudiments of chick embryos. *Exp. Cell Res.* 3: 535–539.

Moscona, A. A. 1961. Rotation-mediated histogenetic aggregation of dissociated cells: A quantifiable approach to cell interaction in vitro. *Exp. Cell Res.* 22: 455–475.

Nagafuchi, A. and Takeichi, M. 1989. Transmembrane control of cadherin-mediated cell adhesion: A 94-kDa protein functionally associated with a specific region of the cytoplasmic domain of E-cadherin. *Cell. Reg.* 1: 37–44.

Nagafuchi, A., Shirayoshi, Y., Okazaki, K., Yasuda, K. and Takeichi, M. 1987. Transformation of cell adhesion properties of exogenously introduced E-cadherin cDNA. *Nature* 329: 341–343.

Nardi, J. B. and Stocum, D. L. 1983. Surface properties of regenerating limb cells: Evidence for gradation along the proximodistal axis. *Differentiation* 25: 27–31.

Nicolson, G. L. 1976. Transmembrane control of the receptors on normal and tumor cells. *Biochim. Biophys. Acta* 457: 57–108.

Nose, A. and Takeichi, M. 1986. A novel cadherin adhesion molecule: Its expression patterns associated with implantation and organogenesis of mouse embryos. *J. Cell Biol.* 103: 2649–2658.

Nose, A., Nagafuchi, A. and Takeichi, M. 1988. Expressed recombinant cadherins mediate cell sorting in model systems. *Cell* 54: 993–1001.

Nuccitelli, R. 1984. The involvement of transcellular ion currents and electric fields in pattern formation. *In* G. M. Malacinski (ed.), *Pattern Formation: A Primer in Developmental Biology*. Macmillan, New York, pp. 23–46.

Nuccitelli, R. and Erickson, C. A. 1983. Embryonic cell motility can be guided by physiological electric fields. *Exp. Cell Res.* 147: 195–201.

Obara, M., Kang, M. S. and Yamada, K. M. 1988. Site-directed mutagenesis of the cell-binding domain of human fibronectin: Separable, synergistic sites mediate adhesive function. *Cell* 53: 649–657.

Orkin, R. W., Knudson, W. and Toole, P. T. 1985. Loss of hyaluronidate-dependent coat during myoblast fusion. *Dev. Biol.* 107: 527–530.

Palade, G. E. 1985. Differentiated microdomains in cellular membranes: Current status. *In* G. M. Edelman and J.-P. Thiery (eds.), *The Cell in Contact*. Wiley, New York, pp. 9–24.

Peracchia, C. and Dulhunty, A. F. 1976. Low resistance junctions in crayfish: Structural changes with functional uncoupling. *J. Cell Biol.* 70: 419–439.

Pierce, M., Turley, E. A. and Roth, S. 1980. Cell surface glycosyltransferase activities. *Int. Rev. Cytol.* 65: 1–47.

Pierschbacher, M. D. and Ruoslahti, E. 1984. The cell attachment activity of fibronectin can be duplicated by small synthetic fragments of the molecule. *Nature* 309: 30–33.

Podleski, T. R., Greenberg, I., Schlessinger, J. and Yamada, K. M. 1979. Fibronectin delays the fusion of L6 myoblasts. *Exp. Cell Res.* 122: 317–326.

Poole, T. J. and Steinberg, M. S. 1982. Evidence for the guidance of pronephric duct migration by a craniocaudally traveling adhesive gradient. *Dev. Biol.* 92: 144–158.

Radice, G. P. 1980. The spreading of epithelial cells during wound closure in *Xenopus laevis*. *Dev. Biol.* 76: 26–46.

Ratner, N., Bunge, R. P. and Glaser, L. 1985. A neuronal cell surface heparan sulfate proteoglycan is required for dorsal root ganglion neuron stimulation of Schwann cell proliferation. *J. Cell Biol.* 101: 744–754.

Rieger, F., Grument, M. and Edelman, G. M. 1985. N-CAM at the vertebrate neuromuscular junction. *J. Cell Biol.* 101: 285–293.

Rosavio, R. A., Delouvée, A., Yamada, K. M., Timpl, R. and Thiery, J.-P. 1983. Neural crest cell migration: Requirements for exogenous fibronectin and high cell density. *J. Cell Biol.* 96: 462–473.

Roseman, S. 1970. The synthesis of complex carbohydrates by multi-glycosyltransferase systems and their potential in intercellular adhesion. *Chem. Phys. Lipids* 5: 270–297.

Roth, S. 1968. Studies on intracellular adhesive selectivity. *Dev. Biol.* 18: 602–631.

Roth, S., McGuire, E. J. and Roseman, S. 1971. An assay for intercellular adhesive specificity. *J. Cell Biol.* 51: 525–535.

Runyan, R. B., Maxwell, G. D. and Shur, B. D. 1986. Evidence for a novel enzymatic mechanism of neural crest cell migration on extracellular glycoconjugate matrices. *J. Cell Biol.* 102: 432–441.

Ruoslahti, E. and Pierschbacher, M. D. 1987. New perspectives in cell adhesion: RGD and integrins. *Science* 238: 491–497.

Rutishauser, U. and Jessell, T. M. 1988. Cell adhesion molecules in vertebrate neural development. *Physiol. Rev.* 68: 819–857.

Rutishauser, U., Hoffman, S. and Edelman, G. M. 1982. Binding properties of a cell adhesion molecule from neural tissue. *Proc. Natl. Acad. Sci. USA* 79: 685–689.

Rutishauser, U., Acheson, A., Hall, A., Mann, D. M. and Sunshine, J. 1988. The neural cell adhesion molecule (N-CAM) as a regulator of cell-cell interactions. *Science* 240: 53–57.

San Antonio, J. D., Winston, B. M. and Tuan, R. S. 1987. Regulation of chondrogenesis by heparan sulfate and structurally related glycosaminoglycans. *Dev. Biol.* 123: 17–24.

Shur, B. D. 1977a. Cell surface glycosyltransferases in gastrulating chick embryos. I. Temporally and spatially specific patterns of four endogenous glycosyltransferase activities. *Dev. Biol.* 58: 23–29.

Shur, B. D. 1977b. Cell surface glycosyltransferases in gastrulating chick embryos. II. Biochemical evidence for a surface localization of endogenous glycosyltransferase activities. *Dev. Biol.* 58: 40–55.

Shur, B. D. 1982. Cell surface glycosyltransferase activities during normal and mutant (T/T) mesenchyme migration. *Dev. Biol.* 91: 149–162.

Shur, B. D., Oettgen, P. and Bennett, D. 1979. UDP galactose inhibits blastocyst formation in the mouse: Implications for the mode of action of T/t-complex mutations. *Dev. Biol.* 73: 178–181.

Singer, S. J. and Nicolson, G. L. 1972. The fluid mosaic model of the structure of cell membranes. *Science* 175: 720–731.

Solowska, J., Guan, J.-L., Marcantonio, E. E., Trevithick, J. E., Buck, C. A. and

Hynes, R. O. 1989. Expression of normal and mutant avian integrin subunits in rodent cells. *J. Cell Biol.* 109: 853–861.

Sorokin, L., Sonnenberg, A., Aumailley, M., Timpl, R. and Ekblom, P. 1990. Recognition of the laminin E8 cell binding site by an integrin possessing the α6 subunit is essential for epithelial polarization in developing kidney tubules. *J. Cell Biol.* 111: 1265–1273.

Spiegel, M. and Spiegel, E. S. 1975. Reaggregation of dissociated embryonic sea urchin cells. *Am. Zool.* 15: 583–606.

Spring, J., Beck, K. and Chiquet-Ehrismann, R. 1989. Two contrary functions of tenascin: Dissection of the active sites by recombinant tenascin fragments. *Cell* 59: 325–334.

Steinberg, M. S. 1964. The problem of adhesive selectivity in cellular interactions. *In* M. Locke (ed.), *Cellular Membranes in Development.* Academic Press, New York, pp. 321–434.

Steinberg, M. S. 1970. Does differential adhesion govern self-assembly processes in histogenesis? Equilibrium configurations and the emergence of a hierarchy among populations of embryonic cells. *J. Exp. Zool.* 173: 395–434.

Stickel, S. K. and Wang, Y. 1988. Synthetic peptide GRGDS induces dissociation of α-actinin and vinculin from the sites of focal contacts. *J. Cell Biol.* 107: 1231–1239.

Stopak, D. and Harris, A. K. 1982. Connective tissue morphogenesis by fibroblast traction. I. Tissue culture observations. *Dev. Biol.* 90: 383–398.

Takeichi, M. 1987. Cadherins: A molecular family essential for selective cell–cell adhesion and animal morphogenesis. *Trends Genet.* 3: 213–217.

Takeichi, M., Ozaki, H. S., Tokunaga, K. and Okada, T. S. 1979. Experimental manipulation of cell surface to affect cellular recognition mechanisms. *Dev. Biol.* 70: 195–205.

Tamkun, J. W., DeSimone, D. W., Fonda, D., Patel, R. S., Buck, C, Horwitz, A. F. and Hynes, R. O. 1986. Structure of integrin, a glycoprotein involved in transmembrane linkage between fibronectin and actin. *Cell* 46: 271–282.

Tan, S.-S., Crossin, K. L., Hoffman, S. and Edelman, G. M. 1987. Asymmetric expression of somites of cytotactin and its proteoglycan ligand is correlated with neural crest cell migration. *Proc. Natl. Acad. Sci. USA* 84: 7977–7981.

Thesleff, I., Vainio, S. and Jalkanen, M. 1989. Cell-matrix interactions in tooth development. *Int. J. Dev. Biol.* 33: 91–95.

Thiery, J.-P., Brackenbury, R., Rutishauser, U. and Edelman, G. M. 1977. Adhesion among neural cells of the chick embryo. II. Purification and characterization of a cell adhesion molecule from neural retina. *J. Biol. Chem.* 252: 6841–6845.

Thiery, J.-P., Delouvée, A., Gallin, W. J., Cunningham, B. A. and Edelman, G. M.

1985. Initial appearance and regional distribution of the neuron–glia cell adhesion molecule of the chick embryo. *J. Cell Biol.* 100: 442–456.

Toole, B. P. 1976. Morphogenetic role of glycosaminoglycans (acid mucopolysaccharides) in brain and other tissues. *In* S. H. Barondes (ed.), *Neuronal Recognition.* Plenum, New York, pp. 276–329.

Tosney, K. W., Watanabe, M., Landmesser, L. and Rutishauser, U. 1986. The distribution of N-CAM in the chick hindlimb during axon outgrowth and synaptogenesis. *Dev. Biol.* 114: 468–481.

Townes, P. L. and Holtfreter, J. 1955. Directed movements and selective adhesion of embryonic amphibian cells. *J. Exp. Zool.* 128: 53–120.

Trinkaus, J. P. 1963. The cellular basis of *Fundulus* epiboly. Adhesivity of blastula and gastrula cells in culture. *Dev. Biol.* 7: 513–532.

Trinkaus, J. P. 1985. Further thoughts on directional cell movement during morphogenesis. *J. Neurosci. Res.* 13: 1–19.

Turley, E. A. and Roth, S. 1979. Spontaneous glycosylation of glycosaminoglycan substrates by adherent fibroblasts. *Cell* 17: 109–115.

Tyler, A. 1946. An auto-antibody concept of cell structure, growth, and differentiation. *Growth* 10 (Symposium 6): 7–19.

Vainio, S., Jalkanen, M., Lehtonen, E. and Bernfeld, M. 1989. Epithelial-mesenchymal interactions regulate stage-specific expression of a cell surface proteoglycan, syndecan, in the development kidney. *Dev. Biol.* 134: 382–391.

Venkatesh, T. R., Zipursky, S. L. and Benzer, S. 1985. Molecular analysis of the development of the compound eye in *Drosophila. Trends Neurosci.* 8: 251–257.

Vuorio, E. 1986. Connective tissue diseases: Mutations of collagen genes. *Ann. Clin. Res.* 18: 234–241.

Warner, A. E., Guthrie, S. C. and Gilula, N. B. 1984. Antibodies to gap junctional protein selectively disrupt junctional communication in the early amphibian embryo. *Nature* 311: 127–131.

Wehrle, B. and Chiquet, M. 1990. Tenascin is accumulated along developing peripheral nerves and allows neurite outgrowth *in vitro. Development* 110: 401–415.

Weiss, P. 1934. In vitro experiments on the factors determining the course of the outgrowing nerve fiber. *J. Exp. Zool.* 68: 393–448.

Weiss, P. 1945. Experiments on cell and axon orientation *in vitro*: The role of colloidal exudates in tissue organization. *J. Exp. Zool.* 100: 353–386.

Weiss, P. 1947. The problem of specificity in growth and development. *Yale J. Biol. Med.* 19: 235–278.

Weiss, P. 1955. Nervous system. *In* B. H. Willier, P. Weiss and V. Hamburger (eds.), *Analysis of Development.* Saunders, Philadelphia, pp. 346–401.

Werb, Z., Tremble, P. and Damsky, C. H. 1990. Regulation of extracellular matrix degradation by cell–extracellular matrix interaction. *Cell Differ. Dev.* [Suppl.]. In press.

Wessel, G. M., Tomlinson, C. R., Lennarz, W, J and Klein, W. H. 1989. Transcription of the *spec1*-like gene of *Lytechinus* is selectively inhibited in response to disruption of the extracellular matrix. *Devel.* 106: 355–365.

Wilcox, M., DiAntonio, A. and Leptin, M. 1989. The functions of the PS integrins in Drosophila wing morphogenesis. *Development* 107: 891–897.

Williams, A. F. and Barclay, A. N. 1988. The immunoglobulin superfamily: Domains for cell surface recognition. *Annu. Rev. Immunol.* 6: 381–405.

Wood, A. and Thorogood, P. 1987. An ultrastructural and morphometric analysis of an *in vivo* contact guidance system. *Development* 101: 363–381.

Woodruff, R. I. and Telfer, W. H. 1974. Electrical properties of ovarian cells linked by intercellular bridges. *N.Y. Acad. Sci.* 238: 408–419.

Yamada, K. M. and Kennedy, D. W. 1985. Amino acid sequence specificities of an adhesive recognition signal. *J. Cell Biochem.* 28: 99–104.

Yamada, K. M., Critchley, D. R., Fishman, P. H. and Moss, J. 1983. Exogenous gangliosides enchance the interaction of fibronectin with ganglioside-deficient cells. *Exp. Cell Res.* 143: 295–302.

Yamada, K. M., Humphries, M. J., Hasegawa, T., Hasegawa, E., Olden, K., Chen, W.-T. and Akiyama, S. K. 1985. Fibronectin: Molecular approaches to analyzing cell interactions with the extracellular matrix. *In* G. M. Edelman and J.-P. Thiery (eds.), *The Cell in Contact*. Wiley, New York, pp. 303–332.

Yelton, D. E. and Scharff, M. D. 1980. Monoclonal antibodies. *Am. Sci.* 68: 510–516.

Yow, H., Wong, J. M., Chen, H. S., Lee, C., Steel, G. D. Jr. and Chen, L. B. 1988. Increased mRNA expression of a laminin-binding protein in human colon carcinoma: Complete sequence of a full-length cDNA encoding the protein. *Proc. Natl. Acad. Sci. USA* 85: 6394–6398.

Zackson, S. L. and Steinberg, M. S. 1986. Cranial neural crest cells exhibit directed migration on the pronephric duct pathway: Further evidence for an *in vivo* adhesion gradient. *Dev. Biol.* 117: 342–353.

Zackson, S. L. and Steinberg, M. S. 1987. Chemotaxis or adhesion gradient? Pronephric duct elongation does not depend on distant sources of guidance information. *Dev. Biol.* 124: 418–422.

Zackson, S. L. and Steinberg, M. S. 1988. A molecular marker for cell guidance information in the axolotl embryo. *Dev. Biol.* 127: 435–442.

Zackson, S. L. and Steinberg, M. S. 1989. Axolotl pronephric duct cell migration is sensitive to phosphatidylinositol-specific phospholipase C. *Development* 105: 1–7.

Zigmond, S. H. 1978. Chemotaxis by polymorphonuclear leukocytes. *J. Cell Biol.* 77: 269–287.

Ziomek, C. A., Schulman, S. and Edidin, M. 1980. Redistribution of membrane proteins in isolated mouse intestinal epithelial cells. *J. Cell Biol.* 86: 849–857.

Zipursky, S. L., Venkatesh, T. R., Teplow, D. B. and Benzer, S. 1984. Neuronal development in the *Drosophila* retina: Monoclonal antibodies as molecular probes. *Cell* 36: 15–26.

16

Proximate tissue interactions: Secondary induction

In dealing with such a complex system as the developing embryo, it is futile to inquire whether a certain organ rudiment is "determined" and whether some feature of its surroundings, to the exclusion of others, "determines" it. A score of different factors may be involved and their effects most intricately interwoven. In order to resolve this tangle we have to inquire into the manner in which the system under consideration reacts with other parts of the embryo at successive stages of development and under as great a variety of experimental conditions as it is possible to impose.

—R. G. HARRISON (1933)

The aspiration to truth is more precious than its assured possession.

—G. E. LESSING (1778)

In the previous chapter, we saw that the cell surface plays an important role in providing spatial information to adjacent cells. In this chapter and the next, we shall discuss the ways in which this information is used by cells to create tissues and organs. In particular, this chapter concerns the events wherein sets of cells influence the behavior of other nearby cell populations during organogenesis. Organs are complex structures composed of numerous types of tissues. If one considers an organ such as the vertebrate eye, for example, one finds that light is transmitted through the transparent corneal tissue; it is focused by the lens tissue, the diameter of which is controlled by muscle tissue; and it eventually impinges upon the tissue of the neural retina. The precise arrangement of tissues in this organ cannot be disturbed without damaging its function. Such coordination in the construction of organs is accomplished by one group of cells changing the behavior of an adjacent set of cells, thereby causing them to change their shape, mitotic rate, or differentiation. This action at close range, or PROXIMATE INTERACTION, enables one group of cells to respond to a second group of cells, and, in changing, often to become able to alter a third set of cells. This phenomenon has been called SECONDARY INDUCTION.

Instructive and permissive interactions

Howard Holtzer (1968) has distinguished two major modes of proximate tissue interactions. One of them is PERMISSIVE INTERACTION. Here, the responding tissue contains all the potentials needed to be expressed, and it only has to have an environment that permits expression of these traits. Many developing tissues need a solid substrate containing fibronectin or laminin in order to develop. The fibronectin or laminin does not alter the type of cell that is to be produced, but only enables its expression.*

The other variety of proximate tissue interaction is INSTRUCTIVE INTERACTION. This kind of interaction changes the cell type of the responding cell. For example, in Chapter 8 we discussed the ability of the notochord to induce the formation of floor plate cells in the neural tube. All neural tube cells were capable of responding to the notochord signal, but only those closest to the notochord were induced. The other cells became non-floor plate cells. Moreover, if one removed the notochord from the embryo, those cells that would normally have become floor plate cells would not differentiate into that type of cell, and if one added an additional notochord laterally to the neural plate, this new notochord would induce a secondary set of floor plate cells. The responding neural tube cells had somehow been told to express a set of genes different from the set they would have expressed had they not been in contact with the notochord. The notochord is said to be an inducing tissue acting *instructively*. Wessells (1977) has proposed four general principles characteristic of most instructive interactions.

1. In the presence of tissue A, responding tissue B develops in a certain way.
2. In the absence of tissue A, responding tissue B does not develop in that way.
3. In the absence of tissue A, but in the presence of tissue C, tissue B does not develop in that way.
4. In the presence of tissue A, a tissue D, which would normally develop differently, is changed to develop like B.

Competence

It should be noted that in the fourth principle, the responding tissue must be competent to respond. COMPETENCE is the ability to respond to an inductive signal (Waddington, 1940). It is not a passive state, but an actively acquired condition. When we detailed the induction of the neural tube in Chapter 8, we discussed the observations that the gastrula ectoderm is capable of being induced by the dorsal blastopore lip or its mesodermal derivatives. Thus, the gastrula ectoderm is said to be *competent* to respond to the inductive stimuli. This competence for neural induction is acquired during late cleavage stages and it is lost during the late gastrula stages. As this competence to respond to dorsal lip induction diminishes, the same ectoderm gains the competence to respond to lens inducers. Still later, the competence to lens inducers is lost, but the ectoderm can respond

---

*It is easy to distinguish permissive and instructive relationships by an analogy with a more familiar situation. This textbook is made possible by permissive and instructive interactions. The reviewers can convince me to change the material in the chapter. This is an *instructive interaction*, as the information is changed from what it would have been. However, the information in the book could not be expressed without *permissive interactions* by the publisher and printer.

to ear placode inducers (Servetnick and Grainger, 1990). Therefore, competence is, itself, a differentiated phenotype that distinguishes cells spatially and temporally.

Although the mechanisms that enable a cell to become competent are not known, it is generally thought that competence can be attained in several ways. First, a cell can become competent by synthesizing a receptor for the inducer molecule. As we will see in Chapter 19, the ability to respond to a hormone such as thyroxine depends upon whether the cell contains that hormone's receptor. Similarly, competence to respond to an inducer depends upon the cell's having a receptor for that inducer. Second, a cell might achieve competence by synthesizing a molecule that allows the receptor to function. Receptors alone may bind the inducer, but that does not mean that they are functional. Often, a receptor acts by sending a signal to the nucleus. As we have seen in Chapter 2, when the sperm receptor of the egg binds its sperm ligand, it initiates a cascade of such signals. If one of the enzymes involved in this cascade is not present, the signal is not transmitted. So a cell might achieve competence by synthesizing a missing link in the signaling pathway.

Third, competence may be acquired by the repression of an inhibitor. If the inhibitor is present, a cell might bind the inducer, send the signal to the nucleus, and still not be able to be become induced. For example, inducers often cause cell shape changes (as in the induction of the neural tube). If the cell were inhibited from changing its shape, it would not be able to respond. This may be the case in the induction of ectodermal structures such as the neural tube, lens, ear, and feathers in the chick embryo. A particular cytoskeletal protein exists in the epiblast of the chick embryo. However, as neurulation begins, this protein is lost in the head and in the presumptive neural plate region, even though it is retained by the presumptive epidermal cells (Charlebois et al., 1990; Grunwald et al., 1990; Figure 1). The epidermal cells lose this protein immediately prior to their forming feathers. Thus, the loss of this cytoskeletal protein correlates with the cell's ability to change shape upon induction.

FIGURE 1
Loss of cytoskeletal protein from those ectodermal cells undergoing shape changes during chick neurulation. (A) Cross section of the epiblast and hypoblast of a 1-day chick embryo stained with fluorescent antibody to a specific cytokeratin protein of the cells. (B) Loss of this protein from those cells that are forming the neural plate. (From Grunwald et al., 1990; photographs courtesy of the authors.)

Epithelio-mesenchymal interactions

Some of the best studied cases of secondary induction are those involving the interactions of epithelial sheets with adjacent mesenchymal cells. These are called EPITHELIO-MESENCHYMAL INTERACTIONS. The epithelium can come from any germ layer whereas the mesenchyme is usually derived from loose mesodermal or neural crest tissue. Examples of epithelio-mesenchymal interactions are listed in Table 1.

(A)

(B)

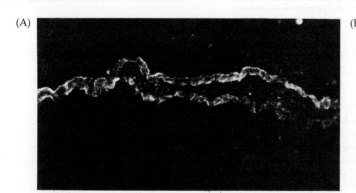
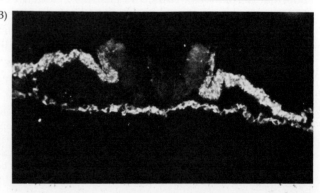

TABLE 1
Some epithelio-mesenchymal interactions

Organ	Epithelial component	Mesenchymal component
Cutaneous structures (hair, feathers, sweat glands, mammary glands)	Epidermis (ectoderm)	Dermis (mesoderm)
Limb	Epidermis (ectoderm)	Mesenchyme (mesoderm)
Gut organs (liver, pancreas, salivary glands)	Epithelium (endoderm)	Mesenchyme (mesoderm)
Pharyngeal and respiratory associated organs (lungs, thymus, thyroid)	Epithelium (endoderm)	Mesenchyme (mesoderm)
Kidney	Ureteric bud epithelium (mesoderm)	Mesenchyme (mesoderm)
Tooth	Jaw epithelium (ectoderm)	Mesenchyme (neural crest)

Regional specificity of induction

Using the induction of cutaneous structures as our examples, we will look at the properties of epithelio-mesenchymal interactions. The first phenomenon is that of the regional specificity of induction. Skin is composed of two main tissues: an outer epidermis derived from ectoderm, and a dermis derived from mesoderm. Chicken skin gives rise to three major cutaneous structures that are made almost entirely of ectodermal cells. These are the broad wing feathers, the narrow thigh feathers, and the scales and claws of the feet. After separating the embryonic epithelium and mesenchyme from each other, one can recombine them in different ways (Saunders et al., 1957). Some of the recombinations are illustrated in Figure 2. As you can see, the mesenchyme is responsible for the specificity of induction in the competent ectoderm. This same type of ectoderm develops according to the region from which the mesoderm was taken. Here, the mesenchyme has an instructive role, calling into play different sets of genes in the responding cells.

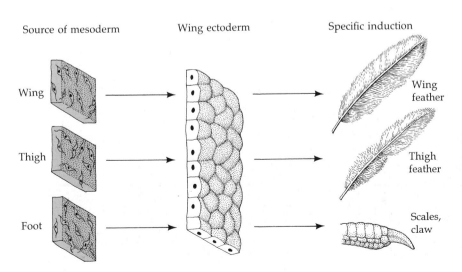

FIGURE 2
Regional specificity of induction. When cells of the dermis (mesoderm) are recombined with the epidermis (ectoderm) in the chick, the type of cutaneous structure made by the ectoderm is determined by the original location of the mesoderm. (Adapted from Saunders, 1980.)

This regional specificity of induction is critical during the development of the digestive system and the respiratory system. In the morphogenesis of the endodermal tubes, the endodermal epithelium is able to respond differently to different regionally specific mesenchymes. This enables the digestive tube and respiratory tube to develop different structures in different regions of the tube (Figure 3). Thus, as the digestive tube meets new mesenchymes, it differentiates into esophagus, stomach, small intestine, and colon (Gumpel-Pinot, et al., 1978; Fukumachi and Takayama, 1980.) This regional specificity of mesenchymal induction is dramatically apparent in the formation of the respiratory system. The respiratory epithelium is not as finicky as kidney and salivary epithelia and it responds to numerous mesenchymes. In the developing mammal, it responds in two distinct fashions. When in the region of the neck, it grows straight, forming the trachea. After entering the thorax, it branches, forming two

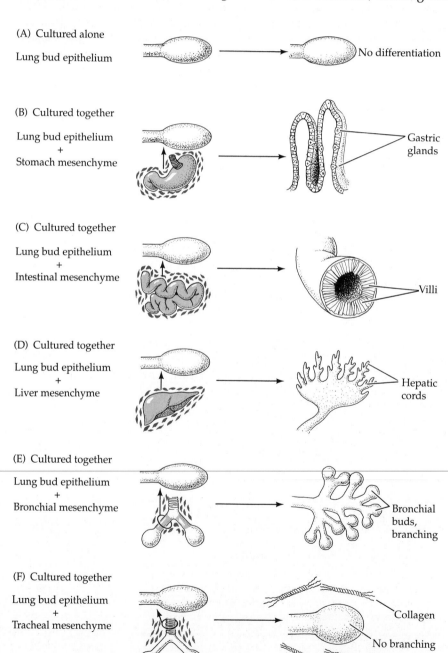

(A) Cultured alone

Lung bud epithelium

No differentiation

(B) Cultured together

Lung bud epithelium
+
Stomach mesenchyme

Gastric glands

(C) Cultured together

Lung bud epithelium
+
Intestinal mesenchyme

Villi

(D) Cultured together

Lung bud epithelium
+
Liver mesenchyme

Hepatic cords

(E) Cultured together

Lung bud epithelium
+
Bronchial mesenchyme

Bronchial buds, branching

(F) Cultured together

Lung bud epithelium
+
Tracheal mesenchyme

Collagen

No branching

FIGURE 3
Ability of presumptive lung epithelium to differentiate with respect to the source of the mesenchyme. (A) Lung epithelium does not differentiate when cultured in the absence of mesenchymal cells. (B–F) Mesenchyme-specific differentiation of epithelium. (Modified from Deuchar, 1975.)

bronchi and then the lungs. The respiratory epithelium can be isolated soon after it has split into two bronchi, and the two sides can be treated differently. Figure 4 shows the result of such an experiment. The right bronchial epithelium retained its lung mesenchyme, whereas the left bronchus was surrounded by tracheal mesenchyme (Wessells, 1970). The right bronchus proliferated and branched under the influence of the lung mesenchyme whereas the left side continued to grow in an unbranched manner. Thus, respiratory epithelium is extremely malleable and can differentiate into many other endodermal structures.

Genetic specificity of induction

Whereas the mesenchyme may instruct the epithelium as to what sets of genes to activate, the responding epithelium can comply with this information only so far as its genome permits. In a classic experiment, Hans Spemann and Oscar Schotté (1932) transplanted flank ectoderm from an early frog gastrula to the region of a newt gastrula destined to become parts of the mouth. Similarly, the presumptive flank ectodermal tissue of newt gastrula was placed into the presumptive oral regions of frog embryos. The structures of the mouth region differ greatly between these salamander larvae and the frog larvae. The *Triturus* salamander larva has club-shaped balancers beneath its mouth, whereas the frog tadpoles produce mucus-secreting glands or suckers (Figure 5). The frog tadpoles also have a horny jaw without teeth, whereas the salamander has a set of calcareous teeth in its jaw. The larvae resulting from the transplants were

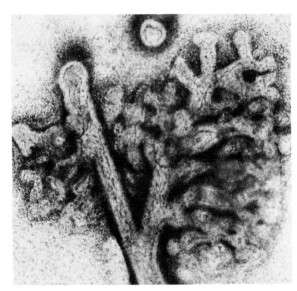

FIGURE 4
Ability of presumptive lung epithelium to differentiate with respect to the source of the inducing mesenchyme. After embryonic mouse lung epithelium has branched into two bronchi, the entire rudiment is excised and cultured. The right bronchus is left untouched, while the tip of the left bronchus is covered with tracheal mesenchyme. The tip of the right bronchus forms the branches characteristic of the lung, whereas no branching occurs in the tip of the left bronchus. (From Wessells, 1970; courtesy of N. Wessells.)

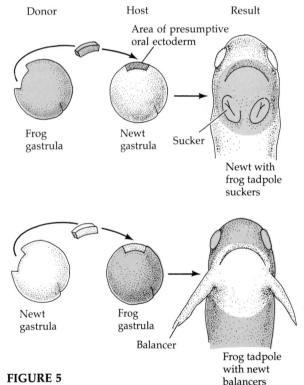

FIGURE 5
Genetic specificity of induction. Reciprocal transplantation between the presumptive oral ectoderm regions of newt and frog gastrulae leads to newt larvae with tadpole suckers and to frog tadpoles with newt balancers. (After Hamburgh, 1970.)

FIGURE 6
Genetic specificity of cutaneous induction. (A) Section of the corneal region of a 17-day chick embryo. At 5 days of incubation, the lens of this eye had been replaced by the flank dermis of an early mouse embryo. A condensation of the mouse embryo cells is located directly beneath the chick epithelium. (B) Feather forming from the corneal epithelium from such a specimen. Mouse cells are present in the feather rudiment. (From Coulombre and Coulombre, 1971; courtesy of A. J. Coulombre.)

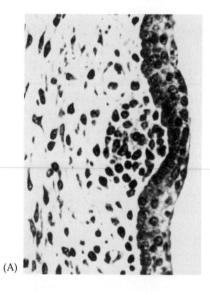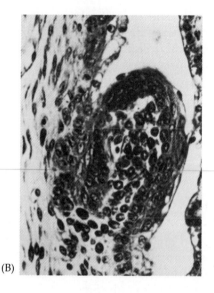

(A) (B)

chimeras. The salamander larvae had froglike mouths and the frog tadpoles had salamander teeth and balancers. In other words, the mesodermal cells instructed the ectoderm to make a mouth, but the ectoderm responded by making the only mouth it "knew" how to make, no matter how inappropriate.*

The same genetic specificity is seen in combinations of chicken skin and mouse skin (Coulombre and Coulombre, 1971). When ectoderm normally destined to become cornea is isolated from chicken embryos and combined with chick skin mesoderm, the ectoderm produces feather buds typical of chick skin. Moreover, when the same tissue—presumptive cornea ectoderm—is combined with *mouse* skin mesoderm, feather buds also appear (Figure 6). The mouse mesoderm has instructed the chick cornea to make a cutaneous structure. This would normally be hair, for the mouse. The competent chick ectoderm, however, does the best it can, developing its cutaneous structures—namely, feathers.

Thus, the instructions sent by the mesenchymal tissue can cross species barriers. Salamanders respond to frog signals, and chick tissue responds to mammalian inducers. The response of the epithelium, however, is species-specific for that epithelium. So, whereas organ-type specificity (feather or claw) is usually controlled by the mesenchyme within a species, species specificity is usually controlled by the responding epithelium.

This observation has more recently been used to generate a long-lost structure: hen's teeth (Kollar and Fisher, 1980). Epithelium from the jaw-forming region of 5-day-old chick embryos (the first and second pharyngeal arches) was isolated and combined with molar mesenchyme of 16- to 18-day-old mouse embryos. These recombined tissues were allowed to adhere to each other and were then cultured within the anterior chamber of a mouse eye. Several recombinations resulted in the formation of complete teeth that were unlike those of mammals (Figure 7). The cells of the

*Spemann is reported to have put it as follows: "The ectoderm says to the inducer, 'you tell me to make a mouth; all right, I'll do so, but I can't make your kind of mouth; I can make my own and I'll do that.'" (Quoted in Harrison, 1933.)

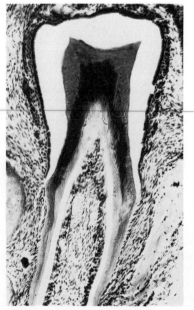

FIGURE 7
"Hen's tooth" formed after combination of chick pharyngeal (presumptive jaw) epithelium and mouse molar mesenchyme. (From Kollar and Fisher, 1980; courtesy of E. J. Kollar.)

chick pharyngeal arches, which have not made teeth for nearly 100 million years, still retained the genetic potential to do so in response to an appropriate inducer.*

Cascades of intercellular interactions: Lens induction

The phenomena of lens induction

Proximate cell interactions provide a mechanism whereby coordinated organ development can occur, for a responding tissue can also become an inducing tissue. Recent studies have shown that secondary induction is a very complex process. Indeed, what we have traditionally been calling "secondary" inductions are usually only the last induction in a cascade that began much earlier in embryogenesis, and many tissues acquire their competence through a previous induction. Although these tissues may look unchanged through a microscope, they have been induced such that they can respond to a new inducer. This is probably true of the epidermal inductions mentioned above, and it is certainly true in the case of that most intensively studied secondary induction, the formation of the lens.

As discussed in Chapter 5, the cells that form the lens are derived from a region of head ectoderm that is contacted by the optic vesicle of the forebrain. This work was pioneered by Hans Spemann, and his review of these studies in 1938 has made lens induction the paradigm of secondary inductive events. The basic experiments were as follows: First, when Spemann (1901) destroyed the optic vesicle primordium of the frog *Rana temporaria*, lenses failed to develop. Thus, Spemann concluded that contact of the optic vesicle with the overlying ectoderm was essential for inducing the formation of the lens. Second, Warren Lewis (1904; 1907) confirmed and extended this conclusion. He removed optic vesicles from late stage neurulae and transplanted them to the head ectoderm of regions that would not usually form lenses. He found that the head ectoderm of this region would then form lens-like structures, and he concluded that the optic vesicle was sufficient to induce the formation of lens tissue in ectoderm that would not have otherwise formed it. It appeared that the optic vesicle was all that was needed to induce lenses in the overlying ectoderm. This model is summarized in Figure 8, and this is the way that lens induction has traditionally been depicted.

However, there were dissenters from this view. Mencl (1908) noticed that certain fish had congenital defects wherein no eye formed. Nevertheless, these fish had lenses in their head ectoderm. More importantly, when King (1905) tried to repeat Spemann's experiments, she found, contrary to her expectations, that lenses still formed even when the optic vesicle rudiments had been obliterated. These and other investigators began to find that lenses could form without contact with optic vesicle.[†]

As more data began accumulating in various species, it seemed that

*This may not be the case for all birds, however. When quail mandibular epithelium is recombined with mouse molar mesenchyme, teeth do not form. Such experiments must be done with remarkable care, because very small numbers of mesenchyme cells can instruct tooth morphogenesis in the epithelium (Arechaga et al., 1983). The importance of induction and competence during evolution will be detailed in Chapter 23.

[†]The interpretation of these experiments has been extremely difficult due to species differences in the induction mechanisms, the temperatures at which maximal induction takes place, and the difficulty in getting uncontaminated pieces of tissue for transplantation. See Jacobson and Sater (1988) and Saha (1989, 1991) for reviews of these dissenters and their experiments.

FIGURE 8
Traditional use of optic vesicle induction of the lens to illustrate instructive interaction. The interaction illustrated is the induction of the lens from head ectoderm by the optic vesicle. (1) Normal induction. (2) In the absence of the optic vesicle, no lens forms. (3) Substitution of another tissue for the optic vesicle gives no induction. (4) Placement of the optic vesicle adjacent to another region of head ectoderm induces lens in a different place. (Modified from Wessells, 1977.)

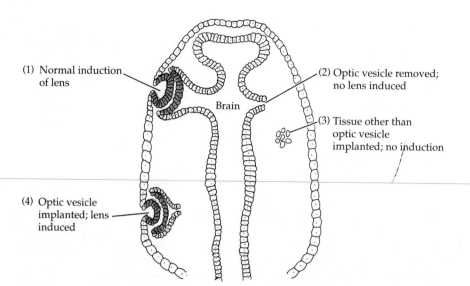

(1) Normal induction of lens

(2) Optic vesicle removed; no lens induced

(3) Tissue other than optic vesicle implanted; no induction

Brain

(4) Optic vesicle implanted; lens induced

there was a great deal of species diversity. Some species appeared to form lenses without the need for optic vesicles, while the lenses of other species appeared to be completely dependent on the optic vesicle contact. Spemann (1938) reconciled these results by arguing that an organism could evolve a margin of safety by developing two ways of forming a particular tissue. Thus, the lens would normally arise by contact with the optic vesicle, but, failing this, could arise separately if it had to. This concept was called the "double assurance" hypothesis. In 1966, Jacobson integrated more data into this model. He noted that the lens-forming ectoderm sequentially comes into contact with presumptive foregut endoderm, presumptive heart mesoderm, and the optic vesicle. Each of these tissues, he predicted, would act in an additive way to induce the formation of lenses in this tissue. In some species, the threshold for lens induction would be low, and contact with the endoderm would be sufficient. In other species, the threshold would be high, and all three inducers would have to be active. Here, lens formation would seem to be dependent upon the optic vesicle, but in actuality, the optic vesicle would be only the last of three inducers (Figure 9).

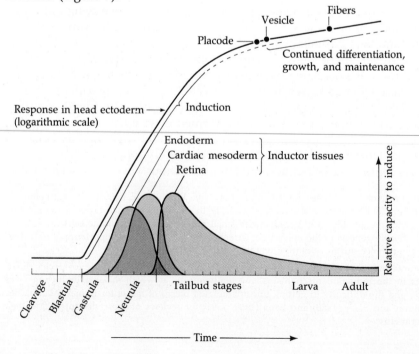

Fibers

Vesicle

Placode

Continued differentiation, growth, and maintenance

Response in head ectoderm → (logarithmic scale)

Induction

Endoderm
Cardiac mesoderm } Inductor tissues
Retina

Relative capacity to induce

Cleavage · Blastula · Gastrula · Neurula · Tailbud stages · Larva · Adult

Time

FIGURE 9
Representation of lens induction in an embryonic salamander to show additive effects of possible inducers. The ability to produce lens tissue is first depicted as coming from the endoderm, then from cardiac mesoderm, and finally from retinal tissue. The competency of the head ectoderm to respond to inducers increases logarithmically from early gastrula through the tailbud larval stages. (Modified from Jacobson, 1966.)

While not ruling out the role of the mesoderm and endoderm, recent studies in *Xenopus* stress the importance of the anterior neural plate as an early inducer of lens ectoderm. These experiments indicate that the presumptive lens ectoderm receives its ability to become lens very early in development (during late-gastrula to mid-neurula stages) and that the optic vesicle merely localizes the differentiation of this already autonomous tissue. In other words, the head ectoderm will form lenses without the contact of the optic cup, but the optic cup is needed for the full differentiation of the lens and its proper positioning with respect to the rest of the eye. This model (Figure 10; Saha et al., 1989; Henry and Grainger, 1990) divides the determination of lens ectoderm into three stages: competence, bias, and final determination. Competence to respond to the initial inducing signal is seen to be an autonomous process within the ectoderm, bias to produce lenses is provided by the anterior neural plate, and final determination is induced by the optic vesicle.

Ectodermal competence and bias

In 1987, Henry and Grainger demonstrated that the determination of lens-forming ability occurs very early in *Xenopus* development. They transplanted ectoderm from various regions of *Xenopus* gastrulae into the lens-forming region of the neurula. Would that ectoderm be able to form a lens when contacted by the optic vesicle? Ectoderm from very young gastrulae did not form lenses. When ectoderm from the same prospective lens ectoderm in later embryos was grafted to the neurula, it was able to respond to the optic vesicle by forming a lens (Table 2). No other tissue responded in this way. Other ectodermal regions from gastrulae also had a limited ability to form lenses, but this ability was lost as development proceeded.

It appeared, then, that the lens-forming ectoderm achieved the ability to form lenses long before the optic vesicle contacted it. When was this lens-forming state achieved, and how was it effected? Experiments by Nieuwkoop (1952) had suggested that a signal from the neural plate could travel through the ectoderm. Could the neural plate induce the presumptive epidermis lateral to it to become lens-forming ectoderm? Henry and Grainger (1990) tested this hypothesis by combining the prospective anterior neural plate region of late gastrula embryos with ectoderm from the region that would eventually become lens. While isolated ectoderm from the potential lens-forming region did not form lens proteins when cultured alone, the same region did form lens proteins when cultured next to the prospective anterior neural plate tissue. Although lens differentiation was often rudimentary, it was very specific. Lens proteins were not made when the gastrula ectoderm was combined with other tissues, including foregut endoderm or cardiac mesoderm. These experiments show that the anterior portion of the prospective neural plate (which contains the future retinal regions) provides a signal that biases this tissue to become lens.

FIGURE 10

A current model for lens induction. (A) During midgastrula, the lens-forming ectoderm becomes competent to respond to lens-inducing signal. (B) During late gastrula, a signal from neuralized ectoderm (most likely the presumptive eye region) induces the presumptive lens-forming ectoderm. (C) At early neurula, this signal is reinforced by induction signal from the anterior lateral mesoderm. (D) At the late neurula stage, the optic vesicle contacts the lens-forming ectoderm, signalling the differentiation of this tissue into lens. (After Saha et al., 1989; Henry and Grainger, 1990.)

(A) Mid gastrula stage

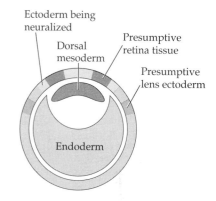

(B) Late gastrula stage

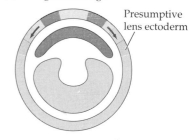

(C) Early neurula stage

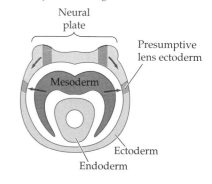

(D) Late neurula stage

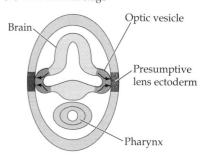

TABLE 2
Increasing responsiveness of prospective lens ectoderm with age

Donor stage	Operation Donor	Host neurula	Number examined	Induced lens (%)	Lens-like body (%)	Ectodermal thickening (%)	Non-lens body (%)	No response (%)	Total positive
Midgastrula			24	0	4	38	8	50	1 (4%)
Late gastrula			21	10	14	42	10	24	5 (24%)
Early neurula			24	75	8	0	4	13	20 (83%)
Late neurula			20	95	5	0	0	0	20 (100%)

Source: Henry and Grainger (1987).

But are all tissues able to respond to the signal coming from the anterior neural plate? Servetnick and Grainger (1990) showed that only mid-late gastrula ectoderm is competent to respond to these signals. They removed animal cap ectoderm from various gastrula stages and then transplanted them into the presumptive lens region of neural plate-stage embryos. Ectoderm from early gastrulae showed little or no competence to form lenses (as assayed by the production of crystallin proteins), but ectoderm from slightly later stages were able to form lenses. By the end of gastrulation, this ability to respond to the neural plate signal had been lost. This competence was seen to be inherent within the ectoderm itself, and was not induced by other surrounding tissues. The animal cap ectoderm from various embryonic stages could be removed, cultured in glass for a certain period of time, and then placed back into neural plate-stage embryos. Such ectoderm showed the same pattern of competence even though it had spent part of its development inside a petri dish. It appears then, that the ectoderm acquires the competence to respond to inducing signals from the anterior neural plate at the early mid-gastrula stages, and during late gastrula, the anterior neural plate induces a lens-forming bias in this tissue. Similar experiments have shown that dorsolateral mesoderm at this stage, while it will not induce the lens-forming bias in the ectoderm, will enhance the lens-forming bias imparted by the neural plate.

Final determination of the lens

Although the optic vesicle does not appear to play a major role in the initial induction of lens formation in *Xenopus*, it does play a role in enabling the complete lens phenotype to become expressed. The lenses that form in the absence of the optic vesicle are usually very rudimentary. It is not known whether the influence of the optic vesicle is a directly positive one, promoting the differentiation of the lens placode into a fully differentiated lens, or if the influence is to remove an inhibitor of lens differentiation. It has been proposed (von Woellwarth, 1961; Henry and Grainger, 1987) that the cranial neural crest cells prevent lens differentiation, and that contact with the optic vesicle shields the lens placode from these inhibiting signals.

Cornea formation

Once the lens placode has invaginated, it becomes covered by two cell layers from the adjacent ectoderm. Now the developing lens can act as an inducer. The ectoderm that is to become the cornea has probably also been determined during an earlier stage of development (Meier, 1977). Now, under the influence of the lens, corneal differentiation takes place. The overlying ectoderm becomes columnar and fills with secretory granules. These granules migrate to the bases of the cells and secrete a PRIMARY STROMA containing about 20 layers of types I and II collagen (Figure 11). Neighboring capillary endothelial cells migrate into this region (upon the primary stroma) and secrete hyaluronic acid into this matrix. The hyaluronic acid causes the matrix to swell and to become a good substrate for the migration of two waves of mesenchymal cells derived from the neural crest. Upon entering the matrix, the second wave of mesenchymal cells remains there, secreting type I collagen and hyaluronidase. The hyaluronidase causes the stroma to shrink. Under the influence of thyroxine from the developing thyroid gland, this SECONDARY STROMA is dehydrated and the collagen-rich matrix of epithelial and mesenchymal tissues becomes the transparent CORNEA.

We can see, then, that "simple" inductive interactions are actually well

FIGURE 11
Corneal development. Under the inductive influence of the lens, the corneal epithelium differentiates and secretes a primary stroma consisting of collagen layers. Endothelial cells then secrete hyaluronic acid into this region, enabling mesenchymal cells from the neural crest to enter. Afterwards, hyaluronidase (secreted either by the mesenchyme or endothelium) digests the hyaluronic acid, causing the primary stroma to shrink. (After Hay and Revel, 1969.)

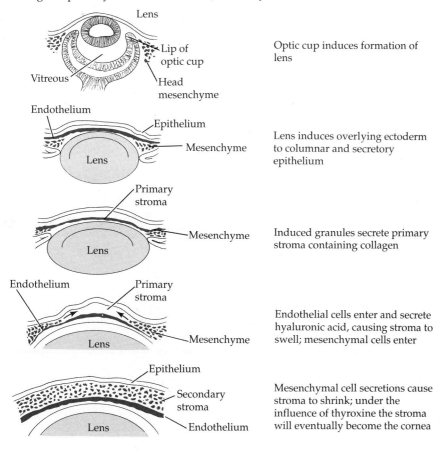

Optic cup induces formation of lens

Lens induces overlying ectoderm to columnar and secretory epithelium

Induced granules secrete primary stroma containing collagen

Endothelial cells enter and secrete hyaluronic acid, causing stroma to swell; mesenchymal cells enter

Mesenchymal cell secretions cause stroma to shrink; under the influence of thyroxine the stroma will eventually become the cornea

coordinated dramas in which the actors must come on stage and speak their lines at the correct times and positions. In acquiring new information, they can also impart information for others to use. With this in mind, we can now study some principles about secondary induction obtained from other developing organs.

Formation of parenchymal organs

Epitheliomesenchymal interactions are also seen in the formation of duct-forming organs such as the kidney, liver, lung, mammary gland, and pancreas.

Morphogenesis of the mammalian kidney

The progression of renal tubules. Like the eye, the mammalian kidney is an exceedingly intricate structure. Its functional unit, the nephron, contains over 10,000 cells and at least twelve different cell types, each cell type located in a particular place in relation to the other cell types along the length of the nephron. The development of the mammalian kidney progresses through three major stages. Early in development (day 22 in humans, day 8 in mice), the pronephric duct arises in the mesoderm just ventral to the anterior somites. The cells in this duct migrate caudally (as discussed for amphibian embryos in Chapter 15), and the anterior region of the duct induces the adjacent mesenchyme to form the PRONEPHRIC KIDNEY TUBULES (Figure 12A). While these pronephric tubules form functioning kidneys in fish and in amphibian larvae, they are not thought to be active in mammalian amniotes. In mammals, the pronephric tubules and the anterior portion of the pronephric duct degenerate, but the more caudal portions of the pronephric duct persist and become "the central component of the excretory system throughout development" (Saxén,

FIGURE 12
General scheme of development in the vertebrate kidney. (A) The original tubules, constituting the pronephric kidney, are induced from the nephrogenic mesenchyme by the pronephric duct as it migrates caudally. (B) As the pronephros degenerates, the mesonephric tubules form. (C) The final mammalian kidney, the metanephros, is induced by the ureteric bud. (D) Section of a mouse kidney showing the initiation of the metanephric kidney (bottom) while the mesonephros is still apparent. The duct tissue is stained with a fluorescent antibody to a cytokeratin found in the pronephric duct and its derivatives. (A–C after Saxén, 1987; D courtesy of S. Vainio.)

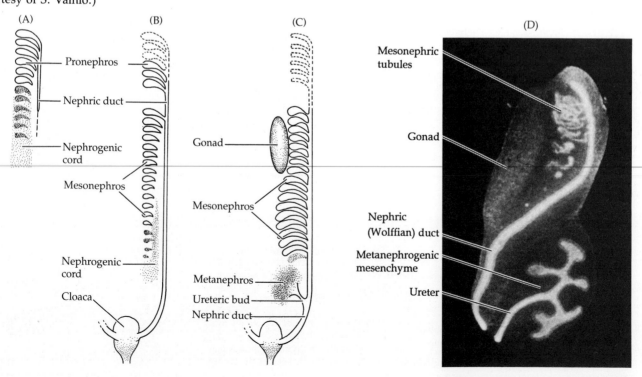

1987). This remaining duct is often referred to as the nephric, or Wolffian, duct.

As the pronephric tubules degenerate, the midportion of the nephric duct initiates a new set of kidney tubules in the adjacent mesenchyme. This set of tubules constitutes the MESONEPHROS, or MESONEPHRIC KIDNEY. In humans, about 30 mesonephric tubules form, beginning around day 25. However, as more tubules are induced caudally, the anterior mesonephric tubules also begin to regress (Figure 12B). In female mammals, this regression is complete. However, as we will discuss in Chapter 21, some of these mesonephric tubules persist in male mammals to become the sperm-carrying tubes (the vas deferens and efferent ducts) of the testes.

The permanent kidney of amniotes, the METANEPHROS, is generated by the same components as the earlier, transient, kidney types. The two nephric ducts, remnants of the original kidney, put forth tubes of epithelium called the URETERIC BUDS. These eventually separate from the nephric duct to become the ureters that take the urine to the bladder. When the ureteric buds emerge from the nephric duct, they enter the regions of kidney-forming mesenchymal cells called the METANEPHROGENIC MESENCHYME. This mesenchymal tissue is induced by the ureteric buds to generate the nephrons of the mammalian kidney (Figure 12C,D).

Reciprocal induction during kidney development. These two mesodermal tissues, the ureteric bud and the metanephrogenic mesenchyme, interact and reciprocally induce each other (Figure 13). The metanephrogenic mesenchyme causes the ureteric bud to elongate and branch. At the tip of these branches, the ureteric bud induces the loose mesenchymal cells to form an epithelial aggregate. Each aggregate of about 20 cells will divide and differentiate into the intricate structure of the renal nephron. First, each node elongates into a "comma" shape and then forms a characteristic S-shaped tube. Soon after the S-shaped tube is formed, the cells of this epithelium begin to differentiate into the regionally specific cell types such

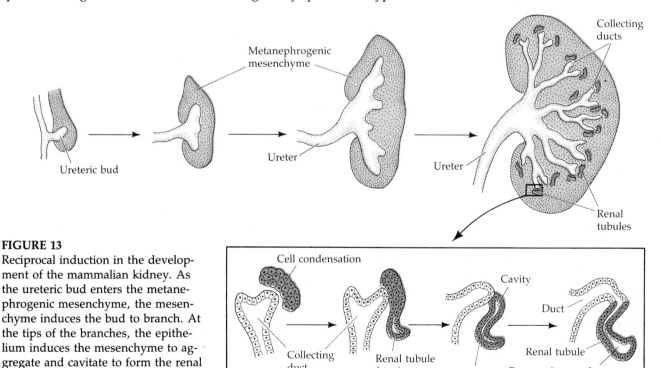

FIGURE 13
Reciprocal induction in the development of the mammalian kidney. As the ureteric bud enters the metanephrogenic mesenchyme, the mesenchyme induces the bud to branch. At the tips of the branches, the epithelium induces the mesenchyme to aggregate and cavitate to form the renal tubules.

as the capsule cells, the podocytes, and the distal and proximal tubule cells. At this time, a connection develops between the ureteric bud and the newly formed tube, thereby enabling material to pass from one to the other. The tubes at the end of the ureteric bud form the secretory nephrons of the functioning kidney, and the branched ureteric bud gives rise to the renal collecting ducts and to the ureter, which drains the urine from the kidney.

Clifford Grobstein (1955, 1956) documented this RECIPROCAL INDUCTION in vitro. He separated the ureteric bud from the mesenchyme and cultured them individually or together. In the absence of mesenchyme, the ureteric bud does not branch. In the absence of the ureteric bud, the mesenchyme does not condense to form tubules. When they are placed together, however, the ureteric bud grows and branches, and tubules form throughout the mesenchyme (Figure 14). Although certain other tissues (notably neural tube) will enable the metanephrogenic mesenchyme to form kidney tubules, the ureteric bud branches only under instructions from the metanephrogenic mesenchyme. Mesenchymes that induce branching in other epithelia (such as salivary gland mesenchyme) will not cause the ureteric bud to branch (Bishop-Calame, 1966). Thus, the ureteric bud induces the metanephrogenic mesenchyme to become the kidney tubules and the mesenchyme acts to cause the epithelial cells to branch.

The first metanephric nephrons, then, are immediately joined to the collecting ducts. These nephrons are carried outward into the metanephrogenic mesenchyme as the ureter continues to grow (Figure 15A). The ends of the ureteric buds, however, retain their ability to induce tubule formation in this mesenchyme, and the result is the formation of tubular arcades (Figure 15B). As the ureteric branch migrates further through the mesenchyme, new nephrons are formed and become joined to the same collecting duct (Figure 15C; Osathanondh and Potter, 1963).

The mechanisms of kidney organogenesis

It is one thing to say that the ureteric bud induces the metanephrogenic mesenchyme to become the epithelium of the nephrons. It is another thing to understand how this process occurs. Like the development of the lens by the optic vesicle, it is thought that induction by the ureteric bud is only the final step that triggers a cascade of events in the competent mesenchyme. Only the metanephrogenic mesenchyme has the ability to respond

FIGURE 14
Kidney induction observed in vitro. (A) An 11-day mouse metanephric rudiment includes both ureteric bud and metanephrogenic mesenchyme. (B) After the first day of culture, tubules can be seen at the tips of the branching ureters. (C) The branching collecting ducts formed by the ureteric bud and the kidney tubules formed by the mesenchymal condensations at the tips of these buds can be clearly seen after eight days of culture. (A and B from Saxén and Sariola, 1987; C from Grobstein, 1955; all photographs courtesy of the authors.)

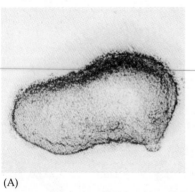
(A)

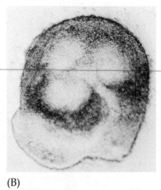
(B)

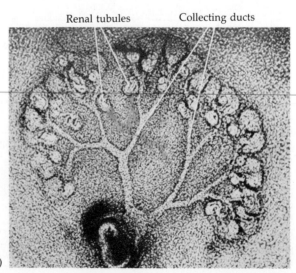
Renal tubules   Collecting ducts
(C)

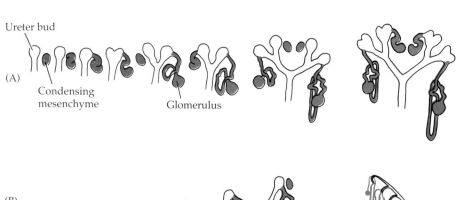

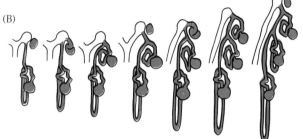

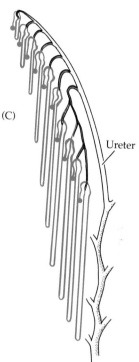

FIGURE 15
Schematic representation of human nephron development. (A) Formation of
early nephrons directly attached to the ureteric bud epithelium. (B) Formation of
nephron arcades where several nephrons are joined to the same collecting duct.
(C) General arrangement of human nephrons at birth. The deeper nephrons constitute an arcade, while the nephrons closer to the surface are directly connected
to the collecting duct of the ureter. (After Osathanondh and Potter, 1963.)

to the ureteric bud to form kidney tubules, and if induced by other epithelia (such as embryonic salivary gland or neural tube), the metanephrogenic mesenchyme will respond by forming kidney tubules and no other
structures (Saxén, 1970; Sariola et al., 1982). Thus, the metanephrogenic
mesenchyme cannot become any other tissue than kidney tubules (if induced) or renal stromal cells (if uninduced). Based on his experiments on
the induction of the chick mesonephros, Etheridge (1968) has suggested
that this determination is the result of interactions with the endoderm
earlier in development.

What, then, is this program of events carried out by the induced
mesenchyme? The conversion of mesenchymal cells into an epithelium is
an uncommon event in organogenesis, occuring in the kidney, the gonads
(where gonadal mesenchyme forms the sex cords), and the mesothelium
(which arises from the mesenchymal cells lining the coelom). As discussed
in Chapter 15, such changes would involve substantial remodeling of the
extracellular matrix, and indeed, the ureteric bud causes dramatic changes
in the extracellular matrix of the metanephrogenic mesenchyme cells. The
uninduced mesenchyme secretes an extracellular matrix consisting largely
of fibronectin and collagen types I and III. Upon induction, these proteins
disappear and are replaced by a basal lamina comprised of laminin and
type IV collagen. The cytoskeleton also changes from one typical of mesenchymal cells to one typical of epithelia (Ekblom et al., 1983, Lehtonen
et al., 1985). In this manner, the loose mesenchymal cells are linked
together as a polarized epithelium on a basal lamina.

Prior to these changes, the newly induced metanephrogenic mesenchyme synthesizes the membrane proteoglycan syndecan (Vainio et al.,
1989a). This cell surface proteoglycan is first seen around the mesenchymal
cells surrounding the ureteric bud as the bud first enters the region of
mesenchyme. As the ureteric bud initiates its first branch, the entire

FIGURE 16
The extracellular matrix proteoglycan syndecan is not synthesized or secreted by mesenchymal cells until after induction. This molecule is probably involved in structuring the new tubule epithelium, and it distinguishes the cells of the tubule from the remaining mesenchyme. (A) Immunological staining of syndecan shows its presence in the newly induced mesenchyme cells (T) that are becoming epithelial. Some staining is also seen on the ureter bud epithelium (U). (B) Intense syndecan staining is seen in the region of the developing tubule that is to become the renal glomerulus (G). (From Vainio et al., 1989a; photographs courtesy of L. Saxén.)

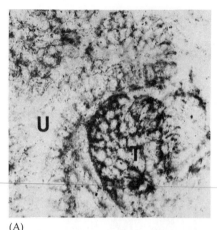

(A)

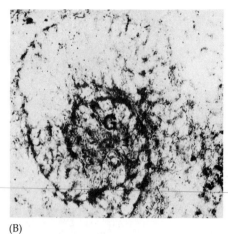

(B)

mesenchymal region around the branch stains positively for syndecan (Figure 16). The cells layers immediately adjacent to the ureteric bud stain most intensely. By using inducers from different species and species-specific antibodies for syndecan, Vainio and co-workers demonstrated that the syndecan was indeed synthesized by the mesenchymal cells. Syndecan is important for shaping epithelia (Repraeger and Bernfield, 1985; Vainio et al., 1989b), and its synthesis may be critical for the aggregation of the mesenchymal cells. If proteoglycan synthesis is inhibited in embryonic kidney rudiments, mesenchymal cells cease to form the epithelial tubules and the ureter fails to branch when it enters the mesenchymal region (Lelongt et al., 1988).

Although it has long been thought that the ureteric bud initiates this cascade of changes by activating and repressing specific genes within the mesenchyme, we still do not know the inducer substance synthesized by the ureteric bud, the receptor for this substance in the responsive mesenchyme, or the genes activated by this interaction. Recently, though, a gene has been identified that is a good candidate for initiating the sequence by which the mesenchymal cells become an epithelium. This gene, *WT2-1*, has been identified as being absent or mutated in patients with Wilm's tumor, a cancer involving embryonic kidney cells. The protein it encodes has the sequence characteristic of a zinc-finger transcription factor, and in situ hybridization shows that this gene is normally expressed in the aggregating renal, gonadal, and mesothelial mesenchymes. In the human kidney, *WT2-1* mRNA is not expressed in either the ureteric bud nor in the uninduced mesenchyme. However, it is readily detected on the condensing aggregate and in the S-shaped epithelium (Pritchard-Jones et al., 1990; van Heyningen et al., 1990).

SIDELIGHTS & SPECULATIONS

Coordinated differentiation and morphogenesis

During the morphogenesis of any organ, numerous dialogues are occurring between the interacting tissues. In epithelial–mesenchymal interactions, the mesenchyme in-

fluences the epithelium; the epithelial tissue, once changed by the mesenchyme, can secrete factors that change the mesenchyme. Such interactions continue until an organ is formed with organ-specific mesenchymal cells and organ-specific epithelia. The identification of the chemicals involved in these intertissue conversations is under way in

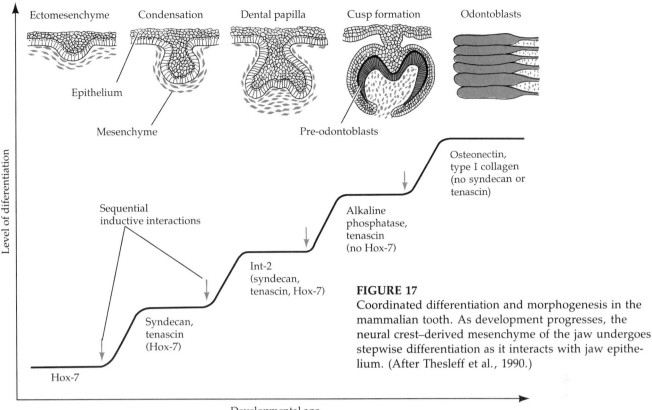

Ectomesenchyme Condensation Dental papilla Cusp formation Odontoblasts

Epithelium

Mesenchyme

Pre-odontoblasts

Level of differentiation

Sequential
inductive interactions

Osteonectin,
type I collagen
(no syndecan or
tenascin)

Alkaline
phosphatase,
tenascin
(no Hox-7)

Int-2
(syndecan,
tenascin, Hox-7)

Syndecan,
tenascin
(Hox-7)

Hox-7

FIGURE 17
Coordinated differentiation and morphogenesis in the mammalian tooth. As development progresses, the neural crest–derived mesenchyme of the jaw undergoes stepwise differentiation as it interacts with jaw epithelium. (After Thesleff et al., 1990.)

Developmental age

several laboratories. Some of the most extensively studied interactions are those that form the mammalian tooth. Here, the jaw epithelium differentiates into the enamel-secreting ameloblasts, while the neural crest-derived mesenchymal cells become the dentin-secreting odontoblasts.

First, the epithelium causes the mesenchyme to aggregate at specific sites. At this time, the epithelium possesses the potential to generate tooth structures out of several types of mesenchymal cells (Mina and Kollar, 1987; Lumsden, 1988). However, this tooth-forming potential soon becomes transferred to the mesenchyme that aggregated beneath it. These mesenchymal cells form the dental papilla and are able to induce tooth morphogenesis in other epithelia (Kollar and Baird, 1970). At this stage, the jaw epithelium has lost its ability to instruct tooth formation in other mesenchymes. At the basement membrane that separates the epithelium from the mesenchyme, the epithelium induces the mesenchyme to become the odontoblasts, while the mesenchyme induces the epithelium to become the ameloblast precursors (Figure 17; Thesleff et al., 1989).

Recent studies have correlated these morphological changes with the differentiation of these two tissues. Figure 17 shows the various stages of mesenchyme differentiation at different stages of morphogenesis. As can be seen, the mesenchyme at one stage is different from the mesenchyme at other stages. Before any induction by the epithelium, the mesenchymal cells express the *Hox-7* gene and secrete a mesenchymal extracellular matrix. As these cells

condense, they are induced to synthesize the membrane protein syndecan and the extracellular matrix protein tenascin. These proteins (which can bind each other) appear at the time the epithelium induces mesenchymal aggregation, and Thesleff and colleagues (1990) have proposed that these two molecules may interact to bring about this condensation.

After the mesenchymal cells have aggregated, they begin to secrete a mitosis-promoting protein, INT-2 (Wilkinson et al., 1989). It is not certain whether this protein is stimulating the growth of the mesenchymal cells, the epithelial cells, or both. As the mesenchymal cells begin to differentiate into odontoblasts, tenascin is induced to be expressed at much higher levels and at the same sites as alkaline phosphatase. Both these proteins have been associated with bone and cartilage differentiation, and they may promote the mineralization of the extracellular matrix (Mackie et al., 1987).

Finally, as the odontoblast phenotype emerges, osteonectin and type I collagen are secreted as components of the extracellular matrix. By this steplike process, the cranial neural crest cells of the jaw can be transformed into the dentin-secreting odontoblasts. These interactions occur at specific times of development and are correlated to the maturation of the epithelium. Under normal circumstances, two independent phenomena—morphogenesis and cytodifferentiation—are coordinated in organ formation.

The nature of proximity in epithelio-mesenchymal inductions

Proximate induction occurs when an inducing tissue is brought near a competent responding tissue. Do these cells have to make physical contact or do they interact over a small distance? Three types of interactions can be postulated: cell–cell contact, cell–matrix contact, and diffusion of soluble signals (Figure 18; Grobstein, 1955; Saxén et al., 1976). In some tissues, cell–cell contact appears to be required. The induction of kidney tubules by the ureteric bud appears to depend upon their intimate contact. This dependency was shown by Lehtonen and Saxén (Lehtonen, 1975; Lehtonen et al., 1975). Modifying Grobstein's technique of growing inducer and mesenchyme on different sides of a porous filter, they showed that tubule formation occurred in the mesenchyme culture opposite the inducer tissue (Figure 19A). In cross section, small projections of inducing tissue were seen to traverse the filter, contacting the mesenchyme (Figure 19B). By changing the porosity of the filter, the investigators were able to correlate the induction with intercellular contact. A similar condition occurs in nature. Kidneys of mice homozygous for *Danforth's short tail* mutation fail to develop. The ureteric bud comes within a cell diameter of the mesenchyme but progresses no further. In the absence of contact, no induction occurs (Gluecksohn-Schoenheimer, 1943). Contacts between inducing and responding cells are also seen in the induction of teeth (Slavkin and Bringas, 1976) and in the induction of the submandibular salivary gland (Cutler and Chaudhry, 1973).

In other organs, the extracellular matrix of one cell type is seen to cause the differentiation of another set of cells. In Figure 43 of Chapter 15, we saw that the extracellular matrix plays an important role in the differentiation of the rat Sertoli cell. The heart and cornea also appear to be induced by extracellular matrices. As discussed earlier, the differentiation of the corneal epithelium depends upon an inductive influence from the capsule of the lens (Hay and Revel, 1969). Unlike kidney induction, where it seems necessary that the inducer tissue be live, dead lens capsule will also work as long as collagen is still present on it, and nearly any source of collagen will induce corneal epithelial cells to synthesize their matrix (Hay and Dodson, 1973; Meier and Hay, 1974). When filters are placed between the lens capsule and the corneal epithelium, the epithelial cells send out processes that cross the filter to contact the capsule (Meier and Hay, 1975). This work suggests that the surface of the corneal epithelial cell receives some instructions from the collagen-rich capsule of the lens. In the developing heart, the myocardium induces endothelial cells in a certain region to become the mesenchyme that will form the heart valves. This induction appears to depend upon fibronectin-containing vesicles that are secreted by the myocardial cells to form the basement membrane in this region (Mjaatvedt and Markwald, 1989).

The extracellular matrix can also provide positional information for secondary inductions. In 5- to 6-day chick embryos, the dermal cells of the skin begin to condense at particular sites. These places are the sites of feather development. The dermal cells are not randomly arranged but follow a precise pattern. A row of condensed dermal cells appears, each FEATHER GERM arising almost simultaneously. The rows adjacent to it form next, each focus of dermal condensation being between those of the first row (Figure 20). It is at these points—and only at these points—that feathers emerge by the interaction of these dermal cells with the overlying ectoderm. Stuart and co-workers (1972) suggested that the dermal con-

FIGURE 18

Three possible ways that inductive interactions might occur. (After Grobstein, 1956.)

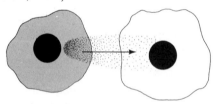

Diffusion of inducers from one cell to another

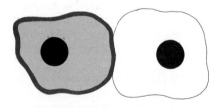

Contact of matrix from one cell to another

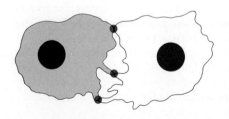

Contact between the inducing and responding cells

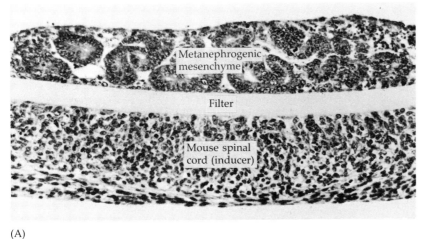

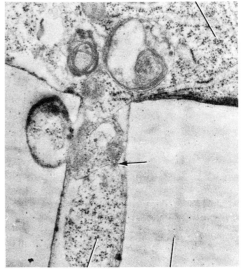

Metanephrogenic
mesenchymal cell

(A)

(B)

Neural tube cell   Filter

FIGURE 19

Transfilter induction of kidney tubules. (A) Metanephrogenic mesenchyme is above the filter and the inducer (in this case, mouse spinal cord, which mimics the effect of the ureteric bud) is placed below. Tubules have been induced in the mesenchyme. (B) Electron micrograph showing cell contact (arrow) through the pore of a filter "separating" the metanephrogenic mesenchyme from the inducing spinal cord. (Photographs courtesy of L. Saxén.)

densations arise from the migration of the mesenchymal cells along a preformed collagen matrix (Figure 20C). By treating back skin with collagenase, they were able to destroy the hexagonal pattern of collagen deposition and inhibit the condensation of dermal cells.

Goetinck and Sekellick (1972) have shown that this hexagonal pattern is lacking in the skin of an embryo carrying the *scaleless* mutation; this embryo is unable to form these condensations. The skin of these mutants makes normal amounts of collagen, but the specific hexagonal lattice is not seen. The ectodermal component of the skin appears to be the source of the mutation, as both lattice and feather germs formed when normal epidermis was combined with *scaleless* dermis, but no lattice or feather germ appeared when mutant epidermis was combined with normal dermis. More recent studies (Mauger et al., 1982, 1983) have shown that during normal skin development, collagen I and III and fibronectin are uniformly distributed at the dermis–epidermis junction. Feather and scale development continue as the collagens disappear from the regions where the feather or scale bud is to form. Conversely, collagen accumulations increase between the interscale regions. Soon afterward, the cell adhesion molecule N-CAM is expressed in the dermis in a punctate fashion, directly under the future feather bud. It is thought (Chuong and Edelman, 1985) that the N-CAM protein specifically aggregates the dermal cells beneath the epidermis and unites them as a separate entity (Chapter 15). It is conceivable, then, that during normal skin development the ectoderm transmits information that enables the dermal cells to regulate their pattern of collagen and fibronectin synthesis, thereby laying down a pattern of collagen fibers upon which the migratory cells travel. Thus, the ectoderm influences the placement of the mesenchyme and then the mesenchyme condenses in particular regions and induces the overlying ectoderm to produce feathers. Thus, although the mechanism for this patterning is not

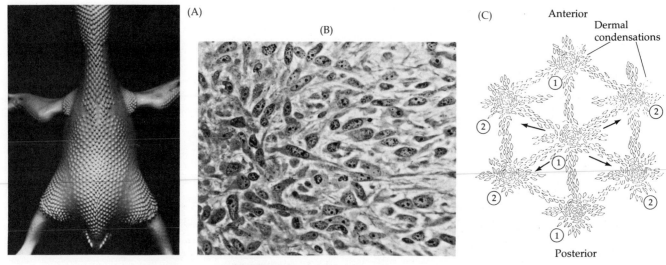

FIGURE 20

Positioning of feather tracts in chick embryo. (A) Feather tracts on the dorsum of a 9-day chick embryo. Note that each feather primordium is located between the primordia of adjacent rows. (B,C) Pattern of dermal condensations giving rise to feather rudiments on the dorsum. (B) Condensation of dermal mesenchyme cells on the left, with nonaggregated cells oriented along the axis connecting the condensation to a neighboring one. (C) Hexagonal pattern of dermal cell aggregation and alignment: (1) primary row of dermal condensations (papillae) and cells aligned along the axis connecting them; (2) secondary rows of papillae and their connections. (A courtesy of P. Sengal; B from Wessells and Evans, 1968, courtesy of N. Wessells; C from Saunders, 1980.)

FIGURE 21

Retention and differentiation of pancreatic cells on Sepharose beads coated with mesenchymal factor. The zymogen granules and microvilli (arrows) characteristic of differentiated pancreatic cells are evident. (From Levine et al., 1973; courtesy of W. J. Rutter.)

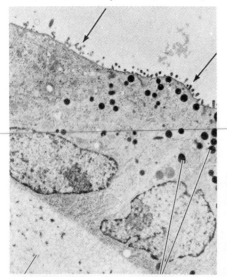

Sepharose   Zymogen
bead   granules

fully understood, the extracellular matrix appears to be very important in determining the site of secondary inductions in the skin.

There are some inductive systems in which contact is not needed. One of these is induction of the neural tube by the action of chordamesoderm cells upon the overlying ectoderm. No contact is seen between inducing and responding cell sets, and this induction can occur through filters separating the two components (Toivonen, 1979).

A similar lack of contact can exist in pancreatic development. Investigators in Rutter's laboratory (Rutter et al., 1978) have isolated from various mesenchymal tissues a glycoprotein that is capable of inducing the differentiation of pancreatic epithelial cells. This MESENCHYMAL FACTOR can be bound to Sepharose (resin) beads. When fragments of pancreatic epithelium adhere by their basal surfaces to these coated beads, they synthesize DNA and produce enzyme-containing secretory granules characteristic of differentiated pancreas cells. Similar tissue fragments adhere to uncoated beads but do not produce granules. The effect is specific for this mesenchymal factor, and other proteins will not induce this differentiation. As shown in Figure 21, the epithelial cells binding to the factor-coated beads will divide and differentiate. Thus, morphogenesis is not necessary for cell differentiation to occur. The cells will produce their particular products irrespective of whether or not they have branched. Although this factor is probably a membrane protein, it may be sloughed off through the normal turnover of membranes. It would not be surprising if induction could occur either by contact of membranes or by the interaction of cells with the sloughed off membrane products of their neighbors. What does seem to be important is that the effects are mediated at the level of the cell surface and that inducing molecules do not travel far from their source of origin.

Mechanism of branching in the formation of parenchymal organs

As seen in the kidney and many other organs, the mesenchyme can interact with an epithelial tube by causing it to branch. Branching occurs when the epithelial outgrowths are divided by clefts, yielding lobules on either side of the cleft. These lobules grow to create branches. The branching of epithelial buds depends upon the presence of the mesenchyme. In some cases, such as the interaction of respiratory epithelium with several different mesenchymes, the interaction is instructive. In most cases, however, these interactions are merely permissive. The buds are prepared to branch and form acini, but they need support from the mesenchyme. It is now thought that the mesenchyme causes cleft formation and branching by splitting the lobule and selectively digesting away part of the epithelial tissue's basal lamina.

The control of cleft formation appears to be a function of collagen molecules. Collagen III fibrils are produced by the mesenchymal cells but accumulate only within the clefts of lobules (Figure 22; Grobstein and Cohen, 1965; Nakanishi et al., 1988). Moreover, the extent of branching can be artificially regulated by preserving or removing collagen molecules (Nakanishi et al., 1986a). Figure 23 shows the branching of a 12-day rudiment of a submandibular salivary gland under conditions that prevent the degradation of collagen fibrils (collagenase inhibitor was added to the medium) or accelerate its degradation (exogenous collagenase was added to the medium). Without the collagen, no clefts are seen, but when the endogenous collagenase is unable to remove excess collagen, supernumerary clefts appear.

The mechanism by which collagen acts to initiate this branching is still unclear. Nakanishi and co-workers (1986b) have proposed that the mesenchymal cells align collagen fibrils by their traction to form ridges that cut into the lobular epithelium to form clefts. These clefts then become clearly defined as more migrating mesenchymal cells further deform the lobule by their traction. This initial cleft formation does not depend on epithelial cell proliferation (Nakanishi et al., 1987). The collagen fibrils may also be responsible for the development of the cleft into distinct branches.

Mesenchymal cell

Epithelial cell

Collagen in cleft between epithelial cells

FIGURE 22

Scanning electron micrograph of collagen fibril accumulation within an early cleft of a 12 day mouse salivary gland. (From Nakanishi et al., 1986b; photograph courtesy of Y. Nakanishi.)

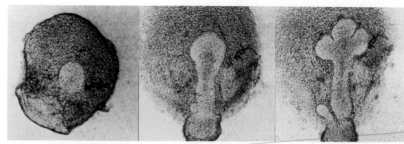

(A) Control

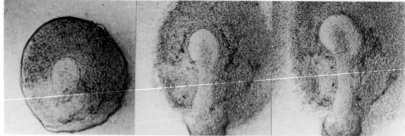

(B) Collagenase added

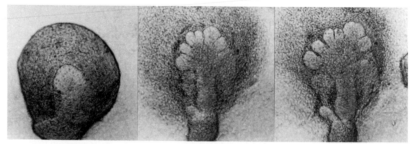

(C) Collagenase inhibitor added

FIGURE 23

Control of epithelial cleft formation by mesenchymal collagen. 12-day mouse salivary gland rudiments were cultured and observed at 1, 18, and 25 hours. (Row A) Normal development, showing three major lobes. (Row B) Growth of lobule but no branching when exogenous collagenase (5 μg/ml) is added to the medium. (Row C) Supernumerary branches when collagenase inhibitor (5 μg/ml) is added to the medium to suppress endogenous collagenase activity. (From Nakanishi et al., 1986a; photograph courtesy of Y. Nakanishi.)

Bernfield and Banerjee (1982) have proposed that collagen can act to protect the basal lamina of the epithelial cells against hyaluronidase secreted by the mesenchymal cells. They showed that the mesenchymal cells do indeed digest glycosaminoglycan from the lobule (Banerjee and Bernfield, 1979) and that the GAGs at the tips are more susceptible than those in the clefts. The breakdown of the basal lamina would enable the expansion of the branch by the increased mitoses stimulated in that area. In this model, shown in Figure 24, the mesenchyme promotes epithelial growth, degrades the GAG, and deposits collagen fibers in the cleft. The epithelium synthesizes the basal lamina materials and stimulates mesenchymal collagen synthesis. The result is a differential breakdown of the basal lamina at the tips of the lobes, thus enabling the dividing cells of the lobe to form branches. Here, the interaction of mesenchymal cells with the extracellular matrix of the epithelium would determine the branching pattern of the organ.

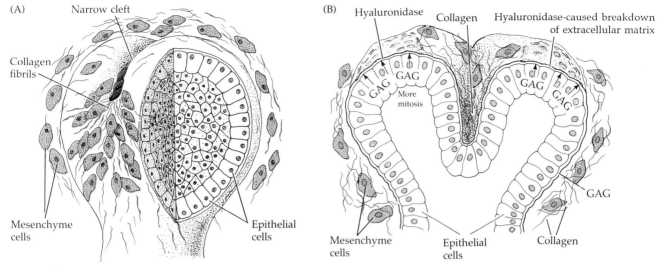

(A) Narrow cleft

Collagen fibrils

Mesenchyme cells

Epithelial cells

(B) Hyaluronidase   Collagen   Hyaluronidase-caused breakdown of extracellular matrix

GAG   GAG   More mitosis   GAG   GAG

Mesenchyme cells   Epithelial cells   Collagen   GAG

FIGURE 24

Possible model for cleft formation and branching in the mouse salivary gland rudiment. (A) A furrow is made in the lobule by contracting a bundle of collagen fibrils (shown here as a twisted, ropelike structure) by the traction of the mesenchymal cells. As is shown in Figure 20, the fibrils extend between two groups of mesenchymal cells. (B) The elongation of the two separated lobules into branches may ensue, as the GAGs at the tips of the lobules are more sensitive to hyaluronidase since they do not have the protection of the collagen fibrils. The stalk of the lobule is stabile, while the increased division (stimulated by the mesenchyme) at the tips pushes the lobule forward. (A from Nakanishi et al., 1986b; B from Wessells, 1977.)

SIDELIGHTS & SPECULATIONS

Induction of plasma cells: Macrophages and helper T cells

It is extremely difficult to study the processes of embryonic induction. The amounts of tissue that one isolates from embryos are painfully small, and the extraction of defined inducing agents from them has not yet been accomplished. However, animals never stop developing, and certain adult cells are constantly differentiating from stem cells. In the adult vertebrate, the differentiation of the antibody-secreting plasma cell has been the subject of intense scrutiny for several decades. As mentioned earlier, the plasma cell is derived from a B lymphocyte. Each B lymphocyte can make a specific type of antibody and place it on its cell membrane as an antigen receptor. When an antigen binds to it, the B lymphocyte divides several times, develops the characteristic structures of a secretory cell (differentiates), and secretes its specific antibody (Chapter 10).

But antigen alone is not enough to trigger this differentiation. In order to proliferate and differentiate, the presence of two other cell types is required. One of these cell types is the MACROPHAGE. This cell presents the antigen to the B lymphocyte. The B lymphocyte does not usually respond well to antigen in solution. Rather the antigen is

first "processed" by the macrophage, which presents the antigen on its cell surface (Mosier, 1967; Unanue and Askonas, 1968).

The second type of cell involved in B lymphocyte differentiation is the T lymphocyte. When T lymphocytes are absent, B lymphocytes do not proliferate or differentiate, even if macrophages and antigens are present. Patients lacking T cells (as happens to people whose thymuses fail to develop) lack the ability to make antibodies to most substances, even though they have a population of B cells. Thus, the induction of the B cell to become an antigen-secreting plasma cell involves antigen, macrophages, and HELPER T CELLS.

How do these other cells interact with the B cell? The studies of Guillet and colleagues (1987) suggest the following model (Figure 25). First, the macrophage binds the antigen (such as a bacterium) and partially digests it. The partially digested fragments are bound by certain membrane glycoproteins (the class II major histocompatibility complex antigens) and are placed on the macrophage cell surface. Second, the T cell recognizes certain antigen fragments on the macrophage cell surface when they are bound to the class II MHC glycoproteins. It effects this recognition via antigen-specific receptors on the T cell surface. These

FIGURE 25

Model for the induction of B cell differentiation by helper (inducer) cells and macrophages. Activation of the helper T cells is effected when macrophages ingest antigen, par-

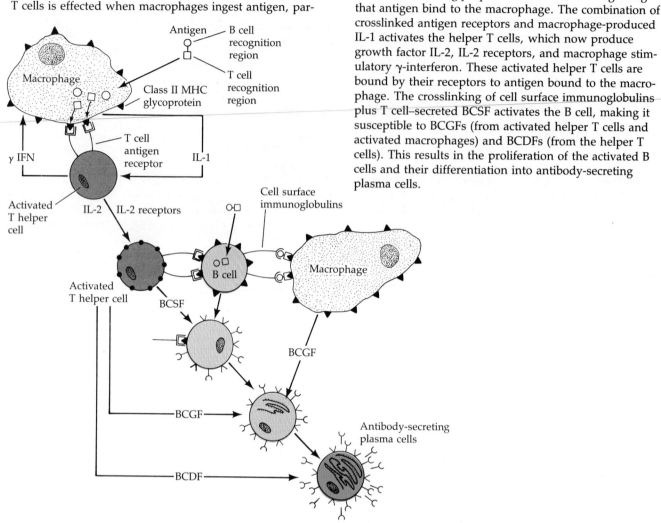

tially digest it, and display it on their cell surfaces bound to a class II MHC glycoprotein. The T cells recognizing that antigen bind to the macrophage. The combination of crosslinked antigen receptors and macrophage-produced IL-1 activates the helper T cells, which now produce growth factor IL-2, IL-2 receptors, and macrophage stimulatory γ-interferon. These activated helper T cells are bound by their receptors to antigen bound to the macrophage. The crosslinking of cell surface immunoglobulins plus T cell–secreted BCSF activates the B cell, making it susceptible to BCGFs (from activated helper T cells and activated macrophages) and BCDFs (from the helper T cells). This results in the proliferation of the activated B cells and their differentiation into antibody-secreting plasma cells.

T cell receptors become clustered together in one portion of the T cell membrane. During this process, the macrophage secretes INTERLEUKIN 1 (IL-1), which activates the helper T cell. The activated helper T cell produces both the T cell growth factor INTERLEUKIN 2 (IL-2) and the receptors for IL-2. (They also produce γ-interferon, which further activates the macrophage.) The B cell has also placed partially digested antigen on its cell surface (bound by class II MHC glycoproteins) and presents it to the activated helper T cells (Lanzavecchia, 1985; Janeway et al., 1987). At the same time, the B cell binds to a specific part of the antigen on the macrophage in such a way that the B cell, helper T cell, and macrophage are all linked together (Inabar et al., 1984; Snow et al., 1983). This binding of B cell to antigen is effected by the cell surface immunoglobulins, and their crosslinking by the macrophage activates the B cell. The B cell now synthesizes the *receptors* for the B cell growth and differentiation factors synthesized by the activated T cells and macrophages (Melchers and Andersson, 1986). These inducing factors include B CELL STIMULATING FACTOR I

(BCSF-I), which enables the resting B cell to resume proliferation (going from G_0 to G_1 of the cell cycle) and to produce more antibody molecules (Lee et al., 1986; Killar et al., 1987), and B CELL GROWTH FACTOR II (BCGF-II), which has been shown to act in the late G_1 stage of the cell cycle (O'Garra et al., 1986). But these proliferating B cells cannot yet secrete their antibodies until acted upon by at least two other B CELL DIFFERENTIATION FACTORS (BCDFs) (Howard and Paul, 1983; Hamaoko and Ono, 1986).

The helper T cell, therefore, is responsible for inducing the B cell to become a set of antibody-secreting plasma cells. Upon binding antigen, the B cells become competent to respond to the inducing molecules of the T cell. These helper T cells also induce the macrophages to synthesize a new set of digestive enzymes that allow them to destroy bacteria more effectively. If the helper T cells are destroyed, these inductions cannot take place and the immune system is severely impaired. This appears to be the case when human T helper lymphocytes are specifically killed by the HUMAN IMMUNODEFICIENCY VIRUS, which attaches to an adhesion molecule (the CD4 glycoprotein) on their cell surfaces* (Klatzmann et al., 1984; Dalgleish et al., 1984). The loss of these inducer T cells causes the ACQUIRED IMMUNE DEFICIENCY SYNDROME (AIDS), which is characterized by the body's inability to combat infectious microorganisms.

During pregnancy, however, it is important to *maintain* a local immune deficiency within the uterus to prevent the fetus from being destroyed (as a graft would be). A placental protein that prevents the proliferation of helper T cells was discussed in Chapter 4; the uterus also contains another population of T cells (called suppressor T cells) that produces factors that block the activity of IL-2 (Daya et al., 1987). Yet another factor, secreted by the kidneys of pregnant women, is able to block the activity of IL-1 (Brown et al., 1986). It appears that the pregnant uterus uses several mechanisms to block the induction of immunocompetent cells that may react against the developing fetus.

We have a system, then, in which one set of cells—the helper T lymphocytes—can induce the differentiation of another group of cells—the B lymphocytes. The B lymphocytes (like pancreatic cells developing), although capable of synthesizing a small amount of differentiated protein, do not secrete it. Upon stimulation by another cell type, cytodifferentiation takes place. The Golgi apparatus and rough endoplasmic reticulum are formed; protein is synthesized and modified in these structures and then secreted from the cell.

B cell induction is thus very similar to inductions seen in the embryo, but is much easier to study. One of the advantages of studying B cell induction is that we can work with clones of adult tumor cells that still retain their ability to make or respond to the inducing molecules. The lymphokines were discovered by using clones of malignant T cells that were secreting certain proteins and by using clones of B cell tumors that could only grow if they were supplied with these factors. These tumors represent the "freezing" of different stages of lymphocyte development. Using these tumor cell lines, researchers can obtain a large number of lymphocytes at any particular stage of differentiation. So far, such stage-specific tumor lines exist only for lymphocytes, but the possibility exists that this will be a more generally used technique (Houghton et al., 1987). For the study of induction and the identifying of the inducer molecules, tumors may become for developmental biologists what mutations have been for geneticists (Auerbach, 1972; Pierce, 1985). Their use has enabled studies of the immune system to provide us with some of our first insights into the molecular nature of vertebrate induction.

---

*In humans, these T cells are called "helper/inducer T cells," a name that recognizes their developmental role. The CD4 glycoprotein is normally involved in mediating nonspecific cell adhesion between the helper/inducer T cells and B lymphocytes (Doyle and Strominger, 1987).

Induction at the single-cell level

Sevenless and *bride of sevenless*

Embryonic induction occurs when interactions between inducing and responding cells bring about changes in the developmental pathway of the responding cell (Jacobson and Sater, 1988). Without the induction, the responding cell would become one cell type; with the induction, it becomes another. Our discussions of induction, however, have usually concerned tissues, not cells. Recent research into the developmental genetics of *Drosophila* and *Caenorhabditis* have shown that induction does indeed occur on the cell-to-cell level. Some of the best studied examples involve the formation of the retinal photoreceptors in the *Drosophila* eye. The retina consists of about 800 units called *ommatidia* (Figure 26). Each ommatidium is composed of twenty cells arranged in a precise pattern. The eye develops in the flat epithelial layer of the eye imaginal disc of the larva. There are no cells directly above or below this layer, so the interactions are confined to neighboring cells. The differentiation of the randomly arranged epithelial cells into retinal photoreceptors and their surrounding lens tissue occurs during the last (third) larval stage. An indentation forms at the posterior margin of the imaginal disc, and this furrow begins to travel forward towards the anterior of the epithelium (Figure 27). As the furrow passes through a region of cells, those cells begin to differentiate in a specific order. The first cell to develop is the central (R8) photoreceptor. (It is not yet known how the furrow instructs certain cells to become R8

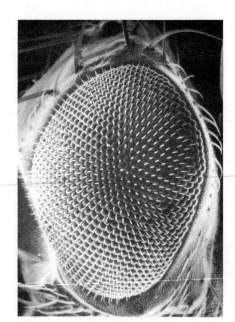

FIGURE 26
Scanning electron micrograph of a compound eye in *Drosophila*. Each facet is a single ommatidium. A sensory bristle projects from each ommatidium. (Photograph courtesy of T. Venkatesh.)

photoreceptors.) The R8 cell is thought to induce the cell above it and the cell below it (with respect to the furrow) to become the R2 and R5 photoreceptors, respectively. The R2 and R5 photoreceptors are functionally equivalent, so the signal from R8 is probably the same to both cells (Tomlinson and Ready, 1987). Signals from these cells induce four more adjacent cells to become the R3, R4, and then the R1 and R6 photoreceptors. Last, the R7 photoreceptor appears. The other cells around these photoreceptors become the lens cells. Lens determination is the "default" condition if the cells are not induced.

A series of mutations has been found that blocks some of the steps of this induction cascade. The *rough* (*ro*) mutation, for instance, blocks the induction of the R3 and R4 photoreceptors. The *sevenless* (*sev*) mutation and the *bride of sevenless* (*boss*) mutation can each prevent the R7 cell from differentiating. (This cell becomes a lens cell instead.) Analysis of these mutations has shown that they are involved in the inductive process. The *sevenless* gene is required in the R7 cell itself. If mosaic embryos are made such that some of the cells of the eye disc are wild-type and some are homozygous for the *sevenless* mutation, the R7 photoreceptor is seen to

FIGURE 27
Differentiation of photoreceptors in the late larval eye imaginal disc. The morphogenetic furrow (arrow) crosses the disc from posterior (left) to anterior (right). Behind the furrow, the photoreceptor cells differentiate in a defined sequence (shown below). The first photoreceptor cell to differentiate is R8. R8 appears to induce the differentiation of R2 and R5, and the cascade of induction continues until the R7 photoreceptor is differentiated (After Tomlinson, 1988; photograph courtesy of T. Venkatesh.)

Antennal portion of disc

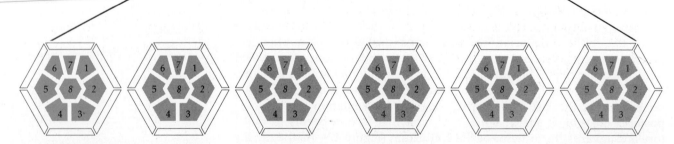

Later differentiation (posterior to morphogenetic furrow)

Early differentiation (entering morphogenetic furrow)

develop only if the R7 precursor cell has the wild-type *sevenless* allele (Basler and Hafen, 1989; Bowtell et al., 1989). Antibodies to this protein find it in the cell membrane, and the sequence of the *sevenless* gene suggests that it is transmembrane protein with a tyrosine kinase site in its cytoplasmic domain (Hafen et al., 1987; Banerjee et al., 1987). This is consistent with the protein's being a receptor for some signal.*

This signal for the R7 precursor to differentiate into the R7 photoreceptor probably comes directly or indirectly from a protein encoded by the wild-type allele of *bride of sevenless*. Flies homozygous for this *boss* mutation also do not have the R7 photoreceptors. Gene mosaic studies wherein some of the cells of the imaginal disc are wild-type and some of the cells are homozygous for the *boss* mutation show that the wild-type *boss* gene is not needed in the R7 precursor cell. Rather, the R7 photoreceptor only differentiates if the wild-type *boss* gene is expressed in the *R8* cell. Thus, the *bride of sevenless* gene is encoding some protein whose existence in the R8 cell is necessary for the differentiation of the R7 cell. The signal produced by the boss protein probably works by cell contact. Wild-type *boss* genes in an R8 cell in one ommatidium will not correct the deficiency of a mutant *boss* allele in the adjacent ommatidia (Reinke and Zipursky, 1988). A summary of the known cell-to-cell inductions in the *Drosophila* retina (Figure 28) shows that individual cells are able to induce other individual cells to create the precise arrangement of cells in particular tissues.

Vulval induction in *Caenorhabditis elegans*

The development of the vulva in *C. elegans* shows several instances of induction on the cellular level. The first instance concerns induction of the gonadal anchor cell. As discussed in Chapter 11, the formation of the anchor cell is mediated by the *lin-12* gene which is thought to encode a cell surface receptor protein. In wild-type hermaphrodites, two adjacent cells, Z1.ppp and Z4.aaa, have the potential to become the gonadal anchor cell. They interact in a manner that causes one of them to be the anchor cell while the other one becomes the precursor of the uterine tissue. In recessive *lin-12* mutants, both cells become anchor cells, while in dominant mutation, both cells become uterine precursors (Greenwald et al., 1983). Studies using genetic mosaics and cell ablations have shown that this decision is made in the second larval stage and that the *lin-12* gene only needs to function in that cell destined to become the uterine precursor cell. The presumptive anchor cell does not need it. Seydoux and Greenwald (1989) speculate that these two cells originally synthesize both the signal for uterine differentiation (which is still unidentified) and the receptor for this module, the lin-12 protein. During a particular time in larval development, the cell which, by chance, is secreting more of this differentiation signal causes its neighbor to cease its production of the signal molecule and to increase its production of lin-12 protein. The cell secreting the signal becomes the gonadal anchor cell, while the cell receiving the signal through its lin-12 protein becomes the ventral uterine precursor cell.

---

*As we saw in Chapter 2 and will detail in Chapter 20, one of the most effective ways of getting developmental information into a cell is through a membrane protein that has a cytoplasmic tyrosine kinase domain. When the extracellular portion of the protein binds to its ligand (whether it is on another cell, a part of the extracellular matrix, or in solution), its tyrosine kinase is activated. This enzymatic domain phosphorylates certain tyrosine residues in particular proteins, thereby activating these proteins. The activated proteins can alter the developmental fate of the cell. (The sperm–egg interaction can even be seen as reciprocal secondary induction on the cell-to-cell level. Secretions from the egg activate the sperm. The activated sperm can then bind and induce changes in the egg.)

FIGURE 28
Summary of genes known to be involved in the induction of *Drosophila* photoreceptors. For development to continue beyond the differentiation of the R8, R2, and R5 photoreceptors, the *rough* gene (*ro*) must be present in both the R2 and R5 cells. For the differentiation of the R7 photoreceptor, the *sevenless* gene (*sev*) has to be active in the R7 precursor cell, while the *bride of sevenless* (*boss*) gene must be active in the R8 photoreceptor. (After Rubin, 1989.)

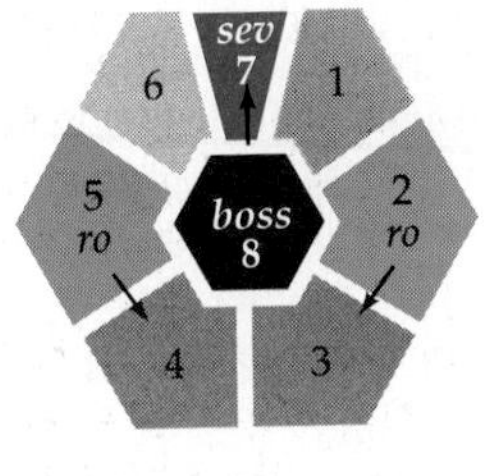

Thus, the two cells are thought to determine each other prior to their respective differentiation events.

Next, the anchor cell determines the fates of six VULVAL PRECURSOR CELLS. Each vulval precursor cell is equipotential and can acquire any one of three possible fates. These fates are determined largely by their proximity to the anchor cell. The anchor cell emits a signal, believed to be a diffusible molecule. This signal acts in a position-dependent manner (Figure 29). The vulval precursor cell closest to the anchor cell (usually P6.p) is given a "primary" fate and is instructed to divide symmetrically three times to form vulval cells. The two cells to the side of this primary cell are given "secondary" fates, instructed by the anchor cell to divide in an asymmetric manner to generate further vulval cells. The other three potential vulval precursors are not instructed to make vulval cells. Their "tertiary" fate is to divide once to provide cells for the hypodermis of the nematode (Ferguson et al., 1987). If the anchor cell is ablated by focused laser beams, all six vulval precursor cells remain uninduced and form hypodermis (Kimble, 1981).

In addition to the signal provided by the anchor cell, a second signal from the primary vulval precursor appears to stabilize the cell fates. This signal acts at very close range to inhibit the secondary vulval precursors from assuming a primary fate. Thus, there is LATERAL INHIBITION of the secondary vulval precursor cells by the primary vulval precursor cell (Sternberg, 1988).

The genetic basis of these inductions is presently being studied. Research by Sternberg and Horvitz (1989) and Seydoux and Greenwald (1989) have shown that at least three sets of genes regulate vulva development. These are *Vulvaless (Vul)*, *Multivulva (Muv)*, and *lin-12*. Nematodes deficient in *Vul* lack vulvas (so that the wild-type gene is necessary for vulva production). All the vulval precursor cells of these mutant nematodes acquire the tertiary cell fate. Conversely, animals lacking *Muv* genes have vulvas made from each of the six potential vulval precursor cells, indicating that the wild-type allele inhibits cells from generating vulval tissue. The

FIGURE 29

Determination of vulval cell lineages in *C. elegans*. Signal from the anchor cell causes the *Vulvaless* genes to be activated in three of the six potential precursor cells. The cell closest to the anchor cell becomes the primary vulval precursor. In the three cells that do not receive the anchor cell stimulus, the *Multivulva* genes repress the *Vulvaless* genes, and these cells generate hypodermis rather than vulval tissue. The precursor cell closest to the anchor cell secretes a short-range signal that induces the neighboring cells to activate the *lin-12* gene. This gene causes them to become secondary vulval precursors.

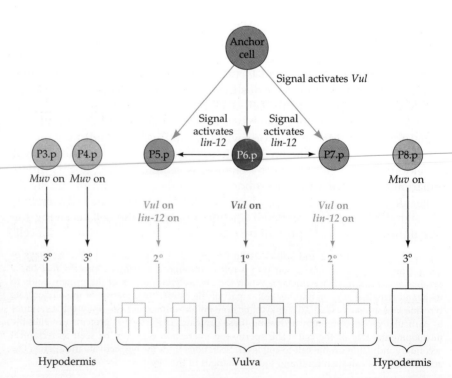

original state of the six vulval precursor cells has the *Multivulva* gene on. The gene product of the *Multivulva* gene is thought to inhibit the *Vulvaless* gene. In this state, the precursor cells do not make vulval cells, but will form hypodermis. However, the signal from the anchor cell abrogates *Muv* repression, allowing the wild-type *Vulvaless* gene to function in those cells receiving the anchor cell stimulus. In this step, then, the anchor cell signal determines that three cells will be primary or secondary vulval precursors and that the other three cells will generate hypodermis. The next step involves the distinction between the primary and the secondary vulval precursor cells. This is accomplished by the primary vulval precursor secreting a signal that activates the *lin-12* gene in its neighboring cells. If the *lin-12* gene is active in the cell, that cell becomes a secondary vulval precursor. In those nematodes lacking a functional *lin-12* gene, their vulval precursor cells acquire either primary or tertiary fates, but have no secondary vulval precursor cells. Thus, the combined actions of two intercellular inductions specifies the three fates of the potential vulval precursors.

The *lin-12* gene is obviously critical in the determination of cell fate and appears to encode a receptor protein. The sequence of the *lin-12* gene demonstrated the existence of a family of membrane proteins that functions to receive differentiation signals from neighboring cells (Yochem and Greenwald, 1989; Austin and Kimble, 1989). As shown in Figure 30, the sequence of lin-12 protein greatly resembles that of two other proteins, glp-1 and Notch. The glp-1 protein of *Caenorhabditis elegans* (as we will detail in Chapter 22) is needed in germ line cells if they are to receive the signal from the gonad to continue mitosis. The Notch protein (as discussed in Chapter 11) is needed for the differentiation of epidermal cells in *Drosophila*. Ectodermal cells lacking this protein become neuroblasts.

All three proteins begin with a stretch of repeated sequences that resemble a portion of the epidermal growth factor molecule; this "opening stretch" is therefore called "EGF-like repeats." Next come three repeated elements that are specific to these membrane proteins. After their transmembrane sections, the three proteins contain a region very similar to the cdc-10 protein of yeast that is involved in regulating cell division. The

FIGURE 30
Structural similarities in lin-12, glp-1, and Notch proteins. These three proteins are aligned by their postulated transmembrane sequences, with the extracellular domains to the left. A fourth member of this family has recently been found (Coffman et al., 1990) by hybridizing a radioactive *Notch* gene probe to a cDNA library from *Xenopus* neurula mRNA. This *Xenopus* gene has a structure extremely similar to that of the *Drosophila Notch* gene, and is also expressed in the ectoderm. (After Yochem and Greenwald, 1989.)

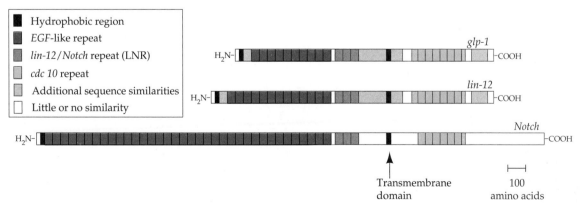

mechanisms by which these proteins receive their respective morphogenetic signals and transmit them to the nucleus is currently being studied. A combination of genetic, molecular, and embryological approaches is starting to delineate the inductive events by which cell fates are regulated, and we are beginning to discover the molecules involved in these processes.

Induction of the chick limb

Some developmental biologists claim that all the processes of development can be seen in the formation of limbs; and ever since R. G. Harrison began the experimental analysis of limb development at the beginning of this century, some of the most important concepts in developmental biology have come from the investigation of these structures. The major features of limb development are common throughout vertebrates. Limb development begins when mesenchymal cells are released from the somatic layer of the lateral plate mesoderm (Figure 31). The cells migrate laterally and accumulate under the epidermal tissue of the neurula. The circular bulge on the surface of the embryo is called the LIMB BUD.

Harrison (1918) discovered that these *mesenchymal* cells play an essential role in limb formation.

1 When the mesenchymal cells are removed, no limb forms.
2 When the limb mesenchyme is grafted to a new site, a new limb forms.
3 When the hindlimb mesenchyme is combined with forelimb ectoderm, a hindlimb forms.
4 When the limb ectoderm is grafted to a nonlimb site, no new limb forms at that new site.

We see, then, that limb mesenchyme is able to induce the formation of limbs. In birds and mammals, the mesoderm is seen to induce the ectodermal cells to elongate and form a special structure, the APICAL ECTODERMAL RIDGE (AER) (Figure 32). Saunders and his co-workers (1976) have also shown that the AER is a self-sustaining population of cells that does not exchange cells with its surroundings as the limb develops. Chick

FIGURE 31
Limb bud formation. Migration of mesodermal cells from the somatic region of the lateral plate mesoderm causes the limb bud in the amphibian embryo to bulge out. (After Balinsky, 1975.)

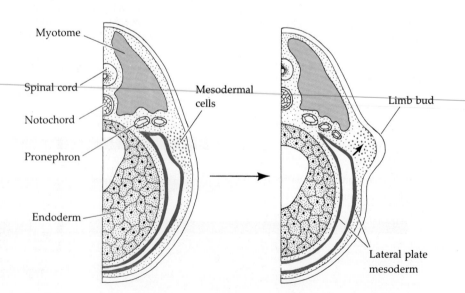

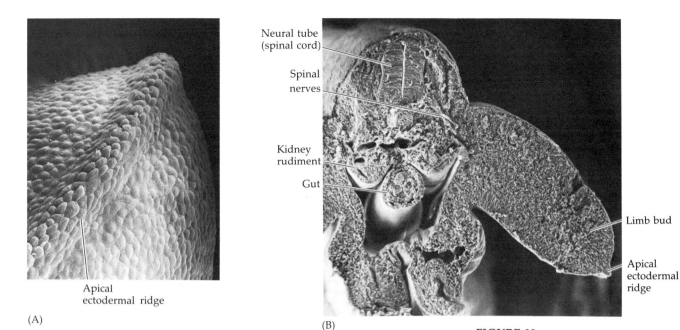

(B)

FIGURE 32
The apical ectodermal ridge. (A) Scanning electron micrograph of an early chick forelimb bud, with its apical ectodermal ridge in the foreground. (B) Scanning electron micrograph of a cross section of the 4-day chick embryo through the region of the limb bud. The AER can be seen at the distal tip of the bud. In the center of the embryo, one can see (starting dorsally) the neural tube, notochord, aorta, lateral kidney rudiments, and gut. Axons from the spinal ganglion can be seen elongating into the limb bud at this stage. (B from Tosney and Landmesser, 1985; photographs courtesy of K. W. Tosney.)

and quail cells can readily be distinguished by their nuclear heterochromatin (Chapter 6), and when a quail AER is placed upon the stump of a chick limb bud, the limb continues to grow. However, the AER of that limb remains composed of quail cells, and the surrounding chick ectoderm is not seen to become part of the AER. Once induced, this AER becomes essential to limb growth and interacts with the mesenchyme (Figure 33).

1 When the AER is removed at any time during limb development, further distal limb development ceases.

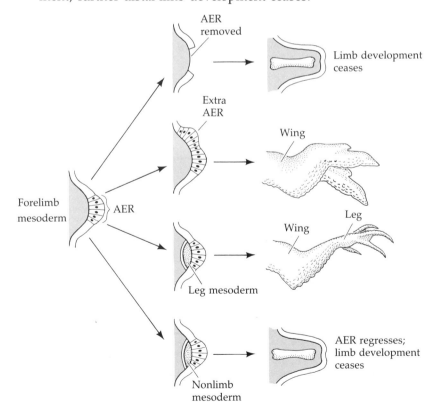

FIGURE 33
Summary of the inductive role of the apical ectodermal ridge (AER) upon the underlying mesenchyme. (Modified from Wessells, 1977.)

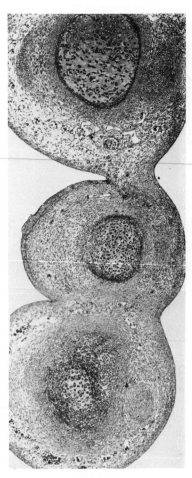

FIGURE 34

Cross section through the distal region of a chick limb 3 days after a wedge of the AER was removed from the area that would form interdigital tissue. Instead of degenerating, the remaining interdigital tissue formed an extra digit. (From Hurle et al., 1989.)

2 When an extra AER is grafted onto an existing limb bud, supernumerary structures are formed, usually toward the distal end of the limb.

3 When leg mesoderm is placed beneath the wing AER, hindlimb structures develop from that point onward.

4 When nonlimb mesoderm is grafted beneath the AER, the AER regresses and limb development ceases.

Thus, although the mesenchymal cells induce and sustain the AER and instruct the AER to produce a certain type of limb, the AER is responsible for the sustained outgrowth and development of the limb (Zwilling, 1955; Saunders et al., 1957; Saunders, 1972). The AER is also important for inhibiting the formation of cartilage in the mesenchymal cells directly beneath it. Solursh and Reiter (1988) have shown that limb ectoderm in general inhibits chondrogenesis in older mesenchyme, and Hurle and co-workers (1989) found that if they cut away a small portion of the AER in a region that would normally fall between the digits of the chick leg, an extra digit emerged at that place (Figure 34). It thus appears that the AER also has the effect of inhibiting induction of cartilage in the mesenchyme cells directly beneath it.

The relationships between the AER and the limb bud mesenchyme can best be seen by two mutations of chick limb development, *polydactylous* and *eudiplopodia*. The *polydactylous* mutation, as the name implies, confers extra digits on each limb. By recombining mutant and wild-type tissues (Table 3), the defect can be traced to mesodermal cells that induce too broad an AER. In the mutant *eudiplopodia* (Greek, meaning "two good feet"), one has not only extra digits, but two complete rows of toes on each hindlimb (Figure 35). Similar reconstitution experiments (Table 3) show that here the defect is in the ectodermal tissue.

It is clear that secondary induction, often of a reciprocal nature as seen in kidney and limb development, is essential for organ formation. But secondary inductions alone are not sufficient for generating an entire organ. There must exist a mechanism for ensuring that a limb does not develop with its palm away from the body, or develop with two thumbs. There must be a mechanism for ensuring that legs and arms occur at the appropriate site in the body and not at other sites as well. Similarly, although the retina can receive images from the incredibly well coordinated eye, there must be a mechanism whereby the nerves from the retina

TABLE 3

Mutations affecting the reciprocal interactions between AER and its underlying mesenchyme[a]

Mesoderm	Epidermis	Result	Conclusion
POLYDACTYLOUS			
Polydactylous	Wild-type	Polydactylous	Mesoderm is affected
Wild-type	Polydactylous	Wild-type	by the mutation
EUDIPLOPODIA			
Eudiplopodia	Wild-type	Wild-type	Ectoderm is affected
Wild-type	Eudiplopodia	Eudiplopodia	by the mutation

[a] By reciprocal transplantation between wild-type and mutant AER and mesenchyme, the aberrant compartment of the induction can be identified.

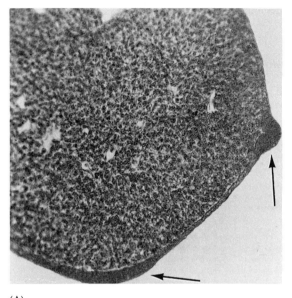
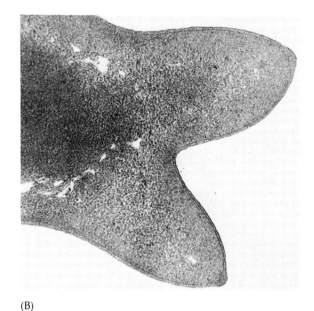

(A)   (B)

FIGURE 35
Cross sections of hindlimb buds from *eudiplopodia* chick embryos. (A) Two AERs on hindlimb bud; extra outgrowth on the dorsal side will form an extra set of toes. (B) Both outgrowth regions are covered by an AER. (From Goetinck, 1964; courtesy of P. Goetinck.)

relay this information to the correct area of the brain. The next chapter will continue the study of organ formation and will focus on pattern formation in the embryo.

LITERATURE CITED

Arechaga, J., Karcher-Djvricic, V. and Ruch, J. V. 1983. Neutral morphogenetic activity of epithelium in heterologous tissue recombinations. *Differentiation* 25: 142–147.

Auerbach, R. 1972. The use of tumors in the analysis of inductive tissue interaction. *Dev. Biol.* 28: 304–309.

Austin, J. and Kimble, J. 1989. Transcript analysis of *glp-1* and *lin-12*, homologous genes required for cell interactions during development of *C. elegans*. *Cell* 58: 565–571.

Balinsky, B. I. 1975. *Introduction to Embryology*, 4th Ed. Saunders, Philadelphia.

Banerjee, S. and Bernfield, M. 1979. Developmentally regulated neutral hyaluronidase activity during epithelial mesenchymal interaction. *J. Cell Biol.* 83: 469a.

Banerjee, U., Renfranz, P. J., Pollock, J. A. and Benzer, S. 1987. Molecular characterization and expression of *sevenless*, a gene involved in neuronal pattern formation in the *Drosophila* eye. *Cell* 49: 281–291.

Basler, K. and Hafen, E. 1989. Ubiquitous expression of *sevenless*: Position-dependent specification of cell fate. *Science* 243: 931–934.

Bernfield, M. and Banerjee, S. D. 1982. The turnover of basal lamina glycosaminoglycan correlates with epithelial morphogenesis. *Dev. Biol.* 90: 291–305.

Bishop-Calame, S. 1966. Étude experimentale de l'organogenese du systéme urogénital de l'embryon de poulet. *Arch. Anat. Microsc. Morphol. Exp.* 55: 215–309.

Bowtell, D. D. L., Simon, M. A. and Rubin, G. M. 1989. Ommatidia in the developing *Drosophila* eye require and can respond to sevenless for only a restricted period. *Cell* 56: 931–936.

Brown, K. M., Muchmore, A. V. and Rosenstreich, D. L. 1986. Uromodulin, an immunosuppressive protein derived from pregnancy urine, is an inhibitor of interleukin 1. *Proc. Natl. Acad. Sci. USA* 83: 9119–9123.

Charlebois, T. S., Spencer, D. H., Tarkington, S. K., Henry, J. J. and Grainger, R. M. 1990. Isolation of a chick cytokeratin cDNA clone which detects regional specialization in early embryonic ectoderm. *Development* 108: 33–45.

Chuong, C.-M. and Edelman, G. M. 1985. Expression of cell adhesion molecules in embryonic induction. I. Morphogenesis of nestling feathers. *J. Cell Biol.* 101: 1009–1026.

Coffman, C., Harris, W. and Kintner, C. 1990. *Xotch*, the *Xenopus* homolog of *Drosophila* Notch. *Science* 249: 1438–1441.

Coulombre, J. L. and Coulombre, A. J. 1971. Metaplastic induction of scales and feathers in the corneal anterior epithelium of the chick embryo. *Dev. Biol.* 25: 464–478.

Cutler, L. S. and Chaudhry, A. P. 1973. Intercellular contacts at the epithelial–mesenchymal interface during prenatal development of the rat submandibular gland. *Dev. Biol.* 33: 229–240.

Dalgleish, A. G., Beverley, P. C. L., Clapham, P. R., Crawford, D. H., Greaves, M. F. and Weiss, R. A. 1984. The CD4 (T4) antigen is an essential component of the receptor for the AIDS retrovirus. *Nature* 312: 763–767.

Daya, S., Rosenthal, K. L. and Clark, D. A. 1987. Immunosuppressor factor(s) produced by decidua-associated suppressor cells: A proposed mechanism for fetal allograft survival. *Am. J. Obstet. Gynecol.* 156: 344–350.

Deuchar, E. M. 1975. *Cellular Interactions in Animal Development*. Chapman and Hall, London.

Doyle, C. and Strominger, J. L. 1987. Interaction between CD4 and class II MHC molecules mediates cell adhesion. *Nature* 330: 256–259.

Ekblom, P., Thesleff, I., Saxén, L., Miettinen, A. and Timpl, R. 1983. Transferrin as a fetal growth factor: Acquisition of responsiveness related to embryonic induction. *Proc. Natl. Acad. Sci. USA* 80: 2651–2655.

Etheridge, A. L. 1968. Determination of the mesonephric kidney. *J. Exp. Zool.* 169: 357–370.

Ferguson, E. C., Sternberg, P. W. and Horvitz, R. 1987. A genetic pathway for the specification of vulval cell lineages of *Caenorhabditis elegans*. *Nature* 326: 259–267.

Fukumachi, H. and Takayama, S. 1980. Epithelial-mesenchymal interaction in differentiation of duodenal epithelium of fetal rats in organ culture. *Experientia* 36: 335–336.

Gluecksohn-Schoenheimer, S. 1943. The morphological manifestations of a dominant mutation in mice affecting tail and urogenital system. *Genetics* 28: 341–348.

Goetinck, P. F. 1964. Studies on limb morphogenesis. II. Experiments with the polydactylous mutant *eudiplopodia*. *Dev. Biol.* 10: 71–91.

Goetinck, P. F. and Sekellick, M. J. 1972. Observations on collagen synthesis, lattice formation, and the morphology of scaleless and normal embryonic skin. *Dev. Biol.* 28: 636–648.

Greenwald, I., Sternberg, P. W. and Horvitz, H. R. 1983. The *lin-12* locus specifies cell fates in *Caenorhabditis elegans*. *Cell* 34: 435–444.

Grobstein, C. 1955. Induction interaction in the development of the mouse metanephros. *J. Exp. Zool.* 130: 319–340.

Grobstein, C. 1956. Trans-filter induction of tubules in mouse metanephrogenic mesenchyme. *Exp. Cell Res.* 10: 424–440.

Grobstein, C. and Cohen, J. 1965. Collagenase: Effect on the morphogenesis of embryonic salivary epithelium in vitro. *Science* 150: 626–628.

Grunwald, G. B., Gilbert, S. F., Brewer, K., Cleland, L. and Kawai, M. 1990. Immunocytochemical analysis of embryonic compartmentation with a monoclonal antibody against a cytokeratin-related antigen. *Histochemistry* 94: 545–553.

Guillet, J.-G. and 7 others. 1987. Immunological self, nonself discrimination. *Science* 235: 865–870.

Gumpel-Pinot, M., Yasugi, S. and Mizuno, T. 1978. Différenciation d'épithéliums endodermiques associés au mésoderme splanchnique. *Comp. Rend. Acad. Sci. (Paris)* 286: 117–120.

Hafen, E., Basler, K., Edstrom, J. E. and Rubin, G. M. 1987. *sevenless*, a cell-specific homeotic gene of *Drosophila*, encodes a putative transmembrane receptor with a tyrosine kinase domain. *Science* 236: 55–63.

Hamaoko, T. and Ono, S. 1986. Regulation of B-cell differentiation. *Annu. Rev. Immunol.* 4: 167–204.

Hamburgh, M. 1970. *Theories of Differentiation.* Elsevier, New York.

Harrison, R. G. 1918. Experiments on the development of the forelimb of *Amblystoma*, a self-differentiating equipotential system. *J. Exp. Zool.* 25: 413–461.

Harrison, R. G. 1920. Experiments on the lens in *Amblystoma*. *Proc. Soc. Exp. Biol. Med.* 17: 199–200.

Harrison, R.G. 1933. Some difficulties of the determination problem. *Am. Nat.* 67: 306–321.

Hay, E. D. and Dodson, J. W. 1973. Secretion of collagen by corneal epithelium. I. Morphology of the collagenous products produced by isolated epithelia grown on frozen-killed lens. *J. Cell Biol.* 57: 190–213.

Hay, E. D. and Revel, J.-P. 1969. Fine structure of the developing avian cornea. *In* A. Wolsky and P. S. Chen (eds.), *Monographs in Developmental Biology*. Karger, Basel.

Henry, J. J. and Grainger, R. M. 1987. Inductive interactions in the spatial and temporal restriction of lens-forming potential in embryonic ectoderm. *Dev. Biol.* 124: 200–214.

Henry, J. J. and Grainger, R. M. 1990. Early tissue interactions leading to embryonic lens formation in *Xenopus laevis*. *Dev. Biol.* 141: 149–163.

Holtzer, H. 1968. Induction of chondrogenesis: A concept in terms of mechanisms. *In* R. Gleischmajer and R. E. Billingham (eds.), *Epithelial–Mesenchymal Interactions*. Williams & Wilkins, Baltimore, pp. 152–164.

Houghton, A., Real, F. X., Davis, L. J., Cordon-Cardo, C. and Old, L. J. 1987. Phenotypic heterogeneity of melanoma. Relation to the differentiation program of melanocyte cells. *J. Exp. Med.* 164: 812–829.

Howard, M. and Paul, W. E. 1983. Regulation of B-cell growth and differentiation by soluble factors. *Annu. Rev. Immunol.* 1: 307–333.

Hurle, J. M., Gañan, Y. and Macias, D. 1989. Experimental analysis of the in vivo chondrogenic potential of the interdigital mesenchyme of the chick limb bud subjected to local ectodermal removal. *Dev. Biol.* 132: 368–374.

Inabar, K., Witmer, M. D. and Steinman, R. M. 1984. Clustering of dendritic cells, helper T lymphocytes, and histocompatible B cells during primary antibody responses in vitro. *J. Exp. Med.* 160: 858–876.

Jacobson, A. G. 1966. Inductive processes in embryonic development. *Science* 152: 25–34.

Jacobson, A. G. and Sater, A. K. 1988. Features of embryonic induction. *Development* 104: 341–359.

Janeway, C. A., Ron, J. and Katz, M. E. 1987. The B cell is the initiating antigen-presenting cell in peripheral lymph nodes. *J. Immunol.* 138: 1051–1055.

Killar, L., MacDonald, G., West, J., Woods, A. and Bottomly, K. 1987. Cloned, Ia-restricted T cells that do not produce IL4

(BSF-1) fail to help antigen-specific B cells. *J. Immunol.* 138: 1674–1679.

Kimble, J. 1981. Alterations in cell lineage following laser ablation of cells in the somatic gonad of *Caenorhabditis elegans*. *Dev. Biol.* 87: 286–300.

King, H. D. 1905. Experimental studies on the eye of the frog embryo. *Wilhelm Roux Arch. Entwicklungsmech. Org.* 19: 85–107.

Klatzmann, D. and 10 others. 1984. Selective tropism of lymphadenopathy associated virus (LAV) for helper–inducer T lymphocytes. *Science* 225: 59–63.

Kollar, E. J. and Baird, G. 1970. Tissue interaction in developing mouse tooth germs. II. The inductive role of the dental papilla. *J. Embryol. Exp. Morphol.* 24: 173–186.

Kollar, E. J. and Fisher, C. 1980. Tooth induction in chick epithelium: Expression of quiescent genes for enamel synthesis. *Science* 207: 993–995.

Lanzavecchia, A. 1985. Antigen-specific interaction between T and B cells. *Nature* 314: 537–539.

Lee, T. and 11 others. 1986. Isolation and characterization of a mouse interleukin cDNA clone that expresses B-cell stimulatory factor 1 activity and T-cell- and mast cell-stimulating activities. *Proc. Natl. Acad. Sci. USA* 83: 2061–2065.

Lehtonen, E. 1975. Epithelio-mesenchymal interface during mouse kidney tubule induction in vivo. *J. Embryol. Exp. Morphol.* 34: 695–705.

Lehtonen, E., Wartiovaara, J., Nordling, S. and Saxén, L. 1975. Demonstration of cytoplasmic processes in Millipore filters permitting kidney tubule induction. *J. Embryol. Exp. Morphol.* 33: 187–203.

Lehtonen, E., Virtanen, I. and Saxén, L. 1985. Reorganization of the intermediate cytoskeleton in induced mesenchyme cells is independent of tubule morphogenesis. *Dev. Biol.* 108: 481–490.

Lelongt, B., Makino, H., Dalecki, T. M. and Kanwar, Y. S. 1988. Role of proteoglycans in renal development. *Dev. Biol.* 128: 256–276.

Levine, S., Pictet, R. and Rutter, W. J. 1973. Control of cell proliferation and cytodifferentiation by a factor reacting with the cell surface. *Nature New Biol.* 246: 49–52.

Lewis, W. 1904. Experimental studies on the development of the eye in amphibia. I. On the origin of the lens, *Rana palustris*. *Am. J. Anat.* 3: 505–536.

Lewis, W. 1907. Experimental studies on the development of the eye in Amphibia. III. On the origin and differentiation of the lens. *Am. J. Anat.* 6: 473–509.

Lumsden, A. G. S. 1988. Spatial organization of the epithelium and the role of neural crest cells in the initiation of the mammalian tooth germ. *Development* 103 [Suppl.]: 155–169.

Mackie, E. J., Thesleff, I. and Chiquet-Ehrismann, R. 1987. Tenascin is associated with chondrogenic and osteogenic differentiation *in vivo* and promotes chondrogenesis *in vivo*. *J. Cell Biol.* 105: 2569–2579.

Mauger, A., Demarchez, M., Herbage, D., Grimaud, J. A., Druguet, M., Hartmann, D. and Sengel, P. 1982. Immunofluorescent localization of collagen types II and III, and of fibronectin during feather morphogenesis in the chick embryo. *Dev. Biol.* 94: 93–105.

Mauger, A., Demarchez, M., Herbage, D., Grimaud, J. A., Druguet, M., Hartmann, D. J., Foidart, J. M. and Sengel, P. 1983. Immunofluorescent localization of collagen types I, III, and IV, fibronectin, and laminin during morphogenesis of scales and scaleless skin in the chick embryo. *Wilhelm Roux Arch. Dev. Biol.* 192: 205–215.

Meier, S. 1977. Initiation of corneal differentiation prior to cornea–lens association. *Cell Tiss. Res.* 184: 255–267.

Meier, S. and Hay, E. D. 1974. Control of corneal differentiation by extracellular materials. Collagen as promoter and stabilizer of epithelial stroma production. *Dev. Biol.* 38: 249–270.

Meier, S. and Hay, E. D. 1975. Stimulation of corneal differentiation by interaction between the cell surface and extracellular matrix. I. Morphometric analysis of transfilter induction. *J. Cell Biol.* 66: 275–291.

Melchers, F. and Andersson, J. 1986. Factors controlling the B cell cycle. *Annu. Rev. Immunol.* 4: 13–36.

Mencl, E. 1908. Neue Tatsachen zur Selbstdifferenzierung der Augenlinse. *Wilhelm Roux Arch. Entwicklungsmech. Org.* 25: 431–450.

Mina, M. and Kollar, E. J. 1987. The induction of odontogenesis in non-dental mesenchyme combined with early murine mandibular archepithelium. *Arch. Oral Biol.* 32: 123–127.

Mjaatvedt, C. H. and Markwald, R. R. 1989. Induction of an epithelial-mesenchymal transition by an *in vivo* adheron-like complex. *Dev. Biol.* 136: 118–128.

Mosier, D. E. 1967. A requirement for two cell types for antibody formation in vitro. *Science* 158: 1573–1575.

Nakanishi, Y., Sugiura, F., Kishi, J.-I. and Hayakawa, T. 1986a. Collagenase inhibitor stimulates cleft formation during early morphogenesis of mouse salivary gland. *Dev. Biol.* 113: 201–206.

Nakanishi, Y., Sugiura, F., Kishi, J.-I. and Hayakawa, T. 1986b. Scanning electron microscopic observations of mouse embryonic submandibular glands during initial branching: Preferential localization of fibrillar structures at the mesenchymal ridges participating in cleft formation. *J. Embryol. Exp. Morphol.* 96: 65–77.

Nakanishi, Y., Morita, T. and Nogawa, H. 1987. Cell proliferation is not required for the initiation of early cleft formation in mouse embryonic submandibular epithelium in vitro. *Development* 99: 429–437.

Nakanishi, Y., Nogawa, H., Hashimoto, Y., Kishi, J.-I. and Hayakawa, T. 1988. Accumulation of collagen III at the cleft points of developing mouse submandibular gland. *Development* 104: 51–59.

Nieuwkoop. P. 1952. Activation and organization of the central nervous system in amphibians. *J. Exp. Zool.* 120: 1–108.

Nieuwkoop, P. 1963. Pattern formation and artificially activated ectoderm. *Dev. Biol.* 7: 255–279.

O'Garra, A., Warren, D. J., Holman, M., Popham, A. M., Sanderson, C. J. and Klaus, G. G. B. 1986. Interleukin 4 (B cell growth factor II/eosinophil differentiation factor) is a mitogen and differentiation factor for preactivated B lymphocytes. *Proc. Natl. Acad. Sci. USA* 83: 5228–5232.

Osathanondh, V. and Potter, E. 1963. Development of human kidney as shown by microdissection. III. Formation and interrelationships of collecting tubules and nephrons. *Arch. Pathol.* 76: 290–302.

Pierce, B. G. 1985. Carcinoma is to embryology as mutation is to genetics. *Am. Zool.* 25: 707–712.

Pritchard-Jones, K. and 11 others. 1990. The candidate Wilms' tumour gene is involved in genitourinary development. *Nature* 346: 194–197.

Reinke, R. and Zipursky, A. L. 1988. Cell-cell interaction in the *Drosophila* retina: The *bride of sevenless* gene is required in photoreceptor cell R8 for R7 cell development. *Cell* 55: 321–330.

Repraeger, A. and Bernfield, M. 1985. Cell surface proteoglycan of mouse mammary epithelial cells: Protease releases a heparin sulfate-rich ectodomain from a putative membrane-anchored domain. *J. Biol. Chem.* 260: 4103–4109.

Roux, W. 1894. The problems, methods, and scope of developmental mechanics. *Lectures of the Marine Biological Laboratory.* Ginn & Co., Boston, pp. 149–190.

Rubin, G. M. 1989. Development of the *Drosophila* retina: Inductive events studied at single cell resolution. *Cell* 57: 519–520.

Rutter, W. J., Wessells, N. K. and Grobstein, C. 1964. Controls of specific synthesis in the developing pancreas. *Natl. Cancer Inst. Monogr.* 13: 51–65.

Rutter, W. J., Pictel, R. L., Harding, J. D., Chirgwin, J. M., MacDonald, R. J. and Przybyla, A. E. 1978. An analysis of pancreatic development: Role of mesenchymal factor and other extracellular factors. *In* J. Papaconstantinou and W. J. Rutter (eds.), *Molecular Control of Proliferation and Differentiation.* Academic Press, New York, pp. 205–277.

Saha, M. S. 1991. Spemann seen through a lens. *In* S. F. Gilbert (ed.), *A Conceptual History of Modern Embryology.* Plenum, New York. In press.

Saha, M. S., Spann, C. L. and Grainger, R. M. 1989. Embryonic lens induction: More than meets the optic vesicle. *Cell Differ. Dev.* 28: 153–172.

Sariola, H., Ekblom, P. and Saxén, L. 1982. Restricted developmental options of the metanephric mesenchyme. *In* M. Burger

and R. Weber (eds.), *Embryonic Development,* Part B: *Cellular Aspects.* Alan R. Liss, New York, pp. 425–431.

Saunders, J. W., Jr. 1972. Developmental control of three-dimensional polarity in the avian limb. *Ann. N.Y. Acad. Sci.* 193: 29–42.

Saunders, J. W., Jr. 1980. *Developmental Biology.* Macmillan, New York.

Saunders, J. W., Jr., Cairns, J. M. and Gasseling, M. T. 1957. The role of the apical ridge of ectoderm in the differentiation of the morphological structure and inductive specificity of limb parts of the chick. *J. Morphol.* 101: 57–88.

Saunders, J. W., Jr., Gasseling, M. T. and Errick, J. E. 1976. Inductive activity and enduring cellular constitution of a supernumerary apical ectodermal ridge grafted to the limb bud of the chick embryo. *Dev. Biol.* 50: 16–25.

Saxén, L. 1970. Failure to demonstrate tubule induction in heterologous mesenchyme. *Dev. Biol.* 23: 511–523.

Saxén, L. and Sariola, H. 1987. Early organogenesis of the kidney. *Pediat. Nephrol.* 1: 385–392.

Saxén, L., Lehtonen, E., Karkinen-Jääskeläinen, M., Nordling, S. and Wartiovaara, J. 1976. Are morphogenetic tissue interactions mediated by transmissible signal substances or through cell contacts? *Nature* 259: 662–663.

Servetnick, M. and Grainger, R. M. 1990. Changes in competence occur autonomously in *Xenopus* ectoderm. (Abstract) Forty-Ninth Annual Symposium of the Society for Developmental Biology. Alan R. Liss, New York. p. 4.

Seydoux, G. and Greenwald, I. 1989. Cell autonomy of *lin-12* function in a cell fate decision in *C. elegans*. *Cell* 57: 1237–1245.

Slavkin, H. C. and Bringas, P., Jr. 1976. Epithelial–mesenchymal interactions during odontogenesis. IV. Morphological evidence for direct heterotypic cell–cell contacts. *Dev. Biol.* 428–442.

Snow, E. C., Noelle, R. J., Uhr, J. W. and Vitetta, E. S. 1983. Activation of antigen-enriched B cells. II. Role of linked recognition in B cell proliferation to thymus-dependent antigens. *J. Immunol.* 130: 614–618.

Solursh, M. and Reiter, R. S. 1988. Inhibitory and stimulatory effects of limb ectoderm on in vitro chondrogenesis. *J. Exp. Zool.* 248: 147–154.

Spemann, H. 1901. Über Correlationen in der Entwicklung des Auges. *Verh. Anat. Ges. 15 Vers. Bonn.* 61–79.

Spemann, H. 1938. *Embryonic Development and Induction.* Yale University Press, New Haven.

Spemann, H. and Schotté, O. 1932. Über xenoplatische Transplantation als Mittel zur Analyse der embryonalen Induktion. *Naturwissenschaften* 20: 463–467.

Sternberg, P. W. 1988. Lateral inhibition during vulval induction in *Caenorhabditis elegans*. *Nature* 335: 551–554.

Sternberg, P. W. and Horvitz, H. R. 1989. The combined action of two intercellular signalling pathways specifies three cell fates during vulval induction in *C. elegans*. *Cell* 58: 679–693.

Stuart, E. S., Garber, B. and Moscona, A. 1972. An analysis of feather germ formation in normal development and in skin treated with hydrocortisone. *J. Exp. Zool.* 179: 97–110.

Thesleff, I., Vainio, S. and Jalkanen, M. 1989. Cell–matrix interaction in tooth development. *Int J. Dev. Biol.* 33: 91–97.

Thesleff, I., Vaahtokari, A. and Vainio, S. 1990. Molecular changes during determination and differentiation of the dental mesenchyme cell lineage. *J. Biol. Bucalle* 18: 179–188.

Toivonen, S. 1979. Transmission problem in primary induction. *Differentiation* 15: 177–181.

Tomlinson, A. 1988. Cellular interactions in the developing *Drosophila* eye. *Development* 104: 183–193.

Tomlinson, A. and Ready, D. F. 1987. Cell fate in the *Drosophila* ommatidium. *Dev. Biol.* 123: 264–275.

Tosney, K. W. and Landmesser, L. T. 1985. Development of the major pathways for neurite outgrowth in the chick hindlimb. *Dev. Biol.* 109: 193–214.

Unanue, E. R. and Askonas, B. A. 1968. The immune response of mice to antigen in macrophages. *Immunology* 15: 287–296.

Vainio, S., Lehtonen, E., Jalkanen, M., Bernfield, M. and Saxén, L. 1989a. Epithelial–mesenchymal interactions regulate the stage-specific expression of a cell surface proteoglycan, syndecan, in the developing kidney. *Dev. Biol.* 134: 382–391.

Vainio, S., Jalkanen, M. and Thesleff, I. 1989b. Syndecan and tenascin expression is induced by epithelial–mesenchymal interactions in embryonic tooth mesenchyme. *J. Cell Biol.* 108: 1945–1954.

van Heyningen, V. and 11 others. 1990. Role for Wilms tumor gene in genital development? *Proc. Natl. Acad. Sci. USA* 87: 5383–5386.

von Woellwarth, V. 1961. Die Rolle des neuralleistenmaterials und der Temperatur bei der Determination der Augenlinse. *Embriologia* 6: 219–242.

Waddington, C. H. 1940. *Organisers and Genes*. Cambridge University Press, Cambridge.

Wessells, N. K. 1970. Mammalian lung development: Interactions in formulation and morphogenesis of tracheal buds. *J. Exp. Zool.* 175: 455–466.

Wessells, N. K. 1977. *Tissue Interaction and Development*. Benjamin, Menlo Park, CA.

Wessells, N. K. and Evans, J. 1968. The ultrastructure of oriented cells and extracellular materials between developing feathers. *Dev. Biol.* 18: 42–61.

Wilkinson, D. G., Bhatt, S. and McMahon, A. P. 1989. Expression of the FGF-related proto-oncogene *int*-2 suggests multiple roles in fetal development. *Development* 105: 131–136.

Yochem, J. and Greenwald, I. 1989. *glp-1* and *lin-12* genes implicated in distinct cell–cell interactions in *C. elegans*, encode similar transmembrane proteins. *Cell* 58: 553–563.

Zwilling, E. 1955. Ectoderm–mesoderm relationship in the development of the chick embryo limb bud. *J. Exp. Zool.* 128: 423–441.

17

Pattern formation

Thus, beyond all questions of quantity there lie questions of pattern which are essential for understanding Nature.

—ALFRED NORTH WHITEHEAD (1934)

Biochemistry and morphology are very shortly going to blend into each other without any difference or inequality. . . . Form is no longer the perquisite of the morphologist, and molecular exactitude no longer the preserve of the chemist.

—JOSEPH NEEDHAM (1967)

Theory without fact is fantasy, but facts without theory is chaos.

—C. O. WHITMAN (1894)

Pattern formation is the activity by which embryonic cells form ordered spatial arrangements of differentiated tissues. The ability to carry out this process is one of the most dramatic properties of developing organisms, and one that has provoked a sense of awe in scientists and laypeople alike. How is it that the embryo is able not only to generate the different cell types of the body, but also to produce them in a way that forms functional tissues and organs? It is one thing to differentiate the chondrocytes and osteocytes that synthesize the cartilage and bone matrices, respectively; it is another thing to produce these cells in a temporal–spatial orientation that generates a functional bone. It is still another thing to make that bone a humerus and not a pelvis or a femur. The ability of cells to sense their relative positions within a limited population of cells and to differentiate with regard to their position has been the subject of intense debate and experimentation.

This chapter will concentrate on two problems of pattern formation. The first involves the formation of the vertebrate limb. How are the cells that differentiate into the cartilage of the embryonic bone specified so as to form digits at one end and a shoulder at the other? (It would be quite a useless appendage if the order were reversed.) Here the cell types are the same, but the patterns they form are different. The second problem concerns how neurons find their target tissues. How is a nerve axon informed that it is to grow from the spinal cord to a fingertip rather than to a toe or a tooth? Once a neuron from the retina reaches the brain, how does it form synapses with the appropriate neurons that interpret vision? The answers to these questions are not fully known. What will be presented here are the theories for pattern formation and evidence to support these views.

THE FORMATION OF THE VERTEBRATE LIMB

Pattern formation during limb development: Specification of axes

The vertebrate limb is an extremely complex organ with an asymmetric pattern of parts. The bones of the forelimb, be it wing, hand, flipper, or fin, consist of a proximal humerus (adjacent to the body wall), a radius and an ulna in the middle region, and the distal bones of the wrist and the digits (Figure 1). Originally, these structures are cartilaginous, but eventually most of the cartilage is replaced by bone. The position of each of the bones and muscles in the limb is precisely organized. Polarity exists in other dimensions as well. In humans, it is obvious that each hand develops as a mirror image of the other. It is possible for other arrangements to exist—such as the thumb developing on the left side of both hands—but this is not generally seen. In some manner, the three-dimensional pattern of forelimb is routinely produced.

The basic "morphogenetic rules" for forming a limb appear to be the same in all tetrapods. Fallon and Crosby (1977) showed that grafted pieces of reptile or mammal limb bud could direct the formation of chick limb development, and Sessions and co-workers (1989) have extended this analysis to frogs and salamanders. Moreover, the regeneration of salamander limbs appears to follow the same rules (Muneoka and Bryant, 1982). But what are these "morphogenetic rules"? Recent research has proposed three types of mechanisms to account for the formation of the pattern of the vertebrate limb. The first type of pattern-forming mechanism is a *gradient*: a homogeneous population of cells is instructed to act in certain ways by the concentration of a soluble molecule. The second pattern-generating mechanism concerns the induction of cell growth when certain cells are adjacent to one another. The third mechanism involves chemical reactions whereby the formation of mesenchymal cell conden-

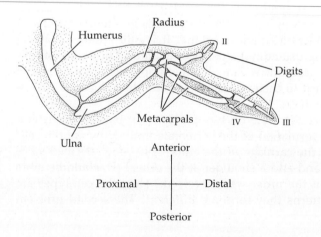

FIGURE 1
Skeletal pattern of the chick wing. Digits are numbered according to convention: II, III, IV. Digits I and V are not found in chick wings. (After Saunders, 1982).

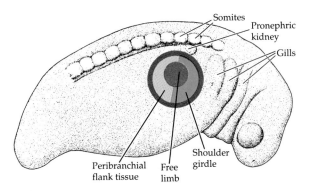

FIGURE 2

Prospective forelimb field of the salamander *Ambystoma maculatum*. The central area contains those cells destined to form the free limb; the cells surrounding the free limb are those that give rise to the peribrachial flank tissue and the shoulder girdle. The cells outside these regions usually are not included in limbs but can regulate to form a limb if the more central tissues are extirpated. (After Stocum and Fallon, 1982.)

sations is promoted in some regions and inhibited in others. But before discussing the models for limb pattern formation, we must first review the biology of the developing limb.

The limb field

The mesodermal cells that give rise to the vertebrate limb can be identified by (1) removing certain groups of cells and observing whether a limb develops in their absence (Detwiler, 1918; Harrison, 1918); (2) transplanting certain groups of cells to new locations and observing whether they form a limb (Hertwig, 1925); and (3) marking groups of cells with dyes or radioactive precursors and observing which descendants of marked cells partake in limb development (Rosenquist, 1971). By these procedures, the PROSPECTIVE LIMB AREA has been precisely localized in many vertebrate embryos. Figure 2 shows the prospective forelimb area in the tailbud stage of the salamander *Ambystoma maculatum*. The center of this disc marks the lateral plate mesoderm cells normally destined to give rise to the limb itself. Adjacent to it are the cells that will form the peribrachial flank tissue and the shoulder girdle. These two regions encompass the classic "limb disc" that will be used in the experiments to be mentioned in this chapter. However, if all these cells are extirpated from the embryo, a limb will still form, albeit somewhat later, from an additional ring of cells that surrounds this area. If this last ring of cells is included in the extirpated tissue, no limb will develop. This larger region, representing all the cells in the area capable of forming a limb, is called the LIMB FIELD. A FIELD can be described as a group of cells whose position and fate are specified with respect to the same set of boundaries (Weiss, 1939; Wolpert, 1977).

The limb field originally has the ability to regulate for lost or added parts. In the tailbud-stage *Ambystoma*, any half of the limb disc is able to regenerate the entire limb when grafted to a new site (Harrison, 1918). This potential can also be shown by splitting the limb disc vertically into two or more segments and placing thin barriers between these segments to prevent their reunion. When this is done, each part develops into a full limb. The regulative ability of the limb bud was recently highlighted by a remarkable experiment of nature. In a pond in Santa Cruz, California, numerous multilegged frogs and salamanders have been found (Figure 3). The presence of these extra appendages has been linked to the infestation of the larval abdomen with parasitic trematode worms. The eggs of these worms apparently split the limb bud in several places while the tadpole was first forming these structures (S. Sessions, unpublished data). Thus, like an early sea urchin embryo, the limb field may represent a "harmonious equipotential system" wherein a cell can be instructed to form any part of the limb.

FIGURE 3

Regulative ability of the limb field, seen when the early hindlimb fields of a *Hyla regila* tadpole were split by numerous trematode eggs. (Photograph courtesy of S. Sessions.)

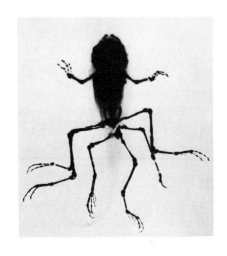

Regeneration of salamander limbs

One of the most impressive instances of regulation is the regeneration of the amphibian limb. When a salamander limb is amputated, the remaining cells are able to reconstruct a complete limb with all its differentiated cells arranged in the proper order. It is remarkable that not only has the limb regenerated, but the remaining cells retain the information specifying their position and the position of the cells that have been removed. In other words, the new cells make only the missing structures and no more; for example, when a wrist is amputated, the salamander forms a new wrist and not a new elbow (Figure 4).

Upon amputation, epidermal cells from the remaining stump migrate to cover the wound surface. This single layer of cells then proliferates to form the APICAL ECTODERMAL CAP. The cells beneath this cap undergo a dramatic dedifferentiation: bone cells, cartilage cells, fibroblasts, myocytes, and neural cells lose their differentiated characteristics and become detached from each other. The well structured limb region at the cut edge of the stump thus forms a proliferating mass of indistinguishable, dedifferentiated cells just beneath the apical ectodermal cap. This dedifferentiated cell mass is called the REGENERATION BLASTEMA, and these cells will continue to proliferate and differentiate to form the new structures of the limb. If the blastema cells are destroyed, no regeneration takes place (Butler, 1935). Moreover, once the cells have dedifferentiated to form a blastema, they have regained their embryonic plasticity. Carlson (1972) has shown that when at least 99 percent of the muscle cells are removed from a newly amputated salamander limb, the regenerated limb contains a normal supply of muscles in their appropriate positions. Thus, other cells in the blastema—cells derived from nonmuscle tissue—must be able to form the muscles of the regenerated limb.

In most instances, however, neural tissue is essential for the formation of the new limb by other cells. Singer (1954) demonstrated that a minimum number of nerve fibers must be present for regeneration to take place. It is thought that the neurons release a mitosis-stimulating factor that increases the proliferation of the blastema cells (Singer and Caston, 1972; Mescher and Tassava, 1975). One candidate for this crucial neural substance is GLIAL GROWTH FACTOR (GGF). This peptide is known to be produced by newt neural cells, is present in the blastema, and is lost upon denervation. When GGF is added to a denervated blastema, the mitotically arrested cells are able to divide again (Brockes and Kinter, 1986).

So we are faced with a situation wherein the adult cells of an organism can return to an "embryonic" condition and begin the formation of the limb anew. Just as in embryonic development, the blastema forms successively more distal structures (Rose, 1962). Thus, the blastema must contain some positional information that directs a blastema on a stump containing a humerus neither to make another humerus nor to start immediately producing digits. Not only does the blastema regenerate those structures beginning at the appropriate proximal–distal level in the limb, but the polarity of the anterior–posterior ("thumb–pinky") and dorsal–ventral ("wrist–palm") axes also correspond to those of the stump.

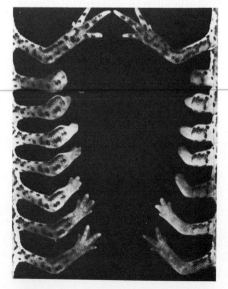

FIGURE 4

Regeneration of a salamander forelimb. Upper photographs show the original limbs. On the left, the amputation was made below the elbow; the amputation shown on the right cut through the humerus. In both cases the correct positional information is respecified. (From Goss, 1969; courtesy of R. J. Goss.)

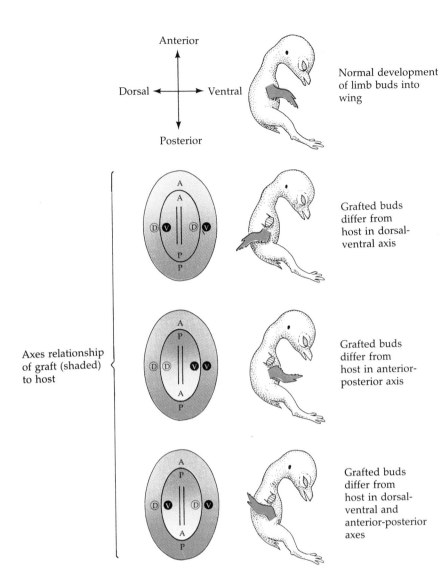

FIGURE 5
Specification of the anterior–posterior and dorsal–ventral axes in the chick wing. The grafted limb bud develops in accordance with its own polarity and does not adopt the polarity of its host. Wings that develop from grafted limb buds are shown in color. For the sake of clarity, the host's normally developed wing is not shown. (After Hamburger, 1938.)

The limb axes

The positional information needed to construct a limb has to function in a three-dimensional coordinate system.* The axes of polarity appear to be determined in the following sequence: anterior–posterior (as in the line between the thumb and the little finger); dorsal–ventral (as in the line between the upper and lower surfaces of the hand); and proximal–distal (as in a line connecting the shoulder and the fingertip).

The self-differentiation of the anterior–posterior axis is the first change from the pluripotent condition. In chicks, this axis is specified long before a limb bud is recognizable. Hamburger (1938) showed that as early as the 16-somite stage, prospective wing mesoderm transplanted to the flank area develops into a limb with the anterior–posterior and dorsal–ventral polarities of the donor graft and not those of the host tissue (Figure 5).

The proximal–distal axis is defined only after the induction of the apical ectodermal ridge (AER) by the underlying mesoderm. The limb elongates by the proliferation of the mesenchymal cells underneath the AER. This region of cell division is called the PROGRESS ZONE. As cells leave the progress zone, they have their proximal–distal values specified.

*Actually, it is a four-dimensional system in which time is the fourth axis. Developmental biologists get used to seeing nature in four dimensions.

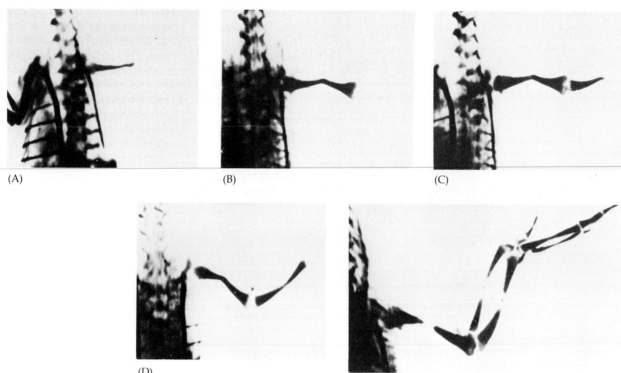

(A)　　　　　　　　　　(B)　　　　　　　　　　(C)

(D)

(E)

FIGURE 6
Dorsal view of chick skeletal pattern after removal of the entire AER from the right wing bud of embryos at various stages. The last picture is of a normal wing skeleton. (From Iten, 1982; courtesy of L. Iten.)

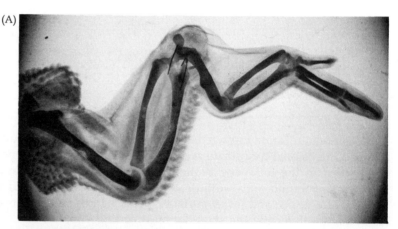

(A)

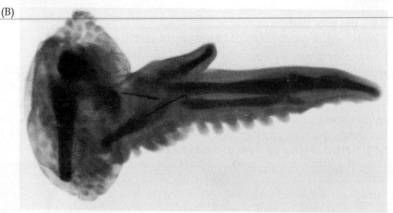

(B)

FIGURE 7
Control of proximal–distal specification by the cells of the progress zone (PZ). (A) Extra set of ulna and radius formed when early bud PZ is transplanted to late wing bud that has already formed ulna and radius. (B) Lack of intermediate structures seen when late limb bud PZ is transplanted to early limb bud. The hinges indicate the location of the grafts. (From Summerbell and Lewis, 1975; courtesy of D. Summerbell.)

The first cells leaving the progress zone form proximal structures; those cells that have undergone numerous divisions in the progress zone become the more distal structures (Saunders, 1948; Summerbell, 1974). Therefore, when the AER is removed from an early-stage wing bud, the cells of the progress zone stop differentiating and only a humerus forms. When the AER is removed slightly later, humerus and radius and ulna form (Figure 6; Rowe et al., 1982).

Proximal–distal polarity (like that of the other two axes) resides in the mesodermal compartment of the limb. If the AER provides the positional information—somehow instructing the undifferentiated mesoderm beneath it as to what structures to make—then older AERs should produce more distal structures when placed on young mesoderm. This was not found to be the case, however (Rubin and Saunders, 1972), as the normal complete sequence of limb development occurred when young mesoderm was combined with any stage AER. But when the entire progress zone, including the mesoderm, from an early embryo was placed on the limb bud of a later-stage embryo, new proximal structures were produced beyond those already present. Conversely, when old progress zones were added to young limb buds, distal structures immediately developed so that digits were seen to emerge from the humerus without the intervening ulna and radius (Figure 7; Summerbell and Lewis, 1975).

Specification of the limb anterior–posterior axis: Gradient models

The zone of polarizing activity

Although the differentiation of the proximal–distal structures is thought to depend upon how many divisions a cell undergoes while in the progress zone, positional information instructing a cell as to its position on the anterior–posterior and dorsal–ventral axes must come from other sources. There are two major models to account for the specificity of the anterior–posterior axis. One of the major hypotheses to explain this polarity involves a gradient of material from one part of the chick limb bud to

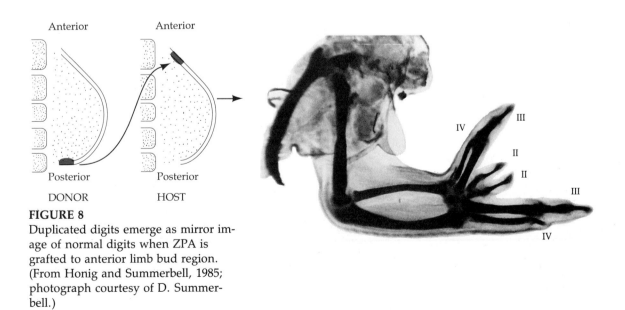

FIGURE 8
Duplicated digits emerge as mirror image of normal digits when ZPA is grafted to anterior limb bud region. (From Honig and Summerbell, 1985; photograph courtesy of D. Summerbell.)

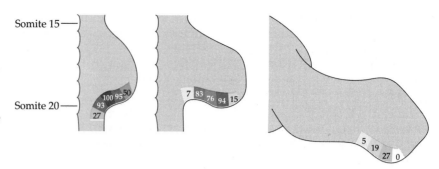

FIGURE 9
Strength of ZPA signaling activity as the limb develops. Different regions of the limb in the ZPA area were excised and grafted into the anterior margin of the early limb bud. If each graft from the area caused the formation of a complete duplication (as seen in Figure 8), that area was scored as 100 percent. As can be seen here, the ability of a tissue to act as ZPA varies with the position in the margin and the stage of development. (After Honig and Summerbell, 1985.)

another. Several experiments (Saunders and Gasseling, 1968; Tickle et al., 1975; Summerbell, 1979) suggest that the anterior–posterior axis is specified by a small block of mesodermal tissue near the posterior junction of the young limb bud and the body wall. When this tissue from a young limb bud is transplanted into a position on the anterior side of another limb bud (Figure 8), the number of digits of the resulting wing is doubled. Moreover, the structures of the extra set of digits are mirror images of the normally produced structures. The polarity has been maintained, but the information is now coming from both an anterior and a posterior direction. This region of the mesoderm has been called the ZONE OF POLARIZING ACTIVITY (ZPA).

The distribution and strength of the ZPA's positional signaling activity in the chick wing and leg buds have been mapped (Honig and Summerbell, 1985; Hinchliffe and Sansom, 1985). As shown in Figure 9, the polarizing activity (measured after grafting the posterior marginal cells into the anterior margin of the limb bud) is highest in a particular region of the posterior margin and tapers off from there. It weakens as development progresses. It has been hypothesized that this ZPA tissue operates by secreting a morphogen, which diffuses from its source to form a concen-

FIGURE 10
Model for a diffusible morphogen secreted by ZPA tissue. (A) Normal limb, wherein the ZPA produces a diffusible compound whose gradient declines anteriorly. As it falls through certain threshold levels, it instructs the cells to make specific digits. The posterior digit is 4, the anterior is 2. (B) When an extra ZPA is placed in the anterior region of the limb bud, the concentration of the postulated morphogen changes, establishing a symmetrical distribution of the morphogen. The threshold conditions still apply, so the result is a symmetrical duplication.

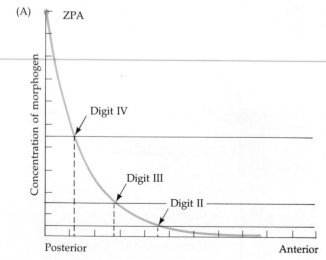

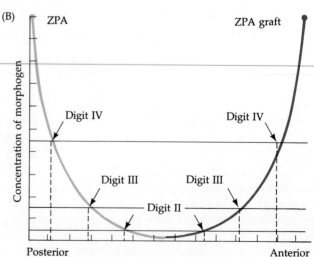

tration gradient from the posterior end to the anterior end of the limb bud. Those cells nearest the ZPA would be exposed to the highest amount of this compound, whereas those farthest from the ZPA would be exposed to a relatively low concentration. The original support for this hypothesis came from a series of experiments (Summerbell, 1979) in which the anterior limb bud was separated from the posterior limb bud by a nonpermeable barrier. In such cases, the anterior structures were no longer formed. These results were interpreted to mean that the ZPA was secreting a morphogen that organized the anterior–posterior gradient in the limb bud and that the anterior limb structures were unable to form because they could not receive this morphogen (Figure 10). The polarizing activity of the ZPA could be attenuated by reducing the number of cells grafted from the ZPA to the anterior margin (Tickle, 1981).

SIDELIGHTS & SPECULATIONS

Gradient models of positional information

How can cells be informed of their position in the embryo and then use that information to differentiate into the appropriate cell type? One major type of explanation proposes GRADIENTS of morphogenic substances (Boveri, 1901; Child, 1941; Wolpert, 1971). In these models, a soluble substance (MORPHOGEN) is posited to diffuse from a "source" (where it is produced) to a "sink" (where it is degraded), establishing a continuous range of concentrations within that region. (Morphogens are thus analogous to chemotactic substances, discussed in Chapter 15, which also operate by establishing a concentration gradient from a source. Like chemotactic substances, morphogens must be readily diffusible and readily degraded or inactivated.) Theoretical considerations (Crick, 1970) suggest that such gradients could only function over relatively small distances, less than 100 cell diameters. We have seen such gradient models used to explain sea urchin pattern formation (Chapter 8).

In gradient models, the concentration of the morphogen changes over distance, the highest concentrations being near the source of the morphogen. The cells would have to have "sensors" that would respond differently to different concentrations of the gradient. As will be seen in the next chapter, this can be done by having enhancer elements that can bind the morphogen at different levels (Figure 11). For example, if one has a morphogen being made at the anterior of the body, the genes responsible for organizing head development may have an enhancer that binds the morphogen poorly. Only when there is a large concentration of the morphogen present will that gene be active. The gene responsible for thorax formation could have an enhancer that binds the morphogen rather well. This would enable it to respond to relatively low levels of that morphogen. The cells of the head would express both of these genes, while the cells of the thorax would express only that gene whose enhancer could bind low amounts of the morphogen. The cells in the posterior portion of the

FIGURE 11
A hypothetical model for gradients establishing positional information. The concentration of morphogen drops from the source. In this diagram, the receptors of the morphogen are enhancer elements of two genes that control cell fate, but the receptors could also be cytoplasmic receptors or cell membrane receptors. One of the receptors (in this case, the enhancer on gene A) needs a high concentration of morphogen to act. The other receptor needs only a little morphogen to act. At high concentrations, both genes are active. In moderate concentrations, just the second gene is active. Where the morphogen is absent, neither gene is active. Between these regions are thresholds at which the cell type would change. (After Wolpert, 1978).

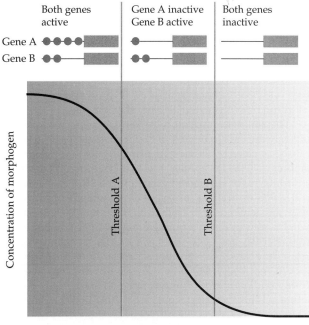

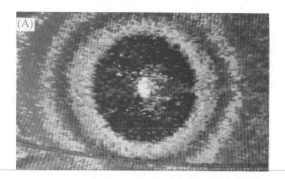
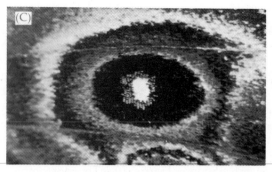
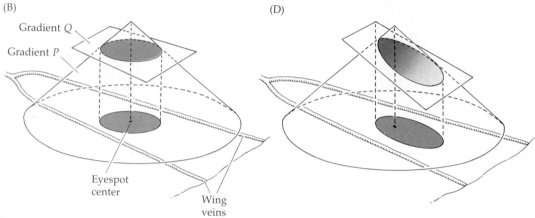

FIGURE 12

Gradient model of positional information proposed to explain the establishment of markings on butterfly wings. (A) Photograph of the eyespot on the wing of the butterfly *Morpho peleides*. (B) Diagram of a two-gradient model that might explain the way in which the eyespot is generated. The three-dimensional shape of a linear gradient (that is, a chemical diffusing equally in all directions) in a cone (gradient P), the height of which reflects the concentration of the signal. The origin of the signal is the apex of the cone, which corresponds to the center of the developing eyespot. Concentration Q represents the level of chemical necessary to reach the threshold sensitivity for the formation of a color in the cells of the wing. If that sensitivity level is the same throughout the wing, it will cut the cone orthogonally, resulting in a circular eyespot. (C) In other species, in this case *Smyrna blomfildia*, the eyespots are elliptical. (D) Different orientations of the sensitivity gradient Q could result in such elliptical eyespot patterns. (After Nijhout, 1981).

body would not see any of this morphogen, and neither of these genes would be activated. In this way, cells can sense the presence of a morphogen and respond differentially, depending on the morphogen concentration. The sensor does not have to be an enhancer; it can just as well be a cell receptor in the cytoplasm or cell membrane.

The ability to respond to certain morphogens in a certain way appears to be part of the properties of cells within a field. A piece of tissue transplanted from one region to another retains its species identity but can differentiate according to its new position within a field. For instance, as detailed in Chapter 8, a piece of presumptive belly skin from a salamander gastrula will form mouth structures when placed in the presumptive oral ectoderm region of a frog gastrula, but these parts will form a salamander jaw, not a frog jaw. The positional information is specified (mouth, not belly), but how the cells use this information to differentiate (into balancers rather than suckers) is inherent within the cells. Similarly, if cells that would normally become the middle of a *Drosophila* leg are removed from a late leg imaginal disc and placed into the center of

an antennal disc (where they would become the end of the antenna), they differentiate into claws (the end of the *foot*) at the tip of the antenna. *Specification* of position (middle or tip) is separate from the *determination* of cell type (leg or antenna).

Gradients have to be "read" by cells, and this interpretation of gradients does not have to be linear. Take, for example, a series of exam grades that stretches uniformly from 100 to 60. In one scheme (a "linear" reading), a grade between 100 and 90 is A, 80 to 89 is B, 70 to 79 is C, and 60 to 69 is D. In another class (using a "curved" reading), 100 to 95 is A, 95 to 85 is B, 84 to 70 is C, and 69 to 60 is D. Nijhout (1981) has used a two-gradient model to explain the development of "eyespot" patterns on butterfly wings. One gradient consists of a linear diffusion of a morphogen. The second gradient involves the interpretation of this morphogen; in other words, the sensitivity threshold of the cells involved differs at different regions of the wings. The existence of the second gradient gives rise to an elliptical spot, not the circular spot that would result if the sensitivity gradient were absent (Figure 12).

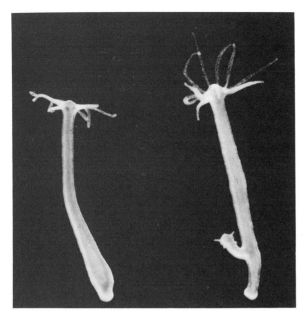

FIGURE 13
Budding in *Hydra pirardi*. The hydra on the left lacks a bud; the presence and polarity of the bud on the right-hand specimen is obvious. (Photograph courtesy of G. E. Lesh-Laurie.)

Gradients in **Hydra**

The development of a hydra illustrates how interacting gradients can determine cell fate as well as the size of an organism. Hydras are capable of sexual reproduction but do so only under adverse circumstances, such as overcrowding or starvation. Under normal conditions, they reproduce asexually by budding off new individuals (Figure

13). The hydra is basically a narrow tube, two cell layers thick, with an extracellular matrix layer sandwiched between the two epithelial sheets. Polarity in the hydra exists along the apical–basal axis. The basal end consists of a mucus-secreting BASAL DISC, which anchors the hydra to its substrate. The head at the apical end consists of the HYPOSTOME, containing the mouth and a ring of tentacles that surrounds the mouth. Between them is the body column. The buds form as outgrowths from the body wall and have the same polarity as their parent.

When a hydra is cut in half, the half containing the basal disc will form a new hypostome and the half containing the hypostome will generate a new basal disc. Moreover, if a hydra is cut into several pieces perpendicular to the body axis, the middle rings of cells will regenerate into complete, proportionally correct animals with both a basal disc and a head. These parts are not replaced by cell division (as happens in regenerating amphibian limbs and imaginal discs) but by the rearrangement and respecification of the existing cells of the adult organism. This type of regeneration without cell division is referred to as MORPHALLAXIS.

Every region of the hydra's stalk can give rise to an entire organism. Yet hypostomes do not form just anywhere; they form only at the apical end of any excised piece of tissue. Again, this situation is reminiscent of the "harmonious equipotential" sea urchin embryo, and the mechanism governing the integration of the totipotent parts into a unified polar whole may be very similar. In both cases, the spatial information appears to be provided by a series of gradients arising from the two poles.

Evidence for morphogenetic gradients in the hydra was obtained by grafting experiments. When hypostome tissue is transplanted into the middle of another hydra, it forms a new apical–basal axis with the hypostome extending outward (Figure 14). When basal disc cells are so

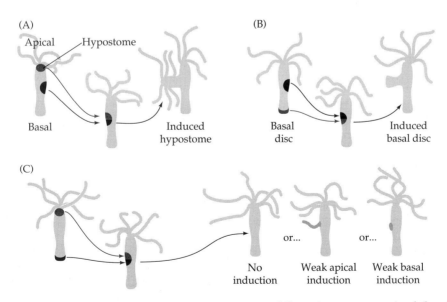

FIGURE 14
Evidence for gradients in *Hydra*. (A) Hypostome plus mid-hydra regions grafted laterally to the host's middle region cause distal (hypostome) induction. (B) Basal disc plus mid-hydra regions grafted laterally to the host's middle region cause proximal (basal disc) induction. (C) Hypostome plus basal disc region grafted laterally cause little or no induction with no distinct polarity. (After Newman, 1974).

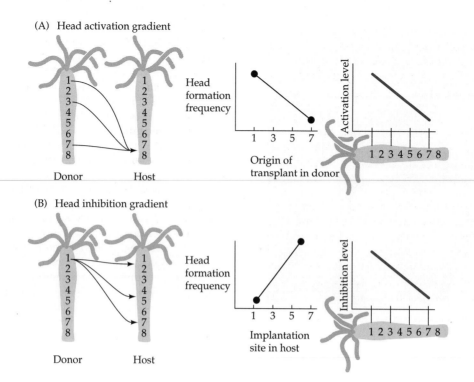

(A) Head activation gradient

Donor   Host

Head formation frequency

Origin of transplant in donor

Activation level

(B) Head inhibition gradient

Donor   Host

Head formation frequency

Implantation site in host

Inhibition level

FIGURE 15

Assaying for head activator and inhibitor activity in *Hydra*. (A) Head activation gradient is demonstrated by placing different regions from a donor hydra into the bottom portion of a host hydra. One assays the ability to form a head structure. (B) Head inhibition gradient is demonstrated by inserting a region from the hypostome of a donor hydra into different places in the host hydra. One assays whether the hypostome tissue can induce secondary head formation. (From Bode and Bode, 1984.)

grafted, a new axis forms, but has the opposite polarity, extending a basal disc. Moreover, when cells from the two poles are grafted together into a host hydra, the resulting bud has no polarity (Browne, 1909; Newman, 1974). Other experiments (Rand et al., 1926) have shown that the normal regeneration of the hypostome can be inhibited when an intact hypostome is grafted adjacent to the amputation site.

These experiments have been interpreted to indicate the existence of both a head activator gradient and a head inhibitor gradient (Webster and Wolpert, 1966; Mac-Williams, 1983a,b). The HEAD ACTIVATOR GRADIENT can be measured by implanting rings of tissue from various levels of the donor hydra into a particular region in the trunk of the host. The higher the level of head activator in the donor tissue, the greater the frequency of implants that form hypostomes (Figure 15). This activating factor is found to be most concentrated in the apical end and decreases linearly toward the base. This gradient appears to be very stable and is probably responsible for the formation of the hypostome at the apical end of the hydra. However, although any part of a hydra can form a head, such formations do not occur regularly. Extra heads fail to form because the existing head produces an inhibitory influence that prevents the formation of other heads. This HEAD INHIBITOR GRADIENT is measured by inserting subhypostomal tissue (having a relatively high concentration of head activator factor) into various regions along the column of host hydras. The more powerful the inhibition at a given level in the host hydras, the fewer grafts that form heads.

The source of this inhibitor gradient is also the hypostome, but the inhibiting factor is probably a very labile and diffusible molecule. It prevents heads from forming elsewhere only while an intact head is present. Therefore, when a head is present, both gradients are functioning, but when the head is removed, the labile inhibitor disappears. The result is a head on the distal end of the piece of hydra (Webster, 1966; Wilby and Webster, 1970). The basal disc has also been seen to be the source of two gradients, one that activates foot development (MacWilliams and Kafatos, 1974; Hicklin and Wolpert, 1973; Grimmelikhuijzen and

FIGURE 16

Head inhibition (——) and foot inhibition (- - -) gradients in newly dropped buds, young adults, and budding adults of *Hydra*.

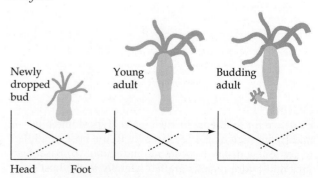

Newly dropped bud

Young adult

Budding adult

Head   Foot

Schaller, 1977) and one that inhibits it (Schmidt and Schaller, 1976). Possible activator and inhibitor substances have been isolated (Schaller et al., 1979; Berking et al., 1979), and the localization of these small molecules is consistent with the gradient hypothesis.

In young adults, the gradients of head and foot inhibitors appear to block bud formation. However, as the hydra grows, the sources of these labile gradients are further separated, creating a region of tissue, about two-thirds down the trunk, where both inhibitory gradients are minimal. Here is where the bud forms (Figure 16; Bode and Bode, 1984). Certain mutants of *Hydra* have defects in their ability to form buds; these bud-forming defects can be traced to defects in their gradients. The *L4* mutant of *H. magnipapillata*, for instance, buds very slowly and only after reaching a size about twice as long as wild-type individuals. The amount of head inhibitory activity in these mutants was found to be much greater than that in wild-type *Hydra* (Takano and Sugiyama, 1983). There appears to be a cor-relation, then, between the size of the animal, the place where buds are formed, and the strengths of the inhibitory gradients.

Although the integration of a small field can theoretically be controlled by diffusible substances that form gradients, there must be some mechanism to ensure that the morphogen does not diffuse into the environment or into the animal's gut. Wakeford (1979) demonstrated that positional information from grafted hypostomes was transferred to the host only if gap junctions had formed between host and donor cells. More recent grafting experiments (Fraser et al., 1987) have shown that antibodies directed against gap junction proteins inhibit the head inhibition gradient. It seems probable, then, that the morphogen is carried from cell to cell by these junctions rather than by diffusing through the primitive gut or circulatory system of the hydra. Although some of these compounds may have been purified, their mechanisms of action remain unknown.

Evidence that retinoic acid may be the ZPA morphogen

Given the evidence that a gradient from the ZPA is responsible for specifying the anterior–posterior axis, what could this morphogen be? Recent evidence implicates retinoids—specifically, retinoic acid*—as the ZPA morphogen.

The teratological effects of retinoic acid are well known and were reviewed in Chapter 5. The injection of vitamin A into pregnant mice leads to dose-dependent limb reduction in the fetuses. After a maternal dose of 10 mg retinoic acid/kg, the level of limb bud retinoic acid increases 50-fold over endogenous levels, and a mild degree of limb abnormalities is seen. When the dose is raised to 100 mg retinoic acid/kg, the level of limb bud retinoic acid increases 300-fold, and all the offspring have a thalido-mide-like syndrome of limb shortening (Kochhar, 1977; Satre and Kochhar, 1989).

Retinoic acid also causes teratogenic effects in regenerating limbs. Immersion of regenerating salamander limb blastemas in high concentrations of vitamin A can cause shortening of the limb as in chick embryos. Lower concentrations of retinoic acid, however, produce dose-dependent proximal–distal duplication of limb structures. In other words, a complete limb (starting with the most proximal bone) grows out of the amputated portion of the leg (Figure 17) (Niazi and Saxena, 1978; Maden, 1982). It appears that the positional value of the blastema cells is proximalized by retinoic acid, causing the cells to "forget" their distal position and regenerate more structures than they normally would. The regenerating cells are most sensitive to retinoids during the time when cells are dedifferentiated and the blastema cells begin to accumulate (Thoms and Stocum, 1984). In regenerating frog and salamander limbs, both the proximal–distal and anterior–posterior axes can be affected, leading to mirror-image duplication of the entire limb starting at the point of amputation (Figure 18).

Tickle and her colleagues (1982, 1985) hypothesized that retinoic acid might be the morphogen or the mimic of a morphogen responsible for

*The morphogenetically active retinoic acids appear to be all-*trans*-retinoic acid (which is used most frequently) and 3,4,-didehydroretinoic acid, both of which are found in the chick limb bud (Thaller and Eichele, 1990).

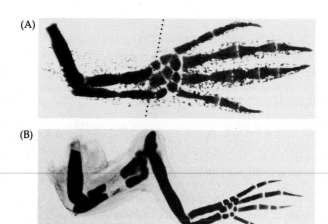

FIGURE 17

Effect of vitamin A on regenerating salamander limbs. (A) Normal regenerated axolotl limb (×9) with its humerus, paired radius and ulna, carpals, and digits. Dotted line shows point of amputation. (B) Regeneration after amputation through the carpal area, but after the limb blastema had been placed in retinol palmitate for 15 days. A new humerus, ulna, radius, carpal set, and digit set emerge (×5). (From Maden, 1982; photographs courtesy of M. Maden.)

specifying the anterior–posterior axis. If so, it would mimic the effect of a ZPA. They then performed a "ZPA graft" but used a bead soaked in all-*trans*-retinoic acid instead of ZPA tissue. Remarkably, inserting a retinoic acid-soaked bead into the anterior margin of the chick limb bud (Figure 19) produced the same mirror-image duplication of digits as the insertion of ZPA tissue. In fact, the retinoic acid-soaked beads acted with the same kinetics as ZPA tissue, creating a gradient across the limb bud. Eichele (1989) has extended this by showing that retinoic acid-soaked beads can substitute for a ZPA in embryos where the ZPA has been removed.

If retinoic acid were actually the morphogen responsible for the anterior–posterior polarity of the chick limb, then four conditions must be met: (1) retinoic acid must be present in the early chick limb bud; (2) it must have a graded distribution from the posterior (ZPA) to anterior margins; (3) retinoic acid-binding proteins must be expressed in the limb bud cells; and (4) retinoic acid must be present at concentrations known to cause digit duplications when added exogenously.

All these conditions seem to occur. In 1987, Thaller and Eichele identified a gradient of naturally occurring retinoic acid across the early chick limb bud; this gradient had its peak at the ZPA. Because biochemical analyses of embryos is a difficult task (due to their size), this study used

FIGURE 18

Complete limb plus its mirror image is regenerated when the regenerating limb has spent 6–9 days in retinol palmitate. (From Maden, 1984; photograph courtesy of M. Maden.)

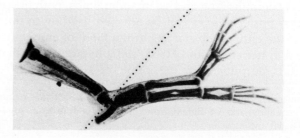

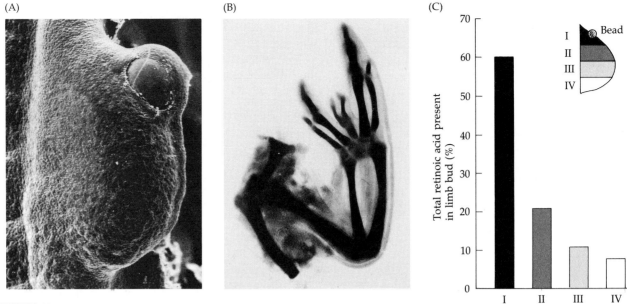

(A)   (B)   (C)

FIGURE 19
Retinoic acid effects in chick limb development. (A) Scanning electron micrograph of retinoic acid-soaked resin bead placed into early limb bud of chick embryo. (B) Duplicated wing resulting when bead soaked with 10 μg/ml retinoic acid is placed into limb bud. (C) Gradient of retinoic acid seen in limb bud 6 hours after placing bead into the limb bud. (From Tickle et al., 1985; photographs courtesy of C. Tickle.)

sections of 5536 limb buds, each bud having a volume of about 0.8 μl. The whole limb bud was found to have a concentration of retinoic acid approximating 25 nM. This concentration is capable of eliciting complete mirror-image duplications of digits when applied exogenously at the anterior margin. Retinoic acid-binding proteins have also been identified in these limb bud cells. The *cellular* retinoic acid-binding protein (which binds retinoic acid in the cytoplasm) is concentrated in the limb bud and neural crest-derived cells that are responsive to retinoic acid (Maden et al., 1989). A specific *nuclear* retinoic acid acid receptor (a protein similar to the steroid-binding transcription factors) has also been isolated. The gene for this protein has been cloned, and radioactive cDNA from the gene is able to locate the mRNA for this nuclear retinoic acid receptor mRNA in the vertebrate limb bud and cranial crest derivatives (Figure 20). In the 11-day mouse embryo, retinoic acid receptor transcripts can be seen throughout the limb bud, eventually becoming localized to the cartilaginous condensations (Ruberte et al., 1990). In regenerating newt limbs, nuclear retinoic acid receptors are detected in the blastema shortly after amputation. They are expressed at high levels during cartilage differentiation and diminish when regeneration is completed (Giguère et al., 1989; Ragsdale et al., 1989). Thus, it is very probable that the vertebrate ZPA secretes retinoic acid that forms a gradient instructing the digit pattern in the limb bud.

The next question becomes: If retinoic acid is the vertebrate ZPA morphogen, what is it doing to the cells? One hypothesis is that different concentrations of retinoic acid can activate different homeobox genes. These homeobox genes are themselves transcription factors and have been known (Chapters 12 and 18) to activate different batteries of genes along an anterior–posterior axis. Dollé and his co-workers have seen that near the ZPA of the mouse embryo, homeotic genes *Hox5.2, 5.3, 5.5,* and *5.6*

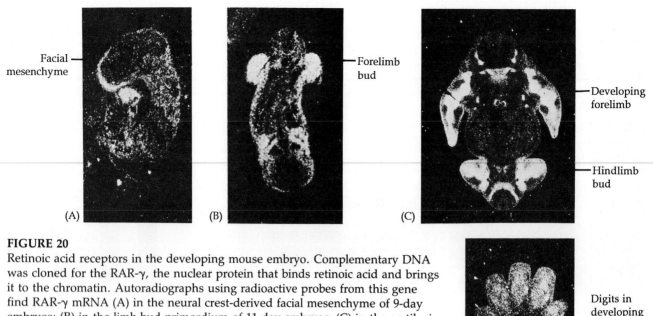

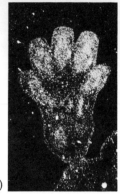

FIGURE 20
Retinoic acid receptors in the developing mouse embryo. Complementary DNA was cloned for the RAR-γ, the nuclear protein that binds retinoic acid and brings it to the chromatin. Autoradiographs using radioactive probes from this gene find RAR-γ mRNA (A) in the neural crest-derived facial mesenchyme of 9-day embryos; (B) in the limb bud primordium of 11-day embryos; (C) in the cartilaginous condensation of the limb buds of 12.5-day embryos; and (D) in the condensing digits of 13.5-day embryos. (From Ruberte et al., 1990; photographs courtesy of P. Chambon).

are all active. Slightly above the ZPA, *Hox5.6* transcripts are no longer detected, and above that, *Hox5.5* mRNA is not seen. As one goes further from the ZPA, only *Hox5.2* remains transcriptionally active (Figure 21). The genes activated by these homeobox transcription factors may be responsible for the different properties of the limb cells at these different regions. One property may involve the cell surface. These cell membrane compounds affected by retinoic acid have not yet been identified.

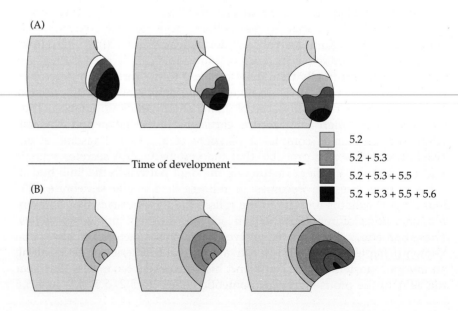

5.2
5.2 + 5.3
5.2 + 5.3 + 5.5
5.2 + 5.3 + 5.5 + 5.6

FIGURE 21
Localization of *Hox5* homeobox gene products in the developing chick limb bud. (A) Progression of *Hox5* gene sequences as the limb bud develops. (B) Hypothetical scheme for the generation of such patterns by the ZPA. (After Dollé et al., 1989).

Specification of limb skeleton by reaction–diffusion processes

As useful as the gradient model has been for explaining pattern formation in the vertebrate limb, it cannot be the entire story. The morphogen gradient model does not explain how some cells in the limb become cartilage and other cells equally distant from the morphogen source become fibroblastic connective tissue. In other words, the gradient model assumes that there is some determination of cells into either a chondrocyte lineage or a fibroblast lineage before limb patterning has begun. Moreover, when an extra ZPA is transplanted to the anterior margin of the limb bud, a secondary humerus fails to form, as would be expected by the ZPA gradient model (Wolpert and Hornbruch, 1987).

Stuart Newman (1988; Newman and Frisch, 1979) proposed a model by which the initially uniform population of mesenchymal cells is progressively determined into cartilage and fibroblastic lineages. In this model, it is the progress zone rather than the ZPA that is important. The mesenchymal cells of the limb bud are derived from two distinct sources. Quail–chick chimeras (Christ et al., 1977; Chevallier et al., 1977) and lineage-specific monoclonal antibodies (George-Weinstein et al., 1988) have shown that the precursor cells for cartilage and fibroblasts arise from the somatopleure, while the muscle cell precursors are somitic in origin. Thus, the muscle cells are already determined before limb development. However, which cells become the cartilage-forming chondrocytes and which form the fibroblasts has not yet been determined. Cartilage formation begins in the proximal region and ends with the digits (Figure 22). At each stage, there is a condensation of mesenchymal cells to form tight clusters of cells. This condensation is probably mediated by fibronectin, whose presence is seen at various foci concomitant with such condensation. Fibronectin may promote the clustering of these cells by linking together the proteoglycans on adjacent cell surfaces (Frenz et al., 1989). Cyclic AMP levels increase in these condensing mesenchymal cells, followed by the modification of chromatin proteins and the transcription of new cartilage-specific mRNAs. Shortly thereafter, the cartilage-specific extracellular matrix molecules are secreted (Solursh, 1984; Newman, 1988).

But which cells are to form these precartilaginous clusters and which cells are to be excluded? Simple gradient models do not suffice. Why should one cartilage condensation (say, the humerus) be formed in the most proximal region, followed by two (say, ulna and radius) in the mid-limb, followed by a whole array of cartilaginous condensations to form the digits? Alan Turing (1952), one of the founders of computer science (and the mathematician who cracked the German "Enigma" code during World War II), proposed a model wherein two homogenously distributed solutions would interact to produce stable patterns during morphogenesis. These patterns would represent regional differences in the concentrations of the two substances. These interactions would produce an ordered structure out of random chaos.

This REACTION–DIFFUSION MODEL involves two substances. One of them, substance S, inhibits the production of the other, substance P. Substance P promotes the production of more substance P as well as more substance S. If S diffuses more readily than P, Turing's mathematics show that sharp waves of concentration differences will be generated for substance P (Figure 23). These waves have been observed in certain chemical reactions (Prigogine and Nicolis, 1967; Winfree, 1974) and can be observed in three dimensions (Figure 24; Welsh et al., 1983).

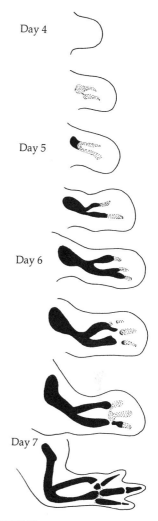

FIGURE 22
Cartilaginous condensations in the chick wing bud between days 4 and 7. Stippled regions indicate new areas of mesenchymal cell condensation; solid black regions indicate definitive cartilage. Areas of cartilage are seen to be foreshadowed by mesenchymal condensations about 12 hours earlier. (From Newman, 1988).

FIGURE 23

Generation of periodic spatial heterogeneity when two reactants, S and P, are mixed together under the conditions that S inhibits P, P catalyzes the formation of both S and P, and S diffuses faster than P. (A) The distributions are initially random, but quickly begin to fluctuate. (B,C) As P increases locally, it produces more S, which inhibits other peaks of P in the near vicinity. The result is a series of P peaks ("standing waves") at regular intervals.

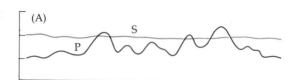

(A)

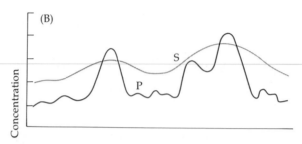

(B)

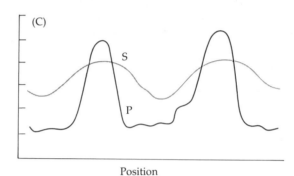

(C)

Position

The reaction–diffusion model predicts alternating areas of high and low concentrations of some substance. When the concentration of such a substance is above a threshold level, a cell (or group of cells) may be instructed to differentiate in a certain way. An important part of Turing's model is that particular chemical wavelengths will be amplified while all others will be suppressed. As local concentrations of P increase, the values of S form a peak centering on the P peak but becoming broader and shallower because of its more rapid diffusion. These S peaks inhibit other P peaks from forming. But which of the many P peaks will survive? This depends on the size and shape of the tissues in which the oscillating reaction is occurring. (This is analogous to the harmonics of vibrating strings, as in a guitar. Only certain resonance vibrations will be permitted, based on the boundaries of the string.)

The mathematics describing which wavelengths are selected at a particular length consist of complex polynomial equations called BESSEL FUNCTIONS. Using these equations (Newman and Frisch, 1979; Oster et al., 1988), we can predict that the size of the original limb bud would allow only one such peak—hence, only one cartilaginous condensation (Figure 25). As the limb bud grows, however, there comes a point at which two waves can fit into the size of the bud. Then, two cartilaginous rods condense from the homogeneous mesenchyme tissue. As growth along the

FIGURE 24

Photograph of a 10-mm diameter test tube wherein standing waves have been generated from a homogeneous solution. In this reaction, malonic acid is oxidized by bromate in the presence of cerium ions. A dye (ferroin) reflects the periodic changes in cerium^{3+}/cerium^{4+} levels. (From Welsh et al., 1983.)

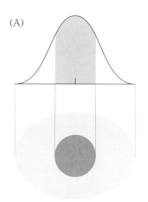

FIGURE 25
The number of standing waves fitting into a region depends upon the size of the region. As the region grows, it becomes possible to fit more complete wavelengths into the region. (A) When the limb cross-section is only able to accommodate a single wavelength, a single condensation emerges. (B) If the domain changes shape so that two complete wavelengths can fit into the cross section, then two condensations arise. (After Oster et al., 1988.)

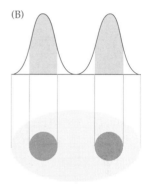

proximal–distal axis continues, the size enlarges, so more reaction–diffusion compounds can fit several complete waves into this region. Here the digits develop.

In this reaction–diffusion hypothesis, the aggregations of cartilage actively recruit more cells from the surrounding area and laterally inhibit the formation of other foci of condensation. The number of foci, then, depends on the geometry of the tissue and the strength of the lateral inhibition. If the inhibition remains the same, tissue volume must increase in order to get two foci forming where one had been allowed before. This model predicts that slight size changes of the distal limb bud alter the number of digits. This is indeed found to be the case, and it may be a very simple way of gaining or losing digits during evolution (see Chapter 23). These standing waves constitute the PREPATTERN of the limb. This reaction–diffusion mechanism does not rule out prelocalized morphogens or gradients (for which there is good evidence), for it can exist as a separate morphogenetic mechanism to structure the embryo.

The molecule forming these standing waves (and therefore responsible for organizing these precartilage condensations) has not yet been identified. Newman and co-workers (1988) suggest that transforming growth factor-B (TGF-β) may be involved. This protein is known to stimulate its own synthesis. Moreover, it has been seen to stimulate the production of fibronectin, which draws the cells together. The reaction–diffusion model and the ZPA model are not mutually exclusive, and both of them may be functioning to provide the positional information needed to create such a complex organ as a limb. Indeed, other pattern-generating mechanisms (see Sidelights & Speculations) may also be active in the creation or maintenance of limb polarity.

SIDELIGHTS & SPECULATIONS

The polar coordinate model of positional information in the developing and regenerating limb

Supernumerary limb elements

Research on developing and regenerating limb buds has also shown another level of pattern formation. When anterior and posterior cells from regeneration blastema or limb buds are juxtaposed, the results are dramatic and strange: the resulting limb contains supernumerary structures, often with two or three sets of digits (Figure 26; Bryant and Iten, 1976; Javois and Iten, 1986).

Moreover, this patterning mechanism appears to be the same for both regenerating and normally developing limbs. When axolotl limb buds were transferred to regenerating axolotl blastema stumps in a way that maintained the original polarity with respect to the stump, normal limbs developed. However, when the polarity of the anterior–posterior axis was reversed with respect to the stump, mirror-image supernumerary digits emerged (Figure 27; Muneoka and Bryant, 1982). These results strongly suggest that the patterning rules are the same for developing and regenerating limbs.

Polar coordinate model

By comparing such results in regenerating vertebrate limbs, insect limbs, and insect imaginal discs, French and his colleagues (1976; Bryant et al., 1981) have proposed a series

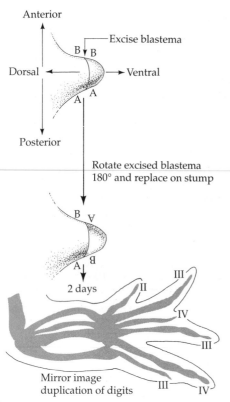

FIGURE 26

Limb bud rotation experiment. When a newly emerged limb bud (or regeneration blastema) is severed at its base and rotated 180 degrees on its stump, the result is a limb with three areas of outgrowth. In this case, the limb bud proceeded to form a radius and ulna with two new ulnas between them. At their tips were digits representing three partial wings. (After Javois and Iten, 1986.)

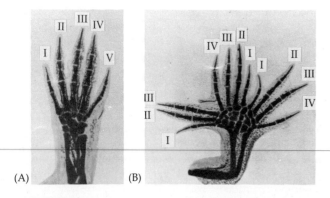

FIGURE 27

Ability of regenerating salamander limb blastema to be controlled by progress zone of developing limb bud. (A) Control showing normal five-digit right hindlimb where right hindlimb bud was placed on right regenerating hindlimb stump without rotation. (B) Limb resulting when a graft from the left hindlimb bud was placed onto a right regenerating hindlimb stump. Here the anterior–posterior axes are reversed between the host and the graft. (From Muneoka and Bryant, 1982; courtesy of K. Muneoka.)

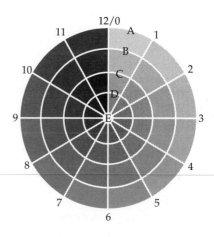

FIGURE 28

Polar coordinate model for the specification of positional information. Each cell has a circumferential value (0–12) specifying the anterior–posterior axis and a radial value (A–E) specifying the proximal (A) to distal (E) axis. (From French et al., 1976.)

of empirical rules that predict the outcome of a wide variety of experimental perturbations with regenerating appendages. Starting from Wolpert's (1969) premise that pattern arises from a cell's recognition of its relative positions in a developing population, they speculate that a cell assesses its physical location in a system of polar (clocklike) coordinates (Figure 28). In this system, each cell has a circumferential value (from 0 to 12) as well as a radial value (from A to E). In regenerating limbs, the outer circle represents the proximal (shoulder) boundary of the limb field; the innermost circle represents the most distal regions.

As we have seen, when normally nonadjacent tissues of a field are juxtaposed, duplications often arise. Yet other transplanted tissue (not of the field) will not cause these duplications. The polar coordinate model has been extremely useful in predicting the extent of these duplicated structures. The SHORTEST INTERCALATION RULE states that when two normally nonadjacent cells are juxtaposed, growth occurs at the junction until the cells between these two points have all the positional values between the original points (Figure 29). The circular sequence, like a clock, is continuous, 0 being equal to 12 and having no intrinsic value in itself. Being circular, however, means that there are two paths by which intercalation can occur between any two points. For example, when cells having the values 4 and 7 are placed next to each other, there are two possible routes between them: 4, 5, 6, 7 and 4, 3, 2, 1, 12, 11, 10, 9, 8, 7. According to this model, the shortest route is taken. The exception, of course, is when the cells have values that fall exactly opposite each other in the coordinate system, so that there is no one shortest route. In this case, all values are formed between the two opposites.

The second rule is the COMPLETE CIRCLE RULE FOR DISTAL TRANSFORMATION. Once the complete circle of positional values has been established on the wound surface, the cells proliferate and produce the more distal structures. The mechanism by which this is thought to occur is outlined in Figure 29 and again involves intercalation of structures between cells having different positional information

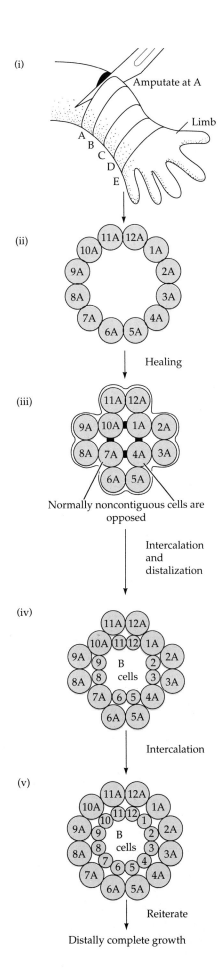

(i)

Amputate at A

Limb

A
B
C
D
E

(ii)

Healing

(iii)

Normally noncontiguous cells are opposed

Intercalation and distalization

(iv)

B cells

Intercalation

(v)

B cells

Reiterate

Distally complete growth

FIGURE 29

Model for distal outgrowth. (i, ii) Limb is cut at position A, proximal to positions B–E. This exposes circumferential positions at A. (iii) Healing leads to the apposing of normally noncontiguous cells (such as 10A and 1A) near the blastema tip. Cells between the newly apposed tissues proliferate and acquire positional specification between the two sites. However, these cells are adjacent to preexisting cells sharing the same circumferential values. By the "distalization rule," such cells acquire a more distal positional value. (iv, v) Intercalation then occurs between these newly specified cells, creating a new surface that contains all the circumferential values. This scheme is repeated until the limb is complete. (After Bryant et al., 1981.)

(Bryant et al., 1981). The predictive value of these rules can be seen when a transplant is made between regeneration blastemas, a transplant in which the anterior and posterior axes are reversed (Figure 30). The result is a limb with three distal portions (Iten and Bryant, 1975). This outcome can be explained by viewing the anterior–posterior axis on the grid as having two opposite numbers—say, 3 and 9. In juxtaposing the values of 3 and 9, one generates a complete circle of values at each of the extreme sites, and a smaller intercalating series at all other sites. The result is three complete circles, which, by the law of distal transformation, will generate three complete limbs from that point on.

The polar coordinate model also predicts the effects of regulation when a portion of the tissue is lost. The newly formed cells would have positional values intermediate between those of the remaining cells and would reconstruct the appropriate part of the tissue. Because the basis of regeneration appears to be the recognition of differences between adjacent tissues, it is probable that epimorphic pattern formation during regeneration and normal pattern formation during embryonic limb development are the result of the proximate interactions between adjacent cells rather than the result of long-range gradients (Bryant et al., 1981).

Retinoic acid and the intercalation

Crawford and Stocum (1988) showed that retinoic acid is able to change the positional values of the regeneration blastema cells in a proximal direction. Normal wrist blastema cells will sort out with other wrist cells in vitro and will migrate to the wrist region when transplanted to an amputated limb (Chapter 15). Wrist blastema cells soaked in retinoic acid sorted out in vitro and in vivo to the forearm region. Not only did the retinoic acid wrist cells sort out to the forearm level, but when retinoic acid-treated wrist blastema were apposed to forearm stumps, no intercalary regeneration was stimulated. Therefore, the cellular mechanism that recognizes the disparities between non-neighboring cells along the proximal–distal axis and initiates intercalary regeneration is proximalized along with the positional "memory."

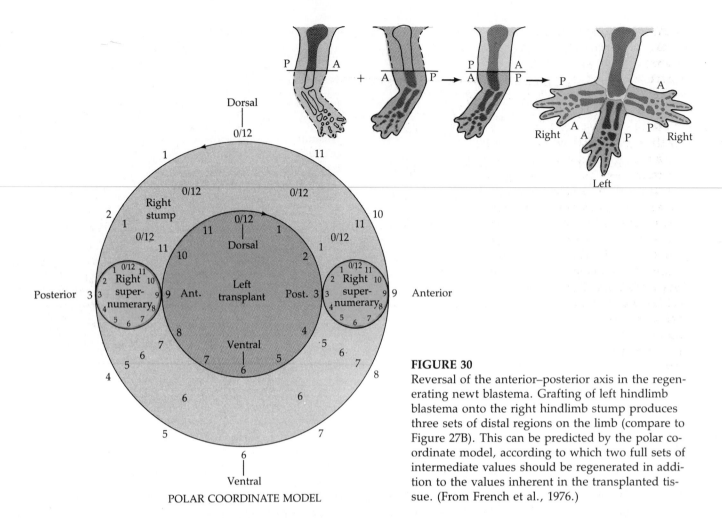

POLAR COORDINATE MODEL

FIGURE 30

Reversal of the anterior–posterior axis in the regenerating newt blastema. Grafting of left hindlimb blastema onto the right hindlimb stump produces three sets of distal regions on the limb (compare to Figure 27B). This can be predicted by the polar coordinate model, according to which two full sets of intermediate values should be regenerated in addition to the values inherent in the transplanted tissue. (From French et al., 1976.)

It is possible that retinoic acid acts as a morphogen to establish the positional values on the cell surfaces of the developing limb and the limb blastema cells. This may have important consequences for the interpretation of the retinoic acid as ZPA morphogen. Wanek and colleagues (S. Bryant, personal communication) have found that anterior chick limb bud cells adjacent to a retinoic acid-soaked bead acquire ZPA-like properties. If the RA-exposed cells are grafted to the anterior margin of a host limb, they act as a ZPA and cause the formation of supernumerary digits. When this experiment is done between retinoic acid-exposed chick cells and host quail embryos, the supernumerary digits are almost exclusively from the quail. Thus, retinoic acid treatment changes cells next to the bead into extreme posterior (ZPA) cells. Since one of the proposed properties of the ZPA is to make retinoic acid, the cells originally exposed to the retinoic acid-soaked bead are now able to produce more of this retinoic acid. This causes a problem with the gradient model, because if retinoic acid causes adjacent cells to become posterior ZPA-like cells, one should find the entire limb bud becoming progressively more ZPA-like and the levels of retinoic acid growing increasingly high, eliminating the gradient.

Bryant and her colleagues interpret these experiments to show that retinoic acid does not provide a gradient of positional information. Rather, retinoic acid is seen as being capable of transforming anterior limb bud tissue into posterior limb bud tissue. The host limbs would now contain "posterior" tissue (the graft of retinoic acid-treated anterior cells) next to anterior tissue (the host limb bud). The result would be intercalary regeneration to restore the positions between the two normally nonapposed tissues. In this interpretation, retinoic acid is an agent that can modify positional values within the limb field.

PATTERN FORMATION IN THE NERVOUS SYSTEM

So far, we have been discussing relatively simple patterns such as the gradients in hydras or chick wing buds. These may be understood within our lifetime. But now we come to the nervous system. The human brain is the most ordered piece of matter known. In most tissues, a given cell may interact in specific ways with fewer than a dozen neighboring cells, but each neuron has the potential to specifically interact with thousands of cells. There are more than 10^{12} neurons in the human brain, and a large neuron (such as a Purkinje cell or motor neuron) can receive input from over 10^5 other cells as well as making its own connections to about a thousand cells (Gershon et al., 1985). This complexity is one of the reasons human and other vertebrate brains are so poorly understood. Most of our knowledge of the nervous system is based on invertebrate nervous systems, which consist of a relatively small number of large, easily identified neurons.

The functioning of the vertebrate brain depends not only on the differentiation and positioning of the neural cells, but also on the specific connections these cells make among themselves. In some manner, nerves from a sensory organ such as the eye must connect to specific neurons in the brain that can interpret visual stimuli, and axons from the nervous system must cross large expanses of tissue before innervating the appropriate target tissue. How does the nerve axon "know" to traverse numerous other potential target cells to make its specific connection? Harrison (1910) suggested that the specificity of axonal growth is due to PIONEER NERVE FIBERS, which go ahead of other axons and serve as guides for them.* This observation simplifies, but does not solve, the problem of how neurons form the appropriate patterns of interconnections. Harrison also noted, however, that axons must grow upon a solid substrate, and he speculated that differences in the embryonic surfaces might allow axons to travel in certain specified directions. The final connections would occur by complementary interactions on the cell surface.

> That it must be a sort of a surface reaction between each kind of nerve fiber and the particular structure to be innervated seems clear from the fact that sensory and motor fibers, though running close together in the same bundle, nevertheless form proper peripheral connections, the one with the epidermis and the other with the muscle. . . . The foregoing facts suggest that there may be a certain analogy here with the union of egg and sperm cell.

Harrison's brilliant experimental work and speculations laid the groundwork for studies of neuronal patterns, just as others of his studies formed the basis for our knowledge of limb pattern formation.

Mechanisms of directing axon growth

The contact guidance hypothesis

One of the first hypotheses to account for the specificity of axonal growth involves CONTACT GUIDANCE (STEREOTROPISM). Here, physical cues in the substratum direct neural growth. Following Harrison's technique of growing axons on clotted blood, Weiss (1955) noted that the growing axons not only needed a solid substrate on which to migrate but also that the migration tended to follow discontinuities in the clot. When the fibers of the

---

*The growth cones of pioneer neurons migrate to their target tissue while embryonic distances are still short and the intervening embryonic tissue is still relatively uncomplicated. Later in development, the other neurons that innervate the target tissue bind (fasciculate) to the pioneer neuron and thereby enter the target tissue. Klose and Bentley (1989) have shown that in some cases, the pioneer neurons die after the other neurons reach their destination. Yet, were that pioneer neuron prevented from differentiating, the other axons would not have reached their target tissue.

blood clot were randomly oriented, the axons followed this random pattern. But when the clot fibers were made parallel by applying tension to the clot, the nerve axons traveled along these fibers, not veering from the straight and narrow (see Figure 9 in Chapter 15). Here, then, is evidence for the involvement of physical factors in contact guidance. Singer and his co-workers (1979) found evidence that such physical factors operated in vivo. They detected large channels between ependymal cells of the developing newt spinal cord through which growing axons migrated. They hypothesized that these channels provided cues for guiding the axons toward the appropriate regions of the brain. Cellular channels have also been detected in the mouse retina (Silver and Sidman, 1980), and these appear to guide the retinal ganglion cell axons into the optic stalk as they develop.

The presence of pre-existing channels is probably not critical for the growth of most axons. The growth cone appears to be capable of digesting its own channels through an extracellular matrix by secreting two types of proteolytic enzymes—plasminogen activator and metalloproteases—into its immediate environment (Pittman, 1985).

The differential adhesive specificity hypothesis

A modified form of the contact guidance hypothesis sees the direction for axonal growth being provided by preferential growth cone binding to different adhesive molecules. The DIFFERENTIAL ADHESIVE SPECIFICITY (HAPTOTAXIS) HYPOTHESIS thus postulates that there are differences in adhesivity across the path of an axon. It is analogous to the model of adhesive specificity mentioned earlier (Chapter 15) to explain migrations of the pronephric duct tissue.

In one form of this hypothesis, a growth cone is presented with a "patchy environment." Given the choice between migrating on one surface or another, it will migrate on the surface that is more adhesive. Growth cones often have to migrate across extracellular matrices, and the molecules within the extracellular matrix can have a profound effect on axon growth. A piece of neural retinal tissue, for instance, will not readily send forth axons onto tissue-culture plastic. However, when the plastic is coated with fibronectin or laminin, long axonal outgrowths are observed. The presence of such molecules may delineate pathways through the embryo (Figure 31; Akers et al., 1981; Gundersen, 1987). Letourneau and co-workers (1988) have shown that the axons of certain spinal neurons travel through the neuroepithelium over a transient laminin-coated surface that precisely marks the path of these axons. Conversely, glycosaminoglycans, another set of molecules associated with extracellular matrices, appear to impede neural outgrowth (Tosney and Landmesser, 1985). Thus, an axon seeing two alternative surfaces might choose one over the other.

Such alternative surfaces exist in the embryo. There is a very good correlation between the elongation of retinal axons and the presence of laminin on the neuroepithelial cells and astrocytes in the embryonic mouse brain (Cohen et al., 1986, 1987; Liesi and Silver, 1988). Punctate laminin deposits are seen on the cell surfaces along the pathway leading from the retina to the optic tectum, whereas adjacent areas where the optic nerve fails to grow lack these laminin deposits. Laminin deposits are not located on the basement membranes, but are found where the nerve growth cone can contact the cell. After the retinal axons have reached the tectum, the glial cells differentiate and lose their laminin. At this point, the retinal ganglial neurons that have formed the optic nerve lose their ability to recognize laminin. Laminin deposits may also be necessary for the regen-

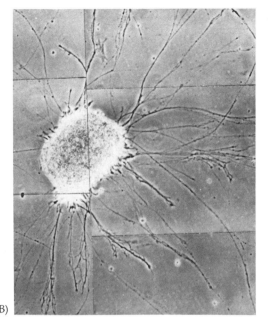

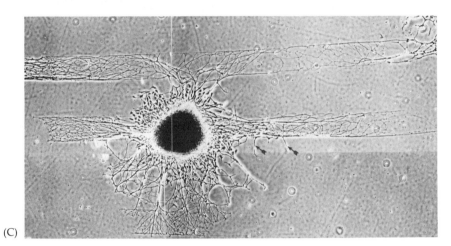

(A)

(B)

(C)

FIGURE 31
Effects of substrate factors on neural outgrowth. (A,B) Effects of fibronectin on neural outgrowth from neural retina aggregates. The aggregate in A was cultured 36 hours on untreated tissue culture plastic. The aggregate in B was cultured on plastic treated with 50 μg/ml fibronectin. (C) Outgrowth of sensory neurons placed on patterned substrate consisting of parallel sripes of laminin applied to a background of type IV collagen. (A and B from Akers et al., 1981; C from Gundersen, 1987. Photographs courtesy of J. Lilien and R. W. Gundersen.)

eration of neural tissue. Astroglial cells can induce regeneration when placed into embryos where the normal neuronal pathways of the corpus callosum have been severed. These astrocytes have punctate laminin on them.

In another form of this model, the extracellular matrix provides a gradient of adhesivity. This would be expected if the extracellular matrix was broad and would thus permit the axon to go in several directions. A gradient would make one direction favored. This type of adhesive gradient has been deduced in certain migrations of insect axons. Nardi (1983) has seen that when sensory neurons from the tips of pupal moth wings grow toward the central nervous system, they are tightly associated with the extracellular matrix of the upper epithelial layer. When he placed grafts along the route of axon development, Nardi found that the neurons adhered more tightly to proximal tissue (tissue closer to the central nervous system) than to distal tissue. Scanning electron micrographs of the extracellular matrix showed that the matrix changed along the route. It became progressively more "bumpy" and fibrillar as it got closer to the central nervous system (Figure 32). Berlot and Goodman (1984) also showed that grasshopper sensory neurons originating in the antennae probably follow a gradient of adhesivity to the central nervous system.

FIGURE 32
Gradients of extracellular matrix fibrils across the developing wing of the moth *Manduca sexta*. Double-headed arrows represent the proximal–distal axis. (Scanning electron micrographs of the three regions from Nardi, 1983; photographs courtesy of J. Nardi.)

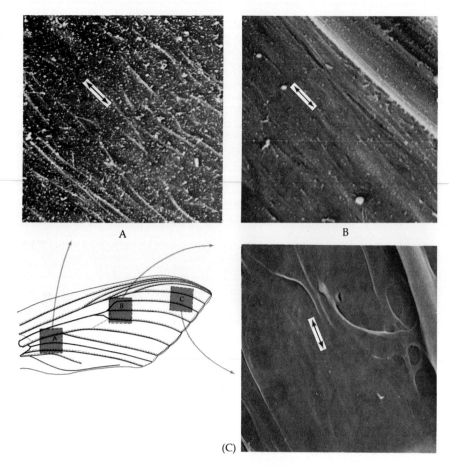

The labeled pathway hypothesis

Extracellular matrix molecules such as laminin and N-CAM can provide only general cues for directionality. They cannot, for instance, tell one nerve axon to travel one way and an adjacent axon to migrate in another direction. Yet in *Drosophila*, grasshoppers, and *Caenorhabditis* (and probably most invertebrates), the patterning of axonal movement is an astoundingly precise process, and adjacent axons are given different migratory instructions by their environment. The embryonic ectoderm of insects contains clusters of cells that become either epidermal or neural tissues. Experiments using laser ablation (Taghert et al., 1984) have shown that the decision to be neural (rather than epidermal) comes about due to cell–cell interactions within these defined clusters, and once the decision has been made, the neural precursor cells divide in a predictable way to generate a family of neurons.

For example, within each segment of the grasshopper, 61 neuroblasts emerge (30 on each side, one in the center). The remaining ectoderm forms the epidermis. One cluster of ectodermal cells is responsible for generating the 7-4 neuroblast (Figure 33). If the entire cluster is destroyed by laser irradiation, no 7-4 neuroblast will form. However, when the cell that is enlarging to form this neuroblast is selectively destroyed, a neighboring cell will take its place to become the 7-4 neuroblast.

The 7-4 neuroblast is a stem cell and gives rise to a family of six neurons, termed C, G, Q1, Q2, Q5, and Q6. This family of neurons is shown in Figure 33 and as the yellow neurons in the color portfolio. The axonal growth cones of these neurons reach their targets by following specific pathways formed by other earlier neurons. Q1 and Q2 follow a straight path together, traversing numerous other cells, until they meet the axon from the dorsal midline precursor neuron (dMP2), which they then follow posteriorly. The other four neurons of the 7-4 family migrate across the dMP axon as if it did not exist. Axons from the C and G neurons progress a long way together, but ultimately C follows nerves X1 and X2 toward the posterior of the segment, while G adheres to the P1 and P2 axons (which go posteriorly) and moves anteriorly on their surfaces (Goodman et al., 1984).

The G growth cone will have encountered over 100 different surfaces to which it could have adhered, but it is specific for the P neurons. If the P neurons are ablated by laser, the G growth cone will act abnormally, its filopodia searching randomly for its proper migratory surface. If any of the other hundred or so axons are destroyed, the G growth cone behaves normally.

This formulation of axonal pathfinding in insects has been called the LABELED PATHWAYS HYPOTHESIS because it implies that a given neuron can specifically recognize the surface of another neuron that has grown out before it. Evidence for this specificity comes from monoclonal antibodies (Bastiani et al., 1987). Neurons aCC and pCC are sibling neurons (both are derived from neuroblast 1-1) that have very different fates. Moreover, different sets of axons adhere to each of them, creating independent bundles of axons, called fascicles. When monoclonal antibodies were tested on the neurons of 10-hour embryos, the antibodies bound specifically to a protein called fasciclin I present on the two aCC neurons of each segment, but they did not bind to the pCC neurons. By hour 11, however, other neurons (but not pCC) began to express this cell surface molecule. The nerves were precisely those (RP1, RP2, U1, U2) that fasciculate with aCC. More recent experiments (Zinn et al., 1988; Harrelson and Goodman,

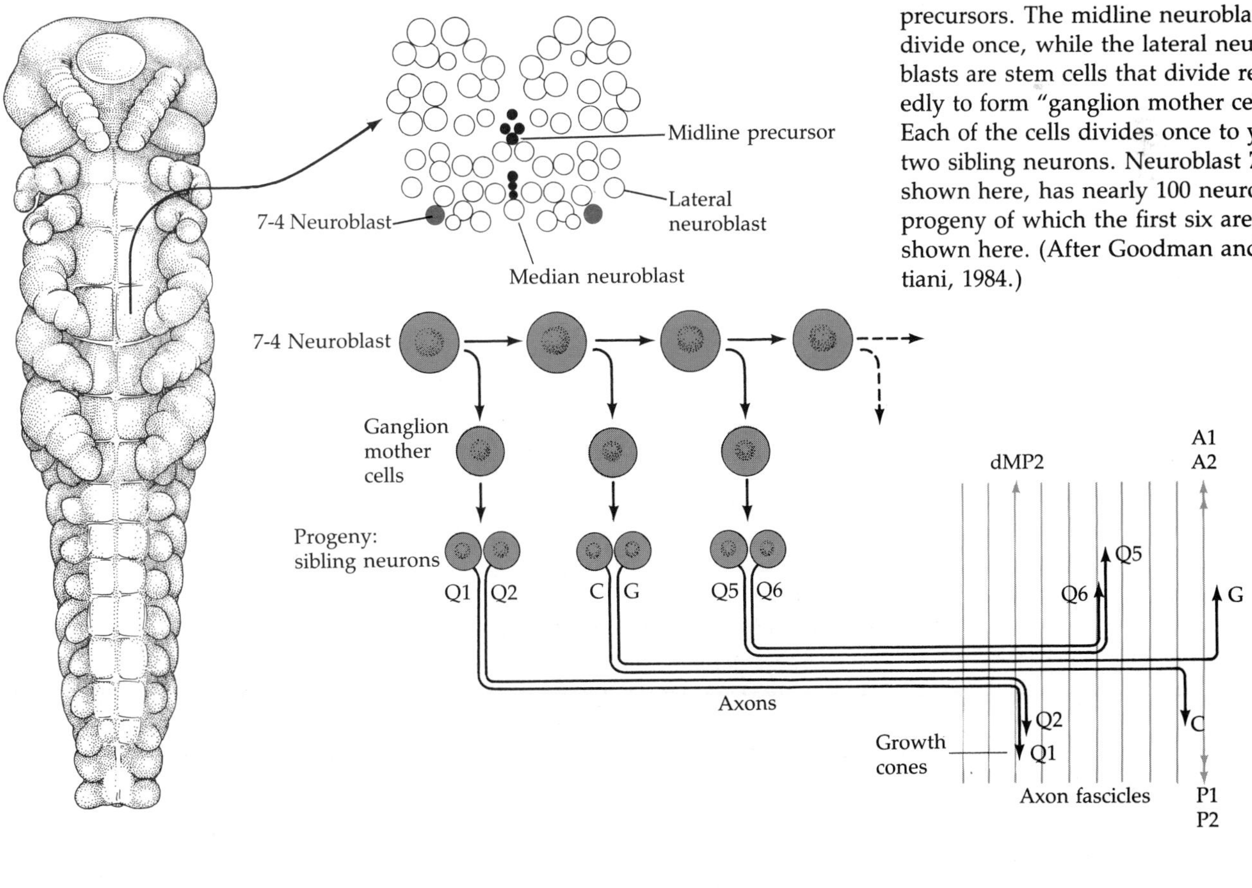

FIGURE 33
Each of the 17 segments of the early grasshopper embryo has the same pattern of neuroblasts. There are 30 lateral neuroblasts on each side, one median neuroblast, and seven midline precursors. The midline neuroblasts divide once, while the lateral neuroblasts are stem cells that divide repeatedly to form "ganglion mother cells." Each of the cells divides once to yield two sibling neurons. Neuroblast 7-4, shown here, has nearly 100 neuronal progeny of which the first six are shown here. (After Goodman and Bastiani, 1984.)

1988) have shown that there are at least three fasciclin molecules expressed on different subsets of neurons, and that each of these molecules allows the grown cones of certain neurons to recognize specifically those axons with which they will fasciculate.

Studies on vertebrates lag far behind those on insects, but recent studies on zebrafish motor neurons suggest that labeled pathways may function here as well. Zebrafish may become the organism of choice in vertebrate neurobiology because their development is very rapid, numerous individuals can be compared, and the embryos are optically clear. This enables neurobiologists to observe the growth of axons in living embryos. Neurons can be identified by injecting fluorescently labeled beads into the neuronal precursors (Kimmel and Law, 1985) and following the axon growth with eye and video recorder. Eisen and her colleagues (1986) watched the axonal elongation of three pioneer motor neurons in these embryos. After the axons left the spinal cord, all three followed the same path along a muscle until they reached a particular site of the embryo. At this point they diverged into three neuron-specific pathways that innervated the appropriate muscles.

The chemotactic hypothesis

The growth of axons from a ganglion to a target tissue may be directed by soluble molecules arising from the target. The axon's growth cone would be expected to migrate up the concentration gradient of these molecules (chemotaxis). Evidence for such a view comes from explanted ganglia placed close to, but not touching, various potential target tissues. Lumsden and Davies (1983, 1986) have dissected the trigeminal ganglia of sensory neurons out of embryonic mice before the neurons contacted their normal target tissue, the whisker pad. When these explanted ganglia are placed in culture medium near a variety of potential target tissues from the embryonic mouse, the trigeminal neurons migrate only toward the whisker pad tissue. In fact, when cultured between whisker pad epithelial tissue and whisker pad mesenchymal tissue, they will only migrate toward the epithelial tissue (Figure 34). This observation can best be explained by postulating the secretion of some chemotactic substance by the whisker pad epithelium. The chemotaxis of the trigeminal nerve to the whisker pad is not only target-restricted but also neuron-restricted and temporally

FIGURE 34
Summary of the experiments of Lumsden and Davies showing chemotaxis between neural tissue (trigeminal ganglion) and its target (whisker pad). The chemoattraction is specific for the target (A) and the epithelial cells of the target (B). Moreover, the chemotactic ability of the whisker pad is specific for the trigeminal neurons (C).

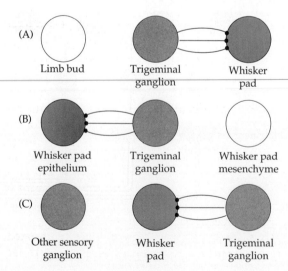

(A) Limb bud   Trigeminal ganglion   Whisker pad

(B) Whisker pad epithelium   Trigeminal ganglion   Whisker pad mesenchyme

(C) Other sensory ganglion   Whisker pad   Trigeminal ganglion

restricted. The whisker pad tissue does not attract the axons from other sensory ganglia, and it will attract the trigeminal neurons only during a specified time.

Another example of chemotaxis is in the spinal commissural neurons. The axons of these neurons begin growing ventrally down the side of the neural tube. However, about two-thirds of the way down, their direction changes, and they project through the motor neuron area of the neural tube toward the floor plate cells (Figure 35). The floor plate cells, the most ventral portion of the neural tube, have been shown to synthesize a diffusible molecule that attracts these axons specifically (Tessier-Lavigne et al., 1988; Placzek et al., 1990). Moving the floor plate cells to different positions will change the direction of the commissural axons, and floor plate explants added to cultured spinal cord explants will direct the migration of commissural axons towards them.

The best characterized neural chemotactic molecule is NERVE GROWTH FACTOR (NGF), a glycoprotein composed of two identical 13,200-Da subunits. NGF is needed for the survival of sympathetic and sensory neurons during development, and exogenously supplied NGF can influence the direction of nerve growth (Angeletti and Vigneti, 1971; Levi-Montalcini, 1976; Ebendal et al., 1982). In 1989, Edwards and his colleagues provided dramatic evidence that naturally produced NGF could lead to innervation of the cells that produced it. They constructed a plasmid containing the mouse NGF gene fused to the upstream regulatory elements of the rat insulin II gene. In other words, they placed the NGF gene under the control of the factors responsible for transcribing insulin in the pancreatic β cells. They then injected this gene into mouse embryos and observed NGF secretion in the resulting transgenic mice. Not only were the pancreatic islet β cells producing large amounts of NGF, but these cells induced spectacular innervation of the islet by sympathetic neurons (Figure 36). The action of NGF appears to be local. New neurons were not created; rather, there was tremendous sprouting and directed growth of neurons that already existed near the area. In normal development, NGF is synthesized by numerous target tissues including glands, whisker pads, and other neurons.

NGF appears to be both a neurotropic factor (i.e., one that can direct influence the direction of neuronal growth) and a neurotrophic factor. Neurotrophic factors are chemicals that are necessary for the continued survival of neurons. During normal development of the central and peripheral nervous systems, there is widespread neuronal death (see Sidelights & Speculations). The removal of target tissues during development often results in neuronal cell death due to the lack of neurotrophic factors, and there appears to be competition for the small amount of factors secreted by the target tissue. The result is that many neurons begin to innervate the target tissue, but after several days usually only one remains. NGF is necessary for the survival of sympathetic and sensory neurons. Treating mouse embryos with anti-NGF antibodies reduces the number of trigeminal and dorsal root ganglia neurons to 20 percent of their control amounts (Levi-Montalcini and Booker, 1960; Pearson et al., 1983). Target tissues for sympathetic neurons have been found to secrete NGF, and removal of these tissues causes the death of the neurons that would have innervated them. There is also a good correlation between the amount of NGF secreted and the survival of neurons that innervate these tissues (Korsching and Thoenen, 1983; Harper and Davies, 1990).

At least two other neurotrophic proteins have been characterized. These proteins—brain-derived neurotrophic factor (BDNF) and neurotro-

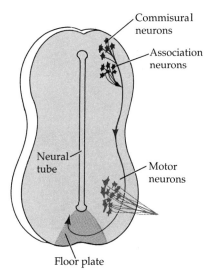

Spinal cord root

Commissural neurons

Association neurons

Motor neurons

Neural tube

Floor plate

FIGURE 35
Trajectory of the commissural axons in the rat spinal cord. Axons from the commissural neurons (color) and association neurons (black) both form in the dorsal region of the spinal cord and begin migrating on embryonic day 11. Association neurons project laterally, while the commissural axons grow ventrally down the lateral margin of the spinal cord. When they approach the region of motor neurons, the commissural axons alter their path and travel towards the floor plate cells. Upon reaching the floor plate, they turn 90 degrees toward the brain. (After Tessier-Lavigne et al., 1988.)

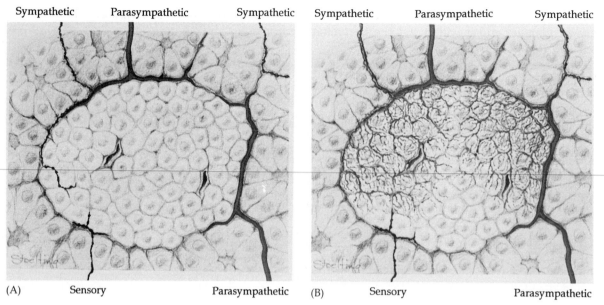

Sympathetic Parasympathetic Sympathetic Sympathetic Parasympathetic Sympathetic

(A) Sensory Parasympathetic (B) Sensory Parasympathetic

FIGURE 36

Nerve growth factor induces the selective outgrowth of sympathetic axons into a new area that otherwise would not have secreted NGF. Transgenic mice were bred such that an exogenous NGF gene was placed under the control of insulin gene regulatory elements. In this way, the β-islet cells of the pancreas (which produce and secrete insulin) were caused to secrete NGF as well. (A) Control mouse showing the positions of the sympathetic, parasympathetic, and sensory neurons around the borders of the islet. (B) Islet of transgenic mouse whose islet cells were secreting NGF. Hyperinnervation of sympathetic neurons is seen. (From Edwards et al., 1989; photographs courtesy of R. H. Edwards.)

phin-3 (NT-3)—share the same basic structure as NGF. However, they favor the survival of slightly different groups of neurons. While some neurons will respond to all three factors, other neurons will respond to only one or two. Liver and skeletal muscle synthesize NT-3, which can support the survival of visceral neurons that are not responsive to NGF. NGF, however, can support the survival of sympathetic ganglion neurons that do not respond to NT-3 (Hohn et al., 1990; Maisonpierre et al., 1990).

One interesting observation is that neurotransmitters released by one nerve's growth cone can inhibit or accelerate the elongation of other growth cones. In the snail *Helisoma*, the pattern of neuronal connection is influenced by such interactions. Neuron 5 secretes serotonin, thereby preventing nerve 19 from synapsing with it (Figure 37). Serotonin has no effect on neuron 5 but is able to completely inhibit the axonal growth of neuron 19 by causing the retraction of that growth cone's filopodia (Haydon et al., 1985; Kater, 1985). Nerves can also alter the growth cones of other neurons by electrical activity. Once electrical potentials are established (as at neuromuscular junctions), the action potential can cause the cessation of axonal outgrowth for other neurons (Cohan and Kater, 1986).

Multiple guidance cues

One of the major mysteries of vertebrate developmental neurobiology concerns the innervation of the limb muscles. Axonal outgrowth of motor neurons occurs very early in development, before the soma of the motor neurons have migrated to their definitive ganglia and before the muscles have condensed out of mesenchyme. This stage can be seen in Figure 32B of Chapter 16. Thus, at the time the axons innervate the limb, neither the neurons, their targets, nor the pathways appear very well defined (Hollyday, 1980; Landmesser, 1978). However, in another sense, they appear very well defined. If one redirects a different source of nerves (from a different ganglion) into the limb, the pattern of branching remains normal. In other words, the limb is able to dictate the pattern of innervation to a set of axons that normally would not enter the limb (Hamburger, 1939; Hollyday et al., 1977). Moreover, if segments of the chick spinal cord are reversed—so that the motor neurons are in new locations—their axons will find their original targets (Figure 38; Lance-Jones and Landmesser,

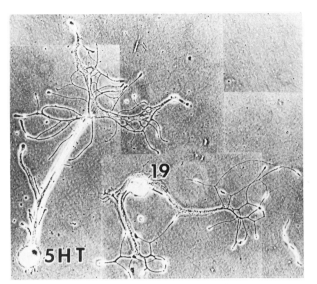

FIGURE 37
The growth cones of neuron 19 turn away from serotonin-producing neuron 5 when the two *Heliosoma* neurons are cultured together. When another two neurons (such as two neurons 19) are brought together, they attract one another. (From Kater, 1985; photograph courtesy of S. B. Kater.)

1980). Yet, when limbs develop with duplicated areas (such as two thighs), the neurons innervating the second thigh will not be "thigh"-specific neurons, but will be neurons that usually innervate the calf (Whitelaw and Hollyday, 1983). These experiments present a paradox that has yet to be resolved: "Particular axons are biased to grow to specific places, yet axons from different motor neuron pools can substitute for one another in the establishment of normal nerve patterns" (Purves and Lichtman, 1985). The most plausible explanation is that there are multiple mechanisms acting simultaneously to ensure that the axons get to their appropriate places.

Multiple guidance cues have also been postulated to explain how individual retinal neurons from *Xenopus laevis* embryos are able to send axons to the appropriate area of the brain even when transplanted away from the optic nerve (Harris, 1986). This ability implies that the guidance cues are not distributed solely along the normal pathway but exist through-

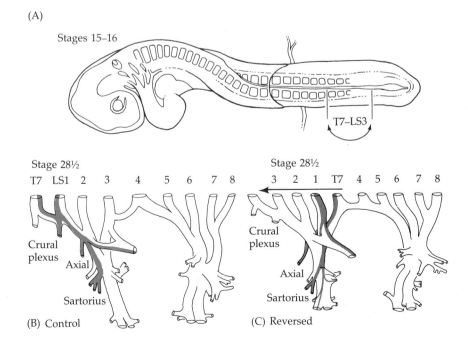

FIGURE 38
Compensation for small dislocations of axonal initiation position in the chick embryo. (A) A length of spinal cord comprising several segments T7–LS3, (seventh thoracic to third lumbarsacral segments) is reversed in the 2.5-day embryo. (B) Normal pattern of axon projection into the different muscles at 6 days. (C) Projection of axons in the reversed segment. The ectopically placed neurons eventually found their proper neural pathways and innervated the appropriate muscles. (From Lance-Jones and Landmesser, 1980.)

out the embryonic brain. Guiding an axon from the nerve cell body to its destination across the embryo is a complex phenomenon, and Spemann (1938) may have been correct when he predicted that several different types of cues might be used simultaneously to assure that the correct connections get established.

Mechanisms of synaptic specificity

Qualitative differences: Gradients

Getting an axon to its target organ does not finish its trek. The core of neural function is the ability to form specific connections from one nerve axon to another cell or group of cells. This is called SYNAPTIC SPECIFICITY, and it is beautifully illustrated in the connections made by axons emerging from the retinal ganglion cells. These retinal ganglion cells send out axons that form the optic nerve. In nonmammalian vertebrates, the axons traverse large expanses of the brain without making a single connection and eventually terminate by forming synapses with cells of the OPTIC TECTUM. Each retinal axon sends its impulse to one specific site (a cell or small group of cells) within the tectum (Sperry, 1951). As shown in Figure 39, there are two optic tecta in the brain. The axons from the right eye enter the left optic tectum and those from the left eye form synapses with the cells of the right optic tectum.

The map of retinal connections to the optic tectum (RETINOTECTAL PROJECTION) was detailed by Marcus Jacobson (Jacobson, 1967). Jacobson defined this map by shining a narrow beam of light on a small, limited region of the retina and noting, by means of a recording electrode in the tectum, which tectal cells were being stimulated. The retinotectal projection of *Xenopus laevis* is shown in Figure 39. As the light moves from the ventral to the dorsal surface of the retina, it stimulates cells from the lateral side to the medial side of the tectum. As the light impinges upon cells from the posterior to the anterior of the retina, the tectal stimulation goes from the caudal (tail) to the rostral (lip) end. Thus, we see that there is a point-for-point correspondence between the cells of the retina and the cells of the tectum. When a group of retinal cells is activated, a very small and specific group of tectal cells is stimulated. We also can observe that the points form a continuum; in other words, adjacent points on the retina project onto adjacent points on the tectum. This arrangement enables the frog to see an unbroken image.

This intricate specificity has given rise to the CHEMOAFFINITY HYPOTHESIS (or the HYPOTHESIS OF NEURONAL SPECIFICITY). According to Sperry (1965),

> The complicated nerve fiber circuits of the brain grow, assemble, and organize themselves through the use of intricate chemical codes under genetic control. Early in development, the nerve cells, numbering in the millions, acquire and retain thereafter, individual identification tags, chemical in nature, by which they can be distinguished and recognized from one another.

Within this framework, however, regulation can still occur. If, at stage 32, a composite retina were constructed consisting of the left half of one retina and the ventral half of another, one would expect there would be no axonal input into that area of the tectum that normally receives neurons from the right dorsal quadrant (and one would expect to see twice as

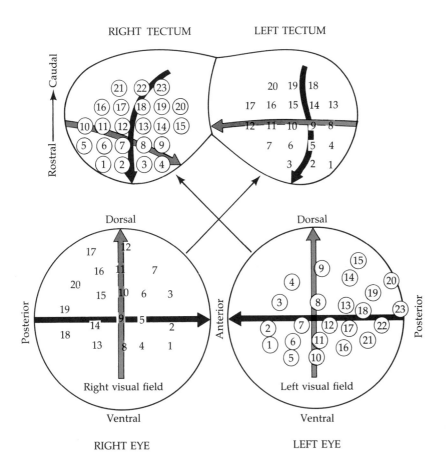

RIGHT TECTUM   LEFT TECTUM

FIGURE 39
Map of the normal retinotectal projection in the adult *Xenopus*. The right eye innervates the left tectum and the left eye innervates the right tectum. The numbers on the visual fields (retina) and the tecta show regions of correspondence; that is, stimulation of spot 15 on the right retina sends electrical impulses to left tectal region 15. The black and colored arrows indicate the direction of light impulses in the eye and the corresponding path of stimulation in the tectum. (From Jacobson, 1967.)

RIGHT EYE   LEFT EYE

many axons entering those regions innervated by the left ventral quadrant, which is represented twice). This does not happen. Rather, the retinotectal projection from such a retina is normal. Similarly, if half the tectum is removed, the entire projection is compressed to fit into the smaller area (Gaze, 1978). This suggests that the retina and tectum form their specificities in a gradient-type pattern of recognition. Rather than having strict codes, with each neuron being qualitatively different, there might be gradients that inform the cells of their relative positions on the dorsal–ventral or anterior–posterior axes. In the developing chick eye, the ability to regulate is extremely well defined. Crossland and co-workers (1974) removed segments of the optic cup at various stages of chick retina development and observed the innervation of the tectum by the remaining neurons. When the retinal segment was removed prior to day 3, the chicken developed a normal tectal projection, the remaining neurons reaching all the areas of the tectum. After day 3, however, the ablation of retinal segments resulted in uninnervated portions of the tectum.

Can the retinal ganglion cells actually distinguish between different cells of the optic tectum? There is good evidence to indicate that retinal ganglial cells can certainly distinguish between regions of the tectum. Cells prepared from the ventral half of chick neural retina preferentially adhere to dorsal halves of the tectum (Figure 40; Roth and Marchase, 1976). Gottlieb and co-workers (1976) found that neurons taken from the most dorsal part of the chick retina adhere preferentially to the most ventral portion of the tectum and that the extreme ventral neurons of the retina preferentially adhere to the most dorsal extremes of the tectum. These results were confirmed in other experimental conditions using axonal tips rather than entire neurons (Halfter et al., 1981).

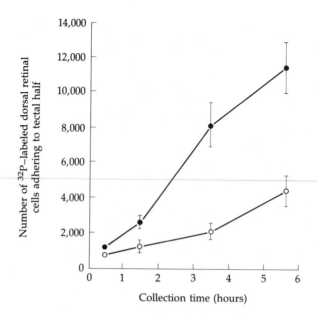

FIGURE 40
Differential adhesion of dorsal radioactive chick retinal cells to dorsal and ventral tectal halves. Radioactive cells from the dorsal half of 7-day chick retinas were added to dorsal (○) and ventral (●) halves of 12-day chick optic tecta. The data show the selective adhesion of the dorsal retina cells to the ventral tectal tissue. (After Roth and Marchase, 1976.)

Gradients of specific compounds have been observed on the cell surface of the neural retina, but their functions are not known. Trisler and his co-workers (1981), using monoclonal antibodies to chick neural retinal cells, found one antibody that localized in a gradient fashion across the retinal surface. The antigen identified by this antibody is 35 times more concentrated in the dorsal region of the retina than it is at the ventral margin (Figure 41). Moreover, the same cell surface molecule is present in the optic tectum, where the gradient is reversed. Because dorsal retinal cells send axons to the ventral tectum (and ventral retinal cells project to the dorsal tectum), this antigen may be involved in orienting specific axonal connections (Trisler and Collins, 1987). When injected into the vitreous humor of 11-day embryonic chick eyes, this antibody inhibits synapse formation (Trisler et al., 1986; Trisler, 1987), further suggesting that this topologically asymmetric antigen may be involved in proper neuronal recognition. In rat embryos, another antigen has been found in a dorsal–ventral gradient across the retina (Figure 41B) (Constantine-Paton et al., 1986). Its biochemical structure does not seem at all related to the antigen on the chick retinal cells.

One gradient that has been identified functionally is a gradient of

FIGURE 41
Gradient of antigens on neural retinal cells. (A) Gradient of antibody binding to a cell surface molecule in the 14-day embryonic chick neural retina. Each retina was divided into eight 45-degree sections and then divided into an inner and an outer segment. A radioactive monoclonal antibody was then bound to it. Numbers represent counts per minute per milligram protein. (B) A gradient across the 13-day embryonic rat neural retina is visualized with a fluorescent monoclonal antibody. (A after Trisler, 1987; B from Constantine-Paton et al., 1986; photograph courtesy of M. Constantine-Paton.)

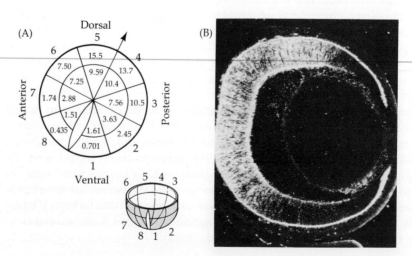

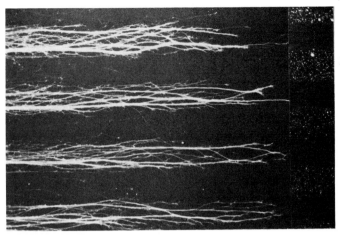

Tectal membranes

Anterior

Posterior

Anterior

Posterior

Anterior

Posterior

Anterior

FIGURE 42
Differential extension of temporal retinal axons on tectal membranes. Alternating stripes of anterior and posterior tectal membranes were absorbed onto filter paper. When neurons from temporal (posterior) retina are plated on such alternating carpets, they preferentially extend axons on the anterior tectal membranes. (From Walter et al., 1987.)

"repulsion" that is highest in the posterior tectum and weakest in the anterior tectum. Walter and colleagues (1987) prepared a "carpet" of tectal membranes having alternating stripes from the posterior and anterior tectum. They then let cells from the nasal (anterior) or temporal (posterior) regions of the retina settle onto this carpet and extend axons. The ganglial cells from the *nasal* portion of the retina extended axons equally well on both the anterior and posterior tectal membranes. The neurons from the *temporal* side of the retina, however, extended axons only when they settled on the anterior tectal membranes (Figure 42). The basis for this specificity appears to be a repulsive factor on the posterior tectal cell membranes. When the growth cone of a temporal retinal axon contacts a posterior tectal cell membrane, the filopodia of the growth cone withdraw, and the growth cone collapses and retracts (Cox et al., 1990). Neither tectal membrane had any effect on the axons from the nasal-region retinal neurons. There appears, then, to be a repulsive factor that steers temporal (posterior) retinal axons away from the posterior tectum but which has no effect on the nasal (anterior) retinal cells. The result would be the directing of the temporal retinal axons to the anterior portion of the tectum. Presumably, the nasal retinal axons could be directed (perhaps by exclusion) to the posterior of the tectum.

Qualitative neuronal differences

The preceding model is a quantitative type of hypothesis, wherein all cells have the same type of glue but the amounts differ. It may be useful for directing axons to a general location, but there is a debate over whether such coding can accurately pinpoint specific locations. There is evidence from other organisms that each neuron might have qualitatively different cell surface molecules and that these molecules might be important in synaptic specificity. The nervous system of the leech consists of 34 paired ganglia containing about 400 neurons each. Individual neurons have been identified, and the functions of many of these neurons are known. Zipser and McKay (1981) injected the leech nervous system into mice and obtained hundreds of monoclonal antibodies, which bound to various regions of the nervous system. Out of the original 300 monoclonal antibodies, 41 of them recognized subsets of the leech nervous system, a finding indicating that the cell surfaces of the neurons differed. In some cases, such differences could be correlated with function. Monoclonal antibody

Lan 3-1 bound specifically to a single pair of neurons in the midbody ganglia and recognized the same right and left neurons in each of the ganglia (Figure 43). These pairs of neurons are known to control the process of penile eversion in mating leeches. Another monoclonal antibody, Lan 3-2, recognized all four neurons that respond to noxious mechanical stimuli. "The situation," according to Zipser and McKay, "seems quite analogous to colour-coded electrical cable containing many wires, where each wire has its own molecule (dye) to facilitate proper recognition and connection at terminals."

Development of behaviors

One of the most fascinating aspects of developmental neurobiology is the correlation of certain neuronal connections with certain behaviors. There are two remarkable aspects of this phenomenon. First there are those cases where complex behavioral patterns are inherently present in the "circuitry" of the brain at birth. The heartbeat of a 19-day chicken embryo will quicken when it hears the distress call, and no other call will evoke this response (Gottlieb, 1965). Furthermore, a newly hatched chick will immediately seek shelter if presented with the shadow of a hawk. The actual hawk is not needed; the shadow cast by a paper silhouette will suffice, and the shadow of no other bird will cause this response (Tinbergen, 1951). There appear, then, to be certain neuronal connections that lead to inherent behaviors in vertebrates.

Equally remarkable are those instances where the nervous system is seen to be so plastic that new experiences can modify the original set of neuronal connections, causing the creation of new neurons or the formation of new synapses between existing neurons. Since neurons, once formed, do not divide, the "birthday" of a new neuron can be identified by treating the organism with radioactive thymidine. Normally, very little radioactive thymidine is taken up into the DNA of a neuron that has already been formed. However, if a new neuron differentiates by cell division during the treatment, it will incorporate radioactive thymidine into its DNA. Such new neurons are seen to be generated when male songbirds learn their songs. Juvenile zebra finches memorize a model song and then learn the pattern of muscle contractions necessary to sing a particular phrase. In this learning and repetition process, new neurons are seen to be generated in the hyperstriatum of the finch's brain. Many of these new neurons send axons to the archistriatum, which is responsible

FIGURE 43

Specific functional neurons stained by monoclonal antibodies to cell surface components. (A) Lan 3-1 antibodies recognize a single pair of neurons in a particular ganglion. These neurons function in penile eversion. (B) A set of neurons recognized by Lan 3-2 antibodies; these neurons respond to noxious stimulants on the leech's skin. (From Zipser and McKay, 1981; courtesy of B. Zipser.)

(A)

(B)

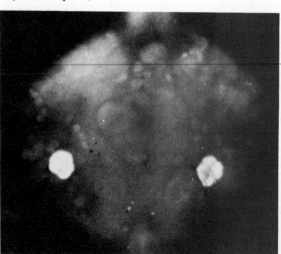
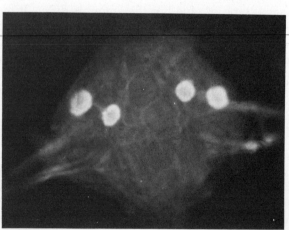

for controlling the vocal musculature (Nordeen and Nordeen, 1988). These changes are not seen in males who are too old to learn the song, nor are they seen in juvenile females (who do not sing these phrases).

Even the adult brain is developing in response to new experiences. When adult canaries learn new songs, new neurons are generated whose axons project from one vocal region of the brain to another (Alvarez-Buylla et al., 1990). Similarly, when adult rats learn to keep their balance on dowels, their cerebellar Purkinje cell neurons develop new synapses (Black et al., 1990). Thus, the nervous system continues to develop in adult life, and the pattern of neuronal connections is a product of inherited patterning and patterning produced by experiences. This interplay between innate and experiential development has been detailed most dramatically in studies on mammalian vision.

Experiential changes in inherent mammalian visual pathways

Some of the most interesting research on mammalian neuronal patterning concerns the effects of sensory deprivation on the developing visual system in kittens and monkeys. The paths by which electrical impulses pass from the retina to the brain in mammals are shown in Figure 44. Axons from the retinal ganglion cells form the two optic nerves, which meet at the optic chiasm. As in *Xenopus* tadpoles, some fibers go to the opposite (CONTRALATERAL) side of the brain, but unlike most other vertebrates, mammalian retinal cells also send inputs into the same (IPSILATERAL) side of the brain. These nerves end at the two LATERAL GENICULATE BODIES. Here the input from each eye is kept separate, the uppermost and anterior layers receiving the neurons from the contralateral eye, the middle of the bodies receiving input from the ipsilateral eye. The situation becomes more complicated as neurons from the lateral geniculate bodies connect with the neurons of the VISUAL CORTEX. Over 80 percent of the neural cells in the cortex receive inputs from both eyes. The result is binocular vision and depth perception. Another remarkable finding is that the retinocortical projection is the same for both eyes. If a cortical neuron is stimulated by light flashing across a region of the left eye 5 degrees above and 1 degree

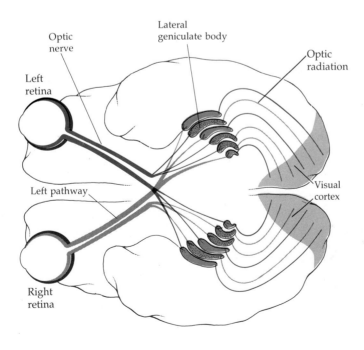

FIGURE 44
Major pathways of the mammalian visual system. In mammals, the optic nerve from each eye branches, sending nerve fibers to a lateral geniculate body on each side of the brain. Fibers from the left side of each retina enter the left lateral geniculate body, and fibers from the right side of each retina enter the right lateral geniculate body. Neurons from each lateral geniculate body enter the visual cortex on the same side.

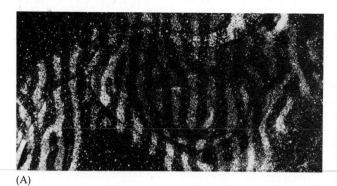

(A)

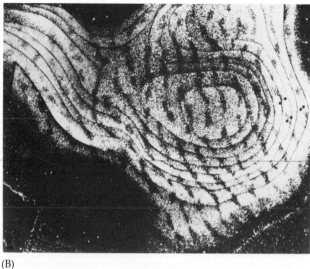

(B)

FIGURE 45
Dark-field autoradiographs of monkey striate cortex 2 weeks after one eye was injected with [³H]proline in the vitreous humor. Each retinal neuron will take up the radioactive label and transfer it to the cells with which it forms synapses. (A) Normal labeling pattern. The white stripes indicate that roughly half the columns took up the label while the other half did not, a pattern reflecting that half the cells were innervated by the labeled eye and half were innervated by the unlabeled eye. (B) Labelling pattern when the unlabelled eye is sutured shut for 18 months. The axonal projections from the normal (labeled) eye take over the regions that would normally have been innervated by the sutured eye. (From Wiesel, 1982; photographs courtesy of T. Wiesel.)

to the left of the fovea,* it will also be stimulated by a light flashing across a region of the right eye 5 degrees above and 1 degree to the left of the fovea. Moreover, the response evoked in the cortical cell when both eyes are stimulated is greater than the response when either retina is stimulated alone.

Hubel and his co-workers (see Hubel, 1967) demonstrated that the development of the nervous system depended to some degree upon the experience of the individual during a critical period of development. In other words, all of neuronal development is not encoded in the genome: some is learned. Experience appears to strengthen or stabilize neuronal connections that are already present at birth. These conclusions come from studies of partial sensory blocks. Hubel and Wiesel (1962, 1963) sewed shut the right eyelids of newborn kittens and left them closed for 3 months. After this time they unsewed the lids of the right eye. The cortical cells of such kittens could not be stimulated by shining light in the right eye. Almost all the inputs into the visual cortex came from the left eye only, and the behavior of the kittens revealed the inadequacy of their right eyes. When only the left eyes of these animals were covered, the kittens became functionally blind. Because the lateral geniculate body neurons appeared to be stimulated from both right and left eyes in these kittens, the physiological defect appeared to be between the lateral geniculate bodies and the visual cortex. In rhesus monkeys, where similar phenomena are observed, the defect has been correlated with a lack of protein synthesis in the lateral geniculate neurons innervated by the covered eye (Kennedy et al., 1981).

Although it would be tempting to conclude that the resulting blindness was due to a failure to learn the proper visual connections, this is not the case. Rather, a kitten is born with all the appropriate neuronal connections in its visual system (Hubel and Wiesel, 1963). However, when one eye is covered early in the kitten's life, those connections into the visual cortex are taken over by the other eye (Figure 45). There is competition occurring, and experience plays a role in strengthening and stabilizing the connections from each lateral geniculate body to the visual cortex. Thus, when both eyes of a kitten are sewn shut for 3 months, the visual system remains

*The fovea is a depression in the center of the retina where only cones are present and the rods and blood vessels are absent. Here it serves as a convenient landmark.

intact. Both eyes become functional. The critical time in kitten development for this validation of neuronal connections is between the fourth and
the sixth week of the kitten's life. Monocular deprivation up to the fourth
week produces little or no physiological deficit, but after 6 weeks it produces all the characteristic neuronal changes.

Two new principles, then, can be seen in the patterning of the mammalian visual system. First, neuronal connections involved in vision are
present even before the animal sees; and second, experience plays an
important role in determining whether or not certain connections remain.*

*The notion that experience can influence neural pathways is also seen in the hypothesis of
the Hebbian synapse. If an axon from neuron A activates neuron B, so that the firing of
neuron B is always associated with the firing of neuron A, then the synapse between neuron
A to neuron B is strengthened. There are many means by which this strengthening could
occur, but most hypotheses focus on changes that would enable calcium ions to more readily
enter neuron B. This type of synapse could explain the phenomenon of long-term potentia-
tion, which is thought to be the basis of correlative memory (where one sensation recalls
others) (Brown et al., 1990). In addition to the competition between neurons seen in the
retina occlusion studies, there may be Hebbian synapse formation in visual cortex as well
(Stent, 1973; Reiter and Stryker, 1988).

SIDELIGHTS & SPECULATIONS

Morphogenesis by specific cell death

In both the developing limb and nervous system, cell death
plays a major role in the final pattern. In most regions of
the developing nervous systems of amphibians, birds, and
mammals, neurons are overproduced and then die in specific regions. This cell death helps produce the patterns of
cells characteristic of the central nervous system tissues. In
addition, there is a drastic reduction in the number of axons
that innervate peripheral tissues.

It had been generally thought that there was competition for neurotrophic factors (Purves and Lichtman, 1980).
Those neurons that received the factors were nourished
while those that did not get them starved and died. New
research, however, suggests that at least some of this neuronal cell death is an active process initiated by a set of
"self-destruct genes" and that the expression of these genes
is inhibited by signals coming from neurotrophic molecules
such as NGF.

When Martin and co-workers (1988) inhibited new
RNA and protein synthesis in cultured sympathetic neurons that had been deprived of NGF, they found that this
inhibition prevented the cells from dying. Thus, it appeared that the death of cells due to NGF deprivation was
not a direct result of nutrient starvation or inhibition of
protein synthesis. Similar results were seen when inhibitors of RNA and protein synthesis were given to chick
embryos in vivo. Oppenheim and colleagues (1990) injected
these drugs into the chorioallantoic membrane blood vessels in doses that did not interfere with the gross anatomical development of the chick. However, the inhibitors of
protein synthesis suppressed nearly all motor neuron degeneration when they shut down protein synthesis to
about 40 percent of the control levels.

The death of neuronal cells thus appears to have some
genetic basis. In *Caenorhabditis elegans*, some of these genes
may have been discovered. Mutations in the *ced-3* and *ced-4*
genes prevent almost all the naturally occurring neuronal
cell death. Cells of neural lineage that normally die in the
wild-type nematodes survive and differentiate into neurons

(A) CHICK LEG PRIMORDIUM
Extensive cell death

(B) DUCK LEG PRIMORDIUM
Minimal cell death

FIGURE 46
Patterns of cell death in leg primordia of chick (A) and
duck (B) embryos. Shading indicates areas of cell death.
(After Saunders and Fallon, 1966.)

in these mutants (Yuan and Horvitz, 1990). Although it is known that the genes are functioning within the cells that die, it is not known whether they are causing the death themselves or if they are responding to signals produced by other cells.

Cell death also plays a role in sculpting the limb, heart, and skeleton. There are obvious cases where differential cell death is genetically programmed and has been selected for during evolution. One such case involves the webbing or nonwebbing of feet. The difference between a duck's foot and that of a chicken is the presence or absence of cell death between the digits (Figure 46). Cell death is also thought to be important in the formation of skeletal joints and in separating the ulna from the radius (Zaleske, 1985). Saunders and his co-workers (1962; Saunders and Fallon, 1966) have shown that after a certain stage, cells destined to die will do so even if transplanted to another region of the embryo or placed into culture. Before that time, however, transplantation will save them. Between the time when the cell's death is determined and when death actually takes place, levels of DNA, RNA, and protein synthesis decrease dramatically (Pollak and Fallon, 1976).

LITERATURE CITED

Akers, R. M., Mosher, D. F. and Lilien, J. E. 1981. Promotion of retinal neurite outgrowth by substratum-bound fibronectin. *Dev. Biol.* 86: 179–188.

Alvarez-Buylla, A., Kirn, J. R. and Nottebohm, F. 1990. Birth of projection neurons in adult avian brain may be related to perceptual or motor learning. *Science* 249: 1444–1446.

Angeletti, P. and Vigneti, E. 1971. Assay of nerve growth factor (NGF) in subcellular fractions of peripheral tissues by microcomplement fixation. *Brain Res.* 33: 601–604.

Bastiani, M. J., Harrelson, A. L., Snow, P. M. and Goodman, C. S. 1987. Expression of fasciclin I and II glycoproteins on subsets of axon pathways during neuronal development in the grasshopper. *Cell* 48: 745–755.

Berking, S. 1979. Analysis of head and foot formation in hydra by means of an endogenous inhibitor. *Wilhelm Roux Arch. Entwicklungsmech. Org.* 186: 189–210.

Berlot, J. and Goodman, C. S. 1984. Guidance of peripheral pioneer neurons in the grasshopper: Adhesive hierarchy of epithelial and neuronal surfaces. *Science* 223: 493–496.

Black, J. E., Issacs, K. R., Anderson, B. J. Alcantara, A. A. and Greenough, W. T. 1990. Learning causes synaptogenesis, whereas motor activity causes angiogenesis, in cerebellar cortex of adult rats. *Proc. Natl. Acad. Sci. USA* 87: 5568–5572.

Bode, P. M. and Bode, H. R. 1984. Patterning in *Hydra*. In G. M. Malacinski and S. V. Bryant (eds.), *Pattern Formation*. Macmillan, New York, pp. 213–241.

Boveri, T. 1901. Über die Polarität des Seeigeleiers. *Eisverh. Phys. Med. Ges. Würzburg* 34: 145–175.

Brockes, J. P. and Kinter, C. R. 1986. Glial growth factor and nerve-dependent proliferation in the regeneration blastema of urodele amphibians. *Cell* 45: 301–306.

Brown, T. H., Kairiss, E. W. and Keenan, C. L. 1990. Hebbian synapses: Biophysical mechanisms and algorithm. *Annu. Rev. Neurosci.*. 13: 475–511.

Browne, E. N. 1909. The production of new hydranths in *Hydra* by the insertion of small grafts. *J. Exp. Zool.* 7: 1–23.

Bryant, S. V. and Iten, L. E. 1976. Supernumerary limbs in amphibians: Experimental production in *Notophthalamus viridescens* and a new interpretation of their formation. *Dev. Biol.* 50: 212–234.

Bryant, S. V, French, V. and Bryant, P. J. 1981. Distal regeneration and symmetry. *Science* 212: 993–1002.

Butler, E. G. 1935. Studies on limb regeneration in X-rayed *Ambystoma* larvae. *Anat. Rec.* 62: 295–307.

Carlson, B. M. 1972. Muscle morphogenesis in axolotl limb regenerates after removal of stump musculature. *Dev. Biol.* 28: 487–497.

Chevallier, A., Kieny, M., Mauger, M. and Sengel, P. 1977. Limb-somite relationships: Origin of limb musculature. *J. Embryol. Exp. Morphol.* 41: 245–258.

Child, C. M. 1941. *Patterns and Problems of Development*. University of Chicago Press, Chicago.

Christ, B., Jacob, H. J. and Jacob, M. 1977. Experimental analysis of the origin of wing musculature in avian embryos. *Anat. Embryol.* 150: 171–186.

Cohan, C. S. and Kater, S. B. 1986. Suppression of neurite elongation and growth cone motility by electrical activity. *Science* 232: 1638–1640.

Cohen, J., Burne, J. F., Winter, J. and Bartlett, P. 1986. Retinal ganglial cells lose responsiveness to lamina with maturation. *Nature* 322: 465–467.

Cohen, J., Burne, J. F., McKinlay, C. and Winter, J. 1987. The role of laminin and the laminin/fibronectin receptor complex in the outgrowth of retinal ganglial cell axons. *Dev. Biol.* 122: 407–418.

Constantine-Paton, M., Blum A. S., Mendez-Otero, R. and Barnstable, C. J. 1986. A cell surface molecule distributed in a dorsoventral gradient in the perinatal rat retina. *Nature* 324: 459–462.

Cox, E. C., Müller, B. and Bonhoeffer, F. 1990. Axonal guidance in chick visual system: Posterior tectal membranes induce collapse of growth cones from temporal retina. *Neuron* 2: 31–37.

Crawford, K. and Stocum, D. L. 1988. Retinoic acid proximalizes level-specific properties responsible for intercalary regeneration in axolotl limbs. *Development* 104: 703–712.

Crick, F. H. 1970. Diffusion in embryogenesis. *Nature* 225: 420–422.

Crossland, W. J., Cowan, W. M., Rogers, C. A. and Kelly, J. 1974. The specification of the retinal-tectal projection in the chick. *J. Comp. Neurol.* 155: 127–164.

Detwiler, S. R. 1918. Experiments on the development of the shoulder girdle and the anterior limb of *Amblystoma punctatum*. *J. Exp. Zool.* 25: 499–538.

Dollé, P., Izpisúa-Belmonte, J.-C., Falkenstein, H., Renucci, A. and Duboule, D. 1989. Coordinate expression of the murine *Hox-5* complex homeobox-containing genes during limb pattern formation. *Nature* 342: 767–772.

Ebendal, T., Hedlund., K.-O. and Norrgran, G. 1982. Nerve growth factor in chick tissues. *J. Neurosci. Res.* 8: 153–164.

Edwards, R. H., Rutter, W. J. and Hanahan, D. 1989. Directed expression of NGF to pancreatic β cells in transgenic mice leads to selective hyperinnervation of the islets. *Cell* 58: 161–170.

Eichele, G. 1989. Retinoic acid induces a pattern of digits in anterior half wing buds that lack the zone of polarizing activity. *Development* 107: 863–867.

Eisen, J., Meyers, P. Z. and Westerfield, M. 1986. Pathway selection by growth cones of identified motoneurones in live zebra fish embryos. *Nature* 320: 269–271.

Fallon, J. F. and Crosby, G. M. 1977. Polarizing zone activity in limb buds of amniotes. In D. A. Ede, J. R. Hinchliffe and M. Balls (eds.), *Vertebrate Limb and Somite Morphogenesis*. Cambridge University Press, Cambridge.

Fraser, S. E., Green, C. R., Bode, H. R. and Gilula, N. B. 1987. Selective disruption of gap junctional communication interferes with a patterning process in hydra. *Science* 237: 49–55.

French, V., Bryant, P. J. and Bryant, S. V. 1976. Pattern regulation in epimorphic fields. *Science* 193: 969–981.

Frenz, D. A., Jaikaria, N. S. and Newman, S. A. 1989. The mechanism of precartilage mesenchymal condensation: A major role for interaction of the cell surface with the amino-terminal heparin-binding domain of fibronectin. *Dev. Biol.* 136: 97–103.

Gaze, R. M. 1978. The problem of specificity in the formation of nerve connections. *In* D. R. Garrod (ed.), *Specificity of Embryological Interactions*. Chapman and Hall, London, pp. 51–96.

George-Weinstein, M., Decker, C. and Horwitz, A. 1988. Combinations of monoclonal antibodies distinguish mesenchyme, myogenic, and chondrogenic precursors of the developing chick embryo. *Dev. Biol.* 125: 34–50.

Gershon, M. D., Schwartz, J. H. and Kandel, E. R. 1985. Morphology of chemical synapses and pattern of interconnections. *In* E. R. Kandel and J. H. Schwartz (eds.), *Principles of Neural Science*, 2nd Ed. Elsevier, New York, pp. 132–147.

Giguère, V., Ong, E. S., Evans, R. M. and Tabin, C. J. 1989. Spatial and temporal expression of the retinoic acid receptor in the regenerating amphibian limb. *Nature* 337: 566–570.

Goodman, C. S. and Bastiani, M. J. 1984. How embryonic nerve cells recognize one another. *Sci. Am.* 251(6): 58–66.

Goodman, C. S., Bastiani, M. J., Doe, C. Q., du Lac, S., Helfand, S. L., Kuwada, J. Y. and Thomas, J. B. 1984. Cell recognition during neuronal development. *Science* 225: 1271–1287.

Goss, R. J. 1969. *Principles of Regeneration*. Academic Press, New York.

Gottlieb, D. I., Rock, K. and Glaser, L. 1976. A gradient of adhesive specificity in developing avian retina. *Proc. Natl. Acad. Sci. USA* 73: 410–414.

Gottlieb, G. 1965. Prenatal auditory sensitivity in chickens and ducks. *Science* 147: 1596–1598.

Grimmelikhuijzen, C. J. P. and Schaller, H. C. 1977. Isolation of a substance activating foot formation in hydra. *Cell Differ.* 6: 297–305.

Gundersen, R. W. 1987. Response of sensory neurites and growth cones to patterned substrata of laminin and fibronectin in vitro. *Dev. Biol.* 121: 423–431.

Halfter, W., Claviez, M. and Schwarz, U. 1981. Preferential adhesion of tectal membranes to anterior embryonic chick retina neurites. *Nature* 292: 67–70.

Hamburger, V. 1938. Morphogenetic and axial self-differentiation of transplanted limb primordia of 2-day chick embryos. *J. Exp. Zool.* 77: 379–400.

Hamburger, V. 1939. The development and innervation of transplanted limb primordia of chick embryos. *J. Exp. Zool.* 80: 347–389.

Harper, S. and Davies, A. M. 1990. NGF mRNA expression in developing cutaneous epithelium related to innervation density. *Development* 110: 515–519.

Harrelson, A. L. and Goodman, C. S. 1988. Growth cone guidance in insects: Fasciclin II is a member of the immunoglobulin superfamily. *Science* 242: 700–708.

Harris, W. A. 1986. Homing behavior of axons in the embryonic vertebrate brain. *Nature* 320: 266–269.

Harrison, R. G. 1910. The outgrowth of the nerve fiber as a mode of protoplasmic movement. *J. Exp. Zool.* 9: 787–848.

Harrison, R. G. 1918. Experiments on the development of the forelimb of *Amblystoma*, a self-differentiating equipotential system. *J. Exp. Zool.* 25: 413–461.

Haydon, P. G., Cohan, C. S., McCobb, D. P., Miller, H. R. and Kater, S. B. 1985. Neuron-specific growth cone properties as seen in identified neurons of *Helisoma*. *J. Neurosci. Res.* 13: 135–147.

Hertwig, G. 1925. Haploidkernige Transplante als Organisatoran diploidkeniger Extremitaten be *Triton*. *Anat. Anz.* [Suppl.] 60: 112–118.

Hicklin, J. and Wolpert, L. 1973. Positional information and pattern regulation in hydra: Formation of the foot end. *J. Embryol. Exp. Morphol.* 30: 727–740.

Hinchliffe, J. R. and Sansom, A. 1985. The distribution of the polarizing zone (ZPA) in the legbud of the chick embryo. *J. Embryol. Exp. Morphol.* 86: 169–175.

Hohn, A., Leibrock, J., Bailey, K. and Barde, Y.-A. 1990. Identification and characterization of a novel member of the nerve growth factor/brain-derived neurotrophic factor family. *Nature* 344: 339–341.

Hollyday, M. 1980. Motoneuron histogenesis and the development of limb innervation. *Curr. Top. Dev. Biol.* 15: 181–215.

Hollyday, M., Hamburger, V. and Farris, J. M. G. 1977. Localization of motor neuron pools supplying identified muscles in normal and supernumerary legs of chick embryos. *Proc. Natl. Acad. Sci. USA* 74: 3582–3586.

Honig, L. S. and Summerbell, D. 1985. Maps of strength of positional signaling activity in the developing chick wing bud. *J. Embryol. Exp. Morphol.* 87: 163–174.

Hubel, D. H. 1967. Effects of distortion of sensory input on the visual system of kittens. *Physiologist* 10: 17–45.

Hubel, D. H. and Wiesel, T. N. 1962. Receptive fields, binocular interaction and functional architecture in the cat's visual cortex. *J. Physiol.* 160: 106–154.

Hubel, D. H. and Wiesel, T. N. 1963. Receptive fields of cells in striate cortex of very young, visually inexperienced kittens. *J. Neurophysiol.* 26: 944–1002.

Iten, L. E. 1982. Pattern specification and pattern regulation in the embryonic chick limb bud. *Am. Zool.* 22: 117–129.

Iten, L. E. and Bryant, S. V. 1975. The interaction between blastema and stump in the establishment of the anterior–posterior and proximal–distal organization of the limb regenerate. *Dev. Biol.* 44: 119–147.

Jacobson, M. 1967. Retinal ganglion cells: Specification of central connections in larval *Xenopus laevis*. *Science* 155: 1106–1108.

Javois, L. C. and Iten, L. E. 1986. The handedness and origin of supernumerary limb structures following 180° rotation of the chick limb bud on its stump. *J. Embryol. Exp. Morphol.* 91: 135–152.

Kater, S. B. 1985. Dynamic regulators of neuronal form and conductivity in the adult snail *Helisoma*. *In* A. I. Selverston (ed.), *Model Neural Networks and Behavior*. Plenum, New York, pp. 191–209.

Kennedy, C., Suda, S., Smith, C. B., Miyaoka, M., Ito, M. and Sokoloff, L. 1981. Changes in protein synthesis underlying functional plasticity in immature monkey visual system. *Proc. Natl. Acad. Sci USA* 78: 3950–3953.

Kimmel, C. B. and Law, R. D. 1985. Cell lineage of zebrafish blastomeres I. Cleavage pattern and cytoplasmic bridges between cells. *Dev. Biol.* 108: 78–101.

Klose, M. and Bentley, D. 1989. Transient pioneer neurons are essential for formation of an embryonic peripheral nerve. *Science* 245: 982–984.

Kochhar, D. M. 1977. Cellular basis of congenital limb deformity induced in mice by vitamin A. *In* D. Bergsma and W. Lenz (eds.), *Morphogenesis and Malformation of the Limb*. Alan R. Liss, New York, pp. 111–154.

Korsching, S. and Thoenen, H. 1983. Nerve growth factor in sympathetic ganglia and corresponding target organs of the rat: correlation with density of sympathetic innervation. *Proc. Natl. Acad. Sci. USA* 80: 3513–3516.

Lance-Jones, C. and Landmesser, L. 1980. Motor neuron projection patterns in chick hindlimb following partial reversals of the spinal cord. *J. Physiol.* 302: 581–602.

Landmesser, L. 1978. The development of motor projection patterns in the chick hindlimb. *J. Physiol.* 284: 391–414.

Letourneau, P., Madsen, A. M., Palm, S. M. and Furcht, L. T. 1988. Immunoreactivity for laminin in the developing ventral longitudinal pathway of the brain. *Dev. Biol.* 125: 135 -144.

Levi-Montalcini, R. 1976. The nerve-growth factor: Its role in growth, differentiation, and formation of the sympathetic adrenergic neuron. *Prog. Brain Res.* 45: 235–258.

Levi-Montalcini, R. and Booker, B. 1960. Destruction of the sympathetic ganglia in mammals by an antiserum to the nerve growth factor protein. *Proc. Natl. Acad. Sci. USA* 46: 384–390.

Liesi, P. and Silver, J. 1988. Is astrocyte laminin involved in axon guidance in the mammalian CNS? *Dev. Biol.* 130: 774–785.

Lumsden, A. G. S. and Davies, A. M. 1983. Earliest sensory nerve fibers are guided to peripheral targets by attractants other than nerve growth factor. *Nature* 306: 786–788.

Lumsden, A. G. S. and Davies, A.M. 1986. Chemotropic effect of specific target epithelium in the developing mammalian nervous system. *Nature* 323: 538–539.

MacWilliams, H. K. 1983a. Hydra transplantation phenomena and the mechanism of hydra head regeneration. I. Properties of head inhibition. *Dev. Biol.* 96: 217–238.

MacWilliams, H. K. 1983b. Hydra transplantation phenomena and the mechanism of hydra head regeneration. II. Properties of head activation. *Dev. Biol.* 96: 239–272.

MacWilliams, H. K. and Kafatos, F. C. 1974. The basal inhibition of hydra may be mediated by a diffusible substance. *Am. Zool.* 14: 633–645.

Maden, M. 1982. Vitamin A and pattern formation of the regenerating limb. *Nature* 295: 672–675.

Maden, M. 1984. Retinoids as probes for investigating the molecular basis of pattern formation. *In* G. M. Malacinski and S. V. Bryant (eds.), *Pattern Formation.* Macmillan, New York, pp. 539–555.

Maden, M., Ong, D. E., Summerbell, D., Chytil, F. and Hirst, E. A. 1989. Cellular retinoic acid-binding protein and the role of retinoic acid in the development of the chick embryo. *Dev. Biol.* 135: 124–132.

Maisonpierre, P. C., Belluscio, L., Squinto, S., Ip, N. Y., Furth, M. E., Lindsay, R. M. and Yancopoulos, G. D. 1990. Neurotrophin-3: A neurotrophic factor related to NGF and BDNF. *Science* 247: 1446–1451.

Martin, D. P., Schmidt, R. E., DiStefano, P. S., Lowry, O. H., Carter, J. G. and Johnson, E. M. Jr. 1988. Inhibitors of protein synthesis and RNA synthesis prevent neuronal cell death caused by growth factor deprivation. *J. Cell Biol.* 106: 829–844.

Mescher, A. L. and Tassava, R. A. 1975. Denervation effects on DNA replication and mitosis during the initiation of limb regeneration in adult newts. *Dev. Biol.* 44: 187–197.

Muneoka, K. and Bryant, S. V. 1982. Evidence that patterning mechanisms in developing and regenerating limbs are the same. *Nature* 298: 369–371.

Nardi, J. B. 1983. Neuronal pathfinding in developing wings of the moth *Manduca sexta. Dev. Biol.* 95: 163–174.

Newman, S. A. 1974. The interaction of the organizing regions in hydra and its possible relation to the role of the cut end in regeneration. *J. Embryol. Exp. Morphol.* 31: 541–555.

Newman, S. A. 1988. Lineage and pattern in the developing vertebrate limb. *Trends Genet.* 4: 329–332.

Newman, S. A. and Frisch, H. L. 1979. Dynamics of skeletal pattern formation in developing chick limb. *Science* 205: 662–668.

Newman, S. A., Frisch, H. L. and Percus, J. K. 1988. The stationary state analysis of reaction-diffusion mechanisms for biological pattern formation. *J. Theoret. Biol.* 134: 183–197.

Niazi, I. A. and Saxena, S. 1978. Abnormal hind limb regeneration in tadpoles of the toad *Bufo andersonii* exposed to excess vitamin A. *Folia Biol.* (Krakow) 26: 3–8.

Nijhout, H. F. 1981. The color patterns of butterflies and moths. *Sci. Am.* 245 (5): 140–151.

Nordeen, K. W. and Nordeen, E. J. 1988. Projection neurons within a vocal pathway are born during song learning in zebra finches. *Nature* 334: 149–151.

Oppenheim, R. W., Prevette, D., Tytell, M. and Homma, S. 1990. Naturally occurring and induced neuronal cell death in the chick embryo *in vivo* requires protein and RNA synthesis: Evidence for the role of cell death genes. *Dev. Biol.* 138: 104–113.

Oster, G. F., Shubin, N., Murray, J. D. and Alberch, P. 1988. Evolution and morphogenetic rules: The shape of the vertebrate limb in ontogeny and phylogeny. *Evolution* 42: 862–884.

Pearson, J., Johnson, E. M., Jr. and Brandeis, L. 1983. Effects of antibodies to nerve growth factor on intrauterine development of derivatives of cranial neural crest and placode in the guinea pig. *Dev. Biol.* 96: 32–36.

Pittman, R. N. 1985. Release of plasminogen activator and a calcium-dependent metalloprotease from cultured sympathetic and sensory neurons. *Dev. Biol.* 110: 91–101.

Placzek, M., Tessier-Lavigne, M., Jessell, T. and Dodd, J. 1990. Orientation of commissural axons in vitro in response to a floor-plate-derived chemoattractant. *Development* 110: 19–30.

Pollak, R. D. and Fallon, J. F. 1976. Autoradiographic analysis of macromolecular synthesis in prospectively necrotic cells of the chick limb bud. II. Nucleic acids. *Exp. Cell Res.* 100: 15–22.

Prigogine, I. and Nicolis, G. 1967. On symmetry-breaking instabilities in dissipative systems. *J. Chem. Phys.* 46: 3542–3550.

Purves, D. and Lichtman, J. W. 1980. Elimination of synapses in the developing nervous system. *Science* 210: 153–157.

Purves, D. and Lichtman, J. W. 1985. *Principles of Neural Development.* Sinauer Associates, Sunderland, MA.

Ragsdale, C. W., Jr., Petkovich, M., Gates, P. B., Chambon, P. and Brockes, J. P. 1989. Identification of a novel retinoic acid receptor in regenerative tissues of the newt. *Nature* 341: 654–657.

Rand, H. W., Board, J. F. and Minnich, D. E. 1926. Localization of formative agencies in hydra. *Proc. Natl. Acad. Sci USA* 12: 565–570.

Reiter, H. O. and Stryker, M. P. 1988. Neural plasticity without postsynaptic action potentials: Less-active inputs become dominant when kitten visual cortical cells are pharmacologically inhibited. *Proc. Natl. Acad. Sci. USA* 85: 3623–3627.

Rose, S. M. 1962. Tissue-arc control of regeneration in the amphibian limb. *In* D. Rudnick (ed.), *Regeneration.* Ronald, New York, pp. 153–176.

Rosenquist, G. C. 1971. The origin and movement of the limb-bud epithelium and mesenchyme in the chick embryo as determined by radioautographic mapping. *J. Embryol. Exp. Morphol.* 25: 85–96.

Roth, S. and Marchase, R. B. 1976. An in vitro assay for retinotectal specificity. In S. H. Barondes (ed.), *Neuronal Recognition.* Plenum, New York, pp. 227–248.

Rowe, D. A., Cairnes, J. M. and Fallon, J. F. 1982. Spatial and temporal patterns of cell death in limb bud mesoderm after apical ectodermal ridge removal. *Dev. Biol.* 93: 83–91.

Ruberte, E., Dollé, P., Krust, A., Zelent, A., Morriss-Kay, G. and Chambon, P. 1990. Specific spatial and temporal distribution of retinoic acid receptor gamma transcripts during mouse embryogenesis. *Development* 108: 213–222.

Rubin, L. and Saunders, J. W., Jr. 1972. Ectodermal–mesodermal interactions in the growth of limbs in the chick embryo: Constancy and temporal limits of the ectodermal induction. *Dev. Biol.* 28: 94–112.

Satre, M. A. and Kochhar, D. M. 1989. Elevations in the endogenous levels of the putative morphogen retinoic acid in embryonic mouse limb-buds associated with limb dysmorphogenesis. *Dev. Biol.* 133: 529–536.

Saunders, J. W., Jr. 1948. The proximal–distal sequence of origin of the parts of the chick wing and the role of the ectoderm. *J. Exp. Zool.* 108: 363–404.

Saunders, J. W., Jr. 1982. *Developmental Biology.* Macmillan, New York.

Saunders, J. W., Jr. and Fallon, J. F. 1966. Cell death in morphogenesis. *In* M. Locke (ed.), *Major Problems of Developmental Biology.* Academic Press, New York, pp. 289–314.

Saunders, J. W., Jr. and Gasseling, M. T. 1968. Ectodermal–mesodermal interactions in the origin of limb symmetry. *In* R. Fleischmajer and R. E. Billingham (eds.), *Epithelial–Mesenchymal Interactions.* Williams & Wilkins, Baltimore, pp. 78–97.

Saunders, J. W., Jr., Gasseling, M. T. and Saunders, L.C. 1962. Cellular death in morphogenesis of the avian wing. *Dev. Biol.* 5: 147–178.

Schaller, H. C., Schmidt, T. and Grimmelikhuijzen, C. J. P. 1979. Separation and specificity of action of four morphogens from hydra. *Wilhelm Roux Arch. Entwicklungsmech. Org.* 186: 139–149.

Schmidt, T. and Schaller, H. C. 1976. Evidence for a foot-inhibiting substance in hydra. *Cell Differ.* 5: 151–159.

Sessions, S. K., Gardiner, D. M. and Bryant, S. V. 1989. Compatible limb patterning mechanisms in urodeles and anurans. *Dev. Biol.* 131: 294–301.

Silver, J. and Sidman, R.L. 1980) A mechanism for guidance and topologic patterning of retinal ganglion cell axons. *J. Comp. Neurol.* 189: 101–111.

Singer, M. 1954. Induction of regeneration of the forelimb of the postmetamorphic frog by augmentation of the nerve supply. *J. Exp. Zool.* 126: 419–472.

Singer, M. and Caston, J. D. 1972. Neurotrophic dependence of macromolecular synthesis in the early limb regenerate of the newt, *Triturus*. *J. Embryol. Exp. Morphol.* 28: 1–11.

Singer, M., Norlander, R. and Egar, M. 1979. Axonal guidance during embryogenesis and regeneration in the spinal cord of the newt: The blueprint hypothesis of neuronal pathway patterning. *J. Comp. Neurol.* 185: 1–22.

Solursh, M. 1984. Ectoderm as a determinant of early tissue pattern in the limb bud. *Cell Diff.* 15: 17–24.

Spemann, H. 1938. *Embryonic Development and Induction*. Yale University Press, New Haven.

Sperry, R. W. 1951. Mechanisms of neural maturation. *In* S. S. Stevens (ed.), *Handbook of Experimental Psychology*. Wiley, New York, pp. 236–280.

Sperry, R. W. 1965. Embryogenesis of behavioral nerve nets. *In* R. L. DeHaan and H. Ursprung (eds.), *Organogenesis*. Holt, Rinehart & Winston, New York, pp. 161–186.

Stent, G. S. 1973. A physiological mechanism for Hebb's postulate of learning. *Proc. Natl. Acad. Sci. USA* 70: 997–1001.

Stocum, D. L. 1986. Retinoids: Probes for understanding pattern regeneration in regenerating amphibian limbs. *In* R. M. Gorczynski (ed.), *Receptors in Cellular Recognition and Developmental Processes*. Academic Press, New York, pp. 165–181.

Stocum, D. L. and Fallon, J. F. 1982. Control of pattern formation in urodele limb ontogeny: A review and a hypothesis. *J. Embryol. Exp. Morphol.* 69: 7–36.

Summerbell, D. 1974. A quantitative analysis of the effect of excision of the AER from the chick limb bud. *J. Embryol. Exp. Morphol.* 32: 651–660.

Summerbell, D. 1979. The zone of polarizing activity: Evidence for a role in abnormal chick limb morphogenesis. *J. Embryol. Exp. Morphol.* 50: 217–233.

Summerbell, D. and Lewis, J. H. 1975. Time, place and positional value in the chick limb bud. *J. Embryol. Exp. Morphol.* 33: 621–643.

Taghert, P. H., Doe, C. Q. and Goodman, C. S. 1984. Cell determination and regulation during development of neuroblasts and neurones in grasshopper embryo. *Nature* 307: 163–165.

Takano, J. and Sugiyama, T. 1983. Genetic analysis of developmental mechanisms in hydra. VIII. Head-activation and head-inhibition potentials of a slow-budding strain (L4. *J. Embryol. Exp. Morphol.* 78: 141–168.

Tessier-Lavigne, M., Placzek, M., Lumsden, A. G. S., Dodd, J. and Jessell, T. 1988.

Chemotropic guidance of developing axons in the mammalian central nervous system. *Nature* 336: 775–778.

Thaller, C. and Eichele, G. 1987. Identification and spatial distribution of retinoids in the developing chick limb bud. *Nature* 327: 625–628.

Thaller, C. and Eichele, G. 1990. Isolation of 3,4-didehydroretinoic acid, a novel morphogenetic signal in the chick limb bud. *Nature* 345: 815–819.

Thoms, S. D. and Stocum, D. L. 1984. Retinoic acid-induced pattern duplication in regenerating urodele limbs. *Dev. Biol.* 103: 319–328.

Tickle, C. 1981. The number of polarizing region cells required to specify additional digits in the developing chick wing. *Nature* 289: 295–298.

Tickle, C., Summerbell, D. and Wolpert, L. 1975. Positional signaling and specification of digits in chick limb morphogenesis. *Nature* 254: 199–202.

Tickle, C., Alberts, B., Wolpert, L. and Lee, J. 1982. Local application of retinoic acid to the limb bud mimics the action of the polarizing region. *Nature* 296: 564–566.

Tickle, C., Lee, J. and Eichele, G. 1985. A quantitative analysis of the effect of all-*trans*-retinoic acid on the pattern of chick wing development. *Dev. Biol.* 109: 82–95.

Tinbergen, N. 1951. *The Study of Instinct*. Clarendon Press, Oxford.

Tosney, K. W. and Landmesser, L. T. 1985. Development of the major pathways for neurite outgrowth in the chick hindlimb. *Dev. Biol.* 109: 193–214.

Trisler, D. 1987. Molecular markers of cell position in avian retina are responsible for synapse formation. *Am. Zool.* 27: 189–206.

Trisler, D. and Collins, F. 1987. Corresponding spatial gradients of TOP molecules in the developing retina and optic tectum. *Science* 237: 1208–1210.

Trisler, D., Schneider, M. D. and Nirenberg, M. 1981. A topographic gradient of molecules in retina can be used to identify neuron position. *Proc. Natl. Acad. Sci. USA* 78: 2145–2149.

Trisler, D., Bekenstein, J. and Daniels, M. P. 1986. Antibody to a molecular marker of cell position inhibits synapse formation in the retina. *Proc. Natl. Acad. Sci. USA* 83: 4194–4198.

Turing, A. M. 1952. The chemical basis of morphogenesis. *Philos. Trans. R. Soc. Lond. B* 237: 37–72.

Wakeford, R. J. 1979. Cell contact and positional communication in hydra. *J. Embryol. Exp. Morphol.* 54: 171–183.

Walter, J., Henke-Fahle, S. and Bonhoeffer, F. 1987. Avoidance of posterior tectal mem-

branes by temporal retinal axons. *Development* 101: 909–913.

Webster, G. 1966. Studies on pattern regulation in hydra. II. Factors controlling hypostome formation. *J. Embryol. Exp. Morphol.* 16: 105–122.

Webster, G. and Wolpert, L. 1966. Studies of pattern regeneration in hydra. *J. Embryol. Exp. Morphol.* 16: 91–104.

Welsh, B. J., Gomatam, J. and Burgess, A. E. 1983. Three-dimensional chemical waves in the Belousov-Zhabotinski reaction. *Nature* 304: 611–614.

Weiss, P. 1939. *Principles of Development*. Holt, Rinehart & Winston, New York.

Weiss, P. A. 1955. Nervous system. *In* B. H. Willier, P. A. Weiss and V. Hamburger (eds.), *Analysis of Development*. Saunders, Philadelphia, pp. 346–402.

Whitelaw, V. and Hollyday, M. 1983. Position-dependent motor innervation of the chick hindlimb following serial and parallel duplications of limb segments. *J. Neurosci.* 3: 1216–1225.

Wiesel, T. N. 1982. Postnatal development of the visual cortex and the influence of the environment. *Nature* 299: 583–591.

Wilby, O. K. and Webster, G. 1970. Experimental studies on axial polarity in hydra. *J. Embryol. Exp. Morphol.* 24: 595–613.

Winfree, A. T. 1974. Rotating chemical reactions. *Sci. Am.* 230(6): 82–95.

Wolpert, L. 1969. Positional information and the spatial pattern of cellular formation. *J. Theoret. Biol.* 25: 1–47.

Wolpert, L. 1971. Positional information and pattern formation. *Curr. Top. Dev. Biol.* 6: 183–224.

Wolpert, L. 1977. *The Development of Pattern and Form in Animals*. Carolina Biological, Burlington, NC.

Wolpert, L. 1978. Pattern formation in biological development. *Sci. Am.* 239(4): 154–164.

Wolpert, L. and Hornbruch, A. 1987. Position signalling and the development of the humerus in the chick limb bud. *Development* 100: 333–338.

Yuan, J. and Horvitz, H. R. 1990. The *Caenorhabditis elegans* genes *ced-3* and *ced-4* act autonomously to cause programmed cell death. *Dev. Biol.* 138: 33–41.

Zaleske, D. J. 1985. Development of the upper limb. *Hand. Clin.* 1985(3): 383–390.

Zinn, K., McAllister, L. and Goodman, C. S. 1988. Sequence analysis and neuronal expression of fasciclin I in grasshopper and *Drosophila*. *Cell* 53: 577–587.

Zipser, B. and McKay, R. 1981. Monoclonal antibodies distinguish identifiable neurones in the leech. *Nature* 289: 549–554.

18

The genetics of pattern formation in *Drosophila*

The Library is limitless and periodic. If an eternal voyager were to traverse it in any direction, he would find, after many centuries, that the same volumes are repeated in the same disorder (which repeated, would constitute an order: Order itself). My solitude rejoices in this elegant hope.

—J. L. BORGES (1962)

Those of us who are at work on Drosophila *find a particular point to the question. For the genetic material available is all that could be desired, and even embryological experiments can be done. . . . It is for us to make use of these opportunities. We have a complete story to unravel, because we can work things from both ends at once.*

—JACK SCHULTZ (1935)

The beginnings of this century saw the emergence of genetics and embryology as two separate fields (Mayr, 1982; Raff and Kaufman, 1983). By 1926, however, the two sciences had their own journals, their own textbooks, and their own jargon. Moreover, each discipline had largely ceased referencing papers in the other two fields. Geneticists studied the transmission of inherited traits and embryologists studied the expression of these traits.

In the late 1930s, several individuals tried to reunite these fields (Goldschmidt, 1938; Just, 1939; Waddington, 1940), but their attempts were premature. In the 1990s, techniques of recombinant DNA and monoclonal antibodies have brought genetics and embryology closer together. One of the major intersections of these areas has concerned the genetics of insect segment identity. This is, in fact, precisely the place where Goldschmidt and Waddington predicted the integration would occur. The study of pattern formation in developing embryos has been greatly aided by the use of mutations. Here *Drosophila* becomes important, because its genetics are among the best known of any eukaryotic organism.

A summary of *Drosophila* development

As discussed in Chapter 3, *Drosophila* embryos develop very rapidly through a series of nuclear divisions that form a syncytial blastoderm. During the ninth division cycle, about five nuclei reach the surface of the posterior pole of the embryo. These nuclei become enclosed by cell membranes and generate the pole cells that give rise to the gametes of the adult. Most of the other nuclei arrive at the periphery of the embryo at cycle 10 and then undergo four more divisions at progressively slower rates. Following cycle 13, cell membranes grow between the nuclei to form a cellular blastoderm of about 6000 cells (Turner and Mahowald, 1977; Foe and Alberts, 1983). At cycle 14, the level of transcription, which had been very low, increases dramatically. At the same time, the cells become motile, and the 2- to 3-hour embryo begins gastrulation (Edgar and Schubiger, 1986).

The first movements of *Drosophila* gastrulation segregate the presumptive mesoderm, endoderm, and ectoderm (Figure 1). The prospective mesoderm—about 1000 cells constituting the ventral midline—folds in to produce the VENTRAL FURROW. This furrow eventually pinches off from the surface to become the ventral tube within the embryo. It then flattens to form a layer of mesodermal tissue beneath the ventral ectoderm. The prospective endoderm invaginates as two pockets at the anterior and posterior ends of the ventral furrow. The pole cells are internalized along with the endoderm. At this time, the embryo bends to form the CEPHALIC FURROW and the ANTERIOR and POSTERIOR TRANSVERSE FOLDS.

The cells remaining on the surface migrate towards the ventral midline to form the GERM BAND. The cells of the germ band migrate posteriorly and then around the dorsal surface so that at the end of germ band formation, the cells destined to form the most posterior larval structures are located immediately behind the future head region. At this time, the body segments begin to appear, dividing the ectoderm and mesoderm. The germ band then contracts, placing the presumptive posterior segments into the posterior tip of the embryo. It is also during this time that the imaginal discs are set aside from the ectoderm (Chapters 9 and 19) and that the nervous system forms from two strips of ectodermal cells along the ventral midline. Unlike vertebrate nervous systems, the insect neural tube is formed ventrally, and it arises from neuroblasts differentiating from the ventral neurogenic ectoderm of each segment and from the procephalic neurogenic ectoderm of the head. Some of the genes involved in this differentiation (such as *Notch*) and patterning have been discussed in Chapters 11 and 17.

THE ORIGINS OF ANTERIOR–POSTERIOR POLARITY

The *Drosophila* egg, embryo, larva, and adult are polar structures with respect to the anterior–posterior axis. The oocyte develops in such a way that one end (the future anterior) connects directly with the cytoplasm of the follicular nurse cells, which transport proteins, ribosomes, and mRNA into the egg. The opposite end of the egg accumulates the pole plasm that gives rise to the germ cells of the adult (Chapter 7). The embryo, larva, and adult fly have a distinct head end and a distinct tail end, between

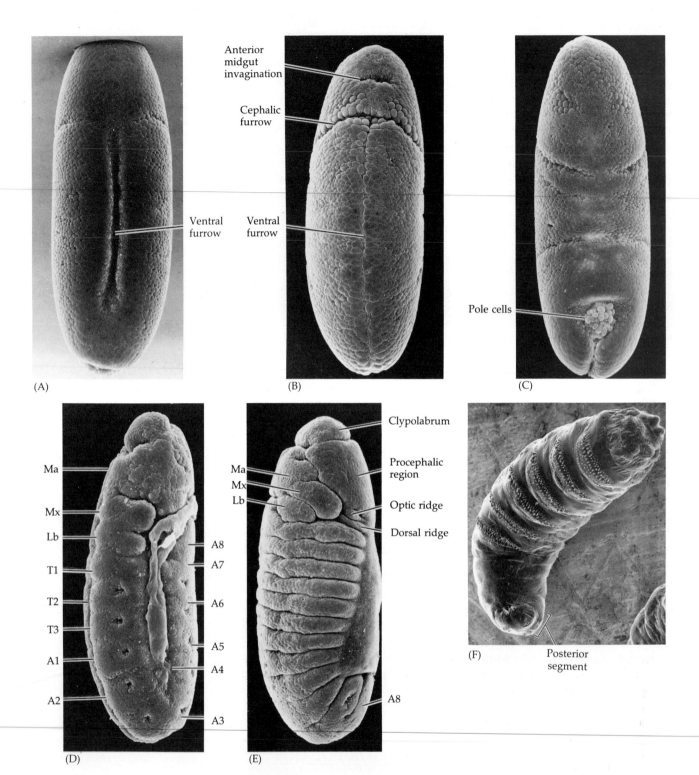

FIGURE 1

Gastrulation in *Drosophila*. (A) Ventral furrow beginning to form as cells migrate toward ventral midline. (B) Invagination of the ventral furrow. (C) Dorsal view of a slightly older embryo showing the migration of the pole cells and posterior endoderm into the embryo. (D) Lateral view showing fullest migration of germ band. Incipient segments of the germ band are labeled: Ma, Mx, and Lb correspond to the mandibular, maxillary, and labial head segments; T1–T3, thoracic segments; A1–A8, abdominal segments. (E) Germ band reversing direction. The clypolabrum, procephalic region, optic, and dorsal ridges of the head are distinguished. (F) Newly hatched first instar larva. (Photographs courtesy of F. R. Turner.)

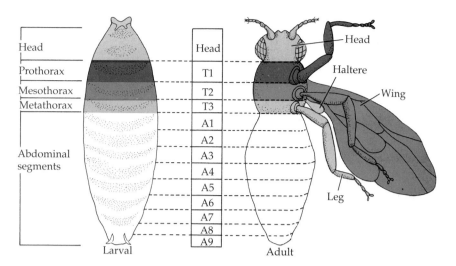

Head

Prothorax

Mesothorax

Metathorax

Abdominal
segments

| Head |
| T1 |
| T2 |
| T3 |
| A1 |
| A2 |
| A3 |
| A4 |
| A5 |
| A6 |
| A7 |
| A8 |
| A9 |

Head

Haltere

Wing

Leg

Larval   Adult

FIGURE 2
Comparison of larval and adult segmentation in *Drosophila*. The three thoracic segments can be distinguished by their appendages: T1 (prothoracic) has legs only; T2 (mesothoracic) has wings and legs; T3 (metathoracic) has halteres and legs.

which are repeating segmental units (Figure 2). Three of these segments form the thorax, while another eight segments form the abdomen. Each segment of the adult fly has its own identity. The first thoracic segment, for example, has only legs; the second thoracic segment contains legs and wings. The third thoracic segment has legs and halteres (balancers). Thoracic and abdominal segments can also be distinguished by their cuticle.

How does the polarity of the *Drosophila* egg give rise to the polarity of the fly body with its repetitive yet individuated segments? During the past decade, a model has emerged that synthesizes many of the data in this field (Figure 3). First, maternal mRNAs residing in various regions of the egg encode transcriptional and translational regulatory proteins that diffuse through the syncytial blastoderm and activate or repress the expression of certain zygotic genes. Second, the zygotic genes regulated by these maternal factors are expressed in certain broad (about three segments wide), partially overlapping domains. These genes are called GAP GENES (because mutations in them cause gaps in the segmentation pattern) and they are among the first genes transcribed in the embryo. Third, the different concentrations of the gap gene proteins cause the PAIR-RULE genes to be transcribed in the primordia of every other segment. The transcription pattern of each of these pair-rule genes gives a striped pattern of seven vertical bands along the anterior–posterior axis. The stripes of the pair-rule gene proteins activate the transcription of the SEGMENT PO-LARITY GENES. Their mRNA and protein products form 14 bands that divide the embryo into segment-wide units. At the same time, proteins of the gap, pair-rule, and segment polarity genes interact to activate another class of genes, the HOMEOTIC GENES, whose transcription determines the developmental fate of each segment.

FIGURE 3
Generalized model of *Drosophila* pattern formation. The pattern is established by maternal effect genes that form gradients and regions of morphogenetic proteins. These morphogenetic determinants activate the gap genes that define broad territories of the embryo. The gap genes enable the expression of pair-rule genes, each of which divides the embryo into regions about two segment primordia wide. The segment polarity genes then divide the embryo into segment-sized units along the anterior–posterior axis. The combination of these genes defines the spatial domains of the homeotic genes that define the identities of each of the segments.

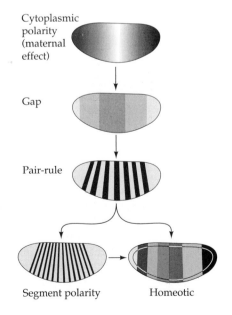

Cytoplasmic polarity (maternal effect)

Gap

Pair-rule

Segment polarity   Homeotic

The maternal effect genes

Embryological evidence of polarity regulation by oocyte cytoplasm

Classical embryological experiments demonstrated that there were at least two "organizing centers" in the insect egg. One was the ANTERIOR ORGA-NIZING CENTER, the other the POSTERIOR ORGANIZING CENTER. Sander (1975) postulated that these two organizing areas formed two gradients, one initiated at the anterior end and the other initiated in the posterior end.

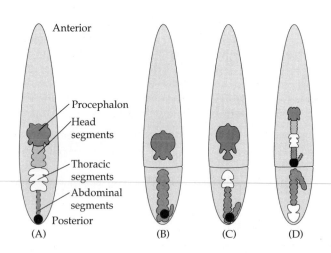

FIGURE 4
Sander's ligature experiments on the embryo of the leafhopper insect *Euscelis*. (A) Normal embryo seen in ventral view. The black ball at the bottom represents a cluster of symbiotic bacteria that mark the posterior pole. (B) After ligating the early embryo, partial embryos form, but the head and thoracic segments are missing from both embryos. (C) When ligated later (at the blastoderm stage), more of the missing segments are formed, but the embryos still lack the most central segments. (D) When the posterior pole cytoplasm is transplanted into an embryo ligated at the blastoderm stage, a small but complete embryo forms in the anterior half, while the posterior half forms an inverted partial embryo. These results can be explained in terms of gradients at the poles of the embryo that turn on one set of structures and repress the formation of others. (After Sander, 1960 and French, 1988.)

Each of these gradients formed its own structures and interacted with the other gradient to form the central portion of the embryo. Sander based this model on experiments that involved ligating the embryo at various times during development and transplanting regions of polar cytoplasm from one region of the egg to another (Figure 4). First, if he were to move cytoplasm from the posterior pole more anteriorly, he obtained a small embryo anterior to the posterior pole plasm, while extra segments, not organized into an embryo, formed behind it. Second, if he were to ligate the egg early in development, separating the anterior from the posterior region, one half developed into an anterior embryo and one half developed into a posterior embryo, but neither half contained the middle segments of the embryo. The later in development the ligature was made, the fewer middle segments were missing. Thus, it appeared that there were gradients emanating from the two poles during cleavage and that these gradients interacted to produce the positional information determining the identity of each segment.

The possibility that mRNA was responsible for generating the anterior gradient was suggested in a series of experiments by Kalthoff and Sander (1968). They found that when the anterior portion of the *Smittia* (midge) egg was exposed to ultraviolet light at wavelengths capable of inactivating RNA (265 and 285 nm), the resulting embryo lacked its head and thorax. Instead, the embryos developed two abdomens with mirror-image symmetry (Figure 5). Further evidence that RNA was important in specifying the anterior portion of the fly embryo was obtained by Kandler-Singer and Kalthoff (1976), who submerged *Smittia* eggs in solutions containing various enzymes and then punctured the eggs in specific regions. Double abdomens resulted when ribonuclease (an enzyme that digests RNA) was permitted to enter the anterior end. Other enzymes did not cause this abnormality, nor did ribonuclease effect this change when it entered other regions of the egg.

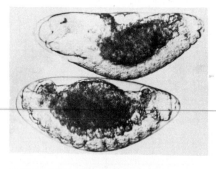

FIGURE 5
Normal and irradiated embryos of the midge *Smittia*. The normal embryo (top) shows a head on the left and abdominal segments on the right. The UV-irradiated embryo has no head region, but has abdominal segments at both ends. (From Kalthoff, 1969; photograph courtesy of K. Kalthoff.)

The anterior organizing center: The gradient of bicoid protein

In 1988, the gradient hypothesis was united with a genetic approach to the study of *Drosophila* embryogenesis. If there were gradients, what were the morphogens whose concentrations changed over space? What were the genes that shaped these gradients? And did these substances act by activating or inhibiting certain genes in the areas where they were concentrated? Christiane Nüsslein-Volhard and others found that certain genes encoded gradient morphogens for the anterior part of the embryo, other genes encoded the morphogens responsible for organizing the abdominal region of the embryo, and a third series of genes encoded proteins that produced the termininal regions at both ends of the embryo (Figure 6; Table 1).

In *Drosophila*, the phenotype of the *bicoid* mutant is very interesting if one is thinking in terms of gradients (see Figure 22 in Chapter 7). Instead of having anterior structures (acron, head, thorax), abdominal structures, and telson, the structure of the *bicoid* mutant is telson, abdomen, telson (Figure 7). It would appear that these embryos lack whatever morphogen is needed for anterior stuctures. Moreover, one could postulate that the substance these mutants lack is necessary to turn on genes for anterior structures and to turn off genes for the telson structures.

Recent evidence has strengthened the view that the product of the wild-type *bicoid* (*bcd*) gene is the morphogen that controls anterior development. First, *bicoid* is a maternal effect gene. The messenger RNA from the mother's *bicoid* genes are placed into the embryo by the mother's ovarian cells (Frigerio et al., 1986; Berleth et al., 1988). As we discussed in Chapter 8, transplantation of purified *bicoid* mRNA or anterior cytoplasm from wild-type eggs into the anterior region of *bicoid*-deficient eggs can restore a near-normal phenotype to the *bicoid*-deficient hosts. Moreover, wherever this anterior cytoplasm is injected becomes the head of the embryo.

The *bicoid* mRNA is strictly localized in the anterior portion of the oocyte (Figure 8A). This would not be compatible with its proposed role as a morphogen whose influence extends as a gradient over half the length of the egg. However, Driever and Nüsslein-Volhard (1988a) have shown that when bicoid protein is translated from this mRNA during early cleavage, it forms a gradient with its highest concentration in the anterior of the egg, and reaches background levels in the posterior third of the egg.

FIGURE 6

Three independent genetic pathways interact to form the anterior–posterior axis of the *Drosophila* embryo. In each case, the initial asymmetry is established during oogenesis, and the pattern is organized by the maternal products soon after fertilization. The realization of the pattern comes about when the localized maternal products activate or repress specific zygotic genes in different regions of the embryo. (After Anderson, 1989.)

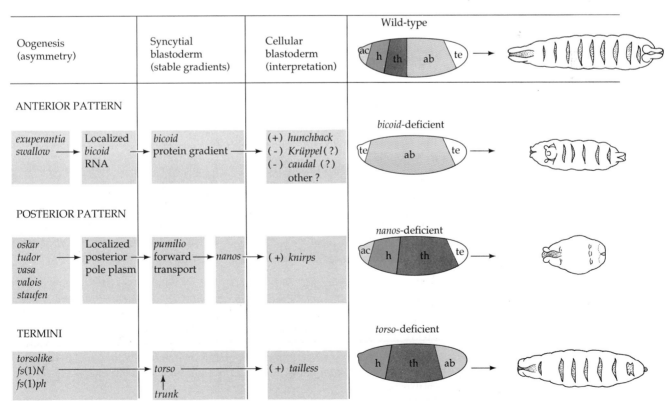

TABLE 1
Maternal effect genes that effect the anterior–posterior polarity of the *Drosophila* embryo

Gene	Phenotype	Proposed function and structure
ANTERIOR GROUP		
bicoid (bcd)	Head and thorax deleted, replaced by inverted telson	Graded anterior determinant; contains homeodomain
exuperantia (exu)	Anterior head structures deleted	Anchors *bicoid* mRNA
swallow (swa)	Anterior head structures deleted	Anchors *bicoid* mRNA
POSTERIOR GROUP		
tudor (tud)	No abdomen, no pole cells	
oskar (osk)	No abdomen, no pole cells	
vasa (vas)	No abdomen, no pole cells; oogenesis defective	
valois (val)	No abdomen, no pole cells; cellularization defect	
staufen (stau)	No abdomen, no pole cells; head defect	
nanos (nos)	No abdomen	Posterior morphogen
pumilio (pum)	No abdomen	Transport posterior signal to abdomen
Bicaudal (Bic)	Double abdomens	Possible attachment site for nanos protein
TERMINAL GROUP		
torso (tor)	No terminals	Possible morphogen for terminals
trunk (trk)	No terminals	
torsolike (tsl)	No terminals	
fs(1)Nasrat[fs(1)N]	No terminals; collapsed eggs	
fs(1)pole hole[fs(1)ph]	No terminals; collapsed eggs	

After Anderson (1989).

Moreover, this protein soon becomes concentrated in the embryonic nuclei in the anterior portion of the embryo (Figures 8B and C, and color portfolio).

Further evidence that bicoid protein is the anterior morphogen came from experiments that altered the steepness of the gradient. Two genes, *exuperantia* and *swallow*, are responsible for keeping the *bicoid* message at the anterior pole of the egg. In their absence, the *bicoid* message diffuses further into the posterior of the egg, and the gradient of bicoid protein is shallower (Driever and Nüsslein-Volhard, 1988b). The phenotype pro-

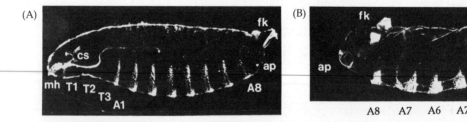

FIGURE 7
Phenotype of a strongly affected embryo from a female deficient in the *bicoid* gene. (A) Wild-type cuticle pattern. (B) *Bicoid* mutant. The head and thorax have been replaced by a second set of posterior telson structures. (fk, filzkörper; ap, anal plates, both telson structures. T1–T3 refer to the thoracic segments, while A1 and A8 represent the two terminal abdominal segments. mh and cs refer to head structures). (From Driever et al., 1990: photographs courtesy of W. Driever.)

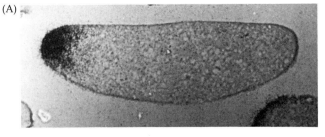

(A)

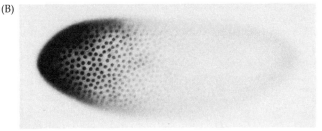

(B)

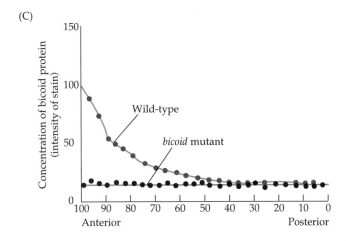

(C)

FIGURE 8

Gradient of bicoid protein in the early *Drosophila* embryo. (A) Localization of *bicoid* mRNA to the anterior tip of the embryo. (B) Gradient of bicoid protein shortly after fertilization. Note that the concentration is greatest anteriorly and trails off posteriorly. Notice also that bicoid protein is concentrated in the nuclei of the embryo. (C) Densitometric scan of the bicoid protein gradient. The upper curve represents the gradient of bicoid protein in wild-type embryos. The lower curve represents bicoid protein in embryos of *bicoid*-deficient mothers. (A from Kaufman et al., 1990; B, C from Driever and Nüsslein-Volhard, 1988a.)

duced by these two mutants is similar to that of bicoid-deficient embryos, but is less severe. These embryos lack their most anterior structures and have an extended jaw and thoracic region. Thus, by altering the gradient of bicoid protein, one correspondingly alters the fate of the embryonic regions.

The next question became: How might a gradient in bicoid protein control the determination of the anterior–posterior axis? One clue was that the bicoid protein was found in nuclei, and it contained a homeodomain. These facts suggested that bicoid protein could bind DNA and regulate gene expression. The earliest zygotic genes expressed are the gap genes. One of them, *hunchback (hb)* lacks mouthparts and the thorax structures. Furthermore, *hunchback* is active in the anterior half of the embryo—the region where bicoid protein was seen. In the late 1980s, two laboratories independently demonstrated that bicoid protein bound to and activated the *hunchback* gene (Driever and Nüsslein-Volhard, 1989; Struhl et al., 1989). The *hunchback* gene is thought to repress abdominal-specific genes, thereby allowing the region of *hunchback* expression (hence, the region of *bicoid* expression) to form the head and thorax. Using DNase footprinting, they found that bicoid protein bound to five sites in the upstream promoter region of the *hunchback* gene. These sites all had the consensus sequence 5′–TCTAATCCC–3′. But binding does not necessarily mean activation. The activation of this gene by bicoid protein was shown by fusing these *hunch-*

back promoter sites to the chloramphenicol acetyltransferase (CAT) reporter gene and injecting these genes into early *Drosophila* embryos. In all cases, bicoid protein was needed to activate the reporter genes. If injected into *bicoid*-deficient embryos, no CAT was produced (Figure 9). It was also shown that while some activation is seen when only one of the five bicoid protein-binding sequences was present, the full expression of the reporter gene (and presumably of *hunchback*) came when three of the five sites were present. Thus, the bicoid protein gradient probably functions by activating *hunchback* gene transcription in the anterior portion of the embryo.

Driever and co-workers hypothesize that another anterior gene besides *hunchback* must be activated by bicoid. First, deletions of *hunchback* produce only some of the defects seen in the *bicoid* mutant phenotype. Second, as we saw in the *swallow* and *exuparentia* experiments, only moderate levels of bicoid protein are needed to activate thorax formation (i.e., *hunchback* gene expression), but head formation needs higher concentrations. Driever et al. (1989) predict that the promoters of such a head-specific gap gene would have low-affinity binding sites for bicoid protein; this gene would be activated only at extremely high concentrations of bicoid protein—that is, near the anterior tip of the embryo. The positional fate of a cell is specified here both by the concentration of morphogen (bicoid protein) and by those of the cell's genes that can bind the morphogen if they see it.

The posterior organizing center: Activating and transporting the *nanos* product

The posterior organizing center was defined by seven maternal effect genes whose absence in the mother causes the absence of the embryonic abdomen (Schüpbach and Wieschaus, 1986; Lehmann and Nüsslein-Volhard, 1986). There are two major ways in which this posterior organizing activity differs from the anterior organizing center. First, the defects involve the differentiation of the abdominal region, not its respecification. In other words, the border of the thorax does not extend into the abdomen in these mutants. Rather, the posterior cells simply fail to form proper abdominal structures. Second, cytoplasmic transplantation experiments show that the posterior organization center is not located in the abdomen itself (where the defect is localized). Rather, the posterior organization center appears to be localized at the posterior pole of the embryo. The abdominal segments in the mutant embryos can be rescued by the injection

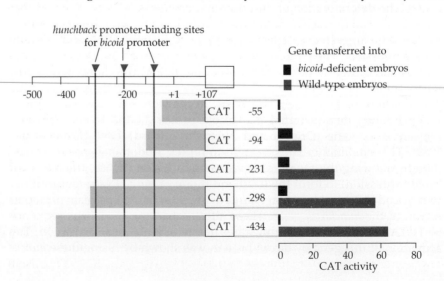

FIGURE 9
Influence of bicoid protein in activating the *hunchback* gene. Different regions of the *hunchback* promoter were fused to the CAT reporter gene and injected into either wild-type embryos or embryos from *bicoid*-deficient mothers. The more bicoid-binding sites were in the promoter region, the more effective was its expression in wild-type embryos. In embryos where bicoid protein was absent, no transcription resulted from any of the *hunchback* promoter-driven genes. (After Driever and Nüsslein-Volhard, 1989.)

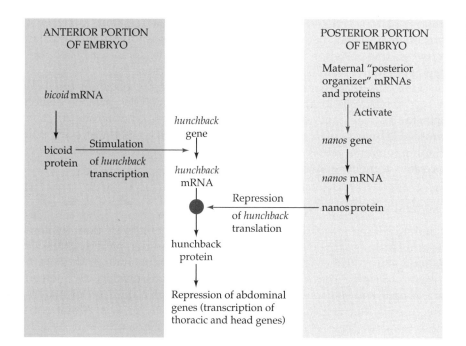

FIGURE 10
Schematic model for the action of bicoid and nanos proteins in specifying the anterior–posterior axis of the embryo. In the anterior region (color), bicoid protein activates *hunchback* transcription, and nanos protein is not there to stop the translation of the *hunchback* transcripts into protein. In the posterior region (gray), nanos represses *hunchback* translation, thereby allowing abdominal genes to be expressed.

ANTERIOR PORTION OF EMBRYO

bicoid mRNA

bicoid protein

Stimulation of *hunchback* transcription

hunchback gene

hunchback mRNA

Repression of *hunchback* translation

hunchback protein

Repression of abdominal genes (transcription of thoracic and head genes)

POSTERIOR PORTION OF EMBRYO

Maternal "posterior organizer" mRNAs and proteins

Activate

nanos gene

nanos mRNA

nanos protein

of wild-type cytoplasm only when this cytoplasm is injected into the abdomen of the defective embryo. However, the wild-type cytoplasm must come from the posterior pole, not from the abdominal region of the wild-type embryo. Thus, some product(s) made in the posterior pole must be transported anteriorly into the abdominal region of the embryo.

It is currently thought that five of the maternal effect genes are needed to activate the wild-type allele of *nanos* (Figure 6). The *nanos* gene product would then regulate gene expression in the abdomen. Unlike the activation of a gene in the anterior (where bicoid protein activates the *hunchback* gene), the formation of the *Drosophila* abdomen seems to be a system in which inhibitors must be inhibited. Tautz (1988) has shown that during normal abdomen formation, the protein product of the *nanos* gene represses the *translation* of *hunchback* mRNA. If the *nanos* gene product is not present, hunchback protein is made, and presumably inhibits the abdomen-generating gap genes such as *knirps* and *Krüppel* (Hülskamp et al., 1989; Struhl, 1989; Irish et al., 1989). The *hunchback* gene, therefore, appears to be the focal point, under the regulation of both the anterior and posterior organization centers long known to exist in insect development The presence of bicoid protein stimulates the transcription of *hunchback* in the anterior portion of the embryo and not in the posterior portion. Hunchback protein turns off the abdomen-generating gap genes such as *knirps* while activating the set of head- and thorax-specific genes. The *nanos* gene product prohibits the translation of any *hunchback* product in the posterior of the egg, thereby enabling the abdomen-generating genes to become expressed there (Figure 10). These studies on *nanos* and *bicoid* expression can now explain why destroying *bicoid* mRNA (for example, with UV light or RNase) allows the formation of a second abdomen, and why ligation procedures stop the formation of the central segments of the fly embryo.

Somehow, however, the *nanos* gene product must get from the posterior pole to the abdominal region. This appears to be the function of the *pumilio* gene product (Lehmann and Nüsslein-Volhard, 1987). If *pole plasm* is transferred from one *pumilio* mutant embryo into the *abdominal region* of another *pumilio* embryo, the host mutant embryo is rescued and forms an

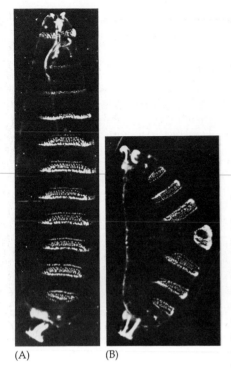

FIGURE 11
The action of the *BicaudalD* gene may be to localize the *nanos* product. (A) Ventral aspect of a wild-type embryo seen in darkfield. (B) Embryo of a *Bicaudal* mutant wherein the head, thorax, and anteriormost abdominal segments are replaced by mirror-images of the posterior abdominal segments and telson. (From Wharton and Struhl, 1989.)

abdomen. Moreover, if one deletes a number of segments between the posterior pole and the abdomen region (thereby shortening the distance between the posterior organizing center and the abdomen), abdominal segmentation is restored in *pumilio* mutant embryos.

Once the *nanos* gene product is moved from the posterior organizing center, it has to be localized. Wharton and Struhl (1989) provide evidence that the *BicaudalD* gene encodes the cytoskeletal binding site for nanos protein. Mutants of *BicaudalD* accumulate nanos protein at both poles, leading to nanos activity at both ends of the embryo and the formation of two mirror-image abdomens (Figure 11).

The terminal gene group

If both the anterior and the posterior organizing centers are nonfunctional, an embryo still can develop some anterior–posterior pattern (Nüsslein-Volhard et al., 1987). When females are made doubly mutant for both the anterior and posterior morphogens, their embryos produce two telsons, one at each end of the embryo. Thus, there exists a third set of maternal effect genes that help to create the anterior–posterior axis. Mutations in these terminal genes result in the loss of the unsegmented extremities of the organism: the anterior acron and the posterior telson. Therefore, the terminal gene set defines the boundaries of the segmented parts of the body. In the absence of these gene products, the segmented portion of the embryo expands to the extremities (Degelmann et al., 1986; Klingler et al., 1988).

The critical gene here appears to be *torso* (Figure 6). In the absence of *torso* product, there is neither acron nor telson. Rather, the segmented portion of the embryo takes up all the egg. Conversely, in a dominant mutation of this gene, the entire anterior half of the embryo is converted into acron and the entire posterior half into telson. Embryos defective in *torso* can be rescued by injecting material from wild-type embryos into both ends of the mutant. Injecting the wild-type cytoplasm into the center of the embryo is not effective. The wild-type cytoplasm, however, can be taken from any part of the embryo, suggesting that the *torso* product is synthesized throughout the embryo, but that it is active only at the tips. Stevens and her colleagues (1990) have shown that torso protein is probably activated by follicle cells at either pole of the embryo. The mutation *torsoless* creates a phenotype almost identical to *torso*, but this gene is expessed in the follicle cells that surround the egg. By somatic mutation, Stevens and colleagues showed that if the follicle cells at the poles of the egg chamber are deficient in the *torsoless* gene (even if the other follicle cells express the wild-type allele of this gene), the resulting embryo will have a phenotype similar to *torso*.

The anterior–posterior axis of the embryo is seen, therefore, to be specified by three sets of genes: those that define the anterior organizing center, those that define the posterior organizing center, and those that define the terminal boundary region. The anterior organizing center is located at the anterior end of the embryo and acts through a gradient of bicoid protein that activates anterior-specific gap genes and suppresses posterior-specific gap genes. The posterior organizing center is located at the posterior pole and acts through the formation of nanos protein, which gets transported into the abdominal region. Here, nanos inhibits the inhibitor of abdominal-specific gene expression. The boundaries of the acron and telson are defined by the *torso* gene product, which is activated at the tips of the embryo. This protein probably acts on the *tailless* and *huckebein* gap genes to create conditions that allow the expression of acron- and telson-specific genes.

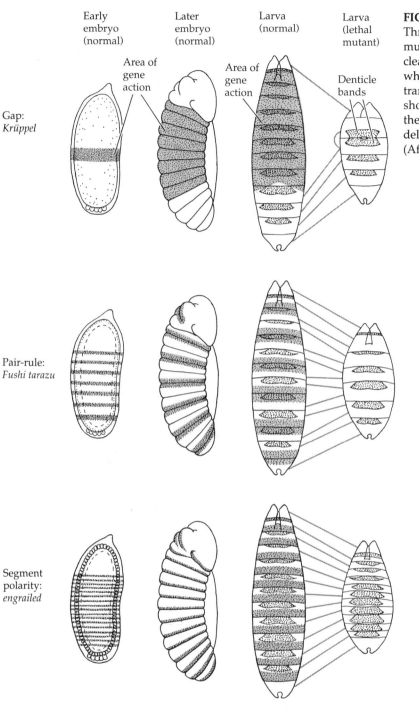

Early embryo (normal) Later embryo (normal) Larva (normal) Larva (lethal mutant)

Gap: *Krüppel*

Area of gene action

Area of gene action

Denticle bands

Pair-rule: *Fushi tarazu*

Segment polarity: *engrailed*

FIGURE 12
Three types of segmentation pattern mutants. The left panel shows the cleavage-stage embryo, with the region where the particular gene is normally transcribed in wild-type embryos shown in color. In the three panels to the right, the areas shown in color are deleted when these mutants develop. (After Mange and Mange, 1990.)

TABLE 2
Major loci affecting segmentation pattern in *Drosophila*

Category	Loci
Gap genes	*Krüppel (Kr)* *knirps (kni)* *hunchback (hb)* *giant (gt)* *tailless (tll)* *huckebein (hkb)*
Pair-rule genes (primary)	*hairy (h)* *even-skipped (eve)* *runt (run)*
Pair-rule genes (secondary)	*fushi tarazu (ftz)* *odd-paired (opa)* *odd-skipped (sdd)* *sloppy-paired (slp)* *paired (prd)*
Segment polarity genes	*engrailed (en)* *wingless (wg)* *cubitus interruptus^D (cb^D)* *hedgehog (hh)* *fused (fu)* *armadillo (arm)* *patched (ptc)* *gooseberry (gsb)*

The segmentation genes

Segmentation genes divide the early embryo into a repeating series of segmental primordia along the anterior–posterior axis. Mutations in segmentation genes cause the embryo to lack certain segments or parts of segments, and these mutations show the existence of three classes of segmentation genes (Table 2). The GAP GENES are activated or repressed by the maternal effect genes and divide the embryo into broad regions containing several segmental primordia. The *Krüppel* gene, for example (Figure 12A), is primarily expressed in parasegments 4–6 in the center of the *Drosophila* embryo (Figure 13), and the absence of this gene causes the

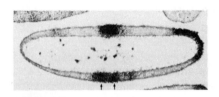

embryo to lack these regions. The term PARASEGMENT is used to define regions of the embryo that are separated by mesodermal thickenings and ectodermal grooves (Martinez-Arias and Lawrence, 1985). These grooves divide the embryo into 14 regions that often coincide with the domains of segmentation gene activity. The parasegments of the embryo do not become the segments of the larva or adult. Rather, they include the posterior part of an anterior segment and the anterior portion of the segment behind it (Figure 14). Next, the PAIR-RULE genes subdivide the broad gap gene domains into segments. Mutations of the pair-rule genes (as in *fushi tarazu*; Figure 12B) usually delete portions of every other segment. Figures 14 and 15 compare the morphology of the wild-type embryo with that of the *fushi tarazu* mutant. Last, the SEGMENT POLARITY genes are responsible for maintaining certain repeated structures within each segment. Mutations in this group of genes cause a portion of each segment to be deleted and replaced by a mirror-image structure of another portion of the segment. For instance, the *engrailed* gene is needed for the maintainance of the anterior–posterior boundary between segments. In *engrailed* mutants (Figure 12C), adjacent segments fuse together.

The regulation of these different genes is structured as a hierarchy, as shown in Figure 16. The cascade is initiated by the maternal effect genes that control the activation of the gap genes. The gap genes interact with one another to control their own transcription, and acting as a group they control the activation of the pair-rule genes. The pair-rule genes interact among themselves to form the repetitive segmental divisions of the body, and they also control the activation of the segment polarity genes. The pair-rule genes and gap genes also interact to regulate the HOMEOTIC genes that determine the structure of each segment. By the end of the cellular blastoderm stage, each segment primordium has a unique identity by its having a unique constellation of gap, pair-rule, and homeotic gene products (Levine and Harding, 1989).

FIGURE 14
Segments and parasegments. A and P represent the anterior and posterior compartments of the segments. The parasegments are shifted one compartment forward. Ma, Mx, and Lb represent the three head segments (mandibular, maxillary, and labial), the T segments are thoracic, and the A segments are abdominal. The parasegments are numbered 1 through 14. Underneath the map are the boundaries of gene expression observed by the in situ hybridization of radioactive cDNA from the pair-rule gene *fushi tarazu* (*ftz*). (After Martinez-Arias and Lawrence, 1985.)

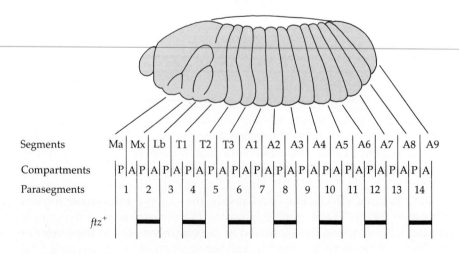

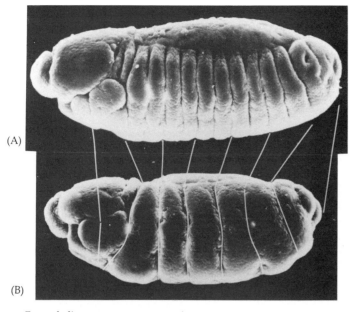

(A)

(B)

FIGURE 15
Defects seen in the *ftz⁻* embryo. (A) Scanning electron micrograph of wild-type embryo seen in lateral view. (B) Same stage of a *ftz⁻* embryo. The white lines connect the homologous portions of the segmented germ band. (C) Diagram of wild-type embryonic segmentation. The shaded regions show the parasegments of the germ band that are missing in the *ftz⁻* embryo. (A,B from Kaufman et al., 1990; C after Kaufman et al., 1990.)

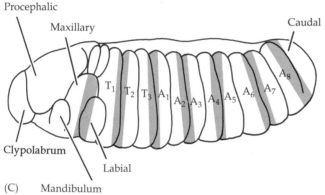

Procephalic

Maxillary

Caudal

T₁ T₂ T₃ A₁ A₂ A₃ A₄ A₅ A₆ A₇ A₈

Clypolabrum

Labial

(C) Mandibulum

FIGURE 16
Summary of the proposed cascade of developmental gene regulation. The straight arrows represent the regulation of one class of genes by the other. The circular arrows indicate regulation between members of the particular group of genes. (After Levine and Harding, 1989.)

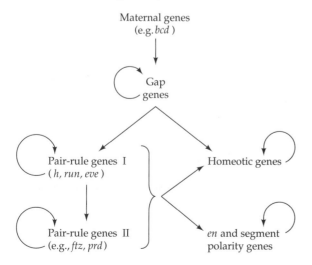

Maternal genes
(e.g. *bcd*)

Gap genes

Pair-rule genes I
(*h, run, eve*)

Homeotic genes

Pair-rule genes II
(e.g., *ftz, prd*)

en and segment polarity genes

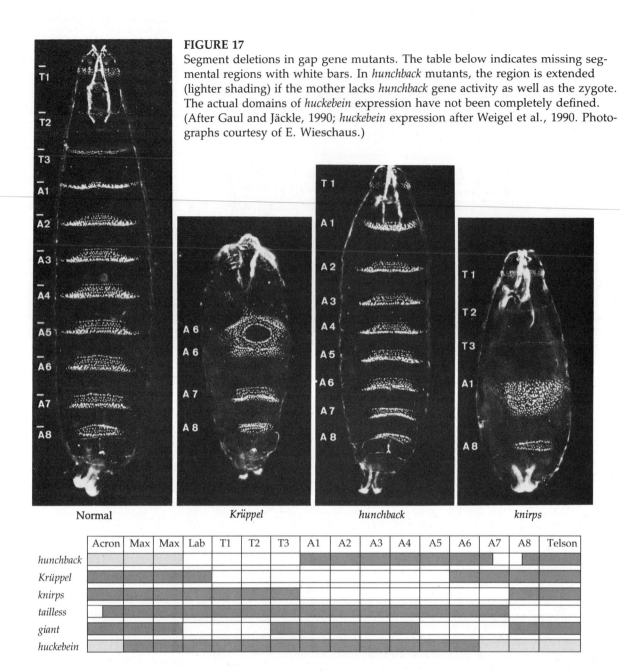

Normal **Krüppel** **hunchback** **knirps**

FIGURE 17
Segment deletions in gap gene mutants. The table below indicates missing segmental regions with white bars. In *hunchback* mutants, the region is extended (lighter shading) if the mother lacks *hunchback* gene activity as well as the zygote. The actual domains of *huckebein* expression have not been completely defined. (After Gaul and Jäckle, 1990; *huckebein* expression after Weigel et al., 1990. Photographs courtesy of E. Wieschaus.)

	Acron	Max	Max	Lab	T1	T2	T3	A1	A2	A3	A4	A5	A6	A7	A8	Telson
hunchback	░	░	░	░	░	░	░	█	█	█	█	█	█	□	█	
Krüppel	█	█	█	█				█	█	█	█	█	█	█		
knirps	█	█	█	█					█	█	█	█	█	█	█	
tailless	□	█	█	█											□	█
giant	█	█	█	█		█							█	█	█	█
huckebein	░	█	█	█	█								█	░		

The gap genes

The gap genes were originally defined by a series of mutations whose embryos lacked groups of consecutive segments (Nüsslein-Volhard and Weischaus, 1980). As shown in Figure 17, deletions caused by the *hunchback* (*hb*), *Krüppel* (*Kr*), and *knirps* (*kni*) genes span the entire segmented region of the *Drosophila* embryo. The *giant* (*gt*) gap gene overlaps with these three, and the phenotypes of the *tailless* and *huckebein* mutants delete portions of the unsegmented termini of the embryo.

The expression of these genes is a dynamic situation. There is usually a low level of transcriptional activity across the entire embryo that becomes defined into discrete regions of high activity as cleavage continues. The *hunchback* gene (*hb*) is expressed strongly in parasegments 1–3 and 13, and in the anterior component of parasegment 14. *Krüppel* (*Kr*) is transcribed in parasegments 4–6 and the posterior compartment of parasegment 14, while *knirps* function is essential for the function of parasegments 7–12

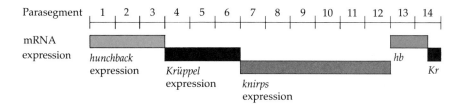

mRNA expression

hunchback expression

Krüppel expression

knirps expression

hb

Kr

Domains of gap gene expression. Non-overlapping domains of gap gene expression based on in situ hybridization of radioactive cDNA from gap genes *hunchback* (hb) and *Krüppel* (Kr) and from the phenotype of *knirps* mutants. (Note that this gene expression map shows where the mRNAs are localized. The proteins appear to diffuse more and the boundaries are not as distinct.) (From Jäckle et. al., 1986.)

(Figure 18; Jäckle et al., 1986). The transcription patterns of the gap genes are initiated by the maternal effect genes, but are stabilized and maintained by interactions between the different gap genes themselves.* As we saw above, the expression of the hunchback protein is stimulated by bicoid protein and inhibited by the nanos protein, thus confining hunchback expression to the anterior half of the embryo. *Krüppel* is also regulated by maternal proteins, but in a different way. Bicoid, nanos, and torso proteins all inhibit *Krüppel* transcription. In *bicoid* mutant embryos, *Krüppel* gene expression reaches to the anterior of the embryo, and in mutants where *bicoid*, *nanos*, and *torso* are all eliminated, *Krüppel* is transcribed by every embryonic cell. However, in the middle of the embryo, bicoid, nanos, and torso proteins are each in low concentrations, thereby allowing the transcription of *Krüppel* (Figure 19).

These initial patterns of transcription are refined and stabilized by the interactions between the different gap genes. For instance, *Krüppel* gene expression is negatively regulated on its anterior boundary by hunchback protein and on its posterior boundary by knirps protein (Jäckle et al., 1986; Harding and Levine, 1988). If *hunchback* activity is lacking, the domain of *Krüppel* expression extends anteriorly. If *knirps* activity is lacking, *Krüppel* gene expression extends more posteriorly.

The boundaries between the regions of gap gene transcription are probably created by mutual repression. Just as the *hunchback* gene can control the anterior boundary of *Krüppel*, so can *Krüppel* determine the posterior boundary of *hunchback* transcription. If an embryo lacks the *Krüppel* gene, *hunchback* transcription continues into the area usually alloted to *Krüppel* (Jäckle et al., 1986). These boundary-forming inhibitions are thought to be directly mediated by the gap gene products, because each of the three major gap genes (*hb*, *Kr*, and *kni*) encodes zinc-finger proteins that appear to act as transcription factors (Knipple et al., 1985; Gaul and Jäckle, 1990).

Moreover, these interactions are highly specific, and the product of one gap gene can bind to the promoters of other gap genes. DNase I footprinting shows that the protein encoded by the wild-type *Krüppel* gene binds to the promoter region of the *hunchback* gene (which it inhibits) and to the promoter region of the *knirps* gene (which it stimulates). The *knirps* promoter region is also recognized by the protein product of the *tailless* gene that inhibits *knirps* transcription. Hunchback protein (in addition to recognizing the *Krüppel* promoter) also recognizes its own promoter, suggesting that *hunchback* is involved in regulating its own expression (Pankratz et al., 1990; Štanojević et al., 1989; Treisman and Desplan, 1989).

The pair-rule genes

The first indication of repeated segmentation in the fly embryo comes when the pair-rule genes are expressed during the thirteenth division cycle. The transcription patterns of these genes are striking in that they

Maternal effect gene control of gap genes *hunchback* and *Krüppel*. (After Gaul and Jäckle, 1990.)

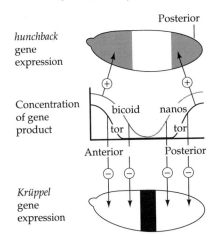

hunchback gene expression

Posterior

Concentration of gene product

bicoid
tor
nanos
tor

Anterior

Posterior

Krüppel gene expression

*These interactions between genes and gene products are facilitated by the fact that all these reactions occur within a syncytium. The cell membranes have not yet formed.

FIGURE 20

The expression of wild-type *fushi tarazu* (*ftz*) gene products in the cellularizing blastoderm of *Drosophila* embryos. The dark bands represent localization of fushi tarazu protein against a background of nuclei stained brightly for DNA. (From Carroll et al., 1988; photograph courtesy of B. Thalley and S. Carroll.)

each divide the embryo into the areas that are the precursors of the segmental body plan. As can be seen in the color portfolio and in Figure 20, one vertical stripe of nuclei (the cells are just beginning to form) will express this gene, then another stripe of nuclei do not express it, and then another stripe of nuclei expresses this gene. The result is a "zebra stripe" pattern along the anterior–posterior axis, dividing the axis into 15 subunits (Hafen et al., 1984). Eight genes are presently known to be capable of dividing the early embryo in this fashion; they are listed in Table 2. It is important to note that not all nuclei express the same pair-rule genes. In fact, in each parasegment, each row of nuclei probably has its own constellation of pair-rule genes that distinguishes it from any other row. Figure 21 shows the relationship between three pair-rule genes, *even-skipped*, *hairy*, and *fushi tarazu*.

How are some nuclei of the *Drosophila* embryo told to transcribe a particular gene while their neighbors are told not to transcribe it? The answer appears to come from the distribution of the protein products of the gap genes. Whereas the *mRNA* of each of the gap genes has a very discrete distribution that defines abutting or slightly overlapping regions of expression, the *protein* products of these genes extend more broadly. In fact, they overlap by at least 8–10 nuclei (which at this stage accounts for about 2–3 segment primordia). This was demonstrated in a striking manner by Štanojević and co-workers (1989). They fixed embryos that were starting to form cells, stained the hunchback protein with an antibody carrying a red dye, and simultaneously stained the Krüppel protein with an antibody carrying a green dye. Cellularizing regions that contained *both* proteins bound both antibodies and were stained bright yellow (color endpaper). Similarly, Krüppel protein overlaps with knirps protein in the posterior region (Pankratz et al., 1990).

Three genes are known to be the PRIMARY PAIR-RULE GENES. These genes—*hairy*, *even-skipped*, and *runt*—are essential for the formation of the periodic pattern, and they are the genes directly controlled by the gap proteins. The promoters of the primary pair-rule genes are recognized by

FIGURE 21

Relationship between three pair-rule genes. (A) The transcription pattern of the *even-skipped* gene. (B) Double-labeling of the *even-skipped* and *fushi tarazu* genes showing that *fushi tarazu* (see Figure 20) is expressed between the domains of *even-skipped* transcription. (C) Diagram of the expression of three pair-rule genes —*even-skipped*, *hairy*, and *fushi tarazu*—in individual nuclei at the cellular blastoderm stage. (After Carroll et al., 1988; photographs courtesy of B. Thalley and S. Carroll.)

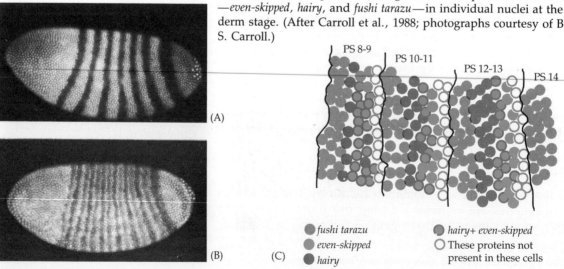

gap gene proteins, and it is thought that the different concentrations of gap gene proteins determine whether the gene is active or not. For example, one particular deletion in the promoter region of the *even-skipped* gene prevents the formation of the seventh *even-skipped* stripe, while a deletion slightly more downstream causes the loss of the second *even-skipped* stripe (Figure 22; Howard et al., 1988). DNase I footprinting of this latter region shows that it contains three binding sites for Krüppel protein and three binding sites for hunchback protein. The region responsible for the third stripe of *even-skipped* transcription contains 20 hunchback protein binding sites and not one site for the Krüppel protein (Štanojević et al., 1989). This situation would enable the site to respond to very low levels of the *hunchback* gene product. A model for the local activation of the second and third *even-skipped* stripe is shown in Figure 23.

It is still uncertain whether individual gap genes activate gene transcription in the stripes or repress the transcription of the genes in the interstripe nuclei. Carroll and Vavra (1989) have shown that when *hb*, *Kr*, and *kni* are all deleted from the embryo, the *hairy* gene is expressed throughout the normally striped territory. In other words, the interstripe nuclei of these mutants were transcribing the *hairy* gene, instead of being repressed. A similar result can be obtained by injecting protein synthesis inhibitors into embryos at the time of pair-rule gene stripe formation. If the gap gene proteins were activators, the pair-rule genes should be repressed. If the gap gene proteins were inhibitors of otherwise active genes, the pair-rule genes should be turned on throughout the embryo. The latter effect was seen (Edgar et al., 1986). Other investigators, however, argue that gap gene proteins can activate the transcription of specific bands and repress the transcription of others. Howard and co-workers (1988) have shown that different regions of the *hairy* promoter are responsible for different stripes. These regions can be separated by restriction enzymes, and, when fused to a new gene, will cause that new gene to be transcribed at a particular region of the *Drosophila* embryo. Pankratz and colleagues (1990) fused these portions of the *hairy* gene promoter to a bacterial β-galactosidase gene and inserted this gene into embryos that were mutant for either *Krüppell*, *knirps*, or *tailless*. By analyzing gap gene mutants, *Krüppel* was found to activate stripe 5 and repress stripe 6, while *knirps* was seen to activate stripe 6 but to repress stripe 7. Tailless protein, however, was found to be essential for activating *hairy* stripe 7. Pankratz et al. have modeled the mechanism by which *Krüppel* and *knirps* (both of which bind to the *hairy* gene promoter) may interact to generate the sixth transcription stripe of *hairy*. It is still uncertain whether the stripe pattern of these primary pair-rule genes is initiated by activation, inhibition, or by the competition between these processes.

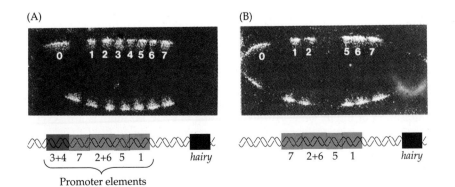

(A)

(B)

FIGURE 22
Specific promoter regions of the *hairy* gene control specific transcription bands in the embryo. (A) Wild-type transcription pattern of seven bands (plus a band in the head region). (B) Loss of bands 3 and 4 when a specific region of the promoter has been deleted. Gene map of the promoter region is printed below each figure, and the numbers correspond to the band that is absent when this particular region is deleted. (From Howard et al., 1988; photographs courtesy of the authors.)

FIGURE 23

Model for the formation of the second and third stripes of *even-skipped*. The hunchback and Krüppel proteins are expressed in broad overlapping areas, and a segment of about eight cells expresses both proteins. Each *eve* stripe contains about four cells. Local concentrations of the hunchback and Krüppel proteins are postulated to stimulate the transcription of the *eve* gene, while other concentrations of these proteins may act to repress it. Stripe 2 coincides with cells having high hunchback concentrations but lacking Krüppel. Stripe 3 occurs in cells expressing low levels of hunchback and high levels of Krüppel (After Štanojević et al., 1989.)

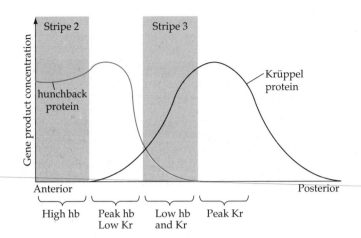

Once initiated by the gap proteins, the transcription pattern of the primary pair-rule genes becomes stabilized by their interactions among themselves (Levine and Harding, 1989). The primary pair-rule genes also form the context that allows or inhibits the expression of the secondary ("late-acting") pair-rule genes. One such secondary pair-rule gene is *fushi tarazu* (*ftz*; Japanese for "too few segments.") Early in cycle 14, *ftz* mRNA and protein are seen throughout the segmented portion of the embryo. However, as the proteins from the primary pair-rule genes begin to interact with the *ftz* promoter, the *ftz* gene is repressed in certain bands of nuclei to create the interstripe regions (Figure 24; Edgar, et al., 1986; Karr and Kornberg, 1989).

Segment polarity genes

The segment polarity genes (Table 2) are transcribed from one band of nuclei in each segment. The pair-rule genes define which nuclei can express a particular segment polarity gene. Perhaps the best studied segment polarity gene is *engrailed* (Figure 25), and it is regulated directly by *fushi tarazu* and *even-skipped*. The *even-skipped* (*eve*) and *ftz* genes are both expressed in seven bands across the anterior–posterior axis, and each band includes 4–5 nuclei. This is the entire length of a parasegmental primordium. Since eve protein inhibits the transcription of the *ftz* gene, the *ftz* gene products alternate with the *eve* gene products. This can be seen in Figure 21. By the beginning of gastrulation, these bands of ftz and eve protein have "sharpened"—each gene is being expressed in only three cells and is separated from the next band by an interstripe region of five cells.

Both eve and ftz proteins bind to the *engrailed* gene promoter and stimulate transcription (Desplan et al., 1985; Hoey and Levine, 1988). However, it seems that eve or ftz protein must be present in large amounts to activate the gene, so the *engrailed* gene is only expressed in the poste-

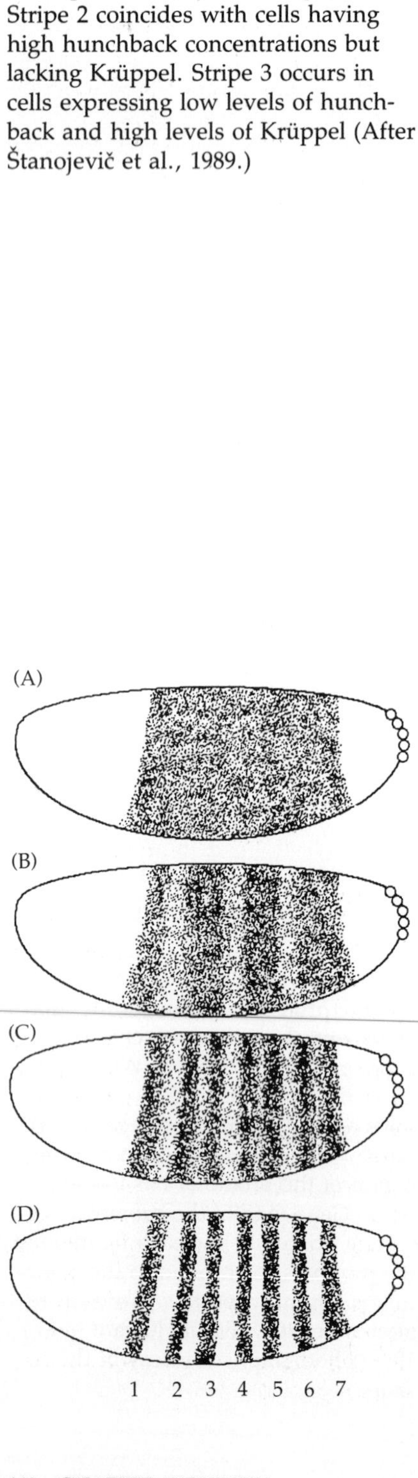

(A)

(B)

(C)

(D)

1 2 3 4 5 6 7

FIGURE 24

Transcription of the *ftz* gene. At the beginning of cycle 14, there is low-level transcription in each of the nuclei in the segmented region of the *Drosophila* embryo. Within the next thirty minutes, the expression pattern alters as *ftz* transcription is enhanced in certain regions (which form the stripes) and repressed in the interstripe regions. (From Karr and Kornberg, 1989.)

(A)

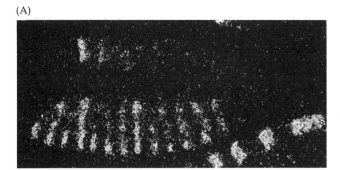

(B)

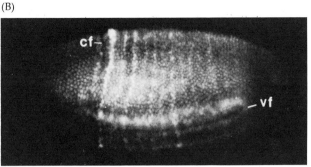

cf→

→vf

riormost cell of each segment (hence, the anteriormost cell of each para-segment, as seen by a comparison of Figures 25 and 21). A model for this situation is depicted in Figure 26. As predicted by this model, embryos lacking *fushi tarazu* gene expression lack the seven even-numbered *engrailed* stripes. Other pair-rule genes are also involved in the activation or main-tenance of *engrailed* transcription, and the other segment polarity genes are probably transcribed due to interactions with another group of pair-rule genes. The *wingless* gene, for instance, is expressed in those regions where *odd-paired* is present, but neither *ftz* nor *even-skipped* is active. This enables *wingless* to be transcribed solely in the cell directly anterior to the cell where *engrailed* is transcribed.

FIGURE 25

Expression of the *engrailed* gene prod-uct in the cellular blastoderm of *Dro-sophila*. (A) Localization of fourteen bands of *engrailed* transcripts in the blastoderm. Note that the bands alter-nate in intensity. Each band spans only one cell. (B) Localization of 14 bands of engrailed protein in such an embryo. The protein bands, like those of the RNA transcripts, are first seen in the anterior part of the embryo, and they are confined to the posterior com-partment of each segment. cf, cephalic fold; vf, ventral furrow. (A from How-ard and Ingham, 1986; courtesy of P. Ingham. B from O'Farrell et al., 1985; courtesy of P. O'Farrell.)

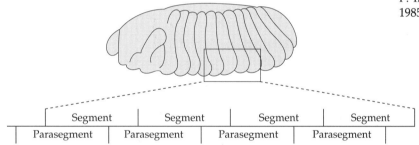

Segment	Segment	Segment	Segment
Parasegment	Parasegment	Parasegment	Parasegment

(A) Wild-type gene expression

Cells:

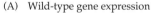

Gene product concentration

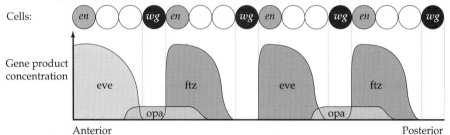

eve   ftz   eve   ftz

opa   opa

Anterior   Posterior

(B) Mutation affecting *ftz* expression affects segment polarity genes

Cells:

Gene product concentration

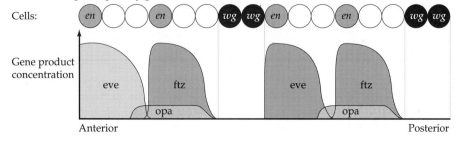

eve   ftz   eve   ftz

opa   opa

Anterior   Posterior

FIGURE 26

Model for the expression of segment polarity genes *engrailed* (*en*) and *wing-less* (*wg*). (A) The *engrailed* gene is ex-pressed when the cells contain high concentrations of either even-skipped or fushi tarazu proteins. The *wingless* gene is transcribed when neither of these two genes is active, but a third gene (probably *odd-skipped*) is present. (B) In certain mutants, the pattern of *fushi tarazu* gene expression is altered, and it is transcribed directly after *even-skipped*. The result is that the *engrailed* genes become active in new cells and that two cells adjacent to each other express *wingless*. (After Levine and Harding, 1989.)

FIGURE 27
The functional domains of the bithorax complex and antennapedia complex genes in *Drosophila*. The bithorax complex (BX-C; top) has been divided into the three lethal complementation groups identified by E. B. Lewis. The antennapedia complex (ANT-C; bottom) genes are *labial* (*lab*), *Deformed* (*Dfd*), *Sex comb reduced* (*Scr*), and *Antennapedia* (*Antp*). (After Harding et al., 1985.)

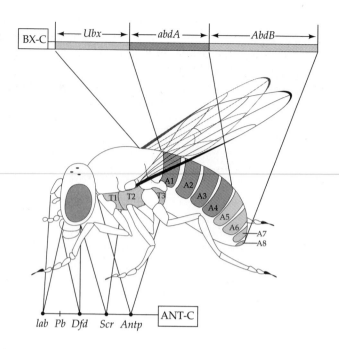

(A)

(B)

The homeotic selector genes

Patterns of homeotic gene expression

After the segmental boundaries have been established, the characteristic structures of each segment are specified. This specification is accomplished by the HOMEOTIC SELECTOR GENES (Lewis, 1978). There are two regions of *Drosophila* chromosome 3 that contain most of these homeotic genes (Figure 27). One region, the ANTENNAPEDIA COMPLEX, contains the homeotic genes *labial* (*lab*), *Antennapedia* (*Antp*), *Sex comb reduced* (*Scr*), *Deformed* (*Dfd*), and *proboscipedia* (*pb*). The *labial* and *Deformed* genes act to specify the head segments, while *Sex comb reduced* and *Antennapedia* contribute to giving the thoracic segments their identities. The *proboscipedia* gene appears to act only in adults, but in its absence, the labial palps of the mouth are transformed into legs (Wakimoto, 1984; Kaufman et al., 1990). The second region of homeotic genes is the BITHORAX COMPLEX (Lewis, 1978). There are three protein-coding genes found in this complex: *Ultrabithorax* (*Ubx*), which is required for the identity of the third thoracic segment; and *abdominal A* (*abdA*) and *Abdominal B* (*AbdB*) which are responsible for the segmental identities of the abdominal segments.

These eight homeotic genes were defined by genetic analysis wherein mutations in two *different* genes were able to complement each other (i.e., give a wild-type phenotype), but mutations within the *same* gene (one on each copy of chromosome 3) gave a mutant phenotype (Sánchez-Herrero et al., 1985). Moreover, the lethal phenotype of the triple-point mutant Ubx^-, $abdA^-$, $AbdB^-$ is identical to that of a deletion of the entire bithorax complex (Casanova et al., 1987), confirming the data that the bithorax

FIGURE 28
(A) Head of a wild-type fly. (B) Head of a fly containing the *Aristopedia* mutation that converts antennae into legs. (From Kaufman et al., 1990; photograph courtesy of T. C. Kaufman.)

complex has only these three transcription units. An additional homeotic gene, *caudal*, is in neither of these regions and is involved in specifying the telson (MacDonald and Struhl, 1986; Mlodzik and Gehring, 1987).

As these genes are responsible for the specification of body parts, mutations in these genes lead to bizarre phenotypes. William Bateson (1894) called these organisms "homeotic mutants," and they have fascinated developmental biologists for decades. The *Antennapedia* gene, for instance, is thought to specify the identity of the second thoracic segment. In the *dominant* mutation of *Antennapedia*, this gene is expressed in the head as well as in the thorax, and the imaginal discs of the head region are specified as thoracic. Thus, legs grow out of the head sockets rather than antennae (Figure 28). In the *recessive* mutant of *Antennapedia*, the gene fails to be expressed in the second thoracic segment, and antennae sprout out of the leg positions (Struhl, 1981; Frischer et al., 1986; Schneuwly et al., 1987). Likewise, when the Ultrabithorax complex is deleted, the third thoracic segment (which is characterized by halteres) becomes transformed into another *second* thoracic segment. The result (Figure 29) is a fly with four wings—an embarassing situation for a classic dipteran.*

These eight major homeotic selector genes have been cloned and their expression analyzed by in situ hybridization (Harding et al., 1985; Akam, 1987). A summary of these experiments is shown in Figure 30. Transcripts from each locus are detected in specific regions of the embryo, and are especially prominent in the central nervous system. In homeotic mutants, this normal expression is altered. For instance, in the dominant *Antennapedia* alleles mentioned above, the *Antennapedia* gene has been inverted on the chromosome, so that it has lost its own promoter and is under the control of a different promoter that is active in the head. This causes the ectopic expression of *Antp* in the head. Similarly, if the *Ultrabithorax* gene is placed on a new promoter and expressed in the head region, the antennae will be transformed into legs (Mann and Hogness, 1990).

---

*Dipterans (two-winged insects such as flies) are thought to have evolved from normal four-winged insects; it is possible that this change arose via alterations in the bithorax complex. Chapter 23 includes more speculation on the relationship between *bithorax* genes and evolution.

FIGURE 29
This four-winged fruit fly was constructed by putting together three mutations in *cis*-regulators of the *Ultrabithorax* gene. These mutations effectively transform the third thoracic segment into another second thoracic segment (i.e., halteres into wings). (Photograph courtesy of E. B. Lewis.)

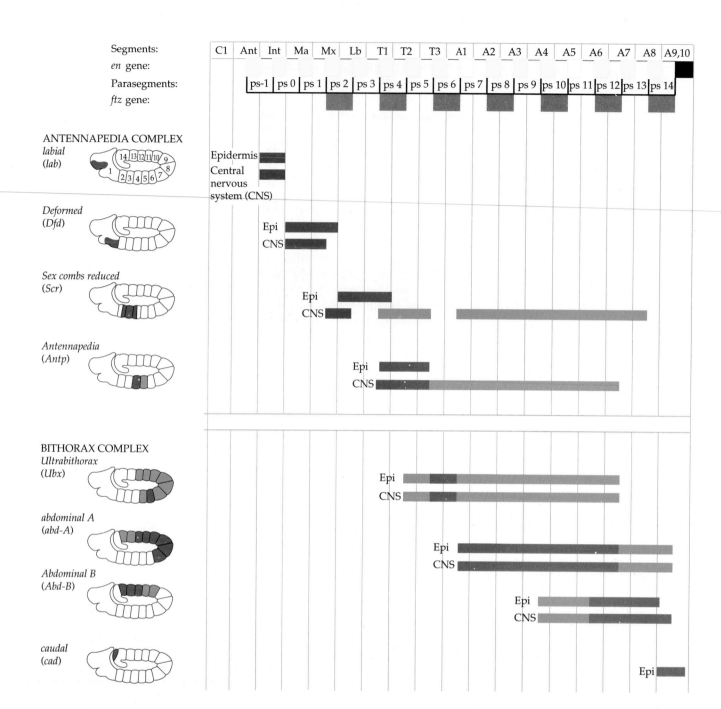

FIGURE 30
Regions of homeotic gene expression (both mRNA and protein) in the blastoderm of the *Drosophila* embryo and (a few hours later) in its central nervous system. The darker shaded areas are those segments or parasegments with the most product. The diagram beside the chart represents gene expression within the parasegmental boundaries. (After Kaufman et al., 1990.)

Initiating the patterns of homeotic gene expression

The initiation of the homeotic gene domains is influenced by the pair-rule genes and the gap genes. In *ftz*⁻ embryos, for instance, the initial transcription rates of *Scr*, *Antp*, and *Ubx* are much lower than in wild-type embryos. The early expression of these genes is usually confined to parasegments 2, 4, and 6, respectively; these are the sites corresponding to *ftz* gene expression bands 1, 2, and 3 (Figure 31; Ingham and Martinez-Arias, 1986). Conversely, ftz protein has a negative effect on *Dfd* transcription. In wild-type embryos, the anterior margin of a ftz protein band forms the posterior border of the *Dfd* territory. In *ftz*⁻ embryos, the posterior of *Dfd* extends posteriorly. Other pair-rule genes, such as *even-skipped* and *odd-paired*, stimulate *Dfd* expression (Jack et al., 1988).

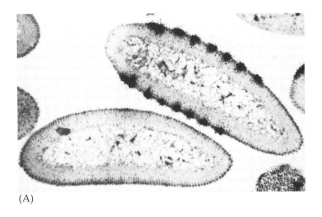

(A)

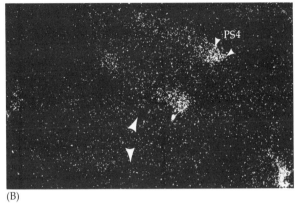

PS4

(B)

FIGURE 31
Dependence of two homeotic genes on the proper expression of the *fushi tarazu* segmentation gene. In situ hybridization of various radioactive DNA probes was performed on adjacent microscopic sections from wild-type (lower) and *ftz⁻* (upper) *Drosophila* embryos. Both embryos are at the cellularizing blastoderm stage. (A) Autoradiograph of section hybridized with radioactive *ftz* DNA. Seven areas of *ftz* RNA accumulation are seen in the wild-type embryo, but none is seen in the *ftz⁻* mutant. (B) Darkfield image of the autoradiograph using radioactive DNA of the *Antennapedia* gene shows wild-type gene expression in the fourth parasegment (PS4 and small arrows) but no transcription in the same region (large arrows) in the *ftz⁻* mutant. (From Ingham and Martinez-Arias, 1986; photographs courtesy of the authors.)

But the pair-rule gene products alone cannot account for the boundaries of the homeotic genes. (Otherwise the homeotic genes would show periodicity across the entire anterior–posterior axis.) Since each homeotic gene is usually expressed in contiguous segments (or parasegments), the gap genes are obvious candidates for establishing the limits of homeotic gene expression. Much evidence has accumulated for this view. The *Krüppel* gene, for instance, activates *Antp* transcription and inhibits *AbdB* transcription (Figure 32). The gap gene *knirps* also is required for delimiting the initial domains of *Antp* and *AbdB*, but it acts to inhibit both genes. Thus, the presence of high concentrations of Krüppel protein in the central region of the embryo enhances the transcription of *Antennapedia* while it keeps the *Abdominal B* gene from being transcribed (Harding and Levine, 1988).

Maintaining the patterns of homeotic gene expression

The expression of homeotic genes is also a dynamic process. The *Antp* gene, for instance, although initially expressed in presumptive parasegment 4, soon appears in parasegment 5. As the germ band expands, *Antp* gene expression is seen in the presumptive neural tube as far posterior as parasegment 12. During further development, the pattern contracts again, and *Antp* transcripts are localized strongly to parasegments 4 and 5. *Antp* transcription is negatively regulated by all the homeotic gene products posterior to it. In other words, each of the bithorax complex genes represses the transcription of *Antennapedia*. If *Ultrabithorax* is deleted, *Antp* activity extends through the region that would normally have expressed *Ubx* and stops where the *Abd* regions begins. (This allows the third thoracic segment to form wings like the second thoracic segment, as in Figure 29.)

FIGURE 32
The initial expression of the homeotic gene *Antennapedia* (B) is predicated upon the prior expression of *Krüppel* (A) in the same area. If the placement of *Krüppel* expression is altered, so is the expression of *Antennapedia*. (From Levine and Harding, 1989; photographs courtesy of the authors.)

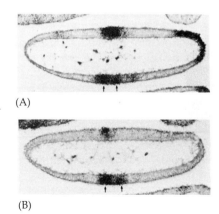

(A)

(B)

If the entire bithorax complex is deleted, *Antp* expression extends throughout the abdomen. (The larva doesn't survive, but the cuticle pattern throughout the abdomen is the same as the second thoracic segment.) In general, the anteriorly expressed homeotic genes are inhibited by those homeotic proteins more posterior to them (Harding and Levine, 1989). Thus, *Antp* is repressed by *Ubx*, *Ubx* is repressed by *abdA*, and *abdA* is repressed by *Abd-B*.*

The molecular structure of homeotic genes

In order to understand how the homeotic genes function, it is necessary to know their structure. The structure of the homeotic genes points toward complex modes of regulation. Some of the genes have multiple promoters and multiple sites for transcription initiation. The *Antennapedia* gene, for instance, has two transcription products, one of them initiating within an intron of the other (Figure 33; Frischer et al., 1986). Transcription from the first promoter is activated by Krüppel protein and repressed by Ultrabithorax protein, while transcription from the second promoter is activated by hunchback and fushi tarazu proteins and inhibited by oskar protein (Boulet and Scott, 1988; Krasnow et al., 1989; Irish et al., 1989). Although the protein produced from both these promoters is the same, the promoters cause it to be expressed in different cells. The transcript of the first promoter is necessary for dorsal thorax development, whereas the transcript of the second promoter is required for ventral thorax (leg) development and for embryonic viability (Bermingham et al., 1990). Hogness and his co-workers (Beachy et al., 1988; Krasnow et al., 1989) used DNA footprinting to show that Ultrabithorax protein binds to a series of sites in the first promoter region of the *Antennapedia* gene. They then showed that this binding represses *Antennapedia* gene activity by fusing the *Antennapedia* first promoter region to the CAT reporter gene and placing the fused gene into cultured cells that do not usually transcribe any of the homeotic genes (Figure 34). The *Antennapedia* gene was activated by the presence of ftz protein (supplied by adding the *ftz* gene on a strong constitutive promoter). However, when a third gene was transfected into the cells—*Ubx* on a strong constitutive promoter—the *Antennapedia* gene was repressed.

Another complication of the homeotic genes is that several of them can produce families of related proteins by alternative RNA splicing. The *Ultrabithorax* gene produces several proteins by such a mechanism, and

*Aficionados of information theory will recognize that the process by which the anterior–posterior information in morphogenic gradients is transferred to discrete domains of homeotic selector genes represents a transition from analogue to digital specification. This enables the transient information of the gradients in the syncitial blastoderm to be stabilized so that it can be utilized much later in development (Baumgartner and Noll, 1990).

FIGURE 33

Transcription of the *Antennapedia* gene is initiated at either promoter (P1 or P2). The actual *Antennapedia* protein product is encoded by exons 5–8 (black) whereas the first four exons make a leader sequence. The homeodomain is encoded in the 3′ terminal exon. In addition, there are two places where poly(A) addition can terminate the transcript (double arrowheads). (After Frischer et al., 1986; Kaufman et al., 1990.)

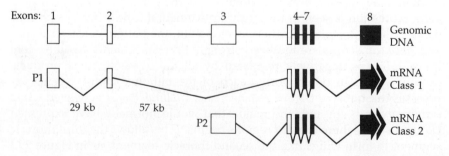

(A) Fusion of *CAT* gene and *Antp* promoter and transfection in *Drosophila* cells

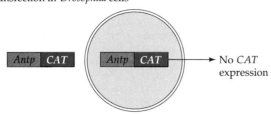

No *CAT* expression

(B) Addition of *ftz* on a strong constitutive promoter

Actin promoter

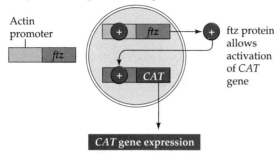

ftz protein allows activation of *CAT* gene

CAT gene expression

(C) Addition of *Ubx* suppresses *CAT* expression again

Actin promoter

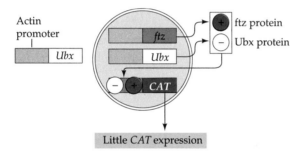

ftz protein
Ubx protein

Little *CAT* expression

FIGURE 34
Regulation of *Antennapedia* expression positively by *fushi tarazu* and negatively by *Ultrabithorax*. (A) Cultured *Drosophila* cells will not express the chloramphenicol acetyltransferase gene (*CAT*) when it is fused with an *Antp* promoter and transfected into the cell. (B) However, when an *ftz* gene is also transfected into these cells and allowed to express its product, the homeotic promoters are activated and the *CAT* gene is expressed. (C) When an active *Ubx* gene is also present, the *CAT* gene is not expressed.

these proteins have different, although overlapping, specificities (Figure 35). All of them will transform antennae into legs if they are expressed at the time of the second-to-third larval molt. However, in the early embryo, one of the Ubx proteins is important in determining the parasegmental identity of the peripheral nervous system, while other Ubx proteins are not (Mann and Hogness, 1990).

Another striking thing about the homeotic genes is that some of them have enormous introns. One of the several *Antennapedia* introns is about 57,000 base pairs long—over 10 times the combined length of all its exons! This extra length may be important in temporal regulation of the gene's expression. In *Drosophila* embryogenesis, events happen quickly. The gap, pair-rule, and homeotic selector genes in the embryo are each active for only about three hours. Since it is estimated that *Drosophila* genes are transcribed at a rate of 1000 nucleotides per minute at 25 degrees Centigrade (Ashburner, 1990), this huge *Antennapedia* intron adds nearly 1 hour to the time lag before the protein is expressed.

Each of the homeotic genes (and some other important developmental regulators such as *bicoid* and *fushi tarazu*) contain the homeobox. This 180-base pair sequence encodes a DNA-binding region, called the homeodomain, in the protein. The homeodomains appear to specify the binding sites for these proteins and are critical in specifying cell fate. For instance, if a chimeric protein is constructed mostly of Antennapedia but with the

FIGURE 35

Ultrabithorax mRNA structures generated by differential RNA processing. (A) *Ultrabithorax* gene. (B) The mRNAs that include two "microexons." The 5' splicing site of the first exon can vary within this group. (C) The mRNA family that includes one or no "microexons." The homeodomain of both sets of Ultrabithorax proteins is encoded in the 3' exon. (After Beachy, 1990.)

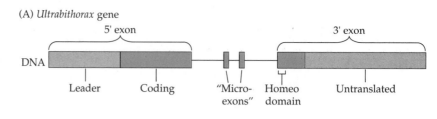

(A) *Ultrabithorax* gene

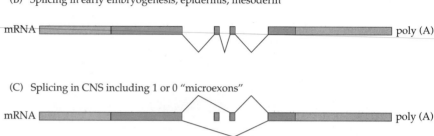

(B) Splicing in early embryogenesis, epidermis, mesoderm

(C) Splicing in CNS including 1 or 0 "microexons"

carboxy terminus (including the homeodomain) of Ultrabithorax, the protein can substitute for Ultrabithorax and specify the appropriate cells as parasegment 6 (Mann and Hogness, 1990). The isolated homeodomain of Antennapedia will bind to the same promoters as the entire Antennapedia protein, indicating that the binding of this protein is dependent upon its homeodomain (Müller et al., 1988). Many homeobox-containing genes have been found in other organisms (see Chapters 12 and 17), and their genetic organization is strikingly like that seen in *Drosophila*.

SIDELIGHTS & SPECULATIONS

Cis-*regulatory elements and the bithorax complex*

The three bithorax complex genes are obviously critical to the fly's development, but how can these three genes establish the segment identities of the the third thoracic segment and the ten abdominal segments? One possibility is that differential concentrations of the three proteins are expressed differently in each of the segments. Another model involves *cis*-regulator sites in these proteins that are segment-specific. This model was proposed on the basis of genetic studies by Lewis (1978, 1985). Lewis noted that when the entire bithorax complex was deleted, all remaining segments of the dying embryo resembled the second thoracic segment. He postulated that this segment was the "baseline" from which all the other segments would be modified.

Lewis then postulated that the bithorax complex should contain at least one gene for each segment below the second thoracic level. In other words, the development of T3 would involve the turning on of all the "ground-level" T2 genes plus the gene(s) responsible for T3 characteristics (Figure 36). A mutation in these T3 genes would return the segment to the T2 state. Similarly, the development of the first abdominal segment would require that the T2 and T3 genes be expressed in addition to the A1

gene. Mutations of the A1 gene would cause the segment to develop T3 structures. This activation of different genes in different segments could be explained by postulating a gradient of a repressor molecule that inhibits the expression of all these genes. The repressor would be concentrated in the anterior segments, where fewer genes are turned on, and would be at a low concentration in the posterior segments. Furthermore, the genes controlling the development of the most posterior segments would have a higher affinity for the repressor molecule. In the anterior segments, therefore, these genes would be more readily turned off. They would be active only in the posterior segments, where repressor concentration is lowest.

But how can Lewis's model work if there are only three protein-encoding genes within the bithorax complex? Lewis and his colleagues (Bender et al., 1983; Karch et al., 1985) have identified other regions of the bithorax complex that produce homeotic transformations. The *anterobithorax* (*abx*) and *bithorax* (*bx*) mutants cause the anterior compartment of the third thoracic segment (anterior balancers) to assume the identity of the anterior compartment of the *second* thoracic segment (anterior wings). Similarly, the *posterobithorax* (*pbx*) and *bithoraxoid* (*bxd*) mutants cause the posterior compartment of the third thoracic segment to resemble that of the second thoracic segment. The combination of *abx*, *pbx*, and *bxd* mutations in a single embryo

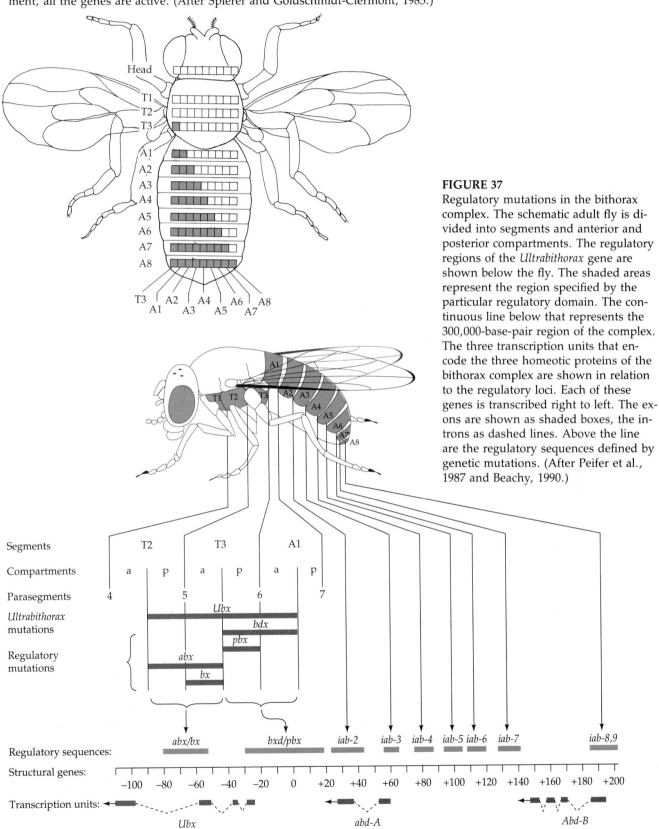

FIGURE 36
Representation of Lewis's model for gene regulation of the bithorax complex. The second thoracic segment is considered the "ground state." For each segment after the second thoracic segment, another gene gets activated. In the last segment, all the genes are active. (After Spierer and Goldschmidt-Clermont, 1985.)

FIGURE 37
Regulatory mutations in the bithorax complex. The schematic adult fly is divided into segments and anterior and posterior compartments. The regulatory regions of the *Ultrabithorax* gene are shown below the fly. The shaded areas represent the region specified by the particular regulatory domain. The continuous line below that represents the 300,000-base-pair region of the complex. The three transcription units that encode the three homeotic proteins of the bithorax complex are shown in relation to the regulatory loci. Each of these genes is transcribed right to left. The exons are shown as shaded boxes, the introns as dashed lines. Above the line are the regulatory sequences defined by genetic mutations. (After Peifer et al., 1987 and Beachy, 1990.)

causes the entire transformation of the third thoracic segment into another second thoracic segment. (The result is the fly shown earlier, in Figure 29.) Whereas these mutations were originally considered to be on separate genes, it now appears that they are mutations of enhancer elements, which enable the tissue-specific expression of the *Ubx* gene (Lewis, 1985; Peifer et al., 1987).

The relationship between the *cis*-regulatory mutations and the three transcription units of the bithorax complex is shown in Figure 37. The protein-encoding regions of the bithorax complex take up less than one-tenth of the DNA in this complex. The regulatory mutations generally map to the flanking regions of these three genes or to introns within the genes. Further evidence that *abx*, *bx*, and *bxd* are *cis*-regulatory elements comes from analyzing specific mutations and deletions. The deletion of the *Ubx* gene results in the homeotic transformation of parasegment 5 (posterior T2 and anterior T3) and parasegment 6 (posterior T3 and anterior A1) into copies of parasegment 4 (posterior T1 and anterior T2). Such a transformation is lethal; the embryo dies before hatching. In *abx* and *bx* mutants, however, only parasegment 5 is transformed into parasegment 4, as *Ubx* expression is reduced in parasegment 5 (Casanova et al., 1985; Peifer and Bender, 1986). Hence the anterior wing emerges in what would otherwise have been anterior haltere. Similarly, the *bxd* mutations reduce *Ubx* expression in parasegment 6 (Peifer et al., 1987). In the abdominal region, the *cis*-regulatory sequences *intraabdominal (iab) 2–9* direct the expression of *abdA* or *AbdB* in the various segments.

Compartmentalization in insect development

The *Drosophila* embryo is formed as a linear array of compartments. It is still not known what genes are activated (or repressed) by the homeotic selector genes, or how the "memory" of these divisions into segments and parasegments is stored in a manner that can be expressed in the adult fly after the larva undergoes metamorphosis. One of the first evidences of compartmentalization is seen shortly after cellularization and the beginnings of gastrulation. About an hour into the fourteenth division cycle, the blastoderm becomes divided into about 25 discrete mitotic domains (Figure 38). Each domain is duplicated bilaterally on the two sides of the embryo, and each domain enters mitosis at different times. The boundaries of these mitotic domains is determined by the expression stripes of engrailed protein. It is thought that the different mitotic domains give rise to functionally different units. Thus, engrailed protein appears to set the boundaries of the functional units of *Drosophila* organogenesis (Foe and Odell, 1989).

These embryonic compartmental boundaries can define specific functional units of adults as well. As we shall see, engrailed protein is also thought to be critical in this specification. This can be shown by inducing genetic changes in an individual cell. The mutant cell is thereby marked as being different from its neighbors, and one can observe where the descendants of this cell are located. In *Drosophila* embryos, such mitotic recombination events can be induced by radiation, and the resulting cells

FIGURE 38

Mitotic domains of *Drosophila* embryos in the fourteenth division cycle. The staining for β-tubulin causes mitotic figures to appear white and interphase cells to appear gray. (A) Tipped dorsal view of an 80-minute embryo showing that domains 1, 3, 4, 5, and 6 are in mitosis. (B) Ventrolateral view of 90-minute embryo wherein domains 9, 11, 14, and 16 have entered mitosis. Domain 2 is in interphase of cycle 15, having already completed mitotic division 14. CF is the cephalic fold. (From Foe and Odell, 1989; photographs courtesy of V. E. Foe.)

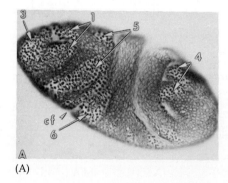
(A)

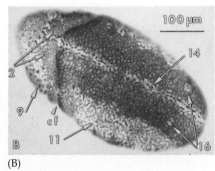
(B)

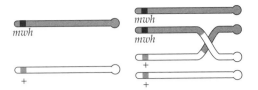
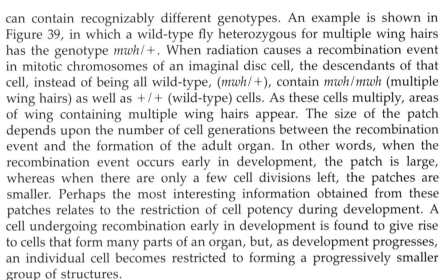

Mutant cells that
can produce
multiple wing hairs

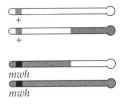

Wild-type wing

Area of multiple
wing hairs

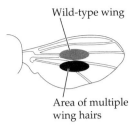

FIGURE 39
Generation of a genetically marked clone of cells by somatic crossing over. Larval *Drosophila* heterozygous for the recessive mutation *multiple wing hair* (*mwh*) have normal wing bristles. In the event of a radiation-induced crossover, daughter cells can get two wild-type chromosomes (+/+) or two chromosomes having *mwh* genes (*mwh*/ *mwh*). Because there is no mechanism for homologous pairing as there is in meiosis, such mitotic recombination is a very rare event. The homozygous wild-type cells are indistinguishable from the heterozygous background, but those wing discs having cells with two *mwh* genes produce wing tissue with mutant bristles.

can contain recognizably different genotypes. An example is shown in Figure 39, in which a wild-type fly heterozygous for multiple wing hairs has the genotype *mwh*/+. When radiation causes a recombination event in mitotic chromosomes of an imaginal disc cell, the descendants of that cell, instead of being all wild-type, (*mwh*/+), contain *mwh*/*mwh* (multiple wing hairs) as well as +/+ (wild-type) cells. As these cells multiply, areas of wing containing multiple wing hairs appear. The size of the patch depends upon the number of cell generations between the recombination event and the formation of the adult organ. In other words, when the recombination event occurs early in development, the patch is large, whereas when there are only a few cell divisions left, the patches are smaller. Perhaps the most interesting information obtained from these patches relates to the restriction of cell potency during development. A cell undergoing recombination early in development is found to give rise to cells that form many parts of an organ, but, as development progresses, an individual cell becomes restricted to forming a progressively smaller group of structures.

This type of experiment has uncovered a remarkable phenomenon: the progeny of cells in the larval imaginal discs do not cross certain distinct boundaries. Rather, determination appears to occur within discrete compartments that do not overlap. In the development of an imaginal disc, groups of cells are determined together. These groups of cells are called POLYCLONES because they are not related by ancestry but are merely a cluster of neighboring cells. Just after blastoderm formation, the 10–50 cells that will constitute the mesothorax imaginal disc are already divided into two polyclones (Figure 40). One of these polyclones generates the anterior portion of the wing and most of the dorsal thorax, whereas the other polyclone produces the posterior part of the wing and the remaining dorsal thorax. A mutant clone will spread in one of these compartments but will not spread into the other. The next determinative event separates the wing disc from the leg disc. The wing disc so derived keeps the anterior–posterior boundaries made earlier. The third determination is the distinction between dorsal and ventral surfaces of the wing, followed by the separation of those central elements (which form the wing proper) from those cells that generate the wing blade. Thus, there is a sequential division of the imaginal disc into the series of compartments, and the compartmental boundaries reflect the commitment of adjacent cells to different fates (Crick and Lawrence, 1975; Garcia-Bellido, 1975; Kauffman et al., 1978). In this way, compartments are established very early in development and are maintained through adult life.

Figure 41 shows the imaginal disc of the wing and the compartmental boundaries that give rise to the adult structure. When the crossover patches were analyzed, they were found not to cross the anterior–posterior border. That is, the homozygous mutant patches would remain on either the anterior or the posterior side of the wing and not venture into the

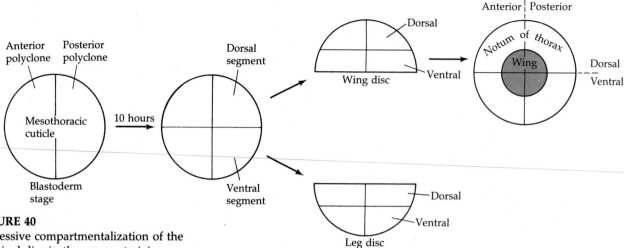

FIGURE 40

Successive compartmentalization of the imaginal disc in the segment giving rise to the wing. The mesothoracic cuticle of that segment is divided into anterior and posterior polyclones at the blastoderm stage. The separation of the dorsal segment (which becomes the wing disc) from the ventral segment (leg disc) occurs by the first 10 hours of development. The wing and leg discs then physically separate; each retains the original anterior–posterior boundary. This boundary is bisected by another division of the respective discs into dorsal and ventral compartments. Each of these compartments of the wing disc is further subdivided into polyclones giving rise to the wing (central) and to the notum of the thorax (peripheral). (After Morata and Lawrence, 1977.)

other. A clone will proceed to the border and then stop. Figure 41B shows a large clone, the posterior border of which runs along the anterior–posterior boundary of the wing. Something must exist that informs the cells of their proper spatial arrangement. That "something" may be the product of the *engrailed* gene. In addition to being involved in segmentation, the *engrailed* gene is also active in the posterior polyclone of each wing imaginal disc (Figure 42A). Posterior compartment cells, induced by somatic crossing over to contain two mutant engrailed genes, fail to respect the anterior–posterior compartment border and invade the space usually "reserved" for the anterior portion of the wing.

It is not known how these boundaries are created, but the formation

FIGURE 41

Compartmental boundaries of the wing disc. (A) Projection of the five compartmental boundaries onto the fate map of the wing imaginal disc. Line 1 separates anterior and posterior compartments. Line 2 separates the dorsal wing surface from the ventral wing surface. Line 3 separates the lateral thorax from the dorsal hinge region (line 3a) and the notal portion of the wing blade from the wing proper (line 3b). Line 4 divides the notum into the scutellum and scutum regions (line 4a). (Another line 4 is postulated to exist in the lateral thorax region; line 4b.) Line 5 separates the free wing surfaces from the hinge regions. (B) Projection of the disc compartmental boundaries along the surface of an adult wing. The color-shaded region represents a large clone of genetically marked cells whose posterior border runs along the anterior–posterior border but does not cross the boundary. (After Lawrence and Morata, 1976.)

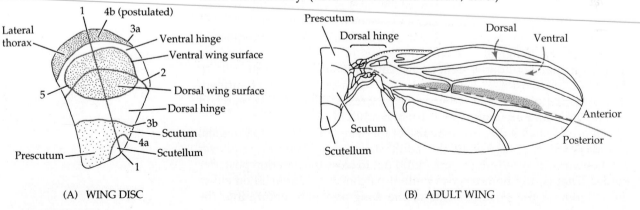

(A) WING DISC

(B) ADULT WING

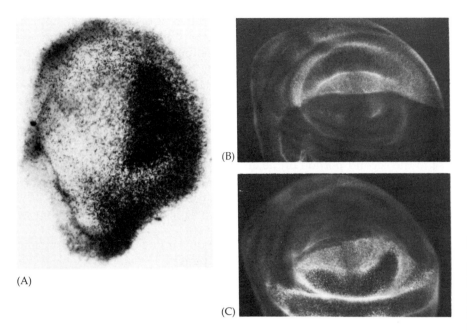

(A)

(B)

(C)

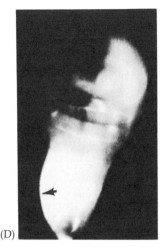

(D)

FIGURE 42

Compartments in imaginal discs. (A) Localization of *engrailed* transcripts by in situ hybridization with radioactive cDNA from the *engrailed* gene. The expression of *engrailed* genes is only seen in those cells generating the posterior compartment. (B) Monoclonal antibodies to the the PS1 antigen localizes this protein in the dorsal (lower) half of the disc, while the PS2 antigen (C) can be seen on the ventral (upper) half of the disc. In both cases, a band of lower intensity can also be seen bisecting the disc along the anterior (left)–posterior (right) border. (D) Fluorescent image of a wing disc 52 minutes after injection of Lucifer Yellow into the notal region (arrow). Dye has moved into the anterior half of the wing region and through the notal cells. Movement into the posterior half has been restricted. (A From Kornberg et al., 1985; B and C from Brower et al., 1985; D from Weir and Lo, 1982; photographs courtesy of the authors.)

of one of the major compartmental boundaries has been correlated with the restriction of integrin proteins to particular cells of the wing imaginal disc. The dorsal–ventral compartmentalization is seen to occur in the wing discs of the third instar larva. At this time, certain cell surface compounds are expressed differentially on the dorsal or ventral cells (Figure 42B,C; Brower et al., 1985). The formation of compartmental boundaries has also been correlated with the borders of sets of communicating cells. These studies involve the injection of a low-molecular-weight fluorescent dye (Lucifer Yellow) into some of the cells of an imaginal disc. This dye can pass through gap junctions and enter all the cells that are so coupled. Weir and Lo (1982) showed that the primary boundaries of dye transfer coincide with the anterior and posterior compartments (Figure 42D). Later the communication of dye becomes restricted to smaller groups of cells, paralleling the formation of other compartmental borders. Thus, restrictions on gap junction communication may play a role in forming or maintaining these boundaries.

The specification of cells: Homologous structures

The final step in the realization of the *Drosophila* phenotype involves the activation of specific batteries of genes (presumably by the homeotic selector proteins). This is an exceedingly complicated event that involves two processes. One is differentiation, wherein cells are instructed as to what proteins to make. But before differentiation occurs, the cells must first be specified as to their position. For instance, is this particular group of cells in the second thoracic segment going to be near the beginning of a leg (such as the coxa or trochanter structures) or at the end of the leg (tarsal or claw structures)? The types of cells in the structures may be the same, but their placement is different.

It appears that the specification of the cells—the commitment to form a particular part of the structure within a morphogenetic field—is quite independent of the determination of the cells. For example, the determination of each imaginal disc is different. Wing discs give rise to wings, leg discs to legs. The specifications of the different discs, however, appear to be identical. This can be seen with certain homeotic mutants such as *Antennapedia*, wherein antennal structures are transformed into those of the legs (Postlethwait and Schneiderman, 1971). Occasionally, the entire antenna becomes an entire leg, but it is more common that only a portion of the antenna becomes leglike. In these latter cases, the replacement is absolutely position-specific. The cells of the antennal disc that normally would have formed the distal tip of the antenna (arista) are transformed into the most distal portion of the leg (claw); those cells specified to give rise to the second portion of the antenna are transformed into the second portion (trochanter) of the leg. The corresponding parts of the two structures are shown in Figure 43. It is apparent, then, that the two differently determined discs must use a common mechanism for the specification of cell fates within the respective discs.

We are beginning to learn how the genome influences the construction of the organism. The genes regulating pattern formation in *Drosophila* operate according to certain principles. First, there is a temporal order wherein the different classes of genes are transcribed. Second, the genes are transcribed in a precise spatial arrangement that is brought about by regulatory interactions with previously acting genes (and gene products) and with other genes of the same class. Third, the spatial patterns of gene expression provide a unique identity to every cell along the anterior–posterior axis. This constellation of gene products is thought to specify the identity of each cell. The mechanisms by which the segmentation genes interpret the constellation of maternal products and specify the activity of homeotic genes, and the mechanisms wherein the homeotic selector genes regulate each other and activate the genes characteristic of particular cell types, are major areas of current investigation.

We are far from a complete understanding of *Drosophila* pattern formation, but we are much more aware of its complexity than we were five years ago. The mutations of *Drosophila* have given us our first glimpses of the multiple levels of pattern regulation in a complex organism and have enabled the isolation of these genes and their gene products. The next decade should further enhance our appreciation of the fruit fly and its development.

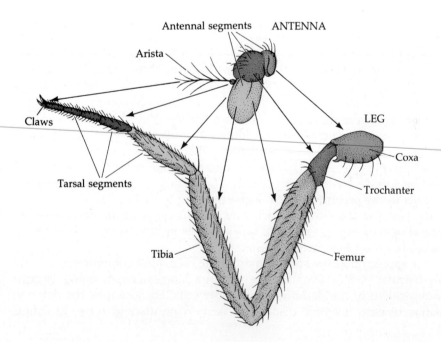

FIGURE 43
Correspondence between portions of the antenna and portions of the leg. In the mutant *Antennapedia*, regions of the antenna are transformed into leg structures. The arrows show the portions of the antenna that form specific corresponding portions of the leg. (After Postlethwait and Schneiderman, 1971.)

LITERATURE CITED

Akam, M. E. 1987. The molecular basis for metameric pattern in the *Drosophila* embryo. *Development* 101: 1–22.

Anderson, K. V. 1989. *Drosophila*: The maternal contribution. *In* D. M. Glover and B. D. Hames (eds.), *Genes and Embryos*. IRL, New York, pp 1–37.

Ashburner, M. 1990. Puffs, genes, and hormones revisited. *Cell* 61: 1–3.

Bateson, W. 1894. *Materials for the Study of Variation*. Macmillan, London.

Baumgartner, S. and Noll, M. 1990. Networks of interaction among pair-rule genes regulating *paired* expression during primordial segmentation of *Drosophila*. *Mech. Devel.* 1: 1–18.

Beachy, P. A. 1990. A molecular view of the *Ultrabithorax* homeotic gene of *Drosophila*. *Trends Genet.* 6(2): 46–51.

Beachy, P. A., Krasnow, M. A., Gavis, E. R. and Hogness, D. S. 1988. An *Ultrabithorax* protein binds sequences near its own and the *Antennapedia* P1 promoters. *Cell* 55: 1069–1081.

Berleth, T. and several others. 1988. The role of localization of bicoid RNA in organizing the anterior pattern of the *Drosophila* embryo. *EMBO J.* 7: 1749–1756.

Bermingham, J. R. Jr., Martinez-Arias, A., Pettit, M. G. and Scott, M. 1990. Different patterns of transcription from the two *Antennapedia* promoters during *Drosophila* embryogenesis. *Development* 109: 553–566.

Boulet, A. M. and Scott, M. P. 1988. Control elements of the P2 promoter of the *Antennapedia* gene. *Genes Dev.* 2: 1600–1614.

Brower, D. L., Piovant, M. and Reger, L. A. 1985. Developmental analysis of *Drosophila* position-specific antigens. *Dev. Biol.* 108: 120–130.

Carroll, S. B. and Vavra, S. H. 1989. The zygotic control of pair-rule gene expression. II. Spatial repression by gap and pair-rule gene products. *Development* 107: 673–683.

Carroll, S. B., Laughton, A. and Thalley, B. S. 1988. Expression, function, and regulation of the hairy segmentation protein in the *Drosophila* embryo. *Genes Dev.* 2: 883–890.

Casanova, J., Sanchez-Herrero, E. and Morata, G. 1985. Prothoracic transformation and functional structure of the *Ultrabithorax* gene of *Drosophila*. *Cell* 42: 663–669.

Casanova, J., Sanchez-Herrero, E., Busturia, A. and Morata, G. 1987. Double and triple mutant combination of the bithorax complex of *Drosophila*. *EMBO J.* 6: 3103–3109.

Crick, F. H. C. and Lawrence, P. A. 1975. Compartments and polyclones in insect development. *Science* 189: 340–347.

Degelmann, A., Hardy, P. A., Perrimon, N. and Mahowald, A. P. 1986. Developmental analysis of the *torso*-like phenotype in *Drosophila* produced by a maternal-effect locus. *Dev. Biol.* 115: 479–489.

Desplan, C., Theis, J. and O'Farrell, P. H. 1985. A *Drosophila* developmental gene, *engrailed*, encodes a sequence-specific DNA binding activity. *Nature* 318: 630–635.

Driever, W. and Nüsslein-Volhard, C. 1988a. A gradient of *bicoid* protein in *Drosophila* embryos. *Cell* 54: 83–93.

Driever, W. and Nüsslein-Volhard, C. 1988b. The *bicoid* protein determines position in the *Drosophila* embryo in a concentration-dependent manner. *Cell* 54: 95–104.

Driever, W. and Nüsslein-Volhard, C. 1989. The *bicoid* protein is a positive regulator of *hunchback* transcription in the early *Drosophila* embryo. *Nature* 337: 138–143.

Driever, W., Thoma, G. and Nüsslein-Volhard, C. 1989. Determination of spatial domains of zygotic gene expression in the *Drosophila* embryo by the affinity of binding sites for the *bicoid* morphogen. *Nature* 340: 363–367.

Driever, W., Siegel, V. and Nüsslein-Volhard, C. 1990. Autonomous determination of anterior structures in the early *Drosophila* embryo by the *bicoid* morphogen. *Development*. 109: 811–820.

Edgar, B. A. and Schubiger, G. 1986. Parameters controlling transcriptional activation during early *Drosophila* development. *Cell* 44: 871–877.

Foe, V. E. and Alberts, B. M. 1983. Studies of nuclear and cytoplasmic behavior during the five mitotic cycles that precede gastrulation in *Drosophila* embryogenesis. *J. Cell Sci.* 61: 31–70.

Foe, V. E. and Odell, G. M. 1989. Mitotic domains partition fly embryos, reflecting early biological consequences of determination in progress. *Amer. Zool.* 29: 617–652.

French, V. 1988. Gradients and insect segmentation. *In* V. French, P. Ingham, J. Cooke and J. Smith (eds.), *Mechanisms of Segmentation*. Company of Biologists, Cambridge, pp. 3–16.

Frigerio, G., Burri, M., Bopp, D., Baumgartner, S. and Noll, M. 1986. Structure of the segmentation gene *paired* and the *Drosophila* PRD gene set as part of a gene network. *Cell* 47: 735-746.

Frischer, L. E., Hagen, F. S. and Garber, R. L. 1986. An inversion that disrupts the *Antennapedia* gene causes abnormal structure and localization of RNAs. *Cell* 47: 1017–1023.

Garcia-Bellido, A. 1975. Genetic control of wing development in *Drosophila*. *Ciba Found. Symp.* 29: 161–182.

Gaul, U. and Jäckle, H. 1990. Role of gap genes in early *Drosophila* development. *Annu. Rev. Genetics* 27: 239–275.

Gehring, W. J. 1985. Homeotic genes, the homeobox, and the genetic control of development. *Cold Spring Harbor Symp. Quant. Biol.* 50: 243–251.

Goldschmidt, R. B. 1938. *Physiological Genetics*. McGraw-Hill, New York.

Hafen, E., Levine, M. and Gehring, W. J. 1984. Regulation of *Antennapedia* transcript distribution by the bithorax complex in *Drosophila*. *Nature* 307: 287–289.

Harding, K. and Levine, M. 1988. Gap genes define the limits of *Antennapedia* and *Bithorax* gene expression during early development in *Drosophila*. *EMBO J.* 7: 205–214.

Harding, K. and Levine, M. 1989. *Drosophila*: The zygotic contribution. *In* D. M. Glover and B. D. Hames (eds.), *Genes and Embryos*. IRL, New York, pp. 38–90.

Harding, K., Wedeen, C., McGinnis, W. and Levine, M. 1985. Spatially regulated expression of homeotic genes in *Drosophila*. *Science* 229: 1236–1242.

Hoey, T. and Levine, M. 1988. Divergent homeo box proteins recognize similar DNA sequences in *Drosophila*. *Nature* 332: 858–861.

Howard, K. and Ingham, P. 1986. Regulatory interactions between the segmentation genes *fushi tarazu*, *hairy*, and *engrailed* in the *Drosophila* blastoderm. *Cell* 44: 949–957.

Howard, K., Ingham, P. and Rushlow, C. 1988. Region-specific alleles of the *Drosophila* segmentation gene *hairy*. *Genes Dev.* 2: 1037–1046.

Hülskamp, M., Schröder, C., Pfeifle, C., Jäckle, H. and Tautz, D. 1989. Posterior segmentation of the *Drosophila* embryo in the absence of a maternal posterior organizer gene. *Nature* 338: 629–632.

Ingham, P. W. and Martinez-Arias, A. 1986. The correct activation of *Antennapedia* and bithorax complex genes requires the *fushi tarazu* gene. *Nature* 324: 592–597.

Irish, V., Lehmann, R. and Akam, M. 1989. The *Drosophila* posterior-group gene *nanos* functions by repressing hunchback activity. *Nature* 338: 646–648.

Irish, V. F., Martinez-Arias, A. and Akam, M. 1989. Spatial regulation of the *Antennapedia* and *Ultrabithorax* genes during *Drosophila* early development. *EMBO J.* 8: 1527–1537.

Jack, T., Regulski, M. and McGinnis, W. 1988. Pair-rule segmentation genes regulate the expression of the homeotic selector gene, *Deformed*. *Genes Dev.* 2: 592–597.

Jäckle, H., Tautz, D., Schuh, R., Seifert, E. and Lehmann, R. 1986. Cross-regulatory interactions among the gap genes of *Drosophila*. *Nature* 324: 668–670.

Just, E. E. 1939. *The Biology of the Cell Surface*. Blakiston, Philadelphia.

Kalthoff, K. 1969. Der Einfluss verscheiedener Versuchparameter auf die Häufigkeit der Missbildung "Doppelabdomen" in UV-bestrahlten Eiern von *Smittia* sp. (Diptera, Chironomidae). *Zool. Anz. Suppl.* 33: 59–65.

Kalthoff, K. and Sander, K. 1968. Der Enwicklungsgang der Missbildung "Doppelabdomen" im partiell UV-bestrahlten Ei von *Smittia parthenogenetica* (Diptera, Chironomidae). *Wilh. Roux Arch. Entwick. Org.* 161: 129–146.

Kandler-Singer, I. and Kalthoff, K. 1976. RNase sensitivity of an anterior morphogenetic determinant in an insect egg (*Smittia* sp., Chironomidae, Diptera. *Proc. Natl. Acad. Sci. USA* 73: 3739–3743.

Karch, F. and 7 others. 1985. The abdominal region of the bithorax complex. *Cell* 43: 81–96.

Karr, T. L. and Kornberg, T. B. 1989. *fushi tarazu* protein expression in the cellular blastoderm of *Drosophila* detected using a novel imaging technique. *Development* 105: 95–103.

Kauffman, S. A., Shymko, R. M. and Trabert, K. 1978. Control of sequential compartment formation in *Drosophila*. *Science* 199: 259–270.

Kaufman, T. C., Seeger, M. A. and Olsen, G. 1990. Molecular and genetic organization of the Antennapedia gene complex of *Drosophila melanogaster*. *Adv. Genet.* 27: 309–362.

Klingler, M., Erdélyi, M., Szabad, J. and Nüsslein-Volhard, C. 1988. Function of *torso* in determining the terminal anlagen of the *Drosophila* embryo. *Nature* 335: 275–277.

Knipple, D. C., Seifert, E., Rosenberg, U. B., Preiss, A. and Jäckle, H. 1985. Spatial and temporal patterns of *Krüppel* gene expression in early *Drosophila* embryos. *Nature* 317: 40–44.

Kornberg, T., Sidén, I., O'Farrell, P. and Simon, M. 1985. The *engrailed* locus of *Drosophila*: In situ localization of transcripts reveals compartment-specific expression. *Cell* 40: 45–53.

Krasnow, M. A., Saffman, E. E., Kornfeld, K. and Hogness, D. S. 1989. Transcriptional activation and repression by *Ultrabithorax* proteins in cultured *Drosophila* cells. *Cell* 57: 1031–1043.

Lawrence, P. A. and Morata, G. 1976. Compartments in the wing of *Drosophila*: A study of the *engrailed* gene. *Dev. Biol.* 50: 321–337.

Lehmann, R. and Nüsslein-Volhard, C. 1986. Abdominal segmentation, pole cell formation and embryonic polarity require the localized activity of *oskar*, a maternal gene in *Drosophila*. *Cell* 47: 141–152.

Lehmann, R. and Nüsslein-Volhard, C. 1987. Involvement of the *pumilio* gene in the transport of the abdominal signal in the *Drosophila* embryo. *Nature* 329: 167–170.

Levine, M. S. and Harding, K. W. 1989. *Drosophila*: The zygotic contribution. *In* D. M. Glover and B. D. Hames (eds.), *Genes and Embryos*. IRL, New York, pp. 39–94.

Lewis, E. B. 1978. A gene complex controlling segmentation in *Drosophila*. *Nature* 276: 565–570.

Lewis, E. B. 1985. Regulation of the genes of the bithorax complex in *Drosophila*. *Cold Spring Harbor Symp. Quant. Biol.* 50: 155–164.

MacDonald, P. M. and Struhl, G. 1986. A molecular gradient in early *Drosophila* embryos and its role in specifying the body pattern. *Nature* 324: 537–545.

Mange, A. P. and Mange, E. J. 1990. *Genetics: Human Aspects*. Sinauer Associates, Sunderland, MA.

Mann, R. S. and Hogness, D. S. 1990. Functional dissection of *Ultrabithorax* proteins in *D. melanogaster*. *Cell* 60: 597–610.

Martinez-Arias, A. and Lawrence, P. A. 1985. Parasegments and compartments in the *Drosophila* embryo. *Nature* 313: 639–642.

Mayr, E. 1982. *The Growth of Biological Thought: Diversity, Evolution, and Inheritance*. Harvard University Press, Cambridge, MA.

Mlodzik, M. and Gehring, W. J. 1987. Expression of the *caudal* gene in the germ line of *Drosophila*: Formation of an RNA and protein gradient during early embryogenesis. *Cell* 48: 465–478.

Morata, G. and Lawrence, P. A. 1977. Homeotic genes control compartments and cell determination in *Drosophila*. *Nature* 265: 211–216.

Müller, M. Affolter, M., Leupin, W., Otting, G., Wüthrich, K. and Gehring, W. J. 1988. Isolation and sequence-specific DNA binding of the *Antennapedia* homeodomain. *EMBO J.* 7: 4299–4304.

Nüsslein-Volhard, C. and Wieschaus, E. 1980. Mutations affecting segment number and polarity in *Drosophila*. *Nature* 287: 795–801.

Nüsslein-Volhard, C., Frohnhöfer, H. G. and Lehmann, R. 1987. Determination of anterioposterior polarity in *Drosophila*. *Science* 238: 1675–1681.

O'Farrell, P. H. and 7 others. 1985. Embryonic pattern in *Drosophila*: The spatial distribution and sequence-specific DNA binding of *engrailed* protein. *Cold Spring Harbor Symp. Quant. Biol.* 50: 235–242.

Pankratz, M. J., Seifert, E., Gerwin, N., Billi, B., Nauber, U. and Jäckle, H. 1990. Gradients of *Krüppel* and *knirps* gene products direct pair-rule gene stripe patterning in the posterior region of the *Drosophila* embryo. *Cell* 61: 309–317.

Peifer, M. and Bender, W. 1986. The *anterobithorax* and *bithorax* mutations of the bithorax complex. *EMBO J.* 5: 2293–2303.

Peifer, M., Karch, F. and Bender, W. 1987. The bithorax complex: Control of segmental identity. *Genes Dev.* 1: 891–898.

Postlethwait, J. H. and Schneiderman, H. A. 1971. Pattern formation and determination in the antenna of the homeotic mutant *Antennapedia* of *Drosophila melanogaster*. *Dev. Biol.* 25: 606–640.

Raff, R. and Kaufman, T. C. 1983. *Embryos, Genes and Evolution: The Developmental–Genetic Basis of Evolutionary Change*. Macmillan, New York.

Sánchez-Herrero, E., Vernós, I., Marco, R. and Morata, G. 1985. Genetic organization of *Drosophila* bithorax complex. *Nature* 313: 108–113.

Sander, K. 1960. Analyse des ooplasmatischen Reaktionssystems von Euscelis plebajus Fall (Circadina) durch Isolieren und Kombinieren von Keimteilen. II. Die Differenzierungsleistungen nach Verlagern von Hinterpolmaterial. *Wilhelm Roux Arch. Entwicklungsmech. Org.* 151: 660–707.

Sander, K. 1975. Pattern specification in the insect embryo. In *Cell Patterning*. *CIBA Foundation Symp.* 29: 241–263.

Schneuwly, S., Kuroiwa, A. and Gehring, W. J. 1987. Molecular analysis of the dominant homeotic *Antennapedia* phenotype. *EMBO J.* 6: 201–206.

Schüpbach, T. and Wieschaus, E. 1986. Maternal effect mutations altering the anterior–posterior pattern of the *Drosophila* embryo. *Roux Arch. Dev. Biol.* 195: 302–317.

Scott, M. P. and Carroll, S. B. 1987. The segmentation and homeotic gene network in early *Drosophila* development. *Cell* 51: 689–698.

Spierer, P. and Goldschmidt-Clermont, M. 1985. La génétique du développement de la mouche. *La Recherche* 16: 452–461.

Štanojević, D., Hoey, T. and Levine, M. 1989. Sequence-specific DNA-binding activities of the gap proteins encoded by *hunchback* and *Krüppel* in *Drosophila*. *Nature* 341: 331–335.

Stevens, L. M., Frohnhöfer, H. G., Klingler, M. and Nüsslein-Volhard, C. 1990. Localized requirement for *torso*-like expression in follicle cells for development of terminal anlagen of the *Drosophila* embryo. *Nature* 346: 660–662.

Struhl, G. 1981. A homeotic mutation transforming leg to antenna in *Drosophila*. *Nature* 292: 635–638.

Struhl, G. 1989. Differing strategies for organizing anterior and posterior body pattern in *Drosophila* embryos. *Nature* 338: 741–744.

Struhl, G., Struhl, K. and Macdonald, P. M. 1989. The gradient morphogen *bicoid* is a concentration-dependent transcriptional activator. *Cell* 57: 1259-1273.

Tautz, D. 1988. Regulation of the *Drosophila* segmentation gene *hunchback* by two maternal morphogenetic centers. *Nature* 332: 281–284.

Treisman, J. and Desplan, C. 1989. The products of the *Drosophila* gap genes *hunchback* and Krüppel bind to the *hunchback* promoters. *Nature* 341: 335–336.

Turner, F. R. and Mahowald, A. P. 1977. Scanning electron microscopy of *Drosophila melanogaster* embryogenesis. *Dev. Biol.* 57: 403–416.

Waddington, C. H. 1940. *Organisms and Genes*. Cambridge University Press, Cambridge.

Wakimoto, B. T., Turner, F. R. and Kaufman, T. C. 1984. Defects in embryogenesis in mutants associated with the *Antennapedia* gene complex of *Drosophila melanogaster*. *Dev. Biol.* 102: 147–172.

Weigel, D., Jürgens, G., Klingler, M. and Jäckle, H. 1990. Two gap genes mediate maternal terminal information in *Drosophila*. *Science* 248: 495–498.

Weir, M. P. and Lo, C. 1982. Gap junction communication compartments in the *Drosophila* wing disc. *Proc. Natl. Acad. Sci. USA* 79: 3232–3235.

Wharton, R. P. and Struhl, G. 1989. Structure of the *Drosophila* BicaudalD protein and its role in localizing the posterior determinant nanos. *Cell* 59: 881–892.

19

Cell interactions at a distance: Hormones as mediators of development

The old order changeth, yielding place to the new.

—ALFRED LORD TENNYSON (1886)

The earth-bound early stages built enormous digestive tracts and hauled them around on caterpillar treads. Later in the life-history these assets could be liquidated and reinvested in the construction of an entirely new organism—a flying-machine devoted to sex.

—CARROLL M. WILLIAMS (1958)

Organ formation in animals is accomplished by the interactions of numerous cell types. In the preceding four chapters, we have seen how developmental interactions can be mediated by adjacent cell populations. In this chapter, we shall discuss the regulation of development by diffusible molecules that travel long distances from one cell type to another. Diffusible regulators of development that travel through the blood to cause changes in the differentiation or morphogenesis of other tissues are called HORMONES.

Metamorphosis: The hormonal redirecting of development

Because minute quantities of hormones are enough to accomplish their actions, it is exceedingly difficult to isolate them from embryos. The most thorough analysis of the hormonal control of development has therefore centered upon the dramatic "reprogramming" of development known as METAMORPHOSIS.

In many species of animals, embryonic development leads to a larval stage with characteristics very different from those of the adult organism. Very often, the larval forms are specialized for some function, such as growth or dispersal. The pluteus larva of the sea urchin, for instance, can travel on ocean currents, whereas the adult urchin leads a sedentary existence. The caterpillar larvae of butterflies and moths are specialized for feeding, whereas their adult forms are specialized for flight and reproduction and often lack the mouth parts necessary for eating. The division of functions between larva and adult is often remarkably distinct (Wald, 1981). Mayflies hatch from eggs and develop for several months. All this development enables them to spend one day as fully developed winged insects, mating quickly before they die. As might be expected from the above discussion, the larval form and the adult form often live in different environments. The adult viceroy butterfly mimics the unpalatable monarch butterfly, but the viceroy caterpillar does not resemble the beautiful larva of the monarch. Rather, the viceroy larva escapes detection by resembling bird droppings (Begon et al., 1986).

During metamorphosis, developmental processes are reactivated by specific hormones, and the entire organism changes to prepare itself for its new mode of existence. These changes are not solely ones of form. In amphibian tadpoles, metamorphosis causes the developmental maturation of liver enzymes, hemoglobin, and eye pigments, as well as the remodeling of the nervous, digestive, and reproductive systems. Thus, metamorphosis is a time of dramatic developmental change affecting the entire organism.

In this chapter, we will focus on three cases in which hormones reactivate the developmental processes after birth: amphibian metamorphosis, insect metamorphosis, and mouse breast development.

Amphibian metamorphosis

The phenomenon of amphibian metamorphosis

In amphibians, metamorphosis is generally associated with those changes that prepare an aquatic organism for a terrestrial existence. In urodeles (salamanders), the changes include the resorption of the tail fin, the destruction of the external gills, and the change of skin structure. In anurans (frogs and toads), the metamorphic changes are most striking, and almost every organ is subject to modification (Table 1). The changes in form are very obvious (Figure 1). Regressive changes include the loss of the tadpole's horny teeth and internal gills, as well as the destruction of the tadpole's tail. At the same time, constructive processes such as limb development and dermoid gland construction are also evident. For locomotion, the paddle tail recedes while the hindlimbs and forelimbs differentiate. The horny teeth constructed for tearing pond plants disappear as the mouth and jaw take a new shape and the tongue muscle develops. Meanwhile the large intestine characteristic of herbivores shortens to suit the more carnivorous diet of the adult frog. The gills regress and the gill arches degenerate. The lungs enlarge, and muscles and cartilage develop for pumping air in and out of the lungs. The sensory apparatus changes, too, as the lateral line system of the tadpole degenerates and the eye and ear undergo further differentiation. In the ear, the middle ear develops, as does the tympanic membrane so characteristic of frogs and toads. In the eye, both nictitating membrane and eyelids emerge. Moreover, the eye pigment changes. In tadpoles, as in freshwater fishes, the major retinal photopigment is PORPHYROPSIN, a complex between the protein opsin and the aldehyde of vitamin A_2 (Figure 2). In adult frogs, the pigment changes to RHODOPSIN, the characteristic photopigment of terrestrial and marine vertebrates. Rhodopsin consists of opsin conjugated to the aldehyde of vitamin A_1 (Wald, 1945, 1981; Smith-Gill and Carver, 1981).

Other biochemical events are also associated with metamorphosis.

FIGURE 1
Sequence of metamorphosis in the frog *Rana pipiens*. (A) Premetamorphic tadpole; (B) prometamorphic tadpole showing hindlimb growth; (C) onset of metamorphic climax as forelimbs emerge; (D,E) climax stages.

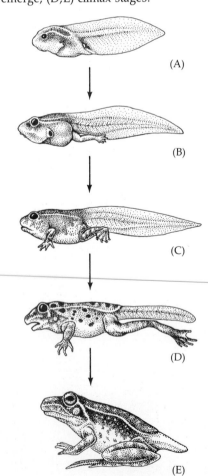

(A)

(B)

(C)

(D)

(E)

TABLE 1
Summary of some metamorphic changes in anurans

System	Larva	Adult
Locomotory	Aquatic; tail fins	Terrestrial; tailless tetrapod
Respiratory	Gills, skin, lungs; larval hemoglobins	Skin, lungs; adult hemoglobins
Circulatory	Aortic arches; aorta; anterior, posterior, and common cardinal veins	Carotid arch; systemic arch; jugular veins
Nutritional	*Herbivorous*: Long spiral gut— intestinal symbionts; small mouth—horny jaws, labial teeth	*Carnivorous*: Short gut— proteases; large mouth— long tongue
Nervous	Lack of nictitating membrane, porphyropsin, lateral line system—Mauthner's neurons	Development of ocular muscles, nictitating membrane, rhodopsin, loss of lateral line system—degeneration of Mauthner's neurons; tympanic membrane
Excretory	Mesonephros: Largely ammonia, some urea (ammonotelic)	Mesonephros: Largely urea, high activity of enzymes of ornithine–urea cycle (ureotelic)

Source: Turner and Bagnara (1976).

Tadpole hemoglobin binds oxygen faster and releases it more slowly than adult hemoglobin (McCutcheon, 1936). Moreover, Riggs (1951) showed that the binding of oxygen by tadpole hemoglobin is independent of pH, whereas frog hemoglobin (as in most other vertebrate hemoglobins) shows increased oxygen binding as the pH rises (Bohr effect). Probably the most striking biochemical change in metamorphosis is the induction of those enzymes necessary for the production of urea. Tadpoles, like most fresh-water fishes, are AMMONOTELIC; that is, they excrete ammonia. Adult frogs, however, like most land vertebrates, are UREOTELIC, excreting urea. During metamorphosis, the liver develops those enzymes necessary to

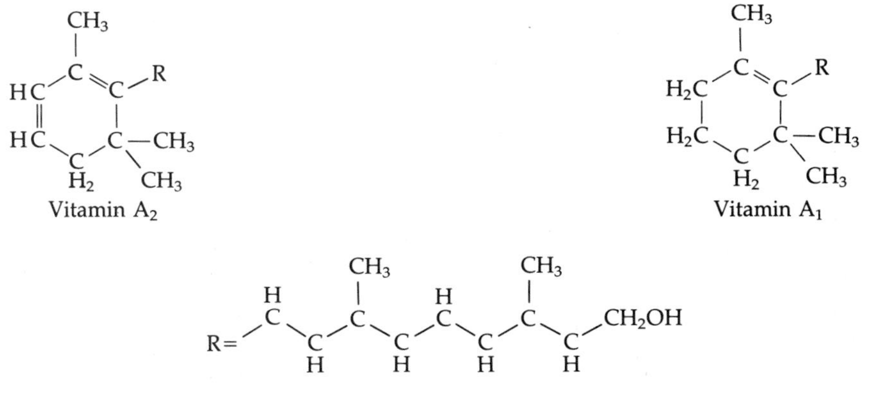

FIGURE 2

Structural formula of the vitamin A portions of the visual pigments of tadpoles and frogs. Tadpole porphyropsin (using vitamin A_2) absorbs light at longer wavelengths than frog rhodopsin (using vitamin A_1).

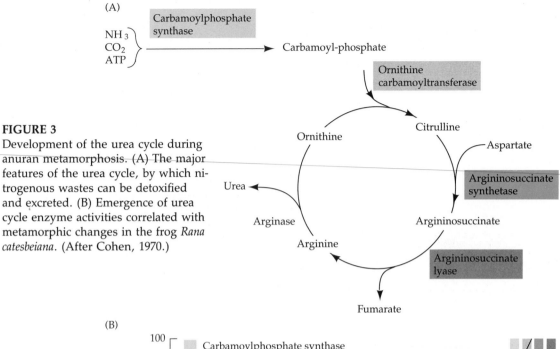

FIGURE 3
Development of the urea cycle during anuran metamorphosis. (A) The major features of the urea cycle, by which nitrogenous wastes can be detoxified and excreted. (B) Emergence of urea cycle enzyme activities correlated with metamorphic changes in the frog *Rana catesbeiana*. (After Cohen, 1970.)

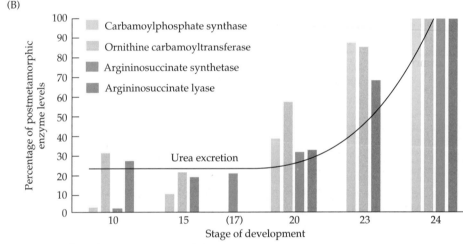

Gilbert 3/E, Figure#19.3, G#468 Black, Focus: 100%

create urea from carbon dioxide and ammonia. These enzymes constitute the UREA CYCLE, and each of them is seen to arise during metamorphosis (Figure 3).

Hormonal control of amphibian metamorphosis

All these diverse changes are brought about by the secretion of the hormones THYROXINE (T_4) and TRIIODOTHYRONINE (T_3) from the thyroid during metamorphosis (Figure 4). It is now believed that T_3 is the active hormone, as it will cause the metamorphic changes in thyroidectomized tadpoles in much lower concentrations than will T_4 (Kistler et al., 1977; Robinson et al., 1977). In both urodeles and anurans, the thyroid produces small amounts of T_3 and T_4 in the larva, but they are overbalanced by the secretion of the hormone PROLACTIN from the anterior pituitary. Prolactin acts as a larval growth hormone and inhibits metamorphosis (Etkin and Gona, 1967; Bern et al., 1967). During metamorphosis, the thyroid hormones T_3 and T_4 increase in concentration, thereby causing the tadpoles to become frogs and stimulating larval newts to become land-dwelling efts. In salamanders, thyroxine is overpowered by prolactin later in de-

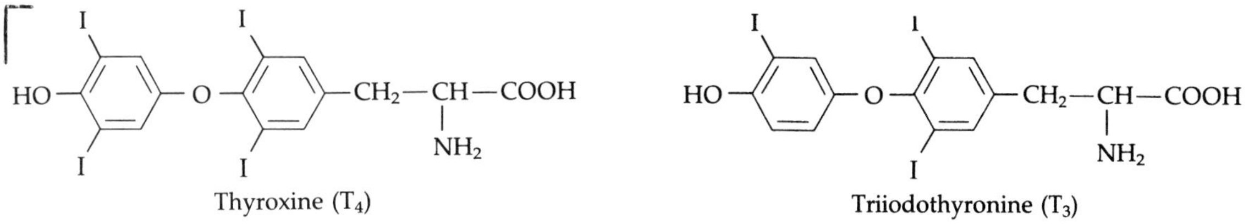

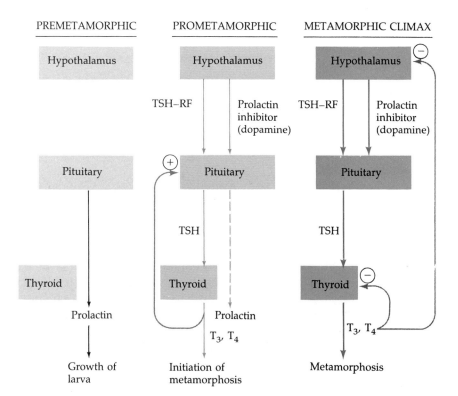

Thyroxine (T₄)

Triiodothyronine (T₃)

FIGURE 4
Formulas of thyroxine (T_4) and triiodothyronine (T_3).

velopment, thereby causing the animals to return to the water to spawn as adults (Grant and Grant, 1958). Thus, in some salamanders, there are two metamorphoses—the first stimulated by thyroxine, the second induced by prolactin. Moreover, even after the second metamorphosis, an injection of thyroxine will cause the adult newt to return to land as a dry-skinned eft (Grant and Cooper, 1964).

The control of metamorphosis by thyroid hormones was shown earlier in this century by Gudernatsch (1912), who found that tadpoles metamorphosed prematurely when fed powdered sheep thyroid gland. Allen (1916) and Hoskins and Hoskins (1917) found that when they removed the thyroid rudiment from early tadpoles, the larvae never metamorphosed, becoming giant tadpoles instead.

The release of T_3 is under the control of the hypothalamic portion of the brain (Etkin, 1968). During the period of larval growth (premetamorphosis), this portion of the brain is underdeveloped, so the brain exerts little control over the anterior pituitary (Figure 5). In the absence of hypothalamic regulation, the pituitary secretes high levels of prolactin but little or no THYROID-STIMULATING HORMONE (TSH). Thus, T_3 levels are low and prolactin levels are high. As the hypothalamus develops, its production of THYROID-STIMULATING HORMONE RELEASING FACTOR (TSH-RF)

FIGURE 5
The hypothalamus–pituitary–thyroid axis during different stages of anuran metamorphosis. As the hypothalamus develops, it stimulates the pituitary to instruct thyroid hormone secretion and inhibit prolactin secretion. T_3, triiodothyronine; T_4, thyroxine; TSH, thyroid-stimulating hormone; TSH-RF, thyroid-stimulating hormone releasing factor.

FIGURE 6
Regression of isolated tail ends under the influence of thyroxine. (A) Control tips from *Xenopus* tadpoles cultured in Holtfreter's salt solution for 6, 8, 10, and 12 days. (B) Treated tail tips at the same ages as controls; thyroxine was added to their salt solutions. The bar represents 1 mm. (From Weber, 1965; courtesy of R. Weber.)

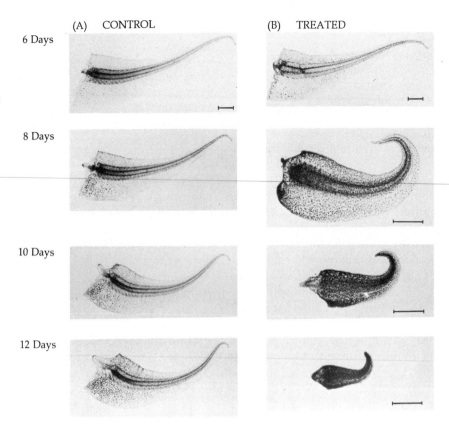

(A) CONTROL   (B) TREATED

6 Days

8 Days

10 Days

12 Days

causes a rise in the level of TSH. This TSH causes the thyroid to increase T_3 and thyroxine synthesis. Thus, T_3 is released from the thyroid and its concentration gradually increases until the first changes of metamorphosis (PROMETAMORPHOSIS) appear. During this time, the hindlimbs start to enlarge. The rising T_3 levels also stimulate the further development of the median eminence of the pituitary, which mediates the flow of TSH-RF to the anterior pituitary. There is, then, a POSITIVE FEEDBACK wherein increased T_3 production causes more T_3 production. In addition, the hypothalamus begins secreting a substance—probably dopamine (see White and Nicoll, 1981)—that inhibits the pituitary synthesis of prolactin. Thus, the ratio of T_3 to prolactin increases enormously. This leads to the METAMORPHIC CLIMAX in which most of the development associated with metamorphosis takes place. One of the effects of metamorphosis is the partial degeneration of the thyroid, and it is also possible that high levels of T_3 have an inhibitory effect on TSH or TSH-RF production (Goos, 1978). In this manner, a new balance of hormones is acquired.

The various organs of the body respond differently to hormonal stimulation. The same stimulus will cause certain tissues to degenerate while causing others to develop and differentiate. For instance, tail degeneration is clearly associated with the increasing levels of thyroid hormones. The degeneration of tail structures is relatively rapid as the bony skeleton does not extend to the tail, which is supported only by the notochord (Wassersug, 1989). This degeneration can be shown in vitro (Weber, 1967) when isolated tail pieces are placed in agar dishes and subjected to chemical treatments (Figure 6). Those tails grown in untreated medium remain healthy, whereas those placed into medium containing thyroid hormones undergo characteristic regression. Moreover, prolactin inhibits the degeneration of the tail induced by thyroid hormones (Brown and Frye, 1969). The regression of the tail is thought to occur in four stages. First, protein synthesis decreases in the striated muscle cells of the tail (Little et al., 1973). Then, there is an increase in the lysosomal enzymes. Concentrations of cathepsin D (a protease), RNase, DNase, collagenase, phosphatase, and glycosidases all rise in the epidermis, notochord, and nerve cord cells (Fox, 1973). Cell death is probably caused by the release of these enzymes into the cytoplasm. The epidermis helps to digest the muscle tissue, probably by releasing these digestive enzymes. If the epidermis is surgically removed from the tail tips, the tips will not regress when cultured in thyroxine (Eisen and Gross, 1965; Niki et al., 1982). After this death, macrophages collect in the tail region, digesting the debris with their own proteolytic enzymes (Kaltenbach et al., 1979). The result is that the tail becomes a large sac of proteolytic enzymes (Figure 7).

The response to thyroid hormones is intrinsic to the organ itself and is not dependent on surrounding tissues. In the epidermis, the response to thyroid hormones depends on which part of the body the epidermis covers. Head and body epidermal cells undergo a small turnover, as expected in skin. T_3 does not change this rate. In the *tail*, however, T_3 causes a rapid increase in the keratinization and death of these cells. It also causes a tail-specific suppression of stem cell divisions that could give rise to more epidermal cells. The result is the death of the tail epidermal cells while the head and body epidermis continues to function (Nishikawa et al., 1989). These local epidermal responses appear to be controlled by the regional specificity of the dermal mesoderm. If tail dermatome cells (that generate the tail dermis) are transplanted into the trunk, the epidermis they contact will degenerate upon metamorphosis. Conversely, when trunk dermatome is transplanted into the tail, those regions of skin persist. Changing the ectoderm does not alter the regional response to thyroid hormones (Kinoshita et al., 1989).

This organ-specific response is dramatically demonstrated when tail tips are transplanted to the trunk region or when eye cups are placed in the tail (Schwind, 1933; Geigy, 1941). The extra tail tip placed into the trunk is not protected from degeneration, but the eye retains its integrity despite the fact that it lies within the degenerating tail (Figure 8). Thus,

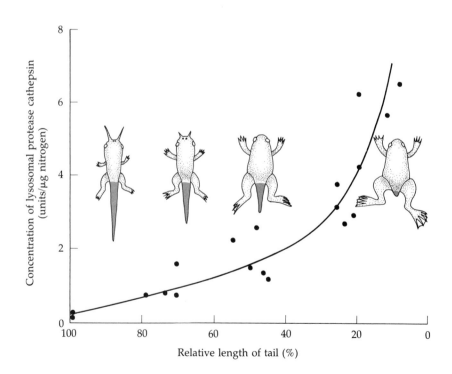

FIGURE 7
Increase of lysosomal protease cathepsin during tail regression in *Xenopus laevis*. The lysosomal enzymes are thought to be responsible for digesting the tail cells. (After Karp and Berrill, 1981.)

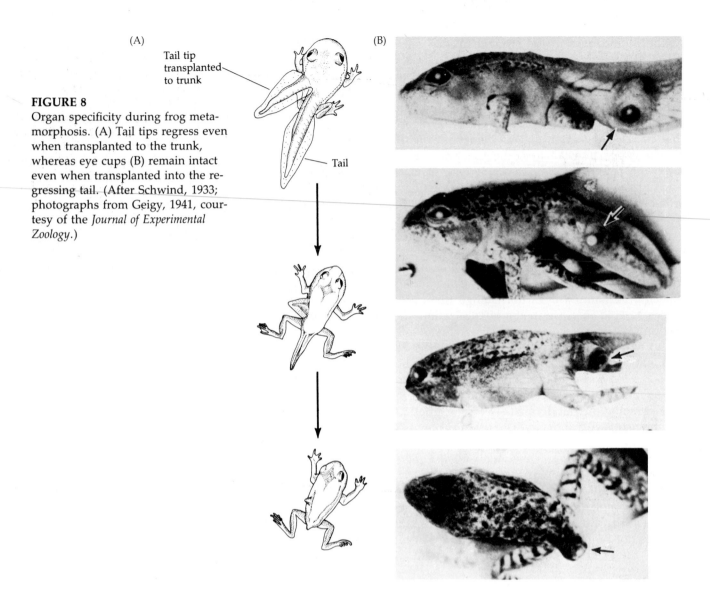

FIGURE 8
Organ specificity during frog metamorphosis. (A) Tail tips regress even when transplanted to the trunk, whereas eye cups (B) remain intact even when transplanted into the regressing tail. (After Schwind, 1933; photographs from Geigy, 1941, courtesy of the *Journal of Experimental Zoology*.)

(A)

Tail tip transplanted to trunk

Tail

(B)

the degeneration of the tail represents a programmed cell death. Only specific tissues die when a signal is given. Such programmed cell deaths are important in molding the body. In humans, such programmed degeneration occurs in the tissues between our fingers and toes, and the degeneration of the human tail during week 4 of development resembles the regression of the tadpole tail (Fallon and Simandl, 1978).

One of the major problems of metamorphosis is the coordination of developmental events. The tail should not degenerate until some other means of locomotion—the limbs—has developed, and the gills should not regress until the animal can utilize its newly developed lung muscles. The means of coordinating the metamorphic events appears to be that different amounts of hormone are needed to produce different specific effects (Kollros, 1961). This model is called the THRESHOLD CONCEPT. As the concentration of thyroid hormones gradually builds up, different events occur at different concentrations of the hormone. If tadpoles are deprived of their thyroids and are placed in a dilute solution of thyroid hormones, the only morphological effects are the shortening of the intestines and accelerated hindlimb growth. However, at higher concentrations of thyroid hormones, tail regression is seen before the hindlimbs are formed. These experiments suggest that as thyroid hormone levels rise, the hindlimbs develop first and then the tail regresses. Thus, the timing of metamorphosis is regulated by the competency of different tissues to respond to thyroid hormones.

But what happens to the nervous system when the animal is constructing a new organism from the old? The adaptational anatomy of a frog certainly differs from that of its tadpole. One readily observed consequence of anuran metamorphosis is the movement of the eyes forward from their originally lateral position* (Figure 9). The lateral eyes of the tadpole are typical of preyed-upon herbivores, whereas the frontally located eyes of the frog befit its more predatory lifestyle. In order to catch its prey, the frog needs to see in three dimensions. That is, it has to acquire a binocular field of vision wherein input from both eyes converges in the brain. In the tadpole, the right eye innervates the left side of the brain and vice versa. There are no ipsilateral (same-side) projections of the retinal neurons. During metamorphosis, however, these additional ipsilateral pathways emerge, enabling input from both eyes to reach the same area of the brain (Currie and Cowan, 1974; Hoskins and Grobstein, 1985a). In *Xenopus*, these new neuronal pathways result not from the remodeling of existing neurons, but from the formation of new neurons that differentiate in response to thyroid hormones (Hoskins and Grobstein, 1985a,b). Both the movement of the eyes to their new positions and the differentiation of new neurons that extend processes ipsilaterally to the brain are thyroid hormone–dependent changes.

Other neurons also undergo profound changes. Some nerve cells, such as those that innervate the tadpole tail muscles, die (Forehand and Farel, 1982). This neuronal death does not seem to be caused by the death of the target tissue but appears to be a separate response to thyroid hor-

*One of the most spectacular movements of the eyes during metamorphosis occurs in flatfish such as flounder. Originally the eyes are on opposite sides of the face. However, as during metamorphosis, one of the eyes migrates dorsally to meet the other at the top of the head, allowing the fish to dwell on the bottom, looking upward (Martin and Drewry, 1978).

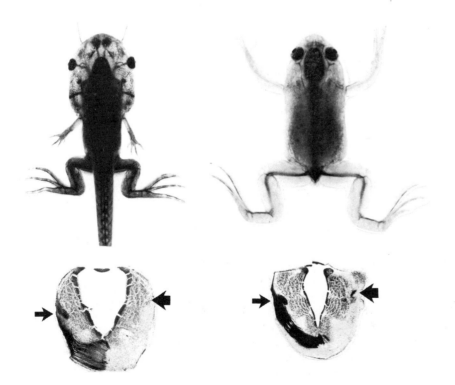

FIGURE 9

Eye migration and associated neuronal changes during metamorphosis of the *Xenopus laevis* tadpole. The eyes of the tadpole are laterally placed, so there is relatively little binocular field. Eyes migrate dorsally and rostrally during metamorphosis, creating a large binocular field for the adult frog. Below the metamorphosing tadpole is a representation of the optic region of its brain. When horseradish peroxidase is injected into the retina, the optic neurons translocate horseradish peroxidase to the contralateral (opposite) side of the brain (small arrow), but not to the ipsilateral side. As metamorphosis continues, the ipsilateral projections (involved in binocular vision) begin to be seen (large arrow). (From Hoskins and Grobstein, 1984; courtesy of P. Grobstein.)

mones. Other neurons, such as certain motor neurons in the tadpole jaw, switch their allegiances from larval muscle to the newly formed adult muscle (Alley and Barnes, 1983). Still other neurons, such as those innervating the tongue (a newly formed muscle not present in the larva), have lain dormant during the tadpole stage and first form connections during metamorphosis (Grobstein, 1987). The brain also undergoes changes in its structure during metamorphosis. Thus, the anuran nervous system undergoes enormous restructuring during metamorphosis. Some neurons die, others are born, and others change their specificity. In these ways, the organism can adapt to its new environmental conditions.

Molecular responses to thyroid hormones during metamorphosis

Thyroid hormones can cause existing tissues to break down or can remold the tissues to their adult function. The cells of the tadpole liver, for instance, are not destroyed and replaced during metamorphosis. Rather, the

(A)

(B)

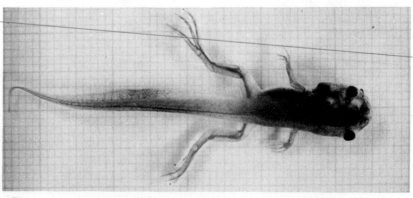

FIGURE 10
Role of new transcription in metamorphosis. (A) *Xenopus laevis* tadpole at the onset of the metamorphic climax. (B) Normal tadpole 7 days later. (C) Tadpole at the same age as in (B); this tadpole was injected at stage (A) with a single dose of actinomycin D. (From Weber, 1967; courtesy of R. Weber.)

(C)

structure of the existing liver cells is changed. This change is accompanied by dramatic increases in ribosomal and messenger RNA synthesis, and the rate of protein synthesis increases nearly 100-fold within 4 hours of thyroid hormone stimulation (Cohen et al., 1978). Many of these new mRNAs are those coding for the new functions of the adult liver. Much of this increase in protein synthesis appears to come from the transcription of new mRNAs, and Weber (1967) has demonstrated that the injection of actinomycin D into normal prometamorphic tadpoles inhibits tail regression and head remodeling (Figure 10).

Experiments hybridizing radioactive mRNA from premetamorphic and metamorphosing bullfrog tadpoles to cloned genes (as dot blots) have demonstrated three types of responses to thyroid hormones. The transcription of one set of genes increases in response to either natural or thyroxine-induced metamorphosis; the transcription of another set of genes is dramatically decreased; while a third set of genes remains unaffected by thyroid hormones (Lyman and White, 1987; Mathison and Miller, 1987). Mori and co-workers (1979) have shown that much of the increase in carbamolyphosphate synthase can be attributed to the increased transcription from that gene. Therefore, metamorphosis is to some extent controlled at the transcriptional level. Other evidence also points to the ability of thyroid hormones to regulate gene activity at the level of transcription. Mammalian T_3-binding proteins have been isolated, and they show remarkable homology to the steroid-binding proteins that bring steroids to the DNA (Chapter 12). Indeed, these T_3-binding proteins contain a region that binds to specific DNA sequences found in the promoter regions of T_3-responsive genes (Weinberger et al., 1986; Crew and Spindler, 1986). This does not mean that transcription is the only level of gene regulation working during metamorphosis, but it is obviously an important one.*

*In general, low-molecular-weight hormones (such as thyroxine and the sex steroids) are able to pass through the cell membrane and bind to specific receptors inside the cell. Peptide hormones, however, usually remain outside the cell and accomplish their activities by binding to specific receptors on the cell membrane. Often, the binding of the peptide hormone to its receptor activates a membrane-bound enzyme, ADENYLATE CYCLASE, which converts ATP into cyclic AMP (cAMP). The cAMP is then able to alter the types of protein synthesized by the cell. Thus, hypothalamic TSH-RF functions by binding to TSH-RF receptors on the set of pituitary cells that make TSH. The binding stimulates the cells' adenylate cyclase to manufacture more cAMP, and the cAMP acts as a "second messenger" to instruct the pituitary cell to synthesize and release TSH. The action of TSH on the thyroid cells is also thought to be mediated by cAMP.

SIDELIGHTS & SPECULATIONS

Heterochrony

Most species of animals develop through a larval phase. These are often highly specialized stages that allow the organism to find and digest specific types of food and to survive in a particular environment. However, some species have modified their life cycles by either greatly extending or shortening their larval period. The phenomenon wherein animals change the relative time of appearance and rate of development for characters already present in their ancestors is called HETEROCHRONY. Here we will discuss three of the extreme types of heterochrony. NEOTENY refers to the retention of juvenile form produced by the retardation of body development with respect to the germ cells and gonads, which achieve maturity at the normal time. PROGENESIS also involves the retention of the juvenile form, but here the gonads and germ line develop at a faster rate than normal, and they become mature while the rest of the body is still in a juvenile phase. In DIRECT DEVELOPMENT, the embryos abandon the stages of larval development entirely and proceed to construct a small adult.

Neoteny

Many salamanders are able to retain their larval form throughout their lives, becoming sexually mature without undergoing metamorphosis. Here, heterochrony involves

(A)   **(B)**

FIGURE 11

Metamorphosis induced in the axolotl. (A) Normal condition of the axolotl. (B) Specimen treated with thyroxine to induce metamorphosis. (Courtesy of G. Malacinski.)

a rate of development of the reproductive system (and germ cells) that is more rapid than that of the rest of the body. Such retention of juvenile characteristics in sexually mature individuals is called *neoteny*. The degree to which metamorphosis occurs differs from species to species. The Mexican axolotl, *Ambystoma mexicanum*, does not undergo metamorphosis in nature because it does not release an active TSH to stimulate its thyroid glands (Prahlad and DeLanney, 1965; Norris et al., 1973; Taurog et al., 1974). Thus, when investigators gave *Ambystoma mexicanum* either thyroid hormones or TSH, they found that the salamander metamorphosed into an adult not seen in nature (Huxley, 1920). Other species, such as *Ambystoma tigrinum*, metamorphose only if given cues from the environment. Otherwise, they become neotenic, successfully mating as larvae. In part of its range, *A. tigrinum* is a neotenic salamander, paddling its way through the cold ponds of the Rocky Mountains. However, in the warmer region of its range, the larval form of *A. tigrinum* is transitory, leading to the land-dwelling tiger salamander. Neotenic populations from the Rockies can be induced to undergo metamorphosis simply by placing them in water at higher temperatures. It appears that the hypothalamus of this species cannot produce TSH-RF at the low temperatures.

Some salamanders, however, are permanently neotenic, even in the laboratory. Whereas thyroxine is able to produce the long-lost adult form of *A. mexicanum* (Figure 11), the neotenic species of *Necturus* and *Siren* remain unresponsive to thyroid hormones. It appears that the target tissues have lost their capacity to respond to thyroid hormones; thus, their neoteny is permanent (Frieden, 1981). The genetic lesions responsible for neoteny in several species are shown in Figure 12.

De Beer (1940) and Gould (1977) have speculated that neoteny is a major factor in the evolution of more complex

taxa. By retarding the development of somatic tissues, natural selection is given a flexible substrate. According to Gould, neoteny would "provide an escape from specialization. Animals can relinquish their highly specialized adult forms, return to the lability of youth, and prepare themselves for new evolutionary directions."

Progenesis

The retention of juvenile form by adult descendants can be brought about in two ways. In neoteny, the development of most somatic tissues is retarded while the gonads mature at their usual rate. In *progenesis*, gonadal maturation is accelerated while the rest of the body develops normally to a certain stage. Progenesis has enabled some salamander species to find new ecological niches. *Bolitoglossa occidentalis* is a tropical salamander that, unlike other members of its genus, lives in trees. This salamander has webbed feet and a small body size that suit it for arboreal existence, the webbed feet producing suction for climbing and the small

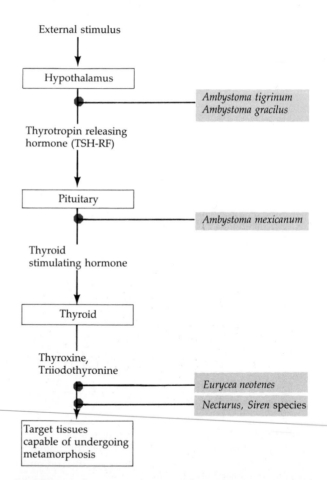

FIGURE 12

Stages along the hypothalamus–pituitary–thyroid axis of salamanders at which various species have blocked metamorphosis. *Eurycea*, *Necturus*, and *Siren* appear to have a receptor defect in the responsive tissues. *Eurycea* will metamorphose when exposed to extremely high concentrations of thyroxine, while *Necturus* and *Siren* do not respond to any dose. (After Frieden, 1981.)

body size making such traction efficient. Alberch and Alberch (1981) have shown that *B. occidentalis* resembles juveniles of the related species *B. subpalmata* and *B. rostrata* (whose young are small, with digits that have not yet grown past their webbing). It is thought that *B. occidentalis* became a sexually mature adult at a much smaller size than its predecessors. This gave it a phenotype that made tree-dwelling a possibility.

Direct development

While some animals have extended their larval period of life, others have "accelerated" their development by abandoning their "normal" larval forms. This latter phenomenon is called *direct development* and is typified by frog species that lack tadpoles and by sea urchins that lack pluteus larvae. Elinson and his colleagues (Elinson, 1987; del Pino and Elinson, 1983) have studied *Eleutherodactylus coqui*, a small frog so common in Puerto Rico that it is one of the most populous animals on the island. Unlike eggs of *Rana* and *Xenopus*, the eggs of *E. coqui* are fertilized while they are still within the female's body. Each egg is about 3.5 mm in diameter (roughly 20 times the volume of *Xenopus* eggs). After the eggs are laid, the male gently sits upon the developing embryos, protecting them from predators and desiccation (Taigen et al., 1984). Early development is like that of most frogs. Cleavage is holoblastic, gastrulation is initiated at a subequatorial position (Figure 13A), and the neural folds become elevated from the surface (Figure 13B). However, shortly after the neural tube closes, limb

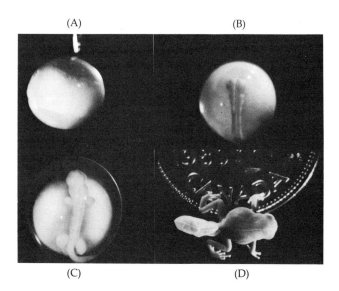

FIGURE 13
Direct development of the frog *Eleuthorodactylus coqui*. (A) Early gastrula showing blastopore lip. (B) Dorsal view of neurula showing raised neural folds. (C) A day after the closure of the neural folds, limb buds are seen. (D) Three weeks after fertilization, a tiny froglet hatches, seen here beside a Canadian penny (the inflation of the tail is an artifact caused by the chemical fixatives used to prepare the specimen). (From Elinson, 1987; courtesy of R. P. Elinson.)

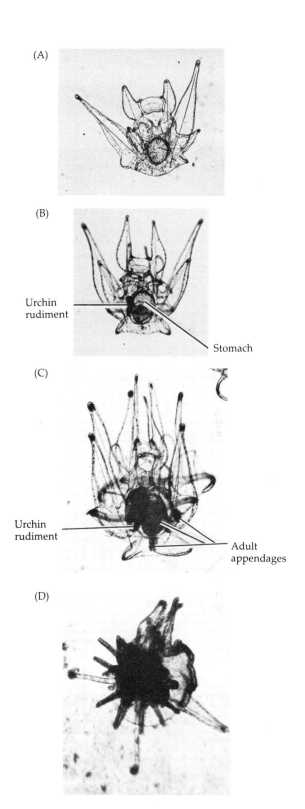

FIGURE 14
Normal metamorphosis of pluteus larva into adult in the sea urchin *Lytechinus pictus*. (A) Pluteus larva 8 days after fertilization. (B) 11-day pluteus larva with urchin rudiment on left coelomic sac. (C) 19-day pluteus with developing sea urchin rudiment. (D) 11 minutes after attaching to substrate, the larval arms are being resorbed. (From Hinegardner, 1969; photographs courtesy of R. T. Hinegardner.)

TABLE 2

Relationship of developmental mode to egg size in sea urchins

Number of species	Egg size range (μm)	Developmental mode
83	60–345	Feeding pluteus larva
1	280	Facultative feeding pluteus
2	300–350	Abbreviated pluteus, nonfeeding
19	400–2000	Pluteus lost; direct development

Source: Raff (1987).

buds appear on the surface (Figure 13C). This early emergence of limb buds is the first indication that development is direct and will not pass through a limbless tadpole stage. Moreover, the emergence of the limbs is not dependent upon thyroid hormones (Lynn and Peadon, 1955). What emerges from the egg jelly three weeks after fertilization is not a tadpole but a little frog (Figure 13D). The froglet has a tail for the first part of its life, but it is used for respiration rather than locomotion. Such direct-developing frogs do not need ponds for their larval stages and can therefore colonize new regions inaccessible to other frogs.

Raff (1987) has studied direct development in sea urchins. In the "typical" sea urchin, the primary mesenchyme cells invaginate and secrete the calcium carbonate skeleton of the pluteus larvae. These larvae feed and grow until coelomic vesicles (also derived from the micromeres) form at the sides of the gut (Pehrson and Cohen, 1986). The left coelom grows further to produce a hydrocoel, which induces the overlying ectoderm to invaginate to form a vestibule. The hydrocoel and vestibule form a rudiment that grows within the larva until it is released at metamorphosis to become a juvenile sea urchin (Figure 14).

Several different sea urchin species have suppressed stages of the pluteus larvae while accelerating the development of the adult rudiment. Like the above-mentioned case of direct development in frogs, direct development in sea urchins also depends upon a large, yolky egg. In fact, Raff has found a correlation between egg volume and the extent of direct development (Table 2). North American and European sea urchins have eggs whose diameters range from 60 to 200 μm. These species undergo indirect development through the pluteus larvae. Eggs in the 300–350 μm range produce partial pluteus larvae that have larval skeletons but no gut (and therefore cannot feed). These species show an accelerated growth of the adult rudiment so a feeding, juvenile urchin is quickly formed.

TABLE 3

Development features in typical and direct-developing sea urchin species

A. Development features between fertilization and initiation of larval development

B. Development features of larvae

	Typical (*Heliocidaris tuberculata*)	Direct-developing (*Heliocidaris erythrogramma*)		Typical (*Heliocidaris tuberculata*)	Direct-developing (*Heliocidaris erythrogramma*)
Egg size (μm)	60–345 (ca. 95)	ca. 450	Initiation of larval skeleton	Yes	No
Fertilization membrane	Yes	Yes	Morphogenesis of typical pluteus arms	Yes	No (no larval skeleton)
Radial cleavage	Yes	Yes	Formation of gut	Yes	No
Fourth cleavage	Small micromeres	Equal cleavage	Formation of coelom	Yes	Yes
Blastula	Smooth	Convoluted	Formation of hydrocoel	Yes	Yes
Active hatching	Yes	Yes	Invagination of vestibule	Yes	Yes
Cilia	Yes	Yes	Formation of adult	Yes	Yes
Primary mesenchyme	32 cells	Large number of cells			
Archenteron invagination	Yes	Yes			
Archenteron elongation	Yes	No			

There are some yolky eggs that reach 2 mm in diameter (about the same volume as *Xenopus* eggs). These embryos develop directly without any pluteus stage. The feeding stage is not needed because nutrition can be provided by the yolk.

Nature has provided an excellent comparison in two species of the Australian sea urchin genus, *Heliocidaris*. *Heliocidaris erythrogramma* and *Heliocidaris tuberculata* are common species that, according to morphological and DNA sequencing data, are very closely related. They live side by side, and both spawn during the same time in the summer. *Heliocidaris erythrogramma*, however, has an egg with a diameter of 425 μm and is a direct developer. *Heliocidaris tuberculata* produces an egg with a diameter of 95 μm and develops through a typical pluteus larva. The comparison between these species (Table 3) reveals that the direct developer has eliminated the larval stages and proceeds directly to coelom formation. The pluteus larva is designed for swimming and feeding (Strathmann, 1971, 1975), using its arms as supports for bands of cilia that sweep food particles into the mouth. The direct developer has neither a functional larval gut nor a larval skeleton. Thus, it can live without this nutritional system and goes about the business of producing the structures of the adult urchin. In the direct-developing sea urchins, one does not see the descendants of the micromere cells invaginating to form the larval skeleton. Rather, these cells are immediately involved in forming the calcareous spine of the young adult.

Thus, we have an interesting paradox. In one sense, the development of larval stages seems highly constrained. Larvae from different classes of echinoderms are very similar, and the tadpoles of different groups of frogs are also very much the same. However, these constraints can be eliminated by abandoning the need for a feeding larval stage. Increasing the amount of yolk available to the embryo seems to make this possible.

Metamorphosis in insects

Eversion and differentiation of the imaginal discs

Whereas amphibian metamorphosis is characterized by the remodeling of existing tissues, insect metamorphosis often involves the destruction of larval tissues and their replacement by an entirely different population of cells. Like amphibian metamorphosis, insect metamorphosis involves the balance between a hormone necessary for continued larval growth and a hormone capable of stimulating the new developmental changes.

There are three major patterns of insect development. A few insects, such as springtails, have no larval stage and undergo direct development. Other insects, notably grasshoppers and bugs, undergo a gradual, HEMI-METABOLOUS metamorphosis (Figure 15A). Here, adult organs are formed without any profound discontinuity. The rudiments of the wing, genital organs, and other adult structures are present at hatching, and they become more mature with each molt. At the last molt, the emerging insect is a winged and sexually mature adult. The larval form of a hemimetabolous insect is called a NYMPH.

In the HOLOMETABOLOUS insects (flies, beetles, moths, and butterflies), there is a dramatic and sudden transformation between the larval and adult stages (Figure 15B). The juvenile larva (caterpillar, grub, maggot) undergoes a series of molts as it becomes larger. The newly hatched insect larva is covered by a hard CUTICLE. To grow, the insect must produce a new, larger, cuticle and shed the old one. Thus, the postembryonic development of these insects consists of a succession of molts. The number of molts before becoming an adult is characteristic for the species, although environmental factors can increase or decrease the number. The stages between these molts are called INSTARS. Here, growth of the larva is the major characteristic. After the last instar stage, the larva undergoes a METAMORPHIC MOLT to become a PUPA. The pupa does not feed, and its energy must come from those foods ingested while a larva.

It is within the pupal cuticle that the transformation of juvenile into adult occurs. Most of the old body of the larva is systematically destroyed as new adult organs develop from undifferentiated nests of cells, the

FIGURE 15
(A) Hemimetabolous (incomplete) metamorphosis. (B) Holometabolous (complete) metamorphosis.

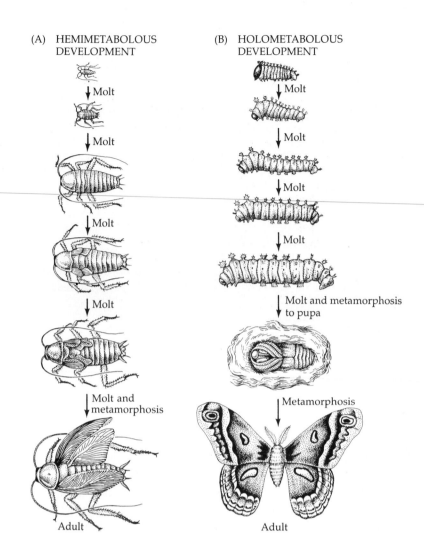

(A) HEMIMETABOLOUS DEVELOPMENT

Molt

Molt

Molt

Molt

Molt

Molt and metamorphosis

Adult

(B) HOLOMETABOLOUS DEVELOPMENT

Molt

Molt

Molt

Molt

Molt and metamorphosis to pupa

Metamorphosis

Adult

IMAGINAL DISCS (and, in some insects, histoblasts). When the adult organism (IMAGO) is developed, the IMAGINAL MOLT results in the shedding of the pupal cuticle, and the mature insect emerges. In holometabolous larvae, then, there are two cell populations: the larval cells, which are used for the functions of the juvenile, and the imaginal cells, which lie in clusters awaiting the signal to differentiate.

We have already encountered imaginal discs in our previous discussions of transdetermination and differential gene transcription. In *Drosophila*, there are ten major pairs of imaginal discs, which reconstruct the entire adult (except for the abdomen), and a genital disc, which forms the reproductive structures. The abdominal epidermis forms from a small group of imaginal cells called HISTOBLASTS, which lie in the region of the larval gut, and other nests of histoblasts located throughout the larva form the internal organs of the adult. The imaginal discs can be seen in the newly hatched larva as local thickenings of the epidermis. In *Drosophila*, these newly hatched eye–antenna, wing, haltere, leg, and genital discs contain 70, 38, 20, 36–45, and 64 cells, respectively (Madhavan and Schneiderman, 1977). Whereas most of the larval cells have a very limited mitotic capacity, the imaginal discs divide rapidly at specific characteristic times (Figure 16). As the cells proliferate, they form a tubular epithelium that folds in upon itself in a compact spiral (Figure 17). The largest disc, that of the wing, contains some 60,000 cells, whereas the leg and haltere discs

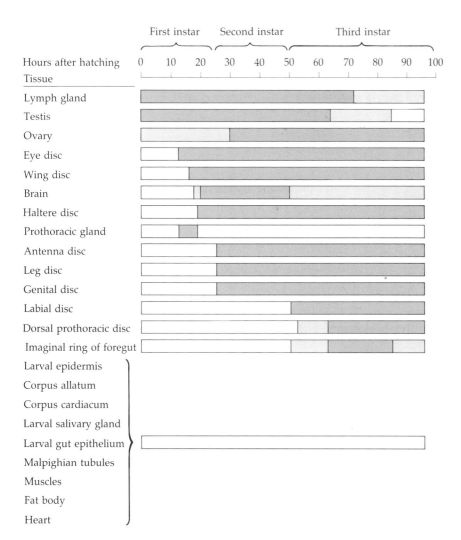

FIGURE 16
Patterns of cell division for several larval and imaginal tissues during the development of *Drosophila* larvae. (After Madhavan and Schneiderman, 1977.)

First instar   Second instar   Third instar

Hours after hatching   0   10   20   30   40   50   60   70   80   90   100

Tissue

Lymph gland
Testis
Ovary
Eye disc
Wing disc
Brain
Haltere disc
Prothoracic gland
Antenna disc
Leg disc
Genital disc
Labial disc
Dorsal prothoracic disc
Imaginal ring of foregut
Larval epidermis
Corpus allatum
Corpus cardiacum
Larval salivary gland
Larval gut epithelium
Malpighian tubules
Muscles
Fat body
Heart

☐ No cells dividing   ▨ Few cells dividing   ▩ Many cells dividing

contain around 10,000 (Fristrom, 1972). At metamorphosis, these cells differentiate and elongate. The fate map and elongation sequence of the leg disc is shown in Figure 18. The cells at the center of the disc telescope out to become the most distal portions of the leg—the claws and the tarsus—and the outside cells become the proximal structures—the coxa and the adjoining epidermis. After differentiating, the cells of the appendages and epidermis secrete a cuticle appropriate for the specific region. Although the discs are primarily composed of epidermal cells, a small number of ADEPITHELIAL cells migrate into the disc early in development. During the pupal period, these cells give rise to the muscles and nerves that serve that structure.

FIGURE 17
Imaginal disc elongation. Scanning electron micrograph of *Drosophila* third instar leg disc (A) before and (B) after elongation. (From Fristrom et al., 1977; courtesy of D. Fristrom.)

(A)

(B)

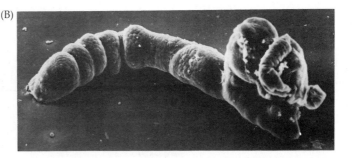

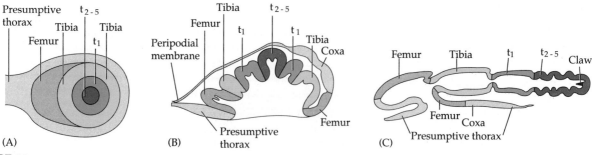

FIGURE 18

Elongation sequence of *Drosophila* leg disc. (A) Surface view of uneverted disc. (B,C) Longitudinal section through elongating and fully everted leg disc. t_1, basitarsus; t_{2-5}, tarsal segments 2–5. (After Fristrom and Fristrom, 1975.)

The process of elongation can be initiated in culture by placing imaginal discs in a solution containing the molting hormone, 20-hydroxyecdysone. Moreover, such eversion can be inhibited by adding any of four sets of drugs: (1) Inhibitors of RNA synthesis and protein synthesis inhibit eversion when added to cultured imaginal discs at the same time as 20-hydroxyecdysone. It is known that RNA and protein syntheses occur prior to elongation and that some of these proteins are needed for this elongation to occur. (2) Cytochalasin B, an inhibitor of microfilament function, also inhibits elongation, thereby indicating a need for actin microfilaments. (3) Protease inhibitors also inhibit elongation (Pino-Heiss and Schubiger, 1989), as cell surface proteases are required for the releasing of constraints upon cell shape. Taken together, these data suggest that the eversion of the imaginal discs requires new protein synthesis, a well developed system of actin microfilaments, and some cellular communication by the cell surface (Fristrom et al., 1977).

Recent studies by Condic and her colleagues (1990) have demonstrated that the elongation of the imaginal disc is due primarily to cell shape change within the disc epithelium. Using fluorescently labeled phalloidin to stain the peripheral microfilaments of the leg disc cells, they showed that the cells of the early third instar discs were tightly compressed along the proximal–distal axis. This compression was maintained through several rounds of cell division. Then, when the tissue began elongating, the compression was removed and the cells "sprung" into their longer state (Figure 19). This conversion of an epithelium of compressed cells into a longer epithelium of noncompressed cells represents a novel mechanism for the extension of an organ during development.

FIGURE 19

Changes in cell shape during the elongation of a *Drosophila* leg imaginal disc. Top panel: Optical sections through the elongating second leg disc. The arrows mark the basitarsal segments, and the calibration bar is 100 μm. Bottom panel: Higher magnification (calibration bar is 10 μm) of cell apices through the basitarsal area. The cell boundaries are stained by fluorescently labeled phalloidin. (A) Beginning prepupa stage. (B) 6-hour prepupa. (C) Leg disc from a beginning prepupa treated with trypsin. The basitarsal cells are initially compressed along the proximal–distal axis. Upon ecdysone treatment or trypsinization, the compression is released and the cells expand to elongate the tissue. (From Condic et al., 1990; photograph courtesy of the authors.)

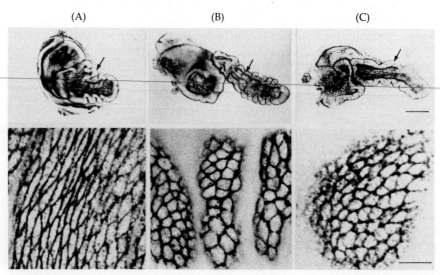

Remodeling of the nervous system

As in anuran metamorphosis, insect metamorphosis causes a major restructuring of the organism's nervous system. Some nerves die, while other nerves take over new functions. In Chapter 15, we saw the development of photoreceptors from the epithelial cells of the eye disc. Here, a new set of neurons is generated to take on a new function. The neurons that have been connected to dying tissues either die with the tissue or are respecified for new functions. The nerve that innervates the proleg muscle of the caterpillar of the *Manduca* moth is independently sensitive to ecdysone and perishes simultaneously with its larval target tissue. However, the motor neuron innervating the second oblique muscle of the larva survives the death of its target to innervate a newly formed adult muscle (the fourth dorsal external muscle) that differentiates during metamorphosis (Truman et al., 1985).

In some instances, larval functions are taken over by different regions in the adult. The larval firefly has its paired lanterns in the eighth (last) abdominal segment. The neurons from the eighth abdominal segment control its luminescence. During pupation, the sixth and seventh segments also develop the light-producing photocytes and the nerves to control the timing of the flash. By the end of pupation, only the sixth and seventh segments have functional lanterns. Moreover, if the larval lanterns are removed, the adult lanterns will still form (Strause et al., 1979). Thus, what had been a neural function of the eighth segment ganglia has become a function of the ganglia of the sixth and seventh segments.

Hormonal control of insect metamorphosis

The hormonal control of insect metamorphosis was shown by the dramatic experiments of Wigglesworth (1934). Wigglesworth studied *Rhodnius prolixus*, a blood-sucking bug that has five instars before undergoing a striking metamorphosis. When a first instar larva of *Rhodnius* was decapitated and fused to a molting fifth instar larva, the minute first instar developed the cuticle, body structure, and genitalia of the adult (Figure 20A). This

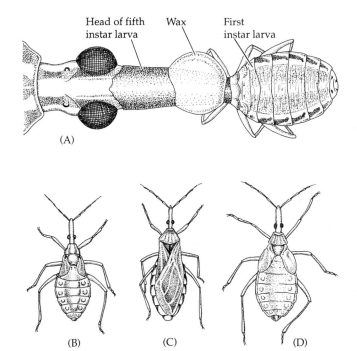

Head of fifth
instar larva   Wax   First
                               instar larva

(A)

(B)   (C)   (D)

FIGURE 20
Demonstration of hormonal control of insect metamorphosis. (A) Technique for producing precocious "adult" from the first instar larva of *Rhodnius* by fusing it to the head of a molting fifth instar larva. (B) Normal fifth instar larva of *Rhodnius*. (C) Normal adult *Rhodnius*. (D) "Sixth instar larva" produced when corpora allata from a fourth instar larva were implanted into the abdomen of a fifth instar larva. (After Wigglesworth, 1939.)

showed that blood-borne hormones were responsible for the induction of metamorphosis. Wigglesworth also showed that the CORPORA ALLATA, near the insect brain, produced a hormone that counteracted this tendency to undergo metamorphosis. If the corpora allata were removed from a third instar larva, the next molt would turn the larva into a precocious adult. Conversely, if the corpora allata from fourth instar larvae were implanted into fifth instar larvae, these larvae would molt into extremely large "sixth instar" larvae rather than into adults (Figure 20B–D).

Transplantation of insect tissues carried out in several laboratories

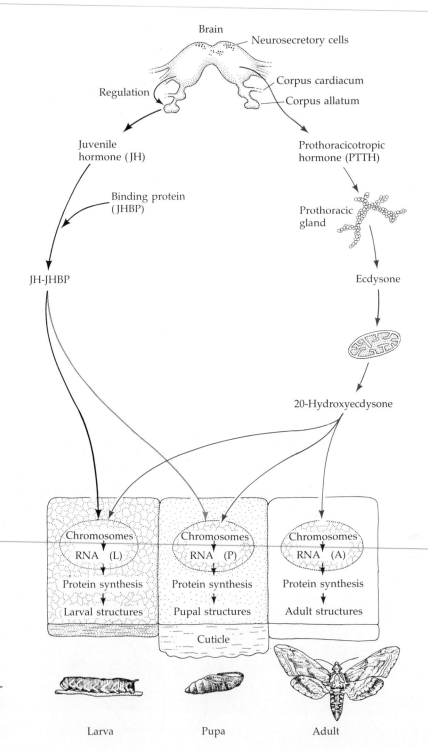

FIGURE 21
Schematic diagram illustrating the control of molting and metamorphosis in the tobacco hornworm moth. (After Gilbert and Goodman, 1981.)

eventually generated an integrated account of how metamorphosis takes place. Although the detailed mechanisms of metamorphosis differ between species, the general pattern of hormone action is usually very similar (Figure 21). Like amphibian metamorphosis, the metamorphosis of insects is regulated by two effector hormones controlled by neurosecretory peptide hormones in the brain (for reviews, see Granger and Bollenbacher, 1981; Gilbert and Goodman, 1981). The molting process is initiated in the brain, where neurosecretory cells release PROTHORACICOTROPIC HORMONE (PTTH) in response to neural, hormonal, or environmental factors. PTTH is a family of peptide hormones with a molecular weight of approximately 40,000, and it stimulates the production of ECDYSONE by the PROTHORACIC GLAND (Figure 22). Ecdysone, however, is not an active hormone; rather it is a prohormone that must be converted into an active form. This conversion is accomplished by a heme-containing oxidase in the mitochondria and microsomes of peripheral tissues such as the fat body. Here, the ecdysone is changed to the active hormone 20-HYDROXYECDYSONE (Figure 23).*

Each molt is occasioned by one or more pulses of 20-hydroxyecdysone. For a molt from a larva, the first pulse produces a small rise in the hydroxyecdysone concentration in the larval hemolymph (blood) and elicits a change in cellular commitment. The second, large pulse of hydroxyecdysone initiates the differentiation events associated with molting. The hydroxyecdysone produced by these pulses commits and stimulates the epidermal cells to synthesize enzymes that digest and recycle the components of the cuticle. In some cases, environmental conditions can control molting, as in the case of the silkworm moth *Hyalophora cecropia*. Here, PTTH secretion ceases after the pupa has formed. The pupa remains in this suspended state, called DIAPAUSE, throughout the winter. If not exposed to cold weather, diapause lasts indefinitely. Once exposed to 2 weeks of cold, however, the pupa can molt when returned to a warmer temperature (Williams, 1952, 1956).

The second major effector hormone in insect development is JUVENILE HORMONE (JH). The structure of a common juvenile hormone active in butterfly and moth caterpillars is shown in Figure 23A. Juvenile hormone is secreted by the corpora allata. The secretory cells of the corpora allata are active during larval molts but are inactive during the metamorphic molt. This hormone is responsible for preventing metamorphosis. As long as juvenile hormone is present, the hydroxyecdysone-stimulated molts result in a new larval instar. In the last larval instar, the medial nerve from the brain to the corpora allata inhibits the gland from producing juvenile hormone, and there is a simultaneous increase in the body's ability to

*Since its discovery in 1954, when Butenandt and Karlson isolated 25 mg of ecdysone from 500 kg of silkworm moth pupae, 20-hydroxyecdysone has gone under several names, including β-ecdysone, ecdysterone, and crustecdysone.

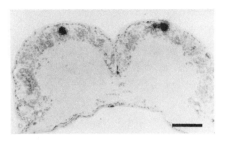

FIGURE 22
Cellular localization of PTTH mRNA in the *Bombyx mori* (silkworm moth) larva. In situ hybridization of a radioactive cloned gene for the 224-amino acid peptide localizes the PTTH mRNA to two neurosecretory cells in the left hemisphere of the brain and two neurosecretory cells in the right hemisphere. In this section, one PTTH-secreting cell can be seen on each side. The bar equals 100 μm. (From Kawakami et al., 1990; photograph courtesy of H. Ishizaki and A. Kawakami.)

FIGURE 23
Structures of a commonly occurring juvenile hormone, of ecdysone, and of the active molting hormone 20-hydroxyecdysone.

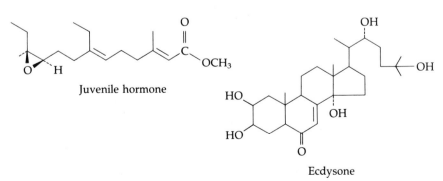

Juvenile hormone

Ecdysone

20-Hydroxyecdysone

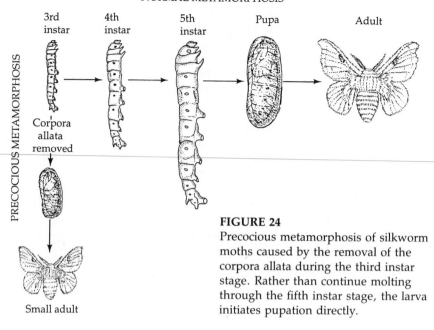

NORMAL METAMORPHOSIS

FIGURE 24
Precocious metamorphosis of silkworm moths caused by the removal of the corpora allata during the third instar stage. Rather than continue molting through the fifth instar stage, the larva initiates pupation directly.

degrade existing juvenile hormone (Safranek and Williams, 1989). Both of these mechanisms cause JH levels drop below a critical threshold value. This triggers the release of PTTH from the brain (Nijhout and Williams, 1974; Rountree and Bollenbacher, 1986). PTTH, in turn, stimulates the prothoracic glands to secrete a small amount of ecdysone. The resulting hydroxyecdysone, in the relative scarcity of JH, commits the cells to pupal development. The subsequent molt, occurring in the relative scarcity of JH, shifts the organism from larva to pupa. During pupation, the corpora allata do not release any juvenile hormone, and the hydroxyecdysone-stimulated pupa will eventually metamorphose into the adult insect. Removal of the corpora allata from larvae will cause the premature metamorphosis of the larvae into small pupae and small adults (Figure 24), while the absorption of exogenously supplied juvenile hormone into last-instar larvae can elicit an extra larval molt and thereby delay the pupal molt. Metamorphosis to the adult, then, occurs when the insect is exposed to hydroxyecdysone in the absence of juvenile hormone. The molecular basis of this involves the transcription of new gene products. Richards (1978) showed that in the absence of juvenile hormone, hydroxyecdysone induces the synthesis of the proteins from the late puffs. When juvenile hormone is present during the initial hydroxyecdysone pulse, the early puffs are produced as usual, but the later puffs are repressed.

SIDELIGHTS & SPECULATIONS

Environmental control over larval form and functions

Childhood's end: Signals to initiate metamorphosis

In 1880, William Keith Brooks, an embryologist at Johns Hopkins University (and thesis advisor to T. H. Morgan, E. B. Wilson, R. G. Harrison, and E. G. Conklin) was asked to help the ailing oyster industry of Chesapeake Bay. For decades, oysters had been dredged from the bay, and there had always been a new crop to take their place. But recently, each year brought fewer oysters. What was responsible for the decline? Experimenting with larval oysters, Brooks discovered that the American oyster (unlike its better studied European cousin) needed a hard substrate on which to metamorphose. For years, oystermen had been throwing the shells back into the sea; but with the advent of suburban sidewalks, the oystermen were selling the shells to the cement factories. Brooks' solution: throw the shells back into the bay. The oyster population responded and the Baltimore wharfs still sell their descendants.

Brooks's study raises an important point. Larvae will not metamorphose under just any set of conditions. In almost all cases, the larvae need a substrate upon which to settle. Cameron and Hinegardner (1974) found that the plutei of two sea urchins, *Arbacia punctulata* and *Lytechinus pictus*, settled only on containers containing a bacterial film on them. The bacteria produced a factor that induced them to settle. This factor could pass through a dialysis membrane that held back any molecule greater than 5000 Da. If the plutei were kept in clean containers, they floated for about 2 months and then deteriorated (Hinegardner, 1969). Other echinoderm plutei settle for different conditions.

The adult sand dollar *Dendraster excentricus* helps its larvae to metamorphose by secreting a chemical (probably a peptide less than 10,000 Da) into the sand. This chemical attracts *Dendraster* larvae and induces their metamorphosis. The result is that larvae tend to associate within or near existing sand dollar beds, causing some areas of a beach to have several hundred adults per square meter, while nearby areas have none. Highsmith (1982) has speculated that the adults in such clusters aid individual survival by protecting the newly metamorphosed *Dendraster* from a particular crustacean predator.

As one might predict, those individuals survive best that can be induced to metamorphose close to protection or potential food sources. It would not be advantageous for a larva to settle where there was no food for it, especially if the adult form were relatively sedentary. In molluscs, there are often very specific cues for settlement (Table 4). Most nudibranch (sea slug) larvae will undergo metamorphosis only if triggered by living adult prey (which differs from species to species). In some cases, the soluble product of the prey that triggers metamorphosis has been identified (Hadfield, 1977). The larva of the shipworm *Teredo navalis* is induced to settle by compounds released by wood, and soluble material eluted from oyster shells induces the settlement of oyster larvae.

The chemical that induces settlement and/or metamorphosis induces a particular set of behaviors in the larvae that receive it. The veliger larvae of the nudibranch mollusc *Onchidoris bilamellata* settle when they encounter chemicals from live barnacles. They cease swimming, sink to the bottom, and begin crawling around the substrate on their larval feet. If they meet a live barnacle within thirty minutes, they begin their metamorphosis. If they do not encounter such a barnacle within that time, they begin swimming again. The settlement cue is received by a particular neural cell on the larval surface. If the larva is exposed to seawater that has had barnacles living in it, this cell is depolarized, sending a neural signal that initiates the behaviors that characterize settlement and the search for the barnacle substrate on which to metamorphose (Arkett et al., 1989).

In insects, the size of the larva plays a major role in initiating metamorphosis. The tobacco hornworm caterpil-

TABLE 4
Specific settlement substrates of molluscan larvae

Molluscan species	Substrate
GASTROPODA (SNAILS, NUDIBRANCHS)	
Nassarius obsoletus	Mud from adult habitat
Philippia radiata	*Porites lobata* (a cnidarian)
Adalaria proxima	*Electra pilosa* (a bryozoan)
Doridella obscura	*Electra crustulenta* (a bryozoan)
Phestilla sibogae	*Porites compressa* (a cnidarian)
Rostanga pulchra	*Ophlitaspongia pennata* (a sponge)
Trinchesia aurantia	*Tubularia indivisa* (a cnidarian)
Elysia chlorotica	Primary film of microorganisms from adult habitat
Haminoea solitaria	Primary film of microorganisms from adult habitat
Aplysia californica	*Laurencia pacifica* (a red alga)
Aplysia juliana	*Ulva* spp. (green algae)
Aplysia parvula	*Chondrococcus hornemanni* (a red alga)
Stylocheilus longicauda	*Lyngbya majuscula* (a cyanobacterium)
Onchidoris bilamellata	Living barnacles
AMPHINEURA (CHITONS)	
Tonicella lineata	*Lithophyllum* sp. and *Lithothamnion* sp. (red algae)
LAMELLIBRANCHIA (BIVALVES)	
Teredo sp.	Wood
Bankia gouldi	Wood
Mercenaria mercenaria	Clam liquor; sand
Placopecten magellanicus	Adult shell; sand; etc.
Mytilus edulis	Filamentous algae; other nonbiological silk material
Crassostrea virginica	Shell liquor; body extract; "shellfish glycogen"
Ostrea edulis	Muscle extract and extrapallial fluid of adult

Source: Hadfield (1977).

lar will not metamorphose until it weighs 3 g, and in some cases it needs an extra larval molt to get there (Safranek and Williams, 1984). It appears that upon reaching this size, the brain releases a substance that inhibits juvenile hormone release (Bhaskaran et al., 1980). In bees, the size of the female larva at its pupal molt determines whether the individual is to be a worker or a queen. A larva fed nutrient-rich "royal jelly" retains the activity of its corpora allata during its last instar stage. The juvenile hormone secreted by this organ delays pupation, thus allowing the resulting bee to emerge larger and (in some species) more specialized in her anatomy (Plowright and Pendrel, 1977; Brian, 1980).

Derangement of metamorphosis by plant compounds: Precocenes and juvenile hormones

When Karel Sláma came from Czechoslovakia to work in Carroll Williams's laboratory at Harvard, he brought with him his chief experimental animal, the European plant bug *Pyrrhocoris apterus*. To the consternation of the entire laboratory, these bugs failed to undergo metamorphosis at the end of the fifth instar stage. Rather, they became large sixth instar larvae—something never before observed in nature or in the laboratory—which ultimately died before becoming adults. After testing many variables, the paper towels lining the dishes were tested for their effect on the larvae. The results were as conclusive as they were surprising: larvae reared on European paper (including pages of the journal *Nature*) underwent metamorphosis as usual, while those larvae reared on American paper (such as shredded copies of the journal *Science*) did not undergo metamorphosis. It was eventually determined that the source of the American paper was the balsam fir, a tree indigenous to the northern United States and Canada. This tree synthesizes a compound that closely resembles juvenile hormone (Sláma and Williams, 1966; Bowers et al., 1966; Williams, 1970), and it probably employs this juvenile hormone analogue to get rid of certain insect predators.

Some other plants have compounds that produce the same effect—the death of insect predators—but do so by eliciting metamorphosis too early. Two compounds that have been isolated from composite herbs have been found to cause the premature metamorphosis of certain insect larvae into sterile adults (Bowers et al., 1976). These compounds are called PRECOCENES and their chemical structures are shown in Figure 25A. When the larvae or nymphs of these insects are dusted with either of these compounds, they will undergo one more molt and then metamorphose into the adult form (Figure 25B). Precocenes accomplish this by causing the selective death of the corpora allata cells in the immature insect (Schooneveld, 1979; Pratt et al., 1980). These cells are responsible for synthesizing juvenile hormone. Thus, without juvenile hormone, the larva commences its metamorphic and imaginal molts. Moreover, juvenile hormone is also responsible for the maturation of the insect egg (Chapter 21). Without this hormone, females are sterile. So the precocenes are able to protect the plant by causing the premature metamorphosis of certain insect larvae into sterile adults.

Plant-induced polyphenism

Plants and animals adapt their life cycles to each other in many ways, and they produce numerous chemicals to induce changes in the other species. In some cases, the morphology of the larva depends on cues from the environment. This ability of an individual to express one phenotype under one set of circumstances and another phenotype under another set of environmental conditions is called POLYPHENISM. One dramatic example of such polyphenism occurs in the moth *Nemoria arizonaria*. This moth has a fairly typical life cycle. Eggs hatch in the spring, and the caterpillars feed on young oak flowers (catkins). These larvae metamorphose in the late spring, mate in the summer, and produce another brood of caterpillars on the oak trees. These caterpillars eat the oak leaves, metamorphose, mate, and their eggs overwinter to start the cycle over again next spring. What is remarkable is that the caterpillars that hatch in the spring look nothing like their progeny that hatch in the summer (color portfolio). The caterpillars that hatch in the spring and eat oak catkins are yellow–brown, rugose, and beaded, resembling nothing else but an oak catkin. They are magnificently camouflaged against predation. But what of the caterpillars that hatch in the summer, after all the catkins are gone? They, too, are well camouflaged, appearing like year-old oak twigs. What controls this? By doing reciprocal feeding experiments, Greene (1989) was able to convert the spring morphs into the summer forms by feeding them oak leaves. The reciprocal experiment did not turn the later morphs into catkin-like caterpillars. Thus, it seems like the catkin form is the "default state" and that something induces the twig-like morphology. This substance is probably a tannin that is concentrated in the oak leaves as they mature. The life cycles of metamorphosing animals are always integrated into that "tangled web of interrelationships" that link one species to the other.

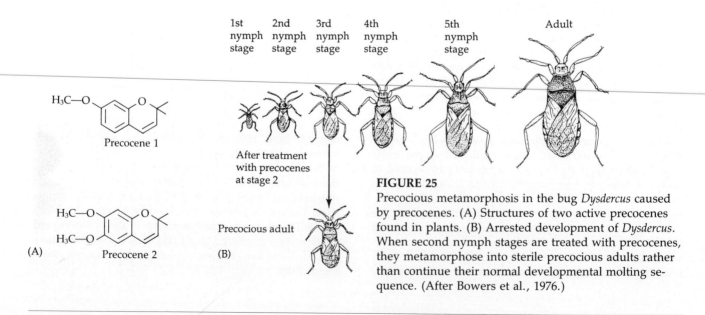

FIGURE 25

Precocious metamorphosis in the bug *Dysdercus* caused by precocenes. (A) Structures of two active precocenes found in plants. (B) Arrested development of *Dysdercus*. When second nymph stages are treated with precocenes, they metamorphose into sterile precocious adults rather than continue their normal developmental molting sequence. (After Bowers et al., 1976.)

1st nymph stage 2nd nymph stage 3rd nymph stage 4th nymph stage 5th nymph stage Adult

Precocene 1

Precocene 2

(A)

After treatment with precocenes at stage 2

Precocious adult

(B)

Multiple hormonal interactions in mammary gland development

Hormones have both constructive and destructive effects in development. In metamorphosis, hormones instruct some cells to die while instructing other cells to form new organs. In breast development, different hormones provide different information to the rudimentary tissue. Mammary development can be divided into three stages: the embryonic stage, the adolescent stage, and the lactating (pregnant) stage. The differentiated products of the mammary glands—casein and other milk proteins—are made only during the final stage (Topper and Freeman, 1980).

Embryonic stage

In the normal development of the female mouse, two bands of raised epidermal tissue appear on both sides of the ventral midline on day 11 of gestation. This tissue is called the MAMMARY RIDGE. Within each ridge, cells collect at centers of concentration and remain there, forming the MAMMARY BUDS (Figure 26). In the mouse, there are five of these buds on each side; in humans, only one per side. In the days immediately prior to birth, the epithelial cells at these places proliferate rapidly, giving rise to the MAMMARY CORD. This cord opens at the skin at one end, forming the nipple, while its other end begins branching into ducts. Here development ceases until puberty.

The development of mammary tissue in male mice is identical to that of females until days 13–15 of gestation. At this time, the mesenchyme condenses around the center of the mammary bud, and the cells of the cord die. Thus, a small cord of epithelial cells is detached from the skin (Figure 27), and the mammary gland does not extend to the surface. No further development occurs.

(A)

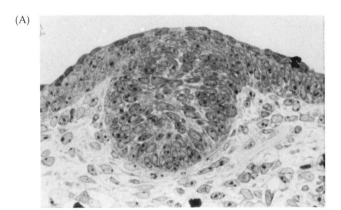

(B)

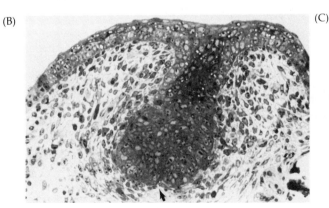

(C)

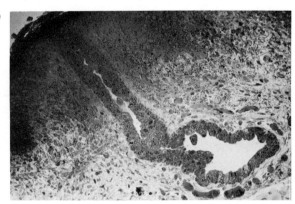

FIGURE 26
Sequence of early mammary gland development in the female mouse. (A) Mammary bud of 12-day fetus. Epithelial ectoderm cells protrude into mesemchyme. (B) Mammary cord of 15-day fetus. A small cleft at the bottom signals the initiation of branching. (C) Cord cavity extending to form a hollow lumen in the 20-day fetus. (From Hogg et al., 1983; photographs courtesy of C. Tickle.)

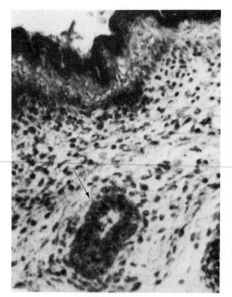

FIGURE 27
Mammary rudiment in a male mouse fetus. The rudiment (arrow) has separated from the epidermis. (From Raynaud, 1961.)

This cell death in the mammary cord of males has been studied by culturing the mouse mammary buds in vitro. Such buds from female mice normally develop lobes connected to the surface (Figure 28). However, when testosterone is added to the culture medium, the buds degenerate. Mammary buds from male mice will also produce lobes, provided they are cultured in the absence of testosterone. Thus, the hormone testosterone prevents mammary development in the male. Testosterone causes this specific cell death by instructing the mesenchymal cells to destroy the epithelial cord. This was shown by a series of recombination experiments. There exists in mice (and in humans as well) a mutation called ANDROGEN INSENSITIVITY SYNDROME, in which chromosomally male (XY) individuals do not make a functional testosterone receptor. Thus, even though these individuals have testes that are actively secreting testosterone, they are unable to respond to it. One of the results is that these individuals have female breast development (see Figure 9 in Chapter 21). Kratochwil and Schwartz (1976) isolated mesenchyme and epithelial cells from normal and mutant mammary buds and cultured them in various combinations. Some cultures were given testosterone and some were not. The results are shown in Figure 29. When both mesenchyme and epithelium were wild-type, the rudiment developed into breast tissue. When testosterone was added, the mesenchyme condensed around the bud and the cord was severed. When normal epithelium was cultured with mutant mesenchyme (which could not respond to testosterone), normal breast development occurred in the presence of testosterone. However, when the mesenchyme was normal and the epithelium was mutant, testosterone was able to cause the degeneration of the mammary cord. Thus, the target of testosterone is the mesenchyme, not the epithelium. The mesenchyme must be responsive to testosterone for its action to occur. In males, the testosterone induces the mammary mesenchyme to destroy its adjacent epithelium. The effect is specific for the organ in that no other mesenchyme will kill

FIGURE 28
Role of testosterone in mediating the detachment of the mammary cord. (A) Female mouse mammary tissue, either in vivo or in culture, will grow downward from the epidermis and branch. (B) When female mouse mammary tissue is cultured in the presence of testosterone, the bud elongates, but mesenchymal cells aggregate around the stalk and the lower portion is cut off, just as in normal male development. (C) When male mouse mammary tissue is cultured in the absence of testosterone, it develops as it would in the female mouse. (After Kratochwil, 1971.)

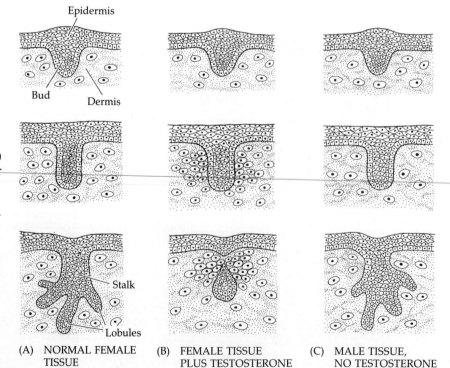

(A) NORMAL FEMALE TISSUE

(B) FEMALE TISSUE PLUS TESTOSTERONE

(C) MALE TISSUE, NO TESTOSTERONE

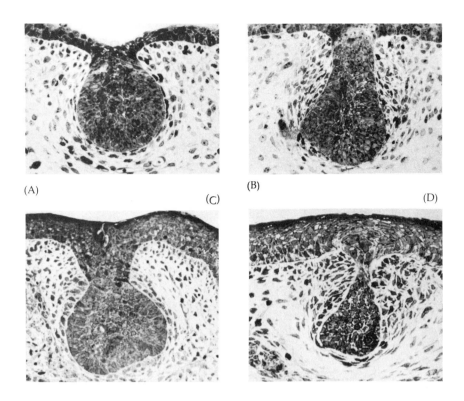

(A)

(B)

(C)

(D)

FIGURE 29
Evidence that the mesenchymal cell is the target of testosterone in the arrest of mammary development. (A) Cultured mammary rudiment from 14-day female embryo. (B) Mammary rudiment from 14-day male embryo beginning its response to testosterone. (C) Recombined mammary bud containing wild-type epithelial cells and androgen-insensitive mesenchyme, cultured with testosterone. No androgen response is seen. (D) Recombined mammary bud containing androgen-insensitive epithelial cells and wild-type mesenchyme, cultured with testosterone. Mesenchyme cells are condensing at the neck of the bud. (From Kratochwil and Schwartz, 1976; courtesy of K. Kratochwil.)

the mammary epithelium and no other epithelium can be destroyed by mammary mesenchyme (Dürnberger and Kratochwil, 1980).*

Adolescence

During adolescence (which in the mouse occurs from week 4 to week 6), the duct system of the mammary gland proliferates extensively. The milk-secreting alveolar cells at the tips of the ducts have not differentiated yet, and no milk is produced. The extensive cell division is under the control of estrogen and growth hormones and appears to be concentrated at the ductal tips. Recent studies by Coleman and her colleagues (1988) implicate epidermal growth factor (EGF) as being responsible for controlling ductal growth during this period. They implanted EGF in slow-release plastic pellets into the mammary glands of 5-week mice. These mice had had their ovaries removed and their mammary development was thus halted. The ducts adjacent to the EGF implant reinitiated their growth and morphological development, while the ducts further away from the implant did not (Figure 30). Moreover, when sections of the mammary gland were incubated with radioactive EGF, the EGF was seen to bind to the tip of the ducts and to be associated with the cells that were undergoing mitosis. It is probable that EGF may act directly to cause the growth of the mammary glands during adolescence.

Pregnancy

Between adolescence and pregnancy, mouse breast cells are mitotically dormant and undifferentiated. This state is changed during the second half of pregnancy. Under the influence of the hormones estrogen and progesterone (the latter from the placenta), new ducts are formed and the

*This phenomenon seems to resemble that of the mesenchymal specificity in anuran metamorphosis that we discussed earlier in this chapter.

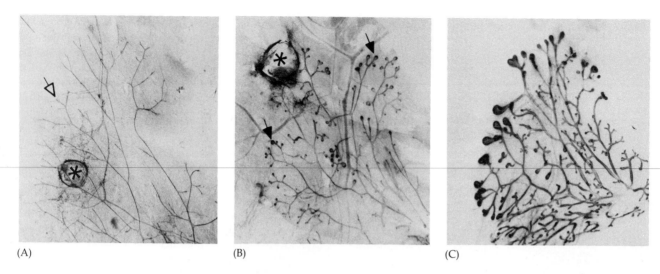

(A) (B) (C)

FIGURE 30
EGF-dependent mammary gland growth in the absence of estrogen. (A) No duct growth or differentiation is seen in estrogen-deficient mice when a pellet of bovine serum albumin (*) is implanted into the mammary gland. (B) When a pellet containing EGF is implanted into the estrogen-deficient mammary gland, nearby ducts increase their size and develop lobular tissue at their tips (arrows). (C) Normal mammary duct development in a control 5-week virgin mouse. (From Coleman et al., 1988; photographs courtesy of S. Coleman.)

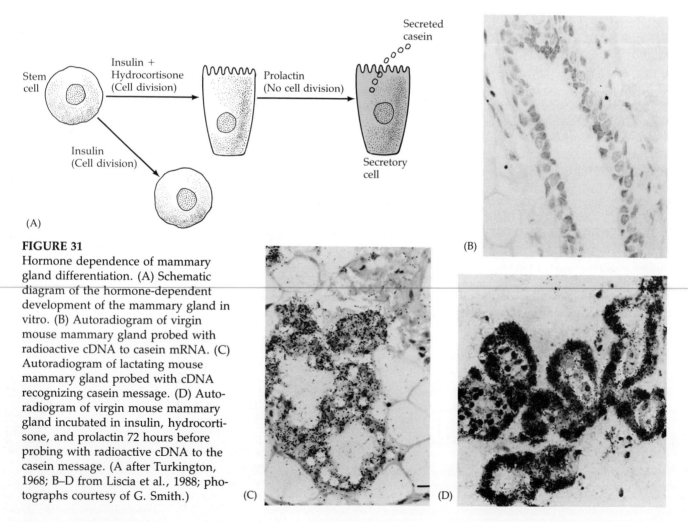

FIGURE 31
Hormone dependence of mammary gland differentiation. (A) Schematic diagram of the hormone-dependent development of the mammary gland in vitro. (B) Autoradiogram of virgin mouse mammary gland probed with radioactive cDNA to casein mRNA. (C) Autoradiogram of lactating mouse mammary gland probed with cDNA recognizing casein message. (D) Autoradiogram of virgin mouse mammary gland incubated in insulin, hydrocortisone, and prolactin 72 hours before probing with radioactive cDNA to the casein message. (A after Turkington, 1968; B–D from Liscia et al., 1988; photographs courtesy of G. Smith.)

distal cells of the ducts begin to develop the characteristics of secretory tissue.

When midpregnancy mammary glands are cultured in vitro, most of the cells have little rough endoplasmic reticulum and Golgi apparatus, and no casein granules. When insulin or another promoter of DNA synthesis is added to these cultures, the cells become responsive to other hormones (Turkington et al., 1965). Glucocorticoids can induce the formation of the rough endoplasmic reticulum, and prolactin can then stabilize the casein mRNA to ensure appropriate milk protein synthesis (Figure 31).

The development of the mammary gland, then, involves a complex interplay of several hormones at three different stages of life: embryonic, adolescent, and pregnant. The mammary gland never develops in normal males and does not become a fully differentiated organ in females until the middle of pregnancy in the adult organism.

SIDELIGHTS & SPECULATIONS

Puberty as a variation on the theme of metamorphosis

In mammals, one of the most striking displays of hormonal control of differentiation occurs during human puberty. If metamorphosis is understood to comprise the processes of dramatic change whereby a juvenile reaches maturity, both sexually and biochemically, then human puberty may be considered a variation on the metamorphic theme. Recent research suggests that the processes of metamorphosis and puberty may in fact be quite similar. A good summary of puberty is given by Styne and Grumbach (1978):

> Puberty is the period of transition between the juvenile state and adulthood; during this stage of development, secondary sex characteristics appear and mature, the adolescent growth spurt occurs, fertility is achieved, and profound psychologic effects are observed. These changes result directly or indirectly from maturation of the hypothalamic-pituitary gonadotropin unit, stimulation of the sex organs, and the secretion of sex steroids.

Certainly, striking changes in body form occur during this time. At the beginning of puberty, boys and girls have the same proportion of muscle mass, skeletal mass, and body fat. By the end of puberty, men have 1.5 times the skeletal and muscle mass of women, whereas women develop twice as much body fat as men (Forbes, 1975). Secondary sex characteristics* also develop at this time, marking the change from the juvenile form to the sexually mature adult. In women, the development of breasts is controlled by a surge of estrogen secreted by the ovaries; and in men, the maturation of the penis and testes is controlled by the testosterone released from the testes. In both sexes, pubic and axillary (underarm) hair development is regulated by testosterone, which is secreted by the testes in males and

*The primary sex characteristic is the presence of ovaries or testes. The development of the primary sex organs will be considered in detail in Chapter 21.

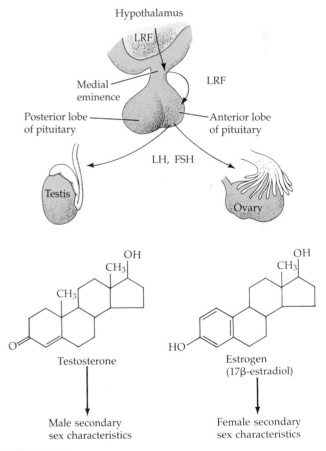

FIGURE 32

Hypothalamus–pituitary–gonadal axis in mammalian sexual development.

by the adrenal gland in females. Men also undergo testosterone-dependent enlargement of the larynx and its associated muscles and cartilage, leading to a deepening of the voice. In both sexes, puberty is the time when major sexual

Sex steroids	Low	Unchanged	Adult level
Feedback	Operative and sensitive	Decreasing in sensitivity	Operative at adult level
Gonadotropins	Low	Increasing	Adult level

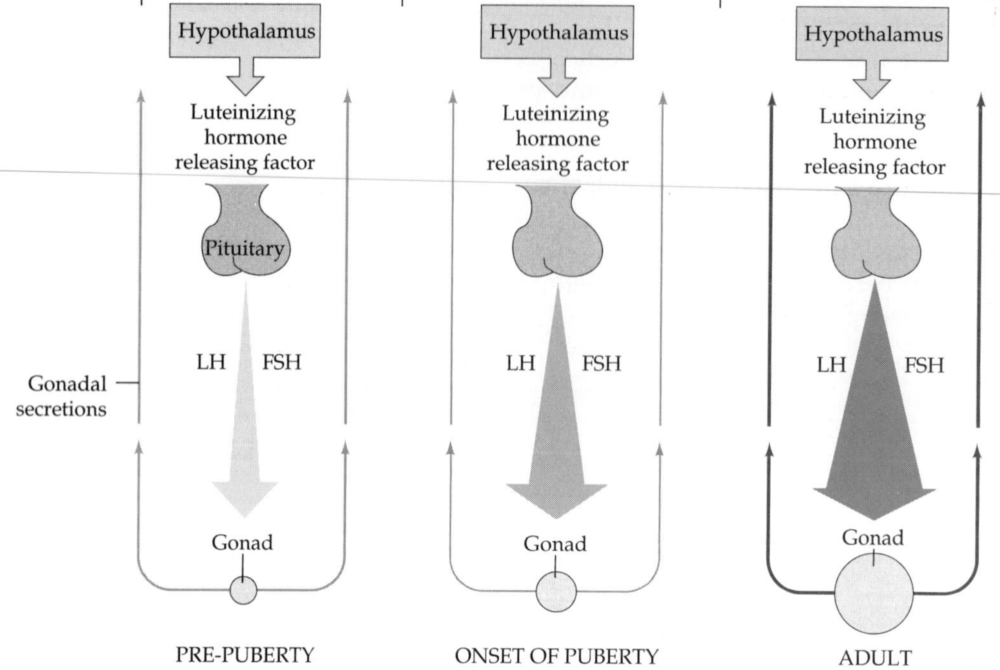

PRE-PUBERTY   ONSET OF PUBERTY   ADULT

FIGURE 33

Proposed mechanism for the induction of puberty in humans. Before puberty, the hypothalamus is sensitive to very small concentrations of sex steroids and stops LRF production, thereby halting further steroid synthesis. Sex steroids are kept at low levels. At the onset of puberty, the hypothalamus becomes progressively less sensitive to the sex steroids, thereby allowing more of them to be synthesized, until finally the adult level of sex steroids is achieved. Relative widths of arrows indicate levels of hormone production. (After Grumbach et al., 1974.)

development takes place. In females, menarche—the first menstrual period—represents the new integration of the hormonal cycles releasing the developing egg from the ovary. In males, fertility is achieved when meiosis begins in the male germ cells and the spermatic ducts hollow out to form a channel for the sperm cells to pass from the testes into the urethra.

The hormonal basis of puberty is thought to be very similar to that of metamorphosis. The metamorphoses of amphibians and insects were both seen to be regulated by hormonal changes that were initiated by the neurohormones of the brain (TSH-RF and PTTH, respectively). The changes of human puberty in both sexes are initiated by LUTEINIZING HORMONE RELEASING FACTOR (LH-RF) from the hypothalamus of the brain* (Figure 32). Like TSH-RF, this factor is released from the hypothalamic neurons to the

MEDIAL EMINENCE (infundibulum) of the pituitary gland. Similarly, the LH-RF is then transported by the blood vessels of the pituitary to the anterior lobes of that gland. Once in the anterior pituitary, the releasing factor causes the release of a tropic hormone. In human puberty, LH-RF releases LUTEINIZING HORMONE (LH) and FOLLICLE-STIMU-LATING HORMONE (FSH). Collectively, these two hormones are called GONADOTROPINS because they stimulate the development of testes in males and ovaries in females. As a result of this stimulation, the gonads secrete the SEX HOR-MONES: testosterone from the testes and estrogen from the ovaries. The various morphological and behavioral changes of puberty are due to the actions of these hormones on the various target tissues. As in metamorphosis, there appears to be a maturation-inhibiting hormone whose activity decreases to permit the reactivation of development. In humans, this hormone is probably melatonin, whose serum concentration decreases as that of LH rises (Waldhauser and Dietzel, 1985).

Although the details concerning the initiation of puberty are not known, Grumbach and co-workers (1974) have proposed the following mechanism (Figure 33). Before puberty, the child secretes a small amount of LH-RF, so that the amounts of circulating LH and FSH are very small. Therefore, the gonad remains immature and secretes little estrogen or testosterone. Moreover, the hypothalamus is

*LH-RF is also known as GONADOTROPIN RELEASING FACTOR (GnRF) since it acts to release both luteinizing hormone and follicle cell stimulating hormone. Until recently, it was assumed that the LH-RF–secreting neurons of the hypothalamus originated in the brain and differentiated there. However, new evidence (Schwanzel-Fukada and Pfaff, 1989; Wray et al., 1989) indicates that these neurons originate in the olfactory placode of the embryonic nose. These cells then migrate into the brain during embryonic development.

highly sensitive to these sex hormones and "turns off" LH-RF production when the circulating sex hormones get anywhere above their very low levels. The onset of puberty is thought to involve the maturation of the hypothalamus. At this time, the hypothalamus becomes less responsive to the negative feedback of testosterone and estrogen. Thus, it takes more of these hormones to turn off LH-RF secretion. (This is analogous to giving a thermostat a higher set point.) Thus, the hypothalamus makes more LH-RF, ultimately causing the further differentiation of the gonads and the release of more testosterone or estrogen. This, in turn, causes the development of those secondary sexual characteristics.

It can readily be seen, then, that puberty is a hormonally controlled reactivation of development leading to sexual maturity and changes in bodily form and physiology. As such, it has many parallels to the metamorphic changes seen elsewhere in the animal kingdom.

We have seen that diffusible regulation of cell–cell interactions is also important in regulating development. In studying the reactivation of development that occurs during metamorphosis, we can identify the roles of hormones in eliciting new patterns of differentiation and morphogenesis. We can also see the interactions between the development of the organism and the ecosystem of which it is part. In the next chapter, we will study the roles of diffusible factors in the processes responsible for the integrated growth of the organism.

LITERATURE CITED

Alberch, P. and Alberch, J. 1981. Heterochronic mechanisms of morphological diversification and evolutionary change in the neotropical salamander *Bolitoglossa occidentalis* (Amphibia: Plethodontidae). *J. Morphol.* 167: 249–264.

Allen, B. M. 1916. Extirpation experiments in *Rana pipiens* larva. *Science* 44: 755–757.

Alley, K. E. and Barnes, M. D. 1983. Birthdates of trigeminal motor neurons and metamorphic reorganization of the jaw myoneural system in frogs. *J. Comp. Neurol.* 218: 395–405.

Arkett, S. A., Chia, F.-S., Goldberg, J. I. and Koss, R. 1989. Identified settlement receptor cells in a nudibranch veliger respond to a specific clue. *Biol. Bull.* 176: 155–160.

Begon, M., Harper, J. L. and Townsend, C. R. 1986. *Ecology: Individuals, Populations, and Communities.* Blackwell Scientific, Oxford.

Bern, H. A., Nicoll, C. S. and Strohman, R. C. 1967. Prolactin and tadpole growth. *Proc. Soc. Exp. Biol. Med.* 126: 518–521.

Bhaskaran, G., Jones, G. and Jones, D. 1980. Neuroendocrine regulation of corpus allatum activity in *Manduca sexta*: Sequential neurohormonal and nervous inhibition in the last instar larva. *Proc. Natl. Acad. Sci. USA* 77: 4407–4411.

Bowers, W. S., Fales, H. M., Thompson, M. J. and Uebel, E. C. 1966. Identification of an active compound from balsam fir. *Science* 154: 1020–1021.

Bowers, W. S., Ohta, T., Cleere, J. S. and Marsella, P. A. 1976. Discovery of insect anti-juvenile hormones in plants. *Science* 193: 542–547.

Brian, M. V. 1980. Social control over sex and caste in bees, wasps and ants. *Biol. Rev.* 55: 379–415.

Brooks, W. K. 1880. The development of the American oyster. *Studies Biol. Lab. Johns Hopkins Univ.* 1: 1–81.

Brown, P. S. and Frye, B. E. 1969. Effect of prolactin and growth hormone on growth and metamorphosis of tadpoles of the frog *Rana pipiens*. *Gen. Comp. Endocrinol.* 13: 139–145.

Butenandt, A. and Karlson, P. 1954. Über die Isolierung eines Metamorphosen-Hormons der Insekten in kristallisierter Form. *Z. Naturforsch., Teil B* 9: 389–391.

Cameron, R. A. and Hinegardner, R. T. 1974. Initiation of metamorphosis in laboratory-cultured sea urchins. *Biol. Bull.* 146: 335–342.

Cohen, P. P. 1970. Biochemical differentiation during amphibian metamorphosis. *Science* 168: 533–543.

Cohen, P. P., Brucker, R. F. and Morris, S. M. 1978. Cellular and molecular aspects of thyroid-hormone action during amphibian metamorphosis. *Horm. Prot. Peptides* 6: 273–381.

Coleman, S., Silberstein, G. B. and Daniel, C. W. 1988. Ductal morphogenesis in the mouse mammary gland: Evidence supporting a role for epidermal growth factor. *Dev. Biol.* 127: 304–315.

Condic, M. L., Fristrom, D. and Fristrom, J. W. 1990. Apical cell shape changes during *Drosophila* imaginal leg disc elongation: A novel morphogenetic mechanism. *Development*. In press.

Crew, M. D. and Spindler, S. R. 1986. Thyroid hormone regulation of transfected rat growth hormone promoter. *J. Mol. Biol.* 261: 5018–5022.

Currie, J. and Cowan, W. M. 1974. Evidence for the late development of the uncrossed retinothalamic projections in the frog *Rana pipiens*. *Brain Res.* 71: 133–139.

De Beer, G. 1940. *Embryos and Ancestors.* Clarendon Press, Oxford.

del Pino, E. M. and Elinson, R. P. 1983. A novel development pattern for frogs: Gastrulation produces an embryonic disk. *Nature* 306: 589–591.

Dürnberger, H. and Kratochwil, K. 1980. Specificity of tissue interaction and origin of mesenchymal cells in the androgen response of the embryonic mammary gland. *Cell* 19: 465–471.

Eisen, A. Z. and Gross, J. 1965. The role of epithelium and mesenchyme in the production of a collagenolytic enzyme and a hyaluronidase in the anuran tadpole. *Dev. Biol.* 12: 408–418.

Elinson, R. P. 1985. Change in developmental patterns: Embryos of amphibians with large eggs. In R. A. Raff and E. C. Raff (eds.), *Development as an Evolutionary Process.* Alan R. Liss, New York, pp. 1–21.

Etkin, W. 1968. Hormonal control of amphibian metamorphosis. In W. Etkin and L. I. Gilbert (eds.), *Metamorphosis: A Problem in Developmental Biology.* Appleton-Century-Crofts, New York, pp. 313–348.

Etkin, W. and Gona, A. G. 1967. Antagonism between prolactin and thyroid hormone in amphibian development. *J. Exp. Zool.* 165: 249–258.

Fallon, J. F. and Simandl, B. K. 1978. Evidence of a role for cell death in the disappearance of the embryonic human tail. *Am. J. Anat.* 152: 111–130.

Forbes, G. B. 1975. Puberty: Body composition. *In* S. R. Berenson (ed.), *Puberty.* Stenfert-Kroese, Leiden, pp. 132–145.

Forehand, C. J. and Farel, P. B. 1982. Spinal cord development in anuran larvae. I. Primary and secondary neurons. *J. Comp. Neurol.* 209: 386–394.

Fox, H. 1973. Ultrastructure of tail degeneration in *Rana temporaria* larva. *Folia Morphol.* 21: 103–112.

Frieden, E. 1981. The dual role of thyroid hormones in vertebrate development and calorigenesis. *In* L. I. Gilbert and E. Frieden (eds.), *Metamorphosis: A Problem in Developmental Biology.* Plenum, New York, pp. 545–564.

Fristrom, D. and Fristrom, J. W. 1975. The mechanisms of evagination of imaginal disks of *Drosophila melanogaster*. I. General considerations. *Dev. Biol.* 43: 1–23.

Fristrom, J. W. 1972. The biochemistry of imaginal disc development. *In* H. Ursprung and R. Nothiger (eds.), *The Biology of Imaginal Discs.* Springer-Verlag, Berlin, pp. 109–154.

Fristrom, J. W., Fristrom, D., Fekete, E. and Kuniyuki, A. H. 1977. The mechanism of evagination of imaginal discs of *Drosophila melanogaster*. *Am. Zool.* 17: 671–684.

Geigy, R. 1941. Die metamorphose als Folge gewebsspezifischer determination. *Rev. Suisse Zool.* 48: 483–494.

Gilbert, L. I. and Goodman, W. 1981. Chemistry, metabolism, and transport of hormones controlling insect metamorphosis. *In* L. I. Gilbert and E. Frieden (eds.), *Metamorphosis: A Problem in Developmental Biology.* Plenum, New York, pp. 139–176.

Goos, H. J. Th. 1978. Hypophysiotropic centers in the brain of amphibians and fish. *Am. Zool.* 18: 401–410.

Gould, S. J. 1977. *Ontogeny and Phylogeny.* Harvard University Press, Cambridge, MA, p. 283.

Graham, C. F. and Wareing, P. F. 1976. *The Developmental Biology of Plants and Animals.* Saunders, Philadelphia.

Granger, N. A. and Bollenbacher, W. E. 1981. Hormonal control of insect metamorphosis. *In* L. I. Gilbert and E. Frieden (eds.), *Metamorphosis: A Problem in Developmental Biology.* Plenum, New York, pp. 105–138.

Grant, W. C., Jr. and Cooper, G. 1964. Endocrine control of metamorphic and skin changes in *Diemictylus viridescens*. *Am. Zool.* 4: 413–414.

Grant, W. C. and Grant, J. A. 1958. Water drive studies on hypophysectomized efts of *Diemyctylus viridescens*. Part I. The role of the lactogenic hormone. *Biol. Bull.* 114: 1–9.

Greene, E. 1989. A diet-induced developmental polymorphism in a caterpillar. *Science* 243: 643–646.

Grobstein, P. 1987. On beyond neuronal specificity: Problems in going from cells to networks and from networks to behavior. *In* P. Shinkman (ed.), *Advances in Neural and Behavioral Development*, Vol. 3, Ablex, pp. 1–58.

Grumbach, M. M., Roth, J. C., Kaplan, S. L. and Kelch, R. P. 1974. Hypothalamic-pituitary regulation of puberty in man: Evidence and concepts derived from clinical research. *In* M. M. Grumbach, G. D. Grave and F. E. Meyer (eds.), *Control of the Onset of Puberty.* Wiley, New York, p. 115–166.

Gudernatsch, J. F. 1912. Feeding experiments on tadpoles. I. The influence of specific organs given as food on growth and differentiation. A contribution to the knowledge of organs with internal secretion. *Wilhelm Roux Arch. Entwicklungsmech. Org.* 35: 457–483.

Hadfield, M. G. 1977. Metamorphosis in marine molluscan larvae: An analysis of stimulus and response. *In* R.-S. Chia and M. E. Rice (eds.), *Settlement and Metamorphosis of Marine Invertebrate Larvae.* Elsevier, New York, pp. 165–175.

Highsmith, R. C. 1982. Induced settlement and metamorphosis of sand dollar (*Dendraster excentricus*) larvae in predator-free sites: Adult sand dollar beds. *Ecology* 63: 329–337.

Hinegardner, R. T. 1969. Growth and development of the laboratory cultured sea urchin. *Biol. Bull.* 137: 465–475.

Hogg, N. A. S., Harrison, D. J. and Tickle, C. 1983. Lumen formation in the mammary gland. *J. Embryol. Exp. Morphol.* 73: 39–57.

Hoskins, E. R. and Hoskins, M. M. 1917. On thyroidectomy in amphibia. *Proc. Soc. Exp. Biol. Med.* 14: 74–75.

Hoskins, S. G. and Grobstein, P. 1984. Thyroxine induces the ipsilateral retinothalamic projection in *Xenopus laevis*. *Nature* 307: 730–733.

Hoskins, S. G. and Grobstein, P. 1985a. Development of the ipsilateral retinothalamic projection in the frog *Xenopus laevis*. II. Ingrowth of optic nerve fibers and production of ipsilaterally projecting cells. *J. Neurosci.* 5: 920–929.

Hoskins, S. G. and Grobstein, P. 1985b. Development of the ipsilateral retinothalamic projection in the frog *Xenopus laevis*. III. The role of thyroxine. *J. Neurosci.* 5: 930–940.

Huxley, J. 1920. Metamorphosis of axolotl caused by thyroid feeding. *Nature* 104: 436.

Kaltenbach, J. C., Fry, A. E. and Leius, V. K. 1979. Histochemical patterns in the tadpole tail during normal and thyroxine-induced metamorphosis. II. Succinic dehydrogenase, Mg- and Ca-adenosine triphosphatases, thiamine pyrophosphatase, and 5' nucleotidase. *Gen. Comp. Endocrinol.* 38: 111–126.

Karp, G. and Berrill, N. J. 1981. *Development.* McGraw-Hill, New York.

Kawakami, A. and 9 others 1990. Molecular cloning of the *Bombyx mori* prothoracicotropic hormone. *Science* 247: 1333–1335.

Kinoshita, T., Takahama, H., Sasaki, F. and Watanabe, K. 1989. Determination of cell death in the developmental process of anuran larval skin. *J. Exp. Zool.* 251: 37–46.

Kistler, A., Yoshizato, K. and Frieden, E. 1977. Preferential binding of tri-substituted thyronine analogs by bullfrog tadpole tail fin cytosol. *Endocrinology* 100: 134–137.

Kollros, J. J. 1961. Mechanisms of amphibian metamorphosis: Hormones. *Am. Zool.* 1: 107–114.

Kratochwil, K. 1971. In vitro analysis of the hormonal basis for sexual dimorphism in the embryonic development of the mouse mammary gland. *J. Embryol. Exp. Morphol.* 25: 141–153.

Kratochwil, K. and Schwartz, P. 1976. Tissue interaction in androgen response of embryonic mammary rudiment of mouse: Identification of target tissue for testosterone. *Proc. Natl. Acad. Sci. USA* 73: 4041–4044.

Liscia, D. S., Doherty, P. J. and Smith, G. H. 1988. Localization of α-casein gene transcription in sections of epoxy resin-embedded mouse mammary tissues by *in situ* hybridization. *J. Histochem. Cytochem.* 36: 1503–1510.

Little, G., Atkinson, B. G. and Frieden, E. 1973. Changes in the rates of protein synthesis and degradation in the tail of *Rana catesbeiana* tadpoles during normal metamorphosis. *Dev. Biol.* 30: 366–373.

Lyman, D. F. and White, B. A. 1987. Molecular cloning of hepatic mRNAs in *Rana catesbeiana* response to thyroid hormone during induced and spontaneous metamorphosis. *J. Biol. Chem.* 262: 5233–5237.

Lynn, W. G. and Peadon, A. M. 1955. The role of the thyroid gland in direct development of the anuran *Eleutherodactylus martinicenis*. *Growth* 19: 263–286.

Madhavan, M. M. and Schneiderman, H. A. 1977. Histological analysis of the dynamics of growth of imaginal disc and histioblast nests during the larval development of *Drosophila melanogaster*. *Wilhelm Roux Arch. Dev. Biol.* 183: 269–305.

Martin, F. D. and Drewry, G. E. 1978. *Development of Fishes of the Mid-Atlantic Bight*, Vol. 6. U. S. Department of the Interior, Washington, D. C.

Mathison, P. M. and Miller, L. 1987. Thyroid hormone induction of keratin genes: A two-step activation of gene expression during development. *Genes Dev.* 1: 1107–1117.

McCutcheon, F. H. 1936. Hemoglobin function during the life history of the bullfrog. *J. Cell. Comp. Physiol.* 8: 63–81.

Mori, M., Morris, S. M., Jr. and Cohen, P. P. 1979. Cell-free translation and thyroxine induction of carbamylphosphate synthetase I messenger RNA in tadpole liver. *Proc. Natl. Acad. Sci. USA* 76: 3179–3183.

Nijhout, H. F. and Williams, C. M. 1974. Control of moulting and metamorphosis in the tobacco hornworm, *Manduca sexta*: Cessation of juvenile hormone secretion as a trigger for pupation. *J. Exp. Biol.* 61: 493–501.

Niki, K., Namiki, H., Kikuyama, S. and Yoshizato, K. 1982. Epidermal tissue requirement for tadpole tail regression induced by thyroid hormone. *Dev. Biol.* 94: 116–120.

Nishikawa, A., Kaiho, M. and Yoshizato, K. 1989. Cell death in the anuran tadpole tail: Thyroid hormone induces keratinization and tail-specific growth inhibition of epidermal cells. *Dev. Biol.* 131: 337–344.

Norris, D. O., Jones, R. E. and Criley, B. B. 1973. Pituitary prolactin levels in larval, neotenic, and metamorphosed salamanders (*Ambystoma tigrinum*). *Gen. Comp. Endocrinol.* 20: 437–442.

Pehrson, J. R. and Cohen, L. H. 1986. The fate of the small micromeres in sea urchin development. *Dev. Biol.* 113: 522–526.

Pino-Heiss, S. and Schubiger, G. 1989. Extracellular protease production by *Drosophila* imaginal discs. *Dev. Biol.* 132: 282-291.

Plowright, R. C. and Pendrel, B. A. 1977. Larval growth in bumble-bees. *Can. Entomol.* 109: 967–973.

Prahlad, K. V. and DeLanney, L. E. 1965. A study of induced metamorphosis in the axolotl. *J. Exp. Zool.* 160: 137–146.

Pratt, G. E., Jennings, R. C., Hammett, A. F. and Brooks, G. T. 1980. Lethal metabolism of precocene-I to a reactive epoxide by locust corpora allata. *Nature* 284: 320–323.

Raff, R. A. 1987. Constraint, flexibility, and phylogenetic history in the evolution of direct development in sea urchins. *Dev. Biol.* 119: 6–19.

Raynaud, A. 1961. Morphogenesis of the mammary gland. *In* S. K. Kon and A. T. Cowrie (eds.), *Milk: The Mammary Gland and Its Secretion*, Vol. 1. Academic Press, New York, pp. 3–46.

Richards, G. 1978. Sequential gene activity in polytene chromosomes of *Drosophila melanogaster*. VI. Inhibition by juvenile hormones. *Dev. Biol.* 66: 32–42.

Riggs, A. F. 1951. The metamorphosis of hemoglobin in the bullfrog. *J. Gen. Physiol.* 35: 23–40.

Robinson, H., Chaffee, S. and Galton, V. A. 1977. Sensitivity of *Xenopus laevis* tadpole tail tissue to the action of thyroid hormones. *Gen. Comp. Endocrinol.* 32: 179–186.

Rountree, D. B. and Bollenbacher, W. E. 1986. The release of the prothoracicotropic hormone in the tobacco hornworm, *Manduca sexta*, is controlled intrinsically by juvenile hormone. *J. Exp. Biol.* 120: 41–58.

Safranek, L. and Williams, C. M. 1984. Critical weights for metamorphosis in the tobacco hornworm, *Manduca sexta*. *Biol. Bull.* 167: 555–567.

Safranek, L. and Williams, C. M. 1989. Inactivation of the corpora allata in the final instar of the tobacco hornworm, *Manduca sexta*, requires integrity of certain neural pathways from the brain. *Biol. Bull.* 177: 396–400.

Schooneveld, H. 1979. Precocene-induced collapse and resorption of corpora allata in nymphs of *Locusta migratoria*. *Experientia* 35: 363–364.

Schwanzel-Fukada, M. and Pfaff, D. W. 1989. Origin of luteinizing hormone releasing hormone (LH-RH) neurones. *Nature* 338: 161–164.

Schwind, J. L. 1933. Tissue specificity at the time of metamorphosis in frog larvae. *J. Exp. Zool.* 66: 1–14.

Sláma, K. and Williams, C. M. 1966. The juvenile hormone. V. The sensitivity of the bug, *Pyrrhocoris apterus*, to a hormonally active factor in American paper-pulp. *Biol. Bull.* 130: 235–246.

Smith-Gill, S. J. and Carver, V. 1981. Biochemical characterization of organ differentiation and maturation. *In* L. I. Gilbert and E. Frieden (eds.), *Metamorphosis: A Problem in Developmental Biology*. Plenum, New York, pp. 491–544.

Strathmann, R. R. 1971. The feeding behavior of planktotrophic echinoderm larvae: Mechanisms, regulation and rates of suspension feeding. *Exp. Mar. Biol. Ecol.* 6: 109–160.

Strathmann, R. R. 1975. Larval feeding in echinoderms. *Am. Zool.* 15: 717–730.

Strause, L. G., DeLuca, M. and Case, J. F. 1979. Biochemical and morphological change accompanying light organ development in the firefly, *Photuris pennsylvanica*. *J. Insect Physiol.* 125: 339–347.

Styne, D. M. and Grumbach, M. M. 1978. Puberty in the male and female: Its physiology and disorders. *In* S. S. C. Yen and R. B. Jaffe (eds.), *Reproductive Endocrinology*. Saunders, Philadelphia, pp. 189–240.

Taigen, T. L., Plough, F. H. and Stewart, M. M. 1984. Water balance of terrestrial anuran (*Eleutherodactylus coqui*) eggs: Importance of paternal care. *Ecology* 65: 248–255.

Taurog, A., Oliver, C., Porter, R. L., McKenzie, J. C. and McKenzie, J. M. 1974. The role of TRH in the neoteny of the Mexican axolotl (*Ambystoma mexicanum*). *Gen. Comp. Endocrinol.* 24: 267–279.

Topper, Y. J. and Freeman, C. S. 1980. Multiple hormone interactions in the developmental biology of the mammary gland. *Physiol. Rev.* 60: 1049–1106.

Truman, J. W., Weeks, J. and Levine, R. B. 1985. Developmental plasticity during the metamorphosis of an insect nervous system. *In* M. J. Cohen and F. Strumwasser (eds.), *Comparative Neurobiology*. Wiley, New York, pp. 25–44.

Turkington, R. W. 1968. Hormone-dependent differentiation of mammary gland in vitro. *Curr. Top. Dev. Biol.* 3: 199–218.

Turkington, R. W., Juergens, W. G. and Topper, Y. J. 1965. Hormone-dependent synthesis of casein *in vitro*. *Biochem. Biophys. Acta* 111: 573–576.

Turner, C. D. and Bagnara, J. T. 1976. *General Endocrinology*, 6th Ed. Saunders, Philadelphia.

Wald, G. 1945. The chemical evolution of vision. *Harvey Lect.* 41: 117–160.

Wald, G. 1981. Metamorphosis: An overview. *In* L. I. Gilbert and E. Frieden (eds.), *Metamorphosis: A Problem in Developmental Biology*. Plenum, New York, pp. 1–39.

Waldhauser, F. and Dietzel, M. 1985. Daily and annual rhythms in human melatonin secretion: Role in puberty control. *Ann. N.Y. Acad. Sci.* 453: 205–214.

Wassersug, R. J. 1989. Locomotion in amphibian larvae (or why aren't tadpoles built like fish). *Am. Zool.* 29: 65–84.

Weber, R. 1965. Inhibitory effect of actinomycin D on tail atrophy in *Xenopus laevis* larvae at metamorphosis. *Experientia* 21: 665–666.

Weber, R. 1967. Biochemistry of amphibian metamorphosis. *In* R. Weber (ed.), *The Biochemistry of Animal Development*, Vol. 3. Academic Press, New York, pp. 227–301.

Weinberger, C., Thompson, C. C., Ong, E. S., Lebo, R., Gruol, D. J. and Evans, R. M. 1986. The c-*erb*-A gene encodes a thyroid hormone receptor. *Nature* 324: 641–646.

White, B. H. and Nicoll, C. S. 1981. Hormonal control of amphibian metamorphosis. *In* L. I. Gilbert and E. Frieden (eds.), *Metamorphosis: A Problem in Developmental Biology*. Plenum, New York, pp. 363–396.

Wigglesworth, V. B. 1934. The physiology of ecdysis in *Rhodnius prolixus* (Hemiptera). II. Factors controlling moulting and metamorphosis. *Q. J. Microsc. Sci.* 77: 121–222.

Wigglesworth, V. B. 1939. *Principles of Insect Physiology*. Chapman and Hall, London.

Williams, C. M. 1952. Physiology of insect diapause. IV. The brain and prothoracic glands as an endocrine system in the *Cecropia* silkworm. *Biol. Bull.* 103: 120–138.

Williams, C. M. 1956. The juvenile hormone of insects. *Nature* 178: 212–213.

Williams, C. M. 1970. Hormonal interactions between plants and insects. *In* E. Sondheimer and J. B. Simeone (eds.), *Chemical Ecology*. Academic Press, New York, pp. 103–132.

Wray, S., Nieburgs, A. and Elkabes, S. 1989. Spatiotemporal cell expression of luteinizing hormone-releasing hormone in the perinatal mouse. Evidence for an embryonic origin in the olfactory placode. *Dev. Brain Res.* 46: 309–318.

20

Patterns of growth and oncogenesis

The curve of the unwinding fern,
And the purple shell in the sea;
These are the spaces of the notes
Of every kind of music.
                    —KENNETH REXROTH (1952)

The agreement of so many species of animals in a common schema,
which appears to be grounded not only in their skeletal structure
but also in the organization of other parts, whereby a multiplicity
of species may be generated by an amazing simplicity of a funda-
mental plan, through the suppressed development of one part and
the greater articulation of another, the lengthening of now this part
accompanied by the shortening of another, gives at least a glimmer
of hope that the principle of mechanism, without which no science
of nature is possible, may be in a position to accomplish something
here.

                    —IMMANUEL KANT (1790)

THE MATHEMATICS OF ORGANISMAL GROWTH

Growing pains: Physical limits to growth

Unlike new machines, which do not need to function until after they have
left the assembly line, new organisms have to function as they develop.
The embryonic cells have to digest, metabolize, and excrete the entire time
the animal is developing, and these requirements place major constraints
on the ways in which animals can grow. Increasing size is not easily
accomplished, for growth can involve changes incompatible with life. The
ways in which animals grow represent creative compromises between
increasing size and the needs of all cells for oxygen and nutrients.

One of the major constraints on growth is the ratio of surface to
volume. If an animal were to grow larger while retaining its same shape,

then its surface area (needed for the absorption of oxygen and nutrients) decreases relative to its new volume. This happens because area changes as the square of the length (L^2), whereas volume changes as the cube of the length (L^3). If we have an animal whose radius is 4 units, then its surface area ($4\pi r^2$) is 200 square units and its volume ($4/3\pi r^3$) is 268 cubic units. The surface area to volume ratio is 0.74. If this animal were to double its length to a radius of 8 units, the new surface area would expand to 804 square units, whereas the volume would increase to 2143 cubic units. Its new surface area to volume ratio has become 0.37, half that of before. Another doubling would further halve the surface area to volume ratio. Thus, we have a fundamental paradox that becomes central to any growing organism: as the animal doubles its length, its surface to volume ratio is halved.

Put another way, the area:volume relationship states that as a body enlarges and retains the same shape, its volume grows faster than its surface. In any organism (including embryos under construction), the food and oxygen supply to the organism depend upon diffusion along absorptive surfaces. Fick's diffusion law holds that the flow of a given substance into a volume (such as a cell) is directly proportional to the area exposed to it. How, then, can organisms grow? One strategy is to become very broad or very flat, thereby exposing a large area to the source of nutrients and oxygen and letting these compounds enter a volume in close proximity to its surface. This strategy is used by the aptly named flatworms, which (as in the case of tapeworms) may grow 12 meters long, but never more than a few millimeters thick. Another strategy is to form a hollow tube, thereby keeping all the cells adjacent to the nutrient source. This strategy is employed by many cnidarians such as coral.

The most successful growth strategy to circumvent the surface:volume problem is invagination. The lungs and intestines are invaginated regions of the outer surface that have become convoluted to provide a huge area for the absorption of oxygen and nutrients, respectively. This modification of organism shape is often accompanied by another strategy: internal circulation. Here, nutrients are brought within the body so that diffusion can occur inside, rather than outside, the organism. This process is accomplished by the development of a circulatory system. But even a circulatory system runs into problems of size. According to the laws of fluid movement, the most effective transport of fluids is performed by large tubes. As the radius of the blood vessel gets smaller, resistance to flow increases as r^{-4} (Poiseuille's law). A blood vessel that is half as wide as another has a resistance to flow 16 times greater. So here is another paradox. The constraints of diffusion mandate that the vessels be small, while the laws of hydraulics mandate that the vessels be large. Living organisms have solved this paradox by evolving circulatory systems with a hierarchy of vessel sizes (LaBarbera, 1990). This hierarchy is formed very early in development, as can be seen in the 3-day chick embryo. In dogs, blood in the large vessels (aorta and vena cava) flows over 100 times faster than it does in the capillaries. By having large vessels specialized for transport and small vessels specialized for diffusion (where the blood spends most of its time), nutrients and oxygen can reach the individual cells of the growing organism.

But this is not the entire story either. If fluid, under constant pressure, moves directly from a large-diameter pipe into a small-diameter pipe (as in a hose nozzle), the fluid velocity increases. The solution to this problem was the emergence of many smaller vessels branching out from a larger one, such that the collective cross-sectional area of all the smaller vessels is greater than that of the larger vessel. This relationship (known as

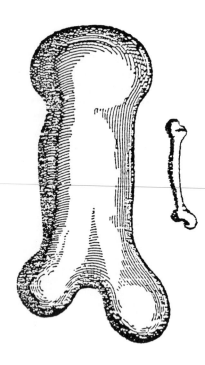

FIGURE 1
Galileo's illustration of the same bone (femur) from animals of different sizes. Whereas the lengths of the bones differ by about 2.5 times, the width of the bones differ over tenfold.

Murray's law) is that the cube of the radius of the parent vessel approximates the sum of the cubes of the radii of the smaller vessels. In this manner, the physical properties of matter have constrained the ways in which animals can develop. If animals do not develop a circulatory system, they must develop either a flat or a tubular surface. If a circulatory system does develop, its shape is constrained by the laws of hydraulics, causing the vessels to have different sizes arranged as branches.

There are other developmental constraints imposed by increasing size. The ability of a bone to support a heavy structure is proportional to the bone's cross-sectional area (L^2). Weight, however, increases as the volume of the organism (L^3). Thus, an animal that grows twice as large as it once was must develop limbs capable of bearing 8 times the weight. If it grows 4 times as large as it was, the limbs would have to support 64 times the original weight. If the animals developed in such a way that the shapes of their bones remained constant (ISOMETRIC GROWTH), the resulting skeletons would be unable to support their growing bodies. This was first recognized by Galileo in his *Discoursi* (written while under house arrest in 1638). He argued that the bones of large animals must become thicker relative to their length, compared with the thin bones of smaller animals (Figure 1). Therefore, as the organism grows, the ratio of height:girth decreases. This differential growth rate (in this case, the cross-sectional area of the bone growing faster relative to the height of the animal) is called ALLOMETRIC GROWTH.

As the animal grows, its absorptive surfaces, circulatory system, and skeleton must all undergo extensive remodeling. J. B. S. Haldane was undoubtedly correct when he wrote that "comparative anatomy is largely the story of the struggle to increase surface in proportion to volume."

Isometric growth

Brooks's law of linear growth

Most organisms grow by incorporating new material into the existing tissues of the body. In this way, the organism increases its volume and retains the same proportions. On the basis of the above discussion, one can determine that, theoretically, an animal that increases its weight twofold will increase its length by only 1.26 times (as $1.26^3 = 2$). This growth rate is seen frequently in nature. Brooks (1886) noted that the length of certain deep-sea arthropods collected by the *Challenger* expedition increased by about 1.25 times between molts. In 1904, Przibram and his colleagues let mantises hatch in captivity and weighed them after each successive molt. They found that the weight of any individual animal doubled from one molt to another. The shed exoskeletons of these animals were also collected. When the lengths of specific segments (such as the length of the tibia) were compared, the mean coefficient of size increase between molts was found to be 1.26. Even the hexagonal eye facets of this insect (which grow by expansion and not by cell division) increase their diameters by 1.26 times between molts. Other molting arthropods such as

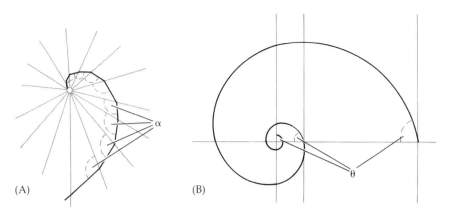

(A)   (B)

FIGURE 2
Equiangular spiral. (A) Equiangular spiral generated by a line cutting evenly spaced radii at the same angle α. (B) Descartes's analysis of an equiangular spiral showing that if the growing curve cuts each radius vector at a constant angle, then the curve can grow continuously without ever changing its shape. (From Thompson, 1942.)

crabs, lobsters, grasshoppers, and lice also follow this rule (known as Brooks's law) that a 1.26-fold increase in length reflects a twofold increase in weight.

The equiangular spiral

For the growth of an animal within a confined space such as a shell, the simplest condition is that it will widen and lengthen in the same proportion throughout its life. This type of growth produces either cones or equiangular (logarithmic) spirals. Such spirals have been defined by Thompson (1942) as "any curve proceeding from a fixed point . . . such that the arc intercepted by any two radii at a given angle to one another is always similar to itself." This is illustrated in Figures 2 and 3. An equiangular spiral can increase in size indefinitely and never change its shape. A small equiangular spiral has the same proportions as a large equiangular spiral in which the original angle is the same. The equation for such a spiral is $r = a^\theta$.

Such a growth pattern is typical of animals in which growth is permitted only at one end. This is the case in certain shells, horns, and claws. (Human nails will curve if grown long enough; and if rodents fail to gnaw constantly, their front teeth can curve back and strangle them). As Thompson said, "This remarkable property of increasing by *terminal* growth but nevertheless retaining unchanged the form of the entire figure, is characteristic of the equiangular spiral and of no other mathematical curve." The equiangular spiral is most readily seen in the relatively planar shells of nautiloid molluscs such as the chambered nautilus, a cross section of which is shown in Figure 3. As the nautilus grows, it builds a new chamber and moves into it. Only the last chamber houses the mollusc; the old compartments are filled with gas and contribute to the buoyancy and mobility of the animal. The remnants of these former living chambers persist as septa, and these septa distinguish nautilus shells from those of other types of molluscs.

Many shells appear not to be spirals, but merely domes. Thompson showed that these shells are actually pieces of equiangular spirals whose angles preclude a 360-degree turn. In a shell such as that found in *Nautilus*, one can measure the angle at which the growing curve intersects the radii. In most nautiloid and gastropod molluscs, this angle is between 80 and 85 degrees. (An angle of 90° would form a circle and the animal would not grow.) The angle of the chambered nautilus is usually 80°. Thompson derived a formula to calculate the ratio of widths of one whorl to the next.* Some of the ratios derived from this formula are shown in Table 1. Using

*As derived in Thompson (1942), the formula is $r = e^{2\pi(\cot\theta)}$.

FIGURE 3
Equiangular spiral growth patterns. The shell of a chambered nautilus and a ram's horn both show equiangular spiral growth. The nautilus shell (below) is cut in cross section. The growth of a ram's horn spirals in two planes. The base is triangular and none of the three sides grows at the same rate. The result is that the horn grows outward as well as backward, each ring representing a specific duration of growth.

TABLE 1
Constant angle of an equiangular spiral determines the ratio of the widths between whorls

Constant angle	Ratio of widths[a]
90°	1.0
89°8′	1.1
86°18′	1.5
83°42′	2.0
80°5′	3.0
75°38′	5.0
69°53′	10.0
64°31′	20
58°5′	50
53°46′	10^2
42°17′	10^3
34°19′	10^4
28°37′	10^5
24°28′	10^6

Source: Thompson (1942).
[a]Ratio of widths is calculated by dividing the width of one whorl by the width of the next larger whorl.

this table, one can calculate that if one whorl were 1 inch in breadth at one point on a radius and the angle of the spiral was 80°, the next whorl would have a width of 3 inches on the same radius. If the angle were 60°, the next whorl would be 4 feet wide on the same radius, and if the angle were 17° the next whorl would occupy a distance of 15,000 miles. Such low-angle curvatures conforming to sections of equiangular spirals are seen in some shells (mainly bivalves) and are very common in teeth and claws.

Using computer-generated curves, Raup (1962, 1966) showed that the growth of the more complex shell types seen in gastropod (snail) and pelecypod (bivalves such as clams) molluscs can be explained by four parameters:

w. The rate of whorl expansion, described above.
t. The rate of whorl expansion along the *y*-axis. This parameter indicates the proportion of the height that is covered as the curve moves on its helical pattern.
s. The shape of the growing curve. This value expresses the outline of the growing edge of the shell and represents a cross section of the hollow tube.
d. The position of the curve with relation to its axis. This value is largely a function of the angle at which the curve is growing.

In general, the shape of the spiral is defined by the *w* and *t* parameters.* Figure 4 shows five hypothetical snail shells drawn by computer-generated cross sections. Only the *w* and *t* parameters have been changed. In this way, it can readily be seen that extremely different changes in shell shapes can be generated by simply changing one or the other of these terms. In this manner, the types of mollusc shell types can be related to one another. Figure 5 shows how a hypothetical snail shell can be changed into types representative of various molluscan groups. In this figure, one begins with a shape that is found in numerous taxonomic groups of shells. If one then increases the rate of whorl translation (from $t = 0$ to $t = 3$), one generates the shapes of common snails. By changing the whorl expansion rate from $w = 3.5$ to $w = 10^4$, one goes from the region of one-shelled gastropods to that of bivalve shapes common to clams and scallops; and by increasing the distance from the coiling axis to the curve (from $d = 0$ to $d = 0.3$), one makes the shells more stretched out.

*If r_0 is the initial distance of point A from the axis, r_θ, the distance after θ revolutions is $r_\theta = r_0 w^{\theta/2\pi}$. It is obvious that this value will depend both on the angle of whorl expansion (*w*) and the relationship between whorl expansion in the *x*-axis and its translation down the *y*-axis (*t*). In nautiloid shells, $t = 0$, whereas in gastropods and pelycipods, *t* becomes important in this equation. Here, $y_\theta = y_0 w^{\theta/2\pi} + rt(w^{\theta/2\pi} - 1)$.

FIGURE 4
Hypothetical snail shells drawn by computer. Only two of the four parameters, the rate of curve enlargement (*w*) and the rate of curve translation (*t*) are varied in this series. (From Raup, 1962.)

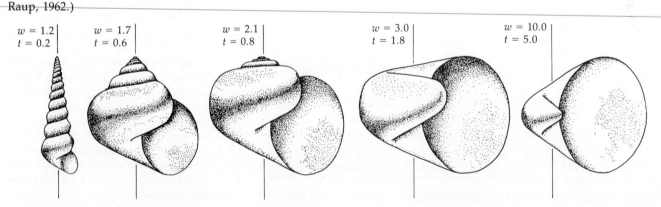

$w = 1.2$ $t = 0.2$	$w = 1.7$ $t = 0.6$	$w = 2.1$ $t = 0.8$	$w = 3.0$ $t = 1.8$	$w = 10.0$ $t = 5.0$

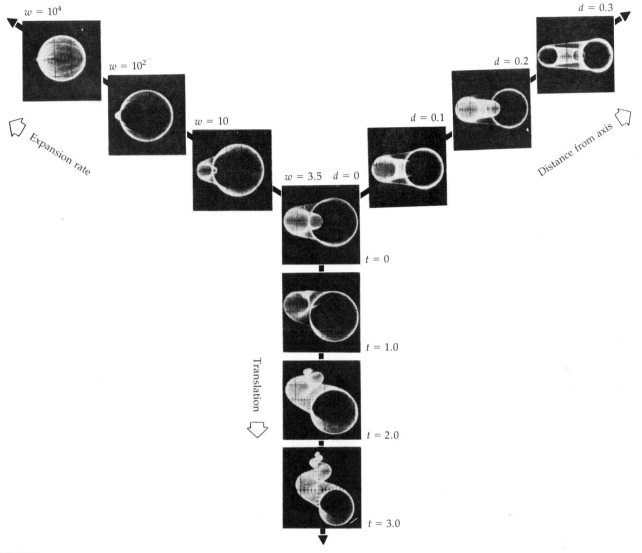

$w = 10^4$

$w = 10^2$

$w = 10$

Expansion rate

$w = 3.5$ $d = 0$

$d = 0.1$

$d = 0.2$

$d = 0.3$

Distance from axis

$t = 0$

$t = 1.0$

Translation

$t = 2.0$

$t = 3.0$

FIGURE 5
Changes in snail shell coiling parameters yield different types of snail shells. By altering the growth ratios of computer-generated shell curves, one can derive different families of molluscan shells. (From Raup, 1966.)

In the above discussion, it was assumed that the growth of the shell remains constant over the animal's lifetime. But this does not have to be the case. In some molluscs, the change is sudden and often corresponds to a major change in the animal's mode of life (as in going from a planktonic, free-swimming existence to a benthic, surface-dwelling mode of life). In other cases, the change is more gradual, allowing the snail to change its morphology as it develops. This strategy can be seen in the "worm-snails" shown in Figure 6. Differential growth rates may be an adaptive strategy, since an organism may encounter different conditions (such as water flow) as it enlarges (Randall, 1964).

By allowing variability in the parameters of their growth, molluscs have evolved a very flexible way of adapting to a large variety of conditions. Indeed, closely related molluscs may have enormously different shell morphologies. Although this plasticity complicates attempts to classify molluscs, it is testimony to the importance of growth patterns to evolutionary change.

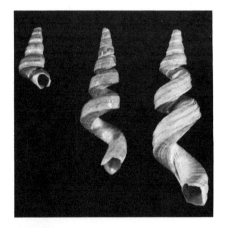

FIGURE 6
Changes of growth parameters during the lifetime of an organism. In these vermetid snail shells, the translation of the whorl about its axis changes as the animal matures.

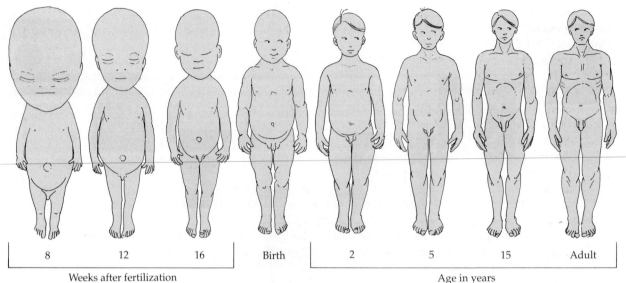

| 8 | 12 | 16 | Birth | 2 | 5 | 15 | Adult |

Weeks after fertilization   Age in years

FIGURE 7

Allometry in humans. The embryo produces a head that is exceedingly large in proportion to the rest of the body. After the embryonic period, the head grows slower than the torso, hands, and legs. Human allometry has been represented in Western art only since the Renaissance. Before that, children resembled little adults. (After Moore, 1983.)

Allometric growth

In many organisms growth is not a uniform phenomenon. It is obvious that there are some periods in an organism's life during which growth is more rapid than in others. Physical growth during the first ten years of a person's existence is much more dramatic than in the ten years following college graduation. Moreover, not all parts of the body grow at the same rate. This phenomenon of the different growth rates of parts within the same organism is called ALLOMETRY. Human allometry is depicted in Figure 7. Our arms and legs grow at a faster rate than our torso and head, such that adult proportions differ markedly from those of infants.

Julian Huxley (1932) has likened allometry to putting money in the bank at two different continuous interest rates. The formula for allometric growth (or for relating moneys invested at two different interest rates) is $y = bx^{a/c}$, where a and c are the growth rates of two body parts, and b is the value of y when $x = 1$. If $a/c > 1$, then that part of the body represented by a is growing faster than that part of the body represented by c. In logarithmic terms (which are much easier to graph), $\log y = \log b + (a/c)\log x$.

One of the most vivid examples of allometric growth is seen in the male fiddler crab, *Uca pugnax*. In small males, the two claws are of equal weight, each constituting about 8 percent of the total crab weight. As the crab enlarges, the size of its chela (the large crushing claw) grows even more rapidly, eventually constituting about 38 percent of the crab's weight (Figure 8; Table 2). When these data are plotted on double logarithmic plots (the body mass on the *x*-axis, the chela mass on the *y*-axis), one obtains a straight line whose slope is the *a*:*c* ratio. In the male *Uca pugnax* (whose name is derived from the huge claw), the *a*:*c* ratio is 6:1. This means that the mass of the chela increases six times faster than the mass of the rest of the body. In females of the species, the claw remains about 8 percent of the body weight throughout growth. It is only in the males (who use the claw for defense and display) that this allometry occurs.

Another important example of allometric growth is found in social insects. In ant colonies (as in those of bees and wasps), the society is composed almost entirely of females, and the structure of the society is based on the morphological differences between its female members. Com-

FIGURE 8
Specimens of the fiddler crab, *Uca pugnax*. The allometric growth occurs only in one of the male claws. In females (not shown), both claws retain isometric growth.

TABLE 2
Relative dry weights of large claw and body in fiddler crabs (*Uca pugnax*)

Weight of body without claw (mg)	Weight of claw (mg)	Percent of total weight in claw
58	5	8.6
80	9	11.2
200	38	19.0
300	78	26.0
420	135	32.1
536	196	36.6
618	243	39.3
743	319	42.9
872	418	47.9
1080	537	49.7
1212	617	50.9
1363	699	51.3
1449	773	53.7
1808	1009	55.8
2233	1380	61.7

Source: Huxley (1932).

plex ant societies contain several morphologically distinct types of sterile "worker" females and a fertile "queen," whose function is reproductive. The larger worker ("soldier") ants are distinguished by their large size, broad head, and large mandibles. In smaller workers ("minors"), the legs predominate. Within any species, the different sizes and shapes of the worker groups fit on a single allometric curve (Huxley, 1932; Wilson, 1953). This relationship is shown in Figure 9.

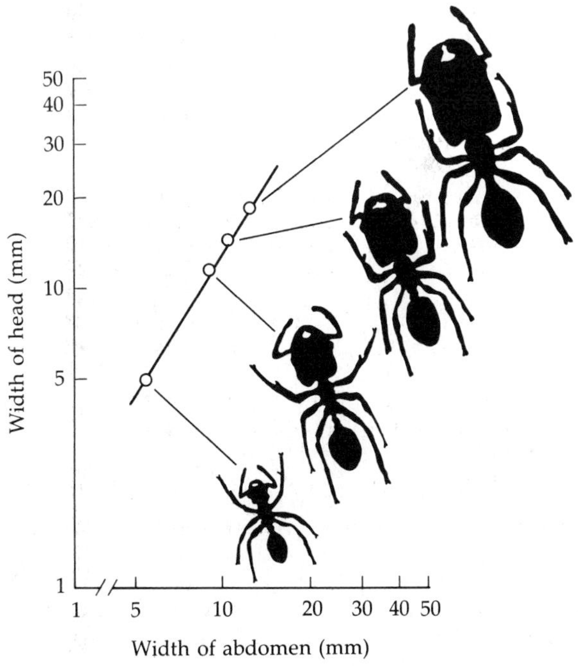

FIGURE 9
Allometric relationship between head size and body size in the ant *Pheidole instabilis*. Note the logarithmic axes of the graph. (After Raff and Kaufman, 1983.)

Newborn

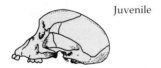

Juvenile

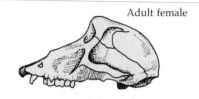

Adult female

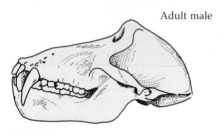

Adult male

FIGURE 10
Allometry in the baboon head. Skulls from individuals of various ages show the enormous increase of facial structures relative to those of the head. (After Huxley, 1932.)

The regulator of growth (and therefore of caste determination) in social insects appears to be juvenile hormone. It is known that juvenile hormone is important in determining which of the females becomes the queen in ant and bee colonies. Wheeler and Nijhout (1981) have also shown that the same hormone plays a decisive role in determining whether a larva is to become a soldier or a minor worker. Worker larvae of the ant *Pheidole bicarinata* undergo three larval molts between hatching and metamorphosis. The last larva-to-larva molt occurs when the larvae are approximately 0.6 mm long. Most of these larvae undergo the pupal molt 9 days later, when they are about 1.3 mm long. They metamorphose into small workers. A small percentage of the third instar larvae, however, are fed a more enriched diet. These larvae do not undergo a pupal molt for 15 days after their last molt, at which time they are approximately 1.8 mm long. These larvae metamorphose into soldiers. Wheeler and Nijhout were able to artificially prolong the duration of time spent as third instar larvae by applying juvenile hormone analogues to the larvae just before they reached 1.3 mm. The treated larvae continued to grow and metamorphosed into soldiers. Wheeler and Nijhout postulate that the rich diet fed to the soldier larvae elevates juvenile hormone levels, and that pupal molting will not take place until the juvenile hormone is eliminated from larval circulation. In the well-fed larvae, this does not take place for an extra 6 days, during which time the larvae have continued to grow. When these larvae metamorphose, the allometric growth produces the large jaw structures characteristic of soldier workers.

Allometric growth is also seen in nonhuman vertebrates. In baboons, the jaw and other facial structures have a growth rate about 4.25 times that of the skull (Figure 10). The more protruding face of the male baboon is merely the result of his being larger than the female. A similar instance of allometry is seen in the fossils of that most majestic of deer, *Megalocerus giganteus*, the so-called Irish elk. These enormous deer stood nearly 10 feet tall and had antlers that spanned up to 12 feet (Figure 11). They perished

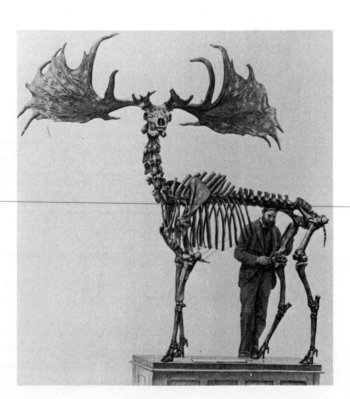

FIGURE 11
Positive allometry of antlers with respect to the skull of *M. giganteus*. Skeleton of the "Irish elk" (with *H. sapiens* for size comparison). Gould argues that antler positioning would favor a display function rather than a combative one. (From Gould, 1974.)

about 11,000 years ago when the climate of Europe became colder. Gould (1974) has analyzed 74 fossil skulls and antlers from this species and has shown a positive allometry for their antlers, which grow 2.5 times faster than the skull.

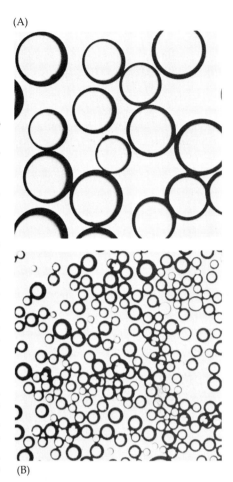

(A)

(B)

THE PHYSIOLOGY OF ORGAN GROWTH

Growth hormones and mitosis

There are two ways of increasing the size of an organ. The first is to retain a constant number of cells but to have the cells increase in volume. This is called HYPERTROPHY. The second mechanism is to increase the number of cells (each of which can remain the same size). This is called HYPERPLASIA (Goss, 1978). Hypertrophy is often found in those tissues that can no longer undergo mitosis. Adipose tissue in which the cytoplasm of the adipocytes is occupied by fatty lipid inclusions and skeletal muscle whose multinucleated cells have ordered arrays of actin and myosin are in this category. So are the larval cells of insects, which expand by hypertrophy while their neighboring imaginal discs expand by hyperplasia. In humans, the precursors of adipocytes divide through the first year of age (Brook, 1972; Adebonojo, 1975). After that, their number is constant, although there is some evidence of new fat cells appearing during adolescence. The growth of fat tissue can be induced by giving animals high-fat diets, which increase the size of the individual adipocytes without changing their number (Stern and Greenwood, 1974), or by prolonging the time during which the fat cell precursors can divide. This extra division time appears to be responsible for the genetic obesity seen in certain inbred strains of animals (Johnson et al., 1971). Figure 12 shows the rhythm of adipocyte hypertrophy in the yellow-bellied marmot, *Marmota flaviventris*. These animals acquire enormous fat deposits, which enable them to hibernate over the winter. Upon waking in the spring, their adipocytes are fat-depleted, and the marmot emerges from its den a leaner animal (Ward and Armitage, 1981). The hypertrophy of human skeletal musculature (and blood vessels) can be seen in Figure 13. Such growth of particular sets of muscles is often seen in shotputters and carpenters.

In some organisms, cell hypertrophy is seen in cells where there are more than two sets of chromosomes per nucleus. For example, in diploid newt larvae, the pronephric tubule is usually formed by a ring consisting of five cells. In pentaploid larvae (containing five haploid chromosome sets), one cell can curve around to make the entire ring. Similarly, the brain cells of these pentaploid larvae are much larger than those of diploid salamanders, but their are many fewer cells (Frankhauser, 1952).

Most tissue growth in vertebrates occurs by mitotic division. The functional units of the kidney and lung—the nephrons and alveoli, respectively—are fixed in their numbers early in infancy. Their growth is due to the addition of more cells to the existing structures. Similarly, growing skin tissue expands by the addition of more cells and not by the enlargement of existing ones. Some tissues, such as the liver, retain a population of cells that not only can divide but also can form new structures such as lobules (Goss, 1966). Figure 14 shows the times during which functional units may be added to growing structures in the human body.

What is it that controls the rate of cell division? Some cells (such as neurons, fat cells, and skeletal muscle cells) hardly divide at all during

FIGURE 12
Hypertrophy of adipose cells in the marmot. Dispersed adipocytes (fat cells) of the yellow-bellied marmot (A) before hibernation and (B) after hibernation. (Photographs courtesy of G. Florant.)

FIGURE 13

Hypertrophy of human musculature seen in the deltoid and biceps muscles of a professional "body-builder." This growth occurs long after the muscle cells have lost their ability to divide. Note also the hypertrophy of the blood vessels servicing these muscles. (Photograph courtesy of J. Watson.)

growth. Other cells (such as fibroblasts, bone cells, and kidney cells) have a limited and precisely regulated pattern of cell division. Still other cells (such as the stem cells for blood cells and sperm) divide continuously during the lifetime of the organism. There appear to be two major levels of control over cell division. The first is *extrinsic* control, wherein the growth of an organ depends upon factors derived from other tissues, and the second is *intrinsic* control, wherein a tissue or organ regulates its own growth.

Extrinsic growth control is seen when one organ influences the growth of another. In most vertebrates, growth control over the entire body is regulated by the synthesis of SOMATOTROPIN, or GROWTH HORMONE (GH). Its most obvious effects are exerted over the long bones of the limb (Chapter 6), and lack of growth hormone produces proportional dwarfism in humans. In order to coordinate bodily growth, however, growth hormone must increase the size of other organs as well, and responsiveness to this hormone has also been seen in liver, kidney, bone marrow, pancreas, mammary glands, gonads, thymus, adipose tissue, and hypothalamus. Fetal tissue appears to be more sensitive to somatotropin than adult tissue, and somatotropin is produced in the human fetus as early as 70 days after fertilization (Martin, 1985).

The ability of growth hormone to coordinate bodily growth was shown in a dramatic experiment using recombinant plasmids. In 1982, Doehmer and co-workers created a "library" of rat chromosomal DNA by cutting the rat genome with low concentrations of restriction enzymes such as *Eco*RI and cloning these fragments into a plasmid vector, pBR322. By using

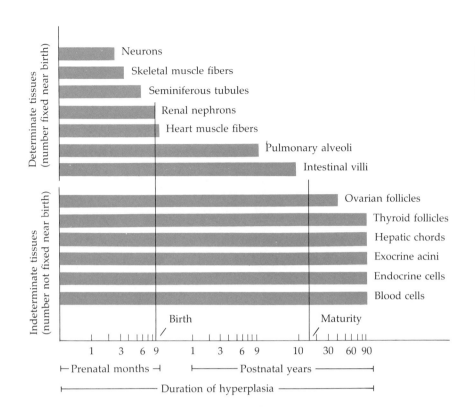

FIGURE 14
Duration of mitotic phase of growth for various human tissues. (After Goss, 1966.)

low concentrations of the enzyme, many pieces of DNA would include an *Eco*RI site that was not cleaved. This method increased the chances of isolating relatively long fragments that contain an entire gene. A radioactive cDNA probe was made from rat growth hormone mRNA by reverse transcriptase (Chapter 10). This probe was used to identify recombinant DNA clones containing the rat growth hormone gene. One such clone contained a 7600-base pair stretch of DNA that appeared to include the entire rat growth hormone gene. When this gene was transfected into cultured mouse fibroblast cells, the gene was incorporated into the cells' DNA and was able to synthesize rat growth hormone.

Palmiter and co-workers (1982) then isolated the 7600-base pair gene from the recombinant plasmid and removed the promoter sequence from the rat growth hormone gene. This sequence contains the *cis*-regulatory sequences that govern its normal transcription. Into this space they substituted the *cis*-regulatory sequence of another gene, that of mouse metallothionein I *(MT-I)*, a small protein involved in regulating serum zinc levels. This hybrid gene is shown in Figure 15A. The *MT-I* gene can be induced by the presence of heavy metals such as zinc or cadmium, and the sequences responsible for this induction are on the 5′ flanking region of the gene. By fusing this metallothionein promoter region to the rat growth hormone gene *(rGH)*, the rat growth hormone gene is placed under the control of the metallothionein promoter. In this case, rat growth hormone should be made when the promoter is activated by the presence of zinc or cadmium.

The triple recombinant plasmid (pBR322 bacterial plasmid-mouse metallothionein promoter-rat growth hormone gene) was grown in bacteria, and the *MT-I-rGH* piece was isolated. About 600 copies of this fragment were injected into pronuclei of recently fertilized mouse eggs. Of the 170 zygotes that were injected and implanted into the uteri of pseudopregnant mice, 21 developed to term. DNA hybridization showed that these mice had incorporated numerous copies of the rat growth hormone gene into

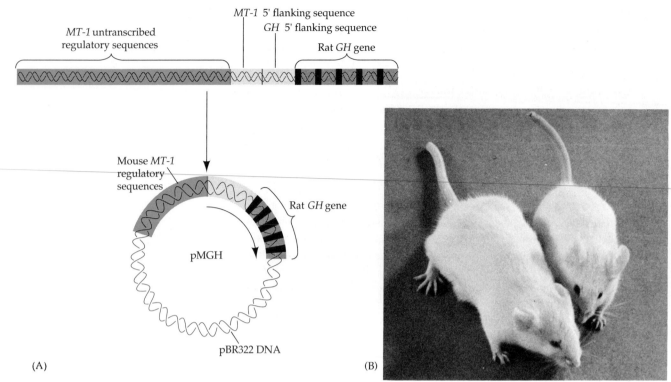

FIGURE 15

Growth control in transgenic mice. (A) Recombinant plasmid containing rat growth hormone structural gene, mouse metallothionein regulatory region, and bacterial plasmid pBR322. The plasmid, pmGH, was injected into the mouse oocytes. The dark boxes on the injected plasmid correspond to the exons of the GH gene. The direction of transcription is indicated by an arrow. Above the circular plasmid is an enlargement of the junction between the metallothionein regulatory region and the rat GH gene. (B) A mouse derived from the eggs injected with pMGH (left) and a normal littermate (right). (From Palmiter et al., 1982; photograph courtesy of R. L. Brinster.)

their chromosomes. These transgenic mice (mice having a gene from another species—in this case, the rat growth hormone gene) were then fed a diet supplemented with zinc. The zinc induced large amounts of rat growth hormone to be made from the livers of these mice. (The liver is where metallothionein is usually made. Growth hormone is usually secreted from the pituitary gland.) The amount of growth hormone secreted correlated with the size of these mice. The transgenic mice became enormous, up to 80 percent larger than their normal littermates (Figure 15B). The "supermice" are all normally proportioned, as each organ has been appropriately enlarged. The only difference between the large mice and their normal littermates was the presence of rat growth hormone genes, so it appears that the growth hormone is responsible for the coordinated regulation of mammalian growth.*

Growth hormone probably acts directly on some cells and indirectly on others. Isaksson and co-workers (1982) postulated a direct effect of growth hormone on bone growth. After removing the pituitary glands from a group of young rats, they added growth hormone directly to the limb growth plates on one side of the animals and not on the other side. The treated limbs grew, while the untreated limbs did not. Growth hormone binds with high affinity to cartilage cells and stimulates DNA synthesis in cultured chondrocytes (Eden et al., 1983; Madsen et al., 1983). The indirect effects of growth hormone are thought to be effected by compounds called SOMATOMEDINS. These peptides are secreted by the liver in response to growth hormone, and they are low in those people with growth hormone insufficiency and are elevated when these individuals are given growth hormone. Somatomedins are also elevated in adolescents

*Two other things grew with this experiment. The first is our potential ability to cure genetic disease by fertilizing eggs in vitro and injecting a normal gene into a pronucleus. These eggs can begin their development and then be returned to the woman's uterus. The second thing that grew was our responsibility (which usually is proportional to our power, whether we like it or not).

undergoing their "growth spurt" and in people with growth hormone-secreting tumors (Vassilopoulou-Sellin and Phillips, 1982; Luna et al., 1983). The somatomedins include INSULIN-LIKE GROWTH FACTOR I (IGF-I; somatomedin C) and INSULIN-LIKE GROWTH FACTOR II (IGF-II; somatomedin A).

These growth factors work at specific stages of the cell cycle. As summarized earlier, the somatic cell cycle is divided into mitosis (M), a "gap" (G_1), a period of DNA synthesis (S), and another "gap" (G_2) between S and M. Cells that are no longer dividing are said to be in a quiescent, G_0, state. This G_0 differs from G_1 in that a cell in G_0 synthesizes certain proteins different from those synthesized by the same cell when it enters the cell cycle at G_1 (Baserga, 1985). Cells starting in G_0 are stimulated by growth factors to a state where they are "competent" to enter the cell cycle. Here, they are essentially equivalent to those cycling cells that have just finished mitosis. About halfway through this G_1 period, they reach a point where a decision is made to divide. Before this point, inhibitors of transcription or translation could have stopped mitosis. After this point, S, G_2, and M proceed independent of external stimulation (Figure 16; Yang and Pardee, 1986).

IGF-I works by allowing cells to pass from the G_1 phase of the cell cycle and enter S phase (Leof et al., 1982). Nilsson and colleagues (1986) have recently shown that this protein is important in growth hormone-regulated limb growth. When growth hormone is added to the tibial growth plate of young hypophysectomized mice (as in the above-mentioned experiment), it stimulates the formation of IGF-I from the chondrocytes in the proliferative zone (Figure 17). The combination of growth hormone and IGF-I may provide an extremely strong mitotic signal. Growth hormone may stimulate the division of stem cells into cells that are responsive to IGF-I. These new cells would undergo further replication in response to the somatomedin (Zezulak and Green, 1986). The pygmies of the Ituri Forest of Zaire have normal GH and IGF-I levels until puberty. However, at puberty, the pygmies' IGF-I levels fall to about one-third that of other adolescents. It appears that IGF-I is essential for the normal pubertal growth spurt (Merimee et al., 1987).

As their alternate names imply, the somatomedins have amino acid sequences similar to that of insulin, which is also a growth-promoting peptide (as in the case of the developing mammary gland in mice; Chapter 19). Growth hormone itself is also homologous to insulin, as are placental

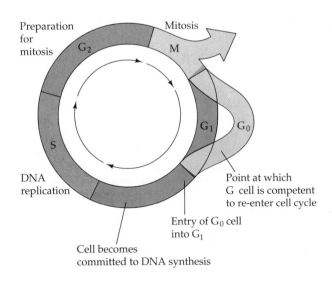

FIGURE 16

Cell cycle of normal mammalian cells, showing points where G_0 cells become competent to enter the cell cycle and where G_1 cells become committed to undergo DNA synthesis.

FIGURE 17

Effect of growth hormone (GH) on cartilage growth and insulin-like growth factor I (IGF-I) synthesis. Growth hormone was injected into the growth plate of the rat tibia on one side of a hypophysectomized rat (A) and not in the other (B). Microscopic sections were stained with an antibody recognizing IGF-I. Arrows indicate the width of the epiphyseal growth plate. The micrographs show that GH induces growth and the production of IGF-I. (From Nilsson et al., 1986; photographs courtesy of O. Isaksson.)

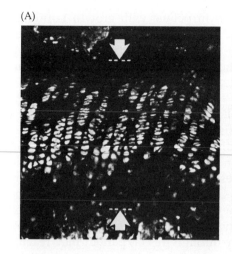

(A)

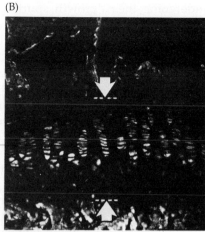

(B)

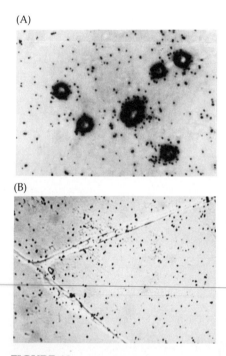

(A)

(B)

FIGURE 18

Fibroblast growth factor (FGF) binding during muscle development. (A) Binding of radioactive FGF to dividing muscle cells shows the presence of membrane receptors of FGF. (B) These receptors are lost when the cells stop dividing to form myotubes. (From Olwin and Hauschka, 1988; photographs courtesy of the authors.)

lactogen (a hormone that is made by the trophoblast and that probably affects fetal growth altering the glucose metabolism in the mother) and nerve growth factor (whose functions were detailed in Chapter 5). Thus, insulin, growth hormone, somatomedins, placental lactogen, and nerve growth factor are a family of related growth factors.

Other important extrinsic growth factors have also been identified. The PLATELET-DERIVED GROWTH FACTOR (PDGF) stimulates the division of smooth muscle cells, fibroblasts, and glial cells (Ross and Vogel, 1978). It triggers the nondividing cell in the G_0 state to enter the G_1 phase of the cell cycle (Stiles et al., 1979). It is probably the growth factor responsible for healing blood vessel tears, as it increases in those cells exposed to the free edge of a cut. PDGF reacts with its cell-surface receptor to induce protein synthesis from the nucleus (Olashaw and Pledger, 1982). Another important growth factor, EPIDERMAL GROWTH FACTOR (EGF), promotes division in many tissues, including mammary epithelium and epidermis. It is necessary for the formation of the hard palate and the epithelial linings of the esophagus, mouth, and cornea. EGF functions both as a growth factor and as a differentiation-promoting hormone. When injected into newborn mice, it accelerates the opening of the eyelids and the eruption of the teeth from the gums (Savage and Cohen, 1972). Whereas PDGF enables nondividing cells to enter the cell cycle (the G_0 to G_1 competency), EGF is able to get cells to go from G_1 into S phase (Pledger et al., 1984).

The ability to respond to particular growth factors can be an extremely important developmental event. In Chapter 6 we saw how different types of blood cell precursors became able to respond differentially to hematopoietic growth factors, and in Chapter 16 we discussed how B cells acquire the receptors for the growth factors derived from helper T cells. Skeletal myoblasts divide in the presence of either EGF or FIBROBLAST GROWTH FACTOR. When they fuse together to form myotubes, they lose the receptors for these mitosis-stimulating proteins (Figure 18; Lim and Hauschka, 1984; Olwin and Hauschka, 1988). There are numerous different growth-promoting hormones; some of these can stimulate division in a wide variety of cells, while others have very specific targets for their activities.

The insulin-like family of growth factors, which include PDGF, EGF, and at least one hematopoietic growth factor—colony-stimulating factor I (CSF-I)—all appear to operate by the same general mechanism. In each case, the receptor for these hormones is a protein that spans the cell membrane and can be divided into three functional parts: (1) an outer domain that recognizes and binds with the specific growth factor; (2) a

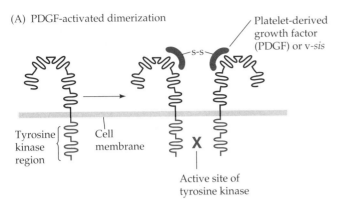

(A) PDGF-activated dimerization

Platelet-derived growth factor (PDGF) or v-*sis*

Tyrosine kinase region

Cell membrane

Active site of tyrosine kinase

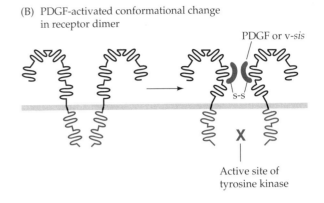

(B) PDGF-activated conformational change in receptor dimer

PDGF or v-*sis*

Active site of tyrosine kinase

FIGURE 19

Interaction of the platelet-derived growth factor and its receptor. The receptor for platelet-derived growth factor spans the cell membrane. Its external domain can contains five immunoglobulin-like domains that can bind PDGF. Within the cytoplasm are two regions that together constitute a tyrosine kinase. The constraints on protein structure suggest that the active tyrosine kinase is actually a dimer of two receptors. In model (A), the PDGF protein is active in causing these dimers to occur. In model (B), PDGF interacts with pre-existing dimers to create the active tyrosine kinase by conformational change. (After Williams, 1989.)

domain of hydrophobic residues that spans the plasma membrane; and (3) a domain within the cytoplasm that can be activated into an ATP–TYROSINE PHOSPHOKINASE. This enzymatic ability (usually just called tyrosine kinase activity) is relatively inactive until the receptor binds its particular growth factor. When that occurs, a conformational change enables the internal domain to catalyze the transfer of the terminal phosphate of ATP to tyrosine residues on adjacent proteins (Cohen et al., 1980; Ullrich et al., 1985; Ebina et al., 1985). This phosphorylation of tyrosine residues is a remarkable event. Almost all known protein phosphorylations (over 95 percent) occur at serine and threonine residues. The phosphorylation of tyrosine residues appears to be a major signal instructing several kinds of cells to initiate DNA synthesis and cell division (Figure 19).

SIDELIGHTS & SPECULATIONS

Spotted mice and stem cell growth factor

Hormones such as somatotropin work in an endocrine fashion. They are synthesized in one organ and travel through the blood to reach their target tissues. *Growth factors*, however, usually function in a PARACRINE fashion. They are synthesized and secreted by cells adjacent to the ones they affect. Therefore, instead of circulating through the blood, most growth factors diffuse locally and act in a small area. These growth factors appear to be crucial for morphogenesis, and they are often seen to be synthesized transiently, just before or during times when neighboring tissues are proliferating. For instance, transforming growth factor-β (TGF-β, which we discussed as a possible mesoderm inducer in Chapter 8) is seen to be synthesized in the tooth epithelium during the time that the neighboring tooth mesenchyme is proliferating (Pelton et al., 1990; Vaahtokari et al., 1990).

Recently, two mutations in the mouse led to the discovery of an important growth factor whose synthesis is necessary for the proliferation of three types of stem cells. Mice having mutations at the dominant *White spotting* (W) and *Steel* (Sl) loci have deficiencies involving three independent cell lineages. Viable homozygotes of the mutants have a severe reduction in their number of melanocytes, hematopoietic stem cells, and primordial germ cells. As a result, these mutants are white, anemic, and sterile (Sarvella and Russell, 1956; Mintz and Russell, 1957). Heterozygotes of these mutants show fewer than normal cells of these lineages, as is evident from their coat colors. *Steel* heterozygotes are grayish due to their deficiency of melanocytes, while *White spotting* heterozygotes have a white belly spot due to the inability of the melanocytes to migrate from the neural crest into the ventral ectoderm (Figure 20).

Transplantation of neural crests and bone marrow cells from wild-type embryos into *White spotting* embryos re-

FIGURE 20
Ventral surface of a mouse heterozygous for the *White spotting* mutation. The mouse has reduced numbers of blood cells, germ cells, and melanocytes. The white belly spot is a result of the diminished number of melanocytes.

sulted in the proliferation and migration of the wild-type melanoblasts and hematopoietic stem cells. However, neural crests and bone marrow cells from wild-type embryos did not proliferate or migrate in the *Steel* hosts. Similarly, these stem cells from *White spotting* embryos did not proliferate in wild-type embryos, whereas the *Steel* stem cells did (Mayer and Green, 1968; Bernstein et al., 1968).

These studies concluded that the *W* gene is expressed in these three stem cell lineages, while the *Sl* gene affects the microenvironment in which these cells divide and migrate.

In 1988, the wild-type *W* locus was found to encode a transmembrane tyrosine kinase that had the structure of a growth factor receptor (Chabot et al., 1988; Geissler et al., 1988). This tyrosine kinase, c-*kit*, has been found in the cell membranes of hematopoietic stem cells, primordial germ cells, and presumptive subepidermal melanocytes. (Orr-Urteger et al., 1990). Because the product of the wild-type *W* allele looked like a growth factor receptor, the search began for its ligand, a presumed growth factor. In 1990, several laboratories (see Witte, 1990) found a new growth factor that specifically bound to the *W*-encoded tyrosine kinase. Moreover, it was encoded by the wild-type *Steel* allele. This protein, stem cell factor, was able to correct the inherited defect of the *Steel* mice.

The cells secreting stem cell factor are associated with the migratory pathways of these three cell lineages (Matsui et al., 1990). Messenger RNA for this protein is seen (a) along the hindgut, mesentery, and genital ridge cells over which the primordial germ cells migrate; (b) in the ectoderm and dorsal region of the somites through which the melanoblasts migrate; and (c) in the yolk sac, embryonic liver, and developing bone marrow that provide the microenvironments for hematopoietic stem cells. In addition, there are regions of the nervous system that express the stem cell growth factor, suggesting that this growth factor may stimulate other cells as well. The identification of the *Steel* product as a growth factor for stem cells and the identification of the *White spotting* product as its receptor solves an embryological enigma over 40 years old. This identification also helps us understand how the selective proliferation of these three cell lineages is so well coordinated in the mammalian embryo.

Growth inhibitory factors

In addition to factors that stimulate cell growth, there are also proteins that inhibit it. Two of these proteins have been identified as β-INTERFERON (β-IFN) and transforming growth factor-β (TGF-β). The effect of TGF-β on cells depends on the target cell type and the presence of other growth factors in its environment. Indeed, its ambiguous name reflects its dual capacity. One laboratory (Roberts et al., 1981, 1983) isolated this factor by its ability to transform normal cells into malignant ones. Normal fibroblasts will not grow without a solid substrate to support them. Tumor cells, on the other hand, can grow in media such as soft agarose. When treated with TGF-β, some normal cells began to exhibit the growth properties of tumor cells. Another laboratory (Holley et al., 1980) isolated the same molecule by assaying for its ability to inhibit the division of cultured kidney cells. They found that when present at only 10 ng/ml, this protein inhibited division by 80 percent. It is now known that a major effect of TGF-β is to inhibit cell growth (Proper et al., 1982; Lawrence et al., 1984; Roberts et al., 1985) and that the effects of this protein can be changed by adding other factors. Several tumors (including lung carcinoma, mammary carcinoma, and melanoma cell lines) are inhibited by TGF-β (Roberts et al., 1985). The growth inhibitory effects of TGF-β can be seen on lymphocytes and on many types of epithelial cells, including epidermal cells. Normal retina cells are very sensitive to TGF-β, but retinoblastoma cells have lost the receptor for this protein and are resistant to this factor (Kimchi et al., 1988). TGF-β, however, is a growth factor for fibroblast cells and embryonic capillary endothelial cells (Hsu and Wang, 1986; Heine et al., 1987). In 11- to 18-day mouse embryos, TGF-β is seen predominantly in those areas of the mesenchyme where there is angiogenesis.

β-Interferon also inhibits cell proliferation and blocks cells between G_0 and G_1 (Einat et al., 1985). Hunter (1986) and Resnitzky and co-workers (1986) have reported that certain growth factors (including EGF, CSF-I, and PDGF) induce the synthesis of β-IFN as one of their later effects. This finding establishes an important feedback loop that might act to stop runaway cell proliferation (Figure 21). β-Interferon inhibits cell proliferation by preventing cells from entering S phase (Sokawa et al., 1977). This is the same site where the growth factor PDGF is thought to work, and recent evidence suggests that β-interferon blocks PDGF activity (Lin et al., 1986). When PDGF is added to a plate of mouse fibroblasts that are in G_0 as a result of cell crowding, the growth factor will cause the cells to enter S phase. However, when β-IFN is added simultaneously, the cells do not respond to the PDGF. Moreover, the β-IFN-treated cells are inhibited from synthesizing the specific proteins that are associated with PDGF stimulation. Thus, β-IFN appears to counteract the action of PDGF, thereby preventing cells from entering S phase of the cell cycle.

Cell division appears to be controlled by the interplay of growth-promoting factors and growth-suppressing factors. Moreover, these factors can act in concert to regulate cell growth in such a way that mitosis not only can occur but can also be stopped before the cells expand beyond the genetically defined limits.

FIGURE 21
Feedback control of growth during blood cell differentiation. (A) Model of the regulation of cell proliferation by β-interferon. Growth factors—in this instance, colony-stimulating factor 1 (CSF-1)—instruct cells to divide and at the same time initiate the synthesis and secretion of β-interferon (β-IFN). The cell surface receptor for β-IFN (on the same cell) binds the β-IFN, causing (by an as-yet-unknown mechanism) the degradation of c-*myc* RNA. (B) The expression of c-*myc* genes during the terminal differentiation of mouse myeloid cells as seen in Northern blots of cellular RNA. The c-*myc* RNA is seen while the cells are proliferating (day 0). When they cease dividing and are completely differentiated (day 3), the c-*myc* transcripts disappear. However, if the cell's own β-IFN is inactivated by adding anti-IFN antibodies to the medium, the cells continue to grow and c-*myc* RNA is seen. (From Resnitzky et al., 1986; photograph courtesy of D. Resnitzky.)

(A)

CSF-I receptors bind CSF-I

↓

Tyrosine phosphorylation

↓

Cytoplasmic signals to divide

Activation of β-IFN

↓

Secretion of β-IFN

↓

Binding of β-IFN by cell surface β-IFN receptors

Destruction of c-*myc* RNA

Activation of c-*myc*

↓

Accumulation of c-*myc* mRNA and protein

↓

Commitment of cell to divide

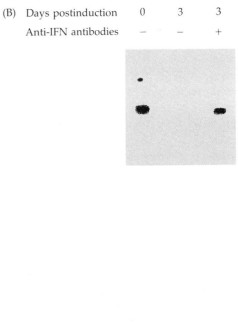

(B)

Days postinduction	0	3	3
Anti-IFN antibodies	−	−	+

ONCOGENES AND CELL GROWTH

Viral oncogenes

During the past decade, research into the physiology of cell growth has become united with research coming from a very different perspective— virology. The mechanism by which normal cells become cancer cells is called ONCOGENESIS and has been the object of speculation for centuries. At the turn of the century, Theodor Boveri speculated that cancers arose from defects in chromatin structure, but he had no way of proving that idea. At about the same time, Peyton Rous, a young investigator at the Rockefeller Institute, claimed that viruses were responsible for causing cancers. But until recently neither claim could be rigorously tested. In the 1980s, however, the techniques of gene cloning and monoclonal antibody production have made it possible to analyze the mechanisms of oncogenesis both within the nucleus and at the cell surface. The result has been a remarkably stimulating union of the viral and cellular theories of oncogenesis. It now appears that cancers arise as defects in the developmental growth pattern of cells and that the viral cancer-causing genes are actually the descendants of growth control genes that once resided within cells.

Rous (1910) isolated a virus capable of producing connective tissue tumors (SARCOMAS) in chickens. In 1958, Temin and Rubin grew chick cells in culture and were able to show that the Rous sarcoma virus could infect these cells and transform them into tumor cells. Normal fibroblasts display contact inhibition of movement and cell division. When they touch one another, they will move away in the opposite direction. If surrounded on all sides, the cell will neither move nor divide. Thus, normal cells form a single layer when placed into culture. Malignant cells, however, have lost these inhibitory properties. They can continue to grow even when touched on all sides, and their shape changes drastically from what it had been. Cells that have become malignant (TRANSFORMED) in culture can be recognized by their characteristic formation of mounds of dividing cells (FOCI) on an otherwise flat layer of cells (Figure 22). Because certain viruses were able to transform normal cells, it was hypothesized that there might be a gene within the virus that was responsible for transforming the cells. These hypothesized genes were called ONCOGENES (Gk. *onkos*, "cancer"), but it was not until 1979 that Copeland and co-workers were able to transform chick fibroblasts with a restriction enzyme fragment of Rous sarcoma virus DNA that contained a single gene. Not only did this gene, called *src*, produce the transformed phenotype when incorporated into cultured cells, but it also caused tumors when injected into otherwise healthy chicks. This *src* gene was found to encode a 60,000-Da protein (Brugge and Erikson, 1977; Beemon and Hunter, 1977).

As more tumor-causing viruses and their oncogenes became isolated, a strange pattern presented itself: the oncogenes were not related to the other genes of the virus and were totally dispensable for viral function. They were merely hitchhikers in the virus (see Bishop, 1985). Temin (1971) proposed that the viral oncogenes were normal cellular genes that had mutated and become packaged into viral particles. The most convincing evidence for the conversion of normal cellular genes into viral oncogenes came from the work of Hanafusa and co-workers (1977), who showed that a virus could pick up an oncogene when it infected a cell. They used a mutant of Rous sarcoma virus that lacked most of the *src* gene and was therefore unable to transform normal cells into tumor cells. This defect did not prohibit the virus from infecting cells and replicating within them.

FIGURE 22

Assay for malignancy. Normal cells (A) cease movement and DNA synthesis when they make contact with other cells. This leads to a single layer of cells on the surface of a petri dish. Malignant cells (B) show no such restraints, so they produce mounds of cells atop each other. (Photographs courtesy of G. S. Martin.)

(A)

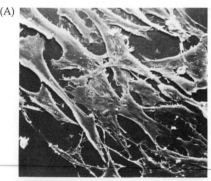

(B)

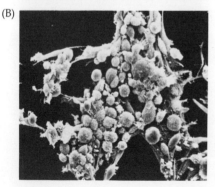

After several rounds of infection, some virus particles regained the ability to produce tumors when injected into chickens. When these new oncogenic viruses were analyzed, Hanafusa and his colleagues found that an entire new *src* gene was present. Thus, the virus had been able to acquire the *src* gene from a normal cell.

By making radioactive cDNA to the viral oncogenes, it became possible to demonstrate the existence of similar genes in the normal cell genome (Stehelin et al., 1976). This can be done by Southern blotting using the radioactive viral DNA. Genomic DNA is cut by restriction enzymes, separated by electrophoresis, and transferred to nitrocellulose paper. After being separated into single strands by treatment with alkali, it is incubated with radioactive DNA derived from the tumor virus (Figure 23). If a cellular DNA sequence is homologous to the viral oncogene, it will bind the radioactive cDNA and can be detected by autoradiography. Indeed, such sequences have been identified, and some of them are listed in Table 3. These cellular genes are called PROTO-ONCOGENES or CELLULAR ONCOGENES.

FIGURE 23

Presence of cellular homologue to the v-*fms* oncogenes of the feline sarcoma virus. The DNA from normal human cells was digested with restriction enzyme *Eco*RI, separated by electrophoresis, and transferred to filter paper by the Southern blot technique. The presence of a human gene homologous to v-*fms* proto-oncogenes is readily seen. (The lower band is a control probe binding to the gene for dihydrofolate reductase, which is not located on chromosome 5.) Individuals lacking part of the large arm of chromosome 5 (lanes 2 and 4) show only one-half the binding seen in the controls, a result suggesting that c-*fms* resides in the long arm of human chromosome 5. (From Nienhaus et al., 1985; photograph courtesy of A. W. Nienhaus.)

Oncogenes and cellular growth

Given that proto-oncogenes exist in normal cells, three related questions emerge: What are the normal products of these genes? How do they function in the normal life of the cell? And how do these oncogenes give rise to cancerous cells? Recent studies show that many of these proto-oncogenes are the genes responsible for controlling DNA synthesis and cell division (Heldin and Westermark, 1984).

Growth factor variants

Cancer can be caused by the malregulation of genes that are the normal constituents of the cellular division pathway. A hypothetical pathway is outlined in Figure 24. A peptide growth factor in the environment of the cell is bound by a growth factor receptor in the plasma membrane of the cell. This growth factor receptor undergoes a conformational change that causes its cytoplasmic region to acquire tyrosine kinase activity. Alternatively, it can induce tyrosine kinase activity in other membrane molecules. This tyrosine kinase activity then activates a component of the phosphatydylinositol pathway that we discussed in Chapter 2. As a result of the cellular changes brought about by this pathway, certain nuclear proteins are induced that bring about a change in the transcriptional activity of the cell and initiate DNA synthesis.

The malregulation leading to cancer can occur anywhere along this pathway. In some cases, a viral oncogene seems to have once been a gene encoding a growth factor. Such is the case with the v-*sis* oncogene of the SIMIAN SARCOMA VIRUS. (The letter "v" prefaces viral oncogenes, while the letter "c" distinguishes their cellular counterparts.) One of the pivotal discoveries in cancer research has been the realization that the v-*sis* oncogene encodes one of the subunits of platelet-derived growth factor (Doolittle et al., 1983; Waterfield et al., 1983). Simian sarcoma virus produces connective tissue tumors in monkeys, and these tumors are found only in those cell types (fibroblasts, glial cells, and myoblasts) that have receptors for PDGF. When the virus infects these cells, the v-*sis* gene is transcribed and then becomes translated into PDGF subunits. This subunit is recognized by the cells' own PDGF receptors and is sufficient for signaling the cell to divide (Kelly et al., 1985). The result is AUTOCRINE

TABLE 3
Viral and cellular oncogenes and their possible cellular functions

Viral oncogene	Retrovirus source	Cellular proto-oncogene	Protein identity, homology, or size	Cell location and distribution	Activities	Possible roles
I. GROWTH FACTOR MIMIC						
v-sis	Simian sarcoma virus	c-sis	PDGF B chain 14K	Secreted protein	homo- or heterodimer; binds to PDGF-R	Mitogenesis Wound healing Early embryonic growth factor (?)
II. TYROSINE KINASES AND RELATED C-ONC						
v-erb-B	Avian erythro-blastosis virus	c-erb-B	EGF receptor 170K	Plasma membrane of mesodermal, ectodermal, and endodermal cells	EGF binding TGF binding Tyrosine kinase	Signal transduction for mitogenesis and differentiation Stimulation of tooth eruption, eye-opening, lung development
v-fms	McDonough feline sarcoma virus	c-fms	CSF-1 receptor 140K	Plasma membrane of macrophage and extraembryonic cells	CSF-1 binding Tyrosine kinase	Signal transduction for mitogenesis and differentiation
v-src	Rous sarcoma virus	c-src	60K	Cytoplasmic face of membranes Adhesion plaques	Tyrosine kinase Phosphorylates many cellular proteins, e.g., vinculin, vimentin	Neuron and platelet development; signal transduction
v-mos	Moloney murine sarcoma virus	c-mos	37K	Cytoplasmic Embryonic testis and ovary	Serine/threonine kinase	$G_2 \rightarrow M$ regulation
v-abl	Abelson murine leukemia virus	c-abl	150K	Plasma membrane	Tyrosine kinase Phosphorylates vinculin	B cell differentiation or proliferation
v-fes	Feline sarcoma virus	c-fes	92K	Cytoplasm Plasma membrane	Tyrosine kinase	White blood cell development
v-fps	Fujinami sarcoma virus	c-fps	98K	Cytoplasm Plasma membrane	Tyrosine kinase	Macrophage and granulocyte development
III. GTPASES						
v-ras^Ki	Kirsten murine sarcoma virus	c-Ki-ras-2, c-H-ras-1, N-ras }	21K	Membrane cytoplasmic face in a wide variety of cells	GTP + GDP binding GTPase	Signal transduction
v-ras^Ha	Harvey murine sarcoma virus					

IV. NUCLEAR PRODUCTS

v-*fos*	FBJ osteosarcoma virus	c-*fos*	55K; subunit of AP-1 transcription factor	Nucleus of extraembryonic tissues, hematopoietic cells, and macrophages. All other cells at lower level.	DNA binding with an accessory protein	G_0 to G_1 transition; Differentiation
v-*myc*	Avian myelocytomatosis virus	c-*myc*, N-*myc*, L-*myc*	62K–66K	Nuclear matrix of most cells; in some tumors and embryonic tissues	DNA binding	Proliferation; Regulation of DNA synthesis
v-*myb*	Avian myeloblastosis virus	c-*myb*	75K	Nucleus of hematopoietic cells	DNA binding	Differentiation of hematopoietic cells
v-*jun*	Avian sarcoma virus-17	c-*jun*	47K; subunit of AP-1 transcription factor	Nucleus of most cells	DNA binding	Activation of gene expression (?)

V. OTHERS

v-*erb*-A	Avian erythroblastosis virus	c-*erb*-A	T_3 receptor; homology to glucocorticoid receptor	Cytoplasm and nucleus	Binds thyroxine	Metabolic regulation (?)

Source: Primarily from Adamson (1987).

FIGURE 24
Schematic diagram of the cell division control network. Growth factors (or their oncogenetic relatives) bind to their specific receptor. This receptor can act as a tyrosine kinase in its own right or activate other molecules to function as tyrosine kinases. These kinases activate the phosphatidylinositol cycle that activates phosphokinase C and elevates the concentration of free calcium ions in the cytoplasm. These factors act upon the nuclear oncogenes to regulate DNA synthesis. (After Bishop, 1989.)

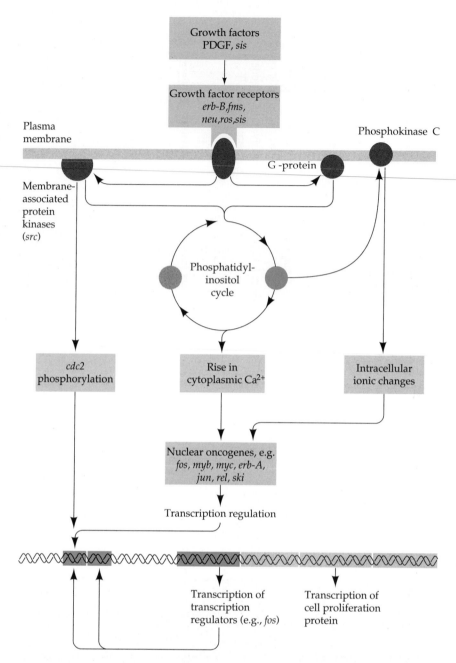

STIMULATION,* whereby the cells make their own growth factor and respond to it. Betscholtz and co-workers (1984) have evidence suggesting that the interaction between the v-*sis* growth factor and its receptor does not even wait until they are on the cell surface. Rather, the interaction and subsequent stimulation of mitosis can occur when both the factor and its receptor are being processed in the endoplasmic reticulum (Betscholtz et al., 1984; Fleming et al., 1989). Autocrine stimulation is not usually

*Physiologists have pointed to three levels wherein a cell receives external chemical signals. In the *endocrine* manner, molecules travel long distances in the blood from one cell type to another. An example is estrogen from the ovary being received by uterine cells. In the *paracrine* manner, one set of cells influences the behavior of its neighbors through diffusible molecules that need not go through the blood. This, as we have seen, is the situation when growth factors from one group of cells stimulate the division of an adjacent group of cells or when a diffusible signal from the anchor cell causes the differentiation of nearby vulval precursor cells (Chapter 16). In *autocrine* stimulation, a set of cells makes and receives its own signal molecules.

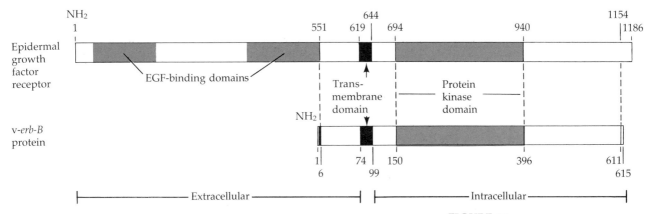

found in developing organisms. However, there is one exceptional case in which the autocrine stimulation by PDGF may be very important. This is in the mammalian cytotrophoblast. Not only can these placental cells secrete PDGF, but they can also bind it (Goustin et al., 1985). Thus, the enormous expansion of placental tissue may be the result of autocrine stimulation wherein the cytotrophoblast cells are capable of both making and responding to PDGF.

Growth factor receptor variants

Other oncogenes are known to code for proteins that mimic growth factor receptors. The viral oncogene v-erb-B of the avian erythroblastosis virus codes for a protein that is almost identical to residues 551 to 1154 of the human receptor for epidermal growth factor (Figure 25; Yamamoto et al., 1983; Downward et al., 1984). This protein represents the cytoplasmic and transmembrane domains and only a small portion of the part of the protein that actually binds EGF. Usually the EGF receptor has a small amount of tyrosine kinase activity, which is greatly stimulated when it binds EGF. This tyrosine kinase activity has been seen to phosphorylate phospholipase C, thereby activating the phosphatidyl inositol pathway toward cell division (Margolis et al., 1989; Meisenhelder et al., 1989). It is hypothesized (see Bell, 1986, and Figure 26) that growth factor receptors, such as the EGF receptor, activate the enzyme phospholipase C either directly or indirectly through a G-protein. Phospholipase C splits phosphatidylinositol 4,5-bisphosphate into diacylglycerol and inositol trisphosphate. The first of these products can activate a sodium–proton pump to raise the intracellular pH, while the inositol trisphosphate releases calcium ions stored in the endoplasmic reticulum. The combination of elevated pH and free intracellular calcium is known to stimulate cell division (as we discussed in Chapter 2).

The "truncated" EGF receptor produced by the v-erb-B oncogene appears to have a tyrosine kinase activity that is always high (Hayman et al., 1983). This ability to phosphorylate protein tyrosine residues is known to be a signal for mitosis and has been correlated with transformation in many cell types (Sefton et al., 1980; Klarlund, 1985; Martin-Zanca et al., 1986). It appears that the EGF-binding domain of the usual receptor inhibits its tyrosine kinase activity and that this inhibition is lost when EGF is bound. The v-erb-B protein lacks this binding domain and remains in its stimulated condition. Thus, cells containing v-erb-B appear to make a growth factor receptor that continuously stimulates the cell to divide, even in the absence of that growth factor.

Another oncogene that mimics a growth hormone receptor is the v-fms

FIGURE 25
The v-erb-B protein as a truncated EGF receptor. The v-erb-B protein (bottom) has six amino acids from a virally encoded gene (*gag*) joined to 609 amino acids encoded by what appears to be a gene for carboxyl half of a EGF receptor (top). This part includes the protein kinase domain, the transmembrane domain, and a small portion of the extracellular domain. (From Hunter, 1985.)

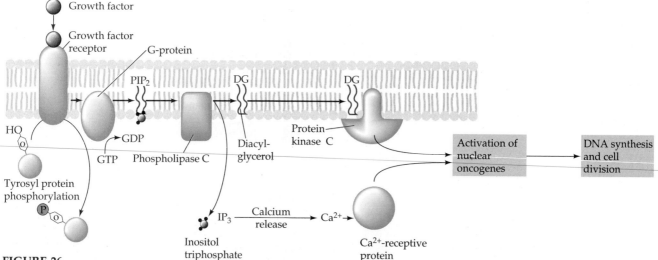

FIGURE 26

Composite model of some of the changes leading to cell division. Binding of growth factor to receptor leads to tyrosine kinase activity of receptor and the activation of phospholipase C. Phospholipase C breaks phosphatidylinositol 4,5-bisophosphate (PIP₂) into inositol trisphosphate (IP₃) and diacylglycerol (DAG). IP₃ mobilizes calcium stored in the endoplasmic reticulum and DAG activates protein kinase C. Protein kinase C phosphorylates proteins such as the Na⁺/H⁺ transport pump. The ionic and metabolic changes affected by these events are able to induce the expression of nuclear oncogenes such as c-*fos* and c-*myc*.

oncogene of the McDonough feline sarcoma virus. It mimics the receptor for CSF-1, a growth factor specific for macrophages and their precursors. The v-*fms* product can bind CSF-1 and has a tyrosine kinase activity when it does (Sacca et al., 1986). The human proto-oncogene for the CSF-1 receptor can be recognized by cDNA made from the v-*fms* gene and maps to the long arm of human chromosome 5. Interestingly, individuals who have deletions in the long arm of chromosome 5 have lost a CSF-1 receptor gene and have abnormal blood cell maturation (Nienhuis et al., 1985). Extra copies of the v-*fms* gene in the bone marrow cells of mice cause the malignant growth of many hematopoietic cell lineages (Heard et al., 1987).

Some other viral oncogenes encode proteins with tyrosine kinase activities, although they are not known to be homologous to any known growth factor receptor (Figure 27). One of these is the v-*src* oncogene of the Rous sarcoma virus. It resides on the inner surface of the plasma membrane of infected cells, and if the tyrosine kinase activity is deficient (as it is in some mutants of the virus), the gene does not transform the cells (Erikson et al., 1979; Levinson et al., 1980). The c-src protein can be activated by a site-specific tyrosine phosphorylation. This phosphorylation is accomplished by the activated PDGF receptor (Piwnica-Worms et al., 1987; Kmiecik and Shalloway, 1987). The product of the c-*src* gene (complete with its tyrosine kinase activity) appears naturally during the course of embryonic development and is not usually seen in adults (Schartl and Barnekow, 1984). It is especially prominent in the development of the neural retina, where it is found on neurons and platelets as they differentiate (Sorge et al., 1984).

The src protein appears to phosphorylate tyrosine residues on several important protein substrates. One important substrate for tyrosine phosphorylation is the src protein itself. This is called AUTOPHOSPHORYLATION, and it is seen in a number of receptors with tyrosine kinase activities. This autophosphorylation is important in increasing the tyrosine phosphorylation activity toward other substrates (Weber et al., 1984; Downward et al., 1984). Two other substrates phosphorylated by src protein are involved in cell attachment. The src protein is also capable of phosphorylating integrin, and phosphorylated integrin is seen shortly after cells are infected with Rous sarcoma virus (Hirst et al., 1986). When infection occurs, this receptor loses its ability to bind both fibronectin (on the external side of the membrane) and talin (on the internal side). This inhibition of receptor function following phosphorylation probably contributes to the adhesive and morphological changes seen in transformed cells. The v-*src* gene also appears to phosphorylate vinculin, one of the proteins at the adhesion plaques where cells contact their substrates (Sefton et al., 1981).

The src protein may also phosphorylate the molecules involved directly in stimulating cell division. Draetta and his colleagues (1988) have demonstrated that the src protein can phosphorylate the cdc2 (p34) subunit of MPF. As shown in Chapter 3, p34 is itself an important kinase that induces the entry of the cell into mitosis. It becomes active only when phosphorylated, and it then allows the cell to proceed from G_2 into M. Thus, *src* may play a major role in the normal pathway of cell division, and its aberrant expression may inappropriately stimulate cell division.

Signal transmitter variants. Some oncogenes appear to mimic elements that take the signal to divide from the membrane and transmit it to the nucleus. This is one of the most controversial and exciting areas of cancer and cell growth research, because these oncogene products may help identify these as-yet-unknown components. The v-*ras* oncogene of the Harvey and Kirsten sarcoma viruses codes for a 21,000-Da membrane-bound protein that binds GTP and GDP. GTP-binding proteins have been implicated in cell growth. When they bind GTP, the protein becomes active and controls some effector function such as opening ion channels in the cell membrane or activating protein kinase C (Figure 26). Then they hydrolyze the GTP to GDP and become inactive until stimulated again by a hormone receptor. Lacal and co-workers (1986a,b) have demonstrated that GTP binding is essential for the oncogenic activity of the v-ras protein, but that the levels of GTP hydrolysis do not correlate with the malignant transformation of cells. The viral *ras* oncogene is not as efficient in GTP hydrolysis as is the normal c-ras protein.

Nuclear factor variants. The last classes of oncogenes work after the signal to divide has been transduced from the membrane through the cytoplasm and received by the nucleus. These oncogenes encode proteins that give the nuclear responses to the cytoplasmic stimuli calling for mitosis. Progression through the cell cycle is usually regulated during G_1. During this time, cells make new mRNAs that are essential for the synthesis of proteins required for cell division, and it is estimated that about 5 percent of the mRNA types of dividing cells are absent in the cytoplasm of nondividing cells (Williams and Penman, 1975; Lau and Nathans, 1985).

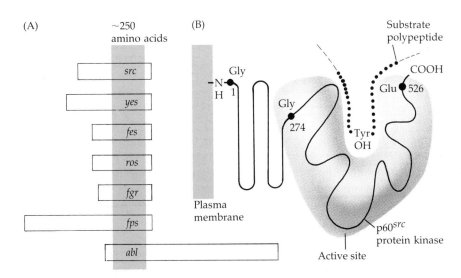

FIGURE 27

Tyrosine kinase regions of oncogene proteins. (A) Seven proteins encoded by different oncogenes. Although the proteins vary greatly in size, they each contain a region of about 250 amino acids catalyzing the tyrosine kinase reaction. (B) The *src* protein kinase (p60^*src*) consists of 526 amino acids bound to the cell membrane by its amino-terminal residue. The tyrosine kinase site is near the carboxyl end. (After Avers, 1985.)

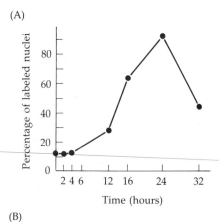

(A)

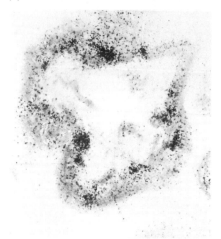

(B)

FIGURE 28

Induction of c-*myc* oncogenic protein in human placental cells. (A) Induction of c-*myc* expression in nuclei of cultured trophoblast cells after stimulation with PDGF at time 0. (B) The expression of c-*myc* in the proliferating chorion villi of a human placenta as seen by in situ hybridization with radioactive c-*myc* DNA. The center of the figure contains the embryo. The areas darkened by radioactivity are primarily the cytotrophoblastic cells of the villi. (A from Goustin et al., 1985; B from Pfeifer-Ohlsson et al., 1984; photograph courtesy of the authors.)

If these mitosis-inducing proteins are encoded by oncogenic viruses, the cell can get these proteins constantly during the time it is infected with the virus. This seems to be the situation with the v-*myc* and v-*fos* oncogenes. The c-*fos* proto-oncogene is expressed for a very brief period of time when G_0 cells enter the cell cycle at G_1, as when serum-starved fibroblasts are refed (Greenberg and Ziff, 1984; Kruijer et al., 1984). The c-*fos* gene can be activated by PDGF, interleukin-3, nerve growth factor, or CSF-1 within five minutes of their binding to the cell surface (Gonda and Metcalf, 1984; Kruijer et al., 1985; Conscience et al., 1986). The fos protein is found in the nucleus of these stimulated cells and binds to the DNA (Sambucetti and Curran, 1986). In fact, c-*fos* dimerizes with another proto-oncogene, c-*jun*, to create the human transcription factor AP-1 (Halazonetis et al., 1988; Rauscher et al., 1988). The v-*fos* oncogene of the FBJ mouse osteosarcoma virus appears to provide the cell with the same information. If cells are infected with the v-*fos* oncogene, they no longer have to rely on growth factors for their proliferation.

The product of the avian myelocytomatosis virus oncogene v-*myc* binds to nuclear chromatin (Donner et al., 1982). The cellular homologues to the v-*myc* gene (there are several genes that are very closely related to it and constitute a family of proto-oncogenes) synthesize very short-lived mRNA and protein products when stimulated by a variety of growth factors (Figure 28; Kelly et al., 1983). Like the c-*fos* proto-oncogenes, the c-*myc* gene products appear suddenly as cells are induced from the G_0 state into G_1.

The *myc* gene products, however, are not sufficient in themselves to give the cell the ability to keep growing or to become malignant. The easiest cells for assaying the oncogenic potential of a virus or gene sequence are often NIH3T3 CELLS. These cultured mouse fibroblasts are easy to grow and will keep dividing as long as they do not touch other cells on all sides and as long as they have a supply of growth factors that are provided by the serum in which they grow. The 3T3 cell has escaped senescence and become potentially immortal. However, these cells are not malignant. They have contact inhibition of movement and division, they will not grow in suspension, and they usually will not form tumors when injected into mice. But the insertion of a *ras* oncogene into these cells will cause them to become malignant. If the *myc* or *ras* oncogenes are inserted into *primary* (mortal) mouse fibroblasts, however, the fibroblasts will not be transformed. Land and co-workers (1983) showed that transformation of normal fibroblasts can occur if *both* the *ras* and *myc* oncogenes are inserted into the same cells. It appears that *ras* and *myc* cooperate with other oncogene or proto-oncogene products to effect malignant transformation and that the NIH3T3 cell already has the "*myc*" function.

How cellular proto-oncogenes go wrong

Viruses do not seem to be responsible for the formation of most human tumors. Rather, there are other mechanisms by which proto-oncogenes gain the ability to transform a cell. These mechanisms include mutation, gene amplification, and chromosomal rearrangement.

Oncogenesis by mutation

It has been known since the early 1960s that mutagens (mutation-causing agents) are also carcinogens (cancer-causing agents). Ames (1979) demonstrated that the mutagenic ability of a given compound was directly proportional to its capacity to induce tumors. Somehow, molecules that can damage DNA (as shown by their ability to induce mutations in bacteria) can generate tumors when given to mice.

In 1982, Shih and Weinberg transfected NIH3T3 cells with DNA from human bladder tumor cells (Figure 29). Some of the mouse fibroblasts picked up some human DNA, and some of these picked up the human gene that had transformed the bladder cells. The NIH3T3 cells that contained this gene grew rapidly into foci of transformed cells and were isolated. The DNA from these transformed cells was then isolated and fractionated, and it was used to transfect another group of NIH3T3 cells. Again, some cells took up the human gene that conferred the transformed phenotype to the cells. The DNA from these cells was cloned, and those clones containing human DNA were detected by DNA hybridization. In this manner, the human gene that caused the malignant growth properties was isolated. This gene was found to be the human cellular homologue of the *ras* oncogene of the Harvey rat sarcoma viruses. Moreover, it differed

FIGURE 29

Isolation of mutant *ras* gene from human bladder carcinoma. Human DNA has characteristic short repetitive DNA segments called *Alu* sequences throughout its genome. The DNA from a human bladder cancer was extracted, cut with restriction enzymes, and added to cultured mouse fibroblasts (NIH/3T3 cells). Those mouse cells that incorporated the oncogene became malignant, producing foci of cells. The DNA of these foci was extracted, cleaved, and added to another dish of mouse fibroblasts. When foci emerged, the DNA of these cells was cut with restriction enzymes and incorporated into phage vectors. The phage multiplied on *E. coli* and when the cells burst, the DNA was placed onto nitrocellulose paper and hybridized to the radioactive *Alu*-sequence DNA. Those clones having an *Alu* sequence are detected by autodiography. This sequence indicates a human (rather than mouse) origin. Because they were selected for malignancy, there is a high probability that they also contain the oncogene responsible for the malignant transformation. (After Weinberg, 1983.)

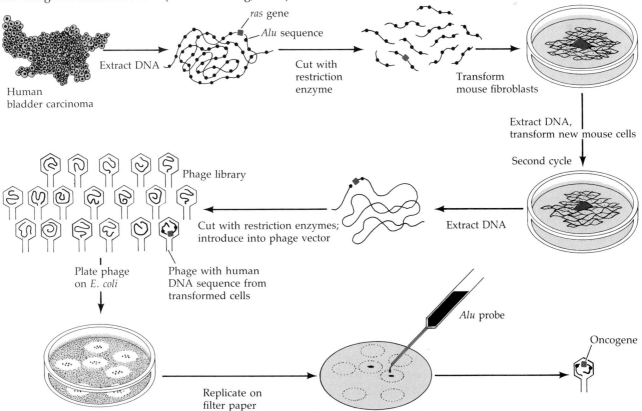

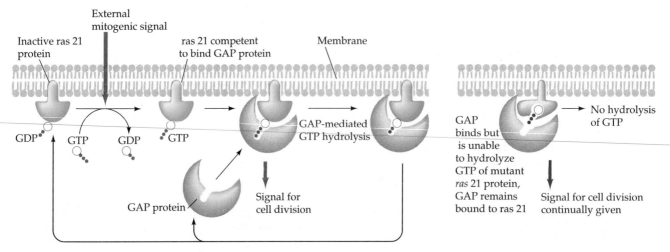

(A) Wild-type ras protein

External mitogenic signal

Inactive ras 21 protein

ras 21 competent to bind GAP protein

Membrane

GDP   GTP   GDP   GTP

GAP-mediated GTP hydrolysis

GAP protein

Signal for cell division

(B) Mutant ras protein

No hydrolysis of GTP

GAP binds but is unable to hydrolyze GTP of mutant *ras* 21 protein, GAP remains bound to ras 21

Signal for cell division continually given

FIGURE 30

Mechanism for oncogenesis by mutant ras proteins. (A) In cells with wild-type ras proteins, an external mitogenic signal causes the phosphorylation of GDP bound to inactive ras protein. The GTP-bound ras protein is able to complex the GAP protein and transduce its signal for cell division. The GAP protein also hydrolyzes the GTP back to GDP and leaves the ras protein, which is now inactive. (B) In mutant ras proteins, the GAP protein binds the GTP-bound form of ras and allows the mitogenic signal to be given. However, it cannot hydrolyze the GTP to GDP, so it remains bound to the ras protein, a situation causing the continuous emission of its mitogenic signal. (After McCormick, 1989.)

from the normal human gene by only one nucleotide (Tabin et al., 1982; Reddy et al., 1982; Capon et al., 1983). The codon encoding the twelfth amino acid had changed from GTC (glycine) to GGC (valine). Studies using other human tumor lines have shown that amino acid substitutions at positions 12, 13, 59, 61, or 63 of the 189-amino acid protein can activate its transforming activity (Der et al., 1986; Levinson, 1986). Barbacid and co-workers (Zarbl et al., 1985), fed rats mutagens such as nitrosomethylurea (which causes point mutations in which guanine is replaced by adenine). The tumors caused by these agents were removed and their DNA was isolated. In the case of the breast tumors induced by nitrosomethylurea, the twelfth codon of the c-*ras* gene had been changed by the mutation of a G to an A.*

The mutations that make the *ras* protein oncogenic are involved in inhibiting its GTPase activity. Normally, the ras protein binds GTP and catalyzes it to GDP, which remains bound to it. This catalysis is greatly stimulated by the normal complexing of the ras protein to the GTPASE ACTIVATING PROTEIN (GAP). This 120,000-Da protein increases its GTP hydrolyzing activity over 100-fold (Trahey and McCormick, 1987; Gibbs et al., 1988). Thus, most of the time, the ras protein has GDP, not GTP, bound to it. The GDP-bound form of the ras protein is inactive, while the GTP-bound form of ras protein activates other proteins. The mutations of the *ras* gene that make it oncogenic all inhibit the GTPase activity of the GAP protein. The GAP protein remains bound to the ras protein, and the ras protein, with its GTP, remains active in transmitting a mitogenic signal to the nucleus (Figure 30; Cales et al., 1988; McCormick, 1989).

Oncogenesis by gene amplification

Another mechanism by which proto-oncogenes can be converted into oncogenes is gene amplification. Schwab and co-workers (1983) found that human neuroblastoma (neural tumor) cells contained many copies of the

---

*The transforming activities of other proto-oncogenes have also been found to be activated by mutations in the protein-coding region of the genes. This would also explain the origin of leukemia by radioactive fallout. Strontium-90, a long-lived radioactive isotope produced by atomic fission, mimics calcium and can become incorporated into growing bones. It is thereby placed near the bone marrow stem cells for lymphocytes and blood cells. These dividing cells are susceptible to mutagenesis by the high-energy electrons emitted by the isotope.

DNA sequence related to the v-*myc* oncogene (Figure 31). They called this gene N-*myc* (the N standing for nerve). The amplification of the N-*myc* proto-oncogene is seen to be associated with rapidly proliferating neuroblastomas (Seeger et al., 1985). It is thought that the amplified genes transcribe more mRNA for the protein. This overabundance of *myc* message would enable the protein to be present in larger amounts and therefore some N-myc protein would remain when it would normally be degraded. The result would be the continued presence of the protein. Because the c-myc protein is usually synthesized in response to growth signals coming from the cell surface, the result is that the nucleus perceives that it has received these growth signals when in reality it has not.

Oncogenesis by insertion mutagenesis

There are other ways by which a proto-oncogene can be deregulated and enabled to overproduce its message (and hence its protein product). For instance, viruses can place a strong promoter or enhancer next to or within a growth control gene, causing it to be expressed in higher than normal levels. Conversely, viruses or mutations can occur within negative regulatory regions, again causing the overproduction of the gene product. In chickens, some tumor viruses do not carry their own oncogenes. Rather, they have strong promoters that can be inserted upstream from the proto-oncogene, thereby placing a proto-oncogene under viral control. This mechanism is called PROMOTER INSERTION (Hayward et al., 1981). The avian leukosis virus (ALV) increases c-*myc* gene transcription when it inserts its promoter near the gene. When this occurs, the chromatin structure of the c-*myc* gene changes into the DNAse I-sensitive conformation typical of highly active genes (Linial et al., 1985). In this manner, the c-*myc* gene transcribes far more mRNA than it would normally synthesize (Figure 32).

Oncogenesis by chromosomal rearrangement

In human B lymphocyte tumors, the c-*myc* gene is often seen to be deregulated by CHROMOSOMAL REARRANGEMENT (Croce, 1987). In 1979, the three human immunoglobulin genes were mapped to chromosomes 14, 22, and 2. These genes are only active during the maturation of the B lymphocyte and plasma cell. In Burkitt lymphoma, a tumor of human B cells, a chromosomal translocation has transferred a segment of chromosome 8 (a small portion of the long arm, close to the end of the chromosome) to the region of these genes (Figure 33). Klein (1983) hypothesized that the gene on chromosome 8 might be involved in growth regulation and that it came under the control of the immunoglobulin gene promoter or enhancers when it was translocated into these regions of the genome. Three pieces

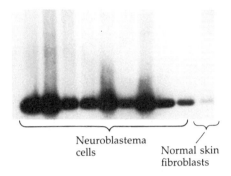

FIGURE 31

Amplification of the c-*myc* DNA in neuroblastoma cells. DNAs from nonmalignant human skin fibroblasts human neuroblastomas were digested with restriction enzymes, transferred to nitrocellulose filters, hybridized with radioactive *myc* genes. (From Schwab et al., 1983; photograph courtesy of the authors.)

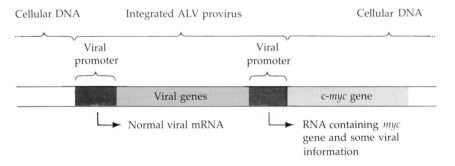

FIGURE 32

Amplification of c-*myc* gene by the promoter of avian leukosis virus (ALV). ALV integrates randomly into the host genome, duplicating its promoter as it does so. If the virus integrates into the region of DNA adjacent to the c-*myc* gene, the c-*myc* becomes controlled by the "right-hand" viral promoter. (After Hayward et al., 1981.)

Normal chromosomes

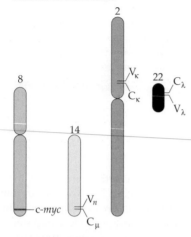

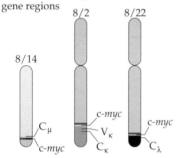

Chromosomal translocations moving
c-*myc* to immunoglobin
gene regions

FIGURE 33
Three chromosomal translocations that
independently give rise to Burkitt
lymphoma of B cells each transfers the
c-*myc* gene (on chromosome 8) to one
of the three immunoglobulin gene regions (on chromosomes 14, 2, or 22.)
(After Croce, 1985.)

of evidence now support this hypothesis. First, the human c-*myc* oncogene, recognized by in situ hybridization, is located at the site of chromosome 8 that is translocated to the immunoglobulin gene regions. Second, when DNA is isolated from Burkitt lymphoma cells and is cleaved
by restriction enzymes, the fragments containing the c-*myc* DNA are found
together with the immunoglobulin genes (Leder et al., 1983). Third, the
amount of c-*myc* mRNA transcribed from these translocated chromosomes
correlates with the activation of the immunoglobulin genes. If the immunoglobulin genes are turned off (as can be done by fusing a Burkitt
lymphoma cell with a fibroblast), the translocated chromosome no longer
transcribes the c-*myc* gene. Similarly, no c-*myc* message seems to be transcribed from the normal chromosome 8 in Burkitt lymphoma cells (Nishikura et al., 1983; Croce et al., 1984). This activation of c-*myc* can be
brought about by separating c-*myc* from its normal negative regulatory
elements or by placing c-*myc* near the immunoglobulin gene enhancer. In
mice, translocations have been seen to bring the *myc* gene into proximity
of the immunoglobulin gene enhancer elements. The enhancers stimulate
the transcription of these *myc* genes, thus causing the lymphocytes
(wherein the enhancer is able to function) to become malignant (Adams
et al., 1985; Fahrlander et al., 1985).

Proto-oncogenes can therefore be transformed into oncogenes by at
least four methods:

1 Point mutations or deletions can create proteins whose functions
  give them the ability to deregulate cell growth. (This is seen in
  ras and *erbB*.)
2 Amplification of proto-oncogenes can create too much of the
  protein product. (This is seen in N-*myc*.)
3 The insertion by a virus of a new promoter or enhancer can deregulate oncogene message synthesis by placing it under viral
  control.
4 The translocation of a proto-oncogene into a different region of
  the genome can place the proto-oncogene under the control of a
  different cellular promoter or enhancer.

Oncogenesis by loss of tumor suppressor genes

In addition to the above four mechanisms of oncogenesis, there is a fifth
mechanism that involves the loss of an inhibitor of cell division. It has
been known for decades that there are families with predispositions toward certain types of tumors (Knudson, 1971, 1976). One of these cancers,
retinoblastoma, is a tumor of the retina and has both a sporadic and a
hereditary form. Several families with susceptibility to retinoblastoma had
members who lacked a certain region of the long arm of chromosome 13
(Benedict et al., 1983). It was proposed that the gene for retinoblastoma
resistance was located in that region, and that members of these families
inherited only one normal allele for retinoblastoma resistance, while most
people had two such genes per cell. A somatic mutation that destroyed
the function of the one remaining wild-type allele could cause that individual cell to continue dividing and become a tumor (Figure 34).

DNA samples from children with this tumor showed that nearly all of
their cells were heterozygous for this region (i.e., that they had only one
functional allele per cell), but that there were no functional alleles in the
actual tumor cells (Cavanee et al, 1983; Friend et al., 1986; Horowitz et al.,
1990). The loss of these retinoblastoma suppressor genes, called *RB1*, could
be caused by point mutation or by deletion. A cell appears to need at least

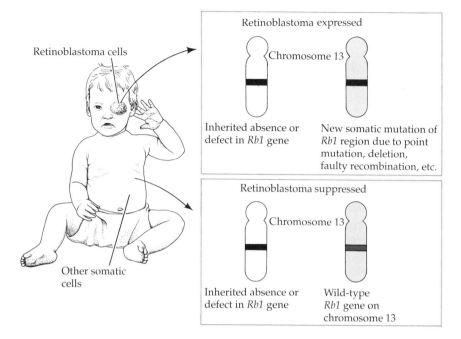

Retinoblastoma cells

Other somatic cells

Retinoblastoma expressed

Chromosome 13

Inherited absence or defect in *Rb1* gene

New somatic mutation of *Rb1* region due to point mutation, deletion, faulty recombination, etc.

Retinoblastoma suppressed

Chromosome 13

Inherited absence or defect in *Rb1* gene

Wild-type *Rb1* gene on chromosome 13

FIGURE 34
Mechanism for the hereditary tendency towards retinoblastoma in certain families. A person can inherit a genotype where one of the two retinoblastoma suppressor (*RB1*) genes is absent or defective. During the life of this individual, a somatic mutation of this one remaining *RB1* gene in a retinal cell can cause the absence of the *RB1* product and allow the uncontrolled proliferation of the cell.

one copy of this gene in order to suppress cell division. Thus, *RB1* is considered to be a tumor suppressor gene whose complete loss triggers oncogenesis.

The protein encoded by this gene is a 105-kDa phosphoprotein that can bind DNA and whose phosphorylation is dependent upon the stage of the cell cycle (Lee et al., 1987). The RB protein is located in the nucleus and is relatively unphosphorylated in resting (G_0 or G_1) cells. However, during S, almost all the RB protein becomes phosphorylated at several serine and threonine sites (DeCaprio et al., 1989; Buchkovich et al., 1989; Chen et al., 1989). It is possible that the dephosphorylated form of the RB protein blocks the entry of the cell into S phase, and that the phosphorylation of the protein takes away this inhibition. (Similarly, the absence of the protein would take away this inhibition.) This model relates directly to the hypothesis that cell division is controlled by protein kinases. The RB gene has been cloned, and its insertion into cultured retinoblastoma cells slows their growth rate and restores normal growth properties (Huang et al., 1988). Abnormalities of this gene have also been found in several other tumors besides retinoblastoma. Recent evidence (Robbins et al., 1990) suggests that RB protein functions by inhibiting the transcription of the c-*fos* gene. The c-*fos* gene encodes a transcription factor that is necessary to take cells from G_0 to G_1. Thus, RB protein would keep cells in their nondividing state.

The importance of RB protein is also seen in that it is a target for the transforming proteins of various tumor viruses. The large T protein of SV40, the E1A protein of adenovirus, and possibly the E7 protein of human papilloma virus each binds to the nonphosphorylated (presumably active) RB protein (Cooper and Whyte, 1989). These tumor virus proteins may operate by functionally depleting the cell of RB protein, thereby enabling the continuous cycling of these cells.

The malignant phenotype is often the result of many mutations. In cultured cells, as we have described earlier in this chapter, at least two genes (a gene encoding a cytoplasmic protein such as *ras* and a gene encoding a nuclear factor such as *myc*) need to malfunction. In certain cancers such as retinoblastoma, perhaps only a single suppressor gene

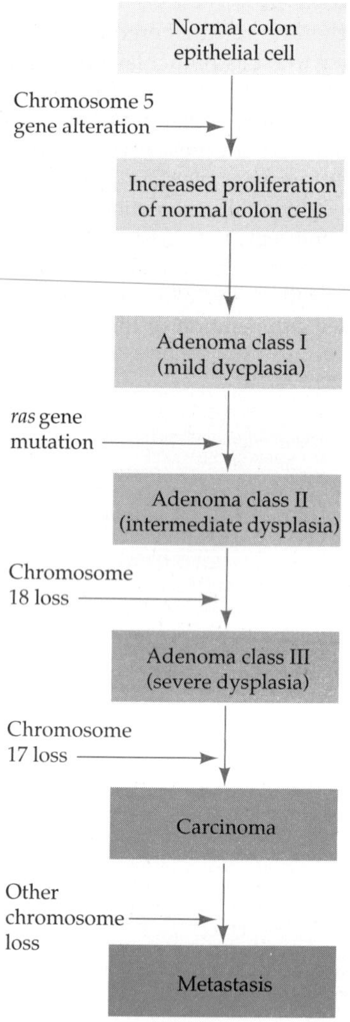

Normal colon epithelial cell
Chromosome 5 gene alteration
Increased proliferation of normal colon cells
Adenoma class I (mild dycplasia)
ras gene mutation
Adenoma class II (intermediate dysplasia)
Chromosome 18 loss
Adenoma class III (severe dysplasia)
Chromosome 17 loss
Carcinoma
Other chromosome loss
Metastasis

FIGURE 35
Model for the progression of colon cancer through a series of intermediate cell phenotypes characterized by the loss of tumor suppressor genes on chromosomes 5, 18, and 17, and the activation of the *ras* oncogene. (After Marx, 1989.)

pair needs to be eliminated for the cell to become malignant. But in most cells of the body, the tumor formation is more complex, and malignancy may involve both the activation of certain oncogenes and the elimination of tumor suppressor genes such as *RB1*. This process has been observed in slow-growing tumors that progress through many stages. Vogelstein and his coworkers (1989) have tracked chromosomes through the progression of colon cancer. Here, cells that lose a portion of chromosome 5 have an increased growth rate and become nonmalignant adenomas. The progression from adenoma to carcinoma (when the cell becomes malignant and can penetrate the colon wall) coincides with the activation of *ras* by mutation and the loss of certain parts of chromosomes 18 and 17 (Figure 35). Since both copies of a tumor suppressor gene have to be deleted, there may be as many as a dozen chromosomal changes in the transformation of a normal cell into a malignant tumor cell.

The products of these tumor suppressor genes may be related to the phenotypes expressed by the transformed cell. The tumor suppressor gene on chromosome 18 has been identified (Fearon et al., 1990) and appears to be related to the known cell adhesion molecules N-CAM and fasciclin II. This may explain the disorganized morphology of adenoma and carcinoma tissue, the dedifferentiation of these cells, and the disruption of normal cell contacts. A tumor suppressor gene found in hamster kidney cells appears to be associated with repressing angiogenesis (see Chapter 6). When deleted, the cell acquires the ability to secrete a factor that induces blood vessel formation (Rastinejad et al., 1989).

The same tumor suppressor genes are not always active in the same cell. The tumor suppressor gene on chromosome 17 has been isolated and is found to be missing in colon cancers and in small-cell carcinoma of the lung. The *RB1* tumor suppressor gene isn't involved in these tumors, but it does seem to be important in stopping osteosarcomas. The correlation of different chromosomal events with the progression of a cell towards malignancy has great potential in diagnosing the stage of tumor development and predicting the likelihood of cells' becoming malignant.

SENESCENCE

The development of an organism takes it from a single-celled zygote through maturity. But what happens afterwards? Is SENESCENCE—the process of aging—also a developmentally programmed event? There is no unanimity of opinion on this question, but a consensus is emerging that developmentally regulated factors do play a major role in the aging process. Senescence involves progressive and irreversible loss of functions that increase the likelihood of death. These changes can be gradual (as in humans) or sudden and dramatic (as in mayflies or salmon). Moreover, not all senescent changes have to happen at the same time. Most animals do not live long enough to have their bodies undergo senescence. It is a privilege reserved for those on top of the food pyramids. Field mice (*Mus musculus*) can live to a ripe old age of 3 or 4 years in the laboratory, but this condition is never observed in the wild. Over 99 percent of field mice die before celebrating their first birthday. Mice "grow up fast," reaching sexual maturity at about 8 weeks and their maximum size within 3 months. Even their gestation period is only 20 days long.

The timing of senescence appears to be a developmentally regulated event. This pattern is easily seen in mayflies and salmon, who die almost

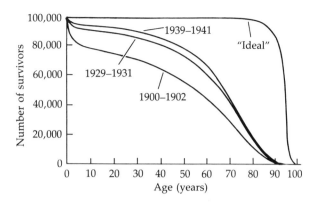

FIGURE 36
Survival curves for white male U.S. citizens at different periods of history. The top curve is a postulated ideal wherein all deaths due to accident or disease are eliminated and individuals would die only of physiological deficits associated with normal aging. The three curves show that while the average health of this group has increased from 1900 to 1941, the longevity of the oldest individual has not. (After Hayflick, 1980.)

immediately upon mating, but it can also be seen in other organisms as well. In human populations, public sanitation and medical science have greatly increased the number of years a person may expect to live (the LIFE EXPECTANCY), but they have not increased the MAXIMUM LONGEVITY of the species. A person is not expected to live far beyond 95 years. Figure 36 shows the life expectancy curves for different populations in the U.S. In all cases, maximum life span is between 90 and 100 years. The healthiest populations, where infectious disease as a major factor in mortality is suppressed, approach an ideal "rectangular" curve, wherein all the individuals live as long as biologically possible. In less healthy environments, more individuals die before reaching this age. Throughout most of human history, the average life span has hovered between 35 and 40 years of age. (This was also the condition of some communities in the U.S. as recently as 1902.) In this century, the average life expectancy in the U.S. has risen to 74.6 years, but the *maximum* longevity has not changed.

Medawar and others (see Cutler, 1979) have suggested that evolutionary pressures would select for a life span that would enable an organism to breed, bear, and in some cases raise its young, but that after this time, random degeneration would take place. The evolutionary conditions would differ from animal to animal. Table 4 shows the maximum life spans of primates. Humans (probably the longest-lived mammal ever on Earth) live about 14 times longer than the shortest-lived primate. The genetic control of longevity is also seen in certain mutations in which the patients age prematurely (Brown, 1985). In PROGERIA (Hutchinson-Gilford syndrome), aging is apparent by the child's first year. When these children die (usually of heart failure at around age 12), they have the skin, hair pattern, and fat distribution characteristic of the elderly.

Many changes are seen during senescence, and even though it is easy to establish age-related correlations, it is very difficult to establish which phenomena are causes of aging and which are merely the effects. Some phenomena appear to be organismal in nature and the result of the accumulation of toxic materials over time. This is probably the case in atherosclerosis ("hardening of the arteries"), where the accumulation of cholesterol can block the passage of blood. Other aging phenomena are primarily cellular, as in the altered immune responsiveness of elderly organisms. Elderly people and aged mice are both prone to infection (where the immune response does not work to destroy the foreign organisms) and to autoimmune diseases (where the immune system malfunctions, attacking its own body). Both of these conditions can be traced to the deterioration of the thymus. As the body ages, more of the thymic tissue is replaced by adipose tissue. The levels of thymic hormones decrease and are essentially undetectable by age 60. This change is consistent with the decline in the number of T cells seen in elderly individuals

TABLE 4
Maximum life span potentials for some primates

Species	Maximum life span potential (years)
Tree shrew	7
Marmoset	15
Squirrel monkey	21
Rhesus monkey	29
Baboon	36
Gibbon	32
Orangutan	50
Gorilla	40
Chimpanzee	45
Human	95

Source: Cutler (1979).

(Bender, 1985). Moreover, those older T cells cannot respond as well as T cells from younger individuals. When T cells from older, healthy individuals are stimulated by antigens or by mitogens that usually cause T cell division, their ability to divide is only half that of the T cells from younger subjects (Weksler, 1983). Thus, older people appear to have fewer T cells, and these T cells are less active. Since T cells function both to destroy foreign organisms and to stop any immune response against the body itself, the loss of T cells may be an important factor in human senescence.

B cells—those cells capable of synthesizing antibodies—also show a decrease in both number and potency as the body ages. The highest incidence of tetanus, for example, is in the elderly population, and it is thought that this is due to the lack of anti-tetanus antibodies in the sera of these people. Kishimoto and co-workers (1982) have supported this notion by co-culturing B cells from young or old adults with T cells from a person who had recently received a "booster" injection to raise the anti-tetanus antibody titre. The "old" B cells produced far less anti-tetanus antibody than did the "young" B cells. Further studies showed that this is in part due to a fivefold decrease in the number of B cells able to respond to the tetanus antigen.

The defect of T cells and B cells probably involves cell division and the ability to respond to extrinsic growth factors. Antigen-stimulated T cells from old mice and elderly humans have fewer receptors for the T cell growth factor interleukin-2 and less ability to cap their cell surface receptors (Noronha et al., 1980; Brohee et al., 1982). Those "old" T cells that do divide have a longer division time and often do not complete the first division (Hefton and Weksler, 1981; Hefton et al., 1980).

Because senescence may be largely due to such cellular phenomena, one of the major ways to study senescence has been to observe changes in cultured fibroblasts after they have been isolated from the organism. While most nonmalignant epithelial, neuronal, and lymphocyte cells will not grow in culture, fibroblasts divide about once every day. The tissue culture environment can be made the same for these cells at all times, and one can ask whether they display senescent changes when separated from an aging immune system, cardiovascular system, and endocrine system. Hayflick and Moorhead (1961) found that fibroblasts in culture do indeed undergo a type of senescence. They demonstrated that fibroblasts cultured from a human embryo had a definite, limited, life span that could be divided into three stages (Figure 37). First there is a period of slow growth as the cells become acclimated to the conditions of tissue culture. This lag phase is followed by logarithmic growth, during which time the cells divide about once every 21 hours. After 40–50 divisions, the cells begin to show a longer cell cycle and have a larger volume. Thereafter, the cells

FIGURE 37

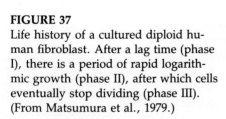
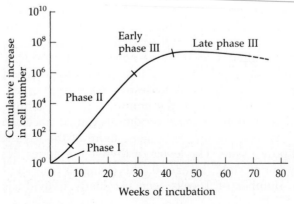

Life history of a cultured diploid human fibroblast. After a lag time (phase I), there is a period of rapid logarithmic growth (phase II), after which cells eventually stop dividing (phase III). (From Matsumura et al., 1979.)

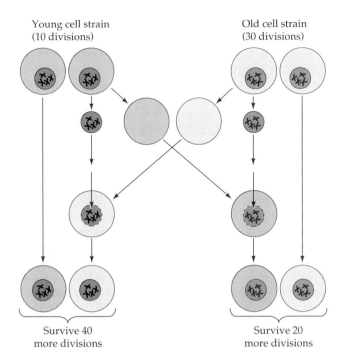

Young cell strain
(10 divisions)

Old cell strain
(30 divisions)

Survive 40
more divisions

Survive 20
more divisions

FIGURE 38
Nuclear control of aging. When old nuclei are transplanted into young cells and vice versa, the age of the nucleus, not of the cytoplasm, controls the number of subsequent cell divisions. (From Hayflick, 1980.)

stop dividing and appear to be blocked in the G_1 phase of the cell cycle. Some cells "escape" this limit by an as-yet-unknown mechanism to become "immortal" cell lines, most of which (with the exception of lines such as the NIH3T3) are malignant.

As might be expected, fibroblasts from older individuals divide fewer times than those from younger subjects (Schneider and Mitsui, 1976; Martin et al., 1970; Bruce et al., 1986) and have a longer cell cycle (about 24.3 hours). Moreover, the fibroblasts appear to "remember" how many divisions they have undergone. If human fibroblasts are frozen for years after their twentieth division, they will still divide about 30 more times. The "clock" for these senescent events appears to reside in the nucleus. Muggleton-Harris and Hayflick (1976) transplanted nuclei from "young" (10 divisions) fibroblasts into the enucleated cytoplasm of "old" (30 divisions) fibroblasts and vice versa. The number of further cell divisions was controlled by the nuclei of these hybrid cells (Figure 38).

What causes this lack of division? Phillips and co-workers (1984) demonstrated that as fibroblasts undergo repeated cell divisions, they lose the ability to respond to the normal physiological stimulators of fibroblast mitosis—PDGF, EGF, insulin, and transferrin. Although the number of EGF receptors does not appear to change with age, their ability to phosphorylate themselves and other protein substrates is dramatically reduced (Carlin et al., 1983; Philips et al., 1983). The critical point may be the induction of the fos protein. This transcription factor is usually induced by PDGF and is expressed during the transition from G_0 to G_1. In senescent cells, *fos* is not transcribed and the cell is irreversibly in G_0 (Seshadri and Campisi, 1990). Fibroblasts from patients with certain premature aging syndromes also show reduced responsiveness to EGF and PDGF (Bauer et al., 1986). It appears that much of senescence may involve the inability to respond to extrinsic factors that are needed for the growth and maintenance of the body.

Patterns of organismal growth may be studied at any level from the molecular through the organismal. Because different tools and conceptual ideas have kept these areas separated from each other, physiologists, anatomists, and cell biologists have all had different views of what "growth" was about. Developmental biologists have to deal with all these different conceptions of growth and synthesize them as much as possible. In the past decade, the study of oncogenes has begun to integrate notions of growth as studied by the cell biologists and physiologists. It may soon be possible to explain allometric growth rates and other organismal phenomena by these new ideas as well. Joseph Needham's statement of 1951 is as true today as then:

> Until we can find the links between the superimposed levels of organization, there remains a certain meaninglessness about the genetics of size and shape or the mathematics of spirals and polyhedra. A unified science of life must inevitably seek to know how one level is connected with the others.

LITERATURE CITED

Adams, J. M. and 7 others. 1985. The c-*myc* oncogene by immunoglobulin enhancers induces lymphoid malignancy in transgenic mice. *Nature* 318: 533–538.

Adamson, E. 1987. Oncogenes on development. *Development* 99: 449–471.

Adebonojo, F. O. 1975. Studies on human adipose cells in culture: Relation of cell size and cell multiplication to donor age. *Yale J. Biol. Med.* 48: 9–16.

Ames, B. 1979. Identifying environmental chemicals causing mutations and cancer. *Science* 104: 587–593.

Avers, C. M. 1985. *Molecular Cell Biology.* Benjamin/Cummings, Menlo Park, CA.

Baserga, R. 1985. *The Biology of Cell Reproduction.* Harvard University Press, Cambridge, MA.

Bauer, E. A., Silverman, N., Busick, D. F., Kronbergan, A. and Deuel, T. F. 1986. Diminished response of Werner's Syndrome fibroblast to growth factors PDGF and EGF. *Science* 234: 1240–1243.

Beemon, K. and Hunter, T. 1977. In vitro translation yields a possible Rous sarcoma virus src gene product. *Proc. Natl. Acad. Sci. USA* 74: 3302–3306.

Bell, R. M. 1986. Protein kinase C activation by diacylglycerol second messengers. *Cell* 45: 631–632.

Bender, B. S. 1985. B lymphocyte function in aging. *Rev. Biol. Res. Aging* 2: 143–154.

Benedict, W. F., Murphree, A. L., Banerjee, A., Spina, C. A., Sparkes, M. D. and Sparkes, R. S. 1983. Patient with 13 chromosome deletion: evidence that the retinoblastoma gene is a recessive cancer gene. *Science* 219: 973–975.

Bernstein, S. E., Russell, E. S. and Keighley, G. H. 1968. Two hereditary mouse anemias (S1/S1^d and W/W^v) deficient in response to erythropoietin. *Ann. N.Y. Acad. Sci.* 149: 475–485.

Betscholtz, C., Westermark, B., Ek, B. and Heldin, C. H. 1984. Coexpression of a PDGF-like growth factor and PDGF receptors in a human osteosarcoma cell line: Implications for autocrine receptor activation. *Cell* 39: 447–457.

Bishop, J. M. 1985. Viral oncogenes. *Cell* 42: 23–38.

Bishop, J. M. 1989. Viruses, genes, and cancer. *Am. Zool.* 29: 653–666.

Brohee, D., Kennes, B. and Neve, P. 1982. Increased stability of E-rosettes and restricted capping of sheep erythrocytes by lymphocytes of aged humans. *Mech. Aging Dev.* 18: 47–52.

Brook, C. G. D. 1972. Evidence for a sensitive period in adipose-cell replication in man. *Lancet* 2: 624–627.

Brooks, W. K. 1886. Report on the Stomatopoda collected by H. M. S. Challenger. *Challenger Reports* 16: 1–114.

Brown, W. T. 1985. Genetics of human aging. *Rev. Biol. Res. Aging* 2: 105–114.

Bruce, S. A., Deamond, S. F. and Tso, P. O. P. 1986. In vitro senescence of Syrian hamster mesenchymal cells of fetal to adult origin: Inverse relationship between in vivo donor age and in vitro proliferative capacity. *Mech. Aging Dev.* 34: 151–173.

Brugge, J. S. and Erikson, R. L. 1977. Identification of a transformation-specific antigen induced by an avian sarcoma virus. *Nature* 269: 346–348.

Buchkovich, K., Duffy, L. A. and Harlow, E. 1989. The retinoblastoma gene is phosphorylated during specific phases of the cell cycle. *Cell* 58: 1097–1105.

Cales, C., Hancock, J. F., Marshall, C. J. and Hall, A. 1988. The cytoplasmic protein GAP is implicated as a target for regulation by the ras gene product. *Nature* 332: 548–551.

Capon, D. J., Chen, E. Y., Levinson, A. D., Seeburg, P. H. and Goeddel, D. V. 1983. Complete nucleotide sequences of the T24 human bladder carcinoma oncogene and its normal homologue. *Nature* 302: 33–37.

Carlin, C. R., Phillips, P. D., Knowles, B. B. and Cristofalo, V. J. 1983. Diminished in vitro tyrosine kinase activity of senescent human fibroblasts. *Nature* 306: 617–620.

Cavanee, W. K. and 8 others. 1983. Expression of recessive alleles by chromosomal mechanisms in retinoblastoma. *Nature* 305: 779–784.

Chabot, B., Stephenson, D. A., Chapman, V. M., Besmer, P. and Bernstein, A. 1988. The proto-oncogene c-kit encoding a transmembrane tyrosine kinase receptor maps to the mouse W locus. *Nature* 335: 88–89.

Chen, P.-L., Scully, P., Shew, J.-Y., Wang, J. Y. J. and Lee, W.-H. 1989. Phosphorylation of the retinoblastoma gene product is modulated during the cell cycle and cellular differentiation. *Cell* 58: 1193–1198.

Cohen, S., Carpenter, G. and King, L, Jr. 1980. Epidermal growth factor–receptor protein kinase interaction. *J. Biol. Chem.* 255: 4834–4842.

Conscience, J.-F., Verrier, B. and Martin, G. 1986. Interleukin-3-dependent expression of the c-myc and c-fos proto-oncogenes in hemopoietic cells. *EMBO J.* 5: 317–323.

Cooper, J. A. and Whyte, P. 1989. RB and cell cycle: Entrance or exit? *Cell* 58: 1009–1011.

Copeland, N. G., Zelenetz, A. D. and Cooper, G. M. 1979. Transformation of NIH /3T3 cells by DNA of Rous sarcoma virus. *Cell* 17: 993–1002.

Croce, C. M. 1987. Role of chromosome translocations in human neoplasia. *Cell* 49: 155–156.

Croce, C. M., Thierfelder, W., Erikson, J., Nishikura, K., Finan, J., Lenoir, G. M. and Nowell, P. C. 1984. Transcriptional activation of an unarranged and untranslocated c-myc oncogene by translocation of a Cλ locus in Burkitt lymphoma cells. *Proc. Natl. Acad. Sci. USA* 80: 6922–2926.

Cutler, R. G. 1979. Evolution of human longevity: A critical overview. *Mech. Aging Dev.* 9: 337–354.

DeCaprio, J. A. and 7 others. 1989. The product of the retinoblastoma susceptibility gene has properties of a cell cycle regulatory element. *Cell* 58: 1085–1095.

Der, C. J., Finkel, T. and Cooper, G. M. 1986. Biological and biochemical properties of human ras^H genes mutated at codon 61. *Cell* 44: 167–176.

Doehmer, J., Barinaga, M., Yale, W., Rosenfeld, M. G., Verma, I. M. and Evans, R. M. 1982. Introduction of rat growth hormone gene into fibroblasts via a retroviral DNA vector: Expression and regulation. *Proc. Natl. Acad. Sci. USA* 79: 2268–2272.

Donner, P., Greiser-Wilka, I. and Moelling, K. 1982. Nuclear localization and DNA binding of the transforming gene product of avian myelocytomatosis virus. *Nature* 296: 262–266.

Doolittle, R. F., Hunkapiller, M. W., Hood, L. E., Devare, S. G., Robbins, K., Aaronson, S. A. and Antoniades, H. N. 1983. Simian sarcoma virus *onc* gene, v-*sis*, is derived from the gene (or genes) encoding a platelet- derived growth factor. *Science* 221: 275–277.

Downward, J. and 8 others. 1984. Close similarity of epidermal growth factor receptor and v-*erb*-B oncogene protein sequences. *Nature* 307: 521–527.

Draetta, G., Piwnica-Worms, H., Morrison, D., Druker, B., Roberts, T. and Beach, D. 1988. Human *cdc*2 protein kinase is a major cell-cycle regulated tyrosine kinase substrate. *Nature* 336: 738–744.

Ebina, Y. and 8 others. 1985. The human insulin receptor cDNA: The structural basis for hormone-activated transmembrane signalling. *Cell* 40: 747–758.

Eden, S., Isakksson, O. G. P., Madsen, K. and Friberg, U. 1983. Specific binding of growth hormone to isolated chondrocytes from rabbit ear and epiphyseal plate. *Endocrinology* 112: 1127–1129.

Einat, M., Resnitzky, D. and Kimsky, A. 1985. Close link between reduction of c-*myc* expression by interferon and G_0/G_1 arrest. *Nature* 313: 597–600.

Erikson, R. L., Collett, M. S., Erikson, E. L. and Purchio, A. F. 1979. Evidence that avian sarcoma virus transforming gene product is a cyclic AMP-independent protein kinase. *Proc. Natl. Acad. Sci. USA* 76: 6260–6264.

Fahrlander, P. D., Sümegi, J., Yang, J.-Q., Weiner, F., Marcu, K. B. and Klein, G. 1985. Activation of the c-*myc* oncogene by the immunoglobulin heavy chain gene enhancer after multiple switch region-mediated chromosome rearrangements in a murine plasmacytoma. *Proc. Natl. Acad. Sci. USA* 82: 3746–3750.

Fearon, E. R. and 10 others. 1990. Identification of a chromosome 18q gene that is altered in colonorectal cancers. *Science* 247: 49–56.

Fleming, T. P., Matusi, T., Molloy, C. J., Robbins, K. C. and Aaronson, S. 1989. Autocrine mechanism for v-*sis* transformation requires cell surface localization of internally activated growth factor receptor. *Proc. Natl. Acad. Sci. USA* 86: 8063–8067.

Frankhauser, G. 1952. Nucleo-cytoplasmic relations in amphibian development. *Int. Rev. Cytol.* 1: 165–193.

Friend, S. H., Bernards, R., Rogelj, S., Weinberg, R. A., Rapaport, J. M., Albert, D. M. and Dryja, T. P. 1986. A human DNA segment with properties of the gene that predisposes to retinoblastoma and osteosarcoma. *Nature* 323: 643–646.

Geissler, E. N., Ryan, M. A. and Housman, D. E. 1988. The dominant white-spotting (*W*) locus of the mouse encodes the c-*kit* proto-oncogene. *Cell* 55: 185–192.

Gibbs, J. B., Scaber, M. D., Allard, W. J., Sigal, I. S. and Scolnick, E. M. 1988. Purification of *ras* GTPase activating protein from bovine brain. *Proc. Natl. Acad. Sci. USA* 85: 5026–5030.

Gonda, T. J. and Metcalf, D. 1984. Expression of *myb*, *myc*, and *fes* proto- oncogenes during the differentiation of a murine myeloid leukemia. *Nature* 310: 249–251.

Goss, R. J. 1966. Hypertrophy versus hyperplasia. *Science* 153: 1615–1620.

Goss, R. J. 1978. *The Philosophy of Growth*. Academic Press, New York.

Gould, S. J. 1974. The origin and function of "bizarre" structures: Antler size and skull size in the "Irish Elk," *Megalocerus giganteus*. *Evolution* 28: 191–220.

Goustin, A. S. and 9 others. 1985. Coexpression of the *sis* and *myc* proto- oncogenes in developing human placenta suggests autocrine control of trophoblast growth. *Cell* 41: 301–312.

Greenberg, M. E. and Ziff, E. B. 1984. Stimulation of 3T3 cells induces transcription of c-*fos* proto-oncogene. *Nature* 311: 433–438.

Halazonetis, T. D., Georgopoulos, K., Greenberg, M. E. and Leder, P. 1988. c-*Jun* dimerizes with itself and with c-*Fos*, forming complexes with different DNA binding affinities. *Cell* 55: 917–924.

Hanafusa, H., Halpern, C. C., Bucchaga, D. L. and Kauvai, S. 1977. Recovery of avian sarcoma virus from tumors induced by transformation- defective mutants. *J. Exp. Med.* 146: 1735–1747.

Hayflick, L. 1980. The cell biology of human aging. *Sci. Am.* 242 1): 58–65.

Hayflick, L. and Moorhead, P. S. 1961. The limited in vitro lifetime of human diploid cell strains. *Exp. Cell Res.* 25: 585–621.

Hayman, M. J., Ramsay, G. M., Savin, K., Kitchener, G., Graf, T. and Beug, H. 1983. Identification and characterization of the avian erythioblastosis virus *erb*-B gene product as a membrane glycoprotein. *Cell* 32: 579–588.

Hayward, W., Neel, B. G. and Astrin, S. 1981. Activation of a cellular *onc* gene by promoter insertion on ALV-induced lymphoid leukosis. *Nature* 290: 475–480.

Heard, J. M., Roussel, M. F., Rettenmeier, C. W. and Sherr, C. J. 1987. Multilineage hematopoietic disorders induced by transplantation of bone marrow cells expressing the v-*fms* oncogene. *Cell* 51: 663–673.

Hefton, J. M. and Weksler, M. E. 1981. The study of immune function in aged humans. *In* W. H. Adler and A. A. Nordin (eds.), *Immunological Techniques Applied to Aging Research*. CRC Press, Boca Raton, FL.

Hefton, J. M., Darlington, G., Casazza, B. A. and Weksler, M. E. 1980. Immunological studies of aging. V. Impaired proliferation of PHA-responsive human lymphocytes in culture. *J. Immunol.* 125: 1007–1010.

Heine, U. I. and 7 others. 1987. Role of transforming growth factor β in the development of the mouse embryo. *J. Cell Biol.* 105: 2861–2876.

Heldin, C.-H. and Westermark, B. 1984. Growth factors: Mechanism of action and relation to oncogenes. *Cell* 37: 9–20.

Hirst, R., Horowitz, A., Buck, C. and Rohrschneider, L. 1986. Phosphorylation of the fibronectin receptor complex in cells transformed by oncogenes that encode tyrosine kinases. *Proc. Natl. Acad. Sci. USA* 83: 6470–6474.

Holley, R. W., Bohlen, P., Fava, R., Baldwin, J. H., Kleeman, G. and Armour, R. 1980. Purification of kidney epithelial cell growth inhibitors. *Proc. Natl. Acad. Sci. USA* 77: 5989–5992.

Horowitz, J. M. and 8 others. 1990. Frequent inactivation of the retinoblastoma anti-oncogene is restricted to a subset of human tumor cells. *Proc. Natl. Acad. Sci. USA* 87: 2775–2779.

Hsu, Y.-M. and Wang, J. L. 1986. Growth control in cultured 3T3 fibroblasts. V. Purification of an M_r 13,000 polypeptide responsible for growth inhibitory activity. *J. Cell Biol.* 102: 362–369.

Huang, H.-J. S. and 7 others. 1988. Suppression of the neoplastic phenotype by replacement of the RB gene in human cancer cells. *Science* 242: 1563–1566.

Hunter, T. 1985. Oncogenes and growth control. *Trends Biochem. Sci.* 10: 275–280.

Hunter, T. 1986. Cell growth control mechanisms. *Nature* 322: 14–16.

Huxley, J. S. 1932. *Problems of Relative Growth*. Dial Press, New York.

Isaksson, O. G. P., Jansson, J.-O. and Gauss, I. A. M. 1982. Growth hormone stimulates longitudinal bone growth directly. *Science* 216: 1237–1239.

Johnson, P. R., Zucker, L. M., Cruce, J. A. and Hirsch, J. 1971. Cellularity of adipose deposits in the genetically obese Zucker rat. *J. Lipid Res.* 12: 706–714.

Kelly, J. D., Raines, E. W., Ross, R. and Murray, M. J. 1985. The β chain of PDGF is sufficient for mitogenesis. *EMBO J.* 3: 3399–3405.

Kelly, K., Cochrane, B. H., Stiles, C. D. and Leder, P. 1983. Cell-specific regulation of the c-*myc* gene by lymphocyte mitogens and platelet-derived growth factor. *Cell* 35: 603–610.

Kimchi, A., Wang, X.-F., Weinberg, R. A., Cheifetz, S. and Massagué, J. 1988. Absence of TGF-β receptors and growth inhibitory responses in retinoblastoma cells. *Science* 240: 196–198.

Kishimoto, S., Tomino, S., Mitsuya, H. and Nishimura, H. 1982. Age-related decrease in frequencies of B-cell precursors and specific help on T-cells involved in the 19G anti-tetanus toxoid antibody protection in humans. *Clin. Immunol. Immunopathol.* 25: 1–10.

Klarlund, J. D. 1985. Transformation of cells by an inhibitor of phosphatases acting on phosphotyrosine in proteins. *Cell* 41: 707–717.

Klein, G. 1983. Specific chromosomal translocations and the genesis of B-cell derived tumors in mice and men. *Cell* 32: 311–315.

Kmiecik, T. E. and Shalloway, D. 1987. Activation and suppression of pp60^{c-src} transforming ability by mutation of its primary sites of tyrosine phosphorylation. *Cell* 49: 65–73.

Knudson, A. G. 1971. Mutation and cancer: Statistical study of retinoblastoma. *Proc. Nat. Acad. Sci. USA* 68: 820–823.

Knudson, A. G., Meadows, A. G., Nichols, W. W. and Hill, R. 1976. Chromosome deletion and retinoblastoma. *N. Engl. J. Med.* 295: 1120–1123.

Kruijer, W., Cooper, J. A., Hunter, T. and Verma, I. M. 1984. Platelet- derived growth factor induces rapid but transient expression of the c-*fos* gene and protein. *Nature* 312: 711–716.

Kruijer, W., Schubert, D. and Verma, I. M. 1985. Induction of the proto-oncogene *fos* by nerve growth factor. *Proc. Natl. Acad. Sci. USA* 82: 7330–7334.

LaBarbera, M. 1990. Principles of design of fluid transport systems in zoology. *Science* 249: 992–1000.

Lacal, J. C., Srivastava, S. K., Anderson, P. S. and Aaronson, S. A. 1986a. *ras* p21 proteins with high or low GTPase activity can efficiently transform NIH/3T3 cells. *Cell* 44: 609–617.

Lacal, J. C., Anderson, P. S. and Aaronson, S. A. 1986b. Deletion mutants of Harvey *ras* p21 protein reveal absolute requirement of at least two distant regions for GTP-binding and transforming activities. *EMBO J.* 5: 679–687.

Land, H., Parada, L. F. and Weinberg, R. A. 1983. Tumorigenic conversion of primary embryo fibroblasts requires at least two cooperating oncogenes. *Nature* 304: 596–602.

Lau, L. F. and Nathans, D. 1985. Identification of a set of genes expressed during the G_0/G_1 transition of cultured mouse cells. *EMBO J.* 4: 3145–3151.

Lawrence, D. A., Pircher, R., Kryceve-Martinerie, C. and Jullien, P. 1984. Normal embryo fibroblasts release transforming growth factors in a latent form. *J. Cell. Physiol.* 121: 184–188.

Lee, E. and 7 others. 1987. The retinoblastoma susceptibility gene encodes a nuclear phosphoprotein associated with DNA binding activity. *Nature* 329: 642–645.

Leder, P. and 7 others 1983. Translocations among antibody genes in human cancer. *Science* 222: 765–771.

Leof, E. B., Wharton, W., Van Wyk, J. J. and Pledger, W. J. 1982. Epidermal growth factor and somatomedin C regulate G_1 progression in competent BALB/c-3T3 cells. *Exp. Cell Res.* 141: 107–115.

Levinson, A. D. 1988. Normal and activated *ras* oncogenes and their encoded products. *Trends Genet.* 2: 81–85.

Levinson, A. D., Opperman, H., Varmus, H. E. and Bishop, J. M. 1980. The purified product of the transforming gene of avian sarcoma virus phosphorylates tyrosine. *J. Biol. Chem.* 255: 1973–1980.

Lim, R. W. and Hauschka, S. D. 1984. A rapid decrease in EGF-binding capacity accompanies the terminal differentiation of mouse myoblasts in vitro. *J. Cell Biol.* 98: 739–747.

Lin, S. L., Kikuchi, T., Pledger, W. J. and Tamm, I. 1986. Interferon inhibits the establishment of competence in G_0/S-phase transition. *Science* 233: 356–359.

Linial, M., Gunderson, N. and Groudine, M. 1985. Enhanced transcription of c-*myc* in bursal lymphoma cells requires continuous protein synthesis. *Science* 230: 1126–1132.

Luna, A. M., Wilson, D. M., Wibbelsman, C. J., Brown, R. C., Nagashima, R. J., Hintz, R. L. and Rosenfeld, R. D. 1983. Somatomedins in adolescence: A cross-sectional study of the effect of puberty of plasma insulin-like growth factor I and II levels. *J. Clin. Endocrinol. Metab.* 304: 545–547.

Madsen, K., Friberg, U., Roos, P., Eden, S. and Isaksson, B. 1983. Growth-hormone stimulates the proliferation of cultured chondrocytes from rabbit ear and rat rib growth cartilage. *Nature* 304: 545–547.

Margolis, B. and 8 others. 1989. EGF induces tyrosine phosphorylation of phospholipase C-II: A potential mechanism for EGF receptor signalling. *Cell* 57: 1101–1107.

Martin, C. R. 1985. *Endocrine Physiology.* Oxford University Press, New York.

Martin, G. M., Sprague, C. A. and Epstein, C. J. 1970. Replicative life span of cultured human cells: Effects of donor's age, tissue, and genotype. *Lab. Invest.* 23: 86–92.

Martin-Zanca, D., Hughes, S. H. and Barbacid, M. 1986. A human oncogene formed by the fusion of truncated tropomyosin and protein tyrosine kinase sequences. *Nature* 319: 743–748.

Marx, J. 1989. Many gene changes found in cancer. *Science* 246: 1386–1388.

Matsui, Y., Zsebo, K. M. and Hogan, B. L. M. 1990. Embryonic expression of a maematopoietic growth factor encoded by the *Sl* locus and the ligand for c-kit. *Nature* 347: 667–669.

Matsumura, T., Zerrudo, Z. and Hayflick, L. 1979. Senescent human diploid cells in culture: survival, DNA synthesis, and morphology. *J. Gerontol.* 34: 328–334.

Mayer, T. C. and Green, T. C. 1968. An experimental analysis of the pigment defect caused by mutations at the *W* and *Sl* loci in mice. *Dev. Biol.* 18: 62–75.

McCormick, F. 1989. *ras* GTPase activating protein: Signal transmitter and signal terminator. *Cell* 56: 5–8.

Meisenhelder, J., Suh, P-G., Rhee, S. G. and Hunter, T. 1989. Phospholipase C-g is a substrate for the PDGF and EGF receptor protein-tyrosine kinase in vivo and in vitro. *Cell* 57: 1109–1122.

Merimee, T. J., Zapf, J., Hewlett, B. and Cavalli-Sforza, L. L. 1987. Insulin-like growth factors in pygmies. The role of puberty in determining final stature. *N. Engl. J. Med.* 316: 906–911.

Mintz, B. and Russell, E. S. 1957. Gene-induced embryological modifications of primordial germ cells in the mouse. *J. Exp. Zool.* 134: 207–237.

Moore, K. L. 1983. *Before We Are Born: Basic Embryology and Birth Defects*, 2nd Ed. Saunders, Philadelphia.

Muggleton-Harris, A. L. and Hayflick, L. 1976. Cellular aging studied by the reconstruction of replicating cells from nuclei and cytoplasms isolated from normal human diploid cells. *Exp. Cell Res.* 103: 321–330.

Needham, J. 1951. Biochemical aspects of form and growth. *In* L. L. Whyte (ed.), *Aspects of Form.* Lund Humphries, London, pp. 77–86.

Nienhaus, A. W., Bunn, H. F., Turner, P. H., Gopal, T. V., Nash, W. G., O'Brien, S. J. and Sherr, C. J. 1985. Expression of the human c-*fms* proto-oncogene in hematopoietic cells and its deletion in the 5q$^-$ syndrome. *Cell* 42: 421–428.

Nilsson, A., Isgaard, J., Lindahl, A., Dahlström, A., Skottner, A. and Isaksson, O. G. P. 1986. Regulation by growth hormone of number of chondrocytes containing IGF-I in rat growth plate. *Science* 233: 571–574.

Nishikura, K., Rushdim, A., Erikson, J., Watt, R., Rovera, G. and Croce, C. M. 1983. Differential expression of the normal and of the translocated human c-*myc* oncogenes in B-cells. *Proc. Natl. Acad. Sci. USA* 80: 4822–4286.

Noronha, A. B. C., Antel, J. P., Roos, R. P. and Arnason, B. G. W. 1980. Changes in concanavalin: A capping of human lymphocytes with age. *Mech. Aging Dev.* 12: 331–337.

Olashaw, N. E. and Pledger, W. J. 1982. Association of platelet-derived growth factor-induced protein with nuclear material. *Nature* 306: 272–274.

Olwin, B. B. and Hauschka, S. D. 1988. Cell surface fibroblast growth factor and epidermal growth factor receptors are permanently lost during skeletal muscle terminal differentiation in culture. *J. Cell Biol.* 107: 761–769.

Orr-Urteger, A., Avivi, A., Zimmer, Y., Givol, D., Yarden, Y. and Lonai, P. 1990. Developmental expression of c-*kit*, a proto-oncogene encoded by the *W* locus. *Development* 104: 911–924.

Palmiter, R. D., Brinster, R. L., Hamm, R. E., Trumbauer, M. E., Rosenfeld, M. G., Birnberg, N. C. and Evan, H. 1982. Dramatic growth of mice that develop from eggs microinjected with metallothionein growth hormone fusion genes. *Nature* 300: 611–615.

Pelton, R. W., Dickinson, M. E., Moses, H. L. and Hogan, B. L. M. 1990. In situ hybridization analysis of TGF-β3 RNA expression during mouse development: Comparative studies with TGF-β1 and β2. *Development* 110: 609–620.

Pfeifer-Ohlsson, S., Goustin, A. S., Rydnert, J., Wahlström, T., Bjersing, L., Stehelin, D. and Ohlsson, R. 1984. Spatial and temporal pattern of cellular *myc* oncogene expression in developing human placenta: Implications for embryonic cell proliferation. *Cell* 38: 585–596.

Phillips, P. D., Kuhne, E. and Cristofalo, V. J. 1983. ^{125}I-EGF binding ability is stable throughout replicative life span of WI-38 cells. *J. Cell. Physiol.* 114: 311–316.

Phillips, P. D., Kaji, K. and Cristofalo, V. J. 1984. Progressive loss of the proliferative response of senescing WI-38 cells to PDGF, EGF, insulin, transferrin, and dexamethasone. *J. Gerontol.* 39: 11–17.

Piwnica-Worms, H., Saunders, K. B., Roberts, T. M., Smith, A. E. and Chang, S. H. 1987. Tyrosine phosphorylation regulates the biochemical and biological properties of pp60^{c-src}. *Cell* 49: 75–82.

Pledger, W. J., Estes, J. B., Howe, P. H. and Leof, E. B. 1984. Serum factor requirements for the initiation of cell proliferation. *In* J. P. Mather (ed.), *Mammalian Cell Culture.* Plenum, New York, pp. 1–15.

Proper, J. A., Bjornson, C. L. and Moses, H. L. 1982. Mouse embryos contain polypeptide growth factors capable of inducing a reversible neoplastic phenotype in nontransformed cells in culture. *J. Cell. Physiol.* 110: 169–174.

Przibram, H. 1904. Reported in Przibram, H. 1931. *Connecting Laws in Animal Morphology.* University of London Press, London.

Raff, R. A. and Kaufman, T. C. 1983. *Embryos, Genes, and Evolution.* Macmillan, New York.

Randall, J. E. 1964. Contributions to the biology of the queen conch, *Strombos gigas.* *Bull. Mar. Sci.* 14: 246–295.

Rastinejad, F., Polverini, P. J. and Bouck, N. 1989. Regulation of the activity of a new inhibitor of angiogenesis by a cancer suppressor gene. *Cell* 56: 345–355.

Raup, D. M. 1962. Computer as aid in describing form in gastropod snails. *Science* 138: 150–152.

Raup, D. M. 1966. Geometric analysis of shell coiling: General problems. *J. Paleontol.* 40: 1178–1190.

Rauscher, F. J. III, Sambucetti, L. C., Curran, T., Distel, R. J. and Spiegelman, B. M. 1988. Common DNA binding site for Fos protein complexes and transcription factor AP-1. *Cell* 52: 471–480.

Reddy, E. P., Reynolds, R. K., Santos, E. and Barbacid, M. 1982. A point mutation is responsible for the acquisition of transforming properties by the *T24* human bladder cancer oncogene. *Nature* 300: 149–152.

Resnitzky, D., Yarden, A., Zipori, D. and Kimchi, A. 1986. Autocrine β-related interferon controls c-*myc* suppression and growth arrest during hematopoietic cell differentiation. *Cell* 46: 31–40.

Robbins, P. D., Horowitz, J. M. and Mulligan, R. C. 1990. Negative regulation of human c-*fos* expression by the retinoblastoma gene product. *Nature* 346: 668–671.

Roberts, A. B., Anzano, M. A., Lamb, L. C., Smith, J. M. and Sporn, M. B. 1981. New class of transforming growth factors potentiated by epidermal growth factor: Isolation from non-neoplastic tissues. *Proc. Natl. Acad. Sci. USA* 78: 5339–5343.

Roberts, A. B., Frolik, C. A., Anzano, M. A. and Sporn, M. B. 1983. Transforming growth factor for neoplastic and non-neoplastic tissues. *Fed. Proc.* 42: 2621–2626.

Roberts, A. B., Anzano, M. A., Wakefield, L. L., Roche, N., Stern, D. F. and Sporn, M. B. 1985. Type β transforming growth factor: A bifunctional regulator of cellular growth. *Proc. Natl. Acad. Sci. USA* 82: 119–123.

Ross, R. and Vogel, A. 1978. The platelet-derived growth factor. *Cell* 14: 203–210.

Rous, P. 1910. A transmissible avian neoplasm (sarcoma of the common fowl). *J. Exp. Med.* 12: 696–705.

Sacca, R., Stanley, E. R., Sherr, C. J. and Rettenheimer, C. W. 1986. Specific binding of the mononuclear phagocyte colony stimulating factor CSF-1 to the product of the *fms* oncogene. *Proc. Natl. Acad. Sci. USA* 83: 3331–3335.

Sambucetti, L. C. and Curran, T. 1986. The *fos* protein complex is associated with DNA in isolated nuclei and binds to DNA cellulose. *Science* 234: 1417–1419.

Sarvella, P. A. and Russell, L. B. 1956. *Steel,* a new dominant gene in the mouse. *J. Hered.* 47: 123–128.

Savage, C. R. and Cohen, S. 1972. Epidermal growth factor and a new derivative. *J. Biol. Chem.* 247: 7609–7611.

Schartl, M. and Barnekow, A. 1984. Differential expression of the cellular *src* gene during vertebrate development. *Dev. Biol.* 105: 415–422.

Schneider, E. L. and Mitsui, Y. 1976. The relationship between in vitro cellular aging and in vivo human age. *Proc. Natl. Acad. Sci. USA* 3584–3588.

Schwab, M. and 8 others. 1983. Amplified DNA with limited homology to *myc* oncogene is shared by human neuroblastoma cell lines and a neuroblastoma tumour. *Nature* 305: 245–248.

Seeger, R. C., Brodeur, G. M., Sather, H., Dalton, A., Siegel, S. E., Wong, K. Y. and Hammond, D. 1985. Association of multiple copies of the N-*myc* oncogene with rapid progression of neuroblastomas. *N. Engl. J. Med.* 313: 1111–1116.

Sefton, B. M., Hunter, T., Beemon, K. and Eckhart, W. 1980. Evidence that the phosphorylation of tyrosine is essential for cellular transformation by Rous sarcoma virus. *Cell* 20: 807–816.

Sefton, B. M., Hunter, T. J., Ball, E. H. and Singer, S. J. 1981. Vinculin-A cytoskeletal target of the transforming protein of Rous sarcoma virus. *Cell* 24: 165–174.

Seshadri, T. and Campisi, J. 1990. Repression of c-*fos* transcription and an altered genetic program in senescent human fibroblasts. *Science* 247: 205–208.

Shih, C. and Weinberg, R. A. 1982. Isolation of a transforming sequence from a human bladder carcinoma cell line. *Cell* 29: 161–169.

Sokawa, Y., Watanabe, Y. and Kawade, Y. 1977. Interferon suppresses the transaction of quiescent 3T3 cells to a growing state. *Nature* 268: 236–238.

Sorge, L. K., Levy, B. T. and Maness, P. F. 1984. pp60^{c-src} is developmentally regulated in the neural retina. *Cell* 36: 249–257.

Stehelin, D., Varmus, H. E., Bishop, J. M. and Vogt, P. K. 1976. DNA related to transforming gene(s) of avian sarcoma viruses is present in normal avian DNA. *Nature* 260: 170–173.

Stern, J. S. and Greenwood, M. R. C. 1974. A review of development of adipose cellularity in man and animals. *Fed. Proc.* 33: 1952–1955.

Stiles, C. D., Capone, G. T., Scher, C. D., Antoniades, H. N., Van Wyk, J. J. and Pledger, W. J. 1979. Dual control of cell growth by somatomedins and platelet-derived growth factor. *Proc. Natl. Acad. Sci. USA* 76: 1279–1283.

Tabin, C. J. and 8 others. 1982. Mechanism of activation of a human oncogene. *Nature* 300: 143–149.

Temin, H. M. 1971. The protovirus hypothesis. *J. Natl. Cancer Inst.* 46: iii–vii.

Temin, H. M. and Rubin, H. 1958. Characteristics of an assay for Rous sarcoma cells in culture. *Virology* 6: 669–688.

Thompson, D. W. 1942. *On Growth and Form.* Cambridge University Press, Cambridge.

Trahey, M. and McCormick, F. 1987. A cytoplasmic protein stimulates normal N-*ras* p21 GTPase, but does not affect oncogenic mutants. *Science* 238: 542–544.

Ullrich, A. and 9 others. 1985. Human insulin receptor and its relationship to the tyrosine kinase family of oncogenes. *Nature* 313: 756–761.

Vaahtokari, A., Sahlberg, C., Vainio, S. and Thesleff, I. 1990. Spatial and temporal changes in transforming growth factor β1 RNA expression during tooth development. *Sixteenth EMBO Symposium Abstracts,* p. 110.

Vassilopoulou-Sellin, R. and Phillips, L. S. 1982. Extraction of somatomedin activity from rat liver. *Endocrinology* 110: 582–589.

Vogelstein, B. and others. 1989. Genetic alterations accumulate during colorectal tumorigenesis. *In* W. Cavanee, N. Hastle and E. Stanbridge (eds.), *Recessive Oncogenes and Tumor Progression.* Cold Spring Harbor Laboratories, Cold Spring Harbor, NY.

Ward, J. M. and Armitage, K. B. 1981. Circannual rhythms of food consumption, body mass, and metabolism in yellow-bellied marmots (*Marmota flaviventris*). *Comp. Biochem. Physiol.* 69[A]: 621–626.

Waterfield, M. D. and 9 others. 1983. Platelet-derived growth factor is structurally related to the putative transforming protein P28^{sis} of simian sarcoma virus. *Nature* 304: 35–39.

Weber, W., Bertics, P. J. and Gill, G. N. 1984. Immunoaffinity purification of the epidermal growth factor receptor. *J. Biol. Chem.* 259: 14631–14636.

Weinberg, R. A. 1983. A molecular basis of cancer. *Sci. Amer.* 249(5): 102–116.

Weksler, M. E. 1983. Senescence of the immune system. *Med. Clin. North Am.* 67: 263–272.

Wheeler, D. and Nijhout, H. F. 1981. Soldier determination in ants and a new role for juvenile hormones. *Science* 213: 361–363.

Williams, J. G. and Penman, S. 1975. Messenger RNA sequences in growing and resting mouse fibroblasts. *Cell* 6: 197–206.

Williams, L. T. 1989. Signal transduction by the platelet-derived growth factor receptor. *Science* 243: 1564–1570.

Wilson, E. O. 1953. The origin and evolution of polymorphism in ants. *Q. Rev. Biol.* 28: 136–156.

Witte, O. N. 1990. *Steel* locus defines new multipotent growth factor. *Cell* 63: 5–6.

Yamamoto, T. and 7 others. 1983. Similarity of protein encoded by the c- *erb*B2 gene to epidermal growth factor receptor. *Nature* 319: 230–234.

Yang, H. C. and Pardee, A. B. 1986. Insulin-like growth factor I regulation of transcription and replicating enzyme induction necessary for DNA synthesis. *J. Cell Physiol.* 127: 410–416.

Zarbl, H., Sukumar, S., Arthur, A. V., Martin-Zanka, D. and Barbacid, M. 1985. Direct mutagenesis of Ha-*ras*-1 oncogenes by *N*-nitroso-*N*-methylurea during initiation of mammary carcinogenesis in rats. *Nature* 315: 382–385.

Zezulak, K. M. and Green, H. 1986. The generation of insulin-like growth factor-I-sensitive cells by growth hormone action. *Science* 233: 551–553.

21

Sex determination

Sexual reproduction is . . . the masterpiece of nature.

—ERASMUS DARWIN (1791)

It is quaint to notice that the number of speculations connected with the nature of sex have well-nigh doubled since Drelincourt, in the eighteenth century, brought together two-hundred and sixty-two "groundless hypotheses," and since Blumenbach caustically remarked that nothing was more certain than that Drelincourt's own theory formed the two hundred and sixty-third.

—J. A. THOMSON (1926)

The mechanisms by which an individual's sex is determined has been one of the great questions of embryology since antiquity. Aristotle, who collected and dissected embryos, claimed that sex was determined by the heat of the male partner during intercourse. The more heated the passion, the greater the probability of male offspring. (Aristotle counseled elderly men to conceive in the summer if they wished to have male heirs.) Since that time, the environment—heat and nutrition, in particular—was believed to be important in determining sex. In 1890, Geddes and Thomson summarized all available data on sex determination and came to the conclusion that the "constitution, age, nutrition, and environment of the parents must be especially considered" in any such analysis. They argued that those factors favoring the storage of energy and nutrients predisposed one to have female offspring, whereas those factors favoring the utilization of energy and nutrients influenced one to have male offspring.

This environmental view of the determination of sex remained the only major scientific theory until the rediscovery of Mendel's work in 1900 and the rediscovery of the sex chromosome by McClung in 1902. Based on his knowledge of Mendelism, Correns speculated that the 1:1 sex ratio of most species could be achieved if the male was heterozygous and the female homozygous for some sex-determining factor. It was not until 1905, however, that the correlation (in insects) of the female sex with XX sex chromosomes and the male sex with XY or XO chromosomes was established (Stevens, 1905; Wilson, 1905). This suggested strongly that a specific nuclear component was responsible for directing the development of the sexual phenotype. Thus, evidence accumulated that sex determination occurs by nuclear inheritance rather than by environmental happenstance.

Today, we find that both environmental and internal mechanisms of sex determination can operate in different species. We shall first discuss the chromosomal mechanisms of sex determination and then consider the ways by which the environment regulates the sexual phenotype.

Chromosomal sex determination in mammals

In mammals, the determination of sex is strictly chromosomal and is not usually influenced by the environment. In most cases, the female is XX and the male is XY. The Y chromosome is the crucial inherited factor determining sex in mammals. Even if a person were to have five X chromosomes and one Y chromosome, he would be male. Furthermore, an individual with only a single X chromosome and no Y develops as a female.

When Jost (1953) removed fetal rabbit gonads before they had differentiated, he found that the resulting rabbits were female, regardless of whether they were XX or XY. They all had oviducts, uteruses, and vaginas, and they lacked penises and male accessory structures. In the absence of gonads, female development ensued. Thus, development in mammals is in the female direction unless acted upon by some product made, or regulated, by the Y chromosome.

The scheme of mammalian sex determination is shown in Figure 1. If there is no Y chromosome present, the gonadal primordia develop into ovaries. The estrogenic hormones produced by the ovary enable the development of the Müllerian duct into the vagina, cervix, uterus, and oviducts. If a Y chromosome is present, it produces or regulates the

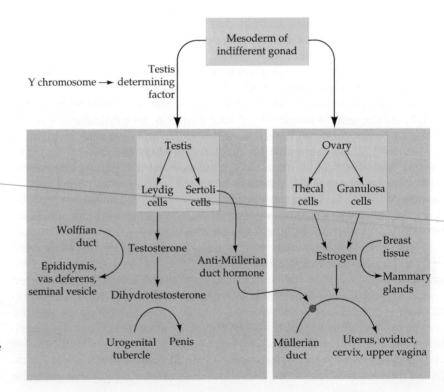

FIGURE 1

Sequence of events leading to the formation of the sexual phenotypes in mammals. In the presence of the Y chromosome, the indifferent gonad is converted into a testis. The testis cells secrete the hormones that cause differentiation of the body in the male direction. In the absence of the Y chromosome, ovarian genes act to form the ovary and develop the female phenotype.

formation of a (still uncharacterized) testis-determining factor. The testis-determining factor organizes the developing gonad into a testis instead of an ovary. Once this testis forms, it secretes two major hormones. The first hormone—ANTI-MÜLLERIAN DUCT HORMONE (often referred to as anti-Müllerian hormone [AMH] or Müllerian inhibiting substance [MIS])—destroys the tissue that would otherwise become the uterus, oviduct, cervix, and upper portion of the vagina. The second hormone—TESTOSTERONE—will masculinize the fetus, stimulating the formation of the penis, scrotum, and other portions of the male anatomy, as well as inhibiting the development of the breast primodia. Thus, the body would have the female phenotype if it is not changed by the two hormones elaborated from the fetal testes. We will now take a more detailed look at these events.

The developing gonads

The development of gonads is a unique embryological situation. All other organ rudiments can normally differentiate into only one type of organ. A lung rudiment can become only a lung, and a liver rudiment can develop only into a liver. The gonadal rudiment, however, has two normal options. When it differentiates, it can develop into either an ovary or a testis. The type of differentiation taken by this rudiment determines the future sexual development of the organism. Before this decision is made, the mammalian gonad first develops through an INDIFFERENT STAGE, during which time it has neither female nor male characteristics. In humans, the gonadal rudiment appears in the intermediate mesoderm during week 4 and remains sexually indifferent until week 7. During this indifferent stage, the epithelium of the genital ridge proliferates into the loose connective mesenchymal tissue above it (Figure 2A,B). These epithelial layers form the SEX CORDS, which will surround the germ cells that migrate into the human gonad during week 6. In both XY and XX gonads, the sex cords remain connected to the surface epithelium.

If the fetus is XY, then the sex cords continue to proliferate through the eighth week, extending deeply into the connective tissue. These cords fuse with each other, forming a network of internal (medullary) sex cords and, at its most distal end, the thinner RETE TESTIS (Figure 2C,D). Eventually, the testis cords lose contact with the surface epithelium and become separated from it by a thick extracellular matrix, the TUNICA ALBUGINEA. Thus, the germ cells are found in the cords *within* the testes. During fetal life and childhood, these cords remain solid. At puberty, however, the cords hollow out to form the seminiferous tubules, and the germ cells begin sperm production. The sperm are transported from the inside of the testis through the rete testis, which joins the EFFERENT DUCTS. These efferent tubules are the remnants of the mesonephric kidney, and they link the testis to the Wolffian duct. This duct used to be the collecting tube of the mesonephric kidney. In males, the Wolffian duct differentiates to become the VAS DEFERENS, the tube through which the sperm pass into the urethra and out of the body. Meanwhile, during fetal development the interstitial mesenchymal cells of the testes have differentiated into LEYDIG CELLS, which make testosterone. The cells of the testis cords differentiate into SERTOLI CELLS, which nurture the sperm and secrete the anti-Müllerian duct hormone.

In females, the germ cells will reside near the outer surface of the gonad. Unlike the male sex cords, which continue their proliferation, the initial sex cords of XX gonads degenerate. However, the epithelium soon produces a new set of sex cords, which do not penetrate deeply into the

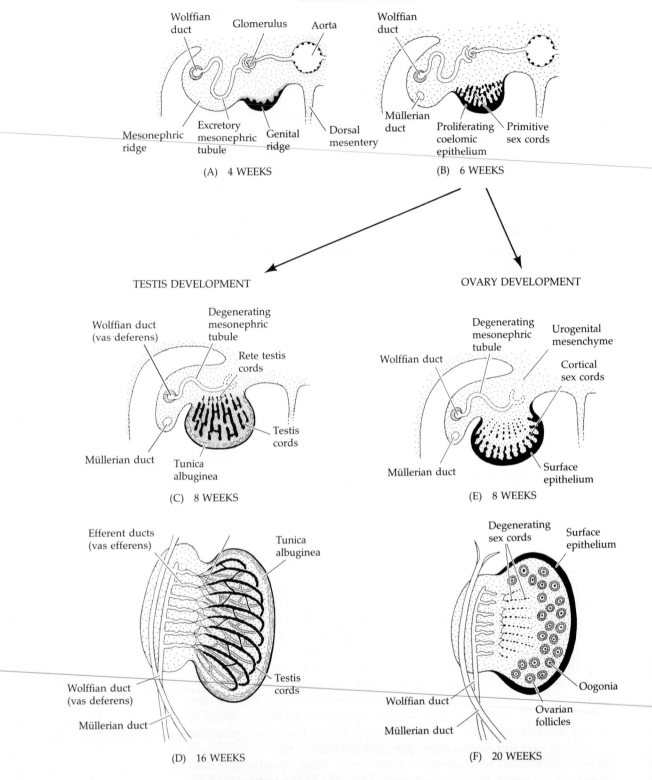

INDIFFERENT GONADS

(A) 4 WEEKS

Wolffian duct — Glomerulus — Aorta

Mesonephric ridge — Excretory mesonephric tubule — Genital ridge — Dorsal mesentery

(B) 6 WEEKS

Wolffian duct

Müllerian duct — Proliferating coelomic epithelium — Primitive sex cords

TESTIS DEVELOPMENT

(C) 8 WEEKS

Wolffian duct (vas deferens) — Degenerating mesonephric tubule

Rete testis cords

Müllerian duct — Tunica albuginea — Testis cords

(D) 16 WEEKS

Efferent ducts (vas efferens) — Tunica albuginea

Wolffian duct (vas deferens) — Testis cords

Müllerian duct

OVARY DEVELOPMENT

(E) 8 WEEKS

Degenerating mesonephric tubule — Urogenital mesenchyme

Wolffian duct — Cortical sex cords

Müllerian duct — Surface epithelium

(F) 20 WEEKS

Degenerating sex cords — Surface epithelium

Wolffian duct — Oogonia

Müllerian duct — Ovarian follicles

mesenchyme but stay near the outer surface (cortex) of the organ. Thus, they are called CORTICAL SEX CORDS. These cords are split into clusters, each cluster surrounding one or more germ cells (Figure 2E,F). The germ cells will become the OVA and the surrounding epithelial sex cords will differentiate into the GRANULOSA CELLS. The mesenchymal cells of the ovary differentiate into the THECAL CELLS. Together, the thecal and granulosa cells form the FOLLICLES that envelop the germ cells and secrete

Differentiation of human gonads shown in transverse section. (A) Genital ridge
of a 4-week embryo. (B) Genital ridge of a 6-week indifferent gonad showing
primitive sex cords. (C) Testis development in the eighth week. The sex cords
lose contact with the cortical epithelium and develop the rete testis. By the sixteenth week of development (D), the testis cords are continuous with the rete
testis and connect with the Wolffian duct. (E) Ovary development in an 8-week
human embryo, as primitive sex cords degenerate. (F) The 20-week human
ovary does not connect to the Wolffian duct, and new cortical sex cords surround the germ cells that have migrated into the genital ridge. (After Langman,
1981.)

steroid hormones. Each follicle will contain a single germ cell. In females,
the Müllerian duct remains intact, and it will differentiate into the oviducts, uterus, cervix, and upper vagina; the Wolffian duct, deprived of
testosterone, degenerates. A summary of the development of mammalian
reproductive systems is shown in Figure 3.

Control of mammalian sex determination:
Y-chromosomal genes

Several genes have been found whose function is necessary for normal
sexual differentiation. In *Drosophila* and *Caenorhabditis elegans*, genetic studies have identified numerous genes in the pathways leading to the sexual
phenotypes. In mammals, where genetic methods are more difficult to
employ, clinical studies have also identified several genes active in determining whether the organism is male or female. The reasons for our
knowledge (meager though it is) concerning mammalian sex-determining
genes is twofold. First, while mutations in most organs are lethal, mutations of the reproductive organs usually produce sex reversal or sterility.
Second, in sex determination, the determining element is known. In the
fruit fly and nematode, it is the ratio of X chromosomes to autosomes. In
mice and humans, it is the presence of a portion of the Y chromosome.
In most developmental pathways, the initiating elements have not yet
been identified.

However, even though the Y chromosome appears to be the major
determinant of sex determination in mammals, many large questions remain. Which portion of the Y chromosome is necessary for the determination of testes? Is this gene sufficient, or are there other gonadal determining genes in addition? How is the switch made between ovarian
determination and testicular determination?

The differentiation of the indifferent gonad into the testicular or ovarian pathways appears to depend on the expression of Y chromosomal
genes in the epithelial sex cord cells. This was shown by making chimeric
mice from XX and XY blastomeres. When such XX/XY chimeras are made,
they are usually fertile males (McLaren et al., 1984). The Leydig cells of
these testes are composed of both XX and XY cells (showing that XX cells
can form testicular structures), but the Sertoli cells are exclusively derived
from the XY embryo (Burgoyne et al., 1988). Thus, the critical event in the
differentiation of testes appears to be the expression of the Y chromosome
sex-determining genes in the Sertoli cell lineage. Since the Sertoli cells are
the first type of testis cell to differentiate (Magre and Jost, 1980), it is
possible that the Y chromosome sex-determining gene functions solely in
these cells, and all other events of testis formation follow as a consequence.

In humans, the gene for the testis-determining factor (*TDF*) is believed

FIGURE 3
Summary of the development of gonads and their ducts in mammals. Note that both the Wolffian and Müllerian ducts are present at the indifferent-gonad stage. The regional development of the Wolffian ducts depends upon the mesenchyme that they encounter. Lower portions of the Wolffian duct that would normally form the epididymis will form seminal vesicle tissue if cultured with mesenchyme associated with the upper (seminal vesicle) portions of the duct (Higgins et al., 1989).

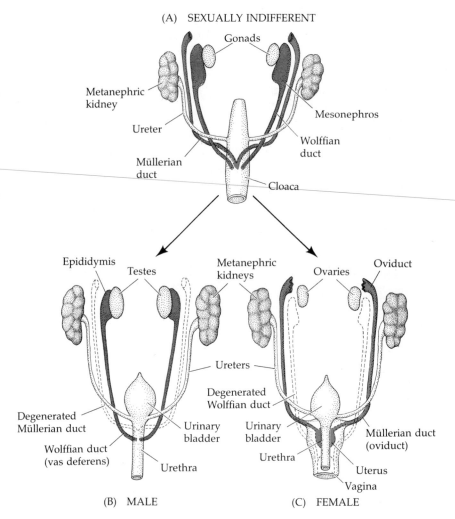

GONADS		
Gonadal type	Testis	Ovary
Sex cords	Medullary (internal)	Cortical (external)
DUCTS		
Remaining duct for germ cells	Wolffian	Müllerian
Duct differentiation	Vas deferens, epididymis, seminal vesicle	Oviduct, uterus, cervix, upper portion of vagina

to be on the short arm of the Y chromosome. Individuals who are born with the short arm but not the long arm of the Y chromosome are male, while individuals born with the long arm of the Y chromosome but not the short arm are female. Since 1985, the region of the short arm containing the *TDF* gene has been identified by combining clinical studies with DNA hybridization. DNA specific for the Y chromosome can be isolated from recombinant DNA libraries made from male cells by screening of DNA prepared from XX and XY cells. Y-specific sequences will bind only to DNA prepared from XY cells (Figure 4). By screening such Y-specific sequences on DNA from individuals missing parts of the Y chromosome, one can determine the portion of the Y chromosome from which that particular sequence is derived (Page, 1985).

In the human population, there exist both XX males (about one out of every 20,000 males) and XY females. The DNA hybridization results (Page et al., 1985; Vergnaud et al., 1986) demonstrated that XX males had Y-

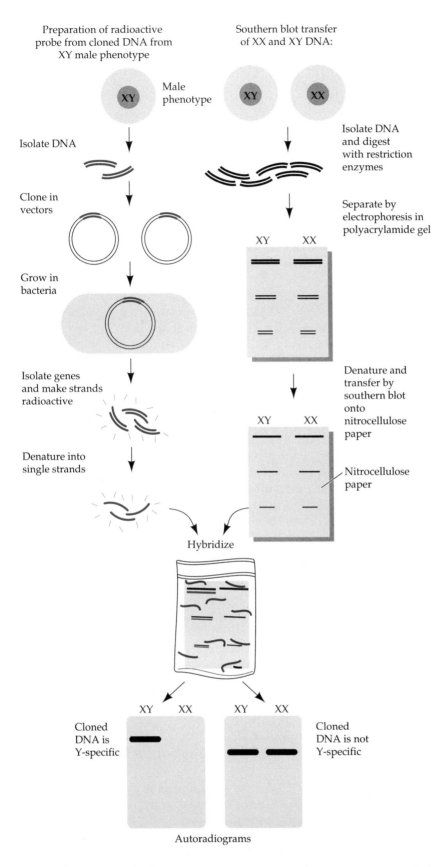

Preparation of radioactive probe from cloned DNA from XY male phenotype

Southern blot transfer of XX and XY DNA:

Male phenotype

Isolate DNA

Isolate DNA and digest with restriction enzymes

Clone in vectors

Grow in bacteria

Separate by electrophoresis in polyacrylamide gel

XY   XX

Isolate genes and make strands radioactive

Denature into single strands

Denature and transfer by southern blot onto nitrocellulose paper

XY   XX

Nitrocellulose paper

Hybridize

XY   XX   XY   XX

Cloned DNA is Y-specific

Cloned DNA is not Y-specific

Autoradiograms

FIGURE 4

Protocol for screening Y-specific DNA clones. DNA from male cells is cloned into viruses and grown in *E. coli*. Each clone is separately cultured and its human DNA is isolated and made radioactive. Each DNA fragment is denatured into single strands and hybridized with DNA from male or female cells that has been electrophoresed and Southern blotted. The Y-specific DNA fragments should bind specifically to the DNA from XY cells but not to XX cells.

specific DNA from Y chromosome region 1 translocated onto one of their chromosomes. Moreover, all the XY females tested were missing this region of the Y chromosome (Figure 5). It seems most likely, then, that the gene for the testis-determining factor maps to region 1 of the Y chro-

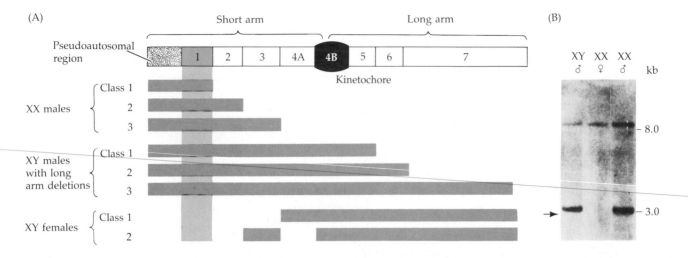

FIGURE 5

Deletion mapping of the gene for the testis-determining factor(s) in humans. (A) The Y chromosome has been divided into nine intervals. Interval 4B is the kinetochore and the stippled region is the "pseudoautosomal" portion, which contains DNA sequences in common with the X chromosome (and allows it to pair with the X during meiosis). XX males all have DNA from region 1 whereas XY females have a deletion of this material from their Y chromosomes. As these regions are often too small to be seen microscopically, the presence or absence of the region was determined by DNA hybridization (as in Figure 4). Results of such a hybridization are shown in B where the radioactive DNA was from region 1 of the Y chromosome. This Y-specific DNA is thought to contain the testis-determining factor since the autoradiogram of the Southern blot demonstrates its presence in XY males and XX males, but not in XX females. (A after Page, 1985; B from Vergnaud et al., 1986; photograph courtesy of D. Page.)

mosome, near the tip of the short arm. Similar evidence from mice also suggested that this region of DNA is the testis-determining gene of the Y chromosome (Singh and Jones, 1982; Eicher et al., 1983).

By analyzing the DNA of XX men and XY women, the position of this testis-determining gene has been further narrowed down to a 35,000-base-pair region of the Y chromosome located just before the pseudoautosomal end (Figure 6). In this region, Sinclair and colleagues (1990) found a sequence of male-specific DNA that could encode a peptide of 223 amino acids. This DNA was found in XY males and in the rare XX males, and it was absent from XX and XY females. If this gene (called *SRY*, for "sex-determining region Y") were actually to encode the major testis-determining factor, one would expect that it would act in the genital ridge immediately before or during testis differentiation. This prediction has been met in studies of the analogous gene found in mice. The mouse gene (called *Sry*) is expressed in the somatic cells of the indifferent gonad immediately before or during its differentiating into a testis. This is the only time and place this gene is seen to be expressed in the embryo (Gubbay et al., 1990; Koopman et al., 1990). Moreover, some XY females have been found to have point or frameshift mutations in this gene (Berta et al., 1990; Jäger et al., 1990).

The *SRY* (*Sry*) gene is currently the best candidate for the mammalian testis-determining gene on the Y chromosome. However, the conclusive proof that this gene is the major testis-determining factor would come from inserting the *Sry* DNA into the genome of a normal XX mouse egg. If such a transgenic XX mouse developed into a male, the *Sry* gene would be shown to be of major importance in the determination of gonadal sex.

Control of mammalian sex determination: Autosomal genes

The testis-determining gene(s) of the Y chromosome are necessary but not sufficient for the development of the mammalian testis. While it is not yet known what genes are being regulated by the testis-determining gene on the Y chromosome, studies on mice have shown that the Y chromosome gene has to be coordinated with certain autosomal genes. Eicher and Washburn (1983) discovered that when the Y chromosome of *Mus domesticus* was bred into an inbred strain of *Mus musculus*, ovarian tissue developed in the XY individuals (Figure 7). Half of these XY mice were

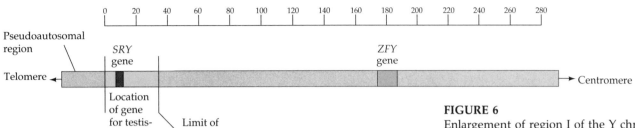

Kilobases DNA

Pseudoautosomal region

Telomere ← | | *SRY* gene | | *ZFY* gene | | → Centromere

Location of gene for testis-determining factor

Limit of breakpoint in XX males

FIGURE 6
Enlargement of region I of the Y chromosome. *SRY* represents the gene found in the 35,000 base pairs of DNA that carries the testis-determining factor. *ZFY* represents another gene, once thought to be the testis-determining factor but which was shown to be separate from it by deletion mapping. *ZFY* appears to be active during sperm formation.

completely sex reversed (having only ovarian tissue as gonads), and the other half of this group contained both testicular and ovarian tissue (Eicher and Washburn, 1983; 1986). The testis-determining gene of the Y chromosome (*Tdy*) of each strain of mouse can cooperate with its own autosomal testis-determining gene (*Tda-1*) but not with the other.

Another autosomal sex-determining gene is the *T-associated sex reversal gene* (*Tas*). This gene segregates with the *T/t* gene complex on chromosome 17 of *Mus musculus,* and different alleles of this gene have been found in different inbred strains of mice (Washburn and Eicher, 1983). The C57 strain of *M. musculus* and the AKR strain of *M. musculus* both form males about 50 percent of the time. However, the C57 Y chromosome is unable to direct testis morphogenesis if chromosome 17 comes from the AKR strain. In such cases, normal testicular determination is disrupted and each gonad contains a mixture of ovarian and testicular tissue.

McLaren (1988) has proposed that one of the genes regulated by the testis-determining factor is the gene for anti–Müllerian duct hormone. This hormone is produced by the Sertoli cells of the testis and is one of the first products of testicular tissue. In addition to its well known ability to cause the degeneration of Müllerian duct tissue (to be discussed at greater length later in this chapter), this protein has also been observed to cause

(A) (B)

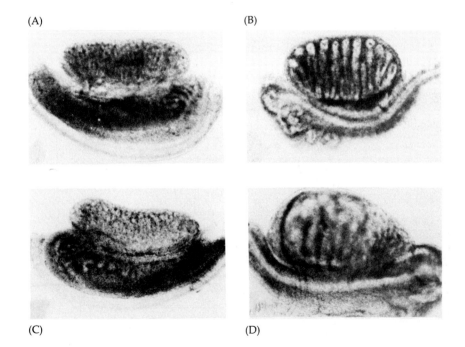

(C) (D)

FIGURE 7
Ovarian tissue in XY gonads where a *Mus domesticus* Y chromosome is in the genome of the C57BL strain of *Mus musculus.* Gonads and attached mesonephros were dissected from 14.5 to 16-day-old fetuses. (A) Normal ovary from XX fetus. (B) Testis from normal C57BL male fetus. (C) Ovary from XY female. (D) "Ovotestis" of XY fetus containing areas of ovarian and testicular tissue. (From Eicher et al., 1982; photographs courtesy of E. Eicher and L. Washburn.)

cultured XX ovarian cells to differentiate into testicular tissue (Vigier et al., 1987). Vigier and co-workers suggest that the anti–Müllerian duct hormone may therefore be responsible for producing the freemartin phenomenon. FREEMARTINS are sterile XX calves whose gonads have been partially or completely masculinized. These individuals are found when the affected cow is one of a pair of twins, the other being XY. The placental blood circulation unites these twins, and it has long been thought (Lillie, 1917; Wachtel et al., 1980) that some hormone travels through the blood from the male fetus to masculinize the female's gonads. Since anti–Müllerian duct hormone can cause rat ovarian cells to organize into structures resembling the seminiferous tubules, it is possibly the cause of both the freemartin phenomenon and the normal morphogenesis of the testes.

A model of primary mammalian sex determination

At the moment, there is no consensus on the mechanism of mammalian sex determination. Figure 8 presents one model, synthesizing the work of Eicher and Washburn (1986) with that of Burgoyne (1989).

This model starts from the premise that the undifferentiated mam-

FIGURE 8
A genetic model of mammalian sex determination. The *Tdy* gene, when present in the gonadal cell lineage, initiates testicular development and prevents the germ cells from producing a signal that would induce the expression of ovary-forming genes in the gonadal tissue. In the absence of the *Tdy* gene, the *Od* gene in the germ cells induces ovarian development at a later time. The autosomally inherited sex reversals seen in mice may be due to timing differences in the testicular pathway genes. (After Eicher and Washburn, 1986, and Burgoyne, 1989.)

(A) *Tdy* gene initiates testicular development

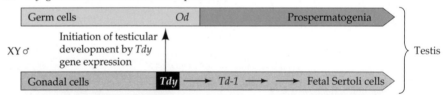

(B) In the absence of *Tdy*, *Od* gene induces ovarian development

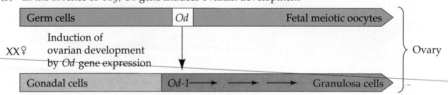

(C) Expression of *Od* and repression of *Tdy* leads to sex reversal

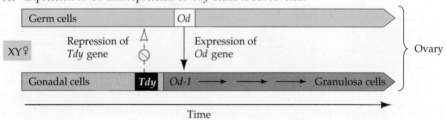

malian fetal gonad is bipotential and that *both* male and female development are active, gene-directed processes. (That is, female development is not a passive "default condition.") Thus, there are two parallel developmental pathways—one for ovarian development, one for testicular development. This model also acknowledges that there are inductive interactions occurring between the germ line cells and the gonadal somatic tissue that they enter. The first evidence of this induction is that germ cells usually become the type of gamete that is produced by the type of gonad they enter. XX germ cells initiate spermatogenesis if they enter a testis. The reciprocal interaction is shown by the observation that the Y chromosome of the XY gonadal epithelium does not transcribe the *ZFY* gene unless induced by incoming germ cells (Koopman et al., 1989).

According to this model, the first gene in the ovarian developmental pathway is the hypothetical gene *Od* (*ovarian-determining*) located on the X chromosome or on an autosome. It would function to activate the next gene on the ovarian pathway *Od-1*. (Both these genes are presently hypothetical, but nobody had been actively looking for them because they were not needed by models of sex determination prior to Eicher and Washburn's hypothesis.) This gene could be a gene active in the germ cell line to induce the activation of ovarian development in the gonadal cells.

The first gene on the testicular pathway is the testis-determining factor of the Y chromosome (*Tdy* in mice, *TDF* in humans). This gene functions *earlier* in development than the *Od* gene and serves two functions. It activates the next gene in the testicular pathway (*Td-1*), and it suppresses the activation of *Od*. In this manner, testes are formed if the *Tdy* gene is present (as it is in XY individuals) and ovaries are formed when it is absent. *Tda-1* and *Tas* are probably two of the genes on the testicular determination pathway. This model suggests that the inherited sex reversal situations involving these loci are due to lack of coordination between *Tdy* and these alleles. This lack of coordination may concern the timing of gene expression. The C57 *Tdy* allele may be activated earlier than the *Mus domesticus Tdy* allele. It is possible that it does not have the opportunity to completely suppress the *Od* gene, thereby enabling ovarian tissue to form.

Although remarkable progress has been made in recent years, the analysis of primary sex determination in mammals remains (as it has since prehistory) one of the great unsolved problems of developmental biology.

Secondary sex determination in mammals

Hormonal regulation of the sexual phenotype

Primary sex determination involves the formation of either an ovary or a testis from the indifferent gonad. This, however, does not give the complete sexual phenotype. Secondary sex determination concerns the development of the female and male phenotypes from the hormones elaborated by the ovaries and testes. Both female and male secondary sex determination have two major temporal components. The first occurs within the embryo during organogenesis; the second (as discussed in Chapter 19) occurs during adolescence.

As mentioned earlier, if the indifferent gonads are removed from an embryonic animal, the female phenotype is realized. The Müllerian ducts develop while the Wolffian duct decays. This is also seen in certain humans

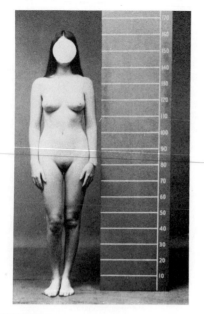

FIGURE 9

XY individual with androgen insensitivity syndrome. Despite the XY karyotype and the presence of testes, the individual develops female secondary sex characteristics. Internally, however, the woman lacks the Müllerian duct derivatives. (Photograph courtesy of C. B. Hammond.)

who are born without functional gonads. Individuals whose cells have only one X chromosome (and no Y chromosome) originally develop ovaries, but these ovaries atrophy before birth and the germ cells die before puberty. Under the influence of estrogen derived from the mother and from the placenta, these infants are born with a female genital tract (Langman and Wilson, 1982).*

The formation of the male phenotype involves the secretion of testicular hormones that promote Wolffian duct development and cause the Müllerian duct to atrophy. The first of these hormones is the anti-Müllerian duct hormone, the Sertoli cell hormone that causes the degeneration of the Müllerian duct. The second of these hormones is the steroid testosterone, which is secreted from the testicular Leydig cells. This hormone causes the Wolffian duct to differentiate into the epididymis, vas deferens, and seminal vesicles, and it causes the urogenital swellings and sinus to develop into the scrotum and penis. The existence of these two independent systems of masculinization is demonstrated by people having androgen insensitivity syndrome. These XY individuals have the testis-determining factor gene and so have testes that make testosterone and AMH. However, these people lack the cytoplasmic testosterone-binding protein and therefore cannot respond to the testosterone made in their testes (Meyer et al., 1975). They are able to respond to estrogen made in their adrenal glands, so they are distinctly female in appearance (Figure 9). However, despite their female appearance, these individuals do have testes, and even though they cannot respond to testosterone, they do respond to AMH. Thus, their Müllerian ducts degenerate. These people develop as normal but sterile women, lacking uteruses and oviducts.

So there are two distinct masculinizing hormones, testosterone and AMH. There is evidence, though, that testosterone might not be the active

*The mechanisms by which estrogen promotes the differentiation of the Müllerian duct are not well understood. During embryonic development, the duct is extremely sensitive to estrogenic compounds, as is known from the teratogenic effects of *diethylstilbesterol* (DES). This compound is a synthetic estrogen that was given to women from the 1940s through the 1960s to help maintain pregnancies. Daughters born to women who used this drug have a high incidence of Müllerian duct anomalies, including malformation of the vaginal and cervical epithelia, structural abnormalities of the oviducts and uterus, and a higher than normal incidence of vaginal cancer (Robboy, 1982; Bell, 1986).

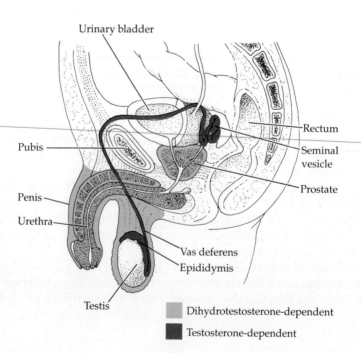

FIGURE 10

Testosterone- and dihydrotestosterone-dependent regions of the fetal human male genital system. (After Imperato-McGinley et al., 1974.)

hormone in certain tissues. In 1974, Siiteri and Wilson showed that testosterone is converted to 5α-DIHYDROTESTOSTERONE in the urogenital sinus and swellings, but not in the Wolffian duct. In the Dominican Republic, Imperato-McGinley and her colleagues (1974) found a small community in which several inhabitants had a genetic deficiency of the enzyme 5α-ketosteroid reductase, the enzyme that converts testosterone to dihydrotestosterone. Although these XY individuals have functioning testes, they have a blind vaginal pouch and an enlarged clitoris. They appear to be girls and are raised as such. Their internal anatomy, however, is male—Wolffian duct development and Müllerian duct degeneration. Thus, it appears that the formation of external genitalia is under the control of dihydrotestosterone, whereas the Wolffian duct differentiation is controlled by testosterone itself (Figure 10). Interestingly enough, the external genitalia become responsive to testosterone at puberty, causing obvious masculinization in a person originally thought to be a girl.

The anti-Müllerian duct hormone is a 560-amino acid glycoprotein (Cate et al., 1986a) made in the Sertoli cells (Tran et al., 1977). When fragments of fetal testes or isolated Sertoli cells are placed adjacent to cultured segments containing portions of the Wolffian and Müllerian ducts, the Müllerian duct atrophies even though no change occurs in the Wolffian duct (Figure 11). This atrophy is caused both by cell death and by the epithelial cells of the duct becoming mesenchymal and migrating away (Trelstad et al., 1982). The Müllerian duct is sensitive to the action of this hormone for only a brief period of time (Figure 12), and the testes stop making this factor at or near the time of birth.*

We see, then, that once the testes are formed, they elaborate two hormones that cause the masculinization of the fetus. One of these hormones—testosterone—may be converted into a more active form by the tissues that create the external genitalia. In females, estrogen secreted from the fetal ovaries appears sufficient to induce the differentiation of the Müllerian duct into the uterus, oviducts, and cervix. In such a manner, the sex chromosomes control the sexual phenotype of an individual.

*AMH may also become a potent anti-tumor drug. Donahoe and her colleagues (Fuller et al., 1984; Cate et al., 1986b) have shown that AMH can kill tumors of the mouse and human female reproductive tracts. Clinical trials await the production of large quantities of AMH from cloned genes.

FIGURE 11
Assay for anti-Müllerian duct hormone activity in the anterior segment of a 14.5-day fetal rat reproductive tract, after 3 days in culture. (A) Both the Müllerian duct (arrow at left) and Wolffian duct (arrow at right) are open. (B) The Wolffian duct (arrow) is open, but the Müllerian duct has degenerated and closed. (Photographs courtesy of N. Josso.)

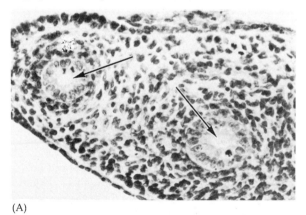

(A)

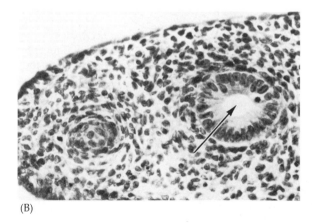

(B)

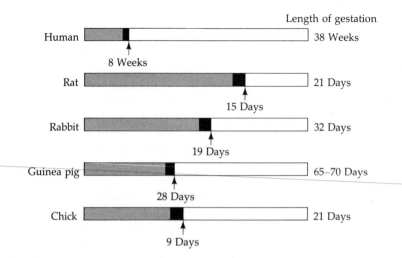

The central nervous system

One of the most controversial areas of secondary sex determination involves the development of sex-specific behaviors. In songbirds, testosterone is seen to regulate the growth of male-specific neuronal clusters in the brain. Male canaries and zebra finches sing eloquently, whereas the females sing little, if ever. These songs are used to mark territories and to attract mates. The ability to sing is controlled by six different clusters of neurons (NUCLEI) in the avian brain (Figure 13). Neurons connect each of these regions to one another. In male canaries, these nuclei are several times larger than the corresponding cluster of neurons in female canaries; and in zebra finches, the females may lack one of these regions entirely (Arnold, 1980; Konishi and Akutagawa, 1985).

Testosterone plays a major role in song production. In adult male zebra finches, Pröve (1978) demonstrated a linear correlation between the amount of song and the concentration of serum testosterone. It has been shown that the seasonal changes in testosterone levels are correlated with the seasonal singing patterns of these birds. When testosterone levels are low, there is not only a decrease in bird song but also a decrease in the size of the male-specific brain nuclei (Nottebohm, 1981). In adult chaffinches, castration eliminates song, but injection of testosterone will induce such birds to sing even in November, when they are normally silent (Thorpe, 1958). In several species of birds, the females can be induced to sing by injecting them with testosterone (Nottebohm, 1980). The four song-controlling regions of the brain will grow 50–69 percent in such birds, whereas other brain regions show no such growth. Autoradiographic studies (Arnold et al., 1976) have shown that the neurons of the song-

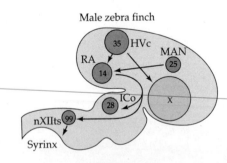

FIGURE 13

Sexual dimorphism in the avian brain. Schematic diagram indicates the major neural areas thought to be involved in the production of bird song in the zebra finch. Circles represent specific brain areas; the size of each circle is proportional to the volume occupied by that region. Circles with dashed lines are estimated volumes. The numbers within each circle represent the percentage of cells therein that incorporate radiolabeled testosterone. The volume differences in three of these regions (HVc, RA, and nXIIts) are significant between the sexes, and area X has not been observed in the brains of female finches. The differences in testosterone binding in regions HVc and MAN are significant, and no sex differences in steroid hormone binding have been observed in other regions of the brain. The arrows indicate the axonal paths connecting the regions in the male finch. (After Arnold, 1980.)

controlling nuclei will incorporate radioactive testosterone, whereas other regions of the brain will not. It is apparent, then, that gonadal hormones can play a major role in the development of the regions of the nervous system that generate sex-specific behaviors.

In mammals, the situation is not as clear, for there are fewer behaviors that exclusively characterize one sex. In rats, penile thrusting is one such behavior, and it is controlled by motor neurons to the levator ani and bulbocavernosus muscles. Both these neurons originate from a spinal nucleus that can specifically concentrate testosterone. In female rats, these muscles are vestigial, and the volume of the controlling neurons is greatly reduced (Breedlove and Arnold, 1980). Testosterone appears to cause two types of changes in these responsive neurons. In fetal and newborn rats, testosterone acts to prevent a "normally" occurring death of neurons in this region. Female rats lose up to 70 percent of the neurons of this spinal nucleus, while male rats lose only 25 percent. In adult rats, testosterone acts upon this nucleus to maintain the size of the nerve cells and their dendrites. The soma area and dendrite length of this spinal nucleus are reduced by half when an adult rat is castrated. This reduction is reversed by the injection of testosterone (Nordeen et al., 1985; Kurz et al., 1986).

Testosterone is not the only steroid capable of mediating behavior. In the mammalian brain, estrogen-sensitive neurons are also seen. These neurons are located (Figure 14) at positions in neural circuits that are known to mediate reproductive behavior: the hypothalamus, the pituitary, and the amygdala (McEwen, 1981). Pfaff and McEwen (1983) demonstrated that estrogen alters the electrical and chemical features of those hypothalamic neurons capable of binding estrogen in their chromatin. Terasawa and Sawyer (1969) previously had found that the electrical activity of these neurons varies during the seasonal estrogen cycle of the rat, becoming elevated at the time of ovulation. Moreover, estrogen appears to stimulate those neurons in the regions that induce female reproductive behavior. Ovariectomized rats given estrogen injections directly to the hypothalamus displayed LORDOSIS, a position that stimulates the male to mount them, whereas control ovariectomized rats did not show that behavior (Barfield and Chen, 1977; Rubin and Barfield, 1980). The mechanism by which estrogen causes the specific neuronal activity at these times is thought to involve increasing the permeability of these neurons to potassium (Nabekura et al., 1986).

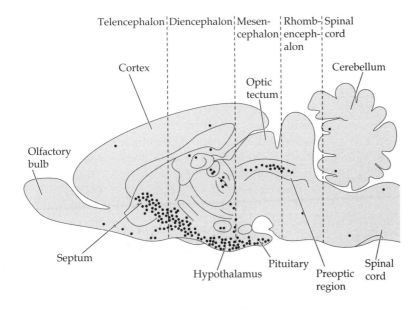

FIGURE 14
Diagrammatic representation of the estrogen-binding regions in a female rat brain. (After Kandel and Schwartz, 1985.)

SIDELIGHTS & SPECULATIONS

The development of sexual behaviors

Exposure of fetal rats and mice to large amounts of hormones has been found to influence their adult behaviors. When newborn male rats are castrated, thereby depriving them of testosterone during the neonatal period, they undergo cyclic gonadotropin release (characteristic of female rats) and display female sexual behaviors such as lordosis when they mature (Tiefer, 1970). Conversely, when newborn female rats are given a single dose of testosterone during the neonatal period, they develop masculine endocrine and sexual behaviors. Thus, each newborn rat is born with the potential to display either male or female behaviors.

What is surprising, however, is that masculine behavior patterns can be permanently induced by injecting newborn female rats with a single dose of the female sex hormone estradiol* (Doughty et al., 1975). This ability of estradiol to help masculinize rat sexual behavior has given rise to the CONVERSION HYPOTHESIS. According to this model, the female pattern of sexual behavior is "intrinsic" to the mammalian brain. However, if the brain receives estradiol during a critical stage in its development (immediately before or after birth, depending upon the species), defeminization will occur. Brain cells can obtain estradiol in two ways. They can receive it directly from the circula-

*The terms estrogen and estradiol are often used interchangeably. However, estrogen refers to a class of steroid hormones responsible for establishing and maintaining specific female characteristics. Estradiol is one of these hormones, and in most mammals (including humans), it is the most potent of the estrogens.

tion or they can synthesize estradiol from circulating testosterone. To prevent defeminization in female mammals, estrogen-binding proteins in the serum eliminate freely circulating estradiol. In males, however, testosterone would not be bound by these proteins and would thus be able to enter the brain and be converted to estradiol. In the newborn rat, this conversion has been seen in the cells of the hypothalamus and the limbic system, two regions of the brain known to regulate hormonal and reproductive behaviors (Reddy et al., 1974).

Neonatal estradiol appears to be responsible for "defeminizing" the brain, whereas the actual masculinization of behavior is probably due to testosterone or dihydrotestosterone. When the conversion of testosterone to estradiol is chemically inhibited, males show both male and female behavior patterns (McEwen et al., 1977; Vreeburg et al., 1977). It is possible, then, that, like the formation of the male genital system, the development of the male nervous system involves both defeminizing and masculinizing steps.

Extrapolating from rats to humans is a risky business, as no stereotypic sex-specific human behavior has yet been identified. What is "masculine" behavior in one society may be "feminine" behavior in another (much to the embarrassment of naive travelers). Although the popular press publicizes research that finds biologically determined differences in skills or behaviors between the sexes, these papers have been widely criticized (Kolata, 1980; Jacklin, 1981; Bleier, 1984; Fausto-Sterling, 1985). Moreover, the overlap in behaviors precludes these from being considered sex-specific.

Chromosomal sex determination in *Drosophila*

The sexual development pathway

Both mammals and insects such as *Drosophila* use an XX/XY system of sex determination. However, the sex-determining mechanisms in these two groups are very different. In mammals, the Y chromosome plays a pivotal role in determining the male sex. Thus, XO mammals are *females*, having ovaries, uteruses, and oviducts (but usually very few, if any, ova). In *Drosophila*, sex determination is achieved by a balance of female determinants on the X chromosome and male determinants on the autosomes (non-sex chromosomes). If there is but one X chromosome in a diploid cell (1X:2A), the organism is male. If there are two X chromosomes in a diploid cell (2X:2A), the organism is female (Bridges, 1921; 1925). Thus, XO *Drosophila* are sterile *males*. Table 1 shows the different X:autosome ratios and the resulting sex.

In *Drosophila*, and in insects in general, one can observe GYNANDROMORPHS—animals in which certain regions are male and other regions are female (Figure 15; color portfolio). This can happen when an X chro-

TABLE 1
Ratios of X chromosomes to autosomes in different sexual phenotypes in
Drosophila melanogaster

X chromosomes	Autosome sets (A)	X:A ratio	Sex
3	2	1.50	Metafemale
4	3	1.33	Metafemale
4	4	1.00	Normal female
3	3	1.00	Normal female
2	2	1.00	Normal female
2	3	0.66	Intersex
1	2	0.50	Normal male
1	3	0.33	Metamale

Source: After Strickberger (1968).

mosome is lost from one embryonic nucleus. The progeny of that nucleus, instead of being XX (female), are XO (male). Because there are no sex hormones in insects to modulate such events, each cell makes its own sexual "decision." The XO cells display male characteristics, whereas the XX cells display female traits. This situation provides a beautiful example of the association between X chromosomes and sex. As can be seen from this example, the Y chromosome plays no role whatever in *Drosophila* sex determination. Rather, it is necessary only to ensure fertility in males. The Y chromosome is active only late in development, during sperm formation.

Any theory of *Drosophila* sex determination must explain how the X:A ratio is read and how this information is transmitted to the genes controlling the male or female phenotypes. Although we do not yet know the intimate mechanisms by which the X:A ratio is made known to the cells, research in the past five years has revolutionized our view of *Drosophila* sex determination. Much of this research has concerned the identification and analysis of genes that are necessary for sexual differentiation and the placing of those genes in a developmental sequence.

As in homeotic genes (Chapter 18) and lineage switch genes (Chapter 11), the sex-determination genes mediate the choice between two alternative developmental pathways. Mutations in most of these genes—*Sex-lethal* (*Sxl*), *transformer* (*tra*), and *transformer-2* (*tra-2*)—transform XX individuals into males. Such loss-of-function mutations have no effect on male sex determination. Homozygosity of the *intersex* (*ix*) gene causes XX flies to develop an intersex phenotype having portions of male and female

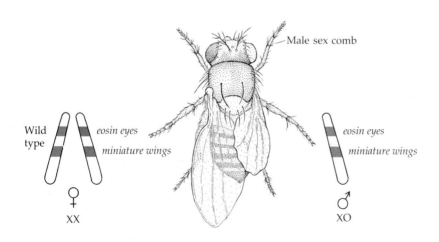

FIGURE 15
Gynandromorph of *Drosophila melanogaster* in which the left side is female (XX) and the right side is male (XO). The male side has lost an X chromosome bearing the wild-type alleles of eye color and wing shape, thereby allowing the expression of the recessive alleles *eosin eye* and *miniature wing* on the remaining X chromosome. (After Morgan, 1919.)

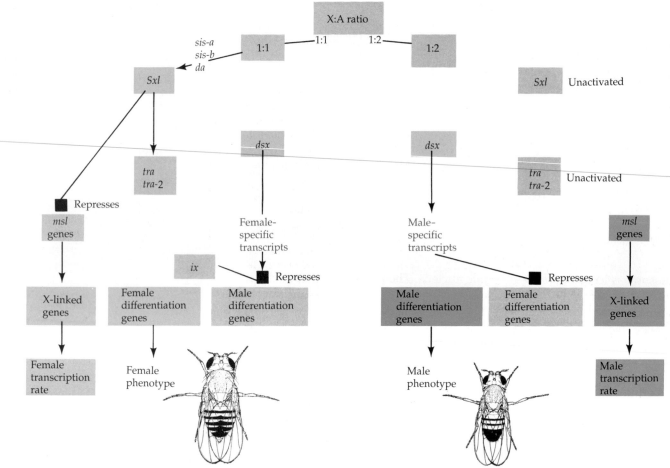

FIGURE 16

Proposed regulation cascade for *Drosophila* somatic sex determination. Arrows represent activation while a block at the end of a line indicates suppression. The *msl* loci, under the control of the *Sxl* gene, regulate the dosage compensatory transcription of the male X chromosome. (After Baker et al., 1987.)

tissue in the same organ. The *doublesex* (*dsx*) gene is important for the sexual differentiation of both sexes. Homozygous recessive *dsx* alleles will turn both XX and XY flies into intersexes (Baker and Ridge, 1980; Belote et al., 1985a).

The position of these genes in a developmental pathway is based on (1) the interpretation of genetic crosses resulting in flies bearing two or more of these mutations, and (2) the determination of what happens when there is a complete absence of the products of one of these genes. Such studies have generated the model of this regulatory cascade seen in Figure 16.

The *Sex-lethal* gene as pivot for sex determination

The first phase of *Drosophila* sex determination involves reading the X:A ratio. What elements on the X chromosome are "counted" and how is this information used? It appears that high values of the X:A ratio are responsible for activating the feminizing switch gene *Sex-lethal* (*Sxl*). At low values (males), *Sxl* remains inactive (Cline, 1983; Salz et al., 1987). In XX *Drosophila*, *Sxl* is activated during the first two hours after fertilization, and this gene transcribes a particular embryonic type of *Sxl* mRNA that is only found for about two more hours (Salz et al., 1989). Once it is activated, the *Sxl* gene remains on despite any further changes in the X:A ratio (Sánchez and Nöthiger, 1983). *Sxl* function is necessary for XX embryos to initiate the female developmental pathway and to maintain an appropriate level of transcription from the two X chromosomes. In XX individuals where *Sxl* is not functional, the male phenotype emerges and the X chromosomes transcribe more rapidly (as would a single X in a normal male embryo, which has to produce as much gene product as the two X chromosomes of a normal female). In XX embryos this overstimulation of X chromosome transcription is lethal.

This sex-specific transcription is thought to be stimulated by "NUMERATOR ELEMENTS" on the X chromosome that constitute the "X" part of the X:A ratio. Cline (1988) has demonstrated that two of these numerator elements are the X chromosome genes *sisterless-a* and *sisterless-b*. In this study, he discovered a variant *Drosophila* autosome that contained two regions of the X chromosome inserted into it. This chromosome was lethal if it was present in males, for males who received this chromosome activated their *Sxl* gene just as if they had inherited two X chromosomes. The activated *Sxl* gene caused the dosage compensation of the X chromosome to be female (that is, to have less transcription), but as the male had only one X chromosome, this proved to be lethal. Cline thus had an assay for the numerator genes. These would be the genes on the X chromosome-derived portion of this autosome whose presence activated the *Sxl* gene. By this technique, he found the *sisterless-a* and *sisterless-b* genes. Having two copies of these genes proved lethal for XY embryos, and the simultaneous deletion of one copy of each *sis* locus proved lethal to females (whose embryos did not activate the *Sxl* gene even though they were XX). Therefore, the presence of two copies of each of the *sis* genes is equivalent to having two copies of the X chromosome. The *sis* genes are probably not the only numerator elements, as Cline has documented cases where the genetic background of the wild-type male fly is not sensitive to double doses of these genes.

The *Sxl* gene does not seem to sense these numerator elements without the presence of a maternal gene whose product is present in the egg cytoplasm. Without the product of the *daughterless (da)* gene, *Sxl* will not be activated. The *daughterless* gene is necessary in the mothers of females. If this gene is not active in the ovaries, the eggs she produces will lack this substance. This lack of daughterless protein prevents the activation of *Sxl*. This does not affect XY embryos (since they do not activate *Sxl* anyway), but it is lethal in female embryos, since the mechanism for dosage compensation causes the two X chromosomes to be transcribed at the faster, male, rate (Cline, 1986; Cronmiller and Cline, 1987), hence, the name of the mutation. Thus, shortly after fertilization, the *sis-a*, *sis-b*, and *da* genes enable *Sxl* to be transcriptionally active only in female embryos.

Shortly after this transcription has taken place, a second promoter on the *Sex-lethal* gene is activated, and this gene is transcribed in both males and females. However, analysis of the cDNA from *Sxl* mRNA shows that the *Sxl* mRNA of males differs from the *Sxl* mRNA of females (Bell et al., 1988). This is the result of differential RNA processing. Moreover, the *Sxl* protein appears to bind to its own mRNA precursor to splice it in the female manner. Since males do not have any available *Sxl* protein, their new *Sxl* transcripts are processed in the male manner. The male *Sxl* mRNA is nonfunctional. While the female-specific *Sxl* message encodes a protein of 354 amino acids, the male-specific *Sxl* transcript contains a translation termination codon (UGA) after amino acid 48. The differential RNA processing that puts this termination codon into the male-specific mRNA is shown in Figure 17. In males, the nuclear transcript is spliced in a manner that yields three exons, and the termination codon is within the central exon. In females, RNA processing yields only two exons, and the male-specific central exon is now spliced out as a large intron. Thus, the female-specific mRNA lacks the termination codon.

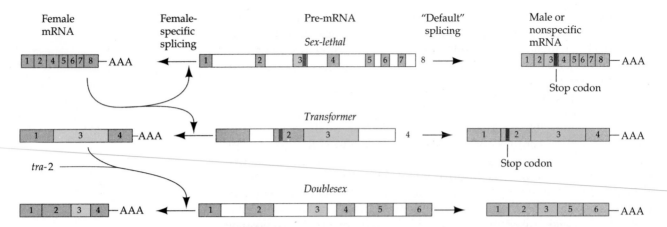

FIGURE 17

The pattern of sex-specific RNA splicing in three major *Drosophila* sex-determining genes. The pre-mRNAs are located in the center of the diagram and are identical in both the male and female nuclei. In each case, the female-specific transcript is shown at the left, while the default transcript (whether male or nonspecific) is shown to the right. Exons are numbered, and the positions of the termination codons and poly(A) sites are marked. (After Baker, 1989.)

The protein made by the female-specific *Sxl* transcript can be predicted from its nucleotide sequence. This protein would contain two regions that are important for binding to RNA. These regions are shared with nuclear RNA-binding proteins such as those in snRNPs. Bell and colleagues (1988) have proposed that there are two targets for the RNA-binding protein encoded by *Sxl*. One of these targets is the pre-mRNA of *Sxl* itself. This would be the mechanism that would maintain the female state of the pathway after the initial activating event has passed. The second target of the female-specific *Sxl* protein would be the pre-mRNA of the next gene on the pathway, *transformer*.

The *transformer* genes

The *Sxl* gene regulates somatic sex determination by controlling the processing of the *transformer* gene transcript. As we saw in Chapter 13, the *transformer* gene (*tra*) is alternatively spliced in males and females. There is a female-specific mRNA and also a nonspecific mRNA that is found in both females and males. Like the male *Sxl* message, the nonspecific *tra* mRNA contains a termination codon early in the message, making the protein nonfunctional (Boggs et al., 1987). In *tra*, the second exon of the non-specific mRNA has the termination codon. This exon is not utilized in the female-specific message (Figure 17). How is it that the females should make a different transcript than the males? It is thought that the female-specific protein from the *Sxl* gene binds to some of the *tra* pre-mRNA, causing them to be processed in a way that splices out the second exon. (See Chapter 13 for details.) The result would be a female-specific protein. This protein is critical in female sex determination. If the female-specific transcript is artificially produced in XY flies, those flies become female. The nonspecific transcript has no effect on either males or females (McKeown et al., 1988).

The female-specific *tra* product acts in concert with the *transformer-2* (*tra-2*) gene to help generate the female phenotype. (The *tra-2* gene is not needed for male sex determination, although it is needed for male spermatogenesis later in development.) The *tra-2* gene is constitutively active and makes the same protein product in both males and females. This tra-2 protein, like that of the female-specific Sxl protein, contains an RNA-binding domain (Amrein et al., 1988; Goralski et al., 1988). It is proposed that the *tra-2* gene can bind to the transcript of the *doublesex* gene, but only in the presence of the female-specific tra protein (Baker, 1989).

Doublesex: The switch gene of sex determination

The *doublesex* gene is active in both males and females, but its primary transcript is processed in a sex-specific manner (Baker et al., 1987; Figure 17 in Chapter 13). Male and female transcripts are identical through the first three exons. The 3' exons differ in a way reminiscent of the two types of IgM molecules discussed in Chapter 13. What is an exon for the female-specific transcripts is part of the untranslated 3' end of the male-specific message. Moreover, molecular analyses of the dominant *dsx* mutations reveal them to be insertions in the female-specific exon. If a dominant *dsx* allele exists in an XX individual, the fly becomes male.

The alternative RNA processing appears to be the result of the *transformer* genes (Figure 17). If only *tra-2* is active, the *doublesex* transcript is spliced in the male-specific manner. This transcript makes a protein that represses the genes for female cell characteristics. On the other hand, if *tra* is making its active, female-specific, protein, it causes a different type of processing to be done. This creates the female-specific transcript whose product inhibits the genes for male-specific characteristics. According to this model (Baker, 1989), the sexual determination cascade comes down to what type of mRNA is going to be processed from the *doublesex* transcript. If the X:A ratio is 1, then *Sxl* makes a female-specific splicing factor that causes the *tra* gene transcript to be spliced in a female-specific manner. This female-specific protein interacts with the *tra-2* splicing factor to cause the *doublesex* pre-mRNA to be spliced in a female-specific manner. If the *doublesex* transcript is not acted upon in this way, it will be processed in a "default" manner to make the male-specific message.

Target genes for the sex determination cascade

There are numerous proteins in *Drosophila* that are present in one sex and not the other. In females, these include yolk proteins and eggshell (chorion) proteins. In males, numerous semen proteins have been found that cause behavioral changes in the female after mating and that stabilize sperm in the female reproductive tract (Wolfner, 1988). These genes are regulated by the cascade of sex determination. By using temperature-sensitive (*ts*) alleles of the *tra-2* gene, the sexual development pathways in *Drosophila* were shown to be active from the late larval stages through the adult period. The *tra-2^{ts}* allele is a temperature-sensitive allele in which the female phenotype is expressed at permissive (colder) temperatures and the male phenotype is expressed at nonpermissive (warmer) temperatures. During late larval and pupal stages, raising the temperature from permissive to nonpermissive will cause an XX larva or pupa to develop into a male. Moreover, when adult mutants are kept at low temperatures, the adult fat body makes yolk proteins that will enter the oocyte. When shifted to a higher, nonpermissive temperature, transcription of the yolk protein genes ceases (Belote et al., 1985b). One remarkable finding has been that if adult XX *tra-2^{ts}* flies are kept at the nonpermissive temperature for several days, they begin to exhibit male courtship behaviors (Belote and Baker, 1987). Although our understanding of *Drosophila* sex determination has increased enormously in the past decade, the above model is incomplete. The mechanisms by which the X:A ratio is read, the *doublesex* gene product spliced, and the sexual phenotypes of the different cells generated are still unknown. The sequencing of these genes and the isolation of their products should give us more insights into the determination of sex in *Drosophila*.

Hermaphroditism

The nematode *Caenorhabditis elegans* usually has two sexual types: hermaphrodite and male. Most individuals of this species are hermaphroditic,* having both testes and ovaries. As larvae, they make sperm, which is stored as the organism develops (Figure 18). The adult ovary produces eggs that become fertilized as they migrate to the uterus. (*C. elegans* hermaphrodites do not copulate with themselves, as some other hermaphroditic organisms are known to do.) Self-fertilization almost always produces more hermaphrodites. Only 0.2 percent of the progeny are males. These males, however, can fertilize hermaphrodites; and because their sperm has a competitive advantage over endogenous hermaphroditic sperm, the sex ratio resulting from such matings is about 50 percent hermaphrodites and 50 percent males (Hodgkin, 1985).

In *C. elegans*, the hermaphrodite is XX and the male is XO. As in *Drosophila*, sex is determined by the ratio of X chromosomes to autosomes. In closely related species of nematodes, XX females are found, suggesting that the hermaphrodites evolved from females. Somatically, the females and hermaphrodites are identical, the only difference being that the hermaphrodites make sperm during their early development before switching over to egg production. In *C. elegans*, there even exists a dominant mutation (*tra-1^D*) that transforms XX or XO individuals into fertile females. In colonies with such an allele, three sexes are possible and functioning (Hodgkin, 1980).

As in *Drosophila*, sex determination in *C. elegans* involves several autosomal genes that read and respond to the X:A ratio. Some of these are essential for hermaphrodite development (*tra* genes), whereas others are necessary for the expression of the male phenotype (*her* genes). Homozygous mutant *tra* genes will transform XX individuals into males, while the mutant *her* and *fem* alleles will transform males into hermaphrodites.

*Hermaphrodites are named after the son of Hermes (Mercury) and Aphrodite (Venus). Having inherited the beauty of both parents, he excited the love of the nymph of the Salmacis fountain. As he bathed in this fountain, she embraced him, praying to the gods that they might forever be united. She got her wish in the most literal of fashions.

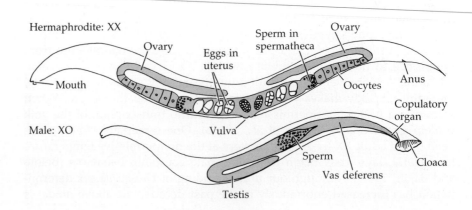

FIGURE 18
Schematic diagrams of the hermaphrodite and male *Caenorhabditis elegans* emphasizing their reproductive systems. (From Hodgkin, 1985.)

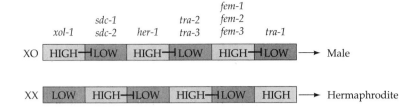

FIGURE 19

Schematic model of somatic sex determination in *C. elegans*. The *sdc-1* gene is postulated to be involved in transmitting the X/A ratio. It controls X chromosome dosage compensation as well as suppressing the *her-1* gene if the ratio is 1. The high/low designation reflects functional gene activity. The activity of the *sdc* genes controls the hermaphroditic (XX) mode of both sex determination and dosage compensation. The activity of the *xol-1* gene controls the male (XO) mode of these processes. (After Hodgkin, 1985, and Miller et al., 1988.)

By creating genotypes carrying different combinations of these mutations, Hodgkin (1980) was able to construct a model for this developmental pathway (Figure 19). He found, for example, that the *tra* mutations all suppressed the *her-1* mutation, indicating that *her-1* was earlier in the pathway.

The crucial gene in the pathway for sex determination appears to be *tra-1*. If the wild-type allele is active, the individual is a hermaphrodite. If the allele is not functional, the individual is a male. The other genes appear to regulate this single switch gene. The first known gene in this pathway is *xol-1* (*XO-lethal*) which can turn off the pathway for hermaphroditic development, thereby turning the animal into a male. It appears to accomplish this by repressing the *sdc* genes whose activities are needed for turning on *tra-1* and making the animal a hermaphrodite (Miller et al., 1988). The *sdc-1* and *sdc-2* genes may be analogous to the *Sxl* gene of *Drosophila* in that they are X-linked and affect both sex determination and X chromosome dosage compensation. It is proposed (Villeneuve and Meyer, 1987; Nusbaum and Meyer, 1989) that the *sdc* products act to repress *her-1* activity in XX animals. When *her-1* activity is high, *tra-2* and *tra-3* activities are low. This combination of activities is thought to allow the expression of the *fem* genes, which will turn off *tra-1*. At low levels of *tra-1* product, the somatic cells of *C. elegans* become male. Conversely, if *her-1* activity is low, *tra-1* activity is high. This leads to the hermaphroditic developmental pattern.

One of the most interesting problems of *C. elegans* is its hermaphroditism. How did such a condition arise in an organism that probably had a male/female sex system? What gene changes arose, and were there other solutions that could have been used? The sex-determining genes of the closely related species *C. ramanei* (with male and female individuals) are now being identified so that such questions may be answered.

While hermaphroditism is not uncommon in worms and insects, it is rarely seen in vertebrates. In birds and mammals, hermaphroditism is a pathological condition causing infertility. The most common vertebrate hermaphrodites are fishes, which display several types of hermaphroditism (Yamamoto, 1969). Some fishes are GONOCHORISTIC; that is, they have a chromosomally determined sex that is either male or female. The hermaphroditic fish species can be divided into three groups. The first are the SYNCHRONOUS hermaphrodites, in which ovaries and testicular tissues exist at the same time and in which both sperm and eggs are produced. One such species is *Servanus scriba*. In nature and in aquaria, these fish form spawning pairs. As soon as one of the fish spawns its eggs, the other fish fertilizes them. Then the fish reverse their roles and the fish that was formerly male spawns its eggs so that they can be fertilized by the sperm of its partner (Clark, 1959).

In other hermaphroditic species, an animal undergoes a genetically programmed sex change during its development. In these cases, the gonads are dimorphic, having both male and female areas. One or the other is predominant during a certain phase of life. In PROTOGYNOUS ("female-

FIGURE 20
Gonadal changes in the hermaphroditic fish *Sparus auratus*. Section through the gonad of the male phase (A), the transitory phase (B), and the final, female phase (C). (Photographs through the courtesy of the family of T. Yamamoto.)

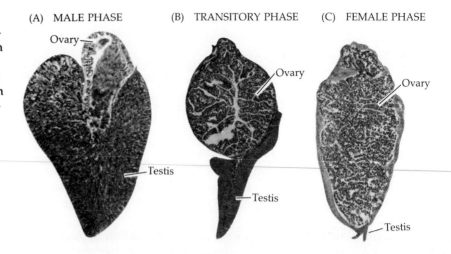

(A) MALE PHASE   (B) TRANSITORY PHASE   (C) FEMALE PHASE

Ovary

Ovary

Ovary

Testis

Testis

Testis

first") hermaphrodites, an animal begins its life as a female, but later becomes male. The reverse is the case in PROTANDROUS ("male-first") species. Figure 20 shows the gonadal changes of the protandrous hermaphroditic fish *Sparus auratus*. At first, testicular tissue predominates, and later, after a transition period during which both testicular and ovarian tissues are seen, the ovarian cells take over.

SIDELIGHTS & SPECULATIONS

Why are there males?

Hermaphrodites appear to have many of the advantages that population biologists look for in fit individuals or successful species. First, nearly all the individuals of a population are capable of bearing young. (Males are considered an extravagance by population biologists.) Second, one individual can colonize a new area. A single nematode or aphid can give rise to a large population very rapidly. Third, the hermaphroditic organism spends hardly any time or energy on courtship and mating (Maynard Smith, 1979).

What, then, keeps the males around? Males are thought to be needed for supplying new combinations of genes. Without males, the gene pool of the species would be exceedingly limited. Van Valen (1973) has proposed that

species cannot remain stable. They must change or they become extinct. This follows from the idea that one's predators and parasites are evolving to better eat their prey and that if the prey does not evolve, the predators could wipe out the species. This proposal is called the RED QUEEN HYPOTHESIS (after the character in *Through the Looking Glass* who claimed that you have to run as fast as you can to remain in the same place). Hermaphroditic species are usually found in those habitats where there are few predators (Glesener and Tilman, 1978), which suggests they have limited interactions with other organisms. The new combinations of genes generated by sexual reproduction coupled with the unpredictability of the environment appear to favor the vagaries of bisexual reproduction over the efficiency of hermaphroditism (Maynard Smith, 1979; Weinshall, 1986).

Environmental sex determination

Temperature-dependent sex determination in reptiles

While the sex of most snakes and most lizards is determined by sex chromosomes at the time of fertilization, the sex of most turtles and all species of crocodilians is determined by the environment after fertilization. In these reptiles, the temperature of the eggs during a certain period of

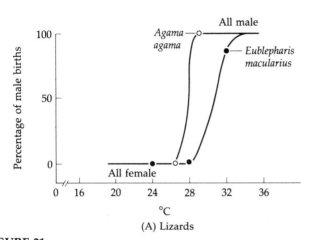

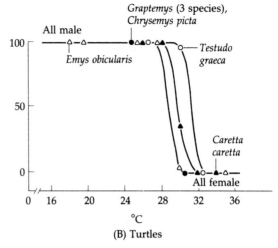

FIGURE 21

Relationship between sex ratio and incubation temperature in reptiles. (A) Two species of lizards in which higher temperatures result in the generation of male offspring. (B) Seven species of turtles in which higher temperatures result in female offspring. (After Bull, 1980.)

development is the deciding factor in determining sex (Bull, 1980), and small changes of temperature can cause dramatic changes in the sex ratio. Generally, eggs incubated at low temperatures (22–27°C) produce one sex, whereas eggs incubating at higher temperatures (30°C and above) produce the other. There is only a small range of temperatures that permits both males and females to hatch from the same brood of eggs. Figure 21 shows the abrupt temperature-induced change in sex ratios for certain species of turtles. If eggs are incubated below 28°C, all the turtles hatching from them will be male. Above 32°C, every egg will give rise to a female. Thus, each brood of eggs usually gives rise to individuals of the same sex. Variations on this theme also exist. Snapping turtle eggs, for instance, become female at either cold (less than or equal to 20°C) or hot (greater than or equal to 30°C) temperatures. In between these extremes, males predominate.

The developmental period during which sex determination occurs can be studied by incubating eggs at the male-producing temperature for a certain amount of time and then shifting the eggs to an incubator of the female-producing temperature. In map turtles and snapping turtles, the middle third of development appears to be the time when the temperature exerts its effect. Ferguson and Joanen (1982) have studied sex determination in the Mississippi alligator, both in the laboratory and in the field, and they have concluded that sex is determined between 7 and 21 days of incubation. Eggs raised at 30°C or below produce female alligators, whereas those incubated at 34°C or above produce all males. Moreover, nests constructed on levees (34°C) give rise to males, whereas those built in wet marshes (30°C) produce females. Thus, the sex of many turtle and alligator species is based on the temperature of the egg's environment.

Ferguson and Joanen (1982) have speculated that this method of temperature-dependent sex determination may have been used by another closely related reptile group—the dinosaurs. If dinosaurs had temperature-dependent sex determination (with a somewhat different threshold temperature for the switch in sex), then a slight change in temperature may have created conditions where only males or females hatched from their eggs. This would explain the sudden and selective extinction of this reptile order.

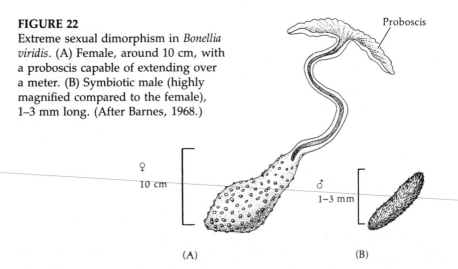

FIGURE 22
Extreme sexual dimorphism in *Bonellia viridis*. (A) Female, around 10 cm, with a proboscis capable of extending over a meter. (B) Symbiotic male (highly magnified compared to the female), 1–3 mm long. (After Barnes, 1968.)

Proboscis

♀
10 cm

♂
1–3 mm

(A)

(B)

Location-dependent sex determination in *Bonellia viridis* and *Crepidula fornicata*

The sex of the echiuroid worm *Bonellia* depends upon where a larva settles. The female *Bonellia* is a marine, rock-dwelling animal, the body of which is about 10 cm long (Figure 22). It has a proboscis, however, that can extend to over a meter in length. This proboscis serves two functions. First, it sweeps food from the rocks into the digestive tract of the female *Bonellia*. Second, should a larva land on the proboscis, it will enter the mouth of the female, migrate to the uterus, and differentiate into a 1- to 3-mm long symbiotic male. Thus, when a larva settles on a rocky surface, it becomes a female, but should that same larva settle on the proboscis of a female, it becomes a male. The male spends the rest of its life within the body of the female, fertilizing eggs.

Baltzer (1914) demonstrated that when larvae were cultured in the absence of adult females, about 90 percent of them became females. However, when these larvae were cultured in the presence of an adult female or its isolated proboscis, 70 percent of them adhered to the proboscis and developed the male structures. These results have been more recently confirmed by Leutert (1974) (Figure 23).

The molecule(s) responsible for masculinizing the larvae can be extracted from the proboscis of adult females. When larvae are cultured in normal seawater in the absence of adult females, most become females. When cultured in seawater containing aqueous extracts of proboscis tissue, most become either male or an intermediate form, neither completely male

FIGURE 23
In vitro analysis of *Bonellia* differentiation. Larval *Bonellia* were placed either in normal seawater or in seawater containing fragments of the female proboscis. A majority of the animals cultured in the presence of the proboscis fragments became males, whereas normally they would have become females. (After Leutert, 1974.)

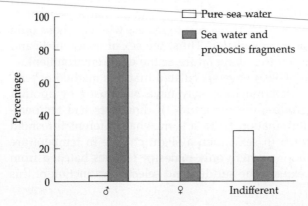

☐ Pure sea water

■ Sea water and proboscis fragments

Percentage

100

80

60

40

20

0

♂   ♀   Indifferent

nor completely female (Nowinski, 1934; Agius, 1979). The purification of the compound(s) that attract the larvae to the proboscis and cause its masculinization is presently being conducted.

Another example in which sex determination is affected by the position of the organism is that of the slipper snail *Crepidula fornicata*. Here, individuals will pile up on each other to form a mound (Figure 24). Young individuals are always male. This phase is followed by the degeneration of the male reproductive system and a period of lability. The next phase can be either male or female; and this depends upon position in the mound. If the snail is attached to a female, it will become male. If such a snail is removed from its attachment, it will become female. Similarly, the presence of large numbers of males will cause some of the males to become females. However, once the individual becomes female, it will not revert to being male (Coe, 1936).

Nature has provided many variations on her masterpiece. In some species, sex is determined solely by chromosomes, whereas in other species, sex is a matter of environmental conditions. Within these two large categories, numerous variations also exist. A complete catalog of known sex-determining mechanisms would take a separate (but very interesting) volume. But the sex-determining mechanism functions as the part of the somatic physiology necessary for the maintenance and propagation of the germ cells. The gonads will die with the body, but the germ cells that resided within them have the potential to renew life. It is to these germ cells that we will now turn our attention.

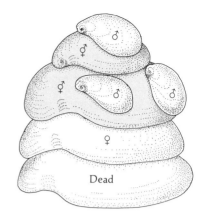

FIGURE 24

Cluster of *Crepidula* snails. Two individuals are changing from male to female. After these molluscs become female, they will be fertilized by the male above them. (After Coe, 1936.)

LITERATURE CITED

Agius, L. 1979. Larval settlement in the echiuran worm *Bonellia vividis*: Settlement on both the adult proboscis and body trunk. *Mar. Biol.* 53: 125–129.

Amrein, H., Gorman, M. and Nöthiger, R. 1988. The sex-determining gene *tra-2* of *Drosophila* encodes a putative RNA binding protein. *Cell* 55: 1025–1035.

Arnold, A. P. 1980. Sexual differences in the brain. *Am. Sci.* 68: 165–173.

Arnold, A. P., Nottebohm, F. and Pfaff, D. W. 1976. Hormone concentrating cells in vocal control and other brain regions of the zebra finch (*Poephila guttata*). *J. Comp. Neurol.* 165: 487–512.

Baker, B. S. 1989. Sex in flies: the splice of life. *Nature* 340: 521–524.

Baker, B. S. and Ridge, K. A. 1980. Sex and the single cell. I. On the action of major loci affecting sex determination in *Drosophila melanogaster*. *Genetics* 94: 383–423.

Baker, B. S., Nagoshi, R. N. and Burtis, K. C. 1987. Molecular genetic aspects of sex determination in *Drosophila*. *BioEssays* 6: 66–70.

Baltzer, F. 1914. Die Bestimmung und der Dimorphismus des Geschlechtes bei *Bonellia*. *Sber. Phys.-Med. Ges. Würzb.* 43: 1–4.

Barfield, R. J. and Chen, J. J. 1977. Activation of estrous behavior in ovariectomized rats by intracerebral implants of estradiol benzoate. *Endocrinology* 101: 1716–1725.

Barnes, R. D. 1968. *Invertebrate Zoology*. Saunders, Philadelphia.

Bell, L. R., Maine, E. M., Schedl, P. and Cline, T. W. 1988. *Sex-lethal*, a *Drosophila* sex determination switch gene, exhibits sex-specific RNA splicing and sequence similarity to RNA binding proteins. *Cell* 55: 1037–1046.

Bell, S. E. 1986. A new model of medical technology development: A case study of DES. *Sociol. Health Care* 4: 1–32.

Belote, J. M. and Baker, B. S. 1987. Sexual behavior: Its genetic control during development and adulthood in *Drosophila melanogaster*. *Proc. Natl. Acad. Sci. USA* 84: 8026–8030.

Belote, J. M., McKeown, M. B., Andrew, D. J., Scott, T. N., Wolfner, M. F. and Baker, B. S. 1985a. Control of sexual differentiation in *Drosophila melanogaster*. *Cold Spring Harbor Symp. Quant. Biol.* 50: 605–614.

Belote, J. M., Handler, A. M., Wolfner, M. F., Livak, K. L. and Baker, B. S. 1985b. Sex-specific regulation of yolk protein gene expression in *Drosophila*. *Cell* 40: 339–348.

Berta, P., Hawkins, J. R., Sinclair, A. H., Taylor, A., Griffiths, B. L., Goodfellow, P. N. and Fellous, M. 1990. Genetic evidence equating SRY and the testis-determining factor. *Nature* 348: 448–450.

Bleier, R. 1984. *Science and Gender*. Pergamon, New York, pp. 80–114.

Boggs, R. T., Gregor, P., Idriss, S., Belote, J. M. and McKeown, M. 1987. Regulation of sexual differentiation in *D. melanogaster* via alternative splicing of RNA from the *transformer* gene. *Cell* 50: 739–747.

Breedlove, S. M. and Arnold, A. P. 1980. Hormone accumulation in a sexually dimorphic motor nucleus of the rat spinal cord. *Science* 210: 565–566.

Bridges, C. B. 1921. Triploid intersexes in *Drosophila melanogaster*. *Science* 54: 252–254.

Bridges, C. B. 1925. Sex in relation to chromosomes and genes. *Am. Nat.* 59: 127–137.

Bull, J. J. 1980. Sex determination in reptiles. *Q. Rev. Biol.* 55: 3–21.

Burgoyne, P. S., Buehr, M., Koopman, P., Rossant, J. and McLaren, A. 1988. Cell-autonomous action of the testis-determining gene: Sertoli cells are exclusively XY in XX/XY chimaeric mouse testes. *Development* 102: 443–450.

Cate, R. L. and 18 others. 1986a. Isolation of the bovine and human genes for Müllerian inhibiting substance and expression of the human gene in animal cells. *Cell* 45: 685–698.

Cate, R. L., Ninfa, E. G., Pratt, D. J., MacLaughlin, D. T. and Donahoe, P. K. 1986b. Development of Müllerian inhibitory substance as an anti-cancer drug. *Cold Spring Harb. Symp. Quant. Biol.* 51: 641–647.

Clark, E. 1959. Functional hermaphroditism and self-fertilization in a serranid fish. *Science* 129: 215–216.

Cline, T. W. 1983. The interaction between *daughterless* and *Sex-lethal* in triploids: A novel sex-transforming maternal effect linking sex determination and dosage compensation in *Drosophila melanogaster*. *Dev. Biol.* 95: 260–274.

Cline, T. W. 1986. A female-specific lethal lesion in an X-linked positive regulator of the *Drosophila* sex determination gene, *Sex-lethal*. *Genetics* 113: 641–663.

Cline, T. W. 1988. Evidence that *sisterless-a* and *sisterless-b* are two of several discrete "numerator elements" of the X/A sex determination signal in *Drosophila* that switch Sxl between two alternative stable expression states. *Genetics* 119: 829–862.

Coe, W. R. 1936. Sexual phases in *Crepidula*. *J. Exp. Zool.* 72: 455–477.

Cronmiller, C. and Cline, T. W. 1987. The *Drosophila* sex determination gene *daughterless* has different functions in the germ line versus the soma. *Cell* 48: 479–487.

Doughty, D., Booth, J. E., McDonald, P. G. and Parrott, R. F. 1975. Effects of oestradiol-17β oestradiol benzoate, and synthetic oestrogen RU2858 on sexual differentiation in the neonatal rat. *J. Endocrinol.* 67: 419–424.

Eicher, E. M. and Washburn, L. L. 1983. Inherited sex reversal in mice: Identification of a new primary sex-determining gene. *J. Exp. Zool.* 228: 297–304.

Eicher, E. M. and Washburn, L. L. 1986. Genetic control of primary sex determination in mice. *Annu. Rev. Genet.* 20: 327-360.

Eicher, E. M., Washburn, L. L., Whitney, J. B. III and Morrow, K. E. 1982. *Mus poschiavinus* Y chromosome in the C57Bl/6J murine genome causes sex reversal. *Science* 217: 535–537.

Eicher, E. M., Phillips, S. J. and Washburn, L. L. 1983. The use of molecular probes and chromosomal rearrangements to partition the mouse Y chromosome into functional regions. *In* A. Messer and I. H. Porter (eds.), *Recombinant DNA and Medical Genetics*. Academic Press, New York, pp. 57–71.

Fausto-Sterling, A. 1985. *Myths of Gender*. Basic Books, New York.

Ferguson, M. W. J. and Joanen, T. 1982. Temperature of egg incubation determines sex in *Alligator mississippiensis*. *Nature* 296: 850–853.

Fuller, A. F., Budzik, G. P., Krane, I. M. and Donahoe, P. K. 1984. Müllerian inhibitory substance inhibition of a human endometrial carcinoma cell line xenografted into nude mice. *Gynec. Oncol.* 17: 124–132.

Geddes, P. and Thomson, J. A. 1890. *The Evolution of Sex*. Walter Scott, London.

Glesener, R. R. and Tilman, D. 1978. Sexuality and the components of environmental uncertainty: Clues from geographic parthenogenesis in terrestrial animals. *Am. Nat.* 112: 659–673.

Goralski, T. J., Edström, J.-E. and Baker, B. S. 1988. The sex determination locus *transformer-2* of *Drosophila* encodes a polypeptide with similarity to RNA binding proteins. *Cell* 56: 1011–1018.

Gubbay, J. and 8 others. 1990. A gene mapping to the sex-determining region of the mouse Y chromosome is a member of a novel family of embryonically expressed genes. *Nature* 346: 245–250.

Higgins, S. J., Young, P. and Cunha, G. R. 1989. Induction of functional cytodifferentiation in the epithelium of tissue recombinants II. Instructive induction of Wolffian duct epithelia by neonatal seminal vesicle mesenchyme. *Development* 106: 235–250.

Hodgkin, J. 1980. More sex-determination mutants of *Caenorhabditis elegans*. *Genetics* 96: 649–664.

Hodgkin, J. 1985. Males, hermaphrodites, and females: Sex determination in *Caenorhabditis elegans*. *Trends Genet.* 1: 85–88.

Imperato-McGinley, J., Guerrero, L., Gautier, T. and Peterson, R. E. 1974. Steroid 5 α reductase deficiency in man: An inherited form of male pseudohermaphroditism. *Science* 186: 1213–1215.

Jacklin, D. 1981. Methodological issues in the study of sex-related differences. *Dev. Rev.* 1: 266–273.

Jäger, R. J., Anvret, M., Hall, K. and Scherer, G. 1990. A human XY female with a frame shift mutation in the candidate testis-determining gene *SRY*. *Nature* 348: 452–454.

Josso, N., Picard, J.-Y. and Tran, D. 1977. The anti-Müllerian hormone. *Recent Prog. Horm. Res.* 33: 117–167.

Jost, A. 1953. Problems of fetal endocrinology: The gonadal and hypophyseal hormones. *Recent Prog. Horm. Res.* 8: 379–418.

Kandel, E. R. and Schwartz, J. H. 1985. *Principles of Neural Science*, 2nd Ed. Elsevier, New York.

Kolata, G. 1980. Math and sex: Are girls born with less ability? *Science* 210: 1234–1235.

Konishi, M. and Akutagawa, E. 1985. Neuronal growth, atrophy, and death in a sexually dimorphic song nucleus in the zebra finch brain. *Nature* 315: 145–147.

Koopman, P., Münsterberg, A., Capel, B., Vivian, N. and Lovell-Badge, A. 1990. Expression of a candidate sex-determining gene during mouse testis differentiation. *Nature* 348: 450–452.

Kurz, E. M., Sengelaub, D. R. and Arnold, A. P. 1986. Androgens regulate the dendritic length of mammalian motoneurons in adulthood. *Science* 232: 395–397.

Langman, J. 1981. *Medical Embryology*, 4th Ed. Williams & Wilkins, Baltimore.

Langman, J. and Wilson, D. B. 1982. Embryology and congenital malformations of the female genital tract. *In* A. Blaustein (ed.), *Pathology of the Female Genital Tract*, 2nd Ed. Springer-Verlag, New York, pp. 1–20.

Leutert, T. R. 1974. Zur Geschlechtsbestimmung und Gametogenese von *Bonellia viridis* Rolando. *J. Embryol. Exp. Morphol.* 32: 169–193.

Lillie, F. R. 1917. The free-martin: A study in the action of sex hormones in the fetal life of cattle. *J. Exp. Zool.* 23: 371–452.

Magre, S. and Jost, A. 1980. Initial phases of testicular organogenesis in the rat. An electron microscope study. *Arch. Anat. Micr. Morph.* 69: 297–318.

Maynard Smith, J. 1979. *The Evolution of Sex*. Cambridge University Press, London.

McClung, C. E. 1902. The accessory chromosome—sex determinant? *Biol. Bull.* 3: 72–77.

McEwen, B. S. 1981. Neural gonadal steroid actions. *Science* 211: 1303–1311.

McEwen, B. S., Leiberburg, I., Chaptal, C. and Krey, L. C. 1977. Aromatization: Important for sexual differentiation of the neonatal rat brain. *Horm. Behav.* 9: 249–263.

McKeown, M., Belote, J. M. and Boggs, R. T. 1988. Ectopic expression of the female *transformer* gene product leads to female differentiation of chromosomally male *Drosophila*. *Cell* 53: 887–895.

McLaren, A. 1988. Sex determination in mammals. *Trends Genet.* 4: 153–157.

McLaren, A., Simpson, E., Tomonari, K., Chandler, P. and Hogg, H. 1984. Male sexual differentiation in mice lacking the H-Y antigen. *Nature* 312: 552–555.

Meyer, W. J., Migeon, B. R. and Migeon, C. J. 1975. Locus on human X chromosome for dihydrotestosterone receptor and androgen insensitivity. *Proc. Natl. Acad. Sci. USA* 72: 1469–1472.

Miller, L. M., Plenefisch, J. D., Casson, L. P. and Meyer, B. 1988. *xol-1*: A gene that controls the male mode of both sex determination and X chromosome dosage compensation in *C. elegans*. *Cell* 55: 167–183.

Morgan, T. H. 1919. *The Physical Basis of Heredity*. Lippincott, Philadelphia.

Nabekura, J., Oomura, Y., Minami, T., Mizuno, Y. and Fukuda, A. 1986. Mechanism of the rapid effect of 17β-estradiol on medial amygdala neurons. *Science* 233: 226–228.

Nordeen, E. J., Nordeen, K. W., Sengelaub, D. R. and Arnold, A. P. 1985. Androgens prevent normally occurring cell death in a sexually dimorphic spinal nucleus. *Science* 229: 671–673.

Nottebohm, F. 1980. Testosterone triggers growth of brain vocal control nuclei in adult female canaries. *Brain Res.* 189: 429–436.

Nottebohm, F. 1981. A brain for all seasons: Cyclical anatomical changes in song control nuclei of the canary brain. *Science* 214: 1368–1370.

Nowinski, W. 1934. Die vermännlichende Wirkung fraktionierter Darmextrakte des Weibchens auf die Larven der *Bonellia viridis*. *Pubbl. Staz. Zool. Napoli* 14: 110–145.

Nusbaum, C. and Meyer, B. J. 1989. The

Caenorhabditis elegans gene *sdc-2* controls sex determination and dosage compensation in XX animals. *Genetics* 122: 579–593.

Page, D. C. 1985. Sex-reversal: Deletion mapping of the male-determining function of the human Y chromosome. *Cold Spring Harbor Symp. Quant. Biol.* 51: 229–235.

Page, D. C., de la Chappelle, A. and Weissenbach, J. 1985. Chromosome Y-specific DNA in related human XX males. *Nature* 315: 224–226.

Pfaff, D. W. and McEwen, B. S. 1983. The actions of estrogens and progestins on nerve cells. *Science* 219: 808–814.

Pröve, E. 1978. Courtship and testosterone in male zebra finches. *Z. Tierpsychol.* 48: 47–67.

Reddy, V. R., Naftolin, F. and Ryan, K. J. 1974. Conversion of androstenedione to estrone by neural tissues from fetal and neonatal rats. *Endocrinology* 94: 117–121.

Robboy, S. J., Young, R. H. and Herbst, A. L. 1982. Female genital tract changes related to prenatal diethylstilbesterol exposure. *In* A. Blaustein (ed.), *Pathology of the Female Genital Tract*, 2nd Ed. Springer-Verlag, New York, pp. 99–118.

Rubin, B. S. and Barfield, R. J. 1980. Priming of estrus responsiveness by implants of 17b-estradiol in the ventromedial hypothalamic nuclei of female rats. *Endocrinology* 106: 504–509.

Salz, H. K., Cline, T. W. and Schedl, P. 1987. Functional changes associated with structural alterations induced by mobilization of a P element inserted into the *Sex-lethal* gene of Drosophila. *Genetics* 117: 221–231.

Salz, H. K., Maine, E. M., Keyes, L. N., Samuels, M. E., Cline, T. W. and Schedl, P. 1989. The *Drosophila* female-specific sex-determination gene, *Sex-lethal*, has stage-, tissue-, and sex-specific RNAs suggesting multiple modes of regulation. *Genes Dev.* 3: 708–719.

Sánchez, L. and Nöthiger, R. 1983. Sex determination and dosage compensation in *Drosophila melanogaster*: Production of male clones in XX females. *EMBO J.* 1: 485–491.

Siiteri, P. K. and Wilson, J. D. 1974. Testosterone formation and metabolism during male sexual differentiation in the human embryo. *J. Clin. Endocrinol. Metab.* 38: 113–125.

Sinclair, A. H. and 9 others. 1990. A gene from the human sex-determining region encodes a protein with homology to a conserved DNA-binding motif. *Nature* 346: 240–244.

Singh, L. and Jones, K. W. 1982. Sex reversal in the mouse (*Mus musculus*) is caused by a recurrent nonreciprocal crossover involving the X and an aberrant Y chromosome. *Cell* 28: 205–216.

Stevens, N. M. 1905. Studies in spermatogenesis with especial reference to the "accessory chromosome." *Carnegie Inst. Rep. Washington* 36.

Terasawa, E. and Sawyer, C. H. 1969. Changes in electrical activity in rat hypothalamus related to electrochemical stimulation of adenohypophyseal function. *Endocrinology* 85: 143–149.

Thorpe, W. H. 1958. The learning of song patterns by birds with especial reference to the song of the chaffinch, *Fringilla coelebs*. *Ibis* 100: 535–570.

Tiefer, L. 1970. Gonadal hormones and mating behavior in adult golden hamster. *Horm. Behav.* 1: 189–202.

Tran, D., Meusy-Dessolle, N. and Josso, N. 1977. Anti-Müllerian hormone is a functional marker of foetal Sertoli cells. *Nature* 269: 411–412.

Trelstad, R. L., Hayashi, A., Hayashi, K. and Donahoe, P. K. 1982. The epithelial–mesenchymal interface of the male rate Müllerian duct: Loss of basement membrane integrity and ductal regression. *Dev. Biol.* 92: 27–40.

Van Valen, L. 1973. A new evolutionary law. *Evol. Theor.* 1: 1–30.

Vergnaud, G. and 8 others. 1986. A deletion map of the human Y chromosome based on DNA hybridization. *Am. J. Hum. Genet.* 38: 109–124.

Vigier, B., Watrin, F. Magre, S., Tran, D. and Josso, N. 1987. Purified bovine AMH induces a characteristic freemartin effect in fetal rat prospective ovaries exposed to it in vitro. *Development* 100: 43–55.

Villeneuve, A. M. and Meyer, B. J. 1987. *sdc-1*: A link between sex determination and dosage compensation in *C. elegans*. *Cell* 48: 25–37.

Vreeburg, J. T. M., van der Vaart, P. D. M. and van der Schoot, P. 1977. Prevention of central defeminization but not masculinization in male rats by inhibiting neonatal estrogen biosynthesis. *J. Endocrinol.* 74: 375–382.

Wachtel, S. S., Hall, J. L., Müller, U. and Chaganti, R. S. K. 1980. Serum-borne H-Y antigen in the fetal bovine freemartin. *Cell* 21: 917–926.

Washburn, L. L. and Eicher, E. M. 1983. Sex reversal in XY mice caused by a dominant mutation on chromosome 17. *Nature* 303: 338–340.

Weinshall, D. 1986. Why is a two-environment system not rich enough to explain the evolution of sex? *Am. Nat.* 128: 736–750.

Wilson, E. B. 1905. The chromosomes in relation to the determination of sex in insects. *Science* 22: 500–502.

Wolfner, M. F. 1988. Sex-specific gene expression in somatic tissues of *Drosophila melanogaster*. *Trends Genet.* 4: 333–337.

Yamamoto, T.-O. 1969. Sex differentiation. *In* W. S. Hoar and D. J. Randall (eds.), *Fish Physiology*, Vol. 3. Academic Press, New York, pp. 117–175.

22

The saga of the germ line

And the end of all our exploring
Will be to arrive where we started
And know the place for the first time.

—T. S. ELIOT (1942)

We began our analysis of animal development by discussing fertilization, and we will finish by investigating gametogenesis, the collection of processes by which the sperm and the egg are formed. Germ cells provide the continuity of life between generations, and the direct mitotic ancestors of our own germ cells once resided in the gonads of reptiles, amphibians, fishes, and invertebrates. In many animals, such as insects and vertebrates, there is a clear and early separation of germ cells from somatic cell types. In many animal phyla (and in the entire plant kingdom), this division is not as well established. In these species (which include cnidarians, flatworms, and tunicates) somatic cells can readily become germ cells even in adult organisms. The zooids, buds, and polyps of many invertebrate phyla testify to the ability of somatic cells to give rise to new individuals.

In those organisms where there is an established germ line that separates early in development, the germ cells do not arise from within the gonad itself. Rather, their precursors—the PRIMORDIAL GERM CELLS (PGCs)—migrate into the developing gonads. The first step in gametogenesis, then, involves getting the primordial germ cells into the genital ridge as the gonad is forming.

Germ cell formation

Germ cell migration in amphibians

In anuran amphibians—frogs and toads—the germ plasm is readily marked by granules in the vegetal hemisphere of the fertilized egg. During the first two cell divisions, small clusters of germ plasm are brought together by microtubules to form a group of large patches overlying the vegetal pole (Ressom and Dixon, 1988). However, this plasm does not

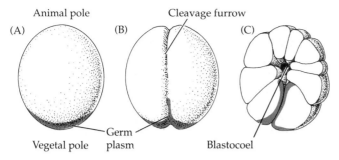

Animal pole

(A)

(B)

Cleavage furrow

(C)

Germ
plasm

Vegetal pole

Blastocoel

FIGURE 1

Changes in the position of germ plasm (color) in an early frog embryo. Originally located near the vegetal pole of the uncleaved egg (A), the germ plasm advances along the cleavage furrows (B) until it becomes localized at the floor of the blastocoel (C). (After Bounoure, 1934.)

remain long in the vegetal region. During cleavage, this material is brought upward through the yolky cytoplasm, and the RNA-rich granules become associated with the cells lining the floor of the blastocoel (Figure 1; Bounoure, 1934). As mentioned in Chapter 7, Blackler traced the migration of these cells into the gonad by transferring blocks of cells from the neurulae of one strain of *Xenopus laevis* into another. The primordial germ cells of this frog are seen to move laterally from the endoderm of the gut to the dorsal mesentery, which connects the gut to the region where the mesodermal organs are forming. They migrate up this tissue until they reach the developing gonads (Figure 2). *Xenopus* PGCs move by extruding a single filopodium and then streaming their yolky cytoplasm into the filopodium while retracting their tail. Contact guidance in this migration seems very likely as the cells over which they migrate are oriented in the direction of that migration (Wylie et al., 1979). Furthermore, PGC adhesion and migration can be inhibited if the mesentery is treated with antibodies against *Xenopus* fibronectin (Heasman et al., 1981). Thus, the pathway for germ cell migration in these frogs appears to be composed of fibronectin-containing extracellular matrix whose fibrils are oriented in the direction of cell migration. The fibrils over which the PGCs travel lose this polarity soon after migration has ended.*

Factors intrinsic to the cell are also important to the specific migration of the primordial germ cells. When the vegetal germ plasm of frog eggs is irradiated with ultraviolet light (which destroys RNA and the disulfide bonds of proteins), the migration of PGCs to the genital ridge is inhibited or delayed significantly (Smith, 1966; Züst and Dixon, 1977; Ikenishi and Kotani, 1979).

The primordial germ cells of urodele amphibians (salamanders) have a different origin, which has been traced by reciprocal transplantation experiments to the regions of the mesoderm that involute through the ventrolateral lips of the blastopore. Moreover, there does not seem to be any particular localized "germ plasm" in salamander eggs. Rather, the interaction of the dorsal endoderm cells and animal hemisphere cells creates the conditions needed to form germ cells in the particular areas that involute through the ventrolateral lips (Sutasurya and Nieuwkoop, 1974). So in salamanders, the primordial germ cells are formed by induction within the mesodermal region and presumably follow a different path into the gonad.

*This does not necessarily hold true for all anurans. In the frog *Rana pipiens*, the germ cells follow a similar route but may be passive travelers along the mesentery rather than actively motile cells (Subtelny and Penkala, 1984).

FIGURE 2

Migration of primordial germ cells in a frog. This phase-contrast photomicrograph of a section through the body wall and dorsal mesentery of a *Xenopus* embryo shows the migration of two large primordial germ cells (arrows) along the dorsal mesentery. (From Heasman et al., 1977; photograph courtesy of the authors.)

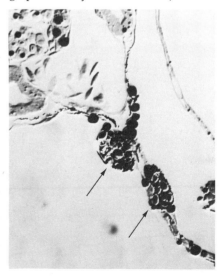

FIGURE 3
Pathway for the migration of mammalian primordial germ cells. (A) Primordial germ cells seen in the yolk sac near the junction of the hindgut and allantois. (B) Migration through gut and, dorsally, up the dorsal mesentery into the genital ridge. (After Langman, 1981.)

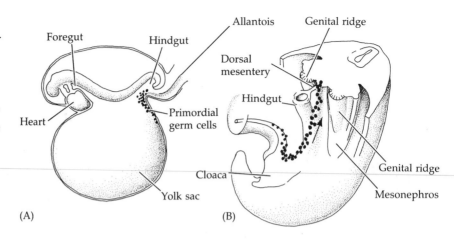

Germ cell migration in mammals

Mammalian germ cells are not morphologically distinct during early development. However, by using monoclonal antibodies that recognize cell surface differences between the PGCs and their surrounding cells, Hahnel and Eddy (1986) showed that mouse PGCs originally reside in the epiblast of the gastrulating embryo. Ginsburg and her colleagues (1990) localized this region to the extraembryonic mesoderm just posterior to the primitive streak of the 7-day mouse embryo. Here about eight large, alkaline phosphatase-staining cells are seen. If this area is removed, the remaining embryo becomes devoid of germ cells, while the isolated segment develops a large number of primordial germ cells. The germ cell precursors in the extraembryonic mesoderm then migrate back into the embryo, first to the mesoderm of the primitive streak, and then to the endoderm by way of the allantois. The route of the mammalian PGC migration from the allantois (Figure 3) resembles that of the anuran PGC migration. After collecting

FIGURE 4
Mouse primordial germ cells at different stages of their migration. (A) PGCs in hindgut of mouse embryo (near the allantois and yolk sac). Four large PGCs (dark circles) stain positively for high levels of alkaline phosphatase. (B) PGCs, stained for alkaline phosphatase, can be seen migrating up the dorsal mesentery and entering the genital ridges. (A from Heath, 1978. B from Mintz, 1957; photographs courtesy of the authors.)

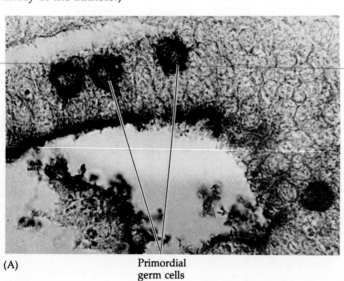

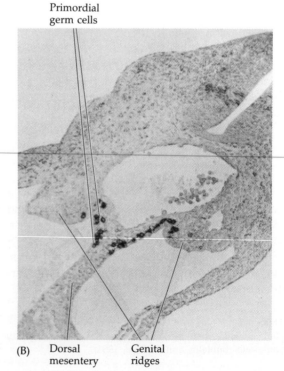

at the allantois by day 7.5 (Chiquoine, 1954; Mintz, 1957), the mammalian PGCs migrate to the adjacent yolk sac (Figure 4). By this time, they have already split into two populations that will migrate to either the right or the left genital ridge. The PGCs then move caudally from the yolk sac through the newly formed hindgut and up the dorsal mesentery into the genital ridge (Figure 4B). Most of the PGCs have reached the developing gonad by the eleventh day after fertilization. During this trek, they have proliferated from an initial population of 10–100 cells to the 2500–5000 PGCs present in the gonads by day 12. Like the PGCs of *Xenopus*, mammalian PGCs appear to be closely associated with the cells over which they migrate, and they move by extending filopodia over the underlying cell surfaces. These cells are also capable of penetrating cell monolayers and migrating through the cell sheets (Stott and Wylie, 1986).

The mechanism by which the primordial germ cells know the route of this journey is still unknown. Godin and colleagues (1990) have in vitro evidence that the genital ridges of 10.5-day mouse embryos secrete diffusible molecules that are capable of attracting and inducing mitosis in 8.5-day mouse primordial germ cells. Whether the genital ridge is able to provide the cues in vivo has still to be tested.

Germ cell migration in birds and reptiles

In birds and reptiles, PGCs are derived from epiblastic cells that migrate from the central region of the area pellucida to a crescent-shaped zone in the endodermal layer at the anterior border of the area pellucida (Figure 5; Eyal-Giladi et al., 1981; Ginsburg and Eyal-Giladi, 1987). This extra-embryonic region is called the GERMINAL CRESCENT, and the primordial germ cells multiply in this region. Unlike the PGCs in amphibians and mammals, germ cells in birds and reptiles migrate primarily by means of the bloodstream. The PGCs of the germinal crescent appear to enter the blood vessels by DIAPEDESIS (Figure 6A), a type of movement common to lymphocytes and macrophages and that enables cells to squeeze in and out of small blood vessels. The PGCs thus enter the embryo by being transported in the blood (Pasteels, 1953; Dubois, 1969). The PGCs must also "know" to get out of the blood when they reach the developing gonad (Figure 6B). When the germinal crescent of a chick embryo is removed, and the circulation of that embryo is joined with that of a normal chick embryo, the primordial germ cells from the normal embryo will migrate into both sets of gonads (Simon, 1960). It is not known what causes this attraction for the genital ridges. One possibility is that the developing

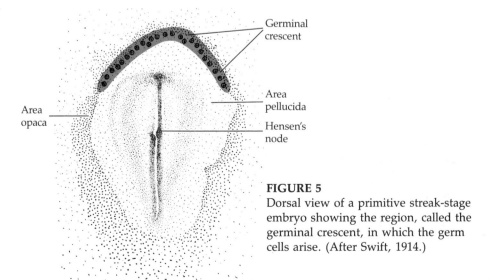

Area opaca

Germinal crescent

Area pellucida

Hensen's node

FIGURE 5
Dorsal view of a primitive streak-stage embryo showing the region, called the germinal crescent, in which the germ cells arise. (After Swift, 1914.)

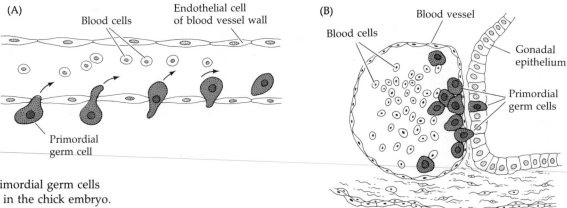

FIGURE 6
Diapedesis of primordial germ cells (shown in color) in the chick embryo. (A) Diagrammatic representation of diapedesis, a process by which PGCs can enter the vitelline blood vessels. (B) Transverse section through the prospective gonadal region of a chick embryo. Several PGCs within the blood vessel cluster next to the genital ridge epithelium. One PGC is crossing through the blood vessel endothelium and another PGC is already located within the gonadal epithelium. (After Romanoff, 1960.)

gonad produces a chemotactic substance that attracts PGCs and retains them in the capillaries bordering the gonad (Rogulska, 1969). (Such substances are known to be secreted by lymphocytes at the sites of infection in order to attract macrophages to that area and to permit them to pass through the capillary wall by diapedesis.) Another possibility is that the endothelial cells of the gonadal capillaries have a cell surface compound that causes the PGCs to adhere there specifically. Using monoclonal antibodies that recognize different cell surface molecules, Auerbach and Joseph (1984) have shown that the endothelial cells of several capillary networks have different cell membrane components and that the endothelial cells from ovarian capillaries differ from all others tested. Whatever the attracting factor is, it is not species specific. The chicken gonad will attract circulating PGCs from the turkey and even the mouse (Reynaud, 1969; Rogulska et al., 1971).

SIDELIGHTS & SPECULATIONS

Cell surface address markers and lymphocyte migration

How can the primordial germ cells of reptiles and chicks, traveling through the blood, "know" to exit the capillaries only in the genital ridge? While we do not know the answer, one possible solution is given us by work in immunology. Mature lymphocytes are bloodborne cells that circulate to the lymphoid organs, notably to the spleen, peripheral lymph nodes, and gut-associated lymphoid tissue. In some fashion, lymphocytes have to recognize where and when to leave the bloodstream (extravasate) and to enter a lymphoid organ. Moreover, certain lymphocytes migrate only to certain lymphoid organs and not to others The mechanism for this "homing" and organ specificity involves the lymphocytes' ability to specifically adhere to the blood vessel endothelial cells in these organs. There is an endothelium in lymphoid organs called the HIGH ENDOTHELIAL VENULES (HEV). These endothelial cells are found immediately after the capillary network of these or-

gans. Monoclonal antibodies, made against these blood vessels, have found differences between the high endothelial venules of different lymphoid organs. Peripheral lymph node endothelial cells contain a molecule in their cell membrane that is essential for the binding and extravasation of those lymphocytes that can recognize it. The high endothelial venules in gut-associated lymphoid tissue have a different molecule in their cell membranes, and this protein is required for certain lymphocytes to adhere to the blood vessel wall. (Streeter et al., 1988a,b; Nakamache, 1989). These adhesive proteins in the endothelial cell membranes are called ADDRESSINS. For each addressin on the high endothelial venules, there is probably a complementary molecule on the lymphocytes that can recognize it (Gallatin et al., 1983, 1986). Addressins may also be important in tissue-specific metastasis, since different tumors recognize and extravasate from different capillary networks (Auerbach et al., 1987).

The migration of cells to a specific organ has therefore been correlated with a specific cell–cell interaction.

Meiosis

Once in the gonad, PGCs continue to divide mitotically, producing millions of potential gametes. The PGCs of both male and female gonads are then faced with the necessity of reducing their chromosome number from the diploid to the haploid condition. In the haploid condition, each chromosome is represented only once, whereas diploid cells have two copies of each chromosome. To accomplish this reduction, the male and female germ cells undergo a type of cell division called MEIOSIS.

After the last mitotic division, a period of DNA synthesis occurs, so that the cells initiating meiosis have twice the normal amount of DNA in their nuclei. In this state, each chromosome consists of two sister CHROMATIDS attached at a common centromere. Meiosis entails two cell divisions (Figure 7). (In other words, although diploid, the cell contains four copies of each chromosome, but the chromosomes are seen as two chro-

FIGURE 7
Chromosomes during the stages of meiosis. The stages leptotene through diakinesis are substages of the first meiotic prophase.

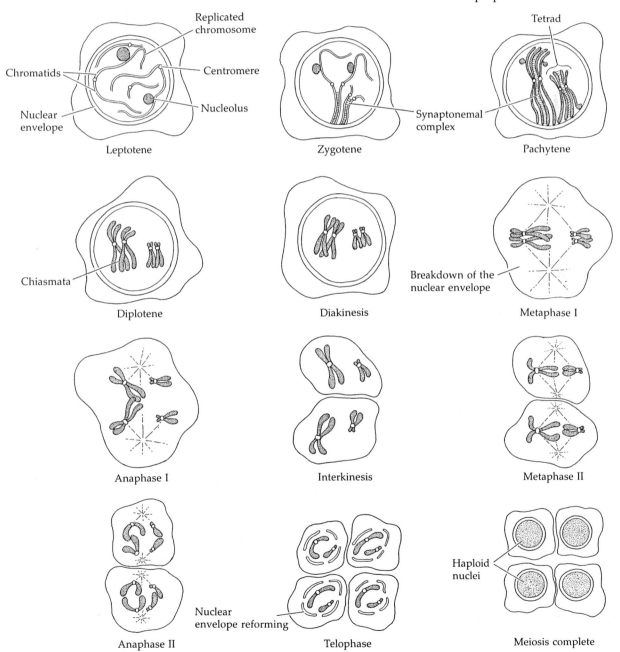

matids bound together). In the first division, homologous chromosomes (for example, the chromosome 3 pair in the diploid cell) come together and are then separated into different cells. Hence, the first meiotic division separates homologous chromosomes into the two daughter cells such that each cell has only one copy of each chromosome. But each of the chromosomes has already replicated. The second meiotic division then separates the two sister chromatids from each other. Consequently, each of the four cells produced by meiosis has a single (haploid) copy of each chromosome.

The first meiotic division begins with a long prophase, which is subdivided into five parts. During the LEPTOTENE (Gk., "thin thread") stage, the chromatin of the chromatids is stretched out very thinly so that it is not possible to identify individual chromosomes. DNA replication has already occurred, however, and each chromosome consists of two parallel chromatids. At the ZYGOTENE (Gk., "yoked threads") stage, homologous chromosomes pair side-by-side. This pairing is called SYNAPSIS, and it is characteristic of meiosis. Such pairing does not occur during mitotic divisions. Although the mechanism whereby each chromosome recognizes its homologue is not known, pairing seems to require the presence of the nuclear membrane and the formation of a proteinaceous ribbon called the SYNAPTONEMAL COMPLEX. This complex is a ladderlike structure with a central element and two lateral bars (von Wettstein, 1984). The chromatin is associated with the two lateral bars and the chromatids are thus joined together (Figure 8). Examinations of the meiotic cell nuclei with the electron microscope (Moses, 1968; Moens, 1969) suggest that paired chromosomes are bound to the nuclear membrane, and Comings (1968) has suggested that the nuclear envelope aids in bringing together the homologous chromosomes. The configuration formed by the four chromatids and the synaptonemal complex is referred to as a TETRAD or a BIVALENT.

During the next stage of meiotic prophase, the chromatids thicken and shorten. This stage has therefore been called the PACHYTENE (Gk.,

(A)

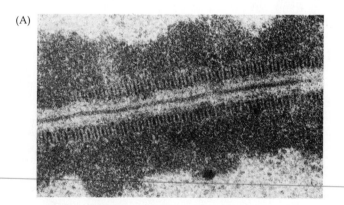

(B)

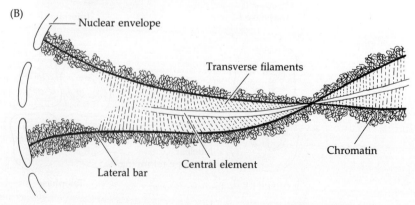

FIGURE 8
The synaptonemal complex. (A) Homologous chromosomes held together at the pachytene stage during meiosis in the *Neottiella* oocyte. (B) Interpretive diagram of the synaptonemal complex structure. (A from von Wettstein, 1971; photograph courtesy of D. von Wettstein. B after Moens, 1974.)

"thick thread") stage. Individual chromatids can now be distinguished under the light microscope, and crossing-over may occur. Crossing-over represents exchanges of genetic material whereby genes from one chromatid are exchanged with homologous genes from another chromatid. This crossing-over continues into the next stage, the DIPLOTENE (Gk., "double threads"). Here, the synaptonemal complex breaks down and the two homologous chromosomes start to separate. Usually, however, they are seen to remain attached at various places called CHIASMATA, which are thought to represent regions where crossing-over is occurring (Figure 9). The diplotene stage is characterized by a high level of gene transcription. In some species, the chromosomes of both male and female germ cells take on the "lampbrush" appearance characteristic of chromosomes that are actively making RNA (Chapter 11). During the next stage, DIAKINESIS (Gk., "moving apart"), the centromeres move away from each other and the chromosomes remain joined only at the tips of the chromatids. This last stage of meiotic prophase ends with the breakdown of the nuclear membrane and the migration of the chromosomes to the metaphase plate.

During anaphase I, homologous chromosomes are separated from each other in an independent fashion. This stage leads to telophase I, during which two daughter cells are formed, each cell containing one partner of the homologous chromosome pair. After a brief INTERKINESIS, the second division of meiosis takes place. During this division, the centromere of each chromosome divides during anaphase so that each of the new cells gets one of the two chromatids; the final result being the creation of four haploid cells. Note that meiosis has also reassorted the chromosomes into new groupings. First, each of the four haploid cells has a different assortment of chromosomes. In humans, where there are 23 different chromosome pairs, there can be 2^{23} (nearly 10 million) different types of haploid cells formed from the genome of a single person. In addition, the crossing-over that occurs during the pachytene and diplotene

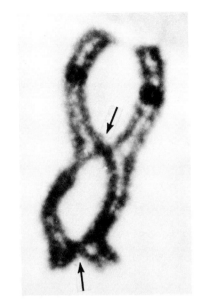

FIGURE 9
Chiasmata in diplotene bivalent chromosomes of salamander oocytes. Centromeres are visible as darkly staining circles; arrows point to the two chiasmata. (Photograph courtesy of J. Kezer.)

FIGURE 10
Sen and Gilbert hypothesis of homologue pairing during meiosis. (A) Four DNA strands may form stable tetrads through hydrogen bonding among four guanosine residues. (B) Diagram of such a stabilized complex showing regions of guanosine tetrads (colored squares). In the region of these tetrads, normal hydrogen bonding between complementary strands is broken. (C) Diagram of the four associated chromatids on the nuclear envelope. (After Sen and Gilbert, 1988.)

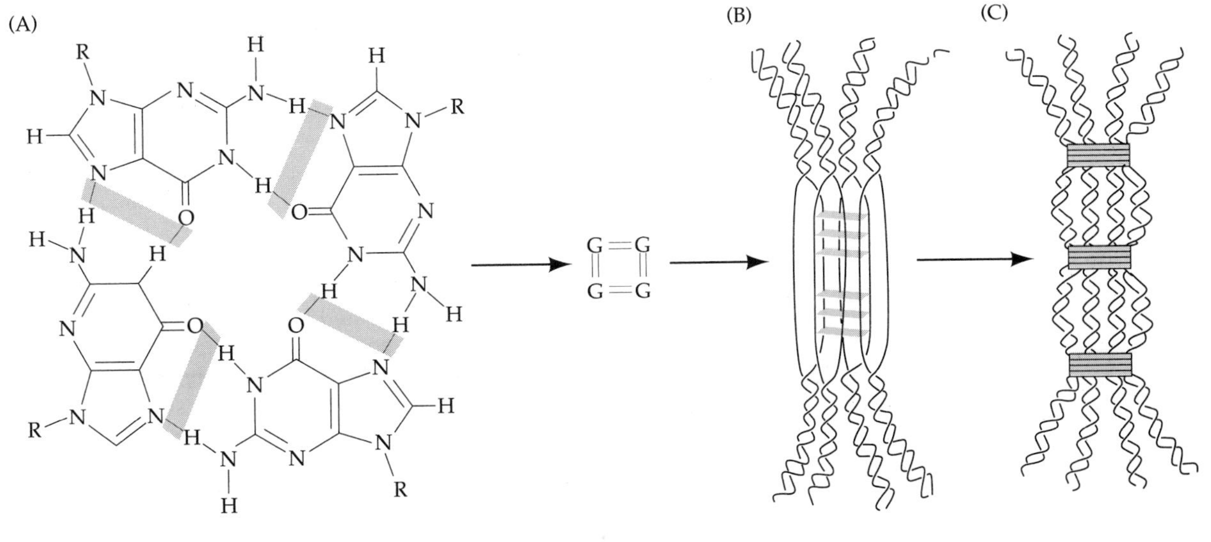

stages of prophase I further increases genetic diversity and makes the number of different gametes incalculable.

The mechanism for synapsis is unknown. One hypothesis is that certain regions of DNA may not replicate during the premeiotic S phase. The chromosomes would be coupled through the base pairs at these sites. Hotta and co-workers (1984) found just such sequences in the lily, where certain single-stranded regions remain in the DNA throughout the formation of the synaptonemal complex and disappear as the complex disassembles. This differentially replicating DNA is rich in guanosine and cytosine, and this pattern of late replication is not seen in the germ cells before or after this period, nor is it seen in any nongerm cell. Sen and Baltimore (1988) have extended these observations into a model where the meiotic chromatids are held together by these late replicating regions of the DNA helices. Following up an observation that certain DNA fragments would not migrate properly in a gel, Sen and Baltimore found that these guanosine-rich sequences self-associated into four-stranded structures at physiological conditions. Chemical analysis showed that these sequences were running parallel to each other, not anti-parallel as in the normal double helices. Such 4-stranded parallel helices would be possible because of strong hydrogen bonding between the guanosines (Figure 10). According to this model, similar G-rich regions at the telomeres would bring homologous chromatids together at the nuclear envelope. (This is where synapsis is often seen to be initiated.) Once joined at the telomeres, the "zipping up" of the homologous chromatids is accomplished by matching the internal G-rich regions together. The G-rich regions may be accessible due to local unwinding that happens when large stretches of G are on one strand of a helix. The G-rich regions would act as the "velcro" for the zipper. In this way, the interaction of the four double helices at meiosis would be caused by the interaction of G-rich regions on four of the strands.

SIDELIGHTS & SPECULATIONS

Big decisions: Mitosis/meiosis and sperm/egg

In many species, the germ cells migrating into the gonad are bipotential and can differentiate into sperm or ova, depending on their gonadal environment. When the ovaries of salamanders are transformed into testes (by grafting a mature testis into the female), the resident germ cells cease their oogenic differentiation and begin developing as sperm (Burns, 1930; Humphrey, 1931). Similarly, in the housefly and mouse, the gonad is able to direct the differentiation of the germ cell (Inoue and Hiroyoshi, 1986; McLaren, 1983).

What genes are involved in this "transdetermination" event? This question can best be studied in hermaphroditic animals, where the change from sperm production to egg production is a naturally occurring physiological event. Using *Caenorhabditis elegans*, Judith Kimble and her colleagues have identified two "decisions" that presumptive germ cells have to make. The first involves whether to enter meiosis or to remain a mitotically dividing stem cell. The second decision concerns whether the meiotic cell is to

become an egg or a sperm. The mitotic/meiotic decision is controlled by a single nondividing cell at the end of each gonad, the DISTAL TIP CELL. The germ cell precursors near this cell divide mitotically, forming the pool of germ cells; but as these cells get further away from the distal tip cell, they enter meiosis. If the distal tip cells are destroyed by focused laser beams, *all* the germ cells enter meiosis, and if the distal tip cell is placed into a different location in the gonad, germ-line stem cells are generated near this new position (Figure 11; Kimble, 1981; Kimble and White, 1981). It appears that the distal tip cells secrete some substance that maintains these cells in mitosis and inhibits their meiotic differentiation.

Although we do not yet know how the distal tip cell interacts with the presumptive germ cells, Austin and Kimble (1987) have isolated a mutation that mimics the phenotype obtained when the distal tip cells are removed. All the germ cell precursors of nematodes homozygous for the recessive mutation *glp-1* initiate meiosis, leaving no mitotic population. Instead of the 1500 germ cells usually found in the fourth larval stage of hermaphroditic development, these mutants produce 5–8 sperm cells. When genetic chi-

(A) Intact gonad

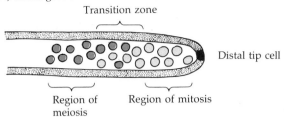

(B) Distal tip cell ablated

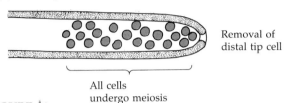

FIGURE 11

Regulation of the mitosis/meiosis decision by the distal tip cell of the *C. elegans* ovotestis. (A) Intact gonad early in development with regions of mitosis (shaded cells) and meiosis. (B) Gonad after laser ablation of the distal tip cell. All germ cells enter meiosis.

meras are made in which wild-type germ cell precursors are found within a mutant larva, the chimeras are able to respond to the distal tip cells and undergo mitosis. However, when mutant germ cell precursors are found within wild-type larvae, they each enter meiosis. Thus, the *glp-1* gene appears to be responsible for enabling the germ cells to respond to the distal tip cell's signal.*

After the cells begin their meiotic divisions, they still must become either sperm or ova. Generally, in each ovotestis, the most proximal germ cells produce sperm, the most distal (near the tip) become eggs (Hirsh et al., 1976). The genetics of this switch is currently being analyzed. As discussed in the previous chapter, the genes for sex determination generate either a female body that is functionally hermaphroditic or a male body. Certain mutations can alter these phenotypes. For example, *tra-1* homozygotes develop as sperm-producing males and *fem-1* homozygotes develop as egg-producing females (Figure 12). The double mutants homozygous for both *tra-1* and *fem-1* have a unique phenotype. They are somatically male but are female in the germ line (Doniach and Hodgkin, 1984). This suggests that *tra-1* is the primary sex-determining gene of the somatic tissues, but that the *fem* genes are responsible for the sperm/oocyte decision.

Hodgkin's laboratory (1985) and Kimble's laboratory (1986) have recently isolated several genes needed for germ cell pathway selection. Figure 13 presents a scheme for how these genes might function. The *fem* genes act together to activate *fog-1*. This gene, when homozygous in the germ

*The *glp*-1 gene appears to be involved in a number of inductive interactions in *C. elegans*. You will no doubt recall that *glp*-1 is also needed by the AB blastomere to receive inductive signals from the EMS blastomere to form pharyngeal muscles (Chapter 7).

(A) Wild-type: Sperm and oocytes

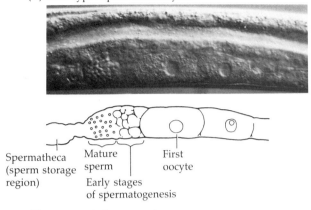

(B) Feminized: Ooctyes only

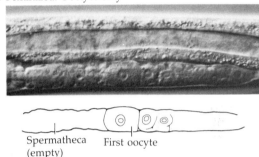

(C) Masculinized: Sperm only

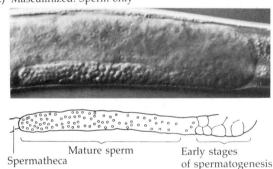

FIGURE 12

Gonads of wild-type and mutant *Caenorhabditis elegans*. (A) Wild-type hermaphrodite first producing sperm and then eggs. (B) Female animal produced by the *fem-1* mutation. Only eggs are produced. (C) Males produced by the *tra-1* mutation, containing only sperm. (Photographs courtesy of J. Kimble.)

cells, eliminates spermatogenesis in XX hermaphrodites (hence its name, feminization of germ line). The wild-type allele of *fog-1* activates the genes needed for the germ cell's maturation as sperm and suppresses the activity of wild-type *mog-3* (whose absence causes masculinization of the germ line). If, however, *tra-1* is active (as would be expected in hermaphrodites), the *fog-1* genes are inactivated, thereby allowing the expression of the oocyte-specific *mog-3* gene. In this way, germ cell sex is usually correlated with somatic sex, and the wild-type *Caenorhabditis* germ cell can become either sperm or egg.

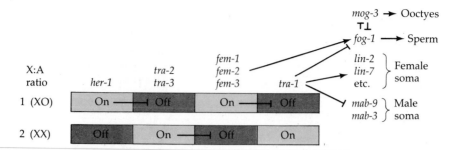

X:A ratio				fem-1 fem-2 fem-3					
	her-1		tra-2 tra-3				tra-1			
1 (XO)	On ——⊣ Off			On ——⊣ Off						
2 (XX)	Off		On ——⊣ Off				On			

mog-3 → Ooctyes
⊤⊥
fog-1 —→ Sperm

lin-2
lin-7 } Female soma
etc.

mab-9 } Male soma
mab-3

FIGURE 13

Model of gonadal sex determination in the nematode *C. elegans*, based on analysis of mutations. Arrows indicate positive regulation; barred arrows show negative regulation. If the *fem* genes are active, *fog-1* is expressed in the germ cells. This suppresses the oocyte activator, *mog-3*, and enables the germ cells to become sperm. Alternatively, when *tra-1* is activated, *fog-1* is suppressed, allowing the expression of oocyte-specific genes. (After Hodgkin et al., 1985.)

Spermatogenesis

Once the vertebrate primordial germ cells arrive at the genital ridge of male embryos, they become incorporated into the sex cords. They remain there until maturity, at which time the sex cords hollow out to form the seminiferous tubules, and the epithelium of the tubules differentiates into the Sertoli cells. These Sertoli cells nourish and protect the developing sperm cells, and SPERMATOGENESIS—the meiotic divisions giving rise to the sperm—occurs in the recesses of the Sertoli cells (Figure 14). The process by which the PGCs generate sperm has been studied in detail in several organisms, but we shall focus here on spermatogenesis in mammals. After reaching the gonad, the PGCs divide to form TYPE A₁ SPERMATOGONIA. These cells are smaller than the PGCs and are characterized by an ovoid nucleus that contains chromatin associated with the nuclear membrane. The A₁ spermatogonia are found adjacent to the outer basement membrane of the sex cords. At maturity, these spermatogonia are thought to divide so as to make another type A₁ spermatogonium as well as a second, paler type of cell, the TYPE A₂ SPERMATOGONIUM. Thus, each type A₁ spermatogonium is a stem cell capable of regenerating itself as well as of producing a new cell type. The A₂ spermatogonia divide to produce the A₃ spermatogonia, which then beget the type A₄ spermatogonia, which beget the INTERMEDIATE SPERMATOGONIA. These intermediate spermatogonia divide to form the TYPE B SPERMATOGONIA, and these cells divide mitotically to generate the PRIMARY SPERMATOCYTES—the cells that enter meiosis.

Looking at Figure 15, we find that during the spermatogonial divisions, cytokinesis is not complete. Rather, the cells form a syncytium whereby each cell communicates to the other via cytoplasmic bridges about 1 μm in diameter (Dym and Fawcett, 1971). The successive divisions produce clones of interconnected cells, and because ions and molecules readily pass through these intercellular bridges, each cohort matures synchronously.

Each primary spermatocyte undergoes the first meiotic division to yield a pair of SECONDARY SPERMATOCYTES, which complete the second division of meiosis. The haploid cells formed are called SPERMATIDS, and they are still connected to each other through their cytoplasmic bridges.

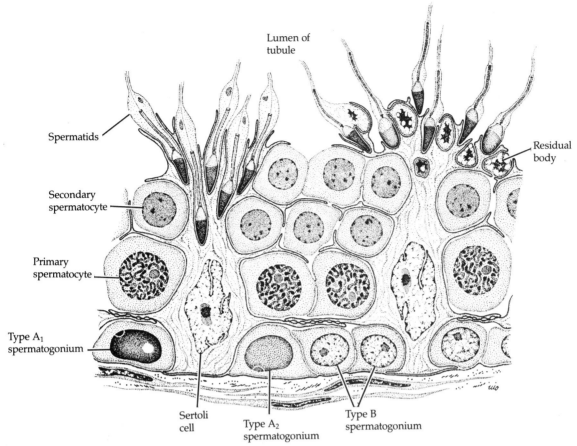

Lumen of
tubule

Spermatids

Secondary
spermatocyte

Primary
spermatocyte

Type A₁
spermatogonium

Sertoli
cell

Type A₂
spermatogonium

Type B
spermatogonium

Residual
body

FIGURE 14
Drawing of a section of the seminiferous tubule, showing the relationship between Sertoli cells and the developing sperm. As cells mature, they progress towards the lumen of the seminiferous tubule. (After Dym, 1977.)

The spermatids that are connected in this manner have haploid nuclei, but are functionally diploid, since the gene product made in one cell can readily diffuse into the cytoplasm of its neighbors (Braun et al., 1989a). During the divisions from type A₁ spermatogonium to spermatid, the cells move farther and farther away from the basement membrane of the seminiferous tubule and closer to its lumen (Figure 14). Thus, each type of cell can be found in a particular layer of the tubule. The spermatids are located at the border of the lumen, and here they lose their cytoplasmic connections and differentiate into sperm cells.

Spermiogenesis

The haploid spermatid is a round, unflagellated cell that looks nothing like the mature vertebrate sperm. The next step in sperm maturation, then, is SPERMIOGENESIS (or SPERMATELIOSIS), the differentiation of the sperm cell. In order for fertilization to occur, the sperm has to meet and bind with the egg, and spermiogenesis differentiates the sperm for these functions of motility and interaction. The processes of mammalian sperm differentiation can be seen in Figure 2 in Chapter 2. The first steps involve the construction of the acrosomal vesicle from the Golgi apparatus. The acrosome forms a cap that covers the sperm nucleus. As the cap is formed, the nucleus rotates so that the acrosomal cap will then be facing the basal membrane of the seminiferous tubule. This rotation is necessary because

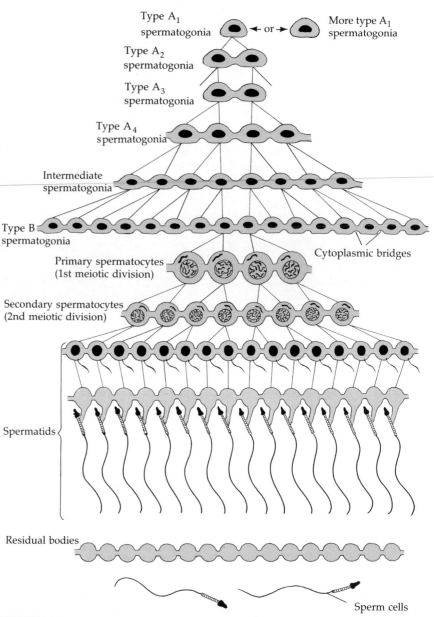

Type A₁ spermatogonia — or → More type A₁ spermatogonia

Type A₂ spermatogonia

Type A₃ spermatogonia

Type A₄ spermatogonia

Intermediate spermatogonia

Type B spermatogonia

Cytoplasmic bridges

Primary spermatocytes (1st meiotic division)

Secondary spermatocytes (2nd meiotic division)

Spermatids

Residual bodies

Sperm cells

FIGURE 15

The formation of syncytial clones of the male germ cells. (After Bloom and Fawcett, 1975.)

the flagellum is beginning to form from the centriole on the other side of the nucleus, and this flagellum will extend into the lumen. During the last stage of spermiogenesis, the nucleus flattens and condenses, the remaining cytoplasm (the "cytoplasmic droplet") is jettisoned, and the mitochondria form a ring around the base of the flagellum. The resulting sperm then enter the lumen of the tubule.

In the mouse, the entire development from stem cell to spermatozoan takes 34.5 days. The spermatogonial stages last 8 days, meiosis lasts 13 days, and spermiogenesis takes up another 13.5 days. In humans, spermatic development takes 74 days to complete. Because the type A₁ spermatogonia are stem cells, spermatogenesis can occur continuously. Each hour, some 100 million sperm are made in each human testicle, and each ejaculation releases 200 million sperm. Unused sperm is either resorbed or passed out of the body in urine.

Gene expression during sperm development

Gene transcription during spermatogenesis takes place predominantly during the diplotene stage of meiotic prophase. This transcription has been observed in many organisms, but the best documented case is probably that of Y chromosome transcription in *Drosophila hydei*. Here, RNA transcripts originating from the Y chromosome are seen to be essential for controlling spermiogenesis. When we recall the function of the Y chromosome in *Drosophila*, this is not surprising, for the Y chromosome is not involved in sex determination here. Rather, it is needed for the formation of viable sperm. The difference between XY *Drosophila* and XO *Drosophila* is that the latter are sterile. Both are male. In *Drosophila hydei*, the Y chromosome extends five prominent loops of DNA (Figure 16). If any of these loops is deleted, the organization of the sperm tail will be abnormal. All the component parts of the sperm will be present, but not properly organized (Hess, 1973). It appears, then, that Y-specific RNA made during meiotic prophase is utilized later during spermiogenesis.

The genes that are transcribed specifically during spermatogenesis are often those whose products are necessary to sperm motility or binding to the egg. In *Drosophila melanogaster*, one of the sperm-specific genes transcribed is for β2-tubulin. This isoform of β-tubulin is only seen during spermatogenesis, and it is responsible for forming the meiotic spindles, the axoneme, and the microtubules associated with the lengthening mitochondria.* Hoyle and Raff (1990) have shown that another β-tubulin isoform, β3-tubulin (which is normally expressed in mesodermal cells and epidermis) cannot substitute for the β2-tubulin. When they fused the 5' regulatory region from the β2-tubulin gene and fused it to the coding sequences of the β3-tubulin gene, the β3-tubulin gene was able to be expressed in the developing sperm. When this gene was expressed in the absence of the β2-tubulin gene, the resulting germ cells failed to undergo meiosis, axoneme assembly, or nuclear shaping. Only the mitochondrial elongation occurred. This indicates that the formation of the meiotic spindles and axoneme of sperm cells cannot be accomplished by just any β-tubulin and that the transcription of sperm-specific isoforms is important.

*Making the sperm axoneme in *Drosophila* is a large undertaking. The sperm tail is 2 mm long—as long as the entire male fly! Even more remarkable, the egg incorporates the entire sperm (Karr, 1988).

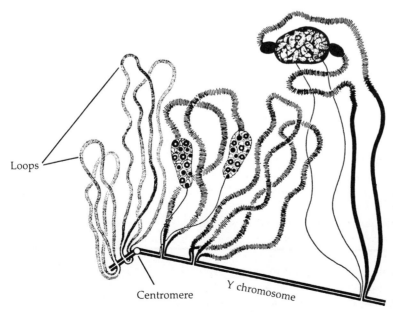

Loops

Centromere

Y chromosome

FIGURE 16
Diagram of the Y chromosome of *Drosophila hydei* in the lampbrush chromosome state. Five loops are readily observed. (After Hess, 1973.)

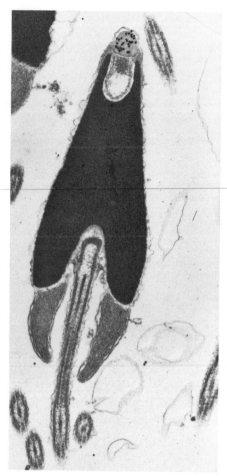

FIGURE 17
Localization of bindin in sperm acrosome by gold-labeled anti-bindin antibodies. The gold atoms enable the antibodies to show up as black dots in the electron micrograph. These sperm are still inside the sea urchin testis. (Photograph courtesy of D. Nishioka.)

Those genes whose products are necessary for the binding of the sperm and the extracellular matrices of the egg are also transcribed during spermatogenesis. The gene for sea urchin bindin is transcribed relatively late in spermatogenesis and its mRNA is translated into bindin shortly after being made (Nishioka et al., 1990). The bindin accumulates in vesicles that fuse together to form the single acrosomal vesicle of the mature sea urchin sperm. Figure 17 shows the localization of the bindin protein in the acrosomal vesicle of the sperm while it is still in the testis.

Like the oocyte, the spermatid can store mRNA for later use. In mammals and birds, a specific form of lactate dehydrogenase—LDH-X— is made during spermatogenesis. (This protein enables the developing sperm to utilize pyruvic acid as an alternative energy source.) The message for this protein can be identified in spermatocyte cytoplasm during meiotic prophase (Blanco, 1980). Similarly, in many species, small proteins called PROTAMINES appear in the nucleus during the final stages of spermiogenesis. These proteins contain about 32 amino acids, all but four or five of which are arginine residues. They replace the nuclear histones and cause the DNA to form a compact, almost crystalline, array (Marushige and Dixon, 1969). DNA complementary to trout protamine mRNA can detect protamine message sequences in the primary spermatocyte. However, these messages are stored in ribonucleoprotein particles. Only during the spermatid stage, approximately a month after its synthesis, is the protamine message translated into protein (Iatrou et al., 1978). The regulation of this mRNA appears to be controlled by its 3' untranslated region. If placed onto another message, this 3' untranslated region will give the new message the same translational regulation as the protamine mRNA (Braun et al., 1989b). We find, then, that just like the oocyte, which can synthesize RNA in the diplotene stage and store it for later use, the sperm can also package messages for later translation.

In addition to gene transcription during meiotic prophase, there is also evidence that certain genes are transcribed in the spermatids (reviewed in Palmiter et al., 1984). This evidence for HAPLOID GENE EXPRESSION comes from studies involving heterozygous mice in which two different populations of sperm are seen to exist—one population expressing the mutant phenotype and one population expressing the wild-type trait. If the synthesis of the RNA or protein were to occur while the cells were still diploid, all the sperm would show the same phenotype.

In some species, sperm provides important developmental information that cannot be compensated for by the egg. We have discussed the imprinting of mammalian chromosomes wherein the sperm and egg DNA differ in their methylation patterns (Chapters 2 and 12). There are also cases of paternal effect genes. Here, homozygous recessive alleles in the male cause abnormal development in the embryo even if the female is

FIGURE 18
Immunofluorescence photomicrographs of mitotic spindles in a first cleavage *C. elegans* embryo when sperm is (A) from a wild-type male, or (B) from a male homozygous for the paternal effect gene *spe-11*. In (B), three microtubule organizing centrioles can be seen instead of the usual two poles of mitosis. (From Hill et al., 1989; photographs courtesy of S. Strome.)

(A)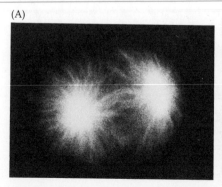

(B)

homozygous for the wild-type allele, while the reciprocal cross, where the father is wild-type and the mother is homozygous for the mutant allele, leads to normal embryos. One such paternal effect gene is *spe-11* in *C. elegans*. The sperm containing mutant alleles at this locus are unable to direct chromosomal movements that orient the mitotic spindle of the embryo, suggesting that the mutation affects the microtubule organizing regions such as centrioles (Figure 18; Hill et al., 1989).

Eventually, the haploid genome is condensed as the histones are replaced by protamines or by specifically modified histones. Many of the sperm histones become modified in the late spermatid stage during spermiogenesis. These modifications (such as dephosphorylating the *N*-terminal regions of certain histones) causes the chromatin to condense. Condensation results in severely reduced transcription. Thus, transcription from the male genome is not detected again until it is reactivated sometime during development (Poccia, 1986; Green and Poccia, 1988).

Oogenesis

Oogenic meiosis

Oogenesis—the differentiation of the ovum—differs from spermatogenesis in several ways. Whereas the gamete formed by spermatogenesis is essentially a motile nucleus, the gamete formed by oogenesis contains all of the factors needed to initiate and maintain metabolism and development. Therefore, in addition to forming a haploid nucleus, oogenesis also builds up a store of cytoplasmic enzymes, messages, organelles, and metabolic substrates. So while the sperm becomes differentiated for motility, the oocyte develops a remarkably complex cytoplasm. Unlike the sperm, which becomes differentiated after its meiotic divisions, the egg grows primarily in an extended period of meiotic prophase.

The mechanisms of oogenesis vary more than do those of spermatogenesis. This difference should not be surprising, since the patterns of reproduction vary so greatly among species. In some species, such as sea urchins and frogs, the female routinely produces hundreds or thousands of eggs at a time, whereas in other species, such as humans and most mammals, only a few eggs are produced during the lifetime of an individual. In those species that produce thousands of ova, the oogonia are self-renewing stem cells that endure for the lifetime of the organism. In the more parsimonious species, the oogonia divide to form a limited number of egg precursor cells. In humans, the thousand or so oogonia divide rapidly from the second to the seventh month of gestation to form roughly 7 million germ cells (Figure 19). After the seventh month of embryonic development, however, the number of germ cells drops precipitously. Most oogonia die during this period, the remaining oogonia entering prophase of the first meiotic division (Pinkerton et al., 1961). These latter cells, called the PRIMARY OOCYTES, progress through the first meiotic prophase until the diplotene stage, at which point they become arrested in the first meiotic prophase. Here they are arrested until puberty. With the onset of adolescence, groups of oocytes periodically resume meiosis. Thus, in the human female, the first part of meiosis is begun in the embryo, and the signal to resume meiosis is not given until roughly 12 years later. In fact, some oocytes are arrested in meiotic prophase for nearly 50 years. As Figure 19 indicates, primary oocytes continue to die even after birth.

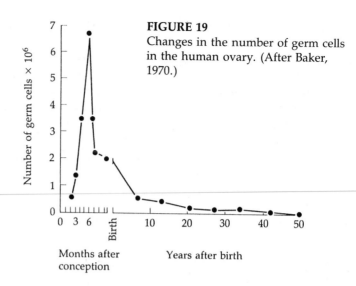

FIGURE 19
Changes in the number of germ cells in the human ovary. (After Baker, 1970.)

Number of germ cells × 10⁶

Months after conception   Years after birth

FIGURE 20
Polar body formation in the oocyte of the whitefish, *Coregonus*. (A) Anaphase of first meiotic division, showing the first polar body pinching off with its chromosomes. (B) Metaphase (within the oocyte, arrow) of the second meiotic division, with the first polar body still in place. The first polar body may or may not divide again. (From Swanson et al., 1981; photographs courtesy of C. P. Swanson.)

Of the millions of primary oocytes present at birth, only about 400 mature during a woman's lifetime.

Oogenic meiosis also differs from spermatogenic meiosis in its placement of the metaphase plate. When the primary oocyte divides, the nuclear membrane of the oocyte, called the GERMINAL VESICLE, breaks down and the metaphase spindle migrates to the periphery of the cell. At telophase, one of the two daughter cells contains hardly any cytoplasm, whereas the other cell has nearly the entire volume of cellular constituents (Figure 20). The smaller cell is called the FIRST POLAR BODY and the larger cell is referred to as the SECONDARY OOCYTE. During the second division of meiosis, a similar unequal cytokinesis takes place. Most of the cytoplasm is retained by the mature egg (OVUM), and a SECOND POLAR BODY receives little more than a haploid nucleus. Thus, oogenic meiosis serves to conserve the volume of oocyte cytoplasm in a single cell rather than splitting it equally among four progeny.

In some species of animals, meiosis is severely modified such that the resulting gamete is diploid and need not be fertilized to develop. Such animals are said to be PARTHENOGENETIC. In the fly *Drosophila mangabeirai*, one of the polar bodies acts as a sperm and "fertilizes" the oocyte after

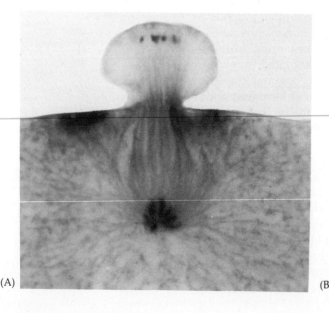

(A)

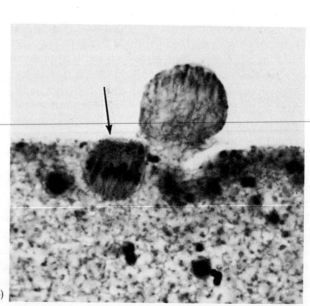

(B)

the second meiotic division. In other insects (such as *Moraba virgo*) and the lizard *Cnemidophorus uniparens*, the oogonia double their chromosome number before meiosis, so that the halving of chromosomes restores the diploid number. The germ cells of the grasshopper *Pycnoscelus surinamensis* dispense with meiosis altogether, forming diploid ova by two mitotic divisions (Swanson et al., 1981). In the above examples, the species consist entirely of females. In other species, haploid parthenogenesis is widely used, not only as a means of reproduction, but also as a mechanism of sex determination. In the Hymenoptera (bees, wasps, and ants), unfertilized eggs develop into males whereas fertilized eggs, being diploid, develop into females. The haploid males are able to produce sperm by abandoning the first meiotic division, thereby forming two sperm cells through second meiosis.

Maturation of the oocyte in amphibians

The egg is responsible for initiating and directing development, and in some species (as seen earlier) fertilization is not even necessary. The accumulated material in the oocyte cytoplasm includes energy sources and organelles (the yolk and mitochondria), the enzymes and precursors for DNA, RNA, and protein syntheses, stored messenger RNAs, structural proteins, and morphogenic determinants. A partial catalog of the materials stored in the oocyte cytoplasm is shown in Table 1. Most of this accumulation takes place during meiotic prophase I, and this stage is often subdivided into PREVITELLOGENIC (Gk., "before yolk formation") and VITELLOGENIC (yolk-forming) phases.

Eggs of fishes and amphibians are derived from a stem cell oogonial population, which can generate a new cohort of oocytes each year. In the frog *Rana pipiens*, oogenesis lasts 3 years. During the first 2 years, the oocyte increases its size very gradually. During the third year, however, the rapid accumulation of yolk material in the oocytes causes them to swell to their characteristically large size (Figure 21). Eggs mature in yearly batches, the first cohort maturing shortly after metamorphosis; the next group matures a year later.

Vitellogenesis occurs when the oocyte reaches the diplotene stage of meiotic prophase. This is also the stage when the lampbrush chromosomes of the nucleus are seen to be actively synthesizing RNA (Chapter 11). Yolk is not a single substance, but a mixture of materials used for embryonic

TABLE 1
Cellular components stored in the mature oocyte of *Xenopus laevis*

Component	Approximate excess over amount in larval cells
Mitochondria	100,000
RNA polymerases	60,000–100,000
DNA polymerases	100,000
Ribosomes	200,000
tRNA	10,000
Histones	15,000
Deoxyribonucleoside triphosphates	2,500

Source: Laskey (1974).

FIGURE 21
Growth of oocytes in the frog. During the first three years of life, three cohorts of oocytes are produced. The drawings follow the growth of the first-generation oocytes. (After Grant, 1953.)

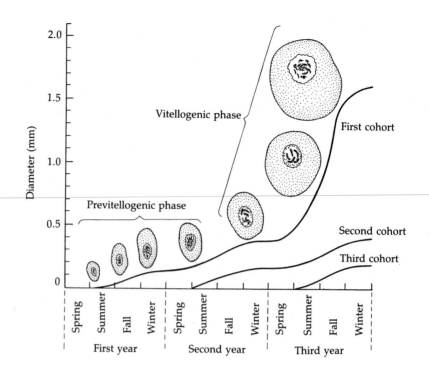

nutrition. The major yolk component is a 470,000-Da protein called VITEL-LOGENIN. It is not made in the frog oocyte (as are the major yolk proteins of organisms such as annelids and crayfish), but it is synthesized in the liver and carried by the bloodstream to the ovary (Flickinger and Rounds, 1956). This large protein passes between the follicle cells of the ovary and is incorporated into the oocyte by MICROPINOCYTOSIS, the pinching off of membrane-bounded vesicles at the base of microvilli (Dumont, 1978). In the mature oocyte, vitellogenin is split into two smaller proteins: the heavily phosphorylated PHOSVITIN and the lipoprotein LIPOVITELLIN. These two proteins are packaged together into membrane-bounded YOLK PLATELETS (Figure 22). Glycogen granules and lipochondrial inclusions store the carbohydrate and lipid components of the yolk, respectively.

Most eggs are highly asymmetric, and it is during oogenesis that the animal–vegetal axis of the egg is specified. Danilchik and Gerhart (1987) have shown that although the concentration of yolk in *Xenopus* oocytes increases nearly tenfold as one goes from the animal to the vegetal poles of the mature egg, vitellogenin uptake is uniform around the surface of the oocyte. What differs is its movement *within* the oocyte, and this de-

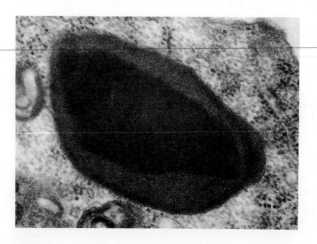

FIGURE 22
An amphibian yolk platelet. (Photograph courtesy of L. K. Opresko.)

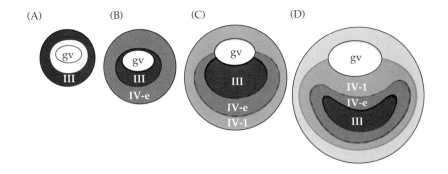

FIGURE 23

Establishment of animal–vegetal polarity of yolk platelets in the *Xenopus* oocyte. (A) In the late stage III (600-μm) oocyte, yolk platelets enter the cell equally at all points of the surface. As the oocyte grows (B–C), the platelets at the future animal pole are displaced towards the vegetal pole, while those at the vegetal pole remain there. More yolk still enters the egg on all sides. By the end of vitellogenesis (D), the earliest platelets (III) are all in the vegetal hemisphere, which has now concentrated roughly 75 percent of the oocyte yolk. Time of entry of yolk into oocyte platelets is indicated by shading and Roman numerals: III, stage III platelets; IV-e, early stage IV platelets; IV-l, late stage IV platelets; V, stage V platelets; gv, germinal vesicle. (After Danilchik and Gerhart, 1987.)

pends upon where the yolk proteins enter. When yolk platelets are formed in the future animal hemisphere, they move inward toward the center of the cell. The vegetal yolk platelets, however, do not actively move but remain at the periphery for long periods of time, enlarging as they stay there. They are slowly displaced from the cortex as new yolk platelets come in from the surface. As a result of this differential intracellular transport, the amount of yolk steadily increases in the vegetal hemisphere, until the vegetal half of a mature *Xenopus* oocyte contains nearly 75 percent of the yolk (Figure 23). The mechanism of this translocation remains unknown.

As the yolk is being deposited, the organelles also form asymmetric arrangements. The cortical granules begin to form from the Golgi apparatus and are originally scattered randomly through the oocyte cytoplasm. They later migrate to the periphery of the cell. The mitochondria replicate at this time, dividing to form millions of organelles that will be apportioned to the different cells during cleavage. (In *Xenopus*, new mitochondria are not formed until after gastrulation is initiated.) As vitellogenesis nears an end, the oocyte cytoplasm becomes stratified. The cortical granules, mitochondria, and pigment granules are found at the periphery of the cell, creating the oocyte cortex. Within the inner cytoplasm, distinct gradients emerge. While the yolk platelets become more heavily concentrated at the vegetal pole of the oocyte, the glycogen granules, ribosomes, lipochondria, and endoplasmic reticulum are found more toward the animal pole. Even specific messenger RNAs stored in the cytoplasm become localized to certain regions of the oocyte (Chapter 7).

While the precise mechanisms for establishing these gradients remain unknown, studies using inhibitors have shown that the cytoskeleton is critically important in localizing specific RNAs and morphogenetic factors. Yisaeli and his co-workers (1990) have shown that the Vg1 mRNA is translocated into the vegetal cortex in a two-step process (see Figure 9 in Chapter 7). In the first phase, microtubules are needed to bring the Vg1 mRNA into the vegetal hemisphere. In the second phase, microfilaments are responsible for anchoring the Vg1 message to the cortex.

In *Xenopus*, the leptotene stage of meiosis lasts only 3–7 days, zygotene takes from 5 to 9 days, and pachytene persists for roughly 3 weeks. The diplotene stage, however, can last years. Even so, vitellogenesis occurs only in part of the diplotene, and the signal for the breakdown of the nucleus (the germinal vesicle) occurs after vitellogenesis is completed. The regulation of these events is controlled by the hormonal interactions between the hypothalamus, pituitary gland, and follicle cells of the ovary (Figure 24). When the hypothalamus receives the cues that the mating season has arrived, it secretes gonadotropin releasing hormone, which is received by the pituitary. The pituitary responds by secreting the gonadotropic hormones into the blood. These hormones stimulate the follicle

(A) VITELLOGENESIS AND
OOCYTE DIFFERENTIATION

Hypothalamus

(B) EGG MATURATION
AND OVULATION

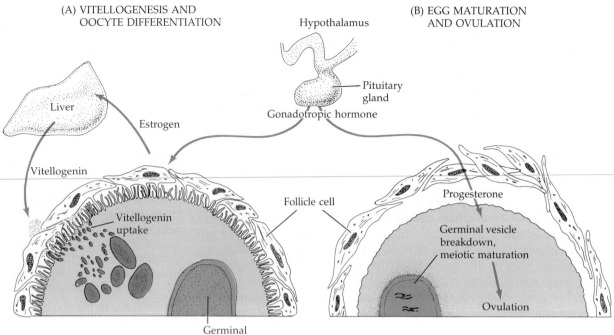

Pituitary
gland

Gonadotropic hormone

Liver

Estrogen

Vitellogenin

Vitellogenin
uptake

Follicle cell

Progesterone

Germinal vesicle
breakdown,
meiotic maturation

Ovulation

Germinal
vesicle

FIGURE 24

Control of amphibian oocyte growth and egg maturation by estrogen and progesterone. (A) Gonadotropic hormone stimulates the follicle cells to produce estrogen, which instructs the liver to secrete vitellogenin. This protein is absorbed by the oocyte. (B) After vitellogenesis, again under the influence of gonadotropic hormone, the follicle cells secrete progesterone. Within 6 hours of progesterone stimulation, the germinal vesicle breaks down, initiating the reactions leading to ovulation. (After Browder, 1980.)

cells to secrete estrogen, and the estrogen instructs the liver to synthesize and secrete vitellogenin. Upon estrogen stimulation, the liver cells change drastically (Figure 25). These changes are usually brought about during the annual mating season, but when estrogen is injected into adult frogs (either male or female) at any time, these changes can be induced and vitellogenin secreted (Skipper and Hamilton, 1977).

Estrogen induces vitellogenin at both the transcriptional and translational levels. Before estrogen release, there are no detectable vitellogenin messenger RNAs in the liver. Following hormone administration, each cell contains some 50,000 vitellogenin mRNA molecules, constituting

FIGURE 25

Effects of estrogen (estradiol) on the ultrastructure of *Xenopus laevis* liver cells. (A) Hepatocytes prior to estrogen injection. (B) After estrogen injection, hepatocytes show enormous expansion of the rough endoplasmic reticulum and Golgi apparatus needed for production and secretion of vitellogenin. (From Skipper and Hamilton, 1977; photographs courtesy of the authors.)

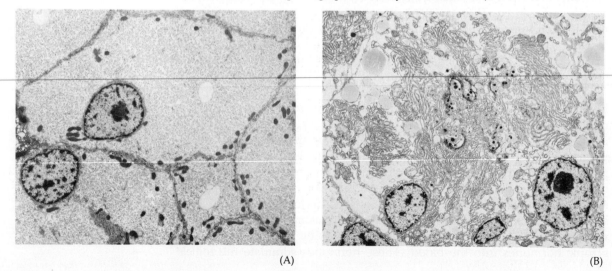

(A)

(B)

roughly half the total cellular mRNA. The transcriptional activation of the vitellogenin genes would normally generate only about 1500 of these mRNA molecules per cell. However, estrogen also specifically stabilizes that message, somehow increasing its half-life from 16 hours to 3 weeks. Because these effects occur in the absence of new protein synthesis, they appear to be direct consequences of estrogen. Thus, estrogen controls the accumulation of vitellogenin by transcriptional and translational gene regulation (Brock and Shapiro, 1983).

Maintenance and release from meiotic arrest

Amphibian oocytes can stay years in the diplotene stage of meiotic prophase. Resumption of meiosis in the amphibian primary oocyte requires progesterone. This hormone is secreted by follicle cells in response to the gonadotropic hormones secreted by the pituitary gland. Within 6 hours of progesterone stimulation, GERMINAL VESICLE BREAKDOWN (GVBD) occurs, the microvilli retract, the nucleoli disintegrate, and the lampbrush chromosomes contract and migrate to the animal pole to begin division. Soon afterward, the first meiotic division occurs, and the mature ovum is released from the ovary by a process called OVULATION. The ovulated egg is in second meiotic metaphase when it is released.

How does progesterone enable the egg to break its dormancy and resume meiosis? To understand the mechanisms by which this activation is accomplished, it is necessary to briefly review the model for early blastomere division that was presented in Chapter 3. Maturation promoting factor (MPF) is responsible for the resumption of meiosis. Its activity is cyclic, being high during cell division and undetectable during interphase. MPF is a protein kinase and is regulated by a cyclin protein. The cyclin is translated during interphase and is rapidly degraded upon cell division. The cyclin activates the MPF. If the cyclin is not degraded at cell division, the MPF remains active, and the cell is blocked in metaphase. Once the cyclin is degraded, the MPF becomes inactive and the cell returns to interphase. Masui (1974) found that cytoplasm from an ovulated amphibian egg would cause the division of other cells to be halted in metaphase. He concluded that the cytoplasm of the ovulated eggs contained a CYTOSTATIC FACTOR (CSF) that blocked cell division at metaphase.

This cytostatic factor has recently been isolated. Studies by Sagata and colleagues (1988) show that progesterone reinitiates meiosis by causing the egg to translate a particular maternal mRNA that has been stored in its cytoplasm. This stored message is encoded by the *mos* protooncogene,[*] and it translates into a 39,000-Da phosphoprotein, $pp39^{mos}$. This protein is detectable only during oocyte maturation and is destroyed quickly upon fertilization. Yet, during its brief lifetime, it plays a major role in releasing the egg from its dormancy. If the translation of $pp39^{mos}$ is inhibited (by injecting *mos*-antisense mRNA into the oocyte), $pp39^{mos}$ does not appear, and germinal vesicle breakdown and the renewal of oocyte maturation does not occur. Having stimulated the reinitiation of meiosis, it then keeps the chromosomes at metaphase until fertilization. In further experiments, Sagata and co-workers (1989) demonstrated that if c-*mos* mRNA were injected into one of the first two blastomeres of a newly cleaved frog egg, the injected blastomere would cease its mitosis in metaphase. Moreover, when antibodies to the $pp39^{mos}$ protein were incubated with homogenates from frog eggs, these homogenates lost their CSF activity. Therefore, CSF is probably the short-lived $pp39^{mos}$ protein.

*This is the *Xenopus* cellular homologue of the viral oncogene of the Moloney murine sarcoma virus, *v-mos* (see Chapter 20).

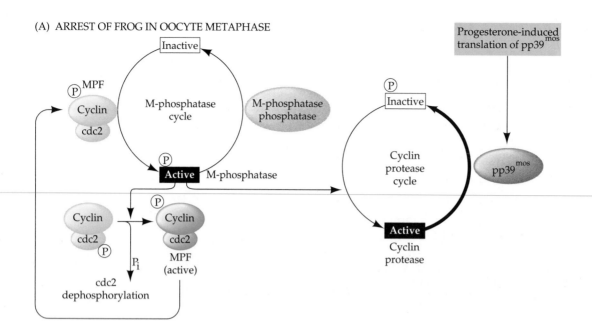

(A) ARREST OF FROG IN OOCYTE METAPHASE

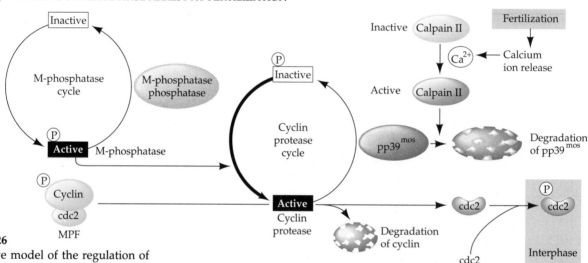

(B) BREAKING OF METAPHASE ARREST AT FERTILIZATION

FIGURE 26
Speculative model of the regulation of meiotic arrest in frog oocytes and its release at fertilization. (A) During meiotic metaphase, progesterone induces the translation of the pp39^{mos} protein (right-hand side of diagram). This inactivates the cyclin protease, preventing the degradation of cyclin. Cyclin remains bound to the cdc2 protein and MPF remains active. The cell is thereby kept in metaphase. (B) On fertilization, calcium ions are released which activate calpain II, a protease that specifically cleaves pp39^{mos}. When pp39^{mos} is degraded, the cyclin protease can be activated by the M-phosphatase. This activated cyclin protease degrades the cyclin, thereby abolishing MPF activity. The cell returns to interphase. It would appear that the cell is controlled by Rube Goldberg devices. (After Hunt, 1989.)

The next question concerns the mechanism by which pp39^{mos} and cyclin causes the renewal of meiosis and its cessation in metaphase. The main theory is that pp39^{mos} acts by inhibiting the degradation of the cyclin component of MPF. It is thought to do this by phosphorylating the specific protease that would otherwise degrade cyclin. Since the cyclin would not be degraded, MPF is stabilized (Figure 26). The cell thus enters metaphase. But until the MPF is destroyed, the cell remains in metaphase.

The degradation of MPF occurs at fertilization, when pp39^{mos} is degraded. But how is the pp39^{mos} degraded upon fertilization? The answer appears to be the activation of a protease which specifically cleaves pp39^{mos} (Watanabe et al., 1989). Upon fertilization, calcium ions are released from the endoplasmic reticulum. These free calcium ions activate the calcium-dependent protease CALPAIN II, which specifically attacks pp39^{mos}. Figure 27 shows the rise of pp39^{mos} upon receiving progesterone and its rapid decline once calpain II is activated. In this way, it is thought that progesterone reinitiates meiosis which can then be completed at fertilization.

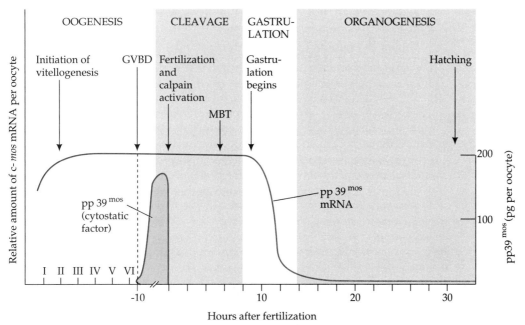

FIGURE 27

Expression of c-*mos* mRNA, pp39[mos], and calpain II during the early development of *Xenopus laevis*. (After Watanabe et al., 1989.)

Gene transcription in oocytes

The changes in oocyte gene transcription have been discussed earlier (Chapters 11 and 12). The patterns of rRNA and tRNA transcription are shown in Figure 28. Transcription appears to begin in early (stage I, 25–40 μm) oocytes, during the diplotene stage of meiosis. At this time, all the ribosomal and transfer RNAs needed for protein synthesis until the midblastula stage are made, and all the maternal mRNA for early development is transcribed. This stage lasts months in *Xenopus*. After reaching a certain size, the chromosomes of the mature (stage VI) oocyte condense and the genes are not actively transcribing. This "mature oocyte" condition can also be months long. Upon hormonal stimulation, the oocyte completes its first meiotic division and is ovulated. The mRNAs stored by the oocyte now join with the ribosomes to initiate protein synthesis. Within hours, the second meiotic division has begun and the secondary oocyte

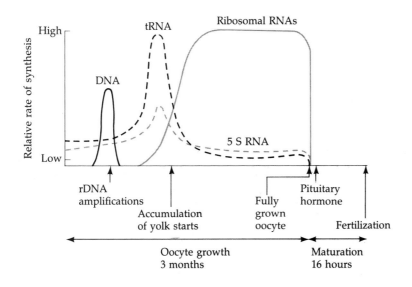

FIGURE 28

Relative rates of DNA, tRNA, and ribosomal RNA synthesis in amphibian oogenesis during the last three months before ovulation. (From Gurdon, 1976.)

has been fertilized. The genes will not begin active transcription until the midblastula transition (Davidson, 1986).

Unlike the situation in sea urchin oocytes, in which the amount and the complexity of the mRNA population are both seen to increase with oocyte maturation (Hough-Evans et al., 1977), the amount and the complexity of frog oocyte mRNA appears to stay relatively constant (Perlman and Rosbash, 1978; Dolecki and Smith, 1979). The amount of increase seems to be balanced by that of degradation, even during the lampbrush stage. Progesterone also initiates the translation of stored messenger RNAs in the oocyte cytoplasm. Thus, whereas in sea urchins the translation of maternal messages is initiated by fertilization, in frogs the signal for such translation is initiated by progesterone as the egg is about to be ovulated.

Fox and co-workers (1989) have shown that, just as in the case of certain spermatic messages, some frog oocyte messages can be activated for translation by the addition of poly(A) residues to their 3' ends. Shortly after progesterone-induced germinal vesicle breakdown, a set of cytoplasmic mRNAs acquire poly(A) stretches from 50 to 300 bases long. These messages are then translated and decay. The signal for this poly(A) addition is (a) the canonical AAUAAA poly(A) addition sequence and (b) a specific UUUUUAU sequence in the 3' untranslated region. This latter sequence can confer on any message the ability to become polyadenylated and translated during progesterone-induced oocyte maturation. A similar situation may also exist during the maturation of mammalian eggs. The message for plasminogen activator, a protease that is probably important for releasing the mammalian egg during ovulation, is stored in an untranslatable state in the meiotically arrested primary oocyte. Following the resumption of meiosis, the 3' end of this mRNA is polyadenylated and the message is translated (Vassalli et al., 1989).

Meroistic oogenesis in insects

Certain insects undergo MEROISTIC oogenesis, wherein cytoplasmic connections remain between the cells produced by the oogonium. In *Drosophila*, each oogonium divides four times to produce a clone of 16 cells connected to each other through RING CANALS. The production of these interconnected cells (called CYSTOCYTES) involves a highly ordered array of cell divisions (Figure 29). Only those two cells having four interconnections

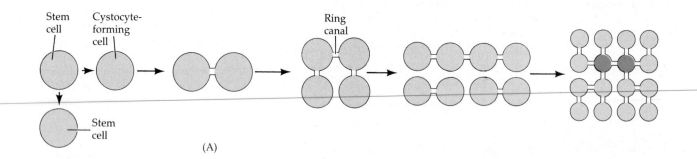

(A)

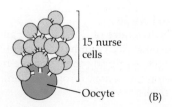

(B)

FIGURE 29
The formation of 16 interconnected cystocytes in *Drosophila*. (A) The cells are represented schematically as dividing in a single plane. The stem cell divides to produce another stem cell plus a cell that is committed to form the cystocytes. Only one of the 16 cystocytes will become an oocyte; the others become nurse cells, connected to the oocyte by ring canals (cytoplasmic bridges). (B) A three-dimensional representation of the oocyte and its 15 nurse cells. (A after Koch, 1967.)

(A)

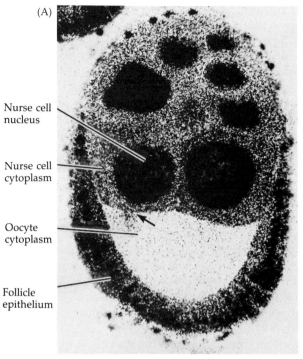

Nurse cell
nucleus

Nurse cell
cytoplasm

Oocyte
cytoplasm

Follicle
epithelium

(B)

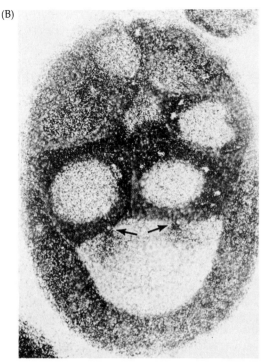

(C)

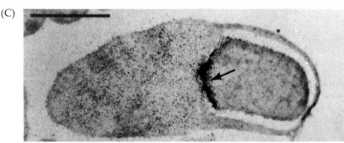

FIGURE 30
Transport of mRNA from nurse cells
into fly oocytes. (A–B) Autoradiographs of the follicle cell of the housefly, *Musca domestica*, after incubation
with [³H]cytidine. (A) Egg chamber
fixed immediately after label was introduced. The nuclei of the nurse cells
are heavily labeled, indicating that
they are synthesizing new RNA. The
oocyte remains unlabeled except
where some RNA is escaping into the
oocyte through the cytoplasmic connection between it and a nurse cell (arrow). (B) A similar egg chamber fixed
five hours later. Label is gone from the
nurse cell nuclei but has moved into
the cytoplasm. Moreover, radioactive
RNA can be seen passing into the oocyte cytoplasm through the two channels between the nurse cells and the
oocyte. (C) Autoradiograph of *Drosophila* egg chamber (arrow) stained by a
radioactive probe to the *bicoid* mRNA.
This message is transported from the
nurse cells (left) and remains in the
anteriormost portion of the oocyte
(smaller, darker oval). (A–B from Bier,
1963; courtesy of D. Ribbert; C from
Stephanson, et al., 1988; courtesy of
E. C. Stephanson.)

are capable of developing into oocytes, and of these two, only one prevails.
The other begins meiosis but does not complete it. Thus, only one of the
16 cystocytes can become an ovum. All the other cells become NURSE
CELLS. As it turns out, the cell destined to become the oocyte is that cell
residing at the most posterior tip of the egg chamber that encloses the 16-
cell clone.

The oocytes of meroistic insects do not pass through a transcriptionally
active stage, nor do they have lampbrush chromosomes. Rather, autoradiographic evidence shows that RNA synthesis is largely confined to the
nurse cells and that the RNA made by these cells is actively transported
into the oocyte cytoplasm. This can be seen in Figure 30. When the egg
chambers of a housefly are incubated in radioactive cytidine, the nuclei of
the nurse cells show intense labeling. When the labeling is stopped and
the cells are incubated for 5 more hours in nonradioactive media, the
labeled RNA is seen to enter the oocyte from the nurse cells (Bier, 1963).
Oogenesis takes place in only 12 days, so the nurse cells are very metabolically active during this time. They are aided in their transcriptional
efficiency by becoming polytene. Instead of having two copies of each
chromosome, they replicate their chromosomes until they have produced
512 copies. The 15 nurse cells are known to pass both ribosomal and
messenger RNAs into the oocyte cytoplasm, and entire ribosomes may be
transported as well. The mRNAs do not associate with polysomes, which
suggests that they are not immediately active in protein synthesis (Paglia
et al., 1976; Telfer et al., 1981).

The meroistic ovary confronts us with some interesting problems. If all the cells are connected so that proteins and RNAs shuttle freely between them, why should they have different developmental fates? Why should one cell become the oocyte while the others become "RNA synthesizing factories"? Why is the flow of protein and RNA in one direction only? One answer to these problems is that within the egg chamber an electrical gradient exists and prevents the diffusion of certain molecules while facilitating the diffusion of others. Woodruff and Telfer (1980; Woodruff et al., 1988b) have shown that such a gradient exists and that it inhibits the movement of most proteins while enhancing the directional movement of acidic macromolecules. This "in vivo electrophoresis" is shown in Figure 31. When an acidic enzyme is labeled with a fluorescent marker and injected into nurse cells, the protein is seen to cross from the nurse cell into the oocyte cytoplasm. However, when that same labeled protein is made basic by small chemical modifications, it will remain in the nurse cell into which it was injected. In this way, developmental cues are not exchanged between the connected cells, but certain proteins and nucleic acids can be transported in a unidirectional fashion.

The three major yolk proteins in *Drosophila* are made in the fat body and ovary, but not in the oocyte itself (Bownes, 1982; Brennen et al., 1982). The hormonal control of yolk synthesis is controlled by juvenile hormone, ecdysone, and a neurosecretory hormone from the brain. It is thought that the brain hormone, responding to environmental cues,* stimulates the corpora allata to secrete juvenile hormone (Figure 32). Juvenile hormone (1) regulates the uptake of the yolk peptides at the surface of the oocyte, (2) stimulates the synthesis of ovarian yolk proteins (which are each identical to those made by the fat body), and (3) causes the ovarian follicles and other abdominal cells to secrete ecdysone. Ecdysone is metabolized into its active form—20-hydroxyecdysone—and stimulates the fat body

*In *Drosophila*, the environmental cue appears to be the photoperiod. In the common mosquito, the cue is the blood meal. Only female mosquitos bite, and they make no vitellogenin before this meal. Some factor in the blood stimulates the mosquito's brain to release juvenile hormone and the corpus cardiacum stimulating factor. The latter factor causes the release of the egg development neurosecretory hormone (EDNH). EDNH stimulates the ovary to secrete ecdysone, which, in concert with juvenile hormone, stimulates the fat body to synthesize vitellogenin (Hagedorn, 1983; Borovsk et al., 1990).

FIGURE 31
Directional movement of macromolecules in egg chamber cells. (A) Egg chamber of the moth *Hyalophora cecropia*, showing the oocyte and four adjacent nurse cells. The micrograph section cuts through the bridge linking one of the nurse cells to the oocyte. (B) Fluorescence photomicrograph of an early vitellogenic egg chamber fixed one hour after one of its nurse cells was injected with methylcarboxylated and fluorescein-labeled lysozyme. This modified protein crossed the intercellular bridge into the oocyte (arrow). (C) The same as (B), except that the basic form of the fluorescein-labeled lysozyme (nonmethylcarboxylated) was injected. No transport of the protein into the oocyte is seen. (From Telfer et al., 1981; photographs courtesy of W. H. Telfer.)

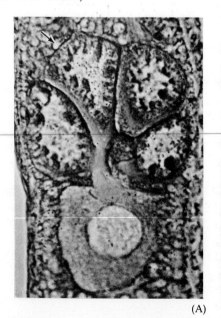
(A)

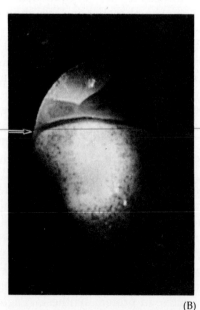
(B)

(C)

to produce yolk proteins just as estradiol stimulated the amphibian liver to do so. Similarly, the administration of ecdysone to adult males will cause their fat bodies to secrete yolk proteins as well (Postlethwait et al., 1980). Although yolk is centrally located in *Drosophila* eggs, polarity exists in these eggs, as well. As discussed earlier (Chapters 7 and 18), the anterior–posterior axis of the embryo is formed by maternal products placed into the oocyte cytoplasm. The *bicoid* gene, for example, has to be expressed in the mother if her offspring are to have heads. (The offspring of mothers lacking this gene have tails at both ends.) The *bicoid* gene is transcribed during oogenesis by the nurse cells and its message accumulates at the anterior end of the oocyte (Figure 30C; Frigerio et al., 1986; Stephanson et al., 1988). Leakage of anterior cytoplasm from wild-type eggs leads to embryos with reduced head structures.

Since there is such polarity in *Drosophila* eggs, the next question concerns how this polarity is maintained. What keeps the *bicoid* message from diffusing throughout the oocyte? What localizes polar granule components to the posterior of the egg? The *cis* component of bicoid localization is present in the *bicoid* message. The *bicoid* mRNA appears to stick to the anterior components of the egg via its 3′ untranslated region. This end of the message undergoes base pairing within itself to generate a secondary structure characteristic of the message. If other mRNAs are given this sequence, they, too, localize to the anterior pole of the *Drosophila* egg (MacDonald, 1990).

Frohnhöfer and Nüsslein-Volhard (1987) and Manseau and Schübach (1989) have shown that two other maternal genes—*swallow* and *exuperantia*—are necessary to keep *bicoid* mRNA in the anterior pole and to keep polar granule components in the posterior. These genes provide the *trans* components of bicoid localization. In eggs produced by mothers with either of these mutations, *bicoid* message extends into more posterior positions in the egg, and the *vasa* gene product, a polar granule component, is not localized to the posterior pole. The *Drosophila* ovary must not only make a cytoplasmically polarized product but must also synthesize some of the oocyte components that provide this polarity.

Oogenesis in humans

The maturation and ovulation of the mammalian egg follows one of two basic patterns, depending upon the species. One type of ovulation is stimulated by the physical act of intercourse itself. Physical stimulation of the cervix triggers the release of gonadotropins from the pituitary. These gonadotropins signal the egg to resume meiosis and initiate the events expelling the ovum from the ovary. This method ensures that most copulations lead to fertilized ova, and animals that utilize this method of ovulation—rabbits and minks—have a reputation for procreative success.

Most mammals, however, have a periodic type of ovulation. The female ovulates only at specific times of the year, called ESTRUS (or its English equivalent, "heat"). In these cases, environmental cues, most notably the amount and type of light during the day, stimulate the hypothalamus to release gonadotropin-releasing factor. This factor stimulates the pituitary to release its gonadotropins—follicle stimulating hormone (FSH) and luteinizing hormone (LH)—which cause the follicle cells to proliferate and to secrete estrogen. The estrogen subsequently binds to certain neurons and evokes the pattern of mating behavior characteristic of the species. Gonadotropins also stimulate follicular growth and the initiation of ovulation. Thus, estrus and ovulation occur close together.

Humans have a variation on the theme of periodic ovulation. Although

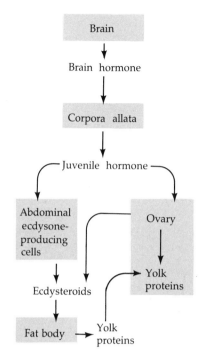

FIGURE 32

Model for the hormonal regulation of yolk peptide synthesis in *Drosophila melanogaster*. In response to a brain hormone, the corpora allata produce juvenile hormone (JH), which causes the ovary to make yolk proteins and ecdysteroids. JH also induces ecdysteroid synthesis in abdominal cells. These ecdysteroids cause the fat body to produce yolk proteins that are transported to the ovary. (After Bownes, 1982.)

human females have a cyclic ovulation (averaging about 29.5 days) and no definitive yearly estrus, most of human reproductive physiology is shared with other primates. The characteristic primate periodicity in maturing and releasing ova is called the MENSTRUAL CYCLE because it entails the periodic shedding of blood and cellular debris from the uterus. The menstrual cycle represents the integration of three very different activities: (1) the ovarian cycle, the function of which is to mature and release an oocyte; (2) the uterine cycle, the function of which is to provide the appropriate environment for the developing blastocyst to implant; and (3) the cervical cycle, the function of which is to allow sperm to enter the female reproductive tract only at the appropriate time. These three functions are integrated through the hormones of the pituitary, hypothalamus, and ovary.

The majority of the oocytes within the adult human ovary are arrested in the prolonged diplotene stage of the first meiotic prophase (often referred to as the DICTYATE state). Each oocyte is enveloped by a PRIMORDIAL FOLLICLE consisting of a single layer of epithelial GRANULOSA CELLS and a less-organized layer of mesenchymal THECAL CELLS (Figure 33). Periodically a group of primordial follicles enter a stage of follicular growth. During this time, the oocyte undergoes a 500-fold increase in volume (corresponding to an increase in oocyte diameter from 10 μm in a primor-

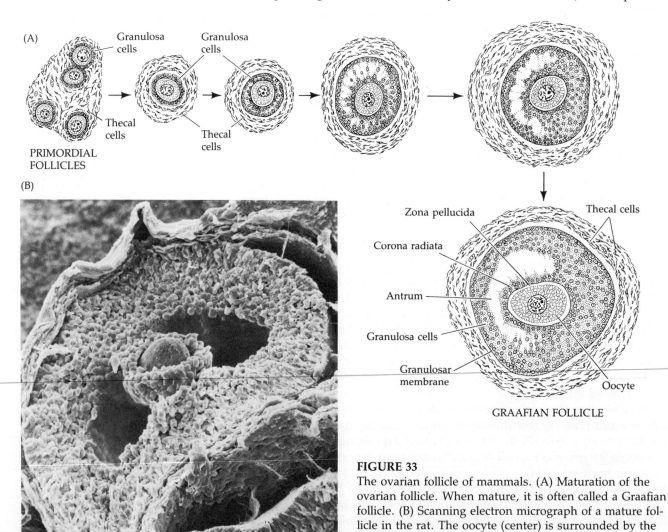

FIGURE 33
The ovarian follicle of mammals. (A) Maturation of the ovarian follicle. When mature, it is often called a Graafian follicle. (B) Scanning electron micrograph of a mature follicle in the rat. The oocyte (center) is surrounded by the smaller granulosa cells that will make the corona. (A after Carlson, 1981; B courtesy of P. Bagavandoss.)

dial follicle to 80 μm in a fully developed follicle). Concomitant with oocyte growth is an increase in the numbers of follicular granulosa cells, which form concentric layers about the oocyte. Throughout this growth period, the oocyte remains arrested in the dictyate stage. The fully grown follicle thus contains a large oocyte surrounded by several layers of granulosa cells. Many of these cells will stay with the ovulated egg, forming the CUMULUS, which surrounds the egg in the oviduct. In addition, during the growth of the follicle, an ANTRUM (cavity) forms, which becomes filled with a complex mixture of proteins, hormones, cyclic AMP, and other molecules. At any given time, a small group of follicles will be maturing. However, after progressing to a more mature stage, most oocytes and their follicles will die. In order to survive, the follicle must find a source of gonadotropic hormones and, "catching the wave" at the right time, must ride it until it peaks. Thus, for oocyte maturation to occur, the follicle needs to be at a certain stage of development when the waves of gonadotropin arise.

Day 1 of the menstrual cycle is considered to be the first day of "bleeding" (Figure 34). This bleeding from the vagina represents the sloughing off of extrauterine tissue and blood vessels that would have aided the implantation of the blastocyst. In the first part of the cycle (called the PROLIFERATIVE or FOLLICULAR phase), the pituitary gland starts secreting increasingly large amounts of FSH. The group of maturing follicles, which have already undergone some development, respond to this hormone by further growth and cellular proliferation. FSH also induces the formation of LH receptors on the granulosa cells. Shortly after this period of initial follicle growth, the pituitary begins secreting LH. In response to LH, the meiotic arrest is broken. The nuclear membranes of competent oocytes break down, and the chromosomes assemble to undergo the first meiotic division. One set of chromosomes is kept inside the oocyte and the other is given to the small polar body. Both are encased by the zona pellucida, which has been synthesized by the growing oocyte. It is in this stage that the egg will be ovulated.

The two gonadotropins, acting together, cause the follicle cells to produce increasing amounts of estrogen, which has at least five major activities in regulating the further progression of the menstrual cycle.

1 It causes the uterine mucosa to begin its proliferation and to become enriched with blood vessels.
2 It causes the cervical mucus to thin, thereby permitting sperm to enter the inner portions of the reproductive tract.
3 It causes an increase in the number of FSH receptors on the follicular granulosa cells (Kammerman and Ross, 1975) while causing the pituitary to lower its FSH production. It also stimulates the granulosa cells to secrete the peptide hormone inhibin, which also suppresses pituitary FSH secretion (Rivier et al., 1986; Woodruff et al., 1988a).
4 At low concentrations, it inhibits LH production, but at high concentrations, it stimulates it.
5 At very high concentrations and over long durations, estrogen interacts with the hypothalamus, causing it to secrete gonadotropin-releasing factor.

Thus, as estrogen levels increase as a result of follicular production, FSH levels decline. LH levels, however, continue to rise as more estrogen is secreted. As estrogens continue to be made (days 7 to 10), the granulosa cells continue to grow. Starting at day 10, estrogen secretion rises sharply.

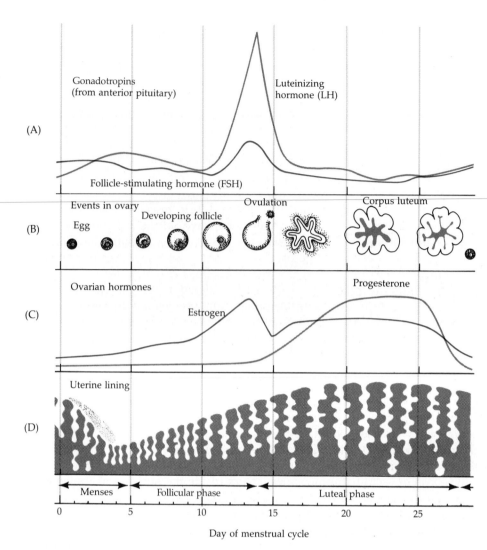

(A)

Gonadotropins
(from anterior pituitary)

Luteinizing
hormone (LH)

Follicle-stimulating hormone (FSH)

(B)

Events in ovary   Ovulation   Corpus luteum

Egg   Developing follicle

(C)

Ovarian hormones   Progesterone

Estrogen

(D)

Uterine lining

Menses   Follicular phase   Luteal phase

0   5   10   15   20   25

Day of menstrual cycle

FIGURE 34

The human menstrual cycle. The coordination of ovarian (B) and uterine (D) cycles is controlled by the pituitary (A) and ovarian (C) hormones. During the follicular phase, the egg matures within the follicle and the uterine lining is prepared to receive the embryo. The mature egg is released around day 14. If an embryo does not implant into the uterus, the uterine wall will begin to break down, leading to menses.

This rise is followed at midcycle by an enormous surge of LH and a smaller burst of FSH. Experiments with female monkeys have shown that exposure of the hypothalamus to greater than 200 pg of estrogen per milliliter of blood for more than 50 hours results in the hypothalamic secretion of gonadotropin-releasing factor. This factor causes the subsequent release of FSH and LH from the pituitary. Within 10–12 hours after the gonadotropin peak, the egg is ovulated (Figure 35; Garcia et al., 1981). Although the detailed mechanism of ovulation is not yet known, the physical expulsion of the mature oocyte from the follicle appears to be due to an LH-induced increase in collagenase, plasminogen activator, and PROSTAGLANDIN within the follicle (Lemaire et al., 1973). The mRNA for plasminogen activator has been dormant in the oocyte cytoplasm. LH causes this message to be polyadenylated and translated into this powerful protease (Huarte et al., 1987). Prostaglandins may cause localized contractions in the smooth muscles in the ovary and may also increase the flow of water

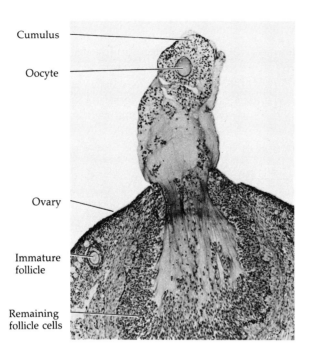

Cumulus

Oocyte

Ovary

Immature
follicle

Remaining
follicle cells

FIGURE 35
Ovulation in the rabbit. The ovary of a living, anesthetized rabbit was exposed and observed. When the follicle started to ovulate, the ovary was removed, fixed, and stained. (Photograph courtesy of R. J. Blandau.)

from the ovarian capillaries (Diaz-Infante et al., 1974; Koos and Clark, 1982). If ovarian prostaglandin synthesis is inhibited, ovulation does not take place. In addition to the prostaglandin-induced pressure, collagenases and the plasminogen activator protease loosen and digest the extracellular matrix of the follicle (Beers et al., 1975; Downs and Longo, 1983). The result of the luteinizing hormone, then, would be increased follicular pressure coupled with the degradation of the follicle wall. A hole would be digested through which the ovum could burst.

Following ovulation, the LUTEAL PHASE of the menstrual cycle begins. The remaining cells of the ruptured follicle under the continued influence of LH become the CORPUS LUTEUM. (They are able to respond to this LH because the surge in FSH stimulates them to develop even more LH receptors.) The corpus luteum secretes some estrogen, but its predominant secretion is PROGESTERONE. This steroid hormone circulates to the uterus, where it completes the job of preparing the uterine tissue for blastocyst implantation, stimulating the growth of the uterine wall and its blood vessels. Blocking the progesterone receptor with the synthetic steroid mifepristone (RU 486) stops the uterine wall from thickening and prevents the implantation of the blastocyst into the uterus* (Couzinet et al., 1986). Progesterone also inhibits the production of FSH, thereby preventing the maturation of any more follicles and ova. (For this reason such a combination of estrogen and progesterone has been used in birth control pills. The growth and maturation of new ova are prevented so long as FSH is inhibited.)

*RU 486 is thought to compete for the progesterone receptor inside the nucleus. In the absence of progesterone, the DNA binding site of the receptor is "capped" by a protein called hsp90. The binding of progesterone to the progesterone receptor causes a change in the shape of the receptor, releasing the hsp90 and allowing the hormone-bound receptor to bind to hormone-responsive elements in the gene promoters and enhancers. RU 486 can bind to the progesterone site in the receptor, but it does not cause the conformational change that releases hsp90. Instead, it prevents progesterone from entering this site and causing this change. The result is that the progesterone receptor cannot bind DNA and progesterone cannot cause its physiological effects (Baulieu, 1989). In Europe, RU 486 has become a widely used alternative to surgical abortion (Palka, 1989).

If the ovum is not fertilized, the corpus luteum degenerates, progesterone secretion ceases, and the uterine wall is sloughed off. With the decline in serum progesterone, the pituitary secretes FSH again, and the cycle is renewed. However, if fertilization occurs, the trophoblast will secrete a new hormone, LUTEOTROPIN, which causes the corpus luteum to remain active and serum progesterone levels to remain high. Thus, the menstrual cycle enables the periodic maturation and ovulation of human ova and allows the uterus to periodically develop into an organ capable of nurturing a developing organism for nine months.

SIDELIGHTS & SPECULATIONS

The maintenance and breaking of meiotic arrest in mammalian oocytes

If numerous follicles are capable of maturing when follicle-stimulating hormone (FSH) is secreted, how is it that usually only one follicle and its oocyte prevail? It appears that the follicle capable of producing the most estrogen in response to FSH is the one that matures, while all the others die. Those sets of follicles initially receiving FSH not only begin to proliferate, they also produce new luteinizing hormone receptors on their thecal cells (Figure 36). The reception of luteinizing hormone causes these thecal cells to initiate estrogen production. As we have seen, estrogen has two disparate effects involving the future reception of FSH. At one level, it turns down the pituitary secretion of FSH, while at another level, it increases the FSH receptors on the follicle cells. Thus, the more estrogen a follicle produces, the more FSH receptors it has while the less FSH remains in circulation. As FSH concentrations get progres-

sively lower, only one follicle can bind the available FSH. Only this follicle can still grow, and the other follicles die.

What does LH do that causes the breaking of meiotic arrest? To address this question, the nature of the meiotic arrest has been intensely studied. Like amphibian oocytes, the arrested (dictyate) stage is extremely important, because that is the time during which oocytes grow, differentiate the structures specific to oocytes, and acquire the ability to resume meiosis (Sorensen and Wassarman, 1976). Early experiments demonstrated that follicle-enclosed oocytes, in vivo or in vitro, do not undergo maturation unless exposed to gonadotropins, whereas oocytes removed from the follicle will spontaneously resume meiosis even without the hormonal stimulus (Pincus and Enzmann, 1935).

It appears, then, that meiosis is normally inhibited by the follicle cells and that it can be reinitiated by gonadotropins. This hypothesis—that the follicle cells are important regulators of meiosis—is strengthened by observations that granulosa cells communicate with the oocyte through processes extending to the oocyte through the zona. These processes have gap junctions that enable small molecules to pass between the oocyte and the granulosa cells of the follicle (Figure 37; Anderson and Albertini, 1976; Gilula et al., 1978)

Because the elevation of cAMP levels inhibits oocyte maturation (Cho et al., 1974), it has been proposed that the meiotic arrest is maintained by the transfer of cAMP through the gap junctions from the follicular granulosa cell to the oocyte (Dekel and Beers, 1978, 1980). The luteinizing hormone surge could trigger maturation by terminating the gap junction communication, thereby inhibiting the transfer of cAMP into the oocyte. Several lines of evidence now support this hypothesis. First, the decline of cAMP appears critical for the resumption of meiosis. Germinal vesicle breakdown can be prevented by inhibiting the degradation of cAMP in follicle-free eggs or by providing such eggs directly with cAMP (Bornslaeger et al., 1986). The decline in oocyte cAMP concentration occurs immediately prior to the resumption of meiosis (Schultz et al., 1983). While the mechanism by which this decline in cAMP causes the resumption of meiosis is not known, it is thought that the falling cAMP level causes a decline in the cAMP-dependent protein kinase enzyme. This protein kinase would be responsible for phosphorylating particular proteins that, in

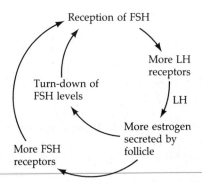

FIGURE 36
Positive feedback cycle in mammalian follicle cells. Reception of follicle stimulating hormone (FSH) leads to the production of more luteinizing hormone (LH) receptors. The follicle cells secrete estrogen when stimulated by LH; the estrogen causes both an increase in the number of FSH receptors and a decrease in the amount of pituitary FSH production. Eventually, very few follicles are able to receive the small amounts of FSH produced, thereby amplifying their ability to receive LH. These few follicles are able to mature.

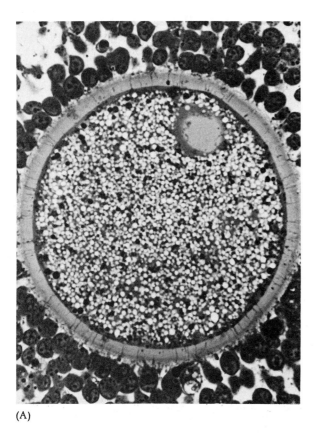

(A)

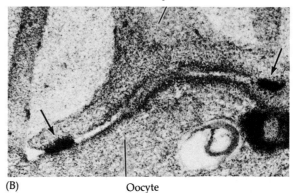

(B)   Oocyte

FIGURE 37
Communication between oocyte and granulosa cells. (A) Sheep oocyte surrounded by the zona pellucida and follicle cells. The granulosa cells of the follicle are extending processes through the zona pellucida and touching the oocyte. (B) Electron micrograph of follicle cell processes establishing gap junction connections with a rhesus monkey oocyte. Gap junctions (arrows) are stained with ionic lanthanum. (A from Moor and Cran, 1980; photograph courtesy of the authors. B from Anderson and Albertini, 1976; photograph courtesy of D. Albertini.)

their phosphorylated states, would maintain meiotic arrest. If inhibitors of this protein kinase are injected into the mouse oocyte, meiosis can resume, even in the presence of cAMP (Figure 38; Bornslaeger et al., 1986).

Second, the gonadotropins can cause the loss of com-

munication between the follicle cells and the oocyte. The follicle cells appear to be important sources of oocyte cAMP, and changes of cAMP concentration in the follicle cells are reflected in oocyte cAMP levels (Racowsky, 1985; Bornslaeger and Schultz, 1985). This observation explains why oocytes remain in meiotic arrest when they are surrounded by the follicle cells but resume meiosis when they are removed from them.

The gonadotropin surge, however, can elevate the follicle cell cAMP concentrations to new levels. In response to this elevation, the mature follicle cells synthesize hyaluronic acid, which causes a physical disruption of the contact between the follicle cell processes and the oocyte (Eppig, 1979; Larsen et al., 1986). The bridges through which cAMP flows from follicular granulosa cell to oocyte have thereby been removed, allowing the mammalian oocyte to resume meiosis (Dekel and Sherizly, 1985; Racowsky and Satterlie, 1985).

Like amphibian oocytes, the ovulated mouse oocyte is

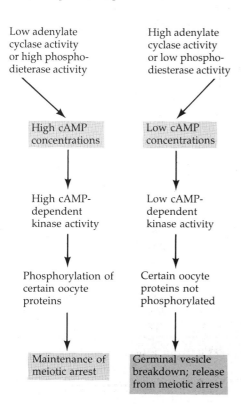

Low adenylate cyclase activity or high phospho-diesterase activity	High adenylate cyclase activity or low phospho-diesterase activity
High cAMP concentrations	Low cAMP concentrations
High cAMP-dependent kinase activity	Low cAMP-dependent kinase activity
Phosphorylation of certain oocyte proteins	Certain oocyte proteins not phosphorylated
Maintenance of meiotic arrest	Germinal vesicle breakdown; release from meiotic arrest

FIGURE 38
Summary of proposed mechanism whereby oocyte cyclic AMP level regulates resumption of meiosis by the oocyte. The cAMP levels within the oocyte are provided, at least in part, by cAMP from the follicle cells. Cyclic AMP cannot cross cell membranes, but it can enter the oocyte through gap junctions connecting the oocyte with its follicle cells. When the connections are released, oocyte cAMP levels decline, leading to the release of meiotic arrest.

arrested in the second meiotic metaphase and is fertilized in that state. Paules and his co-workers (1989) have shown that maturing mouse oocytes also contain the pp39^{mos} cytostatic factor. If the translation of the c-mos gene that makes this factor is specifically interrupted, the oocyte does not release its first polar body and fails to undergo normal maturation. Evidently, similar events must take place for the maturation of the amphibian and mammalian oocytes.

The egg and sperm will both die if they do not meet. We are now back where we began—the stage is set for fertilization to take place. As F. R. Lillie recognized in 1919, "The elements that unite are single cells, each on the point of death; but by their union a rejuvenated individual is formed, which constitutes a link in the eternal process of Life."

LITERATURE CITED

Anderson, E. and Albertini, D. F. 1976. Gap junctions between the oocyte and companion follicle cells in the mammalian ovary. *J. Cell Biol.* 71: 680–686.

Auerbach, R. and Joseph, J. 1984. Cell surface markers on endothelial cells: A developmental perspective. *In* E. A. Jaffe (ed.), *The Biology of Endothelial Cells.* Nijhoff, The Hague.

Auerbach, R., Lu, W. C., Pardon, E., Gumkowski, F., Kaminska, G. and Kaminska, M. 1987. Specificity of adhesion between murine tumor cells and capillary endothelium: an *in vitro* correlate of preferential metastasis *in vivo. Cancer Res.* 47: 1492–1496.

Austin, J. and Kimble, J. 1987. glp-1 is required in the germ line for regulation of the decision between mitosis and meiosis in *C. elegans. Cell* 51: 589–599.

Baker, T. G. 1970. Primordial germ cells. *In* C. R. Austin and R. V. Short (eds.), *Reproduction in Mammals,* Vol. 1: *Germ Cells and Fertilization.* Cambridge University Press, Cambridge, pp. 1–13.

Baulieu, E.-E. 1989. Contragestion and other clinical applications of RU 486, an antiprogesterone at the receptor. *Science* 245: 1351–1357.

Beers, W. H., Strickland, S. and Reich, E. 1975. Ovarian plasminogen activator: Relationship to ovulation and hormonal regulation. *Cell* 6: 387–394.

Bier, K. 1963. Autoradiographische Untersuchungen über die Leistungen des Follikelepithels und der Nährzellen bei der Dottbildung und Eiweisssynthese im Fliegenova. *Wilhelm Roux Arch. Entwicklungsmech Org.* 154: 552–575.

Blanco, A. 1980. On the functional significance of LDH-X. *Johns Hopkins Med. J.* 146: 231–235.

Bloom, W. and Fawcett, D. W. 1975. *Textbook of Histology,* 10th Ed. Saunders, Philadelphia.

Bornslaeger, E. A. and Schultz, R. M. 1985. Regulation of mouse oocyte maturation: Effect of elevating cumulus cell cAMP on oocyte cAMP levels. *Biol. Reprod.* 33: 698–704.

Bornslaeger, E. A., Mattei, P. and Schultz, R. M. 1986. Involvement of cAMP-dependent protein kinase and protein phosphorylation in regulation of mouse oocyte maturation. *Dev. Biol.* 114: 453–462.

Borovsk, D., Carlson, D. A., Griffin, P. R., Shabanowitz, J. and Hunt, D. F. 1990. Mosquito oostatic factor: A novel decapeptide modulating trypsin-like enzyme biosynthesis in the midgut. *FASEB J.* 4: 3015–3020.

Bounoure, L. 1934. Recherches sur lignée germinale chez la grenouille rousse aux premiers stades au développement. *Ann. Sci. Zool. Ser.* 17, 10: 67–248.

Bownes, M. 1982. Hormonal and genetic regulation of vitellogenesis in *Drosophila. Q. Rev. Biol.* 57: 247–274.

Braun, R. E., Behringer, R. R., Peschon, J. J., Brinster, R. L. and Palmiter, R. D. 1989a. Genetically haploid spermatids are phenotypically diploid. *Nature* 337: 373–376.

Braun, R. E., Peschon, J. J., Behringer, R. R., Brinster, R. L., and Palmiter, R. D. 1989b. Protamine 3′ untranslated sequences regulate temporal translational control and subcellular localization of growth hormone in spermatids of transgenic mice. *Genes Dev.* 3: 793–802.

Brennen, M. D., Weiner, A. J., Goralski, T. J. and Mahowald, A. P. 1982. The follicle cells are a major site of vitellogenin synthesis in *Drosophila melanogaster. Dev. Biol.* 89: 225–236.

Brock, M. L. and Shapiro, D. J. 1983. Estrogen stabilizes vitellogenin mRNA against cytoplasmic degradation. *Cell* 34: 207–214.

Browder, L. 1980. *Developmental Biology.* Saunders, Philadelphia.

Burns, R. K., Jr. 1930. The process of sex-transformation parabiotic *Amblystoma.* I. Transformation from female to male. *J. Exp. Zool.* 55: 123–169.

Chiquoine, A. D. 1954. The identification, origin, and migration of the primordial germ cells in the mouse embryo. *Anat. Rec.* 118: 135–146.

Cho, W. K., Stern, S. and Biggers, J. D. 1974. Inhibitory effect of dibutyryl cAMP on mouse oocyte maturation in vitro. *J. Exp. Zool.* 187: 383–386.

Comings, D. E. 1968. The rationale for an ordered arrangement of chromatin in the interphase nucleus. *Am. J. Hum. Genet.* 20: 440–460.

Couzinet, B., Le Strat, N. Ulmann, A., Baulieu, E. E., and Schaison, G. 1986. Termination of early pregnancy by the progesterone antagonist RU486 (Mifepristone). *N. Eng. J. Med.* 315: 1565–1570.

Danilchik, M. V. and Gerhart, J. C. 1987. Differentiation of the animal–vegetal axis in *Xenopus laevis* oocytes: Polarized intracellular translocation of platelets establishes the yolk gradient. *Dev. Biol.* 122: 101–112.

Davidson, E. 1986. *Gene Activity in Early Development,* 3rd Ed. Academic Press, Orlando, FL.

Dekel, N. and Beers, W. H. 1978. Rat oocyte maturation in vitro: Relief of cyclic cAMP inhibition by gonadotropins. *Proc. Natl. Acad. Sci. USA* 75: 4369–4373.

Dekel, N. and Beers, W. H. 1980. Development of the rat oocyte in vitro: Inhibition and induction of maturation in the presence or absence of the cumulus oophorus. *Dev. Biol.* 75: 247–254.

Dekel, N. and Sherizly, I. 1985. Epidermal growth factor induces maturation of rat follicle-enclosed oocytes. *Endocrinology* 116: 406–409.

Diaz-Infante, A., Wright, K. H. and Wallach, E. E. 1974. Effects of indomethacin and PGF$_{2\alpha}$ on ovulation and ovarian contraction in the rabbit. *Prostaglandins* 5: 567–581.

Dolecki, G. J. and Smith, L. D. 1979. Poly(A) RNA metabolism during oogenesis in *Xenopus laevis. Dev. Biol.* 69: 217–236.

Doniach, T. and Hodgkin, J. 1984. A sex-determining gene, *fem-1,* required for both male and hermaphroditic development in *C. elegans. Dev. Biol.* 106: 223–235.

Downs, S. and Longo, F. J. 1983. Prostaglandins and preovulatory follicular maturation in mice. *J. Exp. Zool.* 228: 99–108.

Dubois, R. 1969. Le mécanisme d'entrée des cellules germinales primordiales dans le réseau vasculaire, chez l'embryon de poulet. *J. Embryol. Exp. Morphol.* 21: 255–270.

Dumont, J. N. 1978. Oogenesis in *Xenopus laevis*. VI. Route of injected tracer transport in follicle and developing oocyte. *J. Exp. Zool.* 204: 193–200.

Dym, M. 1977. The male reproductive system. *In* L. Weiss and R. O. Greep (eds.), *Histology*, 4th Ed. McGraw-Hill, New York, pp. 979–1038.

Dym, M. and Fawcett, D. W. 1971. Further observations on the number of spermatogonia, spermatocytes, and spermatids connected by intercellular bridges in the mammalian testis. *Biol. Reprod.* 4: 195–215.

Eppig, J. J. 1979. FSH stimulates hyaluronic acid synthesis by oocyte–cumulus cell complexes from mouse preovulatory follicles. *Nature* 281: 483–484.

Eyal-Giladi, H., Ginsburg, M. and Farbarou, A. 1981. Avian primordial germ cells are of epiblastic origin. *J. Embryol. Exp. Morphol.* 65: 139–147.

Flickinger, R. A. and Rounds, D. E. 1956. The maternal synthesis of egg yolk proteins as demonstrated by isotopic and serological means. *Biochim. Biophys. Acta* 22: 38–72.

Fox, C. A., Sheets, M. D. and Wickens, M. P. 1989. Poly(A) addition during maturation of frog oocytes: distinct nuclear and cytoplasmic activities and regulation by the sequence UUUUUAU. *Genes Dev.* 3: 2151–2162.

Frigerio, D., Burri, M., Bopp, D., Baumgartner, S. and Noll, M. 1986. Structure of the segmentation gene *paired* and the *Drosophila* PRD gene set as part of a gene network. *Cell* 47: 735–746.

Frohnhöfer, H. G. and Nüsslein-Volhard, C. 1987. Maternal genes required for the anterior localization of *bicoid* activity in the embryo of *Drosophila*. *Genes Dev.* 1: 880–890.

Gallatin, W. M., Weissman, I. L. and Butcher, E. C. 1983. A cell surface molecule involved in organ-specific homing of lymphocytes. *Nature* 304: 30–35.

Gallatin, W. M., St. John, T. P., Siegelman, M., Reichert, R., Butcher, E. C. and Weissman, I. L. 1986. Lymphocyte homing receptors. *Cell* 44: 673–680.

Gilula, N. B., Epstein, M. L. and Beers, W. H. 1978. Cell-to-cell communication and ovulation. A study of the cumulus–oocyte complex. *J. Cell Biol.* 78: 58–75.

Ginsburg, M. and Eyal-Giladi, H. 1987. Primordial germ cells of the young chick blastoderm originate from the central zone of the area pellucida irrespective of the embryo-forming process. *Development* 101: 209–219.

Ginsburg, M., Snow, M. H. L. and McLaren, A. 1990. Primordial germ cells in the mouse embryo during gastrulation. *Development* 110: 521–528.

Godin, I., Wylie, C. and Heasman, J. 1990. Genital ridges exert long-range effects on primordial germ cell numbers and direction of migration in culture. *Development* 108: 357–363.

Grant, P. 1953. Phosphate metabolism during oogenesis in *Rana temporaria*. *J. Exp. Zool.* 124: 513–543.

Green, G. R. and Poccia, D. L. 1988. Interaction of sperm histone variants and linker DNA during spermiogenesis in the sea urchin. *Biochemistry* 27: 619–625.

Gurdon, J. B. 1976. *The Control of Gene Expression in Animal Development*. Harvard University Press, Cambridge, MA.

Hagedorn, H. H. 1983. The role of ecdysteroids in the adult insect. *In* G. Downer and H. Laufer (eds.), *Endocrinology of Insects*. Alan R. Liss, New York, pp. 241–304.

Hahnel, A. C. and Eddy, E. M. 1986. Cell surface markers of mouse primordial germ cells defined by two monoclonal antibodies. *Gamete Res.* 15: 25–34.

Heasman, J., Mohun, T. and Wylie, C. C. 1977. Studies on the locomotion of primordial germ cells from *Xenopus laevis* in vitro. *J. Embryol. Exp. Morphol.* 42: 149–162.

Heasman, J., Hynes, R. D., Swan, A. P., Thomas, V. and Wyle, C. C. 1981. Primordial germ cells of *Xenopus* embryos: The role of fibronectin in their adhesion during migration. *Cell* 27: 437–447.

Heath, J. K. 1978. Mammalian primordial germ cells. *Dev. Mammals* 3: 272–298.

Hess, D. 1973. Local structural variations in the Y chromosome of *Drosophila hydei* and their correlation to genetic activity. *Cold Spring Harbor Symp. Quant. Biol.* 38: 663–672.

Hill, D. P., Shakes, D. C., Wards, S. and Strome, S. 1989. A sperm-supplied product essential for initiation of normal embryogenesis in *Caenorhabditis elegans* is encoded by the paternal effect embryonic-lethal gene, *spe-11*. *Dev. Biol.* 136: 154–166.

Hirsh, D., Oppenheim, D. and Klass, M. 1976. Development of the reproductive system of *Caenorhabditis elegans*. *Dev. Biol.* 49: 200–219.

Hodgkin, J., Doniach, T. and Shen, M. 1985. The sex determination pathway in the nematode *Caenorhabditis elegans*: Variations on a theme. *Cold Spring Harbor Symp. Quant. Biol.* 50: 585–593.

Hotta, Y., Tabata, S. and Stern, H. 1984. Replicating and nicking of zygotene DNA sequences: control by a meiosis-specific protein. *Chromosoma* 90: 243–253.

Hough-Evans, B. R., Wold, B. J., Ernst, S. G., Britten, R. J. and Davidson, E. H. 1977. Appearance and persistence of maternal RNA sequence in sea urchin development. *Dev. Biol.* 60: 258–277.

Hoyle, H. D. and Raff, E. C. 1990. Two *Drosophila* β-tubulin isoforms are not functionally equivalent. *J. Cell Biol.* 111: 1009–1026.

Huarte, J., Belin, D., Vassalli, A., Strickland, S. and Vassalli, J.-D. 1987. Meiotic maturation of mouse oocytes triggers the translation and polyadenylation of dormant tissue-type plasminogen activator mRNA. *Genes Dev.* 1: 1201–1211.

Humphrey, R. R. 1931. Studies of sex reversal in *Amblystoma*. III. Transformation of the ovary of *A. tigrinum* into a functional testis through the influence of a testis resident in the same animal. *J. Exp. Zool.* 58: 333–365.

Hunt, T. 1989. Embryology: Under arrest in the cell cycle. *Nature* 342: 483–484.

Iatrou, K., Spira, A. and Dixon, G. H. 1978. Protamine messenger RNA: Evidence for early synthesis and accumulation during spermatogenesis in rainbow trout. *Dev. Biol.* 64: 82–98.

Ikenishi, K. and Kotani, M. 1979. Ultraviolet effects on presumptive and primordial germ cells (pPGCs) in *Xenopus laevis* after the cleavage stage. *Dev. Biol.* 69: 237–246.

Inoue, H. and Hiroyoshi, T. 1986. A maternal-effect sex-transformation mutant of the housefly, *Musca domestica* L. *Genetics* 112: 469–481.

Kammerman, S. and Ross, J. 1975. Increase in numbers of gonadotropin receptors on granulosa cells during follicle maturation. *J. Clin. Endocrinol.* 41: 546–550.

Karr, T. L. 1988. Persistence of the sperm tail during early embryonic development in *Drosophila* visualized using sperm-specific monoclonal antibodies. *J. Cell Biol.* 107: 173a.

Kimble, J. E. 1981. Strategies for control of pattern formation in *Caenorhabditis elegans*. *Philos. Trans. R. Soc. Lond.* [B] 295: 539–551.

Kimble, J. E. and White, J. G. 1981. Control of germ cell development in *Caenorhabditis elegans*. *Dev. Biol.* 81: 208–219.

Kimble, J., Barton, M. K., Schedl, T. B., Rosenquist, T. A. and Austin, J. 1986. Controls of postembryonic germ line development in *Caenorhabditis elegans*. *In* J. Gall (ed.), *Gametogenesis and the Early Embryo*. Alan R. Liss, New York, pp. 97–110.

Koch, E. A., Smith, P. A. and King, R. C. 1967. The division and differentiation of *Drosophila* cystocytes. *J. Morphol.* 121: 55–70.

Koos, R. D. and Clark, M. R. 1982. Production of 6-keto-prostaglandin $F_{1\alpha}$ by rat granulosa cells in vitro. *Endocrinology* 111: 1513–1518.

Langman, J. 1981. *Medical Embryology*, 4th Ed. Williams & Wilkins, Baltimore.

Larsen, W. J., Wert, S. E. and Brünner, G. D. 1986. A dramatic loss of cumulus cell gap junctions is correlated with germinal vesicle breakdown in rat oocytes. *Dev. Biol.* 113: 517–521.

Laskey, R. A. 1974. Biochemical processes in early development. *In* A. T. Bull, J. R. Lagnado, J. O. Thomas and K. F. Tipton (eds.), *Companion to Biochemistry*, Vol. 2. Longman, London, pp. 137–160.

Lemaire, W. J., Yang, N. S. T., Behram, H. H. and Marsh, J. M. 1973. Preovulatory changes in concentration of prostaglandin in rabbit graafian follicles. *Prostaglandins* 3: 367–376.

Lillie, F. R. 1919. *Problems of Fertilization.* University of Chicago Press, Chicago.

MacDonald, P. M. 1990. *bicoid* mRNA localization signal: Phylogenetic conservation of function and RNA secondary structure. *Development* 110: 161–171.

Manseau, L. J. and Schübach, T. 1989. *cappuccino* and *spire*: two unique maternal-effect loci required for both the anteroposterior and dorsoventral patterns of the *Drosophila* embryo. *Genes Dev.* 3: 1437–1452.

Marushige K. and Dixon, G. H. 1969. Developmental changes in chromosome composition and template activity during spermatogenesis in trout testis. *Dev. Biol.* 19: 397–414.

Masui, Y. 1974. A cytostatic factor in amphibian: Its extraction and partial characterization. *J. Exp. Zool.* 187: 141–147.

McLaren, A. 1983. Does the chromosomal sex of a mouse cell affect its development? *Symp. Brit. Soc. Dev. Biol.* 7: 225–227.

Mintz, B. 1957. Embryological development of primordial germ cells in the mouse: Influence of a new mutation. *W. J. Embryol. Exp. Morphol.* 5: 396–403.

Moens, P. B. 1969. The fine structure of meiotic chromosome polarization and pairing in *Locusta migratoria. Chromosoma* 28: 1–25.

Moens, P. B. 1974. Coincidence of modified crossover distribution with modified synaptonemal complexes. *In* R. F. Grell (ed.), *Mechanisms of Recombination.* Plenum, New York, pp. 377–383.

Moor, R. M. and Cran, D. G. 1980. Intercellular coupling of mammalian oocytes. *Dev. Mammals* 4: 3–38.

Moses, M. J. 1968. Synaptonemal complex. *Annu. Rev. Genet.* 2: 363–412.

Nakamache, M., Berg, E. L., Streeter, P. R. and Butcher, E. C. 1989. The mucosal vascular addressin is a tissue-specific endothelial cell adhesion molecule for circulating lymphocytes. *Nature* 337: 179–181.

Paglia, L. M., Berry, J. and Kastern, W. H. 1976. Messenger RNA synthesis, transport, and storage in silkmoth ovarian follicles. *Dev. Biol.* 51: 173–181.

Palka, J. 1989. The pill of choice? *Science* 245: 1319–1323.

Palmiter, R. D., Wilkie, T. M., Chen, H. Y. and Brinster, R. L. 1984. Transmission distortion and mosaicism in an unusual transgenic mouse pedigree. *Cell* 36: 869–877.

Pasteels, J. 1953. Contributions à l'étude du developpement des reptiles. I. Origine et migration des gonocytes chez deux Lacertiens. *Arch. Biol.* 64: 227–245.

Paules, R. S., Buccione, R., Moscel, R. C., Vande Woude, G. F. and Eppig, J. J. 1989. Mouse *mos* protoncogene product is present and functions during oogenesis. *Proc. Nat. Acad. Sci. USA* 86: 5395–5399.

Perlman, S. and Rosbash, M. 1978. Analysis of *Xenopus laevis* ovary and somatic cell polyadenylated RNA by molecular hybridization. *Dev. Biol.* 63: 197–212.

Pincus, G. and Enzmann, E.V. 1935. The comparative behavior of mammalian eggs in vivo and in vitro. I. The activation of ovarian eggs. *J. Exp. Med.* 62: 665–675.

Pinkerton, J. H. M., McKay, D. G., Adams, E. C. and Hertig, A. T. 1961. Development of the human ovary: A study using histochemical techniques. *Obstet. Gynecol.* 18: 152–181.

Poccia, D. 1986. Remodeling of nucleoproteins during gametogenesis, fertilization, and early development. *Int. Rev. Cytol.* 105: 1–65.

Postlethwait, J. H., Brownes, M. and Jowett, T. 1980. Sexual phenotype and vitellogenin synthesis in *Drosophila melanogaster. Dev. Biol.* 79: 379–387.

Racowsky, C. 1985. Effect of forskolin on the spontaneous maturation and cyclic AMP content of hamster and oocyte–cumulus complexes. *J. Exp. Zool.* 234: 87–96.

Racowsky, C. and Satterlie, R. A. 1985. Metabolic, fluorescent dye and electrical coupling between hamster oocytes and cumulus cells during meiotic maturation in vivo and in vitro. *Dev. Biol.* 108: 191–202.

Reynaud, G. 1969. Transfert de cellules germinales primordiales de dindon à l'embryon de poulet par injection intravasculaire. *J. Embryol. Exp. Morphol.* 21: 485–507.

Ressom, R. E. and Dixon, K. E. 1988. Relocation and reorganization of germ plasm in *Xenopus* embryos after fertilization. *Development* 103: 507–518.

Rivier, C., Rivier, J., and Vale, W. 1986. Inhibin-mediated feedback control of follicle-stimulating hormone secetion in the female rat. *Science* 234: 205–208.

Rogulska, T. 1969. Migration of chick primordial germ cells from the intracoelomically transplanted germinal crescent into the genital ridge. *Experientia* 25: 631–632.

Rogulska, T. Ozdzenski, W. and Komer, A. 1971. Behavior of mouse primordial germ cells in chick embryo. *J. Embryol. Exp. Morphol.* 25: 155–164.

Romanoff, A. L. 1960. *The Avian Embryo.* Macmillan, New York.

Sagata, N., Oskarsson, M., Copeland, T., Brumbaugh, J., and Vande Woude, G. F. 1988. Function of *c-mos* proto-oncogene product in meiotic maturation in *Xenopus* oocytes. *Nature* 335: 519–525.

Sagata, N., Watanabe, N., Vande Woude, G. F. and Ikawa, Y. 1989. The *c-mos* proto-oncogene product is a cytostatic factor responsible for meiotic arrest in vertebrate eggs. *Nature* 342: 512–518.

Schultz, R. M., Montgomery, R. R. and Belanoff, J. R. 1983. Regulation of mouse oocyte maturation: Implications of a decrease in oocyte cAMP and protein dephosphorylation in commitment to resume meiosis. *Dev. Biol.* 97: 264–273.

Sen, D. and Gilbert, W. 1988. Formation of parallel four-stranded complexes by guanine-rich motifs in DNA and its implications for meiosis. *Nature* 334: 364–366.

Simon, D. 1960. Contribution à l'étude de la circulation et du transport des gonocytes primaires dans les blastodermes d'oiseau cultivé in vitro. *Arch. Anat. Micr. Morph. Exp.* 49: 93–176.

Skipper, J. K. and Hamilton, T. H. 1977. Regulation by estrogen of the vitellogenin gene. *Proc. Natl. Acad. Sci. USA* 74: 2384–2388.

Smith, L. D. 1966. The role of a "germinal plasm" in the formation of primordial germ cells in *Rana pipiens. Dev. Biol.* 14: 330–347.

Sorensen, R. and Wassarman, P. M. 1976. Relationship between growth and meiotic maturation of the mouse oocyte. *Dev. Biol.* 50: 531–536.

Stephanson, E. C., Chao, Y.-C. and Fackenthal, J. D. 1988. Molecular analysis of the *swallow* gene of *Drosophila melanogaster. Genes Dev.* 2: 1655–1665.

Stott, D. and Wylie, C. C. 1986. Invasive behaviour of mouse primordial germ cells in vitro. *J. Cell Sci.* 86: 133–144.

Streeter, P. R., Berg, E. L., Rouse, B. T. N., Bargatze, R. F. and Butcher, E. C. 1988a. A tissue-specific endothelial cell molecule involved in lymphocyte homing. *Nature* 331: 41–46.

Streeter, P. R., Rouse, B. T. N. and Butcher, E. C. 1988b. Immunohistologic and functional characterization of a vascular addressin involved in lymphocyte homing into peripheral lymph nodes. *J. Cell Biol.* 107: 1853–1862.

Subtelny, S. and Penkala, J. E. 1984. Experimental evidence for a morphogenetic role in the emergence of primordial germ cells from the endoderm of *Rana pipiens. Differentiation* 26: 211–219.

Sutasurya, L. A. and Nieuwkoop, P. D. 1974. The induction of primordial germ cells in the urodeles. *Wilhelm Roux Arch. Entwicklungsmech. Org.* 175: 199–220.

Swanson, C. P., Merz, T. and Young, W. J. 1981. *Cytogenetics: The Chromosome in Division, Inheritance and Evolution.* Prentice-Hall, Englewood Cliffs, NJ.

Swift, C. H. 1914. Origin and early history of the primordial germ-cells in the chick. *Am. J. Anat.* 15: 483–516.

Telfer, W. H., Woodruff, R. I. and Huebner, E. 1981. Electrical polarity and cellular differentiation in meroistic ovaries. *Am. Zool.* 21: 675–686.

Vassalli, J.-D. and 8 others. 1989. Regulated polyadenylation controls mRNA translation during meiotic maturation of mouse oocytes. *Genes Dev.* 3: 2163–2171.

von Wettstein, D. 1971. The synaptonemal complex and four-strand crossing over. *Proc. Natl. Acad. Sci. USA* 68: 851–855.

von Wettstein, D. 1984. The synaptonemal complex and genetic segregation. *In* C. W. Evans and H. G. Dickinson (eds.), *Controlling Events in Meiosis.* Cambridge University Press, Cambridge, pp. 195–231.

Watanabe, N., Vande Woude, G. F., Ikawa, Y. and Sagata, N. 1989. Specific proteolysis of the *c-mos* proto-oncogene product by calpain on fertilization of *Xenopus* eggs. *Nature* 342: 505–517.

Woodruff, R. I. and Telfer, W. H. 1980. Electrophoresis of proteins in intercellular bridges. *Nature* 286: 84–86.

Woodruff, R. I., Kulp, J. H., and LaGaccia, E. D. 1988. Electrically mediated protein movement in *Drosophila* follicles. *Roux Arch. Dev. Biol.* 197: 231–238.

Woodruff, T. K., D'Agostino, J., Schwartz, N. B., and Mayo, K. E. 1988. Dynamic changes in inhibin messenger RNAs in rat ovarian follicles during the reproductive cycle. *Science* 239: 1296–1299.

Wylie, C. C., Heasman, J., Swan, A. P. and Anderton, B. H. 1979. Evidence for substrate guidance of primordial germ cells. *Exp. Cell Res.* 121: 315–324.

Yisaeli, J. K., Sokol, S. and Melton, D. A. 1990. A two-step model for the localization of a maternal mRNA in *Xenopus* oocytes: Involvement of microtubules and microfilaments in translocation and anchoring of Vg1 mRNA. *Development* 108: 289–298.

Züst, B. and Dixon, K. E. 1977. Events in the germ cell lineage after entry of the primordial germ cell into the genital ridges in normal and UV-irradiated *Xenopus laevis. J. Embryol. Exp. Morphol.* 41: 33–46.

23

Developmental mechanisms of evolutionary change

How does newness come into the world? How is it born? Of what fusions, translations, conjoinings is it made? How does it survive, extreme and dangerous as it is? What compromises, what deals, what betrayals of its secret nature must it make to stave off the wrecking crew, the exterminating angel, the guillotine?

—SALMAN RUSHDIE (1988)

The first Bird was hatched from a Reptile's egg.

—WALTER GARSTANG (1922)

"Unity of Type and Conditions of Existence"

Charles Darwin was the heir to centuries of speculation concerning the origins of the diversity of animal life. Darwin's own education was steeped in the British tradition of natural theology that held that God's wisdom, omnipotence, and benevolence could be seen in the works of His creation. The dominant part of this tradition was an account of Creation proclaiming that the species were intricately designed works of the Creator. The fingers of the human hand were seen as exquisitely (some said perfectly) designed contrivances that allowed humans mastery of their environment. The shovel-like claw of the mole was, again, perfectly adapted for its "office of existence," as were the wings of birds and the fins of fish. A more sophisticated form of natural theology, championed in Britain by anatomist and embryologist Richard Owen, held that adaptations were merely of secondary importance. What was really important was the idea that the human hand, the mole's claw, the bird's wing, and the fish's fin were based on the same plan. In abstracting the plan of the limb, we could determine the grand design upon which God had constructed all vertebrate appendages. To Owen (1848), the homologies underlying animal diversity were what counted, not the secondary adaptations of these basic unities.

Charles Darwin's synthesis

Darwin acknowledged his debt to these earlier debates when he wrote (1859), "It is generally acknowledged that all organic beings have been formed on two great laws—Unity of Type, and Conditions of Existence." Darwin went on to explain that his theory would explain unity of type by descent. The changes creating these types and causing the marvelous adaptations to the conditions of existence, moreover, were explained by natural selection. Darwin called this "descent with modification." After reading Johannes Müller's summary of von Baer's laws in 1842, Darwin saw that embryonic resemblances would be a very strong argument in favor of the genetic connectedness of different animal groups. "Community of embryonic structure reveals community of descent," he would conclude in *Origin of Species*.

Larval forms had been used for taxonomic classification even before Darwin. J. V. Thompson, for instance, had demonstrated that larval barnacles were almost identical to larval crabs, and he therefore counted barnacles as arthropods, not molluscs (Figure 1; Winsor, 1969). Darwin, an expert on barnacle taxonomy, celebrated this finding: "Even the illustrious Cuvier did not perceive that a barnacle is a crustacean, but a glance at the larva shows this in an unmistakable manner." Darwin's evolutionary interpretation of von Baer's laws set a paradigm that was to be followed for many decades, namely, that relationships between groups can be discovered by finding common larval forms. Kowalevsky (1871) would soon make a similar discovery (publicized in Darwin's *Descent of Man*) that tunicate larvae had notochords and formed their neural tube and other organs in a manner very similar to that of the primitive chordate amphioxus. The tunicates, another enigma of classification schemes (usually placed, along with barnacles, as a mollusc), had thereby found a home with the chordates. Darwin also noted that embryonic organisms sometimes make structures that are inappropriate for their adult form but that show their relatedness to other animals. He pointed out the existence of eyes in embryonic moles, pelvic rudiments in embryonic snakes, and teeth in embryonic baleen whales. In this book, we noted that mammalian embryos form a rudimentary yolk sac, send blood vessels to this yolk sac, and undergo gastrulation in a manner resembling birds and reptiles, whose development is constrained by the yolk.

Darwin also argued that adaptations that departed from the "type" and that allowed an organism to survive in its particular environment developed late in the embryo. He noted that the differences between species and genera were, as predicted by von Baer's laws, produced only late in development; he even chloroformed pigeons (with great reluctance) to prove to himself that this was indeed the case. Thus, Darwin recognized two ways of looking at "descent with modification." One could emphasize the common *descent* by pointing out embryonic homologies between two or more groups of animals, or one could emphasize the *modifications* by showing how development was altered to produce structures that enabled animals to adapt to particular conditions.

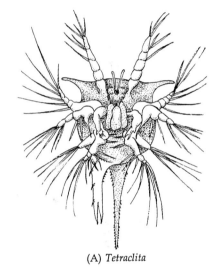

(A) *Tetraclita*

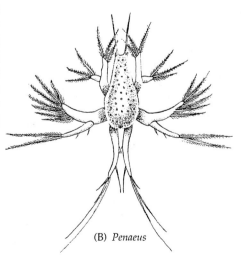

(B) *Penaeus*

FIGURE 1
Nauplius larvae of (A) a barnacle (*Tetraclita*, seen in ventral view) and (B) a shrimp (*Penaeus*, seen in dorsal view). The shrimp and barnacle share a similar larval stage despite their radical divergence in later development. (After F. Müller, 1864.)

Darwin did not attempt to construct complete phylogenies from embryological data, but his work influenced many of his contemporaries to do so. One of the first scientists to realize the evolutionary importance of von Baer's studies was Elie Metchnikoff. Metchnikoff appreciated that evolution consisted of modifying embryonic organisms, not adult ones. He wrote (1891):

> Man appeared as a result of a one-sided, but not total, improvement of organism, by joining not so much adult apes, but rather their unevenly developed fetuses. From the purely natural historical point of view, it would be possible to recognize man as an ape's 'monster,' with an enormously developed brain, face and hands.

Thus, organisms were seen to evolve through changes in their embryonic development.

E. B. Wilson and F. R. Lillie

If changes in embryonic development effected evolutionary changes, how did these developmental changes take place? During the late 1800s, many investigators attempted to link development to phylogeny through the analysis of cell lineages. They meticulously observed each cell in developing embryos and compared the ways that different organisms formed their tissues. In 1898, two eminent embryologists gave cell lineage lectures at the Marine Biology Laboratories at Woods Hole, Massachusetts, and their lectures served to emphasize the two ways that embryology was being used to support evolutionary biology. The first lecture, presented by E. B. Wilson, was a landmark in the use of embryonic homologies to establish phylogenetic relationships. Wilson had observed the spiral cleavage patterns of flatworms, molluscs, and annelids, and he had discovered that, in each case, the same organs came from the same groups of cells. For him this meant that these phyla all had a common ancestor. The various groups of cleavage-stage cells in flatworms, molluscs, and annelids

> . . . show so close a correspondence both in origin and in fate that it seems impossible to explain the likeness save as a result of community of descent. The very differences, as we shall see, give some of the most interesting and convincing evidence of genetic affinity; for processes which in the lower forms play a leading role in the development are in the higher forms so reduced as to be no more than vestiges or reminiscences of what they were, and in some cases seem to have disappeared as completely as the teeth of birds or the limbs of snakes.

The next lecturer was F. R. Lillie, who had also done his research on the development of mollusc embryos and on modifications of cell lineage. He stressed the modifications, not the similarities, of cleavage. His research on *Unio*, a mussel whose cleavage is altered to produce the "beartrap" larva that enables it to survive in flowing streams, was highlighted in Chapter 3. Lillie argued that "modern" evolutionary studies would do better to concentrate on changes in embryonic development that allowed for survival in particular environments than to focus on ancestral homologies that united animals into lines of descent.

In 1898, then, the two main avenues of approach to evolution and development were clearly defined: finding underlying unities that unite disparate groups of animals, and detecting the differences in development that enable species to adapt to particular environments. (These same lines of argument characterized the two types of natural theologies before Darwin.) Darwin had thought these to be temporally distinguished (i.e., that one would find underlying unities in the earliest stages while the later stages would diverge to allow specific adaptations [see Ospovat, 1981]). However, Wilson and Lillie were both discussing the cleavage stage of embryogenesis. These two ways of characterizing evolution and development are still the major approaches today.

The evolution of early development: *E Pluribus Unum*

The emergence of embryos

In the evolution and the development of living organisms, one sees the emergence of multicellularity from single-celled organisms. A new whole is formed from component cellular parts. This is a fundamental step in the emergence of a new level of complexity. The Volvocaceans and the Dictyostelids mentioned in Chapter 1 represent but two of the 17 types of protists in which multicellularity was attained (Buss, 1987). However, only three groups (those that generated fungi, plants, and animals) evolved the ability to form multicellular aggregates that could differentiate into particular cell types—that is, an embryo.

The first embryos had to solve a fundamental problem. Since each of the component cells had the genetic apparatus and the cytoplasmic architecture needed to divide, why shouldn't each cell continue its own proliferation? What would cause them to sacrifice their proliferative capacity to form a collective individual? There may have been more than one solution. Buss suggests that in these early embryos, there was a sharp dichotomy between proliferation and differentiation and that our protist ancestors never learned the trick of dividing once they had differentiated cilia. While

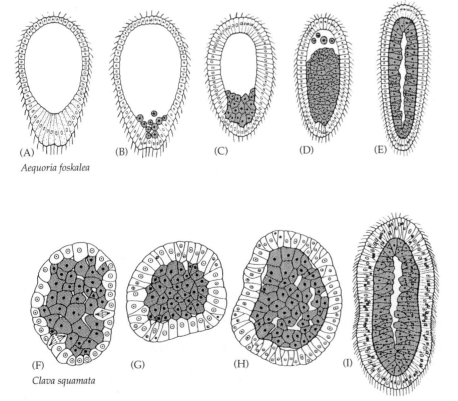

(A)

Aequoria foskalea

(B) (C) (D) (E)

(F) (G) (H) (I)

Clava squamata

FIGURE 2
Gastrulation in two hydroid cnidarians. (A–E) Gastrulation in *Aequoria foskalea*, wherein a ciliated blastula is formed. Cells at the vegetal pole lose their cilia and migrate into the blastocoel to form a mitotically dividing population. (F–I) Gastrulation in *Clava squamata*, wherein a cell-filled stereoblastula is formed and the outer layer then becomes ciliated. Both plans converge on the ciliated "planula" larva characteristic of cnidarians. (Epiboly of a nonciliated ectoderm is not seen in embryos that are free-swimming.) (After Buss, 1987.)

some other protist groups (especially ciliates) could make more microtubule organizing centers, our ancestors could not. To this day, no ciliated metazoan cell divides (although ciliated metazoan cells can lose their cilia and then divide). Buss speculates that the ancestors of today's metazoans stopped their cellular proliferation by differentiating into a blastula of ciliated cells. The early embryos of the first metazoan phyla—sponges and cnidarians—are characterized as balls of ciliated cells, just like the sea urchin embryos discussed in Chapter 3. These ciliated blastulae could move, but it would appear that all their development had ceased, for ciliated cells do not divide, nor do they become any other differentiated cell type. To develop into an organism, this dilemma had to be solved.

This problem was solved by retaining or producing a population of nonciliated cells. These nonciliated cells could proliferate new cells, while the ciliated cells allowed the embryo to move. But these dividing cells could not go just anywhere. They could not grow on top of the ciliated cells or movement would cease. They could not grow into the water or they would be a drag on the embryo's movement. Rather, they would have to migrate *inside* the blastocoel (Figure 2). This movement and proliferation of cells is thought to be the origin of gastrulation. Thus, the blastula arose as a means of joining autonomous cells into a federation. The gastrula arose as a compromise within this federation that allowed the embryo to develop while moving* (Buss, 1987).

The earliest embryos probably developed in this mosaic fashion. However, induction provided a second mechanism to ensure that totipotent blastomeres remained together to form a single individual. Here, each cell sacrificed its autonomy to create a coherent community. Henry and coworkers (1989) have found that whereas individual sea urchin blastomeres can be totipotent, aggregates made of these same cells are not. Rather, each cell restricts the potency of its neighbor (Chapter 8). This restrictive regulation is also seen in allophenic mice (Chapter 3), wherein mammalian blastomeres combine to form a single chimeric mouse rather than two individual mice. There appear to be very important restrictions on cell potency once cells are brought together. Moreover, once the inside population can interact with the outside population of cells and with other parts of the inside population, inductive events can occur to give rise to new organs.

Whichever way this community of cells has been formed, their integration into a unified embryo is accomplished by maternal input into the egg cytoplasm. It is this set of instructions that causes cells to cleave in certain arrangements, to stick to one another, and to differentiate at particular times. As we saw in Chapter 14, the sea urchin embryo becomes a ciliated blastula even in the absence of nuclear transcription. Only at gastrulation does the nucleus begin to regulate development. Thus, selection at the level of cell propagation (which had been the rule of survival among the protists) has been superceded by selection at the level of the individual multicellular organism.

*This is a modification of the theory originally proposed by Metchnikoff (1886) to account for the origin of multicellular organisms. Using hydroid and sponge embryos, Metchnikoff pointed out that certain cells from the wall of the blastula "draw in their flagellum, become amoeboid and mobile, multiply by division, fill the cavity of the blastula, and become capable of digesting." This embryonic state, he felt, "is therefore entitled to be considered the prototype of multicellular beings." Metchnikoff attempted to make a phylogeny of all organisms on the basis of their germ layers, and he believed that all mesodermal cells could be characterized by their ability to phagocytize foreign substances. His discoveries in comparative embryology eventually allowed him to formulate the conceptual foundations of a new science, immunology. (For details of Metchnikoff's theory of multicellular origins, see Chernyak and Tauber, 1988, 1991.)

Formation of the phyla: Modifying developmental pathways

Only about 33 animal body plans are presently being used on this planet (Margulis and Schwartz, 1988). These constitute the animal phyla. This is not to say that these body plans are the only possible ones. The Burgess Shale, a repository of early Cambrian soft-body fossils, is interpreted as containing representatives of 20 phyla that never evolved descendants in the upper strata (Figure 3). In addition, this small band of sediment, about the size of a city block, contains about a dozen previously unknown classes of arthropods. These animals are not "primitive" members of existing phyla or classes, but are specialized examples of their own groups (Whittington, 1985; Gould, 1989). There are also two specimens in the Burgess Shale that may be related to ancestral forms of existing phyla. One is a peripatus-like animal that may be close to the ancestral form of insects; the other appears to be a well preserved chordate (called *Pikaia gracilens*) that may be related to the ancestral chordates (Figure 3C). This latter fossil has several features that recommend its being classified in our phylum: it appears to have a notochord, and the zigzag bands along its sides look very much like the somite-derived musculature found in *Amphioxus* (Conway Morris and Whittington, 1979). Thus, all the known metazoan phyla (and many heretofore unknown ones) may have been formed about 570 million years ago.

If the interpretations of the fossil record are correct, eukaryotic cells emerged some 1.4 billion years ago, the phyla all evolved during the next 800 million years, and no new morphological pattern (*Bauplan*) has been added in the past half-billion years. How did these basic body plans come into existence in such a relatively short time, and why has no other phylum emerged since then? Kauffman (in press) has a mathematical model that predicts that any evolving system (whether it be a phylum, species, automobile, or religion) displays this pattern of divergence followed by the locking in on a particular subset of the original diversity. Kauffman uses the metaphor of a rugged fitness landscape wherein there are mountains and valleys of fitness, and all organisms start out with the same average fitness value. If they take large jumps, they have a 50 percent chance of becoming fitter organisms. Eventually, the chance of finding a fitter body plan decreases if one takes a jump far from where one is situated. Long jumps become risky, and the chances are that these higher peaks are already occupied. Instead, small jumps (on the same peak) may make one somewhat fitter than the surrounding population. What we see then is a diversification around a few successful models. In general, the duration between successful long jumps doubles with each attempt. During the early Cambrian, it is possible that the genome had not become stabilized into the sets of interactions that we see today. Moreover, in many invertebrate groups, there is an alternation of generations, wherein a sexual form generates an asexual form (zooid, polyp, bud) that then gives rise to the sexual form again. In such cases, somatic mutations in the asexual form can enter into the body of the sexual form and be propagated very rapidly (Figure 4; Buss, 1987).

How, then, can one modify one *Bauplan* to create another *Bauplan*? The first way would be to modify the earliest stages of development. According to von Baer (Chapter 5), animals of different species but of the same genus diverge very late in development. The more divergent the species are from one another, the earlier one can distinguish their embryos. Thus, embryos of the snow goose are indistinguishable from those of the blue goose until the very last stages. However, snow goose development diverges from chick embryos a bit earlier, and goose embryos can be

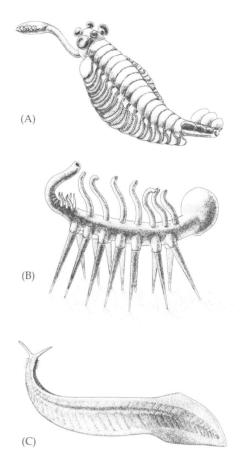

(A)

(B)

(C)

FIGURE 3
Reconstructions of three fossil organisms found in the Cambrian Burgess Shale. (A) *Opabina*, an organism having five eyes on its head, a frontal appendage with a terminal claw, body segments with dorsal gills, and a three-segment tail piece. (B) The aptly named *Hallucigenia* with its seven pairs of struts and dorsal tentacles. (C) *Pikaia gracilens*, a possible chordate. (From Gould, 1989.)

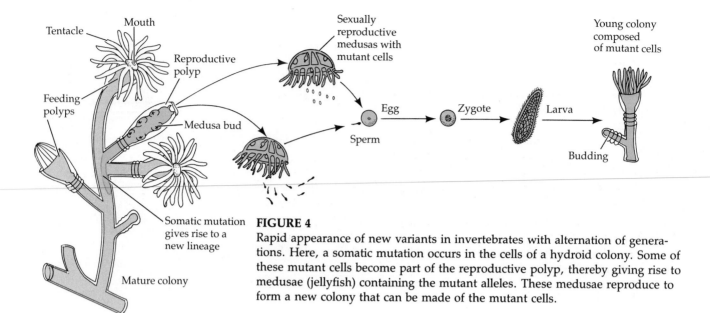

FIGURE 4

Rapid appearance of new variants in invertebrates with alternation of generations. Here, a somatic mutation occurs in the cells of a hydroid colony. Some of these mutant cells become part of the reproductive polyp, thereby giving rise to medusae (jellyfish) containing the mutant alleles. These medusae reproduce to form a new colony that can be made of the mutant cells.

distinguished from lizard embryos at even earlier stages. It appears, then, that mutations that create new *Baupläne* could do so by altering the earliest stages of development.

These early developmental changes can be effected by changing the localization of cytoplasmic determinants, changing the rate of cell division of one cell or group of cells relative to the others, or changing the positions of the cells as they divide. In Chapter 3, we saw that a modification of molluscan cleavage can give the bulk of cytoplasm to the ectodermal cells that form the larval shell. This was due to changing the manner in which the blastomeres divide and apportion cytoplasm. In annelid worms, the differences between polychaetes and oligochaetes stem from differences in the cytoplasmic localization of morphogens within the egg (Figure 5). While they both undergo spiral cleavage, they apportion their morphogens into different cells. Polychaetes undergo a relatively standard spiral cleavage to give rise to the trochophore larva. Oligochaetes, however, put most of their cytoplasm into those cells destined to form adult, rather than larval, structures. This group then skips the larval stage. If a mutation were to place a certain cytoplasmic morphogen into one region of the egg instead of another, or if a mutation caused a change in the axis of cell division so that different sets of cells acquired these determinants, then a radically different phenotype could be produced. As E. G. Conklin wrote in 1915, "We are vertebrates because our mothers were vertebrates and produced eggs of the vertebrate pattern."

Another way of evolving new phyla may involve modifying the larva. Darwin and others thought that similarities of larval form signified common descent. However, this can be reinterpreted to mean that changes that give rise to different phyla may occur in larvae. Snails, echiuroids, and polychaetes have very similar patterns of division and form trochophore larvae (Figure 6). In fact, the placement of the newly discovered phylum Vestimentifera (the bright red, gutless invertebrates found in the deep ocean trenches) near the annelids was made in part on the basis of vestimentiferans' having trochophore larvae (Jones and Gardiner, 1989). Thus, one of the principal mechanisms for establishing new phyla and classes may be the rearrangement of development during the larval stage so that metamorphosis brings about new types of organization. Garstang

(A)   *Podarke*

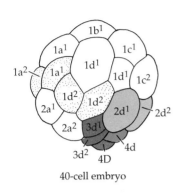

40-cell embryo

Fate map

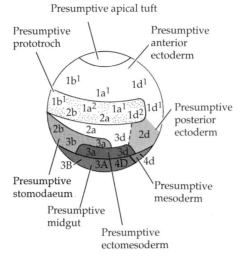

Presumptive apical tuft

Presumptive
prototroch

Presumptive
anterior
ectoderm

Presumptive
posterior
ectoderm

Presumptive
mesoderm

Presumptive
stomodaeum

Presumptive
midgut

Presumptive
ectomesoderm

Trochophore larva

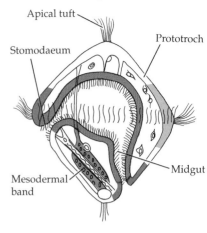

Apical tuft

Prototroch

Stomodaeum

Mesodermal
band

Midgut

(B)   *Tubifex*

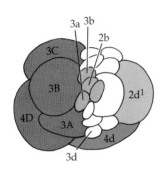

Cleaving embryo

Fate map

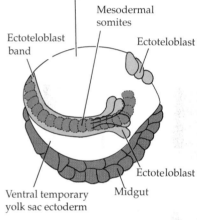

Presumptive
stomodaeum

Presumptive
yolk sac
ectoderm

Presumptive
ectoteloblast
ectoderm

Presumptive
mesoderm

Presumptive
midgut

Gastrulation

Dorsal temporary
yolk sac ectoderm

Mesodermal
somites

Ectoteloblast

Ectoteloblast
band

Ectoteloblast

Midgut

Ventral temporary
yolk sac ectoderm

FIGURE 5
Comparison of the development of two classes of annelid worms, (A–C) the polychaete *Podarke* and (D–F) the oligochaete *Tubifex*. Their cleaving embryos, blastula fate maps, and products of gastrulation are seen. In *Podarke*, gastrulation leads to the formation of a trochophore larva. In *Tubifex*, there is no larval stage, and the embryo develops directly into a segmented body. (After Raff and Kaufman, 1983.)

(1928) showed how the veliger larva of certain snails could have arisen by mutation and then been selected because the new arrangement of head and shell allowed the head to retract beneath the shell for safety. He also framed the hypothesis that chordates arose from ancestral tunicate larvae that had become neotenic. Unfortunately, soft-bodied larvae rarely fossilize, so we know very little about the mechanisms by which chordates and other phyla may have arisen from early Cambrian larvae.*

* Larval forms are often used to bridge the gap between the different adult forms. The larval form is seen as either being ancestral to two groups or as "breaking away" by neoteny and forming a different type of organism. This has often been hypothesized as the mechanism by which chordates emerged from invertebrates and vertebrates arose from chordates. The tornaria larva of hemichordates is formed in a deuterostome manner similar to that of echinoderm larvae and looks enough like echinoderm larvae to have been originally mistaken for them. This would link echinoderms and chordates. Garstang (1928) and Berrill (1955) hypothesized that the larvae of certain tunicates could have evolved into chordates such as amphioxus by neotenic development. In this way, the tunicates would keep the notochord, larval musculature, and feeding apparatus of the larval tunicate while becoming sexually mature. There are, in fact, neotenic free-swimming tunicates (such as *Larvacea*). Modifications of this view (using a different protochordate stock) have recently been suggested by Jefferies (1986). The origin of chordates remains a difficult problem.

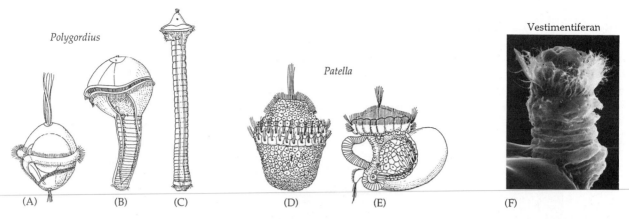

FIGURE 6

Divergence of development after the trochophore larval stage. (A–C) Metamorphosis of the polychaete annelid *Polygordius* from its free-swimming trochophore larva shows the formation of a segmented trunk. Eventually, the larval structures shrink at the anterior end as the head forms. (D,E) Metamorphosis of the prosobranch (clam) mollusc *Patella*. After the trochophore stage, it develops a molluscan foot, shell gland, and visceral hump. (F) Scanning electron micrograph of the trochophore larva of a vestimentiferan. (A–E after Grant, 1978; F from Jones and Gardner, 1989, photograph courtesy of the authors.)

The developmental–genetic mechanisms of evolutionary changes

Isolation

Recent research is attempting to circumvent this gap in our understanding by studying the genes that control embryonic and larval development. One way developmental changes can bring about evolutionary change is to cause the reproductive isolation of a group of organisms within a population. Such isolation is thought to be essential for the formation of new species. The snail shell coiling mutations discussed in Chapter 3 are mutations that act during early development to change the position of the mesodermal organs. Mating between left-coiling and right-coiling snails is mechanically very difficult, if not impossible, in some species (Clark and Murray, 1969). As this mutation is inherited as a maternal effect gene, a group of related snails would emerge that could mate with one another but not with other members of the original population. These reproductively isolated snails could expand their range and, by the accumulation of new mutations, form a new species (Alexandrov and Sergievsky, 1984).

Homeosis

Major developmental changes can occur when the interactions between genes are changed. This can give rise to new cellular phenotypes, and, in some instances, to new organismal phenotypes as well. We have seen in Chapters 12 and 20 that when a structural gene is set next to a new enhancer, it becomes expressed at the time or place specified by the enhancer. A globin gene will be expressed at midblastula transition if placed next to certain *Xenopus* enhancers, and a proto-oncogene will lead to uncontrolled lymphocyte growth when translocated near an immunoglobulin gene enhancer in a B cell precursor. One possible way of changing a cell's differentiated phenotype would be to alter a gene's enhancer, either by mutation or by translocation.

Rabinow and Dickinson (1986) are testing this hypothesis on Hawaiian *Drosophila*. These flies have undergone an extremely rapid rate of speciation, and several species have changes in the tissue specificity of gene expression. The larva of *D. hawaiiensis*, for instance, expresses its alcohol dehydrogenase (ADH) in its fat bodies and carcass, whereas the larva of *D. formella* expresses its alcohol dehydrogenase only in its fat bodies. The two ADHs can be distinguished electrophoretically. These species can mate with one another to generate hybrid larvae. When the fat bodies of these larvae were examined, both *D. hawaiiensis* and *D. formella* ADHs were seen. However, when the carcasses were analyzed, only the *D. hawaiiensis* protein was observed. This is the pattern one would expect if the ADH genes were controlled by *cis*-regulatory elements and if they differed between the two closely related species. (Were expression based on *trans*-regulatory elements, both genes would either be on or off.) It appears, then, that enhancers may change during evolution, thereby causing different products to be synthesized in different tissues.

If cellular phenotypes can change when the *cis*-regulatory elements of structural genes are altered, the alterations can be even more far-reaching when the *trans*-regulatory factors are altered. In Chapters 12 and 18, we discussed homeotic transformations in *Drosophila*. These mutations are caused when certain *trans*-regulatory enhancer-binding factors are altered. In 1940, Richard Goldschmidt and Conrad H. Waddington pointed out that these homeotic mutations provided the means to link together development, genetics, and evolution. Waddington (1940) stressed that homeotic genes were embryonic switches, analogous to switches of railroad yards that directed trains into one path rather than another. Goldschmidt (1940) saw homeotic genes as generating several large changes in the embryo simultaneously, thereby causing the formation of new species. In focusing attention on homeotic transformations, they were far ahead of the science of their day.

In many ways, the phenotypes of the homeotic deletions resemble the postulated earlier stages of *Drosophila* evolution (Figure 7). *Drosophila*, a two-winged (dipteran) fly, is thought to have arisen from normal four-winged insects (pterygotes). This condition can be achieved by deleting the *Ubx* portion of the bithorax complex. Winged insects are thought to have evolved from nonwinged (apterygote) insects. This situation is mimicked by the deletion of the *Antennapedia* gene, in whose absence all the thorax develops like the wingless first thoracic segment. The apterygote insects are believed to have arisen from a myriapod-like creature (millipedes, centipedes), a situation similar to the phenotype generated when the entire bithorax complex is missing and the entire posterior of the

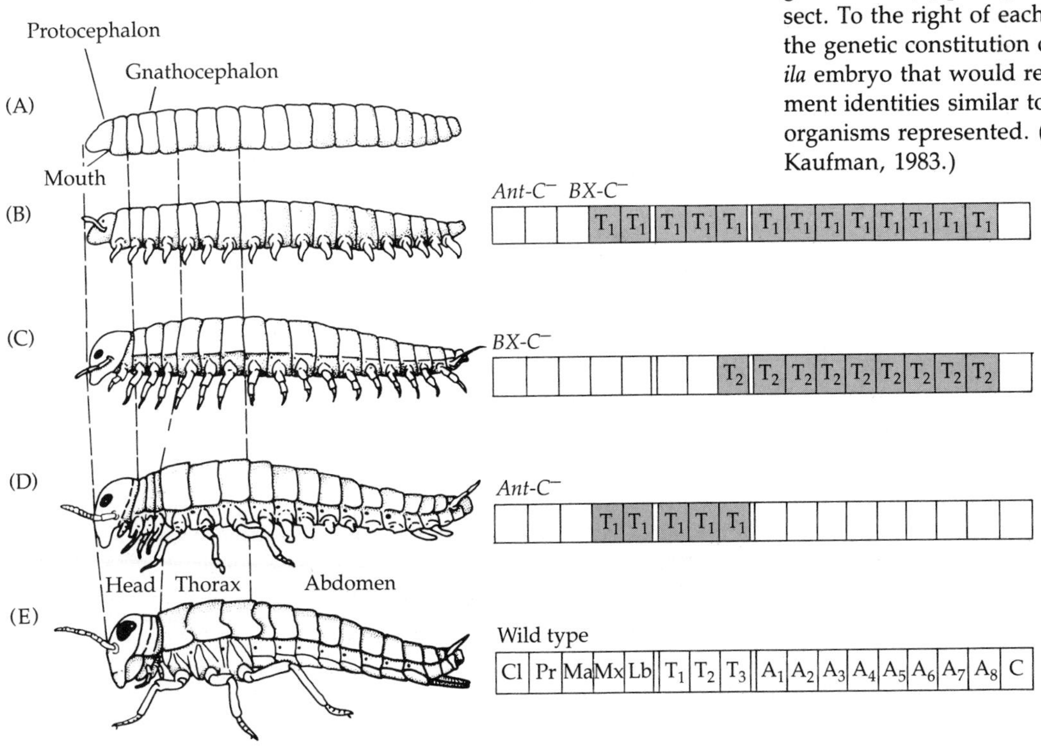

FIGURE 7
Insect evolution as it might be occasioned by the addition of homeotic gene functions. (A) Annelid, (B) onychophoran, (C) myriapod, (D) apterogote insect, (E) pterogote (winged) insect. To the right of each drawing is the genetic constitution of the *Drosophila* embryo that would result in segment identities similar to those of the organisms represented. (After Raff and Kaufman, 1983.)

embryo develops like a legged thorax segment. Finally, if both sets of homeotic genes (the *Antennapedia* and *Bithorax* complexes) are removed, the result is a condition similar to that of the most primitive known arthropods, the onychophorans. As Raff and Kaufman (1983) have said, "Thus, by sequential deletion of a relatively small amount of genetic material we have managed to traverse a rather large amount of evolutionary ground."*

New combinations of genes could also create new cell types. One of the major changes in phyla has been the occurrence of cell types with new characteristics. Kauffman (in press) has mathematically modeled the generation of new cell types from a random genome consisting of 10,000 genes, each regulated by two other genes. In such cases, he finds only 100 stable states (out of $2^{10,000}$ possible states). Each of these stable states represents a differentiated cell type. In some cases, the mutation of a regulatory gene is sufficient for the restructuring of the interactions, and a new cell type is created. Most genes, however, remain unaffected by this new arrangement. The creation of a new cell type is a rare event in nature and can often change the nature of the beast. Neural crest cells, for instance, distinguish vertebrates from the protochordates and invertebrates (Figure 8). The protochordates have a dorsal neural tube and notochord, but no real "head." The cranial neural crest cells are largely responsible for the creation of the face, skull, and the branchial arches. It is thought that the development of the head originally allowed for more efficient predation, placing the sensory structures adjacent to the prey-capturing jaws (Gans and Northcutt, 1983; Langille and Hall, 1989). As Figure 8 shows, the vertebrates are thought to have arisen from invertebrates in several steps that involved the formation and modification of new cell types.

Development and evolution within established *Baupläne*

Developmental constraints

We have alluded to the fact that there are relatively few *Baupläne*, and that one can easily imagine types of animals that do not exist within existing phyla. Why aren't there more major body types among the animals? To answer this, we have to consider the constraints imposed upon evolution. There are three major classes of constraints upon morphogenetic evolution. First, as mentioned in Chapter 20, there are PHYSICAL CONSTRAINTS on the construction of the organism. These constraints of diffusion, hydraulics, and physical support allow only certain mechanisms of devel-

---

*Raff and Kaufman also warn us that these interesting analogies should not be taken too literally. When the *Antennapedia* genes are deleted from the *Drosophila* genome, one gets a highly aberrant fly larva, not a millipede. Similarly, the four-winged flies created by mutations in *Ultrabithorax* have two sets of forewings, not a forewing-hindwing combination. However, recent studies point to some remarkable similarities between segmentation in *Drosophila* and that in vertebrates. For instance, in both vertebrates and *Drosophila*, the order of homeotic genes on the chromosome corresponds to their expression on the anterior–posterior axis of the embryo (*Drosophila*) or spinal cord (vertebrates). This can be seen by comparing Figure 27 in Chapter 18 with Figure 22 in Chapter 12. The antennapedia and bithorax complex genes on *Drosophila* chromosome 3 show the same order of anterior–posterior expression as the *Hox-2* genes on mouse chromosome 11 (see Holland, 1990, for speculations on the evolution of these pattern-forming genes). The recent finding that chick hindbrain segments have cell lineage boundary restrictions (like the compartmental boundaries in *Drosophila*) also strengthens hypotheses postulating a common ancestor to metameric patterns in insects and vertebrates (Fraser et al., 1990).

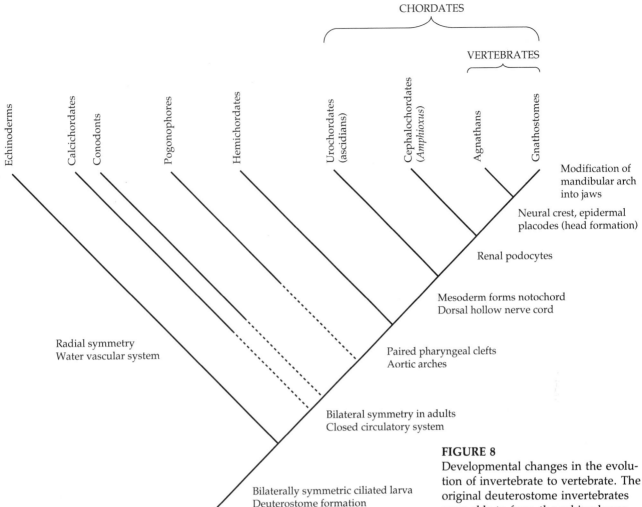

CHORDATES

VERTEBRATES

Echinoderms

Calcichordates

Conodonts

Pogonophores

Hemichordates

Urochordates (ascidians)

Cephalochordates (*Amphioxus*)

Agnathans

Gnathostomes

Modification of mandibular arch into jaws

Neural crest, epidermal placodes (head formation)

Renal podocytes

Mesoderm forms notochord
Dorsal hollow nerve cord

Paired pharyngeal clefts
Aortic arches

Radial symmetry
Water vascular system

Bilateral symmetry in adults
Closed circulatory system

Bilaterally symmetric ciliated larva
Deuterostome formation
Enterocoelous mesoderm

FIGURE 8

Developmental changes in the evolution of invertebrate to vertebrate. The original deuterostome invertebrates were able to form the echinoderms and those other organisms that eventually gave rise to the vertebrate lineage. The ability of the mesoderm to form a notochord and its overlying ectoderm to become a neural tube, separated the chordates from the remaining invertebrates. The development of neural crest cells and the epidermal placodes that give rise to the sensory nerves of the face distinguish the vertebrates from the protochordates. (After Langille and Hall, 1989; Gans, 1989.)

opment to occur. One cannot have a vertebrate on wheeled appendages (of the sort that Dorothy saw in Oz) because blood cannot circulate to a rotating organ; this entire possibility of evolution has been closed off. Similarly, structural parameters and fluid dynamics forbid the existence of five-foot-tall mosquitos.

Second, there are those constraints involving MORPHOGENETIC CONSTRUCTION RULES (Oster et al., 1988). Bateson (1894) noted that when organisms depart from their normal development, they do so in only a limited number of ways. Research in this area attempts to find the architectural parameters upon which organisms are constructed and seeks to show how these parameters can be modified during evolution. Some of the best examples of these types of constraints come from the analysis of limb formation in vertebrates. Holder (1983) pointed out that although there have been many modifications of the vertebrate limb over 300 million years, some modifications are not found (Figure 9). Moreover, analyses of natural populations suggest that there are a relatively small number of ways that limb changes can occur (Wake and Larson, 1987). If a longer limb is favorable in a given environment, the humerus may become elongated. One never sees two smaller humeri joined together in tandem, although one could imagine the selective advantages that such an arrangement might have. This indicates a construction scheme that has certain rules.

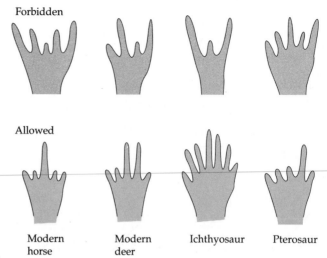

Forbidden

Allowed

| Modern horse | Modern deer | Ichthyosaur | Pterosaur |

FIGURE 9

Some "forbidden" and some permissible tetrapod limb structures. (From Thomson, 1988.)

The principal rules of vertebrate limb formation have been summarized by Oster and his co-workers (1988). They find that a reaction–diffusion mechanism can explain the known morphologies of the limb and can explain why other morphologies are forbidden. As discussed in Chapter 17, this model posits that the aggregations of cartilage actively recruit more cells from the surrounding area and laterally inhibit the formation of other foci of condensation. The number of foci depends on the geometry of the tissue and the strength of the lateral inhibition. If the inhibition remains the same, the size of the tissue volume must increase in order to get two foci forming where one had been allowed before. At a certain threshold (called a bifurcation threshold), this size is reached, and the limb can branch into two foci.

Evidence for this mathematical model comes from experimental manipulations and comparative anatomy. When an axolotl limb bud is treated with the antimitotic drug colchicine, the dimensions of the limb are reduced. In such limbs, there is not only a reduction of digits, but a reduction of certain digits in a certain order, as expected by the mathematical model and from the "forbidden" morphologies. Moreover, these reductions of specific digits are very similar to those limbs of progenetic salamanders, those species that reach maturity at a smaller stage than did their ancestors and whose limbs develop from smaller limb buds (Figure 10; Alberch and Gale, 1983, 1985). Thus, the use of reaction–diffusion mechanisms to construct limbs may constrain the possibilities that can be generated during development, because only certain types of limbs are possible using these rules.

The third set of constraints upon the evolution of new types of structures are PHYLETIC CONSTRAINTS (Gould and Lewontin, 1979). These are the historical restrictions based on the genetics of an organism's development. For instance, once a structure is generated by inductive interactions, it is difficult to start over again. The notochord, which is still functional in adult protochordates (Berrill, 1987), is considered vestigial in adult birds and mammals. Yet it may be transiently necessary in the embryo to specify the neural tube. Similarly, Waddington (1938) noted that although the pronephric kidney of the chick embryo is considered vestigial (since it has no ability to concentrate urine), it is the source of the ureteric bud that induces the formation of a functional kidney during chick development.

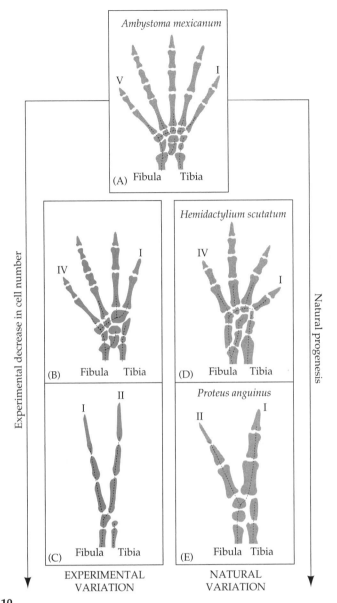

FIGURE 10
Relationship between cell number and number of digits in salamanders. (A) The hindlimb of an axolotl (*Ambystoma mexicanum*) with its five symmetrical digits. (B, C) Digits on axolotl hindlimb after hindlimb bud is incubated in colchicine to reduce cell number. (D, E) Two wild salamanders formed by progenesis, each having a smaller limb bud. (D) *Hemidactylium scuatum* and (E) *Proteus anguinus*. The parallels between the experimental variation and the natural variation can be seen, and the common denominator is reduced cell numbers in the limb buds. (After Oster et al., 1988.)

This type of phyletic constraint has recently been reviewed by Raff and colleagues (1991). Until recently in was thought that the earliest stages of development would be the hardest to change, because altering them would either destroy the embryo or generate a radically new phenotype. But recent work (and the reappraisal of older work) has shown that alterations can be made to early cleavage without upsetting the final form. Modifications of morphogens in mollusc embryos can give rise to new types of larvae that still metamorphose into molluscs, and changes in sea urchin cytoplasmic morphogens can generate sea urchins that develop without larvae but still become sea urchins. In fact, looking at vertebrates,

FIGURE 11
The bottleneck at the "pharyngula" stage of vertebrate development. The bottom of this chart is the standard depiction of von Baer's law (as shown in Chapter 5), demonstrating the divergence of vertebrate classes after a common embryonic stage. The top of this chart represents the divergent beginnings of development. von Baer himself (1886) was aware of this bottleneck. (After Elinson, 1987.)

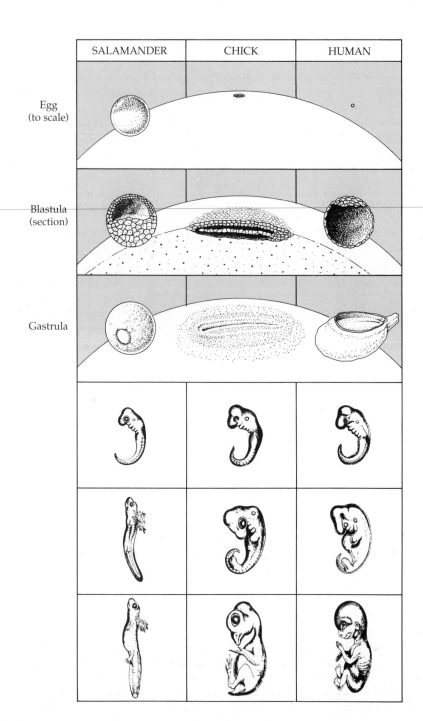

one can see there is an entire history leading up to the famous diagram of von Baer's law shown in Chapter 5. All the vertebrates arrive at this particular stage of development (called the "pharyngula"), but they can do so by different means (Figure 11). Birds, reptiles, and fish arrive there after meroblastic cleavages of different sorts; amphibians get to the same stage by way of radial holoblastic cleavage; and mammals reach the same place after constructing a blastocyst, chorion, and amnion. The earliest stages of development, then, appear to be extremely plastic. Similarly, the later stages are very different, as the different phenotypes of mice, sunfish, snakes, and newts amply demonstrate. There is something in the *middle* of development that appears to be invariant.

Raff argues that the formation of new *Baupläne* is inhibited by the need for global sequences of induction during the neurula stage (Figure 12).

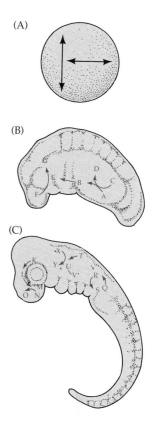

FIGURE 12
Mechanism for the bottleneck at the pharyngula stage of vertebrate development. In the cleaving embryo (A) there are global interactions, but very few of them, mainly to specify the axes of the organism. (B) At the neurula to pharyngula stages, there are many global interactions. After the pharyngula stage (C), there are even more inductive interactions, but these are primarily local in effect, confined to their own fields. (After R. A. Raff, personal communication.)

Before that stage, there are few inductive events. After that period, there are a great many inductive events, but almost all of them are local. During early organogenesis, however, there are several inductive events occurring simultaneously that are global in nature. In vertebrates, to use von Baer's example, the earliest stages involve specifying axes and undergoing gastrulation. Induction has not happened on a large scale. Moreover, as Raff and colleagues have shown (Henry et al., 1989), there is a great deal of regulative ability here, so small changes in morphogen distribution or the position of cleavage planes can be accommodated. After the major body plan is fixed, inductions occur all over the body but are compartmentalized into discrete organ-forming systems. The lens induces the cornea, but if it fails to do so, only the eye is affected. Similarly, there are inductions in the skin that form feathers, scales, or fur. If they do not occur, the skin or patch of skin may lack these structures. But during early organogenesis, the interactions are more global (Slack, 1983). Failure to have the heart at a certain place can affect the induction of eyes (Chapter 16). Failure to induce the mesoderm in a certain region leads to malformations of the kidneys, limbs, and tail. It is this stage that constrains evolution and that typifies the vertebrate phylum. Thus, once a vertebrate, it is difficult to evolve into anything else.

Inductive interactions and the generation of novel structures

In addition to constraining development, inductive interactions can be altered to produce evolutionary novelties. This was first suggested by Walter Garstang. During the first decades of this century, the dominant way of relating embryology to evolution was through Ernst Haeckel's "biogenetic law" (see Sidelights & Speculations). Garstang delivered a stunning critique of this "law" in 1922, showing that it was incompatible with Mendelian genetics and with the observations of embryology. Garstang stood the biogenetic law on its head: ontogeny did not recapitulate phylogeny, it created phylogeny. The animals that arose late in evolutionary history did not arise through a terminal addition onto an existing phylogeny. Rather, they arose by mutations that affected the interactions of modules already existing in the *Bauplan* of the organism: "A house is not a cottage with an extra story on the top. A house represents a higher grade in the evolution of a residence, but the whole building is altered—foundation, timbers, and roof—even if the bricks are the same."

Thus, when we say that the contemporary one-toed horse evolved from a five-toed ancestor, we are saying that hereditable changes occurred in the differentiation of limb mesoderm into chondrocytes during embryogenesis in the horse lineage. In this perspective, evolution is the result of hereditary changes affecting development. This is the case whether the mutation is one that changes the reptilian embryo into a bird or one that changes the color of *Drosophila* eyes.

Therefore, even after the 30 or so contemporary *Baupläne* had been established, changes during development could still effect drastic changes in evolution. There are many ways by which changes in induction can lead to diverse morphogenic phenomena. Changes in induction can

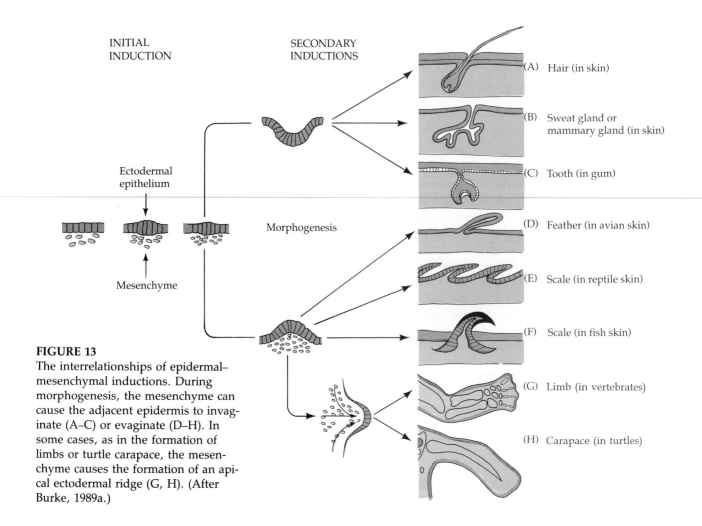

INITIAL
INDUCTION

SECONDARY
INDUCTIONS

Ectodermal
epithelium

Mesenchyme

Morphogenesis

(A) Hair (in skin)

(B) Sweat gland or
      mammary gland (in skin)

(C) Tooth (in gum)

(D) Feather (in avian skin)

(E) Scale (in reptile skin)

(F) Scale (in fish skin)

(G) Limb (in vertebrates)

(H) Carapace (in turtles)

FIGURE 13
The interrelationships of epidermal–mesenchymal inductions. During morphogenesis, the mesenchyme can cause the adjacent epidermis to invaginate (A–C) or evaginate (D–H). In some cases, as in the formation of limbs or turtle carapace, the mesenchyme causes the formation of an apical ectodermal ridge (G, H). (After Burke, 1989a.)

change scales into feathers (as in the case of bantam chicks) and are responsible for such remarkable adaptations as the avian lung, the ruminant stomach, and the tusks of elephants (modified incisors) and walruses (modified upper canine teeth). The secretory glands of the epidermis are each modifications of the same type of induction—mammary glands are embryologically modified sweat glands. Likewise, the fearsome rows of sharks' teeth are modifications of its body scales. Indeed, Burke (1989a) has proposed that all secondary inductions have a common ancestry (Figure 13).

One interesting point in Burke's diagram concerns the mechanism used to create limbs (an apical ectodermal ridge accompanied by a malleable cartilage-forming mesoderm). She finds that one family of vertebrates uses this same mechanism in another region of the body. The carapace shell of the turtle results from such a placement of an epithelial–mesenchymal interaction in a new area. As can be seen in Figure 14, the "carapacial ridge" is formed in a manner analogous to the apical ectodermal ridge of limbs. The patterns of cell proliferation and the distribution of adhesive glycoproteins such as N-CAM and fibronectin also show similarity to limb formation (Burke, 1989b). Induction is also responsible for the creation of "extra" arms in echinoderms. The formation of the five arms of a starfish are caused by the induction of the epidermis by the five protrusions of the hydrocoel. However, in the six-armed starfish, *Leptasterias hexactis*, a sixth protrusion of the hydrocoel occurs, and nine protrusions cause the formation of the nine-armed *Solaster* sea stars (Gemmill, 1912).

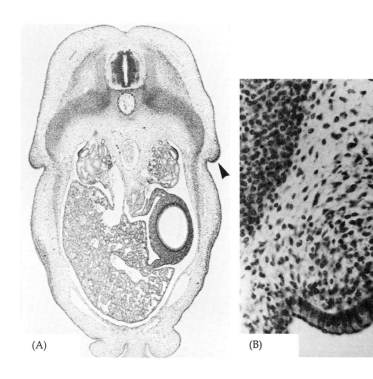

FIGURE 14
Midtrunk cross section through the
embryo of the turtle *Chelydra serpen-
tina*. (A) The carapacial ridge (arrow) is
formed at the boundary of the somitic
and lateral plate mesoderm and now
represents the dorsal–ventral boundary. The thickened mesodermal bands
extending from the center into the carapace area are the rib condensations.
(B) Higher magnification of the carapacial ridge. (From Burke, 1989b; photographs courtesy of the author.)

(A)

(B)

SIDELIGHTS & SPECULATIONS

Ernst Haeckel and the biogenetic law

A disastrous union of embryology and evolutionary biology
was forged in the last half of the nineteenth century by the
German embryologist and philosopher Ernst Haeckel.
Based on the assumption that the laws by which species
arose on this planet (phylogeny) were identical to the laws
by which the individuals of the species developed (ontogeny), he viewed adult organisms as the embryonic stages
of more advanced organisms. This view was summarized
by his "Biogenetic Law": ontogeny recapitulates phylogeny. In other words, development of advanced species was
seen to pass through stages represented by adult organisms
of more primitive species.

In this view, the creation of new phyla is a step towards the completion of human development. In earlier
epochs, only the initial stages of this development occurred, producing protists and cnidarians. Later, more
stages are added sequentially until a human being has
evolved. According to Haeckel, three laws sufficed to explain how this advancing ontogeny could generate new
species. First, there was the law of correspondence. The
human zygote, for instance, was represented by the
"adult" stage of the protists; the colonial protists represented the advancement of development to the blastula
stage; the gill slit stage of human embryos was represented
by adult fish. Haeckel even postulated an extinct organism,
Gastraea, a two-layered sac corresponding to the gastrula,
which he considered the ancestor of all metazoan species.
Second, there was the law of terminal addition. The embryo evolved new species by adding a step at the end of

the previous ones. In such a view, humans evolved when
the embryo of the next highest ape added on a new stage.
This provided a linear, not a branching, phylogeny. There
was also the law of truncation, which held that preceding
development could be foreshortened. This law was needed
to prevent gestation time from being enormous. It also was
needed since embryologists did not observe all these stages
in all animals.

This notion of ontogeny recapitulating phylogeny was
not Darwinism. In fact, Haeckel's synthesis was an attempt
to fuse the works of Darwin, Lamarck, and Goethe. In
Darwinism, contemporary species are seen as having a
common ancestor. The result is a multibranched "bush."
(A tree metaphor has also been used, but trees have a
central axis on which scientists have frequently placed the
lineage leading up to *H. sapiens*.) Humans are not "higher"
than chimps, but have an ancestor from which both groups
diverged. In Haeckel's scheme, animals advance to new
levels by adding stages to existing embryonic development.

Interestingly, von Baer (1828) had disproven the "biogenetic law" before Haeckel ever invented it. In ridiculing
the pre-evolutionary forms of this law, von Baer fantasized
what would happen if birds were writing the embryology
textbooks.

Let us imagine that birds had studied their own development
and that it was they who investigated the structure of the
adult mammal and of man. Wouldn't their physiological textbooks teach the following? "Those four and two-legged animals bear many resemblances to our own embryos, for their
cranial bones are separated, and they have no beak, just as

we do in the first five or six days of our incubation; their extremities are all very much alike, as ours are for about the same period; there is not a true feather on their body, rather only thin feather-shafts, so that we, as fledglings in the nest, are more advanced than they will ever be . . . And these mammals that cannot find their own food for such a long time after birth, that can never rise freely from the earth, want to consider themselves more highly organized than we?"

By observing development, von Baer (Chapter 5) noted that embryos never pass through the adult stages of other animals. However, there are stages that related embryos do share. All vertebrate embryos pass through a stage in which there are embryonic gill slits. Fish elaborate them into true gills, while the slits become part of the jaw or ear apparatus in other vertebrates. But a frog or human embryo never passes through a stage in which it has the structures of an adult fish. However, even though von Baer and others had discredited the recapitulation notion, it became one of the most popular notions in biology.*

*Even more than in biology, Haeckel's "biogenetic law" was adapted uncritically by many of the newly forming social sciences. Early anthropologists espoused the view that other cultures were "primitive" in the embryological sense in that their development had stopped short of our own. Indeed, the word "underdeveloped" is still used to define such a culture. Haeckel also felt that within the human species, races could be ranked in a linear fashion and thereby gave scientific "validity" to the racial prejudices that culminated in the genocidal biopolicy of the Third Reich. (For details, see Gasman, 1971; Gould, 1977a, b; Stein, 1988.) Gould shows that recapitulation has a limited value in looking at the formation of related species but that it is not a universal phenomenon.

Correlated progression

When a drastic remodeling of morphology takes place during evolution, one has to account for the developmental changes that have caused it. As early as 1871, St. George Mivart pointed out that large evolutionary changes were not due to the simple alteration of one structure of an organism. Rather, an entire group of structures changed. Raissa Berg (1960) has called these co-varying constellations of characters "correlation pleiades." The mechanism that enables such changes to occur has been called CORRELATED DEVELOPMENT and was a major feature of embryology at the turn of the century. Spemann (1901) wrote a major work on this topic that formed the context for his work on lens formation. Lenses, he said, form only when induced by other tissues (such as optic vesicle tissue) that form eye structures. Similarly, skeletal cartilage informs the placement of muscles, and muscles induce the placement of nerve axons. Thomson (1988) has used this concept to explain how one part of an organism can change without causing the entire structure to become uncoordinated. If one structure changes, it will induce the other structures to change with it. This is correlated progression.

One of the best examples of such correlated progression involves the formation of the vertebrate jaw (see Gould, 1990, for a review). The first vertebrates had no jaws. When gill slits became supported by cartilaginous elements, the first set of gill supports appears to have surrounded the mouth to form the jaw. There is ample evidence that jaws are modified gill supports. In both cases, these bones are made from neural crest cells rather than from mesodermal tissue as are most bones. Second, both structures form from upper and lower bars that bend forward and are hinged in the middle. Third, the jaw musculature seems to be homologous to the original gill support musculature. Thus, the first transformation of the first branchial arch cartilage was from gill apparatus to jaw apparatus. But the story does not end here.

The second branchial arch supporting the gill becomes the hyomandibular bone of jawed fishes. This element supports the skull and links the jaw to the cranium (Figure 15). As we saw in Chapter 5, this hyomandibular bone functions in mammals as part of the middle ear bones. But fish do not use this bone for hearing, so how did a bone used for gill support and then for cranial support become part of the auditory appa-

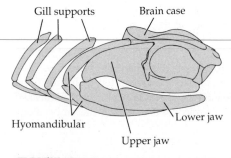

FIGURE 15
Homologies of the jaws and the gill arches as seen in the skull of the paleozoic shark *Cobeledus aculentes*. (After Zangerl and Williams, 1975.)

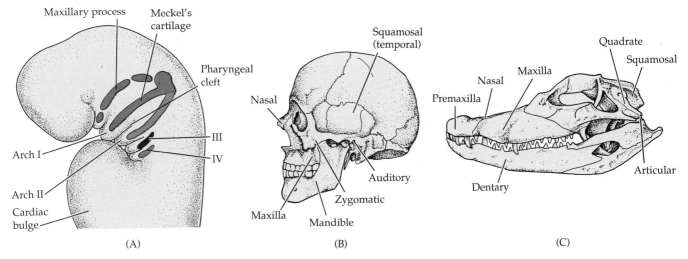

FIGURE 16
The branchial arches and their derivatives. (A) Lateral view of a 4-week human embryo showing the pharyngeal arches and their cartilaginous derivatives. (B) Lateral view of the human skull showing the junction of the lower jaw with the squamosal (temporal) region of the skull. (C) Lateral view of an alligator skull. The articular portion of the lower jaw articulates with the quadrate bone of the skull. In mammals, the quadrate becomes internalized to form the incus of the middle ear. The articular bone retains its contact with the quadrate, becoming the malleus of the middle ear. (A after Langman, 1981.)

ratus? As fish came up onto land, they had a new problem—how to hear in a medium as thin as air. The hyomandibular bone happens to be near the otic capsule, and bony material is excellent for transmitting sound. Thus, while still functioning as a cranial brace, the hyomandibular bone also began functioning as a sound transducer (Clack, 1989). As the terrestrial vertebrates altered their locomotion, jaw structure, and posture, the cranium became firmly attached to the rest of the skull and did not need the hyomandibular brace. It then seems to have become specialized into the stapes bone of the middle ear.

The original jaw bones changed also. The first branchial arch generates the jaw apparatus. In amphibians, reptiles, and birds, the posterior portion of this cartilage forms the quadrate bone of the upper jaw and the articular bone of the lower jaw. These bones connect to one another and are responsible for articulating the upper and lower jaws. However, in mammals, this articulation occurs at another region (the dentary and squamosal bones), thereby "freeing" these bony elements to acquire new functions. The quadrate bone of the reptilian upper jaw evolved into the mammalian incus bone, and the articular bone of the reptile's lower jaw has become our malleus. This latter process was first described by Reichert in 1837 when he observed in the pig embryo that the mandible (jawbone) ossifies on the side of Meckel's cartilage, while the posterior region of Meckel's cartilage ossifies, detaches from the rest of the cartilage, and enters into the region of the middle ear to become the malleus (Figure 16).

These dramatic changes in bone arrangement from agnathans to jawed fish, from jawed fish to amphibians, and from reptiles to mammals were coordinated with changes in jaw structure, jaw musculature, tooth deposition and shape, and modifications of the cranial vault and ear (Kemp, 1982; Thomson, 1988). It therefore seems that one particular change may co-vary with several others.

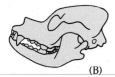
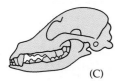
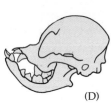

FIGURE 17
Skulls of various breeds of dogs. (A) Borzoi, (B) St. Bernard, (C) Windspiel, and (D) Japanese spaniel. The jaw musculature and facial skeleton develop in a correlated fashion. (After Waddington, 1956.)

Evidence for correlated progression

These changes in jaw and ear formation are the products of development and natural selection that occurred hundreds of millions of years ago. Can one breed for such changes in existing organisms and see if correlated progression occurs? Fortunately for us, humans have a great talent for selecting hereditary variants involving those neural crest cells that form the frontonasal and mandibular processes. Figure 17 shows the skull profiles from different breeds of dogs. Each variation is genetically determined, and it is important to note that each represents a harmonious rearrangement of the different bones with each other and with their muscular attachments. In some cases, such as that of bull dogs, the breed is selected for a wide face with very little angle between head and jaw. Other breeds, such as the collie, are selected for narrow snouts with a long jaw protruding away from the head. All breeds can move their jaws, shake their heads, and bark, despite the differences in the ways their bones are shaped or positioned. As the skeletal elements were selected, so were the muscles that moved them, the nerves that controlled these movements, and the blood vessels that fed them.*

Correlated progression has also been shown experimentally. Repeating earlier experiments of Hampé (1959), Gerd Müller (1989) inserted barriers of gold foil into the prechondrogenic hindlimb buds of a 3.5-day chick embryo. This barrier separated the regions of tibia formation and fibula formation. The results of these experiments are twofold. First, the tibia is shortened and the fibula bows and retains its connection to the fibulare. Such relationships between the tibia and fibula are not usually seen in birds, but they are characteristic of reptiles (Figure 18). Second, the *musculature* of the hindlimb undergoes parallel changes with the bones. Three of the muscles that attach to these bones now show characteristic reptilian patterns of insertion. It seems, therefore, that experimental manipulations that alter the development of one part of the mesodermal limb-forming field also alter the development of other mesodermal components as well. As with the correlated progression seen in face development, these changes all appear to be due to interactions within a field, in this case, the chick hindlimb field. These are not global effects and can occur independently of the other portions of the body.

In some cases, correlated progression may be a pleiotropic effect having a single origin. This could be a "key adaptation" that alters other tissues as a consequence (Larson et al., 1981). For instance, if the bones of the limb or mouth are changed, the mesodermal regions that become bone-associated muscle are simultaneously altered. This situation could be caused either by a single inducer that affects both cell types, or by a cascade of inductions such that the muscles that insert in the upper portion of the bone are induced to insert there by the bone itself. In other cases, correlated progression might be selected for at the level of gene translocations. One such case involves Batesian mimicry in the swallowtail butterfly *Papilio memnon* (Clarke et al., 1968). This butterfly is a tasty morsel for several bird species, but it survives by mimicking two relatively unpalatable butterflies, *Atrophaneura coon* and *A. aristolochiae* (Figure 19). *P. memnon* carries a "supergene" complex containing at least five closely linked genes. One locus controls the presence (*T*) or absence (*t*) of tails. A second gene controls whether the abdomen color is yellow (B^Y), yellow-tipped (b^y), or black (*b*). A third gene controls the hindwing color pattern

---

*This coordination is not quite universal, however. In dogs with greatly shortened faces (such as bulldogs), the skin has not coordinated its development with the bones and therefore hangs in folds from the head (Stockard, 1941).

(A)

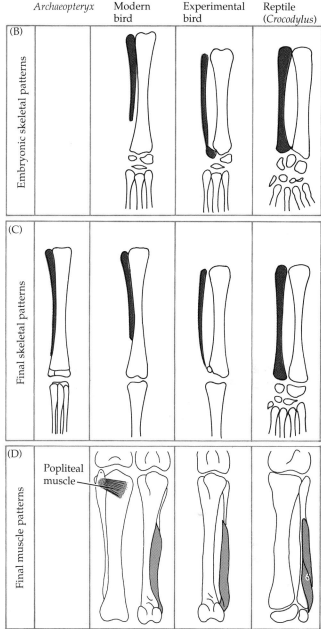

FIGURE 18

Experimental "atavisms" produced by altering embryonic fields in the limb. (A) Fossil *Archaeopteryx* in limestone. Imprints of feathers can be clearly seen. Were it not for the feathers, this toothed organism would probably have been classified as a reptile. (B–D) Results of Müller's experiments wherein gold foil split the chick hindlimb field. B and C show the embryonic and final bone pattern, indicating that the fibulare structure was retained by the experimental chick limb, as it is in extant reptiles and as it is thought to be in *Archaeopteryx*. D shows some of the muscle changes in the experimental chick embryos. The popliteal muscle is present in the chick, but is absent from reptile limbs and from the experimental limb. The fibularis brevis muscle, which normally originates from both the tibia and fibula in chicks, takes on the reptilian pattern of originating solely from the fibula in the operated limbs. (Photograph courtesy of B. A. Miller/Biological Photo Service; B–D after Müller, 1989.)

mimicking the wings of *A. coon* (W^{al}) or *A. aristolochiae* (W^d). A fourth and fifth gene in this cluster control forewing color pattern and the color of the spot at the base of the forewings. In each population of *P. memnon* studied, there are different alleles at this locus that allow the butterfly to mimic the distasteful model. For example, in Hong Kong, *P. memnon* has the gene constellation $TW^{al}B^Y$, which forms a tailed butterfly with abdomen and hindwing colors similar to the *A. aristolochiae* populations found there. This constellation of characters, however, is inherited as a unit because its genes have been selected to be so closely linked together that recombinations are relatively rare events. This enables the butterfly to mimic the model completely. Mimicking only part of the animal is not as good. Thus, correlated variation in some species (such as dogs) probably comes from interactions between correlated components during development, while in other species (such as the swallowtail butterflies), correlated variation can be at the genetic level.

FIGURE 19
Mimicry in *P. memnon*. (A) Unpalatable model, *A. coon*, and its mimic, *P. memnon*. (B) Unpalatable model, *A. aristolochia*, with its mimic, another genetic form of *P. memnon*. The ability of *P. memnon* to mimic either species depends on alleles at a gene complex containing several linked loci. (After Clarke et al., 1968.)

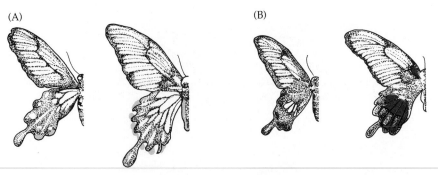

(A)

(B)

| Unpalatable | Mimic | Unpalatable | Mimic |
| *A. coon* | *P. memnon* | *A. aristolochia* | *P. memnon* |

Heterochrony and allometry

Heterochrony is a shift in the relative timing of two developmental processes during embryogenesis from one generation to the next. We have come across this concept in our discussions of neoteny and progenesis in salamanders (Chapter 19). Heterochrony can be caused in different ways. In salamander heterochronies, where the larval stage is retained, heterochrony is caused by gene mutations in the induction–competence system. Other heterochronic phenotypes, however, are caused by the heterochronic expression of certain genes. The direct development of the adult sea urchin rudiment (Chapter 19) involves the early activation of adult genes and the suppression of larval gene expression (Raff and Wray, 1989). As discussed earlier in the text, heterochrony can affect evolution in several ways. It can "return" an organism to a larval state, free from the specialized adaptations of the adult. Heterochrony can also give larval characteristics to an adult organism, as in the small size and webbed feet of arboreal salamanders or the fetal growth rate of human newborn brain tissue.

Allometry occurs when different parts of the organism grow at different rates. As seen in Chapter 20, allometry can be very important in forming variant body plans within a *Bauplan*. Such differential growth changes can involve altering a target cell's sensitivity to growth factors or altering the amounts of growth factors produced. Again, the vertebrate limb can provide a useful illustration. Local differences in chondrocytes cause the central toe of the horse to grow at a rate 1.4 times that of the lateral toes (Wolpert, 1983). This means that as the horse grew larger during evolution, this regional difference caused the five-toed horse to become a one-toed horse. A particularly dramatic example of allometry in evolution comes from skull development. In the very young (4- to 5-mm) whale embryo, the nose is in the usual mammalian position. However, the enormous growth of the maxilla and premaxilla (upper jaw) pushes over the frontal bone and forces the nose to the top of the skull (Figure 20). This new position of the nose (blowhole) allows the whale to have a large and highly specialized jaw apparatus, and to breathe while parallel to the water's surface (Slijper, 1962).

Allometry can also generate evolutionary novelty by small incremental changes that eventually cross some developmental threshold (sometimes called a bifurcation point). Eventually, a change in quantity becomes a change in quality when such a threshold is crossed. This type of mechanism has been postulated to have produced the external fur-lined "neck" pouches of pocket gophers and kangaroo rats that live in deserts. External pouches differ from internal ones in that (1) they are fur-lined, and (2)

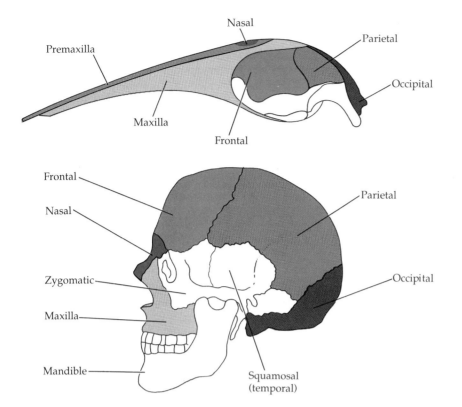

FIGURE 20
Allometric growth in the whale head. The jaw has pushed forward, causing the nose to move to the top of the skull. A human skull is shown for comparison. (The premaxilla is present in the early human fetus, but it fuses with the maxilla by the end of the third month of gestation. The human premaxilla was discovered by Wolfgang Goethe, among others, in 1786.) (After Slijper, 1962.)

they have no internal connection to the mouth. They are very useful in that they allow these animals to store seeds without running the risk of desiccation. Brylski and Hall (1988) have dissected the heads of pocket gopher and kangaroo rat embryos and have looked at the way that the external cheek pouch is constructed. When data from these animals were compared with data from animals that form internal cheek pouches (such as hamsters), the investigators found that the pouches are formed in very similar manners. In both cases, the pouches are formed within the embryonic cheeks by outpocketings of the cheek (buccal) epithelium into the facial mesenchyme (Figure 21). In animals with internal cheek pouches, these evaginations stay within the cheek. However, in animals that form external pouches, the elongation of the snout draws up the outpocketings into the region of the lip. As the lip epithelium rolls out of the oral cavity, so do the outpockets. What had been internal becomes external. The fur lining is probably derived from the external pouches' coming in contact with dermal mesenchyme, which can induce hair to form in epithelia (see Chapter 16). Such a pouch has no internal opening to the mouth. Indeed, the transition from internal to external pouch is one of threshold. The placement of the evaginations anteriorly or posteriorly determines whether the pouch is internal or not. There is no "transition stage" having two openings, one internal and one external.* One could envision this exter-

---

*The lack of transition forms is often used as a criticism of evolution by Creationists. For instance, in the transition from reptiles to mammals, discussed above, three of the bones of the reptilian jaw become the incus and malleus, leaving only one bone (the dentary) in the lower jaw. Gish (1973), a Creationist, says that this is an impossible situation, since no fossil has been discovered showing two or three jaw bones and two or three ear ossicles. Such an animal, he claims, would have dragged its jaw on the ground. However, such a specific transition form (and there are over a dozen documented transition forms between reptile and mammalian skulls) need never have existed. Hopson (1966) has shown on embryological grounds how the bones of the jaw could have divided and been used for different functions, and Romer (1970) has found reptilian fossils wherein the new jaw articulation was already functional while the older bones were becoming useless.

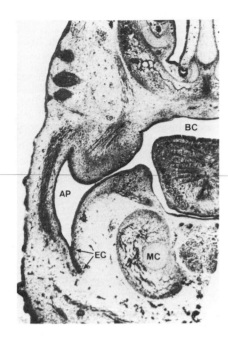

FIGURE 21

Transverse section through the anterior region of the pocket gopher (*Thomomys*) embryo showing the anterior opening of the pouch (AP) and the continuity between the pouch at this stage and the buccal cavity (BC) across the developing lip area. MC, Meckel's cartilage; T, tongue. (From Brylski and Hall, 1988; photograph courtesy of the authors.)

nalization occurring by a chance mutation that shifted the outpocketing to a slightly more anterior location. Such a trait would be selected for in the desert. As Van Valen reflected in 1976, evolution can be defined as "the control of development by ecology."

Transfer of competence

Another way that induction can be used to create developmental changes that may be selected for during evolution is by TRANSFER OF COMPETENCE. This idea, popularized by C. H. Waddington and I. I. Schmalhausen, enables the selection of new phenotypes that are favored for survival. In 1936, Waddington, Needham, and Brachet made an unexpected discovery. They had been hoping to find a specific factor from the notochord that would induce neural plates but instead found that a large variety of natural and artificial compounds were able to cause this induction. This led them to reinterpret their data and to suggest that the actual neuralizing factor lies dormant in the competent ectoderm and that a variety of factors were able to release the neuralizing factor from its inhibitor. Waddington then began to focus on the competent cells rather than on the inducing cells. He thought that many things could act as inducers, but the competent tissue had to have something within it that allowed it to respond to these chemicals. The inducer, he wrote (1940), was only the push. It was the competence that was genetically controlled and that was responsible for the details of the development.

Since competence could be achieved independently from an inducer, and since different compounds could induce the same developmental process in this competent tissue, Waddington proposed that a given competent tissue could transfer its ability to respond from one inducing stimulus to another. Moreover, these inducers could be either internal or external. External inducers were already known, and Waddington used the case of sex determination in *Bonnelia* (Chapter 21) as such an example. He also noted that the ability to form calluses on those areas of skin that abrade the ground was such an example. Here, the skin cells had the ability to form a callus if induced by friction. The genes could respond by causing the proliferation of cells to form the callus structure. While such examples of environmentally induced callus formation are widespread, the ostrich is *born* with them. Waddington hypothesized that since the skin cells were already competent to be induced by friction, they could be induced by other things as well. As ostriches evolved, a mutation appeared that enabled the skin cells to respond to a substance within the embryo. In this way, a trait that had been induced by the environment became part of the genome of the organism and could be selected. He called this phenomenon "genetic assimilation."

Proving transfer of competence is extremely difficult since it means that one has to first establish that a tissue could respond in a particular way to environmental inducers and then that it evolved a receptor for an internal stimulus that would move the cell into the same pathway of differentiation. As in much of evolutionary theory, one has to use circumstantial evidence for such evolution. There are, however, some candidates for evolution by transfer of competence.

Possible transfer of competence in ant populations. One place where competence may be transferred from an environmental stimulus to a genomic stimulus is in the evolution of caste separation in ant species. Ant colonies are predominantly female, and the females can be extremely polymorphic. The two major types of females are the worker and the gyne. The gyne is a potential queen. In more specialized species, a larger worker, the soldier, is also seen. These castes are determined by the levels of juvenile hormone seen by the developing larva. If the larva has more juvenile hormone, the larva grows, and this allometric growth forms larger jaws or allows the development of reproductive organs. The determination of the amount of juvenile hormone in the larva can be nutritionally controlled or regulated internally through maternal hormones that act during embryogenesis.

The developmental mechanisms of caste determination have been analyzed by Diana Wheeler (1986) and are summarized in Figure 22. In most species, ant larvae are bipotential until near pupation. In *Myrmica rubra*, only larvae that overwinter remain bipotential. After winter, the queen stimulates workers to *under*feed the last instar larvae. This means that as long as there is a queen, no new queens can result. If the larvae are fed, they can becomes gynes. Thus, larvae remain bipotential until late in their last instar.

In *Pheidole pallidula*, the queens appear to control caste determination during embryogenesis. There are no bipotential larvae in this species, caste determination being decided completely within the embryo. Gynes are produced from those eggs laid just after hibernation has ended, so it is assumed that eggs laid at different times are biochemically different. Although there is controversy over how the gyne-eggs and the worker-eggs differ, we have here an example of caste determination occurring within the embryo.

These ants show that what had been an environmental type of determination can be internalized. Notice, too, that in *Pheidole* there is polymorphism in the worker caste. Soldiers—large workers—have been formed. Wilson (1971) has pointed out that of the 260 known genera of ants, only seven have complete dimorphism of the workers. If having polymorphic workers is advantageous (Oster and Wilson, 1978), why don't more genera have them? Wheeler answers that worker polymorphism is only seen when the gyne determination has occurred very early. Once that has been determined, the nutritional variations can be used to create different worker castes.

Possible transfer of competence in fish sex determination. If competent tissues can change their developmental triggers from environmental stimuli to internal inducers encoded by the genome, then it would stand to reason that the reverse could also occur. Under some selective conditions, one would expect genetic induction to be replaced by environmental induction. This phenomenon might be the case in certain types of environmental sex determination. Environmental sex determination is found among certain invertebrates and among vertebrates such as fish, amphibians, and reptiles (Chapter 21). Charnov and Bull (1977) have argued that environmental sex determination would be adaptive in certain habitats characterized by patchiness and having certain regions in which it is more

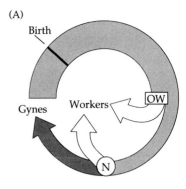

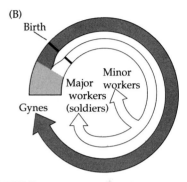

FIGURE 22
Possible transfer of competence in gyne (queen) formation in ants. Lightly colored areas represent bipotentiality to become either workers or gynes. The circled N represents a nutritional switch controlled by the larva's environment. (A) *Myrmica rubra*, wherein only the larvae that overwinter (OW) remain bipotential. In the last instar, the nutritional switch determines caste. (B) *Pheidole pallidula*, wherein the queen controls gyne determination through hormones that act during embryogenesis. (After Wheeler, 1986.)

FIGURE 23
Relationship between temperature and sex ratio F/(F + M) during the period of sex determination in the fish *Menidia menidia*. In those fish collected from the northernmost portion of its range (Nova Scotia), temperature had little effect on sex determination. When embryos were collected from fish at more southerly locations (especially from Virginia through South Carolina), the environment had a large effect. (From Conover and Heins, 1987.)

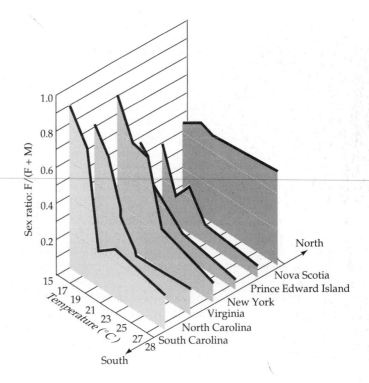

advantageous to be male and other regions in which it is more advantageous to be female. Conover and Heins (1986) provide evidence that in certain fish, females benefit from being larger, since size translates into higher fecundity. It is an advantage to be born early in the breeding season if you are a female *Menidia*, since you would have a longer feeding season and would grow larger. In the males, size is of no importance. Conover and Heins showed that in the southern range of *Menidia*, females are indeed born early in the breeding season. Temperature appears to play a major role. However, in the northern reaches of its range, the same species shows no environmental sex determination. Rather, a 1:1 ratio is generated at all temperatures (Figure 23). The authors speculate that this is because the more northern populations have a very short feeding season so that there is no advantage for a female to be born earlier. Thus, this species of fish has environmental sex determination in those regions where it is adaptive, and genotypic sex determination in those regions where it is not. Here again, one sees that the environment can induce sexual phenotype, or sexual phenotype can be a property of the genome, as it is with most mammals.

Delayed implantation

One probable example of competence being transferred from one inducer to another involves a species of which some subspecies have their inductive stimuli internal and other subspecies have the same phenomenon under the control of another inducer. This phenomenon can be seen in the implantation of blastocysts into the uterus of the spotted skunk. The eastern subspecies of the spotted skunk *Spirogale putorius* has normal implantation. It mates in April, has a gestation period of about two months, and the young are born in June or July when food is abundant. The western subspecies, however, lives at higher altitudes and is not able to mate in April. Rather, the western skunks mate in September. If the embryos developed at the same rate and had a two-month gestation, the young would be born in November and December and would starve.

Instead, the western subspecies have evolved DELAYED IMPLANTATION. Although fertilization and early cleavage start in September, implantation of the blastocyst is delayed until April (the same time as implantation in the eastern subspecies). The timing of the delay is probably regulated by day length, since blinded skunks have prolonged periods of delayed implantation. (In other animals with delayed implantation, the photoperiod is sometimes perceived by the pineal gland.) At the end of the period of delay, progesterone is made, which enables the uterus to receive a blastocyst that has been waiting some 210–230 days (Renfree, 1982). Delayed implantation is part of the life cycle of numerous mammals.

SIDELIGHTS & SPECULATIONS

Transfer of competence by transfer of receptors

According to Waddington, the competent cell evolved a genetically controlled pathway by which it could respond to an inducer. When the competency of the cell was transferred from one inducer to another, the receptor for the signal changed, but the pathway distal to the signal remained the same. We can see this in the competencies that different cells have to mitogenic agents. Let us recall (from Chapter 16) the division of a competent B-lymphocyte. Upon binding antigen through its surface immunoglobulins, the B cell acquires the cell surface receptors for growth factors. In other words, specific competence is provided by the antigen receptor. If a different antigen is presented, a different set of B cells becomes competent to divide and differentiate into antibody-secreting plasma cells. Thus, the binding of antigen to a B cell antigen receptor triggers into action a *preset program* for cell division and differentiation.

We also see transfer of competence when we look at the cell division pathway of the B cell. A G-protein that is bound to the cell surface immunoglobulin activates phospholipase C. Phospholipase C splits phosphatidylinositol bisphosphate (PIP_2) into inositol trisphosphate, which releases calcium from the endoplasmic reticulum, and into diacylglycerol, which activates protein kinase C with a resultant elevation in intracellular pH. Together, these signals activate the nuclear proto-oncogenes that initiate cell division. Interestingly enough, these are the same reactions that happen when antigens bind to the T cell receptor. So it appears that lymphocytes, both B and T, use the same pathway for cell division. Similarly, several neurohormones, such as serotonin and acetylcholine, function by activating G-proteins and initiating the same pathway. The reactions are the same. All that changes is the cell surface receptor.

But this should be an already familiar story. Indeed, lymphocytes and neurons are latecomers to this pathway, for this pathway is the series of reactions by which every cell in the body divides. A more general scheme is known in which a growth factor binds to the growth factor receptor (Chapter 20). This, in turn, activates the G-protein, which activates phospholipase C, which starts the process in motion. But neither is this the entire story. The same pathway of division is seen in the reactions that activate the fertilized egg (Chapter 2). Here, the activator is not an antigen or a peptide growth factor; it is a sperm. One can see that the pathway is independent of the stimulus that initiates the cell division. Different receptors can be inserted into the membrane, causing the pathway to become activated by new compounds. Indeed, one can experimentally add the mRNA for acetylcholine and serotonin receptors and activate *Xenopus* eggs when one adds these neurotransmitters (Kline et al., 1988). Here an internal inducer (neurotransmitter) has substituted for an external inducer (sperm) merely by changing the receptor. Therefore, by changing the cell surface receptor, the competence of the cell is changed, and the cell can respond to a different inducer.

A new evolutionary synthesis

One of the major events in evolutionary theory has been the "modern synthesis" of evolutionary biology and Mendelian genetics (Mayr and Provine, 1980). One outcome of this hard-won merger is that evolution has been redefined to mean changes in gene frequencies in a population over time. "Since evolution is a change in the genetic composition of populations," wrote Dobzhansky (1937), "the mechanisms of evolution constitute problems of population genetics." This definition became largely agreed upon in the United States, England, and the Soviet Union, and thus morphology and development were seen to play little role in modern evolutionary theory (Adams, 1991). The large morphological changes seen during evolutionary history could be explained by the accumulation of small genetic changes. In other words, macroevolution (the large morphological changes seen between species, classes, and phyla) could be explained by the mechanisms of microevolution, the "differential adaptive values of genotypes or deviations from random mating or both these factors acting together" (Torrey and Feduccia, 1979).

However, this view has had its critics (its heretics, as some would say). Perhaps the foremost of these was Richard Goldschmidt. Goldschmidt began his book, *The Material Basis of Evolution* (1940), with a challenge to the modern synthesis. It may be able to explain the *survival* of the fittest, but not the *arrival* of the fittest.

> I may challenge the adherents of the strictly Darwinian view, which we are discussing here, to try to explain the evolution of the following features by accumulation and selection of small mutants: hair in mammals, feathers in birds, segmentation in arthropods and vertebrates, the transformation of the gill arches in phylogeny including the aortic arches, muscles, nerves, etc.; further, teeth, shells of molluscs, ectoskeletons, compound eyes, blood circulation, alternation of generations, statocysts, ambulacral systems of echinoderms, pedicellaria of the same, cnidocysts, poison apparatus of snakes, whalebone, and finally chemical differences like hemoglobin vs. hemocyanin.

He claimed that new species did not arise from the mechanisms of microevolution, and that population genetics was unable to explain new types of structures that involve several components changing simultaneously. Such macroevolutionary change "requires another evolutionary method than that of sheer accumulation of micromutations." Goldschmidt saw homeotic mutants as "macromutations" that could change one structure into another and possibly create new structures or new combinations of structures. These mutations would not be in the structural genes but in the regulatory genes. A new species, Goldschmidt asserted, would start as a "hopeful monster" (a rather unfortunate phrase having its antecedent in Metchnikoff's prose).

At the same time, Waddington was attempting to find developmental mechanisms for producing such new species. He, too, looked at homeotic mutations in flies as models for drastically new phenotypes, and he formulated the notion of competence transfer ("genetic assimilation") to explain certain aspects of morphological evolution. Few scientists paid attention to Goldschmidt or Waddington because they were not writing in the population genetics paradigm of the modern synthesis and their scientific programs were suspect. (Goldschmidt did not believe in Morgan's notion of the gene as a particulate entity, and Waddington's work was misinterpreted as supporting the inheritance of acquired traits.) However, in the 1970s, events in paleontology (the punctuated equilibrium theory), events in society (the Creationists giving the microevolutionary contest to the biologists, but contesting macroevolution), and events in molecular biology (notably King and Wilson's 1975 paper showing that chimps and humans have DNA that is greater than 99 percent identical) prompted scientists to consider seriously the view that mutations in regulatory genes could create large changes in morphology.

A new developmental synthesis is emerging that retains the best of the microevolution-yields-macroevolution model and the macroevolution-as-separate-phenomenon model. From the latter it derives the concept that mutations in regulatory genes can create "jumps" from one phenotype to another without necessary intermediate steps. From the former, it derives the notion that genetic mutations can account for such variants and that selection acts upon them to delete them or retain them in populations. It also retains a multiplicity of paradigms. In some instances (such as the creation of neural crest cells) a qualitative change occurs, whereas in other cases (such as the formation of the pocket gopher pouch) quantity becomes quality when a threshold is passed.

We are at a remarkable point in our understanding of nature, for a synthesis of developmental genetics with evolutionary biology may transform our appreciation of the mechanisms underlying evolutionary change and animal diversity. Such a synthesis is actually a return to a broader-based evolutionary theory that fragmented at the turn of the past century (Figure 24). In the late 1800s, evolutionary biology contained the sciences that we now call evolutionary biology, systematics, ecology, genetics, and

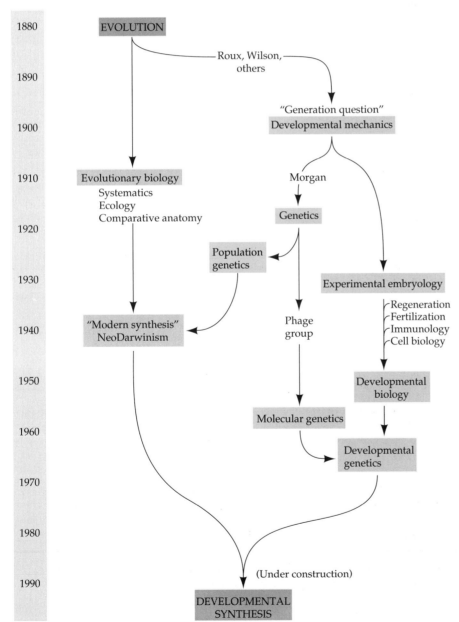

FIGURE 24
Disciplinary road map of the evolutionary side of biology, 1880 to the present. For sake of clarity, other paths (such as those from genetics to human genetics or from evolution to immunology) have not been shown.

development. By the turn of the century, the "question of heredity," i.e., genetics and development, separated from the rest of evolutionary biology. Genetics eventually split into (among other rubrics) population genetics and molecular genetics, while embryology became developmental biology (Gilbert, 1978, 1988). During the mid-twentieth century, population genetics merged with evolutionary biology to produce the evolutionary genetics of the modern synthesis, while molecular genetics merged with developmental biology to produce developmental genetics. These two vast areas, developmental genetics and evolutionary genetics, are on the verge of a merger that may unite these long-separated strands of biology and that may produce a developmental genetic theory capable of explaining macroevolution.

When Wilhelm Roux (1894) announced the creation of "developmental mechanics," he did not fully break with evolutionary biology. Rather, he stated that "an *ontogenetic* and a *phylogenetic developmental mechanics* are to be perfected." He noted further that the developmental mechanics of embryos (the ontogenetic branch) would proceed faster than the phylogenetic studies, but posited that "in consequence of the intimate causal connections between the two, many of the conclusions drawn from the investigation of ontogeny [would] throw light on the phylogenetic processes." We are now at the point where we can attend to the second of Roux's developmental mechanics and create a unified theory of evolution.

LITERATURE CITED

Adams, M. 1991. Soviet perspectives on evolutionary theory. *In* L. Warren and M. Meselson (eds.), *New Perspectives in Evolution.* Liss/Wiley, New York.

Alberch, P. and Gale, E. 1983. Size dependency during the development of the amphibian foot. Colchicine induced digital loss and reduction. *J. Embyol. Exp. Morphol.* 76: 177–197.

Alberch, P. and Gale, E. 1985. A developmental analysis of an evolutionary trend. Digit reduction in amphibians. *Evolution* 39: 8–23.

Alexandrov, D. A. and Sergievsky, S. O. 1984. A variant of sympatric speciation in snails. *Malacol. Rev.* 17: 147.

Ambros, V. 1989. A hierarchy of regulatory genes controls a larva-to-adult developmental switch in *C. elegans. Cell* 57: 49–57.

Bateson, W. 1894. *Materials for the Study of Variation.* Cambridge University Press, Cambridge.

Berg, R. L. 1960. Evolutionary significance of correlation pleiades. *Evolution* 14: 171–180.

Berrill, N. J. 1955. *The Origins of the Vertebrates.* Oxford University Press, New York.

Berrill, N. J. 1987. Early chordate evolution. I. Amphioxus, the riddle of the sands. *Int. J. Invert. Repr. Dev.* 11: 1–27.

Brylski, P. and Hall, B. K. 1988. Ontogeny of a macroevolutionary phenotype: The external cheek pouches of geomyoid rodents. *Evolution* 42: 391-395.

Burke, A. C. 1989a. Epithelial-mesenchymal interactions in the development of the chelonian Bauplan. *Fortschr. Zool.* 35: 206–209.

Burke, A. C. 1989b. Development of the turtle carapace: Implications for the evolution of a novel bauplan. *J. Morphol.* 199: 363–378.

Buss, L. W. 1987. *The Evolution of Individuality.* Princeton University Press, Princeton, NJ.

Charnov, E. L. and Bull, J. J. 1977. When is sex environmentally determined? *Nature* 266: 828–830.

Chernyak, L. and Tauber, A. I. 1988. The birth of immunology: Metchnikoff, the embryologist. *Cell. Immunol.* 117: 218–233.

Chernyak, L. and Tauber, A. I. 1991. *From Metaphor to Theory: Metchnikoff and the Origin of Immunology.* Oxford University Press, New York.

Clack, J. A. 1989. Discovery of the earliest known tetrapod stapes. *Nature* 342: 425–427.

Clark, B. and Murray, J. 1969. Ecological genetics and speciation in land snails of the genus *Partula. Biol. J. Linn. Soc.* 1: 31–42.

Clarke, C. A., Sheppard, P. M. and Thornton, I. W. B. 1968. The genetics of the mimetic butterfly *Papilio memnon. Philos. Trans. R. Soc. Lond.* [B] 254: 37–89.

Conklin, E. G. 1915. *Heredity and Environment in the Development of Men.* Princeton University Press, Princeton, NJ.

Conover, D. O. and Heins, S. W. 1987. Adaptive variation in environmental and genetic sex determination in a fish. *Nature* 326: 496–498.

Conway Morris, S. and Whittington, H. B. 1979. The animals of the Burgess Shale. *Sci. Am.* 240(1): 122–133.

Darwin, C. 1859. *The Origin of Species.* John Murray, London.

Dobzhansky, T. G. 1937. *Genetics and the Origin of Species.* Columbia University Press, New York

Elinson, R. P. 1987. Changes in developmental patterns: Embryos of amphibians with large eggs. *In* R. A. Raff and E. C. Raff (eds.), *Development as an Evolutionary Process.* Alan R. Liss, New York; pp. 1–21.

Fraser, S., Keynes, R. and Lumsden, A. 1990. Segmentation in the chick embryo hindbrain is defined by cell lineage restrictions. *Nature* 344: 431–435.

Gans, C. 1989. Stages in the origin of vertebrates: Analysis by means of scenarios. *Biol. Rev.* 64: 221–268.

Gans, C. and Northcutt, R. G. 1983. Neural crest and the origin of vertebrates: A new head. *Science* 220: 268-274.

Garstang, W. 1922. The theory of recapitulation: a critical restatement of the biogenetic law. *J. Linn. Soc. Zool.* 35: 81–101.

Garstang, W. 1928. Presidential address to the British Association for the Advancement of Science, Section D. Republished in *Larval Forms and Other Zoological Verses.* 1985. University of Chicago Press, Chicago, pp. 77–98.

Gasman, D. 1971. *The Scientific Origins of National Socialism: Social Darwinism in Ernst Haeckel and the German Monist League*. Mac-Donald, London.

Gemmill, J. F. 1912. The development of the starfish *Solaster endica* (Forbes). *Trans. Zool. Soc. Lond.* 20: 1–72.

Gilbert, S. F. 1978. The embryological origins of the gene theory. *J. Hist. Biol.* 11: 307–351.

Gilbert, S. F. 1988. Cellular politics: Ernest Everett Just, Richard B. Goldschmidt, and the attempt to reconcile embryology and genetics. *In* R. Ranger, K. R. Benson and J. Maienschein (eds.), *The American Development of Biology*. University of Pennsylvania Press, Philadelphia, pp. 311–346.

Gish, D. T. 1973. *Evolution? — The Fossils Say No!* Creation-Life Publishers, San Diego.

Goldschmidt, R. B. 1940. *The Material Basis of Evolution*. Yale University Press, New Haven.

Gould, S. J. 1977a. *Ever Since Darwin*. Norton, New York.

Gould, S. J. 1977b. *Ontogeny and Phylogeny*. Harvard University Press, Cambridge, MA.

Gould, S. J. 1989. *Wonderful Life*. Norton, New York.

Gould, S. J. 1990. An earful of jaw. *Natural History* 1990(3): 12–23.

Gould, S. J. and Lewontin, R. C. 1979. The spandrels of San Marcos and the Panglossian paradigm: A critique of the adaptationist program. *Proc. R. Soc. London* [B] 205: 581–598.

Grant, P. 1978. *Biology of Developing Systems*. Holt, Rinehart and Winston, New York.

Haeckel, E. 1900. *The Riddle of the Universe*. Harper & Bros., New York.

Hampé, A. 1959. Contribution à l'étude du développement et al régulation des déficiences et excédents dans la patte de l'embryon de poulet. *Arch. Anat. Micr. Morph. Exp.* 48: 347–479.

Henry, J. J., Amemiya, S., Wray, G. A. and Raff, R. A. 1989. Early inductive interactions are involved in restricting cell fates of mesomeres in sea urchin embryos. *Dev. Biol.* 136: 140–153.

Holder, N. 1983. Developmental constraints and the evolution of vertebrate limb patterns. *J. Theor. Biol.* 104: 451–471.

Holland, P. W. H. 1990. Homeobox genes and segmentation: co-option, co-evolution, and convergence. *Sem. Dev. Biol.* 1: 135–145.

Hopson, J. A. 1966. The origin of the mammalian middle ear. *Am. Zool.* 6: 437–450.

Jefferies, R. P. S. 1986. *The Ancestry of the Vertebrates*. British Museum of Natural History, London.

Jones, M. L. and Gardiner, S. L. 1989. On the early development of the vestimentiferan tube worm *Ridgeia* sp. and observations on the nervous system and trophosome of *Ridgeia* sp. and *Riftia pachyptila*. *Biol. Bull.* 177: 254–276.

Kauffman, S. A. *Origins of Order*. Oxford University Press, New York. In press.

Kemp, T. S. 1982. *Mammal-Like Reptiles and the Origin of Mammals*. Academic Press, New York.

King, M.-C. and Wilson, A. C. 1975. Evolution at two levels in humans and chimpanzees. *Science* 188: 197–116.

Kline, D., Simoncini, L., Mandel, G., Maue, R. A., Kado, R. T. and Jaffe, L. A. 1988. Fertilization events induced by neurotransmitters after injection of mRNA in *Xenopus* eggs. *Science* 241: 464–467.

Kowalevsky, A. 1871. Weitere Studien II. die Entwicklung der einfachen Ascidien. *Arch. Micr. Anat.* 7: 101–130.

LaBarbera, M. 1990. Principles of design of fluid transport systems in zoology. *Science* 249: 992–1000.

Langille, R. M. and Hall, B. K. 1989. Developmental processes, developmental sequences and early vertebrate phylogeny. *Biol. Rev.* 64: 73–91.

Langman, J. 1981. *Medical Embryology*. William and Wilkins, Baltimore.

Larson, A., Wake, D. B., Maxson, L. R. and Highton, R. 1981. A molecular phylogenetic perspective on the origins of morphological novelties in the salamander of the tribe plethodontini (Amphibia, Plethodontidae). *Evolution* 35: 405–422.

Lillie, F. R. 1898. Adaptation in cleavage. *Biological Lectures from the Marine Biological Laboratories, Woods Hole, Massachusetts*. Ginn, Boston, pp. 43–67.

Margulis, L. and Schwartz, K. V. 1988. *The Five Kingdoms*, 2nd Ed. W. H. Freeman, San Francisco.

Mayr, E. 1963. *Animal Speciation and Evolution*. Harvard University Press, Cambridge.

Mayr, E. and Provine, W. 1980. *The Evolutionary Synthesis: Perspectives on the Unification of Biology*. Harvard University Press, Cambridge.

Metchnikoff, E. 1886. *Embryologische Studien an Medusen. Ein Beitrage zur Genealogie der Primitive-Organe*. Vienna.

Metchnikoff, E. 1891. Zakon zhizni. Po povodu nektotorykh proizvedenii gr. L. Tolstogo. *Vest. Evropy* 9: 228–260. Quoted and translated in Chernyak and Tauber, 1990, *From Metaphor to Theory: Metchnikoff and the Origin of Immunology*. Oxford University Press, New York.

Mivart, St. G. 1871. *On the Genesis of Species*. Macmillan, London.

Müller, F. 1864. *Für Darwin*. Engelmann, Leipzig.

Müller, G. B. 1989. Ancestral patterns in bird limb development: A new look at Hampé's experiment. *J. Evol. Biol.* 1: 31–47.

Ospovat, D. 1981. *The Development of Darwin's Theory*. Cambridge University Press, Cambridge.

Oster, G. F. and Wilson, E. O. 1978. *Caste and Ecology in the Social Insects*. Princeton University Press, Princeton, N.J.

Oster, G. F., Shubin, N., Murray, J. D. and Alberch, P. 1988. Evolution and morphogenetic rules: The shape of the vertebrate limb in ontogeny and phylogeny. *Evolution* 42: 862–884.

Owen, R. 1848. *On the Archetype and Homologies of the Vertebrate Skeleton*. London.

Rabinow, L. and Dickinson, W. J. 1986. Complex *cis*-acting regulators and locus structure of *Drosophila* tissue-specific ADH variants. *Genetics* 112: 523–537.

Raff, R. A. and Kaufman, T. C. 1983. *Embryos, Genes, and Evolution*. Macmillan, New York.

Raff, R. A. and Wray, G. A. 1989. Heterochrony: Developmental mechanisms and evolutionary results. *J. Evol. Biol.* 2: 409–434.

Raff, R. A., Wray, G. A. and Henry, J. J. 1991. Implications of radical evolutionary changes in early development for concepts of developmental constraint. *In* L. Warren and M. Meselson (eds.), *New Perspectives in Evolution*. Liss/Wiley, New York.

Renfree, M. B. 1982. Implantation and placentation. *In* C. R. Austin and R. V. Short (eds.), *Reproduction in Mammals 2: Embryonic and Fetal Development*, Cambridge University Press, Cambridge.

Romer, A. S. 1970. The Chanares (Argentina) Triassic reptile fauna VI. A chiniquodontid cynodont with an incipient squamosal-dentary jaw articulation. *Breviora* 344: 1–18.

Roux, W. 1894. The problems, methods, and scope of developmental mechanics. *Biological lectures of the Marine Biology Laboratory, Woods Hole*. Ginn, Boston., pp. 149–190.

Schmalhausen, I. I. 1949. *Factors of Evolution: The Theory of Stabilizing Selection*. Blakiston, Philadelphia.

Slack, J. M. W. 1983. *From Egg to Embryo: Determinative Events in Early Development*. Cambridge University Press.

Slijper, E. J. 1962. *Whales*. (Translated by A. J. Pomerans.) Basic Books, New York.

Spemann, H. 1901. Über Correlationen in der Entwicklung des Auges. *Verh. Anat. Ges.* [Vers. Bonn] 15: 61–79.

Stein, G. J. 1988. Biological science and the roots of Nazism. *Am. Sci.* 76: 50–58.

Stockard, C. R. 1941. The genetic and endocrine basis for differences in form and behaviour as elucidated by studies of contrasted pure-line dog breeds and their hybrids. *Am. Anat. Memoirs* 19.

Thomson, K. S. 1988. *Morphogenesis and Evolution*. Oxford University Press, New York.

Torrey, T. W. and Feduccia, A. 1979. *Morphogenesis of the Vertebrates*. Wiley, New York.

Van Valen, L. M. 1976. Energy and evolution. *Evol. Theor.* 1: 179–229.

von Baer, K. E. 1828. *Entwicklungsgeschichte der Thiere: Beobactung und Reflexion*. Bornträger, Koningsberg.

von Baer, K. E. 1886. *Autobiography of Dr. Karl Ernst von Baer.* (Translated by H. Schneider.) Science History Publications, Canton, MA, pp.261–262.

Waddington, C. H. 1938. The morphogenetic function of a vestigial organ in the chick. *J. Exp. Biol.* 15: 371–376.

Waddington, C. H. 1940. *Organisers and Genes.* Cambridge University Press, Cambridge.

Waddington, C. H. 1956. *Principles of Embryology.* Allen and Unwin, London.

Waddington, C. H., Needham, J. and Brachet, J. 1936. Studies on the nature of the amphibian organization centre. III. The activation of the evocator. *Proc. R. Soc. London* [B] 120: 173-198.

Wake, D. B. and Larson, A. 1987. A multidimensional analysis of an evolving lineage. *Science* 238: 42–48.

Wheeler, D. 1986. Developmental and physiological determinants of caste in social hymenoptera: Evolutionary implications. *Am. Nat.* 128: 13–34.

Whittington, H. B. 1985. *The Burgess Shale.* Yale University Press, New Haven.

Wilson, E. B. 1898. Cell lineage and ancestral reminiscence. *Biological Lectures from the Marine Biological Laboratories, Woods Hole, Massachusetts.* Ginn, Boston, pp. 21-42.

Wilson, E. O. 1971. *The Insect Societies.* Harvard University Press, Cambridge, MA.

Winsor, M. P. 1969. Barnacle larvae in the nineteenth century: A case study in taxonomic theory. *J. Hist. Med. Allied Sci.* 24: 294–309.

Wolpert, L. 1983. Constancy and change in the development and evolution of pattern. *In* B. C. Goodwin, N. Holder and C. C. Wylie (eds.), *Development and Evolution.* Cambridge University Press, Cambridge.

Zangerl, R. and Williams, M. E. 1975. New evidence on the nature of the jaw suspension in Paleozoic anacanthus sharks. *Paleontology* 18: 333–341.

Sources for chapter-opening quotations

Borges, J. L. (1962). "The Library of Babel" in *Ficciones*. Grove Press, New York, pp. 87–88.

Boveri, T. (1904). *Ergebnisse über die Konstitution der chromatischen Substanz des Zelkerns*. Jena (G. Fischer), p 123.

Claude, A. (1974). The coming of age of the cell. Nobel lecture, reprinted in *Science* 189 (1975): 433–435.

Conrad, J. (1920). *The Rescue: A Romance of the Shallows*. Doubleday, Page, and Co., Garden City, NJ (1924), p. 447.

Darwin, C. (1871). *The Descent of Man*. Murray, London, p. 893.

Darwin, E. (1791). Quoted in Ghiselin, M. T. (1974). *The Economy of Nature and the Evolution of Sex*. University of California Press, Berkeley, p. 49.

Doyle, A. C. (1891). "A Case of Identity" in *The Adventures of Sherlock Holmes*. Reprinted in *The Complete Sherlock Holmes Treasury* (1976). Crown, New York, p. 31.

Eliot, T. S. (1936). "The Hollow Men," Part V, in *Collected Poems 1909–1962*. Harcourt, Brace and World, New York, pp. 81–82. Copyright by T. S. Eliot.

Eliot, T. S. (1942). "Little Gidding " in *Four Quartets*. Harcourt, Brace and Company, New York (1943), p. 39. Copyright by T. S. Eliot.

Garstang, W. (1922). The theory of recapitulation: A critical restatement of the biogenetic law. *J. Linn. Soc. Zool.* 35: 81–101.

Hardin, G. (1968). *Exploring New Ethics for Survival: The Voyage of the Spaceship Beagle*. Viking Press, New York, p. 45.

Harrison, R. G. (1933). Some difficulties of the determination problem. *Am. Nat.* 67: 306–321.

Huxley, T. (1895). Quoted in Mitchell, P. L. (1900), *Thomas Henry Huxley: A Sketch of His Life and Work*. Putnam, New York, p. 279.

Jacob, F. and Monod, J. (1963). Genetic repression, allosteric inhibition, and cellular differentiation. In M. Locke (ed.), *Cytodifferentiation and Macromolecular Synthesis*. Academic, New York, p. 31.

Just, E. E. (1939). *The Biology of the Cell Surface*. Blakiston, Philadelphia, p. 288.

Kant, I. (1790). *Critique of Judgement*. Translated by J. H. Bernard. Hafner, New York, pp. 267–268.

Lessing, G. E. (1778). Eine Duplik. In F. Muncker (ed.), *Sämtliche Schriften* 13. Göschen, Leipzig (1897), p. 23.

Levi-Montalcini, R. (1988). *In Praise of Imperfection*. Basic Books, New York, p. 90.

Monod, J. and Jacob, F. (1961). Teleonomic mechanism in cellular metabolism, growth, and differentiation. *Cold Spring Harbor Symp. Quant. Biol.* 26: 389–401.

Muller, H. J. (1922). Variation due to change in the individual gene. *Am. Nat.* 56: 32–50.

Needham, J. (1967). *Order and Life*. M.I.T. Press, Cambridge, MA, p. xv.

Pander, C. (1817). *Beiträge zur Entwickelungsgeschichte des Hühnchens im Eye*. Wuerzburg.

Rexroth, K. (1952). "Golden Section" in *Natural Numbers*. New Directions, New York.

Rostand, J. (1960). *Carnets d'un biologiste*. Librairie Stock, Paris.

Rostand, J. (1962). *The Substance of Man*. Doubleday, New York, p. 181.

Roux, W. (1894). The problems, methods, and scope of developmental mechanics. *Biol. Lect. Woods Hole* 3: 149–190.

Rushdie, S. (1989). *The Satanic Verses*. Viking, New York, p. 8.

Schultz, J. (1935). Aspects of the relation between genes and development in *Drosophila*. *Am. Nat.* 69: 30–54.

Shelley, M. W. (1831). *Frankenstein, or The Modern Prometheus*. Oxford University Press (1969), p. 36.

Spemann, H. (1943). *Forschung und Leben*. Quoted in Horder, T. J., Witkowski, J. A. and Wylie, C. C. (1986), *A History of Embryology*, Cambridge University Press, Cambridge, p. 219.

Tennyson, A. (1886). *Idylls of the King*. Macmillan, London (1958), p. 292.

Thomas, L. (1979). "On Embryology" in *The Medusa and the Snail*. Viking Press, New York, p. 157.

Thomson, J. A. (1926). *Heredity*. Putnam, New York, p. 477.

Virchow, R. (1858). *Die cellularpathologie in ihre Begrundung auf physiologische und pathologische Geweblehre*. Berlin, p. 493.

Virgil (37 B.C.). *Georgics II*: 490.

Waddington, C. H. (1956). *Principles of Embryology*. Macmillan, New York, p. 5.

Weiss, P. (1960). Ross Granville Harrison, 1870–1959. Memorial minute. *Rockefeller Inst. Quarterly*, p. 6.

Whitehead, A. N. (1919). *The Concept of Nature*. University of Michigan Press, Ann Arbor (1957), p. 163.

Whitehead, A. N. (1934). *Nature and Life*. Cambridge University Press, Cambridge, p. 41.

Whitman, C. O. (1894). Evolution and epigenesis. *Biol. Lect. Woods Hole* 3: 205–224.

Whitman, W. (1855). "Song of Myself." In S. Bradley (ed.), *Leaves of Grass and Selected Prose*. Holt, Rinehart & Winston, New York (1949), p. 25.

Whitman, W. (1867). "Inscriptions." In S. Bradley (ed.), *Leaves of Grass and Selected Prose*. Holt, Rinehart & Winston, New York (1949), p. 1.

Williams. C. M. (1958). Quoted in J. A. Miller (1983), A brain for all seasons. *Science News* 123: 268–269.

Wilson, E. B. (1925). *The Cell in Development and Heredity*, 3rd Ed. Macmillan, New York, p. 1112.

Wolpert, L. (1983). Quoted in J. M. W. Slack (1983), *From Egg to Embryo: Determinative Events in Early Development*. Cambridge University Press, Cambridge, p. 1.

Author Index

Blastopore, 122ff
chick, 145ff
twins and, 130–131
Blastopore lip *see* Dorsal lip of blastopore
Blastula, 4, 78ff, 117ff
amphibian, 82, 124–138
chick, 139ff
formation of, 80–83
gastrulation and, 118ff
hatching of, 82
junctions in, 80, 82, 83
primary mesenchyme in, 116–117
sea urchin, 78–82, 116–124
Blood cell
in chick, 223
development of, 229–237
Blood flow, changes at birth, 225–227
Blood island, 161, 222
Blood vessel, formation of, 222–225
Blotting, DNA and RNA, 353–355
Body cavity, formation of, 30
Body fold, 216–217
lateral, 161
Boltenia, 259
Bone
formation, 209–215
growth, 731
osteogenesis, 209–215
Bone marrow
hematopoiesis in, 229
stroma, 233
Bone morphogenic protein, 210ff
Bonellia viridis, sex determination, 784–785
boss gene, 597
Bottle cells, 20
amphibian gastrulation and, 125ff
endoderm formation and, 131–133
Brain
development, 165–167
POU expression and, 432
Brain-derived neurotrophic factor (BDNF),
635–636
Branch site, 470–473
Branchial arch, derivatives of, 845
Branchiostoma, egg, 39
bride of sevenless gene, 595–597
Brine shrimp, translational regulation in, 502
Bronchus, 240
Brooks's law of linear growth, 720–721
Buccinum undatum, 262
Burkitt's lymphoma, 747–748
Bursin, 234
Burst forming unit (BFU), 231

c-*abl*, 738
C blastomere, 84–86, 261ff
c-*erb*-A, 739
c-*erb*-B, 738
c-*fes*, 738
c-*fms*, 738
c-*fos*, 739, 744, 749
translational control of, 489–490
c-*fps*, 738
c-H-*ras*-1, 738
c-*jun*, 739
c-Ki-*ras*-2
c-*mos*, 738
cyclin and, 104
meiotic arrest and, 809–811, 822
oocyte maturation and, 809–810
c-*myb*, 738
c-*myc*, 739, 744, 747, 747–748
c-*onc* (cellular oncogenes), summary of,

738–739
c-*ras*
bladder carcinoma and, 745–746
C region *see* Constant region
c-*sis*, 738
c-*src*, 738, 742
Ca^{2+} *see* Calcium ion
CACCC box, 426
Cadherin, 544–546
Caenorhabditis elegans, 264–269
cell death in, 645–646
determination in, 400–401
as experimental organism, 264–265
glp-1 gene, 268–269
hermaphroditism in, 780–781
induction in, 597–600
lineage chart, 265
meiosis in, 796
paternal effect gene, 802–803
sex determination in, 763, 780–781
temperature-sensitive loci of, 268–269
Calcitonin
differential RNA processing of, 469
cells, origin of, 182
Calcium ion (Ca^{2+})
A23187 and, 56
acrosome reaction, 42–43, 49
aequorin, 56
bone and, 213ff
in cell cycle control, 70
cortical granules and, 55–58
cyclin and, 104–105
cytostatic factor and, 70
in egg activation, 64, 66–67
in endoplasmic reticulum, 56–57
epinephrine, 49
G-protein, 49
growth factor, 49
lymphocyte activation, 49
procaine, 56
Callus, 340
Calpain II, 810–811
cell cycle control and, 70
cytostatic factor and, 70
CAM *see* Cell adhesion molecule
cAMP
chemical structure, 24
chemotaxis and, 23–27
Dictyostelium, 23–24, 26
neural induction and, 317–318
Cancer
causes of, 737–741
oncogenes and, 736
see also Oncogene; Oncogenesis;
Transformation; Virus
Cap (of mRNA), 408–409
structure, 411
translational regulation, 497–498
Cap-binding protein, 483
Cap sequence, 408, 409ff
Capacitation, 46, 49
Capillary formation, 223–224
Capping (of membrane protein), 536–538
Carcinogen, 744ff
Cardia bifida, 221
Cardinal vein, 224
Carotid body, 182
Carrot, 340
Cartilage, 205
in bone formation, 210ff
in head, 182, 190 .LM 21
model, 211ff
Casein, mRNA stability, 501

CAT gene *see* chloramphenicol
acetyltransferase gene
caudal gene, 655
–658, 672
Cavitation, 92
CD blastomere, 261–264
CD4 glycoprotein, 545
cdc2
p34 and, 104
src protein and, 743
cdc13, 104
cdc25, cyclin and, 105–106
cDNA library, 347–350
cDNA *see* DNA, complementary
Cecropia, galvanotaxis in, 530
ced-3, 645–646
ced-4, 645–646
Cell adhesion molecules, 538, 543–553
addressins, 792
cadherin, 544–546
cell sorting and, 544–546
compaction, 93–94
Dictyostelium, 22ff, 26–27
distribution, 547–549
in fertilization, 44–50
in gastrulation, 120ff
glycosyltransferase and, 551–553
immunoglobulin superfamily CAM, 544–545
junctions and, 563–565
L-CAM, 545, 547–549
N-CAM, 319–322
in neurulation, 160ff
saccharide CAM, 544
saccharide-mediated cell adhesion, 549–553
see also Cell recognition; Neural cell
adhesion molecule
Cell affinity
differential, 524–528
gastrulation and, 120ff
Cell–cell interaction, of germ cells, 792
Cell cycle
biphasic, 103
calpain II and, 70
control of *see* Cell cycle control
cyclin and, 70
cytoskeleton in, 107–109
cytostatic factor and, 70
Drosophila, 103
growth factors and, 734–735
maternal mRNA and, 499
maturation promoting factor and, 103–104
midblastula transition, 124
oncogenes and, 743
platelet-derived growth factor (PDGF) and,
732
regulation of, 70, 103–106, 504–505, 731
Xenopus, 103
see also Cell cycle control; Growth control;
Growth factor; Oncogene; Oncogenesis;
Transformation
Cell cycle control
epidermal growth factor and, 753
insulin and, 753
oncogenes and, 743–744
platelet-derived growth factor and, 753
retinoblastoma and, 749–750
transferrin and, 753
tumor suppressors and, 748–750
see also Growth control; Growth factor;
Oncogene; Oncogenesis; Transformation
Cell death, programmed, 645–646
in limb bud, 645–646
mammary gland, 710

Volvox, 20–21
Cell division, 7
  cleavage, 70–71, 108–109
  control of, 727–733, 734–735
  cyclin and, 70
  cytokinesis, 70–71, 108–109
  cytoskeleton in, 107–109
  eukaryotic, 7–8
  karyokinesis, 108–109 microfilaments in, 40,
    107–109
  microtubules in, 107–109
  nuclear division, 107
  prokaryotic, 7–8
  regulation of, 70–71, 103–111
  see also Cell cycle; Cleavage; Meiosis;
    Mitosis
Cell fate
  in amphibians, 300–301
  in *C. elegans*, 400–401
  in *Drosophila*, 401–402
  in mosaic development, 260ff
  in regulative development, 300–301
  in vertebrates, 402–403
Cell fusion, 359–360
  during fertilization, 51–58
  in muscle development, 206–209
Cell junction, 538
  desmosome, 563–565
  gap, 88, 92, 563–565
  in morphogenesis, 563–565
  tight, 80, 82, 83, 563–565
Cell lineage, 400–403
  in *C. elegans*, 265
  evolutionary implications of, 828–829
  in sea urchin, 296–297
Cell membrane
  in cell division, 110–111
  cell migration and, 536–541
  in cell recognition, 15
  cytoskeleton and, 536–538
  developmental changes in, 539–541
  fluid nature of, 536–538
  see also Plasma membrane
Cell migration
  of ectoderm, 145–146
  glycosyltransferase in, 551–553
  laminin in, 558
  mechanisms of, 528–536
  of neural crest, 183
  polyspermy and, 53–58
  positional information and, 538–541
  somite formation and, 204
  sorting out and, 524–527
  tenascin in, 558
  thermodynamic model of, 532–536
  see also Cell movement
Cell movement
  in amphibians, 125ff
  archenteron and, 131–133
  in avian gastrulation, 138ff
  extracellular matrix and, 553–563
  in gastrulation, 118ff, 134–137, 147ff
  in sea urchin, 118ff
  sorting out and, 524–527
  see also Cell migration
Cell recognition
  in *Dictyostelium*, 23ff
  in fertilization, 42–50
  in sexual reproduction, 15
Cell signaling, in *Dictyostelium*, 23ff
Cell sorting
  positional information, 524–527
  role of cadherin, 544–546

Cell surface
  cell migration and, 536–541
  compaction, role in, 93–94
  in preimplantation development, 93–94
  role in signaling, 8
  see also Cell membrane
Cellular blastoderm, 100ff, 276
Cellular differentiation *see* Differentiation
Cellular slime mold *see Dictyostelium*
Cellular oncogene, 737
  summary, 738–739
Cellular regulation, in cancer, 737–740
  see also Cell cycle; Cell cycle control;
    Oncogene; Oncogenesis
Central nervous system, organization of,
  167–174
  see also Nervous system; Neuron; specific
    part of Nervous system
Central promoter element, 396–397
Centriole, 52
Centrolecithal egg, 77
Centromere, 793
Cephalic flexure, 165–167
Cephalic furrow, 651
Cerebellum, 166, 168ff
Cerebral aqueduct, 166
Cerebral hemisphere, 166
Cerebratulus, egg, 39
Cerebrum, 168ff
Cervical flexure, 165–167
Cervical cycle, 816
Cervix
  menstrual cycle, 817
  ovulation and, 815
CFU *see* Colony forming unit
CGRP neuropeptide, differential RNA
  processing of, 469
Chaetopterus, egg, 39
Chemoaffinity hypothesis, 638–641
Chemotaxis, 23, 528
  axon growth and, 634–637
  cAMP in, 23–27 cell migration and, 528–529
  of sperm, 41–42
Chick
  allantois, 218
  axon growth, 636–637
  eye, 177
  gastrulation, 139–146
  limb, 600–603, 608
  mesoderm, 203, 217
  myoblast, 207–209
  neural crest, 181–184
  neurulation, 160ff
Chimera
  germ line, 272–273
  mouse, 241–244
  yolk sac, 236
Chironomus, 329
  chromosome puff, 367–371
Chiton, metamorphosis, 707
Chlamydomonas, 14–16, 17ff
  gametes, 18
  isogamy in, 18
  reproduction in, 18
Chloramphenicol acetyltransferase (CAT) gene,
  420
Cholesterol, in sperm membrane, 46
Chondrocyte, 205, 211ff, 250
  bone formation, 210ff
  hypertrophic, 211ff
Chondrogenesis, 211ff
Chondroitin sulfate, 205, 250, 555
Chordamesoderm, 126ff, 201

  organizer, 303
Chordate, evolutionary scheme of, 29
Chorioallantoic membrane, 218
Chorion, 91, 97, 149, 151, 152, 217
Chorion (of insect), gene amplification in, 383
  Drosophila, 388
Chorionic gonadotropin, 150
  see also Luteotropin
Chorionic somatomammotropin, 150–151
Chorionic villi, 149ff, 224
Chromatid, 7, 793ff
  meiosis, 15
Chromatin
  activation of, 440–446
  condensation, 802–803
  heterochromatin, 377–383
  nuclear matrix and, 447–452
  structure, 437–439
Chromomere, 328
Chromosomal rearrangement
  in meiosis, 15
  oncogenesis and, 747–748
Chromosome diminution, 274ff
Chromosome
  eukaryotic, 7–8
  histones and, 7
  lampbrush, 805–806
  polyspermy, 52
  polytene, 328–329, 367–373, 813–815
  puff, 367–373, 389–392
  prokaryotic, 7–8
  sex determination and, 760–763
  X inactivation, 379–383
Chromosome puff, 367–373, 389–390
  ecdysone and, 389–390
  metamorphosis, 390–392
  see also Polytene chromosome
Chymotrypsin gene, 420
Cidaris, hybrids from, 490
Cigarettes, as teratogens, 194
Cilia
  in blastula, 82
  cell division and, 830
  immotile, 35–36
  Kartagener's triad, 35–36
Ciona, 258
Circulatory system, 219–225
  changes at birth, 225–227
Cis-regulatory elements, 412–417, 418ff
  bithorax complex and, 676–678
  RNA processing and, 475–476
Class switching, 363ff
Clear cytoplasm, in tunicate egg, 253ff
Cleavage, 4, 63
  in amphibians, 82–83
  asynchrony of, 124
  bilateral, 77, 88–89
  in birds, 98–99, 138ff
  cell junctions in, 80, 82, 83, 88
  compaction, 91–94
  contractile ring in, 107ff
  discoidal, 77, 98–99
  equatorial, 78
  in fish, 98–99
  gap junctions in, 88
  holoblastic, 77, 78–83, 88–91
  initiation of, 70
  in insects, 100–102
  left-handed, 85–86
  mammalian, 89–91
  maternal mRNA and, 499
  mechanisms of, 103–111
  membrane formation during, 110–111

amphibian, 127ff, 137–138
   chick, 143ff, 145–146
   derivatives, 127, 157ff, 202
   extraembryonic, 140
   formation, 137–138, 145–146
Ectopic pregnancy, 96
EDNH *see* Egg development neurosecretory
   hormone
Efferent duct (of testis), 761–762
EGF *see* Epidermal growth factor
EGF-like repeat, 599–600
Egg, 37–41
   activation, 62–71
   amniote, 31
   cleavage type, 76–77
   cortex, 107ff
   cortical granule, 39–41
   cytoplasmic rearrangement, 67–70
   diacylglycerol in, 64, 66–67
   G-protein, 49
   hamster, 40
   ionic composition of, 53
   mammalian, 39–40
   maturation, 39
   membrane potential, 53–54
   mRNA, 38
   peroxidase, 55
   pH, 63, 64–67
   plasma membrane, 39, 53–54
   polyspermy, 52–58
   pronucleus, 58, 59–60
   proteins, 38
   response to sperm, 62–71
   RNA, 38
   sea urchin, 37
   signal transduction in, 64, 66–67
   sperm receptor on, 45–46, 48, 49
   structure of, 37–41
   Styela, 88–89
   tRNA, 38
   Volvox, 21–22
   yolk distribution in, 77
Egg development neurosecretory hormone
   (EDNH), 814–815
Egg jelly, 41
   and acrosome reaction, 42–43
EGTA, and egg activation, 63, 67
eIF (eukaryotic initiation factor) *see* Initiation
   factor
Electrical current *see* Galvanotaxis
Electrical gradient, and oocyte maturation, 814
Electrophoresis, 351
Electroporation, 353
Eleutherodactylus, yolk in, 77
Elongation factor (of translation), 483–484
Emboitment, 250
Embryo, 3
   rejection by mother, 91
Embryogenesis, 4, 5
   mammalian, 89–98
Embryoid body, 241ff
Embryology, 3
Embryonal carcinoma, 241ff
Embryonic genome, maternal mRNA and,
   499–500
Embryonic induction, 250, 301–303
   see also Induction; Instructive interaction;
      Permissive interaction; Primary induction;
      Proximate interaction; Secondary
      induction
Emetine, 492
EMS blastomere, 268
Encapsulation, 251

Endocardium, 220ff
Endochondral bone formation, 210ff
Endocrine, 733, 740
Endoderm, 5, 157, 237–241
   chick, 141ff
   derivatives, 127, 202
   extraembryonic, 147ff
   formation, 125ff, 131–133
   yolk sac, 147ff
Endometrium, 149, 151
Endoplasmic reticulum
   calcium ions in, 56–57
   egg activation, 66–67
Endothelial cell, 223ff
Energids, 100–101
engrailed gene, 431, 661, 662, 668–669, 672,
   680
Enhancer, 412, 417–426
   interaction with promoters, 429–420
Entactin, 562
Entelechy, 288
Enterocoelomate, 29, 30
Entwicklungsmechanik, 286–287
Enucleation, maternal regulation and, 490–491
Enzyme isoform, 216
eosin eye gene, 775
Eosinophil, origin, 232
Ependyma, 168ff
Ependymal cell, 172
Epi 1, 319–322
Epiblast, 98–99, 138ff
   embryonic, 147ff
   mammalian, 147ff
Epiboly, 115, 135ff, 829
   amphibian, 126ff
   chick, 145–146
   ectoderm, 137–138
   mesoderm formation and, 135–137
Epidermal growth factor (EGF), 732–733, 734
   differentiation-promoting hormone, 732
   mammary gland differentiation and, 711,
      oncogene and, 741
Epidermis
   epidermal growth factor and, 732
   origin of, 191–193
Epigenesis, 251–252
Epimyocardium, 220ff
Epiphyseal plate, 211ff
Epiphysis, 166
Epithelial integrin, 560
Epithelio-mesenchymal induction, 588–590
Epithelio-mesenchymal interaction, 572–577,
   582–586
   evolution and, 841–842
   examples of, 573
Epithelium
   epidermal growth factor and, 732
   morphogenesis of, 591–593
Epsilon constant region, 363ff
Equatorial cleavage, 78
Equiangular spiral, 721–723
Equipotential harmonious system, 288–289
Equivalence group, of *C. elegans*, 268
ER *see* Endoplasmic reticulum
erbB, 748
Erythroblast, 231ff
Erythrocyte, 250
   globin genes in, 395–396
   origin of, 230
Erythrophore, 182
Erythropoietin, 231ff
Escherichia coli, 8, 258
   gene regulation, 412–413

lac operon, 412–413
   use in molecular biology, 345ff
Esophagus, 238, 241
Esterase, of *C. elegans*, 267
Estrogen receptor, 423–426
Estrogen
   bone growth and, 213
   mammary gland differentiation and, 711–713
   oocyte maturation and, 808–809
   ovalbumin regulation and, 392–394
   in puberty, 713–715
   sexual behavior and, 774
   vitellogenin synthesis and, 808–809
Estrus, 815
Euchromatin, 377
eudiplopodia gene, 602
Eudorina, 16, 18
Eukaryote, 6
   cell division, 7–8
   chromatin structure, 437–439
   colonial, 16–27
   DNA and RNA in, 6–8
   gene expression in, 6–8
   proteins, 7–8
   translation, 482–484
   unicellular, 8–16
Eukaryotic initiation factor (eIF) *see* Initiation
   factor
Euscelis, 654
Eustachian tube, 237
even-skipped gene, 431, 661, 663, 666–669
Evolution
   development and, 827ff, 834–836
   differentiation, 16–18
   embryology and, 155–157
   invertebrate to vertebrate, 837
   metazoan, 28–31
   modern synthesis, 853–856
   of multicellularity, 16–27
Evolutionary divergence, 29–31
Exocytosis
   acrosome reaction, 42
   calcium ions and, 42, 55–58
   cortical granule, 55–58
   in fertilization, 55–58
   insulin release and, 42
   of neurotransmitter, 42
Exogastrulation, 318
Exon, 407–408
   RNA processing and, 471–473
Explant experiments, 310–311
Extension, gastrulation, 134–137
Extracellular matrix, 190, 553–556
   axon growth and, 629–638
   and bone formation, 210ff
   cell migration and, 529–530
   cell movement and, 553–563
   differentiation and, 562–563
   fibronectin in, 556–558
   heart formation and, 220–221
   in induction, 588–590
   integrin in, 558–561
   laminin in, 558
   in neural crest migration, 185–188
   tenascin in, 558
Extraembryonic coelom, 149ff, 161
Extraembryonic endoderm, 147ff
Extraembryonic membrane, 216–219
Extraembryonic mesoderm, 147ff, 151
extramacrochaetae gene, 432
Extrinsic control (of cell division), 728
exuperantia gene, 655–658
   egg polarity and, 815

Eye
cornea, 180–181
development of, 175–181
Drosophila, 595–597
lens, 176, 180–181
pigment, 686
retina of, 177–179
eyeless mutation, 176

fa(1)pole holes gene, 655, 656
Facial nerve, 166, 182
Facultative heterochromatin, 377–378
FAS *see* Fetal alcohol syndrome
Fasciclin, 549
Fasciculation, 549
Fast block to polyspermy, 53–55
Fat body, yolk production and, 814–815
Fate map, 124ff
avian, 140
chick, 140
sea urchin, 296
tunicate, 254
Xenopus laevis, 125ff
see also Gastrulation
FBJ sarcoma virus (*fos*), 739 Feline sarcoma
virus (*fes*), 738
fem gene, 780–781, 797
fem-1 gene, 797
Ferritin mRNA, translational regulation of, 502
Fertilization, 4, 33
ammonia activation, 65–66
calcium in, 63, 65
in *Chlamydomonas*, 18
cortical granules and, 55–58
egg activation in, 62–71
external vs. internal, 46
function of, 33
human, 90
inositol trisphosphate and, 64, 66–67
in mammals, 46–50, 89–90
membrane potential in, 53–54
metabolic activation after, 62–71
metaphase arrest and, 809–811
pH in, 63, 64–67
polyspermy in, 52–58
pronuclear fusion after, 59–60
protein synthesis after, 63, 64–66
response of egg to, 62–71
in sea urchin, 41–46, 55ff
signal transduction during, 49, 64, 66–69
summary, 33
in Volvocales, 18
Fertilization envelope, 55
see also Vitelline envelope; Vitelline layer
Fertilization cone, 51, 52
Fertilization tube, 15
Fetal alcohol syndrome (FAS), 196–197
FGF *see* Fibroblast growth factor
Fibroblast
contact guidance of, 531–532
life cycle of, 752–753
platelet-derived growth factor and, 732
Fibroblast growth factor (FGF), 190, 732
mesoderm induction and, 313–314, 316–317
Fibroin, 388–389
Fibronectin, 121ff, 556–558
in amphibian gastrulation, 135–137
differential RNA processing in, 469
extracellular matrix and, 555, 556–558
in gastrulation, 121ff
germ cell migration and, 789
in heart development, 220
integrin and, 558–561

neural crest migration and, 187–188
somite formation and, 204ff
Filopodia, 118
ectoderm formation, 145–146
gastrulation, 121ff
neuronal, 173–174
Fish
cleavage, 98–99
gonochorism, 781
sex determination in, 851–852
Fission, 13
in *Dictyostelium* 22ff
Flagellum, 11–12, 35–37
Flatworm, cleavage in, 83
Flexure, of neural tube, 165–167
Fluid mosaic model (of membrane structure),
536–537
Focus (of dividing cells), 736
fog-1 gene, 797
Follicle, 816–820
Follicle cell, 39, 47, 388, 762–763
meiotic arrest and, 820–822
nurse cell, 813–815
oocyte maturation and, 807 yolk
accumulation and, 806
Follicle-stimulating hormone (FSH), 714, 815,
817
Follicular phase (of menstrual cycle), 817
Footprinting (of DNA), 398
Foramen ovale, 226
Forebrain, 165–167
Foregut, formation of, 126ff
fos, aging and, 753
Founder cell, sea urchin, 295–297
Fovea, 644
Fraternal twins, 97
Freemartin, 768
Frog, 286
cell division, 109
cleavage, 82–83
cytoplasmic rearrangement, 68–70
developmental cycle, 5
germ cell migration, 788–789
mesoderm, 217
mosaic qualities, 309
neurulation, 158–159
oocyte maturation, 805ff
RNA complexity, 812
Fruiting body, of *Dictyostelium*, 23ff
fs(1)Nasrat gene, 655, 656
FSH *see* Follicle-stimulating hormone
Fujinami sarcoma virus (*fps*), 738
fused gene, 661
fushi tarazu gene, 431, 661ff, 668, 672, 673
Fusion, calcium-mediated, 42
Fusion protein, and polyspermy, 54
Fusogenic proteins, in sperm–egg fusion, 52

G_0 phase, 731
G_1 phase (prereplication gap), 103, 731
G_2 phase (premitotic gap), 103, 731
G6PD *see* Glucose-6-phosphate dehydrogenase
GAG *see* Glycosaminoglycan
Galactose, zona pellucida, 48
β–Galactosidase, stability of, 506–507
Galactosyltransferase
in sperm, 545
from mouse embryo, 545
Gallbladder, 239, 240
Galvanotaxis, 530–531
Gamete, 6, 33
imprinting, 435–436
structure, 34–41

unicellular eukaryotes, 14–15
Volvocales, 17, 18–22
Gamete recognition *see* Sperm–egg interaction
Gametic imprinting, 435–436
Gametogenesis, 5, 6, 36
see also Oogenesis; Spermatogenesis
Ganglion
acoustic, 175
dorsal root, 183, 189ff
parasympathetic, 182
spinal, 181, 182
sympathetic, 181ff, 188ff
GAP *see* GTPase activating protein
Gap gene, 653, 661–662, 664–665
Gap junction, 88, 92, 563–565
in ovarian follicle, 821–822
Gastropoda
metamorphosis, 707
sex determination in, 784–785
see also Snail; other Mollusc class names;
individual species name
Gastrulation, 4, 114–152
amphibian, 124–138
avian, 138–146
blastocoel and, 83
bottle cell and, 125ff, 132–133 cell movement
during, 131ff
convergent extension in, 134–137, 138
cytoplasmic rearrangements and, 68–70,
128–131
deuterostome, 29–31
dorsal lip and, 125ff
in *Drosophila*, 651, 652
endoderm formation, 131–133
extension, 134–137
fate map, 124ff
genomic expression, 152
glycoproteins in, 121–123, 135–138
glycosyltransferase in, 551–553
mammalian, 146–152
marginal zone, 132–133
mesoderm formation, 134–137
patterns of, 115–116
protostome, 29–31
sea urchin, 117–124
twinning and, 130–131
Gene
accessibility of, 445–446
leader sequence, 408
sex determination and, 766–774
sperm-specific, 801–803
stability, 359–360
structure, 407–411
Gene amplification, 383–387
mechanism of, 384–386
oncogenesis and, 746–747
Gene cloning, 344–352
Gene expression, 417–426, 811–812
coordinate regulation of, 423–426
determination, 400–403
differential, 327–328, 353–373, 400–403
DNA methylation and, 432
in *E. coli*, 412–413
enhancers in, 417–426
in gastrulation, 152
growth factors and, 313–314
haploid, 802–803
HMG proteins and, 446
in induction, 311–312
in mammalian embryo, 91
in mesoderm, 311–312
ovalbumin, 392–394
posttranslational regulation of, 503–515

enzymes of, 96
human, 90
Imprinting, gametic, 435–436
In situ hybridization, 352
Incus, 185
Independent development, 300ff
Indifferent stage (of sexual differentiation), 761
Induction, 143, 250, 301–303
  in *C. elegans*, 597–600
  cellular, 595–600
  competence and 322–323
  of cornea, 581–582
  dorsal lip of blastopore and, 301ff
  of *Drosophila* eye, 595–597
  evolution and, 841–843
  gene expression in, 311–312
  genetic specificity of, 575–577
  instructive, 571, 573–575, 578
  in kidney development, 582–586
  of lens, 577–580
  mechanisms of, 305–308, 318–322
  of mesoderm, 299ff, 310–311
  molecules, 312–314
  negative, 294–295
  neural, 317–322
  neurulation and, 158
  of pancreas, 590
  permissive, 571, 573–577, 578
  and polarity, 305–306, 307ff
  primary embryonic, 301–303
  proximate, 570–582
  reciprocal, 583–584
  regional specificity of, 573–575
  secondary, 322–323, 570
  species specificity of, 575–576
  specificity, 303–305
  see also Embryonic induction
Infertility, 50
Informosome, 496
Infundibulum, 237ff
  of oviduct, 392, 393
Ingression, 116ff
  chick, 142ff
Inhibin, 817
Initiation (of translation), 482–483
Initiation codon, 408
  see also Transcription; Translation
Initiation complex, 483–484
Initiation factor (of translation; eIF), 482–483
Inner cell mass (ICM), 92, 94–96, 146ff
Inositol 1,4,5-trisphosphate, 49
  in compaction, 94
  in fertilization, 64, 66–67
Insect
  cleavage, 100–102
  germ cells, 275–279
  meroistic, 812–815
  metamorphosis, 390–392, 699–706, 707
  oogenesis, 812–815
  pattern formation in, 651–682
  polytene chromosome, 389–390, 391
  RNA genes, 383ff
  see also Drosophila; Hymenoptera;
    individual species name
Insertion mutagenesis, 747
Instar, 699ff
Instructive interaction, 571, 573–575, 578
  see also Induction
Insulin, 731
  gene, 420
  mammary gland differentiation and, 713
  mRNA stability, 501
  posttranslational modification of, 504

Insulin–like growth factors (IGF-I, IGF-II), 731–732
Int-2 protein, 587
Integrin, 538, 558–561
  extracellular matrix and, 558–561
  oncogenes and, 742
Intercalation, 626–628
Interferon, 734–735
  growth factors and, 735
Interkinesis (of meiosis), 793, 795
Interleukin-1, 594–595
Interleukin-2, 234, 594–595
Interleukin-3, 233, 234
Interleukin-4, 234
Intermediate mesoderm, 201ff
Intermediate spermatogonium, 798
Interphase, 7
  meiotic, 15
intersex gene, 775–776
Intervening sequence
  mechanism of processing and, 470–473
  see also DNA structure; Intron; RNA
    processing
Intervillous space, 149, 151
Intestine, 239
Intramembranous bone formation, 210ff
Intron, 407–408, 411
  see also Intervening sequences; RNA
    processing
Invagination, 115
  amphibian, 125ff
  gastrulation, 122ff
  sea urchin, 121ff
Invertebrate metamorphosis, 707
Involuting marginal zone, 134–137
Involution, 116
  amphibian, 126ff, 132–133
  gastrulation, 134–137
  marginal zone, 132–133
Ionizing radiation, teratogenic effect, 193
IP$_3$ *see* Inositol 1,4,5-trisphosphate
IRE-BP (iron-responsive protein), 502–503
Iridophore, 182
Iris, 181
Isocitrate dehydrogenase, myotube, 208–209
Isogamy, 18
Isolecithal egg, 77
Isometric growth, 721–723
Isthmus (of oviduct), 392, 393

J region (of antibody genes), 362ff
JH *see* Juvenile hormone
Junction, cell *see* Cell junction
Juvenile hormone, 702, 703–706
  precocenes and, 708
  yolk synthesis in, 814–815

K$^+$ *see* Potassium ion
Kartagener's triad, 35–36
Karyokinesis, 107ff
  regulation of, 108–109
Keratan sulfate, 555
Keratin, 192
Keratinocyte, 192, 250
Kidney, 215
  morphogenesis of, 582–586
Kirsten murine sarcoma virus (Ki-*ras*), 743
Klinefelter syndrome, 194
knirps gene, 655, 661, 658–660, 664–665
Krüppel (*Kr*) gene, 655–658, 660ff, 664–665, 673

L-CAM, 545, 547–549
L-*myc*, 739

Labeled pathway hypothesis (of axon growth), 632–634
labial (*lab*), 670–676
Labyrinth, inner ear, 175
lac operon, 412
Lactate dehydrogenase (LDH)
  sperm-specific, 802
  stability of, 507–508
Lambda phage, 348ff
Lamellibranchia, metamorphosis in, 707
Lamellipodium, 35, 532
Lamin, phosphorylation of, 104
Laminin, 204ff, 558
  integrin and, 558–561
  neural crest, 187–188
  saccharide-mediated cell adhesion and, 552
Laminin–heparan sulfate receptor, 545
Lampbrush chromosome, 389–390, 392
  in oogenesis, 805–806
  and Y chromosome, 801
Lampsilis ventricola, cleavage pattern in, 87
Lanugo, 193
LAR *see* Locus activating region
Lariat (in RNA), 470–473
  see also RNA processing
Larva
  glochidium, 87
  planuloid, 29
  pluteus, 117ff, 287, 291ff
  prism, 116–117, 296ff
  nauplius, 827
  stereoblastula, 84, 829
  tornaria, 29
Laryngotracheal groove, 240–241
Lateral body fold, 161
Lateral geniculate body, 643–645
Lateral inhibition, 598
Lateral plate mesoderm, 140, 201–203, 215–216
LDH *see* Lactate dehydrogenase
LDH-X *see* Lactate dehydrogenase, sperm-specific
Leader sequence, 408
  transcribed, 384
Lecithin, as surfactant, 241
Lens, 180–181, 250
  crystallin, 181
  development, 176
  induction of, 577–580
  placode, 176, 180
  regeneration, 332–333
Leptotene, 793, 794, 807
Lesch–Nyhan syndrome, 380
Leucan integrin, 560
Leucine zipper, 432
Leucocyte *see* B cell; Basophil; Eosinophil;
  Hematopoiesis; T cell
Leydig cell, 250, 761–762, 770–771
LFA-1, 545
LH *see* Luteinizing hormone
LH-RF *see* Luteinizing hormone releasing
  factor
Life cycle
  Chlamydomonas, 14
  Dictyostelium, 22–27
  frog, 5
  Naegleria, 11–12
  Volvox, 16–22
Life expectancy, 751
Life span potential, 751
Light chain (of antibody), 362ff
Limb
  bud, 600–601
  development of, 609–628

growth, 731
induction of, 600–603
mutations of, 602
pattern formation in, 608–628
polar coordinate model, 625–628
supernumerary, 625
Limb field, 609
Limbus, 227
Limnae peregra, 86
lin-12 gene, 400–401, 597–600
Lineage *see* Cell fate; Cell lineage
Lipochondria, 806, 807
Lipovitellin, 806
Lithium chloride, 293, 297, 315
Liver, 238ff
hematopoiesis in, 234–237
yolk synthesis and, 806, 808
Locus activating region (LAR), 420
Longevity, 751
Lordosis, 773
Lucifer yellow, as cell marker, 88
Lung
bud, 238, 241
differentiation of, 574–575
Luteal phase (of menstrual cycle), 819
Luteinizing hormone (LH), 713–715, 815, 817
meiotic arrest and, 820–822
Luteinizing hormone releasing factor (LH-RF),
714–715
Luteotropin, 820
see also Chorionic gonadotropin
Lymphocyte
development of, 593–595
division in, 853
genes, 360–366
see also B cell; T cell
Lymphocyte homing receptor, 545
Lymphoid cell, origin, 231, 232
Lymphokine, 250
Lymphoma, Burkitt's, 747–748
Lysin
sperm, 48, 52
see also Sperm, protease
Lysozyme, 393
Lytechinus
blastula of, 119
cleavage in, 79
egg jelly, 43
fertilization events, 63
gastrulation in, 118, 122, 123
hybrids from, 490 metamorphosis, 707
polyspermy in, 54
RNA from, 355

M phase, 731
see also Cell cycle; Mitosis
Macromere, 79ff
gastrulation and, 119ff
Macronucleus, 13–14
Macrophage, 593–595
origin of, 231ff
Magnum (of oviduct), 392, 393
Major histocompatibility antigen, in placenta,
152
Malleus, 184
Malpighian layer, 191–192
Mammal
allophenic mouse, 95, 98
blastocyst of, 92ff
capacitation in, 46, 49
cleavage in, 89–91
compaction in, 91–94
cortical granule reaction in, 55

egg of, 39–40, 47ff
embryo rejection in, 91 fertilization in, 46–52
gastrulation in, 146–152
germ cell migration in, 790–791
inner cell mass in, 92ff, 94–96
neural crest in, 185
polyspermy in, 58
pronuclear fusion in, 60–62
sex determination in, 760–774
sperm in, 43, 46ff
spermatogenesis in, 798–799
spermiogenesis in, 799, 802
twinning in, 97–98
yolk sac of, 149
see also individual species name
Mammary buds, 709
Mammary gland, 193
differentiation of, 709–713
Mammary ridge, 709
Mantle zone, 167ff
Marginal zone (of amphibian), 126, 131ff, 168ff
involuting, 134–137
noninvoluting, 135ff
Marginal zone cell, deep involuting, 132–133
Marmota, hypertrophy in, 727
Marsupium, of clams, 87
Masculinization, 770–771
Masked mRNA, 495–497
Maternal effect gene, 270, 653–660
Maternal genome, in mammal, 152
Maternal mRNA, 490–499
embryonic genome and, 499–500
longevity of, 499–500
Mating types, 13–15
Matrix-associated region (of DNA), 448
Matrix vesicle (in bone formation), 213
Maturation promoting factor (MPF), 103–104
cyclin and, 104–106
meiotic arrest and, 809–811
posttranslational modification of, 505
regulation of, 104
src protein and, 743
Maturation-inhibiting hormone, 714
Maximum longevity, 751
MBT *see* Midblastula transition
McDonough feline sarcoma virus (*fms*), 738,
742
Meckel's cartilage, 845
Median eminence (of pituitary), 713, 714
Medulla, adrenal, 181
Medulla oblongata, 166, 168ff
Megaloblast, 235
Meiosis, 15, 793–796
oocyte maturation and, 809–811
oogenesis, 803–805
reinitiation at fertilization, 809–811
spermatogenesis, 798–803
stored mRNA and, 809ff
summary of, 793
Meiotic arrest, 803, 809–811
human, 816, 817
mammalian oocyte, 820–822
Melanin, 250
Melanocyte, 182, 250
origin, 184
α-Melanocyte stimulating hormone (α-MSH),
posttranslational modification of, 504
Melanoma, 229–8ff
Melatonin, 714
Membrane, extraembryonic, 216–219
Membrane fusion
in abalone sperm, 52
and acrosome reaction, 42–43

calcium-mediated, 42
of cortical granules, 55–58
influenza virus and, 52
in mammalian sperm, 52
myoblast fusion, 206–209
sea urchin sperm, 52
specializations for, 52
sperm–egg, 51–52
Membrane potential, of egg, 53, 64
Membrane protein, movement of, 536–537
Menarche, 714
Menstrual cycle, 816–820
Mental retardation, 193, 197
Mercury, teratogenic effect of, 194
Meridional cleavage, 77, 78, 98–102
Merogone, 290, 491
Meroistic oogenesis, 812–815
Mesencephalon, 165–167
Mesenchymal factor, 590
Mesenchyme
avian, 141ff
chick, 144–145
head, 181ff
metanephrogenic, 583–584
migrations of, 118ff
primary, 116–117, 118ff
secondary, 116–117, 118ff, 123
skeletogenic, 296–297
see also Mesoderm
Mesentoblast, 264
Mesoderm, 5, 157
activin and, 313–314
amphibian, 125ff
appearance in evolution, 29ff
chick, 141ff
D blastomere and, 262
derivatives, 202
dorsal, 201–205
extraembryonic, 147ff
formation, 134–137, 201–205
gene expression in, 311–312
growth factors and, 312–314
head, 126
induction, 310–311, 312–314, 314–317
intermediate, 201ff, 215–216
lateral plate, 201–203, 215–216
paraxial, 203
parietal, 215–216
regionalization, 314–317
sea urchin, 121ff
somatic, 161, 205, 215–216
splanchnic, 161, 205, 215–216, 221
unsegmented, 203
Xhox3 gene and, 314–317
see also Mesenchyme
Mesodermal pouch, 30–31
Mesolecithal egg, 77
Mesomere, 79ff
gastrulation and, 119ff
Mesonephros, 583
Messenger RNA *see* mRNA
Messenger RNP *see* Ribonucleoprotein particle
Metabolic activation, of egg, 62–71
Metallothionein I (MT-I), 729
Metamorphic climax, 690
Metamorphic molt, 699ff
Metamorphosis (insect), 390–392
Metamorphosis, 685
amphibian, 686–695
and direct development, 695, 697–699
ecdysone in, 702, 703–706
environmental control of, 706–708
hemimetabolous, 699ff

holometabolous, 699ff
hormonal control of, 703–706
insect, 699–706
invertebrate, 707
neoteny, 695–696
precocious, 708
progenesis, 695, 696–697
protandry, 782
Metanephrogenic mesenchyme, 583–584
Metanephros, 583–584
Metaphase, 7
Metaphase arrest, mechanism of, 809–811
Metaphase, meiotic, 15, 793
Metaplasia, 332
Metastasis, 228ff
Metazoa, 28–31
    developmental pattern, 28–31
    evolution of, 28–31
Metencephalon, 166–167
Methylation (of DNA), 398–399, 432–437
    DNase hypersensitivity and, 444–446
    gametic imprinting and, 435–436
    site, identification of, 436
    Z-DNA and, 450–452
    see also DNA methylation
7-Methylguanosine (m⁷G) cap, 483
Microenvironment, hematopoietic, 233
Microfilament
    in acrosome reaction, 42
    in axons, 173–174
    capping and, 536–538
    in cell division, 40, 107–109
    in cleavage, 101
    in compaction, 94
    of contractile ring, 107–109
    cytochalasin B and, 107
    in cytokinesis, 107–109
    in egg, 40
    in insect egg, 101, 102
    in neurulation, 162ff
    in oocyte, 807
    spectrin and, 162
    sperm entry, role in, 40
β₂-Microglobulin, chemotactic factor, 529
Microinjection (of DNA), 353
Micromere, 79ff
    gastrulation and, 119ff
Micronucleus, 14
    migratory, 14
    stationary, 14
Micropinocytosis, 806
Microspike, 173–174
Microtubule, 11–12
    assembly of, 513
    in asters, 108–109
    in axon, 173–174
    in axoneme, 35–37
    capping and, 536–538
    in cell division, 107–109
    in cleavage, 101
    colchicine and, 69, 107
    cytoplasmic rearrangement and, 68, 69
    cytoplasmic rotation and, 129
    in insect egg, 101, 102
    in mitosis, 107–109
    in neurulation, 162ff
    nocodazole and, 107
    in oocyte, 807
    protofilament of, 35–37
    structure, 35–36
Microvilli
    egg, 51–52
    preimplantation embryos, 94

uvomorulin, 94
Midblastula transition (MBT), 102, 124
    enhancers in, 418–419
    gene expression, 372–373
Midbrain, 165–167
Middle ear, 237
Midgut, 161
Mifepristone (RU 486), 819
Mimicry, Batesian, 846–848
miniature wing, 775
Mitochondria
    oogenesis, 807
    protein targeting to, 511–512
    source for embryo, 59
    sperm, 35–37, 43
    sperm-derived, 59
Mitosis, 103ff
    asters, 108–109
    cyclin and, 70
    cytokinesis and, 108–109
    cytoskeleton in, 107–109
    duration of, 729
    karyokinesis, 107ff, 108–109
    membrane formation in, 110–111
    microfilaments and microtubules in, 107–109
    nuclear control of, 104–106
    in polyspermy, 52–53
    spindle, 107–109
    summary of, 7
    see also Cell division; Cell cycle; Cell cycle control
Mitosis-inducing phosphoprotein, 104
Molecular biology, techniques, 342–357
Mollusc, 260ff
    cleavage, 83, 84–85
    evolution, 29
    sperm chemotaxis in, 41
Moloney murine sarcoma virus (mos), 738
    meiotic arrest and, 809–811
MOM19 protein, 512
Monoclonal antibody, 539–540
Monocyte, origin of, 232
Monospermy, 52
Monozygotic twins, 97
Morphallaxis, 617
Morphogen, 615–619, 654–655
    Dictyostelium, 25–26
    differentiation-inducing substance (DIF), 25–26
    egg, 38
    retinoic acid as, 619–622
Morphogenesis, 4
    in Acetabularia, 8–11
    cadherin in, 544–546
    cell adhesion molecules in, 543–553
    cell junctions in, 563–565
    cell death in, 645–646
    coordinate, 586–587
    in Dictyostelium, 22–27
    epithelial, 591–593
    extracellular matrix in, 553–563
    fibronectin in, 558–561, 555, 556–558
    glycosyltransferase in, 551–553
    gradients in, 615–619, 623–627
    integrin in, 558–561
    key issues of, 523–524
    nuclear control of, 9–11
    protists, 8–12
    in Stylonichia, 9
    of unicellular eukaryotes, 8–12
Morphogenetic construction rules, 837–838
Morphogenetic factor, cytoskeleton and, 807
Morphogenic determinant, 253, 260–264

cytoskeleton and, 259–260
    in egg cytoplasm, 38
    gradients of, 615–619
    nature of, 257–259
    in oocyte, 805
    and cytoplasmic rearrangement, 67–70
    in tunicate, 257–259
Morphogenic factors, polarity determination and, 269–273
Morphogenic gradients, 615–619
Morpho peleides, 616
Morula, 82
    mammalian, 91ff
mos proto-oncogene, meiotic arrest and, 809
Mosaic development, 252, 253ff, 260ff, 267–269, 286–287
    cell interaction in, 259–260, 267–269
    and frog embryos, 309
    protostome vs. deuterostome, 30–31
Mosaic mouse, 241ff
Mosaic X chromosome inactivation, 380
Mouse
    agouti, 241ff
    allophenic, 95
    blastocyst, 96
    cleavage, 91ff
    clones, 339–340
    compaction, 95ff
    embryogenesis, 91ff
    implantation, 96
    measles, 359
    nuclear transplantation, 339–340
    preimplantation, 93
    sex determination gene, 766
    T locus, 767
    transgenic, 729–730
    X chromosome inactivation in, 382–383
MPF see Maturation promoting factor
mRNA
    5′ cap, 408–409
    compared to nRNA, 465
    complexity of, 459–461
    cytoplasmic storage of, 10–11
    degradation of, 475–477, 488–490, 501
    in Dictyostelium differentiation, 25
    in early development, 490–499
    in egg, 38
    heterogeneous nuclear RNA and, 459
    hormones and, 501
    inactive, 10–11
    leader sequence, 408
    localization of, 259–260, 494–495
    long-lived, 10–11
    masking of, 495–497
    maternal, 490–499
    morphogenesis and, 10–11
    polyadenylation of, 475–477
    processing of, 461–475
    prokaryotic vs. eukaryotic, 8
    sequestration of, 498
    stability of, 501
    synthesis of, 12
    trailer sequence in, 475–476
    transport of, 477–479
    utilization, 10–11
    see also Stored mRNA
mRNP see mRNA; Ribonucleoprotein
α-MSH see α-Melanocyte stimulating hormone
MT-I (metallothionein I), 729
Mucopolysaccharide, in cortical granule, 40–41, 55
Müller cell see Glia

egg, 127
  genetics of, 269–273
  insect egg, 814–815
  in limb development, 611–62
  specification of, 127–131, 269–273, 312, 314–317
  transcriptional control of, 270
  translational control of, 270–273
  yolk distribution and, 806–807
Polarization, embryo, 93–94
Pole (animal and vegetal), 76
Pole cell, 100ff, 276
Pole plasm, 276
Poly(A) polymerase, 476
Polyadenylation (of mRNA), 348, 408, 410ff
  mRNA degradation and, 475–477, 488–490, 812
polydactylous mutant, 602
Polymerase chain reaction (PCR), 356–357
Polyphenism, 708
Polyribosome, 484–485
Polysome *see* Polyribosome
Polyspermy, 52–58
Polytene chromosome, 328–329, 367–373
  in nurse cells, 813–815
Pons, 166
Porifera, 28–29
Porphobilinogen, 486
Porphyropsin, 686
Positional information, 301
  cell sorting and, 524–527
  cell migration and, 534–536
  determination and, 616
  frog embryos, 299–309
  genetic control of, 314–317
  gradient models, 615–619
  polar coordinate model, 625–628
  role of cell membrane, 538–541
  specification of, 616
  see also Axis; Polarity
Posterior intestinal portal, 238
Posterior marginal zone, 139ff
Posterior organizing center, 655, 658–660
Posterior transverse folds, 651
posterobithorax gene, 676–678
Posttranslational regulation, 503–515
  of morphogenesis, 12
  protein targeting and, 508–512
Potassium ion (K$^+$)
  membrane potential, 53
  polyspermy and, 53–54
Potential difference, in egg, 53–54
POU domain, 431
Pouch
  mesodermal, 30–31
  pharyngeal, 237ff
Precocenes, 708
Preformation, 34, 251
Pregnancy
  ectopic, 96
  immune deficiency in, 595
  mammary gland differentiation in, 711–713
  tubal, 96
Preimplantation
  blastocyst in, 92ff
  compaction in, 91–94
  human, 90
  mammalian, 89–98, 94–96
  Na$^+$/K$^+$-dependent ATPase in, 96
Premature infants, 241
Premitotic gap (G$_2$), 103
Preproinsulin, 504
Prereplication gap (G$_1$), 103

Previtellogenic phase (of oocyte maturation), 805, 806
Primary embryonic induction, 301–303
Primary mesenchyme *see* Mesenchyme, primary
Primary oocyte, 803–805
Primary spermatocyte, 798
Primary villus (of placenta), 149, 150–151
Primate
  germ layer derivatives, 147
  placenta, 150
Primer (DNA), 343ff
Primitive endoderm (mammalian), 147
Primitive groove, 139, 140ff
Primitive knot *see* Hensen's node
Primitive pit, 161
Primitive streak, 139, 140ff
  mammalian, 147ff
  as organizer, 303
Primordial follicle, 816
Primordial germ cells, 788
Prism larva, 116–117, 296ff
Proacrosin, 50, 545
Probe, in molecular biology, 349
proboscipedia gene, 670–676
Procaine, 56
Procreation, 34
Proctodeum, 238
Proerythroblast, 231
Progenesis, 695, 696–697
Progesterone, 150, 819
  mammary gland differentiation and, 711–713
  in meiotic maturation, 103
  meiotic arrest and, 809–811
  oocyte maturation and, 807–808
  translation and, 812
Progress zone, 611–613
Proinsulin, 504
Prokaryotes, 6–8
  cell division in, 7–8
  DNA and RNA in, 6–8
  proteins, 7–8
  transcription in, 8
  translation in, 8
Prolactin
  casein production and, 501
  gene expression of, 423
  mammary gland differentiation and, 713
  metamorphosis, 688
Proliferative phase (of menstrual cycle), 817
Prometamorphosis, 690
Promoter, 408, 412–417
  of 5 S rRNA gene, 396–397
  hormone-responsive, 424–426
  interaction with enhancers, 429–430
  sequence of, 415
  see also Enhancer; Transcription
Promoter-binding protein, 415–417
Promoter insertion, 747
Pronephric duct, 582–583
Pronephros, 215, 582–583
Pronucleus
  female, 39, 59–60
  fusion of, 58, 59–60
  male, 59–60
  migration of, 58–60
  nonequivalence of, 61–62
Prophase, 7
  meiotic, 15
Prosencephalon, 165–167
Prospective limb area, 609
Prostaglandin, ovulation and, 818–819
Protamine, 802, 803

Protandrous, 782
Protease
  and polyspermy, 55–58
  of sperm, 48, 52
Protein
  egg, 38
  prokaryotic vs. eukaryotic, 7–8
Protein degradation, regulatory mechanism of, 506–508
Protein inactivation, differential, 506–507
Protein kinase
  cell cycle control and, 749–750
  maturation promoting factor, 103–104
  meiotic arrest and, 809–811, 820–922
  oncogenes and, 737, 740, 741, 743
  translational control and, 486–488, 502
Protein kinase C
  and B cell division, 853
  compaction, 94
  egg activation, 64, 66–67
  neural induction, 317–318
  oncogenes and, 743–744
Protein phosphorylation *see* Phosphorylation, protein
Protein stability, 506–507
Protein synthesis *see* Translation
Protein targeting, 508–512
Proteoglycan, 553–554 Prothoracic gland, 705
Prothoracicotropic hormone (PTTH), 705–706
Protists
  development in, 8–12
  differentiation, 8–12
  evolutionary scheme, 29
  fission in, 13–14
  gametes, 14–15
  metazoan evolution and, 28–31
  morphogenesis in, 8–28
  sexual reproduction in, 13–16
Protofilament (of microtubules), 35–37
Protogynous, 781–782
Proto-oncogene, 736, 737
  cellular, 744–750
Protoporphyrin, 486
Protostome, 28–31
  development of, 29–31
  egg activation, 63
  evolution, 837
  mosaic development in, 30–31
Protozoa *see* Protists
Proximate interaction, 570–582
Pseudoplasmodium, of *Dictyostelium* 22ff
Psoriasis, 192
Pterygote insects, 835–836
PTTH *see* Prothoracicotropic hormone
Puberty, human, 713–715
pumilio (*pum*) gene, 655, 656, 658–660
Pupa, 699ff
Purkinje neuron, 169, 171

Quail, myoblast, 207–209
Quinine, as teratogen, 194

Rabbit, polyspermy, 58
Radial cleavage, 78–83
Radial symmetry, 29
Radiata, 29
Rana, 289ff, 333ff
  cytoplasmic rearrangement, 68
  germ cell migration, 789
  metamorphosis in, 686
  oogenesis, 805
RAP3/74 protein *see* Transcription factor TFIIF

ras oncogene, 744–746
  colon cancer and, 750
Rat, myoblast, 207–209
Rathke's pouch, 237ff
RB1 (retinoblastoma) gene, 748–750
Reaction diffusion model, 623–625
Receptor protein, 538
Reciprocal induction, 583–584
Recombinant DNA, 345ff
  see also DNA, recombinant; other DNA
    subheadings
Recombinant phage, 347ff
Recombinant plasmid, 345ff
Recombinase, 366
Red Queen hypothesis, 782
Regeneration, 332
  blastema, 610
  nuclear, 9
  salamander limb, 332, 333, 610
Regional specificity, 314–317
Regulative development, 287ff
  Driesch's theory, 287–289
  human, 98
  model, 293–294
  protostome vs. deuterostome, 30–31
Release factor (of translation), 484
Renal tubule, morphogenesis of, 582–583
Replication, after fertilization, 63
Reporter gene, 420
Repressor protein, of *E. coli*, 412–413
Reproduction, 4, 13
  see also Asexual reproduction; Sexual
    reproduction
Reptile
  cleavage in, 98
  germ cell migration in, 791–792
  sex determination in, 782–783
Resact, 41–42
Respiratory system, 240–241
Resting membrane potential *see* Membrane
    potential
Restriction enzyme (endonuclease), 345ff
Rete testis, 761–762
Reticular lamina, 553
Reticulocyte, 231ff
Retina
  axon growth, 637–638
  axon recognition in, 549
  cell aggregation, 533–536
  morphogenesis, 540–541
  N-CAM in, 543–544
  neural, 176, 177–179, 742
  pigmented, 176, 177–179
  synaptic specificity, 638–641
Retinoblastoma, tumor suppression and,
    748–750
Retinoic acid, 194–195
  intercalation of, 627–628
  limb development and, 619–622
  teratogenic potential, 194, 619
Retinotectal projection, 639–641
Retroviral vector, 353
Retrovirus, 353
Reverse transcriptase, 343, 343
Rhesus monkey, implantation in, 93
Rhizoid, 9ff
Rhodopsin, 686
Rhombencephalon, 165–167
Rhombomere, 166–167
  gene expression in, 431
Ribonuclease, 10
Ribonucleic acid *see* RNA
Ribonucleoprotein particles (RNP)

protamine, 802, 803
purification of, 496
sequestration of, 498
spermiogenesis, 802, 803
translational control and, 495–497
  see also snRNP
Ribonucleotide reductase, stored mRNA, 494
Ribosomal RNA (rRNA)
  amplification of, 383–387
  genes, 383–388
  RNA polymerase I, 387–388
  transcription, 387–388
  5 S rRNA, 396–399
Ribosome
  early development, 496–497
  efficiency of, 498–499
  egg, 38
  oocyte, 807
  prokaryotic vs. eukaryotic, 8
  see also Ribosomal RNA, genes;
    Translation; Translational control
Ring canal (in insect oocyte), 812
RNA (ribonucleic acid)
  complexity of, 459–461, 812
  $C_o t$ curve, 460–461
  cytoskeleton and, 807
  heterogeneous nuclear, 459
  nuclear, 408, 459–465
  oogenesis, 805–806
  polyadenylation and, 812
  prokaryotes vs. eukaryotes, 6–8
  ribonucleoprotein, 495, 497
  snRNP, 472–473
  trailer sequence, 475–476
  transport of, 411, 477–479
  Y-chromosome specific, 801
RNA blotting, 353–355
RNA–DNA hybridization, 342–343ff
RNA editing, 502
RNA polymerases, 8, 387–388, 414
  cell cycle control and, 104
  phosphorylation of, 104
  regulation of, 397–399
RNA processing, 461–475
  biochemistry of, 470–472
  differential, 466–469
  regulation of, 475–477
  sex determination and, 473–475
  Sex-lethal gene and, 473–475
  sex-specific, 777–778
  snRNPs and, 472–473
  transformer gene and, 473–475
  transport, 477–479
  U1 RNA and, 471–473
  U7 snRNP and, 477
RNA synthesis, chromosome puff, 367–373
  see also Transcription
RNP *see* Ribonucleoprotein
Robert syndrome, 193
Rods (of retina), 177
Rolling circle model (of DNA replication), 386
Rotational cleavage, 89–91
rough gene, 596–597
Rous sarcoma virus (*src*), 420, 736–737, 738,
    742
rRNA *see* Ribosomal RNA
RU 486 (mifepristone), 819
Rubella, as teratogen, 194
runt gene, 661, 667

S (synthesis) phase, 103, 731
S1 nuclease, 489
Saccharide CAM (cell adhesion molecule), 544

see also Cell adhesion molecules
Saccharide-mediated cell adhesion, 549–553
Saccocirrus, egg, 39
Salamander
  gastrulation, 125ff, 136
  heterochrony in, 848
  neural crest, 186–187
  neurulation, 160, 164
  parthenogenesis, 62
  regeneration in, 332, 333, 610
  regulative development in, 297ff
Sanger technique (for DNA sequencing),
    351–352
Sarcoma, 736
  see also Avian sarcoma virus-17; FBJ
    sarcoma virus; Fujinami sarcoma virus;
    Harvey murine sarcoma virus; Kirsten
    murine sarcoma virus; McDonough feline
    sarcoma virus; Moloney murine sarcoma
    virus; Rous sarcoma virus; Simian sarcoma
    virus
Sarcoplasmic reticulum, 58
scaleless mutation, 589
Schizocoel, formation of, 30
Schwann cell, 172, 174, 176
Scientific method, 27–28
Sclerotome, 205
Sea urchin, 287ff
  blastula stage, 79ff
  calcium ions, 55–58
  cell cycle, 103
  cleavage, 79–82, 107ff
  cortical granule, 55
  cyclin, 499
  differential cell affinity, 526–527
  DNA complexity, 461
  egg, 37ff, 107, 108
  egg peroxidase, 55
  egg activation, 63–67
  fertilization, summary, 63
  gastrulation in, 117–124
  hybrids, 490
  larval stages, 116–117, 287, 291ff, 296ff
  membrane potential, 53–54
  mesenchyme, 118ff
  metamorphosis, 707
  pluteus larva, 117ff, 287, 291ff
  polyspermy block, 52–58
  prism larva, 116–117, 296ff
  pronuclear fusion, 60
  replication, 103
  RNA complexity, 462–465
  sperm, 42ff
  sperm–egg binding, 44–46
  sperm–egg fusion, 52
  stored mRNA in, 493–494
  see also individual species name
Sebaceous glands, 192–193
Secondary induction, 322–323, 570
  see also Induction; Instructive interaction;
    Permissive interaction; Primary induction;
    Proximate interaction
Secondary mesenchyme, 116–117, 11844, 123
Secondary oocyte, 804–805
Secondary spermatocyte, 798
Secondary villus (of placenta), 149, 150–151
Secretory phase (of menstrual cycle) *see* Luteal
    phase
Segment polarity gene, 653, 661, 662, 668–669
Segmental plate, 203
Segmentation gene (in *Drosophila*), 661–669
Selective cell affinity, 525
Selective entry and retention, 510

Styela, 258, 259
  cleavage, 88–89
  cytoplasmic determinants in, 253ff
  cytoplasmic rearrangement, 67–68
  germ layers, 89
Stylonichia, 9
Subcephalic space, 161
Subgerminal cavity, 98
Substance P, differential RNA processing of, 469
Substrate adhesion, 538
Substrate adhesion molecule, 553
Sulcus limitans, 168
Superficial cleavage, 77, 100–102
Supramolecular assembly, 513–514
Surfactant, 241
SV40 *see* Simian virus 40
swallow gene, 655–658
  egg polarity and, 815
Sweat gland, 193
Switch gene, 400–403
Symmetry
  of amphibian embryo, 127–131
  bilateral, 29, 68
  cleavage, 70–71
  dorsal–ventral, 69–70
  establishment of, 127–131
  radial, 29
Sympathetic ganglia, 181ff, 188ff
Synapsis, 794
Synapta, cleavage in 78–79
Synaptic specificity, 638–641
Synaptonemal complex, 793
Syncytial blastoderm, 100ff
Syncitial cables, 118ff
Syncitiotrophoblast, 147ff, 149, 151
Syncitium, 276
Syndecan, 554–555
  kidney morphogenesis, 585–586

T_3 *see* Triiodothyronine
T_4 *see* Thyroxine
T cell (T lymphocyte), 237, 250
  aging and, 751–752
  division pathway, 853
  helper, 593–595
  migration, 529
  origin of, 232
thy-1 gene, 420
T locus, 767
T lymphocyte *see* T cell
T protein (of SV40), 749
TAA codon, 408, 409
TAF *see* Tumor angiogenesis factor
TAG-1, 545
tailless gene, 655, 660, 661, 664–665
Talin, 538
Targeting (of proteins), 508–512
TATA box, 412–414, 415
tc^R, 345–346
Td-1 gene, 768–769
Tda-1 gene, 767
TDF *see* Testis–determining factor
TDF gene, 769
Tdy gene, 765–766, 767, 768–769
Telencephalon, 165–167
Telolecithal egg, 77
Telophase, 7, 793
  meiotic, 15, 795
Temperature, sex determination and, 777–778
Temperature-sensitive mutant, 268–269
Tenascin, 187–188, 558
Teratocarcinoma, 93, 241–244

X chromosome inactivation, 382–383
Teratogen, 193ff
Teratology, 193–197, 252–253
Terminal gene (in *Drosophila*), 655, 660
Termination codon, 408 *see also* Translation; Transcription
Termination factor (of translation), 483–484
Tertiary villus (of placenta), 149, 150–151
Testis, 34, 798–803
  development of, 761–762
Testis-determining factor (TDF), 760, 763–766, 769
Testis-determining gene (*Tdy*), 765–766, 767
Testosterone, 250
  in bone growth, 213
  mammary gland differentiation and, 710
  in puberty, 713–715
  sex determination and, 761, 770–771, 772–773
Tetracycline, 346ff
Tetrad, 793, 794
Tetraparental *see* Allopheny
TFIIA, TFIIB, TFIIC, TFIID, TFIIE, TFIIF *see* Transcription factors
Thalamus, 166
Thalassema, egg, 39
Thalassemia, 409, 410–411, 478–479, 483
Thalidomide, 195–196
Thecal cell (of ovary), 762, 816ff
Thermodynamic model of cell migration, 532–536
Thermus aquaticus, 357
Threshold concept (in metamorphosis), 692–693
thy-1 gene, 420
Thymidine kinase, 352
Thymopoietin, 234
Thymus, 237
  aging and, 751
  chemotactic factors from, 529
Thyroid, 237, 238
Thyroid-stimulating hormone releasing factor (TSH-RF), 689–690
  and metamorphosis, 688–689
Thyroxine (T_4), 688–695
Tight junction, 563–565
Tobacco, 340
Toll gene, 271–272
Tonsils, 237
Tooth, morphogenesis of, 587
Topoisomerase II, 449–450
Tornaria larva, 29
torpedo mutation, 272–273
torso gene, 655, 656, 660, 665
torsoless gene, 660
torso-like gene, 660
Totipotency, 328–329, 334ff
  plant cells, 340
  restriction of, 332–338
  vs. pluripotency, 338
Toxoplasma gondii, as teratogen, 194
Toxopneustes lividus, 35
tra gene, 775–776, 778–779
tra-1, 797
Trachea, 239, 241
Trailer sequence (of mRNA), 475–476
Trans–acting element, 412–417, 418ff, 432ff, 440–446
  families, 430–432
  RNA processing and, 475–476
Transactivation domain, 425–426
Transcription
  actinomycin D and, 12, 492
  amplified genes and, 386–387
  and cap, 408ff

  and cell cycle control, 104
  chorion gene, 389
  chromosome puff, 367–373
  cis–regulator of, 412–417
  Dictyostelium morphogenesis and, 24
  differential, 400–403
  DNase hypersensitivity and, 442–446
  enhancers in, 412, 417–426
  euchromatin and, 377–378
  heterochromatin and, 377–378
  HMG proteins and, 446
  inactivation of, 434–435
  inhibition of, 12, 492
  initiation of, 408, 409, 414–415
  during mesoderm induction, 311–312
  midblastula transition and, 102
  morphogenesis and, 10–11
  morphogenic determinants and, 257–258
  nuclear matrix and, 447–452
  of ovalbumin, 392–394
  in prokaryotes, 8
  promoters in, 412–417
  regulatory elements of, 412–417
  ribosomal RNA gene, 383–387
  RNA polymerase I and, 387388
  rRNA, 383–388, 811
  in spermatogenesis, 801–803
  TATA box and, 412–414
  trans–regulator of, 412–417
  translation (concurrent) and, 8
  tRNA, 483–484, 811
  Y-chromosome, 801–803
Transcription factors, 396–399, 412–417
  5 S RNA gene and, 396–399
  enhancers and, 417–426
  ribosomal RNA genes, 387–388
Transcription unit, 384
Transcriptional control
  in *Acetabularia*, 8–11
  of morphogenesis, 10–12
  in *Naegleria*, 11–12
  of sex determination, 777–778
  see also Transcription
Transdetermination, 329ff, 331, 332, 796
Transfection, 352–353
Transfer of competence, 850–852, 853
Transfer RNA *see* tRNA
Transferrin receptor, 502
Transformation, 736ff
  in *Naegleria*, 11–12
  see also Oncogene; Oncogenesis; Virus
transformer gene, 473–475, 775–776, 778–779
transformer-2, 775–776, 778–779
Transforming growth factor-α, 192
Transforming growth factor-β (TGF-β), 733, 734–735
  mesoderm induction and, 313–314
Transgenic mouse
  growth in, 729–730
Translation
  activation after fertilization, 63, 64
  cap (of mRNA), 408–409
  and egg activation, 64–66
  efficiency of, 498–499
  eukaryotic, 6
  after fertilization, 63, 64
  initiation codon of, 408
  mechanism of, 482–484
  midblastula transition and, 102
  morphogenesis and, 10–11
  nick, 355–356
  in prokaryotes, 6, 8
  protein targeting and, 508–512

sexual reproduction, 18–22
von Baer's Laws, 155–157
Vulval induction (in *C. elegans*), 597–600
Vulvaless gene, 598–600

Wachtiella persicariae, germ cell, 275
weaver mutation, 171
white, 276, 367
white apricot mutant, 367
White blood cells *see* B cell; Basophil;
 Eosinophil; Hematopoiesis; T cell
white spotting, 733–734
White matter, 168ff
White blood cell, origin of, 230ff
wingless (wg) gene, 661, 669
Wolffian duct, 764, 770–771
 derivatives of, 760, 761–762
Wolffian regeneration, 332, 333

X chromosome
 Barr body and, 379–382
 dosage compensation and, 379–382
 inactivation, 379–383
 methylation and, 435
 timing of inactivation, 379–383
Xanthophore, 182
Xenopus, 260, 306–308, 336ff
 blastula, 82
 cell cycle, 103
 cleavage in, 110–111
 cortical endoplasmic reticulum in, 58
 cytoplasmic rearrangements in, 69
 DNA, 385
 determination in, 402–403
 DNA library from, 347ff
 5 S rRNA genes from, 396–399
 galvanotaxis in, 530
 gastrulation in, 124ff
 germ cell migration in, 780
 mating in, 336
 maturation promoting factor in, 103ff
 metamorphosis in, 690ff
 midblastula transition in, 102, 124, 418–419
 mitochondria in, 807
 neural induction in, 318–322

nucleolus of, 385
oocyte components in, 805
oocyte transcription in, 811–812
polarity in, 313–314
retina, 637–638
retinotectal projections, 639–641
ribosomal RNA, 383–384
stage-specific gene expression in, 372–373
yolk accumulation, 806
Xenopus borealis, 307, 308
Xhox3 gene, 314–317
XO-lethal gene, 781
XTC cell line, 313–314, 316–317

Y chromosome
 Drosophila, 801
 sex determination and, 763–766
yellow mutation, 276
Yellow crescent, 68, 89, 253ff, 257, 259–260
 Yolk, 31
 accumulation of, 805–807
 in amphibian egg, 82
 cleavage pattern and, 76, 98–102
 function of, 77
 and gastrulation, 124ff, 138–146
 patterns, 77
 storage, 806–807
Yolk duct, 219
Yolk plug, 127ff
Yolk platelet, 806
Yolk protein, 421–422
 Drosophila, 814–815
 gene expression of, 421–422

Yolk sac, 140, 149, 219
 endoderm, 147ff
 gut, 239–240
 hematopoiesis in, 233–237
 human, 150
yp1 gene, 422
yp2 gene, 422

Z-DNA, 450–452
Zebra fish, 99
zerknüllt gene, 270, 431
Zinc finger (of DNA), 425
Zona pellucida, 39ff, 47–50
 blastocyst, 96
 composition, 47
 glycosyltransferase and, 552
 hatching from, 96
 penetration, 50, 51
 polyspermy and, 55–58
 species-specificity and, 47
Zona reaction 55
Zone of polarizing activity (ZPA), 613–615
ZP1, 47, 50
ZP2, 47, 50
ZP3, 47–50, 67
 calcium ion and, 49
 G-protein and, 49
 polyspermy and, 55
ZPA *see* Zone of polarizing activity
Zygote, 3
 centriole in, 52
 gene expression in, 91
 nucleus, 59–60
 unicellular eukaryotes, 14
 Volvox 21–22
Zygotene, 793, 794, 807

ABOUT THE BOOK

Editor: Andrew D. Sinauer
Project Editor: Carol J. Wigg
Production Manager: Joseph J. Vesely
Book Production: Janice Holabird
Illustration Program: J/B Woolsey Associates
Book and Cover Design: Rodelinde Graphic Design
Copy Editor: Christine Decker
Composition: DEKR Corporation
Prepress: Jay's Publishers Services, Inc.
Cover Manufacture: New England Book Components, Inc.
Book Manufacture: Courier Westford, Inc.